D1748169

Macromolecular Engineering

Edited by
Krzysztof Matyjaszewski,
Yves Gnanou,
and Ludwik Leibler

1807–2007 Knowledge for Generations

Each generation has its unique needs and aspirations. When Charles Wiley first opened his small printing shop in lower Manhattan in 1807, it was a generation of boundless potential searching for an identity. And we were there, helping to define a new American literary tradition. Over half a century later, in the midst of the Second Industrial Revolution, it was a generation focused on building the future. Once again, we were there, supplying the critical scientific, technical, and engineering knowledge that helped frame the world. Throughout the 20th Century, and into the new millennium, nations began to reach out beyond their own borders and a new international community was born. Wiley was there, expanding its operations around the world to enable a global exchange of ideas, opinions, and know-how.

For 200 years, Wiley has been an integral part of each generation's journey, enabling the flow of information and understanding necessary to meet their needs and fulfill their aspirations. Today, bold new technologies are changing the way we live and learn. Wiley will be there, providing you the must-have knowledge you need to imagine new worlds, new possibilities, and new opportunities.

Generations come and go, but you can always count on Wiley to provide you the knowledge you need, when and where you need it!

William J. Pesce
President and Chief Executive Officer

Peter Booth Wiley
Chairman of the Board

Macromolecular Engineering

Precise Synthesis, Materials Properties, Applications

Edited by
Krzysztof Matyjaszewski, Yves Gnanou, and Ludwik Leibler

Volume 1
Synthetic Techniques

WILEY-VCH Verlag GmbH & Co. KGaA

The Editors

Prof. Dr. Krzysztof Matyjaszewski
Carnegie Mellon University
Department of Chemistry
4400 Fifth Ave
Pittsburgh, PA 15213
USA

Prof. Dr. Yves Gnanou
Laboratoire de Chimie des Polymères Organiques
16, ave Pey-Berland
33607 Pessac
France

Prof. Dr. Ludwik Leibler
UMR 167 CNRS-ESPCI
École Supérieure de Physique
et Chimie Industrielles
10 rue Vauquelin
75231 Paris Cedex 05
France

■ All books published by Wiley-VCH are carefully produced. Nevertheless, authors, editors, and publisher do not warrant the information contained in these books, including this book, to be free of errors. Readers are advised to keep in mind that statements, data, illustrations, procedural details or other items may inadvertently be inaccurate.

Library of Congress Card No.: applied for

British Library Cataloguing-in-Publication Data
A catalogue record for this book is available from the British Library.

Bibliographic information published by the Deutsche Nationalbibliothek
The Deutsche Nationalbibliothek lists this publication in the Deutsche Nationalbibliografie; detailed bibliographic data are available in the Internet at http://dnb.d-nb.de.

© 2007 WILEY-VCH Verlag GmbH & Co. KGaA, Weinheim, Germany

All rights reserved (including those of translation into other languages). No part of this book may be reproduced in any form – by photoprinting, microfilm, or any other means – nor transmitted or translated into a machine language without written permission from the publishers. Registered names, trademarks, etc. used in this book, even when not specifically marked as such, are not to be considered unprotected by law.

Composition K+V Fotosatz GmbH, Beerfelden
Printing betz-druck GmbH, Darmstadt
Bookbinding Litges & Dopf GmbH, Heppenheim
Cover Grafik-Design Schulz, Fußgönheim
Wiley Bicentennial Logo Richard J. Pacifico

Printed in the Federal Republic of Germany
Printed on acid-free paper

ISBN 978-3-527-31446-1

Contents

Preface XXV

List of Contributors XXVII

Volume 1
Synthetic Techniques

1	**Macromolecular Engineering** *1*	
	Krzysztof Matyjaszewski, Yves Gnanou, and Ludwik Leibler	
2	**Anionic Polymerization of Vinyl and Related Monomers** *7*	
	Michel Fontanille and Yves Gnanou	
2.1	Introduction *7*	
2.2	General Features of Anionic Polymerization *8*	
2.2.1	Polymerizability of Vinyl and Related Monomers *9*	
2.2.2	Various Parameters Influencing the Structure and Reactivity of Active Centers *11*	
2.2.2.1	Influence of the Type of Monomer *12*	
2.2.2.2	Influence of the Nature of Solvent *13*	
2.2.2.3	Influence of Additives *15*	
2.2.2.4	Influence of the Counterion *16*	
2.2.3	Experimental Constraints Related to Anionic Polymerization *17*	
2.3	Initiation of Anionic Polymerizations *17*	
2.3.1	Initiation by Electron Transfer *18*	
2.3.2	Initiation by Nucleophilic Addition to the Double Bond *19*	
2.3.2.1	In Polar Solvents *19*	
2.3.2.2	In Nonpolar Solvents *21*	
2.3.2.3	Bi- and Multifunctional Initiators *25*	
2.3.3	Initiation by Alkoxides and Silanolates *29*	
2.3.4	Initiation of the Polymerization of Alkyl (Meth)acrylates by Group Transfer *29*	
2.4	Propagation Step *30*	
2.4.1	Kinetics of the Propagation Step *31*	

Macromolecular Engineering. Precise Synthesis, Materials Properties, Applications.
Edited by K. Matyjaszewski, Y. Gnanou, and L. Leibler
Copyright © 2007 WILEY-VCH Verlag GmbH & Co. KGaA, Weinheim
ISBN: 978-3-527-31446-1

2.4.1.1	Kinetics of Polymerization in Non-polar Solvents 31
2.4.1.2	Polymerizations Carried Out in Polar Media 34
2.4.2	Anionic Polymerization of (Meth)acrylic Monomers 36
2.4.2.1	General Characteristics 36
2.4.2.2	Propagation by Group Transfer 37
2.4.3	Anionic Copolymerization 38
2.4.4	Regio- and Stereoselectivity in Anionic Polymerization 39
2.4.4.1	Cases of Conjugated Dienes 39
2.4.4.2	Case of Vinyl and Related Monomers 40
2.5	Persistence of Active Centers 42
2.5.1	Case of Polystyrenic Carbanions 43
2.5.2	Case of Polydiene Carbanions 44
2.5.3	Case of (Meth)acrylic Polymers 44
2.6	Application of Anionic Polymerization to Macromolecular Synthesis 45
2.6.1	Prediction of Molar Masses and Control of Their Dispersion 46
2.6.2	Functionalization of Chain Ends 46
2.6.3	Synthesis of Graft and Block Copolymers 47
2.6.4	Star Polymers 48
2.6.5	Macrocyclic Polymers 49
	References 50
3	**Carbocationic Polymerization** 57
	Priyadarsi De and Rudolf Faust
3.1	Introduction 57
3.2	Mechanistic and Kinetic Details of Living Cationic Polymerization 58
3.3	Living Cationic Polymerization 60
3.4	Monomers and Initiating Systems 61
3.5	Additives in Living Cationic Polymerization 61
3.5.1	Isobutene (IB) 62
3.5.2	β-Pinene 64
3.5.3	Styrene (St) 65
3.5.4	p-Methylstyrene (p-MeSt) 65
3.5.5	p-Chlorostyrene (p-ClSt) 66
3.5.6	2,4,6-Trimethylstyrene (TMeSt) 66
3.5.7	p-Methoxystyrene (p-MeOSt) 66
3.5.8	α-Methylstyrene (αMeSt) 67
3.5.9	Indene 67
3.5.10	N-Vinylcarbazole 68
3.5.11	Vinyl Ethers 68
3.6	Functional Polymers by Living Cationic Polymerization 70
3.6.1	Functional Initiator Method 70
3.6.2	Functional Terminator Method 72
3.7	Telechelic Polymers 74

3.8	Macromonomers	76
3.8.1	Synthesis Using a Functional Initiator	76
3.8.2	Synthesis Using a Functional Capping Agent	78
3.8.3	Chain-end Modification	80
3.9	Block Copolymers	80
3.9.1	Linear Diblock Copolymers	81
3.9.2	Linear Triblock Copolymers	84
3.9.2.1	Synthesis Using Difunctional Initiators	84
3.9.2.2	Synthesis Using Coupling Agents	85
3.9.3	Block Copolymers with Nonlinear Architecture	86
3.9.4	Synthesis of A_nB_n Hetero-arm Star-block Copolymers	87
3.9.5	Synthesis of AA′B, ABB′ and ABC Asymmetric Star-block Copolymers Using Furan Derivatives	87
3.9.6	Block Copolymers Prepared by the Combination of Different Polymerization Mechanisms	89
3.9.6.1	Combination of Cationic and Anionic Polymerization	89
3.9.6.2	Combination of Living Cationic and Anionic Ring-opening Polymerization	90
3.9.6.3	Combination of Living Cationic and Radical Polymerization	92
3.10	Branched and Hyperbranched Polymers	93
3.10.1	Surface-initiated Polymerization: Polymer Brushes	94
3.11	Conclusions	94
	References	95

4	**Ionic and Coordination Ring-opening Polymerization**	**103**
	Stanislaw Penczek, Andrzej Duda, Przemyslaw Kubisa, and Stanislaw Slomkowski	
4.1	Introduction	103
4.2	Thermodynamics of Ring-opening Polymerization	106
4.2.1	Equilibrium Monomer Concentration – Ceiling/Floor Temperatures	106
4.2.2	Recent Results Related to Thermodynamics of Ring-opening Polymerization	107
4.2.2.1	Thermodynamics of γ-Butyrolactone (Co)polymerization	107
4.2.2.2	Copolymerization of Lactide at the Polymer–Monomer Equilibrium	108
4.3	Basic Mechanistic Features of Ring-opening Polymerization	109
4.3.1	Anionic and Coordination Ring-opening Polymerization of Cyclic Ethers and Sulfides	109
4.3.1.1	Initiators and Initiation	109
4.3.1.2	Active Centers – Structures and Reactivities	110
4.3.1.3	Controlled Anionic and Coordination Polymerization of Oxiranes	111
4.3.1.4	Stereocontrolled Polymerization of Chiral Oxiranes	113

4.3.2	Controlled Synthesis of Aliphatic Polyesters by Anionic and Coordination Ring-opening Polymerization 114
4.3.2.1	Initiators and Active Centers – Structures and Reactivities 114
4.3.2.2	Controlled Polymerization of Cyclic Esters with "Multiple-site" Metal Alkoxides and Carboxylates 118
4.3.2.3	Controlled Polymerization of Cyclic Esters with "Single-site" Metal Alkoxides 121
4.3.2.4	Poly(β-hydroxybutyrate)s by Carbonylation of Oxiranes 121
4.3.2.5	Stereocontrolled Polymerization of Chiral Cyclic Esters 122
4.3.2.6	Stereocomplexes of Aliphatic Polyesters 126
4.3.3	Controlled Synthesis of Aliphatic Polycarbonates by Anionic and Coordination Ring-opening Polymerization 127
4.3.4	Controlled Synthesis of Branched and Star-shaped Polyoxiranes and Polyesters 129
4.3.4.1	Anionic Polymerization of Oxiranes 129
4.3.4.2	Coordination Polymerization of Cyclic Esters 131
4.3.5	Controlled Synthesis of Polyamides by Anionic and Coordination Ring-opening Polymerization 132
4.3.5.1	Polymerization of Lactams 132
4.3.5.2	Polymerization of N-Carboxyanhydrides of α-Amino Acids (NCAs) 134
4.3.6	Cationic Ring-opening Polymerization 136
4.3.6.1	Propagation in Cationic Ring-opening Polymerization 137
4.3.6.2	Chain Transfer to Polymer in Cationic Ring-opening Polymerization 138
4.3.6.3	Activated Monomer Mechanism in Cationic Ring-opening Polymerization of Cyclic Ethers and Esters 140
4.3.6.4	Branched and Star-shaped Polymers Prepared by Cationic Ring-opening Polymerization 144
4.3.7	Cationic Polymerization of Cyclic Imino Ethers (Oxazolines) 145
4.4	Dispersion Ring-opening Polymerization 146
4.5	Conclusion 149
	References 150

5	**Radical Polymerization** 161
	Krzysztof Matyjaszewski and Wade A. Braunecker
5.1	Introduction 161
5.2	Typical Features of Radical Polymerization 162
5.2.1	Fundamentals of Organic Radicals 162
5.2.2	Elementary Reactions and Kinetics 163
5.2.3	Copolymerization 165
5.2.4	Monomers 166
5.2.5	Initiators for RP 166
5.2.6	Additives 168
5.2.7	Typical Conditions for RP 168

5.2.8	Commercially Important Polymers by RP	*169*
5.2.8.1	Polyethylene	*169*
5.2.8.2	Polystyrene	*169*
5.2.8.3	Poly(vinyl chloride) (PVC)	*170*
5.2.8.4	Poly(meth)acrylates	*170*
5.2.8.5	Other polymers	*170*
5.3	Controlled/Living Radical Polymerization	*171*
5.3.1	General Concepts	*171*
5.3.2	Similarities and Differences Between RP and CRP	*172*
5.4	SFRP and NMP Systems – Examples and Peculiarities	*173*
5.4.1	Monomers and Initiators	*174*
5.4.2	General Conditions	*175*
5.4.3	Controlled Architectures	*175*
5.4.4	Other SFRP Systems	*175*
5.5	ATRP – Examples and Peculiarities	*176*
5.5.1	Basic ATRP Components	*177*
5.5.1.1	Monomers	*177*
5.5.1.2	Initiators	*178*
5.5.1.3	Transition Metal Complexes as ATRP Catalysts	*179*
5.5.2	Conditions	*180*
5.5.3	Controlled Architectures	*182*
5.6	Degenerative Transfer Processes and RAFT	*182*
5.6.1	Monomers and Initiators	*184*
5.6.2	Transfer Agents	*185*
5.6.3	Controlled Architectures	*185*
5.7	Relative Advantages and Limitations of SFRP, ATRP and DT Processes	*185*
5.7.1	SFRP	*186*
5.7.2	ATRP	*186*
5.7.3	RAFT and Other DT Processes	*186*
5.8	Controlled Polymer Architectures by CRP: Topology	*187*
5.8.1	Linear Chains	*188*
5.8.2	Star-like Polymers	*188*
5.8.3	Comb-like Polymers	*189*
5.8.4	Branched and Hyperbranched Polymers	*191*
5.8.5	Dendritic Structures	*192*
5.8.6	Polymer Networks and Microgels	*192*
5.8.7	Cyclic Polymers	*192*
5.9	Chain Composition	*193*
5.9.1	Statistical Copolymers	*193*
5.9.2	Segmented Copolymers (Block, Grafts and Multisegmented Copolymers)	*193*
5.9.2.1	Block Copolymers by a Single CRP Method	*193*
5.9.2.2	Block Copolymers by Combination of CRP Methods	*194*

5.9.2.3	Block Copolymerization by Site Transformation and Dual Initiators	195
5.9.2.4	Multisegmented Block Copolymers	196
5.9.2.5	Stereoblock Copolymers	197
5.9.3	Graft Copolymers	197
5.9.4	Periodic Copolymers	199
5.9.5	Gradient Copolymers	199
5.9.6	Molecular Hybrids	199
5.9.7	Templated Systems	200
5.10	Functional Polymers	200
5.10.1	Polymers with Side Functional Groups	201
5.10.2	End-group Functionality: Initiators	202
5.10.3	End-group Functionality Through Conversion of Dormant Chain End	202
5.11	Applications of Materials Prepared by CRP	203
5.11.1	Polymers with Controlled Compositions	204
5.11.2	Polymers with Controlled Topology	204
5.11.3	Polymers with Controlled Functionality	204
5.11.4	Hybrids	205
5.12	Outlook	205
5.12.1	Mechanisms	205
5.12.2	Molecular Architecture	206
5.12.3	Characterization	207
5.12.4	Structure–Property Relationship	207
	Acknowledgments	207
	References	208
6	**Coordination Polymerization: Synthesis of New Homo- and Copolymer Architectures from Ethylene and Propylene using Homogeneous Ziegler–Natta Polymerization Catalysts**	**217**
	Andrew F. Mason and Geoffrey W. Coates	
6.1	Introduction, Historical Perspective and Scope of Review	217
6.2	Primer on the Homogeneous Coordination Polymerization of Olefins	218
6.2.1	Nature of the Active Species and Mechanism of Initiation	218
6.2.2	Mechanism of Propagation	219
6.2.3	Mechanisms of Termination and Chain Transfer	220
6.3	Ethylene-based Polymers	222
6.4	Propylene-based Polymers	224
6.4.1	Atactic, Isotactic and Syndiotactic Polypropylene	224
6.4.2	Hemiisotactic Polypropylene	226
6.4.3	Stereoblock Polypropylene	226
6.4.4	Graft and Star Polypropylene	233
6.5	Ethylene–Propylene Copolymers	235
6.5.1	Random Ethylene–Propylene Copolymers	235

6.5.2	Alternating Ethylene–Propylene Copolymers	235
6.5.3	Ethylene–Propylene Block Copolymers	238
6.5.4	Ethylene–Propylene Graft Copolymers	241
6.6	Summary and Outlook	242
	References	243

7 Recent Trends in Macromolecular Engineering 249
Damien Quémener, Valérie Héroguez, and Yves Gnanou

7.1	Introduction	249
7.2	The March Towards Well-defined/Selective Catalysts for ROMP	250
7.2.1	Discovery of Olefin Metathesis and its Mechanism	250
7.2.2	Development of Well-defined ROMP Initiators	252
7.3	Macromolecular Engineering Using ROMP	257
7.3.1	Block Copolymers by ROMP and Combination of ROMP with Other "Living" Polymerizations	258
7.3.2	Graft Copolymers by ROMP with Other "Living" Polymerizations	262
7.3.2.1	ROMP/ROMP	262
7.3.2.2	Anionic Polymerization/ROMP	263
7.3.2.3	Cationic Polymerization/ROMP	265
7.3.2.4	ATRP/ROMP	265
7.3.2.5	ROP/ROMP	266
7.3.2.6	ROMP/ROP/ATRP	267
7.3.3	Star Polymers by ROMP	268
7.4	ROMP in Dispersed Medium	273
7.4.1	Emulsion ROMP	274
7.4.2	Dispersion ROMP	276
7.4.3	Suspension ROMP	277
7.4.4	Miniemulsion ROMP	278
7.5	Advanced Materials by ROMP	280
7.5.1	Liquid Crystalline Polymers	280
7.5.2	Conjugated and Electroactive Polymers	281
7.5.3	Monolithic Supports	283
7.5.4	Supported Catalysts	284
7.5.5	Biological Materials	286
7.5.6	Hybrid Materials/Particles	290
7.6	Conclusion	290
	References	290

8 Polycondensation 295
Tsutomu Yokozawa

8.1	Monomer Reactivity Control (Stoichiometric-imbalanced Polycondensation)	295
8.1.1	Polycondensation of a,a-Dihalogenated Monomers	296
8.1.2	Pd-catalyzed Polycondensation	298

8.1.3	Crystallization Polycondensation	299
8.1.4	Nucleation–Elongation Polycondensation	302
8.2	Sequence Control	303
8.2.1	Sequential Polymers from Symmetrical and Unsymmetrical Monomers	304
8.2.2	Sequential Polymers from Two Unsymmetrical Monomers	308
8.2.3	Sequential Polymers from Two Symmetrical Monomers and One Unsymmetrical Monomer	309
8.2.4	Sequential Polymers from Two Symmetrical Monomers and Two Unsymmetrical Monomers	310
8.3	Molecular Weight and Polydispersity Control	310
8.3.1	Transfer of Reactive Species	311
8.3.2	Different Substituent Effects Between Monomer and Polymer	316
8.3.2.1	Resonance Effect (Polymerization of para-Substituted Monomers)	316
8.3.2.2	Inductive Effect (meta-Substituted Monomers)	321
8.3.3	Transfer of Catalyst	322
8.4	Chain Topology and Polymer Morphology Control	324
8.4.1	Cyclic Polymers	324
8.4.2	Hyperbranched Polymers	326
8.4.2.1	Polyphenylene	326
8.4.2.2	Polyester	326
8.4.2.3	Polyamide	328
8.4.2.4	Polyether	329
8.4.2.5	Poly(Ether Ketone) and Poly(Ether Sulfone)	330
8.4.2.6	Poly(Ether Imide)	331
8.4.2.7	Polyurethane and Polyurea	332
8.4.3	Polymer Morphology Control	332
8.5	Condensation Polymer Architecture	334
8.5.1	Block Copolymers	334
8.5.1.1	Block Copolymers of Condensation Polymers	334
8.5.1.2	Block Copolymers of Condensation Polymers and Coil Polymers	337
8.5.2	Star Polymers	343
8.5.3	Graft Polymers	344
	References	345
9	**Supramolecular Polymer Engineering**	**351**
	G. B. W. L. Ligthart, Oren A. Scherman, Rint P. Sijbesma, and E. W. Meijer	
9.1	Introduction	351
9.2	General Aspects of Supramolecular Polymers	352
9.3	Non-covalent Interactions	356
9.3.1	Hydrogen Bonds	356
9.3.2	Solvophobic and Coulombic Interactions	360
9.4	Supramolecular Polymers	361

9.4.1	Small Building Blocks	*361*
9.4.1.1	Supramolecular Polymers Based on Liquid Crystalline Monomers	*362*
9.4.1.2	Supramolecular Polymers in Isotropic Solution	*363*
9.4.2	Large Building Blocks	*373*
9.4.2.1	Main-chain Supramolecular Polymers	*374*
9.4.2.2	Supramolecular Block Copolymers	*378*
9.4.2.3	Side-chain Supramolecular Polymers	*380*
9.4.2.4	Applications Based on Supramolecular UPy Materials	*385*
9.5	Ring–Chain Equilibria in Supramolecular Polymers	*387*
9.6	Conclusions and Outlook	*392*
	References	*393*

10 **Polymer Synthesis and Modification by Enzymatic Catalysis** *401*
Shiro Kobayashi and Masashi Ohmae

10.1	Introduction	*401*
10.2	Characteristics of Enzymatic Catalysis	*402*
10.3	Synthesis of Poly(aromatic)s Catalyzed by Oxidoreductases	*404*
10.3.1	Synthesis of Polymers from Phenolic Compounds	*405*
10.3.1.1	Polymers from Unsubstituted Phenol	*406*
10.3.1.2	Polymers from Substituted Phenols	*408*
10.3.1.3	Polymerization of Phenols Catalyzed by Enzyme Model Complexes	*413*
10.3.2	Synthesis of Polymers from Polyphenols	*414*
10.3.2.1	Polymers from Catechol Derivatives	*414*
10.3.2.2	Polymers from Flavonoids	*417*
10.3.3	Synthesis of Polyaniline and Its Derivatives	*418*
10.4	Synthesis of Vinyl Polymers Catalyzed by Oxidoreductases	*420*
10.5	Synthesis of Polysaccharides Catalyzed by Hydrolases	*422*
10.5.1	Synthesis of Polysaccharides via Polycondensation	*422*
10.5.1.1	Cellulose and Its Derivatives	*422*
10.5.1.2	Xylan	*427*
10.5.1.3	Amylose Oligomers	*427*
10.5.1.4	Hybrid Polysaccharides	*428*
10.5.1.5	Oligo- and Polysaccharide Synthesis by Mutated Enzymes	*429*
10.5.2	Synthesis of Polysaccharides via Ring-opening Polyaddition	*430*
10.5.2.1	Chitin and its Derivatives	*431*
10.5.2.2	Glycosaminoglycans	*435*
10.5.2.3	Unnatural Hybrid Polysaccharides	*437*
10.6	Synthesis of Polyesters Catalyzed by Hydrolases, Mainly by Lipases	*439*
10.6.1	Polyesters via Ring-opening Polymerization	*439*
10.6.1.1	Ring-opening Polymerization of Lactones	*439*
10.6.1.2	Ring-opening Polymerization of Other Cyclic Monomers	*452*
10.6.2	Polyesters via Polycondensation	*454*

10.6.2.1	Polycondensation of Dicarboxylic Acids and Their Derivatives with Glycols 454	
10.6.2.2	Polycondensation of Oxyacid Derivatives 458	
10.6.2.3	Synthesis of Functional Polyesters 459	
10.7	Modification of Polymers by Enzymatic Catalysis 461	
10.7.1	Modification of Polysaccharides 462	
10.7.2	Modification of Other Polymers 465	
10.8	Conclusion 466	
	References 467	
11	**Biosynthesis of Protein-based Polymeric Materials** 479	
	Robin S. Farmer, Manoj B. Charati, and Kristi L. Kiick	
11.1	Protein Polymers That Mimic Natural Proteins 481	
11.1.1	Silk 481	
11.1.2	Elastin 484	
11.1.2.1	Elastin-like Polypeptide Copolymers 486	
11.1.2.2	Silk–Elastin-like Polypeptides 487	
11.1.2.3	Applications of ELPs 488	
11.1.3	Collagen 489	
11.1.4	Other Naturally Occurring Proteins 491	
11.1.4.1	Resilin 491	
11.1.4.2	Mussel Adhesive Plaque Protein and Glutenin 491	
11.2	Protein Polymers of *De Novo* Design 492	
11.2.1	β-Sheet-forming Protein Polymers 492	
11.2.2	Liquid Crystals 493	
11.2.3	Coiled Coils 494	
11.2.4	Helical Protein Polymers 495	
11.2.5	Non-structured Protein Polymers 495	
11.3	Proteins Containing Non-natural Amino Acids 496	
11.3.1	Synthetic Methodologies 496	
11.3.1.1	Chemical Synthesis 496	
11.3.1.2	*In Vitro* Suppression Strategies 498	
11.3.1.3	*In Vivo* Suppression Strategies 498	
11.3.2	Multisite Incorporation of Non-natural Amino Acids into Protein Polymers *In Vivo* 499	
11.3.3	Types of Chemically Novel Amino Acids Incorporated into Protein Polymers 503	
11.3.3.1	Halide-functionalized Side-chains 503	
11.3.3.2	Azide-functionalized Side-chains 505	
11.3.3.3	Ketone-functionalized Side-chains 507	
11.3.3.4	Alkyne- and Alkene-functionalized Side-chains 507	
11.3.3.5	Photoreactive Side-chains 508	
11.3.3.6	Unsaturated and Structural Amino Acid Analogues 508	
11.4	Prospects for Protein-based Polymers 510	
	References 512	

12 Macromolecular Engineering of Polypeptides Using the Ring-opening Polymerization-Amino Acid N-Carboxyanhydrides 519
Harm-Anton Klok and Timothy J. Deming

12.1 Introduction 519
12.2 Polymerization of α-Amino Acid N-Carboxyanhydrides 520
12.2.1 Conventional Methods 520
12.2.2 Transition Metal-mediated NCA Polymerization 522
12.2.3 Other NCA Polymerization Methods 522
12.3 Block Copolymers 525
12.3.1 Block Copolypeptides 525
12.3.1.1 Conventional NCA Polymerization 525
12.3.1.2 Controlled NCA Polymerization 525
12.3.2 Hybrid Block Copolymers 526
12.3.2.1 Conventional NCA Polymerization 526
12.3.2.2 Controlled NCA Polymerizations 528
12.4 Star Polypeptides 531
12.4.1 Conventional NCA Polymerization 531
12.4.2 Controlled NCA Polymerizations 533
12.5 Graft and Hyperbranched Polypeptides 533
12.6 Summary and Conclusions 537
References 539

13 Segmented Copolymers by Mechanistic Transformations 541
M. Atilla Tasdelen and Yusuf Yagci

13.1 Introduction 541
13.2 Direct Transformations 545
13.3 Indirect Transformation 546
13.3.1 Transformations Involving Condensation Polymerization 546
13.3.1.1 Condensation Polymerization to Conventional Radical Polymerization 547
13.3.1.2 Condensation Polymerization to Controlled Radical Polymerization 549
13.3.1.3 Macrocyclic Polymerization to Condensation Polymerization 551
13.3.1.4 Condensation Polymerization to Anionic Coordination-Insertion Polymerization 552
13.3.1.5 Transformations Involving Suzuki and Yamamoto Polycondensations 553
13.3.2 Transformation of Anionic Polymerization to Radical Transformation 555
13.3.2.1 Anionic Polymerization to Conventional Radical Polymerization 555
13.3.2.2 Anionic Polymerization to Controlled Radical Polymerization 556
13.3.3 Transformation of Cationic Polymerization to Radical Polymerization 563
13.3.3.1 Cationic Polymerization to Conventional Radical Transformation 563

13.3.3.2	Cationic Polymerization to Controlled Radical Transformation	568
13.3.4	Transformation of Radical Polymerization to Anionic Polymerization	572
13.3.4.1	Conventional Radical Polymerization to Anionic Polymerization	572
13.3.4.2	Controlled Radical Polymerization to Anionic Polymerization	573
13.3.5	Transformation of Radical Polymerization to Cationic Polymerization	574
13.3.6	Transformations Involving Anionic and Cationic Polymerizations	578
13.3.7	Transformations Involving Activated Monomer Polymerization	583
13.3.8	Transformations Involving Metathesis Polymerization	585
13.3.9	Transformations Involving ZieglerNatta Polymerization	586
13.3.10	Transformations Involving Group Transfer Polymerization	589
13.4	Coupling Reactions and Concurrent Polymerizations	590
13.5	Conclusions	593
	References	594

14	**Polymerizations in Aqueous Dispersed Media**	**605**
	Bernadette Charleux and François Ganachaud	
14.1	Introduction	605
14.2	Aqueous Suspension Polymerization	606
14.2.1	Conventional Free Radical Polymerization	606
14.2.2	Controlled/Living Free Radical Polymerization	607
14.2.2.1	Nitroxide-mediated Controlled Free Radical Polymerization (NMP)	607
14.2.2.2	Atom-transfer Radical Polymerization (ATRP)	607
14.2.2.3	Other Controlled Free Radical Polymerization Methods	608
14.2.3	Ring-opening Metathesis Polymerization (ROMP)	608
14.2.4	Ionic Polymerizations	608
14.2.4.1	Oil-in-Water Processes	608
14.2.4.2	Water-in-Oil Processes	610
14.2.5	Polycondensation/Polyaddition	610
14.2.5.1	Phase-transfer Catalysis	610
14.2.5.2	Preparation of Microcapsules	612
14.3	Aqueous Miniemulsion Polymerization	613
14.3.1	Conventional Free Radical Polymerization	613
14.3.2	Controlled/Living Free Radical Polymerization	615
14.3.2.1	Nitroxide-mediated Controlled Free Radical Polymerization	615
14.3.2.2	Atom-transfer Radical Polymerization	617
14.3.2.3	Control via Reversible Chain Transfer	619
14.3.2.4	Other Controlled Free Radical Polymerization Methods	620
14.3.3	Ring-opening Metathesis Polymerization	620
14.3.4	Catalytic Polymerization of Ethylene and Butadiene	621
14.3.5	Ionic Polymerization	622
14.3.5.1	Irreversible Deactivation of the Chain Ends by Water	622

14.3.5.2	Reversible Deactivation of the Chain Ends by Water	623
14.3.6	Polycondensation/Polyaddition Reactions	625
14.4	Aqueous Emulsion Polymerization	626
14.4.1	Conventional Free Radical Polymerization	626
14.4.2	Controlled/Living Free Radical Polymerization	629
14.4.2.1	Nitroxide-mediated Controlled Free Radical Polymerization	629
14.4.2.2	Atom-transfer Radical Polymerization	630
14.4.2.3	Control via Reversible Addition–Fragmentation Chain Transfer	630
14.4.3	Ring-opening Metathesis Polymerization	632
14.4.4	Catalytic Polymerization of Ethylene	632
14.4.5	Ionic Polymerization	632
14.5	Polymerization in Surfactant Templates	633
14.5.1	Microemulsion Polymerization	633
14.5.1.1	Conventional Free Radical Polymerization	633
14.5.1.2	Controlled Free Radical Polymerization	634
14.5.1.3	Ionic Polymerization	635
14.5.1.4	Polycondensation	635
14.5.2	Polymerization in Vesicles	635
14.6	Conclusions	636
	References	636
15	**Polymerization Under Light and Other External Stimuli**	**643**
	Jean Pierre Fouassier, Xavier Allonas, and Jacques Lalevée	
15.1	Introduction	643
15.2	Background	643
15.2.1	Photopolymerization Reactions	643
15.2.2	Light Sources	644
15.2.3	Absorption of Light	645
15.2.4	Initiation Step	645
15.3	Photoinitiators and Photosensitizers	645
15.3.1	Photoinitiators	646
15.3.1.1	Direct Production of Reactive Species	646
15.3.1.2	Radical Photoinitiators	646
15.3.1.3	Cationic Photoinitiators	646
15.3.1.4	Anionic Photoinitiators	647
15.3.1.5	Photoacid and Photobase Generators	647
15.3.1.6	Absorption Spectra	647
15.3.1.7	Excited State Reactivity	648
15.3.2	Photosensitizers	648
15.3.2.1	Processes	648
15.3.2.2	Examples	648
15.3.3	Properties of Photoinitiators and Photosensitizers	649
15.4	Monomers and Oligomers	649
15.4.1	Various Systems	650
15.4.1.1	Radical Monomers and Oligomers	650

15.4.1.2	Cationic Monomers and Oligomers	650
15.4.2	Current Developments	650
15.4.3	General Properties	651
15.5	Brief Overview of Applications in UV Curing	652
15.6	Photochemical/Chemical Reactivity and Final Properties	653
15.6.1	Different Aspects of Photopolymerization Reactions	653
15.6.1.1	Examples	653
15.6.1.2	Functional Properties	655
15.6.1.3	Some Typical Reactions of Industrial Interest in the UV Curing Area	657
15.6.2	Kinetics and Efficiency of the Photopolymerization Reaction	661
15.6.2.1	Overall Processes	661
15.6.2.2	Monitoring of the Photopolymerization Reaction	661
15.6.2.3	Kinetics of Photopolymerization	662
15.6.2.4	Photochemical and Chemical Reactivity	663
15.7	Electron Beam, Microwave, Gamma Rays, Plasma and Pressure Stimuli Compared with Temperature and Light	664
15.8	Conclusion	666
	References	667

16 Inorganic Polymers with Precise Structures 673
David A. Rider and Ian Manners

16.1	Metal-containing Polymers	673
16.1.1	Introduction	673
16.1.2	Chain Growth Polymerizations	674
16.1.3	Living Polymerizations and Controlled Polymerizations	675
16.2	Use of Nitroxide-mediated Radical Polymerization	676
16.2.1	Introduction	676
16.2.2	Nitroxide-mediated Radical Polymerization Routes to Ligand Functional Homopolymers	677
16.2.3	Nitroxide-mediated Radical Polymerization Routes to Ligand Functional Block Copolymers	678
16.2.4	Nitroxide-mediated Radical Polymerization of Metallomonomers	680
16.2.5	Nitroxide-mediated Radical Polymerization of Metallomonomers to Prepare Block Copolymers	680
16.3	Use of Atom-transfer Radical Polymerization (ATRP)	681
16.3.1	Introduction	681
16.3.2	Atom-transfer Radical Polymerization to Ligand Functional Homopolymers	682
16.3.3	Atom-transfer Radical Polymerization Routes to Ligand Functional Block Copolymers	682
16.3.4	Atom-transfer Radical Polymerization of Metallomonomers	684
16.3.5	Use of Atom-transfer Radical Polymerization of Metallomonomers to Prepare Block Copolymers	685

16.4	Use of Reversible Addition Fragmentation Termination (RAFT) Polymerization *686*	
16.4.1	Introduction *686*	
16.4.2	Reversible Addition Fragmentation Termination Polymerization to Prepare Ligand Functional Block Copolymers *687*	
16.5	Use of Living Cationic Polymerization *688*	
16.6	Use of Living Anionic Polymerization *690*	
16.6.1	Introduction *690*	
16.6.2	Living Anionic Polymerization Routes to Ligand Functional Homopolymers *691*	
16.6.3	Living Anionic Polymerization Routes to Ligand Functional Block Copolymers *691*	
16.6.4	Living Anionic Polymerization of Metallomonomers *692*	
16.6.5	Living Anionic Polymerization of Metallomonomers to Prepare Block Copolymers *694*	
16.7	Use of Metathesis Polymerization *696*	
16.7.1	Introduction *696*	
16.7.2	Metathesis Polymerization Routes to Ligand Functional Block Copolymers *697*	
16.7.3	Metathesis Polymerization of Metallomonomers *697*	
16.7.4	Use of Metathesis Polymerization of Metallomonomers to Prepare Block Copolymers *702*	
16.8	Indirect Sequential Polymerization Routes to Metal-containing Block Copolymers *704*	
16.9	Routes to Metallo-linked Block Copolymers *707*	
16.10	Routes to Metal-centered Star- and Star-block Copolymers *709*	
16.11	Applications of Metal-containing Polymers with Precise Structures *712*	
16.12	Polymers Based on Main Group Elements *717*	
16.12.1	Introduction *717*	
16.12.2	Polysiloxanes *718*	
16.12.3	Polysilanes *720*	
16.12.4	Polyphosphazenes *724*	
16.13	Conclusions and Outlook *726*	
	References *727*	

Volume 2
Elements of Macromolecular Structural Control

1. **Tacticity** *731*
   Tatsuki Kitayama

2. **Synthesis of Macromonomers and Telechelic Oligomers by Living Polymerizations** *775*
   Bernard Boutevin, Cyrille Boyer, Ghislain David, and Pierre Lutz

3. **Statistical, Alternating and Gradient Copolymers** *813*
   Bert Klumperman

4. **Multisegmental Block/Graft Copolymers** *839*
   Constantinos Tsitsilianis

5. **Controlled Synthesis and Properties of Cyclic Polymers** *875*
   Alain Deffieux and Redouane Borsali

6. **Polymers with Star-related Structures** *909*
   Nikos Hadjichristidis, Marinos Pitsikalis, and Hermis Iatrou

7. **Linear Versus (Hyper)branched Polymers** *973*
   Hideharu Mori, Axel H. E. Müller, and Peter F. W. Simon

8. **From Stars to Microgels** *1007*
   Daniel Taton

9. **Molecular Design and Self-assembly of Functional Dendrimers** *1057*
   Wei-Shi Li, Woo-Dong Jang, and Takuzo Aida

10. **Molecular Brushes – Densely Grafted Copolymers** *1103*
    Brent S. Sumerlin and Krzysztof Matyjaszewski

11. **Grafting and Polymer Brushes on Solid Surfaces** *1137*
    Takeshi Fukuda, Yoshinobu Tsujii, and Kohji Ohno

12. **Hybrid Organic Inorganic Objects** *1179*
    Stefanie M. Gravano and Timothy E. Patten

13. **Core–Shell Particles** *1209*
    Anna Musyanovych and Katharina Landfester

14	Polyelectrolyte Multilayer Films – A General Approach to (Bio)functional Coatings *1249* *Nadia Benkirane-Jessel, Philippe Lavalle, Vincent Ball, Joëlle Ogier, Bernard Senger, Catherine Picart, Pierre Schaaf, Jean-Claude Voegel, and Gero Decher*
15	Bio-inspired Complex Block Copolymers/Polymer Conjugates and Their Assembly *1307* *Markus Antonietti, Hans G. Börner, and Helmut Schlaad*
16	Complex Functional Macromolecules *1341* *Zhiyun Chen, Chong Cheng, David S. Germack, Padma Gopalan, Brooke A. van Horn, Shrinivas Venkataraman, and Karen L. Wooley*

Volume 3
Structure-Property Correlation and Characterization Techniques

1	Self-assembly and Morphology Diagrams for Solution and Bulk Materials: Experimental Aspects *1387* *Vahik Krikorian, Youngjong Kang, and Edwin L. Thomas*
2	Simulations *1431* *Denis Andrienko and Kurt Kremer*
3	Transport and Electro-optical Properties in Polymeric Self-assembled Systems *1471* *Olli Ikkala and Gerrit ten Brinke*
4	Atomic Force Microscopy of Polymers: Imaging, Probing and Lithography *1515* *Sergei S. Sheiko and Martin Moller*
5	Scattering from Polymer Systems *1575* *Megan L. Ruegg and Nitash P. Balsara*
6	From Linear to (Hyper) Branched Polymers: Dynamics and Rheology *1605* *Thomas C. B. McLeish*
7	Determination of Bulk and Solution Morphologies by Transmission Electron Microscopy *1649* *Volker Abetz, Richard J. Spontak, and Yeshayahu Talmon*

Contents

8 **Polymer Networks** *1687*
   Karel Dušek and Miroslava Dušková-Smrčková

9 **Block Copolymers for Adhesive Applications** *1731*
   Costantino Creton

10 **Reactive Blending** *1753*
    Robert Jerome

11 **Predicting Mechanical Performance of Polymers** *1783*
    Han E. H. Meijer, Leon E. Govaert, and Tom A. P. Engels

12 **Scanning Calorimetry**
    René Androsch and Bernhard Wunderlich

13 **Chromatography of Polymers** *1881*
    Wolfgang Radke

14 **NMR Spectroscopy** *1937*
    Hans Wolfgang Spiess

15 **High-throughput Screening in Combinatorial Polymer Research** *1967*
    Michael A. R. Meier, Richard Hoogenboom, and Ulrich S. Schubert

Volume 4
Applications

1 **Applications of Thermoplastic Elastomers Based on Styrenic Block Copolymers** *2001*
   Dale L. Handlin, Jr., Scott Trenor, and Kathryn Wright

2 **Nanocomposites** *2033*
   Michaël Alexandre and Philippe Dubois

3 **Polymer/Layered Filler Nanocomposites: An Overview from Science to Technology** *2071*
   Masami Okamoto

4 **Polymeric Dispersants** *2135*
   Frank Pirrung and Clemens Auschra

5 **Polymeric Surfactants** *2181*
   Henri Cramail, Eric Cloutet, and Karunakaran Radhakrishnan

6	**Molecular and Supramolecular Conjugated Polymers for Electronic Applications** *2225* Andrew C. Grimsdale and Klaus Müllen	
7	**Polymers for Microelectronics** *2263* Christopher W. Bielawski and C. Grant Willson	
8	**Applications of Controlled Macromolecular Architectures to Lithography** *2295* Daniel Bratton, Ramakrishnan Ayothi, Nelson Felix, and Christopher K. Ober	
9	**Microelectronic Materials with Hierarchical Organization** *2331* G. Dubois, R. D. Miller and James L. Hedrick	
10	**Semiconducting Polymers and their Optoelectronic Applications** *2369* Nicolas Leclerc, Thomas Heiser, Cyril Brochon, and Georges Hadziioannou	
11	**Polymer Encapsulation of Metallic and Semiconductor Nanoparticles: Multifunctional Materials with Novel Optical, Electronic and Magnetic Properties** *2409* Jeffrey Pyun and Todd Emrick	
12	**Polymeric Membranes for Gas Separation, Water Purification and Fuel Cell Technology** *2451* Kazukiyo Nagai, Young Moo Lee, and Toshio Masuda	
13	**Utilization of Polymers in Sensor Devices** *2493* Basudam Adhikari and Alok Kumar Sen	
14	**Polymeric Drugs** *2541* Tamara Minko, Jayant J. Khandare, and Sreeja Jayant	
15	**From Biomineralization Polymers to Double Hydrophilic Block and Graft Copolymers** *2597* Helmut Cölfen	
16	**Applications of Polymer Bioconjugates** *2645* Joost A. Opsteen and Jan C. M. van Hest	
17	**Gel: a Potential Material as Artificial Soft Tissue** *2689* Yong Mei Chen, Jian Ping Gong, and Yoshihito Osada	

18   **Polymers in Tissue Engineering** 2719
      Jeffrey A. Hubbell

      **IUPAC Polymer Terminology
      and Macromolecular Nomenclature** 2743
      R. F. T. Stepto

      Subject Index 2747

Preface

Macromolecular Engineering: From Precise Macromolecular Synthesis to Macroscopic Materials Properties and Applications aims to provide a broad overview of recent developments in precision macromolecular synthesis and in the design and applications of complex polymeric assemblies of controlled sizes, morphologies and properties. The contents of this interdisciplinary book are organized in four volumes so as to capture and chronicle best, on the one hand, the rapid advances made in the control of polymerization processes and the design of macromolecular architectures (Volumes I and II) and, on the other, the noteworthy progress witnessed in the processing methods – including self-assembly and formulation – to generate new practical applications (Volumes III and IV).

Each chapter in this book is a well-documented and yet concise contribution written by noted experts and authorities in their field. We are extremely grateful to all of them for taking time to share their knowledge and popularize it in a way understandable to a broad readership. We are also indebted to all the reviewers whose comments and remarks helped us very much in our editing work. Finally, Wiley-VCH deserves our sincere acknowledgements for striving to keep the entire project on time.

We expect that specialist readers will find *Macromolecular Engineering: From Precise Macromolecular Synthesis to Macroscopic Materials Properties and Applications* an indispensable book to update their knowledge, and non-specialists will use it as a valuable companion to stay informed about the newest trends in polymer and materials science.

November 2006
Pittsburgh, USA   Krzysztof Matyjaszewski
Bordeaux, France   Yves Gnanou
Paris, France   Ludwik Leibler

Macromolecular Engineering. Precise Synthesis, Materials Properties, Applications.
Edited by K. Matyjaszewski, Y. Gnanou, and L. Leibler
Copyright © 2007 WILEY-VCH Verlag GmbH & Co. KGaA, Weinheim
ISBN: 978-3-527-31446-1

List of Contributors

Volker Abetz
GKSS Research Centre Geesthacht GmbH
Institute of Polymer Research
Max-Planck-Straße 1
21502 Geesthacht
Germany

Basudam Adhikari
Indian Institute of Technology
Materials Science Centre
Polymer Division
Kharagpur 721302
India

Takuzo Aida
University of Tokyo
School of Engineering
Department of Chemistry
and Biotechnology
7-3-1 Hongo, Bunkyo-ku
Tokyo 113-8656
Japan

Michaël Alexandre
Materia Nova Research Centre asbl
Parc Initialis
1 avenue Nicolas Copernic
7000 Mons
Belgium

Xavier Allonas
University of Haute Alsace, ENSCMu
Department of General
Photochemistry, UMR 7525 CNRS
3 Alfred Werner street
68093 Mulhouse Cedex
France

Denis Andrienko
Max Planck Institute for Polymer Research
Ackermannweg 10
55128 Mainz
Germany

René Androsch
Martin Luther University
Halle-Wittenberg
Institute of Materials Science
06099 Halle
Germany

Markus Antonietti
Max Planck Institute of Colloids
and Interfaces
Colloid Department
Research Campus Golm
14424 Potsdam
Germany

Clemens Auschra
CIBA Specialty Chemicals, Inc.
Research and Development
Coating Effects
Schwarzwaldallee 215
4002 Basel
Switzerland

Ramakrishnan Ayothi
Cornell University
Department of Materials Science
and Engineering
Bard Hall
Ithaca, NY 14853-1501
USA

Vincent Ball
Institut National de la Santé
et de la Recherche Médicale
INSERM Unité 595
11 rue Humann
67085 Strasbourg Cedex
France
and
Université Louis Pasteur
Faculté de Chirurgie Dentaire
1 place de l'Hôpital
67000 Strasbourg
France

Nitash P. Balsara
University of California
Department of Chemical Engineering
and Materials Sciences
and Environmental Energy
Technologies Divisions
Lawrence Berkeley National
Laboratory
Berkeley, CA 94720
USA

Nadia Benkirane-Jessel
Institut National de la Santé
et de la Recherche Médicale
INSERM Unité 595
11 rue Humann
67085 Strasbourg Cedex
France
and
Université Louis Pasteur
Faculté de Chirurgie Dentaire
1 place de l'Hôpital
67000 Strasbourg
France

Christopher W. Bielawski
The University of Texas at Austin
Department of Chemistry
and Biochemistry
Austin, TX 78712
USA

Hans G. Börner
Max Planck Institute of Colloids
and Interfaces
Colloid Department
Research Campus Golm
14424 Potsdam
Germany

Redouane Borsali
Université Bordeaux 1
CNRS, ENSCPB
Laboratoire de Chimie des Polymères
Organiques
16 avenue Pey-Berland
33607 Pessac
France

Bernard Boutevin
Ingénierie et Architectures
Macromoléculaires
Institut Gerhardt, UMR 5253
Ecole Nationale Supérieure Chimie
de Montpellier
8 rue de l'Ecole Normale
34296 Montpellier
France

Cyrille Boyer
Ingénierie et Architectures
Macromoléculaires
Institut Gerhardt, UMR 5253
Ecole Nationale Supérieure Chimie
de Montpellier
8 rue de l'Ecole Normale
34296 Montpellier
France

Daniel Bratton
Cornell University
Department of Materials Science
and Engineering
Bard Hall
Ithaca, NY 14853-1501
USA

Wade A. Braunecker
Carnegie Mellon University
Department of Chemistry
4400 Fifth Avenue
Pittsburgh, PA 15213
USA

Cyril Brochon
Laboratoire d'Ingénierie des
Polymères pour les Hautes
Technologies
UMR 7165 CNRS
Ecole Européenne Chimie Polymères
Matériaux
Université Louis Pasteur
25 rue Becquerel
67087 Strasbourg
France

Manoj B. Charati
University of Delaware
Department of Materials Science
and Engineering
201 DuPont Hall
and
Delaware Biotechnology Institute
15 Innovation Way
Newark, DE 19716
USA

Bernadette Charleux
Université Pierre et Marie Curie
Laboratoire de Chimie des Polymères
4, Place Jussieu, Tour 44, 1er étage
75252 Paris Cedex 5
France

Yong Mei Chen
Hokkaido University
Section of Biological Sciences
Faculty of Science
Laboratory of soft & wet matter
North 10, West 8
060-0810 Sapporo
Japan

Zhiyun Chen
Washington University in Saint Louis
Center for Materials Innovation and
Department of Chemistry
One Brookings Drive
St. Louis, MO 63130-4899
USA

Chong Cheng
Washington University in Saint Louis
Center for Materials Innovation and
Department of Chemistry
One Brookings Drive
St. Louis, MO 63130-4899
USA

Eric Cloutet
Université Bordeaux 1
Laboratoire de Chimie des Polymères
Organiques
Unité Mixte de Recherche
(UMR 5629) CNRS
ENSCPB
16 avenue Pey-Berland
33607 Pessac Cedex
France

Geoffrey W. Coates
Cornell University
Department of Chemistry
and Chemical Biology
Baker Laboratory
Ithaca, NY 14853
USA

Helmut Cölfen
Max Planck Institute of Colloids
and Interfaces
Colloid Chemistry
Research Campus Golm
Am Mühlenberg 1
14476 Potsdam-Golm
Germany

Henri Cramail
Université Bordeaux 1
Laboratoire de Chimie des Polymères
Organiques
Unité Mixte de Recherche
(UMR 5629) CNRS
ENSCPB
16 avenue Pey-Berland
33607 Pessac Cedex
France

Costantino Creton
Laboratoire PPMD
ESPCI
10 rue Vauquelin
75231 Paris
France

Ghislain David
Ingénierie et Architectures
Macromoléculaires
Institut Gerhardt, UMR 5253
Ecole Nationale Supérieure Chimie
de Montpellier
8 rue de l'Ecole Normale
34296 Montpellier
France

Priyadarsi De
University of Massachusetts Lowell
Polymer Science Program
Department of Chemistry
One University Avenue
Lowell, MA 01854
USA

Gero Decher
Institut Charles Sadron
(C.N.R.S. UPR 022)
6 rue Boussingault
67083 Strasbourg Cedex
France
and
Université Louis Pasteur
Faculté de Chimie
1 rue Blaise Pascal
67008 Strasbourg Cedex
France

Alain Deffieux
Université Bordeaux 1
CNRS, ENSCPB
Laboratoire de Chimie des Polymères
Organiques
16 avenue Pey-Berland
33607 Pessac
France

Timothy J. Deming
University of California, Los Angeles
Department of Bioengineering
420 Westwood Plaza
7523 Boelter Hall
Los Angeles, CA 90095
USA

G. Dubois
IBM Almaden Research Center
650 Harry Road
San Jose, CA 95120
USA

Philippe Dubois
Université de Mons-Hainaut
Matériaux Polymères et Composites
Place du Parc 20
7000 Mons
Belgium

Andrzej Duda
Center of Molecular and
Macromolecular Studies
Polish Academy of Sciences
Department of Polymer Chemistry
Sienkiewicza 112
90-363 Łodz
Poland

Karel Dušek
Academy of Sciences
of the Czech Republic
Institute of Macromolecular
Chemistry
Heyrovského nám. 2
162 06 Praha
Czech Republic

Miroslava Dušková-Smrčková
Academy of Sciences
of the Czech Republic
Institute of Macromolecular
Chemistry
Heyrovského nám. 2
162 06 Praha
Czech Republic

Todd Emrick
University of Massachusetts Amherst
Department of Polymer Science
and Engineering
120 Governors Drive
Amherst, MA 01003
USA

Tom A. P. Engels
Eindhoven University of Technology
Department of Mechanical
Engineering
P.O. Box 513
5600 MB Eindhoven
The Netherlands

Robin S. Farmer
University of Delaware
Department of Materials Science
and Engineering
201 DuPont Hall
and
Delaware Biotechnology Institute
15 Innovation Way
Newark, DE 19716
USA

Rudolf Faust
University of Massachusetts Lowell
Polymer Science Program
Department of Chemistry
One University Avenue
Lowell, MA 01854
USA

Nelson Felix
Cornell University
Department of Materials Science
and Engineering
Bard Hall
Ithaca, NY 14853-1501
USA

Michel Fontanille
Université Bordeaux 1
Laboratoire de Chimie des Polymères
Organiques
ENSCPB
16 avenue Pey-Berland
33607 Pessac
France

Jean Pierre Fouassier
University of Haute Alsace,
ENSCMu
Department of General
Photochemistry, UMR 7525 CNRS
3 Alfred Werner street
68093 Mulhouse Cedex
France

Takeshi Fukuda
Kyoto University
Institute for Chemical Research
Uji, Kyoto 611-0011
Japan

François Ganachaud
Ecole Nationale Supérieure de Chimie
de Montpellier
Laboratoire de Chimie
Macromoléculaire
8 rue de l'Ecole Normale
34296 Montpellier Cedex 5
France

David S. Germack
Washington University in Saint Louis
Center for Materials Innovation and
Department of Chemistry
One Brookings Drive
St. Louis, MO 63130-4899
USA

Yves Gnanou
Université Bordeaux 1
Laboratoire de Chimie des Polymères
Organiques
ENSCPB
16 avenue Pey-Berland
33607 Pessac
France

Jian Ping Gong
Hokkaido University
Section of Biological Sciences
Faculty of Science
Laboratory of soft & wet matter
North 10, West 8
060-0810 Sapporo
Japan

Padma Gopalan
University of Wisconsin – Madison
Department of Materials Science
and Engineering
1117 Engineering Research Building
1500 Engineering Drive
Madison, WI 53706
USA

Leon E. Govaert
Eindhoven University of Technology
Department of Mechanical
Engineering
P.O. Box 513
5600 MB Eindhoven
The Netherlands

Stefanie M. Gravano
University of California, Davis
Department of Chemistry
One Shields Avenue
Davis, CA 95616-5295
USA

Andrew C. Grimsdale
Nanyang Technological University
School of Materials Science and
Engineering
50 Nanyang Avenue
Singapore 639798

Nikos Hadjichristidis
University of Athens
Department of Chemistry
Panepistimiopolis Zografou
15771 Athens
Greece

Georges Hadziioannou
Laboratoire d'Ingénierie des
Polymères pour les Hautes
Technologies
UMR 7165 CNRS
Ecole Européenne Chimie Polymères
Matériaux
Université Louis Pasteur
25 rue Becquerel
67087 Strasbourg
France

Dale L. Handlin, Jr.
Kraton Polymers
700 Milam
North Tower
Houston, TX 77002
USA

James L. Hedrick
IBM Almaden Research Center
650 Harry Road
San Jose, CA 95120
USA

Thomas Heiser
Université Louis Pasteur
Institut d'Electronique du Solide
et des Systèmes
UMR 7163, CNRS
23 rue du Loess
67087 Strasbourg
France

Valérie Héroguez
Université Bordeaux 1
Laboratoire de Chimie des Polymères
Organiques
ENSCPB
16 avenue Pey-Berland
33607 Pessac
France

Richard Hoogenboom
Eindhoven University of Technology
and Dutch Polymer Institute (DPI)
Laboratory of Macromolecular
Chemistry and Nanoscience
P.O. Box 513
5600 MB Eindhoven
The Netherlands

Jeffrey A. Hubbell
Ecole Polytechnique Fédérale
de Lausanne
Institute of Bioengineering
1015 Lausanne
Switzerland

Hermis Iatrou
University of Athens
Department of Chemistry
Panepistimiopolis Zografou
15771 Athens
Greece

Olli Ikkala
Helsinki University of Technology
Department of Engineering Physics
and Mathematics
and Center for New Materials
P.O. Box 2200
02015 Hut
Espoo
Finland

Woo-Dong Jang
The University of Tokyo
School of Engineering
Department of Chemistry
and Biotechnology
7-3-1 Hongo, Bunkyo-ku
Tokyo 113-8656
Japan

Sreeja Jayant
Rutgers, The State University
of New Jersey
Department of Pharmaceutics
160 Frelinghuysen Road
Piscataway, NJ 08854-8020
USA

Robert Jerome
University of Liège
Center for Education and Research
on Macromolecules (CERM)
Sart-Tilman, B6a
4000 Liège
Belgium

Youngjong Kang
Massachusetts Institute of Technology
Department of Materials Science
and Engineering and
Institute for Soldier Nanotechnologies
Cambridge, MA 02139
USA

Jayant J. Khandare
Rutgers, The State University
of New Jersey
Department of Pharmaceutics
160 Frelinghuysen Road
Piscataway, NJ 08854-8020
USA

Kristi L. Kiick
University of Delaware
Department of Materials Science
and Engineering
201 DuPont Hall
and
Delaware Biotechnology Institute
15 Innovation Way
Newark, DE 19716
USA

Tatsuki Kitayama
Osaka University
Department of Chemistry
Graduate School of Engineering
Toyonaka, Osaka 560-8531
Japan

Harm-Anton Klok
Ecole Polytechnique Fédérale
de Lausanne (EPFL)
Institut des Matériaux
Laboratoire des Polymères
STI – IMX – LP, MXD 112
(Bâtiment MXD), Station 12
1015 Lausanne
Switzerland

Bert Klumperman
Eindhoven University of Technology
Laboratory of Polymer Chemistry
P.O. Box 513
5600 MB Eindhoven
The Netherlands

Shiro Kobayashi
Kyoto Institute of Technology
R & D Center for Bio-based Materials
Matsugasaki, Sakyo-ku
Kyoto 606-8585
Japan

Kurt Kremer
Max Planck Institute for Polymer
Research
Ackermannweg 10
55128 Mainz
Germany

Vahik Krikorian
Massachusetts Institute of Technology
Department of Materials Science
and Engineering and
Institute for Soldier Nanotechnologies
Cambridge, MA 02139
USA

Przemyslaw Kubisa
Center of Molecular and
Macromolecular Studies
Polish Academy of Sciences
Department of Polymer Chemistry
Sienkiewicza 112
90-363 Łodz
Poland

Alok Kumar Sen
Indian Institute of Technology
Materials Science Centre
Polymer Division
Kharagpur 721302
India

Jacques Lalevée
University of Haute Alsace, ENSCMu
Department of General
Photochemistry, UMR 7525 CNRS
3 Alfred Werner street
68093 Mulhouse Cedex
France

Katharina Landfester
University of Ulm
Department of Organic Chemistry III
– Macromolecular Chemistry
and Organic Materials
Albert-Einstein-Allee 11
89081 Ulm
Germany

Philippe Lavalle
Institut National de la Santé
et de la Recherche Médicale
INSERM Unité 595
11 rue Humann
67085 Strasbourg Cedex
France
and
Université Louis Pasteur
Faculté de Chirurgie Dentaire
1 place de l'Hôpital
67000 Strasbourg
France

List of Contributors

Nicolas Leclerc
Université Louis Pasteur
Laboratoire d'Ingénierie des
Polymères pour les Hautes
Technologies
UMR 7165 CNRS
Ecole Européenne Chimie Polymères
Matériaux
25 rue Becquerel
67087 Strasbourg
France

Young Moo Lee
Hanyang University
School of Chemical Engineering
College of Engineering
Seoul 133-791
Korea

Ludwik Leibler
Matière Molle et Chimie
UMR 167 CNRS
ESPCI
10 rue Vauquelin
75005 Paris
France

G. B. W. L. Ligthart
DSM Campus Geleen
Performance Materials
PO Box 18
6160 MD Geleen
The Netherlands

Wei-Shi Li
ERATO-SORST Nanospace Project
Japan Science and
Technology Agency (JST)
National Museum of Emerging
Science and Innovation
2-41 Aomi, Koto-ku
Tokyo 135-0064
Japan

Pierre Lutz
Institut Charles Sadron
6 rue Boussingault
67083 Strasbourg Cedex
France

Ian Manners
University of Bristol
Department of Chemistry
Cantock's Close
Bristol BS8 1TS
UK

Andrew F. Mason
IBM Almaden Research Center
650 Harry Road
San Jose, CA 95120
USA

Toshio Masuda
Kyoto University
Department of Polymer Chemistry
Graduate School of Engineering
Katsura Campus
Kyoto 615-8510
Japan

Krzysztof Matyjaszewski
Carnegie Mellon University
Department of Chemistry
4400 Fifth Avenue
Pittsburgh, PA 15213
USA

Thomas C. B. McLeish
University of Leeds
IRC in Polymer Science
and Technology
Polymers and Complex Fluids
Department of Physics
and Astronomy
Leeds LS2 9JT
UK

Michael A. R. Meier
Eindhoven University of Technology
and Dutch Polymer Institute (DPI)
Laboratory of Macromolecular
Chemistry and Nanoscience
P.O. Box 513
5600 MB Eindhoven
The Netherlands

E. W. Meijer
Eindhoven University of Technology
Laboratory of Macromolecular
and Organic Chemistry
P.O. Box 513
5600 MB Eindhoven
The Netherlands

Han E. H. Meijer
Eindhoven University of Technology
Department of Mechanical
Engineering
P.O. Box 513
5600 MB Eindhoven
The Netherlands

R. D. Miller
IBM Almaden Research Center
650 Harry Road
San Jose, CA 95120
USA

Tamara Minko
Rutgers, The State University
of New Jersey
Department of Pharmaceutics
160 Frelinghuysen Road
Piscataway, NJ 08854-8020
USA

Martin Moller
RWTH Aachen
Institut für Technische
und Makromolekulare Chemie
Pauwelsstraße 8
52056 Aachen
Germany

Hideharu Mori
Yamagata University
Faculty of Engineering
Department of Polymer Science
and Engineering
4-3-16, Jonan
Yonezawa 992-8510
Japan

Klaus Müllen
Max Planck Institute
for Polymer Research
Ackermannweg 10
55128 Mainz
Germany

Axel H. E. Müller
University of Bayreuth
Macromolecular Chemistry II
95440 Bayreuth
Germany

Anna Musyanovych
University of Ulm
Department of Organic Chemistry III
– Macromolecular Chemistry
and Organic Materials
Albert-Einstein-Allee 11
89081 Ulm
Germany

Kazukiyo Nagai
Meiji University
Department of Applied Chemistry
1-1-1 Higashi-mita, Tama-ku
Kawasaki 214-8571
Japan

Christopher K. Ober
Cornell University
Department of Materials Science
and Engineering
310 Bard Hall
Ithaca, NY 14853-1501
USA

Joëlle Ogier
Institut National de la Santé
et de la Recherche Médicale
INSERM Unité 595
11 rue Humann
67085 Strasbourg Cedex
France
and
Université Louis Pasteur
Faculté de Chirurgie Dentaire
1 place de l'Hôpital
67000 Strasbourg
France

Masashi Ohmae
Kyoto University
Department of Materials Chemistry
Graduate School of Engineering
Katsura, Nishikyo-ku
Kyoto 615-8510
Japan

Kohji Ohno
Kyoto University
Institute for Chemical Research
Uji, Kyoto 611-0011
Japan

Masami Okamoto
Advanced Polymeric Materials
Engineering
Graduate School of Engineering
Toyota Technological Institute
2-12-1 Hisakata, Tempaku
Nagoya 468-8511
Japan

Joost A. Opsteen
Radboud University Nijmegen
Institute for Molecules and Materials
Toernooiveld 1
6525 ED Nijmegen
The Netherlands

Yoshihito Osada
Hokkaido University
Section of Biological Sciences
Faculty of Science
Laboratory of soft & wet matter
North 10, West 8
Sapporo 060-0810
Japan

Timothy E. Patten
University of California, Davis
Department of Chemistry
One Shields Avenue
Davis, CA 95616-5295
USA

Stanislaw Penczek
Centre of Molecular
and Macromolecular Studies
Polish Academy of Sciences
Department of Polymer Chemistry
Sienkiewicza 112
90-363 Łodz
Poland

Catherine Picart
Université de Montpellier II
Laboratoire de Dynamique
des Interactions Membranaires
Normales et Pathologiques
(C.N.R.S. UMR 5235)
Place Eugène Bataillon
34095 Montpellier Cedex
France

Frank Pirrung
CIBA Specialty Chemicals, Inc.
Research and Development
Coating Effects
Schwarzwaldallee 215
4002 Basel
Switzerland

Marinos Pitsikalis
University of Athens
Department of Chemistry
Panepistimiopolis Zografou
15771 Athens
Greece

Jeffrey Pyun
University of Arizona
Department of Chemistry
1306 E. University Boulevard
Tucson, AZ 85721
USA

Damien Quémener
Université Bordeaux 1
Laboratoire de Chimie des Polymères
Organiques
ENSCPB
16 avenue Pey-Berland
33607 Pessac
France

Karunakaran Radhakrishnan
University of Akron
Institute of Polymer Science
170, University Avenue
Akron, OH 44325
USA

Wolfgang Radke
Deutsches Kunststoff-Institut
Darmstadt
Schlossgartenstraße 6
65289 Darmstadt
Germany

David A. Rider
University of Toronto
Department of Chemistry
80 St. George Street
M5S 3H6 Toronto, Ontario
Canada

Megan L. Ruegg
University of California
Department of Chemical Engineering
Berkeley, CA 94720
USA

Pierre Schaaf
Institut Charles Sadron
(C.N.R.S. UPR 022)
6 rue Boussingault
67083 Strasbourg Cedex
France
and
Ecole Européenne de Chimie
Polymères et Materiaux
25 rue Bequerel
67087 Strasbourg Cedex 2
France

Oren A. Scherman
University of Cambridge
Department of Chemistry
Lensfield Road
Cambridge CB2 1EW
UK

Helmut Schlaad
Max Planck Institute of Colloids
and Interfaces
Colloid Department
Research Campus Golm
14424 Potsdam
Germany

Ulrich S. Schubert
Eindhoven University of Technology
and Dutch Polymer Institute (DPI)
Laboratory of Macromolecular
Chemistry and Nanoscience
P.O. Box 513
5600 MB Eindhoven
The Netherlands

Bernard Senger
Institut National de la Santé
et de la Recherche Médicale
INSERM Unité 595
11 rue Humann
67085 Strasbourg Cedex
France
and
Université Louis Pasteur
Faculté de Chirurgie Dentaire
1 place de l'Hôpital
67000 Strasbourg
France

Sergei S. Sheiko
University of North Carolina
at Chapel Hill
Department of Chemistry
Chapel Hill, NC 27599-3290
USA

Rint P. Sijbesma
Eindhoven University of Technology
Laboratory of Macromolecular
and Organic Chemistry
P.O. Box 513
5600 MB Eindhoven
The Netherlands

Peter F. W. Simon
GKSS Research Centre
Geesthacht GmbH
Institute of Polymer Research
Max-Planck-Straße
21502 Geesthacht
Germany

Stanislaw Slomkowski
Center of Molecular and
Macromolecular Studies
Polish Academy of Sciences
Department of Polymer Chemistry
Sienkiewicza 112
90-363 Łodz
Poland

Hans Wolfgang Spiess
Max-Planck-Institute for Polymer
Research
Spectroscopy
P.O. Box 3148
55021 Mainz
Germany

Richard J. Spontak
North Carolina State University
Departments of Chemical and
Biomolecular Engineering and
Materials Science and Engineering
Raleigh, NC 27695
USA

R. F. T. Stepto
The University of Manchester
School of Materials
Materials Science Centre
Polymer Science
and Technology Group
Grosvenor Street
Manchester M1 7HS
UK

Brent S. Sumerlin
Southern Methodist University
Department of Chemistry
3215 Daniel Avenue
Dallas, TX 75240-0314
USA

Yeshayahu Talmon
Technion – Israel Institute
of Technology
Department of Chemical Engineering
32000 Haifa
Israel

M. Atilla Tasdelen
Istanbul University
Department of Chemistry
Maslak
Istanbul 34469
Turkey

Daniel Taton
Université Bordeaux 1
Laboratoire de Chimie des Polymères
Organiques
LCPO CNRS
ENSCPB
16 avenue Pey-Berland
33607 Pessac
France

Gerrit ten Brinke
University of Groningen
Laboratory of Polymer Chemistry
Materials Science Centre
Nijenborgh 4
9747 AG Groningen
The Netherlands

Edwin L. Thomas
Massachusetts Institute of Technology
Department of Materials Science
and Engineering
and
Institute for Soldier Nanotechnologies
Cambridge, MA 02139
USA

Scott Trenor
Kraton Polymers
Houston, TX 77002
USA

Constantinos Tsitsilianis
University of Patras
and FORTH/ICEHT
Department of Chemical Engineering
Karatheodori 1
26504 Patras
Greece

Yoshinobu Tsujii
Kyoto University
Institute for Chemical Research
Uji, Kyoto 611-0011
Japan

Jan C. M. van Hest
Radboud University Nijmegen
Institute for Molecules and Materials
Toernooiveld 1
6525 ED Nijmegen
The Netherlands

Brooke A. van Horn
Washington University in Saint Louis
Center for Materials Innovation and
Department of Chemistry
One Brookings Drive
St. Louis, MO 63130-4899
USA

Shrinivas Venkataraman
Washington University in Saint Louis
Center for Materials Innovation and
Department of Chemistry
One Brookings Drive
St. Louis, MO 63130-4899
USA

Jean-Claude Voegel
Institut National de la Santé
et de la Recherche Médicale
INSERM Unité 595
11 rue Humann
67085 Strasbourg Cedex
France
and
Université Louis Pasteur
Faculté de Chirurgie Dentaire
1 place de l'Hôpital
67000 Strasbourg
France

C. Grant Willson
The University of Texas at Austin
Departments of Chemistry
and Biochemistry
and Chemical Engineering
Austin, TX 78712
USA

Karen L. Wooley
Washington University in Saint Louis
Center for Materials Innovation and
Department of Chemistry
One Brookings Drive
St. Louis, MO 63130-4899
USA

Kathryn Wright
Kraton Polymers
700 Milam
North Tower
Houston, TX 77002
USA

Bernhard Wunderlich
University of Tennessee
Department of Chemistry
200 Baltusrol Road
Knoxville, TN 37992-3707
USA

Yusuf Yagci
Istanbul University
Department of Chemistry
Maslak
Istanbul 34469
Turkey

Tsutomu Yokozawa
Kanagawa University
Department of Material
and Life Chemistry
Rokkakubashi, Kanagawa-ku
Yokohama 221-8686
Japan

1
Macromolecular Engineering

Krzysztof Matyjaszewski, Yves Gnanou, and Ludwik Leibler

The science and industry dealing with synthetic polymers have been growing and developing with amazing vigor and speed. One may be astounded to remember that 100 years ago even the concept of a polymer, a long macromolecular chain, did not exist and one could not find a single chemical plant to synthesize even 1 kg of a polymer. Today, thousands of publications containing the word polymer in their title are published each year and around 2×10^8 tonnes of polymers are produced annually, i.e., 0.5 kg of polymer per capita worldwide every week!

Polymers are preferred materials in applications for industrial practice and we as a society use them daily. Their growth accompanied and pushed economic expansion and industrial revolutions throughout the 20th century, even during the dark and difficult years. Many modern-day technologies emerged due to polymer science and its rapid progress. The aircraft and space industries, movies and music with its superabundance of vinyl records, CDs and DVDs, sport and outdoor equipment and modern packaging are only a fraction of examples being used today. Adhesives are made from polymers and so are paints. Could you imagine life without polymer photoresists that enable lithography to produce all the tiny electric circuits in our laptops, cell phones, Ipods or Blackberries? The interior of every automobile is essentially entirely made from polymers, but they are also used for body parts and for under-the-hood applications. Although we could continue to cite and marvel over all of these practical achievements with polymers, the interested reader can refer to Volume IV of this book, which tells the story of the growth of many applications along with those that are expected to grow provided that critical scientific, industrial and economic bottlenecks and challenges are overcome. Indeed, the world of polymers results from a unique chain of knowledge which we call *macromolecular engineering*. The latter relies on and produces some very fundamental and sometimes even abstract science that finds its way to applications, but at the same time it more often refers to applications and processing for motivation and for inspiration. Many examples will be given throughout the four volumes of *Macromolecular Engineering: From Precise Macromolecular Synthesis to Macroscopic Materials Properties and Applications*.

Polymer science exhibits a typical frontier character by the pioneering nature of its development. Nourished from many areas of chemistry and physics, both fundamental and applied, polymer science generously contributes to these disciplines. For example, statistical physics and small-angle neutron scattering brought very firm and useful bases to build an understanding of polymer solutions, melts and self-assembly of macromolecular systems. At the same time, the Flory approximation, initially intended to predict the swelling of a macromolecule in a good solvent, and Edwards' replica trick introduced the understanding of rubber elasticity; both are now the bread and butter of statistical physicists to help solve many problems. De Gennes' reptation concept, the basis of polymer rheology, is a part of modern physics in its own right. In chemistry, the discovery of Ziegler–Natta catalysis fundamentally changed the nascent polymer industry of the 1950s, introducing novel polyolefin-based thermoplastic polymers through highly selective processes and laying the foundations of an industry of commodity plastics. To give an idea of the vitality of that industry, it took just 5 years to transfer isotactic polypropylene from the laboratory to industrial production. In the same decade, the development of living polymerization by Michael Szwarc provided a deeper insight into the structure and reactivity of ions, ion pairs, their aggregates and dormant species. Very efficient catalysts for ring-opening metathesis polymerization and atom-transfer radical polymerization are now being used by many organic chemists in the highly selective metathesis cyclization or atom-transfer radical addition reactions. In a similar way, metallocene and post-metallocene catalysts had a tremendous impact on coordination and organometallic chemistry. The entire field of dendrimers is shared by organic and polymer chemists, as also is the area of supramolecular chemistry.

During the first enthusiastic decades of polymer chemistry and industrial practice, various polymerization techniques were developed and an amazing number of new polymers, each made from different monomers, were introduced into the market to answer fast-growing and increasingly demanding applications. However, only a limited number of commodity polymers have proved themselves as workhorse materials since the 1960s, impossible to displace thanks to their low cost and versatility. During the last 20 years, these commodity polymers have mutated to truly high-tech materials, with marvelous science, both chemistry and physics, and technology contributing to this metamorphosis. Nevertheless, they are still called commodity because of their price and volume. In the 1970s, fashion, future projections and research efforts were focused on other polymers, with more noble pedigrees, but which often did not meet expectations. During the last 30 years not many new technical thermoplastics have been introduced and the production of those that did appear often ceased soon after their introduction. It is believed that the development of polymers for new classical applications will rely on better synthetic control of the molecular structure, stereoregularity and architecture of chains. It may also be supposed that better control of organization at various scales ranging from nanometers to millimeters will be a constant demand. Hence macromolecular engineering is about obtaining such control on a laboratory, pilot or plant scale, and on designing new polymers for new applications in areas of today's concerns, such as biomedicine and energy economy.

Achieving such precise control of every fragment of a macromolecule resembles in some respects the total synthesis of natural products. Thus, macromolecular synthesis often reaches an extremely high degree of chemoselectivity; in order to prepare ultra-high molecular weight polyethylene with $M_n > 20 000 000$, chemoselectivity of chain growth must exceed 99.9999%! The same is true for regioselectivity, which for many vinyl polymerizations leads to exclusive head-to-tail polymers (i.e. regioselectivity exceeds 99.99%). Also, stereoselectivity in some coordination polymerizations (albeit not in radical and carbocationic systems) may exceed 99%. These values are unmatched by most organic synthetic methodologies. Indeed, polymer chemistry often selects the best organic reactions and implements them to macromolecular synthesis. Thus, only reactions with selectivities >90% can be used in macromolecular synthesis. Then, by adopting protocols from physical organic chemistry, by adjusting the solvent, temperature, pressure, pH, ionic strength and many other parameters, the selectivities are further enhanced to yield precisely controlled high molecular weight polymers.

Macromolecular Engineering: From Precise Macromolecular Synthesis to Macroscopic Materials Properties and Applications is a book dedicated to these aspects and is organized in four volumes. *Volume I* presents essentially all of the synthetic techniques used to prepare well-defined macromolecules. They include carbanionic polymerization, which was the first process to provide living polymers, a pathway to nanostructured materials and many commercially important products. Other ionic processes based on carbocationic polymerization and ionic ring-opening polymerizations also afforded high molecular weight, well-defined (co)polymers with several commercial products.

Free radical polymerization, with the concurrent initiation, propagation and termination steps and thus its chaotic character, is totally unsuited to the preparation of complex architectures, although nearly 50% of all polymeric materials are currently obtained by this means. Recently, however, new controlled/living radical processes, such as atom-transfer radical polymerization (ATRP), stable free radical polymerization (SFRP) and degenerative transfer techniques (mediated by alkyl iodides or dithio esters, i.e. RAFT or MADIX), where the growing radicals are prevented from early termination, have been developed, opening up many avenues to well-defined (co)polymers with polar groups and functional moieties.

Coordination polymerization has industrial relevance similar to radical polymerization. New catalytic/initiating systems based on single-site catalysts expanded the range of polyolefin-based materials and opened up routes to specialty products. Ring-opening metathesis polymerization, recognized by the Nobel Prize awarded to Chauvin, Grubbs and Schrock in 2005, is another stellar example of the creation of a powerful tool of macromolecular engineering combining high levels of structural fidelity and functional group compatibility.

Chain growth addition polymerizations are not the only means available for the preparation of well-defined polymers. Recent advances in polycondensations have permitted the synthesis of low-polydispersity polymers and block copolymers. Enzymatic catalysis and protocols inspired by biotechnologies are two emerging techniques that afford uniform polymers, such as polypeptides, polysaccharides and nucleic acids. Another increasingly prevalent method of design of complex functional systems is through the accurate placement of non-covalent interactions at precise locations within self-assembling building blocks. Some monomers can be polymerized by only one mechanism, therefore a selective transformation between polymerization mechanisms opens up the possibility of preparing unusual polymers, including hybrid materials.

Many polymerizations are carried out under homogeneous conditions without any external stimuli. However, heterogeneous polymerization, especially in dispersed media that employ compartmentalization, provides an avenue to materials that cannot be prepared under homogeneous conditions. In a similar way, external stimuli can provide additional control and improve or alter selectivities in macromolecular synthesis.

Hence the chapters included in Volume I present the state-of-the-art in macromolecular synthesis and describe not only organic, but also inorganic polymers.

The main mission of *Volume II* is to describe various elements of macromolecular architecture. They include microstructure control, in terms of tacticity and composition, leading to statistical, alternating, gradient or block copolymers. Also, functionalities, especially end-functionalities, are very important since they provide unusual building blocks for macromolecular architecture.

Non-linear topologies include cyclic polymers and various degrees of branching, including hyperbranched polymers, stars, microgels, molecular brushes and – ultimately – dendrimers. Advances in polymer synthesis allow the preparation of previously unavailable materials, namely organic–inorganic hybrids, core–shell particles and very dense brushes with compositionally different compressibilities, lubricities and other properties.

Another approach is to combine synthetic polymers with natural products, yielding alternative hybrids or chimera materials. Other approaches using bio-inspired synthesis, including polyelectrolytes, along with a combination of various synthetic techniques yield functional macromolecules with complex shapes and architecture.

The properties of polymeric materials, although encoded in molecular structure, also depend on how they self-assemble to morphologies of the size of micrometers rather than nanometers. The organization on mesoscopic scales is governed by both thermodynamics and processing conditions. Self-assembly can be used to impart original macroscopic properties to the systems, including optical, electrical and transport through polymer matrices. These processes, including formulations and processing and techniques used to characterize polymers, are covered in *Volume III*.

The basis of polymer self-assembly is treated, including developments in simulations and modeling and the means to characterize the assembly features of nanostructured polymer materials are described. These characterization methods include proximal probe techniques such as atomic force microscopy, but also various scattering methods and microscopy.

Since the nanostructured morphologies are not static, dynamic phenomena, i.e. processing, are critically important in many cases and are discussed in Volume III. Further, because many properties will depend on thermal history, mechanical stresses, solvent removal, etc., the polymeric materials need to be precisely characterized at different stages by calorimetric techniques, chromatography or NMR spectroscopy. High-throughput screening allows the use of smaller amounts of polymers to obtain relevant information.

Some polymeric materials are eventually converted to polymer networks; they can be used as various sealants and adhesives and can be designed to react further during a blending process.

The ultimate goals of macromolecular engineering are to target special applications and some of the most important among them are described in *Volume IV*.

Perhaps the most stellar example of products prepared by macromolecular engineering is Kraton®, a landmark material made by living anionic polymerization. This thermoplastic elastomer was originally designed to be used in the tire industry, but the first applications came in footwear and later in other compounding applications, including automotive, wire and cable, medical, soft touch overmolding, cushions, thermoplastic vulcanizates, lubricants, gels, coatings and in flexographic printing and road marking. It is anticipated that materials made by other controlled/living processes will lead to more applications, with an even larger market impact.

However, many other materials benefiting from macromolecular engineering are already commercially available. They include nanocomposites, especially based on layered structures, but also dispersants and surfactants.

It is anticipated that the electronics industry, constantly moving to smaller dimensions and nanometer precision, will present a great need for well-defined polymers. They include many conjugated polymers for molecular electronics, but also many other polymers for microelectronics, including dielectrics and resists. They often require special hierarchical organization and can find applications in soft lithography, magnetic and optoelectronic areas and even as materials for membranes and fuel cells.

Equally important are well-defined copolymers for biomedical applications. The 21st century will be a century of polymeric drugs which will deliver, at a programmed rate, a drug to a specific location and release it at the desired rate. Biomedical applications also include gels for tissue and bone engineering. The concept of biomineralization is expanding to the synthesis of new inorganic crystals with tunable nanostructured morphology. Covalent bonding of well-defined organic polymers with natural polymers can dramatically alter the action of enzymes and other natural products.

The promise of accurately controlling every facet of polymeric architectures through very selective processes and the diversity of available tools – including the programmed inter- and intramolecular assembly of building blocks – holds considerable potential and heralds limitless sophistication in macromolecular design. For this potential to be efficiently harnessed and translated into applications, understanding and optimizing the properties of the (nano)structures formed are instrumental.

Summarizing, the main objectives of this four-volume book. *Macromolecular Engineering: From Precise Macromolecular Synthesis to Macroscopic Materials Properties and Applications,* are to provide a state-of-the-art description of:
- synthetic tools used to control precisely various aspects of macromolecular structure, including chain composition, microstructure, functionality and topology;
- modern characterization techniques at molecular and macroscopic levels used to determine quantitatively various properties of well-defined (co)polymers in solution, bulk and at surfaces;
- correlation of molecular structure with macroscopic properties which can be additionally affected by processing;
- some emerging applications for the (co)polymers.

The book is directed towards chemists and polymer scientists interested in affecting macroscopic properties via precisely designed macromolecular structure and controlled processing. Each chapter provides background information, comparative advantages and limitations and the most recent advances of various synthetic approaches, characterization techniques and intended applications.

2
Anionic Polymerization of Vinyl and Related Monomers

Michel Fontanille and Yves Gnanou

2.1
Introduction

Vinyl, vinylidene, (meth)acrylic monomers and conjugated dienes belong to the broad category of monomers that can undergo polymerization upon repeated nucleophilic attacks and additions on to their carbon–carbon double bond. All these monomers exhibit a strongly negative free energy of polymerization and are free of molecular groups that could interfere with the propagation step and thwart the growth of carbanionic chains.

Even though the anionic polymerization of butadiene was exploited in the industrial production of the so-called Buna synthetic rubbers as early as 1938 [1] and the persistence of the corresponding carbanionic species discovered even earlier in 1936 [2], this type of polymerization has been little investigated until the work of Szwarc, first published in 1956 [3, 4]. He discovered that an anionic polymerization can be conducted to complete conversion without occurrence of any reaction of transfer and termination. Szwarc rightly contended that propagating active centers can be preserved indefinitely from deactivation well beyond the propagation step and coined the term "living" for polymerization that meets this criterion. In such a process, chain growth is proportional to the amount of monomer consumed and propagation can be discontinued temporarily or definitively or even resumed at will.

In addition, when initiation is complete within an interval shorter than that of propagation, the polymerization is called "controlled", as the molar mass of the sample can be precisely predicted and its dispersity is narrow. Besides this, the "living" reactive centers carried by the chain ends can be utilized for purposes of end-functionalization or chain extension.

The papers published by Szwarc and coworkers unleashed the race to describe these systems comprehensively and utilize them to engineer well-defined polymeric architectures. Because of both the persistence of carbanionic centers and their relatively high concentration, it was possible to identify and closely study the various structures taken by these propagating species, to measure

Macromolecular Engineering. Precise Synthesis, Materials Properties, Applications.
Edited by K. Matyjaszewski, Y. Gnanou, and L. Leibler
Copyright © 2007 WILEY-VCH Verlag GmbH & Co. KGaA, Weinheim
ISBN: 978-3-527-31446-1

their intrinsic reactivity and establish the relations between them. In spite of its complexity, anionic polymerization is certainly the chain process that is the most accurately known among all the methods of polymerization. As a result, mechanistic investigations have declined in intensity and emphasis is now more than ever placed on macromolecular engineering using anionic polymerization.

Because of the very high reactivity of the propagating species, anionic polymerization has to be carried out in the total absence of impurities or compounds that are sensitive to the attack of nucleophilic reagents. Initially, it had been thought that such constraints would thwart the development of anionic polymerization, but through rationalizations and simplifications anionic processes now afford materials at a competitive price, even sometimes lower than that of polymers produced by conventional polymerizations. It is beyond the scope of this chapter to describe comprehensively the multiple facets of anionic polymerization and accordingly aspects related in particular to the structure and reactivity of active centers will not be covered. This chapter will focus instead on macromolecular engineering and only review systems that exhibit a "controlled" – if not "living" – character. Further information on anionic polymerization and certain of its aspects are available in the many books and reviews published in the last four decades [5].

2.2
General Features of Anionic Polymerization

Anionic polymerization can be represented by one of the two reactions shown in Scheme 2.1, which depict a propagation step and the insertion of one monomer molecule (M) between the charged or negatively polarized atom carried by the growing chain and its associated counterion (Met$^+$), generally metallic, positively charged or polarized.

In the polymerization of vinyl and related monomers, the active carbanionic species formed are extremely reactive and can add on to vinylic double bonds via a nucleophilic addition. The overall reactivity is thus mainly determined by that of the monomer and that of the carbanionic species. The nucleophilic addition reaction shown in Scheme 2.2 hardly occurs without activation of the nucleophilic reagent and/or of the monomer double bond.

The mechanism illustrated in Scheme 2.2 features the prior coordination of the electron-rich double bond of a monomer to the electron-deficient Met$^+$ counterion before its insertion. Such a coordination was hypothesized to account for certain experimental phenomena such as the decrease in the reactivity of active

$$\sim\!\!\sim\!\!\sim\!\!\sim M_n^-\!,Met^+ + M \longrightarrow \sim\!\!\sim\!\!\sim\!\!\sim M_{(n+1)}^-\!,Met^+$$

$$\sim\!\!\sim\!\!\sim\!\!\sim M_n^{\delta-}\!,Met^{\delta+} + M \longrightarrow \sim\!\!\sim\!\!\sim\!\!\sim M_{(n+1)}^{\delta-}\!,Met^{\delta+}$$

Scheme 2.1

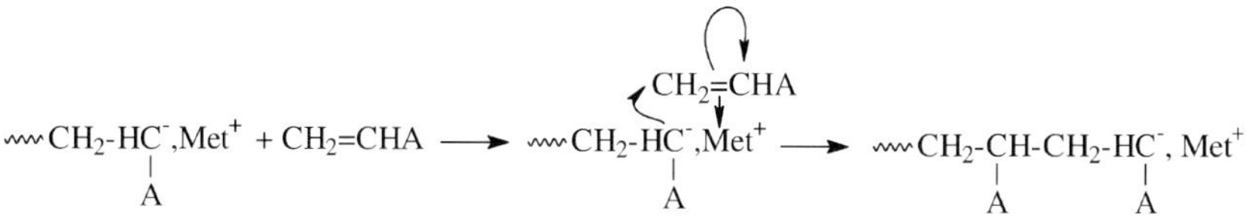

$\text{wwCH}_2\text{-HC}^-,\text{Met}^+ + \text{CH}_2=\text{CHA} \longrightarrow \text{wwCH}_2\text{-HC}^-,\text{Met}^+ \xrightarrow{\text{CH}_2=\text{CHA}} \text{wwCH}_2\text{-CH-CH}_2\text{-HC}^-, \text{Met}^+$
         | | | |
         A   A   A   A

Scheme 2.2

centers placed in the presence of electron donors [6–8], as recently corroborated by calculations [9].

The kinetic study performed by Szwarc et al. to determine the rate constants of addition of polystyryl carbanions to various substituted styrenes in THF showed that these rate constants obey the Hammett relation [10], the nucleophilic attack of the monomer being the rate-determining step of the process. A similar study was carried out by Busson and Van Beylen [11] on the addition of living polystyrene to various substituted diphenylethylenes in benzene. They observed that, in contrast to the reaction in THF, in benzene the monomer very likely coordinates with the cation – at least in the case of Li^+.

2.2.1
Polymerizability of Vinyl and Related Monomers

To undergo polymerization by nucleophilic addition, vinylic monomers must satisfy two types of requirements.

The first is a thermodynamic requirement that applies to any kind of reaction and polymerization. The free energy of polymerization ($\Delta G_{pol} = \Delta H_{pol} - T\Delta S_{pol}$) should take a negative value for polymerization to occur; this condition is generally satisfied due to the high value of the enthalpy of polymerization ($-\Delta H_{pol} \approx 40-80$ kJ mol^{-1}) [12]. However, with certain 1,1-disubstituted monomers such as α-methylstyrene, the rigidity of the chains formed is responsible for a strongly negative entropy of polymerization, implying a low ceiling temperature for this kind of monomer [13].

In the case of "living" polymerizations, the equilibrium

$$P_n^* \rightleftarrows P_{(n-1)}^* + M \tag{1}$$

between M, $P_{(n-1)}^*$ and P_n^* the active chains of degree of polymerization (n–1) and n, respectively, can be instantaneously driven toward one side or another by modification of either the temperature or the monomer concentration and whole chains are concerned. Thus, for a concentration in monomer equal to 1.0 mol L^{-1} at a temperature of 20 °C, the "living" oligo(α-methylstyrene) exist in the form of tetrameric species in equilibrium with the monomer, whereas at 0 °C their degree of polymerization ($\overline{DP}_n$) increases to 100. For an initial molar concentration in monomer of 2.5, $\overline{DP}_n$ is approximately equal to 80 at 20 °C, whereas conversion is practically complete at 0 °C [14].

The second type of requirement pertains to the sensitivity of the monomer to a nucleophilic attack. To undergo such an attack, the monomer must be activated through the positive polarization of its double bond, which can be induced by an electron-withdrawing substituent (in the case of acrylic monomers) and/or by a resonance effect associated with the approach of a strongly nucleophilic entity (in the case of styrenics). By polarizing positively the carbon atom at position 2, electron-withdrawing substituents lower the electron density of the double bond and thus increases its reactivity. Steric effects could limit this activation and in certain cases even prevent propagation. Ethylene is a monomer without any activating substituent, but it can be anionically polymerized or oligomerized because of the extremely high reactivity of the carbanions formed [15].

The polymerizability of monomers – but not their reactivity – can be estimated through the value taken by the rate constant of homopropagation. It is not an intrinsic characteristic of the monomer because, by means of additives, it is possible to activate or deactivate either of the two reactants that participate in the addition process, namely the monomer double bond and the carbanionic species. However, the intrinsic reactivity of the monomer is closely related to that of the active center from which it derives. Indeed, the same substituent either activates or stabilizes both the monomer double bond and the corresponding carbanionic species carried by the growing chain. The more the monomer substituent increases the reactivity of the double bond, the more the charge density and the reactivity of the corresponding carbanionic site decrease, but the increase in reactivity of the monomer prevails over the decrease in reactivity of the active center formed. In other words, for a given family of monomers, their polymerizability is proportional to their intrinsic reactivity whereas the reactivity of the resulting active centers is inversely proportional.

A more rigorous assessment of the reactivity of monomers can be obtained through the determination of their rate constants of cross-addition on carbanionic centers generated from reference monomers [10] or from the determination of reactivity ratios [16].

The main limitation on the extensive utilization of carbanionic polymerization stems from the very high nucleophilicity of the corresponding species and their propensity to react indiscriminately with the monomer double bond and also with molecular groups/substituents carried by the monomer/polymer and sensitive to nucleophilic attack. This is the main cause of the small number of monomers that can be polymerized by anionic polymerization under "living" and controlled conditions. Monomers carrying acidic groups, those whose pK_a is lower than about 30 (alcohols, amines, alkynes, etc.) or functions that undergo nucleophilic addition (ketones, esters, nitriles, etc.), are unsuited to "living" anionic polymerization. By reacting with the growing carbanions, these functional groups generate species of lower reactivity that cannot propagate further, discontinuing the entire polymerization.

2.2.2
Various Parameters Influencing the Structure and Reactivity of Active Centers

The high reactivity of carbanionic active centers can be accounted for by considering that these species are the conjugate bases of the corresponding protonated molecules (conjugated acids). Following Wooding and Higginson [17], who observed, as early as 1952, that the reactivity of anionic active centers is parallel to their basicity, the reactivity of polystyryl carbanions, for instance, can be estimated through the pK_a of toluene (model molecule of the polystyryl chain end). Its very low acidity ($pK_a \approx 41$) [18] indicates a very strong basicity of the corresponding carbanion.

Another way to rank the intrinsic reactivity of carbanionic propagating species is to compare the values of the free energy of ionization of molecules to form anions that are models of the chain ends. Table 2.1 gives an idea of the relative reactivities of totally ionized initiators or of growing carbanionic active centers. It is worth stressing that these values were established in the gas phase and do not take into account the solvation of carbanionic species. The comparison between these values is valid for one type of bond, here –C–H, and reflects the relative reactivities of carbanionic entities associated with alkali metals other than lithium. Their reactivity is extremely high and therefore these species lack stability. In the case of lithiated species, there is no parallel between relative reactivities and free enthalpies of ionization because of the partial covalent character of the carbon–lithium bond.

Such a high reactivity of carbanionic species is utilized to propagate polymerization through repeated nucleophilic additions but, unfortunately, it is also responsible for many side-reactions, which will be discussed later (Sections 2.5.1 and 2.5.2).

Depending on the type of monomer, the counterion, the solvent and the presence of possible additives, the temperature of the reaction medium and their concentrations, active or potentially active species can exist in a wide variety of

Table 2.1 Values of the gas-phase free energy of ionization (ΔG) [19] of model molecules that can be ionized by H^+ elimination.

Organometallic species to be evaluated	Model molecule to be ionized	Ionized model entity	ΔG (kJ mol^{-1})
$CH_3-CH_2-CH_2-CH_2-Met$	$CH_3-CH_2-CH_2-CH_2-H$	$CH_3-CH_2-CH_2-H_2C^-$	1700
$C_6H_5^-, Met^+$	C_6H_6	$H_5C_6^-$	1655
$C_6H_5-CH_2^-, Met^+$ (or PS^-, Met^+)	$C_6H_5-CH_3$	$C_6H_5-CH_2^-$	1555
PAN^-, Met^+	$N\equiv C-CH_3$	$NC-H_2C^-$	1522
$PMMA^-, Met^+$	$CH_3-CO_2-CH_3$	$CH_3-CO_2-H_2C^-$	1513
$(C_6H_5)_2-H_2C^-, Met^+$ and $Met^+, ^-D-D^-, Met^+$	$(C_6H_5)_2-CH_2$	$(C_6H_5)_2-H_2C^-$	1501
$Fluorenyl^-, Met^+$	Fluorene	$Fluorenyl^-$	1438

ionic forms in equilibrium with one another in the reaction medium. Hence their relative concentrations vary with the experimental conditions. Each of these ionic forms reacts with the monomer according to its own reactivity and the polymerization kinetics can be extremely complex due to their multiplicity. Therefore, it is essential to know beforehand how various factors influence the nature of the propagating species formed and their reactivity.

2.2.2.1 Influence of the Type of Monomer

Depending upon the type of substituent(s) carried by the terminal monomeric unit, the charge density on the carbanionic species will be more or less significant. All other parameters being equal, the charge density is maximum when there is no possibility of delocalization of the carbanionic electron pair. For instance, the anionic oligomerization of ethylene occurs only with Li^+ as a counterion [20] because, with this metal, the partial covalency of the C–metal bond stabilizes the species and compensates for the absence of delocalization. With other alkali metal counterions, the reactivity of the carbanionic centers would be higher and such an instability results in

$$\sim\sim CH_2\text{-}CH_2\text{-}H_2C^-,Met^+ \longrightarrow \sim\sim CH_2\text{-}HC=CH_2 + H\text{-}Met \qquad (2)$$

With styrene and diene monomers, the delocalization of the electron pair into the substituent of the double bond is significant and the reactivity is therefore strongly dependent on the nature of the substituting side-group. An analogous study to that performed by Szwarc on substituted styrenes [10] was carried out on the reactivity of the corresponding active centers [21]. Obviously, the substituents of the phenyl ring influence the reactivity of the corresponding carbanionic species: the higher the donor effect of the substituent in the para position, the higher is the reactivity of the polystyryl carbanion. Conversely, the reactivity of the species formed in the polymerization of (meth)acrylic monomers is considerably lower than that of the above-mentioned monomers because the presence of an ester substituent induces a delocalization of the electron pair in the form of an enolate [22], as represented in Scheme 2.3 for the methacrylic propagating center.

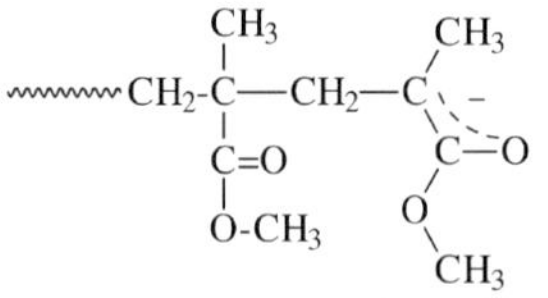

Scheme 2.3

2.2.2.2 Influence of the Nature of Solvent

Because of the very high reactivity of active centers formed in anionic polymerization, the most appropriate solvents are basic or neutral and those exhibiting an acidic character are to be avoided.

The role of a solvent is multifold and, depending on its structure, it can have one, two or three functions.

The first function is that of a diluent. The simultaneous generation of all the propagating centers may cause an uncontrolled elevation of the temperature of the reaction medium, which has to be curbed: the presence of a solvent, by decreasing the viscosity of the reaction medium and permitting its stirring, is therefore essential. Solvents preferred for this task are always aliphatic or aromatic hydrocarbons. They modify little or not at all the structure of the active centers and just dilute the concentration of the various species according to the law of mass action. In hydrocarbons, organolithium compounds are aggregated with degrees of aggregation varying with the nature of the carbanion and, sometimes, with the range of concentration. Thus, polystyryllithium ion pairs are aggregated into dimers [23] as illustrated in Scheme 2.4, whereas polydienyllithiums are aggregated into dimers or tetramers depending on their concentration in the reaction medium [24].

It should be stressed that aggregation constants (K_{ag}) [K_{ag} refers to the aggregation of ion pairs and thus to the formation of weak interactions between them, whereas K_{diss} (see Eq. 4) refers to the dissociation of ion pairs into free ions] for both systems, polystyryl- and polydienyllithium, are very high [25] and only non-aggregated species in equilibrium with aggregated species are reactive [23] (Eq. 3). Such an aggregation into dimers was also found with polystyrylsodium in benzene and in cyclohexane [26, 27], but was not observed with alkali metal cations of larger ionic radius.

$$\left(\sim\sim\sim\sim PS^-, Li^+\right)_2 \underset{K_{ag}}{\rightleftharpoons} 2 \sim\sim\sim\sim PS^-, Li^+ \tag{3}$$
$$\text{inactive} \qquad\qquad\qquad \text{active}$$

The second function of a solvent can be that of solvating agent. Due to their Lewis base character, solvents such as ethers (tetrahydrofuran, dimethoxyethane, dioxane, etc.) or tertiary amines coordinate with the metal cations associated with carbanions and can therefore be used to this end. They can solvate ion pairs either externally or increase the interionic distance in solvent-separated

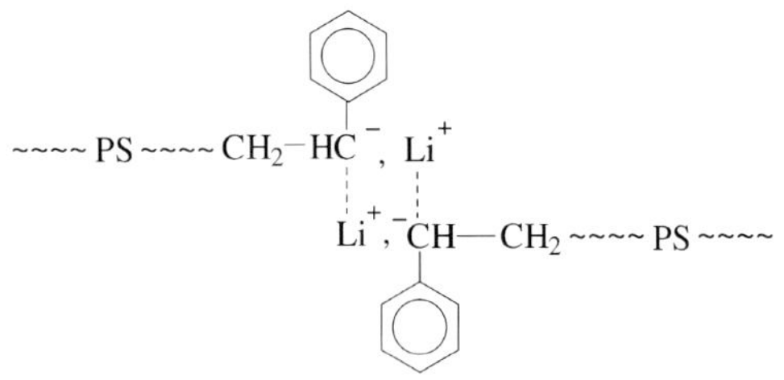

Scheme 2.4

(loose) ion pairs: for instance, PS^-,Na^+ mainly exists in the form of loose ion pairs in THF and in the form of externally solvated ion pairs in dioxane.

Lastly, when the permittivity of the solvent is high enough, it can play the role of a dissociating agent, this effect combining with the solvating effect. The high dielectric constant can induce total separation of electric charges and cause partial dissociation of ion pairs into free ions. The greater the interionic distance, the easier is the dissociation and the higher the dissociation constant:

$$\sim\sim\sim\sim\sim M_n^-,S_x,Met^+ \underset{}{\overset{K_{diss}}{\rightleftarrows}} \sim\sim\sim\sim\sim M_n^- + S_x,Met^+ \qquad (4)$$

where S_x represents x molecules of solvent (S) coordinated to the metal cation. There is a relation between the equilibrium constant of dissociation (K_{diss}) and the permittivity of the medium (ε):

$$-\ln K_{diss} = -\ln K_{diss}^0 + \frac{e^2}{(r_1 + r_2)\varepsilon kT} \qquad (5)$$

where K_{diss}^0 represents the dissociation constant of ion pairs in a medium of infinite permittivity, r_1 and r_2 are the ionic radii of the cation and the anion, respectively, and e is the electron charge. Equation 5 indicates that dissociation is made easier by the solvating effect, which increases the apparent radius of the cation and causes stretching of the ionic bond. Most of the solvents that possess high permittivity also exhibit strong solvating ability. The reactivity of free ions resulting from the dissociation of ion pairs is extremely high and even a low concentration of them can dramatically affect the overall kinetics of polymerization, as shown in Section 2.4.1.1. The dissociation constants of ion pairs into free ions (K_{diss}) can be determined either directly by measurement of the electrical conductance of the corresponding solutions [28] or from kinetic studies of polymerizations performed in the presence of soluble common-ion salts. From the knowledge of the dissociation constant of such salts and of their effect on the polymerization kinetics, K_{diss} for polymeric ion pairs can be determined [29–31].

In contrast to the case of radical polymerization, the same monomer can generate carbanionic propagating species of very different structures depending on the nature of the cation (Met^+) and of its environment. The various forms taken by active carbanionic species are as follows, ranked in order of their increasing reactivity (from highly covalent entities to free ions):

$R-Met < (R^{\delta-}-Met^{\delta+})_x < (R^-,Met^+)_x < R^{\delta-}-Met^{\delta+} < R^-, Met^+S_y < R^-,Met^+ < R^-,S_y Met^+ < R^- + Met^+S_y$

where R represents the macromolecular chain and S a solvent molecule. Even though the reactivity of externally solvated ion pairs is intrinsically higher than that of non-solvated contact ion pairs, the latter bring about faster rates of polymerization due to their capability to coordinate monomer.

2.2.2.3 Influence of Additives

Additives can exhibit a high solvating power (crown ethers, cryptands, tertiary pluriamines, etc.). When added in a small proportion to a hydrocarbon solvent, they can induce the solvation of ion pairs without appreciably modifying the permittivity of the reaction medium. Depending on their geometry and the nature of the cation to be solvated, they give rise to more or less "loose" or externally solvated ion pairs. Generally, they cause the disaggregation of aggregated ion pairs and are used to bring about fast initiation or to accelerate the rate of polymerization of monomers of low polymerizability.

Polystyryllithium ion pairs that are externally solvated by tetramethylethylenediamine (TMEDA) [6], as illustrated in Scheme 2.5, undergo stretching of the carbon–metal bond on addition of a crown tertiary amine, as shown schematically in Scheme 2.6 [32].

Depending on the size of the cation and the geometry of the solvating agent, solvation is more or less effective. The stretching of ion pairs increases their reactivity considerably because it facilitates the insertion of the entering monomer molecule between the anion and the cation. On the other hand, it prevents the coordination of the monomer with the cation (π-complex) and thus its activation. When solvation by an additive is only external, the interionic distance remains unchanged and the obstacle to the monomer coordination that follows results in a decrease in the intrinsic reactivity of the system [6]. Under certain conditions, the overall rate of polymerization decreases and/or side-reactions can be reduced in the presence of σ-complexes [33, 34].

In any case, the most effective method to reduce the global reactivity consists in forming "ate" complexes [35], whose composition and therefore reactivity can be easily fine-tuned [36]. These complexes are formed by addition of Lewis acids such as R_2Mg, R_3Al, R_2Zn, $(RO)_2Mg$ and $(RO)_3Al$ to the reaction medium. These Lewis acids interact with carbanionic species (or enolates in the case of acrylic polymers) to generate entities of reduced reactivity. The molecular structure of "ate" complexes was established by NMR spectroscopy [37] and X-ray diffractometry [38]. Their structure can be represented as shown in Scheme 2.7 for

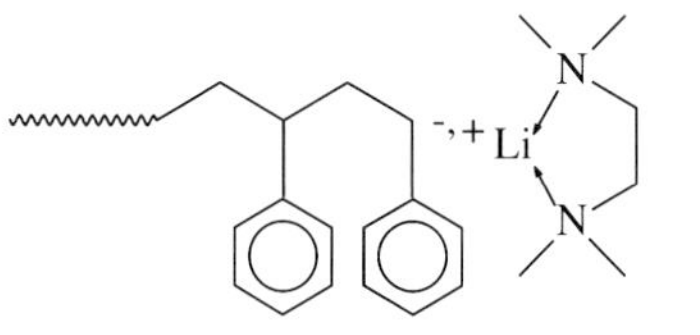

Scheme 2.5

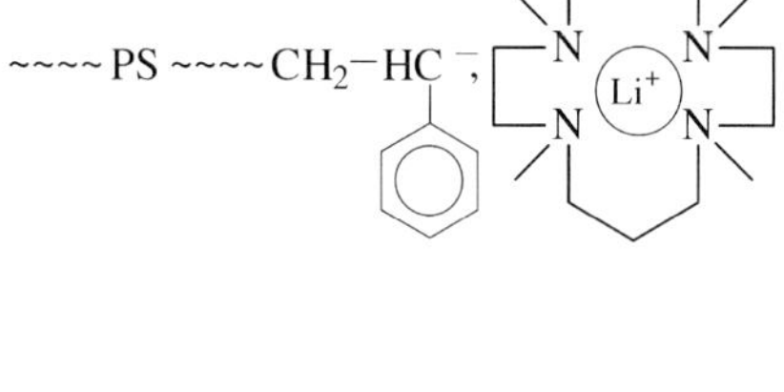

Scheme 2.6

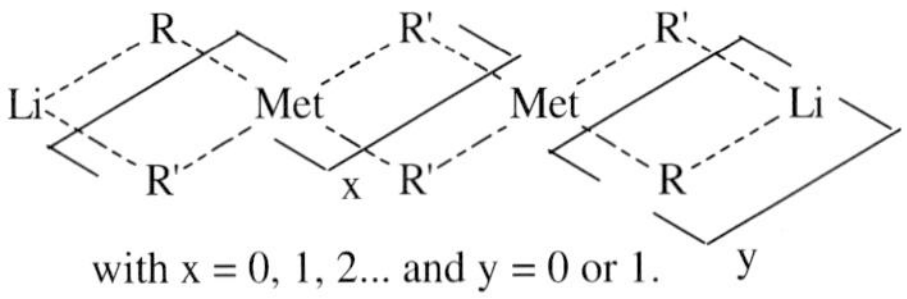

Scheme 2.7

with x = 0, 1, 2... and y = 0 or 1.

a complex formed of an alkyllithium (or a polymer–lithium) (RLi) and an alkyl or alkoxy derivative comprising a divalent or a trivalent metal (R′₂Met). Because the electron pairs are being shared between several sites, the reactivity is considerably lower than that of alkyllithiums considered alone (see Section 2.4.1.1).

2.2.2.4 Influence of the Counterion

Anionic polymerizations are generally initiated by means of organo-alkali metal compounds. Among them, butyllithium (*n*-, *sec*- or *tert*-) is by far the most often utilized. The reason is the partial covalency of the carbon–lithium bond, which entails its organo-solubility and the possibility of carrying out polymerizations in hydrocarbon media. When the lithium cation is associated with a delocalized negative charge, the carbon–lithium bond is fully ionized; given the small ionic radius of this cation, the energy of the ionic carbon–lithium bond is thus very high and the reactivity of the corresponding active centers is reduced to the same extent.

The aptitude of this cation to coordinate to monomer molecules and additives and give π-, σ- and μ-complexes, is one of its main characteristics. Complexation of monomer molecules increases their reactivity and induces regio- and stereoregulating effects, in the case of dienes in particular (see Section 2.4.4). The competing coordination of polar solvents and additives to this cation thwarts these effects, leading to behaviors similar to those observed with cations of larger ionic radius.

With other alkali metal cations, the ionic radius is larger and the ionicity of the carbon–cation bonds is more pronounced, which confers higher reactivity. The aptitude of these cations to be solvated also decreases as the ionic radius increases. If Na^+ and K^+ can be easily solvated by many solvents and additives, the solvation of Rb^+ needs favorable thermodynamic conditions and Cs^+-based ion pairs are generally unsolvated except when the structure of the solvating agent is appropriate for that cation [39].

With alkaline earth metal cations, the situation is similar to that described for alkali metal cations but with a definitely lower reactivity of active species. Thus, certain regio- and stereoregulating effects can be observed with Mg^{2+} cation [40], as for Li^+. Non-metallic cations such as R_4N^+ and R_4P^+, whose ionic radii are particularly large, are sometimes utilized for the polymerization of (meth)acrylic monomers. Due to their steric hindrance, the side-reactions on carbonyl groups are reduced [41, 42] (see Section 2.5.2).

As for other types of cations, they will be discussed when describing the reaction mechanisms they induce.

2.2.3
Experimental Constraints Related to Anionic Polymerization

The high reactivity of propagating active centers implies experimental constraints, in particular the absence of any acidic impurity in the reaction medium. Most of the atmospheric components such as H_2O, O_2 and CO_2 must also be removed and the purification of the solvents and reagents has to be carried out either under an inert atmosphere or under high vacuum. Because active centers are highly reactive and the totality of them are present at the onset of polymerization, conversion is generally fast and its control requires operation in solution.

Whether used for dilution, solvation and/or dissociation purposes, these solvents must not carry functional groups sensitive to nucleophilic reagents. For the investigations of the structure of active species and the determination of their reactivity, it is preferable that all the operations be carried out under high-purity conditions in order to avoid the presence and influence of compounds resulting from an untimely termination of organometallic species. For that purpose, it is advisable to operate under high vacuum in an entirely sealed apparatus equipped with breakseals for easy transfer of the various components. The purification of reagents and solvents is achieved by treating them with strong bases or through contact with alkali metal mirrors. A detailed description of these techniques was first published by Szwarc [43] and reviewed recently [234].

For the preparation of materials, it is more practical to operate under an inert atmosphere, even if these operations have to be carried out with the greatest care. The possible presence of compounds formed during the purification of the reagents generally causes only a minor modification of the reactivity of active centers, without influencing the structure of the polymers obtained.

2.3
Initiation of Anionic Polymerizations

For the purpose of polymer synthesis, initiators have to be selected with care to bring about a controlled polymerization, that is, with an efficiency close to one and a short period of initiation compared with that of propagation. This last criterion does not mean that the rate of initiation ($r_i = k_{i,app}[I][M]$) must be much higher than that of propagation, but it has to be in the same range of magnitude so that the totality of initiator molecules is consumed in the first moments of the chain growth.

Two different mechanisms are used to initiate anionic polymerizations, as described below.

2.3.1
Initiation by Electron Transfer

Metals whose ionization potential is not too high (lower than ~ 6.2 eV for alkali and alkaline earth metals) are able to transfer external electrons to those organic molecules that can accept them and eventually generate a radical anion due to their electron affinity [44]. In solvents with strong solvating capacity and high permittivity, spontaneous ionization can also occur, which results from the simultaneous solvation of both the cation and the electron. Thus, potassium dissolved – in small amounts – in tetrahydrofuran (THF) gives a blue solution of solvated electrons [45]:

$$K^0 \xrightarrow{\text{THF}} K^+, x\text{THF} + e^-, y\text{THF} \tag{6}$$

However, these solutions are too unstable and not sufficiently concentrated to be used for the initiation of polymerizations.

In the presence of a monomer such as styrene (S), the direct electron transfer from alkali metals to this organic molecule occurs in spite of the rather low electron affinity of the latter (~ -0.25 eV) because the transfer equilibrium is shifted towards the dimerization of radical ions ($S^{-\bullet}$) into bicarbanions (^-S–S^-) [46]:

$$2(\text{Met} + S) \rightleftharpoons 2(S^{-\bullet}, \text{Met}^+) \to \text{Met}^+, ^-S - S^-, \text{Met}^+ \tag{7}$$

The second reaction in this process, tail-to-tail radical coupling, is extremely fast, leaving no time for addition of a styrene molecule. This process was studied by Szwarc with 1,1-diphenylethylene, a non-polymerizable model molecule [47].

Although electron transfer from lithium metal to conjugated dienes is difficult to obtain in hydrocarbon solution, this reaction was utilized in industry to produce polyisoprene in bulk [48], a polymer similar to natural rubber in its molecular structure. The same method was utilized before 1940 for the preparation of Buna rubber from sodium metal [1].

The insolubility of alkali metals is a major drawback for their utilization as initiators: the initiation time is prolonged and may overlap with that of propagation. It is therefore preferable to use an intermediate electron transfer agent which permits the accumulation, isolation and stabilization of the radical ions after reaction with the metal. Polycyclic aromatic hydrocarbons are good acceptors of electrons and could be utilized for this purpose. For example, naphthalene and biphenyl, whose electron affinities are in the range 0.10–0.20 eV [49, 50], accept easily the external electron of alkali metals.

This reaction is easier to carry out in solution, in a solvent – or in the presence of an additive – with strong solvating power. The solvation of the metal cation lowers the energy level of the final state. Tetrahydrofuran and dimethoxyethane are solvents that solvate cations efficiently and thus facilitate electron

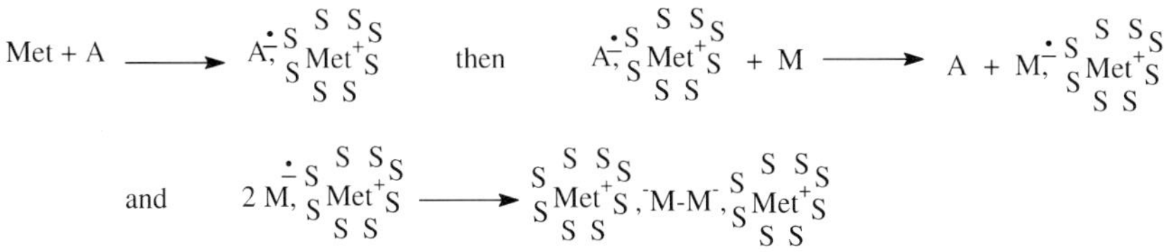

2.3 Initiation of Anionic Polymerizations

Scheme 2.8

transfer. The rate of reaction is directly proportional to the contact surface area between the solution and the metal; it is therefore preferable to use the latter in a highly divided form. With respect to macromolecular synthesis, electron transfer from the aromatic radical anion to the monomer and the dimerization of the resulting monomeric radical anion can be considered to be instantaneous. The reaction pathway is shown in Scheme 2.8, where A is the aromatic hydrocarbon, S the solvent and M the monomer. The resulting bicarbanionic species then bring about the propagation of a polymeric chain at its two ends.

2.3.2
Initiation by Nucleophilic Addition to the Double Bond

This type of initiator is the most often utilized because experiments can be carried out in either polar or nonpolar solvents. For many reasons, butyllithium in its various isomeric forms (*n*-, *sec*-, *tert*-) is the preferred initiator operating by nucleophilic addition.

2.3.2.1 In Polar Solvents

In polar solvents, butyllithium may give rise to side-reactions which may limit its utilization. In THF, for example, it can react with the solvent to give nucleophilic species of lesser reactivity [51, 52] (Scheme 2.9) or also as shown in Scheme 2.10.

The nucleophilicity of butyllithium is very high, which leads to side-reactions with certain monomers such as (meth)acrylic monomers (see Sections 2.4.2 and

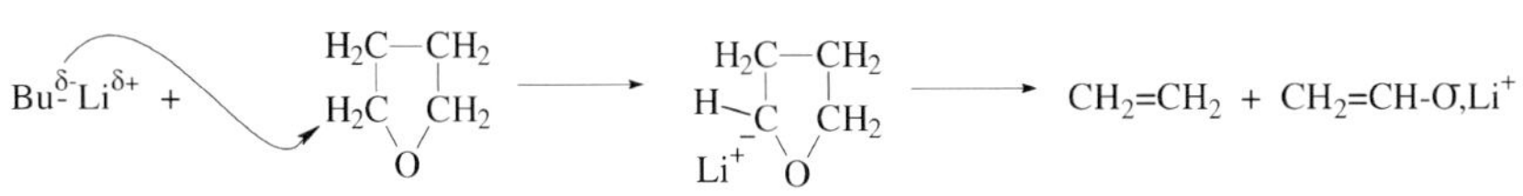

Scheme 2.9

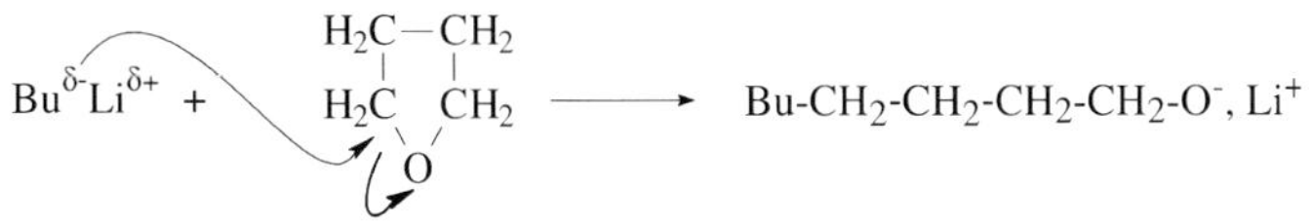

Scheme 2.10

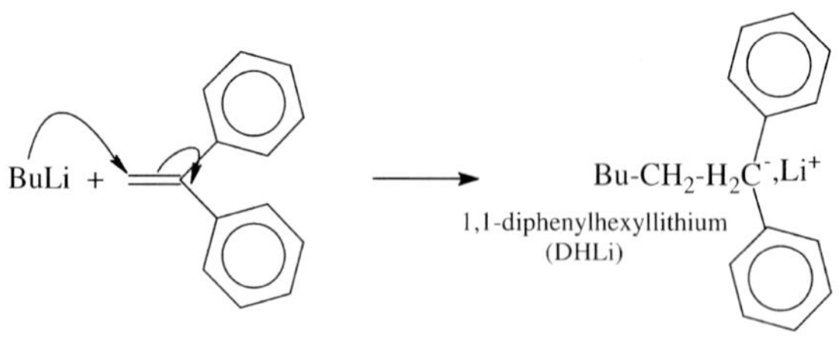

Scheme 2.11

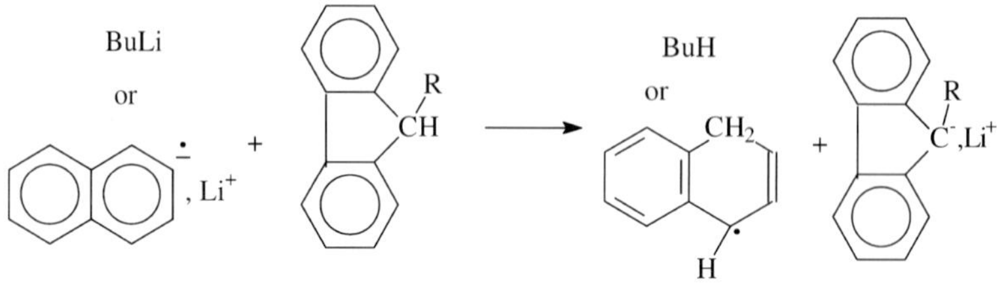

Scheme 2.12

2.5.2). This initiator can be best replaced by the product of its reaction with a non-polymerizable monomer such as 1,1-diphenylethylene, which gives 1,1-diphenylhexyllithium [53–55] (Scheme 2.11), or with molecules such as triphenylmethane or fluorenes, which possess an acidic site. In the last case, butyllithium can be replaced by naphthalene associated with various alkali metal cations. Thus, alkylfluorenyllithium can be obtained as shown in Scheme 2.12.

The resulting carbon–lithium bond is more ionized than that in BuLi but the carbanionic electron pair is strongly delocalized and the reactivity is lowered. In order to limit the side-reactions occurring during initiation of the polymerization of (meth)acrylic monomers, alkali metal enolates such as alkyl α-lithioisobutyrates – which are models of the growing chains – can be used as initiator [56]. However, they are strongly aggregated, which considerably limits their efficiency [57]; finally, it is preferable to use carbanionic species of oligo-α-methylstyryllithium in the presence of a lithium alkoxide [58].

For methacrylic monomers, initiators with very bulky non-metallic counterions and of lesser nucleophilicity are also suitable; the latter are obtained by abstraction of the acidic hydrogen of various molecules using tetrabutylammonium hydroxide [59]. Initiation by triphenylmethyl carbanions associated with tetraphenylphosphonium cation gives similar results [60]. Certain monomers of very high reactivity such as alkyl cyanoacrylates can be polymerized by the sole effect of hydroxy groups resulting from the self-ionization of atmospheric water [61].

Generally, initiators based on alkali metals other than lithium and reacting by nucleophilic addition are less often used than lithium-based initiators. However, it can be interesting to utilize monovalent initiators associated with sodium,

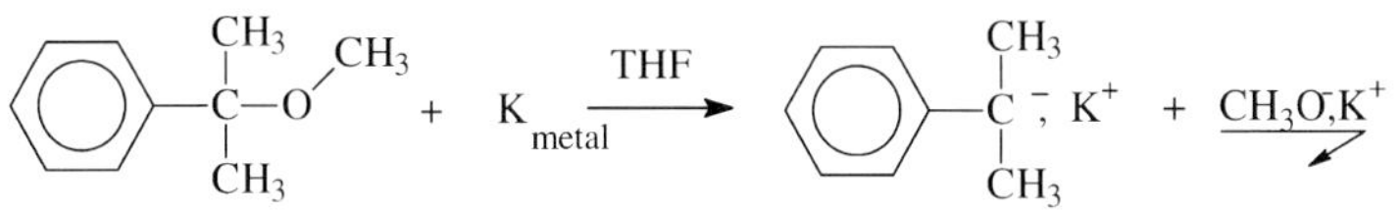

Scheme 2.13

potassium or cesium counterion; the latter are differently solvated by polar solvents because of their varying ionic radius. For instance, Cs^+ is practically unsolvated by THF.

Cumyl carbanions associated with various alkali metal cations are very reactive and therefore often used as monofunctional initiators in polar media. For instance, cumylpotassium is prepared by simple contact of methyl cumyl ether in THF solution with a potassium mirror [62] (Scheme 2.13). The potassium methoxide formed precipitates in the reaction medium and is eliminated by filtration, thus affording pure cumylpotassium.

2.3.2.2 In Nonpolar Solvents

The various isomers of butyllithium are the most commonly utilized initiators in nonpolar solvents (toluene, ethylbenzene, aliphatic or cycloaliphatic hydrocarbons, etc.). Generally, other alkaline compounds are insoluble in nonpolar solvents.

BuLi exists as aggregates whose degree of aggregation depends on the isomer considered (especially steric hindrance due the hydrocarbon moiety) and to a lesser extent on the nature of the nonpolar solvent. Thus, n-butyllithium is hexameric [63], whereas sec-butyllithium [64] and tert-butyllithium [65] are tetrameric and can be purified under high vacuum by distillation or sublimation, respectively.

Kinetic studies investigating the initiation step by organolithium compounds in hydrocarbon media show a complex behavior. Indeed, the first active centers produced after initiation form mixed aggregates with the residual initiator. The corresponding degree of aggregation, the aggregation constant and the reactivity of these aggregates are different from those of the initial initiator. They strongly influence the reaction behavior of the whole system. Due to the lack of detailed information relating to the composition of these mixed aggregates, their extent, their degree of aggregation and their potential reactivity, it is difficult to predict the behavior of these species during initiation. Only experiments and comparison with similar systems can help in making the appropriate choice of initiators with optimal reactivity. Finally, it should be stressed that because of the presence of various impurities in some commercial organolithium compounds, controversial results were published concerning the relative reactivities of alkyllithium isomers [66–68].

To give some qualitative indications, one can consider that, whatever the hydrocarbon solvent used, the order of reactivity of the main alkyllithium compounds is as follows [69]:

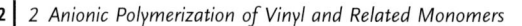

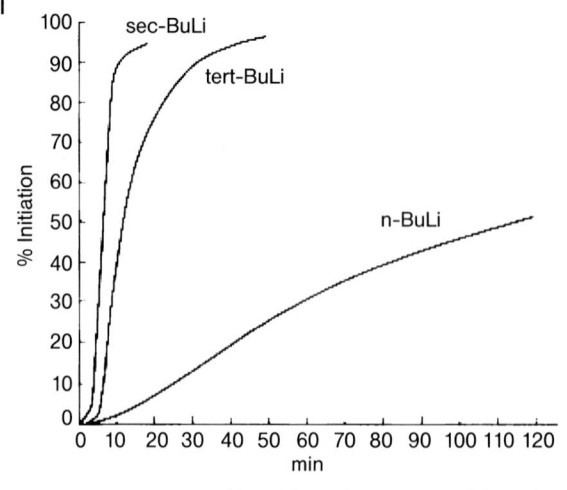

Fig. 2.1 Consumption of butyllithium by initiation of the polymerization of styrene in cyclohexane at 40 °C versus time [70].

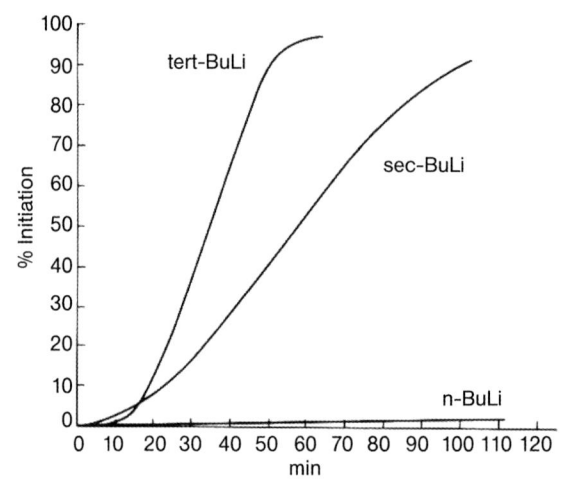

Fig. 2.2 Influence of the nature of the isomer on the consumption of butyllithium when used to initiate the polymerization of isoprene in cyclohexane at 30 °C versus time [5 e].

$$\text{menthylLi} > \textit{sec}\text{-BuLi} \approx \textit{tert}\text{-BuLi} \gg \textit{n}\text{-BuLi}$$

As an example, the results of the polymerization of styrene initiated by the three isomers of BuLi in cyclohexane solution are given in Fig. 2.1. Those corresponding to the polymerization of isoprene in hexane are shown in Fig. 2.2. Obviously the monomer influences the apparent reactivity of the initiators, causing an inversion of the relative reactivities, probably through the formation of mixed

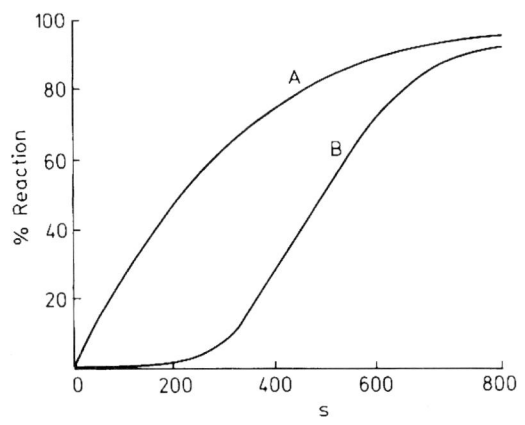

Fig. 2.3 Influence of the nature of the solvent on the consumption of butyllithium when used to initiate the polymerization of styrene as a function of time. A, benzene; B, cyclohexane [5 e].

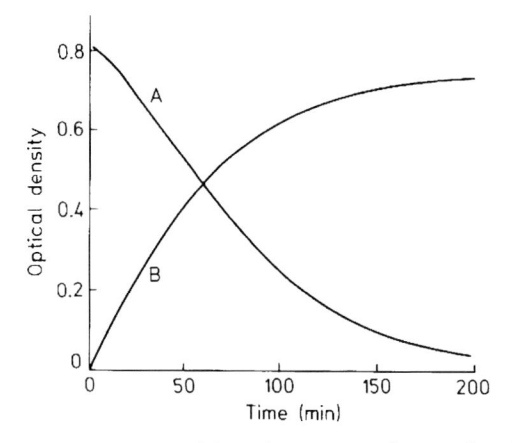

Fig. 2.4 Initiation of the polymerization of styrene by n-butyllithium in benzene solution, monitored by UV-visible spectrophotometry. A, disappearance of the monomer; B, appearance of PSLi [5 e].

aggregates of different stability and/or through the possible coordination of styrene with the lithium cation.

The nature of the hydrocarbon solvent is an essential parameter influencing the capacity of the dispersing medium to favor or not the deaggregation of organometallic species. Interactions between the π-electrons of aromatic solvents and Li$^+$ cation favor disaggregation, which cannot be the case with aliphatic solvents. Thus, the rate of initiation of the polymerization of isoprene by sec-BuLi at $[C^*] \approx 10^{-3}$ M is approximately 10^3 times faster in benzene than in hexane

[71], in spite of the competitive coordination of the monomer and the aromatic solvent to Li$^+$. With styrene, initiation in cyclohexane occurs after an induction period, which is not observed in the presence of benzene (Fig. 2.3) [72].

From the various examples above, it can be seen that depending on both the experimental conditions and the type of BuLi isomer, total initiation may be difficult to achieve before complete monomer consumption [73] (Figs. 2.4 and 2.5). The lack of control that results is mirrored in a broad distribution of molar masses. It is therefore preferable to initiate polymerizations by a solution of sec-BuLi in an aromatic solvent.

However, for polymerizations that do not require stereoregulation, fast and total initiation by n-BuLi can be obtained by adding a small amount of a solvating agent (tertiary ethers or amines) to disaggregate the organolithium even in hydrocarbon medium. For instance, the anionic oligomerization of ethylene [74] and, a fortiori, the controlled polymerization of styrene and dienes can be initiated by associating tetramethylethylenediamine (TMEDA) to n-BuLi.

However, the presence of such additives is not desirable for the polymerization of dienes because the stereoregulated propagation that affords a high content of 1,4-cis-type monomeric units in their absence is disrupted.

With certain highly reactive monomers such as vinylpyridines and (meth)-acrylic monomers, initiation by organomagnesium compounds in the absence of any solvating additive can be resorted to. With such initiators, marked stereo-regulating effects were observed and the polymeric materials formed were found to exhibit enhanced thermomechanical properties. The first attempts were made by Natta et al. [75], who initiated the polymerization of 2-vinylpyridine (2-VP) in a hydrocarbon solvent with desolvated Grignard reagents obtained by heating under vacuum. The insolubility of the initiator in this solvent prevented the control of the molar masses but the polymerization was found to be isospecific and afforded highly isotactic polyvinylpyridine.

Soluble R$_2$Mg initiators were prepared in hydrocarbon solvents by reacting organomercury compounds with a magnesium mirror [76]:

$$R_2Hg \xrightarrow{\text{Mg mirror}} R_2Mg + Hg \quad (8)$$

When R is an alkyl group, the corresponding dialkylmagnesium initiates the polymerization of highly reactive monomers [vinylpyridines and (meth)acrylic monomers] but with low efficiency [77], probably due to the partial covalency of the alkyl–Mg bond. On the other hand, dibenzylmagnesium prepared under the same conditions exhibits complete efficiency, only one chain being generated

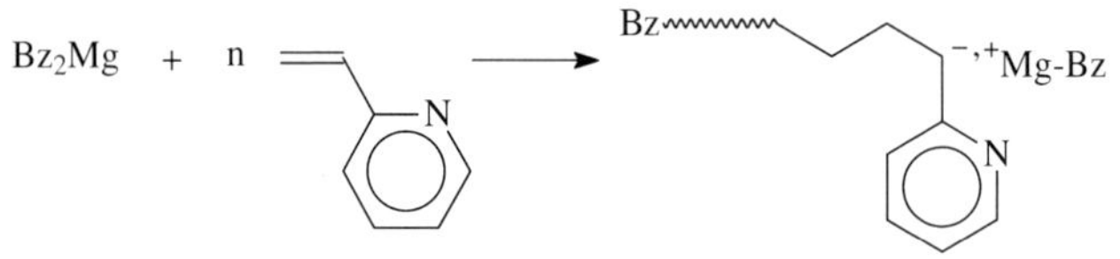

Scheme 2.14

per molecule of initiator. For instance, in the case of 2-VP, only one benzylic carbanion is active, the second remaining inactivate after insertion of the first monomeric unit (Scheme 2.14). This inactivity of one of the two carbanionic entities was confirmed by NMR spectroscopy. A similar behavior was observed with methyl methacrylate as a monomer [78].

Whatever the reaction medium, alkali metal hydrides are also able to initiate the anionic polymerization of most monomers that can be polymerized by nucleophilic addition. However, because they are strongly aggregated, they are hardly soluble in the reaction medium and their efficiency is extremely low. Recent work by Deffieux and coworkers [79] showed that when complexed with alkylaluminum derivatives they form species that are soluble in hydrocarbon solvents and able to initiate the polymerization of styrene at high temperature [80].

2.3.2.3 Bi- and Multifunctional Initiators

The synthesis of bi- and multifunctional initiators capable of initiating the polymerization of vinyl and related monomers in more than one direction has attracted much interest as a means to prepare star and multiple block copolymers.

For reactions carried out in polar solvents, bifunctional initiation is easily obtained from solutions of radical anions associated with various alkali metal cations, as mentioned in Section 2.3.1. Moreover, bifunctional initiators developed for hydrocarbon media and described in this section can also be used in polar media.

The access to initiators of higher and well-defined functionality is more difficult. Using a reaction similar to that yielding cumylpotassium, attempts to synthesize a trifunctional initiator from the reaction of potassium amalgam with 1,3,5-tris(α-methoxybenzyl)benzene in THF solution in the presence of diglyme gave disappointing results [81]. Indeed, the use of the species formed, 1,3,5-tris(benzylpotassium)benzene, as initiator of the polymerization of α-methylstyrene resulted in a mixture of stars and linear chains. In the same way, the metallation of 1,3,5-tris[2-(2'-pyridyl)ethyl]benzene by the dipotassium dimer of α-methylstyrene gave rise to a trifunctional initiator of low efficiency [82]. The compound resulting from the trismetallation of 1,3-cycloheptadiene by a Lochmann's base (n-BuLi/$tert$-BuOK) [237] might be a satisfactory initiator but no convincing characterization of the polymers formed was reported.

Satisfactory results were obtained by Quirk and Tsai [83], who resorted to higher homologs of 1,1-diphenylethylene as multifunctional precursors. The reaction is shown in Scheme 2.15 for a trifunctional initiator. This initiator was synthesized in a nonpolar solvent but required the addition of a small amount of a σ-solvating additive to bring about a fast and efficient initiation of styrene or butadiene polymerization [84].

Initiators with higher functionality were prepared by Vasilenko et al. [85] from dendrimers carrying 8 or 16 internal unsaturations and 16 or 32 decyl groups at their periphery. In the presence of a polar additive, sec-BuLi could add to the un-

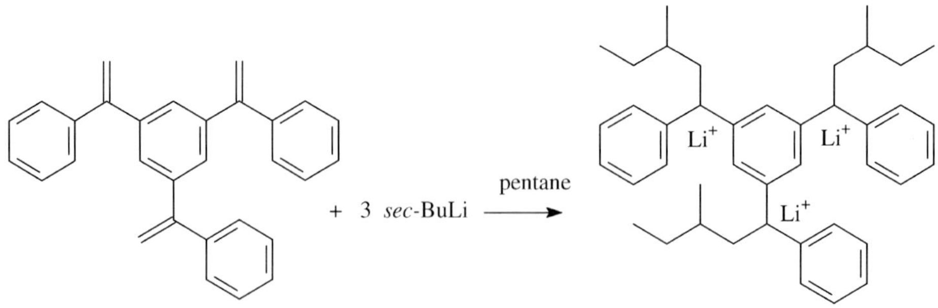

Scheme 2.15

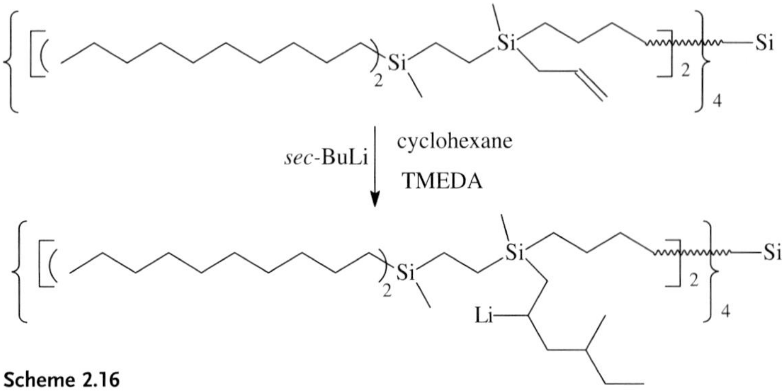

Scheme 2.16

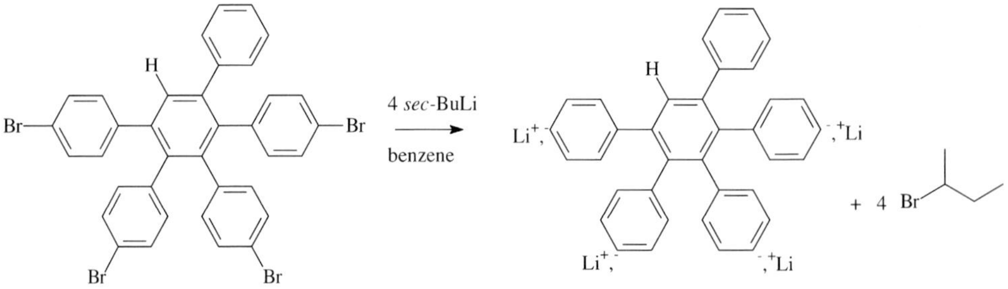

Scheme 2.17

saturations and give a dendrimer comprising 8 or 16 carbanionic entities able to initiate the polymerization of styrene. An example is shown in Scheme 2.16. Such an octafunctional system was found to be soluble in many solvents because of the combined solubilizing and hindering effects of peripheral decyl groups.

In a recent addition, aryl plurihalides were successfully lithiated by *sec*-BuLi to obtain bi-, tri- and tetrafunctional initiators soluble in hydrocarbons in the presence of σ- or μ-coordinating additives. The preparation of this tetrafunc-

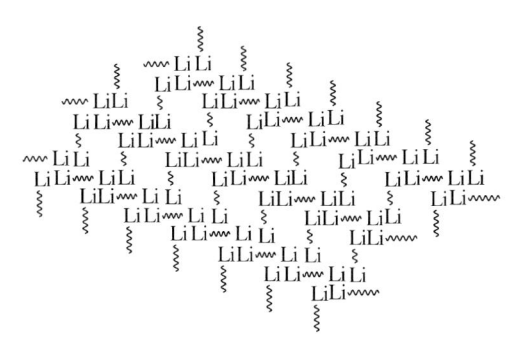

Fig. 2.5 Representation of the aggregation of a dilithium derivative with a degree of aggregation of 4.

tional derivative is illustrated in Scheme 2.17. The lithiated derivatives of this family initiate the polymerization of styrene and conjugated dienes, as demonstrated by the characterization of the resulting polymers [86].

Bi- and multifunctional lithium-based initiators that could be utilized in hydrocarbon solvents in the absence of any solvating additive are of strong interest. Indeed, they are useful for the preparation of telechelics and of triblock copolymers containing a 1,4-*cis*-polydiene central block and intended for application as thermoplastic elastomers. For the stereoregulation of the polydiene block to occur, it is essential that the medium does not contain any additive that could compete with the diene monomer for the complexation to the Li^+ cation. In the absence of such additives, organolithium derivatives with a functionality ≥ 2 tend to aggregate and form a kind of macromolecular three-dimensional network (Fig. 2.5), hence their insolubility in hydrocarbons.

Such a tendency to undergo aggregation can be disrupted, however, by steric and/or solubilizing effects due to the organic part of the organodilithium derivative. Indeed, when the free enthalpy of solubilization is higher than that of aggregation, the organometallic molecule becomes soluble in the reaction medium. The energy of aggregation can be lowered by increasing the relative distance between lithium atoms (steric effects, dissymmetry, etc.) and the energy of solubilization can be increased through either the size of the organic moiety or its interactions with the dispersing medium.

Soluble dilithium initiators were obtained by reaction of *sec*-BuLi with 1,3-diisopropenylbenzene [87–89]. For reasons of steric hindrance, this bifunctional monomer has a low ceiling temperature; it does not homopolymerize and the

Scheme 2.18

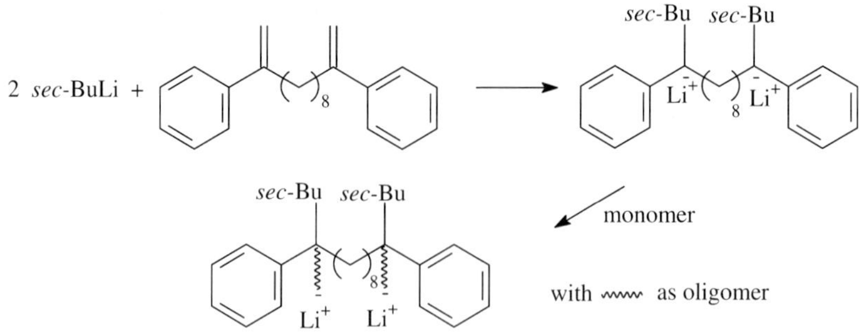

2 Anionic Polymerization of Vinyl and Related Monomers

Scheme 2.19

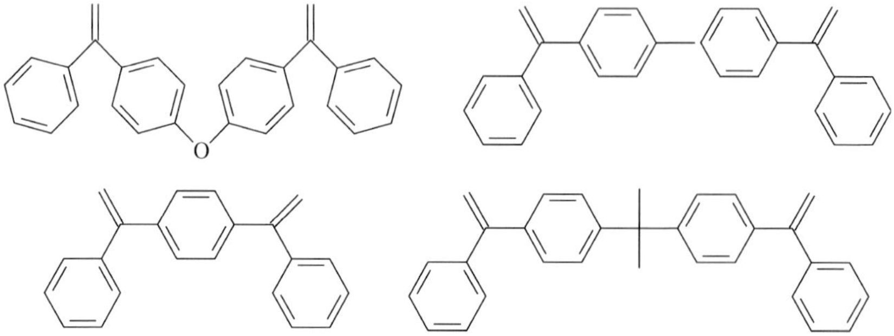

Scheme 2.20

dilithium derivative which is moderately aggregated remains soluble in hydrocarbons. With the meta isomer, isoreactivity of the two ethylenic double bonds was observed (Scheme 2.18). However, the perfect bifunctionality of the corresponding initiators was questioned, probably because of partial aggregation, and the presence of an ether in the reaction medium was shown to be necessary to obtain initiation in two directions [90, 91].

A recent study by Hofmans and Van Beylen [92] showed that addition of π-complexing agents such as durene and tetraphenylethylene induces at least partial deaggregation of the dilithium compounds through weak interactions with the lithium cations, without disturbing the stereoregulating mechanism.

The energy which is necessary to deaggregate the dilithium initiators can also be provided by the free enthalpy of solubilization of small hydrocarbon chains. Thus, dilithium initiators were prepared by reacting sec-BuLi with bifunctional molecules that are higher homologs of α-methylstyrene [93–95]. The resulting dilithium compounds were insoluble and could be purified by filtration. On addition of monomer to finely dispersed particles of such dilithium derivatives, short chains (seeds) could be generated, entailing dissolution of the bifunctional initiator in a hydrocarbon solvent. An example of this type of initiators is shown in Scheme 2.19.

By the same principle, satisfactory results were obtained by reaction of sec-BuLi with difunctional precursors, homologs of 1,1-diphenylethylene [96]. Although their intrinsic reactivity is low, they can be used to initiate the polymerization of main hydrocarbon monomers, dienes in particular (Scheme 2.20).

Starting from a diaryl halide containing a side C_{15}-alkyl chain, Gnanou et al. recently obtained a hydrocarbon-soluble, additive-free, dicarbanionic organolithium initiator by selective halogen–lithium exchange reaction. They used the dilithiated species formed to synthesize poly(styrene-b-butadiene-b-styrene) triblock copolymers with a high content in 1,4-units and excellent mechanical properties [97].

2.3.3
Initiation by Alkoxides and Silanolates

Generally, these species are not reactive enough to add to carbon–carbon double bonds. However, they can be used to release the high ring strain of cyclics such as cyclodisilane derivatives and produce in this way fairly nucleophilic species [98]. Both silanolates and macromolecular alkoxides were reacted with such cyclics. The reaction pathway for silanolates is shown in Scheme 2.21 and for alkoxides in Scheme 2.22. One can thus climb up the ladder of nucleophilic reactivity by this means, but the efficiency of these initiators is only partial.

2.3.4
Initiation of the Polymerization of Alkyl (Meth)acrylates by Group Transfer

Group transfer polymerization (GTP) was discovered in the early 1980s by a team at DuPont de Nemours [99]; when activated by nucleophilic catalysts, it proceeds by repeated nucleophilic additions but applies only to (meth)acrylic monomers. "Living" and controlled polymerization of (meth)acrylic monomers can be achieved at room temperature and under relatively mild experimental conditions [100, 101] by GTP.

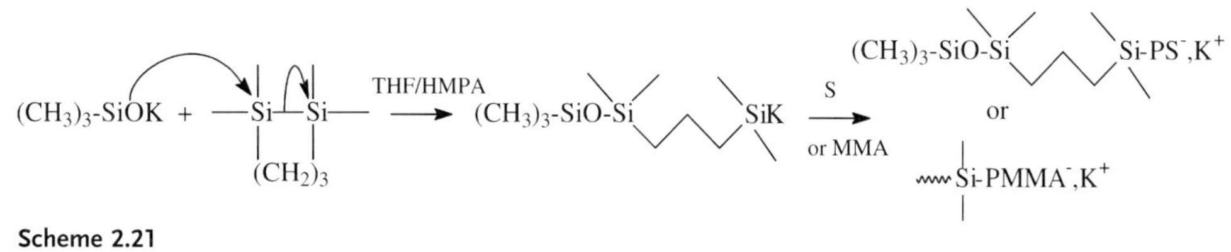

Scheme 2.21

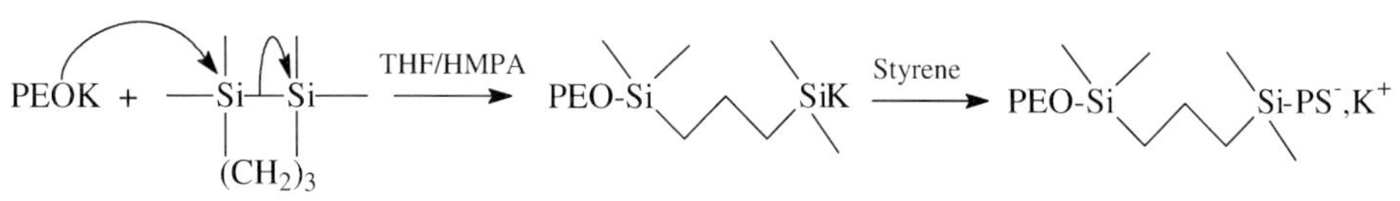

Scheme 2.22

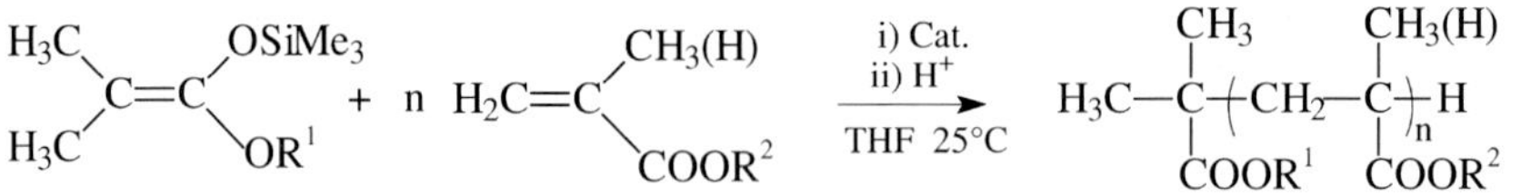

Scheme 2.23

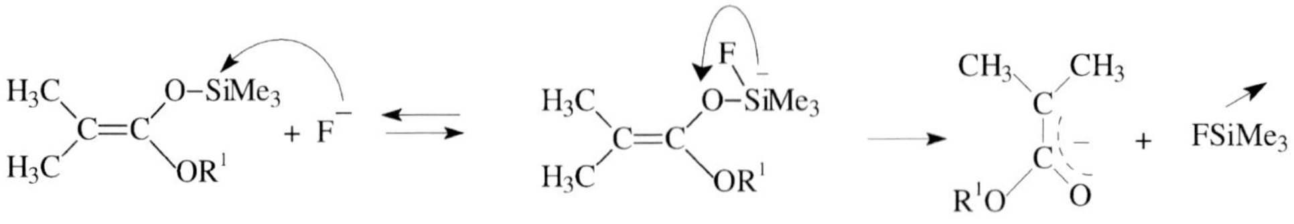

$$H_3C\text{-}C(CH_3)=C(OSiMe_3)(OR^1) + n\ H_2C=C(CH_3(H))(COOR^2) \xrightarrow[\text{THF 25°C}]{\text{i) Cat.}\ \text{ii) H}^+} H_3C-\underset{COOR^1}{\underset{|}{C}}(CH_3)-[CH_2-\underset{COOR^2}{\underset{|}{C}}(CH_3(H))-H]_n$$

Scheme 2.23

Scheme 2.24

$$H_3C\text{-}C(CH_3)=C(O\text{-}SiMe_3)(OR^1) + F^- \rightleftarrows H_3C\text{-}C(CH_3)=C(O\text{-}SiMe_3(F^-))(OR^1) \rightarrow \underset{R^1O}{\overset{CH_3\ CH_3}{C}}\underset{O}{\overset{C^-}{C}} + FSiMe_3$$

Scheme 2.24

The initiator is a silyl dimethylketene acetal [1-methoxy-2-methyl-1-trimethyl-siloxypropene (TMS)], which becomes active only in the presence of a "catalyst". The reaction pathway is shown in Scheme 2.23.

The degree of polymerization obtained is determined primarily by the molar ratio of monomer to initiator (TMS), the "catalyst" concentration influencing only the rate of polymerization. This "catalyst" can be a nucleophilic species, those recommended for their efficiency being fluoride (F$^-$) [102] or bifluoride (HF^{2-}) anions contained in salts such as tris(piperidino)sulfonium bifluoride, [(C$_5$H$_{10}$N)$_3$S$^+$,HF^{2-}] [103], and tetrabutylammonium fluoride (n-C$_4$H$_9$)N$^+$,F$^-$ [104], all soluble in the reaction medium (Scheme 2.24).

Electrophilic "catalysts" of Lewis acid type such as zinc halides and dialkylaluminum chlorides (AlR$_2$Cl) [105] work better for the GTP of acrylates in aromatic hydrocarbon or alkyl chloride solution.

Various mechanisms of propagation that will be discussed in Section 2.4.2.4 have been proposed to account for the experimental observations [106, 107].

From this section on initiation, one can observe that the choice of an initiator is often a compromise between the necessity to achieve a relatively short initiation step with highly reactive organometallic species and the obligation to avoid side-reactions which would deactivate part of the active species by reaction with the monomer or the solvent.

2.4
Propagation Step

With vinyl and related monomers, this step proceeds by nucleophilic addition to their electron-deficient carbon–carbon double bond, after their possible coordination to the metal cation. It is this last sub-step which determines the kinetics of propagation [10].

2.4.1
Kinetics of the Propagation Step

Taking into consideration all possible structures of active centers, the general equation describing the kinetics of propagation can be written as

$$r_p = -\frac{d[M]}{dt} = [M]\sum_i k_{pi}[C_i^*] \tag{9}$$

where k_{pi} represents the absolute rate constant of homopropagation of the monomer M by one structure of active centers (i) whose concentration is $[C_i^*]$. This relation considers that each elementary reaction is bimolecular and first order in monomer and in active centers. It can be adapted as a function of the experimental conditions used to carry out the polymerization.

2.4.1.1 Kinetics of Polymerization in Non-polar Solvents

The plot of $\log(r_p/[M]) = \log[C^*]$ reveals that the polymerization of hydrocarbon monomers initiated by organolithium compounds exhibits a fractional kinetic order with respect to the active centers $[C^*]$. This result can be logically interpreted as being due to the aggregation of active chains as represented in Scheme 2.25, where x is the degree of aggregation and $K_{ag} = [C_x^*]/[C_1^*]^x$ the constant of aggregation of the system considered. Only the reactivity of unimeric species (C_1^*) are to be taken into account.

Aggregated species can be viewed as a reservoir containing "dormant" species in equilibrium with active unimeric entities. The general kinetic equation can thus be written in the form

$$r_p = k_p[M][C_x^*]^{1/x} K_{ag}^x \tag{10}$$

Since the constant of aggregation (K_{ag}) is generally very high, the total concentration in (active or "dormant") chain ends $[C^*]$ is not very different from $x[C_x^*]$, the rate of polymerization can then be written as

$$r_p = k_{p1}[M] K_{ag}^x ([C^*]/x)^{1/x} \tag{11}$$

For example, polystyryllithium-active centers in benzene solution at 30 °C are aggregated into dimeric species with a constant of aggregation estimated as $K_{ag} = [(PS^-,Li^+)_2]/[PS^-,Li^+]^2 \approx 2 \times 10^6$ L mol^{-1}, the rate constant of propagation of unimers being of about 17 L mol^{-1} s^{-1} [108]. For a concentration in active cen-

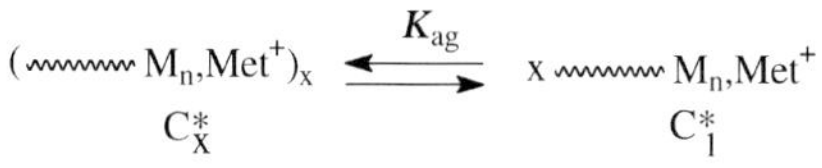

$$(\sim\sim\sim M_n, Met^+)_x \underset{}{\overset{K_{ag}}{\rightleftarrows}} x \sim\sim\sim M_n, Met^+$$
$$C_x^* \hspace{4cm} C_1^*$$

Scheme 2.25

$$(\text{\textasciitilde dienyl}^-, \text{Li}^+)_4 \underset{}{\overset{K_{ag,4}}{\rightleftharpoons}} 2(\text{\textasciitilde dienyl}^-, \text{Li}^+)_2 \underset{}{\overset{K_{ag,2}}{\rightleftharpoons}} 4 \text{\textasciitilde dienyl}^-, \text{Li}^+$$

Scheme 2.26

ters of 1.0×10^{-3} mol L^{-1}, which corresponds to a half-polymerization time of ~60 min, a value much higher than that required for complete initiation by n-, sec- or tert-BuLi.

In the case of dienes, the value of the degree of aggregation of active chains is very controversial [109–111]. The existence of a double equilibrium between tetramers (and possibly hexamers), dimers and unimers can account for most of the phenomena observed [112] (Scheme 2.26).

Thus, at high concentrations in active centers, the species would be mainly in the form of tetrameric aggregates, whereas at the lowest accessible concentrations they would be dimeric. The preponderance of one form over the other probably changes at about 10^{-7} mol L^{-1} in the case of polyisoprenyllithium [113]. Really active unimeric species remain in a very small minority, however. In addition, the equilibrium constants are thought to depend on the nature of the solvent and on the molar mass of the polydiene [114], thus varying progressively with the extent of the polymerization. Kinetic equations are then difficult to establish, and also equilibrium constants and rate constants. It is therefore more judicious to compare half-polymerization times with those of styrene, which turn out to decrease. For instance, for a concentration in polyisoprenyllithium-active centers equal to 1.4×10^{-3} mol L^{-1} – a concentration range where the species are aggregated into tetramers – in benzene solution at 30 °C, the half-polymerization time of isoprene is ~45 min [115].

When faster polymerizations are expected and the stereochemistry of the propagation is not particularly desired, species known to solvate the cation and disaggregate aggregated species are added to the medium. Kinetic studies were carried out to investigate the effect of small amounts of THF [116] or TMEDA [6] on the propagation of polystyryllithium. In the case of TMEDA, there is a total deaggregation of (PSLi)$_2$ for [TMEDA]:[Li]=1, the resulting externally solvated ion pairs being less reactive (k_p=0.15 L mol^{-1} s^{-1} at 25 °C) than non-solvated unimeric ion pairs (k_p=17 L mol^{-1} s^{-1} at 30 °C). The addition of TMEDA thus causes an increase in the overall rate of polymerization at high concentrations in active centers and a decrease at low concentrations. In other words, depending on the concentration range, either the proportion of really active chains or their reactivity can predominate [117]. The situation is more complex when small amounts of THF are added because various degrees of solvation can be obtained. The case of the addition of a π-coordinating agent such as durene is similar to that of a σ-coordinating agent such as THF, but requires the addition of larger amounts than for THF due to its lower coordinating power [7].

If the solvating additive causes stretching of the carbon–metal bond, the reactivity of the corresponding species is strongly enhanced. For instance, tetramethyltetraazacyclotetradecane, a cyclic tertiary tetramine which is a suitable

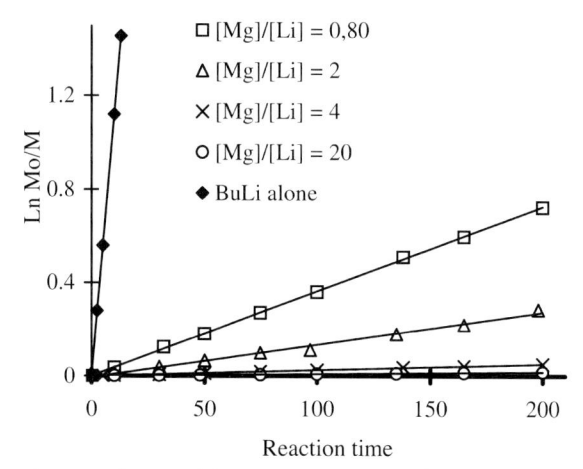

Fig. 2.6 Influence of the ratio $r=[Mg]/[Li]$ on the rate of polymerization of styrene initiated by sec-BuLi/n-, sec-Bu$_2$Mg in cyclohexane solution at 50°C [120].

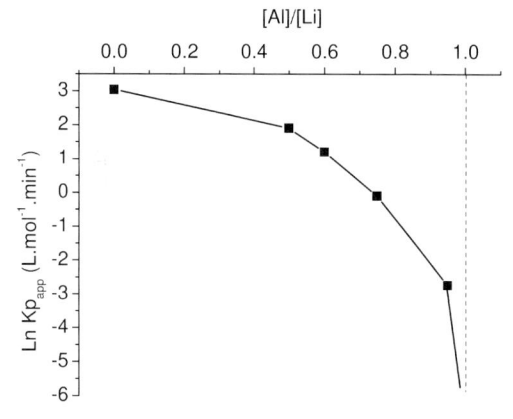

Fig. 2.7 Influence of the ratio $r=[Al]/[Li]$ on the rate of polymerization of styrene initiated by sec-BuLi/i-BuAl$_3$ in cyclohexane solution at 50°C [121].

solvating agent for Li$^+$ cations and is insensitive to the attack of carbanionic species, deaggregates totally polystyryllithium dimers and increases considerably their reactivity (k_p=750 L mol^{-1} s^{-1} in cyclohexane at 30°C) [118]. Also, when larger amounts of THF than mentioned above are added, again solvent-separated or loose ion pairs may be formed and the rate increases [119].

Kinetic studies on the polymerization of styrene initiated by "ate" complexes showed that the reactivity of the propagating species can be dramatically lowered and polymerizations carried out in bulk and at high temperature. Figures 2.7 and 2.8 show the variation of reactivity of the active species as a function of the increasing proportion of Lewis acids in the case of polystyryllithium. Hence it can be seen that with R_2Mg as an additive the reactivity decreases gradually as $r=[Mg]/[Li]$ varies from 0 to 20 [120], whereas with R_3Al the reactivity decreases drastically and reduces to zero for $r=[Al]/[Li]=1$ [121].

2.4.1.2 Polymerizations Carried Out in Polar Media

The combination of solvating and dissociating effects generally results in the partial formation of free ions whose reactivity is much higher than that of ion pairs. For instance, in the case of polystyrylsodium in THF solution, the equilibria shown in Scheme 2.27 coexist.

As the constant of solvation of ion pairs is very high, it can be considered that unsolvated ion pairs exist in negligible proportion. The solvated ion pairs consist of externally solvated contact (or tight) ion pairs of low reactivity in fast equilibrium with solvent separated (or loose) ion pairs, the proportion of which increases with decreasing temperature and whose reactivity is much higher than that of tight ion pairs. Although in very small proportion, free ions cannot be overlooked because of their very high reactivity. For the polymerization of styrene initiated by an organosodium compound at 25 °C in THF solution, the dissociation constant of (solvated) ion pairs into free ions (K_{diss}) was measured by conductometry [122] as equal to 1.5×10^{-7} mol L^{-1}. The kinetic equation of the propagation step for such a system can be written as

$$r_p = [M] \sum_i k_{p,i}[C_i] = [M](k_{p,\pm}[C_\pm] + k_{p,-}[C_-]) \tag{12}$$

where $[C_\pm]$ and $k_{p,\pm}$ represent the concentration of ion pairs and the absolute rate constant of propagation, respectively, and $[C_-]$ and $k_{p,-}$ correspond to the concentration of free ions and their rate constant of propagation, respectively.

As $[C_\pm]$ is not very different from the total concentration of active species $[C^*]$, one obtains

$$[C_-] = K_{diss}^{\frac{1}{2}}[C^*]^{\frac{1}{2}} \tag{13}$$

and

$$r_p = [M][C^*]\underbrace{\left\{ k_{p,\pm} + k_{p,-} K_{diss}^{\frac{1}{2}}[C^*]^{-\frac{1}{2}} \right\}}_{k_{app}} \tag{14}$$

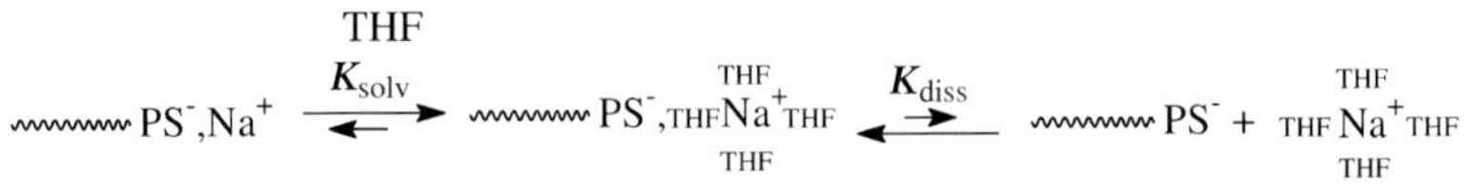

$$\sim\!\!\sim\!\!\sim PS^-,Na^+ \quad \underset{\longleftarrow}{\overset{THF}{\underset{K_{solv}}{\longrightarrow}}} \quad \sim\!\!\sim\!\!\sim PS^-,_{THF}Na^+_{THF} \quad \underset{\longrightarrow}{\overset{K_{diss}}{\longrightarrow}} \quad \sim\!\!\sim\!\!\sim PS^- \;+\; _{THF}Na^+_{THF}$$
(with THF solvation shown around the ion pair and free ion)

Scheme 2.27

Table 2.2 Ion pair/free ion equilibrium constants and the corresponding rate constants for anionic polymerization of styrene initiated by organo-alkali metal derivatives in THF solution [123, 124].

Counterion	K_{diss} (10^7 mol L^{-1})	$k_{p,\pm}$ (L mol^{-1} s^{-1})	k_{p-} (10^{-4} L mol^{-1} s^{-1})
Li^+	1.9	160	–
Na^+	1.5	80	6.5
K^+	0.7	60	–
Cs^+	0.028	22	6.3

Table 2.3 Propagation rate constants for anionic polymerization of styrene initiated by organo-alkali metal derivatives in dioxane solution at 25 °C [126].

Counterion	$k_{p,\pm}$ (L mol^{-1} s^{-1})
Li^+	0.9
Na^+	3.4
K^+	20
Rb^+	21.5
Cs^+	24.6

The plot of the apparent rate constant (k_{app}) versus $[C^*]^{-\frac{1}{2}}$ for various concentrations in active centers, followed by extrapolation to $[C^*]^{-\frac{1}{2}}=0$ of the corresponding straight line, yields $k_{p,\pm}$. From the knowledge of K_{diss} by another means, $k_{p,-}$ can be determined as the slope of this line.

The role of temperature can be complex as it can affect in opposite ways the solvating power of the medium, its dielectric constant and the absolute energy of activation. Due to the exothermic formation of highly reactive loose ion pairs, low in concentration at room temperature but increasing at lower temperature, the decrease in temperature may result in some cases (e.g. in THF) in an increase in the overall rate of polymerization, corresponding to an apparent negative activation energy in part of the temperature domain [235, 236].

Table 2.2 shows the values of the constants measured for the polymerization of styrene in THF, allowing one to calculate the proportion of chains that propagate through free ions. Thus, for $[C^*]=1\times10^{-4}$ mol L^{-1}, the proportion of free ions is about 10% and, due to their high reactivity, they contribute to 90% of the propagation. Hence the effect on the regio- and stereoselectivity of the propagation can be affected.

It is interesting to compare the rate constants of propagation on ion pairs measured in THF with those measured in dioxane. In the latter, there is no loosening of ion pairs and the rate constants are close to those measured in hydrocarbon solvents [125]; they reflect the direct influence of interionic distances (Table 2.3).

2.4.2
Anionic Polymerization of (Meth)acrylic Monomers

2.4.2.1 General Characteristics

This family of monomers exhibit specific problems that are related to their high polymerizability and to the presence of a carbonyl group that is sensitive to the attack by nucleophilic reagents and may cause chain termination [127] (see Section 2.5.2). Hence it is of the utmost importance to lower the reactivity of propagating species in order to achieve control of the propagation [128].

As shown in Scheme 2.28 for the polymerization of MMA, the active propagating centers are enolates whose intrinsic reactivity is considerably lower than that of pure carbanionic species. From kinetic studies carried out in THF solution [129–133], the complex behavior that characterizes these systems could be unraveled: non-aggregated ion pairs are in equilibrium with both free ions present in negligible proportion and dimeric aggregates whose reactivity is negligible. The behavior of these systems is even more complicated when using bifunctional initiators [134]. In spite of the low intrinsic reactivity of propagating species, these systems are characterized by a high rate of polymerization due to the strong reactivity of the monomer [135].

Various methods have been contemplated to improve the control of the polymerization of methyl methacrylate (MMA) in particular, which has been the subject of most of the investigations. All these methods are aimed at decreasing the intrinsic reactivity of both the initiator and of the growing ion pairs and/or their accessibility to the carbonyl groups in order to limit side-reactions. These aspects will be discussed in Section 2.5.2.

An exhaustive compilation pertaining to the control of the polymerization of (meth)acrylic monomers can be found in a recent review by Baskaran [136]. A decrease in the temperature of reaction, increase in the ionic radius of the counterion and of the covalent character of the enolate/counterion bond, addition of σ-, μ- or σ,μ-complexing agents and conversion into "ate" complexes are the solutions that have been explored with more or less satisfactory outcomes to improve the persistence of active centers and control propagation better (see Section 2.5.2). Group transfer polymerization, which is discussed below, was one of the solutions proposed to overcome side-reactions.

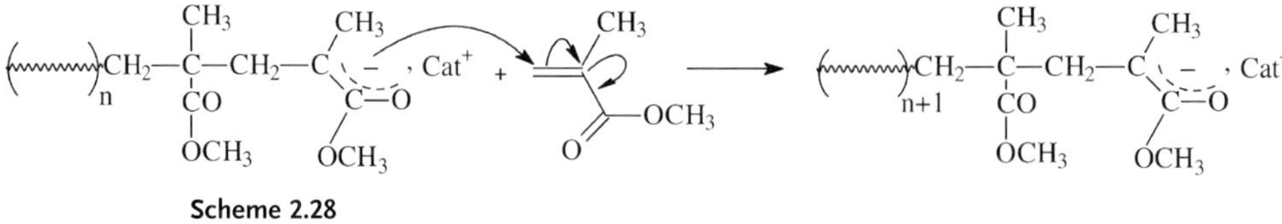

Scheme 2.28

2.4.2.2 Propagation by Group Transfer

When catalyzed by nucleophilic entities (Nu$^-$) such as Bu$_4$N$^+$,F$^-$ (see Section 2.3.3), active centers take the structure of enolates (the same structure as with alkali metal counterions). Because of the nature of the counterion, in particular its size, their reactivity is strongly reduced compared with that of lithium-based systems. Moreover, they are present in lower concentrations because most of the growing chains exist in a "dormant" form in fast exchange with really active enolate centers [238]. The mechanism of propagation can be represented as shown in Scheme 2.29.

It should be noted that the propagation reaction occurs through nucleophilic attack by the carbon atom of the enolate whereas the resilylation of the active chain by intermolecular exchange of the trimethylsilyl group involves the oxygen atom of the same enolate.

This mechanism is called irreversible "dissociative" because the cleavage of (CH$_3$)$_3$–Si is irreversible and the active species formed is completely ionized. A more controversial "reversible dissociative" version postulating the formation of an ester enolate and its resilylation by (CH$_3$)$_3$–Si–Nu was also contemplated.

With weak nucleophiles, an "associative" mechanism was proposed which implies the pentacoordination of silicon and mainly covalent active species (Scheme 2.30).

In the case of catalysis by Lewis acids, which is particularly suitable for the polymerization of acrylates, propagation occurs through activation of the monomer by a different mechanism.

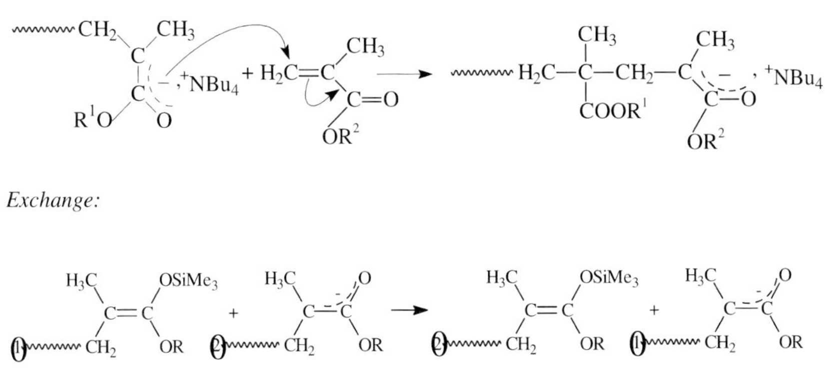

Scheme 2.29

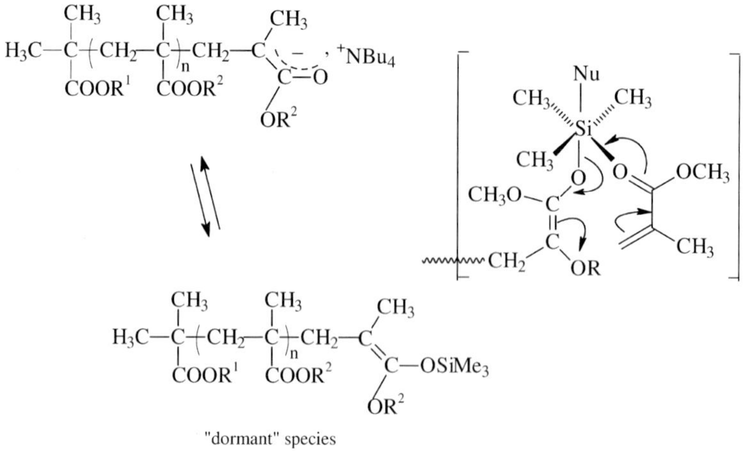

Scheme 2.30

2.4.3
Anionic Copolymerization

In contrast to the case of radical polymerization, monomers that can be polymerized anionically and the propagating centers they generate exhibit large differences with respect to their reactivity. If the Mayo–Lewis method can be applied to predict the structure of statistical copolymers, one has to remember that it is valid only when one type of active center exists at any one time. Similarly to the rate constants of homopropagation, the rate constants of cross-addition (k_{AB}) which appear in the expression of the reactivity ratios are given by $k_{AB} = \sum_i k_{AB,i}([A_i^*]/[A^*])$, where $k_{AB,i}$ is the rate constant of cross-addition of the active species of structure $[A_i^-]$ and concentration $[A_i^*]$ whose proportion is equal to $[A_i^*]/[A^*]$. The multiplicity of structures of active centers which vary with the counterion, the concentration, the temperature and the solvent used prevents an easy prediction of the outcome of such statistical copolymerizations. A few values of reactivity ratios (r_1 and r_2) are shown in Table 2.4 for styrene–butadiene under well-defined experimental conditions. One can observe the marked influence of the experimental conditions.

For the preparation of well-defined block copolymers, the k_{AB}/k_{BB} ratio must be as high as possible to prevent the second block from growing too fast while all the chain-ends of the first block have not been transformed.

However, the large difference in reactivity between monomers and between active centers imposes the requirement to carry out copolymerization according to a precise sequential order even if certain studies show that this rule does not necessarily apply for certain systems [80] (see Section 2.3.4). The cause of such a restriction is twofold. On the one hand, the low reactivity of active centers generated from the most reactive monomers (acrylics, etc.) prevents them from

Table 2.4 Styrene–butadiene anionic copolymerization initiated by organolithium derivatives: influence of the solvent on the reactivity ratios [137].

Solvent	Temperature (°C)	r_1 (styrene)	r_2 (butadiene)
Benzene	25	0.04	11.2
Hexane	25	0.04	10.8
THF	25	4	0.3
THF	−78	0.1	2.8
Diphenyl ether	25	11	0.04

attacking less reactive monomers (styrene, diene, etc.). On the other hand, the high reactivity of species (polystyryl⁻, polybutadienyl⁻, etc.) may cause side-reactions to occur on the chemical groups carried by the most reactive monomers (vinylpyridines, acrylics, etc.).

To obtain well-defined block copolymers by anionic means, monomers have to be polymerized sequentially in order of increasing electroaffinity; for monomers of equivalent reactivity, the order of polymerization matters less (see Section 2.5.2).

2.4.4
Regio- and Stereoselectivity in Anionic Polymerization

2.4.4.1 Cases of Conjugated Dienes

Because of the two conjugated double bonds, there exist (with isoprene in particular) four isomeric enchainments: 4,1-, 1,4-, 1,2- and 3,4-. Whatever the counterion and the nature of the solvent, the allylic active center is delocalized on the three final carbon atoms with an additional negative polarization on the ultimate carbon atom. At first glance, this situation seems favorable to the sole 4,1- (or 1,4-) enchainment [138], but the fact that it is not always formed means that the structure of the inserted monomeric unit is not determined by that of the unit carrying the active center but rather by that of the transition state for the next monomer addition [139]. The latter is very dependent on the nature of the counterion and its environment. Table 2.5 gives an idea of the influence of the polymerization conditions on the molecular structure of polydienes; the configuration of the polydiene formed can also depend on the concentration in active centers.

For 1,4- (or 4,1-) polydienes, geometric isomerism plays an essential role in the structural and thermomechanical properties of polymers. E (or *cis*) isomers are much more interesting than Z (or *trans*) isomers. For polymerizations carried out in hydrocarbon solvents and with lithium as counterion, Table 2.5 shows that 1,4-*cis*-polydienes can be obtained with a high ratio of geometric regularity. The mechanism leading to E (*cis*) isomers is not easy to establish because, more than the geometric configuration of the active centers in the aggre-

Table 2.5 Examples of configuration of polydienes prepared by anionic polymerization as a function of the polymerization conditions.

Polydiene	Counterion	Solvent	T (°C)	Configuration (%): 1,4-cis/1,4-trans/1,2-(3,4-)	Ref.
Polybutadiene	Li⁺	None	20	39/52/9	140
	Li⁺	Hexane	20	30/62/8	140
	Li⁺–TMEDA	Cyclohexane	5	15/85	141
	Li⁺	THF	0	13/87	142
	Li⁺–diglyme (1:1)	Toluene	30	17/83	142
	K⁺	THF	0 to –78	33/67	140
Polyisoprene	Li⁺	Benzene	20	69/25/6	140
	Li⁺	Hexane	20	70/25/5	140
	Li⁺	None	20	77/18/5	140
	Na⁺	None	25	45/55	143
	K⁺	None	25	52/48	142
	Cs⁺	None	25	4/51/45	144

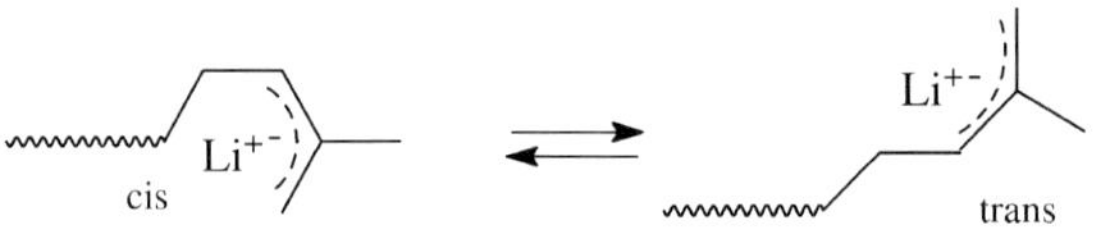

Scheme 2.31

gated form, it is that of the reactive unimeric species which matters and they are in very low proportion. Allylic active centers of polyisoprenyllithium can exist under two forms in fast equilibrium (Scheme 2.31).

NMR studies of the aggregated species in hydrocarbons show that they are mainly in the *trans* form [145, 146] whereas insertion occurs preferentially in the *cis* form. Such an isomerization of chain end competes between *trans* and *cis* forms with monomer addition (*cis* form), the configuration of the penultimate diene unit being fixed after addition of the last monomeric unit. Upon coordination of the monomer to the lithium cation, this configuration is stabilized in the form of a six-membered ring [147].

In the presence of solvating agents, the preceding situation no longer prevails due the competitive coordination of the additive or the solvent, entailing a decrease in the ratio of *cis* monomeric units.

2.4.4.2 Case of Vinyl and Related Monomers

Because of the marked polarization of the double bond, the attack of a carbanionic entity occurs exclusively at C-2, affording regioregular polymers with essentially head-to-tail attachments.

Control of the tacticity of vinyl and related polymers could only be obtained from monomers that carry side-groups able to coordinate to a counterion, e.g. (meth)acrylates, 2-vinylpyridine and o-methoxystyrene. The stereospecific anionic polymerization of (meth)acrylic monomers has been the subject of many studies and a complete review on this topic was published by Hatada et al. [148].

The first stereospecific anionic polymerization of MMA was obtained in 1958 by Fox et al. [149]; it was initiated at low temperature by organolithium compounds in dimethoxyethane solution. The number of studies on these systems increased when a structural characterization by NMR spectroscopy became available [150]. Fowells et al. polymerized various (meth)acrylic esters by using organolithium initiators in toluene solution and observed a marked stereoregulating effect [151]. They suggested the existence of active chains in the form of ion pairs, the two last monomeric units being coordinated to Li^+. Insertion of the monomer was postulated to occur through a "cisoidal" configuration resulting in a threodiisotactic sequence.

Whatever the methacrylic ester considered, polymerizations initiated by organolithiums in hydrocarbons exhibit a marked tendency for isotacticity, this effect being favored with the bulkiness of the ester group [152]. Thus, at $-78\,°C$, the butyllithium-initiated polymerization of triphenylmethyl methacrylate (TPMMA) in toluene yields a polymer with an isotactic content (mm) of 99% [153], whereas a PMMA sample obtained under identical conditions contains only 72–87% of mm triads [154, 155]. Steric hindrance was hypothesized as being responsible for the non-formation of r dyads and the induction of a helicoidal conformation of PTMMA chains, which in turn favored m sequences. This hypothesis was then corroborated by the preparation of chiral poly(triphenylmethyl methacrylate) initiated by an organolithium associated with a chiral ligand such as (−)-sparteine [156].

The polymerization of methyl methacrylate (MMA) carried out in hydrocarbons and initiated by alkali derivatives different from lithium yields samples with lower isotactic content [157] which indicates the influence of the solvation of the cation on the stereoregulating mechanism.

The use of non-solvated organomagnesium compounds to polymerize polar monomers also induces the formation of isotactic sequences, which mirrors the strong capacity of ester groups to solvate the Mg^{2+} cation. The first experiments were carried out in 1960 with MMA as monomer and Grignard reagents as initiators [158], but the multiplicity of the active centers formed [159] prevented the control of the sample tacticity and molar masses, which was achieved later with an isotactic content reaching 97% [160].

Non-solvated symmetrical organomagnesium compounds such as dibenzylmagnesium initiate the polymerization of MMA and yield samples with predominantly isotactic sequences, but the mm triad content is only about 65% [161].

Isoregulation can also be achieved with polar monomers other than (meth)acrylics. For instance, Natta and Mazzanti achieved the isospecific polymerization of 2-vinylpyridine (2-VP) [162] by using a desolvated Grignard reagent in hydrocarbon solution, but could not obtain a "living" character.

A mechanism close to that proposed for PMMA was suggested to account for the isospecific polymerization of o-methoxystyrene [163].

Symmetrical organomagnesium initiators prepared in the total absence of a solvating agent afforded isospecific and "living" polymerization of 2-VP [164]. In apolar solvents, the growing active center is sp^3 hybridized, the isoregulating mechanism resulting from the coordination of the pyridine rings of the ultimate and penultimate units to the metal cation, which in turn favors the formation of m dyads [165].

In polar solvents and/or in the presence of species capable of solvating the cations, syndiotactic sequences are favored in the polymerization of MMA carried out at low temperature. For example, PMMA samples initiated by an organolithium at $-85\,°C$ in THF exhibits an rr triad content of 84% [166]. In those obtained in the same solvent but from vinylbenzylmagnesium chloride at very low temperature ($-110\,°C$), this content rose to 97% [167]. In comparison, radical polymerization of the same monomer at low temperature ($-55\,°C$) yields samples with an mm triad content reaching 78.5% [168]; this was attributed to a greater energetic stability of r dyads compared with m dyads and thus to a rate constant of formation of r dyads (k_r) higher than that of m dyads (k_m).

MMA polymerizations initiated by "ate" complexes resulting from the addition of triethylaluminum to tert-BuLi ([Al]/[Li]≥3) in toluene also yielded samples with high contents of syndiotactic sequences [169]. Comparable results were obtained with acrylates [170].

Aspects pertaining to the stereochemistry of polar vinyl and related polymers have been the subject of several reviews [171].

2.5
Persistence of Active Centers

Anionic polymerizations are generally carried out in solution. Transfer and/or termination reactions that may occur with the solvent and/or with the impurities present in the reaction medium are beyond the scope of this section. Reactions with THF were presented in Section 2.3.2.

Due to the high reactivity of propagating centers, the preservation of the polymerization "livingness" beyond a certain time and above a certain temperature may be problematic. Depending on the substituent of the carbon–carbon double bond, carbanionic species can undergo elimination or react with the sidegroups.

The degradation reactions of "living" polymers was the subject of a review published in 1983 which stressed the contradictions between certain results [172]. Many further reports have appeared since that time.

2.5.1
Case of Polystyrenic Carbanions

As pointed out by Szwarc and coworkers, the polymerization of styrene carried out in THF and with Na^+ as counterion [173, 174] undergoes a degradation reaction as shown in Scheme 2.32. In addition to deactivated polystyrene, an acidic benzylallylic hydrogen is formed that reacts with the living PS^-,Na^+ to form a very stable 1,3-diphenylallylic carbanion unable to reinitiate the polymerization (Scheme 2.33). The latter reaction can be easily followed by UV spectroscopy since PS^-,Na^+ absorbs at 343 nm and the 1,3-diphenylallylic carbanion at 535 nm in THF. The higher the reactivity of the organometallic species (in particular its proportion of free ions), the faster is the degradation/transformation observed. For example, this reaction is instantaneous in pure hexamethylphosphotriamide (HMPA) [175], a solvent of high permittivity.

Independently of transfer reactions that may occur with solvents such as toluene or ethylbenzene [176], polystyryllithium carbanionic species are more stable and more persistent in hydrocarbon solution. Their half-lifetimes depend on the concentration of the species formed, the nature of the solvent, the possible presence of additives and the temperature. For instance, the half-lifetimes of PS^-,Li^+ ($[C^*] \approx 6 \times 10^{-3}$ M) in cyclohexane solution are very long at ambient temperature, $\sim$166 h at 50 °C, 3 h at 100 °C and 3.5 min at 150 °C [177]. The presence of a magnesium phenolate stabilizes the active "ate" species and increases their persistence, the half-lifetime at 150 °C increasing to 2 h [178].

The lithiated species that propagate the polymerization of α-methylstyrene are less stable than PS^-,Li^+. The half-lifetime of poly(α-methylstyryl)lithium in bulk is only 5 h at 25 °C, but stabilization can be obtained by addition of TMEDA to the medium [179]. In addition to the elimination of LiH, the phenyl ring of the penultimate unit can be attacked and indanyl rings formed [180].

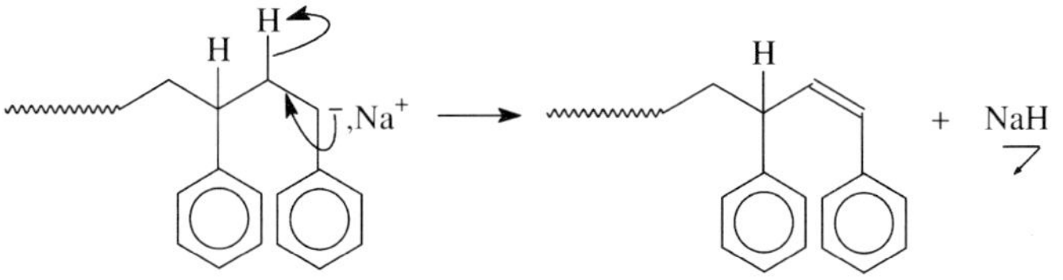

Scheme 2.32

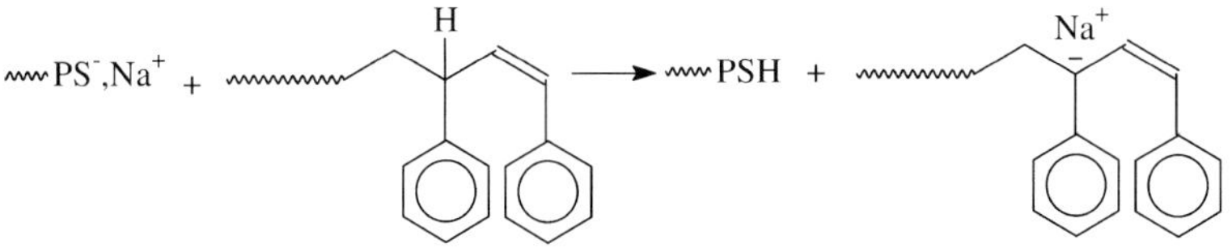

Scheme 2.33

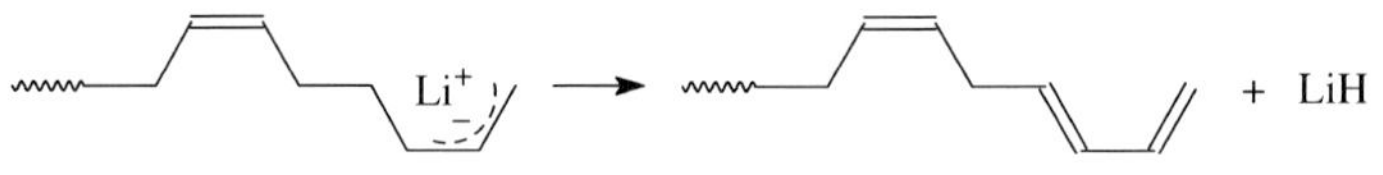

Scheme 2.34

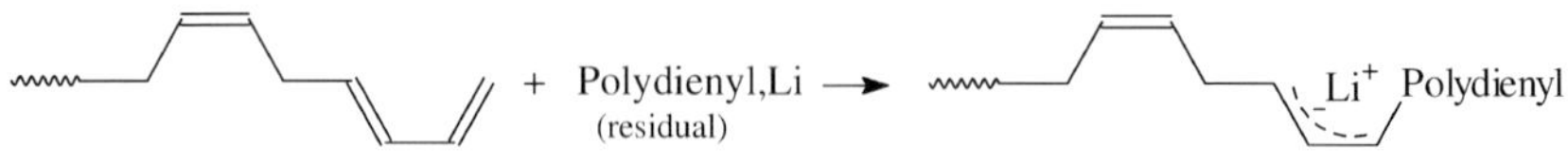

Scheme 2.35

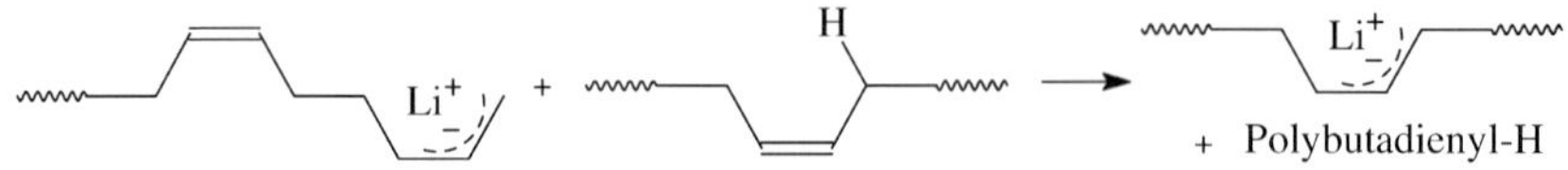

Scheme 2.36

2.5.2
Case of Polydiene Carbanions

The degradations/transformations undergone by these species are faster [181] and of the same type as those of polystyryl carbanions. In polar media, the instability of the organometallic species is also due to its reaction with the solvent, which requires that the solutions be stored at low temperature. Interpretation of the experimental observations is controversial [182].

In hydrocarbon solutions, the stability of polydienyllithium is satisfactory at room temperature, but at high temperature hydride elimination occurs similarly to what is observed with PS^-,Li^+ [183] (Scheme 2.34).

A dienic chain end is then formed which reacts with residual active chains and leads to doubling of the chain length [184]. Addition of a fresh amount of monomer brings about a reinitiation of the polymerization [185] (Scheme 2.35).

Transfer to polymer occurring by abstraction of an allylic hydrogen atom carried by a monomeric unit of the main chain was also contemplated [186] (Scheme 2.36).

2.5.3
Case of (Meth)acrylic Polymers

Independently of the side-reactions that may occur between the initiator and the monomer [187, 188, 239], the predominant termination reaction in this case is the attack of the carbonyl group carried by the antepenultimate monomeric unit of the chain by the growing enolate [189] (Scheme 2.37).

This nucleophilic addition–elimination reaction, first identified on oligomers [190], results from the interaction of Li^+ cation with the lone electron pair of the oxygen atom of the antepenultimate unit. It can also occur randomly between

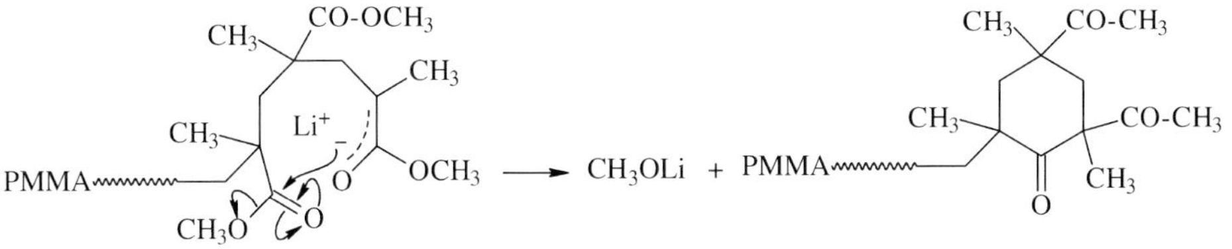

2.6 Application of Anionic Polymerization to Macromolecular Synthesis

Scheme 2.37

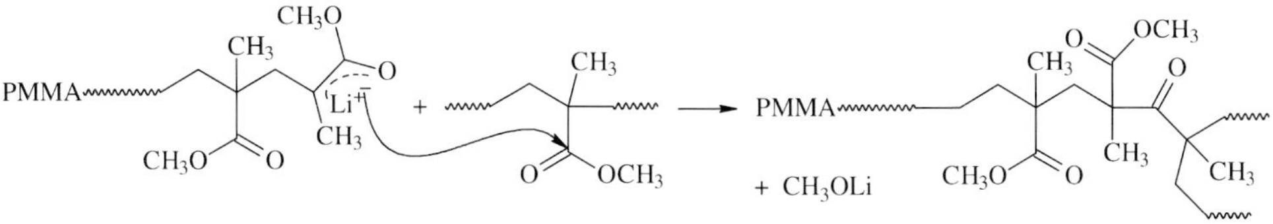

Scheme 2.38

active centers and the carbonyl groups carried either by monomeric units or monomer molecules. Such a reaction gives rise to branching, e.g. Scheme 2.38.

To prevent this termination reaction, two general methods have been contemplated. The first takes advantage of the global decrease in reactivity of active centers which affects both the propagation and the termination steps but to different extents. Such a decrease in reactivity could be obtained by lowering the reaction temperature [191] and/or by generating propagating species of lesser intrinsic reactivity, e.g. enolates associated with Mg^{2+} counterion [192], organic counterions (GTP) [193–195] and "ate" complexes [196–198].

The second method consists in hindering the counterion of the active center and/or saturating its coordination sites in order to decrease its electrophilicity and thus its affinity for carbonyl groups. In this respect, addition of complexing agents entails an increase in the persistence of enolate active centers. In the family of σ-complexing agents, polyethers, crown ethers [199] and cryptands [200], tertiary multiamines [201] and so forth have been used. In that of μ-complexing agents, lithium salts have been successfully utilized [202–204] and also alkoxides [205–207]. σ,μ-Complexing additives such as multi-ether/alkoxides [208] and tertiary multiamines/alkoxides [209] offer the advantage of combining the properties of the two families.

2.6
Application of Anionic Polymerization to Macromolecular Synthesis

It is beyond the scope of this chapter to describe and review exhaustively all macromolecular structures accessible by anionic polymerization. Being the first chain polymerization that gave access to complex and well-defined polymeric structures, some of the original architectures obtained through anionic polymerization are presented in the following section. The latter still remains today the method of choice in macromolecular engineering because of the efficiency of its initiators and the perennial character of its propagating species.

2.6.1
Prediction of Molar Masses and Control of Their Dispersion

When the initiator is suitably chosen, it is entirely and quickly consumed and transformed into carbanionic "living" chains. In the absence of transfer reactions, the number of chains formed is equal to the number of molecules of monofunctional initiator. Since the monomer conversion is generally complete, the number-average degree of polymerization is given by

$$\overline{DP}_n = \frac{[M]}{[I]} \text{ or } \overline{M}_n = \frac{[m]}{[I]} \tag{15}$$

where [I] and [M] are the molar concentrations of initiator and monomer consumed, [m] and $\overline{M}_n$ being the monomer mass concentration and the number-average molar mass, respectively. In the case of initiation by electron transfer, as two molecules of initiator participate in the formation of a bifunctional chain the relation giving $\overline{DP}_n$ is modified:

$$\overline{DP}_n = 2\frac{[M]}{[I]} \tag{16}$$

If all the chains are simultaneously generated, their growth occurs at the same rate and stops when no fresh monomer is available. Samples obtained under such conditions exhibit a "Poisson"-type narrow distribution of molar masses, i.e. which is obtained when one distributes in a random way i objects in j boxes, with $i \gg j$. Polydispersity indices close to 1.0 are thus obtained.

2.6.2
Functionalization of Chain Ends

ω-Functionalized chains can be easily obtained by nucleophilic deactivation of "living" chain ends on to various electrophilic substrates [210]. Examples of "functionalizing" reactions are illustrated in Table 2.6.

a,ω-Difunctional linear chains or telechelics can be obtained through various methods:
- Initiation with bifunctional initiators (see Section 2.3.2.3) followed by deactivation/functionalization as described in Table 2.6. This method could also be utilized to end-functionalize star polymers obtained from multifunctional initiators.
- By coupling/deactivation of two a-functional ω-active chains by means of a bivalent electrophilic molecule [215].

Table 2.6 Examples of functionalization of polystyryllithium chain ends.

Functional group to be generated	Reaction	Ref.
–OH	~~PS⁻,Li⁺ + (epoxide) → ~~PS–CH₂CH₂–OLi →[H⁺] ~~PS–CH₂CH₂–OH	211
–COOH	~~PS⁻,Li⁺ →[CO₂, H⁺ / THF/C₆H₆ (25/75)] ~~PS-COOH	212
–CH=CH₂	~~PS⁻,Li⁺ →[Cl-CH₂-C₆H₄-CH=CH₂ / C₂H₅C(CH₃)(OLi)CH(CH₃)₂] ~~PS-CH₂-C₆H₄-CH=CH₂	213
–NH₂	~~PS⁻,Li⁺ + C₆H₅–CH=N–Si(CH₃)₃ →[H₃O⁺] ~~PS–CH(C₆H₅)–NH₂	214
–COCl	~~PS⁻,Li⁺ + COCl₂ → ~~PS-COCl	215
–CO–CH=CH₂	~~PS–CH₂CH₂–OLi + ClCO–CH=CH₂ → ~~PS–CH₂CH₂–O-CO-CH=CH₂	216
–N(SiR³)₂	~~PS⁻,Li⁺ →[(CH₃)₂SiClH] ~~PS-Si(CH₃)₂-H →[CH₂=CH-CH₂-N[Si(CH₃)₃]₂ / Pt₍₀₎] ~~PS-Si(CH₃)₂-CH₂CH₂CH₂-N[Si(CH₃)₂]₂	240

- By initiating polymerization with a molecule carrying a protected reactive group and functionalizing the growing chain in the ω-position at the end of the polymerization [217].

2.6.3
Synthesis of Graft and Block Copolymers

The method of preparation of block copolymers is derived from that of chain end functionalization, the second block-forming monomer substituting for the functionalizing/deactivating agent as proposed by Szwarc et al. [4]. However, as mentioned in Section 2.4.3, the order of sequential addition/polymerization of monomers must comply with the rule of their increasing electroaffinity and take into account possible side-reactions. The synthesis of block copolymers in-

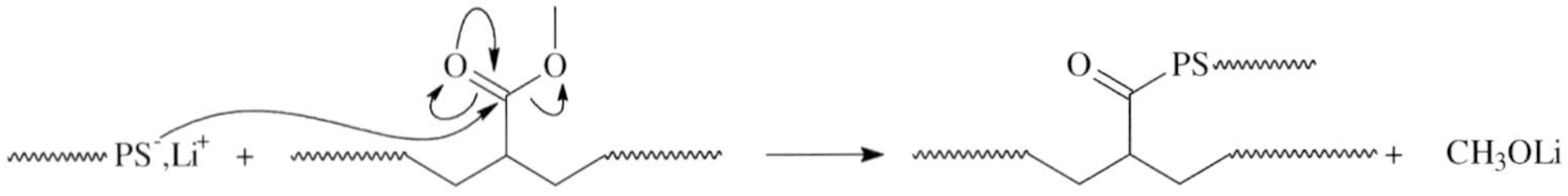

Scheme 2.39

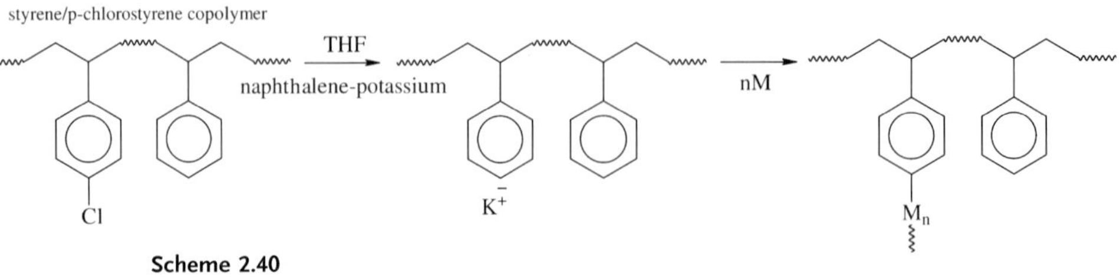

Scheme 2.40

cluding those prepared by anionic polymerization is the subject of Chapter 4 in Volume II.

Anionic polymerization can also be applied to the synthesis of graft copolymers and there are mainly three methods. That resorting to the "living" (co)polymerization of macromonomers [218] prepared by monofunctionalization (as described in Section 2.6.2) requires that the latter do not interfere with the propagation. "Grafting on to" and "grafting from" are the two other methods.

"Grafting on to" methods are based on the deactivation of a "living" anionic polymer on to the electrophilic functions of a pre-existing polymer, thereby generating side-chains from a polymer backbone. For instance, "living" polystyrene can be grafted on to PMMA in this way [219] (Scheme 2.39).

In "grafting from" methods, the side-chains are initiated by active centers carried by a pre-existing polymer backbone [220] (Scheme 2.40).

2.6.4
Star Polymers

The synthesis of star-shaped polymers can be performed in several ways:
- By reacting a "living" precursor with a plurifunctional deactivating reagent. Known as the "convergent" method, it requires a slight excess of the precursor – which subsequently has to be separated – to obtain well-defined stars. For example [221]:

$$4 \sim\!\!\sim PS^-, Li^+ + SiCl_4 \longrightarrow (\sim\!\!\sim PS)_4\text{-}Si + 4\, LiCl \qquad (17)$$

- By means of a multifunctional organometallic initiator, referred to as the "divergent method", its applicability is limited by the insolubility of most multifunctional organometallic compounds. However, a few stars have been obtained by this method [86, 222]:

$$\bigcirc\!\!-\!\!(^-,Li^+)_x \;+\; nM \;\longrightarrow\; \bigcirc\!\!-\!\!(M^-_{n/x}, Li^+)_x \tag{18}$$

- By block copolymerization of a "living" precursor with a suitable divinylic monomer in small amounts. Stars generated by this method are ill-defined and their number of arms cannot be precisely predicted [222–224].

A combination of convergent and divergent methods was used to prepare stars with mixed branches, also called heteroarm stars [225, 226], and even more complex architectures [227].

In a recent addition, dendrimer-like polymers could also be obtained by anionic means, upon repetition of a divergent growth of polymeric arms and their end-derivatization [228].

2.6.5
Macrocyclic Polymers

As with many well-defined polymeric architectures, the first macrocyclic polymers were obtained anionically, by deactivation of a dicarbanionic "living" polymer by a bifunctional electrophile. A recent review on macrocyclic polymers was published by Hogen-Esch [229] and is the subject of Chapter 5 in Volume II.

The formation of macrocyclics requires that the deactivation of the bicarbanionic "living" chains [230] be performed under high dilution and with vigorous stirring, by adding slowly a solution containing the bifunctional electrophile [231]. Under these experimental conditions only the cyclization can be favored against chain extension [232, 233].

A great variety of other tailor-made polymeric architectures have been designed by anionic means and are described in Chapters 2–12 in Volume II.

The purpose of this chapter is to show how anionic polymerization by its "living" character has contributed immensely – and is still contributing – to macromolecular engineering, a research activity that received a decisive impulse from the early work of Szwarc and coworkers in 1956 [3, 4].

The high reactivity of carbanionic active centers may sometimes be an obstacle to its application, but when the conditions of synthesis are adequate with respect to the nature of the monomer, when initiation is fast and quantitative, anionic polymerization remains the most powerful tool for the preparation of polymers with well-defined structures, defined as macromolecular engineering.

References

1 *Ullmann's Encylopedia of Industrial Chemistry*, VCH Publishers, New York, **1993**, A23 (chap. 3).
2 A. Abkin, S. Medvedev, *Trans. Faraday Soc.* **1936**, *32*, 286.
3 M. Szwarc, *Nature*, **1956**, *178*, 1168.
4 M. Szwarc, M. Levy, R. Milkovich, *J. Am. Chem. Soc.*, **1956**, *78*, 2656.
5 (a) M. Szwarc, *Carbanions, Living Polymers and Electron Transfer Processes*, Wiley, New York, **1968**; (b) M. Szwarc (Ed.), *Ions and Ion-Pairs in Organic Reactions*, Wiley-Interscience, New York, Vol. 1, **1972**; Vol. 2, **1974**; (c) S. Bywater, Anionic polymerization of olefins, in *Comprehensive Chemical Kinetics*, Vol. 15, C. H. Bamford, C. F. H. Tipper (Eds.), Elsevier, Amsterdam, **1976**, Ch. 1; (d) J. McGrath (Ed.), *Anionic Polymerization. Kinetics, Mechanism and Synthesis*, ACS Symposium Series, American Chemical Society, Washington, DC, **1981**; (e) M. Szwarc, *Adv. Polym. Sci.*, **1983**, *49*, 1; (f) M. Morton, *Anionic Polymerization: Principles and Practice*, Academic Press, New York, **1983**; (g) S. Bywater, in *Encyclopedia of Polymer Science and Engineering*, Vol. 2, Wiley-Interscience, New York, **1985**, p. 1; (h) T. E. Hogen-Esch, J. Smid (Eds.), *Recent Advances in Polymer Science*, Elsevier, New York, **1987**; (i) G. C. Eastmond, A. Ledwith, S. Russo, P. Sigwalt (Eds.), *Comprehensive Polymer Science*, Vol. 3, *Chain Polymerization*, Pergamon Press, Oxford, **1989**, Ch. 25–30; (j) M. Szwarc, M. van Beylen, *Ionic Polymerization and Living Polymers*, Chapman and Hall, New York, **1993**; (k) S. Bywater, *Prog. Polym. Sci.*, **1994**, *19*, 287; (l) H. L. Hsieh, R. P. Quirk, *Anionic Polymerization – Principles and Practical Applications*, Marcel Dekker, New York, **1996**; (m) D. Baskaran, *Prog. Polym. Sci.*, **2003**, *28*, 521.
6 G. Hélary, M. Fontanille, *Eur. Polym. J.*, **1978**, *14*, 345.
7 G. Wang, M. van Beylen, *Polymer*, **2003**, *44*, 6205.
8 G. M. Wang, K. Janssens, C. van Oosterwijck, A. Yakimansky, M. van Beylen, *Polymer*, **2005**, *46*, 295.
9 P. Verstraete, Structure et réactivité des espèces actives en polymérization anionique du styrène amorcée par les alkyllithiums, PhD Thesis, Université Bordeaux 1, **2002**.
10 M. Shima, D. N. Bhattacharyya, J. Smid, M. Szwarc, *J. Am. Chem. Soc.*, **1963**, *85*, 1306.
11 R. Busson, M. Van Beylen, *Macromolecules*, **1977**, *10*, 1320.
12 J. Léonard, in *Polymer Handbook*, 4th edn, J. Bandrup, E. H. Immergut, E. A. Grulke (Eds.), Wiley, New York, **1999**, Vol. II, p. 363.
13 K. J. Ivin, J. Léonard, *Eur. Polym. J.*, **1970**, *6*, 331.
14 M. Szwarc, *Carbanions, Living Polymers and Electron Transfer Processes*, Wiley, New York, **1968**, p. 135.
15 A. W. Langer, *Adv. Chem. Ser.*, **1974**, 130.
16 H. L. Hsieh, R. P. Quirk, *Anionic Polymerization – Principles and Practical Applications*, Marcel Dekker, New York, **1996**, pp. 244–247.
17 N. S. Wooding, W. C. E. Higginson, *J. Chem. Soc.*, **1952**, 774.
18 E. P. Serjeant, B. Dempsey (Eds.), *Ionization Constants of Organic Acids in Solution*, Chemical Data Series No. 23, Pergamon Press, Oxford, **1979**.
19 H. L. Hsieh, R. P. Quirk, *Anionic Polymerization – Principles and Practical Applications*, Marcel Dekker, New York, **1996**, p. 34.
20 A. W. Langer, *Polym. Prepr. ACS Div. Polym. Chem.*, **1966**, *7*, 132.
21 J. C. Favier, P. Sigwalt, M. Fontanille, *J. Polym. Sci., Polym. Chem.*, **1977**, *15*, 2373.
22 H. Jeuck, A. H. E. Müller, *Makromol. Chem., Rapid Commun.*, **1982**, *3*, 121.
23 D. J. Worsfold, S. Bywater, *Can. J. Chem.*, **1960**, *38*, 1891.
24 S. Bywater, *Macromolecules*, **1998**, *31*, 6010.
25 S. Bywater, *Macromol. Chem. Phys.*, **1998**, *199*, 1217.
26 J. E. Roovers, S. Bywater, *Trans. Faraday Soc.*, **1966**, *62*, 701.
27 J. E. Roovers, S. Bywater, *Can. J. Chem.*, **1968**, *46*, 2711.

28 M. Szwarc, M. van Beylen, *Ionic Polymerization and Living Polymers*, Chapman and Hall, New York, **1993**, p. 58.
29 D. N. Bhattacharyya, C. L. Lee, J. Smid, M. Szwarc, *Polymer*, **1964**, *5*, 54.
30 D. N. Bhattacharyya, C. L. Lee, J. Smid, M. Szwarc, *J. Phys. Chem.*, **1965**, *69*, 608.
31 H. Hostalka, R. V. Figini, G. V. Schultz, *Makrom. Chem.*, **1964**, *71*, 198.
32 G. Hélary, M. Fontanille, *Polym. Bull.*, **1980**, *3*, 159.
33 S. Bywater, D. J. Worsfold, *Can. J. Chem.*, **1962**, *40*, 1564.
34 G. Hélary, M. Fontanille, *Eur. Polym. J.*, **1978**, *14*, 345.
35 G. Wittig, F. J. Meyer, G. Lange, *Liebigs Ann. Chem.*, **1951**, *571*, 167.
36 Ph. Desbois, M. Fontanille, A. Deffieux, V. Warzelhan, S. Lätsch, C. Schade, *Macromol. Chem. Phys.*, **1999**, *200*, 621.
37 L. M. Seitz, T. L. Brown, *J. Am. Chem. Soc.*, **1966**, *88*, 4140.
38 B. Schubert, E. Weiss, *Chem. Ber.*, **1984**, *117*, 366.
39 J. L. Dye, *Pure Appl. Chem.*, **1989**, *61*, 1555.
40 A. Nishioka, H. Watanabe, K. Abe, I. Sono, *J. Polym. Sci.*, **1960**, *48*, 241.
41 A. P. Zagala, T. E. Hogen-Esch, *Macromolecules*, **1996**, *29*, 3038.
42 D. Baskaran, D. Chakrapani, S. Sivaram, T. E. Müller, A. H. E. Hogen-Esch, *Macromolecules*, **1999**, *32*, 2865.
43 M. Szwarc, *Carbanions, Living Polymers and Electron Transfer Processes*, Wiley, New York, **1968**, Ch. 4.
44 D. E. Paul, D. Lipkin, S. I. Weissmann, *J. Am. Chem. Soc.*, **1956**, *78*, 116.
45 J. Jorner, S. A. Rice, *Adv. Chem. Ser.*, **1965**, *50*, 7.
46 E. Megiel, A. Kaim, *J. Polym. Sci., Part A: Polym. Chem.*, **2001**, *39*, 3761.
47 H. C. Wang, G. Levin, M. Szwarc, *J. Am. Chem. Soc.* **1978**, *100*, 6137.
48 F. W. Stavely, *Ind. Eng. Chem.*, **1956**, *48*, 778.
49 P. D. Burrow, J. A. Michejda, K. D. Jordan, *J. Chem. Phys.*, **1987**, *86*, 9.
50 E. C. M. Chen, W. E. Wentworth, *Mol. Cryst. Liq. Cryst.*, **1989**, *171*, 271.
51 A. E. Oberster, T. C. Bouton, J. K. Valaitis, *Angew. Makromol. Chem.*, **1973**, *29/30*, 291.
52 K. S. Das, M. Feld, M. Szwarc, *J. Am. Chem. Soc.*, **1960**, *82*, 1506.
53 D. M. Wiles, S. Bywater, *Trans. Faraday Soc.*, **1965**, *61*, 150.
54 B. C. Anderson, G. D. Andrews, P. Arthur Jr., H. W. Jacobson, L. R. Melby, A. J. Playtis, W. H. Scharkley, *Macromolecules*, **1981**, *14*, 1599.
55 P. Vlcek, L. Lochmann, *Prog. Polym. Sci.*, **1999**, *24*, 793.
56 L. Lochmann, D. Lím, *J. Organomet. Chem.*, **1973**, *50*, 9.
57 L. Lochmann, M. Rodová, J. Trekoval, *J. Polym. Sci., Polym. Chem. Ed.*, **1974**, *12*, 2091.
58 S. K. Varshney, J.-P. Hautekeer, R. Fayt, R. Jérôme, Ph. Teyssié, *Macromolecules*, **1990**, *23*, 2618.
59 M. T. Reetz, S. Hutte, R. Goddard, *J. Phys. Org. Chem.*, **1995**, *8*, 231.
60 D. Baskaran, A. H. E. Müller, *Macromolecules*, **1997**, *30*, 1869.
61 I. C. Eromosele, D. C. Pepper, B. Ryan, *Makromol. Chem.*, **1989**, *190*, 1613.
62 K. Ziegler, H. Dislich, *Chem. Ber.*, **1957**, *90*, 1107.
63 T. Kottke, D. Stalke, *Angew. Chem. Int. Ed. Engl.*, **1993**, *32*, 580.
64 S. Bywater, D. J. Worsfold, *J. Organomet. Chem.*, **1967**, *10*, 1.
65 M. Weiner, C. Vogel, R. West, *Inorg. Chem.*, **1962**, *1*, 654.
66 H. Hsieh, *J. Polym. Sci., A3*, **1965**, 163.
67 L. J. Fetters, M. Morton, *Macromolecules*, **1974**, *7*, 552.
68 J. E. Roovers, S. Bywater, *Macromolecules*, **1975**, *8*, 251.
69 H. L. Hsieh, R. P. Quirk, *Anionic Polymerization – Principles and Practical Applications*, Marcel Dekker, New York, **1996**, p. 147.
70 D. J. Worsfold, S. Bywater, *Can. J. Chem.*, **1964**, *42*, 397.
71 J. E. L. Roovers, S. Bywater, *Macromolecules*, **1968**, *1*, 328.
72 S. Bywater, D. J. Worsfold, *J. Organomet. Chem.*, **1967**, *10*, 1.
73 D. J. Worsfold, S. Bywater, *Can. J. Chem.*, **1960**, *38*, 1891.
74 A. W. Langer, *Adv. Chem. Ser.*, **1974**, 130.
75 G. Natta, G. Mazzanti, P. Longi, G. Dall' Asta, *J. Polym. Sci.*, **1961**, *51*, 487.

76 J.C. Favier, M. Fontanille, *Bull. Soc. Chim. Fr.*, **1971**, 526.
77 A. Soum, M. Fontanille, *Makromol. Chem.*, **1980**, *181*, 799.
78 A. Soum, M. Fontanille, *Makromol. Chem., Rapid Commun.* **1983**, *4*, 471.
79 S. Ménoret, A. Deffieux, Ph. Dubois, *Macromolecules*, **2003**, *36*, 5988.
80 S. Carlotti, S. Ménoret, Ph. Desbois, N. Nissner, V. Warzelhan, A. Deffieux, *Macromolecular Rapid Comm.* **2006**, *27*, 905.
81 T. Fujimoto, S. Tani, K. Takano, M. Ogawa, M. Nagasawa, *Macromolecules*, **1978**, *11*, 673.
82 W. Toreki, M. Takaki, T.E. Hogen-Esch, *Polym. Prepr. ACS Div. Polym. Chem.*, **1988**, *27*, 355.
83 R.P. Quirk, Y. Tsai, *Macromolecules*, **1998**, *31*, 8016.
84 R.P. Quirk, T. Yoo, Y. Lee, J. Kim, B. Lee, *Adv. Polym. Sci.*, **2000**, *153*, 67.
85 N.G. Vasilenko, E.A. Rebrov, A.M. Muzafarov, B. Esswein, B. Striegel, M. Möller, *Macromol. Chem. Phys.*, **1998**, *199*, 889.
86 R. Matmour, A. Lebreton, C. Tsitsilianis, I. Kallitsis, V. Héroguez, Y. Gnanou, *Angew. Chem. Int. Ed.*, **2005**, *44*, 284.
87 G. Beinert, P. Lutz, E. Franta, P. Rempp, *Makromol. Chem.*, **1978**, *179*, 551.
88 P. Lutz, E. Franta, P. Rempp, *Polymer*, **1982**, *23*, 1953.
89 E.A. Mushina, L.S. Muraviova, T.S. Samedova, T.A. Krentsel, *Eur. Polym. J.*, **1979**, *15*, 99.
90 Y.S. Yu, Ph. Dubois, R. Jérôme, Ph. Teyssié, *Macromolecules*, **1996**, *29*, 6090.
91 Y.I. Estrin, *Centr. Eur. J. Chem.*, **2004**, *2*, 52.
92 J. Hofmans, M. van Beylen, *Polymer*, **2005**, *46*, 303.
92 M. Fontanille, P. Guyot, P. Sigwalt, J.-P. Vairon, *French Patent 2313389*, **1975**.
94 P. Guyot, J.C. Favier, M. Fontanille, P. Sigwalt, *Polymer*, **1981**, *22*, 1724.
95 P. Guyot, J.C. Favier, M. Fontanille, P. Sigwalt, *Polymer*, **1982**, *23*, 73.
96 L.H. Tung, G.Y.-S. Lo, D.E. Beyer, *Macromolecules*, **1978**, *11*, 616.
97 R. Matmour, S.A. More, P.P Wadgaonkar, Y. Gnanou, *J. Am. Chem. Soc.*, **2006**, *128*, 8158.
98 T. Zundel, J. Baran, M. Mazurek, J.-S. Wang, R. Jérôme, Ph. Teyssié, *Macromolecules*, **1998**, *31*, 2724.
99 O.W. Webster, W.R. Hertler, D.Y. Sogah, W.B. Farnham, T.V. Ranjanbabu, *J. Am. Chem. Soc.*, **1983**, *105*, 5706.
100 O.W. Webster, in *Encyclopedia of Polymer Science and Engineering*, H. Mark, N.M. Bikales, C.G. Overberger (Eds.) Wiley, New York, **1987**.
101 O.W. Webster, *Adv. Polym. Sci.*, **2004**, *167*, 1.
102 D.Y. Sogah, W.R. Hertler, O.W. Webster, G.M. Cohen, *Macromolecules*, **1987**, *20*, 1473.
103 W. Schubert, H.-D. Sitz, F. Bandermann, *Makromol. Chem.*, **1989**, *190*, 2193.
104 R.P. Quirk, G.P. Bidenger, *Polym. Bull.*, **1989**, *22*, 69.
105 W.R. Hertler, D.Y. Sogah, O.W. Webster, B.M. Trost, *Macromolecules*, **1984**, *17*, 1415.
106 A.H.E. Müller, *Makromol. Chem., Makromol. Symp.*, **1990**, *32*, 87.
107 S. Bywater, *Makromol. Chem., Makromol. Symp.*, **1993**, *67*, 339.
108 S. Bywater, *Fortschr. Hochpolym. Forsch.*, **1965**, *4*, 66.
109 D.J. Worsfold, S. Bywater, *Can. J. Chem.*, **1964**, *42*, 2884.
110 M. Morton, E.E. Bostick, R.A. Livigny, *J. Polym. Sci., Part A1*, **1961**, 1735.
111 H.L. Hsieh, R.P. Quirk, *Anionic Polymerization – Principles and Practical Applications*, Marcel Dekker, New York, **1996**, p. 163.
112 S. Bywater, *Macromol. Chem. Phys.*, **1998**, *199*, 1217.
113 S. Bywater, in *Encyclopedia of Polymer Science and Engineering*, H. Mark, N.M. Bikales, C.G. Overberger (Eds.) Vol. 2, Wiley-Interscience, New York, **1985**, p. 1.
114 D.J. Worsfold, S. Bywater, *Polym. Prepr. ACS Div. Polym. Chem.*, **1986**, *27*, 140.
115 S. Quinebêche, M. Fontanille, Y. Gnanou, to be published.
116 S. Bywater, D.J. Worsfold, *Can. J. Chem.*, **1962**, *40*, 1564.
117 M. Fontanille, G. Hélary, M. Szwarc, *Macromolecules*, **1988**, *21*, 1532.
118 G. Hélary, M. Fontanille, *Polym. Bull.*, **1980**, *3*, 159.

119 D. J. Worsfold, S. Bywater, *J. Phys. Chem.*, **1965**, *69*, 4124.
120 S. Ménoret, S. Carlotti, M. Fontanille, A. Deffieux, Ph. Desbois, Ch. Schade, W. Schrepp, V. Warzelhan, *Macromol. Chem. Phys.*, **2001**, *202*, 3219.
121 Ph. Desbois, M. Fontanille, A. Deffieux, V. Warzelhan, Ch. Schade, *Macromol. Symp.*, **2000**, *157*, 151.
122 D. N. Battacharyya, C. L. Lee, J. Smùid, M. Szwarc, *J. Phys. Chem.*, **1965**, *69*, 612.
123 D. N. Battacharyya, C. L. Lee, J. Smùid, M. Szwarc, *Polymer*, **1964**, *5*, 54.
124 T. Shimomura, K. J. Tölle, J. Smùid, M. Szwarc, *J. Am. Chem. Soc.*, **1967**, *89*, 796.
125 J. E. L. Roovers, S. Bywater, *Can. J. Chem.*, **1968**, *46*, 2711.
126 D. N. Battacharyya, C. L. Lee, J. Smùid, M. Szwarc, *J. Phys. Chem.*, **1965**, *69*, 612.
127 M. Szwarc, A. Rembaum, *J. Polym. Sci.*, **1956**, *22*, 189.
128 A. Roig, J. E. Figueruelo, E. Llano, *J. Polym. Sci., Part B*, **1965**, *3*, 171.
129 G. Löhr, G. V. Schulz, *Makromol. Chem.*, **1973**, *172*, 137.
130 G. Löhr, G. V. Schulz, *Eur. Polym. J.*, **1974**, *10*, 121.
131 V. Warzelhan, G. V. Schulz, *Makromol. Chem.*, **1976**, *177*, 2185.
132 V. Warzelhan, H. Höcker, G. V. Schulz, *Makromol. Chem.*, **1978**, *179*, 2221.
133 C. Tsvetanov, A. E. H. Müller, G. V. Schulz, *Macromolecules*, **1985**, *18*, 863.
134 V. Warzelhan, H. Höcker, G. V. Schulz, *Makromol. Chem.*, **1980**, *181*, 149.
135 H. Jeuck, A. H. E. Müller, *Makromol. Chem., Rapid Commun.*, **1982**, *3*, 121.
136 D. Baskaran, *Prog. Polym. Sci.*, **2003**, *28*, 521.
137 H. L. Hsieh, R. P. Quirk, *Anionic Polymerization – Principles and Practical Applications*, Marcel Dekker, New York, **1996**, Tables 10.1 and 10.2; L. K. Huang, PhD Dissertation, University of Akron, **1979**, pp. 244–247.
138 S. Bywater, D. J. Worsfold, *J. Organomet. Chem.*, **1978**, *159*, 229.
139 E. V. Krystal'nyi, A. A. Arest-Yakubovich, *Polym. Sci., Ser. B*, **2000**, *42*, 39.
140 M. Morton, J. R. Rupert, in *Initiation of Polymerization*, F. E. Bailey (Ed.), ACS Symposium Series, Vol. 212, American Chemical Society, Washington, DC, **1983**, p. 283.
141 S. Bywater, in *Recent Advances in Polymer Science*, T. E. Hogen-Esch, J. Smid (Eds.), Elsevier, New York, **1987**, p. 187.
142 A. A. Arest-Yakubovich, I. P. Golberg, V. L. Zolotarev, V. I. Akzenov, I. I. Ermakova, V. S. Ryakovsky, in *Applications of Anionic Polymerization Research*, ACS Symposium Series, Vol. 193, American Chemical Society, Washington, DC, **1998**, p. 197.
143 A. V. Tobolsky, C. E. Rogers, *J. Polym. Sci.*, **1959**, *40*, 73.
144 F. Foster, J. L. Binder, *Adv. Chem. Ser.*, **1957**, *19*, 26.
145 F. Schué, D. J. Worsfold, S. Bywater, *J. Polym. Sci., Part B*, **1969**, *7*, 821.
146 F. Schué, D. J. Worsfold, S. Bywater, *Macromolecules*, **1970**, *3*, 509.
147 D. J. Worsfold, S. Bywater, *Macromolecules*, **1978**, *11*, 582.
148 K. Hatada, T. Kitayama, K. Ute, *Prog. Polym. Sci.*, **1988**, *13*, 189.
149 T. G. Fox, B. S. Garett, W. E. Goode, S. Gratch, J. F. Kincaid, A. Spell, J. D. Stroupe, *J. Am. Chem. Soc.*, **1958**, *80*, 1718.
150 F. A. Bovey, G. V. D. Tiers, *J. Polym. Sci.*, **1960**, *44*, 173.
151 W. Fowells, C. Schuerch, F. Bovey, *J. Am. Chem. Soc.*, **1967**, *89*, 1396.
152 H. Yuki, K. Hatada, *Adv. Polym. Sci.*, **1979**, *31*, 1.
153 H. Yuki, K. Hatada, T. Niinomi, Y. Kikuchi, *Polym. J.*, **1970**, *1*, 36.
154 H. Yuki, K. Hatada, K. Ohta, Y. Okamoto, *J. Macromol. Sci., Chem.*, **1975**, *A9*, 983.
155 D. M. Wiles, S. Bywater, *Trans. Faraday Soc.*, **1965**, *61*, 150.
156 Y. Okamoto, K. Suzuki, H. Yuki, *J. Polym. Sci., Polym. Chem. Ed.*, **1980**, *18*, 3043.
157 D. Braun, M. Herner, U. Johnsen, W. Kern, *Makromol. Chem.* **1962**, *51*, 15.
158 W. E. Goode, F. H. Owens, R. P. Fellman, W. H. Snyder, J. E. Moore, *J. Polym. Sci.*, **1960**, *46*, 317.

159 K. Matsuzaki, H. Tanaka, T. Kanai, *Makromol. Chem.*, **1981**, *182*, 2905.
160 K. Hatada, K. Ute, K. Tanaka, Y. Okamoto, T. Kitayama, *Polym. J.*, **1986**, *18*, 1037.
161 A. Soum, N. D'Accorso, M. Fontanille, *Makromol. Chem., Rapid Commun.*, **1983**, *4*, 471.
162 G. Natta, G. Mazzanti, *J. Polym. Sci.*, **1961**, *51*, 487.
163 H. Yuki, Y. Okamoto, Y. Kuwae, K. Hatada, *J. Polym. Sci., Part A1*, **1969**, *7*, 1933.
164 A. Soum, M. Fontanille, *Makromol. Chem.*, **1980**, *181*, 799.
165 A. Soum, M. Fontanille, in *Anionic Polymerization, Kinetics, Mechanisms and Synthesis*, J. E. McGrath (Ed.), ACS Symposium Series 166, American Chemical Society, Washington, DC, **1981**, p. 239.
166 H. Jeuck, A. H. E. Müller, *Makromol. Chem., Rapid Commun.*, **1982**, *3*, 121.
167 K. Hatada, H. Nakanishi, K. Ute, T. Kitayama, *Polym. J.*, **1986**, *18*, 581.
168 M. Reinmöller, T. G. Fox, *Polym. Prepr. ACS Div. Polym. Chem.*, **1966**, *7*, 999.
169 T. Kitayama, T. Shinozaki, T. Sakamoto, M. Yamamoto, K. Hatada, *Macromol. Chem., Suppl.*, **1989**, *15*, 167.
170 M. Tabuchi, T. Kawauchi, T. Kitayama, K. Hatada, *Polymer*, **2002**, *43*, 7185.
171 T. E. Hogen-Esch, in *Macromolecular Design of Polymeric Materials*, K. Hatada, T. Kitayama, O. Vögl (Eds.), Marcel Dekker, New York, **1997**, Ch. 10.
172 M. D. Glasse, *Prog. Polym. Sci.*, **1983**, *9*, 133.
173 G. Spach, M. Levy, M. Szwarc, *J. Chem. Soc.*, **1962**, 355.
174 M. Levy, M. Szwarc, S. Bywater, D. J. Worsfold, *Polymer*, **1960**, *1*, 515.
175 M. Fontanille, unpublished results.
176 F. Schué, R. Aznar, J. Sledz, P. Nicol, *Makrom. Chem., Makrom. Symp.*, **1993**, *67*, 213.
177 A. Deffieux, P. Desbois, M. Fontanille, V. Warzelhan, S. Lätsch, C. Schade, in *Ionic Polymerizations and Related Processes*, J. E. Puskas (Ed.), Kluwer, Dordrecht, **1999**.
178 S. Ménoret, M. Fontanille, A. Deffieux, Ph. Desbois, *Macromol. Chem. Phys.*, **2002**, *203*, 1155.
179 D. Adès, M. Fontanille, J. Léonard, *Can. J. Chem.*, **1982**, *60*, 564.
180 D. Margerison, V. A. Nyss, *J. Chem. Soc. C*, **1968**, 3065.
181 S. S. Medvedev, A. R. Gantmakher, *J. Polym. Sci., Part C*, **1963**, *4*, 173.
182 M. D. Glasse, *Prog. Polym. Sci.*, **1983**, *9*, 133.
183 T. A. Antkowiak, *Polym. Prepr. ACS Div. Polym. Chem.*, **1971**, *12*, 393.
184 W. Nentwig, H. Sinn, *Makromol. Chem., Rapid Commun.*, **1980**, *1*, 59.
185 J. N. Anderson, W. J. Kern, T. W. Bethea, H. E. Adams, *J. Appl. Polym. Sci.*, **1972**, *16*, 3133.
186 R. W. Pennisi, L. J. Fetters, *Macromolecules*, **1988**, *21*, 1094.
187 D. M. Wiles, S. Bywater, *Chem. Ind. (London)*, **1963**, 1209.
188 A. H. E. Müller, L. Lochmann, J. Trekoval, *Makromol. Chem.*, **1986**, *187*, 1473.
189 F. J. Gerner, H. Höcker, A. H. E. Müller, G. V. Schulz, *Eur. Polym. J.*, **1984**, *20*, 349.
190 D. L. Glusker, I. Lysloff, E. Stiles, *J. Polym. Sci.*, **1961**, *49*, 315.
191 D. Baskaran, S. Chakrapani, S. Sivaram, *Macromol. Chem. Phys.*, **1995**, *28*, 7315.
192 K. Hatada, K. Ute, K. Tanaka, T. Kitayama, in *Recent Advances in Polymer Science*, T. E. Hogen-Esch, J. Smid (Eds.), Elsevier, New York, **1987**, p. 195
193 M. T. Reetz, *Angew. Chem. (Adv. Mater.)*, **1988**, *100*, 1026.
194 D. Baskaran, D. Chakrapani, S. Sivaram, T. E. Müller, A. H. E. Hogen-Esch, *Macromolecules*, **1999**, *32*, 2865.
195 M. T. Reetz, S. Hütte, H. M. Herzog, R. Goddard, *Macromol. Symp.*, **1996**, *107*, 209.
196 D. Kunkel, A. H. E. Müller, L. Lochmann, M. Janata, *Makromol. Chem., Makromol. Symp.*, **1992**, *60*, 315.
197 T. Kitayama, T. Shinozaki, T. Sakamoto, M. Yamamoto, K. Hatada, *Makromol. Chem., Suppl.*, **1989**, *15*, 167.
198 B. Schmitt, W. Stauf, A. H. E. Müller, *Macromolecules*, **2001**, *34*, 1551.
199 J. S. Wang, R. Jérôme, R. Warin, H. Zhang, Ph. Teyssié, *Macromolecules*, **1994**, *27*, 3377.
200 C. Johan, A. H. E. Müller, *Makromol. Chem., Rapid Commun.*, **1981**, *2*, 687.

201 F. Lavaud, M. Fontanille, Y. Gnanou, *Polymer*, **2002**, *43*, 7195.
202 S. K. Varshney, J. P. Hautekeer, R. Fayt, R. Jérôme, Ph. Teyssié, *Macromolecules*, **1990**, *23*, 2618.
203 D. Baskaran, S. Sivaram, *Macromolecules*, **1997**, *30*, 1550.
204 A. V. Yakimansky, A. H. E. Müller, M. van Beylen, *Macromolecules*, **2000**, *33*, 5686.
205 L. Lochmann, A. H. E. Müller, *Makromol. Chem.*, **1990**, *191*, 1657.
206 J. Kris, J. Dybal, M. Janata, L. Lochmann, P. Vlček, *Macromol. Chem. Phys.*, **1996**, *197*, 1889.
207 P. Vlček, J. Otoupalova, J. Kris, P. Schmidt, *Macromol. Symp.*, **2000**, *161*, 113.
208 P. Bayard, R. Jérôme, Ph. Teyssié, S. Varshney, J. S. Wang, *Polym. Bull.*, **1994**, *32*, 381.
209 J. Marchal, M. Fontanille, Y. Gnanou, *Macromol. Symp.*, **1998**, *132*, 249.
210 R. P. Quirk, in *Comprehensive Polymer Science, 1st Suppl.*, S. L. Aggarwal, S. Russo (Eds.), Pergamon Press, Oxford, **1992**, p. 83.
211 D. H. Richards, M. Szwarc, *Trans. Faraday Soc.*, **1959**, *55*, 1644.
212 R. P. Quirk, J. Yin, L. J. Fetters, *Macromolecules*, **1989**, *22*, 85.
213 R. P. Quirk, J. M. Pickel, M. A. Arnould, K. M. Wollyung, C. Wesdemotis, *Macromolecules*, **2006**, *39*, 1681.
214 A. Hirao, I. Hattori, T. Sadagawa, K. Yamaguchi, S. Nakahama, *Makrom. Chem., Rapid Comm.*, **1982**, *3*, 59.
215 G. Finaz, Y. Gallot, P. Rempp, J. Parrod, *J. Polym. Sci.*, **1962**, *58*, 1363.
216 P. Masson, E. Franta, P. Rempp, *Makrom. Chem., Rapid Commun.*, **1982**, *3*, 499.
217 D. N. Schulz, A. F. Halasa, A. E. Oberster, *J. Polym. Sci.*, **1974**, *12*, 153.
218 P. Rempp, P. Lutz, P. Masson, P. Chaumont, E. Franta, in *Recent Advances in Anionic Polymerization*, T. E. Hogen-Esch, J. Smùid (Eds.), Elsevier, New York, **1987**, p. 353.
219 Y. Gallot, P. Rempp, J. Parrod, *J. Polym. Sci., Polym. Lett. Ed.*, **1963**, *1*, 329.
220 G. Greber, *Makromol. Chem.*, **1967**, *101*, 104.
221 M. Morton, T. F. Helminiak, S. D. Gadkary, F. Bueche, *J. Polym. Sci.*, **1962**, *57*, 471.
222 H. Eschwey, M. L. Hallensleben, W. Burchard, *Makromol. Chem.*, **1973**, *173*, 235.
223 J. G. Zilliox, P. Rempp, J. Parrod, *J. Polym. Sci., Part C*, **1968**, *22*, 145.
224 D. J. Worsfold, J. G. Zilliox, P. Rempp, *Can. J. Chem.*, **1969**, *47*, 3379.
225 C. Tsitsilianis, P. Lutz, S. Graff, J.-P. Lamps, P. Rempp, *Macromolecules*, **1991**, *24*, 5897.
226 C. Tsitsilianis, D. Papanagopoulos, P. Lutz, *Polymer*, **1995**, *36*, 3745.
227 M. Schappacher, A. Deffieux, J. L. Puteaux, R. Lazzaroni, *Macromolecules*, **2003**, *36*, 5776.
228 R. Matmour, Y. Gnanou, to be published.
229 T. E. Hogen-Esch, *J. Polym. Sci., Part A: Polym. Chem.*, **2006**, *44*, 2139.
230 E. F. Cassasa, *J. Polym. Sci., Part A1*, **1965**, *3*, 605.
231 D. Geiser, H. Höcker, *Macromolecules*, **1980**, *13*, 653.
232 G. Hild, H. Kolher, P. Rempp, *Eur. Polym. J.*, **1980**, *16*, 525.
233 R. Yin, T. E. Hogen-Esch, *Macromolecules*, **1993**, *26*, 6952.
234 D. Uhrig, J. W. Mays, *J. Polym. Sci., Part. A: Polym. Chem.*, **2005**, *43*, 6179.
235 T. Shimomura, K. J. Tölle, J. Smùid, M. Szwarc, *J. Am. Chem. Soc.*, **1967**, *89*, 796.
236 I. I. Böhm, M. Chmelir, G. Löhr, B. J. Schmitt, G. V. Schulz, *Adv. Polym. Sci.*, **1972**, *9*, 1.
237 V. Halaska, L. Lochman, D. Lim, *Collect. Czech Chem. Commun.*, **1968**, *33*, 3245.
238 R. P. Quirk, G. P. Bidinger, *Polym. Bull.*, **1989**, *22*, 63.
239 T. Kitayama, N. Fujimoto, K. Hatada, *Makromol. Chem., Macromol. Symp.*, **1993**, *67*, 137.
240 R. P. Quirk, H. Kim, M. J. Polee, C. Wesdemotis, *Macromolecules*, **2005**, *38*, 7895.

3
Carbocationic Polymerization

Priyadarsi De and Rudolf Faust

3.1
Introduction

Macromolecular engineering is defined as the total synthesis and processing of tailor-made macromolecules. The ultimate goal of this technology is to obtain a high degree of control over compositional and structural variables which affect the physical properties of macromolecules, including molecular weight, molecular weight distribution, end-functionality, tacticity, stereochemistry, block sequence and block topology, where the parameters of molecular characterization are well-represented by the ensemble average.

Cationic macromolecular engineering was pioneered by J. P. Kennedy, who stated: "The aim of macromolecular engineering is the synthesis of useful polymers by generating and then exploiting mechanistic understanding of elementary steps of polymerization reactions, i.e. initiation, propagation, chain transfer and termination" [1]. Before the 1970s, the above elementary steps were uncontrolled because traces of moisture initiated the polymerizations, chain transfer to monomer was difficult to avoid and the chemistry of termination was largely unknown. Controlled initiation and termination were discovered during the 1970s that provided head- and end-group control and allowed the synthesis of block and graft copolymers and macromonomers. Controlled chain transfer appeared in the late 1970s when it was discovered that chain transfer to monomer could be avoided by employing certain Lewis acid co-initiators or by using proton traps. Another important step towards the total synthesis of tailor made macromolecules was the development of the *inifer* technique by Kennedy. Controlled chain transfer to the initiator–*transfer* agent allowed the preparation of well-defined telechelic polymers, block copolymers and model networks. The discovery of quasi-living polymerization in 1982, where for the first time reversible termination (activation–deactivation) was postulated, heralded the new era of living cationic polymerization, which arrived a few years later. The early developments have been reviewed [1].

Living polymerizations, which proceed in the absence of termination and chain-transfer reactions, are the best techniques for the preparation of polymers with well-defined structure and, indeed, most of these polymers have been prepared using living polymerization. The resulting model polymers have extensively been used in validation of theories with respect to the properties in solution, melt and solid states [2]. They also have served as excellent standard materials for systematic studies on structure–property relationships of macromolecules, lending an impetus to the major fields of material science and polymer physics. The foregoing activities are made possible by advances in modern synthetic methodologies, combined with state-of-the-art characterization techniques in material science.

The experimental criteria for living polymerizations have been critically reviewed [3]. In general, diagnostic proof for the absence of chain transfer and termination can be obtained from both a linear semilogarithmic kinetic plot ($\ln([M]_0/[M])$ vs. time) and a linear dependence of number-average molecular weight (M_n) vs. monomer conversion (M_n vs. conversion). The linearity of a semilogarithmic kinetic plot implies a constant concentration of the growing center (absence of termination), but it does not guarantee the absence of chain transfer. Additional information for livingness should be obtained from the linearity of molecular weight vs. conversion plot, which proves the absence of chain transfer. Alternatively, according to a combined relation derived by Penczek et al., the linearity of the $\ln\{1-([I]DP_n/[M]_0)\}$ versus time plot (where DP_n is the number-average degree of polymerization, $[I]$ is the concentration of the initiator, $[M]_0$ is the initial concentration of the monomer) proves the simultaneous absence of termination and chain transfer [4]. There are no absolute living systems and the careful control of the experimental conditions (counterion, temperature, solvent) is necessary to obtain sufficient livingness to prepare well-defined polymers, especially when high molecular weights are targeted [5]. The original author of the seminal paper introducing the concept of living polymers concludes that "we shall refer to polymers as living if their end-groups retain the propensity of growth for at least as long a period as needed for the completion of an intended synthesis or any other desired task" [6]. This view has been adopted in this chapter.

3.2
Mechanistic and Kinetic Details of Living Cationic Polymerization

A complete mechanistic understanding of cationic polymerization is fundamental for cationic macromolecular engineering and requires the knowledge of the rate and equilibrium constants involved in the polymerization process. Numerous previous kinetic studies of carbocationic polymerization, however, have generally failed to yield reliable rate constants for propagation (k_p). This is attributed to uncertainties involved in the accurate determination of the active center concentration, a consequence of our incomplete knowledge of the mechanism due to the multiplicity of possible chain carriers (free ions, ion pairs and different solvated species) and the complexity of carbocationic reaction paths. Recently, new methods have been developed for the determination of rate and equilibrium constants in carbocationic polymerizations [7, 8]. These methods have been utilized to determine the rate constant of propagation (k_p), the rate (k_i) and equilibrium constant of ionization (K_i) and deactivation (k_{-i}) for isobutene (IB) [9], styrene (St) [10] and ring-substituted styrenes [11–14]. The results show that previously accepted propagation rate constants [15] are underestimated in some cases by as much as 4–6 orders of magnitude. The k_i values for IB and St have also been determined independently by Storey and coworkers [16–18] from the average number of monomer units added during one initiation–termination cycle (run number). The reported values agreed remarkably well with those published earlier.

The above studies confirmed the results of prior kinetic investigations with model compounds that the propagation rate constant is independent of the nature of Lewis acid and increases moderately with increasing solvent polarity. In agreement with findings of Mayr that fast bimolecular reactions (i.e. $k_p^\pm > 10^7$ L mol^{-1} s^{-1}) do not have an enthalpic barrier [19], $k_p^\pm$ is independent of temperature for IB, p-chlorostyrene (p-ClSt), St and p-methylstyrene (p-MeSt). Although most kinetic investigations have been conducted under conditions where propagation takes place on ion pairs, the propagation rate constant for free ions, k_p^+, for IB is reportedly similar to $k_p^\pm$, suggesting that free and paired cations possess similar reactivity and therefore differentiation between free ions and ion pairs is unnecessary [20].

It is apparent that, due to the extremely rapid propagation, if all chain ends were ionized and grew simultaneously, monomer would disappear at such a high rate that the polymerization would be uncontrollable. In living cationic polymerization, therefore, a dynamic equilibrium must exist between a very small amount of active and a large pool of dormant species. Mechanistically, the equilibrium concentration of cations is provided by the reversible activation/deactivation shown for IB in Scheme 3.1.

From the k_i, k_{-i} and k_p values determined in hexanes–CH$_3$Cl (60:40, v/v) at $-80\,^\circ$C, the sequence of events for an average polymer chain could be ascertained. Using typical concentrations of [TiCl$_4$] = 3.6×10^{-2} mol L^{-1} and [IB] = 1 mol L^{-1}, the following time intervals (τ) between two consecutive events have been calculated:

$\tau_i = 1/k_i[\text{TiCl}_4]^2 = 49$ s;
$\tau_{-i} = 1/k_{-i} = 2.9\times 10^{-8}$ s = 29 ns;
$\tau_p = 1/k_p[\text{IB}] = 1.4\times 10^{-9}$ s = 1.4 ns.

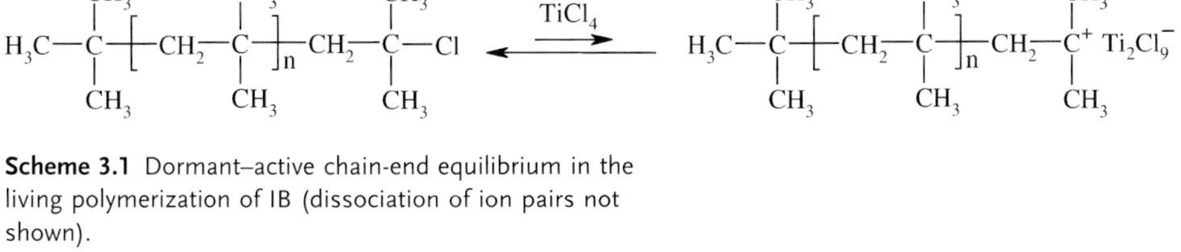

Scheme 3.1 Dormant–active chain-end equilibrium in the living polymerization of IB (dissociation of ion pairs not shown).

Thus, the time interval between two ionizations (activation) is relatively long (49 s). The ionized chain ends stay active for a very short time, only 29 ns, before reversible termination (deactivation) takes place and the polymer end goes back to a dormant, inactive state. Propagation is 20 times faster than deactivation, however (monomer incorporates on average every 1.4 ns), and 20 monomer units are added during one active cycle. This results in a relatively high polydispersity index at the beginning of the polymerization that progressively decreases to the theoretical value at complete conversion [21]:

$$M_w/M_n = 1 + [I]_0 k_p/k_{-i}$$

The above equation, where $[I_0]$ is the total concentration of (active and dormant) chain ends, is valid for unimolecular deactivation (ion pairs); for bimolecular deactivation, the deactivation rate constant should be multiplied by the concentration of the deactivator. The starting [IB] may be decreased to decrease the number of monomer units incorporated during one active cycle and this yields PIB with a lower polydispersity index. For instance, at $[IB] = 0.1$ mol L^{-1}, two monomer units are incorporated during one active cycle even at the onset of the polymerization. At $[TiCl_4] = 3.6 \times 10^{-2}$ mol L^{-1} and $[IB] = 1$ mol L^{-1}, about 4 and 40 min would be necessary for the formation of a PIB with a DP=100 and 1000, respectively.

For a specific monomer, the rate of exchange and the position of the equilibrium and, to some extent, the zero-order monomer transfer constants depend on the nature of the counteranion in addition to temperature and solvent polarity. Therefore, initiator/co-initiator systems that bring about controlled and living polymerization under a certain set of experimental conditions are largely determined by monomer reactivity.

It is important to note that living polymerization does not require the assumption of special growing species such as stretched covalent bonds or stabilized carbocations, as pointed out by Matyjaszewski and Sigwalt [22]. In line with this reasoning, identical propagation rate constants were observed in living and nonliving polymerization, indicating that propagation proceeds on identical active centers [9].

3.3
Living Cationic Polymerization

Since the first reports of living cationic polymerization of vinyl ethers [23] and isobutene [24, 25] in the 1980s, the scope of living cationic polymerization of vinyl monomers has been expanded rapidly in terms of monomers and initiating systems. Compared with anionic polymerization, living cationic polymerization can proceed under much less rigorous and much more flexible experimental conditions. The high-vacuum technique is not indispensable, since alternative routes can consume adventitious moisture without terminating the living chains. Nonetheless, rigorous purification, as in living anionic polymerization, of reagents is still required for the best control.

3.4
Monomers and Initiating Systems

Initiation and propagation take place by electrophilic addition of the monomers to the active cationic sites. Therefore, the monomer must be nucleophilic and its substituents should be able to stabilize the resulting positive charge. As a result, the reactivity of monomers is roughly proportional to the electron-donating ability of their substituents, as can be seen in Fig. 3.1.

In addition to these monomers, the total number of monomers for living cationic polymerization was estimated to be around 100 in 1994 and it appears that this process has a much broader choice of monomers than the living anionic counterpart [26].

With a few exceptions [27, 28], living cationic polymerization is initiated by the initiator/co-initiator (Lewis acid) binary system. Selection of an initiating system for a given monomer is of crucial importance, since there are no universal initiators such as organolithiums in anionic polymerization. For example, whereas weak Lewis acids such as zinc halides may be necessary to effect living polymerization of the more reactive vinyl ethers, they are not effective for the living polymerization of the less reactive monomers, such as IB and St. Detailed inventories of initiating systems for various monomers are well described in recent publications [26, 29, 30].

3.5
Additives in Living Cationic Polymerization

Three main categories of additives have been introduced and extensively utilized in the living cationic polymerization of vinyl monomers; (1) Lewis bases [31] (also called "electron donors" [32] or "nucleophiles" [33]), (2) proton traps [34] and (3) salts [35]. As the different names of the first categories of additives imply, the actual roles of these basic adjuvants and the true mechanisms of enhanced livingness have been longstanding controversies. Higashimura et al.

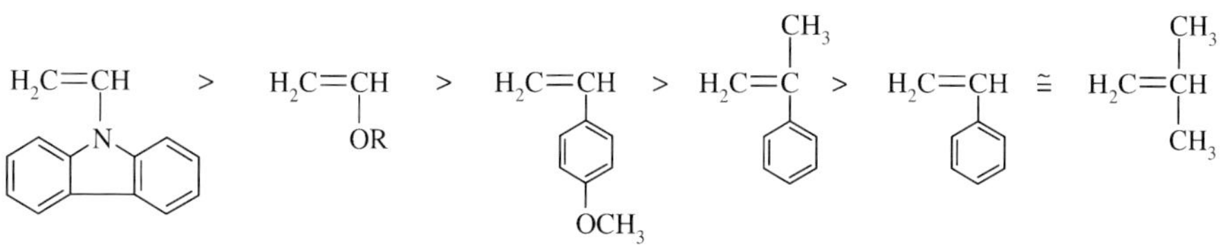

Fig. 3.1 Order of reactivity of monomers.

proposed the theory of carbocation stabilization by nucleophilic additives through weak nucleophilic interaction [36]. A similar opinion was also expressed by Kennedy et al. [29]. In contrast to this view, Matyjaszewski [33] considered that these bases only decrease the concentration of active species by reversible formation of onium ions which do not propagate or by complexing with Lewis acids. It has also been proposed by Penczek [37] that these bases may enhance the rate of equilibrium between dormant and active species via onium ion formation, which provides a thermodynamically more favorable pathway from covalent species to cation and vice versa. Unfortunately, no direct evidence for either nucleophilic interaction or onium ion formation has been provided.

Faust and coworkers demonstrated the living polymerization of IB and styrene co-initiated with $TiCl_4$ or BCl_3 in the *absence* of nucleophilic additives but in the presence of a proton trap (2,6-di-*tert*-butylpyridine), a non-nucleophilic weak base [34, 38]. The addition of nucleophilic additives had no effect on polymerization rates, molecular weights or molecular weight distributions (MWDs). Thus, it was suggested that the major role of added bases and the sole role of the proton trap is to scavenge protogenic impurities in the polymerization system. While supporting view is emerging [39], combination of the first and second additives in one category is still under discussion.

Common ion salts are considered to suppress the ionic dissociation of covalent species and ion-pairs to free ions, which are believed to result in nonliving polymerization [35]. In the light of recent results, which confirmed similar reactivity of free ions and ion pairs, this view may require revision. In addition to the common ion effect, addition of salts can also change the nucleophilicity of counterions by modifying either the coordination geometry [40] or degree of aggregation [41] of Lewis acids or their complex counterions. The former is the case with $SnCl_4$ and the latter with $TiCl_4$ in the presence of tetra-*n*-butylammonium chloride (nBu_4NCl). In both cases, more nucleophilic counterions ($SnCl_6^-$ vs. $SnCl_5^-$ or $TiCl_5^-$ vs. $Ti_2Cl_9^-$) are generated and these are reported to mediate the living cationic polymerizations of styrenic monomers [40, 42] and isobutyl vinyl ether [41].

3.5.1
Isobutene (IB)

IB is the most studied monomer that can only polymerize by a cationic mechanism. The living carbocationic polymerization of IB was first discovered by Faust and Kennedy using an organic acetate–BCl_3 initiating system in CH_3Cl or CH_2Cl_2 solvents at –50 to –10 °C [24, 25]. Living carbocationic polymerizations of IB to date are based on BCl_3, $TiCl_4$ and organoaluminum halide co-initiators. The activity of the BCl_3-based system is greatly solvent dependent, i.e. sufficient activity occurs only in polar solvents. In less polar solvents, the solvation of the counteranion does not promote ion generation and the binary ionogenic equilibrium is shifted strongly to the left. Therefore, the concentration of growing cations is extremely small, resulting in negligible polymerization rates. However, since polyisobutene (PIB) is poorly soluble in polar solvents at low temperatures, the molecular weights are limited with the BCl_3-based initiating systems.

A wide variety of initiators, organic esters, halides, ethers and alcohols have been used to initiate living polymerization of IB at temperatures up to –10 °C. The true initiating entity with ethers and alcohols is the chloro derivative arising by fast chlorination. The polymerization involving the BCl_4^- counteranion is very slow, measured in hours, compared with the fast polymerization by protic impurities and, in the absence of proton scavenger, the monomer is consumed mainly by this process. In the presence of proton trap or electron donors (EDs), similar rates, controlled molecular weights and narrow MWDs (PDI $\approx$ 1.2) have been reported [38]. According to kinetic studies, the polymerization is first order with respect to both monomer and BCl_3 [38]. The absence of a common ion salt effect in polymerizations involving the BCl_4^- counteranion suggests that propagation is mainly via the ion pairs and the contribution of free ions, if any, is negligible [43].

The advantage of the ester initiators is that the nucleophilicity of the counteranions and therefore the activity of the growing centers can be modulated by the nature of the acid. Esters of trichloroacetic, dichloroacetic, acetic, phenylacetic, phenylbutyric, isobutyric, pivalic, benzoic and cinnamic acids have been used as initiators for the living cationic polymerization of IB [44]. The apparent propagation rate constants were found to decrease dramatically (by ~5–6 orders of magnitude) in the same order as decreasing strength of the corresponding acid. Investigations with mono- and difunctional esters of different acids indicate that with stronger acids, not only polymerization rates but also initiation efficiencies are greatly increased [45]. Interestingly, protic initiation is negligible when esters are used as initiators. It has been speculated that the living centers themselves scavenge protic impurities, which may explain the relatively broad MWDs ($M_w/M_n \approx$ 1.4–2.5) [46]. In the presence of the hindered proton trap 2,6-di-*tert*-butylpyridine (DTBP), values of M_w/M_ns $\approx$ 1.1 have been obtained.

Organic esters, halides and ethers have been used to initiate living polymerization of IB at temperatures from –90 up to –40 °C. In conjunction with $TiCl_4$, ethers are converted to the corresponding chlorides almost instantaneously, while the conversion of esters is somewhat slow [34]. According to Chen et al., alcohols are inactive with $TiCl_4$ alone but have been used in conjunction with BCl_3 and $TiCl_4$ [47]. The BCl_3 converts the alcohols to the active chloride, which is activated by $TiCl_4$. In contrast to Chen et al., Puskas and Grassmuller reported chlorination of alcohols and initiation by $TiCl_4$ alone [48].

Under well-dried conditions PIBs with controlled M_ns up to ~60 000 and narrow MWDs could be prepared in the absence of any additives in nonpolar solvent mixtures and low temperatures [34]. PIBs with M_ns up to 150 000 and M_w/M_ns as low as 1.02 have been obtained in the presence of proton trap or Lewis bases. The polymerization is first order in monomer but second order in $TiCl_4$, due to dimeric counteranions [34], although a first-order dependency was reported at $[TiCl_4]<[initiator, I_0]$ [49]. The consequence of the second-order rate dependence is that although excess of $TiCl_4$ over the initiator halide is not required to induce polymerization, at low initiator concentrations to obtain high M_n, acceptable rates are only obtained when high $TiCl_4$ concentrations (16–36 times $[I]_0$) are used. Living polymerization of IB was also reported with the $TiCl_4$–$TiBr_4$ mixed [50] co-initiator that yields mixed $Ti_2Cl_{n+1}Br_{8-n}^-$ counter-anions. By the stepwise replacement of Cl with Br, however, the Lewis acidity decreases, which results in a decreased ionization rate constant and therefore decreasing overall rates of polymerization with decreasing $TiCl_4/TiBr_4$ ratio.

Organoaluminum compounds have also been employed for the living cationic polymerization of IB using 1,4-bis(1-azido-1-methylethyl)benzene–Et_2AlCl–CH_2Cl_2 at –50 °C to produce living polymerization of IB for $M_n < 50000$ where the presence of an electron donor such as DMSO is not necessary [51]. Another polymerization system based on Et_2AlCl and tertiary alkyl halide initiators has been reported but requires the use of an 80:20 (v/v) nonpolar–polar solvent mixture [52]. The first example of Me_2AlCl-catalyzed living polymerizations of IB was presented using conventional tertiary alkyl chloride initiators and 60:40 (v/v) nonpolar–polar solvent mixtures. PIBs were prepared with $M_n = 150000$ and $M_w/M_n = 1.2$ [53] even in the absence of additives such as proton traps or electron donors. The "living" nature of these polymerizations has been demonstrated at –75 to –80 °C in both 60:40 (v/v) hexane–CH_2Cl_2 and hexane–methyl chloride (MeCl) solvent systems. Recently, the living polymerization of IB was also reported using 2-chloro-2,4,4-trimethylpentane (TMPCl)–DTBP–hexanes–MeCl solvent mixtures at –80 °C using Me_2AlCl, $Me_{1.5}AlCl_{1.5}$ or $MeAlCl_2$ [54]. With the latter two co-initiators the polymerization was extremely fast and completed in seconds, which necessitated special considerations for reaction control.

3.5.2
β-Pinene

The first example of living cationic isomerization polymerization of β-pinene was reported with the HCl–2-chloroethyl vinyl ether adduct [$CH_3CH(OCH_2CH_2Cl)Cl$] or 1-phenylethyl chloride–$TiCl_3(OiPr)$ initiating system in the presence of tetra-n-butylammonium chloride (nBu_4NCl) in CH_2Cl_2 at –40 and –78 °C [55, 56]. The polymerization was rather slow even at relatively high initiator (20 mM) and co-initiator (100 mM) concentrations. The much stronger Lewis acid co-initiator $TiCl_4$ induced an extremely rapid polymerization, yielding polymers with controlled molecular weight but with broad molecular weight distributions. The ^1H NMR analysis of the polymers showed a tert-chloride end group and isomerized β-pinene repeat units with a cyclohexene ring. Copolymerization of β-pinene with IB indicated that the two monomers exhibit almost equal reactivity [57].

3.5.3
Styrene (St)

The conventional cationic polymerization of St suffers from side-reactions such as chain transfer by β-proton elimination and inter- and intramolecular Friedel–Crafts alkylation. Hence control of the cationic polymerization of St has been considered difficult. The living carbocationic polymerization of St was first achieved with the 1-(p-methylphenyl)ethyl acetate–BCl$_3$ initiating system in CH$_3$Cl at –30 °C [58]. The MWD was broad (~5–6), most likely because of slow initiation and/or slow exchange between the dormant and active species. Living polymerization of St with controlled molecular weight and narrow MWD was obtained using SnCl$_4$–1-phenylethyl halides as initiating systems in a nonpolar solvent (CHCl$_3$) [59] and solvent mixtures or in a polar CH$_2$Cl$_2$ in the presence of nBu$_4$NCl [60].

Living polymerization was also reported with the TMPCl–TiCl$_4$–methylcyclohexane (MeChx)–MeCl (60:40, v/v) system at –80 °C in the combined presence of an ED and a proton trap [61]. Later studies indicated that the ED is unnecessary and the living nature of the polymerization is not due to carbocation stabilization [62]. The living cationic polymerization of St has also been achieved with TiCl$_3$(OiPr) as an activator, in conjunction with 1-phenylethyl chloride and nBu$_4$NCl in CH$_2$Cl$_2$ at –40 and –78 °C [63]. The molecular weight distributions were narrow throughout the reactions (MWD $\approx$ 1.1). Living St polymerization was also reported with the p-MeSt · HCl adduct (p-MeStCl)–TiCl$_4$–MeChx–MeCl (60:40, v/v) at –80 °C in the presence of a proton trap and it was found that p-MeStCl is a better initiator than TMPCl for St polymerization using TiCl$_4$ in MeChx–MeCl solvent mixture [64]. Recently, living polymerization of St was obtained with the system 1-phenylethyl chloride–TiCl$_4$–Bu$_2$O in a mixture of 1,2-dichloroethane and hexane (55:45, v/v) at –15 °C [65].

3.5.4
p-Methylstyrene (p-MeSt)

Faust and Kennedy [66] reported the living carbocationic polymerization of p-MeSt in conjunction with BCl$_3$ in CH$_3$Cl and C$_2$H$_5$Cl solvents at –30 and –50 °C; however, the MWDs were rather broad (~2–5). TMPCl–TiCl$_4$-initiated living polymerization of p-MeSt in MeChx–MeCl (50:50, v/v) solvent mixture at –30 °C in the presence of nBu$_4$NCl and DTBP has been reported by Nagy et al. [67]. Kojima et al. reported the living cationic polymerization of p-MeSt with the HI–nBu$_4$NCl–ZnX$_2$ system in toluene and CH$_2$Cl$_2$ and obtained polymers of fairly narrow MWD [68]. Lin and Matyjaszewski studied the living cationic polymerization of St and p-MeSt initiated by 1-phenylethyl trichloroacetate–BCl$_3$ in CH$_2$Cl$_2$ [69]. Stöver et al. studied the living cationic polymerization of p-MeSt in the molecular range up to $M_n \approx 5500$, initiated by the 1-phenylethyl bromide–SnCl$_4$ initiating system in CHCl$_3$ or in CHCl$_3$–CH$_2$Cl$_2$ solvent mixtures at –27 °C [70]. The polymerization was very slow even though very high concentrations of SnCl$_4$ (0.23 mol L^{-1}) and initiator, 1-phenylethyl bromide (0.0215 mol L^{-1}), were used. The M_ns of the poly(p-MeSt) obtained were in the agreement with the calculated values, but the polydispersity index (PDI) was relatively high, PDI ≈ 1.5–1.6. Living carbocationic polymerization of p-MeSt was also obtained with 1,1-diphenylethylene-capped TMPCl–TiCl$_4$:Ti(IpO)$_4$ initiating system in the presence of DTBP using hexanes–MeCl or MeChx–MeCl (60:40, v/v) solvent mixture at −80 °C [71]. Recently, the living carbocationic polymerization of p-MeSt was achieved with 1-chloro-1-phenylethane, 1-chloro-1-(4-methylphenyl)ethane and 1-chloro-1-(2,4,6-trimethylphenyl)ethane in conjunction with SnCl$_4$ as Lewis acid and DTBP as proton trap in CH$_2$Cl$_2$ at −70 to −15 °C [14].

3.5.5
p-Chlorostyrene (p-ClSt)

Kennedy and coworkers reported the living carbocationic polymerization of p-ClSt initiated by TMPCl–TiCl$_4$ in the presence of dimethylacetamide as electron donor and DTBP as proton trap in MeCl–MeChx (60:40, v/v) solvent mixture at −80 °C [67, 72]. Kanaoka et al. obtained poly(p-ClSt) of a narrow molecular weight distribution with the 1-phenylethyl chloride–SnCl$_4$ initiating system in CH$_2$Cl$_2$ at −15 to +25 °C in the presence of nBu$_4$NCl [73]. The polymerization was somewhat slow. Controlled cationic polymerization of p-ClSt was also achieved using the alcohol–BF$_3$OEt$_2$ system in the presence of fairly large amount of water [74]. Recently, for the living polymerization of p-ClSt, 1-chloro-1-(4-methylphenyl)ethane and p-ClSt·HCl adduct was used in conjunction with TiCl$_4$–DTBP in MeCl–MeChx (40:60, v/v) solvent mixture at −80 °C [13].

3.5.6
2,4,6-Trimethylstyrene (TMeSt)

In the cationic polymerization of St, one of the major side-reactions is indanic cyclization [75]. Intra- and intermolecular alkylation are absent in the cationic polymerization of TMeSt, which was recognized in an early report on the living polymerization of TMeSt initiated by the cumyl acetate–BCl$_3$ initiating system in CH$_3$Cl at −30 °C [76]. Recently, the living cationic polymerization of TMeSt was initiated by the 1-chloro-1-(2,4,6-trimethylphenyl)ethane, a model propagating end, in CH$_2$Cl$_2$ at −70 °C and yielded polymers with theoretical molecular weights and very low polydispersity (M_w/M_n = 1.02–1.1) [12].

3.5.7
p-Methoxystyrene (p-MeOSt)

The living carbocationic polymerization of p-MeOSt was first reported with the HI–ZnI$_2$ initiating system in toluene at −15 to 25 °C [31, 77]. Living polymerizations were also attained in the more polar solvent CH$_2$Cl$_2$ with the HI–I$_2$ and HI–ZnI$_2$ initiating systems in the presence of nBu$_4$NX (X=Cl, Br, I) [78]. Comparable but less controlled polymerization of p-MeOSt has been reported using iodine as an initiator in CCl_4 [79]. This system gives rise to long-lived but not truly living polymerization. More recently, The p-MeOStCl–$SnBr_4$ initiating system has been used in CH_2Cl_2 at −60 to −20 °C in the presence of DTBP. The M_ns obtained were in good agreement with the calculated values assuming that one polymer chain forms per initiator. Polymers with M_n up to 120 000 were obtained with $M_w/M_n \approx 1.1$ [11]. A report has indicated the controlled cationic polymerization of p-MeOSt with controlled molecular weights and relatively narrow molecular weight distribution (PDI = 1.4) using the p-MeOSt · HCl adduct (p-MeOStCl)–$Yb(OTf)_3$ initiating system in the presence of 2,6-di-*tert*-butyl-4-methylpyridine [80]. The authors also claimed the controlled, albeit very slow, cationic polymerization of p-MeOSt in aqueous media using the p-MeOStCl–$Yb(OTf)_3$ initiating system. Relatively narrow PDIs (~1.4) were observed and the molecular weights increased in proportion to the monomer conversion. Surfactants [81], sulfonic acid-based initiators [82] and various phosphonic acid initiators [83] were also used for the cationic polymerization of p-MeOSt in aqueous medium.

3.5.8
α-Methylstyrene (αMeSt)

The living polymerization of αMeSt was first achieved with the vinyl ether–HCl adduct–$SnBr_4$ initiating system in CH_2Cl_2 at −78 °C [84]. Controlled polymerization of αMeSt was obtained with the cumyl chloride–BCl_3 system at −78 °C in CH_2Cl_2–toluene (1:7, v/v) solvent mixture in the presence of nBu_4NCl [85]. The living polymerization of αMeSt was also studied at −60 °C using iodine in liquid SO_2–CH_2Cl_2 or liquid SO_2–toluene [86]. The living polymerization of αMeSt has also been established in hexanes–MeCl (60:40, v/v) at −60 to −80 °C in the presence of DTBP using 1,1-diphenylethylene (DPE)-capped TMPCl with $SnBr_4$ or $SnCl_4$ [42] and diphenyl alkyl chloride or HCl adduct of αMeSt dimer with BCl_3 or $SnCl_4$ co-initiators [87–89]. Initiation with cumyl chloride, however, is slow relative to propagation, due to the absence of back strain. The living polymerization of p-chloro-α-methylstyrene was achieved using the 1,3-dimethyl-1,3-diphenyl-1-chlorobutane–BCl_3 initiating system in MeChx–MeCl (60:40, v/v) at −80 °C [90].

3.5.9
Indene

Chain transfer to monomer via indane formation (intramolecular alkylation), the most important side-reaction in the polymerization of styrene, cannot take place with indene. Therefore, even conventional initiating systems give high molecular weight and negligible transfer at low temperatures. Cumyl methyl ether [91] or cumyl chloride [92] in conjunction with $TiCl_3OBu$ or with $TiCl_4$ and dimethyl sulfoxide as an ED [93] initiate living polymerization of indene.

Living polymerization is also claimed with the TMPCl–TiCl$_4$ initiating system using hexanes–MeCl (60:40, v/v) [94] and MeChx–MeCl (60:40, v/v) [95] mixed solvents at –80 °C. Thus, polyindene of theoretical molecular weight up to at least 13 000 and with a narrow MWD ($M_w/M_n \approx 1.2$) can be obtained with the cumyl chloride–BCl$_3$ initiating system in MeCl at –80 °C [94].

3.5.10
N-Vinylcarbazole

There are very few reports available on the living cationic polymerization of N-vinylcarbazole, one of the most reactive monomers for cationic polymerization. Living polymerization of N-vinylcarbazole was reported using the toluene or CH$_2$Cl$_2$ solvent system with only HI [96] and I$_2$ as initiator in CH$_2$Cl$_2$ and CH$_2$Cl$_2$–CCl$_4$ (1:1, v/v) [97]. Living cationic polymers of N-vinylcarbazole were synthesized with I$_2$ at –78 °C in CH$_2$Cl$_2$ in the presence of nBu$_4$NI [98].

3.5.11
Vinyl Ethers

Alkyl vinyl ethers are among the most reactive vinyl monomers in cationic polymerization. The pendant alkoxy groups provide the growing vinyl ether carbocation with high stability. Controlled/living cationic polymerization of isobutyl vinyl ether was first discovered with the HI–I$_2$ initiating system [23]. The living cationic polymerization of vinyl ethers (CH$_2$=CHOR, where R=CH$_3$, C$_2$H$_5$, isopropyl, n-butyl, isobutyl, n-hexadecyl, 2-chloroethyl, benzyl, cyclohexyl, etc.) was first developed with the HI–I$_2$ initiating system and has been reviewed extensively. Subsequently other weak Lewis acids, e.g. ZnCl$_2$, ZnBr$_2$ and ZnI$_2$, have also been employed. A review is available [99]. The living cationic polymerization of tert-butyl vinyl ether was achieved with the CH$_3$CH(OiBu)OCOCH$_3$–Et$_{1.5}$AlCl$_{1.5}$ initiating system in the presence of THF as a "Lewis base" at –20 °C [100]. However, the polymerization was slow and close to quantitative yield was reached only after 60 h.

Recently, the living cationic polymerization of tert-butyl vinyl ether [101] and cyclohexyl vinyl ether [102] was accomplished in hexanes–MeCl solvent mixtures at –80 °C using TMPCl capped with 1,1-ditolylethylene as initiator and TiCl$_4$–Ti(OiP)$_4$ as co-initiators. The process involved capping the initiator, TMPCl, with 1,1-ditolylethylene in the presence of TiCl$_4$, followed by fine-tuning of the Lewis acidity with the addition of Ti(OiP)$_4$ to match the reactivity of tert-butyl vinyl ether or cyclohexyl vinyl ether. Both polymers exhibited T_g (88 and 61 °C, respectively) well above room temperature. Poly(vinyl ether)s with a T_g as high as 100 °C have been obtained in the living cationic polymerization of vinyl ethers with a bulky tricyclodecane or tricyclodecene unit using HCl–ZnCl$_2$ in toluene at –30 °C [103]. The living cationic polymerization of 4-[2-(vinyloxy)ethoxy]azobenzene was achieved with various Lewis acids in the presence of an ester as an added base [104]. The fast living cationic polymerization of vinyl ethers with

SnCl$_4$ combined with EtAlCl$_2$ in the presence of an ester as an added base has been reported [105]. Fast living polymerization with the SnCl$_4$–EtAlCl$_2$ initiating system was tested with isobutyl vinyl ether, ethyl 4-[2-(vinyloxy)ethoxy]benzoate and 4-[2-(vinyloxy)ethoxy]azobenzene and yielded polymers with theoretical molecular weights and very low polydispersity ($M_w/M_n = 1.02–1.09$). The cationic polymerization of vinyl ethers with a urethane group, 4-vinyloxybutyl n-butylcarbamate and 4-vinyloxybutyl phenylcarbamate, was studied with the HCl–ZnCl$_2$ initiating system in CH$_2$Cl$_2$ solvent at –30 °C. The results indicated the formation of living polymer from vinyl ether with 4-vinyloxybutyl phenylcarbamate. The difference in living nature between monomers with 4-vinyloxybutyl n-butylcarbamate and 4-vinyloxybutyl phenylcarbamate was attributed to the difference in the electron-withdrawing power of the carbamate substituents, namely n-butyl versus phenyl units of the monomers [106]. The cationic homopolymerization and copolymerization of five vinyl ethers with silyloxy groups, each with a different spacer length, were examined with a cationogen–Et$_{1.5}$AlCl$_{1.5}$ initiating system in the presence of an added base. When an appropriate base was added, the living cationic polymerization of Si-containing monomers became feasible, giving polymers with narrow MWD [107]. A saccharidic vinyl ether-type monomer, 1,2:3,4-di-O-isopropylidene-6-O-(2-vinyloxyethyl)-D-galactopyranose, was polymerized via a living cationic process using acetaldehyde diethylacetal–trimethylsilyl iodide–ZnCl$_2$ as the initiating system [108]. The authors provided references for the living cationic polymerization of other vinyl ether-type saccharidic monomers. The living cationic polymerization route was reported to give poly(vinyl ether)-type polymers with oligooxyethylene carbonate pendant groups [109]. The living nature of the cationic polymerization of butyl vinyl ether in the fluorinated solvent 1,1,2-trichlorotrifluoroethane was studied using the EtAlCl$_2$–1-butoxyethyl acetate initiating system. The polymerization is living at 0 °C or below in the presence of 1,4-dioxane as a stabilizing Lewis base [110]. The cationic polymerization of 2-[4-(methoxycarbonyl)phenoxy]ethyl vinyl ether, a vinyl ether with a pendant benzoate, was reported to proceed with living/long-lived propagating species with an HCl–ZnCl$_2$ initiating system in dichloromethane at –15 °C [111]. Hexa(chloromethyl)melamine–ZnCl$_2$ was found to be a efficient initiating system for the living cationic polymerization of isobutyl vinyl ether in CH$_2$Cl$_2$ at –45 °C [112]. Characterization of the polymers by GPC and ^{1}H NMR showed that initiation was rapid and quantitative and that the initiator is hexafunctional, leading to six-armed star-shaped polymers. A series of aromatic acetals from substituted phenols were employed as initiators in conjunction with Lewis acids such as AlCl$_3$, SnCl$_4$ and SnBr$_4$ for the cationic polymerization of isobutyl vinyl ether [113]. An AlCl$_3$–aromatic acetals initiating system gave polymers with broad molecular weight distributions, whereas the addition of ethyl acetate (10 vol.%) to these systems led to the formation of controlled polymers whose number-average molecular weights were directly proportional to monomer conversion, with relatively narrow distributions (~1.2). The aromatic acetals in conjunction with tin halides gave relatively high meso contents (81%) in toluene at –78 °C.

3.6
Functional Polymers by Living Cationic Polymerization

Functional polymers are of great interest due to their potential applications in many important areas such as surface modification, adhesion, drug delivery, polymeric catalysts, compatibilization of polymer blends and motor oil additives. In addition to the controlled and uniform size of the polymers, living polymerizations provide the simplest and most convenient method for the preparation of functional polymers. However, there are relatively few end-functionalized polymers (polymers with functional groups selectively positioned at the termini of any given polymeric or oligomeric chain) synthesized by living cationic polymerization of vinyl monomers, although varieties of end-functionalized polymers have been successfully synthesized in anionic polymerization. There are two basic methods to prepare functional polymers by cationic polymerization: (a) initiation from functional initiators and (b) termination by functional terminators

3.6.1
Functional Initiator Method

This method involves the use of functional initiators with a protected or unprotected functional group. When the functional group is unreactive under the polymerization conditions, protection is not necessary. Functional vinyl ethers have been used extensively in the living cationic polymerizations of vinyl ethers and these functional poly(vinyl ethers) can be derivatized to desired functionality by simple organic reactions. Vinyl ethers carrying a variety of functional pendant groups, in the general form shown in Fig. 3.2, have been polymerized in living fashion in toluene (or CH_2Cl_2) using $HI-I_2$ or $HI-ZnI_2$ systems [114].

Functionalized initiators have been used extensively for styrene and derivatives to obtain end-functionalized polymers by living cationic polymerization. A series of α-end-functionalized polymers of St and p-MeSt were synthesized by living cationic polymerizations in CH_2Cl_2 at $-15\,°C$ initiated with the HCl adducts of $CH_2=CH(OCH_2CH_2X)$ (X=chloride, benzoate, acetate, phthalimide, methacrylate) carrying pendant substituents X that serve as the terminal functionalities in the products using $SnCl_4$ in the presence of nBu_4NCl. ^1H NMR

$H_2C=CH-OCH_2-CH_2-R$

R = OMe, OEt, OAc, (CH_2CH_2O) Et_n [n = 1, 2, 4]

$OSi(Me)_3$, $OSi(Me)_2iBu$, $C(CO_2Et)_3$, O_2CPh

OCH_2CO_2Et, $OC(O)C(CH_3)H=CH_2$, $OC(O)CH=CH_2$

$CH(CH_2CO_2Et)_2$, etc.

Fig. 3.2 Functional vinyl ethers used for living cationic polymerization.

3.6 Functional Polymers by Living Cationic Polymerization

analysis showed that all the polymers possess controlled molecular weights and end functionalities close to unity, i.e. one terminal functionality X per chain [115]. The living cationic polymerization of aMeSt initiated by the HCl adduct of 2-chloroethyl vinyl ether–SnBr$_4$ initiating system in CH$_2$Cl$_2$ at –78 °C gave terminal functionalities in the products [84]. The living cationic polymerization of p-ClSt induced with CH$_3$CH(OCH$_2$CH$_2$Cl)Cl–SnCl$_4$–nBu$_4$NCl at 0 °C or room temperature gave living polymers with narrow MWDs, which are a kind of end-functionalized polymers. The authors concluded that a variety of vinyl ethers would lead to end-functionalized poly(pClSt) [73]. The –CH$_2$CO$_2$H and –OH end functionalities have been obtained using the functional initiator method for the living cationic polymerization of p-tert-butoxystyrene [116]. A series of end-functionalized polymers of p-MeOSt were synthesized by the functional initiator method initiated with a functional vinyl ether–HI adduct [XCH$_2$CH$_2$OCH-(CH$_3$)I; X=CH$_3$COO, (EtOCO)$_2$CH, phthalimide]–ZnI$_2$ in toluene at –15 °C [117].

Few end-functionalized PIBs have been obtained by using functional initiators. The ester functional initiators 3,3,5-trimethyl-5-chloro-1-hexyl isobutyrate and methacrylate have been successfully employed for the living polymerization of IB [118]. Subsequently the synthesis of α-carboxylic acid functional PIB was also reported by hydrolysis of the product obtained in the living cationic polymerization of IB using a novel aromatic initiator containing the CH$_3$OCO– moiety [119]. While these ester functionalities form a complex with the Lewis acid co-initiator, quenching the polymerization with CH$_3$OH yield the original ester quantitatively. Initiators containing a cationically unreactive vinyl functionality, e.g. 5-chloro-3,3,5-trimethyl-1-hexene and 3-chlorocyclopentene, have been used to prepare PIBs with α-olefin head-groups. Following the discovery that the chlorosilyl functionality is unreactive toward Lewis acids or carbocations, a series of novel chlorosilyl functional initiators have been employed in the living cationic polymerization of IB to synthesize well-defined PIBs carrying a mono-, di- and trimethoxysilyl head-group and a *tert*-chloro end-group in conjunction with TiCl$_4$ in hexanes–MeCl (60:40, v/v) at –80 °C [120]. A class of unique epoxide initiators, e.g. α-methylstyrene epoxide (MSE), 2,4,4-trimethylpentyl 1,2-epoxide (TMPO-1), 2,4,4-trimethylpentyl 2,3-epoxide (TMPO-2) and hexaepoxysqualene, for the living polymerization of IB in conjunction with TiCl$_4$ has been described by Puskas and coworkers [121, 122]. Ring cleavage induced by TiCl$_4$ produces a tertiary cation that initiates the living polymerization of IB. Upon quenching the polymerization with methanol, PIB carrying a primary hydroxyl head-group and *tert*-chloride end-group is obtained. During initiation, however, simultaneous side-reactions take place. Although these side-reactions do not affect the livingness of the polymerization or the functionality of the PIB, they reduce the initiator efficiency to 3% with TMPO-1 and 40% with MSE.

The α-end-functionalized poly(β-pinene) was obtained by living cationic isomerization polymerization in CH$_2$Cl$_2$ at –40 °C using TiCl$_3$(O*i*Pr)–nBu$_4$NCl and HCl adducts of functionalized vinyl ethers [CH$_3$CH(OCH$_2$CH$_2$X)Cl; X=chloride, acetate and methacrylate] as initiators carrying pendant substituents X that serve as terminal functionalities [55].

3.6.2
Functional Terminator Method

The second method involves end quenching of living polymers with appropriate nucleophiles. Although this approach appears to be more attractive than the first one, *in situ* end functionalization of the living ends is limited to nucleophiles that do not react with the Lewis acid co-initiator. Because the ionization equilibrium is shifted to the covalent species, the concentration of the ionic active species is very low. Quantitative functionalization can only be accomplished, therefore, when ionization takes place continuously in the presence of nucleophile. Quenching the polymerization of vinyl ether polymerization with the malonate anion [123], certain silyl enol ethers [124] and silyl ketene acetals [125] has been successfully used to synthesize end-functionalized poly(vinyl ethers). Alkylamines [126], ring-substituted anilines [127], alcohols [128] and water [129] have also been used to quench the vinyl ether polymerization to synthesize end-functionalized poly(vinyl ethers). Functionalizations by the latter nucleophiles, however, most likely do not entail reactions of the living cationic ends but proceed by S_N1 reactions involving the halogen-terminated chain ends.

In the functionalization of living polymers of hydrocarbon olefins, success remained limited up until recently. Although various methods to modify the resulting chloro chain ends are available, they usually involve a number of steps and are rather cumbersome.

Various nucleophiles that do not react (or react very slowly) with the Lewis acid have been used to prepare functional PIBs by *in situ* functionalization of the living ends. Since these terminators are mostly π-nucleophiles, multiple additions should be avoided. This can be accomplished by employing π-nucleophiles that do not homopolymerize, yielding a stable ionic product or a covalent uncharged product either by rapidly losing a cationic fragment, e.g. Me_3Si^+ or H^+, or by fast ion collapse. Accordingly, the rapid and quantitative addition of various 2-substituted furans to living PIB^+ has been observed in conjunction with $TiCl_4$ as Lewis acid in hexanes–CH_2Cl_2 or CH_3Cl (60:40, v/v) at –80 °C and with BCl_3 in CH_3Cl at –40 °C [130]. The formation of the stable allylic cation was confirmed by trapping the resulting cation with tributyltin hydride, which yielded PIB with a dihydrofuran functionality. Quenching with methanol resulted in the quantitative formation of 2-alkyl-5-PIB-furan. The 2-PIB-Fu was obtained in quantitative yields in a reaction of PIB^+ with 2-Bu_3-SnFu in hexanes–CH_3Cl (60:40, v/v) in the presence of $TiCl_4$ at –80 °C. Using unsubstituted furan, coupling of two living chain ends as a side-reaction could not be avoided. However, thiophene- and N-methylpyrrole-terminated PIB could be obtained by employing unsubstituted thiophene and N-methylpyrrole, respectively [131].

Allyl telechelic PIBs have been obtained by end-quenching with allyltrimethylsilane [20, 132], methallyltrimethylsilane [20], tetraallyltin or allyltributyltin [133]. In the capping reaction of living PIB with 1,3-butadiene, a quantitative cross-over reaction followed by instantaneous termination (absence of multiple addition) and selective formation of the 1,4-addition product yielded chloroallyl func-

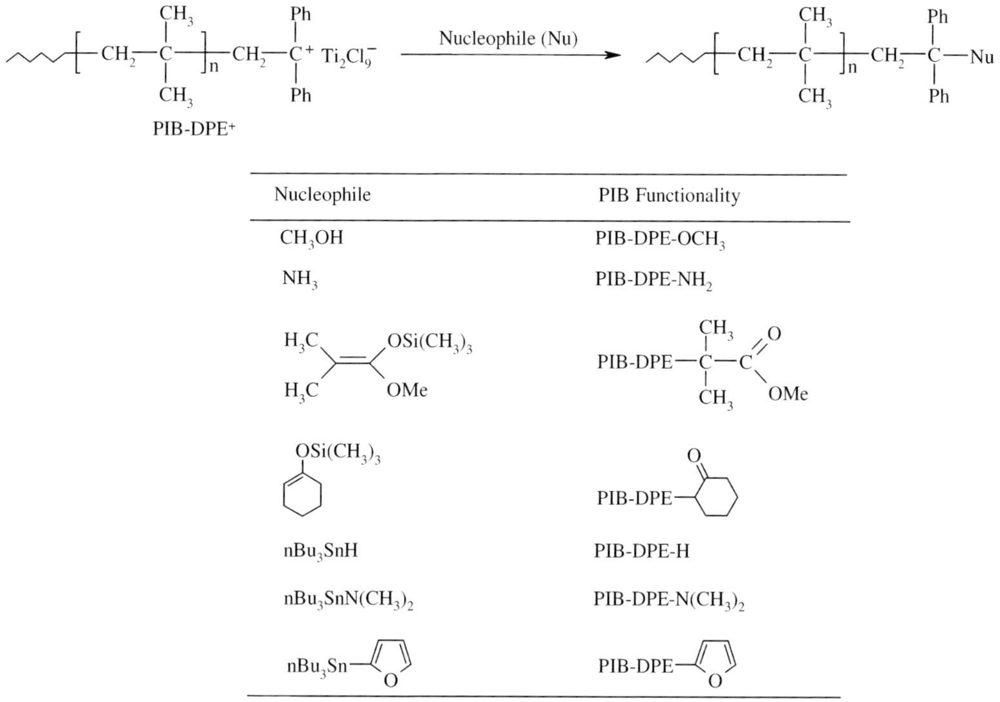

3.6 Functional Polymers by Living Cationic Polymerization

Scheme 3.2 Synthesis of chain-end functionalized PIBs.

Nucleophile	PIB Functionality
CH$_3$OH	PIB-DPE-OCH$_3$
NH$_3$	PIB-DPE-NH$_2$
(H$_3$C)$_2$C=C(OSi(CH$_3$)$_3$)(OMe)	PIB-DPE–C(CH$_3$)$_2$–C(=O)OMe
1-(trimethylsilyloxy)cyclohexene	PIB-DPE–(cyclohexanone)
nBu$_3$SnH	PIB-DPE-H
nBu$_3$SnN(CH$_3$)$_2$	PIB-DPE-N(CH$_3$)$_2$
nBu$_3$Sn–(2-furyl)	PIB-DPE–(2-furyl)

tional PIB [134]. In the functionalization reactions, β-proton abstraction should generally be avoided. Quantitative β-proton abstraction with hindered bases, as reported recently, however, is a valuable method to produce *exo*-olefin-terminated PIB in one pot [135].

A series of end-functionalized polymers of *p*-MeOSt were synthesized by quenching the HI–ZnI$_2$ initiated living poly(*p*-MeOSt)$^+$ cations with a functional alcohol [HOCH$_2$CH$_2$Z; Z=OOCCH$_3$, OOCC(=CH$_2$)CH$_3$, OOCC(=CH$_2$)H] [117].

When all dormant chain ends are converted to active ionic species, as in the capping reaction with diarylethylenes [136], many other nucleophiles, such as NH$_3$ and CH$_3$OH, which also quench the Lewis acid, could be used. *In situ* functionalization of the living ends by a variety of nucleophiles was recently realized via capping with non(homo)polymerizable diarylethylenes or 2-alkylfurans followed by end-quenching [137]. The stable and fully ionized diarylcarbenium ion, obtained in the capping of PIBCl or PStCl with 1,1-diphenylethylene (DPE), is readily amenable to chain-end functionalization by quenching with appropriate nucleophiles, as shown in Scheme 3.2 [138]. Using this strategy, a variety of chain-end functional PIBs, including methoxy, amine, carbonyl and ester end-groups, have been prepared. It is also notable that, when living PIB is capped with DPE, organotin compounds can also be used to introduce new functionalities such as –H, –N(CH$_3$)$_2$ and furan [139].

3.7
Telechelic Polymers

Telechelic or α,β-bifunctional and multifunctional polymers carry functional groups at each terminal. Symmetric telechelic or multifunctional polymers can be readily prepared by employing bi- or multifunctional initiators followed by functionalization of the living end as described above. A bifunctional initiator (in Fig. 3.3) in conjunction with $EtAlCl_2$ was used for the living cationic polymerization of isobutyl vinyl ether in hexane and 2-acetoxyethyl vinyl ether in toluene in the presence of an excess amount of 1,4-dioxane (an added base for cation stabilization) followed by quenching the living ends with the sodiomalonic ester $NaCH(COOC_2H_5)_2$ gives telechelic polymers with malonate groups. The diester terminals could be transformed into the corresponding carboxylic acids by hydrolysis followed by decarboxylation [140].

Symmetric telechelic polymers can also be prepared by coupling of α-functional living polymer chains using any of the recently discovered coupling agents. Bifunctional silyl enol ethers, such as 1,3-bis(p-{1-[(trimethylsilyl)oxy]-vinyl}phenoxy)propane, 1,4-diethoxy-1,4-bis[(trimethylsilyl)oxy]-1,3-butadiene and 2,4-bis[(trimethylsilyl)oxy]-1,3-pentadiene, are efficient bifunctional coupling agents for the living polymers of isobutyl vinyl ether initiated with $HCl-ZnCl_2$ at $-15\,°C$ in CH_2Cl_2 and toluene solvents [124]. The di- (A), tri- (B) and tetra- (C) functional silyl enol ethers (Fig. 3.4) have been used extensively with vinyl ethers to obtain functional polymers and block copolymers [141]. As an exam-

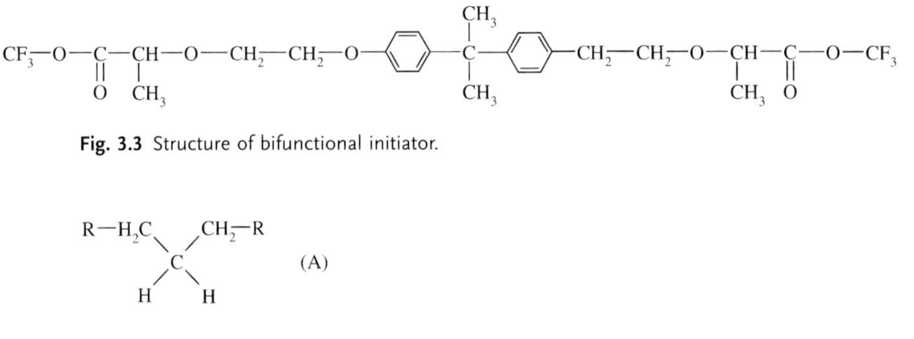

Fig. 3.3 Structure of bifunctional initiator.

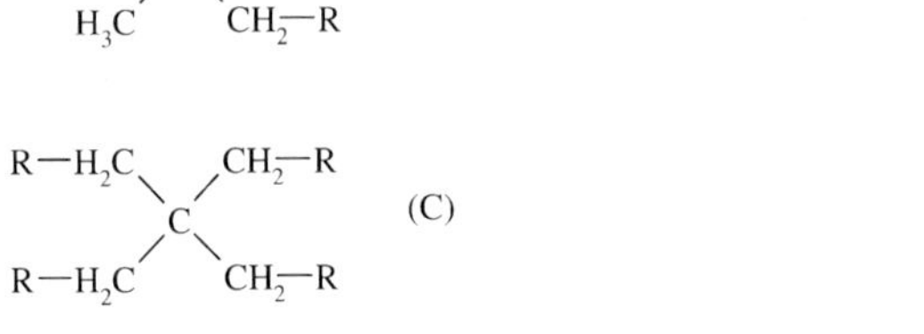

Fig. 3.4 Structures of silyl enol ethers.

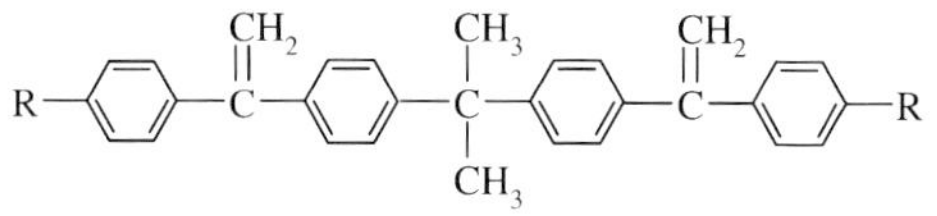

Fig. 3.5 Structures of BDPEP and BDTEP.

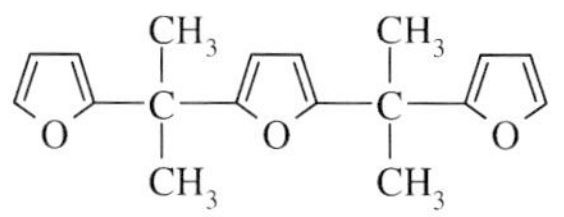

R = H (BDPEP)
 = CH$_3$ (BDTEP)

Fig. 3.6 Structure of BFPF.

ple, telechelic four-arm star polymers were obtained by coupling end-functionalized living poly(isobutyl vinyl ether) with the tetrafunctional silyl enol ethers (C) [142].

Non(homo)polymerizable bis-DPE compounds, such as 2,2-bis[4-(1-phenylethenyl)phenyl]propane (BDPEP) and 2,2-bis[4-(1-tolylethenyl)phenyl]propane (BDTEP) (Fig. 3.5), have been successfully employed in the living coupling reaction of living PIB [143, 144]. It was demonstrated that living PIB reacts quantitatively with BDPEP or BDTEP to yield stoichiometric amounts of bis(diarylalkylcarbenium) ions, as confirmed by the quantitative formation of diarylmethoxy functionalities at the junction of the coupled PIB. Kinetic studies indicated that the coupling reaction of living PIB by BDPEP is a consecutive reaction where the second addition is much faster than the first. As a result, high coupling efficiency was also observed with excess BDPEP.

Since 2-alkylfurans add rapidly and quantitatively to living PIB, yielding stable tertiary allylic cations, the coupling reaction of living PIB was also studied using bis-furanyl compounds [145]. Using 2,5-bis(1-furanyl-1-methylethyl)furan (BFPF) (Fig. 3.6), coupling of living PIB was found to be rapid and quantitative in hexane–MeCl (60:40 or 40:60, v/v) solvent mixtures at –80 °C in conjunction with TiCl$_4$, and also in MeCl at –40 °C with BCl$_3$ as Lewis acid. For instance, *in situ* coupling of living PIB, prepared by haloboration–initiation using the BCl$_3$–MeCl system at –40 °C with BFPF, yielded a,a-telechelic PIB with alkylboron functionality. After oxidation, this telechelic PIB was converted to a,a-hydroxyl PIB. The synthesis of a,a-telechelic PIBs with a vinyl functionality was also achieved by the coupling reaction of living PIB, prepared using 3,3,5-trimethyl-5-chloro-1-hexene as an initiator in the presence of TiCl$_4$ [146].

The a,ω-asymmetric polymers are available by the combination of the functional initiator and functional terminator methods. By the rational combination of haloboration–initiation and capping techniques, a series of a,ω-asymmetrically functionalized PIBs have been prepared [147, 148]. Polymers prepared by haloboration–initiation invariably carry an alkylboron head group [43, 149, 150], which can easily be converted into a primary hydroxy [149] or a secondary amine group [147, 148]. To functionalize the ω-living ends, the functionalization strategy shown in Scheme 3.2 is applicable and has been used to incorporate methoxycarbonyl groups as ω-functionality [151].

3.8
Macromonomers

A macromonomer is a macromolecule containing a (co)polymerizable end functional group. Macromonomers have been synthesized by living cationic polymerization using three different techniques: the functional initiator and functional terminator methods and chain-end modification.

3.8.1
Synthesis Using a Functional Initiator

This technique is the simplest as it generally requires only one step since the polymerizable function is incorporated via the initiator fragment. Most macromonomers have been prepared with a methacrylate end-group [152–154] which is unreactive under cationic polymerization conditions. For instance, the synthesis of the poly(vinyl ether) macromonomer was reported by employing the initiator **1** (Fig. 3.7), which contains a methacrylate ester group and a function able to initiate the cationic polymerization of vinyl ethers. Using initiator **1**, the polymerization of ethyl vinyl ether (EVE) was performed using I_2 as an activator in toluene at −40 °C. ^1H and ^{13}C NMR spectroscopy of the product obtained after complete conversion and quenching with methanol (acetal ω end-group, showed that the number-average end-functionality of the α-methacrylate end-group was very close to 1.

PEVE macromonomers were also prepared using initiator **3** (Fig. 3.8) bearing an allylic function [155]. This reactive group remained intact during the poly-

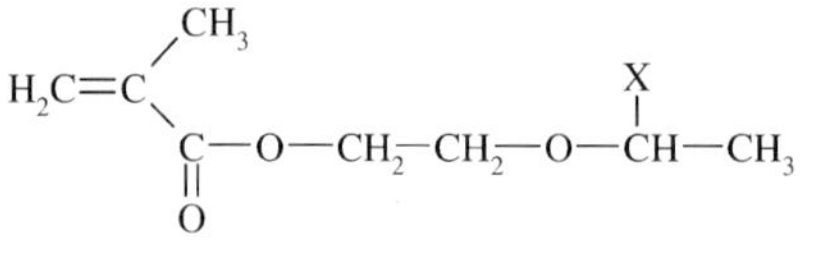

$$H_2C=C(CH_3)-C(=O)-O-CH_2-CH_2-O-CH(X)-CH_3 \quad \begin{array}{l} 1: X = I \\ 2: X = Cl \end{array}$$

Fig. 3.7 Initiators **1** and **2**.

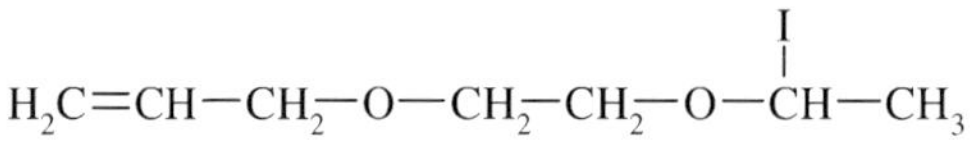

$$H_2C=CH-CH_2-O-CH_2-CH_2-O-CH(I)-CH_3 \quad 3$$

Fig. 3.8 Initiator **3**.

merization and could be transformed into the corresponding oxirane by peracid oxidation.

Other vinyl ethers were also polymerized using initiator 1 under similar conditions [156]. For instance, 2-(*tert*-butyldimethylsilyloxy)ethyl vinyl ether and 2-(trimethylsilyloxy)ethyl vinyl ether provided hydrophobic macromonomers, which were desilylated to obtain the corresponding water-soluble poly(2-hydroxyethyl vinyl ether) without any side-reaction of the methacrylate end-group.

Using functionalized initiator 2 (Fig. 3.7), polystyrene and poly(*p*-methylstyrene) macromonomers bearing a terminal methacrylate [115] could be prepared by living cationic polymerization in CH_2Cl_2 at –15 °C in the presence of $SnCl_4$ and nBu_4NCl. To preserve the end-functionality, mixing of the reagents was carried out at –78 °C. When mixing was performed at –15 °C, the functionality was lower than unity, which was attributed to initiation by protons eliminated following intramolecular alkylation.

A similar procedure was also used for the synthesis of methacrylate functional poly(*a*-methylstyrene) [157]. Thus, 2 was used in conjunction with $SnBr_4$ in CH_2Cl_2 at –78 °C to obtain the macromonomer; however, the M_ns were substantially (~50%) higher than the theoretical value. Allylation of the ω-end of the living polymer was also accomplished by quenching with excess allyltrimethylsilane. In a related development, four-armed poly(*a*-methylstyrene), functionalized with methacrylate end-groups, has been synthesized by a coupling reaction of methacryloxy (head)functional living poly(*a*-MeS), obtained by the procedure above, with tetrafunctional silyl enol ethers [158].

Methacrylate functional PIB macromonomers have been synthesized by living carbocationic polymerization of IB using the 3,3,5-trimethyl-5-chloro-1-hexyl methacrylate (4, Fig. 3.9)–$TiCl_4$ initiating system in hexane–MeCl (60:40, v/v) [118, 159]. By varying the monomer to initiator ratio, PIBs in the molecular weight range 2000–40000 g mol^{-1} were obtained with narrow molecular weight distributions and close to theoretical ester functionality.

The polymerization of β-pinene in conjunction with the 2–$TiCl_3(OiPr)$ initiating system in the presence of nBu_4NCl in CH_2Cl_2 at –40 °C yielded poly(β-pinene) macromonomer with a methacrylate function at the a end and a chlorine atom at the ω end [55]. The macromonomers exhibited narrow MWD and the reported functionality was close to unity.

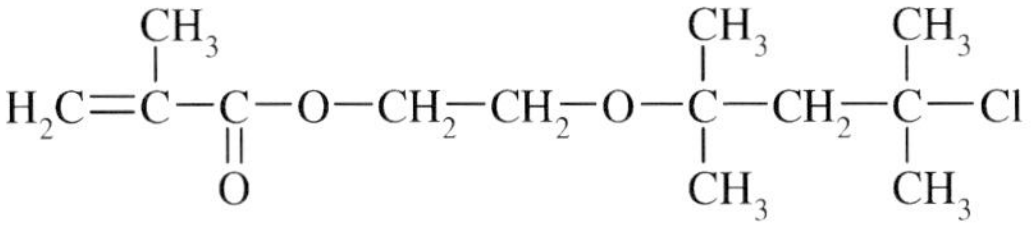

$$H_2C=C(CH_3)-C(=O)-O-CH_2-CH_2-O-C(CH_3)_2-CH_2-C(CH_3)_2-Cl \quad 4$$

Fig. 3.9 Initiator 4.

3.8.2
Synthesis Using a Functional Capping Agent

In this method, the polymerizable group is incorporated at the ω-end of the macromolecule by a reaction between a capping agent and the living polymer end.

The sodium salt of malonate carbanions reacts quantitatively with the living ends of poly(vinyl ether)s to give a stable carbon–carbon bond. This reaction was used to functionalize the ends of living poly(isobutyl vinyl ether) or poly(benzyl vinyl ether) with a vinyl ether polymerizable end-group supported by a malonate ion (end-capping agent **5**, Fig. 3.10) [160].

A hydroxy function is also able to react quantitatively with the living end of poly(vinyl ether)s, but the resulting acetal end-group has poor stability in acidic media, therefore a proton trap should be added in order to scavenge the protons released during the coupling process. Various end-capping agents with a primary alcohol and a polymerizable double bond were used to produce poly(vinyl ether) macromonomers. Most often 2-hydroxyethyl methacrylate (HEMA) [161–164] has been used, but other alcohols with an allylic or olefinic group have also been employed, such as allyl alcohol, 2-[2-(2-allyloxyethoxy)ethoxy]ethanol and 10-undecen-1-ol. Another capping agent with a methacrylate ester group, 2-(dimethylamino)ethyl methacrylate, with a tertiary amine as the coupling nucleophilic function, was also reported. In that case, capping results in the formation of a quaternary ammonium salt.

Well-defined macromonomers of poly[butyl vinyl ether (BVE)], poly[isobutyl vinyl ether (IBVE)] and poly(EVE) with ω-methacrylate end-groups were prepared by reacting the living polymer ends of the corresponding monomers, initiated by trifluoromethanesulfonic acid in CH_2Cl_2 at –30 °C in the presence of thiolane as a Lewis base, with HEMA in the presence of 2,6-lutidine or 2-(dimethylamino)ethyl methacrylate. Quenching the same polymers with allyl alcohol in the presence of the same proton trap yielded poly(vinyl ether)s with an allylic terminal.

Due to the lower stability of growing *p*-alkoxystyrene cations and the possibility of several side-reactions, some end-capping agents that were successfully used for poly(vinyl ether)s, such as sodiomalonic ester and *tert*-butyl alcohol, failed to yield end-functional poly(*p*-alkoxystyrene). In contrast, primary and secondary alcohols underwent quantitative reactions to give stable alkoxy terminals. Thus, 2-hydroxyethyl methacrylate and acrylate were used to introduce a polymerizable group at the ω end [116, 117]. Heterotelechelic poly(*p*-MeOSt)s were

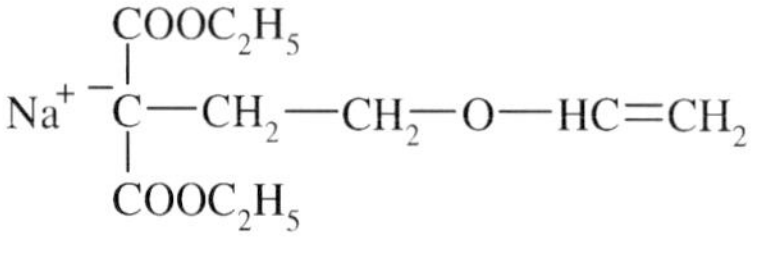

$$Na^+ \quad \underset{\underset{COOC_2H_5}{|}}{\overset{\overset{COOC_2H_5}{|}}{C^-}} - CH_2 - CH_2 - O - HC = CH_2 \quad \mathbf{5}$$

Fig. 3.10 End-capping agent **5**.

also prepared by the combination of the functional initiator method and the functional end-capping method. This allowed the synthesis of a poly(p-MeOSt) macromonomer with one malonate diester at the ω end and one methacrylate group at the ω end.

In contrast to vinyl ethers and p-alkoxystyrenes, quenching the living cationic polymerization of styrene, in conjunction with $SnCl_4$ in the presence of nBu$_4$NCl, with bases such as methanol, sodium methoxide, benzylamine or diethyl sodiomalonate, led to the terminal chloride instead of the specific end-group. This can be explained by the very low concentration of cationic species compared to the dormant C–Cl end-group and by the quenching of $SnCl_4$ with the above Lewis bases. This was overcome by using organosilicon compounds such as trimethylsilyl methacrylate and quantitative functionalization was achieved when the quenching reaction was performed at 0 °C for 24 h, in the presence of a large excess of the quencher and low concentration of the Lewis acid [165].

Allyl-terminated linear and three-arm star PIBs and epoxy and hydroxy telechelics therefrom have been reported by Ivan and Kennedy [132]. Allyl functional PIBs were obtained in a simple one pot procedure involving living IB polymerization using $TiCl_4$ as co-initiator followed by end-quenching with allyltrimethylsilane (ATMS). This method is employed commercially by Kaneka Corp. (Japan) for the synthesis of allyl telechelic PIB, a precursor to moisture-curable PIBs. The procedure was based on earlier reports by Wilczek and Kennedy [166, 167] that demonstrated quantitative allylation of PIB-Cl by ATMS in the presence of Et_2AlCl or $TiCl_4$. Structural characterization by ^{1}H NMR spectroscopy and end-group titration with m-chloroperbenzoic acid demonstrated quantitative allylation. Quantitative hydroboration followed by oxidation in alkaline THF at room temperature resulted in –OH functional PIBs, which were used to form PIB-based polyurethanes [168]. Quantitative epoxidation of the double bonds was also achieved with m-chloroperbenzoic acid in $CHCl_3$ at room temperature, giving rise to macromonomers able to polymerize by ring-opening polymerization.

In a similar development utilizing allylsilanes, the synthesis of α-methylstyryl functional PIB macromonomer was reported by the reaction of 2-phenylallyltrimethylsilane with living PIB [169]. The macromonomer, however, displayed low reactivity in cationic copolymerization with IB, which was ascribed to steric hindrance. In contrast, a reactive and unhindered macromonomer was obtained in the reaction of living PIB with 1-(2-propenyl)-3-[2-(3-trimethylsilyl)propenyl]benzene, where the reactivity of the allylsilyl function is ~1000 times higher than that of the α-methylstyryl function.

Furan telechelic PIB macromonomers with well defined M_ns and narrow MWD were synthesized by end quenching living PIB with 2-tributylstannylfuran or 2,2-difurylpropane [145, 170]. Three-arm star, furan functional PIBs were obtained under identical conditions except that 1,3,5-tricumyl chloride was used as initiator. The resulting telechelic PIBs could be efficiently photocured by UV radiation in the presence of a cationic photoinitiator and a divinyl ether reactive diluent [171]. Due to the lower reactivity of thiophene compared with furan, the reaction of unsubstituted thiophene to living PIB resulted in rapid and quantitative monoaddition and the quantitative formation of 2-polyisobutylenylthiophene [172].

Macromonomers with a terminal non(homo)polymerizable vinylidene group, such as 1,1-diphenylethylene (DPE), have attracted much attention in recent years. One of the most unique and appealing applications of these types of macromonomers is that they can be used as precursor polymers for a variety of block copolymers with controlled architectures such as ABC-type star-block or comb-type graft copolymers. ω-DPE functionalized macromonomers could be prepared by the addition reaction of living cationic polymers to "double" diphenylethylenes such as 1,3-bis(1-phenylethenyl)benzene (or *meta*-double diphenylethylene, MDDPE) or 1,4-bis(1-phenylethenyl)benzene (or *para*-double diphenylethylene, PDDPE) [173]. The addition reaction of living PIB prepared using the 2-chloro-2,4,4-trimethylpentane–$TiCl_4$–DTBP system in hexane–MeCl (60:40, v/v) at $-80\,°C$ with 2 equiv. of PDDPE resulted in the rapid and quantitative formation of PIB–DPE macromonomer, as proved by ^1H and ^{13}C NMR spectroscopy, without the formation of the coupled product. With MDDPE, a larger excess (4 equiv.) was necessary to obtain the monoadduct with negligible amounts of the diadduct.

3.8.3
Chain-end Modification

In this method, the polymerizable function is incorporated by chemical modification of the α- or ω-end-group after isolation of the polymer. Although a wide variety of polymerizable groups can be incorporated in this way, the method generally involves several steps. For example, the synthesis of polyisobutenyl methacrylate reported by Kennedy and Hiza [174] was accomplished by a multistep procedure. First, IB was polymerized using the cumyl chloride–BCl_3 initiating system. Dehydrochlorination followed by hydroboration–oxidation resulted in PIB–CH_2OH, which was subsequently esterified with methacryloyl chloride.

Cyanoacrylate-capped PIB (CA-PIB) has been synthesized by esterification of PIB–CH_2OH with the Diels–Alder adduct of 2-cyanoacryloyl chloride and anthracene followed by deprotection [175, 176].

Vinyl ether-terminated PIBs have been synthesized by Nemes et al. [177]. First PIB-Cl was dehydrochlorinated and metalated in a one-pot procedure, followed by coupling of the resulting PIB anion with 2-chloroethyl vinyl ether. In the second process, phenol was alkylated with PIB-Cl followed by reaction with 2-chloroethyl vinyl ether.

3.9
Block Copolymers

Living polymerization is the most effective and convenient method to prepare block copolymers. Synthetic methodologies however, need to be carefully selected to prepare block copolymers with high structural integrity. In general, block copolymers can be synthesized by sequential monomer addition or by reactions of living or end-functionalized polymer ends. This second method includes the use of a macroinitiator for the polymerization of the second monomer and coupling/linking living and/or end functional polymer ends.

3.9.1
Linear Diblock Copolymers

Living cationic sequential block copolymerization is one of the simplest and most convenient methods to provide well-defined block copolymers. The successful synthesis of block copolymers via sequential monomer addition relies on the rational selection of polymerization conditions such as Lewis acid, solvent, additives and temperature and on the selection of the appropriate order of monomer addition. For a successful living cationic sequential block copolymerization, the rate of crossover to a second monomer (R_{cr}) must be faster than or at least equal to that of the homopolymerization of a second monomer (R_p). In other words, efficient crossover could be achieved when the two monomers have similar reactivities or when crossover occurs from the more reactive to the less reactive monomer. When crossover is from the less reactive monomer to the more reactive one a mixture of block copolymer and homopolymer is invariably formed due to the unfavorable R_{cr}/R_p ratio. The nucleophilicity parameter (N) reported by Mayr's group might be used as the relative scale of monomer reactivity [178].

When the reactivity of the two monomers is similar and steric factors are absent, sequential block copolymerization can be used successfully. Alkyl vinyl ethers have similar reactivity and therefore a large variety of AB- or BA-type diblock copolymers could be prepared by sequential block copolymerization. A review is available [179]. Typical examples are shown [180] in Fig. 3.11.

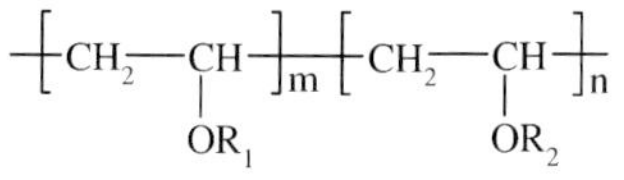

$$\text{+CH}_2\text{-CH+}_m\text{+CH}_2\text{-CH+}_n$$
$$\quad\quad\quad\ \ |\quad\quad\quad\quad\quad |$$
$$\quad\quad\quad\text{OR}_1\quad\quad\quad\quad\text{OR}_2$$

R_1	R_2
$nC_{16}H_{33}$	CH_3, C_2H_5
CH_2CH_2OH	iC_4H_9, nC_4H_9, iC_8H_{17}, $nC_{16}H_{33}$
$CH_2CH_2CH_2COOH$	nC_4H_9, iC_8H_{17}, $nC_{16}H_{33}$
$CH_2CH_2NH_2$	nC_4H_9, $nC_{16}H_{33}$
$CH_2C_6H_5$	iC_4H_9
CH_3 or iC_4H_9	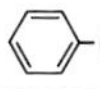 $\langle\bigcirc\rangle$-OMe

Fig. 3.11 Typical examples of AB- and BA-type diblock copolymers prepared by sequential block copolymerization.

Hydrolysis of 2-acetoxyethyl vinyl ether or 2-(vinyloxy)diethyl malonate yielded block segments with pendant hydroxyl or carboxyl groups. Most of these syntheses utilized the $HI-I_2$ or $HX-ZnX_2$ (X=halogen) initiating systems. Within the vinyl ether family, differences in reactivity are relatively small and could be overcome by increasing the concentration of the Lewis acid for the polymerization of the second, less reactive vinyl ether, for instance in the preparation of poly(isobutyl vinyl ether-b-2-acetoxyethyl vinyl ether). Stimuli-responsive diblock copolymers with a thermosensitive segment and a hydrophilic segment have been synthesized via sequential living cationic copolymerization employing $Et_{1.5}AlCl_{1.5}$ as co-initiator in the presence of Lewis base by Aoshima and coworkers [181]. Block copolymers consisting of a poly(vinyl ether) block segment with oxyethylene pendants exhibiting LCST-type phase separation in water and a poly-hydroxyethyl vinyl ether) segment displayed highly sensitive and reversible thermally induced micelle formation and/or physical gelation. Using essentially the same synthetic method, many other vinyl ether-based stimuli-responsive block copolymers (in addition to homo- and random copolymers) have been reported by the same group. A recent review is available [182].

Similarly, block copolymers of vinyl ethers and p-alkoxystyrenes could be prepared by a simple sequential monomer addition [78]. In contrast, the synthesis of poly(methyl vinyl ether-b-styrene) is more difficult. Whereas methyl vinyl ether smoothly polymerized with the $HCl-SnCl_4$ system in the presence of nBu_4NCl at $-78\,°C$, for the second stage polymerization of the less reactive St, an additional amount of $SnCl_4$ and an increase in temperature to $-15\,°C$ were necessary.

Although structurally different, IB and styrene possess similar reactivity and diblock copolymers poly(IB-b-St) [62] as well as the reverse order poly(St-b-IB) [64, 183] could be readily prepared via sequential monomer addition. Moreover, identical co-initiator and reaction conditions could be employed for the living cationic polymerization of both monomers. However, whereas the living PIB chain ends are sufficiently stable under monomer-starved conditions, the living PSt chain ends undergo decomposition at close to ~100% conversion of St [184]. Therefore, IB must be added at ≤95% conversion of St in order to obtain poly(St-b-IB) diblock copolymers with negligible homo-polystyrene contamination [185, 186]. The presence of unreacted St monomer, however, complicates the block copolymerization of IB. The first-order plot of IB, which is linear for homopolymerization, is curved downwards for block copolymerization, indicating decreasing concentration and/or reactivity of active centers with time. This is attributed to the slow formation of –St–IB–Cl chain ends (due to the reactivity ratios, which are both much higher than unity), which are much less reactive than –IB–IB–Cl [43, 148].

In the synthesis of the reverse sequence, poly(IB-b-St), it is important to add St after complete polymerization of IB. When St is added at less the 100% IB conversion the polymerization of St will be slow, which again is due to the formation and low reactivity of –St–IB–Cl chain ends. For instance, when St is added after complete polymerization of IB, St polymerization is complete in

1 h. In contrast, when St is added at 94% IB conversion, St conversion reaches only ~50% in 1 h under otherwise identical conditions.

Living cationic sequential block copolymerization from a more reactive monomer to a less reactive one usually requires a change from a weaker to a stronger Lewis acid. For instance, the living cationic polymerization of aMeSt has been reported using the relatively mild Lewis acids $SnBr_4$, $TiCl_n(OR)_{4-n}$, $SnCl_4$ or BCl_3 as a co-initiators [42, 84, 85, 87, 88, 187]. These Lewis acids, however, are too weak to initiate the polymerization of the less reactive IB, therefore the addition of a stronger Lewis acid, e.g., $TiCl_4$, is necessary to polymerize IB by sequential monomer addition. With $SnBr_4$ or $TiCl_n(OR)_{4-n}$, however, ligand exchange takes place upon addition of $TiCl_4$, which results in mixed titanium halides that are too weak to initiate the polymerization of IB. Ligand exchange is absent with BCl_3, which also induces living cationic polymerization of aMeSt in MeChx–MeCl (60:40, v/v) solvent mixture at –80 °C. Thus BCl_3 is suitable for the synthesis of poly(aMeSt-b-IB) diblock copolymer. Upon addition of IB to the living poly(aMeSt) (PaMeSt) solution, quantitative crossover takes place followed by instantaneous termination (initiation without propagation) and the selective formation of PaMeSt-IB$_1$-Cl [89, 188]. The addition of $TiCl_4$ starts the polymerization of IB.

The living cationic polymerization of p-chloro-a-methylstyrene (pClaMeSt) can also be accomplished under conditions identical with those used for the synthesis of poly(aMeSt-b-IB) copolymer [90, 189]. Using the above method, poly-(pClaMeSt-b-IB) diblock copolymer was also prepared via sequential monomer addition. On the basis of GPC UV traces of the starting PpClaMeSt and the resulting poly(pClaMeSt-b-IB) diblock copolymer, B_{eff} was ~100% and homopolymer contamination was not detected.

Sequential block copolymerization of IB with more reactive monomers such as aMeSt, p-MeSt, isobutyl vinyl ether (IBVE) or methyl vinyl ether (MeVE) as a second monomer invariably leads to a mixture of block copolymer and PIB homopolymer. To overcome the difficulty in the crossover step, a general methodology has been developed for the synthesis of block copolymers when the second monomer is more reactive than the first. It involves the intermediate capping reaction with non(homo)polymerizable monomers such as diarylethylenes and 2-substituted furans.

As shown in Scheme 3.3 [190], this process involves the capping reaction of living PIB with 1,1-diphenylethylene or 1,1-ditolylethylene (DTE), followed by tuning of the Lewis acidity to the reactivity of the second monomer. First, the capping reaction yields a stable and fully ionized diarylcarbenium ion (PIB–DPE$^+$) [191, 192], which has been confirmed using spectroscopic methods (NMR and UV–visible) and conductivity measurements. The capping reaction of living PIB with 1,1-diarylethylenes is an equilibrium reaction, which can be shifted towards completion with decreasing temperature or with increasing Lewis acidity, solvent polarity, electron-donating ability of para substituents or concentration of reactants. The purpose of the Lewis acidity tuning, following the capping reaction, is to generate more nucleophilic counterions, which ensure a

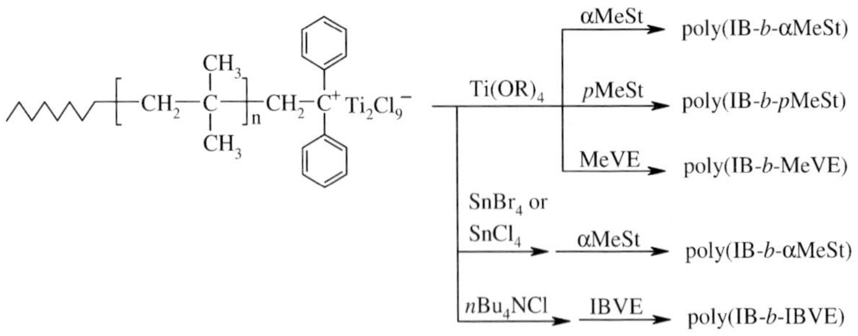

Scheme 3.3 Synthesis of block copolymers via capping reaction of living PIB with DPE, followed by Lewis acidity tuning and sequential monomer addition.

high R_{cr}/R_p ratio and also the living polymerization of a second monomer. This has been carried out using three different methods: (i) by the addition of titanium(IV) alkoxides [Ti(OR)$_4$], (ii) by the substitution of a strong Lewis acid with a weaker one or (iii) by the addition of nBu$_4$NCl.

The first and simplest method has been successfully employed in the block copolymerization of IB with aMeSt [193], p-MeSt [71], MeVE [194, 195], tBuVE [101], $tert$-butyldimethylsilyl vinyl ether [196], cyclohexyl vinyl ether (CHVE) [197] and p-$tert$-butyldimethylsiloxystyrene (tBDMSt) [198].

The substitution of TiCl$_4$ with a weaker Lewis acid (SnBr$_4$ or SnCl$_4$) has also been proven to be an efficient strategy in the synthesis of poly(IB-b-aMeSt) [42, 187] and poly(IB-b-t-BuOSt) [199] diblock copolymers.

The block copolymerization of IB with IBVE was achieved by Lewis acidity tuning using nBu$_4$NCl [41, 200] The addition of nBu$_4$NCl reduces the concentration of free and uncomplexed TiCl$_4$ ([TiCl$_4$]$_{free}$) and mechanistic studies indicated that, when [TiCl$_4$]$_{free}$ > [chain end], the dimeric counterion, Ti$_2$Cl$_9^-$, is converted to a more nucleophilic monomeric TiCl$_5^-$ counterion suitable for the living polymerization of IBVE.

Block copolymerization of IB with MeVE was also carried out using 2-methylfuran or 2-$tert$-butylfuran as a capping agent [130, 201]. However, the crossover efficiency was only ~66% using 2-$tert$-butylfuran and only slightly higher (~75%) when 2-methylfuran was employed as a capping agent under similar conditions.

3.9.2
Linear Triblock Copolymers

3.9.2.1 Synthesis Using Difunctional Initiators
Since soluble multifunctional initiators are more readily available in cationic polymerization than in the anionic counterpart, ABA-type linear triblock copolymers have been almost exclusively prepared using difunctional initiation followed by sequential monomer addition. The preparation and properties of ABA

type block copolymer thermoplastic elastomers (TPEs), where the middle segment is PIB, have been reviewed recently [202].

The synthesis of poly(St-b-IB-b-St) triblock copolymer has been accomplished by many groups [203–209]. The synthesis invariably involved sequential monomer addition using a difunctional initiator in conjunction with TiCl$_4$ in a moderately polar solvent mixture at low temperatures (–70 to –90 °C). As already mentioned at the synthesis of poly(IB-b-St), it is important to add St at ~100% IB conversion. The selection of the solvent is also critical; coupled product that forms in intermolecular alkylation during St polymerization cannot be avoided when the solvent is a poor solvent [e.g. hexanes–MeCl (60:40, v/v)] for polystyrene [210]. The formation of coupled product is slower in nBuCl or in MeChx–MeCl (60:40, v/v) solvent mixture; however, to obtain block copolymers essentially free of coupled product it is necessary to stop the polymerization of St before completion. Detailed morphological and physical properties of poly(St-b-IB-b-St) triblock copolymer have been reported [207, 211–214]. Two step sequential monomer addition method has also been employed to obtain poly(p-chlorostyrene-b-IB-b-p-chlorostyrene) [72, 215], poly(indene-b-IB-b-indene) [216], poly(p-tert-butylstyrene-b-IB-b-p-tert-butylstyrene) [217], poly[(indene-co-p-methylstyrene)-b-IB-b-(indene-co-p-methylstyrene)] [218], poly(p-MeSt-b-IB-b-p-MeSt) [204] and poly(styryl-POSS-b-IB-b-styryl-POSS) [219] copolymers.

When the crossover from the living PIB chain ends is slower than propagation of the second monomer, e.g. aMeSt, p-MeSt and vinyl ethers, the final product is invariably a mixture of triblock and diblock copolymers and possibly homoPIB, which results in low tensile strength and low elongation [220]. This slow crossover can be circumvented by the synthetic strategy shown above, utilizing an intermediate capping reaction of the living PIB with diarylethylenes followed by moderating the Lewis acidity before the addition of the second monomer. This method has been successfully employed for the synthesis of poly(aMeSt-b-IB-b-aMeSt) [221], poly(p-MeSt-b-IB-b-p-MeSt) [222], poly(tBDMSt-b-IB-b-tBDMSt) and poly(p-hydroxystyrene-b-IB-b-p-hydroxystyrene) by subsequent hydrolysis [198], poly(tBuVE-b-IB-b-tBuVE) [101], poly(vinyl alcohol-b-IB-b-vinyl alcohol) by subsequent hydrolysis [223] and poly(CHVE-b-IB-b-CHVE) [102]. The tensile strengths of most of these TPEs as well as triblock copolymers reported above were similar to those obtained with poly(St-b-IB-b-St) and virtually identical with that of vulcanized butyl rubber, indicating failure in the elastomeric domain.

3.9.2.2 Synthesis Using Coupling Agents

Although the synthetic strategy using non(homo)polymerizable monomers has been shown to be highly effective for the synthesis of a variety of di- and triblock copolymers, ABA-type linear triblock copolymers can also be prepared by coupling of living diblock copolymers, a general and useful method in living anionic polymerization.

Several coupling agents for living poly(vinyl ethers) and PaMeSt have been reported in cationic polymerization [124, 224, 225]. Synthetic utilization of non-

(homo)polymerizable diolefins has been first shown for the coupling reaction of living PIB [143, 226]. Using 2,2-bis[4-(1-phenylethenyl)phenyl]propane (BDPEP), 2,2-bis[4-(1-tolylethenyl)phenyl]propane (BDTEP) or 2,5-bis[1-(2-furanyl)-1-methylethyl]-furan as coupling agent, a rapid and quantitative coupling reaction of the living chain end was achieved, independently of the molecular weight of PIB. Kinetic studies indicated that the coupling reaction of living PIB by bis-DPE compounds is a consecutive reaction where the second addition is much faster than the first. As a result, high coupling efficiency was also observed, even when excess BDPEP was used. This coupling agent is therefore best suited for the synthesis of ABA triblock copolymers by coupling of living AB diblock copolymers and has been employed to obtain poly(St-b-IB-b-St) [64] and poly(aMeSt-b-IB-b-aMeSt) [189] triblock copolymers. For the synthesis of poly(St-b-IB-b-St) triblock copolymers, however, the two-step monomer addition method is superior. Since IB must be added at ≤95% St conversion to obtain living poly(St-b-IB) diblocks with negligible PSt homopolymer contamination, the relatively high concentration of unreactive –St–IB–Cl chain ends causes coupling of living poly(St-b-IB) diblocks to be very slow, being incomplete even after 50 h.

3.9.3
Block Copolymers with Nonlinear Architecture

Cationic synthesis of block copolymers with nonlinear architectures has been reviewed [227]. $(AB)_n$-type star-block copolymers, where n represents the number of arms, have been prepared by living cationic polymerization using three different methods: (i) via multifunctional initiators, (ii) via multifunctional coupling agents and (iii) via linking agents.

The synthesis using multifunctional initiators has been the most versatile method due to the abundance of well-defined soluble multifunctional initiators for a variety of monomers. Using trifunctional initiators, many groups have prepared three-arm star block copolymers such as poly(IBVE-b-2-hydroxyethyl vinyl ether)$_3$ [228], poly(IB-b-St)$_3$ [203, 229] and poly(IB-b-p-MeSt)$_3$ [222] star-block copolymers. The synthesis of eight-arm poly(IB-b-St)$_8$ star block copolymers was reported [230] using an octafunctional calix[8]arene-based initiator for the living cationic polymerization of IB followed by sequential addition of St. The synthesis of poly[IB-b-p-(pClSt)]$_8$ star block copolymer was also accomplished using the previously mentioned method [231]. Multi-arm-star block copolymers of poly(IB-b-St) [232] and poly(IB-b-p-tert-butylstyrene) [233] copolymers were synthesized by living cationic polymerization using a hexafunctional initiator, hexaepoxysqualene (HES), which was prepared by a simple epoxidation of squalene.

Novel arborescent block copolymers comprised of rubbery PIB and glassy PSt blocks (arb-PIB-b-PSt) were described by Puskas et al. [234]. The synthesis was accomplished with the use of arb-PIB macroinitiators, prepared by the use of 4-(2-methoxyisopropyl)styrene inimer, in conjunction with TiCl$_4$. Samples with 11.7–33.8 wt% PSt exhibited thermoplastic elastomeric properties with 3.6–8.7 MPa tensile strength and 950–1830% elongation.

Linking reactions of living polymers have been employed as an alternative way to prepare star-block copolymers. The synthesis of poly(St-b-IB) multi-arm star-block copolymers was reported using divinylbenzene (DVB) as a linking agent [186, 235]. The synthesis and mechanical properties of star-block copolymers consisting of 5–21 poly(St-b-IB) arms emanating from cyclosiloxane cores have been published [179, 236]. The synthesis involved the sequential living cationic block copolymerization of St and IB, followed by quantitative allylic chain end-functionalization of the living poly(St-b-IB) and finally linking of these prearms with SiH-containing cyclosiloxanes (2,4,6,8,10,12-hexamethylcyclohexasiloxane) by hydrosilylation. Star-block copolymers of poly(indene-b-IB) have been prepared using the previously mentioned method [237].

3.9.4
Synthesis of A_nB_n Hetero-arm Star-block Copolymers

Bis-DPE compounds such as BDPEP or BDTEP could be useful as "living" coupling agents. It was demonstrated that living PIB reacts quantitatively with these coupling agents to yield stoichiometric amounts of bis(diarylalkylcarbenium) ions. Since diarylalkylcarbenium ions have been shown to be successful for the controlled initiation of reactive monomers such as p-MeSt, aMeSt, IBVE and MeVE, A_2B_2 star-block copolymers may be prepared by this approach.

As a proof of the concept, an amphiphilic A_2B_2 star-block copolymer (A=PIB and B=PMeVE) was prepared by the living coupling reaction of living PIB followed by the chain-ramification polymerization of MeVE at the junction of the living coupled PIB as shown in Scheme 3.4 [144].

While the concept of coupling with ω-furan functionalized PIB as a polymeric coupling agent has been utilized to obtain AB-type block copolymers, it is also apparent that ω-furan functionalized polymers can be used as living coupling polymeric precursors for the synthesis of hetero-arm star-block copolymers. The synthesis of poly(IB_3-star-$MeVE_3$) is shown in Scheme 3.5. Tricumyl chloride is reacted with 2-PIB–furan and after tuning of the Lewis acidity the living linked 2-PIB–Fu initiated the polymerization of MeVE [238].

3.9.5
Synthesis of AA'B, ABB' and ABC Asymmetric Star-block Copolymers Using Furan Derivatives

The strategy for the synthesis of AA'B-type star-block copolymers, where A=PIB(1), A'=PIB(2) and B=PMeVE, is illustrated in Scheme 3.6 [170]. First, quantitative addition of ω-furan functionalized PIB (A'), obtained from a simple reaction between living PIB and 2-Bu_3SnFu, to living PIB (A) could be achieved in hexanes (Hex)–CH_2Cl_2 (40:60, v/v) at –80 °C in conjunction with $TiCl_4$. The resulting living coupled PIB–Fu^+–PIB' was successfully employed for the subsequent chain ramification polymerization of MeVE. This technique is unique in the ability to control A and A' block lengths independently.

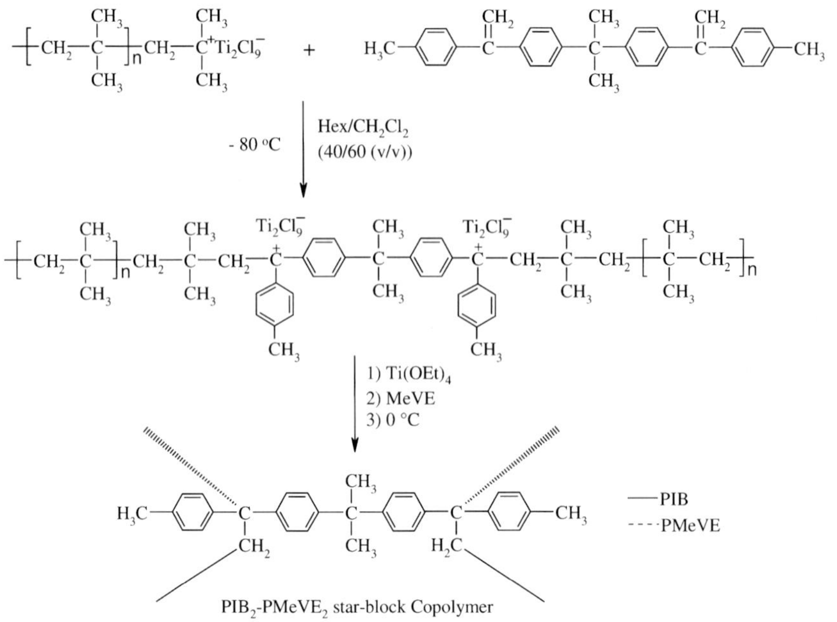

Scheme 3.4 Living coupling reaction of living PIB with BDTEP and chain ramification reaction of MeVE for the synthesis of A_2B_2 star-block copolymer.

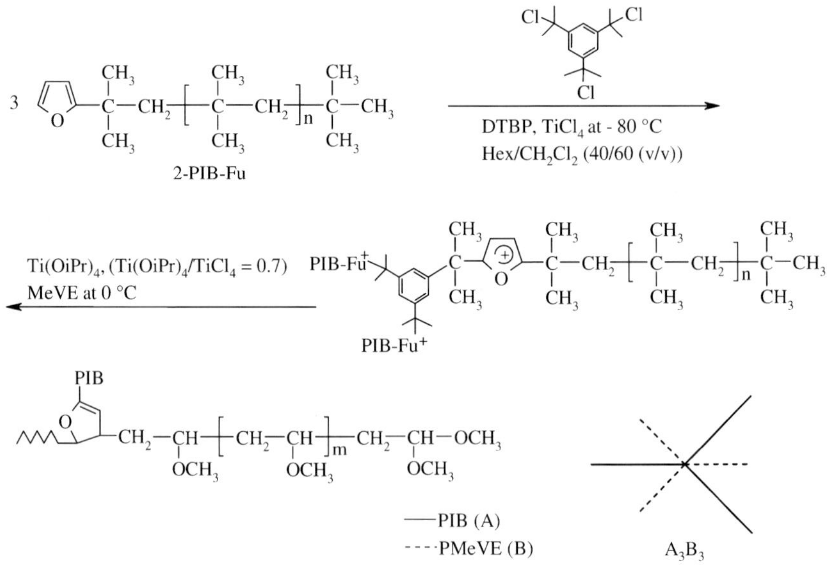

Scheme 3.5 Synthesis of poly(IB_3-star-$MeVE_3$).

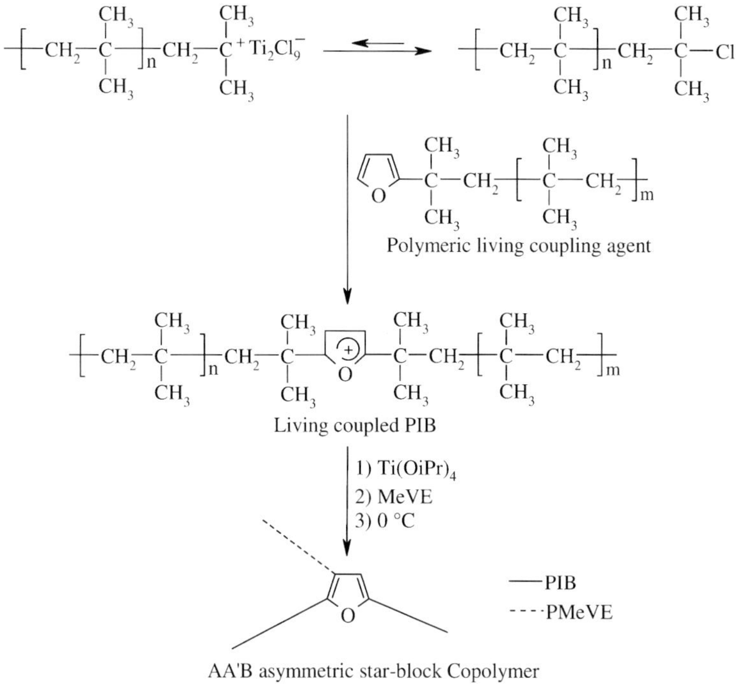

Scheme 3.6 Synthesis of AA'B asymmetric star-block copolymer.

3.9.6
Block Copolymers Prepared by the Combination of Different Polymerization Mechanisms

3.9.6.1 Combination of Cationic and Anionic Polymerization

The combination of living cationic and anionic techniques provides a unique approach to block copolymers not available by a single method. Site transformation and coupling of two homopolymers are convenient and efficient ways to prepare well-defined block copolymers.

Block copolymers of IB and methyl methacrylate (MMA), monomers that are polymerizable only by different mechanisms, can be prepared by several methods. The prerequisite for the coupling reaction is that the reactivities of the end-groups have to be matched and a good solvent has to be found for both homopolymers and copolymer to achieve quantitative coupling. Poly(IB-b-MMA) block copolymers were synthesized by coupling reaction of two corresponding living homopolymers, obtained by living cationic and group transfer polymerization (GTP), respectively [239].

The synthesis of poly(MMA-b-IB-b-MMA) triblock copolymers has also been reported using the site-transformation method, where α,ω-dilithiated PIB was used as the macroinitiator [240]. The site-transformation technique provides a useful alternative for the synthesis of block copolymers consisting of two monomers that are polymerized only by two different mechanisms. In this method, the propagating active center is transformed to a different kind of active center and a second monomer is subsequently polymerized by a mechanism different from the preceding one. The key process in this method is the precocious control of α- or ω-end functionality, capable of initiating the second monomer. Recently a novel site-transformation reaction, the quantitative metalation of DPE-capped PIB carrying methoxy or olefin functional groups, has been reported [241]. This method has been successfully employed in the synthesis of poly(IB-b-tBMA) diblock and poly(MMA-b-IB-b-MMA) triblock copolymers [242]. In this technique, however, metalation of DPE-capped PIB requires Na–K alloy as organolithium compounds are ineffective.

A new synthetic route for the synthesis poly(IB-b-tBMA) developed by combining living carbocationic and anionic polymerizations involves metalation of 2-polyisobutenylthiophene with n-butyllithium in THF at –40 °C. The resulting stable macrocarbanion (PIB-T$^-$,Li$^+$) was successfully used to initiate living anionic polymerization of tBMA yielding poly(IB-b-tBMA) block copolymers [243].

The preparation of poly(IB-b-methyl methacrylate or hydroxyethyl methacrylate) block copolymers has also been accomplished by the combination of living cationic and anionic polymerization. First DPE end-functionalized PIB (PIB–DPE) was prepared from the reaction of living PIB and 1,4-bis(1-phenylethenyl)-benzene (PDDPE), followed by the methylation of the resulting diphenylcarbenium ion with dimethylzinc. PIB–DPE was quantitatively metalated with n-butyllithium in THF at room temperature and the resulting macroinitiator could efficiently initiate the living polymerization of methacrylate monomers at –78 °C, yielding block copolymers with high block efficiency [244].

3.9.6.2 Combination of Living Cationic and Anionic Ring-opening Polymerization

Block copolymers containing crystallizable blocks have been studied not only as alternative TPEs with improved properties but also as novel nanostructured materials with much more intricate architectures compared with those produced by the simple amorphous blocks. Since the interplay of crystallization and microphase segregation of crystalline/amorphous block copolymers greatly influences the final equilibrium ordered states and results in a diverse morphological complexity, there has been a continued high level of interest in the synthesis and characterization of these materials.

Due to the lack of vinyl monomers giving rise to crystalline segments by cationic polymerization, amorphous–crystalline block copolymers have not been prepared by living cationic sequential block copolymerization. Although site transformation has been utilized extensively for the synthesis of block copolymers, only a few PIB–crystalline block copolymers, such as poly(L-lactide-b-IB-b-L-lactide) [245], poly[IB-b-ε-caprolactone (ε-CL)] [246], diblock and poly(ε-CL-b-IB-b-ε-CL) [247] triblock copolymers, have been reported.

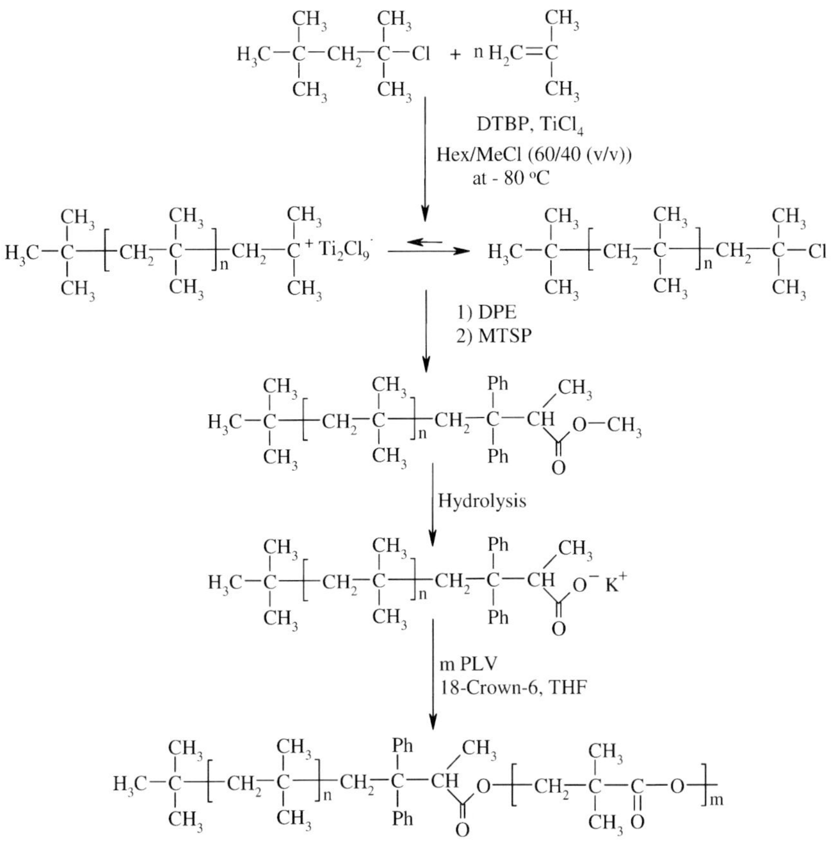

Scheme 3.7 Synthesis of poly(IB-b-PVL) copolymer by site transformation.

The synthesis of poly[IB-b-pivalolactone (PVL)] diblock copolymers has also been accomplished by site transformation of living cationic polymerization of IB to anionic ring-opening polymerization (AROP) of PVL, as shown in Scheme 3.7 [248–250]. First, PIB with ω-carboxylate potassium salt was prepared by capping living PIB with DPE followed by quenching with 1-methoxy-1-trimethylsiloxypropene (MTSP) and hydrolysis of ω-methoxycarbonyl end-groups. The ω-carboxylate potassium salt was successfully used as a macroinitiator for the AROP of PVL in tetrahydrofuran, leading to poly(IB-b-PVL) copolymers. The same methodology as mentioned above was applied for the synthesis of poly(PVL-b-IB-b-PVL) triblock copolymers, except that a difunctional initiator, 5-*tert*-butyl-1,3-bis(1-chloro-1-methylethyl)benzene (*t*BuDiCumCl), was used for the polymerization of IB in the first step.

The preparation of novel glassy (A)-b-rubbery (B)-b-crystalline (C) linear triblock copolymers have been reported, where A block is P*a*MeSt, B block is rubbery PIB and C block is crystalline PPVL. The synthesis was accomplished by living cationic sequential block copolymerization to yield living poly(aMeSt-b-IB) followed by site transformation to polymerize PVL [251]. In the first synthetic step, the GPC traces of poly(aMeSt-b-IB) copolymers with ω-methoxycarbonyl functional group exhibited bimodal distribution in both RI and UV traces and the small hump at higher elution volume was attributed to PaMeSt homopolymer. This product was fractionated repeatedly using hexanes–ethyl acetate to remove homo-PaMeSt and the pure poly(aMeSt-b-IB) macroinitiator was then utilized to initiate AROP of PVL to give rise to poly(aMeSt-b-IB-b-PVL) copolymer.

Complete crossover from living PaMeSt to IB could be achieved by modifying the living PaMeSt chain end with a small amount of pClaMeSt after complete conversion of aMeSt. The poly(aMeSt-b-IB) copolymer carrying ω-carboxylate group, obtained from hydrolysis of the ω-methoxycarbonyl group of the block copolymer, was used to initiate AROP of PVL in conjunction with 18-crown-6 in THF at 60 °C, to give rise to poly(aMeSt-b-pClaMeSt-b-IB-b-PVL) copolymer [252].

Recently, the synthesis of poly(IB-b-ethylene oxide) diblock copolymer has been reported [253]. In the first step, HO-functional PIB was prepared by hydroboration–oxidation of allyl functional PIB, obtained in the reaction of living PIB and allyltrimethylsilane. The ring-opening polymerization of ethylene oxide was initiated by the PIB alkoxide anion in conjunction with the bulky phosphazene t-BuP$_4$.

3.9.6.3 Combination of Living Cationic and Radical Polymerization

The scope of block copolymer synthesis by the combination of two different polymerization techniques has been rapidly expanded with the advent of living/controlled radical polymerization. Although block copolymers such as poly(St-b-IB-b-St) could also be prepared by the combination of cationic and atom transfer radical polymerization [254, 255], the more interesting examples involve monomers that do not undergo cationic polymerization. For instance, the synthesis of poly(IB-b-methacrylic acid) diblocks, poly(methacrylic acid-b-IB-b-methacrylic acid) triblocks and three-arm star block copolymers have been reported by Fang and Kennedy [256], by hydrolysis of the corresponding t-BuMA block copolymers. The hydroxyl functional PIBs, obtained by hydroboration–oxidation of allyl functional PIBs, were reacted with 2-bromoisobutyryl bromide to yield a PIB macroinitiator for the ATRP. Radical polymerization of t-BuMA followed by hydrolysis gave the targeted block copolymers.

Block copolymers containing PMeVE and poly(tert-butyl acrylate), poly(acrylic acid), poly(methyl acrylate) or polystyrene were prepared by Bernaerts and Du Prez [257] by the use of a novel dual initiator, 2-bromo-(3,3-diethoxypropyl) 2-methylpropanoate. In the first step, the living cationic homopolymerization of MeVE is performed with the acetal end-group of the dual initiator as initiating site or by the ATRP homopolymerization of tert-butyl acrylate from the bromoisobutyrate group of the dual initiator. The second step in the preparation of block copolymers well-defined PMeVE-Br and poly-*p*-*t*BA-acetal homopolymers were employed as macroinitiators, respectively, in the ATRP of several monomers and cationic polymerization of MeVE.

3.10
Branched and Hyperbranched Polymers

The synthesis of branched polymers by cationic polymerization of vinyl monomers has been reviewed [227], hence they will be only briefly considered here. Star-shaped or multi-arm star (co)polymers can be prepared three general methods: (a) multifunctional initiator method; (b) multifunctional terminator method; and (c) polymer linking method.

In the first case, the arms are grown from a single core with a given number of potentially active sites or a well-defined multifunctional initiator. In contrast to anionic multifunctional initiators, well-defined soluble multifunctional cationic initiators are readily available. These multifunctional initiators with 3–8 initiating sites have been successfully applied for the synthesis of 3–8-arm star homo- and block copolymers of vinyl ethers, styrene and styrene derivatives and isobutene. For example, six-arm star polystyrenes were prepared using an initiator with six phenylethyl chloride-type functions emanating from a central hexasubstituted benzene ring [258]. By subsequent end-functionalization, a variety of end-functionalized A_n or $(AB)_n$ (see above) star-shaped structures can also be obtained.

In the second and third cases, the arms are first synthesized and then linked together using either a well-defined multifunctional terminator or a bifunctional monomer leading to a cross-linked core. Well-defined star-branched polymers have been obtained by utilizing multifunctional coupling agents with the nucleophilic functions well separated to avoid steric hindrance. For example, high yields were reported in the synthesis of three- or four-arm star polymers by reacting short poly(isobutyl vinyl ether) living polymers with a tri- or tetrafunctional silyl enol ether as multifunctional terminators [259].

Difunctional monomers such as divinylbenzene and divinyl ether have been found to be efficient in the synthesis of star (co)polymers having a cross-linked core from which homopolymer or block copolymer arms radiate outwards. However, so far only the "core last" method has been reported in cationic polymerization. This method is particularly suited to prepare stars with many arms. The average number of arms per molecule is a function of several experimental and structural parameters.

Graft copolymers by cationic polymerization may be obtained by the "grafting from" and "grafting on to" methods and by (co)polymerization of macromonomers. For example, PIB with pendant functionalities could be prepared by copolymerization of IB with a functional monomer such as bromomethylstyrene or chloromethylstyrene in CH_2Cl_2 at $-80\,°C$ with BCl_3. An alternative method to obtain initiating sites along a PIB backbone involves copolymerization with *p*-

MeS followed by selective halogenation [260]. In subsequent initiation of 2-methyl-2-oxazoline, water-soluble amphiphilic graft copolymers have been obtained [261].

Highly branched, so-called "hyperbranched", macromolecules have recently attracted interest, due to their interesting properties, which closely resemble those of dendrimers. Vinyl monomers with a pendant initiating moiety, e.g. 3-(1-chloroethyl)ethenylbenzene, have been reported to give rise to hyperbranched polymers in a process termed "self-condensing vinyl polymerization" [262]. Hyperbranched PIBs have been synthesized by cationic copolymerization of 4-(2-methoxyisopropyl)styrene and IB [263]. Using a similar approach, the preparation of arborescent block copolymers of IB and St (arb-PIB-b-PSt) has also been reported (see above) [234].

3.10.1
Surface-initiated Polymerization: Polymer Brushes

Polymer brushes can be generated by reacting by the "grafting to" or "grafting from" techniques. The "grafting to" technique, where a living polymer or a suitable end-functionalized polymer is reacted with a reactive substrate, yields limited surface grafting density due to steric hindrance. In contrast, surface-initiated polymerization from a self-assembled initiator on the surface results in high grafting density and film thickness that increase linearly with molecular weight. PSt brushes were prepared on flat silicate substrates by cationic polymerization by Zhao and Brittain [264]. The polymerization of St was initiated from self-assembled monolayers (SAMs) of the cationic initiators, 2-[4-(11-triethoxysilylundecyl)phenyl]-2-methoxypropane in conjunction with $TiCl_4$. PIB brushes could also be obtained on flat silica surfaces by a similar method employing SAMs of 3-(1-chlorodimethylsilylmethyl)ethyl-1-(1-chloro-1-methyl)ethylbenzene [265]. Initiation from macroscopic surfaces requires the addition of a sacrificial soluble initiator to control the molecular weight of the polymer brush; however, this results in a large amount of unbound polymer. A sacrificial initiator is not necessary when initiation is from nanoparticles with a large surface area, as demonstrated for the surface-initiated polymerization of IB from SAMs on ~20-nm silica nanoparticles [266]. Due to the high surface area and grafting density (3.3 chains nm^{-1}) approximately 4000 polymer chains of $M_n = 65\,000$ were linked to the each nanoparticle, resulting in a total particle diameter of ~220 nm.

3.11
Conclusions

Cationic polymerization is of great theoretical and practical importance. Worldwide production of polymers by cationic vinyl polymerization is estimated at ~2.5 million metric tons per year [267]. Since the discovery of living cationic systems, cationic polymerization has progressed to a new stage where the synthesis of designed materials is now possible.

The practical importance of cationic macromolecular engineering is wide ranging. Commercialization of new technologies based on living cationic polymerization has already begun. Allyl-telechelic curable PIB elastomers (Epion) are produced by Kaneka (Japan). Kaneka and BASF recently announced the introduction of poly(styrene-b-isobutene-b-styrene) triblock copolymer thermoplastic elastomer to the market. Boston Scientific also commercialized the synthesis of poly(styrene-b-isobutylene-b-styrene) (Translute), which it employs as the polymer drug carrier for the TAXUS^{TM} Express2^{TM} Paclitaxel-Eluting Coronary Stent system.

The rapid advances in this field will lead to useful new polymeric materials and processes that will greatly increase the economic impact of cationic macromolecular engineering.

References

1 J. P. Kennedy, E. Marechal, *Carbocationic Polymerization*, Ch. 9, Wiley-Interscience, New York, **1982**. J. P. Kennedy, *Proc. Robert A. Welch Foundation Conf. Chem. Res.* **1982**, *26*, 70.
2 L. J. Fetters, E. L. Thomas, *Material Science and Technology*, VCH, Weinheim, **1993**, *12*, 1.
3 H. L. Hsieh, R. P. Quirk, *Anionic Polymerization. Principles and Practical Applications*, Ch. 4, Marcel, Dekker, New York, **1996**.
4 S. Penczek, P. Kubisa, R. Szymanski, *Makromol. Chem. Rapid Commun.* **1991**, *12*, 77.
5 K. Matyjaszewski, *Macromolecules* **1993**, *26*, 1787.
6 M. Szwarc, M. V. Beylen, *Ionic Polymerization and Living Polymers*, Chapman and Hall, London, **1993**, p. 12.
7 M. Roth, H. Mayr, *Macromolecules* **1996**, *29*, 6104.
8 H. Schlaad, Y. Kwon, L. Sipos, R. Faust, B. Charleux, *Macromolecules* **2000**, *33*, 8225.
9 L. Sipos, P. De, R. Faust, *Macromolecules* **2003**, *36*, 8282.
10 P. De, R. Faust, H. Schimmel, A. R. Ofial, H. Mayr, *Macromolecules* **2004**, *37*, 4422.
11 P. De, R. Faust, *Macromolecules* **2004**, *37*, 7930.
12 P. De, L. Sipos, R. Faust, M. Moreau, B. Charleux, J.-P. Vairon, *Macromolecules* **2005**, *38*, 41.
13 P. De, R. Faust, *Macromolecules* **2004**, *37*, 9290.
14 P. De, R. Faust, *Macromolecules* **2005**, *38*, 5498.
15 P. H. Plesch, *Prog. React. Kinet.* **1993**, *18*, 1.
16 Q. A. Thomas, R. F. Storey, *Macromolecules* **2005**, *38*, 4983.
17 L. K. Breland, Q. A. Smith, R. F. Storey, *Macromolecules* **2005**, *38*, 3026.
18 R. F. Storey, Q. A. Thomas, *Macromolecules* **2003**, *36*, 5065.
19 H. Mayr, *NATO Sci. Ser., Ser. E* **1999**, *359*, 99.
20 P. De, R. Faust, *Macromolecules* **2005**, *38*, 9897.
21 C. H. Lin, K. Matyjaszewski, *Makromol. Chem. Macromol. Symp.* **1991**, *47*, 221.
22 K. Matyjaszewski, P. Sigwalt, *Polym. Int.* **1994**, *35*, 1.
23 M. Miyamoto, M. Sawamoto, T. Higashimura, *Macromolecules* **1984**, *17*, 265.
24 R. Faust, J. P. Kennedy, *Polym. Bull.* **1986**, *15*, 317.
25 R. Faust, J. P. Kennedy, *J. Polym. Sci., Polym. Chem. Ed.* **1987**, *25*, 1847.
26 K. Matyjaszewski, M. Sawamoto, in *Cationic Polymerization. Mechanism, Synthesis and Applications*, Ch. 4, K. Matyjas-

zewski (Ed.), Marcel Dekker, New York, **1996**.
27 C. G. Cho, B. A. Feit, O. W. Webster, *Macromolecules* **1990**, *23*, 1918.
28 C.-H. Lin, K. Matyjaszewski, *Polym. Prepr.* **1990**, *31*, 599.
29 J. P. Kennedy, B. Ivan, *Designed Polymers by Carbocationic Macromolecular Engineering*, Hanser, Munich, **1991**.
30 F. Faust, in *Polymeric Materials Encyclopedia*, J. C. Salamone (Ed.), CRC Press, Boca Raton, FL, **1996**, *5*, 3816.
31 T. Higashimura, S. Aoshima, M. Sawamoto, *Makromol. Chem., Macromol. Symp.* **1988**, *13/14*, 457.
32 G. Kaszas, J. E. Puskas, C. C. Chen, J. P. Kennedy, *Polym. Bull.* **1988**, *20*, 413; *Macromolecules* **1990**, *23*, 3909.
33 K. Matyjaszewski, *Macromol. Symp.* **1996**, *107*, 53.
34 M. Gyor, H.-C. Wang, R. Faust, *J. Macromol. Sci., Pure Appl. Chem.* **1992**, *A29*, 639.
35 Y. Ishihama, M. Sawamoto, T. Higashimura, *Polym. Bull.* **1990**, *23*, 361; *Polym. Bull.* **1990**, *24*, 201.
36 T. Higashimura, M. Sawamoto, S. Aoshima, Y. Kishimoto, E. Takeuchi, in *Frontiers of Macromolecular Science*, T. Saegusa, T. Higashimura, A. Abe (Eds.), Blackwell, Oxford, **1989**, p. 67.
37 S. Penczek, *Makromol. Chem., Rapid Commun.* **1992**, *13*, 147.
38 L. Balogh, R. Faust, *Polym. Bull.* **1992**, *28*, 367.
39 R. F. Storey, K. R. Choate, Jr., *Polym. Prepr.* **1995**, *36*, 318.
40 K. Matyjaszewski, C.-H. Lin, A. Bon, J. S. Xiang, *Macromol. Symp.* **1994**, *85*, 65.
41 S. Hadjikyriacou, R. Faust, *Macromolecules* **1995**, *28*, 7893.
42 D. Li, S. Hadjikyriacou, R. Faust, *Macromolecules* **1996**, *29*, 6061.
43 L. Balogh, L. Wang, R. Faust, *Macromolecules* **1994**, *27*, 3453.
44 R. Faust, J. P. Kennedy, *J. Macromol. Sci.* **1990**, *A27*, 649.
45 R. Faust, M. Zsuga, J. P. Kennedy, *Polym. Bull.* **1989**, *21*, 125.
46 R. Faust, B. Ivan, J. P. Kennedy, *J. Macromol. Sci.* **1991**, *A28*, 1.
47 C. C. Chen, J. Si, J. P. Kennedy, *J. Macromol. Sci., Pure Appl. Chem.* **1992**, *A29*, 669.
48 J. E. Puskas, M. Grassmuller, *Makromol. Chem., Macromol. Symp.* **1998**, *132*, 117.
49 J. E. Puskas, S. Shaikh, *Macromol. Symp.* **2004**, *215*, 231.
50 M. Tawada, R. Faust, *Macromolecules* **2005**, *38*, 4989.
51 B. Rajabalitar, H. A. Nguyen, H. Cheradame, *Macromolecules* **1996**, *29*, 514.
52 T. D. Shaffer, *US Patent 5 350 819*, **1994**.
53 M. Bahadur, T. D. Shaffer, J. R. Ashbaugh, *Macromolecules* **2000**, *33*, 9548.
54 S. Hadjikyriacou, M. Acar, R. Faust, *Macromolecules* **2004**, *37*, 7543.
55 J. Lu, M. Kamigaito, M. Sawamoto, T. Higashimura, Y.-X. Deng, *J. Polym. Sci., Part A: Polym. Chem.* **1997**, *35*, 1423.
56 J. Lu, M. Kamigaito, M. Sawamoto, T. Higashimura, *Macromolecules* **1997**, *30*, 22.
57 A.-L. Li, W. Zhang, H. Liang, J. Lu, *Polymer* **2004**, *45*, 6533.
58 R. Faust, J. P. Kennedy, *Polym. Bull.* **1988**, *19*, 21.
59 O.-S. Kwon, Y.-B. Kim, S.-K. Kwon, B.-S. Choi, S.-K. Choi, *Makromol. Chem.* **1993**, *194*, 251.
60 T. Higashimura, Y. Ishihama, M. Sawamoto, *Macromolecules* **1993**, *26*, 744.
61 G. Kaszas, J. E. Puskas, J. P. Kennedy, W. G. Hager, *J. Polym. Sci., Part A: Polym. Chem.* **1991**, *29*, 421.
62 Zs. Fodor, M. Gyor, H.-C. Wang, R. Faust, *J. Macromol. Sci., Pure Appl. Chem.* **1993**, *A30*, 349.
63 T. Hasebe, M. Kamigaito, M. Sawamoto, *Macromolecules* **1996**, *29*, 6100.
64 X. Cao, R. Faust, *Macromolecules* **1999**, *32*, 5487.
65 S. V. Kostjuk, F. N. Kapytsky, V. P. Mardykin, L. V. Gaponik, L. M. Antipin, *Polym. Bull.* **2002**, *49*, 251.
66 R. Faust, J. P. Kennedy, *Polym. Bull.* **1988**, *19*, 29.
67 A. Nagy, I. Majoros, J. P. Kennedy, *J. Polym. Sci., Part A: Polym. Chem.* **1997**, *35*, 3341.
68 K. Kojima, M. Sawamoto, T. Higashimura, *J. Polym. Sci. Part A: Polym. Chem.* **1990**, *28*, 3007.

69 C. H. Lin, K. Matyjaszewski, *Polym. Prepr.* **1988**, *29*, 67.

70 M.-L. Yang, K. Li, H. D. Stöver, *Macromol. Rapid Commun.* **1994**, *15*, 425.

71 Zs. Fodor, R. Faust, *J. Macromol. Sci., Pure Appl. Chem.* **1994**, *A31*, 1985.

72 J. P. Kennedy, J. Kurian, *Macromolecules* **1990**, *23*, 3736; J. P. Kennedy, J. Kurian, *Polym. Bull.* **1990**, *23*, 259; J. P. Kennedy, J. Kurian, *J. Polym. Sci., Part A: Polym. Chem.* **1990**, *28*, 3725.

73 S. Kanaoka, Y. Eika, M. Sawamoto, T. Higashimura, *Macromolecules* **1996**, *29*, 1778.

74 K. Satoh, J. Nakashima, M. Kamigaito, M. Sawamoto, *Macromolecules* **2001**, *34*, 396.

75 J. M. Barton, D. C. Pepper, *J. Chem. Soc.* **1964**, 1573; B. Charleux, A. Rives, J.-P. Vairon, K. Matyjaszewski, *Macromolecules* **1996**, *29*, 5777; K. Matyjaszewski, C. Pugh, in *Cationic Polymerizations: Mechanisms, Synthesis and Applications*, K. Matyjaszewski (Ed.), Marcel Dekker, New York, **1996**, p. 22.

76 R. Faust, J. P. Kennedy, *Polym. Bull.* **1988**, *19*, 35.

77 T. Higashimura, K. Kojima, M. Sawamoto, *Polym. Bull.* **1988**, *19*, 7.

78 K. Kojima, M. Sawamoto, T. Higashimura, *Macromolecules* **1990**, *23*, 948; K. Kojima, M. Sawamoto, T. Higashimura, *Polym. Bull.* **1990**, *23*, 149.

79 T. Higashimura, M. Mitsuhashi, M. Sawamoto, *Macromolecules* **1979**, *12*, 178.

80 K. Satoh, M. Kamigaito, M. Sawamoto, *Macromolecules* **1999**, *32*, 3827.

81 K. Satoh, M. Kamigaito, M. Sawamoto, *Macromolecules* **2000**, *33*, 4660; S. Cauvin, F. Ganachaud, V. Touchard, P. Hémery, F. Leising, *Macromolecules* **2004**, *37*, 3214; S. Cauvin, A. Sadoun, R. Dos Santos, J. Belleney, F. Ganachaud, P. Hémery, *Macromolecules* **2002**, *35*, 7919.

82 K. Satoh, M. Kamigaito, M. Sawamoto, *J. Polym. Sci., Part A: Polym. Chem.* **2000**, *38*, 2728.

83 R. F. Storey, A. D. Scheuer, *J. Macromol. Sci., Pure Appl. Chem.* **2004**, *A41*, 257.

84 T. Higashimura, M. Kamigaito, M. Kato, T. Hasebe, M. Sawamoto, *Macromolecules* **1993**, *26*, 2670.

85 K. Matyjaszewski, A. Bon, C.-H. Lin, J. S. Xiang, *Polym. Prepr.* **1993**, *34*, 487.

86 B. G. Soares, A. D. Silva, A. S. Gomes, *Polym. Bull.* **1992**, *29*, 253.

87 Zs. Fodor, R. Faust, *J. Macromol. Sci., Pure Appl. Chem.* **1998**, *35A*, 375.

88 Y. Kwon, X. Cao, R. Faust, *Polym. Prepr.* **1988**, *39*, 494.

89 Y. Kwon, X. Cao, R. Faust, *Macromolecules* **1999**, *32*, 6963.

90 L. Sipos, X. Cao, R. Faust, *Macromolecules* **2001**, *34*, 456.

91 L. Thomas, A. Polton, M. Tardi, P. Sigwalt, *Macromolecules* **1992**, *25*, 5886.

92 L. Thomas, M. Tardi, A. Polton, P. Sigwalt, *Macromolecules* **1993**, *26*, 4075.

93 L. Thomas, A. Polton, M. Tardi, P. Sigwalt, *Macromolecules* **1995**, *28*, 2105.

94 Y. Tsunogae, I. Majoros, J. P. Kennedy, *J. Macromol. Sci., Pure Appl. Chem.* **1993**, *A30*, 253.

95 J. P. Kennedy, S. Midha, B. Keszler, *Macromolecules* **1993**, *26*, 424.

96 M. Sawamoto, J. Fujimori, T. Higashimura, *Macromolecules* **1987**, *20*, 916.

97 T. Higashimura, H. Teranishi, M. Sawamoto, *Polym. J.* **1980**, *12*, 393.

98 T. Higashimura, Y. X. Deng, M. Sawamoto, *Polym. J.* **1983**, *15*, 385.

99 J. E. Puskas, G. Kaszas, *Prog. Polym. Sci.* **2000**, *25*, 403.

100 S. Aoshima, S. K. Sadahito, E. Kobayashi, *Polym. J.* **1994**, *26*, 335.

101 Y. Zhou, R. Faust, S. Chen, S. P. Gido, *Macromolecules* **2004**, *37*, 6716.

102 Y. Zhou, R. Faust, *Polym. Bull.* **2004**, *52*, 421.

103 T. Namikoshi, T. Hashimoto, T. Kodaira, *J. Polym. Sci., Part A: Polym. Chem.* **2004**, *42*, 3649.

104 T. Yoshida, S. Kanaoka, S. Aoshima, *J. Polym. Sci., Part A: Polym. Chem.* **2005**, *43*, 5138.

105 T. Yoshida, T. Tsujino, S. Kanaoka, S. Aoshima, *J. Polym. Sci., Part A: Polym. Chem.* **2004**, *43*, 468.

106 T. Namikoshi, T. Hashimoto, T. Kodaira, *J. Polym. Sci. Part A: Polym. Chem.* **2004**, *42*, 2960.

107 S. Sugihara, K. Hashimoto, Y. Matsumoto, S. Kanaoka, S. Aoshima, *J. Polym. Sci. Part A: Polym. Chem.* **2003**, *41*, 3300.

108 F. D'Agosto, M.-T. Charreyre, F. Delolme, G. Dessalces, H. Cramail, A. Deffieux, C. Pichot, *Macromolecules* **2002**, *35*, 7911.
109 B.-A. Feit, B. Halak, *J. Polym. Sci. Part A: Polym. Chem.* **2002**, *40*, 2171.
110 V. Castelvetro, B. P. Gianluca, F. Ciardelli, *Macromol. Chem. Phys.* **2001**, *202*, 2093.
111 Md. S. Rahman, T. Hashimoto, T. Kodaira, *J. Polym. Sci. Part A: Polym. Chem.* **2000**, *38*, 4362.
112 X. Zhang, E. J. Goethals, T. Loontjens, F. Derks, *Macromol. Rapid Commun.* **2000**, *21*, 472.
113 K. Satoh, M. Kamigaito, M. Sawamoto, *Macromolecules* **2000**, *33*, 748.
114 S. Aoshima, O. Hasegawa, T. Higashimura, *Polym. Bull.* **1985**, *13*, 229; T. Nakamura, S. Aoshima, T. Higashimura, *Polym. Bull.* **1985**, *14*, 515; S. Aoshima, O. Hasegawa, T. Higashimura, *Polym. Bull.* **1985**, *14*, 417; S. Aoshima, T. Nakamura, N. Uesugi, M. Sawamoto, T. Higashimura, *Macromolecules* **1985**, *18*, 2097; T. Higashimura, T. Enoki, M. Sawamoto, *Polym. J.* **1987**, *19*, 515; M. Minoda, M. Sawamoto, T. Higashimura, *Polym. Bull.* **1987**, *17*, 107; T. Hashimoto, H. Ibuki, M. Sawamoto, T. Higashimura, *J. Polym. Sci., Part A: Polym. Chem.* **1988**, *26*, 3361; E. Takeuchi, T. Hashimoto, M. Sawamoto, T. Higashimura, *J. Polym. Sci., Part A: Polym. Chem.* **1989**, *27*, 3303; S. Aoshima, H. Oda, E. Kobayashi, *J. Polym. Sci., Part A: Polym. Chem.* **1992**, *30*, 2407; G. Liu, N. Hu, X. Xu, H. Yao, *Macromolecules* **1994**, *27*, 3892; S. Asthana, S. Varanasi, R. M. Lemert, *J. Polym. Sci., Part A: Polym. Chem.* **1996**, *34*, 1993.
115 K. Miyashita, M. Kamigaito, M. Sawamoto, T. Higashimura, *Macromolecules* **1994**, *27*, 1093.
116 H. Shohi, M. Sawamoto, T. Higashimura, *Makromol. Chem.* **1992**, *193*, 1783.
117 H. Shohi, M. Sawamoto, T. Higashimura, *Macromolecules* **1992**, *25*, 53.
118 L. Balogh, A. Takacs, R. Faust, *Polym. Prepr.* **1992**, *33*, 958.
119 J. Si, J. P. Kennedy, *J. Macromol. Sci., Pure Appl. Chem.* **1993**, *A30*, 863.
120 I.-J. Kim, R. Faust, *J. Macromol. Sci., Pure Appl. Chem.* **2003**, *A40*, 991.
121 J. E. Puskas, L. B. Brister, A. J. Michel, M. G. Lanzendorfer, D. Jamieson, W. G. Pattern, *J. Polym. Sci., Part A: Polym. Chem.* **2000**, *38*, 444.
122 J. Song, J. Bodis, J. E. Puskas, *J. Polym. Sci., Part A: Polym. Chem.* **2002**, *40*, 1005.
123 H. Shohi, M. Sawamoto, T. Higashimura, *Macromolecules* **1992**, *25*, 58.
124 H. Fukui, M. Sawamoto, T. Higashimura, *Macromolecules* **1993**, *26*, 7315.
125 A. Verma, A. Nielsen, J. E. McGrath, J. S. Riffle, *Polym. Bull.* **1990**, *23*, 563; A. Verma, A. Nielsen, J. M. Bronk, J. E. McGrath, J. S. Riffle, *Makromol. Chem. Macromol. Symp.* **1991**, *47*, 239.
126 M. Miyamoto, M. Sawamoto, T. Higashimura, *Macromolecules* **1985**, *18*, 123.
127 M. Sawamoto, T. Enoki, T. Higashimura, *Polym. Bull.* **1987**, *18*, 117.
128 T. Higashimura, M. Miyamoto, M. Sawamoto, *Macromolecules* **1985**, *18*, 611.
129 T. Hashimoto, S. Iwao, T. Kodaira, *Makromol. Chem.* **1993**, *194*, 2323.
130 S. Hadjikyriacou, R. Faust, *Macromolecules* **1999**, *32*, 6393.
131 R. F. Storey, C. D. Stokes, J. J. Harrison, *Macromolecules* **2005**, *38*, 4618.
132 B. Ivan, J. P. Kennedy, *J. Polym. Sci., Part A: Polym. Chem.* **1990**, *28*, 89.
133 R. Faust, B. Ivan, unpublished results.
134 P. De, R. Faust, *Polym. Prepr.* **2005**, *46*, 847.
135 K. L. Simson, C. D. Stokes, J. J. Harrison, R. F. Storey, *Macromolecules* **2006**, *39*, 2481.
136 H. Schlaad, K. Erentova, R. Faust, B. Charleux, M. Moreau, J.-P. Vairon, H. Mayr, *Macromolecules* **1998**, *31*, 8058.
137 Y. C. Bae, S. Hadjikyriacou, H. Schlaad, R. Faust, in *Ionic Polymerization and Related Processes*, J. E. Puskas (Ed.), Kluwer, Dordrecht, **1999**, *359*, 61.
138 S. Hadjikyriacou, Zs. Fodor, R. Faust, *J. Macromol. Sci., Pure Appl. Chem.* **1995**, *A32*, 1137.

139 S. Hadjikyriacou, R. Faust, *PMSE Prepr.* **1997**, *76*, 300.
140 T. Yoshida, M. Sawamoto, T. Higashimura, *Makromol. Chem.* **1991**, *192*, 2317.
141 H. Fukui, M. Sawamoto, T. Higashimura, *Macromolecules* **1994**, *27*, 1297.
142 H. Fukui, M. Sawamoto, T. Higashimura, *J. Polym. Sci., Part A: Polym. Chem.* **1994**, *32*, 2699.
143 Y. C. Bae, Zs. Fodor, R. Faust, *Macromolecules* **1997**, *30*, 198.
144 Y. C. Bae, R. Faust, *Macromolecules* **1998**, *31*, 2480.
145 S. Hadjikyriacou, R. Faust, *Macromolecules* **2000**, *33*, 730.
146 S. Hadjikyriacou, R. Faust, T. Suzuki, *J. Macromol. Sci., Pure Appl. Chem.* **2000**, *A37*, 1333.
147 B. Koroskenyi, R. Faust, *Polym. Prepr.* **1998**, *39*, 492.
148 B. Koroskenyi, R. Faust, *J. Macromol. Sci., Pure Appl. Chem.* **1999**, *A36*, 471.
149 L. Wang, J. Svirkin, R. Faust, *PMSE Prepr.* **1995**, *72*, 173.
150 B. Koroskenyi, L. Wang, R. Faust, *Macromolecules* **1997**, *30*, 7667.
151 B. Koroskenyi, R. Faust, *ACS Symp. Ser.* **1998**, *704*, 135.
152 S. Aoshima, K. Ebara, T. Higashimura, *Polym. Bull.* **1985**, *14*, 425.
153 T. Higashimura, S. Aoshima, M. Sawamoto, *Makromol. Chem., Macromol. Symp.* **1986**, *3*, 99.
154 M. Sawamoto, S. Aoshima, T. Higashimura, *Makromol. Chem., Macromol. Symp.* **1988**, *13/14*, 513.
155 T. Higashimura, S. Aoshima, M. Sawamoto, *Polym. Prepr.* **1988**, *29*, 1.
156 T. Higashimura, K. Ebara, S. Aoshima, *J. Polym. Sci., Part A: Polym. Chem.* **1989**, *27*, 2937.
157 M. Sawamoto, T. Hasebe, M. Kamigaito, T. Higashimura, *J. Macromol. Sci., Pure Appl. Chem.* **1994**, *A31*, 937.
158 H. Fukui, T. Deguchi, M. Sawamoto, T. Higashimura, *Macromolecules* **1996**, *29*, 1131.
159 A. Takacs, R. Faust, *J. Macromol. Sci., Pure Appl. Chem.* **1996**, *A33*, 117.
160 M. Sawamoto, T. Enoki, T. Higashimura, *Polym. Bull.* **1986**, *16*, 117.
161 E. J. Goethals, N. Haucourt, A. M. Verheyen, J. Habimana, *Makromol. Chem., Rapid Commun.* **1990**, *11*, 623.
162 V. Percec, M. Lee, D. Tomazos, *Polym. Bull.* **1992**, *28*, 9.
163 S. S. Lievens, E. J. Goethals, *Polym. Int.* **1996**, *41*, 277.
164 E. J. Goethals, P. Roose, W. Reyntjens, S. Lievens, presented at the International Symposium on Ionic Polymerization, Paris, July 1997.
165 K. Miyashita, M. Kamigaito, M. Sawamoto, *J. Polym. Sci., Part A: Polym. Chem.* **1994**, *32*, 2531.
166 L. Wilczek, J. P. Kennedy, *J. Polym. Sci., Polym. Chem. Ed.* **1987**, *25*, 3255.
167 L. Wilczek, J. P. Kennedy, *Polym. Bull.* **1987**, *17*, 37.
168 M. K. Mishra, J. P. Kennedy, *Desk Reference of Functional Polymers*, Am. Chem. Soc. **1997**, 57.
169 S. Hadjikyriacou, R. Faust, *Polym. Bull.* **1999**, *43*, 121.
170 J.-P. Yun, S. Hadjikyriacou, R. Faust, *Polym. Prepr.* **1999**, *40*, 1041.
171 S. Hadjikyriacou, R. Faust, T. Suzuki, M. Bahadur, *Polym. Prepr.* **2002**, *43*, 1209.
172 J. C. Cho, M. Acar, R. Faust, *Polym. Prepr.* **2003**, *44*, 818.
173 Y. C. Bae, R. Faust, *Macromolecules* **1998**, *31*, 9379.
174 J. P. Kennedy, M. Hiza, *J. Polym. Sci., Polym. Chem. Ed.* **1983**, *21*, 1033.
175 J. P. Kennedy, S. Midha, A. Gadkari, *Polym. Prepr.* **1990**, *31*, 655.
176 J. P. Kennedy, S. Midha, A. Gadkari, *J. Macromol. Sci., Chem.* **1991**, *A28*, 209.
177 S. Nemes, T. Pernecker, J. P. Kennedy, *Polym. Bull.* **1991**, *25*, 633.
178 H. Mayr, in *Cationic Polymerization: Mechanism, Synthesis and Application*, K. Matyjaszewski (Ed.), Marcel Dekker, New York, **1996**, p. 125.
179 W. G. S. Reyntjens, E. J. Goethals, *Polym. Adv. Tech.* **2001**, *12*, 107.
180 M. Sawamoto, in *Cationic Polymerization: Mechanism, Synthesis and Application*, K. Matyjaszewski (Ed.), Marcel Dekker, New York, **1996**, p. 392.
181 S. Sugihara, K. Hashimoto, S. Okabe, S. Satoshi, M. Shibayama, S. Kanaoka, S. Aoshima, *Macromolecules* **2004**, *37*, 336.

182 S. Aoshima, *Springer Ser. Mater. Sci. Macromol. Nanostruct. Mater.* **2004**, *78*, 138.

183 R. Faust, X. Cao, *Polym. Prepr.* **1999**, *40*, 1039.

184 X. Cao, R. Faust, *Polym. Prepr.* **1998**, *39*, 496.

185 J. S. Shim, S. Asthana, N. Omura, J. P. Kennedy, *J. Polym. Sci., Polym. Chem.* **1998**, *36*, 2997.

186 R. F. Storey, K. A. Shoemake, *Polym. Prepr.* **1996**, *37*, 321.

187 D. Li, S. Hadjikyriacou, R. Faust, *Polym. Prepr.* **1996**, *37*, 803.

188 Y. Kwon, X. Cao, R. Faust, *Polym. Prepr.* **1999**, *40*, 1035.

189 X. Cao, L. Sipos, R. Faust, *Polym. Bull.* **2000**, *45*, 121.

190 R. Faust, *Polym. Prepr.* **1999**, *40*, 960.

191 Y. C. Bae, Zs. Fodor, R. Faust, *ACS Symp. Ser.* **1997**, *665*, 168.

192 B. Charleux, M. Moreau, J.-P. Vairon, S. Hadjikyriacou, R. Faust, *Macromol. Symp.* **1998**, *132*, 25.

193 D. Li, R. Faust, *Macromolecules* **1995**, *28*, 1383.

194 S. Hadjikyriacou, R. Faust, *Polym. Prepr.* **1996**, *37*, 345.

195 S. Hadjikyriacou, R. Faust, *Macromolecules* **1996**, *29*, 5261.

196 Y. Zhou, R. Faust, *Polym. Prepr.* **2004**, *45*, 752.

197 Y. Zhou, R. Faust, *Polym. Prepr.* **2004**, *45*, 750.

198 L. Sipos, A. Som, R. Faust, R. Richard, M. Schwarz, S. Ranade, M. Boden, K. Chan, *Biomacromolecules* **2005**, *6*, 2570.

199 R. Faust, et al., unpublished work.

200 S. Hadjikyriacou, R. Faust, *Polym. Prepr.* **1995**, *36*, 174.

201 S. Hadjikyriacou, R. Faust, *Polym. Prepr.* **1998**, *39*, 398.

202 J. P. Kennedy, J. E. Puskas, in *Thermoplastic Elastomers*, 3rd edn, G. Holden, H. Kricheldorf, R. P. Quirk (Eds.), Hanser Publishers, Munich, **2004**, p. 285.

203 G. Kaszas, J. Puskas, J. P. Kennedy, W. G. Hager, *J. Polym. Sci., Polym. Chem.* **1991**, *A29*, 427.

204 H. Everland, J. Kops, A. Nielson, B. Iván, *Polym. Bull.* **1993**, *31*, 159.

205 R. F. Storey, B. J. Chisholm, K. R. Choate, *J. Macromol. Sci., Pure Appl. Chem.* **1994**, *A31*, 969.

206 M. Gyor, Zs. Fodor, H. C. Wang, R. Faust, *Polym. Prepr.* **1993**, *34*, 562.

207 M. Gyor, Zs. Fodor, H. C. Wang, R. Faust, *J. Macromol. Sci., Pure Appl. Chem.* **1994**, *A31*, 2055.

208 Zs. Fodor, R. Faust, *Polym. Prepr.* **1995**, *36*, 176.

209 Zs. Fodor, R. Faust, *J. Macromol. Sci., Pure Appl. Chem.* **1996**, *A33*, 305.

210 R. F. Storey, D. W. Baugh, K. R. Choate, *Polymer* **1999**, *40*, 3083.

211 R. F. Storey, B. J. Chisholm, M. A. Masse, *Polymer* **1996**, *37*, 2925.

212 R. F. Storey, D. W. Baugh, *Polymer* **2001**, *42*, 2321.

213 T. Kwee, S. J. Taylor, K. A. Mauritz, R. F. Storey, *Polymer* **2005**, *46*, 4480.

214 J. E. Puskas, P. Antony, M. El Fray, V. Altstadt, *Eur. Polym. J.* **2003**, *39*, 2041.

215 J. P. Kennedy, J. Kurian, *Polym. Mater. Sci. Eng.* **1990**, *63*, 71.

216 J. P. Kennedy, B. Keszler, Y. Tsunogae, S. Midha, *Polym. Prepr.* **1991**, *32*, 310.

217 J. P. Kennedy, N. Meguriya, B. Keszler, *Macromolecules* **1991**, *24*, 6572.

218 Y. Tsunogae, J. P. Kennedy, *J. Macromol. Sci., Pure Appl. Chem.* **1993**, *A30*, 269.

219 I.-J. Kim, R. Faust, *Polym. Prepr.* **2004**, *45*, 1106.

220 Y. Tsunogae, J. P. Kennedy, *J. Polym. Sci., Polym. Chem.* **1994**, *32*, 403.

221 D. Li, R. Faust, *Macromolecules* **1995**, *28*, 4893.

222 Zs. Fodor, R. Faust, *J. Macromol. Sci., Pure Appl. Chem.* **1995**, *A32*, 575.

223 Y. Zhou, R. Faust, R. Richard, M. Schwarz, *Macromolecules* **2005**, *38*, 8183.

224 H. Fukui, M. Sawamoto, T. Higashimura, *J. Polym. Sci., Polym. Chem.* **1993**, *31*, 1531.

225 H. Fukui, M. Sawamoto, T. Higashimura, *Macromolecules* **1996**, *29*, 1862.

226 Y. C. Bae, S. Coca, P. L. Canale, R. Faust, *Polym. Prepr.* **1996**, *37*, 369.

227 B. Charleux, R. Faust, *Adv. Polym. Sci.* **1999**, *142*, 1.

228 H. Shohi, M. Sawamoto, T. Higashimura, *Polym. Bull.* **1991**, *25*, 529.

229 R. F. Storey, B. J. Chisholm, Y. Lee, *Polymer* **1993**, *34*, 4330.

230 S. Jacob, I. Majoros, J. P. Kennedy, *Polym. Mater. Sci. Eng.* **1997**, *77*,

185; *Rubber Chem. Technol.* **1998**, *71*, 708.

231 S. Jacob, J. P. Kennedy, *Polym. Bull.* **1998**, *41*, 167; S. Jacob, I. Majoros, J. P. Kennedy, *Polym. Prepr.* **1998**, *39*, 198.

232 J. E. Puskas, W. Pattern, P. M. Wetmore, V. Krukonis, *Polym. Mater. Sci. Eng.* **1999**, *80*, 429; *Rubber Chem. Technol.* **1999**, *72*, 559.

233 L. B. Brister, J. E. Puskas, E. Tzaras, *Polym. Prepr.* **1999**, *40*, 141.

234 J. E. Puskas, Y. Kwon, A. Yongmoon, B. Prince, K. Anil, *J. Polym. Sci., Part A: Polym. Chem.* **2005**, *43*, 1811.

235 R. F. Storey, K. A. Shoemaker, *J. Polym. Sci., Polym. Chem.* **1999**, *37*, 1629; S. Asthana, I. Majoros, J. P. Kennedy, *Rubber Chem. Technol.* **1998**, *71*, 949; S. Asthana, J. P. Kennedy, *J. Polym. Sci., Polym. Chem.* **1999**, *37*, 2235.

236 J. S. Shim, J. P. Kennedy, *J. Polym. Sci., Polym. Chem.* **1999**, *37*, 815.

237 J. S. Shim, J. P. Kennedy, *J. Polym. Sci., Polym. Chem.* **2000**, *38*, 279.

238 J. Yun, R. Faust, *Macromolecules* **2002**, *35*, 7860.

239 A. Takács, R. Faust, *Macromolecules* **1995**, *28*, 7266.

240 J. P. Kennedy, J. L. Price, *Polym. Mater. Sci. Eng.* **1991**, *64*, 40; T. Kitayama, T. Nishiura, K. Hatada, *Polym. Bull.* **1991**, *26*, 513; J. P. Kennedy, J. L. Price, K. Koshimura, *Macromolecules* **1991**, *24*, 6567; T. Nishiura, T. Kitayama, K. Hatada, *Polym. Bull.* **1991**, *27*, 615.

241 J. Feldhusen, B. Iván, A. H. E. Müller, *Macromolecules* **1997**, *30*, 6989.

242 J. Feldhusen, B. Iván, A. H. E. Müller, *Macromolecules* **1998**, *31*, 4483; J. Feldhusen, B. Iván, A. H. E. Müller, *ACS Symp. Ser.* **1998**, *704*, 121.

243 N. Martinez-Castro, M. G. Lanzendoerfer, A. H. E. Müller, J. C. Cho, M. H. Acar, R. Faust, *Macromolecules* **2003**, *36*, 6985.

244 J. C. Cho, R. Faust, *Polym. Prepr.* **2004**, *45*, 1099.

245 L. Sipos, M. Zsuga, G. Deák, *Macromol. Rapid Commun.* **1995**, *16*, 935.

246 K. R. Gorda, D. G. Peiffer, T. C. Chung, E. Berluche, *Polym. Commun.* **1990**, *31*, 286.

247 R. F. Storey, J. W. Sherman, L. B. Brister, *Polym. Prepr.* **2000**, *41*, 690; R. F. Storey, L. B. Brister, J. W. Sherman, *J. Macromol. Sci., Pure Appl. Chem.* **2001**, *A38*, 107.

248 Y. Kwon, R. Faust, *Polym. Prepr.* **2000**, *41*, 1597.

249 Y. Kwon, R. Faust, C. X. Chen, E. L. Thomas, *Polym. Mater. Sci. Eng.* **2001**, *84*, 843.

250 Y. Kwon, R. Faust, C. X. Chen, E. L. Thomas, *Macromolecules* **2002**, *35*, 3348.

251 Y. Kwon, M. S. Kim, R. Faust, *Polym. Prepr.* **2001**, *42*, 483.

252 Y. Kwon, R. Faust, *J. Macromol. Sci., Pure Appl. Chem.* **2005**, *A42*, 385.

253 M. Groenewolt, T. Brezesinski, H. Schlaad, M. Antonietti, P. W. Groh, B. Ivan, *Adv. Mater.* **2005**, *17*, 1158.

254 S. Coca, K. Matyjaszewski, *J. Polym. Sci., Part A: Polym. Chem.* **1997**, *35*, 3595.

255 X. Chen, B. Ivan, J. Kops, W. Batsberg, *Polym. Prepr.* **1997**, *38*, 715.

256 Z. Fang, J. P. Kennedy, *J. Polym. Sci., Part A: Polym. Chem.* **2002**, *40*, 3662.

257 K. V. Bernaerts, F. E. Du Prez, *Polymer* **2005**, *46*, 8469.

258 E. Cloutet, J. L. Fillaut, Y. Gnanou, D. Astruc, *J. Chem. Soc., Chem. Commun.* **1994**, 2433.

259 H. Fukui, M. Sawamoto, T. Higashimura, *Macromolecules* **1994**, *27*, 1297.

260 N. A. Merrill, K. W. Powers, H.-C. Wang, *ACS Polym. Prepr.* **1992**, *33*, 962.

261 O. Nuyken, J. R. Sanchez, B. Voit, *Macromol. Rapid Commun.* **1997**, *18*, 125.

262 J. M. J. Frechet, M. Henmi, I. Gitsov, S. Aoshima, M. R. Leduc, R. B. Grubbs, *Science* **1995**, *269*, 1080.

263 C. Paulo, J. E. Puskas, *Macromolecules* **2001**, *34*, 734.

264 B. Zhao, B. J. Brittain, *Macromolecules* **2000**, *33*, 342.

265 I. Kim, R. Faust, *J. Macromol. Sci., Pure Appl. Chem.* **2003**, *A40*, 991.

266 I. Kim, R. Faust, *Polymer Brushes*, Wiley-VCH, Weinheim, **2004**, p. 119.

267 R. J. Pazur, H.-H. Greve, *NATO Sci. Ser., Ser. E* **1999**, *359*, 13.

4
Ionic and Coordination Ring-opening Polymerization

Stanislaw Penczek, Andrzej Duda, Przemyslaw Kubisa, and Stanislaw Slomkowski

4.1
Introduction

There are a large number of cyclic monomers that have been polymerized, depending on the structure of the ring and the ring substituents, by either anionic and/or cationic or the corresponding coordination mechanism. In Fig. 4.1, several examples of (hetero)cyclic monomer structures are given.

The technology of polymers made by ring-opening polymerization (ROP) is usually of a much smaller scale and the total "tonnage" much lower than those of vinyl or olefin polymers. Nevertheless, several technical polymers made by ROP could hardly be replaced by any other synthetic or natural material. The most technologically important polymers made by ROP are collected and briefly characterized in Table 4.1.

In some textbooks [1], a distinction is made between ring-opening polymerization and chain polymerization. This distinction is not correct. According to the preliminary IUPAC document *Terminology of Kinetics, Thermodynamics and Mechanisms of Polymerization*, chain polymerization is "A chain reaction in which the growth of a polymer chain proceeds exclusively by reaction(s) between monomer(s) and reactive site(s) on the polymer chain with regeneration of the reactive site(s) at the end of each growth step" [2]. ROP falls into this category in the same way as the living anionic polymerization of vinyl monomers or atom transfer radical polymerization (ATRP) are chain polymerizations.

ROP has been reviewed in the past in a few monographs and book chapters [3–9], but this review on ROP differs in content and style from the previous reviews. It is not as comprehensive as, for instance, the two-volume review on cationic ROP [5] or several chapters in *Comprehensive Polymer Science* [6], or multichapter monographs in which basic principles of anionic [9] and cationic [8] ROP were reviewed. This difference stems from the restricted size of the present chapter and the availability of the previous reviews.

Among the topical subjects most often studied in recent decades, we review in more detail the polymerization of cyclic ethers and esters, including stereo-

---

Macromolecular Engineering. Precise Synthesis, Materials Properties, Applications.
Edited by K. Matyjaszewski, Y. Gnanou, and L. Leibler
Copyright © 2007 WILEY-VCH Verlag GmbH & Co. KGaA, Weinheim
ISBN: 978-3-527-31446-1

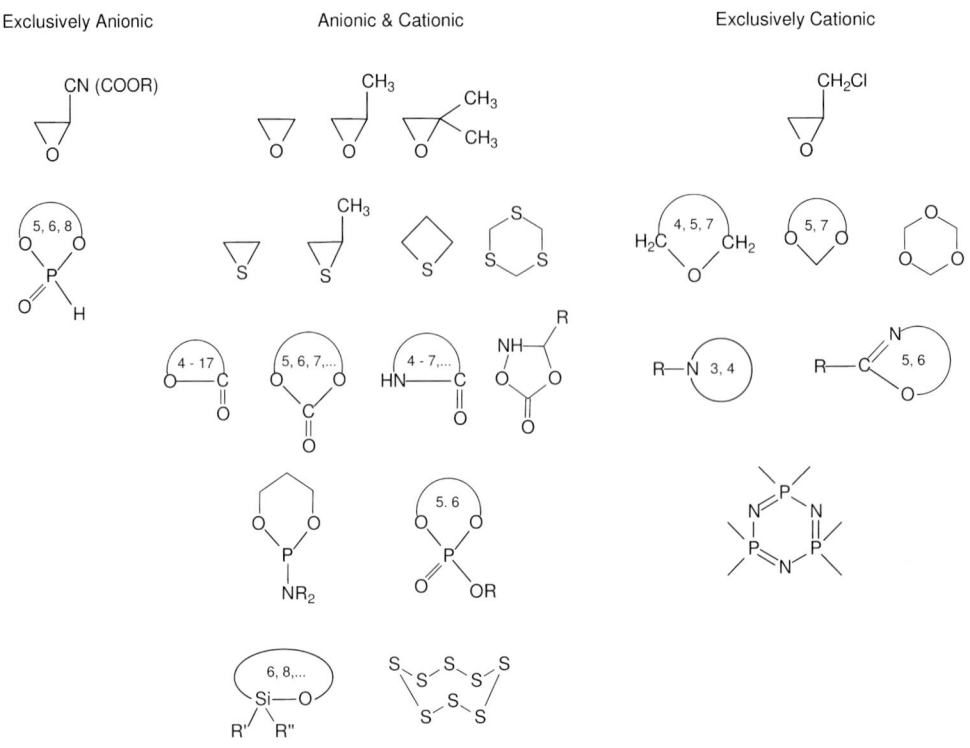

Fig. 4.1 Structures of the (hetero)cyclic monomers known to polymerize by anionic and/or cationic or coordination mechanisms. Inside the rings there are numbers indicating the ring size of the most often studied monomers belonging to a given class.

controlled polymerization, formation of stereocomplexes and novel branched and star-like polymers. Also, novel methods of living polymerization of N-carboxyanhydrides of α-amino acids (NCAs), developed in recent years, are described. Anionic or coordination polymerization of cyclic esters in dispersed media is also treated in more detail.

In the area of cationic polymerization, this chapter mostly discusses recent progress in activated monomer polymerization. Here we discuss the homo- and copolymerization of cyclic ethers and esters and also homopolymerization of monomers having cyclic ether and hydroxyl groups in the same molecules. Resulting from such a polymerization are highly branched polyethers with a larger number of "end-groups" bearing OH groups.

Table 4.1 Most technologically important polymers made by ring-opening polymerization.

Polymer structure	Polymer name	Polymer properties and applications
$-(CH_2-CH_2-O)_n-$	Poly(ethylene oxide) (PEO) [called also poly(ethylene glycol) (PEG)]	Highly hydrophilic and biocompatible. Prepared from low to very high molar masses. PEGylated drugs
$-(CH_2-CH(CH_3)-O)_n-$	Poly(1,2-propylene oxide) (PPO)	Polar rubber. Poly(ethylene oxide-co-propylene oxide)s (Pluronics) – non-ionic surfactants (oligomers)
$-(CH_2-CH(CH_2Cl)-O)_n-$	Poly(a-epichlorohydrin)	Polar rubber. a,ω-Dihydroxytelechelic oligomers as block components
$-((CH_2)_4-O)_n-$	Polytetrahydrofuran (PTHF)	a,ω-Dihydroxytelechelic oligomers as soft blocks in polyurethane, polyester or polyamide elastoplastics
$-[(CH_2O)_3]_n-$	Polytrioxane	Semicrystalline engineering thermoplastics of large-scale production ($\approx 3 \times 10^5$ tons per year) – copolymer with a few mol% of oxyethylene units
$-[C(O)-(CH_2)_5-O]_n-$	Poly(ε-caprolactone) (PCL)	(Bio)degradable thermoplastics a,ω-Dihydroxytelechelic oligomers as block components for polyurethane elastomers
$-(C(O)-CH_2-O)_n-$	Polyglycolide (PGL)	(Bio)degradable and biocompatible thermoplastics. Poly[glycolide-co-(L-lactide)] – bioresorbable surgical sutures
$-(C(O)-CH(CH_3)-O)_n-$	Polylactides (PLAs)	(Bio)degradable and biocompatible thermoplastics from renewable resources (capacity of 1.5×10^5 tons per year from a major producer – Cargill). Commodity polymer and biomedical applications
$-(CH_2-CH_2-NH)_n-$	Polyethylenimine	Water-treatment polymers (flocculants)
$-[C(O)-(CH_2)_m-NH]_n-$	Polylactams	Thermoplastic and fiber-forming polymers (nylons)
$-[(SiR^1R^2O)_m]_n-$	Polysiloxanes	A large family of materials having a wide range of applications, from heat-stable oils to the biomedical area

4.2
Thermodynamics of Ring-opening Polymerization

4.2.1
Equilibrium Monomer Concentration – Ceiling/Floor Temperatures [10, 11]

The thermodynamics of ROP play a much more important role than in vinyl polymerization since the transformation of a double bond into a single bond is accompanied by a much higher energy change (i.e. it is much more exothermic) than ring opening of the majority (three- and four-membered rings are an exception) of heterocyclic compounds. This means that in the ROP propagation–depropagation equilibrium (Eq. 1) is shifted more to the monomer side.

$$\ldots - (m)_n - m^* + M \underset{k_d}{\overset{k_p}{\rightleftharpoons}} \ldots -(m)_{n+1} - m^* \tag{1}$$

where m denotes the polymer repeating unit, m* the active species, M the monomer molecule and k_p and k_d the rate constants of propagation and depropagation, respectively.

ROP, as any other process, takes place spontaneously if the change of the Gibbs energy G_p (also called free enthalpy) is negative, that is

$$\Delta G_p = \Delta H_p - T(\Delta S_p^0 + R \ln [M]) < 0 \tag{2}$$

where ΔH_p and ΔS_p^0 are the enthalpy and standard entropy change, respectively, R the gas constant and T the absolute temperature.

When polymerization comes to equilibrium, the rates of propagation ($k_p[m^*][M]$) and depropagation ($k_d[m^*]$) become equal and macroscopically nothing changes, although both propagation and depropagation proceed all of the time. Under these conditions, $\Delta G_p=0$ and $[M]=[M]_{eq}$. For a given temperature (T_{eq}), monomer and solvent, there is only one equilibrium monomer concentration ($[M]_{eq}$).

Hence there is an infinite number of $[M]_{eq} - T_{eq}$ pairs. When eventually $[M]_{eq}$ reaches at a certain temperature a starting monomer concentration ($[M]_0$) there is no polymer in a system. The particular temperature at which it happens is called the ceiling temperature (T_c) (i.e. the temperature at which and above which polymer is not present in the system) or the floor temperature (T_f) (i.e. the temperature at which and below which polymer is not present in the system). There is only one T_c or T_f for a given starting monomer concentration (in a given solvent if the system is non-ideal and monomer–solvent or polymer–solvent interactions cannot be neglected), although there is an infinite number of pairs of starting monomer concentration and T_c (or T_f). For convenience and for the ability to compare various monomers, T_c (or T_f) is either given for the polymerization in bulk or for $[M]_0 = 1$ mol L^{-1}.

In ROP, the enthalpy change (ΔH_p) is related to the energy that is released upon polymerization (resulting mainly from the release of the ring strain, although steric interactions between substituents may also contribute), whereas ΔS_p is related to the loss of translational entropy and more subtle changes in properties of the chemical bonds in the cyclic and open-chain units, mostly involving vibrational and rotational entropies. These can either increase or decrease depending on the monomer structure. Rings are strained mostly because of the distortion of the angles and related bond lengths. In larger rings there is also possible interaction of substituents or hydrogen atoms across the rings, when they interact in axial positions across the ring. More detailed discussion on ΔH_p and ΔS_p versus various monomer ring sizes can be found, for instance, in [10].

4.2.2
Recent Results Related to Thermodynamics of Ring-opening Polymerization

4.2.2.1 Thermodynamics of γ-Butyrolactone (Co)polymerization

Several polymer chemistry textbooks have claimed that the polymerization of -γbutyrolactone (γBL) and/or its ring opening under ambient conditions is not possible [11–16]. It has recently been checked to what extent these statements are exact [17–19].

According to our recent studies, polymerization (actually oligomerization) of γBL initiated with aluminum tris-isopropoxide [Al(O^{i}Pr)$_3$] gives a series of short-chain oligomers (Scheme 4.1). Both size-exclusion chromatography (SEC) and mass spectrometry showed that oligomers are indeed formed (Fig. 4.2).

These results indicate again that the equilibrium constants for the first few monomer additions (K_0, K_1, ...) are not the same as for addition to a high polymer (K_n). Apparently there is an influence of the head end-group and the conformational flexibility of the macromolecule. Formally, the contribution to the Gibbs energy of the simple concentration term $RT\ln[M]$ is based on the assumed equality $[\ldots\text{-}(m)_n\text{-}\ldots] \approx [\ldots\text{-}(m)_{n+1}\text{-}\ldots]$. For short poly($\gamma$BL) chains, when this equality does not hold, the expression for the Gibbs energy should be

$$\Delta G_n = \Delta H_n - T\left(\Delta S_n^\circ - R\ln\frac{[\text{poly}(\gamma\text{BL})_{n+1}]}{[\gamma\text{BL}][\text{poly}(\gamma\text{BL})_n]}\right) \qquad (3)$$

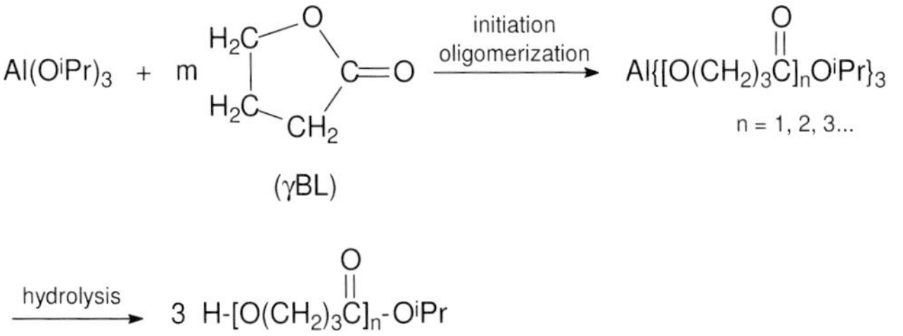

Scheme 4.1 Oligomerization of γ-butyrolactone with Al(OiPr)$_3$.

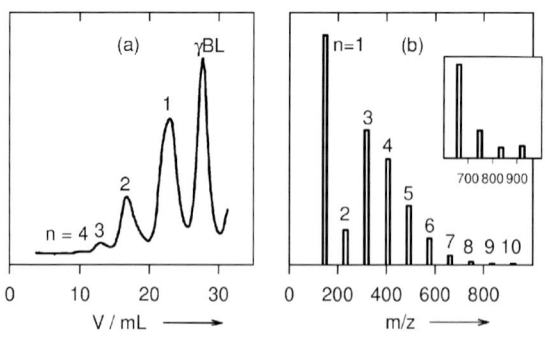

Fig. 4.2 Oligomerization of γ-butyrolactone (γBL) initiated with Al(O^{i}Pr)$_3$. (a) SEC trace; (b) mass spectrum (chemical ionization) of the isolated series of the linear oligomers: H-[O(CH$_2$)$_3$C(O)]$_n$–O^{i}Pr. Conditions of oligomerization: [γBL]$_0$=3.8 mol L^{-1}, [Al(O^{i}Pr)$_3$]$_0$=0.2 mol L^{-1}, THF solvent, 80 °C [19].

For the relatively short chains, the inequality [poly(BL)$_n$] > [poly(γBL)$_{n+1}$] applies and the "entropic term" may outweigh $\Delta H_n \geq 0$, making $\Delta G_n < 0$ and thus allowing shorter chains to be formed.

Further, γBL copolymerizes with other cyclic ester monomers, giving high molar mass copolymers [17, 20, 21]. Obviously, the limit of their proportion in copolymers is close to 50 mol%. This phenomenon of copolymerization of monomers "unable to homopolymerize" has, however, been known for a long time, for example for vinyl monomers [22].

4.2.2.2 Copolymerization of Lactide at the Polymer–Monomer Equilibrium

The situation of a monomer unable to give high polymers in a certain temperature range is similar to a monomer which does homopolymerize, but is reaching its polymer–monomer equilibrium. Under these conditions, no further increase in conversion takes place. Introduction at this moment of another monomer that is able to homopolymerize (and copolymerize) leads to the formation of a block copolymer with a bridge following the first homopolymer block and building of a graded copolymer. This was observed when lactide (LA) was brought to polymer–monomer equilibrium and then ε-caprolactone (CL) was introduced into the system [23].

(LA)   (CL)

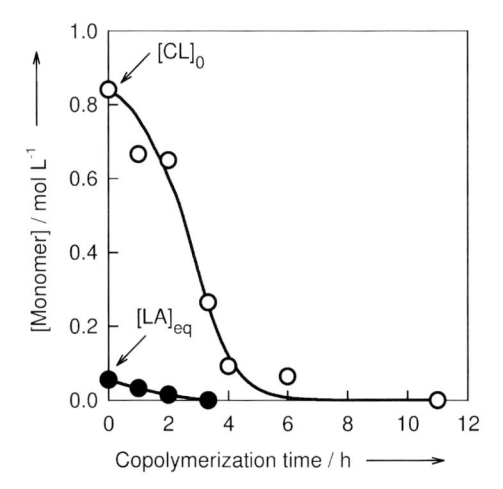

Fig. 4.3 Kinetics of copolymerization of L-lactide (LA) (●) and ε-caprolactone (CL) (○). Conditions: $[CL]_0/[LA]_{eq} = 15.3$, $[CL]_0 = 0.89$ mol L^{-1}; 80 °C, THF as solvent, M_n(living PLA) = 4780 [23].

The actual process is illustrated by the kinetic data in Fig. 4.3. Thus, homopolymerization of LA has been brought to equilibrium and reached $[LA]_{eq} = 0.06$ mol L^{-1}. Then CL was introduced at $[CL]_0 = 0.84$ mol L^{-1} and copolymerization started. LA was virtually consumed when less than 0.2 mol L^{-1} of CL remained and finally homopolymerization of CL proceeded. One could imagine that the proper choice of a monomer and other conditions would give an even lower proportion of comonomer needed for complete conversion of LA. This phenomenon is of interest since it allows a polymer to be produced with complete conversion of a monomer above its T_c.

In the not too distant past, this phenomenon was used for the copolymerization of sulfur (S$_8$) monomer into high polymer below its T_f [24, 25].

4.3
Basic Mechanistic Features of Ring-opening Polymerization

4.3.1
Anionic and Coordination Ring-opening Polymerization of Cyclic Ethers and Sulfides

4.3.1.1 Initiators and Initiation

Our knowledge in this area is mostly based on the anionic ROP of the three-membered cyclic ethers: ethylene oxide (EO) and 1,2-propylene oxide (PO) (oxirane and 2-methyloxirane, respectively), and on anionic ROP and coordination ROP of cyclic esters (e.g. LA and CL discussed in Section 4.3.2). More limited studies have been performed on the anionic ROP of cyclic sulfides, mostly 1,2-propylene sulfide [2-methylthiirane (PS)] and much less on ethylene sulfide [thiirane (ES)].

4 Ionic and Coordination Ring-opening Polymerization

(EO)   (PO)   (ES)   (PS)

Although a very large number of initiators have been used, the simplest and most versatile are alkoxides, carboxylates and thiolates for thiiranes. Alkoxides can initiate practically any polymerization of a cyclic ether or ester, whereas carboxylates can only be used for adequately strained monomers. Another group of initiators are "covalent alkoxides" or "covalent carboxylates", such as the commonly used $Al(OR)_3$ and $Sn[(OC(=O)R]_2$ [3, 4, 6, 9, 10, 26, 27].

Polymerization of cyclic ethers (and esters) can also be initiated by various carbanions and other anions (e.g. amide anions) that are more nucleophilic; however, in many instances the strong bases can induce side-reactions, mostly proton transfer. Therefore, if any new monomer is considered then the proper choice has to be made, taking into account (i) the basicity/nucleophilicity ratio for initiator, (ii) the presence of acidic hydrogen atoms (protons) in the monomer molecule and (iii) steric hindrance in the monomer molecule.

4.3.1.2 Active Centers – Structures and Reactivities

A method has been elaborated that allows the determination of end-groups at concentrations down to less than 10^{-3} mol L^{-1}, which corresponds to $M_n \approx 10^5$ at a typical polymer concentration, using ^{31}P NMR spectroscopy. This method is based on conversion of the active centers into their phosphorus derivatives. The phosphorus atom is now only in the end-groups and the only signals that appear in the spectra are those from the chain ends, so the possible interference from large signals of the polymer backbone (as in ^1H or ^{13}C NMR spectra) is eliminated. Diphenyl chlorophosphate was successfully used as the end-capping agent [28]. The following reaction takes place:

(4)

(counterion omitted)
X = O, S

End-capping with $ClP(=O)(OC_6H_5)_2$ has been applied in the polymerization of oxiranes, thiiranes, lactones, cyclic carbonates and cyclic siloxanes, allowing not only the structure but also the concentration of active centers to be determined.

In the polymerization of substituted oxiranes, it was possible to discriminate between the primary and secondary alkoxide anions comparing the end-capped polymerization product with the corresponding models: $C_6H_5CH_2CH_2OP$

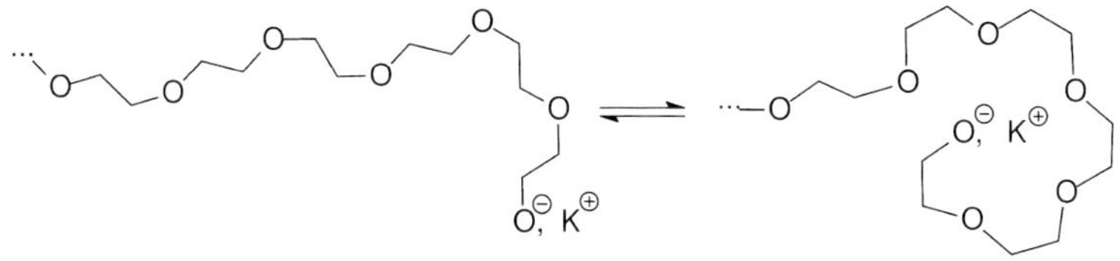

Scheme 4.2 Complexation of the alkoxide active center with the growing poly(ethylene oxide) chain.

(=O)(OC$_6$H$_5$)$_2$ and CH$_3$CH(C$_6$H$_5$)OP(=O)(OC$_6$H$_5$)$_2$ absorbing at $\delta = -12.7$ and -13.1 ppm, respectively. In the P end-capped anionic ROP of styrene a oxide, signal at $\delta = -13.0$ ppm was observed, pointing to the secondary alkoxide anions as the active centers. In the same way, it was established that in polymerization of CL the active centers are alkoxide anions and in the polymerization of β-propiolactone (PL) they are carboxylate anions.

The active centers discussed in this section are either anions or covalent species, leading to coordination polymerization. Anions can exist in various physical forms, such as contact (tight) ion pairs, their aggregates, solvent-separated ion pairs, free ions and some other forms of other aggregated species.

There are only a few heterocyclic monomers, polymerizable anionically, that have been studied quantitatively in the past: EO, PS, PL and CL. These monomers, under certain well-chosen conditions propagate without side-reactions and therefore determination of the pertinent rate constants is possible [9, 10, 29, 30].

The major result of these studies, with regard to oxiranes, can be summarized as follows. Polymerization of EO in THF solvent, initiated in various ways, gives living polymers. Strong aggregation (sometimes called association) of active centers was clearly observed, since polymerization, being first order in active species at low concentration (below 10^{-4} mol L^{-1} for K$^+$ and Cs$^+$ as counterions), becomes of a fractional order at higher concentrations [31]. Some of these results were later analyzed in our group and the degrees of aggregation and the corresponding rate constants of propagation (as elementary reaction) were determined [32]. Polymerization proceeds in THF with auto-acceleration at the beginning of the process (at least for K$^+$ counterion). This is because the polyether chains can specifically solvate the cation and in this way increase the rate constant of propagation on ion pairs, shifting from contact to solvated (separated) species and/or increase the degree of dissociation of ion pairs into ions with conversion (Scheme 4.2).

This is a simplified and hardly realistic picture (the chain on the left-hand side may have different conformations), but shows at least one of the possible ways of complexing a cation by its own chain.

4.3.1.3 Controlled Anionic and Coordination Polymerization of Oxiranes

Anionic ROP of EO can, perhaps, be considered as the first example of living polymerization. Analysis of this process presented by Flory pointed to molar mass control by the consumed monomer to initiator molar ratio and the Poissonian molar mass distribution of the resulting PEO [33]. Despite the apparent simplicity of the anionic polymerization of EO, controlled synthesis of poly(ethylene oxide) (PEO), with regard to its M_n and end-groups structure, requires special precautions (see, e.g., [34] and [35]). PEOs [poly(ethylene glycol)s (PEGs)] bearing end-groups with structures "at will" could simply be prepared with the corresponding metal alkoxides and then, when polymerization is over, terminated, giving the required end-groups. A large number of such PEGs are commercially available [36, 37]. Strictly functionalized PEO oligomers were, nevertheless, prepared by applying phosphazene bases [38–40] or yttrium alkoxides [41] as (co)initiators.

Anionic polymerization of PO is used industrially for the preparation of block copolymers with EO as oligomers, mostly used as nonionic amphiphilic surfactants. High molar mass polymers cannot be prepared in this way, because of an extensive chain transfer to PO, typical for the substituted oxiranes (Scheme 4.3) (see, e.g., [42] and the references cited therein).

The oligomers to be used as surfactants or as starting hydroxyl group-containing precursors of the soft blocks for polyurethanes should have a proper hydroxyl group functionality. This side-reaction can be slightly depressed by counterion complexation with crown ethers or application of cations with larger ionic radii, such as Rb^+ or Cs^+ [43]. Nevertheless, even in such systems, the M_n of poly(1,2-propylene oxide) (PPO) did not exceed 1.5×10^4. Introduction of double metal cyanide complexes, based on $Zn_3[Co(CN)_6]_2$, allows the preparation of PPO with a low degree of unsaturation and better controlled M_n. However, the corresponding polymerization mechanism remains obscure [44].

More recently, Deffieux and coworkers [45] have shown that polymerization of PO initiated with alkali metal alkoxides in hydrocarbon solvents can be substantially accelerated in the presence of more than a threefold molar excess of trialkylaluminum (AlR_3) with respect to the alkoxide initiator. The chain transfer is also strongly reduced, allowing controlled preparation of PPO up to $M_n \geq 2 \times 10^4$. The role of the additive was explained by assuming complexation (activation?) of both PO monomer and alkoxide growing species. In this way, the selectivity of propagation versus transfer was increased. This is shown for the initiation step in Scheme 4.4, based on the authors' proposition.

A similar kind of complexation by a bulky Al Lewis acid, resulting in excellent regiocontrol of PO polymerization, has been reported by Braune and Okuda [46].

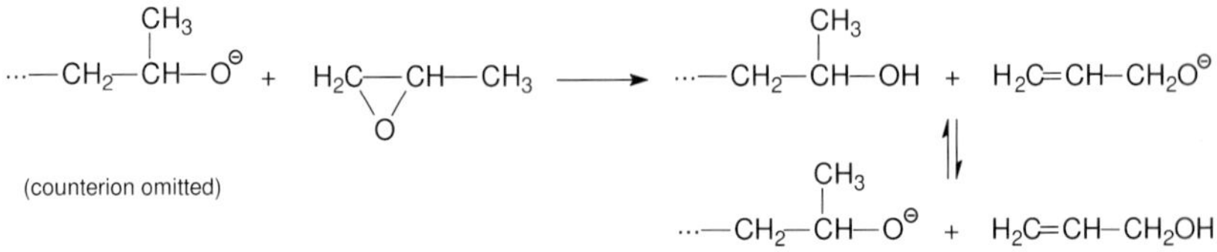

Scheme 4.3 Chain transfer to monomer in the anionic polymerization of 1,2-propylene oxide.

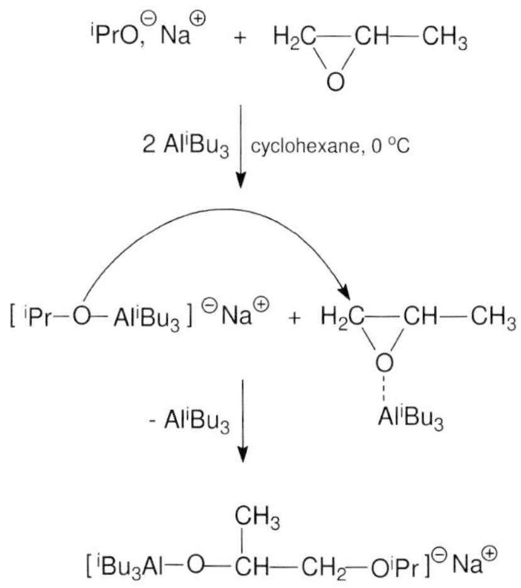

^{i}PrO$^\ominus$ Na$^\oplus$ + H$_2$C—CH—CH$_3$ (epoxide)

Scheme 4.4 Mechanism of propylene oxide polymerization involving complexation of alkoxide active species with Al–Lewis acid.

↓ 2 Al^{i}Bu$_3$ | cyclohexane, 0 °C

[^{i}Pr—O—Al^{i}Bu$_3$]$^\ominus$ Na$^\oplus$ + H$_2$C—CH—CH$_3$ (epoxide···Al^{i}Bu$_3$)

↓ −Al^{i}Bu$_3$

[^{i}Bu$_3$Al—O—CH(CH$_3$)—CH$_2$—O^{i}Pr]$^\ominus$ Na$^\oplus$

These results bear a resemblance to the activation by bulky Al aryloxides of PO polymerization, initiated with Al porphyrins, by Inoue and coworkers [47]. Some polymerizations were reported to be accelerated up to 10^3-fold. Also, detailed studies of PS polymerization by the Paris group revealed the possibility of monomer activation by complexing with the counterion [29].

Up to now, polymers of PS have not found any application, and polymers of ES with properties resembling those of polyacetals, in spite of early hopes, were also hampered for economic reasons and because of the unpleasant nature of some side-products formed upon polymer processing and storage.

4.3.1.4 Stereocontrolled Polymerization of Chiral Oxiranes

Studies of the stereocontrolled polymerization of oxiranes played an important role in development of the understanding of ROP mechanisms. The first synthesis of the isotactic polyoxirane poly(1,2-propylene oxide) (PPO) from the racemic monomer (R,S)-1,2-propylene oxide (rac-PO) is described in a patent by Pruit and Baggett [48] published several months after the pioneering work of Natta on isotactic polypropylene [49].

In spite of the long, more than 50-year, history of stereocontrolled polymerization of three-membered heterocyclics [50, 51], including systematic studies by Tsuruta [52], this process still remains a challenge and further systematic research is required. For example, synthesis of the completely isotactic rac-PPO has been achieved only recently. Coates and coworkers [53] reported that (salph)CoOAc complex (where salph = N,N'-bis(3,5-di-tert-butylsalicylidine-1,2-benzenediamine) initiated polymerization of the rac-PO (Scheme 4.5).

Under the optimized conditions $\{[PO]_0 = 1$ mol L^{-1}, [(salph)CoOAc]$_0 = 2 \times 10^{-3}$ mol L^{-1}, toluene solvent, 0 °C$\}$, rac-PPO was formed in high yield, showing in the ^{13}C NMR spectra more than 99% mm triads. The M_n value for the resulting

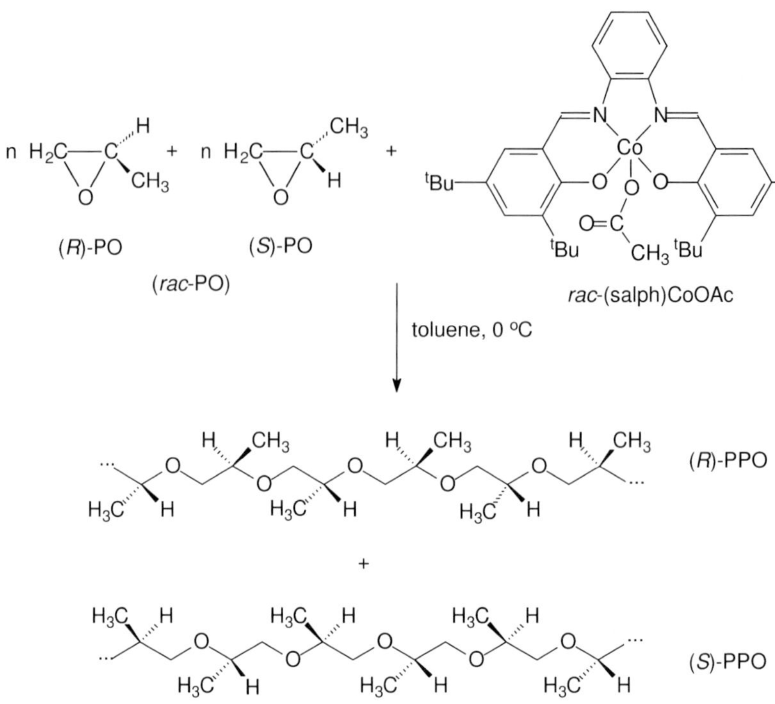

Scheme 4.5 Synthesis of isotactic poly(1,2-propylene oxide) from racemic monomer.

PPO is impressively high (2.87×10^5), it exceeds substantially the M_n predicted from the $([PO]_0 - [PO]_t)/[(salph)CoOAc]_0$ ratio (2.6×10^4).

4.3.2
Controlled Synthesis of Aliphatic Polyesters by Anionic and Coordination Ring-opening Polymerization

Anionic and coordination polymerization of cyclic esters has been reviewed several times [9, 10, 14, 19, 26, 27, 54–63]. Therefore, only the most general phenomena are summarized here.

4.3.2.1 Initiators and Active Centers – Structures and Reactivities

In some processes, structures of active centers can be deduced from the polymer end-groups or direct NMR observation of the growing chain ends, as shown in Scheme 4.6 for the $(C_2H_5)_2AlOC_2H_5$-initiated polymerization of CL [64, 65].

The usual difficulties are well known: the chains are long and the end-groups are at a very low concentration, hence their observation is difficult, since the same atoms are present in the backbones. At M_n higher than $\sim 10^4$, end-groups could hardly be determined by using ^{1}H or ^{13}C NMR spectroscopy. In Section

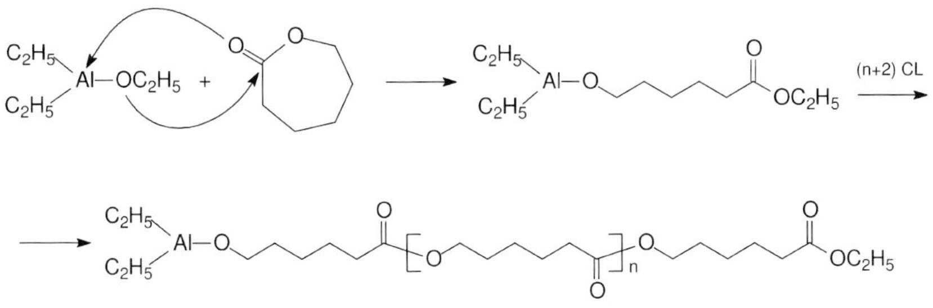

4.3 Basic Mechanistic Features of Ring-opening Polymerization

Scheme 4.6 Coordination polymerization of ε-caprolactone initiated with diethylaluminum ethoxide.

4.3.1.2 (oxiranes), a method for the determination of the structure and concentration of active centers was described, applied to several ROPs, including those of cyclic esters. It was based on the fast reaction of active centers with ClP(=O)(OC$_6$H$_5$)$_2$ followed by ^{31}P NMR measurement. In this way, it was possible to determine several structures and concentrations of the growing ends [28].

The kinetics of the polymerization of lactones and lactides, for both anionic and coordination propagation, have mostly been studied in our group [9, 10, 19, 55, 59]. The first living polymerizations of PL were observed when crowned [66] or cryptated [67] counterions (cations) were used. Apparently, in this way the ratio of the rate constants of propagation and transfer (k_p/k_{tr}) was sufficiently increased that polymers with M_n over 10^5 could be prepared [68, 69]. However, polymerization of the β-substituted four-membered lactones [e.g. β-butyrolactone (γBL)] suffers from the proton abstraction side-reaction, leading to chain transfer to the polymer [70, 71]. Application of Bu$_4$N$^+$ as counterion resulted in further enhancement of the k_p/k_{tr} ratio and allowed the preparation of poly(βBL) with $M_n \leq 2\times10^5$ [72].

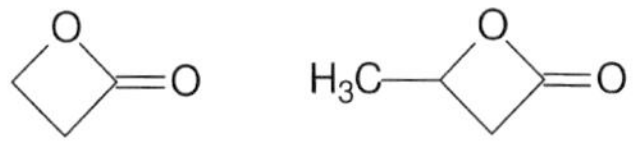

(PL)   (βBL)

Initiation of PL polymerization with alkoxides proceeds in such a way that, independently of the chemistry of the earlier stages, the eventual active species are carboxylate anions [73]. Therefore, polymerization of these four-membered rings can also be initiated by carboxylate anions:

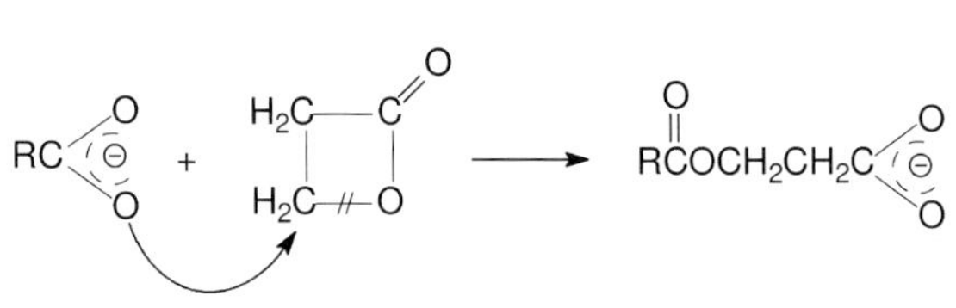

(counterion omitted)   (5)

The observed inversion of configuration, if the substituted carbon is attacked (e.g. in βBL polymerization), is consistent with the S_N2 mechanism [74].

Carboxylate anions are less nucleophilic than alkoxides and therefore are also less efficient as initiators of anionic polymerization. Thus, acetates are known to initiate polymerization of PL but they do not initiate polymerization of less strained lactones (e.g. CL) or lactides.

In contrast to PL, which polymerizes with soft and moderately reactive carboxylate anions, less strained cyclic esters (e.g. CL or LA) require stronger nucleophiles to propagate and in consequence usually are plagued by side-reactions. On the other hand, it has been found that coordination ROP of CL or LA may proceed in a perfectly controlled way due to the high selectivity, as quantitatively expressed by a high k_p/k_{tr} ratio [75–79]. This feature provides a convenient way not only for the synthesis of linear homopolymers, but also for a large variety of macromolecules differing in architecture.

The most often used initiators in coordination ROP of cyclic esters are multivalent metal [e.g. Zn, Sn(II), Al, Sn(IV) and Ti] alkoxides and carboxylates – particularly popular is tin(II) 2-ethylhexanoate [$SnOC(=O)CH(C_2H_5)C_4H_9$], also called tin octoate, $Sn(Oct)_2$]; more exotic alkoxides of, e.g. La, Y and Zr are also used. A large number of metal alkoxides have been studied and it has been shown conclusively that the number of chains growing from one metal atom is equal to the number of alkoxide groups attached to that atom [10, 58].

Initiation of coordination ROP of cyclic esters with dialkylaluminum alkoxides or aluminum trialkoxides is a typical example. This elementary reaction proceeds by a simple monomer insertion, mechanistically identical with the next propagation steps. Whenever the alkoxides exist in different aggregated forms, then it may happen that one form is more reactive than the other. If, in addition, the exchange between these forms is slow, in comparison with the rate of

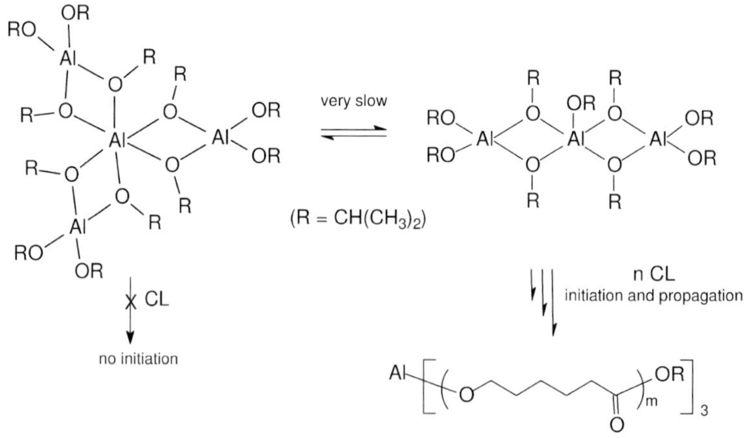

Scheme 4.7 Polymerization of ε-caprolactone initiated with a mixture of tris-aluminum isopropoxide tetramer and trimer.

initiation, then only this reactive form is engaged in initiation, leaving the other form(s) substantially intact when the polymerization is completed. This is the case in initiation of CL with Al(O^{i}Pr)$_3$, consisting of a trimer (reactive) and tetramer (unreactive) at least at ambient temperature (Scheme 4.7) [80–82]. At higher temperatures (e.g. above 100 °C), the rate of trimer–tetramer exchange may approach the rate of initiation, since the temperature coefficient of the rate for exchange is higher than that for initiation [83].

Initiation of CL polymerization with R$_2$AlOR′ was studied by ^{1}H NMR spectroscopy and the structure of the chain-end allowed the elucidation of the corresponding mechanism (cf. Scheme 4.7) [64, 65]. Thus, initiation and then propagation could be classified, at least formally, as the S_N2 nucleophilic attack of the alkoxide moiety on the carbonyl carbon atom with scission of the acyl–oxygen bond. However, the monomer insertion may be preceded by monomer coordination, and the elementary reaction of the Al alkoxide-active species with the CL monomer then proceeds according to the $B_{AC}2$ mechanism [84].

In the group of Al-based initiators, the successful application of aluminum porphyrins in the polymerization of cyclic ethers (including PO) and cyclic esters should be mentioned. The structure of a typical initiator [Al tetraphenylporphyrin (Al-TPP)] is illustrated here.

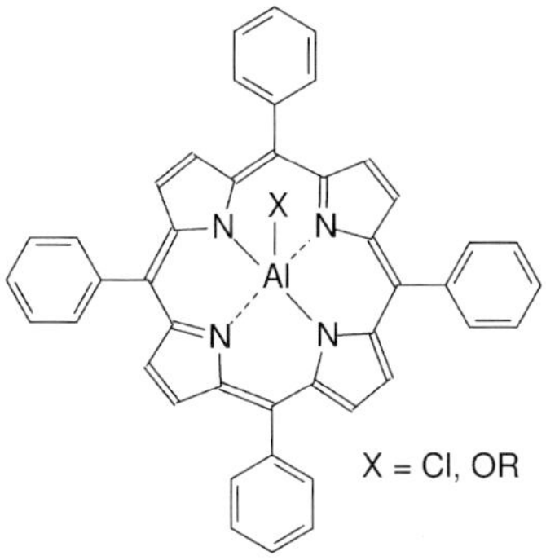

X = Cl, OR

(Al-TPP)

The reason for showing this structure is not only because this class of initiators, developed by Inoue and Aida [30, 85], give living processes, but also because more than 20 years later the structurally related "single-site" initiators became important in the stereocontrolled polymerization of lactides and the discovery of the Japanese researchers has been almost forgotten.

For a long time, the mechanism of initiation with carboxylates was under dispute. Leenslag and Pennings [86] and later Zhang and coworkers [87, 88] proposed a mechanism involving reaction of Sn(Oct)$_2$ with a coinitiator, leading this way to the corresponding alkoxide. The hydroxyl group may be present as an impurity or may be a deliberately added compound in the system [e.g. water, alcohol or hydroxy acid (ROH)] (Scheme 4.8).

Sn(Oct)$_2$ + ROH ⇌ OctSn-OR + OctH

OctSn-OR + ROH ⇌ RO-Sn-OR + OctH

(where OctH stands for 2-ethylhexanoic (octanoic) acid).

Scheme 4.8 Formation of an alkoxide initiator in polymerization carried out in the presence of tin(II) 2-ethylhexanoate.

Then, the resulting tin(II) alkoxide initiates and propagates in the usual manner as with other metal alkoxides. However, there was at that time no direct proof for such a mechanism and several other mechanisms have been proposed [89–93]. The most often cited was the "trimolecular mechanism" [91], in which first the catalyst–monomer complex was formed. This mechanism has been shown conclusively not to operate since it excludes the presence of Sn atoms in the growing macromolecules covalently bound. Recently, we were able to observe by matrix-assisted laser desorption/ionization mass spectrometry (MALDI-TOF-MS) the presence of the tin(II) alkoxides in the growing polyester chains [94]. Moreover, kinetic studies also clearly supported this sequence of events [95–98]. Therefore, these other views [89–93] could be discounted.

In addition to hydroxyl-containing compounds, primary amines can also play the role of the initiator in Sn(Oct)$_2$-catalyzed polymerizations, giving eventually tin(II) alkoxides as the growing species [98] (it should be stressed that the distinction between initiator and catalyst in such binary systems is not always straightforward and those terms are frequently used interchangeably).

There is also a continuous search for metal-free catalysts. Among other nucleophiles, 4-(N,N-dimethylamino)pyridine (DMAP) [99, 100], N-heterocyclic carbenes (NHC) [101–104] and natural amino acids [105] have been particularly explored. It is assumed that metal-free catalysts may give polyesters that can be used advantageously in the biomedical or microelectronic fields. Depending on the initiator structure (ROH), linear and star-shaped homopolymers and block copolymers were prepared with DMAP and NHC in a controlled way. However, control of the number-average degree of polymerization (DP_n), via the monomer to initiator molar ratio, did not exceed the value of $DP_n \approx 200$ and the DP_w/DP_n ratio varied from 1.05 to 1.5, depending on the polymerization conditions. These organocatalytic systems require rather long polymerization times, of the order of 100 h at ambient temperature, to reach $DP_n \approx 200$. Further studies are required to elucidate the DMAP or NHC polymerization mechanism. It has been proposed that at least a certain fraction of the active centers assumes an undesired anionic structure.

4.3.2.2 Controlled Polymerization of Cyclic Esters with "Multiple-site" Metal Alkoxides and Carboxylates

Al and Sn(II) compounds have been investigated in more detail. First, the dialkylaluminum alkoxides (R'$_2$AlOR) were used in both synthetic and mechanistic studies [64, 84, 106–108]. A number of well-defined macromolecules with well-controlled size and end-groups were prepared with these initiators [64, 108]. As can be seen in Fig. 4.4a, M_ns of poly(ε-caprolactone) (PCL) could be controlled up to $\leq 5 \times 10^5$.

After understanding the difference between aluminum Al(O^{i}Pr)$_3$ trimer and tetramer (cf. Scheme 4.7), the isolated trimer has become the most versatile initiator for the controlled polymerization of cyclic esters. It provides fast and quantitative initiation, moderately fast propagation [$k_p = 0.6$ L mol^{-1} s^{-1} compared with 0.039 L mol^{-1} s^{-1} for Et$_2$AlOEt (25 °C, THF)] [81, 106] and relatively good selectivity (with regard to transesterification) [78, 79]. Hence Al(O^{i}Pr)$_3$ in the form of a trimer appears to be ideally suited to the synthesis of aliphatic polyesters, since apart from good selectivity it provides direct control of the degree of polymerization of the resulting polyester, by simply adjusting the ([LA]$_0$ − [LA]$_{eq}$)/3[Al(O^{i}Pr)$_3$]$_0$ ratio. On the other hand, it seems that there is an upper limit of $M_n \approx 3 \times 10^5$ of PLA and PCL which can be obtained with Al(O^{i}Pr)$_3$ [81, 109], whereas with Sn(OBu)$_2$ $M_n \approx 10^6$ has been reached (Fig. 4.4b) [110, 111]. Reasons for such a limitation are not yet well understood, but could be related to the concentration of impurities and/or the intramolecular complexation of Al atoms in the active centers by the acyl oxygen atoms from the growing polyester chain [10, 81].

Sn(Oct)$_2$ is probably the most often used catalyst in the polymerization of cyclic esters. This is mostly due to its commercial availability, physical state (liquid) and higher chemical stability in comparison with alkoxides. DP_n of the polyester formed in the cyclic ester (M)–Sn(Oct)$_2$–initiator systems is given by the ([M]$_0$ −

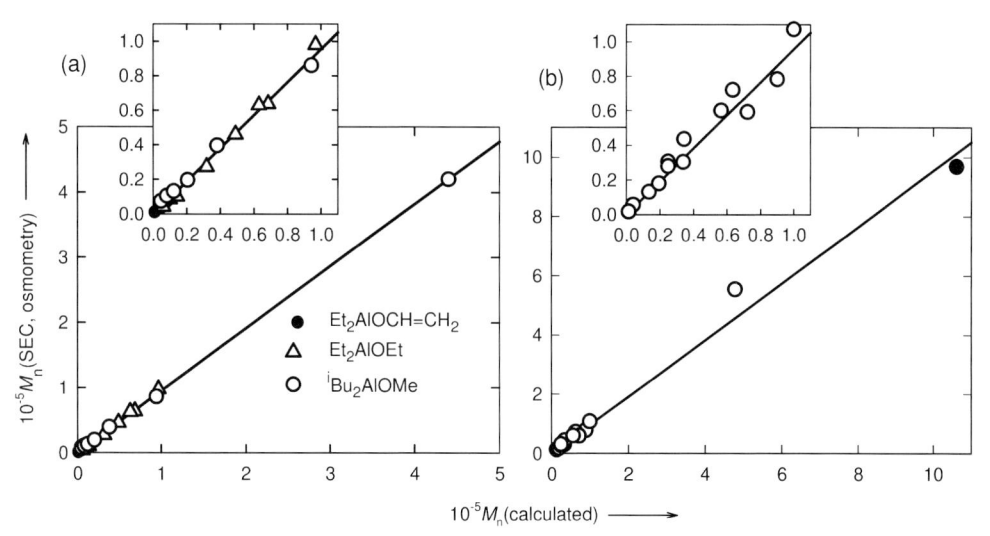

Fig. 4.4 Measured versus calculated (predicted) molar masses of polyesters: (a) polymerization of ε-caprolactone initiated with R$'_2$AlOR, THF, 25 °C [64]; (b) polymerization of L-lactide initiated with Sn(OBu)$_2$: (○) THF, 80 °C and (●) bulk, 120 °C [110].

[M])/[initiator]$_0$ ratio due to the fast initiation and exchange reactions: the chain transfer to water, alcohol or amine and then to the resulting macroalcohol. Using the standard high-vacuum technique and Sn(Oct)$_2$ of 99.0 mol% purity, we were able to obtain both PCL and PLA with M_n up to $\sim 9\times 10^5$ [112]. Thus, $M_n \approx 10^6$ appears to be the limit of M_n of aliphatic polyesters prepared by ROP using the usual techniques and this limit is, most probably, related to the concentration level of impurities. A similar threshold value ($M_v \approx 9\times 10^5$) was reported by Pennings and coworkers about 15 years ago [89]. It is not known whether there is a genuine termination reaction in these polymerizations.

Polymerization of substituted cyclic esters, such as LA or βBL, is a slow process. As a reference value in further discussion, we can use the absolute rate constant $k_p = 5\times 10^{-1}$ mol^{-1} L s^{-1}, determined for the polymerization of CL initiated with Al(O^{i}Pr)$_3$ at 25 °C [81]. At [>Al-O-...] = 10^{-2} mol L^{-1}, this rate constant corresponds to the monomer half-lifetime ($t_{1/2}$) being ~ 2 min. Polymerization of LA and βBL with the same initiator requires much longer times, because k_p^{app}(LA) = 7.5×10^{-5} mol^{-1} L s^{-1} (20 °C) [83] and k_p^{app}(βBL) = 3×10^{-5} mol^{-1} L s^{-1} (50 °C) [113]; the $t_{1/2}$ values are 1.5×10^4 and 3.9×10^4 min, respectively at [>Al-O-...] = 10^{-2} mol L^{-1}.

Therefore, extensive research programs have been started aimed at the elaboration of new, more reactive initiators for LA and βBL polymerization, but devoid of the drawbacks with ionic initiators. The most promising results were obtained with rare earth metal (e.g. La, Sc, Sm, Y, Yb) alkoxides. These derivatives were initially developed by DuPont [114–117] and subsequently in other laboratories [118–128].

Particularly interesting results were obtained by McLain et al. [115] for the polymerization of lactides. For example, in the process initiated by Y(OCH$_2$CH$_2$NMe$_2$)$_3$, k_p^{app} = 5.7×10^{-1} mol^{-1} L s^{-1} was determined at 20 °C, which corresponds to a polymerization rate increase of 7.6×10^3-fold compared with that initiated with Al(O^{i}Pr)$_3$. Similar polymerization rates were also observed by Simic et al. [119, 120]. For the initiating system Y(OAr)$_3$–^{i}PrOH (where Ar = 2,6-di-tert-butylphenyl) applied by Feijen and coworkers, a value of $k_p^{app} \geq 10$ mol^{-1} L s^{-1} (22 °C) can be estimated [124]. This means that the polymerization proceeds with a rate characteristic of the ionic polymerization of cyclic esters.

However, it appears that the possibilities of controlled polymerization initiated with rare earth metal alkoxides are limited. The molar masses reported so far do not exceed $M_n \approx 10^5$. For example, Y(OCH$_2$CH$_2$NMe$_2$)$_3$ taken at a concentration 6.6×10^{-3} mol L^{-1} at 166 °C in an LA melt led to a final monomer consumption of 43%, suggesting the occurrence of chain growth termination [117]. Moreover, MALDI-TOF-MS traces recorded for the LA alkoxide–LA polymerizing mixture revealed the relatively fast formation of an undesired cyclic oligomer fraction [121]. Apparently, in contrast to some other systems, there was not sufficiently efficient kinetic control of propagation.

4.3.2.3 Controlled Polymerization of Cyclic Esters with "Single-site" Metal Alkoxides

More recently, polymerization with the so-called "single-site" catalysts/initiators has been thoroughly explored (see, e.g., [27, 30, 85, 129–131] and the references cited therein). This may be of interest, since it might remove the mechanistic complexity resulting from the aggregation–deaggregation exchange reactions in which multiple-site alkoxides are usually engaged. According to the definition given by Coates et al. [129]: "Single-site catalysts are those polymerization catalysts where enchainment of monomer occurs at a metal center (M, the active site) which is bound by an organic ligand (L). This ancillary ligand remains bound throughout the catalytic reaction, modifying the reactivity of the metal center. Typical single-site catalysts for lactone polymerization are of the form L_nMOR, where the alkoxide group (OR) is capable of propagation."

This concept and definition come from the studies of olefin polymerizations, where the first generations of catalysts were heterogeneous and provided several catalytic sites on the surface of the catalysts. Using the same expression in the polymerization of cyclic esters is somehow ambiguous. Indeed, the dialkylaluminum alkoxides (R'_2AlOR) studied by us [64, 84, 106, 107] belong, in principle, to this category since under properly chosen conditions the R' ligands remain unreactive and the propagation proceeds on the alkoxide group. When R' is a lower alkyl group, aggregation of active centers takes place [84, 106, 107]. However, deaggregation could be observed when simple ligands, such as N,N,N'-trimethylethylenediamine, were attached to the otherwise aggregated species [84]. Further, application of the bulky ancillary ligands L in the catalysts employed hampers chain transfer to macromolecules, slowing both intra- and intermolecular transesterification [76, 78], which manifests in a decrease in the M_w/M_n values for the prepared polyesters. Moreover, some of these derivatives can be successfully applied in the stereocontrolled polymerization of *rac*- and *meso*-LA (cf. Section 4.3.2.5).

The new generation of the single-site catalysts explored so far in CL polymerization do not show any particular advantage over the multiple-site catalysts, such as $Al(O^iPr)_3$, with respect to molar mass control and also molar mass distribution or end-group control in the resulting PCL [132–139]. Results reported for the polymerization of lactides [129, 131, 137, 140] and β-substituted β-lactones [141] initiated with Zn or Mg aminoalkoxide or β-diiminate complexes point to a considerable rate increase of this process compared with polymerizations initiated with $Al(O^iPr)_3$ [83, 113]. For example, for L,L-LA, $k_p = 2.2$ mol^{-1} L s^{-1} at 25 °C ($t_{1/2} = 2.5$ min at $[I]_0 = 10^{-2}$ mol L^{-1}) has been determined [131].

4.3.2.4 Poly(β-hydroxybutyrate)s by Carbonylation of Oxiranes

Syntheses of poly(β-hydroxyalkanoate)s (PHAs) has a long history – particularly of PHAs prepared by polymerization in bacteria [142–144]. These polymers, developed by ICI and Monsanto, are now being further developed industrially by several companies. There have also been attempts to prepare similar polymers in a non-bacterial manner. Rieger and coworkers [145, 146] recently described alternating copolymerization of propylene oxide and CO, catalyzed by a mixture of dicobalto octacarbonyl, [Co$_2$(CO)$_8$], and 3-hydroxypyridine. Copolymerization of rac-PO and CO affords regioregular but atactic poly(β-hydroxybutyrate)s. Application of enantiometrically pure (R)- or (S)-PO results in the formation of isotactic, optically active and crystalline materials. The epoxide units undergo clear regioregular incorporation into the polymer chain with retention of configuration at the stereogenic carbon centers:

$$m \;\triangle\text{-}H_3C\;\;+\;\;m\,CO\;\;\xrightarrow{Co_2CO_8/\,N\text{-}C_5H_4\text{-}OH}\;\;\text{[poly(β-hydroxybutyrate)]}\quad(6)$$

An isotactic polymer of $M_w \leq 6.7 \times 10^3$ melting at 130 °C was prepared, which opened up a new route in PHA synthesis. In contrast to the biotechnological process, the extent of tacticity could be regulated by addition of the racemic PO to the polymerization of one of the PO enantiomers.

Alternatively, carbonylation of the substituted oxiranes leads also to monomeric β-substituted β-lactones. Recently, Coates and coworkers [147], using chromium(III) octaethylporphyrinatotetracarbonycobaltate as catalyst, prepared an impressively broad array of β-lactones bearing aliphatic, cycloaliphatic, ether, ester and amide exocyclic groups. In the majority of the reported examples, yields were quantitative (>99%) and carbonyl insertion proceeded regioselectively. Similar processes, but leading to polycarbonates, are described in Section 4.3.3.

4.3.2.5 Stereocontrolled Polymerization of Chiral Cyclic Esters

There are two major cyclic esters bearing centers of chirality: β-butyrolactones and lactides. The preparation of isotactic, enantiomerically pure polymers from the enantiomerically pure heterocyclic monomers is easy if the corresponding monomers are available. Typically, however, the optically active monomers are not available in a large quantity and their synthesis is much more complicated compared with their racemic counterparts. Until now, the only exception is (S,S)-lactide, a monomer synthesized from (S)-lactic acid, which in turn is produced on an industrial scale by fermentation from carbohydrates of agricultural origin [148, 149]. The resultant high molar mass poly[(S)-lactide] is a crystalline (~60%) polymer melting at 180 °C, whereas racemic poly[(R)-lactide-co-(S)-lactide] with a random distribution of R- and S-units is amorphous.

Initial attempts at stereocontrolled polymerizations [150–157] have shown that in rac-βBL polymerization initiated with Et$_3$Al–H$_2$O, a multiblock stereocopolymer, –{[(R)-βBL]$_x$–[(S)-βBL]$_y$}$_n$–, was formed [150, 151]. However, even the enantiomerically pure initiator Et$_2$Zn–(R)-(CH$_3$)$_3$CCH(OH)CH$_2$OH gave $k_p(S)/k_p(R)$ ratios not higher than 1.7 [157]. Hence in the polymerization of this monomer there is still a lack of effective stereoinitiators.

In contrast, in the stereocontrolled polymerization of lactides, considerable progress has been achieved recently. As is known, LA monomers contain two centers of chirality and therefore they are available in various stereochemical forms: (R,R)-LA, (S,S)-LA, *meso*-(R,S)-LA and racemic, equimolar (R,R)-LA–(S,S)-LA mixture (*rac*-LA) [148].

(R,R)-LA   (S,S)-LA   (R,S)-LA

Thus, the proportion of the R and S repeating units and their distribution along the PLA chains depend not only on the stereospecificity of the active center but also on the stereochemical composition of the monomer. The stereostructure of PLA could also be influenced by racemization. However, this side-reaction in the stereocontrolled synthesis is eliminated when coordination (covalent) initiators – alkoxides of the multivalent metals (e.g. Zn, Al, Sn, Ti) – are used [57, 148].

The PLA actually produced is poly[(S)-LA] [(S)-PLA], which is crystalline and rather rigid. A few percent of R-units is therefore introduced to make the polymer less stiff [148, 158]. On the other hand, the monomer that can be prepared by synthetic chemistry (e.g. from acetaldehyde) is racemic [159]. It provides an interesting model for studying stereocontrolled polymerization. As will be seen later (Section 4.3.2.6), it is important also to have ways of preparing (R)- and (S)-PLA for further stereocomplexation. This could be done either by separate polymerizations of (R,R)- and (S,S)-LA or by stereoselective polymerization of *rac*-LA.

Stereoselective polymerization of *rac*-LA is, however, hampered by the intermolecular chain transfer to a macromolecule side-reaction. Even if initially stereoselective polymerization proceeds and the individual poly[(R)-LA] and poly[(S)-LA] are formed, then transesterification will lead to intermolecular segmental exchange. Eventually, under thermodynamic equilibrium conditions, PLA assumes the form of the poly[(R)-LA-*rand*-(S)-LA] stereocopolymer (Scheme 4.9).

Depression or even elimination of the segmental exchange turned out to be possible by applying metal alkoxide initiators decorated with sterically extended

rac-LA →(stereoselection) ...-RRRRRRRRRR* / *SSSSSSSSSS-...

↓ transesterification

...-SRSSRRSSSR* + ...-SRRRSSRRSR*

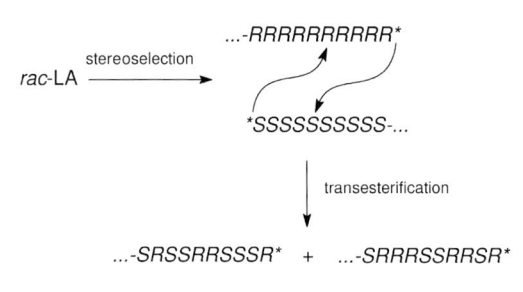

(R, and S stand for (R)- and (S)-(C=O)CH(CH₃)O units, respectively)

Scheme 4.9 Transesterification hampering stereoselective polymerization of racemic lactide.

ligands. Steric hindrance located at the active center decreases the transesterification rate due to the increase in entropy of activation, while the propagation rate remains constant [76].

A single-site, enantioselective (racemate-forming, chirogenic [160]) or enantioselective (asymmetric, enantiomer-differentiating [160]) polymerization of rac-LA has been realized by Spassky and coworkers, who applied achiral and chiral aluminum alkoxides of general structure $SBO_2Al–OR$, bearing Schiff's base ligands [161–163]. The chirality of the derivatives (R)- and (S)-$SBO_2Al–OR$ has its origin in the hindered rotation of the 2- and 2′-substituted binaphthyl moiety.

SBO₂Al-OR   (R)-SBO₂Al-OR   (S)-SBO₂Al-OR

This method, originally developed by Spassky and coworkers, was then followed up by the groups of Baker [164], Coates [165, 166], Feijen [167, 168] and Nomura [169]. It was revealed that polymerization of rac-LA mediated with either achiral or chiral but racemic initiators led to the multiblock stereocopolymers [(S)-PLA-b-(R)-PLA]$_p$, resulting from a chain-end control mechanism (CEM) or an enantiomorphic site-control mechanism (SCM) of the monomer addition, respectively. Initially, formation of the isotactic polymer composed of an equimolar mixture of (R)- and (S)-PLA macromolecules was expected [164]. However, Ovitt and Coates [165, 166], after repeating these experiments, came to the conclusion that the polymer obtained cannot be a mixture of the exclusively homochiral (R)- and (S)-PLA macromolecules (Scheme 4.10).

Comparison of the recorded and simulated ^{1}H NMR spectra, with assumption that polyester chain growth is accompanied by exchange of chiral macromolecules on "foreign" active centers, indicated that the resulting macromolecules were composed of stereoblocks with 11 repeating units derived from LA monomer of a given kind (toluene, 70 °C).

Enantiomerically pure (R)- or (S)-$SBO_2Al–OR$ initiators have been shown to polymerize preferentially one of the enantiomers from rac-LA, i.e. (R,R)- or (S,S)-LA, respectively [162, 170]. This enantiospecific preference was revealed by the stereoselectivity ratio $k_p(RR)/k_p(RS)$ [or $k_p(SS)/k_p(SR)$] equal to 28 [170]. In stereocontrolled polymerization of the lactide monomers, syndio- and heterotactic PLA were also prepared starting from rac- and meso-PLA (see, e.g., [171–173]).

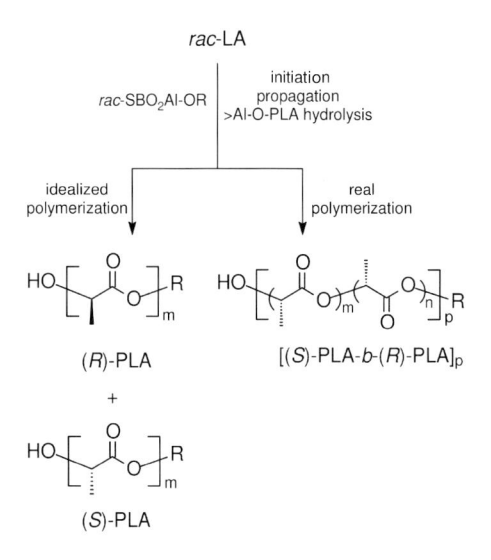

4.3 Basic Mechanistic Features of Ring-opening Polymerization | 125

Scheme 4.10 Idealized versus real stereoselective polymerization of racemic lactide.

One of the most important practical consequences of the results described above, with regard to the mechanical and thermal properties of PLA, is the possibility of stereocomplex formation by sufficiently long stereoblocks of the opposite configuration (cf. Section 4.3.2.6). The T_m of the (R)-PLA–(S)-PLA stereocomplex composed of an equimolar mixture of high molar mass R- and S-macromolecules is 230 °C [158], whereas the highest T_m of the stereocomplex prepared so far by means of the *stereoselective* polymerization of rac-PLA, was 192 °C [169].

More recently, by combining *stereoselective* polymerization with chiral ligand exchange, a stereocomplex with significantly higher $T_m = 210$ °C was obtained [170]. This was realized by two-step polymerization of rac-LA initiated by 2,2'-[1,1'-binaphthyl-2,2'-diylbis(nitrylomethylidyne)]diphenol [(S)-SB(OH)$_2$]–Al(O^{i}Pr)$_3$ trimer mixture, in which the actual initiator (SBO$_2$Al–O^{i}Pr) was formed *in situ*. First, (S)-SB(OH)$_2$–Al(O^{i}Pr)$_3$ mixture was reacted for 24 h in THF as solvent at 80 °C and then rac-LA was introduced. Progress of the polymerization was followed using polarimetry (Fig. 4.5) and by SEC.

Optical rotation (or) readings increased with polymerization time and eventually leveled off. SEC measurement revealed approximately 50 mol% consumption of rac-LA. In the second step, an equimolar quantity of (R)-SB(OH)$_2$, with respect to (S)-SB(OH)$_2$, was introduced. In further polymerization, a gradual decrease in or was observed. Taking into account the determined stereoselectivity ratios [i.e. $k_p(SS)/k_p(SR) = 28$], for the final poly(rac-LA) the gradient poly[(S,S)-LA-*grad*-(R,R)-LA] rather than the pure block copolymer structure was expected. Indeed, homodecoupled ^{1}H NMR spectra showed, apart from the strong signal of the isotactic *mmm* tetrad, also peaks of lower intensity, which can be ascribed to the *mmr*, *rmm* and *rmr* tetrads [170].

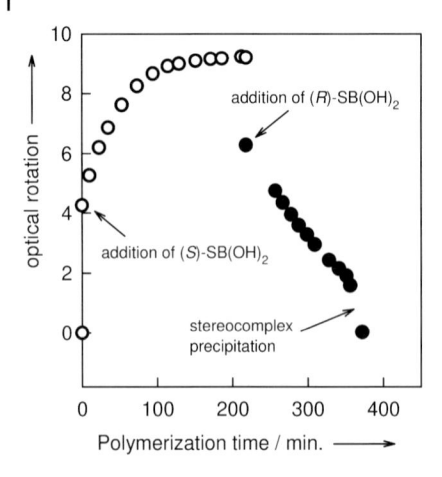

Fig. 4.5 Optical rotation versus time for two-step polymerization of racemic lactide (rac-LA) initiated with Al(O^{i}Pr)$_3$ trimer–(S)-SB(OH)$_2$ mixture and mediated with (R)-SB(OH)$_2$ added consecutively. Conditions: THF, 80 °C, [Al(O^{i}Pr)$_3$]$_0$ = [(S)-SB(OH)$_2$]$_0$ = [(R)-SB(OH)$_2$]$_0$ = 0.023 mol L^{-1}, [rac-LA]$_0$ = 1.5 mol L^{-1} [170].

4.3.2.6 Stereocomplexes of Aliphatic Polyesters

Macromolecules composed of repeating units with complementary attracting sites can form intermolecular complexes. Typical examples are provided by natural macromolecules such as polysaccharides, polypeptides (e.g. collagen) or nucleic acids. Synthetic polymer chemistry, taking a lesson from biology, developed a variety of macro- or supramolecular systems in which complementary interactions give rise to hierarchical structural order. Macromolecules of identical chemical composition but different stereochemical configuration of repeating units, either enantiomeric or diastereomeric, are also able to form intermolecular complexes, called stereocomplexes [174].

Macromolecular stereocomplexes in the solid state usually form crystalline structures melting at higher temperature (T_m) than homochiral components alone, as shown, for example, for polymers derived from the chiral heterocyclic monomers, such as (R)- and (S)-poly(tert-butylthiirane) [175] or (R)- and (S)-poly(α-methyl-α-ethyl-α-propiolactone) [176, 177]. This is also the case with the most frequently studied poly[(R)-lactide]–poly[(S)-lactide] [(R)-PLA–(S)-PLA] stereocomplex, which exhibits T_m up to 50 °C higher than the (R)-PLA and (S)-PLA components alone [162, 168–170, 178–192].

It has been shown recently that the attracting (R)-PLA/(S)-PLA complementary interactions result from weak hydrogen bonding (Scheme 4.11) [187, 188]. The T_m of the high molar mass linear (R)-PLA or (S)-PLA is only ≤180 °C whereas (R)-PLA–(S)-PLA stereocomplexes have T_m reaching 230 °C [178–182]. Unfortunately, for linear high molar mass ($M_n \geq 10^5$) polylactides, there is neither enough memory preserved in preformed stereocomplexes to survive melting nor enough specificity in the formation of stereocomplexes in the poly[(R)-lactide]–poly[(S)-lactide] melt. Eventually, from a melt, a mixture composed of homochiral crystallites and stereocomplex crystallites is formed. However, in the instance of star-shaped high molar mass ($M_n \geq 10^5$) enantiomeric PLA, formation or reformation of the (R)-PLA–(S)-PLA stereocomplex in the

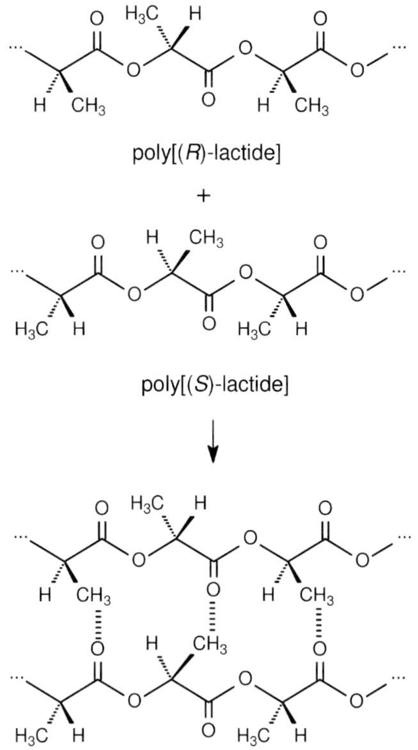

Scheme 4.11 Stereocomplex formation from the homochiral polylactide linear macromolecules.

poly[(R)-lactide]

+

poly[(S)-lactide]

melt is complete and reversible. The authors coined an expression "hardlock-type interactions" in order to explain this difference between linear and star-like stereocomplexes [192].

4.3.3
Controlled Synthesis of Aliphatic Polycarbonates by Anionic and Coordination Ring-opening Polymerization

Mechanistic features of the anionic and coordination ROP of cyclic carbonates, the process by which aliphatic polycarbonates are most often prepared, show a close resemblance to those known for lactones and lactides. Readers who are interested in more detailed information in this area are referred to two authoritative reviews [193, 194] and a few more recent original papers [195–201].

Most often five- and six-membered monomers, such as ethylene carbonate (EC), 1,2-propylene carbonate (1,2-PC), 1,3-propylene carbonate (1,3-PC) and 2,2-dimethyltrimethylene (neopentylene) carbonate (DTC) were studied. Poly(DTC) has been the basis of a fiber developed in an attempt to replace the polyglycolide fibers used for absorbable sutures. This suture on hydrolysis did not cause an inflammatory reaction of the living tissue.

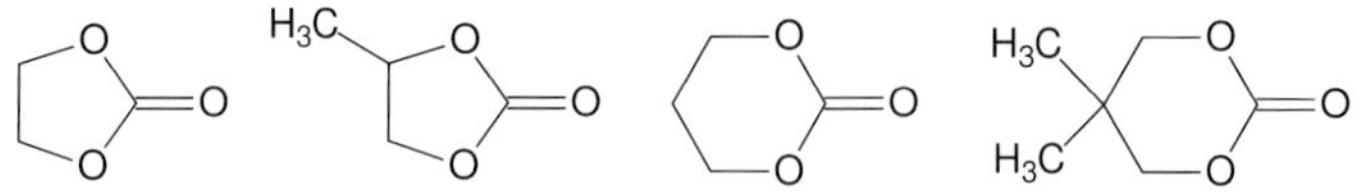

(EC)   (1,2-PC)   (1,3-PC)   (DTC)

Five-membered monomers are able to provide high molar mass polymers, because of thermodynamic restrictions, only via polymerization in which propagation is accompanied by the partial loss of CO_2, exceeding 50 mol% [202]. In contrast, polymerization of the more strained six-membered carbonates gives, in the properly chosen conditions, regular polycarbonate chains without traces of decarboxylation (Scheme 4.12) [193, 198–200].

Anionic polymerization of cyclic carbonates suffers from inter- and/or intramolecular side-reactions and therefore the controlled synthesis of the corresponding polycarbonates has been exclusively realized with the coordination (covalent) initiators/catalysts described in the preceding sections. The elementary reaction of the polycarbonate chain growth proceeds on the alkoxide-active centers with acyl–oxygen bond scission (see [193, 194] and references cited therein). The structure of the active centers was established also by the end-capping method with phosphorus compounds, described earlier in this chapter (Eq. 4) [203].

The environmentally benign methods of polycarbonate synthesis based on the alternating copolymerization of oxiranes (or oxetanes) and CO_2, originally developed by Inoue et al. [204] almost 40 years ago, are still intensively investigated [195–197, 201]. The main drawback of this process is the concomitant formation of thermodynamically stable five-membered cyclic carbonates. Only recently, Coates and coworkers [196] applied optically active (R,R)-(salcy)CoX [where salcy = N,N'-bis(3,5-di-*tert*-butysalicylidene)-1,2-diaminocyclohexane and X = halide or carboxylate] complexes as initiator for 1,2-PO–CO_2 copolymerization. ^{13}C and ^{1}H NMR analyses of the resulting poly(1,2-propylene carbonate) revealed the formation of the regioregular macromolecules containing up to 99 mol% carbonate linkages. In a similar approach, Coates and coworkers [195] used (R)-limonene oxide comonomer, prepared by oxidation of limonene extracted from orange peel, in copolymerization initiated by β-diimidate zinc acetate. The copolymer contained >99 mol% of the carbonate repeating units and M_n values were controlled by the oxirane to initiator molar ratio (cf. Section 4.3.2.4 describing similar processes, but with CO as a comonomer).

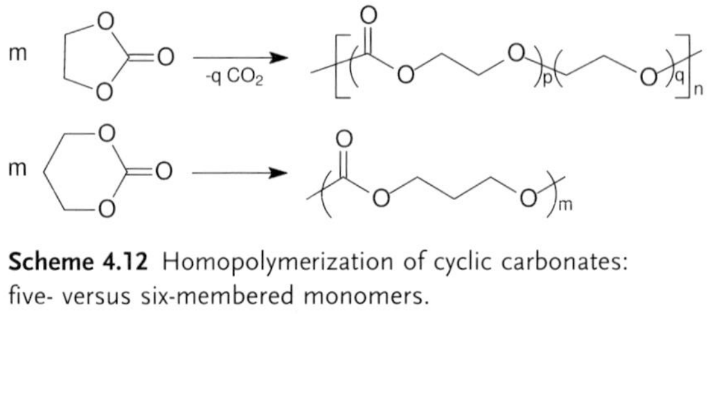

Scheme 4.12 Homopolymerization of cyclic carbonates: five- versus six-membered monomers.

4.3.4
Controlled Synthesis of Branched and Star-shaped Polyoxiranes and Polyesters

4.3.4.1 Anionic Polymerization of Oxiranes

Controlled synthesis of branched and star-shaped PEOs has been systematically explored for more than 20 years (see, e.g., [205–210] and references cited therein). Star-shaped polymers, bearing a strictly defined number of linear arms, provide useful models for studies of this class of polymers. Two different synthetic routes have been explored: "arms first" and "core first" approaches. The latter method is based on initiation of EO polymerization with multifunctional polyols and initiated by alkoxide centers formed therefrom [using, e.g., K metal, KOH (with azeotropic removal of H_2O), NaH or $(C_6H_5)_2$CHK]. Theoretical studies based on models prepared in this way are summarized in the classical paper by Roovers and coworkers [211].

A novel approach to the synthesis of star-like, branched polymers by ROP is based on more recent work in which polymerization of a cyclic monomer is performed in the presence of a core containing two cyclic monomeric rings coupled together. These studies, which started with polymerization of EO in the presence of bis-oxirane compounds, bear a resemblance to the well-known polymerization of vinyl monomers in the presence of divinyl compounds (styrene and divinylbenzene as first described by Rempp et al. [212, 213]). Then, it was realized that the arm precursors can be prepared beforehand and then linked by a bis-oxirane compound. As a consequence, any oligomer with an OH end-group could, in principle, be used as the first arms. Recent work on polystyrene–OH has shown that this method is perhaps one of the most general ways of preparing star-like polymers from any kind of backbones forming arms of the star polymers [206, 214, 215].

It is interesting and important that by properly choosing the length and the ratio of arms (or monomer) to the core-forming compound it has been possible to find conditions when both starting materials reacted quantitatively. When this stage is completed, then on the core there are as many anionic reactive sites as precursors of the first arms were taken for reaction. Therefore, the second generation of arms can be grown from the reactive core. This has been done, first attaching blocked polyglycidol arms and then growing PEO arms from the core. This synthetic route is shown in Scheme 4.13 [209, 210].

Finally, the active sites are transformed into the end-groups of the second generation of arms (A_z in $A_xB_yA_z$), which can further be functionalized.

Another novel method, recently elaborated by Frey and coworkers [216], is based on the anionic polymerization of glycidol leading to polyglycerol. This anionic polymerization has its analogue in the conceptually similar cationic polymerization of this monomer and hydroxymethyloxetanes (Section 4.3.6.3). Frey and coworkers developed then chemistry of several architectural derivatives of the resulting highly branched polyglycerols, building star blocks on the OH groups by ROP and applying other polymerization methods (ATRP) [217].

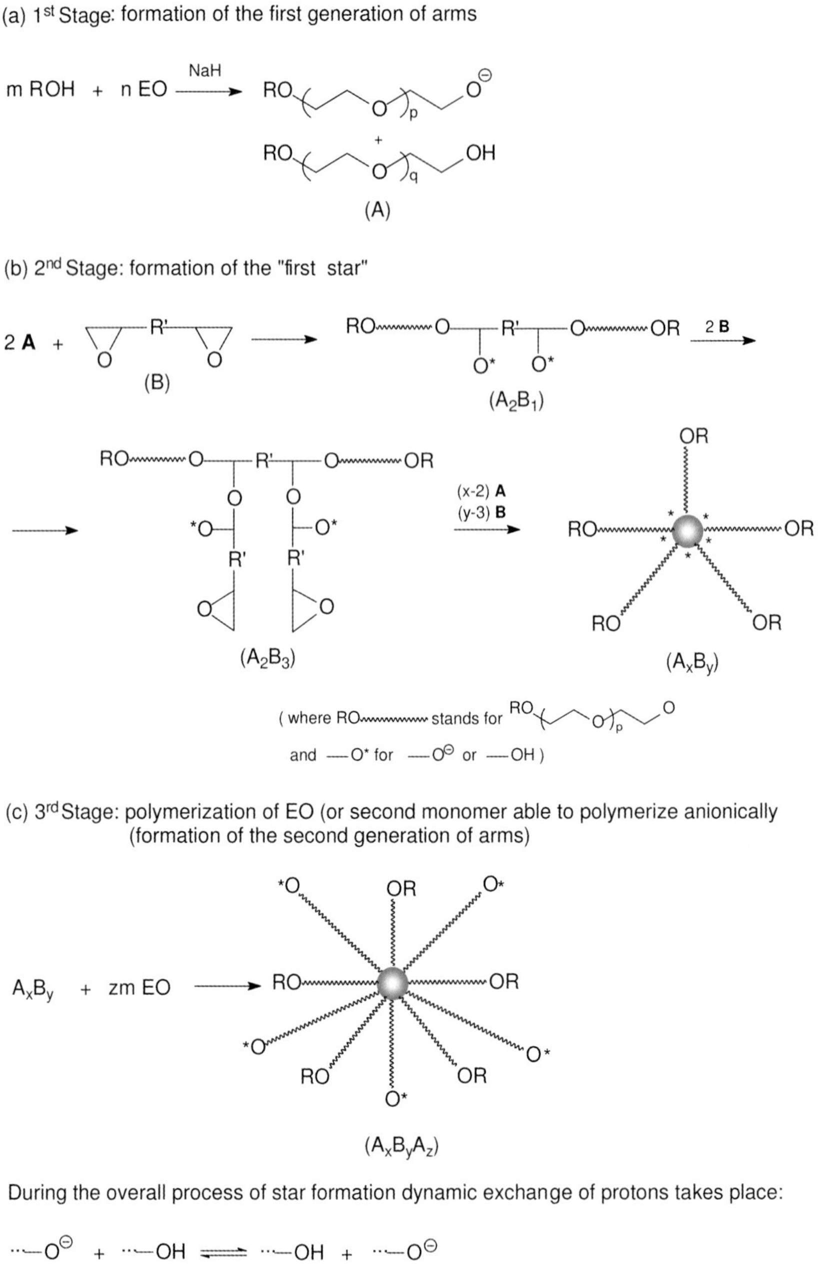

4 Ionic and Coordination Ring-opening Polymerization

(a) 1st Stage: formation of the first generation of arms

$$m\ ROH + n\ EO \xrightarrow{NaH} RO{-}(CH_2CH_2O)_p{-}O^{\ominus} \ + \ RO{-}(CH_2CH_2O)_q{-}OH$$

(A)

(b) 2nd Stage: formation of the "first star"

$$2\ A + (B) \longrightarrow (A_2B_1) \xrightarrow{2\ B} (A_2B_3) \xrightarrow{(x-2)\ A,\ (y-3)\ B} (A_xB_y)$$

(where RO∼∼∼ stands for $RO{-}(CH_2CH_2O)_p{-}O$
and —O* for —O$^{\ominus}$ or —OH)

(c) 3rd Stage: polymerization of EO (or second monomer able to polymerize anionically) (formation of the second generation of arms)

$$A_xB_y + zm\ EO \longrightarrow (A_xB_yA_z)$$

During the overall process of star formation dynamic exchange of protons takes place:

$$\cdots{-}O^{\ominus} + \cdots{-}OH \rightleftharpoons \cdots{-}OH + \cdots{-}O^{\ominus}$$

Scheme 4.13 Synthesis of the star-like branched poly(ethylene oxide).

4.3.4.2 Coordination Polymerization of Cyclic Esters

In a dynamically growing area of biocompatible and biodegradable polymers, the synthesis and characterization of star-shaped aliphatic polyesters are particularly advanced [218–228]. Perhaps the most commonly used method for the synthesis of star-shaped polyesters is based on the "core first" approach in which a multifunctional reagent (a core) plays the role of coinitiator and/or transfer agent. Particularly useful for this purpose is an initiating system employing $Sn(Oct)_2$ (Eq. 7).

$$m \; \text{[cyclic ester]} + n \, R(XH)_p \xrightarrow[80\,°C,\,THF]{Sn(Oct)_2} n \, RX{-}[{-}C(=O){-}O{-}]_{(m/n-1)p}{-}C(=O){-}OH]_p$$

X = NH or O

(7)

The structure of the polyol or polyamine used as a core directly determines the architecture of the polyester macromolecules. The structures shown in Fig. 4.6 give examples of cores applied for the synthesis of branched and star-shaped PCLs and PLAs [98, 229–233].

The M_ns of the resulting polyesters, for all of the monomer–core–$Sn(Oct)_2$ polymerizing systems, were controlled by the ratio of the concentration of the monomer consumed to that of the initiator/transfer agent in the feed, similarly as in the synthesis of linear polyesters. The star shape of the prepared polyester macromolecules has been confirmed by complementary NMR, SEC/RI, SEC/MALLS, LC/CC, double chromatography (SEC/LC/CC), fluorescence spectroscopy and kinetic studies.

Star-like, branched PCLs, with a degradable polyester core based on bis-lactones, were also prepared using an approach similar to that described above for the synthesis of star-like branched polyoxiranes. Two independent papers appeared on this subject recently [234, 235]. However, in both papers the final products still contain unreacted precursors. In the work of Biela and Polanczyk [234] an interesting MALDI-TOF-MS analysis is presented, clearly showing the formation (at a certain stage of reaction) of a population of star-like A_xB_y, where various combinations of $x=1–8$ and $y=1–11$ were observed (where A represents arms and B bis-lactone-derived units). Such data will allow for the first time detailed analysis of star formation processes.

An interesting approach to the synthesis of hyperbranched PLA copolymers has recently been reported by Gottschalk and Frey [236]. A combination of ROP of L,L-lactide catalyzed by $Sn(Oct)_2$ with polycondensation of 2,2-bis(hydroxymethyl)butyric acid as an AB_2 comonomer gave a series of copolymers showing the hydrodynamic volume expressed by the apparent M_ns (SEC with polystyrene standards) in the range from 7×10^2 to 6×10^4 decreasing with the degree of branching.

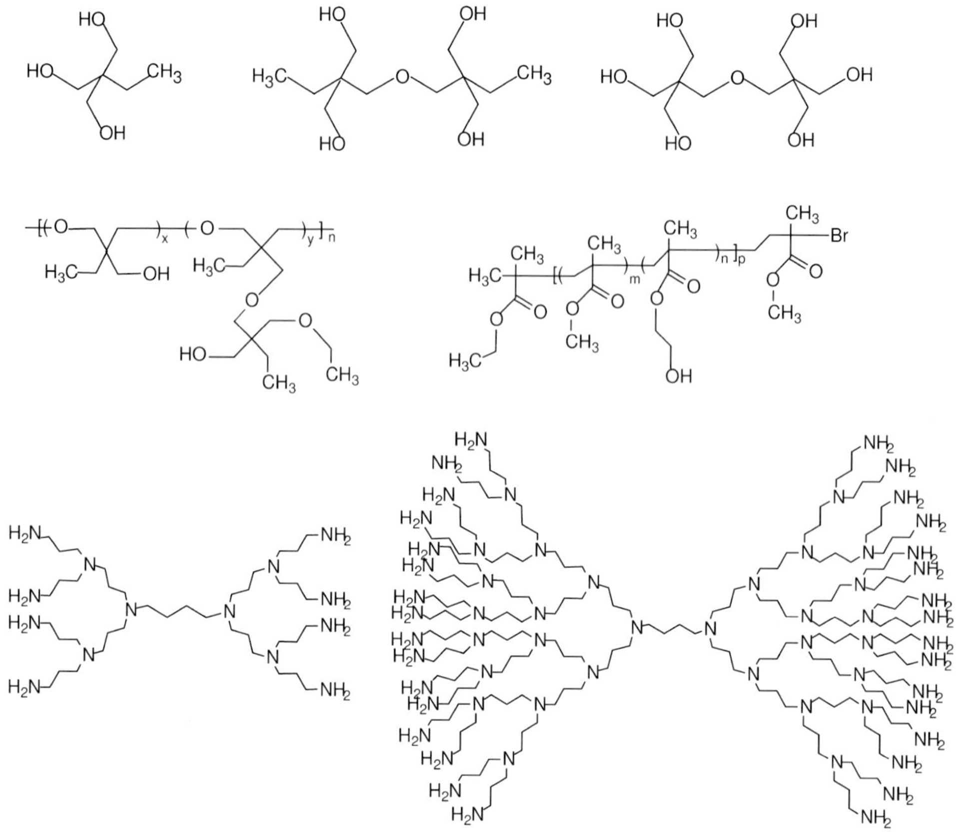

Fig. 4.6 Examples of cores (initiators) employed for star-shaped polyester synthesis [98, 229–233].

4.3.5
Controlled Synthesis of Polyamides by Anionic and Coordination Ring-opening Polymerization

There are two major classes of cyclic monomer polymerization leading to polyamides: lactams and N-carboxyanhydrides of -amino acids (NCAs). The former are employed for the large-scale production of engineering thermoplastic and fiber-forming materials, whereas the latter are mostly used for preparation of polymers mimicking natural proteins.

4.3.5.1 Polymerization of Lactams

ROP of lactams, particularly anionic polymerization, has recently been reviewed by Hashimoto [237]. Previous, comprehensive reviews have been published by the most active researchers in this field, namely Sekiguchi [238] and Sebenda

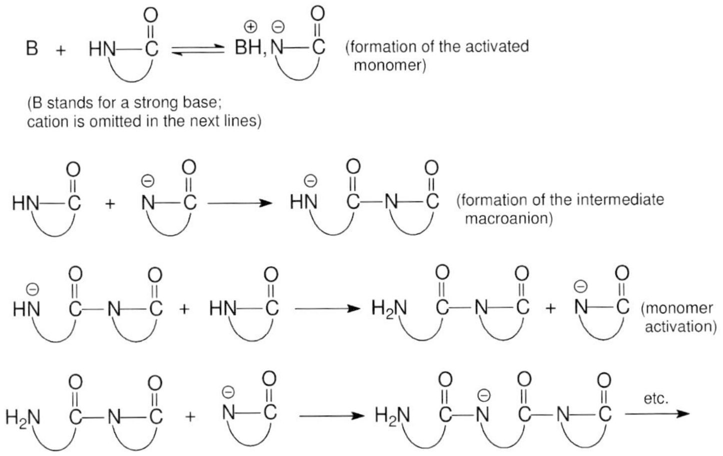

Scheme 4.14 Polymerization of lactams according to the activated monomer mechanism.

[239, 240]. In these reviews, major features of cationic and anionic polymerizations are described. The latter encompasses the active chain end (including hydrolytic polymerization) and activated monomer polymerizations. The expression "activated monomer (AM)" mechanism was coined by Szwarc [241] in order to explain the course of NCA polymerization and has also been adopted for the anionic polymerization of lactams [239]. The AM mechanism in the anionic polymerization of lactams is important, since it allows the fast formation of large objects in bulk. The corresponding mechanism involves several steps, as is concisely shown in Scheme 4.14.

According to the AM mechanism, the anionic active species is located not at the macromolecular growing end but at the deprotonated monomer. The macromolecules bearing anions formed as transient species after addition of the AM are stronger bases than the neutral monomer molecule, therefore the next proton transfer (abstraction) proceeds with the formation again of activated monomer species. Unfortunately, the living polymerization of the unsubstituted lactams, such as seven-membered ε-caprolactam (CLM) or five-membered γ-butyrolactam, have not yet been described. It is difficult to find an appropriate solvent for studying the polymerization of these easily crystallizing polymers. Therefore, the unsymmetrically disubstituted β-butyl-β-methyl-β-propiolactam was used for this purpose [242].

More recent, novel results for lactam polymerization have been described by Loontjens [243]. The bis-lactam [carbonyl-bis-caprolactam (CBC)] introduced by the DSM group appears to be the most efficient among the chain-extending compounds. Major CBC chemistry is outlined in Scheme 4.15.

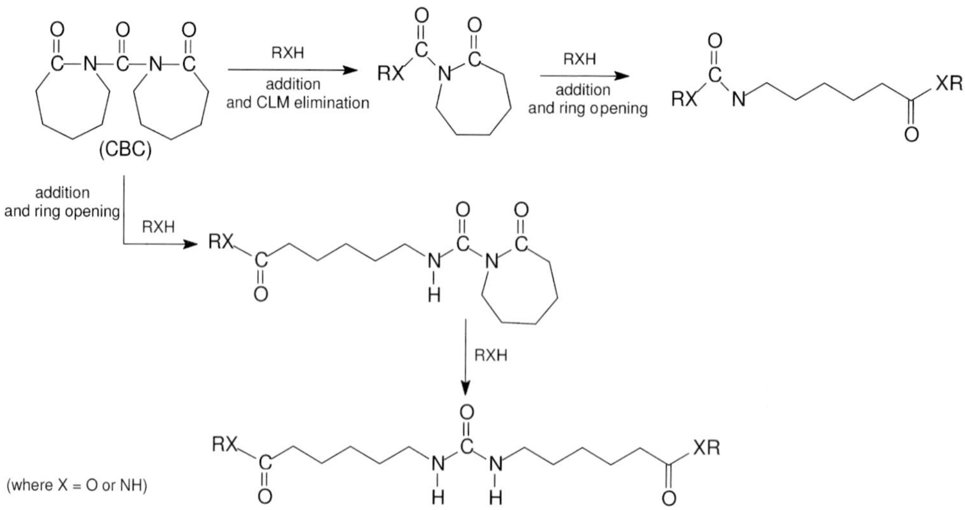

Scheme 4.15 Carbonyl-bis-caprolactam chemistry: reactions with primary amines and alcohols.

(where X = O or NH)

4.3.5.2 Polymerization of N-Carboxyanhydrides of α-Amino Acids (NCAs)

After several decades of discussions on the NCA polymerization mechanism, finally conditions have been found for the controlled polymerization of these monomers. ROP of NCAs is perhaps the most facile and versatile method for the synthesis of model polypeptides (Eq. 8). The availability of more than 200 NCA monomer structures permits the preparation of a vast variety of polypeptide types. Both basic and nucleophilic initiators were able to give high molar mass polymers without detectable racemization at the centers of chirality [244–248].

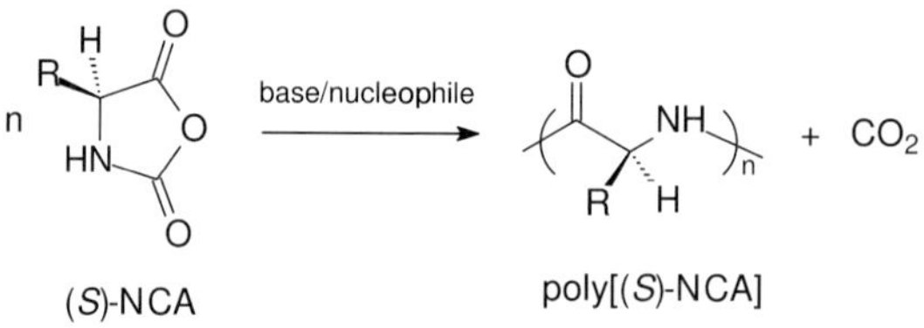

$$(S)\text{-NCA} \xrightarrow{\text{base/nucleophile}} \text{poly}[(S)\text{-NCA}] + CO_2 \tag{8}$$

The synthesis of homopolypeptides by this approach has already been solved. However, the chemical synthesis of copolymers with the required sequence of amino acid residues in the chain is still in its infancy. Here, the pioneering work of Tirrell, using the biotechnological approach, should be noted [249].

In contrast to the considerable progress achieved in the area of the ROP of heterocyclic monomers in general, the possibilities of the controlled polymerization of NCAs have been very limited and remained a challenge since more than 50 years. Chain termination and transfer of various origins precluded molar mass control and led to broad molar mass distributions. In attempts at the synthesis of more complex structures, such as block copolymers, the expected products were contaminated by the undesired homopolymer fraction. This situation was caused by various factors, such as insufficient purity of components of the reacting mixture or interference of different polymerization mechanisms. The latter problem can be illustrated with the example of the competition of amine-activated monomer (AM) mechanisms (Scheme 4.16).

For macromolecular engineering applications, the amine mechanism seems to be the most preferable, since in the absence of side-reactions the degree of polymerization of the resulting polypeptide is strictly controlled by the molar ratio of the NCA consumed to the amine in the feed. The elementary step of propagation in this mechanism is a nucleophilic attack of the primary amine on the carbonyl carbon atom accompanied by acyl–oxygen bond scission and proton transfer, followed finally by CO_2 liberation. Proton abstraction from the NCA molecule leads to the AM pathway and can be considered as a chain transfer to monomer.

Schlaad and coworkers [250–252] have shown recently that the proton abstraction side-reaction might be considerably depressed by initiating the NCA polymerization with a primary amine hydrochloride ($R'NH_3^+Cl^-$) instead of a bare

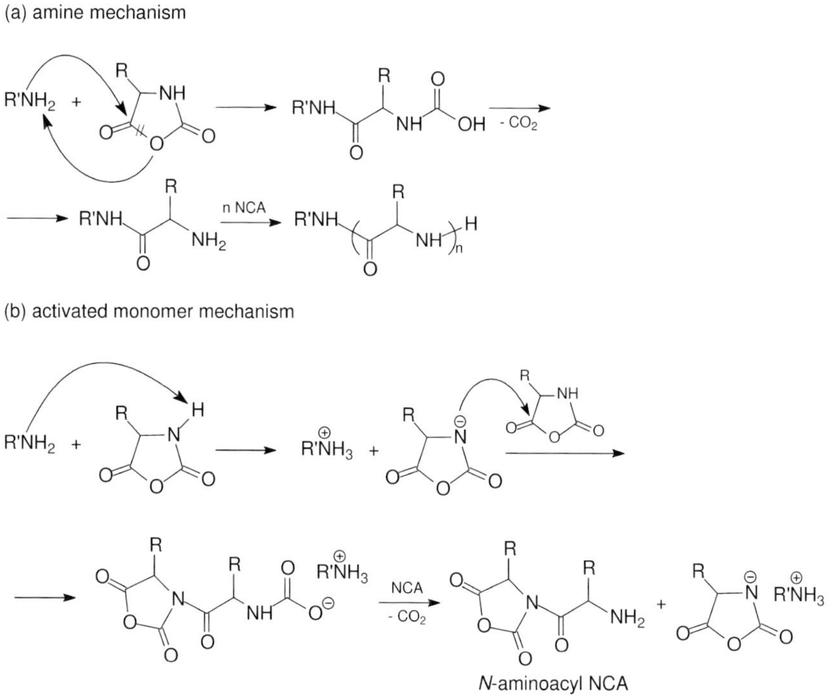

Scheme 4.16 Polymerization of N-carboxyanhydrides of α-amino acids: competition of amine and activated monomer mechanisms.

amine. Most probably R'NH$_3^+$Cl$^-$ plays a double role: as a dormant form of R'NH$_2$ and as a source of protons deactivating the already formed activated monomer (NCA$^-$). Using the amine hydrochloride method, Schlaad and coworkers prepared a number of the well-defined so-called "hybrid block copolymers" or "molecular chimeras" of vinyl polymers and polypeptides with molar mass dispersities as low as $M_w/M_n < 1.03$.

Independently, Hadjichristidis and coworkers [253] applied a high-vacuum technique for the synthesis of NCA and initiator and solvent purification and finally for carrying out polymerizations. Other experimental details are given in [253] and [254]. Finally, they concluded that purity of the reaction mixture is a necessary and sufficient prerequisite for NCA polymerization control, at least for the primary amine-initiated process.

In order to eliminate chain termination and transfer reactions from NCA polymerization, Deming [248] applied coordinate zerovalent nickel or cobalt initiators, instead of primary amine initiators {e.g. bpyNi(COD), where bpy=bipyridine and COD=1,5-cyclooctadiene [255, 256], or (PMe$_3$)$_4$Co [257]}. The M_ns of polypeptides prepared with these Ni and Co zerovalent complexes were predicted by the concentration ratio of the consumed monomer to catalyst in the feed over an impressively wide range of values, from 5×10^2 to 5×10^5. Their molar mass distributions were relatively narrow ($M_w/M_n < 1.2$). Such precise polymerization control also enabled a number of block copolypeptides to be synthesized.

4.3.6
Cationic Ring-opening Polymerization

A larger number of cyclic monomers are polymerized by cationic than by anionic mechanisms. The reason could be understood by comparing the mechanisms in Scheme 4.17.

Thus, in anionic ROP it is the monomer ring that has to be opened whereas in cationic ROP it is the ring in ionic species. The former undoubtedly requires higher energy. The cationic mechanism offers a more general alternative for the polymerization of heterocyclic monomers because some heterocyclic monomers that do not polymerize by the anionic mechanism (such as four- and five-membered cyclic ethers, cyclic acetals and cyclic imines) readily polymerize by the cationic mechanism.

There are several reviews covering the kinetics, mechanisms and synthetic applications of cationic ring-opening polymerization. Although some of those re-

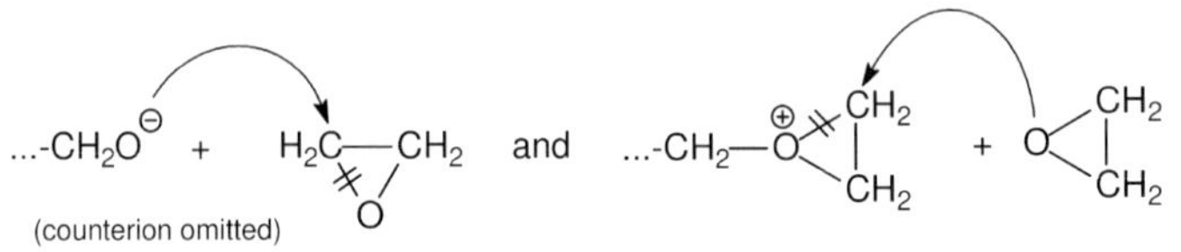

Scheme 4.17 Anionic versus cationic polymerization of oxirane.

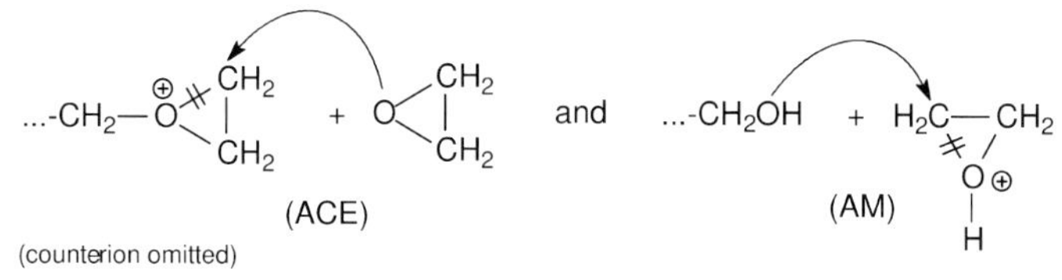

Scheme 4.18 Cationic polymerization of oxirane: active chain end versus activated monomer mechanisms.

views were published more than 20 years ago [5, 8, 258–261], their general picture of the field is still valid in spite of some refinements that have appeared in the more recent literature. A shorter summary has recently been given by Goethals and Verdonck [262]. Our present understanding of cationic ROP is mostly based on extensive studies of the polymerization of tetrahydrofuran (THF) as in anionic ROP, where for the "purely" anionic process the knowledge is built around understanding the polymerization of ethylene oxide and propylene sulfide, in addition to cyclic esters. This is because in the simpler systems, from which additional reactions could be eliminated and the only process left for study is propagation (when initiation is relatively fast), this one, major reaction could be studied in sufficient detail.

There are two different mechanisms by which cationic ROP proceeds: when the ionic active centers are located on the growing chains and propagation involves a non-charged monomer [active chain end (ACE) mechanism] and when the propagation step involves positively charged monomer, adding in this form to the electrically neutral polymer chain [activated monomer (AM) mechanism] (Scheme 4.18).

Both processes belong to S_N2 processes; however, in the ACE mechanism it is a monomer that acts as nucleophile, whereas in the AM mechanism this role is played by the hydroxyl group from a macromolecule. The ACE processes are by far more frequent in cationic ROP. The AM mechanism, however, starts to play an important role in the synthesis of low and medium molar mass reactive oligomers, mostly a,ω-dihydroxytelechelics as blocks for multiblock copolymers. Actually, for reasons explained later, the AM mechanism could hardly lead to high molar mass polymers. The number of monomers is also restricted; as follows from the cationic AM mechanism, the monomer involved should preferably be less nucleophilic than the primary OH groups.

4.3.6.1 Propagation in Cationic Ring-opening Polymerization

Active centers in cationic ROP are onium ions. It has also been assumed that highly stabilized carbenium ions can be present such as oxycarbenium ions ($\ldots-O^+CH_2$) or highly substituted with electron-donating groups [e.g. $\ldots-OCH_2^+C(CH_3)_2$]. Evidence for non-existence of the onium–carbenium ion equilibrium comes, for instance, from the polymerization of THF, from NMR data and from the absence of H^- abstraction that would take place when carbenium ions were present. Nevertheless, and unfortunately, in some textbooks carbenium ions are given as an isomeric form for THF polymerization [263].

Two most important findings in cationic ROP are related to the reactivities of ions and ion pairs, in addition to the observation of the equilibrium between macroions and macroesters, whenever the counterions (anions) are able to form covalent bonds by reacting with their cationic active counterparts. This observation allowed the detailed study of the formation of the less reactive macroesters in a process that was called then (like now for other kinetically similar processes) reversible deactivation of the active species.

The equality of k_p^+ and $k_p^{\pm}$ (and thus reactivities of ions and ion pairs) has been shown for three systems: cyclic sulfides, ethers and imines. For all these systems $k_p^+ \approx k_p^{\pm}$ was observed within experimental error [5, 261]. Further work has shown that $k_p^{\pm}$ does not depend on the anion structure; this result complements the earlier observation of $k_p^+ \approx k_p^{\pm}$. Hence the similar reactivity of ions and ion pairs and the independence of $k_p^{\pm}$ of the structure of anion are the most general features of cationic ROP.

There are two groups of anions (counterions) used in cationic ROP: complex anions, such as AsF_6^-, BF_4^- and SbF_6^-, that are not able to form covalent bonds, since their coordination sphere cannot be further enlarged, and non-complex ions, such as $CH_3SO_3^-$, $CF_3SO_3^-$ and ClO_4^-, that form covalent bonds, using two electrons of the central oxygen atom to form a σ-bond. Therefore, whenever polymerization proceeds with these non-complex anions, reversible ester formation takes place [e.g. for polytetrahydrofuran (PTHF) and $CF_3SO_3^-$] (Eq. 9).

$$...-CH_2OCH_2CH_2CH_2CH_2OSO_2CF_3 \rightleftharpoons ...-CH_2O^{\oplus}\begin{array}{c}CH_2-CH_2\\ |\\ CH_2-CH_2\end{array} \enspace ^{\ominus}OSO_2CF_3 \qquad (9)$$

This particular process was studied extensively, reactivities of covalent and ionic species were determined and the rate constants of interconversion were measured [261, 264]. Polymerization accompanied by formation of esters can be classified as polymerization with reversible deactivation. In such a process, as in the polymerization of THF with triflates or perchlorates, a given macromolecule propagates on its ionic species for a certain time, adding a number of monomer molecules, then the ion pair temporary collapses into the unreactive covalent species. This sequence of events is repeated a number of times during the lifetime of the macromolecule.

4.3.6.2 Chain Transfer to Polymer in Cationic Ring-opening Polymerization

As in anionic or coordination polymerizations, also in the cationic process reaction involving a heteroatom from the chain competes with propagation [5, 258–260, 265]. Intermolecular reaction leads to segmental exchange whereas intramolecular backbiting reaction leads to the formation of cyclic oligomers. The latter is the main factor that limits the synthetic applicability of cationic ROP.

In spite of these phenomena, there are instances known when cationic ROP is applied for manufacturing linear polymers on a technical scale; cationic polymerization of THF leading to polyoxybutylene diols (used, e.g., in polyurethane technology) [266], and cationic copolymerization of 1,3,5-trioxane, leading to engineering plastics, known as polyacetal (polyoxymethylene) [267], are notable examples.

Formation of a cyclic fraction is especially undesired when synthesis of functional polymers (e.g. oligodiols) is attempted, because cyclic macromolecules do not contain functional groups and cannot participate in the chain extension reactions. In the cationic polymerization of THF, essentially linear polymers, free of a cyclic fraction, may be obtained because intramolecular chain transfer to polymer is slow as compared with propagation. Thus, when monomer–polymer equilibrium is reached, the content of the cyclic fraction is still negligible. It has been shown, however, that if the system is kept non-terminated for a longer time, there is a slow build-up of cyclic oligomers [268]. This means that also in cationic polymerization a favorable ratio or rates of propagation and chain transfer to polymer may allow the synthesis of linear polymers, essentially free of a cyclic fraction, when kinetic control is allowed.

Other factors may also limit the formation of a cyclic fraction. In the cationic polymerization of 1,3,5-trioxane, the resulting polymer, being insoluble in common organic solvents, precipitates in the early stages of polymerization and further propagation involves active species immobilized on the surface of growing crystals. The restricted mobility of the polymer chains affects the possibility of intramolecular chain transfer to polymer. Hence in this system, cyclic oligomers cannot be formed by random back-biting. It has been shown, however, that a cyclic fraction is formed in the cationic polymerization of 1,3,5-trioxane (strictly its copolymerization with a few mol% of 1,3-dioxolane or ethylene oxide). The preferred size of the rings formed is, however, governed not by Jacobson–Stockmayer theory [269, 270] but by the size of the crystals, because intramolecular chain transfer to polymer proceeds preferentially at the crystal edges [271].

The significance of cyclization in cationic ROP is best illustrated by the cationic polymerization of three-membered cyclic ethers (oxiranes). Cationic polymerization of oxiranes proceeds smoothly but significant amounts of cyclic oligomers are formed in addition to linear polymer, cyclic dimer in the case of polymerization of EO or cyclic trimers and tetramers in the case of PO. In the polymerization of EO, the content of cyclic dimer (1,4-dioxane) may be as high as 90% [272]. The high tendency of ethylene oxide to form cyclic oligomers in cationic polymerization was ingeniously applied for the synthesis of the crown ethers; polymerization in the presence of alkali metal cations as templates led to preferential formation of well defined cyclic oligomers containing several oxyethylene units in good yields and with size reflecting the ionic radius of the cation template used [273]. Therefore, it may be concluded that reactions involving polymer chain (both intramolecular and intermolecular) are a general feature of cationic ROP.

In order to influence the k_p/k_{tr} ratio (for the particular systems when the relative nucleophilicity of a heteroatom in a monomer molecule and a chain unit depends on the structure and cannot be changed), one would have to devise a reaction mechanism in which those two processes were separated, i.e. other (more reactive) species would participate in propagation and other (less reactive) in chain transfer. Several years ago, we found that such a reaction scheme can indeed be devised if cationic polymerization of cyclic ethers is conducted in the presence of hydroxyl-containing compounds. This development, namely cationic ROP proceeding by an activated monomer mechanism, is described in more detail in a subsequent section.

4.3.6.3 Activated Monomer Mechanism in Cationic Ring-opening Polymerization of Cyclic Ethers and Esters

Cationic polymerization of cyclic ethers in the presence of low molecular weight diols was studied relatively early with the aim of preparing polyether diols [274]. The process was thought to proceed as conventional cationic ring-opening polymerization with diol acting as a chain transfer agent.

Investigating this process in more detail, we observed a striking phenomenon: addition of alcohols to the polymerization of some oxiranes reduced the proportion of cyclics known to be formed by back-biting. Moreover, the lower the instantaneous monomer concentration (i.e. the higher the alcohol to monomer molar ratio), the lower is the proportion of cyclics (see, e.g., [275] and references cited therein).

Back-biting is a unimolecular reaction, therefore any external factor, such as addition of an alcohol, cannot influence its rate and consequently the proportion of cyclics formed. Therefore, we proposed that in the presence of alcohols there should be a mechanism of propagation in which formation of cyclics is either hampered or eliminated until thermodynamic equilibrium is reached. The best candidate for such a mechanism is polymerization in which chain ends in the growing macromolecules are devoid of ions. Then, ions should be located in the monomer molecules. This kind of polymerization could be called "activated monomer (AM)" using the term coined earlier for anionic polymerization of NCAs [241] and lactams [239].

Thus the AM mechanism, in its simplest form, can be described by an equation in which the propagation step proceeds as a nucleophilic attack of the hydroxyl end-group from the macromolecule on the carbon atom, adjacent to the oxonium ion:

$$\cdots\text{O}^{H} + \text{H-O}^{\oplus}\!\!\begin{pmatrix}CH_2\\CH_2\end{pmatrix} \longrightarrow \cdots\text{-O-CH}_2\quad CH_2OH + \text{"H}^{\oplus}\text{"} \tag{10}$$

where "H$^+$" denotes a proton, located on one of the nucleophilic sites in the system, i.e. hydroxyl group, monomer molecule or repeating unit. The AM mechanism may operate effectively, provided that the concentration of the nucleophilic end-groups (reproduced in each propagation step) is high enough to make this route competitive with respect to the active chain end (ACE) mechanism. This may be illustrated by Scheme 4.19, in which both ACE and AM mechanisms can contribute in building of the macromolecules, since both mechanisms can operate simultaneously.

The proof of AM polymerization proceeding was obtained in several ways. The most direct one came from our studies of the polymerization of glycidol (2-hydroxymethyloxirane) (GL) [276], which revealed that the contribution of AM propagation is significant although a contribution of ACE propagation cannot be excluded. In this polymerization, the polymer formed contains both primary and secondary hydroxyl groups. In ACE polymerization, the ring opening should lead to repeating units with pendant CH_2OH groups (Scheme 4.20).

Thus, for both α- and β-ring openings (presumably the α opening should prevail), the polymer backbone is composed of ...–O–C–C–... units with CH_2OH pendant groups. Chain transfer involving hydroxyl groups would not change this structure, because any opening of the tertiary oxonium ion (attack on endocyclic CH_2 or CH groups) would generate a primary hydroxyl group. Secondary hydroxyl groups can, however, be observed only if the AM mechanism operates. Indeed, as shown in Scheme 4.21, in one of the possible modes of attack, namely in α-ring opening, 1,4-units having four atoms (and not three), are formed. The formation of 1,4-units was independently confirmed by analysis of the ^{13}C NMR spectra [277].

The analysis presented above indicates that the concentration of the secondary hydroxyls in the polymer units provides the lowest contribution of the AM mechanism to the chain growth. ^{29}Si NMR spectroscopy was applied after silylation of the polymer and both primary and secondary OH groups were detected [276].

The same reasoning cannot be applied to the cationic polymerization of 3-alkyl-3-hydroxymethyloxetane because the same unit structure is formed independently of the mechanism of propagation and direction of ring opening [278–282]. The two competing propagation pathways in the cationic polymerization

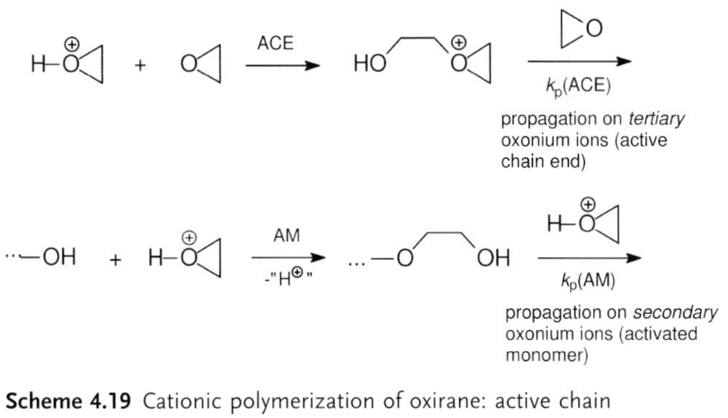

Scheme 4.19 Cationic polymerization of oxirane: active chain end propagation versus activated monomer propagation.

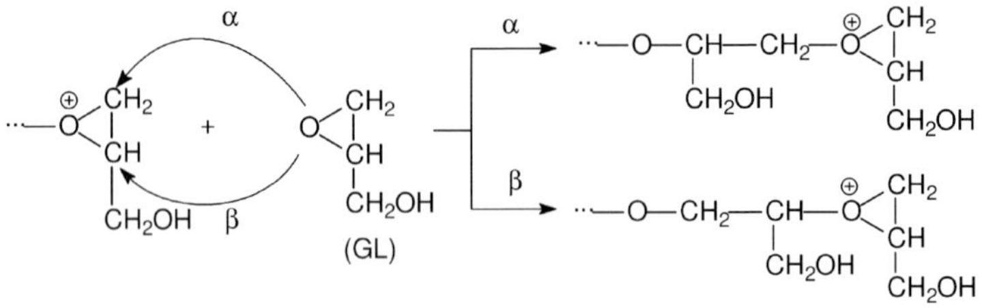

Scheme 4.20 Cationic polymerization of glycidol: propagation according to the active chain end mechanism.

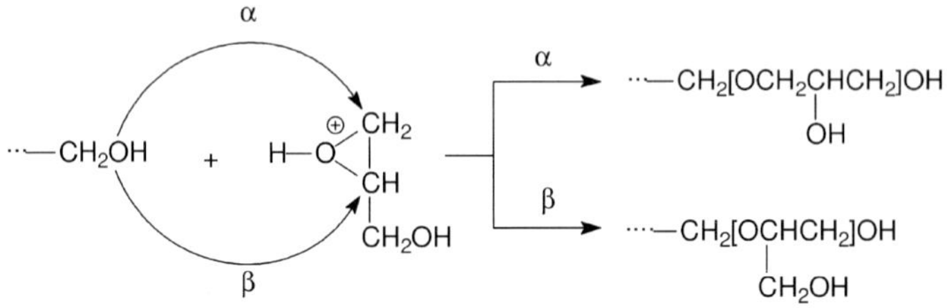

Scheme 4.21 Cationic polymerization of glycidol: propagation according to the activated monomer mechanism.

of 3-ethyl-3-hydroxymethyloxetane (EHMO) can be visualized as shown in Scheme 4.22. Formation of branched units occurs via the AM-type reaction of protonated monomer with any hydroxyl group along the chain.

MALDI-TOF-MS analysis of the product indicated that cyclization cannot be avoided in this process. Cyclization occurs either by intramolecular reaction of tertiary oxonium ion with a hydroxyl group in the same chain (ACE) or by reaction of protonated terminal ring with an OH group (AM). Intramolecular reactions are facilitated by strong intramolecular hydrogen bonding involving OH groups. In contrast to AM polymerization of cyclic ethers in the presence of alcohols (when separate cyclic macromolecules are formed), in this system cycli-

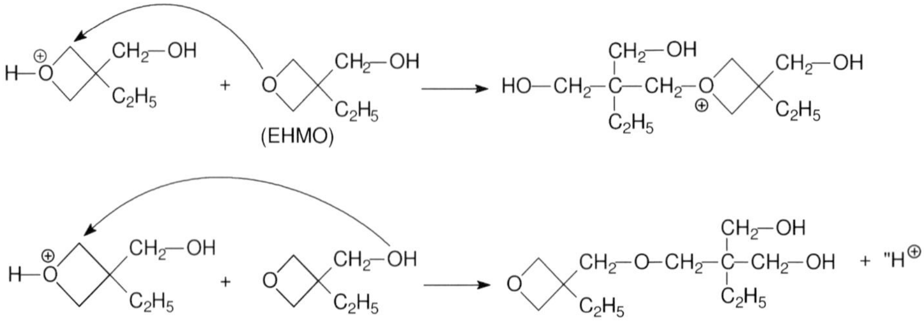

Scheme 4.22 Cationic polymerization of 3-ethyl-3-hydroxymethyloxetane.

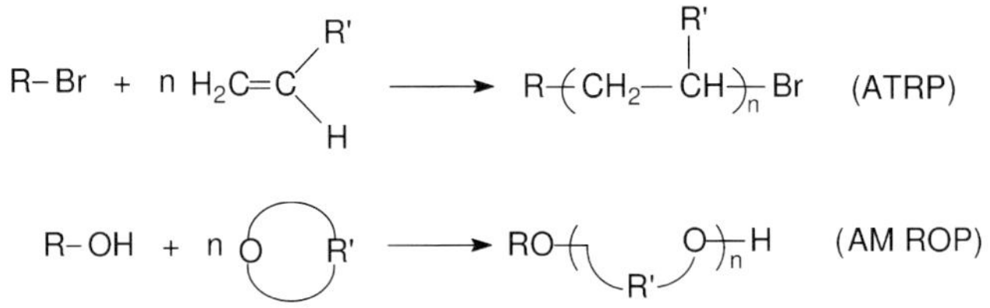

4.3 Basic Mechanistic Features of Ring-opening Polymerization

$$R-Br + n\ H_2C=C\overset{R'}{\underset{H}{\diagdown}} \longrightarrow R-(CH_2-CH(R'))_n-Br \quad (ATRP)$$

$$R-OH + n\ \underset{R'}{\bigtriangleup}O \longrightarrow RO-(CH_2-O)_n-H \quad (AM\ ROP)$$

Scheme 4.23 Controlled polymerization processes: comparison of atom transfer radical polymerization with activated monomer ring-opening polymerization.

zation leads to the formation of the ring which is part of a macromolecule. Because of this process, the molar masses of the products are limited (M_n in the range of a few thousands).

AM polymerization bears a formal resemblance to some processes of controlled polymerization in the sense that both end-groups of the macromolecule that is formed are derived from the initiator (Scheme 4.23). In addition, degrees of polymerization are governed by the monomer to initiator molar ratio and the molar mass distributions are relatively narrow.

Therefore, this type of polymerization is especially well suited for the synthesis of telechelic polymers: oligodiols (with diols as initiators) or macromonomers (with, e.g., $CH_2=CHCOOCH_2CH_2OH$ as an initiator). Such products were successfully prepared by AM polymerization of PO [283], epichlorohydrin [284], functionalized oxiranes such as 1,2-epoxy-3-nitropropane [285], carbazoyl-substituted oxiranes [286] and functionalized oxetanes [287] or copolymerization of EO with THF [288, 289]. The contribution of the AM mechanism may also have a beneficial effect in other systems in which cationic polymerization of oxiranes is employed. It has been found that the rate of photopolymerization of, for instance, epoxidized monomers increases in the presence of alcohols [290].

Lactones, lactides and cyclic carbonates constitute a class of monomers that has recently been extensively studied. Because polyester segments are used as soft blocks in multiblock copolymers (e.g. polyurethanes), there is still a search for convenient methods of synthesis of telechelic polyester diols. Moreover, polymerization by the AM mechanism is interesting since it could provide metal-free polyesters for systematically developed biomedical applications of these polymers. Following investigations of the mechanism of cationic AM polymerization of ethers [276, 283, 284] and lactones [291, 292], a series of reports have appeared in recent years describing the synthesis of telechelic polyesters using this approach. Monomers that were investigated include lactones [293–296], cyclic carbonates [297, 298] and, more recently, L,L-lactide [299].

4.3.6.4 Branched and Star-shaped Polymers Prepared by Cationic Ring-opening Polymerization

One of the first syntheses of three-armed macromolecules was based on the cationic polymerization of THF initiated with the corresponding initiator having three reactive sites (Scheme 4.24) [300].

The "core first" method was applied. The carboxonium initiator started the polymerization of three PTHF chains. It was relatively easy to determine the number of chains (efficiency of initiation), since the core of the star formed was attached to its arms by the relatively labile ester bonds. First, according to NMR, it was established that no oxocations were left, then the molar mass of the star was measured. In the next step, the polymer was hydrolyzed (the ether bonds in arms remained intact) and the molar mass of branches removed from the core was measured again. The degree of polymerization measured agreed with that calculated, taking into account the starting concentration of the initiator, the expected number of arms (equal to the number of active sites) and the starting concentration of the monomer. A similar approach was recently used for the synthesis of θ-shaped polymers. Thus, three-armed PTHF living polymers were coupled with trimesate counterion, giving the desired structure with respectable yield [301].

In another recent example of star-shaped PTHF synthesis, the "arm first" method was employed [302]. Polymers consisting of polypropylenimine dendrimer cores (Astramols®) and linear PTHF were obtained by grafting living cationic PTHF on to the dendrimer (as examples the imine dendrimer structures of third and fifth generation are shown in Fig. 4.6). For PTHF of $M_n \leq 2 \times 10^3$, all primary amino functions of the dendrimers up to 32 primary amino functions could be grafted.

Activated monomer polymerization, starting from polyols of various structures, appears to be a facile method for star-shaped polyester synthesis. For example, Endo and coworkers [303] and then Hult and coworkers [304] applied fumaric acid as an activator in the polymerization of CL and DTC, respectively. As polyols, neopentylene glycol, trimethylolpropane, pentaerythritol and hyperbranched, hydroxyl-terminated polyester (Boltorn H30®) were used.

Understanding of the basic principles of cationic AM polymerization of cyclic ethers stimulated interest in the behavior of monomers that contain both functions, that is, a cyclic ether function and a hydroxyl group in the same molecule

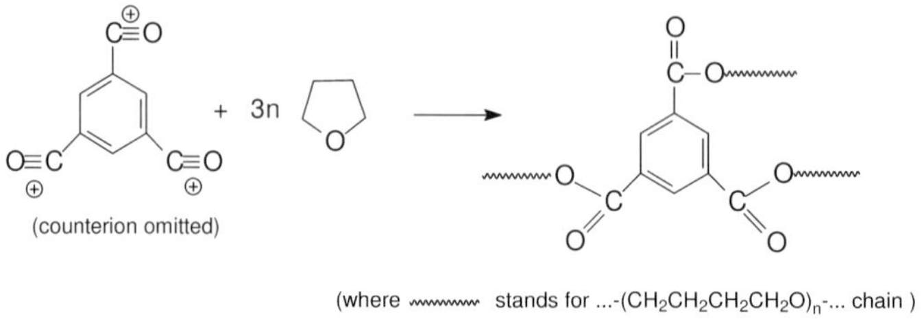

Scheme 4.24 Synthesis of three-armed polytetrahydrofuran.

(e.g. GL or EHMO). As discussed in the preceding section, this kind of polymerization inevitably leads to branched polymer formation. One of the possible structures of poly(EHMO) (PEHMO) prepared in our laboratory is illustrated here [278, 279].

(PEHMO)

Such compact multihydroxy-bearing macromolecules have been used as a core for star polymers, either directly, for example in the coordination polymerization of cyclic esters [230], or after facile transformation of OH groups into groups initiating the ATRP process [305, 306].

4.3.7
Cationic Polymerization of Cyclic Imino Ethers (Oxazolines)

Polymers of imino ethers, mostly 2-oxazolines, have been developed by Seagusa, Kobayashi and coworkers [307]. Recently, a comprehensive paper, in the form of a Highlight, was published by Kobayashi and Uyama [308]. In this paper, not only are polymerization processes discussed, but also numerous applications of polyoxazolines and related polymers are presented. Therefore, in the present chapter on ROP, only the major phenomena are described.

In the polymerization of oxazolines, as in polymerization of THF, both ionic and covalent species could be directly observed by NMR, provided that the anion is not able to form covalent bonds (e.g. BF_4^-). In the polymerization of THF, it has been conclusively shown that propagation on covalent species (e.g. esters, such as triflates or perchlorates) hardly contributes to the chain growth, even if these species prevail in the ester–ion equilibria. In the case of oxazoline, this problem can be visualized as in Scheme 4.25.

Actually, the covalent species could either react with monomer "directly", without an ionic intermediate (this was ruled out in the polymerization of THF), or simply alkylate the next monomer molecule, producing the oxazolinium cation.

More recently, a paper was published giving the quantitative analysis of this situation for at least one system, in which it was shown that ionization proceeds mostly by reaction according to Scheme 4.25 (i.e. unimolecularly) and not by the bimolecular reaction, namely alkylation of the monomer by covalent species [264].

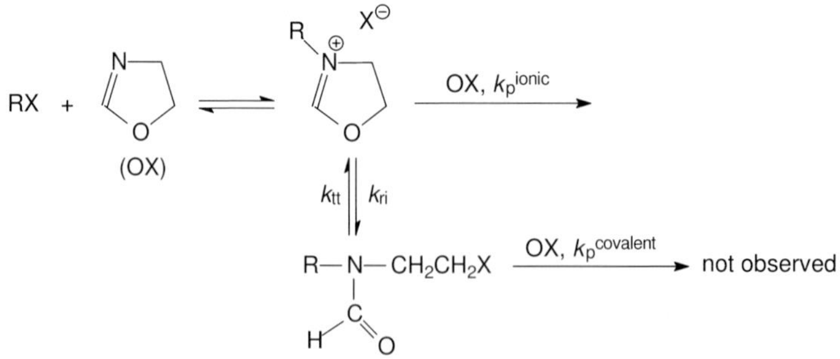

Scheme 4.25 Ester–ion equilibria in polymerization of oxazolines.

The ratio k_{ri}/k_p^c [M] has been measured for the polymerization of 2-methyl-2-oxazoline with Br⁻ counterion (thus X = Br in Scheme 4.25). The monomer concentration at which the rate of intermolecular reinitiation (i.e. resembling covalent growth) would be equal to the rate of intramolecular reinitiation (i.e. giving back ionic growing species) is $\sim 10^3$ mol L^{-1} and exceeds, by almost two orders of magnitude, the monomer concentration in bulk. Thus, ionic centers (formed by unimolecular ionization) are almost exclusively responsible for the chain growth, even if their concentration is only a few percent of the sum of covalent and ionic species.

In the above-cited Highlight, in addition to the analysis of the polymerization process itself, descriptions of block copolymers, macromonomers and telechelics and of major applications of the resulting polymers are presented. The 98 references cited make this Highlight the most authoritative review of the present state of art in this field. On the other hand, the paper by Dworak [264] clarified in more detail the contribution of elementary reactions to the chain growth.

4.4
Dispersion Ring-opening Polymerization

All of the ROP processes described in the preceding sections [with the exception of (co)polymerization of 1,3,5-trioxane] have been studied in homogeneous systems. In this section, we give a short account of ROP conducted in heterogeneous media. This particular process could not compete economically with common industrial processes when monomer taken in bulk is converted into molten polymer (e.g. PEO, PLA, PTHF). However, when the resulting polymer is needed in the form of, for example, microspheres, then dispersion polymerization is a method of choice.

Dispersion polymerization is a polymerization process in which monomer(s), initiator(s) and colloid stabilizer(s) are dissolved in a solvent, forming initially a homogeneous system that produces polymer and results in the formation of polymer particles. There is a large literature on radical dispersion polymerization of vinyl monomers (mostly in water), but only a few reports on dispersion ROP of heterocyclic monomers are available. The first information on dispersion ROP appeared in 1968 [309]. The authors investigated cationic copolymerization of 1,3,5-trioxane and 1,3-dioxolane initiated with $BF_3 \cdot O(C_4H_9)_2$ carried on in heptane in the presence of PEO. This system yielded product in the form of microparticles with number-average diameter ranging from 8 to 12 µm, depending on the concentration of PEO that functioned as a suspension stabilizer and, at the same time, gave block copolymer with polyacetal macromolecules.

A few years later, in 1972, Union Carbide patented the direct synthesis of polyester particles by dispersion polymerization of cyclic esters [310]. Particles were obtained from CL, methyl-ε-caprolactone and δ-valerolactone during polymerization carried out in aliphatic hydrocarbons in the presence of surfactants (vinyl polymers with long aliphatic side-chains). Bu_2Zn and $^iBu_3Al$ were used as initiators. Polymerizations were carried on at temperatures usually exceeding 80 °C, since only under these conditions were the initial solutions of monomers and surface-active agents homogeneous. Diameters of the particles obtained were in the range 1–250 µm, depending on the polymerization conditions.

Teyssié and coworkers noticed that coordination polymerization of GL initiated with ω-Al-alkoxide–PCL macroinitiator and carried out in THF solvent leads to the formation of nanoparticles. In this process, the THF-soluble PCL blocks provided the needed stabilization of the nanoparticles formed [311].

In 1994, the first paper devoted to systematic studies of dispersion ROP of CL and LA was published [312]. Polymerizations were initiated with Et_2AlOEt_2 and carried on in 1,4-dioxane–heptane mixtures (1:4 and 1:9 v/v for LA and CL polymerizations, respectively). Poly(dodecyl acrylate)-g-PCL was used as a particle stabilizer. Polymerizations yielded PCL particles with number-average diameter (D_n) and diameter dispersity factor (D_w/D_n) of 0.63 µm and 1.038, respectively. In a similar system, poly[(R,S)-lactide] microspheres with $D_n = 2.50$ µm and $D_w/D_n = 1.15$ were prepared. Further optimization of the structure of the stabilizer allowed the synthesis of PLA particles with a narrow diameter distribution. $D_w/D_n = 1.03$ was obtained using poly(dodecyl acrylate)-g-PCL with a ratio of M_n of PCL graft to M_n of the poly(dodecyl acrylate)-g-PCL copolymer of 0.25 [313].

Partition of polymerization-active centers between solution and the solid phase (particles) during the initial period of polymerization was monitored [314]. It was established that the concentration of particles does not change after the initial period of about 15% monomer conversion. Then it was also observed that major part (>90%) of monomer conversion occurs when propagation-active centers are localized within particles [315, 316]. On the basis of these observations, the following mechanism of dispersion polymerization of cyclic esters was proposed [317]. All growing chains are initiated at the beginning, during a very short period. After reaching the critical length, they undergo coil-to-globule transition and aggregate into nuclei of microspheres. Once nuclei of microspheres have been formed in a large number, chains from solution are quickly adsorbed on their surface. Thus, all species suitable for reaction with monomer disappear from the liquid phase at the very early stage of polymerization and subsequent propagation proceeds within microspheres swollen with a monomer. Hence, after nucleation, all of the propagation centers are located in microspheres and further polymerization proceeds inside these particles. Propagation stops when the equilibrium monomer concentration is reached within the microspheres. In the absence of particle coalescence, all microspheres formed grow in the same way and their number is constant. The resulting diameter distribution of particles is narrow. Figure 4.7 illustrates mechanism of particle formation and growth.

Measurements of the average weight of the poly[(S)-lactide] microspheres and molar mass of the parent polymer molecules allowed the determination of the average number of polymer chains in one particle [318]. It was found that the average number of chains per microsphere is essentially independent of the concentration of initiator (2,2-dibutyl-2-stanna-1,3-dioxepane) in the range from 8×10^{-4} to 1.0×10^{-2} mol L^{-1} and is equal to 1.8×10^{8}. Later, it was found that the observed independence can be described by a model with the rate of particle nu-

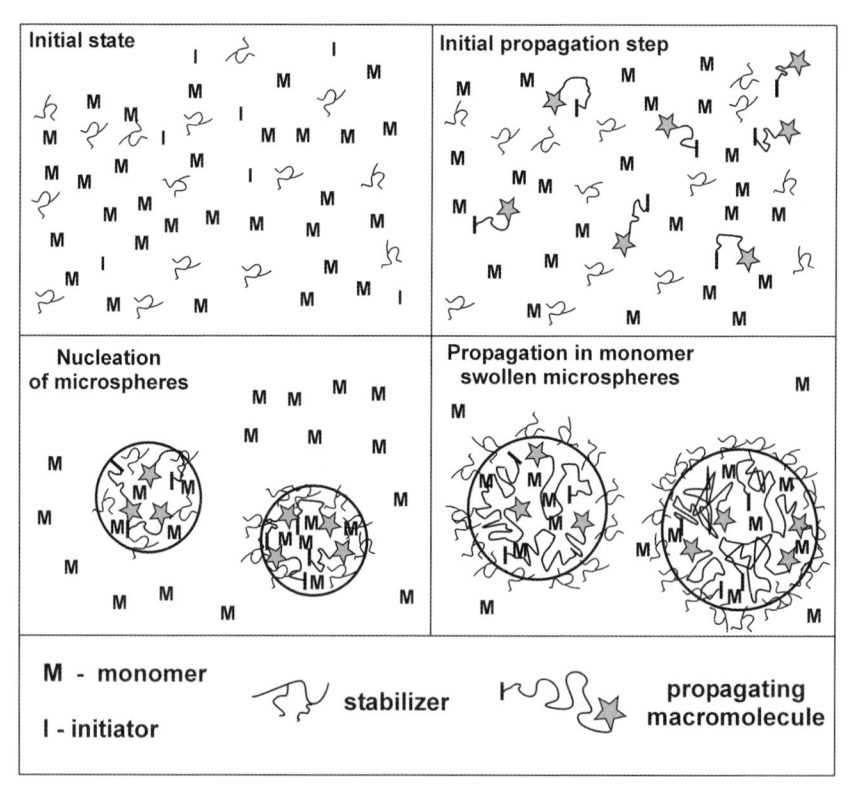

Fig. 4.7 Schematic illustration of particle nucleation and growth in ionic or coordination dispersion ring-opening polymerization.

cleation being of second order in polymer chains with a critical length (chain length at which aggregation begins). The second criterion requires proportionality of the rate of chain adsorption on the already formed particles to the product of concentration of chains in solution, concentration of particles and the average surface area of a particle [319].

The above mechanism of dispersion polymerization of LA and CL has important practical consequences. Since shortly after the beginning of polymerization all propagating species are located inside microspheres and the number of particles does not change during polymerization, the addition of a new monomer portion should lead to an increase in the number-average volume of particles (ΔV_n) according to

$$\Delta V_n = \frac{N_A}{dN} \Delta[M] \tag{11}$$

where N_A, d, N and $\Delta[M]$ denote Avogadro's number, polymer density, concentration of particles (i.e. number of particles per unit volume) and concentration of added monomer converted into polymer, respectively. This relation was observed experimentally [320], confirming the possibility of controlling the diameter of microspheres by gradual addition of the amount of monomer needed (the whole amount of monomer cannot be added at once due to its limited solubility in the polymerization medium).

In model kinetic studies of CL polymerization, it was established that up to $[CL] \approx 2.5 \times 10^{-1}$ mol L^{-1} the concentration of CL in the particles is approximately seven times higher than that in the liquid phase, composed of 1,3-dioxane–heptane (1:9 v/v) [316]. The kinetics of dispersion ROP are discussed in more detail elsewhere [315, 316].

4.5
Conclusion

As stated in the Introduction, in preparing this chapter we had to face obvious restrictions due to the limited space available. Therefore, we (and other chapter authors) had to make a certain subjective choice. Thus, we have deliberately not covered ROP of cyclic siloxanes, silanes and related compounds. A few competent reviews appeared recently [321–324] and a further review is due shortly [325]. We also decided not to cover ROP of phosphorus compounds. Polymerization of cyclic phosphazenes has been reviewed by Allcock, a pioneer of the field [326, 327], and by De Jaeger and Gleria [328].

We took a similar decision regarding the ROP of cyclic esters of phosphoric acid. The corresponding processes were developed in our Center some time ago and summarized several times [329–331]. Since then, there has been not too much progress in ROP itself, although polyesters of phosphoric acid have become popular in biomedicine, as gene carriers, in tissue engineering, drug delivery, etc. (see, e.g., [332–334] and references cited therein). These topics, however, although no less important than ROP itself, are outside of the scope of the present review, in which mostly discussed are the general phenomena and recent results in these areas, which in our judgment are the most active at present.

References

1 G. Odian, *Principles of Polymerization*, 4th edn, Wiley, Hoboken, NJ, **2004**, p. 544.
2 S. Penczek, *J. Polym. Sci., Part A: Polym. Chem.* **2002**, *40*, 1665.
3 K.C. Frish, S.L. Reegen (Eds.), *Ring-Opening Polymerization*, Marcel Dekker, New York, **1969**.
4 K.J. Ivin, T. Saegusa (Eds.), *Ring-Opening Polymerization*, Elsevier Applied Science, London, **1984**.
5 S. Penczek, P. Kubisa, K. Matyjaszewski, *Adv. Polym. Sci.* **1980**, *37*, 1; **1985**, *68/69*, 1.
6 Chain polymerization, Part I, in *Comprehensive Polymer Science. The Synthesis, Characterization, Reactions and Applications of Polymers*, Vol. 3, G.C. Eastmond, A. Ledwith, S. Russo, P. Sigwalt (Eds.), Pergamon Press, Oxford, **1989**, pp. 457–569 (Chapters 31–37), 711–859 (Chapters 54–63).
7 S. Penczek (Ed.), *Models of Biopolymers by Ring-Opening Polymerization*, CRC Press, Boca Raton, FL, **1990**.
8 S. Penczek, P. Kubisa, Cationic ring-opening polymerization, in *Ring-Opening Polymerization. Mechanisms, Catalysis, Structure, Utility*, D.J. Brunelle (Ed.), Hanser, Munich, **1993**, p. 13.
9 S. Slomkowski, A. Duda, Anionic ring-opening polymerization, in *Ring-Opening Polymerization. Mechanisms, Catalysis, Structure, Utility*, D.J. Brunelle (Ed.), Hanser, Munich, **1993**, p. 87.
10 A. Duda, S. Penczek, Mechanisms of aliphatic polyester formation, in *Biopolymers*, Vol. 3b, *Polyesters II. Properties and Chemical Synthesis*, A. Steinbüchel, Y. Doi (Eds.), Wiley-VCH, Weinheim, **2002**, p. 371.
11 H. Sawada, *Thermodynamics of Polymerization*, Marcel Dekker, New York, **1976**, p. 48.
12 M.P. Stevens, *Polymer Chemistry*, Addison-Wesley, Reading, MA, **1975**, p. 265.
13 M. Kucera, *Mechanism and Kinetics of Addition Polymerization*, Academia, Prague/Elsevier Science, Amsterdam, **1992**, p. 35.
14 D.B. Johns, R.W. Lenz, A. Luecke, Lactones, in *Ring-Opening Polymerization*, K.J. Ivin, T. Saegusa (Eds.), Elsevier Applied Science, London, **1984**, Vol. 1, p. 464.
15 R.B. Seymour, Ch.E. Carraher, Jr., *Polymer Chemistry. An Introduction*, 3rd edn, Marcel Dekker, New York, **1992**, p. 261.
16 H.R. Allcock, F.W. Lampe, J.E. Mark, *Contemporary Polymer Chemistry*, 3rd edn, Pearson Education, Upper Saddle River, NJ, **2003**, p. 155.
17 A. Duda, S. Penczek, Ph. Dubois, D. Mecerreyes, R. Jerome, *Macromol. Chem. Phys.* **1996**, *197*, 1273.
18 A. Duda, T. Biela, J. Libiszowski, S. Penczek, Ph. Dubois, D. Mecerreyes, R. Jerome, *Polym. Degrad. Stab.* **1998**, *59*, 215.
19 A. Duda, S. Penczek, *ACS Symp. Ser.* **2000**, *764*, 160.
20 L. Ubaghs, M. Waringo, H. Keul, H. Höcker, *Macromolecules* **2004**, *37*, 6755.
21 S. Agarwal, X. Xie, *Macromolecules* **2003**, *36*, 3545.
22 G.G. Lowry, *J. Polym. Sci.* **1960**, *42*, 463.
23 J. Mosnacek, A. Duda, J. Libiszowski, S. Penczek, *Macromolecules* **2005**, *38*, 2027.
24 S. Penczek, R. Slazak, A. Duda, *Nature* **1978**, *273*, 738.
25 A. Duda, S. Penczek, *Macromolecules* **1982**, *15*, 36.
26 D. Mecerreyes, R. Jerome, Ph. Dubois, *Adv. Polym. Sci.* **1999**, *147*, 1.
27 B.J. O'Keefe, M.A. Hillmyer, W.B. Tolman, *J. Chem. Soc., Dalton Trans.* **2001**, 2215.

28 S. Sosnowski, A. Duda, S. Slomkowski, S. Penczek, *Makromol. Chem., Rapid Commun.* **1984**, *5*, 551.
29 S. Boileau, Anionic Ring-opening polymerization: epoxides and episulfides, in *Comprehensive Polymer Science. The Synthesis, Characterization, Reactions and Applications of Polymers*, Vol. 3, G. C. Eastmond, A. Ledwith, S. Russo, P. Sigwalt (Eds.), Pergamon Press, Oxford, **1989**, p. 467.
30 T. Aida, *Prog. Polym. Sci.* **1994**, *19*, 469.
31 K. S. Kazanskii, A. A. Solovyanov, S. G. Entelis, *Eur. Polym. J.* **1971**, *7*, 1421.
32 A. Duda, S. Penczek, *Macromolecules* **1994**, *27*, 4867.
33 P. J. Flory, *Principles of Polymer Chemistry*, Cornell University Press, Ithaca, NY, **1953**, p. 336.
34 B. Wojtech, *Makromol. Chem.* **1963**, *66*, 180.
35 K. S. Kazanskii, G. Lapienis, V. I. Kuznetsova, L. K. Pakhomova, V. V. Evreinov, S. Penczek, *Vysokomol. Soedin. Ser. A* **2000**, *42*, 915.
36 *Tailored Polymers*, Polymer Source, Dorval, Quebec, **2006**.
37 Shearwater Polymers, *Polyethylene Glycol Derivatives Catalog*, Shearwater Polymers, Huntsville, AL, **2006**.
38 B. Eisswein, M. Moller, *Angew. Chem. Int. Ed. Engl.* **1996**, *35*, 623.
39 B. Eisswein, N. M. Steidl, M. Moller, *Macromol. Rapid Commun.* **1996**, *17*, 143.
40 H. Schlaad, H. Kukula, J. Rudloff, I. Below, *Macromolecules* **2001**, *34*, 4302.
41 Y. K. Choi, W. M. Stevels, M. J. K. Ankone, P. J. Dijkstra, S. W. Kim, J. Feijen, *Macromol. Chem. Phys.* **1996**, *197*, 3623.
42 G.-E. Yu, F. Heatley, C. Booth, T. G. Blease, *J. Polym. Sci., Part A: Polym. Chem.* **1994**, *32*, 1131.
43 R. P. Quirk, G. M. Lizarraga, *Macromol. Chem. Phys.* **2000**, *201*, 1395.
44 Y.-J. Huang, G.-R. Qi, Y.-H. Wang, *J. Polym. Sci., Part A: Polym. Chem.* **2002**, *40*, 1142.
45 C. Billuard, S. Carlotti, P. Desbois, A. Deffieux, *Macromolecules* **2004**, *37*, 4038.
46 W. Braune, J. Okuda, *Angew. Chem. Int. Ed.* **2003**, *42*, 1433.
47 H. Sugimoto, C. Kawamura, M. Kuroki, T. Aida, S. Inoue, *Macromolecules* **1994**, *27*, 2013.
48 M. E. Pruitt, J. M. Baggett (to Dow Chemical), *US Patent 2706182*, **1955**.
49 G. Natta, *J. Polym. Sci.* **1955**, *16*, 143.
50 I. Pasquon, L. Porri, U. Giannini, Stereoregular linear polymers, in *Encyclopedia of Polymer Science and Engineering*, H. Mark, N. M. Bikales, C. G. Overberger, G. Menges, J. I. Kroschwitz (Eds.), Wiley-Interscience, New York, **1989**, Vol. 15, p. 716.
51 Y. Okamoto, T. Nakano, *Chem. Rev.* **1994**, *94*, 349.
52 T. Tsuruta, Anionic ring-opening polymerization: stereospecificity for epoxides, episulfides and lactones, in *Comprehensive Polymer Science. The Synthesis, Characterization, Reactions and Applications of Polymers*, Vol. 3, G. C. Eastmond, A. Ledwith, S. Russo, P. Sigwalt (Eds.), Pergamon Press, Oxford, **1989**, p. 489.
53 K. L. Peretti, H. Ajiro, C. T. Cohen, E. B. Lobkowski, G. W. Coates, *J. Am. Chem. Soc.* **2005**, *127*, 11556.
54 R. D. Lundberg, E. F. Cox, Lactones, in *Ring-Opening Polymerization*, K. C. Frish, S. L. Reegen (Eds.), Marcel Dekker, New York, **1969**, p. 247.
55 R. Jerome, Ph. Teyssié, Anionic ring-opening polymerization: lactones, in *Comprehensive Polymer Science. The Synthesis, Characterization, Reactions and Applications of Polymers*, Vol. 3, G. C. Eastmond, A. Ledwith, S. Russo, P. Sigwalt (Eds.), Pergamon Press, Oxford, **1989**, p. 501.
56 A. Lofgren, A.-Ch. Albertson, Ph. Dubois, R. Jerome, *J. Macromol. Sci. Rev. Macromol. Chem. Phys.* **1995**, *C35*, 379.
57 M. H. Hartmann, High molecular weight polylactic acid polymers, in *Biopolymers from Renewable Resources*, D. L. Kaplan (Ed.), Springer, Berlin, **1998**, p. 367.
58 S. Penczek, T. Biela, A. Duda, *Macromol. Rapid Commun.* **2000**, *21*, 941.
59 S. Penczek, A. Duda, R. Szymanski, T. Biela, *Macromol. Symp.* **2000**, *153*, 1.
60 J. V. Seppala, H. Korhonen, J. Kylma, J. Tuominen, General methodology for chemical synthesis of polyesters, in *Biopolymers, Vol. 3b: Polyesters II. Properties*

and *Chemical Synthesis*, A. Steinbüchel, Y. Doi (Eds.), Wiley-VCH, Weinheim, **2002**, p. 327.
61 A.-Ch. Albertsson, I. K. Varma, *Adv. Polym. Sci.* **2002**, *157*, 1.
62 K. M. Stridsberg, M. Ryner, A.-Ch. Albertsson, *Adv. Polym. Sci.* **2002**, *157*, 41.
63 O. Dechy-Cabaret, B. Martin-Vaca, D. Bourissou, *Chem. Rev.* **2004**, *104*, 6147.
64 A. Duda, Z. Florjanczyk, A. Hofman, S. Slomkowski, S. Penczek, *Macromolecules* **1990**, *23*, 1640.
65 S. Penczek, A. Duda, *Macromol. Symp.* **1996**, *107*, 1.
66 S. Slomkowski, S. Penczek, *Macromolecules* **1976**, *9*, 367.
67 A. Deffieux, S. Boileau, *Macromolecules* **1976**, *9*, 369.
68 S. Slomkowski, S. Penczek, *Macromolecules* **1980**, *13*, 229.
69 S. Slomkowski, *Polymer* **1986**, *27*, 71.
70 H. R. Kricheldorf, N. Scharnagl, *J. Macromol. Sci. Chem.* **1989**, *A26*, 951.
71 A. Duda, *J. Polym. Sci., Part A: Polym. Chem.* **1992**, *30*, 21.
72 P. Kurcok, J. Smiga, Z. Jedlinski, *J. Polym. Sci., Part A: Polym. Chem.* **2002**, *40*, 2184.
73 A. Hofman, S. Slomkowski, S. Penczek, *Makromol. Chem.* **1984**, *185*, 91.
74 Z. Jedlinski, P. Kurcok, R. W. Lenz, *Macromolecules* **1998**, *31*, 6718.
75 A. Hofman, S. Slomkowski, S. Penczek, *Makromol. Chem., Rapid Commun.* **1987**, *8*, 387.
76 S. Penczek, A. Duda, S. Slomkowski, *Makromol. Chem., Macromol. Symp.* **1992**, *54/55*, 31.
77 J. Baran, A. Duda, A. Kowalski, R. Szymanski, S. Penczek, *Macromol. Rapid Commun.* **1997**, *18*, 325.
78 J. Baran, A. Duda, A. Kowalski, R. Szymanski, S. Penczek, *Macromol. Symp.* **1997**, *123*, 93.
79 S. Penczek, A. Duda, R. Szymanski, *Macromol. Symp.* **1998**, *132*, 441.
80 A. Duda, S. Penczek, *Macromol. Rapid Commun.* **1995**, *16*, 67.
81 A. Duda, S. Penczek, *Macromolecules* **1995**, *28*, 5981.
82 A. Duda, *Macromolecules* **1996**, *29*, 1399.
83 A. Kowalski, A. Duda, S. Penczek, *Macromolecules* **1998**, *31*, 2114.
84 A. Duda, S. Penczek, *Makromol. Chem. Macromol. Symp.* **1991**, *47*, 127.
85 T. Aida, S. Inoue, *Acc. Chem. Res.* **1996**, *29*, 39.
86 J. W. Leenslag, A. J. Pennings, *Makromol. Chem.* **1987**, *188*, 1809.
87 X. Zhang, U. P. Wyss, D. Pichora, M. F. A. Goosen, *Polym. Bull.* **1992**, *27*, 623.
88 X. Zhang, D. A. MacDonald, M. F. A. Goosen, K. B. McAuley, *J. Polym. Sci., Part A: Polym. Chem.* **1994**, *32*, 2965.
89 A. J. Nijenhuis, D. W. Grijpma, A. J. Pennings, *Macromolecules* **1992**, *25*, 6419.
90 Y. J. Doi, P. J. Lemstra, A. J. Nijenhuis, H. A. M. van Aert, C. Bastiaansen, *Macromolecules* **1995**, *28*, 2124.
91 H. R. Kricheldorf, I. Kreiser-Saunders, C. Boettcher, *Polymer* **1995**, *36*, 1253.
92 G. Schwach, J. Coudane, R. Engel, M. Vert, *J. Polym. Chem., Part A: Polym. Chem.* **1997**, *35*, 3431.
93 P. J. A. I. Veld, E. M. Velner, P. van de Witte, J. Hamhuis, P. J. Dijkstra, J. Feijen, *J. Polym. Sci., Part A: Polym. Chem.* **1997**, *35*, 219.
94 A. Kowalski, A. Duda, S. Penczek, *Macromolecules* **2000**, *33*, 689.
95 A. Kowalski, A. Duda, S. Penczek, *Macromol. Rapid Commun.* **1998**, *19*, 567.
96 A. Kowalski, A. Duda, S. Penczek, *Macromolecules* **2000**, *33*, 7359.
97 K. Majerska, A. Duda, S. Penczek, *Macromol. Rapid Commun.* **2000**, *21*, 1327.
98 A. Kowalski, J. Libiszowski, T. Biela, M. Cypryk, A. Duda, S. Penczek, *Macromolecules* **2005**, *38*, 8170.
99 F. Nederberg, E. F. Connor, M. Moller, T. Glauser, J. L. Hedrick, *Angew. Chem. Int. Ed.* **2001**, *40*, 2712.
100 F. Nederberg, E. F. Connor, T. Glauser, J. L. Hedrick, *Chem. Commun.* **2001**, 2066.
101 E. F. Connor, G. W. Nyce, M. Myers, A. Mock, J. L. Hedrick, *J. Am. Chem. Soc.* **2002**, *124*, 914.
102 G. W. Nyce, T. Glauser, E. F. Connor, A. Mock, R. M. Waymouth, J. L. Hedrick, *J. Am. Chem. Soc.* **2003**, *125*, 3046.

103 O. Coulombier, A. P. Dove, R. C. Pratt, A. C. Sentman, D. A. Culkin, L. Mespouille, Ph. Dubois, R. M. Waymouth, J. L. Hedrick, *Angew. Chem. Int. Ed.* **2005**, *44*, 4964.
104 O. Coulombier, L. Mespouille, J. L. Hedrick, R. M. Waymouth, Ph. Dubois, *Macromolecules* **2006**, *39*, 4001.
105 J. Liu, L. Liu, *Macromolecules* **2004**, *37*, 2674.
106 A. Duda, S. Penczek, *Macromol. Rapid Commun.* **1994**, *15*, 559.
107 T. Biela, A. Duda, *J. Polym. Sci., Part A: Polym. Chem.* **1996**, *34*, 1807.
108 Ph. Dubois, R. Jerome, Ph. Teyssie, *Polym. Bull.* **1989**, *22*, 475.
109 Ph. Degee, Ph. Dubois, R. Jerome, *Macromol. Chem. Phys.* **1997**, *198*, 1973.
110 A. Kowalski, J. Libiszowski, A. Duda, S. Penczek, *Macromolecules* **2000**, *33*, 1964.
111 R. F. Storey, B. D. Mullen, G. S. Desai, J. W. Sherman, C. N. Tang, *J. Polym. Sci., Part A: Polym. Chem. Ed.* **2002**, *40*, 3434.
112 A. Duda, A. Kowalski, S. Penczek, unpublished results.
113 P. Kurcok, Ph. Dubois, R. Jerome, *Polym. Int.* **1996**, *41*, 479.
114 S. J. McLain, N. E. Drysdale, *Polym. Prepr. Am. Chem. Soc., Div. Polym. Chem.* **1992**, *33*, 174.
115 S. J. McLain, T. M. Ford, N. E. Drysdale, *Polym. Prepr. Am. Chem. Soc., Div. Polym. Chem.* **1992**, *33*, 463.
116 S. J. McLain, T. M. Ford, N. E. Drysdale, J. L. Shreeve, W. J. Evans, *Polym. Prepr. Am. Chem. Soc., Div. Polym. Chem.* **1994**, *35*, 534.
117 S. J. McLain, T. M. Ford, N. E. Drysdale, N. C. Jones, E. F. McCord, *Polym. Prepr. Am. Chem. Soc., Div. Polym. Chem.* **1994**, *35*, 463.
118 A. Le Borgne, C. Pluta, N. Spassky, *Makromol. Chem., Rapid Commun.* **1994**, *15*, 955.
119 V. Simic, N. Spassky, E. G. Hubert-Pfalzgraf, *Macromolecules* **1997**, *30*, 7338.
120 V. Simic, Polymérization et copolymérization d'esters cycliques à l'aide de dérivés de terres rares, PhD Thesis, Université Pierre et Marie Curie Paris VI, **1999**.
121 N. Spassky, V. Simic, M. S. Montaudo, L. G. Hubert-Pfalzgraf, *Macromol. Chem. Phys.* **2000**, *201*, 2432.
122 W. M. Stevels, M. J. K. Ankone, P. J. Dijkstra, J. Feijen, *Macromol. Chem. Phys.* **1995**, *196*, 1153.
123 W. M. Stevels, M. J. K Ankone, P. J. Dijkstra, J. Feijen, *Macromolecules* **1996**, *29*, 3332.
124 W. M. Stevels, M. J. K Ankone, P. J. Dijkstra, J. Feijen, *Macromolecules* **1996**, *29*, 6132.
125 W. M. Stevels, M. J. K. Ankone, P. J. Dijkstra, J. Feijen, *Macromolecules* **1996**, *29*, 8296.
126 Z. Zhong, P. J. Dijkstra, J. Feijen, *Macromol. Chem. Phys.* **2000**, *201*, 1329.
127 M. Save, M. Schappacher, A. Soum, *Macromol. Chem. Phys.* **2002**, *203*, 889.
128 M. Save, A. Soum, *Macromol. Chem. Phys.* **2002**, *203*, 2591.
129 B. M. Chamberlain, M. Cheng, D. R. Moore, T. M. Ovitt, E. B. Lobkovsky, G. W. Coates, *J. Am. Chem. Soc.* **2001**, *123*, 3229.
130 B. J. O'Keefe, L. E. Breyfogle, M. A. Hillmyer, W. B. Tolman, *J. Am. Chem. Soc.* **2002**, *124*, 4384.
131 C. K. Williams, L. E. Breyfogle, S. K. Choi, W. Nam, V. G. Young, Jr., M. A. Hillmyer, W. B. Tolman, *J. Am. Chem. Soc.* **2003**, *125*, 11350.
132 T. Yasuda, T. Aida, S. Inoue, *Bull. Chem. Soc. Jpn.* **1986**, *59*, 3931.
133 M. Endo, T. Aida, S. Inoue, *Macromolecules* **1987**, *20*, 2982.
134 B.-T. Ko, C.-C. Lin, *Macromolecules* **1999**, *32*, 8296.
135 Y.-C. Liu, B.-T. Ko, C.-C. Lin, *Macromolecules* **2001**, *34*, 6196.
136 T.-C. Liao, Y.-L. Huang, B.-H. Huang, C.-C. Lin, *Macromol. Chem. Phys.* **2003**, *204*, 885.
137 H.-Y. Chen, B.-H. Huang, C.-C. Lin, *Macromolecules* **2005**, *38*, 5400.
138 N. Nomura, T. Aoyama, R. Ishii, T. Kondo, *Macromolecules* **2005**, *38*, 5363.
139 S. Matala-Timol, A. Bhaw-Luximon, D. Jhurry, *Macromol. Symp.* **2006**, *231*, 69.

140 H.-Y. Chen, H.-Y. Tang, C.-C. Lin, *Macromolecules* **2006**, *39*, 3745.
141 L. R. Rieth, D. R. Moore, E. B. Lobkovsky, G. W. Coates, *J. Am. Chem. Soc.* **2002**, *124*, 15239.
142 H.-M. Muller, D. Seebach, *Angew. Chem. Int. Ed. Engl.* **1993**, *32*, 477.
143 A. Steinbüchel, Y. Doi (Eds.), *Biopolymers, Vol. 3a, Polyesters I. Biological Systems and Biotechnological Production*, Wiley-VCH, Weinheim, **2002**.
144 K. Sudesh, H. Abe, Y. Doi, *Prog. Polym. Sci.* **2000**, *25*, 1503.
145 M. Allmendinger, R. Eberhardt, G. A. Luinstra, B. Rieger, *J. Am. Chem. Soc.* **2002**, *124*, 5646.
146 M. Allmendinger, R. Eberhardt, G. A. Luinstra, B. Rieger, *Macromol. Chem. Phys.* **2003**, *204*, 564.
147 J. A. R. Schmidt, E. B. Lobkowski, G. W. Coates, *J. Am. Chem. Soc.* **2005**, *127*, 11426.
148 G. B. Kharas, F. Sanchez-Riera, D. K. Severson, Polymers of lactic acid, in *Plastics from Microbes*, D. P. Mobley (Ed.), Hanser, Munich, **1994**, p. 93.
149 P. Gruber, M. O'Brien, Polylactides NatureWorks™ PLA, in *Biopolymers, Vol. 4, Polyesters III. Applications and Commercial Products*, A. Steinbüchel, Y. Doi (Eds.), Wiley-VCH, Weinheim, **2002**, p. 235.
150 R. A. Gross, Y. Zhang, G. Konrad, R. W. Lenz, *Macromolecules* **1988**, *21*, 2657.
151 P. J. Hocking, R. H. Marchessault, *Polym. Bull.* **1993**, *30*, 163.
152 J. E. Kemnitzer, S. P. McCarthy, R. A. Gross, *Macromolecules* **1993**, *26*, 6143.
153 H. R. Kricheldorf, S. R. Lee, N. Scharnagl, *Macromolecules* **1994**, *27*, 3139.
154 K. A. M. Thakur, R. T. Kean, E. S. Hall, J. J. Kolstad, T. A. Lindgren, M. A. Doscotch, J. I. Siepmann, E. J. Munson, *Macromolecules* **1997**, *30*, 2422.
155 A. Le Borne, D. Greiner, R. E. Prud'homme, N. Spassky, P. Sigwalt. *Eur. Polym. J.* **1981**, *17*, 1103.
156 T. Satoh, B. M. Novak, *Polym. Prepr. Am. Chem. Soc., Div. Polym. Chem.* **2000**, *41*, 112.
157 A. Le Borgne, N. Spassky, *Polymer* **1989**, *30*, 2312.
158 A. Sodergard, M. Stolt, *Prog. Polym. Sci.* **2002**, *27*, 1123.
159 C. H. Holten, A. Müller, A. D. Rehbinder, *Lactic Acid*, Chemie, Weinheim, **1971**.
160 K. Hatada, J. Kahovec, M. Baron, K. Horie, T. Kitayama, P. Kubisa, G. P. Moss, R. F. T. Stepto, E. S. Wilkes, *Pure Appl. Chem.* **2002**, *74*, 915.
161 A. Le Borgne, V. Vincens, M. Jouglard, N. Spassky, *Makromol. Chem. Macromol. Symp.* **1993**, *73*, 37.
162 N. Spassky, M. Wisniewski, Ch. Pluta, A. Le Borgne, *Macromol. Chem. Phys.* **1996**, *197*, 2627.
163 M. Wisniewski, A. Le Borgne, N. Spassky, *Macromol. Chem. Phys.* **1997**, *198*, 1227.
164 Ch. P. Radano, G. L. Baker, M. R. Smith, III, *J. Am. Chem. Soc.* **2000**, *122*, 1552.
165 T. M. Ovitt, G. W. Coates, *J. Polym. Sci., Part A: Polym. Chem.* **2000**, *38*, 4686.
166 T. M. Ovitt, G. W. Coates, *J. Am. Chem. Soc.* **2002**, *124*, 1316.
167 Z. Zhong, P. J. Dijkstra, J. Feijen, *Angew. Chem. Int. Ed.* **2002**, *41*, 4510.
168 Z. Zhong, P. J. Dijkstra, J. Feijen, *J. Am. Chem. Soc.* **2003**, *125*, 11291.
169 N. Nomura, R. Ishi, M. Akakura, K. Aoi, *J. Am. Chem. Soc.* **2002**, *124*, 5938.
170 K. Majerska, A. Duda, *J. Am. Chem. Soc.* **2004**, *126*, 1026.
171 M. Bero, P. Dobrzyński, J. Kasperczyk, *J. Polym. Sci., Part A: Polym. Chem.* **1999**, *37*, 4038.
172 T. M. Ovitt, G. W. Coates, *J. Am. Chem. Soc.* **1999**, *121*, 4072.
173 M. Cheng, A. B. Attygale, E. B. Lobkovsky, G. W. Coates, *J. Am. Chem. Soc.* **1999**, *124*, 11583.
174 K. Hatada, T. Kitayama, O. Nakagawa, Stereocomplexes (between enantiomeric and diastereomeric macromolecules), in *Polymeric Materials Encyclopedia*, J. C. Salamone (Ed.), CRC Press, Boca Raton, FL, **1996**, Vol. 10, p. 7950.
175 P. Dumas, N. Spassky, P. Sigwalt, *Makromol. Chem.* **1972**, *156*, 55.
176 D. Greiner, R. E. Prud'homme, *Macromolecules* **1983**, *16*, 302.

177 C. Lavalle, R. E. Prud'homme, *Macromolecules* **1989**, *22*, 2438.
178 Y. Ikada, K. Jamshidi, H. Tsuji, S.-H. Hyon, *Macromolecules* **1987**, *20*, 904.
179 H. Tsuji, F. Hori, S.-H. Hyon, Y. Ikada, *Macromolecules* **1991**, *24*, 2719.
180 H. Tsuji, S.-H. Hyon, Y. Ikada, *Macromolecules* **1991**, *24*, 5651.
181 H. Tsuji, S.-H. Hyon, Y. Ikada, *Macromolecules* **1991**, *24*, 5657.
182 H. Tsuji, Y. Ikada, *Macromolecules* **1993**, *26*, 69186.
183 H. Tsuji, Y. Ikada, *Polymer* **1999**, *40*, 6699.
184 D. Brizzolara, H.-J. Cantow, K. Diederichs, E. Keller, A. J. Domb, *Macromolecules* **1996**, *29*, 191.
185 J. Slager, M. Gladnikoff, A. J. Domb, *Macromol. Symp.* **2001**, *175*, 105.
186 L. Cartier, T. Okihara, B. Lotz, *Macromolecules* **1997**, *30*, 6313.
187 J. Zhang, H. Sato, H. Tsuji, I. Noda, Y. Ozaki, *Macromolecules* **2005**, *38*, 1822.
188 J.-R. Sarasua, N. L. Rodriguez, A. L. Arraiza, E. Meaurio, *Macromolecules* **2005**, *38*, 8362.
189 K. Fukushima, Y. Kimura, *Polym. Int.* **2006**, *55*, 626.
190 L. Li, Z. Zhong, W. H. de Jeu, P. J. Dijkstra, J. Feijen, *Macromolecules* **2004**, *37*, 8641.
191 S. J. de Jong, W. N. E. van Dijk-Wolthuis, J. J. Kettenes-van den Bosch, P. J. W. Shuyl, W. E. Hennink, *Macromolecules* **1998**, *19*, 6397.
192 T. Biela, A. Duda, S. Penczek, *Macromolecules* **2006**, *39*, 3710.
193 G. Rokicki, *Prog. Polym. Sci.* **2000**, *25*, 259.
194 H. Hocker, H. Keul, *Macromol. Symp.* **2001**, *174*, 231.
195 Ch. M. Byrne, S. D. Allen, E. B. Lobkovsky, G. W. Coates, *J. Am. Chem. Soc.* **2004**, *126*, 11404.
196 C. T. Cohen, C. Thu, G. W. Coates, *J. Am. Chem. Soc.* **2005**, *127*, 10869.
197 G. A. Luinstra, G. R. Haas, F. Molnar, V. Bernhart, R. Eberhardt, B. Rieger, *Chem. Eur. J.* **2005**, *11*, 6298.
198 J. Yang, Y. Yu, Q. Li, Y. Li, A. Cao, *J. Polym. Sci., Part A: Polym. Chem.* **2005**, *43*, 373.
199 P. J. Darensbourg, P. Ganguly, D. R. Billodeaux, *Macromolecules* **2005**, *38*, 5406.
200 P. J. Darensbourg, W. Choi, P. Ganguly, C. P. Richers, *Macromolecules* **2006**, *39*, 4374.
201 P. J. Darensbourg, P. Ganguly, W. Choi, *Inorg. Chem.* **2006**, *45*, 3831.
202 L. Vogdanis, B. Martens, H. Uchtman, F. Hentzel, W. Heitz, *Makromol. Chem.* **1990**, *191*, 465.
203 S. Kuhlig, H. Keul, H. Hocker, *Makromol. Chem. Suppl.* **1989**, *15*, 9.
204 S. Inoue, H. Koinuma, T. Tsuruta, *J. Polym. Sci., Part A: Polym. Chem.* **1969**, *7*, 287.
205 D. Taton, M. Saule, J. Logan, R. Duran, S. Hou, E. L. Chaikoff, Y. Gnanou, *J. Polym. Sci., Part A: Polym. Chem.* **2003**, *41*, 1669.
206 R. Francis, D. Taton, J. Logan, P. Masse, Y. Gnanou, R. S. Duran, *Macromolecules* **2003**, *36*, 8253.
207 S. J. Hou, D. Taton, M. Saule, J. Logan, E. L. Chaikoff, Y. Gnanou, *Polymer* **2003**, *44*, 5067.
208 L. M. van Renterghem, X. Feng, D. Taton, Y. Gnanou, F. E. Du Prez, *Macromolecules* **2005**, *38*, 10609.
209 G. Lapienis, S. Penczek, *Macromolecules* **2000**, *33*, 6630.
210 G. Lapienis, S. Penczek, *J. Polym. Sci., Part A: Polym. Chem.* **2004**, *42*, 1576.
211 B. Comanita, B. Noren, J. Roovers, *Macromolecules* **1999**, *32*, 1069.
212 P. Rempp, P. Lutz, E. Franta, *J. Macromol. Sci. Pure Appl. Chem.* **1994**, *31*, 891.
213 P. Rempp, P. Lutz, Star-shaped polymers, in *Polymeric Materials Encyclopedia*, J. C. Salomone (Ed.), CRC Press, Boca Raton, FL, **1996**, p. 7880.
214 Y. Gnanou, P. Lutz, P. Rempp, *Makromol. Chem.* **1988**, *189*, 2885.
215 D. Rein, J. P. Lamps, P. Rempp, P. Lutz, D. Papanagopoulos, C. Tsitsilianis, *Acta Polym.* **1993**, *44*, 225.
216 A. Sunder, R. Hanselmann, H. Frey, R. Mulhaupt, *Macromolecules* **1999**, *32*, 4240.
217 Z. Shen, Y. Chen, E. Barriau, H. Frey, *Macromol. Chem. Phys.* **2006**, *207*, 57.

218 Ch.-M. Dong, K.-Y. Qiu, Zh.-W. Gu, X.-D. Feng, *J. Polym. Sci., Part A: Polym. Chem.* **2001**, *39*, 409.
219 S.-H. Lee, S. H. Kim, Y. K. Han, Y. H. Kim, *J. Polym. Sci., Part A: Polym. Chem.* **2001**, *39*, 973.
220 E. S. Kim, B. Ch. Kim, S. H. Kim, *J. Polym. Sci., Part A: Polym. Chem.* **2004**, *42*, 939.
221 H. Korhonen, A. Helminen, J. V. Seppala, *Polymer* **2001**, *42*, 7541.
222 A. Finne, A.-Ch. Albertsson, *Biomacromolecules* **2002**, *3*, 684.
223 Y. L. Zhao, Q. Cai, J. Jiang, X. T. Shuai, J. Z. Bei, C. F. Chen, F. Xi, *Polymer* **2002**, *43*, 5819.
224 Q. Cai, Y. Zhao, J. Z. Bei, F. Xi, Sh. Wang, *Biomacromolecules* **2003**, *4*, 828.
225 H.-A. Klok, S. Becker, F. Schuch, T. Pakula, K. Muellen, *Macromol. Biosci.* **2003**, *3*, 729.
226 M. S. Hedenquist, H. Yousefi, E. Malmstrom, M. Johanson, A. Hult, U. V. Gedde, M. Trollsas, J. L. Hedrick, *Polymer* **2000**, *45*, 1827.
227 A. Wursch, M. Moller, T. Glauser, L. S. Lim, S. B. Voytek, J. L. Hedrick, C. W. Frank, J. G. Hilborn, *Macromolecules* **2001**, *34*, 6601.
228 M. Trollsas, B. Atthoff, H. Claesson, J. L. Hedrick, *J. Polym. Sci., Part A: Polym. Chem.* **2004**, *42*, 1174.
229 T. Biela, A. Duda, S. Penczek, K. Rode, H. Pasch, *J. Polym. Sci., Part A: Polym. Chem.* **2002**, *40*, 2884.
230 T. Biela, A. Duda, K. Rode, H. Pasch, *Polymer* **2003**, *44*, 1851.
231 I. Ydens, Ph. Degee, Ph. Dubois, J. Libiszowski, A. Duda, S. Penczek, *Macromol. Chem. Phys.* **2003**, *204*, 171.
232 M. Danko, J. Libiszowski, T. Biela, M. Wolszczak, A. Duda, *J. Polym. Sci., Part A: Polym. Chem.* **2005**, *43*, 4586.
233 T. Biela, A. Duda, H. Pasch, K. Rode, *J. Polym. Sci., Part A: Polym. Chem.* **2005**, *43*, 6116.
234 T. Biela, I. Polanczyk, *J. Polym. Sci., Part A: Polym. Chem.* **2006**, *44*, 4214.
235 J. T. Wiltshire, G. G. Ciao, *Macromolecules* **2006**, *39*, 4282.
236 C. Gottschalk, H. Frey, *Macromolecules* **2006**, *39*, 1719.
237 K. Hashimoto, *Prog. Polym. Sci.* **2000**, *25*, 1411.
238 H. Sekiguchi, Lactams and cyclic imides, in *Ring-Opening Polymerization*, K. J. Ivin, T. Saegusa (Eds.), Elsevier Applied Science, London, **1984**, Vol. 2, p. 809.
239 J. Sebenda, Lactams, in *Comprehensive Chemical Kinetics*, C. H. Bamford, C. F. H. Tipper (Eds.), Elsevier Science, Amsterdam, **1976**, Vol. 15, p. 379.
240 J. Sebenda, Anionic ring-opening polymerzation: lactams, in *Comprehensive Polymer Science. The Synthesis, Characterization, Reactions and Applications of Polymers*, Vol. 3, G. C. Eastmond, A. Ledwith, S. Russo, P. Sigwalt (Eds.), Pergamon Press, Oxford, **1989**, p. 511.
241 M. Szwarc, *Adv. Polym. Sci.* **1965**, *4*, 1.
242 J. Sebenda, J. Hauer, *Polym. Bull.* **1981**, *5*, 529.
243 T. Loontjens, *J. Polym. Sci., Part A: Polym. Chem.* **2003**, *41*, 3198.
244 I. Imanishi, *N*-Carboxyanhydrides, in *Ring-Opening Polymerization*, K. J. Ivin, T. Saegusa (Eds.), Elsevier Applied Science, London, **1984**, Vol. 1, p. 558.
245 H. R. Kricheldorf, *α-Aminoacid-N-Carboxyanhydrides and Related Heterocycles*, Springer, New York, **1987**.
246 H. R. Kricheldorf, Anionic ring-opening polymerization: *N*-carboxyanhydrides, in *Comprehensive Polymer Science. The Synthesis, Characterization, Reactions and Applications of Polymers*, Vol. 3, G. C. Eastmond, A. Ledwith, S. Russo, P. Sigwalt (Eds.), Pergamon Press, Oxford, **1989**, p. 531.
247 H. R. Kricheldorf, Polypeptides, in *Models of Biopolymers by Ring-Opening Polymerization*, S. Penczek (Ed.), CRC Press, Boca Raton, FL, **1990**, p. 160.
248 T. J. Deming, *J. Polym. Sci., Part A: Polym. Chem.* **2000**, *38*, 3011.
249 D. A. Tirrell, Artificial polypeptides, in *Encyclopedia of Materials: Science and Technology*, K. H. J. Buschow, R. W. Cahn, M. C. Flemings, B. Ilschner, E. J. Kramer, S. Mahajan (Eds.), Elsevier Science, Oxford, **2001**, p. 347.
250 I. Dimitrov, H. Schlaad, *Chem. Commun.* **2003**, 2944.

251 H. Schlaad, M. Antonietti, *Eur. Phys. J.* **2003**, *E10*, 17.
252 I. Dimitrov, H. Kukula, H. Coelfen, H. Schlaad, *Macromol. Symp.* **2004**, *215*, 383.
253 T. Aliferis, H. Iatrou, N. Hadjichristidis, *Biomacromolecules* **2004**, *5*, 1653.
254 N. Hadjichristidis, H. Iatrou, S. Pispas, M. Pitsikalis, *J. Polym. Sci., Part A: Polym. Chem.* **2000**, *38*, 3211.
255 T. J. Deming, *Nature* **1997**, *390*, 386.
256 T. J. Deming, *J. Am. Chem. Soc.* **1998**, *120*, 4240.
257 T. J. Deming, *Macromolecules* **1999**, *32*, 4500.
258 E. J. Goethals, Cyclic amines, in *Comprehensive Polymer Science. The Synthesis, Characterization, Reactions and Applications of Polymers*, Vol. 2, G. C. Eastmond, A. Ledwith, S. Russo, P. Sigwalt (Eds.), Pergamon Press, Oxford, **1989**, p. 715.
259 E. J. Goethals, Cationic ring-opening polymerization: sulfides, in *Comprehensive Polymer Science. The Synthesis, Characterization, Reactions and Applications of Polymers*, Vol. 3, G. C. Eastmond, A. Ledwith, S. Russo, P. Sigwalt (Eds.), Pergamon Press, Oxford, **1989**, p. 825.
260 E. J. Goethals, Cationic ring-opening polymerization: amines and N-containing heterocycles, in *Comprehensive Polymer Science. The Synthesis, Characterization, Reactions and Applications of Polymers*, Vol. 3, G. C. Eastmond, A. Ledwith, S. Russo, P. Sigwalt (Eds.), Pergamon Press, Oxford, **1989**, p. 837.
261 S. Penczek, *J. Polym. Sci., Part A: Polym. Chem.* **2000**, *38*, 1919.
262 E. J. Goethals, B. Verdonck, Living/controlled cationic polymerization, in *Living and Controlled Polymerization*, J. Jagur-Grodzinski (Ed.), Nova Science, New York, **2005**, p. 131.
263 G. Odian, *Principles of Polymerization*, 4th edn, Wiley, Hoboken, NJ, **2004**, p. 556.
264 A. Dworak, *Macromol. Chem. Phys.* **1998**, *199*, 1843.
265 S. Penczek, S. Slomkowski, Cationic ring-opening polymerization: formation of cyclic oligomers, in *Comprehensive Polymer Science. The Synthesis, Characterization, Reactions and Applications of Polymers*, Vol. 3, G. C. Eastmond, A. Ledwith, S. Russo, P. Sigwalt (Eds.), Pergamon Press, Oxford, **1989**, p. 725.
266 P. Dreyfus, *Polytetrahydrofuran*, Gordon and Breach, New York, **1982**.
267 A. Forschirm, F. D. McAndrew, Acetal resins, in *Polymeric Materials Encyclopedia*, J. C. Salomone (Ed.), CRC Press, Boca Raton, FL, **1996**, Vol. 1, p. 6.
268 G. Pruckmayr, T. K. Wu, *Macromolecules*, **1978**, *11*, 265.
269 H. Jacobson, W. H. Stockmayer, *J. Chem. Phys.* **1950**, *18*, 1600.
270 J. A. Semlyen, *Adv. Polym. Sci.* **1976**, *21*, 43.
271 M. Hasegawa, K. Yamamoto, T. Shiwaku, T. Hashimoto, *Macromolecules* **1990**, *23*, 2629.
272 D. J. Worsfold, A. M. Eastham, *J. Am. Chem. Soc.* **1957**, *79*, 900.
273 J. Dale, K. Daasvatn, *J. Chem. Soc., Chem. Commun.* **1976**, 295.
274 A. I. Kuzaev, O. M. Olkhova, *Vysokomol. Soedin., Ser. A* **1982**, *24*, 2197.
275 P. Kubisa, S. Penczek, *Prog. Polym. Sci.* **1999**, *24*, 1409.
276 R. Tokar, P. Kubisa, S. Penczek, A. Dworak, *Macromolecules* **1994**, *27*, 320.
277 A. Dworak, W. Walach, B. Trzebicka, *Macromol. Chem. Phys.* **1995**, *196*, 1963.
278 M. Bednarek, T. Biedron, J. Helinski, K. Kaluzynski, P. Kubisa, S. Penczek, *Macromol. Rapid Commun.* **1999**, *20*, 369.
279 M. Bednarek, P. Kubisa, S. Penczek, *Macromolecules* **2001**, *34*, 5112.
280 H. Magnusson, E. Malmstrom, A. Hult, *Macromol. Rapid Commun.* **1999**, *20*, 453.
281 H. Magnusson, E. Malmstrom, A. Hult, *Macromolecules* **2001**, *34*, 5786.
282 D. Yan, J. Hou, X. Zhu, J. J. Kosman, H. S. Wu, *Macromol. Rapid Commun.* **2000**, *21*, 557.
283 T. Biedron, K. Brzezinska, P. Kubisa, S. Penczek, *Polym. Int.* **1995**, *36*, 73,
284 T. Biedron, P. Kubisa, S. Penczek, *J. Polym. Sci., Part A: Polym. Chem.* **1991**, *29*, 619.
285 F. Lagarde, L. Reibel, E. Franta, *Makromol. Chem.* **1992**, *193*, 1087.

286 R. Lazauskate, G. Buika, J. V. Grazulavicius, R. Kavaliunas, *Eur. Polym. J.* **1998**, *8*, 1171.

287 T. Fujiwara, U. Makal, K. J. Wynne, *Macromolecules* **2003**, *36*, 9383.

288 M. Bednarek, P. Kubisa, S. Penczek, *Macromolecules* **1999**, *32*, 5237.

289 M. Bednarek, P. Kubisa, *J. Polym. Sci., Part A: Polym. Chem.* **1999**, *37*, 3455.

290 J. V. Crivello, R. A. Ortiz, *J. Polym. Sci., Part A: Polym. Chem.* **2002**, *40*, 2298.

291 Y. Okamoto, *Makromol. Chem., Macromol. Symp.* **1991**, *42*, 117.

292 B. A. Rosenberg, *Makromol. Chem., Macromol. Symp.* **1992**, *60*, 177.

293 Y. Shibasaki, H. Sanada, M. Yokoi, F. Sanda, T. Endo, *Macromolecules* **2000**, *33*, 4316.

294 X. D. Lou, C. Detrembleur, R. Jerome, *Macromolecules* **2002**, *35*, 1190.

295 M. S. Kim, K. S. Seo, G. Khong, H. B. Lee, *Macromol. Rapid Commun.* **2005**, *26*, 643.

296 M. Bednarek, P. Kubisa *J. Polym. Sci., Part A: Polym. Chem.* **2005**, *43*, 3788.

297 Y. Shibasaki, F. Sanda, T. Endo, *Macromol. Rapid Commun.*, **1999**, *20*, 532.

298 T. Endo, Y. Shibasaki, F. Sanda, *J. Polym. Sci., Part A: Polym. Chem.* **2002**, *40*, **1991**.

299 D. Bourissou, B. Martin-Vaca, A. Dumitrescu, M. Graullier, F. Lacombe, *Macromolecules* **2005**, *38*, 9993.

300 E. Franta, L. Reibel, J. Lehmann, S. Penczek, *J. Polym. Sci. Symp.* **1976**, *56*, 139.

301 Y. Tezuka, A. Tsuchitani, Y. Yoshioka, H. Oike, *Macromolecules* **2003**, *36*, 65.

302 L. M. Tanghe, E. J. Goethals, *e-Polymers* **2001**, 017.

303 F. Sanda, H. Sanada, Y. Shibasaki, T. Endo, *Macromolecules* **2002**, *35*, 680.

304 P. Lowenhielm, H. Claesson, A. Hult, *Macromol. Chem. Phys.* **2004**, *205*, 1489.

305 M. Bednarek, P. Kubisa, *J. Polym. Sci., Part A: Polym. Chem.* **2004**, *42*, 608.

306 K. Jankova, M. Bednarek, S. Hvilsted, *J. Polym. Sci., Part A: Polym. Chem.* **2005**, *43*, 3748.

307 T. Saegusa, S. Kobayashi, in *Encyclopedia of Polymer Science and Technology*, H. F. Mark, N. F. Bikales (Eds.), Wiley-Interscience, New York, **1976**, Suppl. 1, p. 220.

308 S. Kobayashi, H. Uyama, *J. Polym. Sci., Polym. Chem. Ed*, **2002**, *40*, 192.

309 S. Penczek, J. Fejgin, W. Sadowska, M. Tomaszewicz, *Makromol. Chem.* **1968**, *116*, 203.

310 R. D. Lubdberg, F. P. D. Gludice (to Union Carbide), *US Patent 3 632 669*, **1972**.

311 I. Barakat, P. Dubois, R. Jérôme, P. Teyssié, M. Mazurek, *Macromol. Symp.* **1994**, *88*, 227.

312 S. Sosnowski, M. Gadzinowski, S. Slomkowski, S. Penczek, *J. Bioact. Compat. Polym.* **1994**, *9*, 345.

313 S. Slomkowski, S. Sosnowski, M. Gadzinowski, *Polym. Prepr. Am. Chem. Soc., Div. Polym. Chem.* **1996**, *37*, 135.

314 M. Gadzinowski, S. Sosnowski, S. Slomkowski, *Macromolecules* **1996**, *29*, 6404.

315 S. Slomkowski, M. Gadzinowski, S. Sosnowski, *Macromol. Symp.* **1998**, *132*, 451.

316 S. Slomkowski, S. Sosnowski, M. Gadzinowski, *Colloids Surf. A* **1999**, *153*, 111.

317 S. Slomkowski, Preparation of biodegradable particles by polymerization processes in colloidal biomolecules, biomaterials and biomedical applications, in *Surfactant Science Series*, A. Elaissari (Ed.), Marcel Dekker, New York, **2003**, Vol. 116, p. 371.

318 S. Sosnowski, S. Slomkowski, A. Lorenc, H. R. Kricheldorf, *Colloid Polym. Sci.* **2002**, *280*, 107.

319 S. Slomkowski, S. Sosnowski, M. Gadzinowski, *Polimery (Warsaw)* **2002**, *7/8*, 485.

320 S. Slomkowski, S. Sosnowski, Diameters and diameter distributions of poly(L,L-lactide) microspheres by ring-opening polymerization of L,L-lactide and from earlier synthesized polymers, in *Polymeric Drug and Drug Delivery Systems*, R. M. Ottenbrite, S. W. Kim (Eds.), CRC Press, Boca Raton, FL, **2001**, p. 261.

321 S. J. Clarson, J. A. Semlyen (Eds.), *Siloxane Polymers*, Ellis Horwood/PTR Prentice Hall, Englewood Cliffs, NJ, **1993**.

322 R. Drake, I. MacKinnon, R. Taylor, Recent advances in the chemistry of si-

loxane polymers and copolymers, in *The Chemistry of Organic Silicon Compounds*, Z. Rappoport, Y. Apeloig (Eds.), Wiley-Interscience, New York, **1998**, p. 2217.
323 R.G. Jones, W. Ando, J. Chojnowski (Eds.), *Silicon-Containing Polymers. The Science and Technology of Their Synthesis and Applications*, Kluwer, Dordrecht, **2000**.
324 J. Chojnowski, M. Cypryk, J. Kurjata, *Prog. Polym. Sci.* **2003**, *28*, 691.
325 M. Mazurek, Silicones, in *Comprehensive Organometallic Chemistry III*, R.H. Crabtree, M.P. Mingos, C. Housecroft (Eds.), Elsevier Science, Oxford, **2006**, in press.
326 H.R. Allcock, Ring-opening polymerization in phosphazene chemistry, in *Ring-Opening Polymerization. Mechanisms, Catalysis, Structure, Utility*, D.J. Brunelle (Ed.), Hanser, Munich, **1993**, p. 217.
327 H.R. Allcock, *Chemistry and Application of Polyphosphazene*, Wiley-Interscience, Hoboken, NJ, **2003**.
328 R. De Jaeger, M. Gleria, *Prog. Polym. Sci.* **1998**, *23*, 179.
329 G. Lapienis, S. Penczek, Cyclic compounds containing phosphorus atoms in the ring, in *Ring-Opening Polymerization*, K.J. Ivin, T. Saegusa (Eds.), Elsevier Applied Science, London, **1984**, p. 919.
330 S. Penczek, P. Klosinski, Synthetic polyphosphates related to nucleic and teichoic acids, in *Models of Biopolymers by Ring-Opening Polymerization*, S. Penczek (Ed.), CRC Press, Boca Raton, FL, **1990**, p. 291.
331 S. Penczek, A. Duda, K. Kaluzynski, G. Lapienis, A. Nyk, R. Szymanski, *Makromol. Chem., Macromol. Symp.* **1993**, *73*, 91.
332 J. Wang, H.-Q. Mao, K.W. Leong, *J. Am. Chem. Soc.* **2001**, *123*, 9480.
333 J. Wen, J.A.G.J.A. Kim, K.W. Leong, *J. Control. Release* **2003**, *92*, 39.
334 D.-P. Chen, J. Wang, *Macromolecules* **2006**, *39*, 473.

5
Radical Polymerization

Krzysztof Matyjaszewski and Wade A. Braunecker

5.1
Introduction

Nearly 50% of all synthetic polymers in the current market are made via radical polymerization (RP) processes. The commercial success of these techniques can largely be attributed to the extensive range of radically polymerizable monomers, their facile copolymerization, the convenient reaction conditions typically employed (room temperature to 100 °C, ambient pressure) and the very minimal requirements for purification of monomers and solvents (which need only be deoxygenated). Additionally, RP is not affected by water and protic impurities and can be carried out in bulk, solution, aqueous suspension, emulsion, dispersion, etc. The range of monomers is larger than for any other chain polymerization because radicals are tolerant to many functionalities, including acid, hydroxy and amino groups. In conventional RP, high molecular weight (MW) polymer is formed at the early stages of the polymerization and neither long reaction times nor high conversions are required, in sharp contrast to step-growth polymerization.

However, RP does have some limitations, especially in comparison with ionic processes. The contribution from chain-breaking reactions such as transfer and termination should be negligible for a polymerization to be considered "living". Whereas transfer is not a major issue in RP when the appropriate conditions are applied, termination is much less avoidable and is one limitation of RP. In contrast to ionic reactions, in which cations or anions do not react via bimolecular termination, radicals terminate with a diffusion-controlled rate. In order to form high-MW polymers, the relative probability of such termination must be minimized. This is accomplished by using very low concentrations (from ppb to ppm) of propagating radicals. The concentrations of growing species are therefore much lower in RP than in conventional anionic or ring-opening polymerizations. In RP, a steady concentration of radicals is established when the rate of termination balances that of initiation. Since the rate of initiation is much slower than the rate of propagation (which is required to keep the radical

Macromolecular Engineering. Precise Synthesis, Materials Properties, Applications.
Edited by K. Matyjaszewski, Y. Gnanou, and L. Leibler
Copyright © 2007 WILEY-VCH Verlag GmbH & Co. KGaA, Weinheim
ISBN: 978-3-527-31446-1

concentration low), the preparation of (co)polymers with controlled architecture is not possible by conventional RP. A propagating radical reacts with monomer every ~1 ms and chains typically terminate after ~1 s, during which time high-MW polymer is formed [degree of polymerization $(DP) \approx 1000$]. At any given moment of the polymerization, the number of growing chains which can be functionalized or chain-extended is very small (<0.1%), unlike in ionic living polymerization, which is essentially carried out without chain-breaking reactions and with instantaneous initiation.

It has not been possible to prepare well-defined (co)polymers via conventional RP. However, new approaches that exploit equilibria between growing radicals and dormant species were recently developed to minimize the proportion of terminated chains in RP. Such controlled/"living" radical polymerizations (CRPs) are mechanistically similar to conventional RP methods and proceed through the same intermediates. However, the proportion of terminated chains in CRP is dramatically reduced from >99.9% to between 1 and 10%. Since termination cannot be entirely avoided in CRP, these systems are never living in the way in which anionic polymerization is living. However, the degree of control is often sufficient to attain many desirable material properties.

In this chapter, the fundamentals of radical polymerization are first covered and are followed by the most recent developments in stable free radical polymerization (SFRP), atom-transfer radical polymerization (ATRP) and reversible addition–fragmentation transfer (RAFT) as an example of a degenerative transfer (DT) process. These CRP systems permit the synthesis of many well-defined copolymers with novel architectures, compositions and functionalities. The possible applications of these materials are also highlighted.

5.2
Typical Features of Radical Polymerization

The major features of conventional RP, which are also typical for CRP, are presented in this section [1–5]. These include basic elementary reactions and typical kinetic parameters. The ranges of polymerizable monomers, initiators, additives and reaction conditions are detailed. Additionally, the most important commercial polymers made by RP are discussed.

5.2.1
Fundamentals of Organic Radicals

Organic radicals (also known as free radicals) are typically sp^2 hybridized intermediates with a very short lifetime. They annihilate in diffusion-controlled biradical termination via disproportionation and coupling reactions. Organic radicals typically have low stereoselectivities due to their sp^2 hybridization. Therefore, polymers formed by RP are usually atactic. However, radicals do have high regioselectivities and add to the less substituted carbon in alkenes; polymers formed by RP thus contain a very high proportion of head-to-tail structures. Exceptions include monomers with either small or weakly stabilizing substituents that result in some regio-errors during polymerization (vinyl acetate, vinyl chloride) or form regio-irregular polymers (vinyl fluoride, trifluorochloroethylene). Radicals can also show high chemoselectivity, as evidenced by the high-MW polymers formed in RP.

5.2.2
Elementary Reactions and Kinetics

RP consists of four elementary steps: initiation, propagation, termination and transfer.

Initiation is usually composed of two processes: the generation of primary initiating radicals (In*) and the reaction of these radicals with monomer (M) (Scheme 5.1). Typically, the former reaction is much slower than the latter and is therefore rate determining with typical values of $k_d \approx 10^{-5}$ s^{-1} and $k_i > 10^4$ M^{-1} s^{-1}.

Propagation occurs by the repetitive addition of the growing radical to the double bond. Rate constants of propagation have a weak chain length dependence with typical values of $k_p \approx 10^{3 \pm 1}$ M^{-1} s^{-1}.

Termination between two growing radicals can occur either by coupling (k_{tc}) or by disproportionation (k_{td}) with rate constants approaching the diffusion-controlled limit, $k_t \approx 10^8$ M^{-1} s^{-1}. Coupling is preferred for radicals with small steric effects (i.e. most monosubstituted radicals). In styrene polymerization, for example, the proportion of coupling approaches 95%. Disproportionation becomes

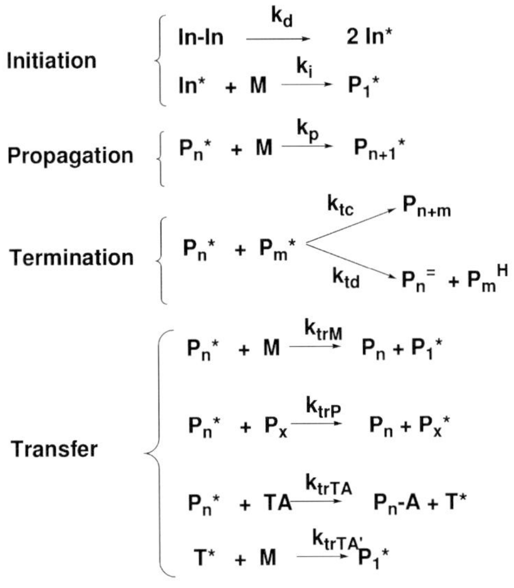

Initiation
- In-In $\xrightarrow{k_d}$ 2 In*
- In* + M $\xrightarrow{k_i}$ P$_1$*

Propagation
- P$_n$* + M $\xrightarrow{k_p}$ P$_{n+1}$*

Termination
- P$_n$* + P$_m$* $\xrightarrow{k_{tc}}$ P$_{n+m}$
- P$_n$* + P$_m$* $\xrightarrow{k_{td}}$ P$_n^=$ + P$_m^H$

Transfer
- P$_n$* + M $\xrightarrow{k_{trM}}$ P$_n$ + P$_1$*
- P$_n$* + P$_x$ $\xrightarrow{k_{trP}}$ P$_n$ + P$_x$*
- P$_n$* + TA $\xrightarrow{k_{trTA}}$ P$_n$-A + T*
- T* + M $\xrightarrow{k_{trTA'}}$ P$_1$*

Scheme 5.1

more important for disubstituted radicals, such as those derived from methyl methacrylate.

Transfer is the fourth basic elementary reaction and transfer to monomer, to polymer or to a number of other transfer agents (TA) can occur. Branching or cross-linking can sometimes result from transfer to polymer. Typically, re-initiation ($k_{trTA'}$) is fast and transfer has no effect on kinetics, only on molecular weights. If re-initiation is slow, then some retardation/inhibition may occur (see below).

Typically, the polymerization rate is first order with respect to monomer and 0.5 order with respect to initiator (Eq. 1). First-order kinetics with respect to monomer are due to the steady concentration of radicals resulting from equal rates of initiation and termination. The rate is a function of the efficiency of initiation (f) and the rate constants of initiation (k_d), propagation (k_p) and termination (k_t) according to Eq. 1. Both rate and MW therefore depend on the $k_p/(k_t)^{1/2}$ ratio. For termination occurring by coupling, MW is defined by Eq. 2. As can be seen by these equations, when polymerization proceeds faster (at higher temperatures or with more initiator), more radicals will be produced. They will also terminate faster, reducing observed polymer molecular weights.

$$R_p = k_p[M](fk_d[I]_0/k_t)^{\frac{1}{2}} \tag{1}$$

$$DP_n = k_p[M]/fk_d[I]_0 k_t)^{-\frac{1}{2}} \tag{2}$$

The absolute values of the rate constants k_p and k_t can be determined under non-stationary conditions, such as those employing a rotating sector or spatially intermittent polymerization. However, pulsed laser photolysis (PLP) techniques are currently the method of choice and provide very precise and reproducible values. A brief analysis of the PLP data in Table 5.1 indicates that alkyl substitu-

Table 5.1 Kinetic rate constants for select monomers obtained by PLP measurements [6, 7].

Monomer [a]	E_p (kJ mol^{-1})	A_p (L mol^{-1} s^{-1})	k_p (M^{-1} s^{-1})	k_t (M^{-1} s^{-1})	Refs.
Dodecyl acrylate	17.0	1.79×10^7	26 100	6.6×10^6	k_p [8], k_t [9]
n-Butyl acrylate	17.9	2.24×10^7	23 100	5.5×10^7	k_p [10], k_t [11]
Methyl acrylate	17.7	1.66×10^7	18 500	2.6×10^7	k_p [8], k_t [9]
Dodecyl methacrylate	21.0	2.50×10^6	786	1.7×10$^{6\,c)}$	k_p [12], k_t [13]
n-Butyl methacrylate	22.9	3.78×10^6	573	5.0×10$^{6\,c)}$	k_p [12], k_t [13]
Methyl methacrylate	22.4	2.67×10^6	490	2.5×10$^{7\,c)}$	k_p [14], k_t [15]
Styrene	32.5	4.27×10^7	162	9.6×10^7	k_p [16], k_t [17]
Butadiene [b]	35.7	8.05×10^7	89	–	k_p [18]

a) Measurements in bulk at 40 °C and ambient pressure, unless stated otherwise.
b) Solvent = chlorobenzene.
c) k_t reported at 1000 bar. E_p and A_p represent the Arrhenius activation energy for propagation and frequency factor for propagation, respectively.

ents in the corresponding (meth)acrylate esters have only a minor effect on kinetics; the rate constants increase very weakly with increasing chain length of the alkyl group. Disubstituted alkenes polymerize more slowly than monosubstituted alkenes. There is also a general reciprocal correlation between the reactivity of monomers and the reactivity of the derived radicals (cf. copolymerization). The rate constants are strongly affected by the radical's structure, meaning that less reactive monomers polymerize faster because they generate more reactive radicals [i.e. $k_{p(butadiene)} < k_{p(styrene)} < k_{p(methacrylate)} < k_{p(acrylate)}$]. On the other hand, rate constants of termination can be much more affected by steric contribution from alkyl substituents; they are similar for dodecyl acrylate and dodecyl methacrylate but are more than an order of magnitude larger for methyl acrylate and methyl methacrylate. Rate coefficients of termination are *not* constant and depend on chain length and conversion (viscosity). Values in Table 5.1 are given for $DP \approx 100$ and low conversion.

5.2.3
Copolymerization

Facile statistical copolymerization is one of the main advantages of free radical polymerization. In contrast to ionic polymerization, the reactivities of many monomers are relatively similar, which makes them easy to copolymerize. There is often a tendency for alternation, especially for monomers with opposite polarities ($r_A r_B < 1$, where $r_A = k_{AA}/k_{AB}$ and $r_B = k_{BB}/k_{BA}$, Scheme 5.2). Electrophilic radicals [i.e. those with –CN, –C(O)OR or –Cl groups] prefer to react with electron-rich monomers (such as styrene, dienes or vinyl acetate), whereas nucleophilic radicals prefer to react with alkenes containing electron-withdrawing substituents. One such example includes styrene and methyl methacrylate ($r_S \approx r_{MMA} \approx 0.5$). An even more dramatic example is the copolymerization of styrene with maleic anhydride ($r_S = 0.02$, $r_{MAH} < 0.001$). For monomers with similar polarities, the reactivity ratios are close to an ideal copolymerization ($r_A = 1/r_B$) [19].

When the reactivity ratios are significantly different, nearly pure homopolymer of the more reactive monomer may be formed at the beginning of the polymerization. As the polymerization progresses and the first monomer is consumed, the second monomer will start to polymerize. In extreme cases, this can lead to the formation of a mixture of two homopolymers. Such compositional drift in monomer feed can be overcome by attenuating the feed with the continuous addition of one or more monomers.

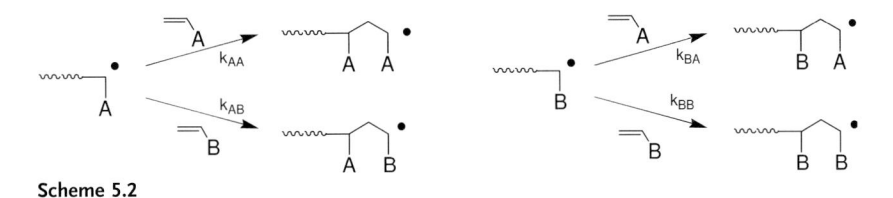

Scheme 5.2

5.2.4
Monomers

Essentially any alkene can be (co)polymerized by radical means. However, the formation of high-MW polymer requires sufficient stability of the propagating radical, sufficient reactivity of the monomer and a small contribution from transfer. Radicals are typically stabilized by resonance. Therefore, styrenes (including vinylpyridines), various substituted methacrylates, acrylates, acrylamides, acrylonitrile, dienes and vinyl chloride are well suited for polymerization by radical means (Scheme 5.3). Although polymerization of less reactive alkenes such as ethylene is more challenging, high-MW polymers are formed commercially in large quantities under special conditions (see below).

Monomers which cannot be successfully homopolymerized radically include simple a-alkenes, isobutene and those monomers with easily abstractable H-atoms (e.g. thiols, allylic derivatives). They have low reactivity and at elevated temperatures participate in extensive transfer.

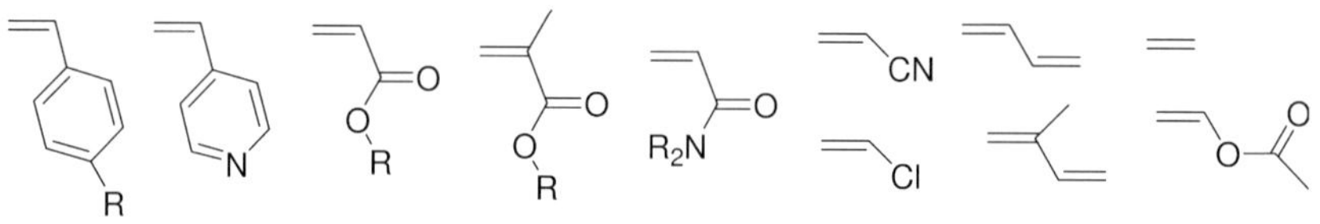

Scheme 5.3

5.2.5
Initiators for RP

The numerous initiators available for RP which can generate radicals capable of addition to double bonds include peroxides, diazenes and various redox systems, and also high-energy sources such as UV light, γ-rays and elevated temperatures. After formation, some radicals will initiate new chains, but others may terminate before escaping the solvent cage and form inactive products. Thus, radical initiation has a fractional efficiency ($0 < f < 1$). It is usually not constant throughout the reaction and decreases with monomer dilution and conversion as [M] becomes lower and viscosity higher.

Peroxy compounds that are used as radical initiators (via homolytic C–O–O–C bond cleavage) include benzoyl and dialkyl peroxides, peresters, percarbonates, hydroperoxides and inorganic peroxides. Some relevant data is listed in Table 5.2. These radicals are generated with very large activation energy and therefore peroxides are stable at room temperature. Polymerization is often carried out at temperatures where the radical half-lifetime ($\tau_{\frac{1}{2}}$) is close to 10 h, assuring a slow and continuous supply of initiating radicals.

Peroxides and hydroperoxides produce very electrophilic radicals that may not only add to double bonds but can also abstract H-atoms and react with aromatic

Table 5.2 Kinetic data for selected peroxides[a].

Peroxide	Temperature range (°C)	k_d (60 °C) (M^{-1} s^{-1})	E_a (kJ mol^{-1})	A (×10^{-15})	$\tau_{1/2}$ = 10 h (°C)
(BPO, dibenzoyl peroxide)	70–90	1.5×10^{-6}	139.0	9.34	78
(di-tert-butyl peroxide)	110–135	2.5×10^{-9}	152.7	2.16	125
(tert-butyl hydroperoxide)	155–175	–	174.2	7.97	168
K$^+$O-S(O)$_2$-O-O-S(O)$_2$-O$^+$K (potassium persulfate)	50–90	4.4×10^{-6}	148.0	7.09	69

a) Data taken from [5].

Table 5.3 Kinetic data for selected azo-initiators[a].

Initiator	Temperature (°C)	k_d (60 °C) (M^{-1} s^{-1})	E_a (kJ mol^{-1})	A (×10^{-15})	$\tau_{1/2}$ = 10 h (°C)
NC-C(CH$_3$)$_2$-N=N-C(CH$_3$)$_2$-CN (AIBN)	50–80	9.6×10^{-6}	131.7	4.31	65
Ph-CH(CH$_3$)-N=N-CH(CH$_3$)-Ph	40–60	1.7×10^{-4}	126.7	12.2	45
(CH$_3$)$_3$C-N=N-C(CH$_3$)$_3$	150–170	5.0×10^{-12}	180.4	91.7	161
(CH$_3$)$_3$C-O-N=N-O-C(CH$_3$)$_3$	30–50	2.2×10^{-4}	119.5	1.17	42

a) Data taken from [5].

compounds. Alkoxy radicals have a stronger affinity towards hydrogen than acyloxy radicals. In the initiation of MMA polymerization with BPO, <1% of H-abstraction is observed compared with >99% addition, whereas initiation with di-(*tert*-butyl) peroxide provides 30% H-abstraction and 70% addition. Additionally, the decomposition of peroxides can be significantly accelerated in the presence of electron-rich compounds such as amines and can also be catalyzed by transition metals.

Diazo compounds form another class of useful radical initiators (Table 5.3). Azobisisobutyronitrile (AIBN) is the most representative and widely used diazene. The useful temperature range for diazenes depends very strongly on the steric effects of its substituents and varies from room temperature to 200 °C. The efficiency of initiation is relatively modest, with cumulative efficiency ~50%

$^-O_3S\text{-}O\text{-}O\text{-}SO_3^- + Fe^{2+} \longrightarrow Fe^{3+} + SO_4^{2-} + SO_4^{-\bullet}$

$^-O_3S\text{-}O\text{-}O\text{-}SO_3^- + S_2O_3^{2-} \longrightarrow {}^\bullet S_2O_3^- + SO_4^{2-} + SO_4^{-\bullet}$

$HO\text{-}OH + Fe^{2+} \longrightarrow Fe^{3+} + HO^- + HO^\bullet$

Scheme 5.4

at higher monomer conversions. The side-products of AIBN initiation are ketenimine and the toxic tetramethylsuccinonitrile.

Other important classes of radical initiators include redox systems (Scheme 5.4) and monomers which undergo thermal self-initiation. For example, styrene participates in the Diels–Alder process forming the labile derivative of tetrahydronaphthalene. This further reacts with another styrene to produce a free radical that can initiate polymer chains. Most monomers do not readily undergo this Diels–Alder process and consequently their rates of self-initiation are much slower than in the case of styrene.

5.2.6
Additives

There are many compounds which can affect either the rate of RP or the MW of the products. Some reagents can inhibit or retard RP. Oxygen is a classic inhibitor which forms relatively stable and unreactive peroxy radicals. Other efficient inhibitors include quinones and phenols (in the presence of oxygen), nitro compounds and some transition metal compounds such as $CuCl_2$ and $FeBr_3$. However, the latter species become very efficient moderators for ATRP in the presence of appropriate ligands, as will be discussed.

Another class of compounds which do not affect the rate but may strongly influence the MW via transfer processes include thiols, disulfides and polyhalogenated compounds. In fact, these transfer agents often react with growing radicals faster than monomers do. If a transfer agent carries some additional functionality, it may result in functional oligomers (telechelics). Some transfer agents may also act as inhibitors and stop radical polymerization by degradative transfer (e.g. α-alkenes).

5.2.7
Typical Conditions for RP

Radical polymerizations can be conducted in bulk, organic or aqueous homogeneous solutions, and also in dispersed media such as suspension, emulsion, miniemulsion, microemulsion, inverse emulsion and even precipitation polymerization. There are also examples of polymerization in the gas phase and from surfaces. Typical polymerizations are carried out between 50 and 100 °C, but many are run at room temperature, at –100 °C (when photoinitiated) or even at 250 °C (in the case of ethylene). The basic prerequisite for RP is deoxygenation of all components; most reactions are run under nitrogen. Inhibitors may be removed by passing the monomer through an alumina column or consumed with excess initiator.

Polymerization is often exothermic and temperature control is important. At high conversion in bulk polymerizations, autoacceleration may occur (known as the Trommsdorff effect) due to a drop in the termination rate as the system becomes more viscous. The evolution of additional heat that ensues can result in faster initiator decomposition which additionally increases the polymerization rate and may ultimately lead to an explosion.

5.2.8
Commercially Important Polymers by RP

5.2.8.1 Polyethylene
Because propagation in ethylene polymerization is very slow, in order to compete successfully with termination, high temperatures (>200 °C) and pressures (>20 000 psi) must be used. Under these conditions, transfer becomes very important. As a result, highly branched polyethylene (PE) is formed with properties very different from the linear analogue made by coordination polymerization and is a very important commercial material. Low-density polyethylene (LDPE) ($T_g \approx -120\,°C$, $T_m \approx 100\,°C$) is flexible, solvent resistant and has good flow properties and good impact resistance. Nearly 4 million tons of LDPE were produced in the USA in 2000. LDPE is used as films for plastic bags, food and beverage containers and plastic wrapping for various other containers (clothing, trash, etc.). LDPE is also extruded in a variety of applications, including toys, plastic containers and lids and for coating electrical wires and cables. Other grades of polyethylene (HDPE, LLDPE) are prepared using coordination polymerization.

5.2.8.2 Polystyrene
Polystyrene is a rigid plastic, with good optical clarity (it is amorphous) and resistance to acids and bases. It can be processed at moderate temperatures ($T_g \approx 100\,°C$) and it is a good electrical insulator. Nearly 4 million tons of polystyrene were produced in the USA in 2000. A drawback of polystyrene is that it is soluble in a variety of solvents (poor solvent resistance) and is relatively brittle. These problems can be overcome by copolymerization of styrene with a variety of other monomers. For example, copolymerization of styrene (S) with acrylonitrile (AN) (~40% AN) imparts good solvent resistance to the polymer and increases its tensile strength. Such SAN copolymers find use in appliances, furniture and electrical applications (battery casings). Nearly 100 000 tons of SAN were produced in the US in 2000. The flexibility of styrene can also be improved by copolymerization with dienes. Elastomeric copolymers of styrene and 1,3-butadiene or SBR rubbers generally contain low amounts of styrene (~25%) and have similar properties to natural rubber. These polymers are primarily

used in the manufacturing of tires, where SBR is vulcanized. Other automotive uses include belts, hoses, shoe soles, flooring materials and electrical insulation. Higher contents of styrene in the copolymers are used in paints. By combining the solvent resistance of SAN and the flexibility of SBR, terpolymers of acrylonitrile, styrene and 1,3-butadiene have been prepared. Generally, styrene and acrylonitrile are copolymerized in the presence of a rubber [poly(1,3-butadiene), SBR or poly(1,3-butadiene-co-acrylonitrile)]. These copolymers are known as ABS. ABS has excellent solvent and abrasion resistance and finds application in housewares (luggage, hairdryers, furniture), construction (bathroom sinks/tubs, piping), electronic housings, automotive parts (light housings) and in recreation (hockey sticks, boat hulls). In 2000, nearly 2 million tons of ABS were produced in the USA.

5.2.8.3 Poly(vinyl chloride) (PVC)

PVC is generally produced by suspension polymerization of gaseous vinyl chloride. Homo- and copolymers of PVC are widely used, with 7 million tons produced in the USA in 2000. Although PVC itself is very brittle and is relatively unstable to heat and light (which results in the loss of HCl), its performance can be greatly improved with the use of additives (i.e. plasticizers and stabilizers). By adding plasticizers (up to 40%), PVC becomes more pliable and thus more useful. PVC is widely used in construction (drainage pipes, vinyl sidings, window frames, gutters, flooring, etc.), packaging (bottles, boxes), wire insulation and biomedical applications (gloves, tubing).

5.2.8.4 Poly(meth)acrylates

Acrylic and methacrylic polymers are primarily used in coatings (automotive, structural and home) and paints. Acrylic paints are generally prepared as emulsions. They are also polymerized in solution for use as oil-based paints. In 2000, the US paint industry sold $20 billion dollars worth of paint, much of it acrylic-based. Acrylic copolymers are also used as adhesives, oil additives and caulks/sealants. Poly(methyl methacrylate) (PMMA) with its excellent optical clarity (also known as Plexiglas) finds application as safety glass. It is also used in light fixtures, skylights, lenses, fiber optics, dentures, fillings and in contact lenses [PMMA, hard contacts; poly(2-hydroxyethyl methacrylate), soft contacts].

5.2.8.5 Other Polymers

There are a variety of other commercial polymers prepared by radical polymerization, but in much smaller quantities than the aforementioned polymers. Polyacrylamides and poly(vinyl alcohol) (PVA) are widely used in aqueous solutions as thickeners, in biomedical applications and in health and beauty products. PVA cannot be prepared by the direct polymerization of vinyl alcohol, but rather is prepared by hydrolysis of poly(vinyl acetate) (PVAc). Polyacrylonitrile is

5.3
Controlled/Living Radical Polymerization

5.3.1
General Concepts

Achieving controlled synthesis over polymeric materials requires that chain-breaking reactions be minimized and the apparent simultaneous growth of all chains be attained. Such a combination of fast initiation and an absence of termination would seem to disagree with the fundamental principles of RP, which proceeds via slow initiation and in which all chains are essentially dead at any given instant. However, such a system was realized with the development of controlled/"living" systems and the intermittent formation of active propagating species. This concept, originally applied to cationic ring-opening polymerization [20], was successfully extended to carbocationic systems [21] and later to anionic and other polymerizations where growing centers are in equilibrium with various dormant species.

Central to all CRP systems is a dynamic equilibrium between propagating radicals and various dormant species [22, 23]. There are two types of equilibria. Radicals may be reversibly trapped ("reversibly terminated") in a deactivation/activation process according to Scheme 5.5, or they can be involved in "reversible transfer", i.e. a degenerative exchange process (Scheme 5.6).

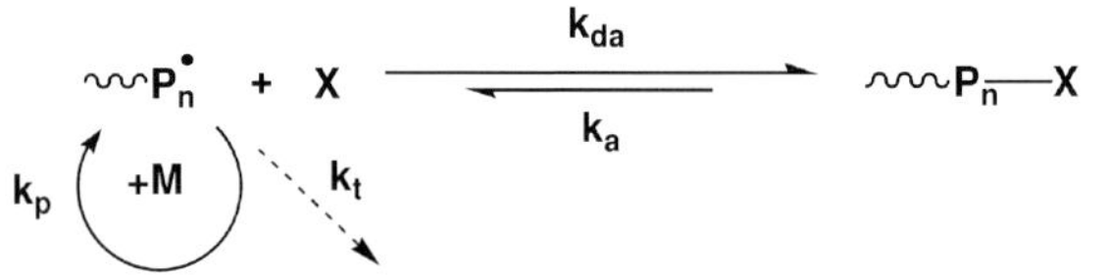

$$\sim\!\!P_n^{\bullet} + X \underset{k_a}{\overset{k_{da}}{\rightleftharpoons}} \sim\!\!P_n\!-\!X$$

with k_p (+M) and k_t

Scheme 5.5

$$\sim\!\!P_n^{\bullet} + \sim\!\!P_m\!-\!X \overset{k_{exch}}{\rightleftharpoons} \sim\!\!P_m^{\bullet} + \sim\!\!P_n\!-\!X$$

with k_p (+M), k_t on both sides

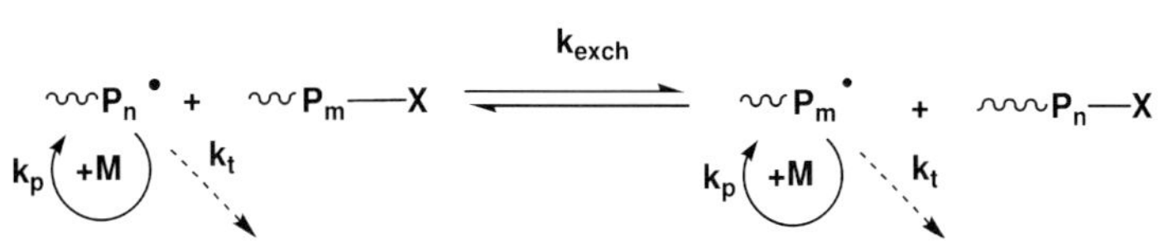

Scheme 5.6

The first approach is more widely used and relies on the persistent radical effect (PRE) [24, 25]. In systems obeying the PRE, newly generated radicals are rapidly trapped in a deactivation process (with a rate constant of deactivation, k_{da}) by species X (which is typically a stable radical such as a nitroxide [26, 27] or an organometallic species such as a cobalt porphyrin [28]). When the dormant species are reactivated (with a rate constant k_a) and re-form growing centers, they can propagate (k_p) but also terminate (k_t). Persistent radicals (X) cannot terminate with each other but only cross-couple with growing species. Every act of radical–radical termination is accompanied by the irreversible accumulation of X. As the concentration of X increases with time, the concentration of radicals decreases (as the equilibrium in Scheme 5.5 becomes shifted towards the dormant species) and the probability of termination is greatly decreased.

In systems based on the PRE, a steady state of growing radicals is established through the activation–deactivation process and *not* through initiation–termination. Because initiation is much faster than termination, this essentially gives rise to the instantaneous growth of all chains. By contrast, systems which employ degenerative transfer are *not* based on the PRE. They follow typical RP kinetics with slow initiation and fast termination. The concentration of transfer agent is much larger than that of radical initiators and thus plays the role of a dormant species.

Good control over molecular weight, polydispersity and chain architecture in all CRP systems requires very fast exchange between active and dormant species. Ideally, a growing species should react with only a few monomer units before it is converted back into a dormant species. Consequently, it would remain active for only a few milliseconds and would return to the dormant state for a few seconds or more. The growth of a polymer chain may consist (for example) of ~1000 periods of ~1 ms activity, interrupted by ~1000 periods of ~1 min dormant states. The lifetime in the active state is thus comparable to the lifetime of a propagating chain in conventional RP (~1000 ms, i.e. ~1 s). However, because the whole propagation process in CRP may take ~1000 min or ~1 day, there exists the opportunity to carry out various synthetic procedures, such as functionalization or chain extension [29].

5.3.2
Similarities and Differences Between RP and CRP

Although CRP and RP proceed via the same radical mechanism, exhibit similar chemo-, regio- and stereoselectivities and can polymerize a similar range of monomers, several important differences exist and are summarized as follows:
1. The lifetime of growing chains is extended from ~1 s in RP to more than 1 h in CRP. This is done by participation of dormant species and intermittent reversible activation.
2. In conventional RP, initiation is slow and free radical initiator is often left unconsumed at the end of the polymerization. In most CRP systems, initiation

is very fast and all chains start growing at essentially the same time, ultimately allowing control over chain architecture.
3. In RP, nearly all chains are dead, whereas in CRP the proportion of dead chains is usually <10%.
4. Polymerization in CRP is often slower than in RP but this does not necessarily have to be the case. Rates may be comparable, if the targeted MW in CRP is relatively small.
5. In RP, a steady-state radical concentration is established with similar rates of initiation and termination, whereas in CRP systems based on the PRE (ATRP, SFRP), a steady radical concentration is reached by balancing the rates of activation and deactivation.
6. In conventional RP, termination usually occurs between long chains and constantly generated new chains. In CRP systems based on the PRE (ATRP and SFRP), at the early stages all chains are short (most probable termination); but as they become much longer, the termination rate decreases significantly. In DT processes (e.g. RAFT), new chains are constantly generated by a small amount of conventional initiator and therefore termination is more likely.

5.4
SFRP and NMP Systems – Examples and Peculiarities

Stable free radical polymerization (SFRP) and, more specifically, nitroxide-mediated polymerization (NMP) systems rely on the reversible homolytic cleavage of a relatively weak bond in a covalent species to generate a growing radical and a less reactive radical (also known as the persistent or stable free radical). As discussed previously, stable free radicals should only react reversibly with growing radicals and should *not* react amongst themselves or with monomers to initiate the growth of new chains and they should *not* participate in side-reactions such as the abstraction of β-H atoms.

Nitroxides were originally employed as stable free radicals in the polymerization of (meth)acrylates [30]. However, only after the seminal paper by Georges et al., who used TEMPO as a mediator in styrene polymerization, did the field of NMP gain momentum (Scheme 5.7) [26]. TEMPO and substituted TEMPO derivatives form relatively strong covalent bonds in alkoxyamines that have very low equilibrium constants in NMP [ratio of rate constants of dissociation (activa-

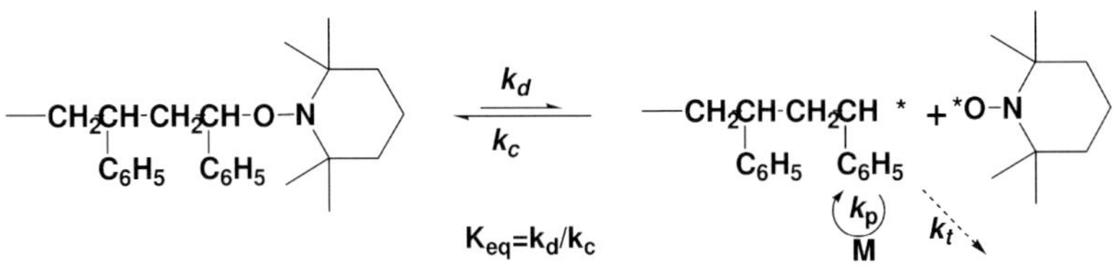

Scheme 5.7

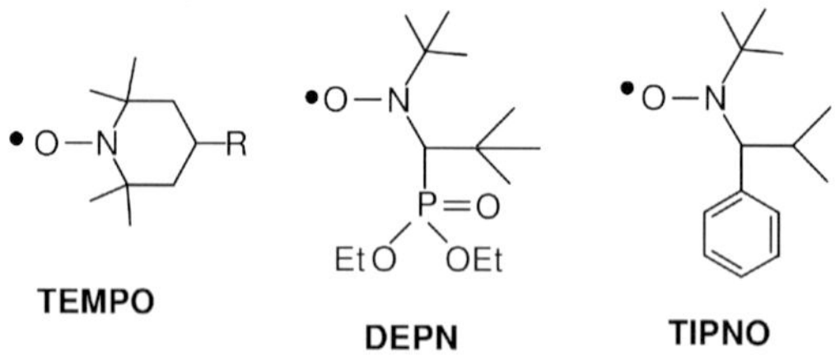

TEMPO **DEPN** **TIPNO**

Scheme 5.8

tion) and coupling (deactivation), $k_d/k_c = K_{eq} = 2.0 \times 10^{-11}$ M at 120 °C for polystyrene] [23] and therefore require high polymerization temperatures.

Subsequently, many other nitroxides were synthesized which provide a range of C–O bond strengths (Scheme 5.8) [27]. DEPN [31] (also known as SG-1) and TIPNO [32] in Scheme 5.8 contain an H-atom at the α-C. Such nitroxides were originally predicted to be very unstable and were expected to decompose quickly by loss of a hydroxyl radical. However, the stability of both DEPN and TIPNO was found sufficient to control the polymerization of many monomers. Steric and electronic effects stabilize these radicals. K_{eq} for DEPN with polystyrene has been estimated as 6.0×10^{-9} M at 120 °C [31], or several hundred times higher than for the TEMPO system. A decrease in the bond dissociation energy of these alkoxyamines actually permits lower polymerization temperatures [33, 34]. Polymerization can also be accelerated in NMP systems with the presence of a constant radical flux, for example by using slowly decomposing free radical initiators [35].

Other systems employing this concept have since been developed utilizing triazolinyl radicals [36, 37], boronyloxy radicals [38], bulky organic radicals such as thermolabile alkanes [39, 40] and also transition metal compounds (see below). Alkyl dithiocarbamates, originally used by Otsu et al., undergo homolytic cleavage when irradiated by UV light and can mediate polymerization following an SFRP mechanism. However, these dithiocarbamyl radicals not only cross-couple but can dimerize and also initiate polymerization [41, 42].

5.4.1
Monomers and Initiators

The range of monomers polymerizable by NMP includes styrene, various acrylates, acrylamides, dienes and acrylonitrile. Monomers forming less stable radicals such as vinyl acetate have not yet been successfully polymerized via NMP due to too low values of K_{eq} (Scheme 5.7). Polymerization of disubstituted alkenes such as methacrylates that form more sterically hindered tertiary radicals cannot yet be controlled owing to very slow deactivation and/or disproportionation. Copolymerization with styrene improves the control [43].

Nitroxides can be employed together with typical radical initiators and used in nearly stoichiometric amounts. A second approach utilizes alkoxyamines, also known as unimolecular initiators [44, 45]. Alkoxyamines can be used in block copolymerizations when present as macromolecular species. Sometimes, an excess of nitroxides can actually enhance block copolymerization efficiency [46].

5.4.2
General Conditions

The bond dissociation energy of most nitroxides and alkoxyamines in NMP is relatively high and therefore high temperatures are generally required. The polymerization of styrene can be successfully mediated with TIPNO and DEPN at >80 °C, but with TEMPO the temperature should be >120 °C. All solvents used in RP can also be employed in NMP. NMP has been successfully used in aqueous systems under homogeneous, emulsion and miniemulsion conditions [47, 48].

5.4.3
Controlled Architectures

NMP has been applied to the synthesis of various homopolymers and statistical, gradient, block and graft copolymers. Star, hyperbranched and dendritic structures have also been prepared [27]. The range of molecular weights attainable in NMP is fairly large, although transfer can limit molecular weights at elevated temperatures. Also, nitroxides may abstract β-H atoms and lead to elimination and a loss of functionality in some systems. Alkoxyamines can be replaced by halogens at the chain end, but functionalization is not very straightforward.

5.4.4
Other SFRP Systems

The SFRP of methyl acrylate with Co porphyrins [28] is one of the most successful SFRP systems to date, leading to well-defined high-MW polyacrylates. More recently, Co(acac)$_2$ derivatives were used to control the polymerization of vinyl acetate and N-vinylpyrrolidone [49, 50]. Co porphyrin and glyoxime derivatives do not control the SFRP of methacrylates but instead lead to a very efficient catalytic chain transfer process (Scheme 5.9). Other transition metal com-

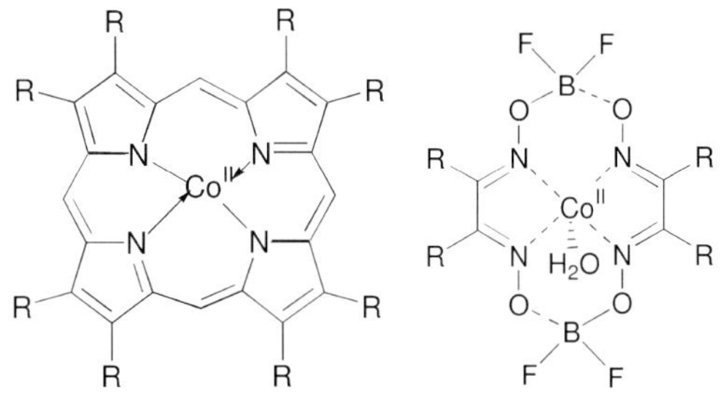

Scheme 5.9

plexes (of Mo [51], Os [52] and Fe [53, 54]) can act as SFRP moderators; furthermore, these complexes may also participate in ATRP equilibria (see below) [55].

5.5
ATRP – Examples and Peculiarities

ATRP, discovered in the last decade [56–60], is currently the most widely used CRP technique. This partially originates from the large range of polymerizable monomers by this technique under a wide range of conditions, but also from the commercial availability of all necessary ATRP reagents, including alkyl halide initiators and transition metal compounds and ligands used as catalysts.

The basic working mechanism of ATRP involves homolytic cleavage of an alkyl halide bond, R–X, by a transition metal complex Mt^n–L (such as $CuBr$–bpy_2) to generate (with the activation rate constant, k_a) the corresponding higher oxidation state metal halide complex X–Mt^{n+1}–L (such as $CuBr_2$–bpy_2) and an alkyl radical $R^{\bullet}$ (a representative example of this process is illustrated in Scheme 5.10) [59, 61]. $R^{\bullet}$ can then propagate with a vinyl monomer (k_p), terminate by coupling and/or disproportionation (k_t) or be reversibly deactivated in this equilibrium by X–Mt^{n+1}–L (k_{da}). Radical termination is diminished as a result of the PRE [24] that ultimately shifts the equilibrium towards the dormant species (activation rate constant ≪ deactivation rate constant). The values of the rate constants in Scheme 5.10 refer to styrene polymerization at 110 °C [62, 63].

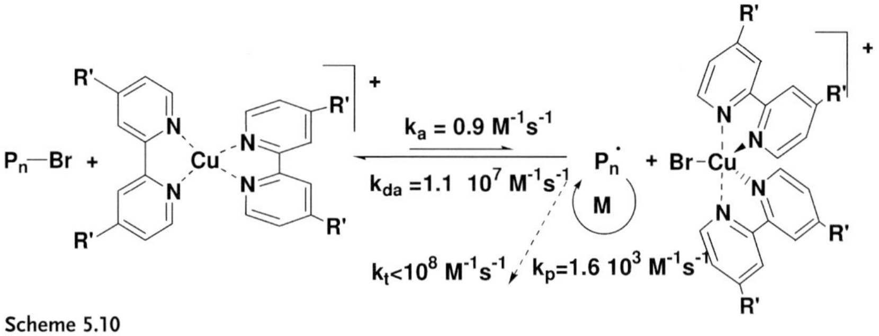

Scheme 5.10

One important difference between SFRP and ATRP is that kinetics and control in ATRP depend not only on the persistent radical (X–Mt^{n+1}–L) but also on the activator (Mt^n–L). Molecular weights are defined by the $\Delta[M]/[RX]_0$ ratio and are not affected by the concentration of transition metal. The polymerization rate increases with initiator concentration (P–X) and is proportional to the ratio of concentrations of activator and deactivator according to

$$R_p = -d[M]/dt = k_p[M][P^*] = k_p[M](k_a[P-X][Mt^n-L])/(k_{da}[X-Mt^{n+1}-L]) \quad (3)$$

The synthesis of polymers with low polydispersities and predetermined molecular weights requires a sufficient concentration of deactivator. Polydispersities decrease with monomer conversion (p) and concentration of deactivator but increase with the k_p/k_{da} ratio. In the absence of any irreversible chain-breaking reactions, polydispersities are also lower for higher targeted MW, i.e. for lower $[RX]_0$ [64, 65]:

$$M_w/M_n = 1 + \{(k_p[RX]_0)/(k_{da}[X-Mt^{n+1}-L])\}/2/p-1) \quad (4)$$

In an ideal case, ATRP proceeds via an inner sphere electron-transfer process, i.e. atom transfer. However, under certain conditions the oxidation or reduction of radicals to carbocations or carbanions by transition metal complexes can occur as a side-reaction [66, 67]. Other side-reactions include the formation of organometallic species (as in SFRP) [51], monomer [68, 69], or solvent coordination to the transition metal catalyst, dissociation of the halide from the deactivator [70], HBr elimination [71–73], etc. Therefore, the appropriate selection of an initiator and catalyst will not only depend on the monomer or targeted molecular architecture but should also be made with respect to plausible side-reactions.

5.5.1
Basic ATRP Components

5.5.1.1 Monomers

The list of monomers successfully homopolymerized by ATRP is fairly extensive and includes various substituted styrenes [74], (meth)acrylates [75–77], (meth)-acrylamides [78, 79], vinyl pyridine [80], acrylonitrile [81], and others [59, 60]. Nitrogen-containing monomers and certain dienes can present challenges in ATRP as monomer or polymer complexation to the catalyst can render it less active. Some nitrogen-containing monomers can also retard polymerization by displacing the terminal halogen of a growing chain. Non-polar monomers such as isobutene [82, 83] and a-alkenes [84, 85] have been successfully copolymerized using ATRP.

ATRP is tolerant to many functional groups such as hydroxy, amino, amido, ether, ester, epoxy, siloxy and others. All of them have been incorporated into (meth)acrylate monomers and successfully polymerized. One exception is the carboxylic acid group, which can protonate amine-based ligands; it must therefore be protected. However, the polymerizations of methacrylic acid at high pH [86], styrenesulfonic acid and acrylamide with sulfonic acid functionality have all been successful in the presence of amines.

Since the ATRP equilibrium constant ($K_{ATRP}=k_a/k_{da}$) depends very strongly on the structure of the monomer and RX, specific catalysts must be used for different monomers. The order of the equilibrium constants in ATRP differs from those in SFRP and RAFT: acrylonitrile > methacrylates > styrene ≈ acrylates > acrylamides >> vinyl chloride > vinyl acetate. This order should be followed to prepare block copolymers efficiently, i.e. polyacrylonitrile should be chain-extended with polyacrylate and not vice versa. However, it is possible to alter this order by using a technique known as halogen exchange whereby a macromolecular alkyl bromide is extended with a less reactive monomer in the presence of a CuCl catalyst [87–89]. The rational behind "halogen exchange" is that the value of the ATRP equilibrium constant for alkyl chloride-type (macro)-initiators is 1–2 orders of magnitude lower than that for the alkyl bromides with the same structure. These C–Cl bonds are activated more slowly and therefore the rate of propagation is decreased with respect to the rate of initiation, which effectively leads to increased initiation efficiency from the added macroinitiator and preparation of a second block with lower polydispersity. In this way, block copolymers have been successfully prepared from poly(n-butyl acrylate) and extended with polyacrylonitrile or poly(methyl methacrylate).

5.5.1.2 Initiators

The broad availability of initiators is perhaps the most significant advantage of ATRP. In fact, many alkoxyamines and RAFT reagents are prepared from activated alkyl halides, i.e. ATRP initiators. Essentially all compounds with halogen atoms activated by α-carbonyl, phenyl, vinyl or cyano groups are efficient initiators. Compounds with a weak halogen–heteroatom bond, such as sulfonyl halides, are also good ATRP initiators [90].

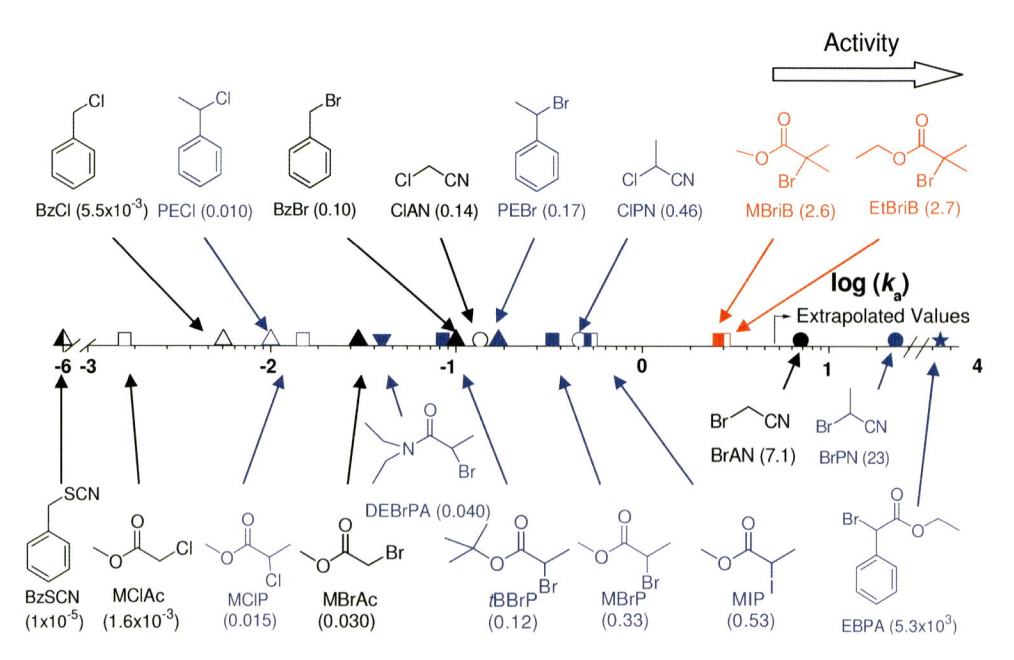

Fig. 5.1 ATRP rate constants of activation for various initiators with CuX–PMDETA in MeCN at 35 °C [92].

The reactivity of initiators depends *reciprocally* on the R–X bond dissociation energy (BDE) [91]. The activation rate constants are typically governed by the degree of initiator substitution [primary (black entries) < secondary (blue) < tertiary (red)], by the radical stabilizing groups [–C(O)NEt$_2$ (inverted triangle) < –Ph (triangle) ≈ –C(O)OR (square) ≪ –CN (circle)] and by the leaving atom/group (e.g. for methyl 2-halopropionates: chloro : bromo : iodo ≈ 1 : 20 : 35). Benzyl bromide is over 10 000 times more reactive than the corresponding thiocyanate. The values of the rate constants of activation for various alkyl halides are shown in Fig. 5.1.

There are many compounds with several activated halogen atoms which have been used to initiate multidirectional growth to form star-like polymers, brushes and related copolymers. Activated halogens can be incorporated at the chain ends of polymers prepared by other techniques such as polycondensation, cationic, anionic, ring-opening metathesis, coordination and even conventional radical processes to form macroinitiators. Chain extensions via ATRP from such macroinitiators were successfully used to form novel diblock, triblock and graft copolymers [93–95]. Inorganic surfaces (wafers, colloids, fibers, various porous structures) have been functionalized with ATRP initiators to grow films of controlled thickness and density [96–98]. Also, several natural products have been functionalized with ATRP initiators to prepare novel bioconjugates [99–101].

5.5.1.3 Transition Metal Complexes as ATRP Catalysts

ATRP has been successfully mediated by a variety of metals, including those from Groups 4 (Ti [102]), 6 (Mo [51, 103, 104]), 7 (Re [105]), 8 (Fe [106–110], Ru [57, 111], Os [52]), 9 (Rh [112], Co [113]), 10 (Ni [114, 115], Pd [116]) and 11 (Cu [56, 59]). Cu is by far the most efficient metal, as determined by the successful application of its complexes as catalysts in the ATRP of a broad range of monomers in diverse media. The transition metal complex is very often a metal halide, but pseudohalides, carboxylates and compounds with non-coordinating triflate and hexafluorophosphate anions have also been used successfully [117].

Ligands include multidentate alkylamines [118, 119], pyridines [120, 121], phosphines [106] or ethers. Ligands serve to adjust the atom transfer equilibrium and to provide appropriate catalyst solubility. Since atom transfer proceeds via a reversible expansion of the metal coordination sphere, the system should be flexible to assure a very fast reversible inner sphere electron transfer process [122]. The atom transfer process should dominate over aforementioned side-reactions. Although some catalysts are suitable for several classes of monomers, other systems may require a specific catalyst and optimized conditions.

Copper complexes with various polydentate N-containing ligands are most often used as ATRP catalysts [59]. This late transition metal does not complex strongly with polar monomers, it has a low tendency to form organometallic species and it typically does not favor hydrogen abstraction. The redox properties of the catalyst can be adjusted over a very broad range (>500 mV, corresponding to a range of ATRP equilibrium constants spanning eight orders of magnitude) [123–125]. More reducing catalysts are characterized by higher

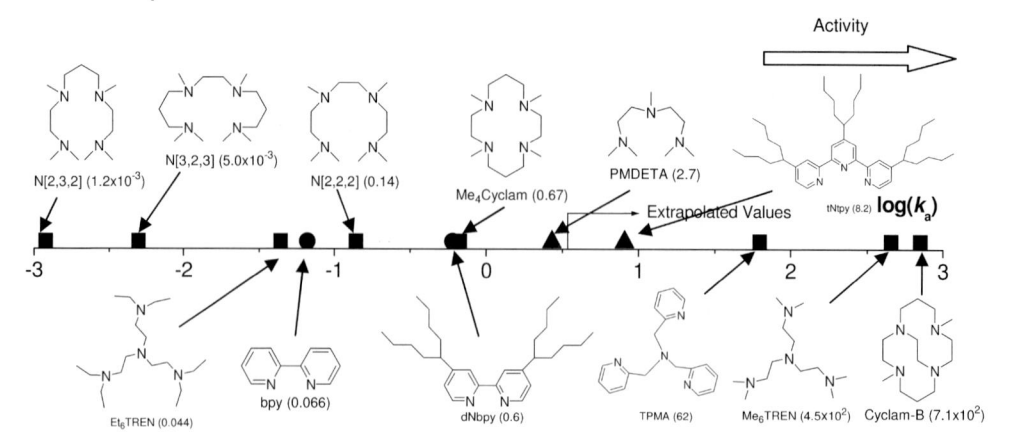

Fig. 5.2 ATRP rate constants of activation for various ligands with ethyl-2-bromoisobutyrate in the presence of CuBr in MeCN at 35 °C [126].

K_{ATRP}. Another parameter which affects equilibrium constants is halogenophilicity, i.e. the affinity of halide anions to the transition metal complexes in their higher oxidation state. For example, since Cu(II) species typically do not form strong complexes with iodides, alkyl iodides are not efficient initiators in Cu-mediated ATRP.

Figure 5.2 illustrates the order of reactivity of various copper complexes with N-base ligands in the activation process [126]. There are a few general rules for catalyst activity: (1) activity depends very strongly on the linking unit between the N atoms (C4 ≪ C3 < C2) and/or the coordination angle; (2) activity depends on the topology of the ligand (cyclic ≈ linear < branched); (3) activity depends on the nature of the N-ligand (arylamine < arylimine < alkylimine < alkylamine ≈ pyridine); (4) steric effects, especially those close to the metal center, are very important (e.g. the Me$_6$TREN complex is 1000 times more active than Et$_6$TREN; 6,6'-diMe-2,2'-bpy is inactive in ATRP in contrast to 2,2'-bpy and 4,4'-disubstituted 2,2'-bpy) [127].

Selection of the catalyst requires attention not only to the appropriate K_{ATRP}, but also to systems with very fast deactivation (k_{da}), as the degree of control over molecular weight distribution in a controlled radical polymerization is defined by the rate of deactivation according to Eq. 4 above.

5.5.2
Conditions

ATRP can be conducted over a very broad temperature range, from sub-zero to >130 °C. Reactions have been successful in bulk, organic solvents, CO_2, water (homogeneous and heterogeneous – emulsion, inverse emulsion, miniemulsion, microemulsion, suspension, precipitation) and even in the gas phase and from solid surfaces.

For very reactive catalysts (which are naturally air sensitive) and for polymerization systems which are difficult to deoxygenate (water, large vessels), irreversible oxidation of the catalyst by residual oxygen can lead to failed *normal* ATRP. Therefore, alternative systems have been developed. *Reverse* ATRPeverse ATRP",4,1> is a convenient method for avoiding such oxidation problems. In this technique, the ATRP initiator and lower oxidation state transition metal activator (Cu^I) are generated *in situ* from conventional radical initiators and the higher oxidation state deactivator (Cu^{II}) [128, 129]. The initial polymerization components are less sensitive to oxygen and yet the same equilibrium between active and dormant species can ultimately be established as in normal ATRP.

However, there are several drawbacks associated with reverse ATRP: (i) because the transferable halogen atom or group is added as a part of the copper salt, the catalyst concentration must be comparable to the concentration of initiator and therefore cannot be independently regulated; (ii) block copolymers cannot be formed; (iii) Cu^{II} complexes are typically much less soluble in organic media than those complexes of Cu^I, often resulting in a heterogeneous (and poorly controlled) polymerization; (iv) very active catalysts must be used at lower temperatures where the gradual decomposition of thermal initiators results in slow initiation and consequently poor control.

In contrast to reverse ATRP, clean block copolymers can be produced when starting from the higher oxidation state transition metal complex if the *activators are generated by electron transfer* (AGET). Reducing agents unable to initiate new chains (rather than organic radicals) are employed in this technique. The principle was demonstrated with a number of Cu^{II} complexes using tin(II) 2-ethylhexanoate [130] and ascorbic acid [131] as the reducing agents, which react with Cu^{II} to generate the Cu^I ATRP activator. Normal ATRP then proceeds in the presence of alkyl halide initiators or macromonomers. The technique has proven particularly useful in miniemulsion systems [132].

Perhaps the most industrially relevant recent development in ATRP was the realization that *activators* could be *regenerated by electron transfer* (ARGET) to decrease the amount of transition metal catalyst necessary to achieve controlled polymerizations (Scheme 5.11) [133, 134]. In principle, the absolute amount of

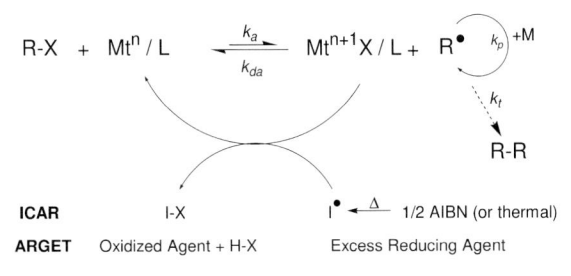

$$R\text{-}X + Mt^n/L \underset{k_{da}}{\overset{k_a}{\rightleftharpoons}} Mt^{n+1}X/L + R^{\bullet} \overset{k_p}{\underset{}{\circlearrowright}} +M$$

$$\downarrow k_t$$
$$R\text{-}R$$

ICAR I-X   I$^\bullet$ ←$^\Delta$— 1/2 AIBN (or thermal)
ARGET Oxidized Agent + H-X   Excess Reducing Agent

Scheme 5.11

copper catalyst can be reduced under normal ATRP conditions without affecting the polymerization rate, a rate which is governed by a ratio of the concentrations of Cu^I to Cu^{II} species according to Eq. 3.

Because the Cu^{II} deactivator accumulates during ATRP due to radical termination reactions, if the initial amount of copper catalyst does not exceed the concentration of those chains which actually terminate, no Cu^I activating species will remain and polymerization will halt. In ARGET ATRP, the relative concentration of catalyst to initiator can be reduced to much lower than under normal ATRP conditions because a reducing agent [such as tin(II) 2-ethylhexanoate or ascorbic acid] is used in excess to regenerate constantly the Cu^I activating species. This technique has effectively reduced the amount of Cu catalyst necessary to achieve control in a polymerization from typical values exceeding 1000 ppm down to approximately 10 ppm (polystyrene with $M_n > 60000$ g mol^{-1} and $M_w/M_n < 1.2$ was produced using this technique with only 10 ppm of Cu catalyst [134]). The economical and environmental implications of this development are obvious; additionally, the catalyst and excess reducing agent can effectively work to scavenge and remove dissolved oxygen from the polymerization system, an added benefit of industrial relevance.

The role of the reducing agent in ARGET can also be played by radical initiators which can regenerate Cu(I) species. In this system, known as *initiators for continuous activator regeneration* (ICAR) ATRP (Scheme 5.11), only very small amounts of radical initiators are used, which remain until the very end of the reaction. ICAR very closely resembles RAFT (see below) [135]. The role of dithio ester in RAFT is played by an alkyl halide in the presence of ppm amounts of copper catalysts. The small amount of radical initiator is responsible for continuing polymerization. If the radical source disappears, RAFT stops and ICAR would also stop very quickly.

5.5.3
Controlled Architectures

ATRP has been successfully used to control topology (linear, stars, cyclic, comb, brushes, networks, dendritic and hyperbranched), composition (statistical, blocks, stereoblocks, multi block copolymers, multisegmented block copolymers, graft, periodic, alternating and gradient copolymers) and functionalities (incorporated into side-chains, end-groups (homo- and heterotelechelics) and in many arms of stars and hyperbranched materials). In addition, natural products and inorganics have been functionalized via ATRP.

5.6
Degenerative Transfer Processes and RAFT

Conventional free radical initiators are used in degenerative transfer (DT) processes and control is assured by the presence of transfer agents (RX) which exchange a group/atom X between all growing chains. This thermodynamically neutral (degenerative) transfer reaction should be fast compared with propagation ($k_{exch} > k_p$). At any instant, the concentration of dormant species P_n–X is much larger than that of the active species P_n^* (by $\sim 10^6$ times). Degrees of polymerization are defined by the ratio of concentrations of consumed monomer to the *sum* of concentrations of the consumed transfer agent and the decomposed initiator:

$$DP_n = \Delta[M]/([TA] + f\Delta[I]) \quad (5)$$

If transfer is fast and the concentration of the decomposed initiator is much lower than that of the transfer agent, good control can be obtained. Polydispersities decrease with conversion and depend on the ratio of the rate constants of propagation to exchange:

$$M_w/M_n = 1 + (k_p/k_{exch})(2/p - 1) \quad (6)$$

DT processes do *not* operate via the PRE. A steady-state concentration of radicals is established via initiation and termination processes. In some cases, the exchange process may occur via some long-lived intermediates which can either retard polymerization or participate in side-reactions such as the trapping of growing radicals (Scheme 5.12).

The simplest example of DT occurs in the presence of alkyl iodides [136]. Unfortunately, the rate constants of exchange in these systems are typically < 3 times larger than the rate constants of propagation for most monomers, resulting in polymers with polydispersities greater than 1.3. Exchange is faster in other DT processes employing derivatives of Te, As or Bi [137–139]; consequently, better control has been obtained with these systems.

Reversible addition–fragmentation transfer (RAFT) polymerization is among the most successful DT processes [140, 141]. While Addition–fragmentation chemistry was originally applied to the polymerization of unsaturated methacrylate esters (prepared by CCT) [142], but offered only limited control. RAFT employs various dithioesters, dithiocarbamates, trithiocarbonates and xanthates as transfer agents, leading to polymers with low polydispersities and various controlled architectures (Scheme 5.13) for a broad range of monomers [143, 144]. The structure of a dithioester has a very strong effect on control. Both R and Z groups must be carefully selected to provide control. Generally, R* should be more stable than P_n^*, in order to be formed and efficiently initiate the polymer-

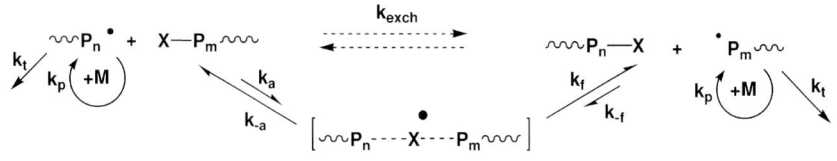

Scheme 5.12

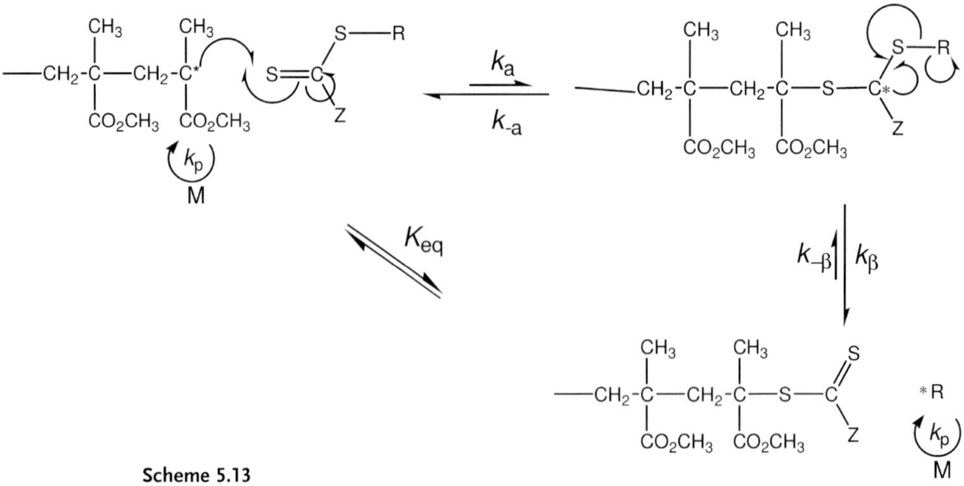

Scheme 5.13

ization. More precisely, the selection of the R group should also take into account stability of the dormant species and rate of addition of R* to monomer. The structure of group Z is equally important. Strongly stabilizing Z groups such as Ph are efficient for styrene and methacrylate but they retard polymerization of acrylates and inhibit polymerization of vinyl esters. On the other hand, very weakly stabilizing groups, such as $-NR_2$ in dithiocarbamates or $-OR$ in xanthates, are good for vinyl esters but less efficient for styrene. However, the proper choice of the Z substituent can dramatically enhance the reactivity of these thiocarbonylthio compounds [145].

The contribution from termination is slightly different in RAFT than it is in SFRP and ATRP. In the last two cases all chains are originally short; as they become longer, the probability of termination decreases. However, in DT systems, there will always be a small proportion of continuously generated chains. They will terminate faster than long chains. This will also make the formation of pure block copolymers via typical DT techniques impossible, since low-MW initiators will always generate new homopolymer chains.

5.6.1
Monomers and Initiators

DT can be applied to any monomer used in RP unless there is some interference with the transfer agent. For example, dithioesters can be decomposed in the presence of primary amines and alkyl iodides may react with strong nucleophiles (including amines). RAFT is also suitable to polymerize hydrophilic monomers directly in aqueous media without any protection group chemistry [146].

In principle, there is no limitation for the radical initiators. However, their concentrations should be lower than that of the transfer agent.

5.6.2
Transfer Agents

As discussed above, transfer agents should be carefully selected for particular monomers in order to provide sufficiently fast addition but also sufficiently fast fragmentation. Alkyl iodides (and derivatives of Te, Sn, Ge, As or Bi) should have a leaving group R which provides a more stable radical. Isobutyrates are good for acrylates, but propionates are not efficient for polymerization of methacrylates. This also has implications for the block copolymerization order.

In the RAFT polymerization of MMA with dithiobenzoates [S=C(Ph)SR], the effectiveness of the RAFT agent (i.e. the leaving group ability of R) decreases in the order $C(alkyl)_2CN \approx C(alkyl)_2Ph > C(CH_3)_2C(=O)OEt > C(CH_3)_2C(=O)NH(alkyl) > C(CH_3)_2CH_2C(CH_3)_3 \approx CH(CH_3)Ph > C(CH_3)_3 \approx CH_2Ph$. In fact, only the first two groups are successful in preparing PMMA with low polydispersity ($M_w/M_n \leq 1.2$) and predetermined molecular weight [147, 148].

Groups Z in RAFT should be more radical stabilizing for monomers forming more stable growing radicals. For the RAFT polymerization of styrene, the chain transfer constants were found to decrease in the series where Z is aryl (Ph) >> alkyl (CH_3) $\approx$ alkylthio (SCH_2Ph, SCH_3) $\approx$ N-pyrrolo >> N-lactam > aryloxy (OC_6H_5) > alkoxy >> dialkylamino [149].

5.6.3
Controlled Architectures

RAFT and other DT systems can be applied to the synthesis of many (co)polymers with controlled architectures. It should be remembered that there will always be some contamination from chains resulting from the radical initiators.

Multifunctional RAFT reagents are usually prepared from multifunctional activated alkyl bromides. RAFT reagents can be attached to polymer chains or surfaces via either the R or Z group. Attachment via the Z group allows polymer chains to grow (and terminate) in solution. This can reduce cross-linking for multifunctional systems [150].

Statistical, block, stereoblock, alternating/periodic and gradient copolymers have been prepared using RAFT. Many functional monomers have been polymerized by RAFT. The dithio ester group can be removed by the addition of a large excess of AIBN [151].

5.7
Relative Advantages and Limitations of SFRP, ATRP and DT Processes

Each CRP system has some comparative advantages and disadvantages. Recent advances in all three techniques have minimized many of the disadvantages, but there will always be some limitations pertinent to each methodology.

5.7.1
SFRP

- *Advantages:* (1) purely organic system for NMP applicable to many monomers including acids; (2) no Trommsdorf effect.
- *Limitations:* (1) relatively expensive moderators must be used in stoichiometric amounts to the chain end; (2) same limitation for potentially toxic transition metal compounds; (3) difficult to control polymerization of disubstituted alkenes such as methacrylates; (4) difficult to introduce end functionality; (5) relatively high temperatures are required for NMP.

5.7.2
ATRP

- *Advantages:* (1) only catalytic amounts of transition metal complexes (commercially available) are necessary; (2) many commercially available initiators including multifunctional and hybrid systems; (3) large range of monomers (except unprotected acids); (4) easy end functionalization; (5) no Trommsdorff effect; (6) large range of polymerization temperatures.
- *Limitations:* (1) transition metal catalyst must often be removed; (2) acidic monomers require protection.

5.7.3
RAFT and Other DT Processes

- *Advantages:* (1) large range of monomers; (2) minimal perturbation to RP systems.
- *Limitations:* (1) transfer agents are not yet commercially available (except some alkyl iodides); (2) dithio ester and other end-groups should be removed due to color, toxicity and potential odor; (3) amine-containing monomers decompose dithio esters; (4) functionalization is usually difficult.

Scheme 5.14 compares SFRP, ATRP and RAFT in the areas related to the synthesis of high- and low-MW polymers, functional polymers, block copolymers, the range of polymerizable monomers, various hybrids, environmental issues and polymerization in water. The left part of the scheme was presented several years ago [152], but many aspects of all three polymerizations have since changed.

All techniques have expanded in all areas and new advances have allowed many of the aforementioned limitations to be overcome.
1. In NMP, MMA polymerization can be controlled in the presence of small amounts of styrene, allowing the formation of block copolymers [43].
2. In ATRP, AGET has allowed the production of clean block copolymers by starting from an oxidatively stable catalyst precursor and employing reducing agents; ARGET and ICAR have lowered the amount of Cu necessary to low ppm levels and allowed higher MW polymers to be synthesized [134, 135];

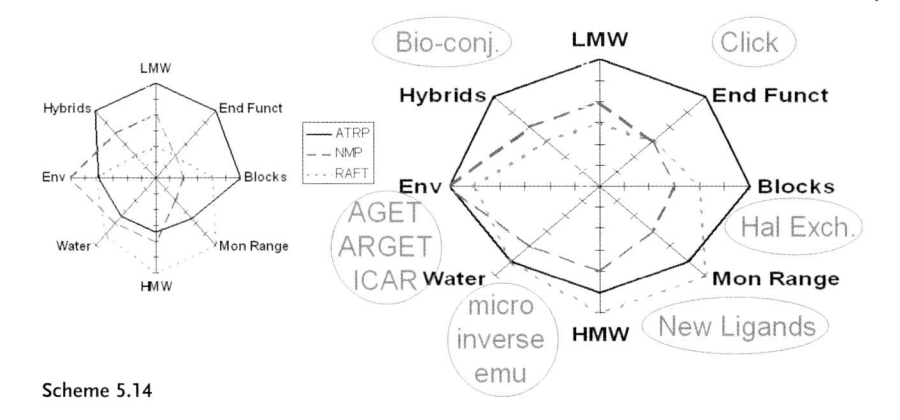

5.8 Controlled Polymer Architectures by CRP: Topology

Scheme 5.14

new ligands have expanded the range of polymerizable monomers to include vinyl acetate and vinyl chloride [110, 153]; simple replacement of a Br end-group by azide has opened up many possibilities for functionalization via click chemistry [154], including new bioconjugates; halogen exchange [87] has allowed the production of many block copolymers with sequences not available via other processes.

3. In RAFT, dithio end-groups could be removed with the simple addition of AIBN [151]; with the rational design of RAFT reagents, retardation and side-reactions can be avoided [155]; the convergent growth of chains in multifunctional system can prevent macroscopic gelation [150].
4. In all systems, a better understanding of the processes in dispersed media has allowed the reduction of the amount of surfactant necessary, has increased colloidal stability and has also resulted in the extension of miniemulsion to true emulsion, microemulsion and inverse miniemulsion systems [47, 131, 156–159].

5.8
Controlled Polymer Architectures by CRP: Topology

CRP allows the preparation of many well-defined polymers with predetermined MW and narrow MWD or even designed MWD. Moreover, polymers with controlled topology, composition and functionality are accessible by CRP. Figure 5.3 presents a few examples of polymers with such controlled topology.

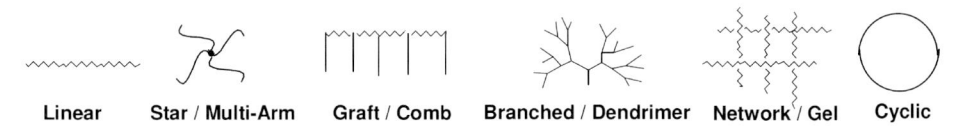

Fig. 5.3 Illustration of polymers with controlled topology prepared by CRP.

Linear   Star / Multi-Arm   Graft / Comb   Branched / Dendrimer   Network / Gel   Cyclic

5.8.1
Linear Chains

All CRP techniques have been used in the synthesis of linear chains. The use of difunctional initiators allowed for the first time simultaneous two-directional growth in a radical process. This gives access to both functional telechelic polymers and polymers with MW double those otherwise attainable under similar conditions. There are many initiators and macroinitiators that contain functional group(s) and therefore add value to the products obtained [160, 161]. Several difunctional activated alkyl halides are commercially available. Many others can be prepared by simple esterification of diols and functional alcohols with, for example, 2-bromoisobutyryl bromide.

5.8.2
Star-like Polymers

Star-like polymers have been prepared via four different approaches [162]:
1. Multifunctional initiators have been used to grow simultaneously several arms via a core-first approach (Scheme 5.15). ATRP is the most frequently used CRP system for this technique due to the availability of many polyols that have been subsequently converted to the initiating core with 3, 4, 6, 12 or more initiating sites [163–165]. However, the synthesis of star-like structures by CRP using a core-first method faces the danger of potential cross-linking and gelation because of radical coupling. Thus, a relatively low concentration of radicals and low conversions are often used to minimize star–star coupling.
2. An coupling-onto approach involves attachment of chains to a functional core (Scheme 5.15). Certain RAFT reagents can be attached to a core via the Z group; therefore, RAFT offers a unique possibility to grow arms in solution [150]. Arms are attached to the core as a dormant species and therefore cannot terminate. Termination of linear chains in solution does not lead to cross-linking. A similar approach involves growing linear monofunctional Br-terminated chains in solution via ATRP and then converting the end-groups to azides. These chains were subsequently "clicked" to a core with several acetylene groups [166, 167].

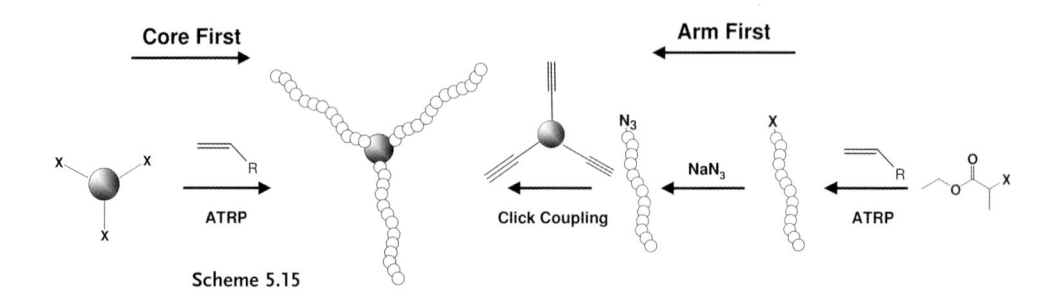

Scheme 5.15

5.8 Controlled Polymer Architectures by CRP: Topology | 189

Scheme 5.16

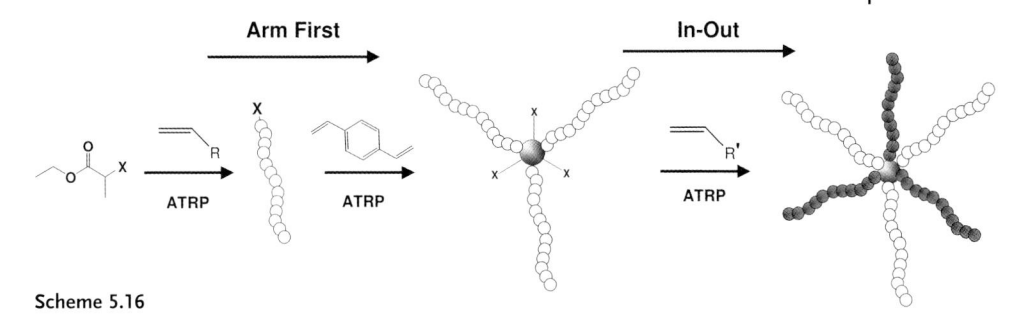

3. Another approach utilizes arms that are cross-linked in the presence of divinyl compounds (Scheme 5.16). In this case, the number of arms in the resulting stars is not precisely controlled. Also, star–star coupling between the cores may occur. This synthetic procedure requires precise manipulation of arm length, the amount of divinyl compound and the reaction time [168–170]. Cross-linker can be added to isolated arms or at a certain moment during the arm synthesis, allowing for control of the core density. All of these parameters affect the resulting star structure. As an example, the synthesis of stars with ~50 arms in >90% yield will require ~10-fold molar excess of cross-linker (vs. arm end-groups) for precursors with $DP \approx 100$.
4. Stars prepared via the arm-first process with divinyl crosslinker (item 3 above) still contain active/dormant species at the core. Hence they can be used to grow a second generation of arms by the so called *in–out* approach (illustrated for ATRP in Scheme 5.16 and also possible for RAFT and NMP). Such miktoarm stars can contain the same monomeric units but can be of different lengths; they can also contain hetero-arms. However, the efficiency of re-initiation is often not very high. By using a degradable cross-linker with an S–S linking unit, it was determined that the efficiency of re-initiation is usually ~30%. Hence, if the weight fractions of the first and second generation arms are similar, the second generation arms will be roughly three times longer. This may be due in part to poor accessibility of the monomer/catalyst to the sterically congested core, but may also result from enhanced termination inside the core [166, 170].

5.8.3
Comb-like Polymers

Comb-like polymers can be prepared using three techniques which are similar to those described for stars [161, 171]. However, instead of a compact core, a long linear backbone is employed. These three techniques correspond to grafting from, onto and through.
1. Grafting from is the most common method and is most similar to the core-first approach for stars. It is often used to synthesize graft copolymers with a different composition of the backbone and the side-chains; it has (for exam-

ple) been successfully conducted from polyethylene [172–174], poly(vinyl chloride) [175, 176], and polyisobutene [177].
2. Grafting onto is very similar to an arm-first approach. Click chemistry with azides and acetylenes was used to attach side-chains efficiently to a backbone [154].
3. In grafting through, monofunctional macromonomers are utilized as comonomers together with a low-MW monomer, which permits the incorporation of macromonomers prepared by other controlled polymerization processes {such as polyethylene [178, 179], poly(ethylene oxide) [180], polysiloxanes [181], poly(lactic acid) [182] and polycaprolactone [183]} into a backbone of polystyrene or poly(meth)acrylate prepared by CRP. The grafting density depends on the proportion of macromonomer and may not always be uniform. Reactivity ratios in copolymerizations of macromonomers depend not only on true chemical reactivity but also on viscosity, compatibility, etc. Controlled/living processes have permitted the preparation of combs of different shapes, including tadpole or dumbbell structures [184].

Other interesting brush topologies with unique properties can be synthesized using different macromonomers in heterografted brush copolymers [185] or block brush copolymers prepared by a combination of grafting from and grafting through [186]. When macromonomers are polymerized by grafting from a macroinitiator, double-grafted brush copolymers can be synthesized where each graft is a brush [187].

Figure 5.4 presents an AFM image of molecular brushes consisting of a methacrylate backbone with poly(n-butyl acrylate) side-chains with a gradient density grafting. These gradient brushes were prepared by copolymerization of MMA with HEMA–TMS using ATRP and a feeding process. Subsequently, TMS groups were converted to 2-bromopropionate groups to initiate growth of side-chains.

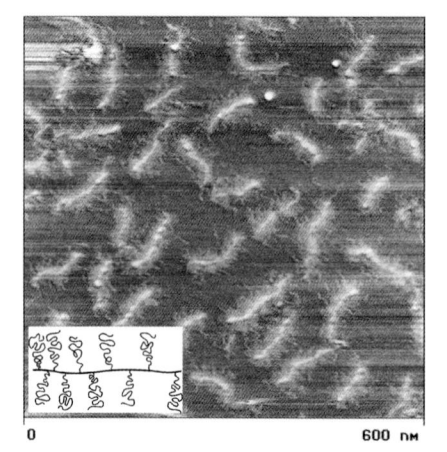

Fig. 5.4 AFM image of gradient molecular brushes. Reprinted with modifications from [188]. Copyright 2002, with permission from ACS.

5.8.4
Branched and Hyperbranched Polymers

Comb-like polymers can be considered as models for regularly branched polymers. Branching occurs in polymers made by conventional RP as a result of transfer to polymer. It may be due to intramolecular transfer (short branches) or intermolecular transfer which would give long chain branching and ultimately gelation. Branching is favored at higher temperatures and for less stabilized radicals. It becomes a very important reaction at high temperatures in the synthesis of polyethylene (resulting in low-density polyethylene). However, CRP is typically run at lower temperatures and it seems that the contribution of branching is reduced in systems with intermittent activation.

CRP can also be used to prepare hyperbranched polymers [77, 189–191]. Monomers that also serve as initiators (so-called inimers) have been successfully used to make such materials (Scheme 5.17). The ratio of reactivities of the initiating site and the active species resulting from the alkene moiety define the de-

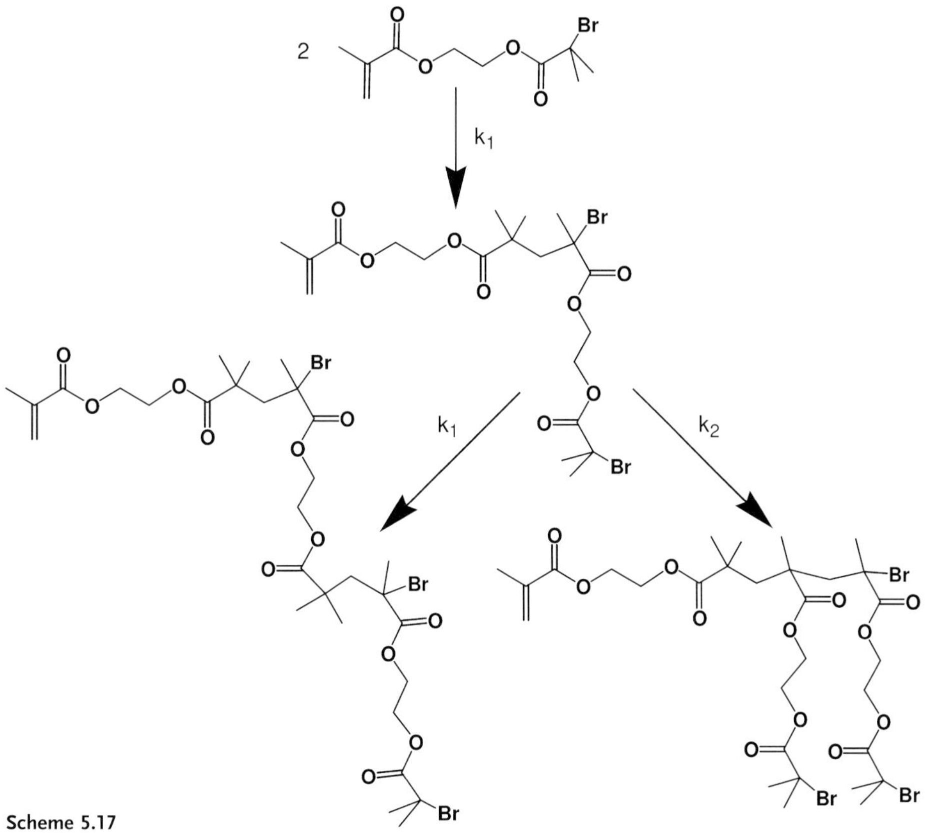

Scheme 5.17

gree of branching (which can approach 50%). It is possible additionally to affect this ratio by varying the concentration of deactivator, as was demonstrated in ATRP. Copolymerization of inimers with regular monomers reduces the degree of branching and increases the branch length [192].

Hyperbranched structures can also be obtained when divinyl monomers are copolymerized in small amounts or in the presence of various compounds (thiols) that inhibit cross-linking and favor the formation of branched products [193].

5.8.5
Dendritic Structures

While hyperbranched systems are irregular, branching is controlled in dendritic structures by the degree of polymerization. Dendritic structures have been prepared by a single CRP method [194, 195] and also with various combinations of CRP and other controlled living polymerization techniques [196–198]. Polymer chains have been grafted from dendrimer surfaces or attached to dendrimers. Single chains have been combined with dendrimer structures. Dendrons have also been functionalized with methacrylate groups and copolymerized to form stiff polymer chains [199–202].

5.8.6
Polymer Networks and Microgels

Most polymer networks have irregular structures. However, CRP can significantly improve network uniformity. When a divinyl compound is copolymerized with a monovinyl compound, the gel point in CRP occurs much later than in RP. This is due to the fact that in CRP, only short chains are formed at the early stages and the number of pendant double bonds in each chain is much lower than in conventional RP [203].

Additionally, it is possible to prepare well-defined polymers with cross-linkable pendant moieties that can subsequently form microgels, even with just a single intramolecularly cross-linked chain. Star polymers with microgel cores have been prepared by ATRP [204, 205]. Water-soluble polymeric nanogels have been synthesized by xanthate-mediated radical cross-linking copolymerization [206]. Degradable gels have been prepared by the copolymerization of styrene or methyl methacrylate and a disulfide-containing difunctional methacrylate [207, 208]. It is also possible to use cross-linkers which can be reversibly cleaved, which leads to the formation of reversible gels [170, 209].

5.8.7
Cyclic Polymers

CRP has no special advantage in forming cyclic polymers. Rather, anionic and/or cationic polymerizations are much better suited to make cyclics by using complementary reagents at very low concentrations. However, cyclization has been reported when click chemistry was used in a reaction of azido- and acetylene-terminated chains [210, 211].

5.9
Chain Composition

CRP is very well suited for preparation of copolymers with precisely controlled composition. CRP combines facile cross-propagation with controlled/living character that can be applied to many monomers. Figure 5.5 presents some examples of copolymers with controlled composition.

5.9.1
Statistical Copolymers

Statistical copolymers are formed in RP when comonomers have similar reactivity. The copolymerization of monomers with different reactivity ratios (e.g. methacrylates and acrylates have reactivity ratios ~3 and ~0.3, respectively) results in copolymers with a spontaneous gradient (cf. 5.9.5) [161, 212]. In general, reactivity ratios in CRP are the same as in RP, although small differences exist which have been ascribed to non-equilibrated systems [19, 213].

5.9.2
Segmented Copolymers (Block, Grafts and Multisegmented Copolymers)

The area of segmented copolymers is perhaps the most prolific in CRP. Due to the tolerance of CRP to many functionalities, essentially an infinite number of possible structures is attainable. The end-groups of polymers prepared by other techniques, including cationic, anionic, coordination and step-growth polymerization, have been converted to CRP initiating sites and used as macroinitiators for CRP. Multisegmented copolymers can also be prepared by employing "click" coupling of complementary macroinitiators.

5.9.2.1 Block Copolymers by a Single CRP Method
Efficient block copolymerization requires an appropriate sequence of blocks. This order generally follows radical stabilities and can be correlated with the proportion of active species. This order may be specific for a particular polymer-

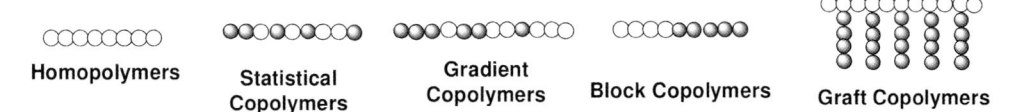

Fig. 5.5 Illustration of polymers with controlled composition.

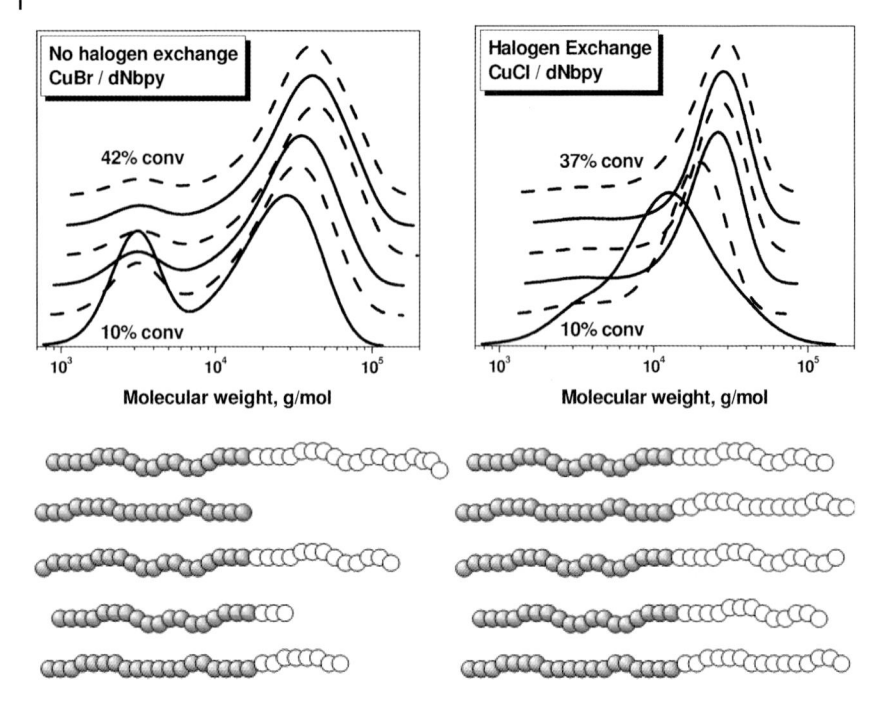

Fig. 5.6 The GPC curves and illustrations show chain extension of a polySty–Br macroinitiator with MMA using (a) CuBr–dNbpy and (b) CuCl–dNbpy as the catalyst. [MMA]:[polySt–Br]:[CuX]:[dNbpy] = 500:1:1:2; 75 °C [216].

ization. In ATRP the order is as follows: acrylonitrile > methacrylates > styrene ≈ acrylates > acrylamides. In RAFT, due to steric effects, methacrylates are most reactive. In ATRP, the block order can be altered by using halogen exchange (as discussed). Figure 5.6 illustrates enhanced control over the chain extension of polystyrene with MMA when halogen exchange is employed. Additionally, thermoplastic elastomers of PMMA–PBA–PMMA were successfully obtained by starting from Br–PBA–Br and using CuCl/L to chain-extend with MMA [87, 88, 214]. ABC, ABCD and ABCBA blocks have also been prepared by ATRP [215].

5.9.2.2 Block Copolymers by Combination of CRP Methods

ATRP macroinitiators have been successfully converted to RAFT and NMP initiators. After site transformation, these initiators served as a starting point for RAFT and NMP-mediated block copolymerization [161].

5.9.2.3 Block Copolymerization by Site Transformation and Dual Initiators

CRP and living carbocationic and carbanionic polymerization

The dormant species in styrene cationic polymerizations and ATRP systems have identical structure. Therefore, Cl-terminated polystyrenes could be efficiently extended via ATRP with polystyrene and also with polyacrylates and poly(meth)acrylates [94]. In principle, the reverse process (starting with ATRP) would also be possible for polystyrene, but carbocationic systems are more sensitive to impurities than ATRP. ATRP has been initiated by acetal-functionalized alkyl halides which were subsequently used to grow vinyl ethers by living cationic polymerization. The reverse process was also successful due to the higher tolerance of vinyl ethers to impurities [217].

Living carbanions are quenched by species serving as initiators for NMP and also for ATRP [218, 219].

CRP and ionic ring-opening polymerization

Cationic polymerization of THF was initiated by 2-bromoisobutyryl bromide and silver triflate. The subsequent addition of ATRP catalyst allowed the controlled polymerization of MMA, BA and styrene from the residual bromo ester moiety [220]. In a similar way, oxonium ions at the chain end of living polyTHF were quenched with 4-hydroxyTEMPO and then used to chain extend the polyTHF with styrene via NMP [221]. Bromo end-groups in ATRP of styrene were used to initiate directly ring-opening polymerization of oxazolines [222].

Hydroxy end-groups in polymers from AROP of ethylene oxide or lactones were directly esterified to bromo esters and used as ATRP initiators. In this way, not only mono- but also di- and multifunctional polymers were synthesized. 2-Hydroxyethyl 2'-bromoisobutyrate is a simple dual initiator capable of initiating both AROP and ATRP [223]. Depending on the initiating/catalytic systems, the rates of both processes can be comparable, allowing the simultaneous presence of two non-interfering processes [224, 225].

CRP and coordination polymerization

Unsaturated chain ends in polyethylene and polypropylene were transformed to bromo esters to initiate ATRP of styrene and (meth)acrylates. Hydroxy functional polyethylene obtained in degenerative transfer polymerization of ethylene was used in a similar way for block copolymerization via ATRP [172, 173, 179, 226]. A similar approach has been used to create graft copolymers. Copolymerization of ethylene and 10-undecen-1-ol in the presence of a trialkylaluminum protecting group can be conducted with a metallocene catalyst. If the hydroxyl functionality is converted into an α-bromoisobutyrate functionality by reaction with excess α-bromoisobutyryl bromide in the presence of triethylamine (Scheme 5.18), the polyethylene multifunctional macroinitiator can be used in a "grafting from" ATRP procedure. In this way, copolymers such as polyethylene-*graft*-poly(methyl methacrylate) have been produced which have found application as compatibilizers for otherwise immiscible polymers [172, 227].

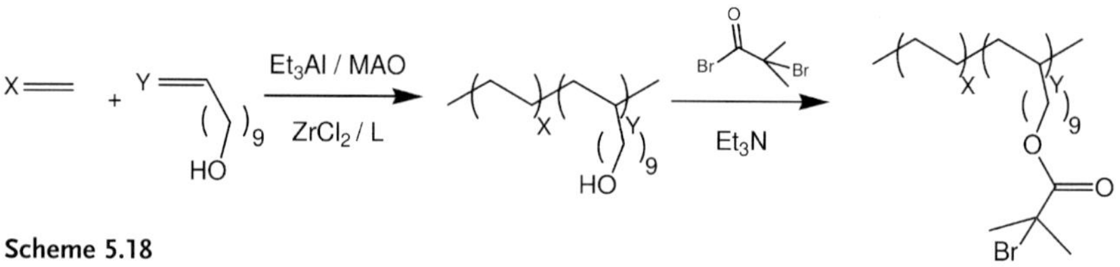

Scheme 5.18

Several methods combine CRP with ROMP. Molybdenum alkylidene chain ends are destroyed by aldehydes. Thus, aldehydes with activated alkyl bromides were used to terminate ROMP of norbornene and subsequently form block copolymers with styrene and (meth)acrylates [93]. Another interesting possibility is the simultaneous growth of polymer chains via ROMP on ruthenium alkylidenes which also serve as ATRP catalysts. Interestingly, an excess of phosphines reduces the rate of ROMP but does not affect the rate of ATRP [228, 229].

CRP and polycondensation polymers

End-groups from many step growth polymers are easily transformed to ATRP macroinitiators. Thus, terminal hydroxy groups from polyesters, polycarbonates or polysulfones were esterified to bromo esters and subsequently chain extended with styrene or (meth)acrylates [95, 230, 231].

Polythiophenes are formed by polycondensation, but the mechanism follows a chain growth rather than a step growth process. It is therefore possible to prepare polythiophenes with low polydispersity. End-groups are well controlled and can be converted to ATRP initiators. Block copolymerizations with this technique have been very efficient and well-defined nanostructured materials have been prepared [232–234].

CRP and RP

CRP has also been combined with polymers prepared by conventional RP. For example, telechelic polymers prepared by RP can either contain ATRP active sites or can be converted to the corresponding macroinitiators and used for controlled ATRP. In such a way, poly(vinyl acetate) was block copolymerized with styrene and acrylates. Conversely, well-defined blocks were used to initiate uncontrolled RP of vinyl acetate to afford another type of block copolymer [235].

5.9.2.4 Multisegmented Block Copolymers

Multisegmented block copolymers are typically prepared by polycondensation processes. ATRP can be used to make ABC or ABCBA segments by sequential polymerization. However, chain coupling can afford multisegmented copolymers in a much simpler way. Such coupling can be done with termination by radical coupling; however, yields in these reactions are rather low.

Another approach involves the aforementioned click chemistry. An ATRP initiator with an acetylene functionality (such as propargyl 2-bromoisobutyrate) can

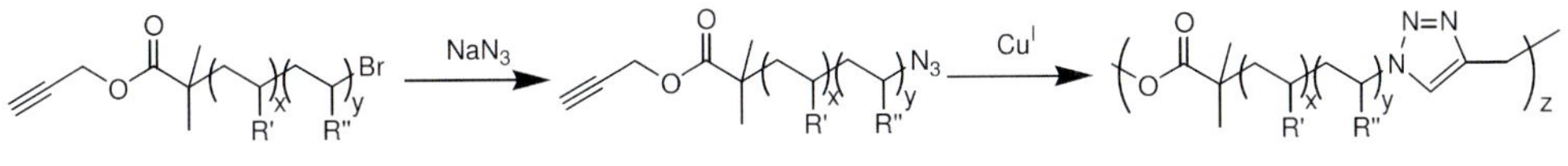

Scheme 5.19

be used to synthesize an α-alkyne-ω-bromo-terminated block copolymer, which can further be reacted with NaN₃ to generate the corresponding α-alkyne-ω-azido-terminated block copolymer. This functional diblock can then be click coupled in a step growth process at room temperature in the presence of Cu^I to generate segmented block copolymers (Scheme 5.19) [210].

5.9.2.5 Stereoblock Copolymers

RP is not stereoselective. However, in the presence of some complexing agents which interact strongly with the pendent groups of the polymer chain end and/or the monomer, the proportion of syndiotactic or isotactic triads can be significantly increased. This has been demonstrated for the polymerization of acrylamides in the presence of catalytic amounts of Yb(OTf)₃ or Y(OTf)₃ [236] and also in the polymerization of vinyl esters in fluoro alcohols [237]. CRP offers the special advantage that atactic segments formed without complexing agent can be extended with stereoregular blocks formed in the presence of such agents. In such a way, stereoblock poly(*atactic*-dimethylacrylamide)-*block*-poly(*isotactic*-dimethylacrylamide) was formed [238–240] (Scheme 5.20).

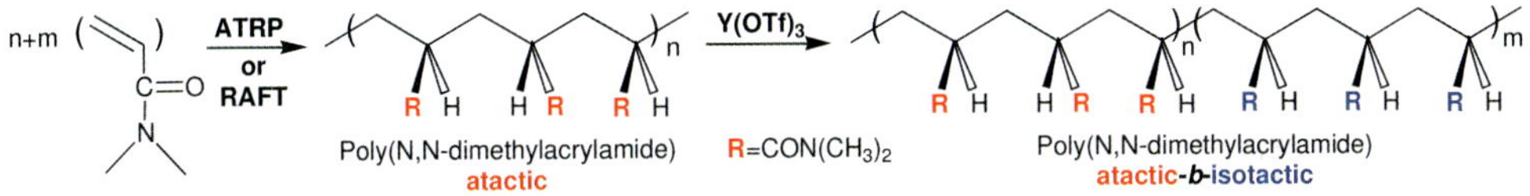

Scheme 5.20

5.9.3
Graft Copolymers

Techniques used for the synthesis of include those applied for the preparation of comb polymers (grafting from, on to and through) and also methods known for block copolymerization, including site transformation.

Polymer backbones containing hydroxy functionalities can be easily esterified to form bromoesters with a controlled amount of ATRP initiating sites for use in "grafting from". In such a way, either very densely grafted molecular brushes or loosely grafted copolymers were prepared. Grafting density can be varied along the chain length (cf. gradients). Various types of block/graft copolymers have also been generated, including core–shell cylinders, double-grafted struc-

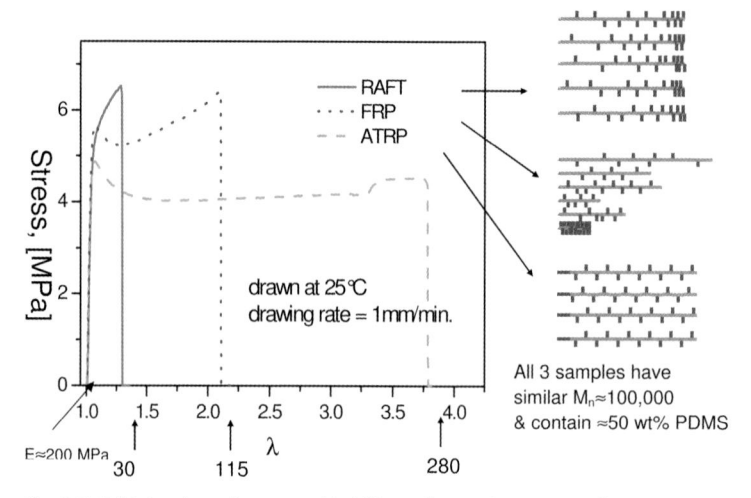

Fig. 5.7 Cold drawing of press-molded films of PMMA–PDMS grafts. Reprinted with modifications from [181]. Copyright 2003, with permission from ACS.

tures, etc. "Grafting onto" is the least explored grafting technique in CRP. Click chemistry or azides and acetylenes and maleic anhydride/phthalimides with amino-terminated polymers have been used for grafting onto. Several macromonomers have been copolymerized by ATRP and other CRP techniques providing graft copolymers with variable density via "grafting through". Copolymerization via CRP provides more uniform grafts because the growth of the backbone is slower and macromonomer has enough time to diffuse to the growing chain [241]. Additionally, it is possible to use macroinitiator with the same chemical composition as macromonomers inherently to compatibilize the graft copolymer.

The mechanical properties of graft copolymers prepared by copolymerization of PDMS macromonomers with MMA using ATRP and RP are dramatically different. Regular graft copolymers with $M_n \approx 100\,000$ and ~50 wt% PDMS ($M_n = 2000$) prepared by ATRP have tensile elongation ~300%, whereas that for irregular grafts is only ~100%. Interestingly, graft copolymers with a gradient composition had tensile elongation only ~30% (Fig. 5.7) [181, 182, 242].

Homopolymerization of macromonomer by CRP is difficult. Only those with relatively "thin" side-chains such as PEO or polyethylene were homopolymerized to high-MW polymers. Macromonomers with "thicker" chains, such as PBA, can be polymerized only to relatively low DP. Homopolymerization of macromonomers by CRP may be more difficult than by RP. Rate constants of propagation of macromonomers are smaller than those of regular monomers, but the rate constants of termination are even less. Thus, in RP, an overall rate of homopolymerization may still be sufficient since it scales with the $k_p/k_t^{1/2}$ ratio. In ATRP, polymerization rate is not affected by k_t and consequently it is much slower for macromonomers than for low-MW monomers.

5.9.4
Periodic Copolymers

Alternating copolymers are the simplest periodic copolymers. There are several comonomer pairs which copolymerize in an alternating fashion by RP. They are typically strong electron donors and strong electron acceptors, such as styrene–maleic anhydride or styrene–maleimides [243]. They were successfully copolymerized by all CRP techniques. It is also possible to use one kinetically reactive but thermodynamically non-homopolymerizable monomer. Maleic anhydride can serve as an example [244]. Another possibility is to use a less reactive monomer in large excess. In such a way, nearly alternating copolymers of acrylates with alkenes were prepared [82, 85]. The reactivity ratios of monomers can also be affected by complexation. For example, in the presence of strong Lewis acids, methyl methacrylate tends to form alternating copolymers with styrene [245].

5.9.5
Gradient Copolymers

In contrast to conventional free radical copolymerization, where differences in comonomer reactivity ratios result in a variation in copolymer composition, in CRP this variation in instantaneous incorporation of the monomers into the copolymer results in a tapering in composition along the main chain of the copolymer. Gradient copolymers can be prepared by either simultaneous copolymerization of monomers with different reactivity ratios (spontaneous gradient) or by continuous feeding of one monomer (forced gradient) [246, 247]. The shape of the gradient depends on concentration of comonomers, on their reactivity ratios and on the feeding process. Gradients can be linear, V-shaped and also curved. Gradient uniformity is a very important parameter. The uniformity depends on the frequency of intermittent activation. If only a few monomer units are added during each activation step, then uniformity should be high and all copolymers should have a similar composition along the chain length. However, if many monomer units are added at each step, then chains would contain long sequences formed at a certain conversion and individual chain composition and microstructure would differ from one another. Additionally, gradient copolymerization can be extended to the formation of graft copolymers with a gradient distribution of grafts (cf. Fig. 5.4) [184, 188].

5.9.6
Molecular Hybrids

Molecular hybrids include synthetic polymers attached covalently to inorganic materials and to natural products [96]. Well-defined organic–inorganic hybrids can be formed by grafting from flat, convex, concave or irregular surfaces [97, 248]. Surfaces of gold, silicon, silica, iron and other inorganic substrates were functionalized with bromo esters to initiate growth of polystyrene, poly(meth)-acrylates, polyacrylonitrile or polyacrylamide chains via ATRP [249]. It is possible to vary the grafting density and reach very densely grafted films, approaching ~0.5 chain nm^{-2} [250, 251]. Such dense grafting is not possible via grafting onto, since attached polymers collapse on the surface and prevent other chains from reaching the surface. Also, high grafting density cannot be achieved without uniform growth. Only when polymer chains grow by adding "about one monomer unit at a time" is high grafting density possible. The resulting hybrids have many new and unusual properties related to their compressibility, lubrication or swelling and interpenetration [252].

It is also possible to graft polymers and block copolymers onto inorganic surfaces. For example, poly(methacrylic acid)-*block*-poly(methyl methacrylate)-*block*-poly(sodium styrenesulfonate) prepared by ATRP was grafted onto ~200-nm diameter iron particles to destroy residual chlorinated solvents [253].

Functional groups in natural products have been converted to ATRP initiators (NH$_2$ or OH groups converted to bromoamides and bromo esters). It is important during polymerization in such systems that the original substrate does not interfere with the ATRP process and that the functionality and properties of the natural products are not destroyed by ATRP. Alternatively, chains prepared by CRP containing functionalities (NH$_2$ or OH) have been used to grow polypeptide or DNA [254]. It is also possible to fuse functionalized natural products and organic polymers together either with click reactions or biotin–avidin chemistry [100, 101, 255–259].

5.9.7
Templated Systems

Templating with CRP can be more precise than with conventional RP. Slower growth may facilitate better control in templated systems. Mesoporous silica has been functionalized with bromoesters and used to grow polyacrylonitrile, polystyrene and poly(meth)acrylate. Growth of polymer chains was followed by TGA, porosity and analysis of detached polymers. Highly uniform growth was achieved on the molecular level (as evidenced by low polydispersity of detached polymers analyzed by GPC) but also macroscopically (overall porosity measurements). The resulting polyacrylonitrile was converted to very regular nanosize carbon filaments ($D \approx 10$ nm, $L > 1$ μm) (Scheme 5.21) [248, 260].

5.10
Functional Polymers

Radical polymerization is tolerant to many polar groups such as –OH, –NH$_2$ or –COOH and can be carried out in water. In CRP, functionalities can be incorporated into specific parts of polymer chains, as shown in Fig. 5.8 [160].

Scheme 5.21

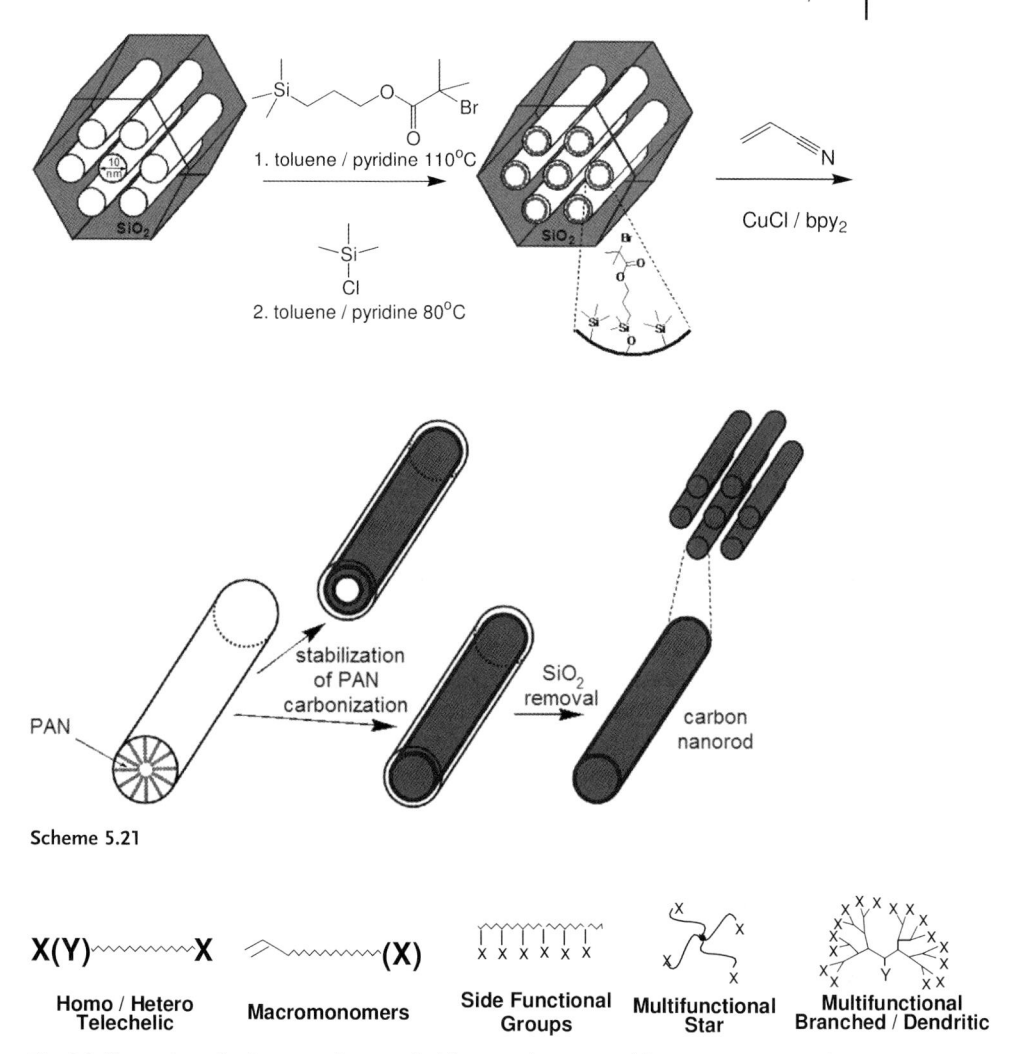

Fig. 5.8 Illustration of polymers with controlled functionality prepared by CRP.

5.10.1
Polymers with Side Functional Groups

Functional monomers have been directly incorporated in polymer backbones and also incorporated in a protected form. For example, hydroxyethyl (meth)acrylate has been homopolymerized and copolymerized by all CRP techniques [261, 262]. Also, trimethylsilyl (TMS) derivatives have been polymerized and subsequently deprotected. The hydroxy groups were then functionalized with 2-bromopropionate groups and extended to form densely grafted molecular brushes [261, 263].

(Meth)acrylic acid has been polymerized directly by RAFT and NMP but less successfully by ATRP [143, 264]. Much better results were obtained when the acid functionality was protected by either acetal, a TMS group, *tert*-butyl or benzyl ester or even converted to a sodium salt [265, 266].

The side functional groups have been subjected to further reactions. In addition to simple deprotection of acetal or ester functionalities, other reactions such as the conversion of cyano groups from acrylonitrile units to tetrazoles and reactions of azides with functional acetylenes have been employed [267].

5.10.2
End-group Functionality: Initiators

End-groups can be incorporated either by using functional initiators or by converting growing chain ends to another functional group [160]. There is strong interest in di- and multifunctional polymers, but functional initiators allow the introduction of functionality to only one chain end. Various functionalized ATRP initiators have been used to prepare hetero-telechelic polymers (Scheme 5.22) [160]. Additionally, atom transfer radical coupling (ATRC) permits the efficient coupling of chains and provides a route to telechelic polystyrenes [268, 269]. ATRC is less efficient for methacrylates (the growing radicals predominantly disproportionate rather than couple) and for acrylates (too low equilibrium constant). However, it is sufficient to add a few styrene units at the acrylate chain end to form polyacrylate telechelics successfully. Multifunctional star polymers have been prepared from functional macroinitiators.

5.10.3
End-group Functionality Through Conversion of Dormant Chain End

The dormant chain end in all CRP processes can be further converted to other functional groups by radical or non-radical processes. For example, dithio esters in RAFT have been converted to thiols, the alkoxyamines in NMP to alkyl chlorides and the alkyl halides in ATRP to allyl, hydroxy, amino, azido and ammonium or phosphonium groups in excellent yields. It is also possible to incorporate one functional monomer at the chain end, if the monomer is much less reactive and forms an unreactive dormant species. Scheme 5.23 presents three

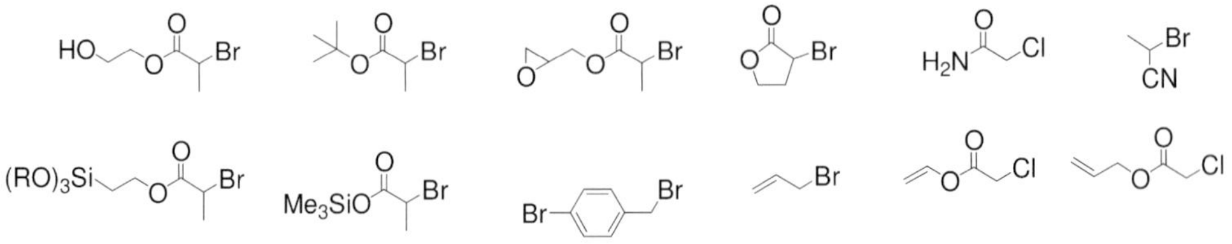

Scheme 5.22

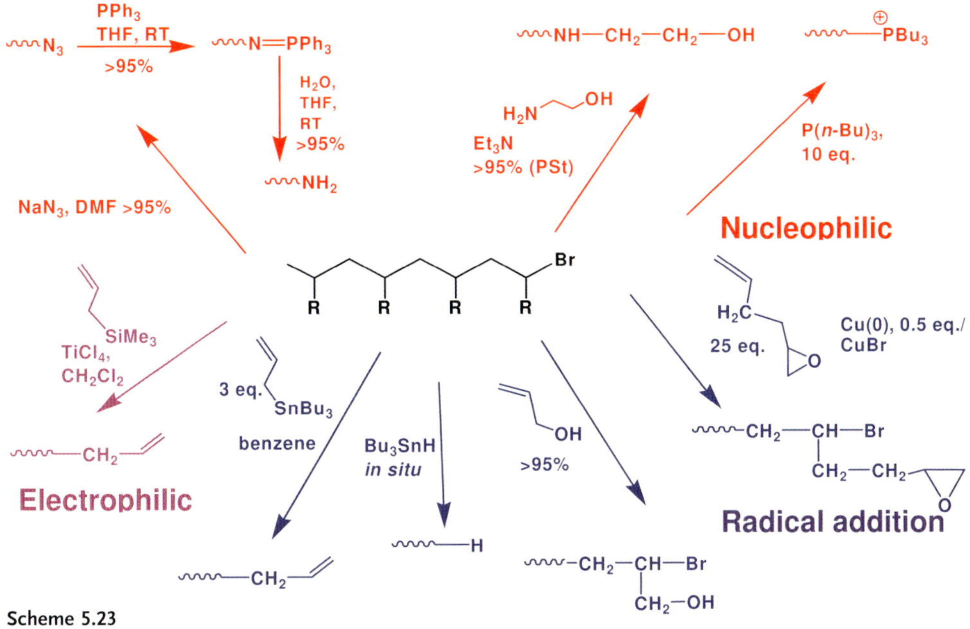

Scheme 5.23

types of reactions employed for end-functionalization of ATRP polymers: nucleophilic substitution (red), electrophilic substitution (magenta) and radical processes (blue).

The conversion of dormant species in chains growing in two directions leads to telechelics and in stars to multifunctional polymers. A similar approach was applied to hyperbranched polymers prepared by (co)polymerization of inimers (Section 5.8.4).

5.11
Applications of Materials Prepared by CRP

Volume IV is devoted to applications of materials prepared by controlled/living polymerization, many of them made by CRP. Although NMP, ATRP and RAFT were invented just in the last decade, they are already finding application in commercial production. It is expected that many more specialty products will become available and be successfully commercialized in the future [270]. A few examples are highlighted in this section.

5.11.1
Polymers with Controlled Compositions

Block copolymers based on acrylates and other polar monomers may find applications as polar thermoplastic elastomers [88] in automotive applications, especially in the presence of hydrocarbons that will not swell. However, they can also be used for much more sophisticated applications such as controlled drug release in cardiovascular stents [271]. Amphiphilic block copolymers with water-soluble segments have been successfully used as very efficient surfactants [272] and have also been used for higher end applications including pigment dispersants, various additives and components of health and beauty products. Segmented copolymers with nanostructured morphologies are promising as microelectronic devices.

Graft copolymers have been used as compatibilizers for polymer blends and may be used in many applications described for block copolymers [172, 273]. Gradient copolymers hold great promise in applications ranging from surfactants to noise- and vibration-damping materials [83].

5.11.2
Polymers with Controlled Topology

Designed branching allows precise control over melt viscosity and polymer processing. Such polymers (and also comb and star polymers) can be used as viscosity modifiers and lubricants.

An ultimate example of controlled topology might be a macromolecular bottle-brush. Such polymers, when lightly cross-linked, result in supersoft elastomers. Materials have been synthesized with moduli of ~1 kPa, in the range attainable by hydrogels. However, hydrogels must be swollen 100-fold by water to reach such low moduli. Molecular brushes are swollen by their own short side-chains that never leach. Thus, applications are predicted ranging from intraocular lenses and other biomedical applications requiring a soft material that does not leach to surrounding tissue, to electronic applications requiring the protection of delicate components by a soft solid. These materials also show very high ionic conductivities, reaching 1 mS cm^{-1} for Li cations at room temperature [274].

5.11.3
Polymers with Controlled Functionality

CRP offers unprecedented control over chain end functionality. End functional polyacrylates are excellent components of sealants for outdoor and automotive applications. Multifunctional low-MW polymers are desirable components in coatings with a low organic solvent content important for VOC reduction. It is also possible to design systems with two types of functionalities, curable by two independent mechanisms. Incorporation of degradable units into the backbone of vinyl polymers allows controlled cleavage and degradation/recycling of such polymers.

5.11.4
Hybrids

Molecular hybrids with a covalent attachment of well-defined functional polymer to either an inorganic component or a natural product are currently being extensively investigated and should lead to numerous materials with previously unattainable properties. Such hybrids allow better dispersibility of inorganic components (pigments, carbon black, carbon nanotubes, nanoparticles), they dramatically enhance the stability of such dispersions and they allow the formation of molecular nanocomposites. Also, dense polymer layers improve lubrication, prevent corrosion and facilitate surface patterning. Precise grafting from chromatographic packing can allow enhanced chromatographic resolution of oligopeptides and prions not previously available [275].

Other potential applications include microelectronics, microfluidics and optoelectronics, in addition to biomedical applications such as components of tissue and bone engineering, controlled drug release and drug targeting, antimicrobial surfaces [276], steering enzyme activity [101] and many others.

5.12
Outlook

CRP is currently the most rapid developing area of synthetic polymer chemistry. It bridges chemistry with physics and processing/engineering in order to rationally design and prepare targeted functional materials. Nevertheless, to reach its full potential, more research in this field will be needed.

5.12.1
Mechanisms

Mechanistic and kinetic studies of all CRP processes are needed. The deeper understanding of structure–reactivity correlation in NMP may lead to satisfactory control over polymerization of methacrylates. A full understanding of structural effects with ATRP catalysts could lead to the development of even more active complexes used in smaller amounts, which in turn could minimize the environmental impact of this chemistry and also expand the range of polymerizable monomers to include acrylic acid and a-alkenes. This may also lead to complexes that can participate in both ATRP and SFRP and potentially in coordination polymerization. In a similar way, the proper choice of dithio esters may decrease retardation effects of RAFT reagents. Model reactions with low-MW analogues and oligomers are increasing our comprehension of penultimate effects. Various additives that can accelerate polymerization and provide enhanced microstructural (tacticity and sequence) control should also be evaluated. Better cross-fertilization between synthetic organic chemistry and polymer chemistry is needed. In the past, achievements in organic chemistry were applied to polymer chemistry. Recently, advances made in polymer chemistry have also benefited organic chemistry, e.g. new catalysts developed for ATRP are used for atom transfer radical addition and cyclization and new nitroxides developed for NMP are used in organic synthesis [34, 277].

It is expected that computational chemistry will start to play a progressively more important role. A deeper insight into the reaction intermediates and energetic pathways will be possible in this way. It can also lead to the development of new mediating systems. However, precise computational evaluation requires a large basis set, since many reaction pathways may become dramatically affected by tiny changes in the structure of the substituents. Thus, continued model studies and their correlation with macromolecular systems are very much needed to understand better and optimize the existing processes.

Although current CRP techniques seem to cover all major mechanistic approaches to equilibria between active and dormant species, it is feasible to imagine more efficient stable free radicals, more active transition metal complexes and better degenerative transfer agents than are currently available. Hence both a serendipitous and rational search for new mediating agents in CRP should continue.

5.12.2
Molecular Architecture

Previous sections described many elements of controlled molecular architectures attainable by CRP. Further studies are needed to prepare higher MW polymers, important for some commercial applications, polymers with better control of end functionality, copolymers with precisely controlled gradients, materials with hierarchical structural control (such as in molecular brushes or other nano-objects with a core–shell) and other nanostructured morphologies.

It will also be important to prepare systems with designed molecular weight distributions, with both symmetrical and asymmetric distributions. Also, the introduction of imperfections, such as missing functionality or a missing arm in a star, may be important for further structure–property evaluation.

Some specialty applications will require systems responding reversibly to external stimuli (pH, temperature, light, pressure, magnetic or electric field). Thus, segmented copolymers with components responding separately to individual stimuli will be needed. Such systems may associate or dissociate under certain conditions and can rearrange, self-repair and be useful for advanced applications in microelectronics and biomedical areas.

The importance of hybrids has already been emphasized and it is expected that with a continued reduction of dimensions for many high-tech applications, the demand for precisely designed multifunctional nanocomposites materials will grow rapidly.

5.12.3
Characterization

Improved synthesis and more complex architectures require more precise characterization techniques to determine quantitatively the level of functionality, detect existing imperfections and precisely describe composition, topology and microstructure. This is needed both at a level of macromolecular assembly and for individual molecules. Therefore, various novel scattering techniques, modulated thermomechanical analysis, more sensitive spectroscopic techniques, multidimensional chromatography and visualization of individual molecules by AFM are just a few examples of techniques needed to characterize the structure of prepared materials.

5.12.4
Structure–Property Relationship

New polymers with precisely controlled architecture are primarily developed to explore novel properties. Such structure–property correlation is very much needed for many applications. Since macromolecular systems are very large and complex, it is difficult to predict all properties by modeling and computational techniques without input from well-defined macromolecules. The detailed correlation of molecular structure with final properties is still not obvious because many properties will also depend on processing conditions, including the solvent used and its removal conditions, mechanical stresses and thermal history. The incorporation of such parameters into simulation will be very beneficial.

Additionally, evaluation of all imperfections and synthetic errors on properties of prepared polymers is needed. This should include the effect of polydispersities and shape of the molecular weight distribution. Polymers with higher polydispersities can allow more flexible processing regimes. In CRP, it will be much more economical to prepare polymers faster, with more termination and with more errors. Therefore, it will be important to understand how these imperfections may affect material properties to optimize the cost/performance ratio.

Acknowledgments

Financial support from the Carnegie Mellon University (CMU) CRP Consortium and NSF grants (CHE-0405627 and DMR-0549353) are greatly appreciated.

References

1. Walling, C. *Free Radicals in Solution*, Wiley, New York, **1957**.
2. Bamford, C. H., Barb, W. G., Jenkins, A. D., Onyon, P. F. *The Kinetics of Vinyl Polymerization by Radical Mechanisms*, Academic Press, New York, **1958**.
3. Bagdasarian, H. S. *Theory of Radical Polymerization*, Izd. Akademii Nauk: Moscow, **1959**.
4. Matyjaszewski, K., Davis, T. P. *Handbook of Radical Polymerization*, Wiley-Interscience, Hoboken, NJ, **2002**.
5. Moad, G., Solomon, D. H. *The Chemistry of Radical Polymerization*, 2nd edn., Elsevier, Oxford, **2006**.
6. Beuermann, S., Buback, M. *Prog. Polym. Sci.* **2002**, *27*, 191–254.
7. Barner-Kowollik, C., Buback, M., Egorov, M., Fukuda, T., Goto, A., Olaj, O. F., Russell, G. T., Vana, P., Yamada, B., Zetterlund, P. B. *Prog. Polym. Sci.* **2005**, *30*, 605–643.
8. Buback, M., Kurz, C. H., Schmaltz, C. *Macromol. Chem. Phys.* **1998**, *199*, 1721–1727.
9. Buback, M., Kuelpmann, A., Kurz, C. *Macromol. Chem. Phys.* **2002**, *203*, 1065–1070.
10. Asua, J. M., Beuermann, S., Buback, M., Castignolles, P., Charleux, B., Gilbert, R. G., Hutchinson, R. A., Leiza, J. R., Nikitin, A. N., Vairon, J.-P., van Herk, A. M. *Macromol. Chem. Phys.* **2004**, *205*, 2151–2160.
11. Beuermann, S., Buback, M., Schmaltz, C. *Ind. Eng. Chem. Res.* **1999**, *38*, 3338–3344.
12. Beuermann, S., Buback, M., Davis, T. P., Gilbert, R. G., Hutchinson, R. A., Kajiwara, A., Klumperman, B., Russell, G. T. *Macromol. Chem. Phys.* **2000**, *201*, 1355–1364.
13. Buback, M., Kowollik, C. *Macromol. Chem. Phys.* **1999**, *200*, 1764–1770.
14. Beuermann, S., Buback, M., Davis, T. P., Gilbert, R. G., Hutchinson, R. A., Olaj, O. F., Russell, G. T., Schweer, J., Van Herk, A. M. *Macromol. Chem. Phys.* **1997**, *198*, 1545–1560.
15. Buback, M., Kowollik, C. *Macromolecules* **1998**, *31*, 3211–3215.
16. Buback, M., Gilbert, R. G., Hutchinson, R. A., Klumperman, B., Kuchta, F.-D., Manders, B. G., O'Driscoll, K. F., Russell, G. T., Schweer, J. *Macromol. Chem. Phys.* **1995**, *196*, 3267–3280.
17. Buback, M., Kuchta, F. D. *Macromol. Chem. Phys.* **1997**, *198*, 1455–1480.
18. Deibert, S., Bandermann, F., Schweer, J., Sarnecki, J. *Makromol. Chem. Rapid Commun.* **1992**, *13*, 351–355.
19. Madruga, E. L. *Prog. Polym. Sci.* **2002**, *27*, 1879–1924.
20. Matyjaszewski, K., Kubisa, P., Penczek, S. *J. Polym. Sci., Polym. Chem. Ed.* **1974**, *12*, 1333–1336.
21. Matyjaszewski, K., Sigwalt, P. *Polym. Int.* **1994**, *35*, 1.
22. Greszta, D., Mardare, D., Matyjaszewski, K. *Macromolecules* **1994**, *27*, 638–644.
23. Goto, A., Fukuda, T. *Prog. Polym. Sci.* **2004**, *29*, 329–385.
24. Fischer, H. *Chem. Rev.* **2001**, *101*, 3581–3610.
25. Tang, W., Tsarevsky, N. V., Matyjaszewski, K. *J. Am. Chem. Soc.* **2006**, *128*, 1598–1604.
26. Georges, M. K., Veregin, R. P. N., Kazmaier, P. M., Hamer, G. K. *Macromolecules* **1993**, *26*, 2987–2988.
27. Hawker, C. J., Bosman, A. W., Harth, E. *Chem. Rev.* **2001**, *101*, 3661–3688.
28. Wayland, B. B., Poszmik, G., Mukerjee, S. L., Fryd, M. *J. Am. Chem. Soc.* **1994**, *116*, 7943–7944.
29. Matyjaszewski, K. *Macromolecules* **1999**, *32*, 9051–9053.
30. Solomon, D. H., Rizzardo, E., Cacioli, P. *Eur. Pat. Appl.* **1985**, 63 p.
31. Benoit, D., Grimaldi, S., Robin, S., Finet, J.-P., Tordo, P., Gnanou, Y. *J. Am. Chem. Soc.* **2000**, *122*, 5929–5939.
32. Benoit, D., Chaplinski, V., Braslau, R., Hawker, C. J. *J. Am. Chem. Soc.* **1999**, *121*, 3904–3920.
33. Studer, A., Harms, K., Knoop, C., Mueller, C., Schulte, T. *Macromolecules* **2004**, *37*, 27–34.
34. Studer, A., Schulte, T. *Chem. Rec.* **2005**, *5*, 27–35.

35 Greszta, D., Matyjaszewski, K. *J. Polym. Sci., Part A: Polym. Chem.* **1997**, *35*, 1857–1861.
36 Steenbock, M., Klapper, M., Muellen, K. *Macromol. Chem. Phys.* **1998**, *199*, 763–769.
37 Klapper, M., Brand, T., Steenbock, M., Mullen, K. *ACS Symp. Ser.* **2000**, *768*, 152–166.
38 Chung, T. C., Janvikul, W., Lu, H. L. *J. Am. Chem. Soc.* **1996**, *118*, 705.
39 Borsig, E., Lazar, M., Capla, M., Florian, S. *Angew. Makromol. Chem.* **1969**, *9*, 89–95.
40 Braun, D. *Macromol. Symp.* **1996**, *111*, 63–71.
41 Otsu, T., Yoshida, M., Tazaki, T. *Makromol. Chem., Rapid Commun.* **1982**, *3*, 133–140.
42 Otsu, T. *J. Polym. Sci., Part A: Polym. Chem.* **2000**, *38*, 2121–2136.
43 Nicolas, J., Dirc, C., Mueller, L., Belleney, J., Charleux, B., Marque, S. R. A., Bertin, D., Magnet, S., Couvreur, L. *Macromolecules* **2006**, *39*, 8274–8282.
44 Hawker, C. J. *J. Am. Chem. Soc.* **1994**, *116*, 11185–11186.
45 Hawker, C. J., Barclay, G. G., Orellana, A., Dao, J., Devonport, W. *Macromolecules* **1996**, *29*, 5245–5254.
46 Tang, C., Kowalewski, T., Matyjaszewski, K. *Macromolecules* **2003**, *36*, 1465–1473.
47 Qiu, J., Charleux, B., Matyjaszewski, K. *Prog. Polym. Sci.* **2001**, *26*, 2083–2134.
48 Nicolas, J., Charleux, B., Guerret, O., Magnet, S. *Macromolecules* **2005**, *38*, 9963–9973.
49 Debuigne, A., Caille, J.-R., Jerome, R. *Angew. Chem. Int. Ed.* **2005**, *44*, 1101–1104.
50 Kaneyoshi, H., Matyjaszewski, K. *Macromolecules* **2006**, *39*, 2757–2763.
51 Le Grognec, E., Claverie, J., Poli, R. *J. Am. Chem. Soc.* **2001**, *123*, 9513–9524.
52 Braunecker, W. A., Itami, Y., Matyjaszewski, K. *Macromolecules* **2005**, *38*, 9402–9404.
53 Claverie, J. P. *Res. Discl.* **1998**, *416*, 1595–1604.
54 Shaver, M. P., Allan, L. E. N., Rzepa, H. S., Gibson, V. C. *Angew. Chem. Int. Ed.* **2006**, *45*, 1241–1244.
55 Poli, R. *Angew. Chem. Int. Ed.* **2006**, *45*, 5058–5070.
56 Wang, J.-S., Matyjaszewski, K. *J. Am. Chem. Soc.* **1995**, *117*, 5614–5615.
57 Kato, M., Kamigaito, M., Sawamoto, M., Higashimura, T. *Macromolecules* **1995**, *28*, 1721–1723.
58 Patten, T. E., Matyjaszewski, K. *Acc. Chem. Res.* **1999**, *32*, 895–903.
59 Matyjaszewski, K., Xia, J. *Chem. Rev.* **2001**, *101*, 2921–2990.
60 Kamigaito, M., Ando, T., Sawamoto, M. *Chem. Rev.* **2001**, *101*, 3689–3745.
61 Matyjaszewski, K. *J. Macromol. Sci., Pure Appl.Chem.* **1997**, *A34*, 1785–1801.
62 Matyjaszewski, K., Patten, T. E., Xia, J. *J. Am. Chem. Soc.* **1997**, *119*, 674–680.
63 Ohno, K., Goto, A., Fukuda, T., Xia, J., Matyjaszewski, K. *Macromolecules* **1998**, *31*, 2699–2701.
64 Matyjaszewski, K., Lin, C.-H. *Makromol. Chem., Macromol. Symp.* **1991**, *47*, 221.
65 Litvinenko, G., Mueller, A. H. E. *Macromolecules* **1997**, *30*, 1253–1266.
66 Matyjaszewski, K. *Macromol. Symp.* **1998**, *134*, 105–118.
67 Matyjaszewski, K. *Macromolecules* **1998**, *31*, 4710–4717.
68 Braunecker, W. A., Pintauer, T., Tsarevsky, N. V., Kickelbick, G., Matyjaszewski, K. *J. Organomet. Chem.* **2005**, *690*, 916–924.
69 Braunecker, W. A., Tsarevsky, N. V., Pintauer, T., Gil, R. R., Matyjaszewski, K. *Macromolecules* **2005**, *38*, 4081–4088.
70 Tsarevsky, N. V., Pintauer, T., Matyjaszewski, K. *Macromolecules* **2004**, *37*, 9768–9778.
71 Matyjaszewski, K., Davis, K., Patten, T. E., Wei, M. *Tetrahedron* **1997**, *53*, 15321–15329.
72 Lutz, J.-F., Matyjaszewski, K. *Macromol. Chem. Phys.* **2002**, *203*, 1385–1395.
73 Lutz, J.-F., Matyjaszewski, K. *J. Polym. Sci., Part A: Polym. Chem.* **2005**, *43*, 897–910.
74 Qiu, J., Matyjaszewski, K. *Macromolecules* **1997**, *30*, 5643–5648.
75 Wang, J.-L., Grimaud, T., Shipp, D. A., Matyjaszewski, K. *Macromolecules* **1998**, *31*, 1527–1534.
76 Davis, K. A., Paik, H.-J., Matyjaszewski, K. *Macromolecules* **1999**, *32*, 1767–1776.
77 Mori, H., Mueller, A. H. E. *Prog. Polym. Sci.* **2003**, *28*, 1403–1439.

78 Teodorescu, M., Matyjaszewski, K. *Macromolecules* **1999**, *32*, 4826–4831.
79 Neugebauer, D., Matyjaszewski, K. *Macromolecules* **2003**, *36*, 2598–2603.
80 Tsarevsky, N. V., Braunecker, W. A., Brooks, S. J., Matyjaszewski, K. *Macromolecules* **2006**, *39*, 6817–6824.
81 Matyjaszewski, K., Jo, S. M., Paik, H.-j., Gaynor, S. G. *Macromolecules* **1997**, *30*, 6398–6400.
82 Coca, S., Matyjaszewski, K. *Polym. Prepr.* **1996**, *37*, 573–574.
83 Lutz, J.-F., Pakula, T., Matyjaszewski, K. *ACS Symp. Ser.* **2003**, *854*, 268–282.
84 Elyashiv-Barad, S., Greinert, N., Sen, A. *Macromolecules* **2002**, *35*, 7521–7526.
85 Venkatesh, R., Klumperman, B. *Macromolecules* **2004**, *37*, 1226–1233.
86 Ashford, E. J., Naldi, V., O'Dell, R., Billingham, N. C., Armes, S. P. *Chem. Commun.* **1999**, 1285–1286.
87 Matyjaszewski, K., Shipp, D. A., Wang, J.-L., Grimaud, T., Patten, T. E. *Macromolecules* **1998**, *31*, 6836–6840.
88 Matyjaszewski, K., Shipp, D. A., McMurtry, G. P., Gaynor, S. G., Pakula, T. *J. Polym. Sci., Part A: Polym. Chem.* **2000**, *38*, 2023–2031.
89 Kowalewski, T., Tsarevsky, N. V., Matyjaszewski, K. *J. Am. Chem. Soc.* **2002**, *124*, 10632–10633.
90 Percec, V., Barboiu, B. *Macromolecules* **1995**, *28*, 7970–7972.
91 Gillies, M. B., Matyjaszewski, K., Norrby, P.-O., Pintauer, T., Poli, R., Richard, P. *Macromolecules* **2003**, *36*, 8551–8559.
92 Tang, W., Matyjaszewski, K. *Polym. Prep.* **2005**, *46*, 211–212.
93 Coca, S., Paik, H.-j., Matyjaszewski, K. *Macromolecules* **1997**, *30*, 6513–6516.
94 Coca, S., Matyjaszewski, K. *Macromolecules* **1997**, *30*, 2808–2810.
95 Gaynor, S. G., Matyjaszewski, K. *Macromolecules* **1997**, *30*, 4241–4243.
96 Pyun, J., Matyjaszewski, K. *Chem. Mater.* **2001**, *13*, 3436–3448.
97 Pyun, J., Kowalewski, T., Matyjaszewski, K. *Macromol. Rapid Commun.* **2003**, *24*, 1043–1059.
98 Luzinov, I., Minko, S., Tsukruk, V. V. *Prog. Polym. Sci.* **2004**, *29*, 635–698.
99 Cunliffe, D., Pennadam, S., Alexander, C. *Eur. Polym. J.* **2003**, *40*, 5–25.
100 Klok, H.-A. *J. Polym. Sci., Part A: Polym. Chem.* **2005**, *43*, 1–17.
101 Lele, B. S., Murata, H., Matyjaszewski, K., Russell, A. J. *Biomacromolecules* **2005**, *6*, 3380–3387.
102 Kabachii, Y. A., Kochev, S. Y., Bronstein, L. M., Blagodatskikh, I. B., Valetsky, P. M. *Polym. Bull.* **2003**, *50*, 271–278.
103 Brandts, J. A. M., van de Geijn, P., van Faassen, E. E., Boersma, J., Van Koten, G. *J. Organomet. Chem.* **1999**, *584*, 246–253.
104 Maria, S., Stoffelbach, F., Mata, J., Daran, J.-C., Richard, P., Poli, R. *J. Am. Chem. Soc.* **2005**, *127*, 5946–5956.
105 Kotani, Y., Kamigaito, M., Sawamoto, M. *Macromolecules* **1999**, *32*, 2420–2424.
106 Matyjaszewski, K., Wei, M., Xia, J., McDermott, N. E. *Macromolecules* **1997**, *30*, 8161–8164.
107 Ando, T., Kamigaito, M., Sawamoto, M. *Macromolecules* **1997**, *30*, 4507–4510.
108 O'Reilly, R. K., Gibson, V. C., White, A. J. P., Williams, D. J. *Polyhedron* **2004**, *23*, 2921–2928.
109 Teodorescu, M., Gaynor, S. G., Matyjaszewski, K. *Macromolecules* **2000**, *33*, 2335–2339.
110 Wakioka, M., Baek, K.-Y., Ando, T., Kamigaito, M., Sawamoto, M. *Macromolecules* **2002**, *35*, 330–333.
111 Simal, F., Demonceau, A., Noels, A. F. *Angew. Chem. Int. Ed.* **1999**, *38*, 538–540.
112 Percec, V., Barboiu, B., Neumann, A., Ronda, J. C., Zhao, M. *Macromolecules* **1996**, *29*, 3665–3668.
113 Wang, B., Zhuang, Y., Luo, X., Xu, S., Zhou, X. *Macromolecules* **2003**, *36*, 9684–9686.
114 Granel, C., Dubois, P., Jerome, R., Teyssie, P. *Macromolecules* **1996**, *29*, 8576–8582.
115 Uegaki, H., Kotani, Y., Kamigaito, M., Sawamoto, M. *Macromolecules* **1997**, *30*, 2249–2253.
116 Lecomte, P., Drapier, I., Dubois, P., Teyssie, P., Jerome, R. *Macromolecules* **1997**, *30*, 7631–7633.
117 Woodworth, B. E., Metzner, Z., Matyjaszewski, K. *Macromolecules* **1998**, *31*, 7999–8004.
118 Xia, J., Matyjaszewski, K. *Macromolecules* **1997**, *30*, 7697–7700.

119 Xia, J., Gaynor, S. G., Matyjaszewski, K. *Macromolecules* **1998**, *31*, 5958–5959.
120 Patten, T. E., Xia, J., Abernathy, T., Matyjaszewski, K. *Science* **1996**, *272*, 866–868.
121 Kickelbick, G., Matyjaszewski, K. *Macromol. Rapid Commun.* **1999**, *20*, 341–346.
122 Pintauer, T., Matyjaszewski, K. *Coord. Chem. Rev.* **2005**, *249*, 1155–1184.
123 Qiu, J., Matyjaszewski, K., Thouin, L., Amatore, C. *Macromol. Chem. Phys.* **2000**, *201*, 1625–1631.
124 Tsarevsky, N. V., Braunecker, W. A., Tang, W., Brooks, S. J., Matyjaszewski, K., Weisman, G. R., Wong, E. H. *J. Mol. Catal. A: Chem.* **2006**, *257*, 132–140.
125 Matyjaszewski, K., Goebelt, B., Paik, H.-j., Horwitz, C. P. *Macromolecules* **2001**, *34*, 430–440.
126 Tang, W., Matyjaszewski, K. *Macromolecules* **2006**, *39*, 4953–4959.
127 Xia, J., Zhang, X., Matyjaszewski, K. *ACS Symp. Ser.* **2000**, *760*, 207–223.
128 Xia, J., Matyjaszewski, K. *Macromolecules* **1997**, *30*, 7692–7696.
129 Moineau, G., Dubois, P., Jerome, R., Senninger, T., Teyssie, P. *Macromolecules* **1998**, *31*, 545–547.
130 Jakubowski, W., Matyjaszewski, K. *Macromolecules* **2005**, *38*, 4139–4146.
131 Min, K., Gao, H., Matyjaszewski, K. *J. Am. Chem. Soc.* **2005**, *127*, 3825–3830.
132 Min, K., Jakubowski, W., Matyjaszewski, K. *Macromol. Rapid Commun.* **2006**, *27*, 594–598.
133 Jakubowski, W., Matyjaszewski, K. *Angew. Chem. Int. Ed. Engl.* **2006**, *45*, 4482–4486.
134 Jakubowski, W., Min, K., Matyjaszewski, K. *Macromolecules* **2006**, *39*, 39–45.
135 Matyjaszewski, K., Jakubowski, W., Min, K., Tang, W., Huang, J., Braunecker, W. A., Tsarevsky, N. V. *Proc. Natl. Acad. Sci. USA* **2006**, *103*, 15309–15314.
136 Gaynor, S. G., Wang, J.-S., Matyjaszewski, K. *Macromolecules* **1995**, *28*, 8051–8056.
137 Grishin, D. F., Moikin, A. A. *Vysokomol. Soedin., Ser. A Ser. B* **1998**, *40*, 1266–1270.
138 Goto, A., Kwak, Y., Fukuda, T., Yamago, S., Iida, K., Nakajima, M., Yoshida, J.-I. *J. Am. Chem. Soc.* **2003**, *125*, 8720–8721.
139 Yamago, S., Ray, B., Iida, K., Yoshida, J., Tada, T., Yoshizawa, K., Kwak, Y., Goto, A., Fukuda, T. *J. Am. Chem. Soc.* **2004**, *126*, 13908–13909.
140 Moad, G., Rizzardo, E., Thang, S. H. *Aust. J. Chem.* **2005**, *58*, 379–410.
141 Perrier, S., Takolpuckdee, P. *J. Polym. Sci., Part A: Polym. Chem.* **2005**, *43*, 5347–5393.
142 Moad, C. L., Moad, G., Rizzardo, E., Thang, S. H. *Macromolecules* **1996**, *29*, 7717–7726.
143 Chiefari, J., Chong, Y. K., Ercole, F., Krstina, J., Jeffery, J., Le, T. P. T., Mayadunne, R. T. A., Meijs, G. F., Moad, C. L., Moad, G., Rizzardo, E., Thang, S. H. *Macromolecules* **1998**, *31*, 5559–5562.
144 Favier, A., Charreyre, M.-T. *Macromol. Rapid Commun.* **2006**, *27*, 653–692.
145 Destarac, M., Charmot, D., Franck, X., Zard, S. Z. *Macromol. Rapid Commun.* **2000**, *21*, 1035–1039.
146 McCormick, C. L., Lowe, A. B. *Acc. Chem. Res.* **2004**, *37*, 312–325.
147 Destarac, M., Brochon, C., Catala, J.-M., Wilczewska, A., Zard, S. Z. *Macromol. Chem. Phys.* **2002**, *203*, 2281–2289.
148 Chong, Y. K., Krstina, J., Le, T. P. T., Moad, G., Postma, A., Rizzardo, E., Thang, S. H. *Macromolecules* **2003**, *36*, 2256–2272.
149 Chiefari, J., Mayadunne, R. T. A., Moad, C. L., Moad, G., Rizzardo, E., Postma, A., Skidmore, M. A., Thang, S. H. *Macromolecules* **2003**, *36*, 2273–2283.
150 Mayadunne, R. T. A., Jeffery, J., Moad, G., Rizzardo, E. *Macromolecules* **2003**, *36*, 1505–1513.
151 Perrier, S., Takolpuckdee, P., Mars, C. A. *Macromolecules* **2005**, *38*, 2033–2036.
152 Matyjaszewski, K. (Ed.) *Advances in Controlled/Living Radical Polymerization (ACS Symp. Ser. 854)*, American Chemical Society, Washington, DC, **2003**.
153 Percec, V., Popov, A. V., Ramirez-Castillo, E., Hinojosa-Falcon, L. A. *J. Polym. Sci., Part A: Polym. Chem.* **2005**, *43*, 2276–2280.
154 Sumerlin, B. S., Tsarevsky, N. V., Louche, G., Lee, R. Y., Matyjaszewski, K. *Macromolecules* **2005**, *38*, 7540–7545.
155 Coote, M. L., Henry, D. J. *Macromolecules* **2005**, *38*, 5774–5779.

156 Min, K., Matyjaszewski, K. *Macromolecules* **2005**, *38*, 8131–8134.
157 Ferguson, C. J., Hughes, R. J., Nguyen, D., Pham, B. T. T., Gilbert, R. G., Serelis, A. K., Such, C. H., Hawkett, B. S. *Macromolecules* **2005**, *38*, 2191–2204.
158 Manguian, M., Save, M., Charleux, B. *Macromol. Rapid Commun.* **2006**, *27*, 399–404.
159 Nicolas, J., Charleux, B., Guerret, O., Magnet, S. *Angew. Chem. Int. Ed.* **2004**, *43*, 6186–6189.
160 Coessens, V., Pintauer, T., Matyjaszewski, K. *Prog. Polym. Sci.* **2001**, *26*, 337–377.
161 Davis, K. A., Matyjaszewski, K. *Adv. Polym. Sci.* **2002**, *159*, 1–166.
162 Matyjaszewski, K. *Polym. Int.* **2003**, *52*, 1559–1565.
163 Matyjaszewski, K., Miller, P. J., Fossum, E., Nakagawa, Y. *Appl. Organomet. Chem.* **1998**, *12*, 667–673.
164 Wang, J.-S., Greszta, D., Matyjaszewski, K. *Polym. Mater. Sci. Eng.* **1995**, *73*, 416–417.
165 Angot, S., Murthy, S., Taton, D., Gnanou, Y. *Macromolecules* **1998**, *31*, 7218–7225.
166 Gao, H., Matyjaszewski, K. *Macromolecules* **2006**, *39*, 3154–3160.
167 Gao, H., Matyjaszewski, K. *Macromolecules* **2006**, *39*, 4960–4965.
168 Zhang, X., Xia, J., Matyjaszewski, K. *Macromolecules* **2000**, *33*, 2340–2345.
169 Xia, J., Zhang, X., Matyjaszewski, K. *Macromolecules* **1999**, *32*, 4482–4484.
170 Gao, H., Tsarevsky, N. V., Matyjaszewski, K. *Macromolecules* **2005**, *38*, 5995–6004.
171 Borner, H. G., Matyjaszewski, K. *Macromol. Symp.* **2002**, *177*, 1–15.
172 Inoue, Y., Matsugi, T., Kashiwa, N., Matyjaszewski, K. *Macromolecules* **2004**, *37*, 3651–3658.
173 Okrasa, L., Pakula, T., Inoue, Y., Matyjaszewski, K. *Colloid Polym. Sci.* **2004**, *282*, 844–853.
174 Kaneyoshi, H., Inoue, Y., Matyjaszewski, K. *Polym. Mater. Sci. Eng.* **2004**, *91*, 41–42.
175 Paik, H. J., Gaynor, S. G., Matyjaszewski, K. *Macromol. Rapid Commun.* **1998**, *19*, 47–52.
176 Percec, V., Asgarzadeh, F. *J. Polym. Sci., Part A: Polym. Chem.* **2001**, *39*, 1120–1135.
177 Hong, S. C., Pakula, T., Matyjaszewski, K. *Polym. Mater. Sci. Eng.* **2001**, *84*, 767–768.
178 Hong, S. C., Jia, S., Teodorescu, M., Kowalewski, T., Matyjaszewski, K., Gottfried, A. C., Brookhart, M. *J. Polym. Sci., Part A: Polym. Chem.* **2002**, *40*, 2736–2749.
179 Kaneyoshi, H., Inoue, Y., Matyjaszewski, K. *Macromolecules* **2005**, *38*, 5425–5435.
180 Neugebauer, D., Zhang, Y., Pakula, T., Sheiko, S. S., Matyjaszewski, K. *Macromolecules* **2003**, 6746–6755.
181 Shinoda, H., Matyjaszewski, K., Okrasa, L., Mierzwa, M., Pakula, T. *Macromolecules* **2003**, *36*, 4772–4778.
182 Shinoda, H., Matyjaszewski, K. *Macromolecules* **2001**, *34*, 6243–6248.
183 Hawker, C. J., Mecerreyes, D., Elce, E., Dao, J., Hedrick, J. L., Barakat, I., Dubois, P., Jerome, R., Volksen, I. *Macromol. Chem. Phys.* **1997**, *198*, 155–166.
184 Lord, S. J., Sheiko, S. S., LaRue, I., Lee, H.-I., Matyjaszewski, K. *Macromolecules* **2004**, *37*, 4235–4240.
185 Gallyamov, M. O., Tartsch, B., Khokhlov, A. R., Sheiko, S. S., Boerner, H. G., Matyjaszewski, K., Moeller, M. *Chem. Eur. J.* **2004**, *10*, 4599–4605.
186 Ishizu, K., Satoh, J., Sogabe, A. *J. Colloid Interface Sci.* **2004**, *274*, 472–479.
187 Neugebauer, D., Zhang, Y., Pakula, T., Matyjaszewski, K. *Polymer* **2003**, *44*, 6863–6871.
188 Boerner, H. G., Duran, D., Matyjaszewski, K., da Silva, M., Sheiko, S. S. *Macromolecules* **2002**, *35*, 3387–3394.
189 Hawker, C. J., Frechet, J. M. J., Grubbs, R. B., Dao, J. *J. Am. Chem. Soc.* **1995**, *117*, 10763–10764.
190 Gaynor, S. G., Edelman, S., Matyjaszewski, K. *Macromolecules* **1996**, *29*, 1079–1081.
191 Matyjaszewski, K., Gaynor, S. G. *Macromolecules* **1997**, *30*, 7042–7049.
192 Gao, C., Yan, D. *Prog. Polym. Sci.* **2004**, *29*, 183–275.
193 Baudry, R., Sherrington, D. C. *Macromolecules* **2006**, *39*, 1455–1460.

194 Lepoittevin, B., Matmour, R., Francis, R., Taton, D., Gnanou, Y. *Macromolecules* **2005**, *38*, 3120–3128.
195 Matmour, R., Lepoittevin, B., Joncheray, T. J., El-Khouri, R. J., Taton, D., Duran, R. S., Gnanou, Y. *Macromolecules* **2005**, *38*, 5459–5467.
196 Angot, S., Taton, D., Gnanou, Y. *Macromolecules* **2000**, *33*, 5418–5426.
197 Gnanou, Y., Taton, D. *Macromol. Symp.* **2001**, *174*, 333–341.
198 Hou, S., Chaikof, E. L., Taton, D., Gnanou, Y. *Macromolecules* **2003**, *36*, 3874–3881.
199 Matyjaszewski, K., Shigemoto, T., Frechet, J. M. J., Leduc, M. *Macromolecules* **1996**, *29*, 4167–4171.
200 Grubbs, R. B., Hawker, C. J., Dao, J., Frechet, J. M. J. *Angew. Chem. Int. Ed. Engl.* **1997**, *36*, 270–272.
201 Liang, C., Frechet, J. M. J. *Prog. Polym. Sci.* **2005**, *30*, 385–402.
202 Frechet, J. M. J. *Prog. Polym. Sci.* **2005**, *30*, 844–857.
203 Ide, N., Fukuda, T. *Macromolecules* **1997**, *30*, 4268–4271.
204 Du, J., Chen, Y. *J. Polym. Sci., Part A: Polym. Chem.* **2004**, *42*, 2263–2271.
205 Connal, L. A., Gurr, P. A., Qiao, G. G., Solomon, D. H. *J. Mater. Chem.* **2005**, *15*, 1286–1292.
206 Taton, D., Baussard, J.-F., Dupayage, L., Poly, J., Gnanou, Y., Ponsinet, V., Destarac, M., Mignaud, C., Pitois, C. *Chem. Commun.* **2006**, 1953–1955.
207 Tsarevsky, N. V., Matyjaszewski, K. *Macromolecules* **2002**, *35*, 9009–9014.
208 Tsarevsky, N. V., Matyjaszewski, K. *Polym. Prepr.* **2005**, *46*, 339–340.
209 Oh, J. K., Tang, C., Gao, H., Tsarevsky, N. V., Matyjaszewski, K. *J. Am. Chem. Soc.* **2006**, *128*, 5578–5584.
210 Tsarevsky, N. V., Sumerlin, B. S., Matyjaszewski, K. *Macromolecules* **2005**, *38*, 3558–3561.
211 Laurent, B. A., Grayson, S. M. *J. Am. Chem. Soc.* **2006**, *128*, 4238–4239.
212 Qin, S., Saget, J., Pyun, J., Jia, S., Kowalewski, T., Matyjaszewski, K. *Macromolecules* **2003**, *36*, 8969–8977.
213 Matyjaszewski, K. *Macromolecules* **2002**, *35*, 6773–6781.
214 Shipp, D. A., Wang, J.-L., Matyjaszewski, K. *Macromolecules* **1998**, *31*, 8005–8008.
215 Davis, K. A., Matyjaszewski, K. *Macromolecules* **2001**, *34*, 2101–2107.
216 Tsarevsky, N. V., Cooper, B. M., Wojtyna, O. J., Jahed, N. M., Gao, H., Matyjaszewski, K. *Polym. Prepr.* **2005**, *46*, 249–250.
217 Bernaerts, K. V., Du Prez, F. E. *Polymer* **2005**, *46*, 8469–8482.
218 Kobatake, S., Harwood, H. J., Quirk, R. P., Priddy, D. B. *Macromolecules* **1998**, *31*, 3735–3739.
219 Acar, M. H., Matyjaszewski, K. *Macromol. Chem. Phys.* **1999**, *200*, 1094–1100.
220 Kajiwara, A., Matyjaszewski, K. *Macromolecules* **1998**, *31*, 3489–3493.
221 Yoshida, E., Sugita, A. *Macromolecules* **1996**, *29*, 6422–6426.
222 Weimer, M. W., Scherman, O. A., Sogah, D. Y. *Macromolecules* **1998**, *31*, 8425–8428.
223 Jakubowski, W., Lutz, J.-F., Slomkowski, S., Matyjaszewski, K. *J. Polym. Sci., Part A: Polym. Chem.* **2005**, *43*, 1498–1510.
224 Mecerreyes, D., Atthoff, B., Boduch, K. A., Trollsaas, M., Hedrick, J. L. *Macromolecules* **1999**, *32*, 5175–5182.
225 Mecerreyes, D., Moineau, G., Dubois, P., Jerome, R., Hedrick, J. L., Hawker, C. J., Malmstrom, E. E., Trollsas, M. *Angew. Chem. Int. Ed.* **1998**, *37*, 1274–1276.
226 Stehling, U. M., Malmstroem, E. E., Waymouth, R. M., Hawker, C. J. *Macromolecules* **1998**, *31*, 4396–4398.
227 Liu, S., Sen, A. *Macromolecules* **2001**, *34*, 1529–1532.
228 Bielawski, C. W., Louie, J., Grubbs, R. H. *J. Am. Chem. Soc.* **2000**, *122*, 12872–12873.
229 Bielawski, C. W., Morita, T., Grubbs, R. H. *Macromolecules* **2000**, *33*, 678–680.
230 Zhang, Y., Chung, I. S., Huang, J., Matyjaszewski, K., Pakula, T. *Macromol. Chem. Phys.* **2005**, *206*, 33–42.
231 Mennicken, M., Nagelsdiek, R., Keul, H., Hoecker, H. *Macromol. Chem. Phys.* **2004**, *205*, 143–153.
232 Yagci, Y., Toppare, L. *Polym. Int.* **2003**, *52*, 1573–1578.

233 Iovu, M.C., Jeffries-El, M., Sheina, E.E., Cooper, J.R., McCullough, R.D. *Polymer* **2005**, *46*, 8582–8586.
234 Kowalewski, T., McCullough, R.D., Matyjaszewski, K. *Eur. Phys. J. E: Soft Matter* **2003**, *10*, 5–16.
235 Paik, H.-j., Teodorescu, M., Xia, J., Matyjaszewski, K. *Macromolecules* **1999**, *32*, 7023–7031.
236 Sugiyama, Y., Satoh, K., Kamigaito, M., Okamoto, Y. *J. Polym. Sci., Part A: Polym. Chem.* **2006**, *44*, 2086–2098.
237 Okamoto, Y., Habaue, S., Isobe, Y., Nakano, T. *Macromol. Symp.* **2002**, *183*, 83–88.
238 Lutz, J.-F., Neugebauer, D., Matyjaszewski, K. *J. Am. Chem. Soc.* **2003**, *125*, 6986–6993.
239 Wan, D., Satoh, K., Kamigaito, M., Okamoto, Y. *Macromolecules* **2005**, *38*, 10397–10405.
240 Ray, B., Isobe, Y., Matsumoto, K., Habaue, S., Okamoto, Y., Kamigaito, M., Sawamoto, M. *Macromolecules* **2004**, *37*, 1702–1710.
241 Roos, S.G., Mueller, A.H.E., Matyjaszewski, K. *Macromolecules* **1999**, *32*, 8331–8335.
242 Shinoda, H., Matyjaszewski, K. *Macromol. Rapid Commun.* **2001**, *22*, 1176–1181.
243 Chen, G.-Q., Wu, Z.-Q., Wu, J.-R., Li, Z.-C., Li, F.-M. *Macromolecules* **2000**, *33*, 232–234.
244 Benoit, D., Hawker, C.J., Huang, E.E., Lin, Z., Russell, T.P. *Macromolecules* **2000**, *33*, 1505–1507.
245 Lutz, J.-F., Kirci, B., Matyjaszewski, K. *Macromolecules* **2003**, *36*, 3136–3145.
246 Matyjaszewski, K., Ziegler, M.J., Arehart, S.V., Greszta, D., Pakula, T. *J. Phys. Org. Chem.* **2000**, *13*, 775–786.
247 Farcet, C., Charleux, B., Pirri, R., *Macromol. Symp.* **2002**, *182*, 249–260.
248 Kruk, M., Dufour, B., Celer, E.B., Kowalewski, T., Jaroniec, M., Matyjaszewski, K. *J. Phys. Chem. B* **2005**, *109*, 9216–9225.
249 Zhao, B., Brittain, W.J. *Prog. Polym. Sci.* **2000**, *25*, 677–710.
250 Tsujii, Y., Ejaz, M., Yamamoto, S., Ohno, K., Urayama, K., Fukuda, T. *Polym. Brushes* **2004**, 273–284.
251 Tsujii, Y., Ohno, K., Yamamoto, S., Goto, A., Fukuda, T. *Adv. Polym. Sci.* **2006**, *197*, 1–45.
252 Yoshikawa, C., Goto, A., Tsujii, Y., Fukuda, T., Yamamoto, K., Kishida, A. *Macromolecules* **2005**, *38*, 4604–4610.
253 Saleh, N., Phenrat, T., Sirk, K., Dufour, B., Ok, J., Sarbu, T., Matyjaszewski, K., Tilton, R.D., Lowry, G.V. *Nano Lett.* **2005**, *5*, 2489–2494.
254 Mei, Y., Beers, K.L., Byrd, H.C.M., VanderHart, D.L., Washburn, N.R. *J. Am. Chem. Soc.* **2004**, *126*, 3472–3476.
255 Becker, M.L., Liu, J., Wooley, K.L. *Chem. Commun.* **2003**, 802.
256 Bontempo, D., Heredia, K.L., Fish, B.A., Maynard, H.D. *J. Am. Chem. Soc.* **2004**, *126*, 15372–15373.
257 Bontempo, D., Li, R.C., Ly, T., Brubaker, C.E., Maynard, H.D. *Chem. Commun.* **2005**, 4702–4704.
258 Bontempo, D., Maynard, H.D. *J. Am. Chem. Soc.* **2005**, *127*, 6508–6509.
259 Joralemon, M.J., Smith, N.L., Holowka, D., Baird, B., Wooley, K.L. *Bioconjug. Chem.* **2005**, *16*, 1246–1256.
260 Kruk, M., Dufour, B., Celer, E.B., Kowalewski, T., Jaroniec, M., Matyjaszewski, K. *Chem. Mater.* **2006**, *18*, 1417–1424.
261 Beers, K.L., Boo, S., Gaynor, S.G., Matyjaszewski, K. *Macromolecules* **1999**, *32*, 5772–5776.
262 Coca, S., Jasieczek, C.B., Beers, K.L., Matyjaszewski, K. *J. Polym. Sci., Part A: Polym. Chem.* **1998**, *36*, 1417–1424.
263 Beers, K.L., Gaynor, S.G., Matyjaszewski, K., Sheiko, S.S., Moeller, M. *Macromolecules* **1998**, *31*, 9413–9415.
264 Couvreur, L., Lefay, C., Belleney, J., Charleux, B., Guerret, O., Magnet, S. *Macromolecules* **2003**, *36*, 8260–8267.
265 Zhang, X., Matyjaszewski, K. *Macromolecules* **1999**, *32*, 7349–7353.
266 Camp, W.V., Du Prez, F.E., Bon, S.A.F. *Macromolecules* **2004**, *37*, 6673–6675.
267 Tsarevsky, N.V., Bernaerts, K.V., Dufour, B., Du Prez, F.E., Matyjaszewski, K. *Macromolecules* **2004**, *37*, 9308–9313.
268 Sarbu, T., Lin, K.-Y., Ell, J., Siegwart, D.J., Spanswick, J., Matyjaszewski, K. *Macromolecules* **2004**, *37*, 3120–3127.

269 Sarbu, T., Lin, K.-Y., Spanswick, J., Gil, R. R., Siegwart, D. J., Matyjaszewski, K. *Macromolecules* **2004**, *37*, 9694–9700.

270 Matyjaszewski, K., Spanswick, J. *Mater. Today* **2005**, *8*, 26–33.

271 Richard, R. E., Schwarz, M., Ranade, S., Chan, A. K., Matyjaszewski, K., Sumerlin, B. *Biomacromolecules* **2005**, *6*, 3410–3418.

272 Burguiere, C., Pascual, S., Coutin, B., Polton, A., Tardi, M., Charleux, B., Matyjaszewski, K., Vairon, J.-P. *Macromol. Symp.* **2000**, *150*, 39–44.

273 Leibler, L. *Prog. Polym. Sci.* **2005**, *30*, 898–914.

274 Zhang, Y., Costantini, N., Mierzwa, M., Pakula, T., Neugebauer, D., Matyjaszewski, K. *Polymer* **2004**, *45*, 6333–6339.

275 McCarthy, P., Tsarevsky, N. V., Bombalski, L., Matyjaszewski, K., Pohl, C. *ACS Symp. Ser.* **2006**, *944*, 252–268.

276 Lee, S. B., Russell, A. J., Matyjaszewski, K. *Biomacromolecules* **2003**, *4*, 1386–1393.

277 Clark, A. J. *Chem. Soc. Rev.* **2002**, *31*, 1–11.

6
Coordination Polymerization: Synthesis of New Homo- and Copolymer Architectures from Ethylene and Propylene using Homogeneous Ziegler–Natta Polymerization Catalysts

Andrew F. Mason and Geoffrey W. Coates

6.1
Introduction, Historical Perspective and Scope of Review

Synthetic polymers have become commonplace materials that can be found in nearly all areas of everyday life. It seems strange to consider that when Leo Baekeland introduced the first synthetic polymer (Bakelite) nearly 100 years ago, there was great debate within the scientific community regarding the molecular structure of these new materials. Efforts to replace natural materials during World War II eventually led to the large-scale industrial production of polymer resins during the second half of the twentieth century. As new plastic materials were introduced, the general public greeted them with an odd mixture of excitement and skepticism. Strong, durable, inexpensive plastic was seen as physical evidence of man's ability to control and improve nature. Cheap, poorly manufactured plastic was seen as leftover chemical waste, a shoddy imitation of the natural materials that it tried to replace. For better or worse, plastic was a tangible example of the difference science could make in one's daily life. Today, this is no longer the case. As polymer-based materials become more and more prevalent, they are noticed less and less. A new material is not noteworthy unless it "does" something new and interesting. Poorly designed plastic materials may still cause frustration, but in many cases more traditional natural materials are no longer available; plastic has become the standard. We may notice that orange juice is sold in plastic jugs instead of wax-lined paper cartons or that prepackaged salads are sold in plastic bags, but these plastics are not seen as exciting innovations even though they may be technically sophisticated. Commodity polymers will undoubtedly continue to become more prevalent yet less visible in our daily lives. In the years to come, polymer-based materials that excite the public imagination will be high-performance, specialty materials having novel molecular architectures [1].

About 60% of the world's thermoplastics are homo- or copolymers made from ethylene and/or propylene. Although some polyethylenes are made using radical-based polymerization methods, coordination polymerization methods dominate their industrial production. There are two classes of catalysts used to produce ethylene and propylene based polymers: (1) heterogeneous Ziegler–Natta catalysts, typically based on supported titanium halides that are activated by aluminum alkyls, and (2) homogeneous catalysts, commonly (but not always) [2, 3] employing activated Group IV metallocenes. At the current time, only about 1% of polyolefins are made using metallocenes [4]. Despite their limited current use for the production of commodity polyolefins, homogeneous coordination polymerization catalysts offer unprecedented opportunities for controlling polyolefin structure at the molecular level (including molecular weight, molecular weight distribution, sequence, topology and tacticity) which is the theme of this book. This review will describe new polymer architectures formed from olefin-based monomers by homogeneous transition metal catalysts. The monomers covered will be limited to ethylene and propylene. Coverage will be as comprehensive as allowed by space limitations, but will focus on the more recent literature, especially those papers regarding new catalysts that allow advances in macromolecular engineering [5–7].

6.2
Primer on the Homogeneous Coordination Polymerization of Olefins

Although the focus of this review is the coordination polymerization of olefins in the context of macromolecular engineering, a brief review of the organometallic mechanisms of these systems is required to understand fully some of their limitations and capabilities. Shortly after the discovery of heterogeneous catalysts, researchers began developing soluble, homogeneous transition metal catalysts. Well-defined molecular species are more amenable to mechanistic study and homogeneous catalysts form polymers with narrower molecular weight distributions, since all the transition metal active sites are more or less identical. In a homogeneous system, all metal centers are potential "active sites" and the overall process is more efficient than the heterogeneous case, where only atoms at the surface engage in polymerization. Little progress was made in developing improved homogeneous polymerization catalysts until the early 1980s, when a new activating agent consisting of partially hydrolyzed trimethylaluminum was reported by Sinn and Kaminsky, now commonly known as methylaluminoxane [8, 9].

6.2.1
Nature of the Active Species and Mechanism of Initiation

Methylaluminoxane (MAO) is an oligomeric material containing $[Al(Me)O]_n$ repeat units. Activation of Group IV metallocene dichlorides with MAO instead of trialkylaluminum reagents greatly increased catalyst activity toward ethylene polymerization [10]. More importantly, MAO allows for the formation of polypropylene and higher polyolefins using a range of homogeneous systems, both me-

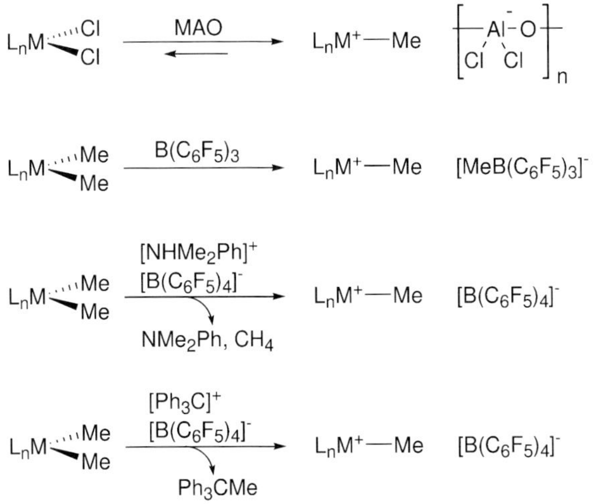

Fig. 6.1 Activation of metallocenes.

tallocene and non-metallocene based. There is a general consensus that MAO activates metal halide complexes by methylation, followed by methyl or halide abstraction to generate a cationic metal methyl of the form L_nM^+–Me, where MAO serves as a counteranion (Fig. 6.1). Metal alkyls can be directly activated by alkyl abstraction by MAO. Owing to the poorly defined nature of MAO, subsequent work has focused on the development of non-coordinating borate counteranions [11–15], a topic that has been thoroughly reviewed [16]. It should be noted that there are examples of neutral metal alkyl complexes that are active for olefin polymerization, such as the metallocene alkyls of scandium and yttrium and also nickel alkyls, that do not require MAO or borate activators [17, 18]. Initiation then begins by insertion of an olefin into the metal alkyl bond (see below). In the case of catalysts that undergo β-hydride elimination during propagation, subsequent chains are initiated by insertion into the metal hydride bond.

6.2.2
Mechanism of Propagation

Propagation occurs by insertion of the olefin into the metal carbon bond of the neutral or cationic active species (Fig. 6.2). This process occurs by a cis-addition process, where the metal and alkyl are added to the same face of the olefin monomer. In the case of α-olefins, the monomer usually inserts by a 1,2 insertion, with the metal adding to the α-carbon of the olefin. In some cases 2,1 insertion occurs; it should be noted that both give polyolefins with substituents on alternating carbons of the main chain [19–21]. It is generally assumed that the olefin coordinates prior to insertion, although this can be experimentally dif-

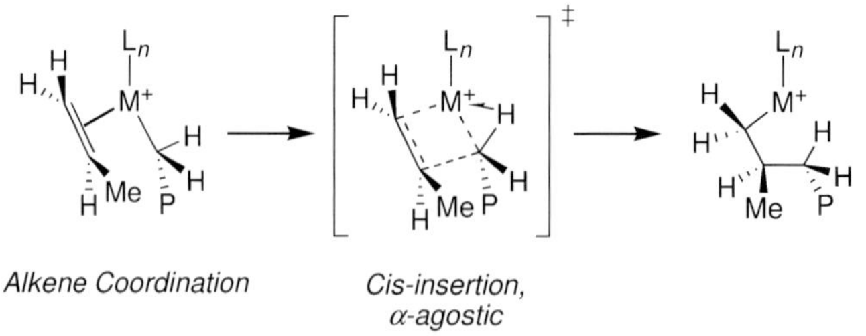

Fig. 6.2 Generalized mechanism of olefin insertion.

ficult to observe [22, 23]. There is experimental and computational evidence in some cases for the existence of an agostic interaction between a hydrogen atom on the α-carbon of the growing polymer chain and the metal during the transition state for olefin insertion [24]. The stereochemistry of the polymer is determined at this stage and this topic will be further addressed in Section 6.4 [6, 25].

6.2.3
Mechanisms of Termination and Chain Transfer

Olefin polymerization complexes typically fall into two categories, those that are living and those that are not. Complexes that can initiate chains rapidly and propagate but do not have appreciable rates for chain termination or transfer are living and can create polymer chains of precise molecular weight and controlled sequence [26]. On the other hand, there are often many possible mechanistic pathways that lead to termination of the chain, and the relative rates of these processes to the rate of propagation will determine the chain length (Fig. 6.3). The most common of these termination processes are: β-hydrogen elimination, β-alkyl elimination, chain transfer to monomer (effectively β-hydrogen elimination) [27], transmetallation and chain transfer to an added agent [28]. Since typical living olefin polymerization systems only produce one polymer chain per metal center, there is a significant amount of interest in developing catalytic systems that can produce multiple chains per metal center. One strategy to accomplish this goal is to add excess amounts of an inexpensive metal complex that will rapidly transmetallate the active catalyst, producing many chains per metal center [29–31]. Metal complexes that exhibit slower rates of transmetallation have the potential to produce block copolymers when two different polymerization catalysts are used [32]. Polymer chains with metal functionality at the end provide the opportunity for subsequent chain-end functionalization [33]. The absence and presence of chain transfer and termination mechanisms are in many cases unique to homogeneous olefin polymerization complexes and can be used to create polyolefins with unprecedented architectures.

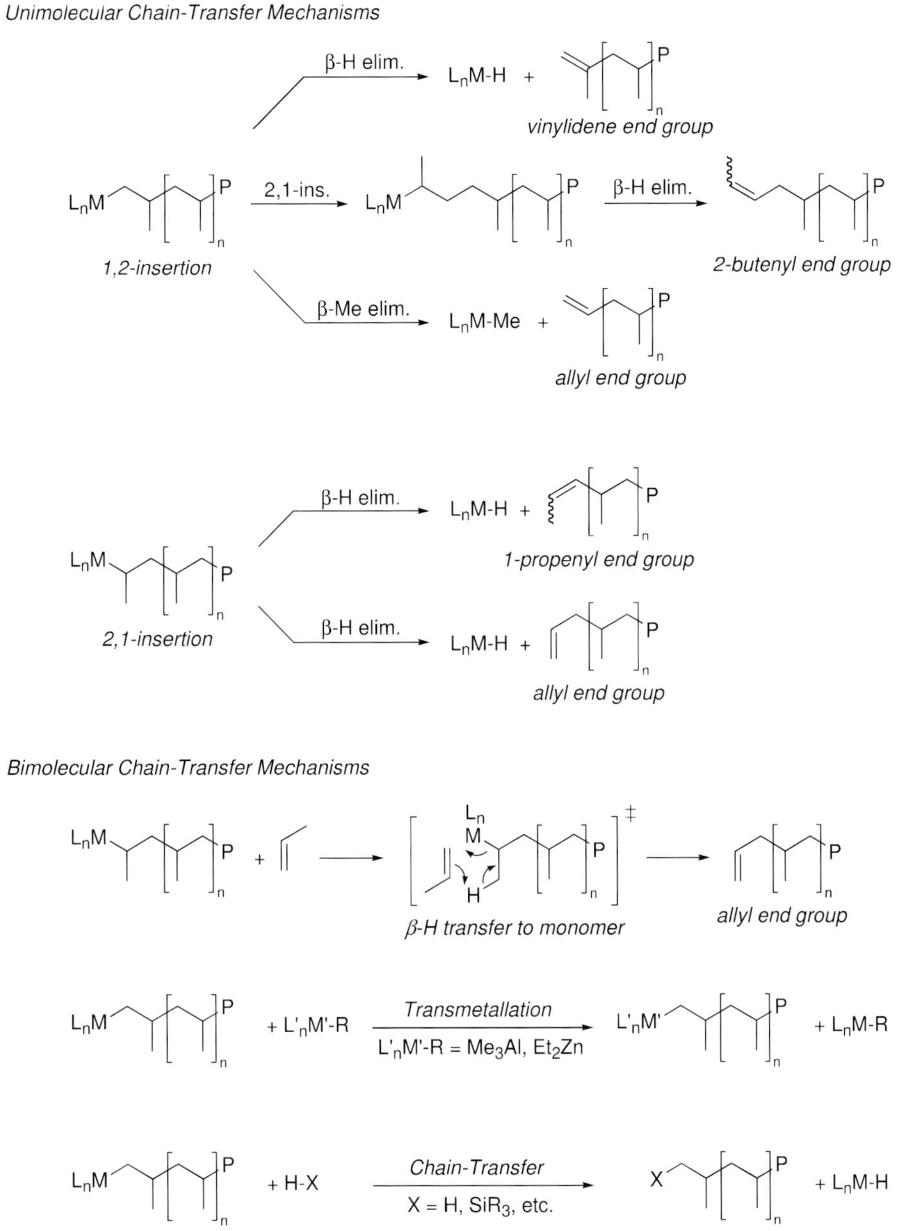

6.2 Primer on the Homogeneous Coordination Polymerization of Olefins

Unimolecular Chain-Transfer Mechanisms

Bimolecular Chain-Transfer Mechanisms

Fig. 6.3 Unimolecular and bimolecular chain transfer mechanisms of olefin polymerization catalysts.

6.3
Ethylene-based Polymers

Low-density polyethylene (LDPE) is synthesized industrially by a free radical process using peroxide initiators at high temperatures and pressures. LDPE has a highly branched structure (Fig. 6.4) due to radical back-biting reactions. The high pressures required for LDPE formation are not cost-effective and LDPE is slowly being replaced in the marketplace by polyethylene formed by transition metal catalysts. Heterogeneous titanium Ziegler–Natta catalysts and supported chromium oxide catalysts [34, 35] are widely used to produce linear high-density polyethylene (HDPE). HDPE has a higher melting-point and is much more crystalline than LDPE. Most homogeneous transition metal catalysts can homopolymerize ethylene to HDPE, but in many cases catalyst performance cannot match that of established heterogeneous systems. Linear low-density polyethylene (LLDPE) is a third major industrially produced material. LLDPE is actually a copolymer of ethylene with short-chain α-olefins (typically C_4–C_8, around 10 wt%). The resulting polymer exhibits good tensile strength due to its long backbone, while the short-chain branches impart toughness. Other industrial polyethylene formulations include ultra-low-density polyethylene (ULDPE), high molecular weight (HMW) HDPE and ultra-high molecular weight (UHMW) HDPE. ULDPE has a higher α-olefin content than LLDPE, while UHMW-HDPE has a molecular weight of over 3×10^6, lacks a melting-point and is commonly compression-molded instead of melt-injected [36].

The range of commercialized ethylene homopolymers does not leave that much room for further innovation. However, there are a number of interesting catalysts that are capable of forming branched LLDPE-like structures from ethylene without adding α-olefin comonomers. Researchers at Dow and Exxon have introduced *ansa*-cyclopentadienylamidotitanium catalysts for olefin polymerization (Fig. 6.5) [37–41]. These constrained-geometry catalysts (CGCs) are particularly useful for the production of LLDPE. In addition to their high thermal stability, CGCs have a more accessible active metal center than metallocene cata-

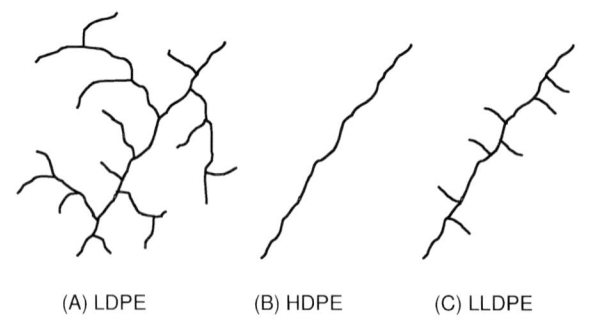

(A) LDPE (B) HDPE (C) LLDPE

Fig. 6.4 Chain branching in common types of polyethylene.

Fig. 6.5 *ansa*-Cyclopentadienylamido constrained geometry catalysts.

R = alkyl, aryl
R' = H, Me
M = Ti, Zr, Hf
X = Cl, Me

lysts and can easily incorporate larger α-olefin comonomers. In the course of characterizing the LLDPE produced by these catalysts, researchers noticed that the polymer samples contained a small amount of long-chain branches (LCBs) that were much longer than any of the branches resulting from α-olefin comonomer incorporation [42, 43]. These LCBs arise from incorporation of "dead" olefin-terminated LLDPE chains by an active catalyst. Conditions that favor LCB formation include high temperature, high conversion and low ethylene or α-olefin concentration [44, 45]. More recently, additional catalysts capable of reinserting long-chain olefins have been reported [46–49]. In some cases, binary catalyst systems are used, where one catalyst oligomerizes ethylene to 1-hexene or 1-octene and a second catalyst incorporates the oligomers into LLDPE chains [50–55].

Late-metal nickel and palladium catalysts can also form branched polyethylene [56]. Developed by Brookhart and coworkers, these catalysts contain bulky diimine ligands (Fig. 6.6) [57]. The degree of branching depends on the metal, ligand structure, temperature and pressure. Nickel catalysts produce polyethylene ranging from nearly linear to moderately branched, while palladium catalysts produce more highly branched chains [58]. These hyperbranched polymers have many more side-chains than LDPE and are often low molecular weight oils. The formation of branched polyethylenes with these systems does not occur via the mechanism described above for early-metal catalysts. Branching occurs through a series of β-hydrogen elimination, olefin rotation and reinsertion events. During this "chain-walking" process, the eliminated olefin remains bound to the active metal center that formed it; unbound α-olefins have not been observed.

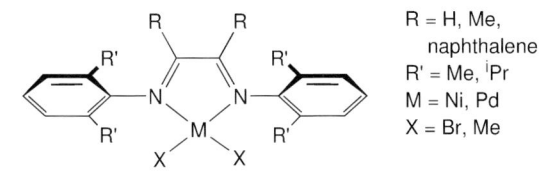

R = H, Me, naphthalene
R' = Me, ^{i}Pr
M = Ni, Pd
X = Br, Me

Fig. 6.6 Late-metal diimine olefin polymerization catalysts.

6.4
Propylene-based Polymers

Over 2.2×10^{10} kg of polypropylene (PP) were made worldwide in 2005 [59]. Although these polypropylenes were almost exclusively isotactic in microstructure, advances in the synthesis of other microstructures and architectures have been reported with increasing frequency. This section will review polypropylene synthesis with a focus on new synthetic capabilities.

6.4.1
Atactic, Isotactic and Syndiotactic Polypropylene

Natta's first propylene polymerization using heterogeneous catalysts produced a mixture of amorphous atactic PP and crystalline isotactic PP that had to be separated by fractionation. Today's state of the art heterogeneous catalysts produce high molecular weight, highly isotactic PP with an *mmmm* pentad >0.99 and a melting-point around 165 °C [60]. Compared with polyethylene (melting-point

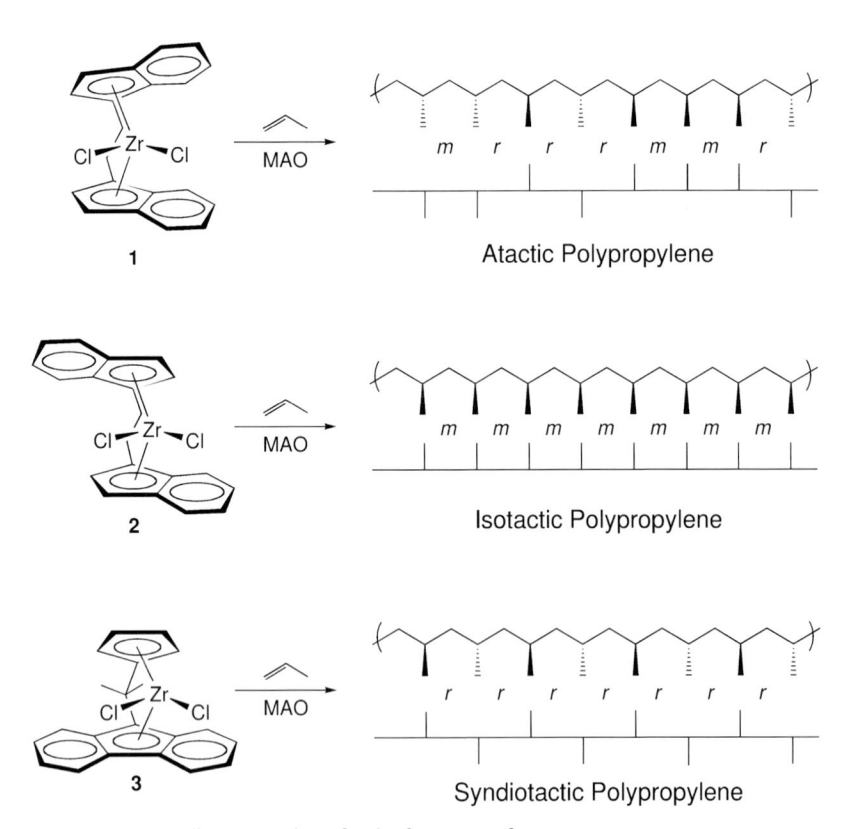

Fig. 6.7 *ansa*-Metallocene catalysts for the formation of atactic, isotactic and syndiotactic polypropylene.

135 °C for HDPE, 110–120 °C for LDPE), isotactic PP is a better material for high-temperature applications. Early homogeneous metallocene catalysts produced only atactic or moderately isotactic PP. Ewen and Brintzinger reported the first stereoregular propylene polymerizations using bridged *ansa*-metallocene systems (Fig. 6.7) [61, 62]. The strapped cyclopentadienyl ligand system restricts ligand rotation, resulting in a rigid steric environment. The bridged bis(indenyl) catalysts demonstrated the importance of ligand structure and ligand geometry in particular. The meso-isomer of the catalyst (**1**) produces atactic PP, but the racemic, C_2-symmetric isomer (**2**) produces isotactic PP ([*mmmm*]=0.78). Ewen et al. later developed a C_s-symmetric zirconocene catalyst bearing a bridged cyclopentadienylfluorenyl ligand (**3**) that produces syndiotactic polypropylene ([*rrrr*]=0.86) [63]. Heterogeneous catalysts are not able to make syndiotactic PP, hence the ability to synthesize a new type of stereoregular polypropylene using homogeneous catalyst systems was a major breakthrough. Syndiotactic PP made using **3** is less crystalline and has a lower melting-point (145 °C) than isotactic PP and is currently produced industrially on a limited scale. Since these early discoveries, the relationship between metallocene ligand structure and catalyst behavior has been studied intensively (Fig. 6.8) [6, 64, 65]. Modern isospecific metallocene catalysts often perform as well as their heterogeneous counterparts. The mechanisms of stereocontrol are also well understood. Chiral metallocenes form isotactic polypropylene by an enantiomorphic site control mechanism, where occasional stereoerrors are immediately corrected by the chiral catalyst site. Syndiospecific metallocene catalysts operate by alternated insertion of propylene with opposite enantiofaces [6].

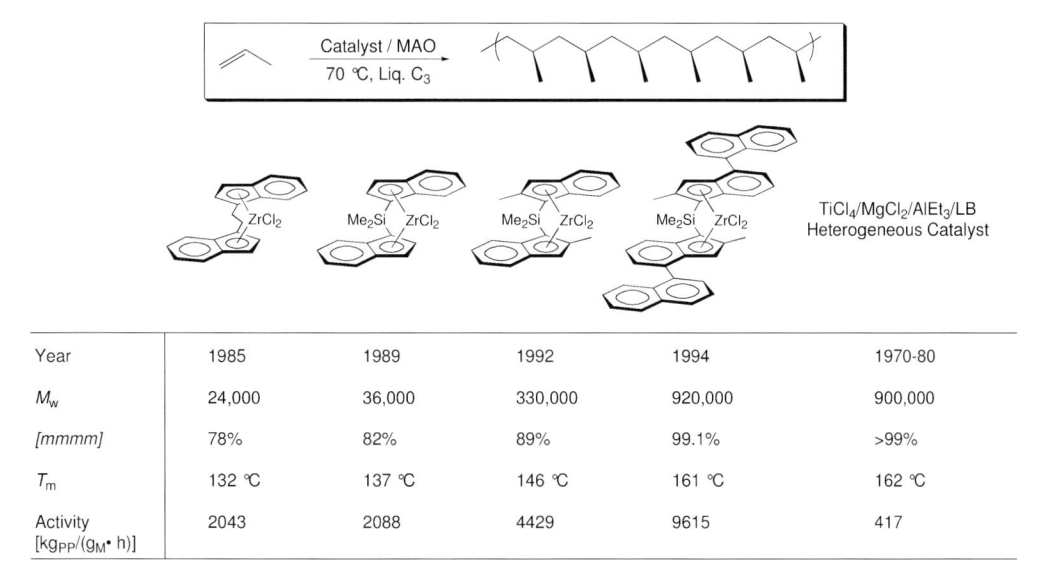

Year	1985	1989	1992	1994	1970-80
M_w	24,000	36,000	330,000	920,000	900,000
[*mmmm*]	78%	82%	89%	99.1%	>99%
T_m	132 °C	137 °C	146 °C	161 °C	162 °C
Activity [kg$_{PP}$/(g$_M$· h)]	2043	2088	4429	9615	417

Fig. 6.8 Advances in isospecific metallocene catalysts.

6.4.2
Hemiisotactic Polypropylene

Accurate predictions regarding polymer tacticity can usually be made when dealing with C_2- or C_s-symmetric catalysts, but asymmetric (C_1 point group) systems are more difficult to understand due to the presence of two different coordination sites. Ewen and coworkers developed an asymmetric catalyst (**4**) for the synthesis of hemiisotactic polypropylene (Fig. 6.9) [66, 67] in which every other methyl group off the main chain has the same stereochemical configuration and the remaining methyl groups have a random configuration. The NMR spectrum of hemiisotactic PP formed by **4** matches that of hemiisotactic PP formed by hydrogenation of poly(*trans*-2-methylpentadiene) [68]. The catalyst has two coordination sites; alternation between isospecific and aspecific sites produces the observed hemiisotactic PP. Stereoerrors appear to be the result of site epimerization (also called chain back-skip) [69] events that move the polymer chain from one site to another without monomer insertion [70–72]. Since hemiisotactic PP is being formed, the rate of site epimerization must be slower than the rate of monomer insertion. Other catalysts for hemiisospecific polymerization have also been reported [73–75].

6.4.3
Stereoblock Polypropylene

Given the variety of possible polypropylene microstructures, propylene block copolymers can be synthesized where the only difference between blocks is polymer stereochemistry [76]. Perhaps the most interesting and useful of these stereoblock polymers combine a crystalline block such as isotactic or syndiotactic PP with an amorphous block such as atactic PP. Polymers with these combinations often behave as thermoplastic elastomers and have properties similar to those of natural rubber [77]. A significant advantage is that, unlike natural rubber, thermoplastic elastomers can be melted and reprocessed (or recycled). Natural rubber is a thermoset plastic containing chemical cross-links; thermoplastic elastomers are physically cross-linked by polymer chains connecting separate crystalline regions of the polymer matrix.

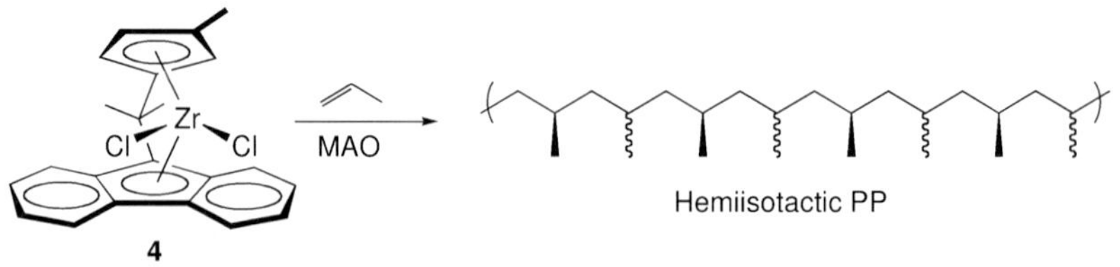

Fig. 6.9 Formation of hemiisotactic polypropylene with asymmetric metallocene catalysts.

Natta's original polypropylene sample contained a small fraction of elastomeric polymer [78], and more recent heterogeneous systems have been developed that produce better yields of this material. Collette and coworkers have developed a catalyst system consisting of Group IV tetraalkyl complexes on an aluminum support [79, 80]. Bulkier alkyl groups such as benzyl, neopentyl and neophyl (2-methyl-2-phenylpropyl) tend to produce better results. Fractionation is still required to isolate the elastomeric material and molecular weight distributions are broad. A thorough analysis of the polymer reveals alternating sequences of isotactic and atactic PP, possibly arising from chain transfer between different catalytic sites [81]. Similar heterogeneous systems were developed using bis(arene) Group IV catalysts [82, 83]. Stereoblock PP formed using these catalysts is usually more crystalline and less elastomeric than PP formed using the tetraalkyl catalysts. More recently, elastomeric PP was obtained using a titanium tetrachloride catalyst supported on magnesium chloride, with dibutyl phthalate and aromatic ether additives [84]. Predominantly syndiotactic elastomers can be synthesized from a complicated catalyst mixture that includes a dialkoxymagnesium compound, a titanium tetrahalide, an organoaluminum activator and an aromatic heterocycle [85].

Chien and coworkers first reported the synthesis of elastomeric PP formed by homogeneous catalysts [14, 86]. Asymmetric titanocenes **5** and **6** (Fig. 6.10) can be activated with MAO at 25 or 50 °C and generate low molecular weight, elastomeric material with an [*mmmm*] pentad range of 0.30–0.40. Compared with heterogeneous catalysts, **5** and **6** produce elastomers with much narrower molecular weight distributions ($M_w/M_n = 1.7–1.9$). A uniform polymer was obtained and the elastomeric material did not have to be isolated by fractionation. Unfortunately, these catalysts suffer from low stability and have relatively short lifetimes.

5, X = Cl
6, X = Me

7, X = C
8, X = Si

9

10

11

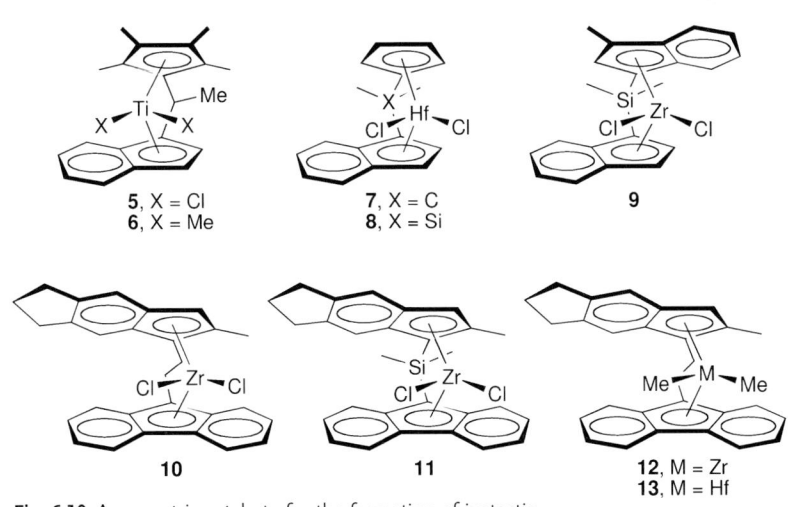

12, M = Zr
13, M = Hf

Fig. 6.10 Asymmetric catalysts for the formation of isotactic–atactic stereoblock polypropylene.

Collins and coworkers later developed a series of related C_1-symmetric hafnocene catalysts [87, 88]. When activated with MAO at 25 °C, **7** and **8** both form elastomeric PP with narrow molecular weight distributions ($M_w/M_n = 1.7$–2.1). Catalyst **8**, with a silicon bridge, produces higher molecular weight PP with greater isotacticity ([mmmm] ≈ 0.50) than carbon-bridged **7** ([mmmm] ≈ 0.30–0.38). Polymer formed by **8** has much better elastomeric properties, suggesting that a molecular weight of at least 50 000 g mol^{-1} is required for physical cross-linking. Zirconocene catalyst **9** also produces elastomeric PP with properties similar to polymer formed by **8** [89].

Bridged indenylfluorenyl metallocenes can also form elastomeric stereoblock polypropylene. Rieger and coworkers identified catalyst **10**, with a 5,6-cyclopentane substituent off the indenyl ring [90]. Replacing the ethylene bridge with a silicon bridge results in a catalyst (**11**) that produces PP with higher molecular weight ($M_w = 158\,000$, $M_w/M_n = 1.9$). Unfortunately, this system rapidly deactivates under polymerization conditions [91]. The behavior of **10** and **11** is highly dependent on the conditions used, resulting in highly tunable catalyst systems. By changing the temperature (20–70 °C) and propylene concentration (0.4–6.1 M), PP tacticities can vary from [mmmm] = 0.20 to 0.72. Isopropyl end-groups are seen in the ^{13}C NMR spectra of these polymers, suggesting that chain transfer to aluminum is a significant source of chain termination. To address this problem, dimethylzirconocene **12** was synthesized and activated with [Ph$_3$C][B(C$_6$F$_5$)$_4$]. Much higher molecular weights ($M_w = 350\,000$ at 10 °C, $M_w/M_n = 3.0$) are obtained in the absence of aluminum chain-transfer agents. Hafnium catalyst **13** performs even better ($M_w = 1\,600\,000$ at 20 °C, $M_w/M_n = 2.6$) and has the highest activity of the series (20 000 s^{-1}) [92]. Compared with **13**, the analogous hafnium dichloride complex has an extremely low polymerization activity when activated with MAO. The ultra-high molecular weight stereoblock elastomers formed by **13** have superior tensile properties and stress–strain profiles to commercial Kraton rubber.

All the asymmetric catalysts in Fig. 6.10 form isotactic–atactic stereoblock polypropylene of similar composition. Structurally similar metallocene catalysts have been described in the patent literature [93, 94]. A number of theories have attempted to explain the formation of elastomeric polymer. Catalyst asymmetry clearly results in two different coordination sites. Isotactic blocks could be formed at the isotactic site and atactic blocks at the atactic site, with occasional site epimerization events interconverting between the two sites [86]. However, based on a migratory insertion mechanism, the polymer chain alternates sites after each insertion. Gauthier and Collins proposed that site epimerization does occur, but at a slower rate than chain propagation. When epimerization does occur from the aspecific site to the isospecific site, the probability of forming an isotactic sequence increases [95]. The tacticity of polypropylene formed by catalysts **7**–**13** is dependent on propylene concentration: an increase in concentration results in a lower [mmmm] pentad value. Increased propylene concentration should increase the rate of monomer insertion without affecting site epimerization. Therefore, less epimerization from aspecific to isospecific should occur, resulting in fewer isotactic sequences. If this is indeed the case, then the chances of forming long isotactic sequences is small. To form a crystalline block, isotactic chain lengths of at least 15–20 are needed [81]. This could explain why high molecular weight polymers are required for good elastomeric properties. Shorter chains do not have enough isotactic sequences of the length required to form a physically cross-linked network.

Elastomeric isotactic–hemiisotactic polypropylene has been synthesized by Miller and Bercaw using a modified version of hemiisospecific catalyst **4**. Replacing the cyclopentadienyl methyl group of **4** with a larger adamantyl substituent produces a catalyst with one highly isospecific site and one slightly isospecific site having an a parameter of 0.60 (a site with $a=0.50$ is completely random and a site with $a=1.00$ is completely isospecific) [96]. Phenyl substituents on the metallocene bridge are required for high molecular weight PP ($M_w=1\,080\,000$ at $0\,°C$, $M_w/M_n=2.33$) and elastomeric behavior. Since isotactic dyad formation is only slightly favored, there are few isotactic sequences of significant length (similar to the isotactic-atactic stereoblocks described above). Ewen et al. recently reported the formation of stereoblock isotactic–hemiisotactic PP using a heterocene catalyst [97].

In most of the cases described above, formation of stereoblock PP elastomers was the unexpected result of various ligand modifications. Waymouth and coworkers used a rational ligand design strategy to form isotactic-atactic stereoblocks using unbridged metallocene **14** (Fig. 6.11) [98, 99]. This catalyst is conformationally flexible and can rotate freely between two isomeric forms. Coates and Waymouth proposed that the C_2-symmetric rotamer produces isotactic blocks and the meso rotamer produces atactic blocks, although a proposal by Busico and coworkers suggests that the atactic block is made when the counteranion allows the ligands to move freely [100, 101]. In the X-ray crystal structure of **14**, both isomers are present in the unit cell. The polymerization behavior of **14** is sensitive to propylene pressure and reaction temperature, with higher pressures and lower temperatures favoring formation of isotactic pentads. Polymerization at $0\,°C$ under 4.4 atm of propylene produces high molecular weight ($M_w=540\,000$, $M_w/M_n=1.7$) elastomeric PP. The effect of ligand structure on

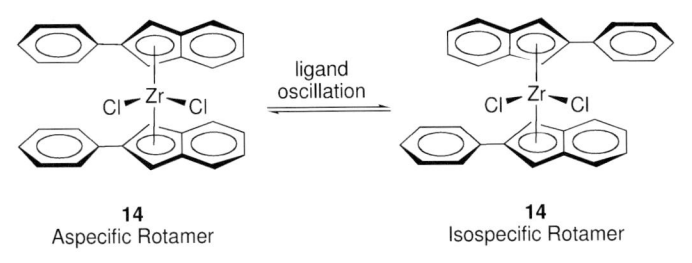

Fig. 6.11 Oscillating metallocene catalyst for the formation of isotactic–atactic stereoblock polypropylene.

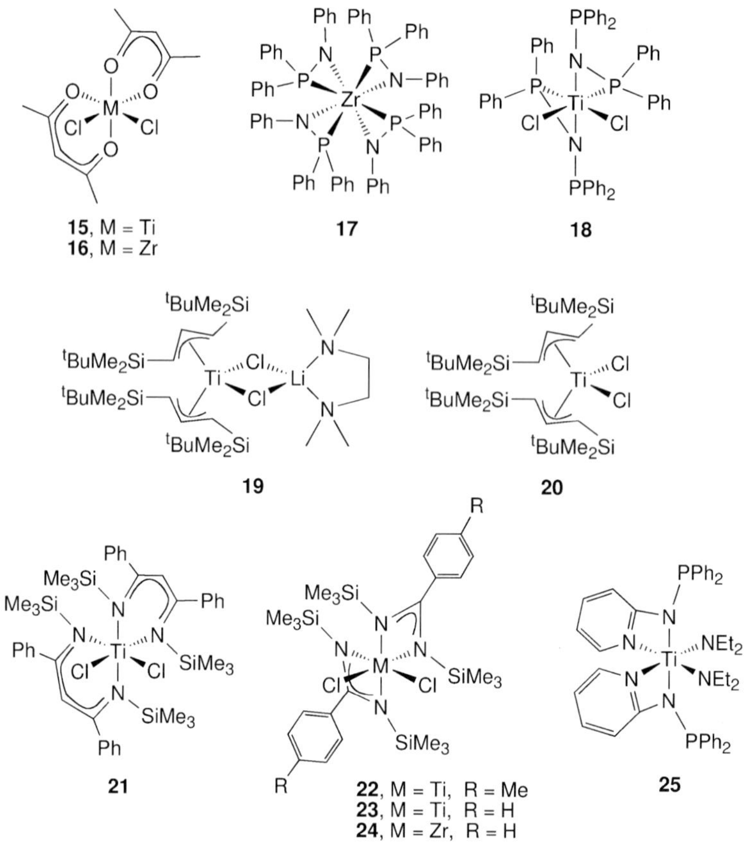

Fig. 6.12 Octahedral non-metallocene catalysts for the formation of isotactic–atactic stereoblock polypropylene.

catalyst behavior and polymer properties for 2-arylindene metallocene catalysts has been extensively studied [102, 103].

Catalysts for the formation of stereoblock polypropylene are not limited to cyclopentadienyl-based ligand frameworks. Eisen and coworkers have identified a number of different non-metallocene Group IV catalysts that form elastomeric PP (Fig. 6.12) [104]. Octahedral bis(acetylacetonate) dichloride complexes **15** and **16** form elastomeric polymers when activated with MAO [105]. Methylene chloride solvent is required for **16** to form stereoblock PP; in toluene solution, highly isotactic PP is formed. Titanium catalyst **15** forms elastomeric polymers in both solvents. Based on NMR data, **15** and **16** are C_2-symmetric in solution and should form isotactic PP based on symmetry arguments. Atactic sequences are caused by epimerization of the growing polymer chain, which inverts the stereochemistry of the last-inserted methyl group [106]. The ability of **15**–MAO to isomerize 1-octene to a mixture of 2- and 3-octene supports this epimerization mechanism. In the case of C_2-symmetric octahedral systems, chain epimerization stereoerrors reduce isotacticity and result in the formation of atactic blocks. This should not be confused with the site epimerization mechanism used to explain stereoblock formation for the assymetric catalysts in Fig. 6.10.

Catalysts **17–24** (Fig. 6.12) also form stereoblock PP as a result of chain epimerization events. The X-ray crystal structure of phosphinoamide complex **17** has four η^2-coordinated ligands, but in solution this structure is in equilibrium with an octahedral C_2-symmetric structure with two η^1-coordinated ligands. It is thought that activation of **17** with MAO generates a cationic bis(η^2-phosphinoamide) active species that carries out the polymerization. Titanium catalyst **18** is almost 20 times faster than **17** [107]. Titanium bis(allyl) complexes **19** and **20** also form elastomeric PP of moderate molecular weight [108]. Titanium(III) **19** can be oxidized directly to **20** with (4-BrC$_6$H$_4$)$_3$NSbCl$_6$. When activated with MAO, **19** and **20** behave similarly, suggesting that **19** is oxidized to titanium(IV) under polymerization conditions. Zirconium versions of **19** and **20** form non-elastomeric isotactic PP. Octahedral β-diketimidinate complex **21**–MAO is an active catalyst for the formation of high molecular weight stereoblock PP ($M_n = 326\,700$ at 60 °C, $M_w/M_n = 2.32$) [109]. Finally, octahedral bis(benzamidinate) complexes **22–24** form elastomeric polypropylene, but only under high propylene pressure (9.2 atm) [110].

Eisen and coworkers have recently found that octahedral, C_2-symmetric catalyst **25** forms isotactic–atactic stereoblock PP when activated with MAO [111]. In this case, stereoblock formation is the result of fluxional ligand behavior, not chain epimerization. NMR studies of **25**–MAO indicate that the pyridine nitrogen donor can bind to an MAO counterion, resulting in a tetrahedral, asymmetric environment around the titanium center. As a result, catalyst stereospecificity changes with ligand binding mode. The elastomeric polymer formed by **25**–MAO does not have any melting transitions due to short isotactic block lengths. Similar ligand dynamics could provide an alternate explanation for stereoblock formation with phosphinoamide complexes **17** and **18**. If the ligand phosphino donors are fluxional under polymerization conditions, then an isospecific octahedral configuration could be in equilibrium with an aspecific tetrahedral configuration.

Binary catalyst mixtures can be used to form stereoblock polypropylene if polymer chains can be exchanged between two different metal centers. Chien et al. combined isospecific metallocene **26** (Fig. 6.13) and aspecific metallocene **27** with iBu$_3$Al–[Ph$_3$C][B(C$_6$F$_5$)$_4$] and formed a polymer mixture containing 10% isotactic–atactic stereoblock PP [112]. The majority of the product was either atactic or isotactic homopolymer. Chain transfer to iBu$_3$Al is the most likely mechanism for stereoblock formation. Syndiospecific metallocene **28** can be used in conjunction with **26**–iBu$_3$Al–[Ph$_3$C][B(C$_6$F$_5$)$_4$] to form syndiotactic–atactic stereoblocks, although once again this material makes up a small fraction of the overall polymer [113]. Lieber and Brintzinger used the combination of **26**–**29**–MAO to form isotactic–atactic stereoblock PP, but no stereoblock formation was observed with **26**–**29**–iBu$_3$Al–[Ph$_3$C][B(C$_6$F$_5$)$_4$] [114]. Isotactic–syndiotactic

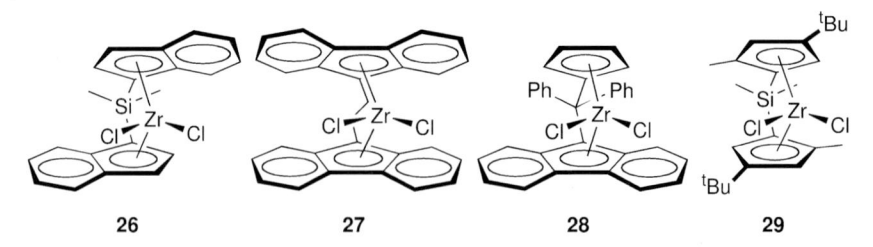

Fig. 6.13 ansa-Metallocene catalysts used as part of binary catalyst mixtures for the formation of stereoblock polypropylene.

stereoblocks have been synthesized using both 2–28–iBu$_3$Al–[Ph$_3$C][B(C$_6$F$_5$)$_4$] and 28–29–MAO catalyst systems [114, 115]. Since both isotactic and syndiotactic PP are crystalline polymers, complete separation of the stereoblock PP is not possible, but NMR data suggest that a stereoblock fraction is formed. A silica-supported binary catalyst system (3–26–MAO–iBu$_3$Al) has also been used to form isotactic–syndiotactic stereoblocks [116]. Synthesizing fewer homopolymer chains with binary systems will require a better understanding of chain-transfer processes and the development of new chain-transfer agents. Nonetheless, a small percentage of stereoblock chains can greatly improve miscibility and lead to useful polymer blends.

Recently, Shiono and coworkers synthesized syndiotactic-atactic diblock polypropylene using a living half-metallocene complex (**30**, Fig. 6.14) [117]. In heptane solvent at 0 °C, **30**–modified MAO (MMAO) produces moderately syndiotactic PP ([rr]=0.72). In chlorobenzene at the same temperature, **30**–MMAO produces atactic PP ([rr]=0.43). Polymer molecular weight distributions are nar-

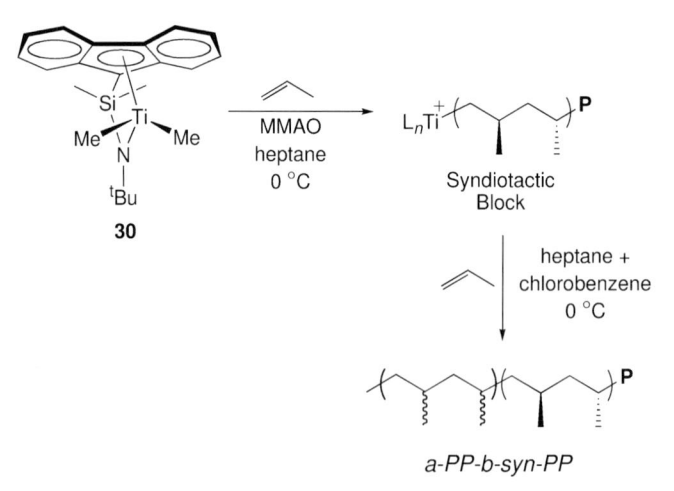

Fig. 6.14 Synthesis of stereoblock polypropylenes by changing solvent polarity.

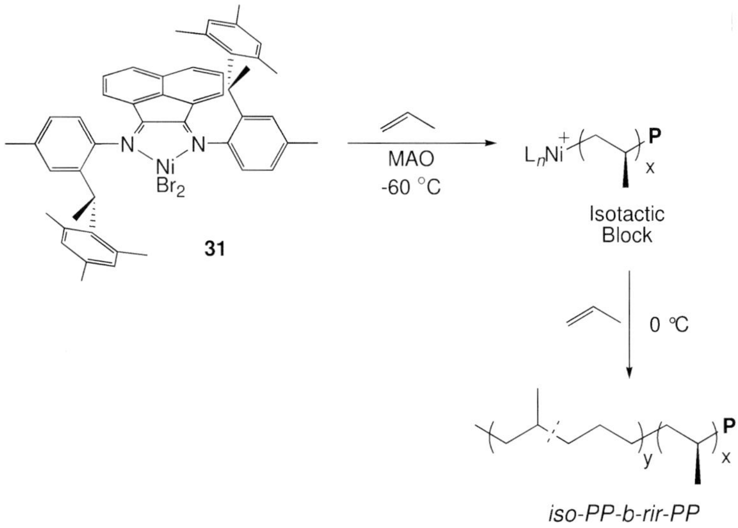

Fig. 6.15 Synthesis of stereoblock polypropylenes by changing temperature.

row (M_w/M_n = 1.14–1.34), suggesting that **30** does not undergo chain transfer or chain termination reactions. Adding chlorobenzene to an active heptane polymerization converts the stereocontrol from syndiotactic to atactic and a diblock polymer is formed. Although the syndiotactic block is not highly syndiotactic, this demonstrates the ability of living catalysts to control polymer architecture.

Coates and coworkers have recently reported propylene polymerization using a living nickel-based complex (**31**) that produces isotactic polypropylene at low temperature and regioirregular polypropylene at ambient temperature (Fig. 6.15). By cooling and warming a propylene polymerization, a diblock poly(-iso-PP-*block*-rir-PP) was made. Triblock copolymers were also made by sequentially changing the temperature of a propylene polymerization [118].

6.4.4
Graft and Star Polypropylene

Graft or comb polypropylene can be synthesized by copolymerizing propylene with long-chain polypropylene macromonomers. Varying the tacticity of the branches relative to the main chain can have a large influence on polymer properties (Fig. 6.16). Incorporation of atactic branches into an isotactic main chain has been accomplished using metallocene catalysts. Vinyl-terminated atactic PP macromonomer was synthesized and isolated using Cp*$_2$ZrCl$_2$–MAO (Cp* = pentamethylcyclopentadienyl) [119]. Vinyl end-groups are required for subsequent macromonomer incorporation and are usually formed by β-methyl elimination chain transfer reactions. Isospecific complex **32** (Fig. 6.17) is used to in-

| Atactic-Branched | Isotactic-Branched | Isotactic-Branched |
| Isotactic PP | Atactic PP | Isotactic PP |

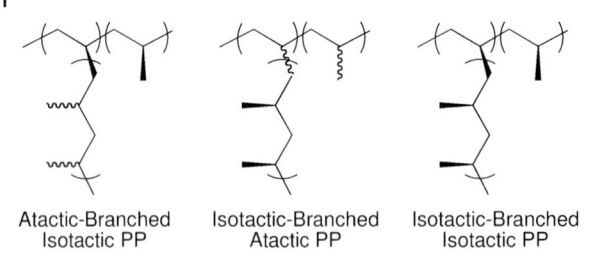

Fig. 6.16 Examples of long-chain branched polypropylene.

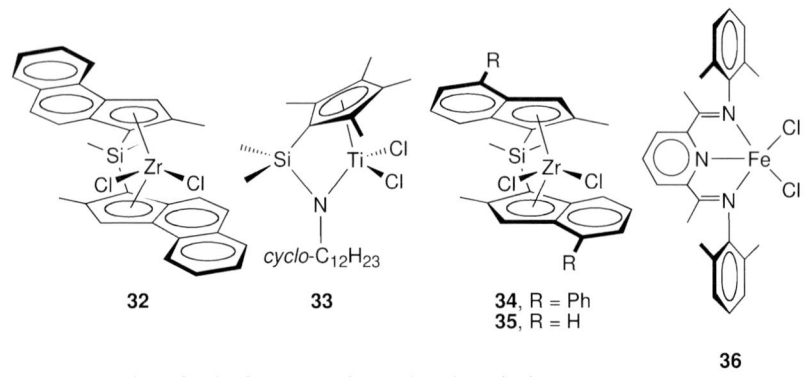

Fig. 6.17 Catalysts for the formation of long-chain branched polypropylene.

corporate macromonomer (about 20% by weight) into the main PP chain. Constrained-geometry catalyst **33** is able to incorporate isotactic PP macromonomers formed by **34** into atactic PP [120]. These isotactic PP macromolecules can also be incorporated into isotactic PP using **34** or **35** [121]. Since **34** is able to form vinyl-terminated chains and incorporate them into growing PP chains, it can be used to form long-chain branched isotactic PP in a one-pot reaction [122]. This method is faster and more efficient, but characterization of the long chain branches is more difficult. Finally, atactic-branched isotactic PP has been synthesized using a binary catalyst system (**32**–**36**–MMAO) [123]. Aspecific iron catalyst **36** inserts propylene via a 2,1-insertion mechanism and produces vinyl-terminated macromonomers by β-hydrogen elimination [19]. To produce the desired branched structure, **36**–MMAO was allowed to form macromonomers for 30–120 min before injection of isospecific catalyst **32**.

Atactic star-like polypropylene was synthesized by Koo and Marks using constrained-geometry catalyst **37** in the presence of 1,3,5-trisilabenzene (Fig. 6.18) [124]. The silane acts as a chain transfer agent during the polymerization. Chain transfer occurs in a stepwise fashion, resulting in polymers having broad molecular weight distributions ($M_w/M_n = 8.7$).

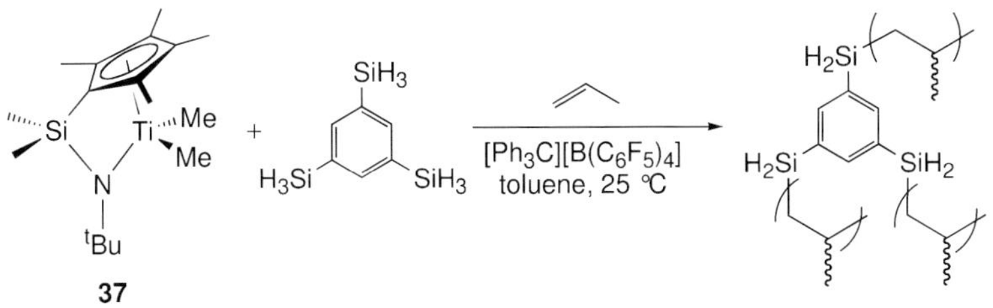

Fig. 6.18 Synthesis of star-like polypropylene using polyfunctional silanes.

6.5 Ethylene–Propylene Copolymers

6.5.1 Random Ethylene–Propylene Copolymers

Random copolymers of ethylene and propylene (EP) are commonly synthesized using either heterogeneous Ziegler–Natta or homogeneous metallocene catalysts [5]. Heterogeneous systems produce chains of varying compositions due to the presence of many different active sites. These catalysts tend to incorporate homopolymer sequences and molecular weights vary with polymer composition [125–127]. Homogeneous systems offer better control over polymer properties and tend to produce more random copolymers. The reactivity ratio for ethylene and propylene can be calculated for a given catalyst and monomer concentrations can be chosen to produce the desired copolymer [128–132]. If an isospecific catalyst only incorporates a small amount of ethylene, a crystalline copolymer is formed that has a lower melting-point than isotactic PP. If an aspecific catalyst is used or if more ethylene is incorporated into the polymer, an amorphous ethylene–propylene rubber (EPR) is formed. EPR generally has a lower glass transition temperature than atactic PP and is a useful material for low-temperature applications. Catalysts that form isotactic or syndiotactic PP by a site control mechanism also control the stereochemistry of propylene units in an EP copolymer.

6.5.2 Alternating Ethylene–Propylene Copolymers

Random EP copolymers containing equal molar amounts of ethylene and propylene can be readily synthesized by choosing appropriate monomer concentrations. Soga's and Waymouth's groups have developed metallocene catalysts that are able to synthesize alternating EP copolymers. The ethylene–propylene polymers were compared with hydrogenated poly(*trans*-1,3-pentadiene) in order to verify their structure [133]. Soga and coworkers used indenylfluorenyl catalyst

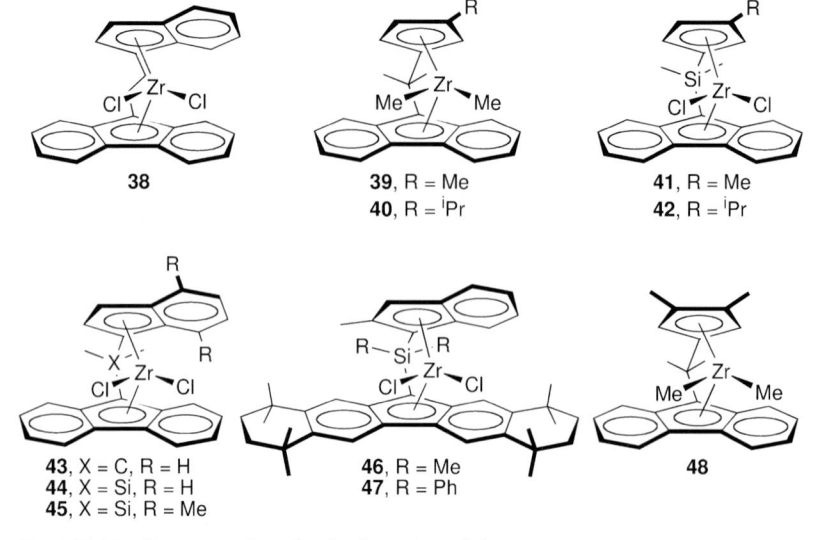

Fig. 6.19 Metallocene catalysts for the formation of alternating ethylene–propylene copolymers.

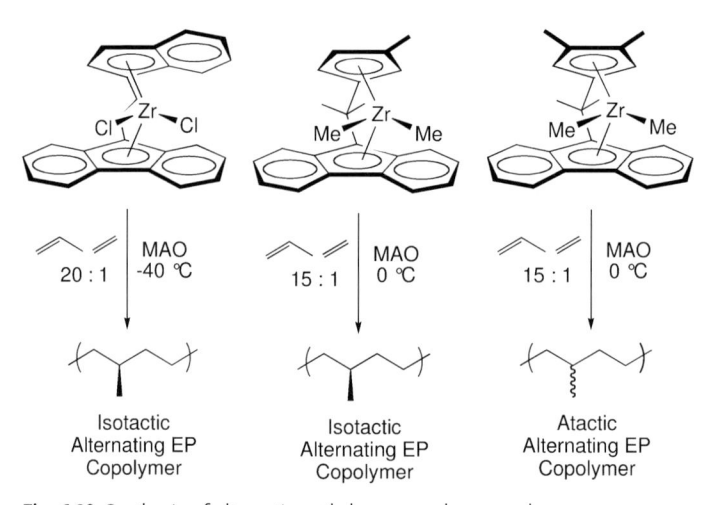

- **38**
- **39**, R = Me
- **40**, R = ^{i}Pr
- **41**, R = Me
- **42**, R = ^{i}Pr
- **43**, X = C, R = H
- **44**, X = Si, R = H
- **45**, X = Si, R = Me
- **46**, R = Me
- **47**, R = Ph
- **48**

Fig. 6.20 Synthesis of alternating ethylene–propylene copolymers using metallocene catalysts.

- 20:1, MAO, −40 °C → Isotactic Alternating EP Copolymer
- 15:1, MAO, 0 °C → Isotactic Alternating EP Copolymer
- 15:1, MAO, 0 °C → Atactic Alternating EP Copolymer

38–MAO at −40 °C to form a polymer containing 93% alternating PEP and EPE sequences by NMR (Figs. 6.19 and 6.20) [134]. Waymouth and coworkers formed a similar polymer containing 70% alternating triads using **39**–MAO at 0 or 20 °C [135]. In both cases, a large excess of propylene monomer (about 15:1 or 20:1 [propylene]:[ethylene]) was used in order to form an alternating microstructure. Steric differences between the two coordination sites are believed to

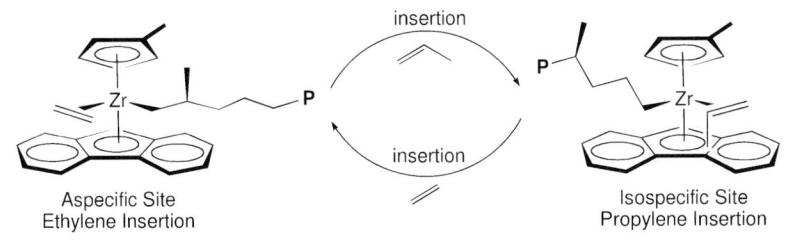

Fig. 6.21 Formation of isotactic alternating EP copolymers using asymmetric metallocene catalysts.

be the source of alternating enchainment (Fig. 6.21). When the growing polymer chain is in the more open coordination site, higher propylene concentrations favor isotactic propylene insertion (analogous to hemiisotactic PP homopolymerization). When the growing chain is at the more crowded site, steric considerations favor ethylene insertion over propylene. Since propylene insertion occurs at an isospecific site, an isotactic alternating polymer is formed. Stereoerrors are the result of site epimerization to the less crowded site resulting in sequential enchainment of propylene units [136]. Kaminsky and coworkers have analyzed these systems using statistical methods to fit experimental NMR data [137].

Waymouth and coworkers have also synthesized a number of structurally similar metallocene catalysts that produce alternating EP copolymers (Fig. 6.19). In addition to **39**, isopropyl-substituted **40** and dimethylsilyl-linked **41** and **42** produce isotactic alternating EP copolymer [136]. The silicon-bridged systems display higher activity and form polymer with higher molecular weight. Indenyl-fluorenyl metallocenes **43–45** are similar to **38** but produce atactic alternating EP copolymer [138, 139]. Heuer and Kaminsky have recently synthesized a series of more complicated derivatives (such as **46** and **47**) that produce similar atactic polymer [140]. Since catalysts **43–47** are all asymmetric, an alternating dual-site mechanism is still proposed with propylene insertion occurring at the aspecific site and ethylene insertion occurring at the isospecific site.

C_s-symmetric **48**–MAO also produces alternating atactic EP copolymers with up to 70% alternating triads (Fig. 6.20) [135]. With this catalyst, the two coordination sites are equivalent and must have equivalent reactivity ratios. Therefore, a different mechanism must be invoked to explain the alternating structure. Presumably, ligand sterics prevent the insertion of propylene into a propylene-terminated chain and high propylene concentrations prevent ethylene insertion into an ethylene-terminated chain. This result indicates that alternating EP copolymers can be synthesized using catalysts with symmetric coordination sites.

6.5.3
Ethylene–Propylene Block Copolymers

Block copolymers made from ethylene and propylene are valuable industrial materials, having potential applications as thermoplastic elastomers and as compatibilizing agents for homopolymer blends. Copolymer properties depend on block microstructure, relative block length and overall molecular weight. ABA triblock copolymers with crystalline A blocks and an amorphous B block are usually required in order to impart elastomeric behavior [141]. Crystalline "hard" blocks include isotactic or syndiotactic polypropylene (iPP or sPP) and linear polyethylene (PE). Amorphous "soft" blocks can be either atactic polypropylene (aPP) or ethylene–propylene copolymer (EPR). Polypropylene stereoblock polymers such as iPP-*block*-aPP have already been described.

There have been many reports in the academic and patent literature describing the synthesis of ethylene–propylene block copolymers using heterogeneous Ziegler–Natta catalysts. Most of these reports claim to synthesize iPP-*block*-EPR diblocks by sequential monomer addition: polymerization under propylene produces an iPP block and subsequent addition of ethylene forms an EPR block. Unfortunately, in most cases it is not clear that the desired block copolymers are actually formed. In the industrial literature, the problem is compounded by frequent use of the term "block" when describing homopolymer blends of iPP, PE and EPR.

One example that illustrates the difficulty in identifying polymer products involves the $TiCl_3$–Et_2AlCl catalyst system, which was reported to produce either iPP-*block*-EPR or PE-*block*-EPR multiblock polymers by adding either ethylene or propylene (respectively) to a propylene or ethylene polymerization for 10–30 s at a time [142, 143]. Attempts to repeat the experiment were not successful and further studies indicate that only homopolymer mixtures are formed, not block copolymers [144]. By reducing the monomer pressure from 1.0 to 0.1 atm, Busico et al. were able to increase chain lifetimes and form the desired block copolymers, but catalyst activities were very low (0.1 g of polymer for every 1.0 g of $TiCl_3$ over 3 h) [145].

Matthews and Nudenberg described a heterogeneous catalyst that produces iPP-*block*-EPR diblock and iPP-*block*-EPR-*block*-iPP triblock copolymers with elastomeric behavior [146]. The catalyst used was $MgCl_2$-supported $TiCl_4$–ethyl benzoate activated with a mixture of Et_3Al and ethyl *p-tert*-butylbenzoate. The amount of homopolymer present in the final product was not determined. Blunt [147] and Lock [148] have developed a $TiCl_3$–$(\eta^5\text{-MeC}_5H_4)_2TiMe_2$ catalyst system for the formation of iPP-*block*-EPR-*block*-iPP thermoplastic elastomers [147]. Molecular weight increases linearly with polymer yield, suggesting that chain-transfer reactions do not occur with this system. Molecular weight distributions are very broad ($M_w/M_n = 4.0$–5.0), presumably due to the presence of multiple catalytic sites. Fractionation studies indicate that most of the EPR chains are bound to iPP blocks, but the number of triblocks could not be determined. The EPR blocks are not homogeneous and contain semicrystalline segments that reduce the melting-point of the overall polymer [148].

The problem of homopolymer formation resulting from chain transfer reactions can be partially addressed by keeping reaction times very short. If the reaction time is shorter than the average lifetime of a growing chain, then chain transfer should be minimized. Heterogeneous and metallocene catalysts have short chain lifetimes of a few seconds. Synthesizing block copolymers in such a short period of time requires a custom-made apparatus. Ver Strate et al. used a narrow tubular reactor to synthesize PE-*block*-EPR diblock copolymers using a VCl$_3$–Et$_3$Al$_2$Cl$_3$ catalyst system [149]. Reactions were completed in about 10 s. The polymer product contained about 70% diblock copolymer, which had a relatively high molecular weight ($M_n = 100\,000$) and a narrow distribution ($M_w/M_n = 1.70$). A third PE block could not be added to the diblock because excess propylene could not be removed from the reactor. Terano and coworkers have developed a stopped-flow polymerization reactor that can be used to form iPP-*block*-EPR diblock copolymers with a MgCl$_2$-supported TiCl$_4$–ethyl benzoate–Et$_3$Al catalyst system [150–154]. Reactions can be completed in fractions of a second. This process generally forms low molecular weight polymers ($M_n < 50\,000$) with broad distributions ($M_w/M_n \approx 3.0$).

Living systems are the most attractive option for the synthesis of well-defined block copolymers. These systems do not undergo chain transfer or chain termination, so homopolymer formation is not a concern. Block length can be easily controlled based on reaction time or monomer concentration. Doi and Ueki reported the first synthesis of a block copolymer using a homogeneous, living polymerization complex [155]. When activated with Et$_2$AlCl and anisole at –78 °C, V(acac)$_3$ (acac = acetylacetonate) can produce sPP-*block*-EPR-*block*-sPP triblock polymers. Moderate chain-end control syndiotacticity ($[r] = 0.81$) results in a non-elastomeric polymer with a very narrow molecular weight distribution ($M_w/M_n = 1.22$). Catalyst activity is low and living behavior is observed only at temperatures below –65 °C.

Turner et al. used Cp$_2$HfMe$_2$ activated with [HMe$_2$NC$_6$H$_5$][B(C$_6$F$_5$)$_4$] at 0 °C to synthesize block copolymers [156]. This system is not living under these conditions and probably produces a significant number of homopolymer chains. Nonetheless, PE-*block*-aPP-*block*-PE and PE-*block*-EPR-*block*-PE triblock copolymers were reported. At lower temperatures, simple metallocene complexes can exhibit living behavior. When activated with nOct$_3$Al–B(C$_6$F$_5$)$_3$ at –75 or –50 °C, Cp$_2$MMe$_2$ (M = Zr, Hf) can form aPP-*block*-EPR diblock copolymers with very narrow molecular weight distributions ($M_w/M_n = 1.07$–1.16) [157]. Once again, these catalysts have low activities at such low temperatures.

Living complexes bearing non-metallocene ligands offer the ability to synthesize block copolymers at more reasonable temperatures. Coates and coworkers [158] and Fujita and coworkers [159] have independently developed bis(phenoxyimine)titanium complexes (Fig. 6.22) for the living polymerization of propylene and ethylene. These C_2-symmetric systems form highly syndiotactic polypropylene by a chain-end control mechanism [160]. Ligand isomerization between monomer insertion events is believed to be the cause of this unusual stereocontrol [161]. Fujita and coworkers have used mono-*tert*-butyl complex **49**–MAO to

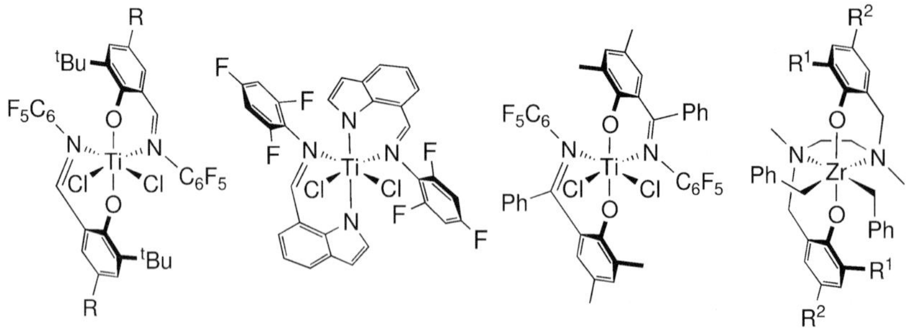

49, R = H
50, R = ^{t}Bu

51

52

53, $R^1 = R^2 = {}^{t}Bu$
54, R^1 = adamantyl, R^2 = Me

Fig. 6.22 Non-metallocene catalysts for the formation of ethylene–propylene block copolymers.

synthesize a variety of block copolymers at 25 °C having narrow molecular weight distributions [159, 162, 163]. sPP-*block*-EPR, PE-*block*-sPP and PE-*block*-EPR diblock copolymers were formed, in addition to PE-*block*-EPR-*block*-sPP and PE-*block*-EPR-*block*-PE triblock copolymers. Coates and coworkers have used di-*tert*-butyl complex **50**–MAO at 0 °C to synthesize a number of sPP-*block*-EPR and sPP-*block*-PE diblock copolymers with varying block lengths and varying propylene content in the EPR block [158, 164]. Ligand modification studies indicate that complexes with *ortho*-fluorinated *N*-aryl groups display living behavior [163, 165], expanding the number of complexes that can form sPP-based block copolymers.

Related ligand frameworks have also yielded living systems. Complex **51**, bearing two indolide-imine ligands, has been used by Fujita and coworkers to synthesize a PE-*block*-EPR diblock copolymer having a narrow molecular weight distribution ($M_w/M_n = 1.17$) [166]. Mason and Coates found that adding a substituent off the imine carbon atom of a phenoxyimine ligand can alter the stereoselectivity of these systems without loss of living behavior. Whereas **50**–MAO forms highly syndiotactic PP by a chain-end control mechanism, **52**–MAO forms moderately isotactic PP by an enantiomorphic site control mechanism [167]. An iPP-*block*-EPR diblock with narrow molecular weight distribution ($M_w/M_n = 1.10$) has been synthesized with **52**–MAO. Busico and coworkers have developed a system that produces highly isotactic PP. Although not rigorously living, at short reaction times zirconium dibenzyl complexes **53** and **54** can be used to form iPP-*block*-PE diblock copolymers. The benzyl complexes are activated with $[HMe_2NC_6H_5][B(C_6F_5)_4]$ in the presence of 2,6-di-*tert*-butylphenol and triisobutyl aluminium [168, 169].

6.5.4
Ethylene–Propylene Graft Copolymers

Various types of ethylene–propylene graft copolymers have been synthesized through incorporation of long-chain macromonomers (Fig. 6.23). In most cases, crystalline branches are incorporated into an amorphous backbone or amorphous branches are incorporated into a crystalline backbone. Polymers with "soft–hard" combinations usually have interesting properties and can be easily separated from unreacted macromonomer.

Shiono and coworkers have incorporated atactic polypropylene macromonomer into a polyethylene main chain using a variety of catalysts [170, 171]. Vinyl-terminated aPP macromonomers containing 15–20 repeat units were synthesized using $Cp_2^*ZrCl_2$–MAO. These were copolymerized with ethylene using either Cp_2ZrCl_2, Cp_2TiCl_2, **55** or **56** (Fig. 6.24) activated with MAO. Based on IR data, constrained-geometry catalyst **56** incorporates the most macromonomer into the main PE chain (up to 5 mol%). Weng and coworkers have synthesized a graft copolymer containing two crystalline subunits. Vinyl-terminated PE macromonomer was synthesized using either Cp_2ZrCl_2 or constrained-geometry catalyst **33**. Incorporation into an iPP main chain was accomplished with either **35**–MAO or **57**–$[HMe_2NC_6H_5][B(C_6F_5)_4]$ [121, 172]. Macromonomers of ethylene–propylene copolymer have been used by Zhu and coworkers to incorporate amorphous chains into crystalline backbones. EPR macromonomers were synthesized using constrained-geometry catalyst **37**–$B(C_6F_5)_3$/MMAO. Equal

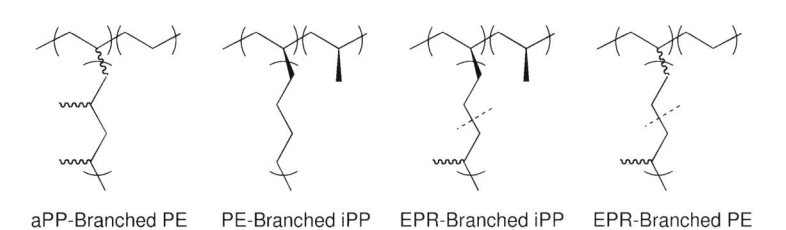

aPP-Branched PE   PE-Branched iPP   EPR-Branched iPP   EPR-Branched PE

Fig. 6.23 Ethylene–propylene long-chain branched copolymers.

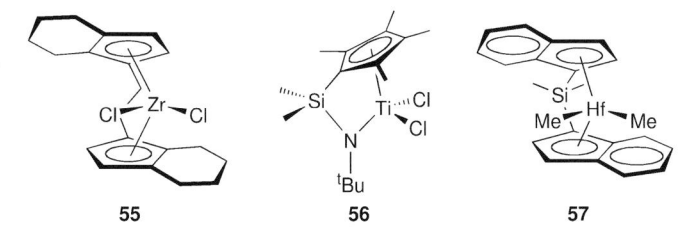

55   **56**   **57**

Fig. 6.24 Catalysts used for the formation of ethylene–propylene long-chain branched copolymers.

amounts of vinyl- and vinylidene-terminated chains were formed, but only the vinyl-terminated macromonomers were copolymerized with propylene using isospecific catalyst **32**–MMAO [173]. Alternatively, the vinyl-terminated EPR macromonomer can be copolymerized with ethylene using **37**–B(C_6F_5)$_3$–MMAO [174]. Using a PE backbone results in the formation of higher molecular weight polymer and a greater number of macromonomers per chain.

6.6
Summary and Outlook

The recent development of well-defined, homogeneous olefin polymerization catalysts has allowed the synthesis of polymeric materials having unique microstructures. Control over polymer structure can take a number of different forms. Polymer stereochemistry (tacticity) is often connected to the steric environment around the catalytic active site. Appropriate ligand design can result in a catalyst that is able to form a "new" polymer from readily available monomer feedstocks. More complex catalyst systems can switch between different stereocontrol mechanisms and can form new types of stereoblock polyolefins. Catalysts can insert different monomers in a controlled fashion to synthesize polyolefin copolymers. If a catalyst has different monomer-selective catalytic sites, then an alternating copolymer can be synthesized. Living catalyst systems can be used to synthesize easily uniform block copolymers by sequential addition of monomers. Given the range of olefin feedstocks available, endless stereo- and regioregular copolymer structures can be designed. The difficulty is finding the right catalyst to execute the desired design.

The controlled polymerizations covered in this review represent a starting point, with possible expansions in any number of directions. Most studies involving homogeneous systems have used metallocene catalysts and have focused on homopolymerization reactions. In the past few years, the expanding scope of research has identified numerous living non-metallocene catalysts that are able to form block copolymers that cannot be synthesized by metallocene-based systems. The range of monomers that can be homopolymerized by these new catalysts has not yet been fully explored and their ability to form copolymers has barely been touched upon.

Despite this great potential, identifying new catalysts remains a challenging task and the ability to design a specific catalyst for a specific task remains more of a dream than a reality. In many instances, new catalysts are screened using a single monomer. The effect of ligand modification is usually judged by how well a modified catalyst performs under a standard set of conditions. This approach to catalyst design tends to result in "specialized" catalysts that are very good at forming one specific type of polymer microstructure. Clearly, one of the challenges for the future is the development of catalysts that are broadly applicable to a wide range of monomers; this will allow for the development of new, revolutionary olefinic polymers.

References

1 J. L. Meikle *American Plastic: a Cultural History*. Rutgers University Press, New Brunswick, NJ, **1995**.
2 V. C. Gibson, S. K. Spitzmesser *Chem. Rev.* **2003**, *103*, 283–315.
3 G. J. P. Britovsek, V. C. Gibson, D. F. Wass *Angew. Chem. Int. Ed.* **1999**, *38*, 428–447.
4 N. Kashiwa *J. Polym. Sci. Part A: Polym. Chem.* **2004**, *42*, 1–8.
5 H. H. Brintzinger, D. Fischer, R. Mülhaupt, B. Rieger, R. Waymouth *Angew. Chem. Int. Ed. Engl.* **1995**, *34*, 1143–1170.
6 L. Resconi, L. Cavallo, A. Fait, F. Piemontesi *Chem. Rev.* **2000**, *100*, 1253–1345.
7 For some recent edited monographs on olefin polymerization, see W. Kaminsky *Metalorganic Catalysts for Synthesis and Polymerization*, Springer, Berlin, **1999**; G. Fink, R. Mülhaupt, H. H. Brintzinger *Ziegler Catalysts*, Springer, Berlin, **1995**.
8 H. Sinn, W. Kaminsky, H. Vollmer, R. Woldt *Angew. Chem. Int. Ed. Engl.* **1980**, *19*, 390–392.
9 E. Zurek, T. Ziegler *Prog. Polym. Sci.* **2004**, *29*, 107–148.
10 H. Sinn, W. Kaminsky *Adv. Organomet. Chem.* **1980**, *18*, 99–149.
11 R. F. Jordan, C. S. Bajgur, R. Willett, B. Scott *J. Am. Chem. Soc.* **1986**, *108*, 7410–7411.
12 R. Taube, L. Krukowka *J. Organomet. Chem.* **1988**, *347*, C9–C11.
13 X. Yang, C. L. Stern, T. J. Marks *J. Am. Chem. Soc.* **1991**, *113*, 3623–3625.
14 J. C. W. Chien, G. H. Llinas, M. D. Rausch, G. Lin, H. H. Winter *J. Am. Chem. Soc.* **1991**, *113*, 8569–8570.
15 H. W. Turner *Eur. Pat. Appl.* 0277004, Exxon, **1988**.
16 E. Y. Chen, T. J. Marks *Chem. Rev.* **2000**, *100*, 1391–1434.
17 E. B. Coughlin, J. E. Bercaw *J. Am. Chem. Soc.* **1992**, *114*, 7606–7607.
18 T. Younkin, E. Conner, J. Henderson, S. Friedrich, R. Grubbs, D. Bansleben *Science* **2000**, *287*, 460–462.
19 B. L. Small, M. Brookhart *Macromolecules* **1999**, *32*, 2120–2130.
20 J. Saito, M. Mitani, M. Onda, J. Mohri, S. Ishii, Y. Yoshida, T. Nakano, H. Tanaka, T. Matsugi, S. Kojoh, N. Kashiwa, T. Fujita *Macromol. Rapid Commun.* **2001**, *22*, 1072–1075.
21 P. D. Hustad, J. Tian, G. W. Coates *J. Am. Chem. Soc.* **2002**, *124*, 3614–3621.
22 Z. Wu, R. F. Jordan, J. L. Petersen *J. Am. Chem. Soc.* **1995**, *117*, 5867–5868.
23 C. P. Casey, S. L. Hallenbeck, D. W. Pollock, C. R. Landis *J. Am. Chem. Soc.* **1995**, *117*, 9770–9771.
24 R. H. Grubbs, G. W. Coates *Acc. Chem. Res.* **1996**, *29*, 85–93.
25 G. W. Coates *Chem. Rev.* **2000**, *100*, 1223–1252.
26 G. W. Coates, P. D. Hustad, S. Reinartz *Angew. Chem. Int. Ed.* **2002**, *41*, 2236–2257.
27 A. Cherian, E. Lobkovsky, G. Coates *Macromolecules* **2005**, *38*, 6259–6268.
28 L. Resconi, I. Camurati, O. Sudmeijer *Top. Catal.* **1999**, *7*, 145–163.
29 H. Kaneyoshi, Y. Inoue, K. Matyjaszewski *Macromolecules* **2005**, 5425–5435.
30 M. Mitani, J. Mohri, R. Furuyama, S. Ishii, T. Fujita *Chem. Lett.* **2003**, *32*, 238–239.
31 M. van Meurs, G. Britovsek, V. Gibson, S. Cohen *J. Am. Chem. Soc.* **2005**, *127*, 9913–9923.
32 For some examples of block copolymer synthesis using polymer exchange pathways, see J. C. W. Chien, Y. Iwamoto, M. D. Rausch, W. Wedler, H. H. Winter *Macromolecules* **1997**, *30*, 3447–3458; J. C. W. Chien, Y. Iwamoto, M. D. Rausch *J. Polym. Sci. Part A: Polym. Chem.* **1999**, *37*, 2439–2445; S. Lieber, H. H. Brintzinger *Macromolecules* **2000**, *33*, 9192–9199; K. C. Jayaratne, L. R. Sita *J. Am. Chem. Soc.* **2001**, *123*, 10754–10755.
33 T. C. Chung *Functionalization of Polyolefins*. Academic Press, London, **2002**.
34 J. P. Hogan, R. L. Banks *US Patent 2825721*, Phillips Petroleum, **1958**.
35 J. P. Hogan *J. Polym. Sci. Part A1: Polym. Chem.* **1970**, *8*, 2637–2652.

36 H. Ulrich *Introduction to Industrial Polymers*, 2nd edn. Hanser, New York, **1993**.
37 A. L. McKnight, R. Waymouth *Chem. Rev.* **1998**, *98*, 2587–2598.
38 J. A. M. Canich *US Patent 5026798*, Exxon Chemical, **1991**.
39 J. A. M. Canich, G. F. Licciardi *US Patent 5057475*, Exxon Chemical, **1991**.
40 J. C. Stevens, D. R. Neithamer *US Patent 5064802*, Dow Chemical, **1991**.
41 J. C. Stevens, D. R. Neithamer *US Patent 5132380*, Dow Chemical, **1992**.
42 S. Lai, J. R. Wilson, G. W. Knight, J. C. Stevens, P. S. Chum *US Patent 5272236*, Dow Chemical, **1993**.
43 S. Lai, J. R. Wilson, G. K. Knight, J. C. Stevens *US Patent 5278272*, Dow Chemical, **1994**.
44 J. C. Stevens *Stud. Surf. Sci. Catal.* **1996**, *101*, 11–20.
45 W. Wang, D. Yan, S. Zhu, A. E. Hamielec *Macromolecules* **1998**, *31*, 8677–8683.
46 C. Pellecchia, D. Pappalardo, G. Gruter *Macromolecules* **1999**, *32*, 4491–4493
47 L. Izzo, L. Caporaso, G. Senatore, L. Oliva *Macromolecules* **1999**, *32*, 6913–6916.
48 E. Kokko, A. Malmberg, P. Lehmus, B. Löfgren, J. V. Seppälä *J. Polym. Sci. Part A: Polym. Chem.* **2000**, *38*, 376–388.
49 E. Kolodka, W. Wang, P. A. Charpentier, S. Zhu, A. E. Hamielec *Polymer* **2000**, *41*, 3985–3991.
50 Z. J. A. Komon, G. C. Bazan *Macromol. Rapid Commun.* **2001**, *22*, 467–478.
51 R. F. de Souza, O. L. Casagrande Jr. *Macromol. Rapid Commun.* **2001**, *22*, 1293–1301.
52 T. M. Pettijohn, W. K. Reagen, S. J. Martin *US Patent 5331070*, Phillips Petroleum, **1994**.
53 R. W. Barnhart, G. C. Bazan *J. Am. Chem. Soc.* **1998**, *120*, 1082–1083.
54 D. Beigzadeh, J. B. P. Soares, T. A. Duever *Macromol. Rapid Commun.* **1999**, *20*, 541–545.
55 R. Quijada, R. Rojas, G. Bazan, Z. J. A. Komon, R. S. Mauler, G. B. Galland *Macromolecules* **2001**, *34*, 2411–2417.
56 S. D. Ittel, L. K. Johnson, M. Brookhart *Chem. Rev.* **2000**, *100*, 1169–1203.
57 L. K. Johnson, C. M. Killian, S. D. Arthur, J. Feldman, E. F. McCord, S. J. McLain, K. A. Kreutzer, M. A. Bennett, E. B. Coughlin, S. D. Ittel, A. Parthasarathy, D. J. Tempel, M. S. Brookhart *WO Patent 96/23010*, Du Pont, **1996**.
58 L. K. Johnson, C. M. Killian, M. Brookhart *J. Am. Chem. Soc.* **1995**, *117*, 6414–6415.
59 *Chem. Eng. News*, **2006**, *84*(28), 59–68.
60 P. Galli *Macromol. Symp.* **1995**, *89*, 13–26.
61 J. A. Ewen *J. Am. Chem. Soc.* **1984**, *106*, 6355–6364.
62 W. Kaminsky, K. Külper, H. H. Brintzinger, F. R. W. P. Wild *Angew. Chem. Int. Ed. Engl.* **1985**, *24*, 507–508.
63 J. A. Ewen, R. L. Jones, A. Razavi, J. D. Ferrara *J. Am. Chem. Soc.* **1988**, *110*, 6256–6258.
64 W. Spaleck, M. Aulbach, B. Bachmann, F. Küber, A. Winter *Macromol. Symp.* **1995**, *89*, 237–247.
65 W. Spaleck, F. Küber, A. Winter, J. Rohrmann, B. Bachmann, M. Antberg, V. Dolle, E. F. Paulus *Organometallics* **1994**, *13*, 954–963.
66 J. A. Ewen *US Patent 5036034*, Fina Technology, **1991**.
67 J. A. Ewen, M. J. Elder, R. L. Jones, L. Haspeslagh, J. L. Atwood, S. G. Bott, K. Robinson *Macromol. Symp.* **1991**, *48/49*, 253–295.
68 M. Farina, G. Di Silvestro, P. Sozzani *Macromolecules* **1982**, *15*, 1451–1452.
69 E. J. Arlman, P. Cossee *J. Catal.* **1964**, *3*, 99–104.
70 N. Herfert, G. Fink *Macromol. Symp.* **1993**, *66*, 157–178.
71 M. Farina, G. Di Silvestro, P. Sozzani *Macromolecules* **1993**, *26*, 946–950.
72 G. Guerra, L. Cavallo, G. Moscardi, M. Vacatello, P. Corradini *Macromolecules* **1996**, *29*, 4834–4845.
73 A. Razavi, J. L. Atwood *J. Organomet. Chem.* **1995**, *497*, 105–111.
74 R. Kleinschmidt, M. Reffke, G. Fink *Macromol. Rapid Commun.* **1999**, *20*, 284–288.
75 A. Yano, T. Kaneko, M. Sato, A. Akimoto *Macromol. Chem. Phys.* **2000**, *200*, 2127–2135.
76 N. M. Bravaya, P. M. Nedorezova, V. I. Tsvetkova *Russ. Chem. Rev.* **2002**, *71*, 49–70.
77 G. Müller, B. Rieger *Prog. Polym. Sci.* **2002**, *27*, 815–851.

78 G. Natta *J. Polym. Sci.* **1959**, *34*, 531–549.
79 J. W. Collette, C. W. Tullock *US Patent 4 335 225*, Du Pont, **1982**.
80 J. W. Collette, C. W. Tullock, R. N. MacDonald, W. H. Buck, A. C. L. Su, J. R. Harrell, R. Mülhaupt, B. C. Anderson *Macromolecules* **1989**, *22*, 3851–3858.
81 J. W. Collette, D. W. Ovenall, W. H. Buck, R. C. Ferguson *Macromolecules* **1989**, *22*, 3858–3866.
82 C. W. Tullock, R. Mülhaupt, S. D. Ittel *Makromol. Chem., Rapid Commun.* **1989**, *10*, 19–23.
83 C. W. Tullock, F. N. Tebbe, R. Mülhaupt, D. W. Ovenall, R. A. Setterquist, S. D. Ittel *J. Polym. Sci. Part A: Polym. Chem.* **1989**, *27*, 3063–3081.
84 R. Ohnishi, S. Yukimasa, T. Kanakazawa *Macromol. Chem. Phys.* **2002**, *203*, 1003–1010.
85 R. C. Job *US Patent 5 270 410*, Shell Oil, **1993**.
86 D. T. Mallin, M. D. Rausch, Y. Lin, S. Dong, J. C. W. Chien *J. Am. Chem. Soc.* **1990**, *112*, 2030–2031.
87 W. J. Gauthier, J. F. Corrigan, N. J. Taylor, S. Collins *Macromolecules* **1995**, *28*, 3771–3778.
88 W. J. Gauthier, S. Collins *Macromol. Symp.* **1995**, *98*, 223–231.
89 A. M. Bravakis, L. E. Bailey, M. Pigeon, S. Collins *Macromolecules* **1998**, *31*, 1000–1009.
90 U. Dietrich, M. Hackmann, B. Rieger, M. Klinga, M. Leskelä *J. Am. Chem. Soc.* **1999**, *121*, 4348–4355.
91 J. Kukral, P. Lehmus, T. Feifel, C. Troll, B. Rieger *Organometallics* **2000**, *19*, 3767–3775.
92 B. Rieger, C. Troll, J. Preuschen *Macromolecules* **2002**, *35*, 5742–5743.
93 A. R. Siedle, D. K. Misemer, V. V. Kolpe, B. F. Duerr *US Patent 6 265 512*, 3M, **2001**.
94 C. C. Meverden, S. Nagy *US Patent 6 541 583*, Equistar Chemical, **2003**.
95 W. J. Gauthier, S. Collins *Macromolecules* **1995**, *28*, 3779–3786.
96 S. A. Miller, J. E. Bercaw *Organometallics* **2002**, *21*, 934–945.
97 J. A. Ewen, R. L. Jones, M. J. Elder, I. Camurati, H. Pritzkow *Macromol. Chem. Phys.* **2004**, *205*, 302–307.
98 G. W. Coates, R. M. Waymouth *Science* **1995**, *267*, 217–219.
99 R. M. Waymouth, E. Hauptman, G. W. Coates, *US Patent 5 969 070*, Stanford University, **1999**.
100 V. Busico, V. V. A. Castelli, P. Aprea, R. Cipullo, A. Segre, G. Talarico, M. Vacatello *J. Am. Chem. Soc.* **2003**, *125*, 5451–5460.
101 V. Busico, R. Cipullo, W. P. Kretschmer, G. Talarico, M. Vacatello, V. V. A. Castelli *Angew. Chem. Int. Ed.* **2002**, *41*, 505–508.
102 S. Lin, R. M. Waymouth *Acc. Chem. Res.* **2002**, *35*, 765–773.
103 S. Mansel, E. Pérez, R. Benavente, J. M. Pereña, A. Bello, W. Röll, R. Kirsten, S. Beck, H. H. Brintzinger *Macromol. Chem. Phys.* **1999**, *200*, 1292–1297.
104 V. Volkis, M. Shmulinson, E. Shaviv, A. Lisovskii, D. Plat, O. Kühl, T. Koch, E. Hey-Hawkins, M. S. Eisen In *Beyond Metallocenes: Next-Generation Polymerization Catalysts*, A. O. Patil, G. G. Hlatky (Eds.). American Chemical Society, Washington, DC, **2003**, p. 46–61.
105 M. Shmulinson, M. Galan-Fereres, A. Lisovskii, E. Nelkenbaum, R. Semiat, M. S. Eisen *Organometallics* **2000**, *19*, 1208–1210.
106 V. Busico, R. Cipullo, L. Caporaso, G. Angelini, A. L. Segre *J. Mol. Catal. A: Chem.* **1998**, *128*, 53–64.
107 O. Kühl, T. Koch, F. B. Somoza Jr., P. C. Junk, E. Hey-Hawkins, D. Plat, M. S. Eisen *J. Organomet. Chem.* **2000**, *604*, 116–125.
108 B. Ray, T. G. Neyroud, M. Kapon, Y. Eichen, M. S. Eisen *Organometallics* **2001**, *20*, 3044–3055.
109 E. Shaviv, M. Botoshansky, M. S. Eisen *J. Organomet. Chem.* **2003**, *683*, 165–180.
110 V. Volkis, E. Nelkenbaum, A. Lisovskii, G. Hasson, R. Semiat, M. Kapon, M. Botoshansky, Y. Eishen, M. S. Eisen *J. Am. Chem. Soc.* **2003**, *125*, 2179–2194.
111 E. Smolensky, M. Kapon, J. D. Woollins, M. S. Eisen *Organometallics* **2005**, *24*, 3255–3265.
112 J. C. W. Chien, Y. Iwamoto, M. D. Rausch, W. Wedler, H. H. Winter *Macromolecules* **1997**, *30*, 3447–3458.

113 J.C.W. Chien *US Patent 6121377*, Academy of Applied Science, **2000**.
114 S. Lieber, H.H. Brintzinger *Macromolecules* **2000**, *33*, 9192–9199.
115 J.C.W. Chien, Y. Iwamoto, M.D. Rausch *J. Polym. Sci. Part A: Polym. Chem.* **1999**, *37*, 2439–2445.
116 C. Przybyla, G. Fink *Acta Polym.* **1999**, *50*, 77–83.
117 K. Nishii, T. Shiono, T. Ikeda *Macromol. Rapid Commun.* **2004**, *25*, 1029–1032.
118 A.E. Cherian, J.M. Rose, E.B. Lobkovsky, G.W. Coates *J. Am. Chem. Soc.* **2005**, *127*, 13770–13771.
119 T. Shiono, S.M. Azad, T. Ikeda *Macromolecules* **1999**, *32*, 5723–5727.
120 W. Weng, A.H. Dekemzian, E.J. Markel, D.L. Peters *US Patent 6184327*, Exxon Chemical, **2001**.
121 W. Weng, E.J. Markel, A.H. Dekmezian *Macromol. Rapid Commun.* **2001**, *22*, 1488–1492.
122 W. Weng, W. Hu, A.H. Dekmezian, C.J. Ruff *Macromolecules* **2002**, *35*, 3838–3843.
123 Z. Ye, S. Zhu *J. Polym. Sci. Part A: Polym. Chem.* **2003**, *41*, 1152–1159.
124 K. Koo, T.J. Marks *J. Am. Chem. Soc.* **1999**, *121*, 8791–8802.
125 K. Soga, T. Sano, R. Ohnishi, T. Kawata, K. Ishii, T. Shiono, Y. Doi *Stud. Surf. Sci. Catal.* **1986**, *25*, 109–122.
126 M. Avella, E. Martuscelli, G.D. Volpe, A. Segre, E. Rossi, T. Simonazzi *Makromol. Chem.* **1986**, *187*, 1927–1943.
127 V. Busico, P. Corradini, C. De Rosa, E. Di Benedetto *Eur. Polym. J.* **1985**, *21*, 239–244.
128 J.A. Ewen *Stud. Surf. Sci. Catal.* **1986**, *25*, 271–292.
129 J.C.W. Chien, D. He *J. Polym. Sci. Part A: Polym. Chem.* **1991**, *29*, 1585–1593.
130 T. Uozumi, K. Soga *Makromol. Chem.* **1992**, *193*, 823–831.
131 M. Galimberti, F. Piemontesi, O. Fusco, I. Camurati, M. Destro *Macromolecules* **1998**, *31*, 3409–3416.
132 M. Galimberti, F. Piemontesi, N. Mascellani, I. Camurati, O. Fusco, M. Destro *Macromolecules* **1999**, *32*, 7968–7976.
133 H. Chien, D. McIntyre, J. Cheng, M. Fone *Polymer* **1995**, *36*, 2559–2565.

134 J. Jin, T. Uozumi, T. Sano, T. Teranishi, K. Soga, T. Shiono *Macromol. Rapid Commun.* **1998**, *19*, 337–339.
135 M.K. Leclerc, R.M. Waymouth *Angew. Chem. Int. Ed.* **1998**, *37*, 922–925.
136 W. Fan, M.K. Leclerc, R.M. Waymouth *J. Am. Chem. Soc.* **2001**, *123*, 9555–9563.
137 M. Arndt, W. Kaminsky, A. Schauwienold, U. Weingarten *Macromol. Chem. Phys.* **1998**, *199*, 1135–1152.
138 W. Fan, R.M. Waymouth *Macromolecules* **2001**, *34*, 8619–8625.
139 W. Fan, R.M. Waymouth *Macromolecules* **2003**, *36*, 3010–3014.
140 B. Heuer, W. Kaminsky *Macromolecules* **2005**, *38*, 3054–3059.
141 G. Holden, H.R. Kricheldorf, R.P. Quirk *Thermoplastic Elastomers*, 3rd edn. Hanser, Munich, **2004**.
142 P. Prabhu, A. Schindler, M.H. Theil, R.D. Gilbert *J. Polym. Sci. Polym. Lett.* **1980**, *18*, 389–394.
143 P. Prabhu, A. Schindler, M.H. Theil, R.D. Gilbert *J. Polym. Sci. Polym. Chem.* **1981**, *19*, 523–537.
144 V. Busico, P. Corradini, P. Fontana, V. Savino *Makromol. Chem., Rapid Commun.* **1984**, *5*, 737–743.
145 V. Busico, P. Corradini, P. Fontana, V. Savino *Makromol. Chem., Rapid Commun.* **1985**, *6*, 743–747.
146 D.N. Matthews, W. Nudenberg *US Patent 4491652*, Uniroyal, **1985**.
147 H.W. Blunt *US Patent 4408019*, Hercules, **1983**.
148 G.A. Lock In *Advances in Polyolefins*, R.B. Seymour, T. Cheng (Eds.) Plenum Press, New York, **1985**, 59–74.
149 G. Ver Strate, C. Cozewith, R.K. West, W.M. Davis, G.A. Capone *Macromolecules* **1999**, *32*, 3837–3850.
150 H. Mori, M. Yamahiro, K. Tashino, K. Ohnishi, K. Nitta, M. Terano *Macromol. Rapid Commun.* **1995**, *16*, 247–252.
151 M. Yamahiro, H. Mori, K. Nitta, M. Terano *Macromol. Chem. Phys.* **1999**, *200*, 134–141.
152 M. Yamahiro, H. Mori, K. Nitta, M. Terano *Polymer* **1999**, *40*, 5265–5272.
153 H. Mori, M. Yamahiro, V.V. Prokhorov, K. Nitta, M. Terano *Macromolecules* **1999**, *32*, 6008–6018.

154 M. Terano *European Patent 703 253*, Research Development Corporation of Japan, **1998**.

155 Y. Doi, S. Ueki *Makromol. Chem., Rapid Commun.* **1982**, *3*, 225–229.

156 H. W. Turner, G. G. Hlatky, H. W. Yang, A. C. Gadkari, G. F. Licciardi *US Patent 5 391 629*, Exxon Chemical, **1995**.

157 Y. Fukui, M. Murata *Appl. Catal. A: Gen.* **2002**, *237*, 1–10.

158 J. Tian, P. D. Hustad, G. W. Coates *J. Am. Chem. Soc.* **2001**, *123*, 5134–5135.

159 S. Kojoh, T. Matsugi, J. Saito, M. Mitani, T. Fujita, N. Kashiwa *Chem. Lett.* **2001**, *30*, 822–823.

160 J. Tian, G. W. Coates *Angew. Chem. Int. Ed.* **2000**, *39*, 3626–3629.

161 G. Milano, L. Cavallo, G. Guerra *J. Am. Chem. Soc.* **2002**, *124*, 13368–13369.

162 J. Saito, M. Mitani, J. Mohri, Y. Yoshida, S. Matsui, S. Ishii, S. Kojoh, N. Kashiwa, T. Fujita *Angew. Chem. Int. Ed.* **2001**, *40*, 2918–2920.

163 M. Mitani, J. Mohri, Y. Yoshida, J. Saito, S. Ishii, K. Tsuru, S. Matsui, R. Furuyama, T. Nakano, H. Tanaka, S. Kojoh, T. Matsugi, N. Kashiwa, T. Fujita *J. Am. Chem. Soc.* **2002**, *124*, 3327–3336.

164 J. Ruokolainen, R. Mezzenga, G. H. Fredrickson, E. J. Kramer, P. D. Hustad, G. W. Coates *Macromolecules* **2005**, *38*, 851–860.

165 A. F. Mason, J. Tian, P. D. Hustad, E. B. Lobkovsky, G. W. Coates *Isr. J. Chem.* **2002**, *42*, 301–306.

166 T. Matsugi, S. Matsui, S. Kojoh, Y. Takagi, Y. Inoue, T. Nakano, T. Fujita, N. Kashiwa *Macromolecules* **2002**, *35*, 4880–4887.

167 A. F. Mason, G. W. Coates *J. Am. Chem. Soc.* **2004**, *126*, 16326–16327.

168 V. Busico, R. Cipullo, N. Friederichs, S. Ronca, M. Togrou *Macromolecules* **2003**, *36*, 3806–3808.

169 V. Busico, R. Cipullo, N. Friederichs, S. Ronca, G. Talarico, M. Togrou, B. Wang *Macromolecules* **2004**, *37*, 8201–8203.

170 T. Shiono, Y. Moriki, K. Soga *Macromol. Symp.* **1995**, *97*, 161–170.

171 T. Shiono, Y. Moriki, T. Ikeda *Macromol. Chem. Phys.* **1997**, *198*, 3229–3237.

172 E. J. Markel, W. Weng, A. J. Peacock, A. H. Dekmezian *Macromolecules* **2000**, *33*, 8541–8548.

173 E. Kolodka, W. Wang, S. Zhu, A. E. Hamielec *Macromolecules* **2002**, *35*, 10062–10070.

174 E. Kolodka, W. Wang, S. Zhu, A. E. Hamielec *Macromol. Rapid Commun.* **2003**, *24*, 311–315.

7
Recent Trends in Macromolecular Engineering using Ring-opening Metathesis Polymerization

Damien Quémener, Valérie Héroguez, and Yves Gnanou

7.1
Introduction

While attempting to polymerize propylene [1], researchers at DuPont de Nemours discovered in the 1950s that olefins can exchange their groups around their double bonds in the presence of certain transition metal compounds. This discovery, also referred to as olefin metathesis to reflect the swapping of groups between two acyclic olefins, resulted in several other outcomes such as the closure of large rings (ring-closing metathesis), the polymerization of cycloolefins [ring-opening metathesis polymerization (ROMP)] and the polymerization of acyclic dienes. In recognition of their outstanding contribution to "the development of the metathesis method in organic synthesis", the Nobel Prize in Chemistry was awarded in 2005 to Yves Chauvin [2], who first postulated the existence of metal carbenes and the formation of metallacyclobutane to account for the mechanism of metathetical reactions, and to Richard R. Schrock [3] and Robert H. Grubbs [4], who first obtained well-defined metathesis-active cocatalysts.

Olefin metathesis nowadays stands at the forefront of organic synthesis and polymer chemistry, allowing the synthesis of a large variety of (macro)molecules from drugs to electroactive polymers through very efficient processes, which can be directly correlated with the improvements made over the last two decades in the design of well-defined metal carbene catalysts/initiators.

Through focused research programs going back to the early 1980s, R. Schrock and R. Grubbs developed such catalysts. In the mid-1980s, Schrock came up with highly reactive catalysts based on tungsten and molybdenum, whereas Grubbs later introduced ruthenium-based catalysts of excellent functional group tolerance and yet of high reactivity towards carbon–carbon double bonds. The ready access to selective catalysts has totally changed the prospects of ring-opening metathesis polymerization of cycloolefins, which has since then experienced phenomenal development: several polymers such as polyoctenamers, polynorbornenes and polydicyclopentadienes obtained by ROMP are found in commercial applications and many cycloolefins can now be polymerized under truly

Macromolecular Engineering. Precise Synthesis, Materials Properties, Applications.
Edited by K. Matyjaszewski, Y. Gnanou, and L. Leibler
Copyright © 2007 WILEY-VCH Verlag GmbH & Co. KGaA, Weinheim
ISBN: 978-3-527-31446-1

"living" conditions. As a result, macromolecular engineering based on ROMP is currently an active field that has generated a number of complex and yet well-defined architectures, including block copolymers and stars, and also functional materials. This review intends to cover the contribution of ROMP of cycloolefins to macromolecular engineering – thus excluding other aspects of metathesis reactions such as alkyne polymerization and ring-closing – and is organized in four parts. After an introductive review of all alkylidene initiators now available, the various complex structures obtained through ROMP are described, and then the synthesis of ROMP-made polymer particles/lattices by various heterogeneous processes is outlined. The final part reviews access to advanced functional materials via ROMP.

7.2
The March Towards Well-defined/Selective Catalysts for ROMP

7.2.1
Discovery of Olefin Metathesis and its Mechanism

"Metathesis" originates from the Greek words *meta* (change) and *tithemi* (place) and commonly means the transposition of phonemes in a word or a group of words. In chemistry, an olefin metathesis reaction can be described as a reaction in which carbon–carbon double bonds in olefins are cut and then rearranged in a statistical fashion (Scheme 7.1).

Early work carried out at DuPont Nemours [1, 5, 6] and then at Standard Oil [7] showed that propylene can be converted into ethylene and butene when heated in the presence of molybdenum–trialkylaluminum catalysts, through a reaction initially called "disproportionation" of acyclic olefins. In 1967, Calderon's group [8–10] at Goodyear realized that olefin "disproportionation" and the polymerization of cycloolefins by the same catalyst are two facets of the same reaction and coined the term "olefin metathesis" to account for the process occurring with both cyclic and acyclic olefins. They performed experiments with butene and deuterated 2-butene in the presence of a homogeneous catalyst [9] (Scheme 7.2) and demonstrated that the double bonds break and swap their alkylidene groups.

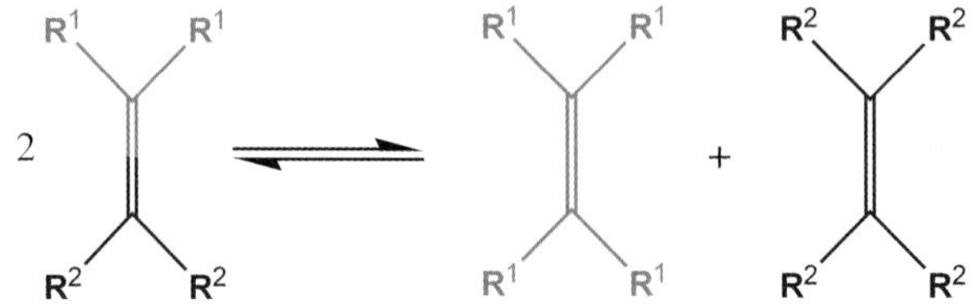

Scheme 7.1 Olefin metathesis.

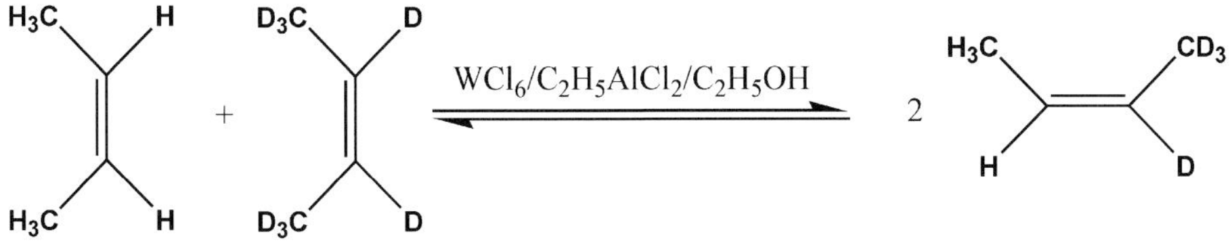

Scheme 7.2 Acyclic olefin metathesis via deuterated compounds.

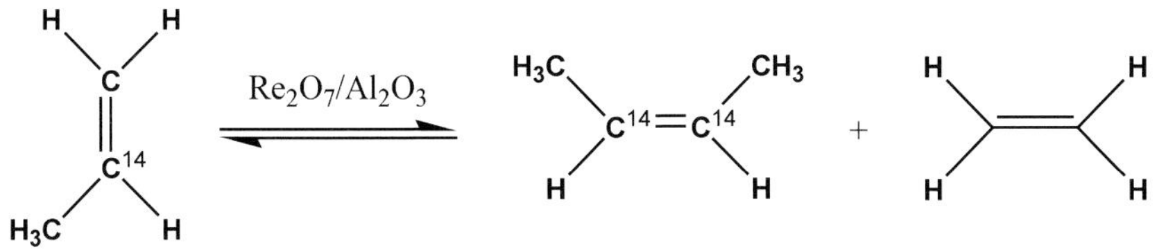

Scheme 7.3 Acyclic olefin metathesis via C-14-labeled compounds.

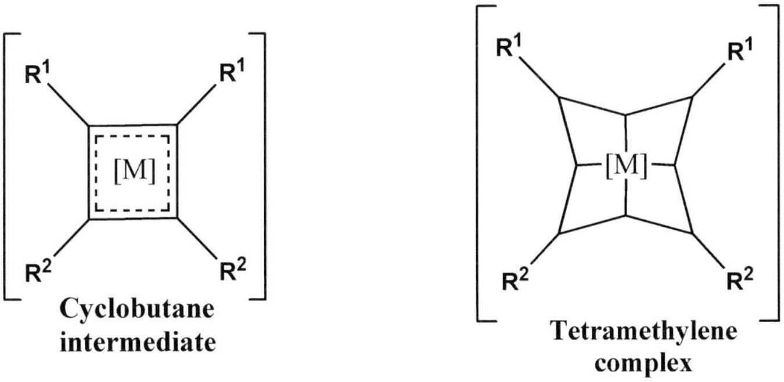

Scheme 7.4 Various intermediates of olefin metathesis.

A similar observation was reported at the same time by Mol et al. [11], with propene and carbon-14-labeled propene in the presence of a heterogeneous catalyst (Scheme 7.3).

Stimulated by the work of the groups of Calderon and Mol, the race to elucidate the mechanism of this unusual reaction lasted about 10 years. At first various mechanisms postulating for the formation of a cyclobutane intermediate [9] and the multicentered species [12] drawn in Scheme 7.4 and the transformation shown in Scheme 7.5 [13] were proposed, but were all subsequently dismissed, none of them being found correct.

From the observation that metal carbenes and four-membered metal-containing rings [14] are known species, Hérisson and Chauvin [15] at Institut Français du Pétrole proposed an alternative mechanism which did not require the unusual theoretical explanations for the previous ones. In this mechanism, a metal

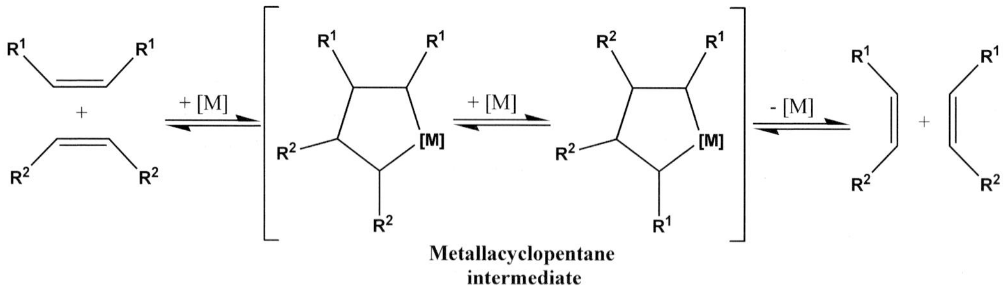

Scheme 7.5 Mechanism of olefin metathesis proposed by Grubbs et al.

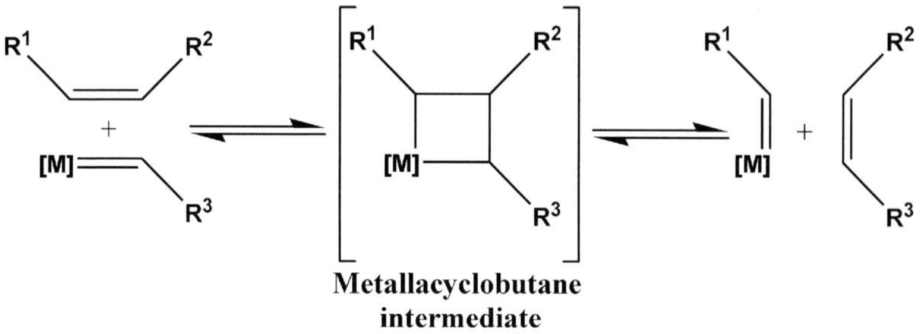

(Metallacyclopentane intermediate)

Scheme 7.6 Mechanism of olefin metathesis proposed by Chauvin et al.

(Metallacyclobutane intermediate)

carbene reacts with an olefin and forms an unstable metallacyclobutane intermediate, which eventually affords a new olefin and a new metal carbene which propagates the reaction (Scheme 7.6).

Despite experimental evidence that accounted for all the facts observed in olefin metathesis, the Chauvin mechanism received full support only after Katz and McGinnis [16] and Grubbs et al. [17] provided additional evidence regarding the metal carbene mechanism. Finally, Chauvin's mechanism was definitely accepted following the work of the groups of Schrock [18–20] and Grubbs [21], who subsequently developed well-defined metal carbenes and used these to catalyze olefin metathesis and initiate ROMP.

7.2.2
Development of Well-defined ROMP Initiators

Over the last 30 years, ROMP initiators have evolved from poorly defined, heterogeneous mixtures to well-defined, single-component metallacycles and metal carbenes (or alkylidenes) [22, 23]. A wide range of catalyst systems based on transition metals (Ti, V, Nd, Ta, Cr, Mo, W, Re, Co, Ir, Ru, Os) are able to trigger the ROMP.

Deriving from Ziegler–Natta catalysts, the first generation of ROMP initiators were ill-defined metallic salts [22, 24] (WCl_6, $Mo(CO)_6$, $RuCl_3$, etc.) that usually required a cocatalyst (Me_4Sn, Et_3Al, n-butanol, etc.). The actual nature of the active site at the metal center was not well known and its *in situ* generation usually proceeded with low yields. Since the propagation rates in the presence of these catalysts were very high compared with the slow initiation rate, very poor control of the molar mass distribution could be achieved in the final polymers. The discovery by Fischer and Maasböl [14] in 1964 of the first well-defined metal carbenes prompted Chauvin to consider their participation in the metathesis mechanism, the major difficulty being to find a way to prove it. After it had been established that metal carbenes are the species responsible olefin metathesis, a large number of them were synthesized and many of them found active in ROMP. Some of the well-characterized catalysts include titanium and tantalum cyclobutanes, tungsten and molybdenum alkylidenes and ruthenium carbenes.

In 1986, Guillom and Grubbs [25, 26] prepared bis(cyclopentadienyl)titanacyclobutanes, the first well-defined metathesis initiators found to bring about "living" polymerizations (Scheme 7.7).

Such complexes trigger ROMPs of norbornene at about 60 °C without any side-reactions, affording polymers with narrow molar mass distributions ($M_w/M_n = 1.08$), but the initiation and propagation were found to depend strongly on temperature (Scheme 7.8).

ROMP of norbornene was also carried out with a 2,6-diisopropylphenoxide-based tantalacycle [27, 28] (Scheme 7.9). Polymers with narrow molar mass dis-

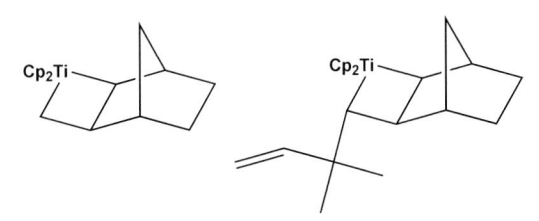

Scheme 7.7 First well-defined titanacyclobutanes active in ROMP.

Scheme 7.8 Mechanism of titanacyclobutane initiation and propagation.

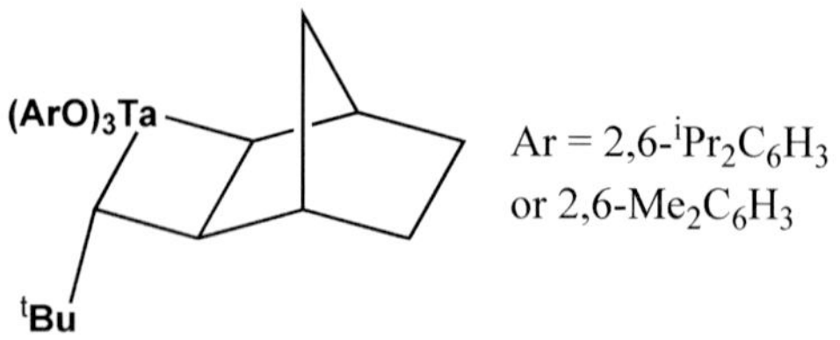
(ArO)3Ta
tBu
Ar = 2,6-iPr2C6H3
or 2,6-Me2C6H3

Scheme 7.9 Tantalacyclobutane catalysts.

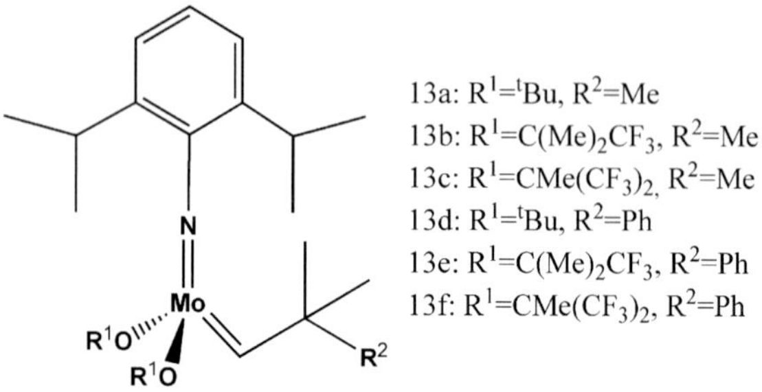
N
Mo
R1O
R1O
R2
13a: R1=tBu, R2=Me
13b: R1=C(Me)2CF3, R2=Me
13c: R1=CMe(CF3)2, R2=Me
13d: R1=tBu, R2=Ph
13e: R1=C(Me)2CF3, R2=Ph
13f: R1=CMe(CF3)2, R2=Ph

Scheme 7.10 Imidotungsten complexes for ROMP.

tributions ($M_w/M_n = 1.04$) were obtained for polymerizations discontinued after 75% monomer conversion. However, if the propagating species were left in solution for an extended period, an increase of the trans double-bond content from 55 to 65% was observed due to secondary reactions, attesting to the non-"living" character of polymerizations initiated by tantalacyclobutane complexes.

Living ROMP of norbornene was also achieved using imidotungstenalkylidene complexes [29] at 25 °C (Scheme 7.10). Metathesis activity was found to depend strongly on alkoxide ligands. Indeed, with the complex **12a**, ROMP of norbornene proceeded without any side-reactions whereas an intermolecular metathesis reaction (chain transfer) was detected during polymerizations with imidotungstenalkylidenes **12b** and **12c**.

ROMP of cyclobutene [30] with imidotungsten complex **12a** was controlled in the presence of an additional amount of a donor component such as PMe_3 or phenyldimethylphosphine. Analogues of poly(1,4-butadiene) were thus obtained with polydispersity indexes as low as 1.1.

On the basis of the results obtained with well-defined imidotungstenalkylidenes, Schrock's group [31, 32] reported in the 1990s the synthesis of molybdenum analogues (Scheme 7.11). Being much more tolerant of functional groups, these molybdenum complexes are currently preferred over tungsten complexes.

These complexes, commonly named "Schrock catalysts", are known to polymerize norbornene, 7-oxanorbornene and their derivatives [33–36] such as 2,3-dicarboxylic acid dimethyl ester, 5-norbornene-*exo-cis*–2,3-diol diacetate and 5-norbornene-2-carbonitrile.

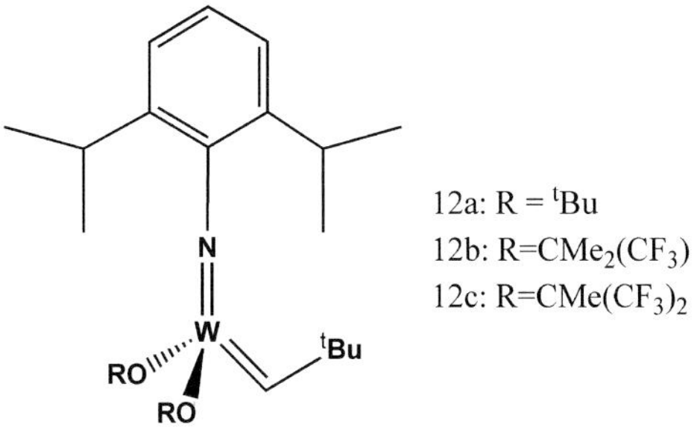

Scheme 7.11 Imidomolybdenum complexes for ROMP.

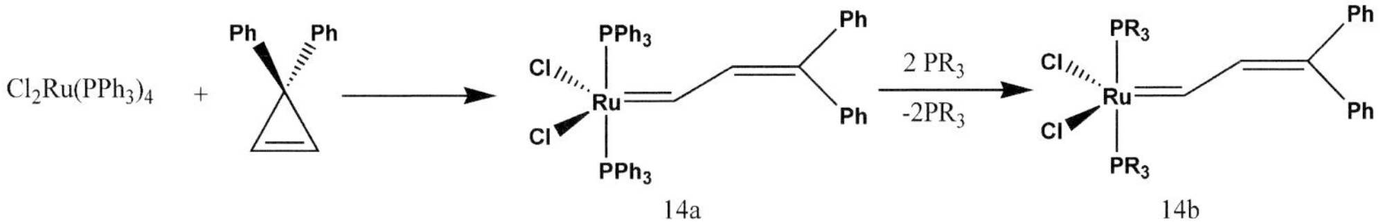

12a: R = ^{t}Bu
12b: R=CMe$_2$(CF$_3$)
12c: R=CMe(CF$_3$)$_2$

Scheme 7.12 Synthesis of the first ruthenium carbene complexes.

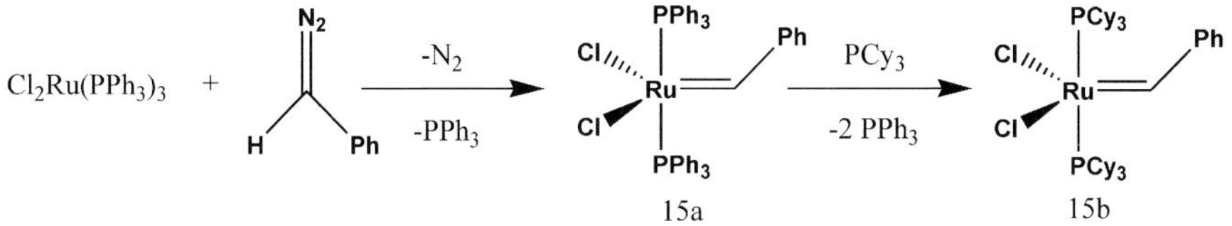

Scheme 7.13 Powerful route to synthesize the first-generation Grubbs catalyst.

In the continuity of efforts to design metal alkylidene complexes more tolerant of polar functions, Grubbs' group [21] described in 1992 the synthesis of the first well-defined ruthenium alkylidene (**14a** in Scheme 7.12). This complex, which combines a 16-electron metal center with one empty coordination site for olefin binding, has a higher preference for carbon–carbon double bonds than for alcohol, amide, aldehyde and carboxylic acid groups. It brings about the "living" polymerization of norbornene [36] but is not appropriate for less strained cycloolefins.

Upon substituting ligands with higher electron-donating character such as tricyclohexylphosphine or triisopropylphosphine for PPh$_3$, Grubbs and coworkers [37] successfully used **14b** to polymerize less strained cycloolefin ligands (Scheme 7.12). The same team described an alternative easier pathway [36] to produce ruthenium alkylidenes through the use of diazoalkane compounds (Scheme 7.13).

The complex **15b**, also known as a first-generation Grubbs catalyst, exhibits a high metathesis activity towards various cyclic olefins such as norbornene [36], 7-oxanorbornene [38] and cyclobutene [36]. Selected variants of complexes **15a**

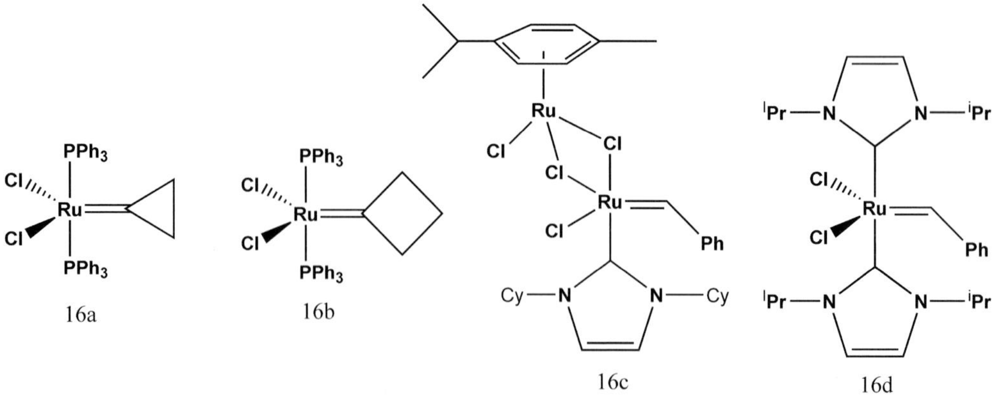

Scheme 7.14 Various ruthenium alkylidene catalysts.

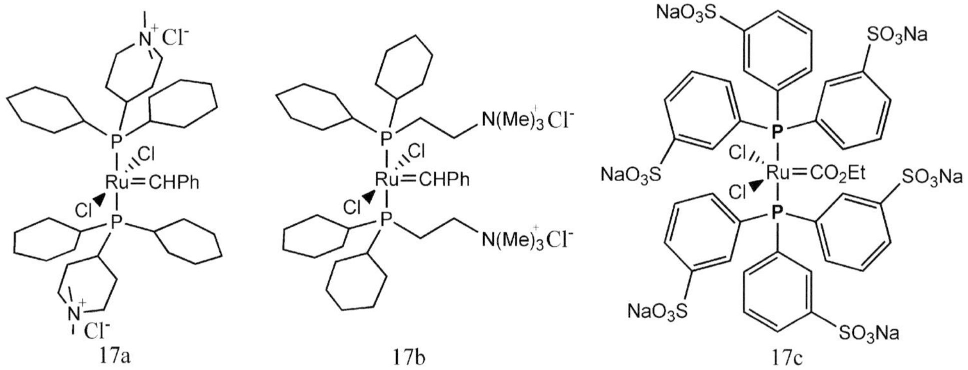

Scheme 7.15 Water-soluble ruthenium alkylidenes.

and **15b** have been prepared by modifying either the carbene or the ligand (Scheme 7.14), affording new complexes with fine-tuned reactivities. For example, **16a** and **16b** bring about the ROMP of norbornene in a "living" fashion [39].

Dramatic improvements in activity could be observed with complexes **16c** and **16d**, using a five-membered N-heterocyclic carbene (NHC) ligand, due to their better -donor character compared with that of phosphines. For example, the bis-(ruthenium) complex **16c** brings about the "living" ROMP of norbornadiene-2,3-dicarboxylic ester [40] ($M_w/M_n = 1.1$) whereas **16d** cannot for a series of functionalized 5-substituted norbornene derivatives [40] ($M_w/M_n = 1.4$–2.3).

With the advent of the discovery of ruthenium alkylidenes, ROMPs in water in their presence became possible [41] and water-soluble complexes were prepared by the modification of the phosphine ligands (Scheme 7.15).

Complexes **17a** and **17b** were shown to bring about the "living" aqueous ROMP of norbornene and 7-oxanorbornene derivatives. In a later contribution,

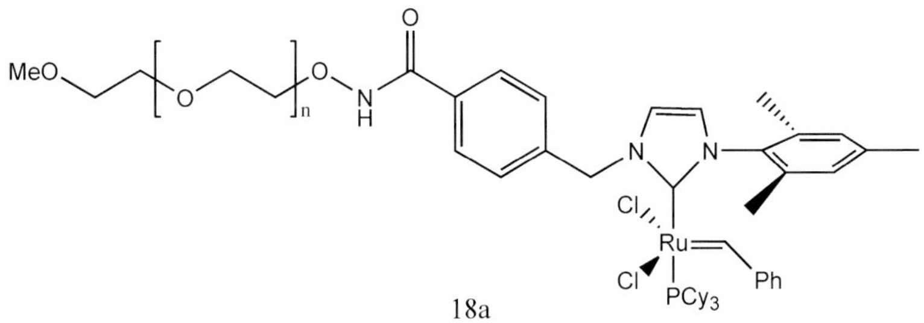

Scheme 7.16 PEO-based water-soluble ruthenium alkylidene.

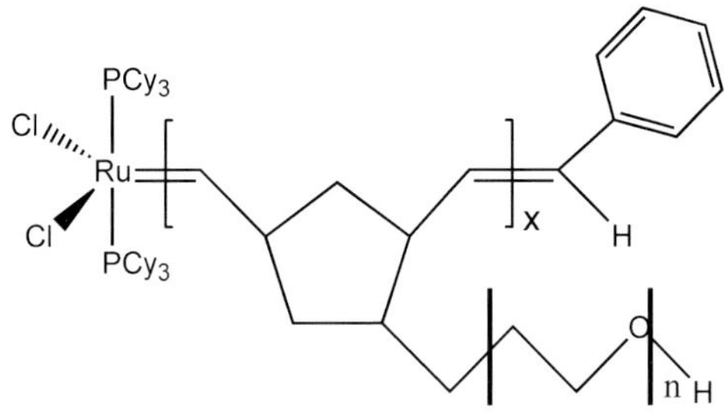

Claverie et al. [42] synthesized **17c**, a complex generated *in situ* by reaction of ethyl diazoacetate with bis[tris(3-sulfonatophenyl)phosphine, sodium salt]. Emulsion ROMP of norbornene was performed with success, but polymerizations of 1,5-cyclooctadiene and cyclooctene failed to occur. In a recent contribution, Grubbs' group [43] synthesized the water-soluble complex **18a**, whose NHC ligand carried a side poly(ethylene oxide) (PEO) chain (Scheme 7.16). ROMPs of norbornene derivatives and cyclooctene could be achieved in water and methanol. Around the same time, our group reported the synthesis of another PEO-based ruthenium carbene [44] (**18b**), made water soluble through the oligomerization of ω-norbornenyl-PEO macromonomer. Carbene **18b** allowed the successful miniemulsion ROMPs of norbornene with relatively good control over molar masses and molar mass distributions ($M_w/M_n = 1.4$).

7.3
Macromolecular Engineering Using ROMP

With metal alkylidene initiators at hand that bring about the "living" polymerization of cyclic olefins – even the less strained ones and those carrying polar functions – in apolar and polar media, all the necessary tools were made available to engineer polyalkenamers of all sorts and all kinds of macromolecular structures including block, graft and star (co)polymers by ROMP. In addition to these, in this chapter polymeric architectures obtained by combination of ROMP with other polymerization processes will also be reviewed.

7.3.1
Block Copolymers by ROMP and Combination of ROMP with Other "Living" Polymerizations

The most common method to prepare block copolymers is to polymerize successively comonomers through a controlled or "living" process. Early ill-defined catalysts were unsuited to the synthesis of block copolymers because of the non-"living" character of the polymerization that they trigger. The situation changed with the discovery of well-defined carbene and alkylidene complexes. The first significant work on block copolymers was published in 1988 by Cannizzaro and Grubbs [45], who described the synthesis of di- and tri-block copolymers (A–B and A–B–A) of narrow molar mass distribution ($M_w/M_n < 1.1$) resulting from the successive ROMPs of norbornene derivatives (norbornene, benzonorbornadiene, 6-methylbenzonorbornadiene and dicyclopentadiene) in the presence of titanacyclobutane initiator (Scheme 7.17).

Grubbs' group [46] showed in 1993 that polynorbornene initiated with this titanacyclobutane catalyst could be quenched with a trace amount of ethanol, a reaction that converted the metal carbene into a metal–alkyl bond; after activation of the latter by addition of ethylaluminum dichloride Ziegler–Natta polymerization of ethylene could be carried on and poly(ethylene)-b-polynorbornene copolymer produced (Scheme 7.18).

Titanacyclobutane end-groups could be also deactivated by aldehyde functions through a Wittig-type reaction which permitted the subsequent preparation of polynorbornene-b-poly(silyl vinyl ether) copolymer by changing from ROMP to group transfer polymerization (Scheme 7.19) [47, 48].

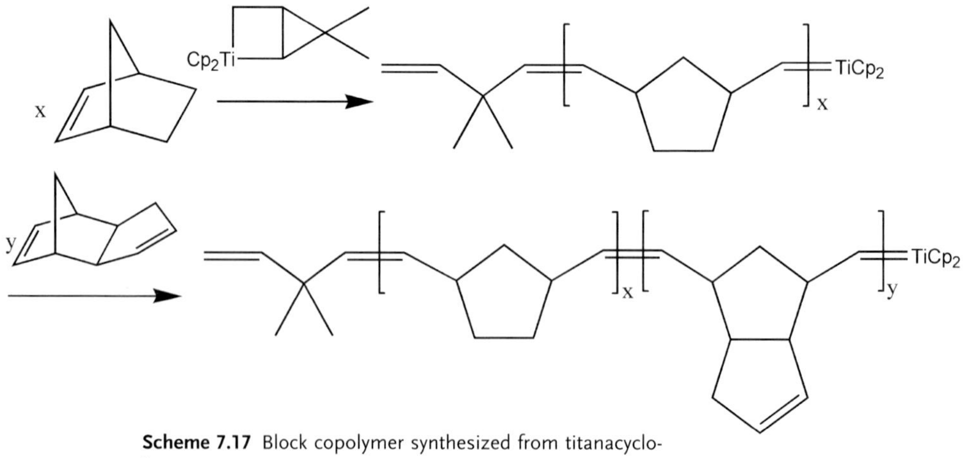

Scheme 7.17 Block copolymer synthesized from titanacyclobutane initiator.

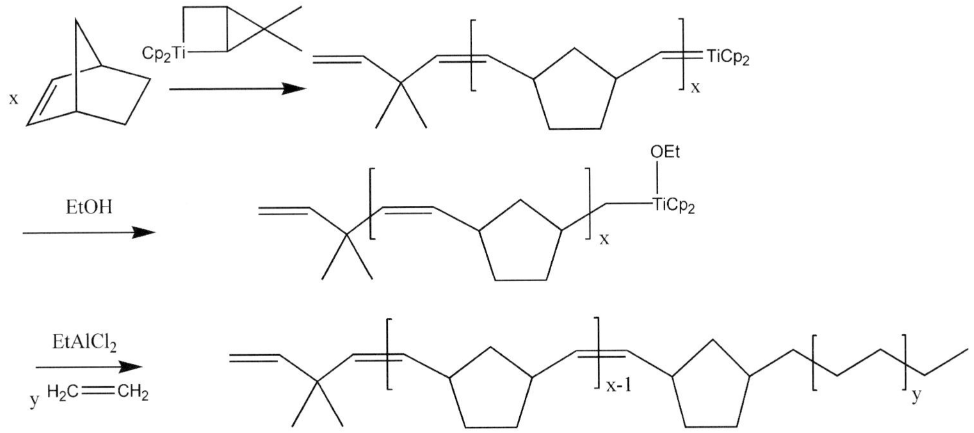

7.3 Macromolecular Engineering Using ROMP | 259

Scheme 7.18 Block copolymer from ROMP and Ziegler–Natta polymerization.

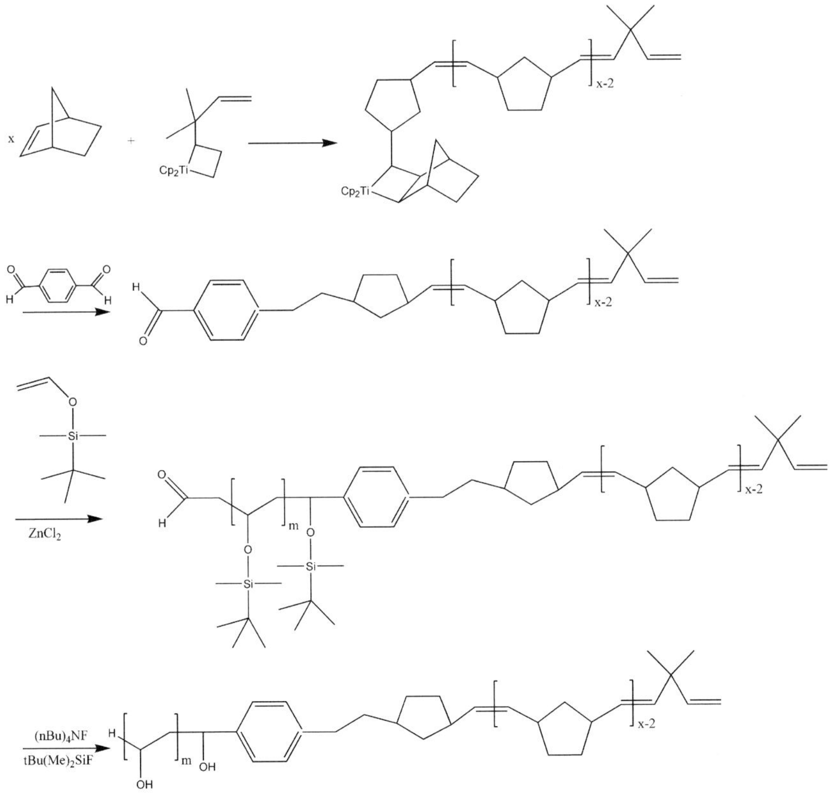

Scheme 7.19 Synthesis of polynorbornene-*b*-poly(silyl vinyl ether) copolymer.

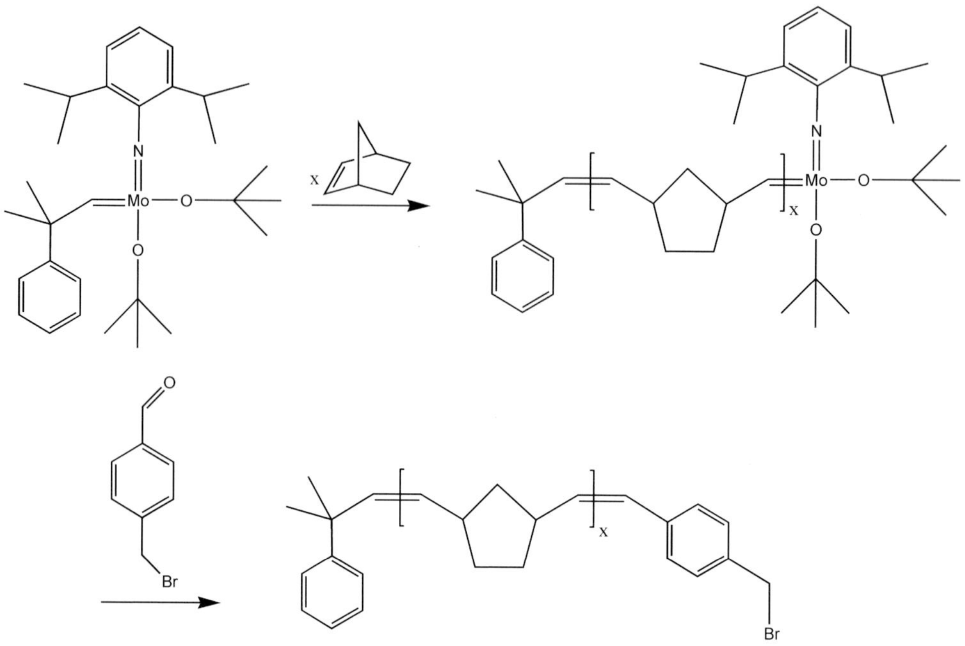

Scheme 7.20 Combination of ROMP with ATRP to synthesize diblock copolymers.

In 1997, Matyjaszewski's group [49] reported the synthesis of block copolymers by combination of "living" ROMP with controlled atom transfer radical polymerization (ATRP) (Scheme 7.20). Upon deactivation of polynorbornene chain ends by a Wittig-type reaction, an ATRP macroinitiator was obtained which was used to polymerize styrene and methyl acrylate under controlled/"living" conditions ($M_w/M_n < 1.2$, $M_n = 5000–100\,000$ g mol^{-1}).

In 2000, a similar ROMP–ATRP tandem reaction was investigated by Grubbs' group [50] (Scheme 7.21). They first synthesized an α,ω, dichloro telechelic of 100% 1,4-polybutadiene by ROMP of 1,5-cyclooctadiene (COD) in the presence of an appropriate chain transfer agent which they used to produced by ATRP triblock copolymers with either PS or polyacrylate as external blocks [poly(styrene-b-butadiene-b-styrene)], $M_w/M_n = 1.2–1.7$, $M_n = 4000–40\,000$ g mol^{-1}).

Mahanthappa et al. [51] also utilized the ruthenium-catalyzed ROMP of cycloolefins in the presence of chain transfer agents, which they combined with RAFT as a means to produce ABA triblock copolymers. Upon completion of the ROMP of 1,5-cyclooctadiene with a chain transfer agent containing latent reversible addition fragmentation chain transfer (RAFT) initiating sites (trithiocarbonate group), they polymerized styrene and butyl methacrylate and eventually obtained poly(styrene-b-butadiene-b-styrene) and poly(tert-butylacrylate-b-butadiene-b-tert-butylacrylate) of varying compositions and monomodal molar mass distributions.

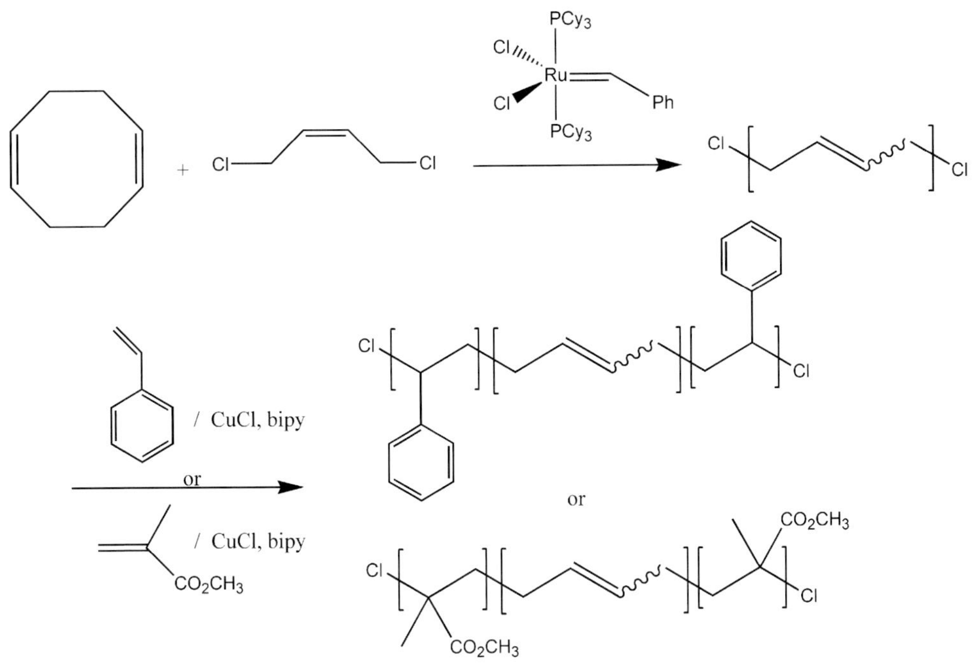

Scheme 7.21 ABA triblock copolymers synthesized by combining ROMP and ATRP.

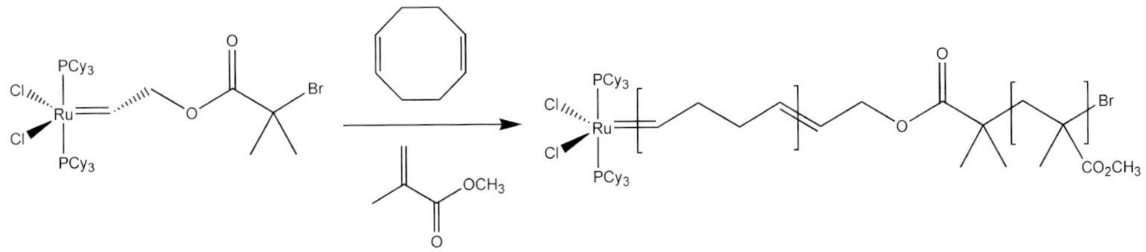

Scheme 7.22 Block copolymer synthesis using ATRP and ROMP bifunctional initiator.

In the same year a new bifunctional ruthenium catalyst was reported [52] that was used to initiate in a one-pot reaction the ROMP of 1,5-cyclooctadiene and the ATRP of MMA (Scheme 7.22). Block copolymers were obtained with fairly broad molar mass distributions ($M_w/M_n=1.5$, $M_n=6000–20000$ g mol^{-1}). Subsequent hydrogenation of the copolymer led to well-defined poly(ethylene-b-methyl methacrylate).

Khosravi's group [53] in 2004 prepared AB-type block copolymers by combining successively "living" anionic polymerization and "living" ROMP (Scheme 7.23). First, ethylene oxide was polymerized anionically using diphenylmethylpotassium as initiator and was terminated with vinylbenzyl chloride. After an al-

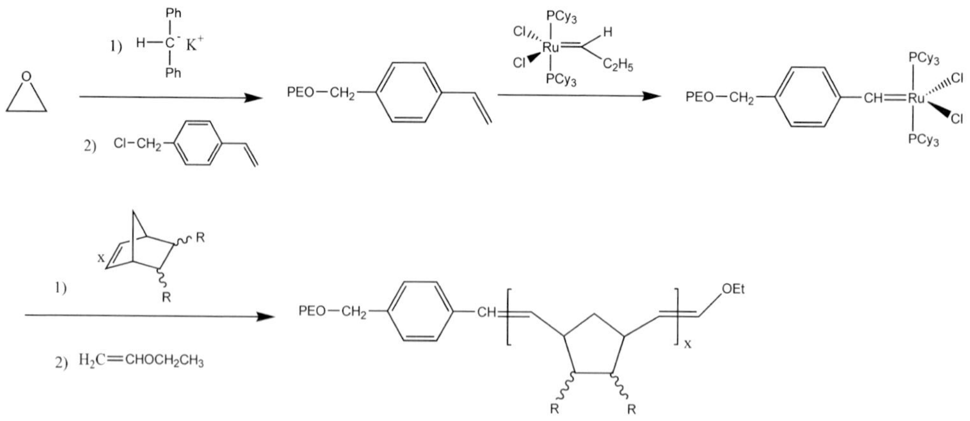

Scheme 7.23 Synthesis of PEO–PNB block copolymer.

kylidene exchange reaction, the PEO-based ROMP macroinitiator of PEO served to polymerize norbornene derivatives.

7.3.2
Graft Copolymers by ROMP with Other "Living" Polymerizations

Graft copolymers in which side-chains or grafts are chemically different from the backbone [54] are found in several applications such as impact-resistant materials, thermoplastic elastomers, compatibilizers and emulsifiers. In as much as ROMP is concerned, they were synthesized by the "grafting from" method or the macromonomer route by combining either ROMP/ROMP, anionic/ROMP, cationic/ROMP, ATRP/ROMP or ring-opening polymerization (ROP)/ROMP. The "macromonomer" method is well known and gives rise to regular multibranch polymers that are highly compact. Nevertheless, one major limitation associated with this method is the difficulty of polymerizing quantitatively the macromonomer due to its lack of reactivity. Provided that certain conditions are fulfilled, such as to work with long-lived active species and macromonomers end-fitted with a norbornene unsaturation, their ROMP was shown to occur under "living" conditions and to proceed to complete conversion.

7.3.2.1 ROMP/ROMP

Only one study relied on ROMP for the growth of both the backbone and the grafts: most of the polymers produced by ROMP exhibiting similar macroscopic characteristics, and there was little incentive to associate in a graft copolymer structure two polymers obtained by ROMP. However, Imanishi's group [55] reported in 2001 the synthesis of various polymacromonomers comprising a polynorbornene backbone and polynorbornene side-chains (Scheme 7.24). Since

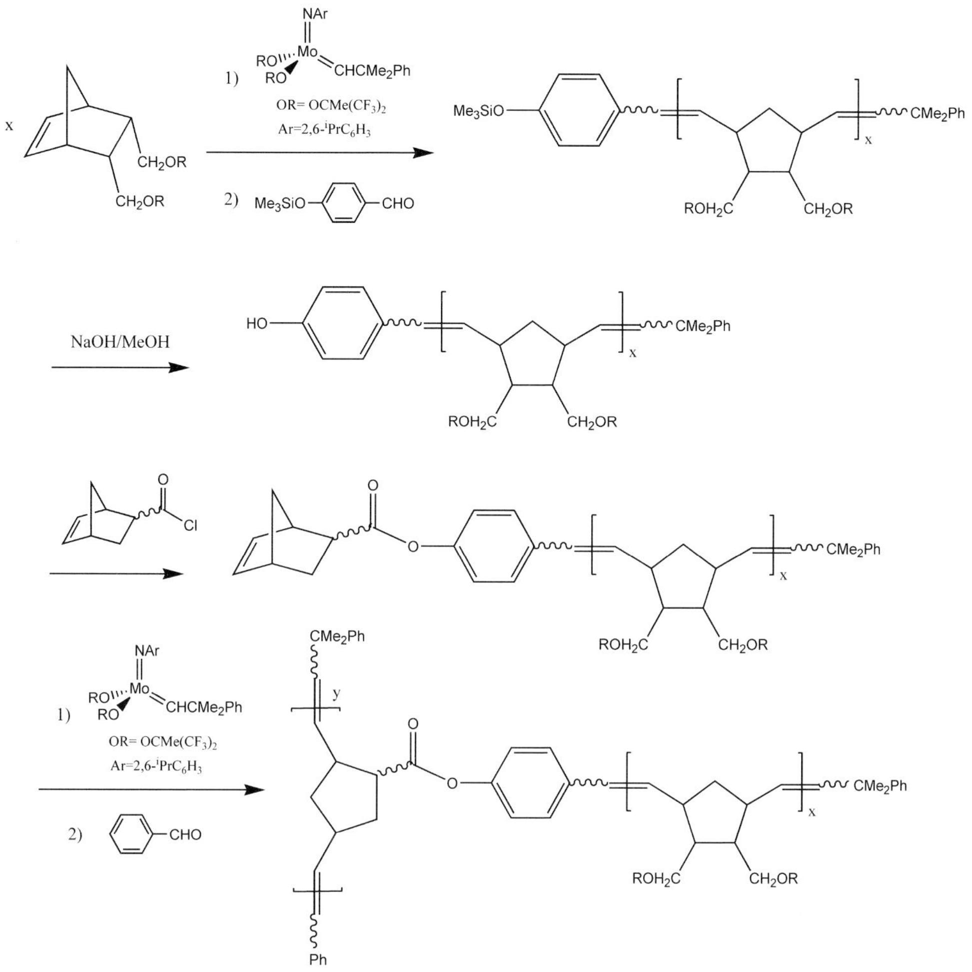

Scheme 7.24 Graft copolymer from ROMPs.

ROMP proceeded in a "living" fashion in most cases, well-defined graft copolymers were synthesized with polydispersity as low as 1.1.

7.3.2.2 Anionic Polymerization/ROMP

Several well-defined graft copolymers have been synthesized by combining anionic polymerization and ROMP. Generally, a macromonomer prepared by anionic polymerization and carrying a terminal cycloolefin was subjected to ROMP along with a comonomer to obtain the expected graft copolymers. For example, Khosravi's group reported the synthesis of a graft copolymer by ROMP of the anionically obtained 5-norbornene-2,3-*trans*-bis(polystyrylcarboxylate) macromo-

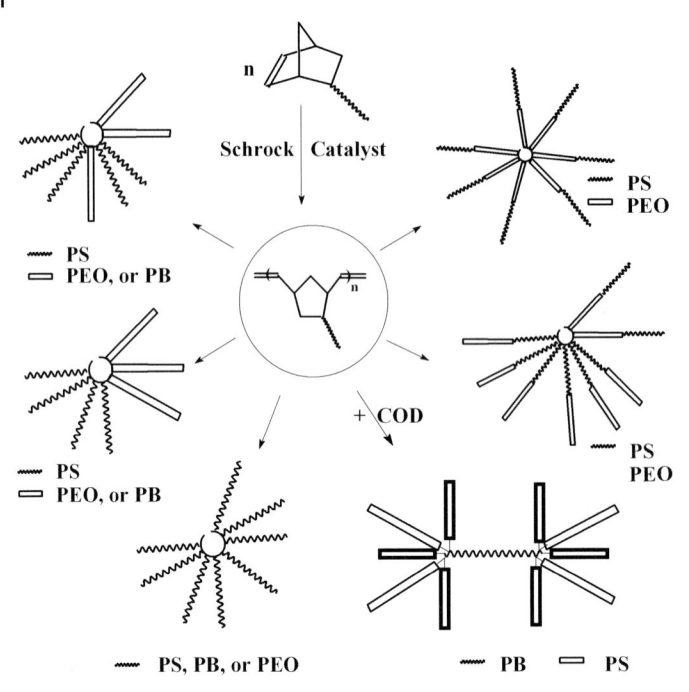

Scheme 7.25 Graft copolymers obtained by ROMP of macromonomer anionically prepared.

nomers. By virtue of the selectivity of the Schrock molybdenum initiator [Mo(N-2,6-iPr$_2$C$_6$H$_3$)(OCMe$_3$)$_2$(CHR), R=CMe$_3$ or CMe$_2$Ph] [56, 57], the graft copolymers exhibited a well-defined structure. However, due to steric crowding around the norbornenyl unsaturation of the macromonomer, it was not possible to polymerize the latter to high DP$_n$. The same group subsequently used a macromonomer containing only one polystyrene graft on each norbornene unit [58] and successfully polymerized the latter to high conversion.

At around the same time, Héroguez and coworkers described the synthesis of various graft copolymers with unusual topologies by ROMP of anionically prepared macromonomers. Sphere-type, Janus-type, bottle-brush-type polymers and structures with dumbbell-like and palm-tree-like shapes were prepared with excellent control over topological features even though they did not exhibit a precise number of arms/grafts. The difficulty in such an endeavor lies more in the synthesis of the appropriate macromonomer than in its ROMP. Macromonomers of poly(ethylene oxide) (PEO) [59], polystyrene (PS) [60] and polybutadiene (PB) [61] and also PS-b-PEO macromonomers end-fitted by norbornenyl unsaturation were prepared and successfully subjected to ROMP [62, 63] (Scheme 7.25). Complete conversions were achieved, regardless of the degree of polymerization targeted and the polymacromonomers isolated exhibited low polydispersity indices and experimental molar masses in good agreement with the expected values. As long as the backbone of the polymacromonomer remains small, it adopts a spherical shape that progressively shifts towards a less compact bottle-brush shape as $\overline{DP}_n$ increases [64, 65].

7.3.2.3 Cationic Polymerization/ROMP

Graft copolymers comprising a polynorbornene backbone and polyphosphazene grafts were synthesized by coupling cationic polymerizations and ROMP [66] (Scheme 7.26). These copolymers find applications as fire-resistant materials and as solid electrolyte polymers. Mono- and dinorbornenylpolyphosphazenes were synthesized by means of "living" cationic polymerization which afforded macromonomers with well-defined molar masses and narrow polydispersities. Homo- and co-ROMPs of these macromonomers with norbornene were carried out using the Grubbs ruthenium carbene. Monofunctionalized polyphosphazene macromonomers produced soluble graft copolymers with a polydispersity ranging from 1.29 to 2.07 and bifunctionalized ones resulted in cross-linked materials.

7.3.2.4 ATRP/ROMP

Poly(norbornene-g-methylmethacrylate) copolymers were synthesized by the "grafting from" method [52] (Scheme 7.27). In a first step, ROM copolymerization

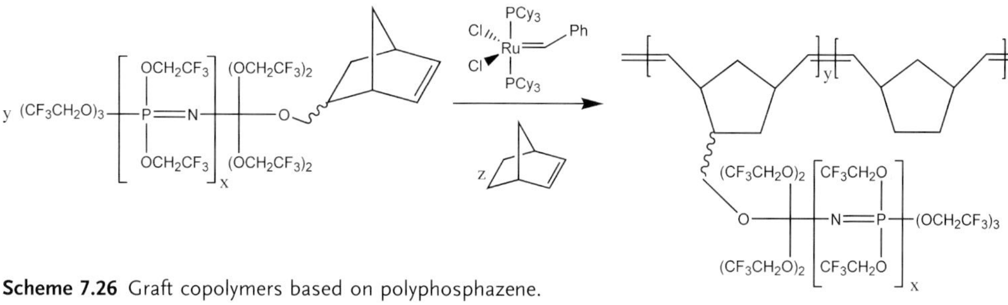

Scheme 7.26 Graft copolymers based on polyphosphazene.

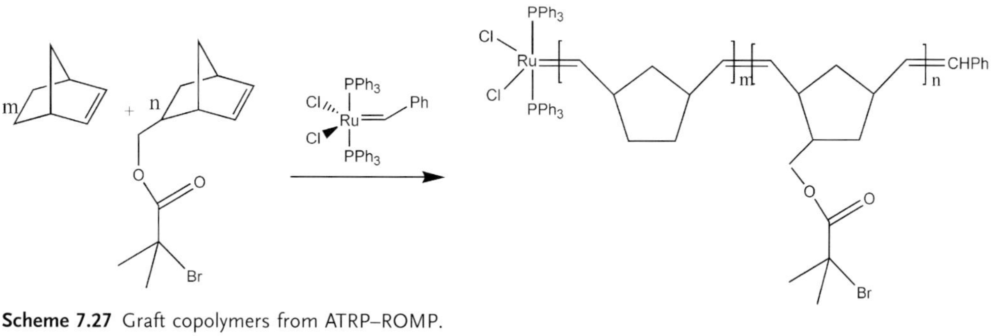

Scheme 7.27 Graft copolymers from ATRP–ROMP.

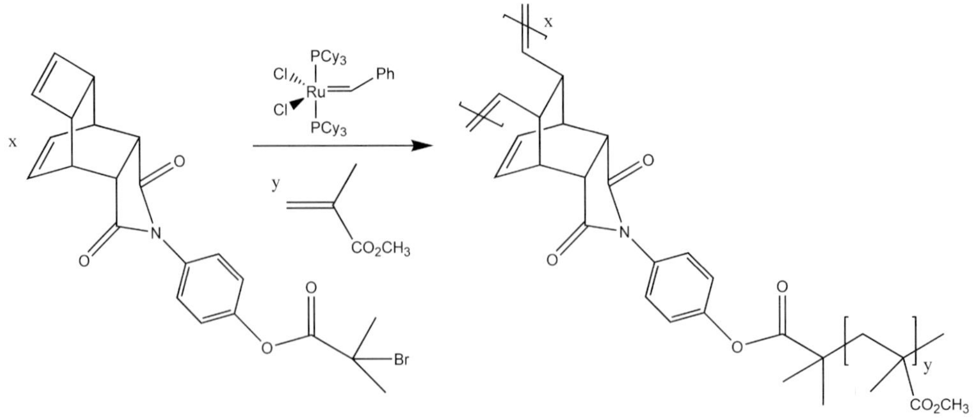

Scheme 7.28 One-pot synthesis of graft copolymers through tandem polymerizations.

of NB with a norbornene derivative functionalized with an initiating ATRP site, i.e. a 2-bromo-2-methylpropionate group, were carried out in the presence of $RuCl_2(CHPh)(PPh_3)_2$. The copolymer obtained was used, in a second step, as a macroinitiator for ATRP of MMA using the ruthenium ethoxycarbene $RuCl_2$-$(CHOEt)(PPh_3)_2$ as catalyst. The resulting graft copolymers were obtained in high yields and with narrow polydispersities.

A few years later, Demonceau's group [67, 68] showed that a ruthenium phenylcarbene, $RuCl_2(CHPh)(PCy_3)_2$, can trigger ATRP of MMA. Inspired by this work, Charvet and Novak [69] investigated the one-pot synthesis of graft copolymers and used for that a single catalyst that was able to initiate both the ROMP of a cyclobutene derivative end-functionalized with 2-bromo-2-methylpropionate and catalyze the ATRP of MMA through activated bromide pendant groups (Scheme 7.28).

A similar strategy was followed by Weck's group [70], who obtained graft copolymers comprising polynorbornene backbone and poly(*tert*-butyl acrylate) or poly(acrylic acid) grafts. The viscosity of these polymers could be finely tuned through the hydrolysis of the *tert*-butyl ester groups as a function of pH, temperature and polymer composition.

7.3.2.5 ROP/ROMP

Jérôme's group [71] reported the synthesis of graft copolymers by ROMP of poly(ε-caprolactone) macromonomers (Scheme 7.30). In a first step, macromonomers in the range 2000–10 000 g mol^{-1} were synthesized by ROP of ε-caprolactone in the presence of 2-hydroxymethyl-5-norbornene and triethylaluminum. In a second step, the copolymerization of these macromonomers with either norbornene or norbornene acetate by ROMP was initiated by the catalyst system (*p*-cymene)ruthenium(II) chloride dimer–tricyclohexylphosphine–(trimethylsilyl)-

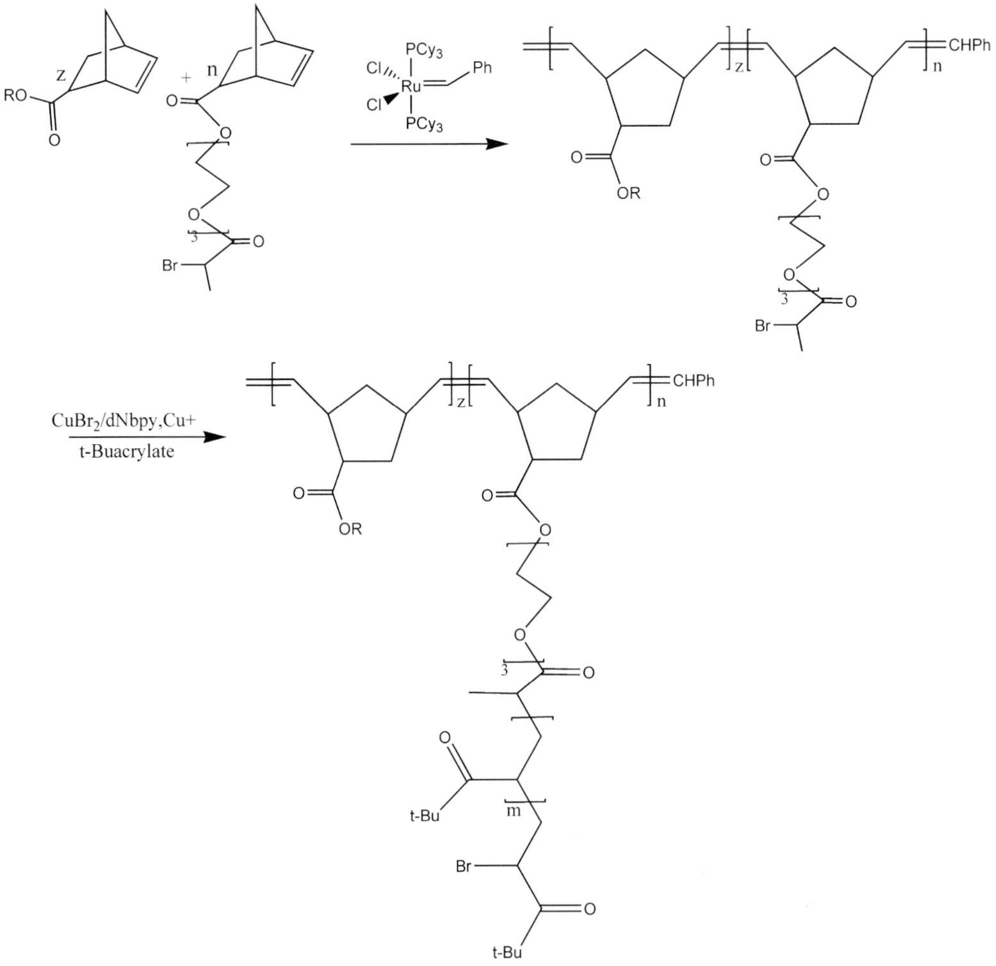

Scheme 7.29 Synthesis of graft copolymer with tunable properties.

diazomethane. Due to the non-"living" character of these polymerizations, the graft copolymers obtained exhibited a broad molar mass distribution (M_w/M_n = 1.25–2.40).

7.3.2.6 ROMP/ROP/ATRP

Block copolymers carrying PS and polylactide grafts were generated from polynorbornene backbones [72]. The latter were obtained in a first step by copolymerization under "living" conditions of two *exo*-norbornene ester (**A**, **B**, Scheme 7.31) macromonomers, one carrying an ATRP initiating site and the other a hydroxy group in the ω-position. Second, these copolymers were quantitatively

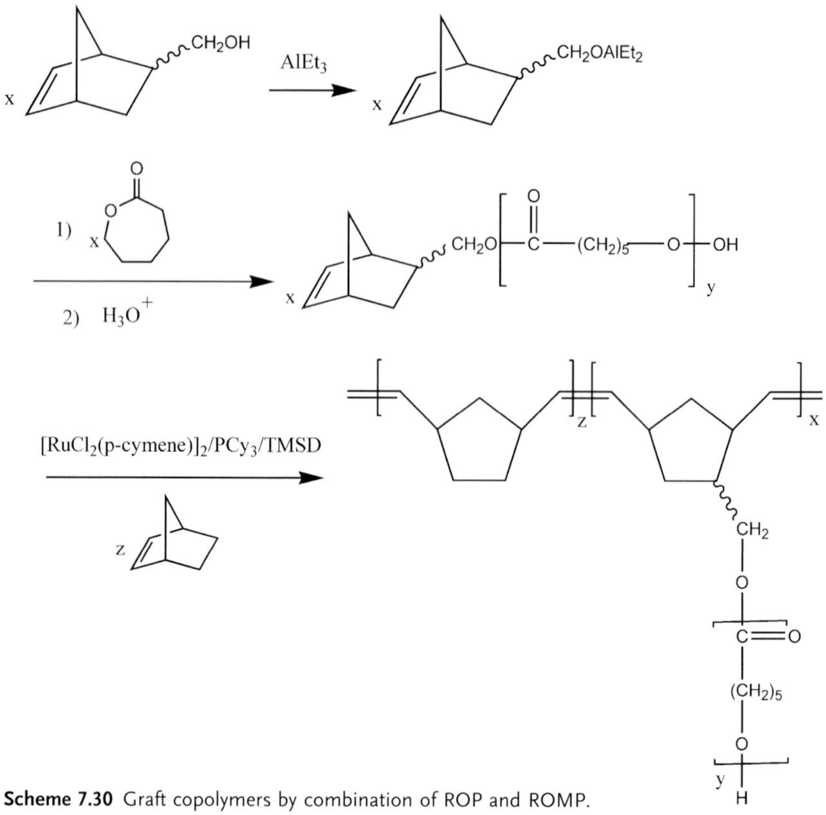

Scheme 7.30 Graft copolymers by combination of ROP and ROMP.

hydrogenated in the presence of 10 equiv. of the Grubbs catalyst, silica gel and H_2. Comb block copolymers with molar masses up to 63×10^6 g mol^{-1} were eventually isolated after growing the PS grafts by ATRP and the polylactide grafts by RO of lactide. New applications in nanofluidics and photonics are contemplated for these polymers.

7.3.3
Star Polymers by ROMP

Polymers with a star-type architecture can be obtained by either of these methods, neutralization of "living" chains on to a multifunctional deactivating agent (arm-first method), branch growth from a multifunctional initiator (core-first method) and copolymerization of "living" chains with a bifunctional – here diunsaturated – cyclic monomer. In 1991, Bazan and Schrock [73] resorted to the latter method to prepare star polymers: W(CHR)(NAr)(O-tBu)$_2$ and Mo(CHR)(NAr)(O-tBu)$_2$ were the two initiators and norbornadiene was the diunsaturated monomer that was copolymerized with "living" polynorbornene

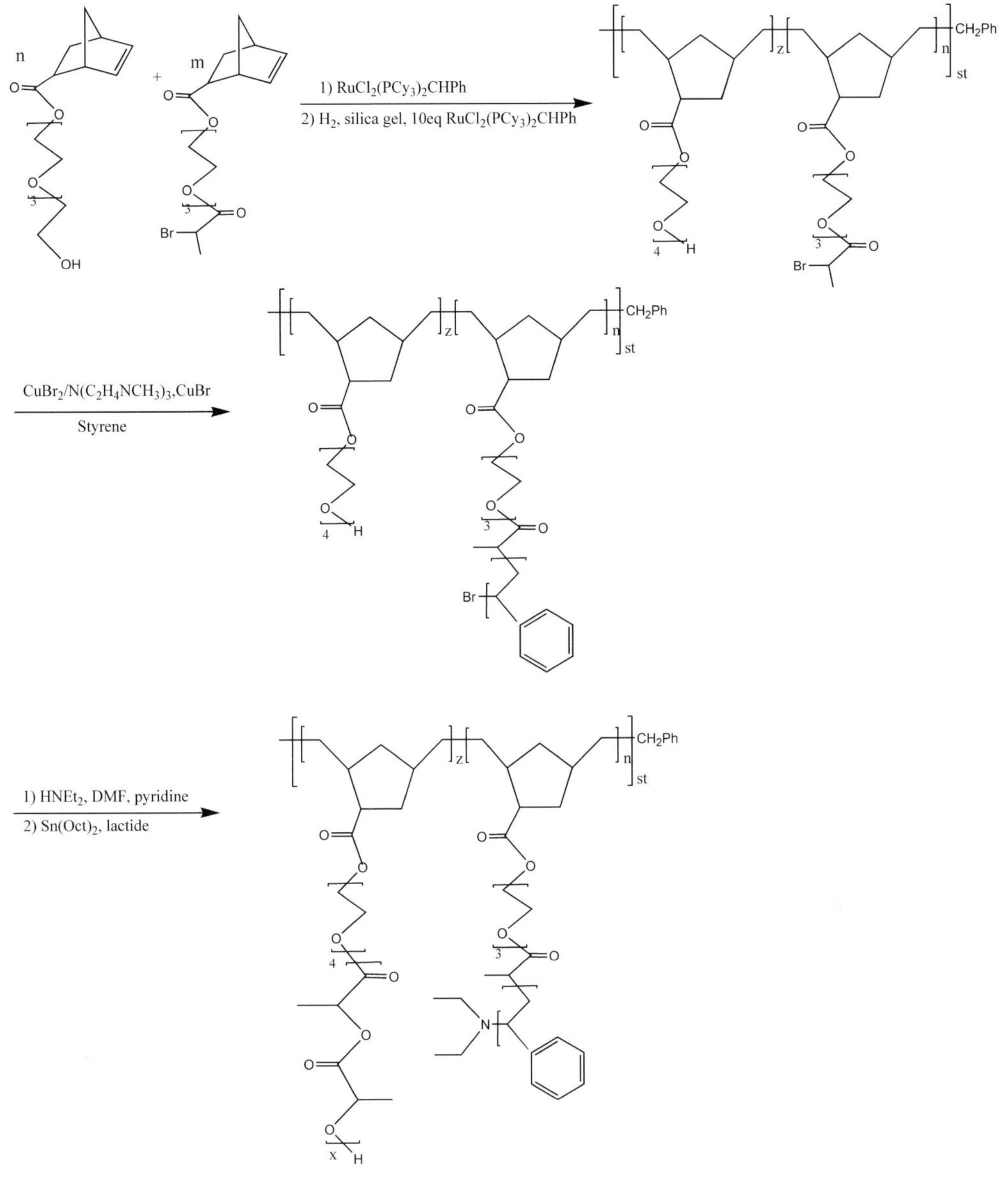

Scheme 7.31 Block copolymer by combination of ROMP, ATRP and ROP.

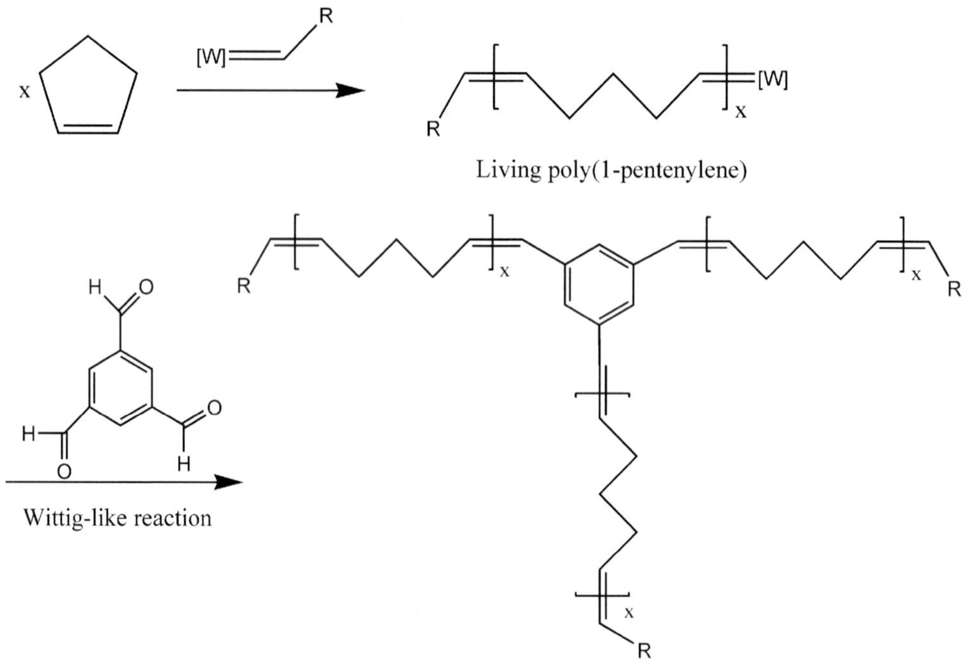

Scheme 7.32 Poly(1-pentenylene) star polymer.

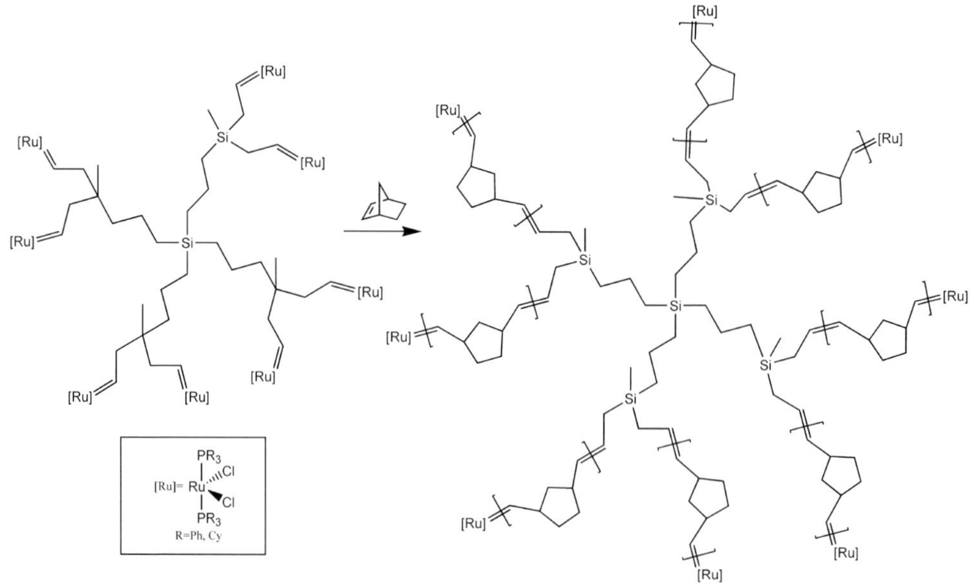

Scheme 7.33 Eight-branch star polymers from dendritic Ru complexes.

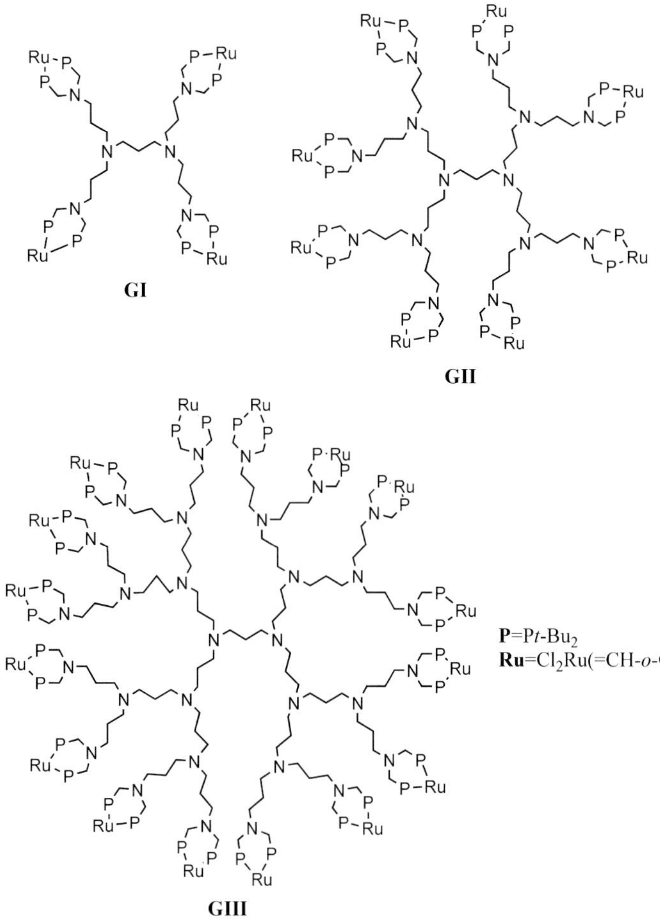

P=Pt-Bu_2
Ru=$Cl_2Ru(=CH$-o-O-$i$$PrC_6H_4)$

Scheme 7.34 Three generations of dendritic ruthenium benzylidene complexes.

chains. Such stars were rather ill-defined, with fluctuations in the number of "arms". Schrock's group also prepared amphiphilic star diblock copolymers [74] by the same strategy.

Examples of three-arm stars obtained by the arm-first method were first described by Dounis and Feast [75]. They generated poly(1-pentenylene) stars by deactivating "living" chains with a trifunctional aldehyde in a Wittig reaction (Scheme 7.32). As these end-capping reactions were extremely sensitive to the stoichiometry between the aldehyde and the "living" chain ends, the final sample exhibited a rather broad distribution (1.61). Subsequent hydrogenation gave rise to polyethylene star polymers.

Stars grown from a multifunctional core carrying a precise number of metal alkylidene sites are more recent. In 2002, Verpoort's group [76] described how

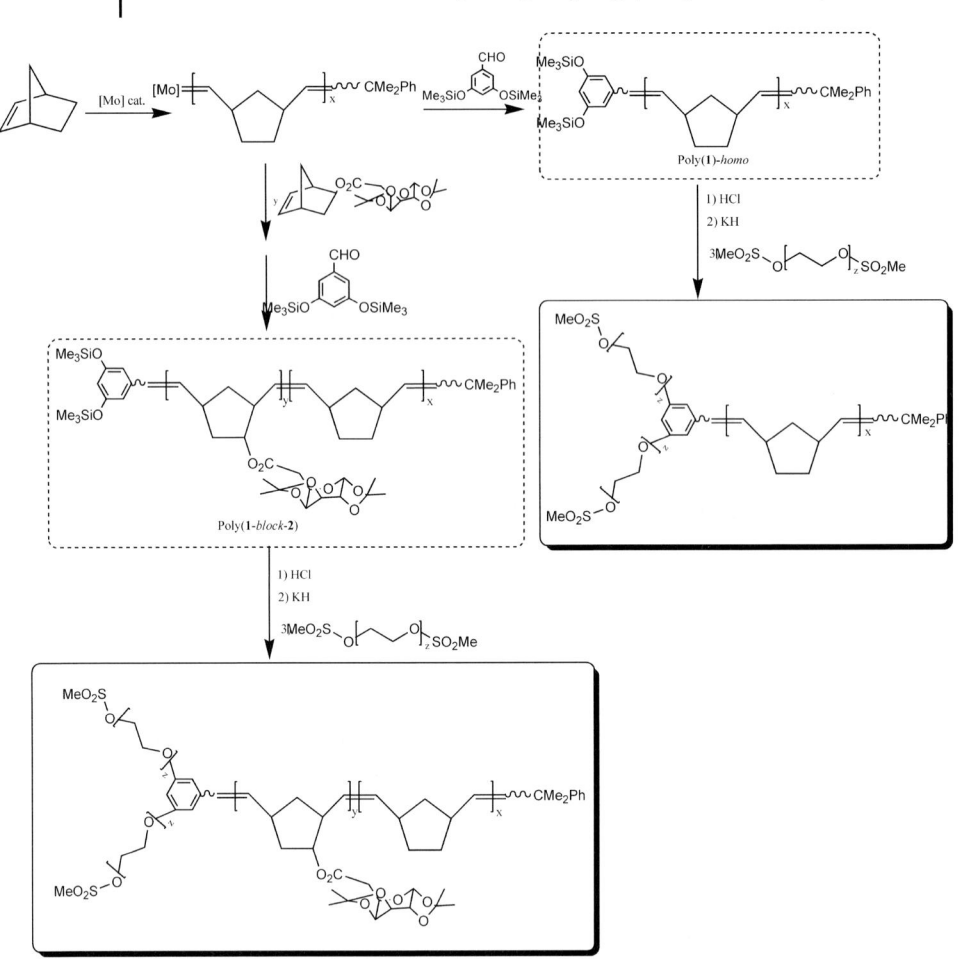

Scheme 7.35 Amphiphilic heteroarm star copolymers.

to modify the dendrimer tips with ruthenium complexes and efficiently initiate the ROMP of norbornene (Scheme 7.33). Star polymers with four and eight branches were synthesized in this way. Monomodal molar mass distributions were obtained with molar masses in good accordance with the expected value. Depending on the phosphine ligand used (PPh$_3$ or PCy$_3$), the sample polydispersity ranged from 1.17 to 2.11.

Astruc and coworkers [77, 78] also reported the use of dendrimers modified with ruthenium complexes (Scheme 7.34) to synthesize polynorbornene stars. Interestingly, they found that the metallodendrimers formed catalyze a faster ROMP of norbornene than molecular complexes. A facile decoordination of the phosphine ligand in the dendrimers was shown to be responsible for this effect.

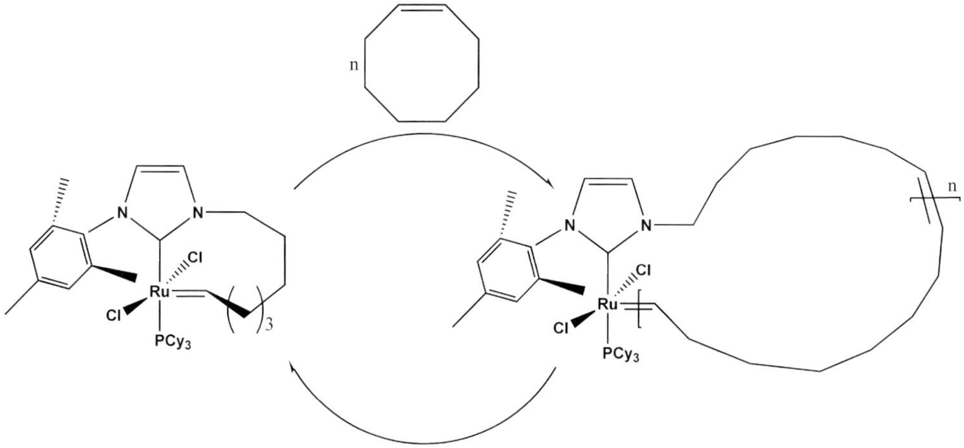

Scheme 7.36 Synthesis of cyclic polymers using ROMP.

As expected, the catalytic efficiency decreases on increasing the dendrimer generation due to steric hindrance.

Nomura's group [79] reported in 2005 the synthesis of various three-armed star-shaped amphiphilic copolymers (Scheme 7.35). Such a synthesis involved three steps. First, homopolynorbornene or diblock copolymers with one of the two blocks functionalized with a sugar moiety were prepared using a molybdenum alkylidene initiator. Through Wittig-type reactions these ROMP (co)polymers were end-functionalized with two trimethylsilyl groups. The latter were, in a second step, transformed into hydroxyl group, which then served to attach PEO chains. Monomodal and narrow molar mass distributions were observed ($M_w/M_n = 1.11–1.12$), attesting to the efficiency of this "grafting to" approach.

Grubbs' group [50] developed in 2002 a new synthetic route to obtain cyclic polymers through the ROMP of cyclooctene initiated by cyclic ruthenium-based catalysts (Scheme 7.36). After hydrogenation, cyclic polyethylenes were eventually isolated. In this approach, there is no need to work under high dilution, which is generally the case with other synthetic methods for macrocycles.

7.4
ROMP in Dispersed Medium

Polymerization under heterogeneous dispersed conditions offers many advantages, such as easier heat removal, better viscosity control and easier recovery of the polymers formed. Moreover, today's environmental concerns can be better addressed through the use of environmentally friendly solvents, ideally water. As ROMP requires transition metal catalysts, most studies have so far been carried out in bulk or in organic, generally apolar, media under anhydrous homogeneous conditions. By comparison, fewer than 20 papers in the last 40 years have been devoted to ROMP performed in polar dispersed media. The discovery by Grubbs of ruthenium-based carbene complexes that can retain high activity even in the presence of water has opened unparalleled perspectives for the implementation of ROMP in dispersed media.

7.4.1
Emulsion ROMP

> In an emulsion polymerization, the water-insoluble monomer is dispersed in the continuous aqueous phase by addition of a surfactant. The majority of the monomer (>95%) is dispersed in the form of droplets (1–10 µm). Introduced in excess, surfactant molecules stabilize the monomer droplets (10^{10}–10^{11} L^{-1}) and contribute to the formation of a large number of micelles (10^{17}–10^{18} L^{-1}). Polymerization starts upon introduction of the water-soluble initiator in the medium. The latter reacts with the few monomer molecules solubilized in water and the oligoinitiator then formed diffuses into the micelles where the polymerization proceeds.

The first reports or ROMP in emulsion date back to 1965 with the work of Rinehart and Smith [80, 81] and Michelotti and Keaveney [82]. They both employed ill-defined catalyst systems based on hydrosoluble metal salts. Activated by a reducing agent, ROMPs of norbornene and other norbornenyl derivatives were carried out in the presence of an ionic surfactant [sodium dodecyl sulfate (SDS)]. Although these early studies demonstrated that strained cycloolefins could well be polymerized in water under heterogeneous conditions, the yields obtained were too low for industrial development. Indeed, the best catalytic systems allowed only 5.4 g of polymer to be produced from 1 g of catalyst and the stability of the emulsion formed could not be clearly proved.

Through better catalyst design, significant improvements could by made, with a report by Wache [83] of a quantitative emulsion ROMP of norbornene in the presence of two novel water-soluble catalysts based on Ru(IV) (Scheme 7.37).

SDS was used to stabilize the particles formed but no further colloidal data were provided. A high content in cis double bonds (>85%) was reported, whereas ruthenium-based catalysts commonly give high yields of trans products.

In the same period, well-defined ruthenium alkylidene complexes tolerant of protic functions and yet highly active were synthesized by Grubbs' group, as mentioned previously. This permitted the ROMP of hydrophobic and hydrophilic norbornene derivatives (**A** and **B**, Scheme 7.38) to be carried out in the presence of dodecyltrimethylammonium bromide (DTAB) [38], an ionic surfactant. No colloidal study was reported and the surfactant was added to maintain the monomer dispersed in the water. The initiators being insoluble in water and

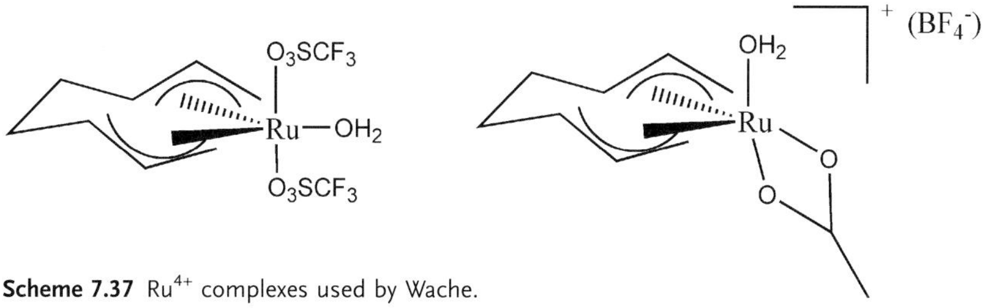

Scheme 7.37 Ru^{4+} complexes used by Wache.

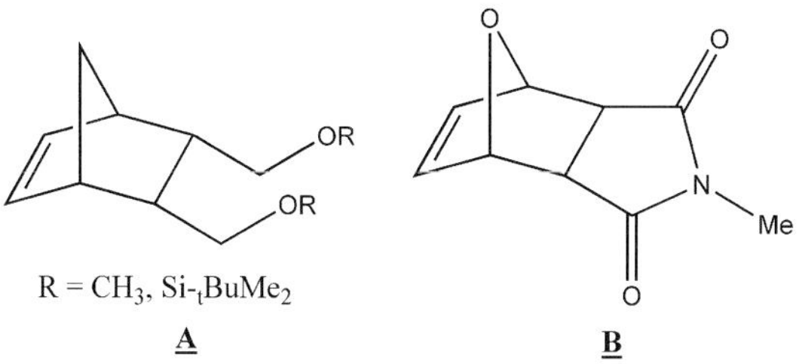

Scheme 7.38 Hydrophobic and hydrophilic norbornene derivatives.

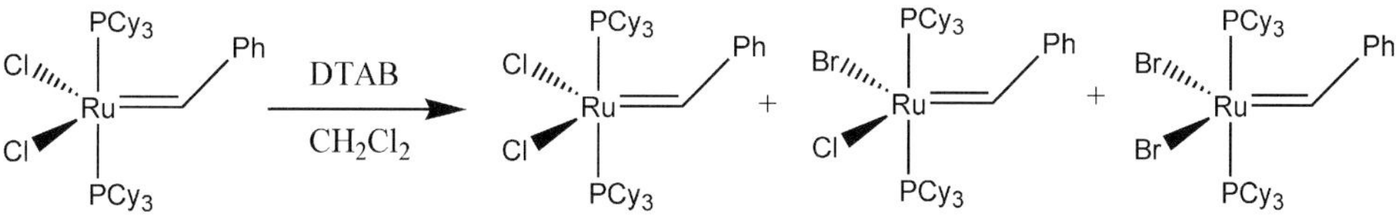

Scheme 7.39 Anionic ligand-exchange reaction.

the monomer **B** hydrophilic, it was inappropriate to term the reactions "emulsion polymerization", but through this contribution, Grubbs and colleagues described the first example of a truly "living" ROMP in water.

In 2000, two water-soluble ruthenium carbenes bearing cationically functionalized ligands were synthesized (**1** and **2**, Scheme 7.15), allowing Grubbs' team to work in rigorous emulsion conditions [84]. Dodecyltrimethylammonium chloride (DTAC), a chloride analogous to DTAB, was used as a stabilizer to prevent an exchange of the halide ligand that could have affected the reactivity of the carbene complexes (Scheme 7.39). The initiation step was fast and the monomer conversion complete. No colloidal study was reported, however, but the high surfactant concentration used to stabilize the particles indicated a microemulsion-type process, in which all the monomer was solubilised into micelles.

One year later, Claverie et al. [42] reported a well-characterized colloidal study of norbornene ROMP in emulsion carried out in the presence of two water-soluble ruthenium carbenes, **1** and **3** (Scheme 7.15). Synthesis of **1** was cumbersome whereas **2** could be easily prepared *in situ* by reaction of ethyl diazoacetate with bis[tris(3-sulfonatophenyl)phosphine sodium salt]ruthenium dichloride. With these two complexes, fast polymerization rates and quantitative yields were observed. Stabilized by an ionic surfactant (SDS or Dowfax 3B2), the polynorbornene particles formed exhibited a monomodal size distribution with diameters ranging from 50 to 100 nm. Attempts to polymerize less reactive cycloolefins such as 1,5-cyclooctadiene and cyclooctene with these complexes resulted in low yields (<10%).

7.4.2
Dispersion ROMP

> *A dispersion polymerization starts as a homogeneous solution containing the monomer, the initiator and the stabilizer, which eventually turned into a colloidal dispersion upon formation of the polymer.*

The first example of dispersion ROMP was reported in 1993 by Booth's group [85], who polymerized *exo,exo*-2,3-bis(methoxymethyl)-7-oxanorbornene in water in the presence of a steric stabilizer [poly(ethylene oxide)-*b*-poly(propylene oxide)-*b*-poly(ethylene oxide)]. Initiated by $RuCl_3 \cdot xH_2O$, the polymerization was not controlled and the catalyst efficiency was low. Particles with diameters ranging from 40 to 330 nm and stable for more than 5 months were obtained. Since relatively high concentrations of surfactant (9–27 g L^{-1}) were used, polymerization could have occurred in monomer-swollen micelles.

On the other hand, Héroguez and coworkers showed that norbornenyl-PEO macromonomer can efficiently stabilize latex particles synthesized by ROMP in dispersed medium. Unlike classical ionic and steric stabilizers, these reactive steric stabilizers have the advantage of being covalently bound to the particle surface, preventing their desorption under stress. Macromonomers thus appeared as efficient stabilizers that improved significantly the colloidal stability of latex particles. In a first contribution, latex particles of PNB and polybutadiene (PB) were synthesized by dispersion ROMP in a dichloromethane–ethanol mixture using a norbornenyl macromonomer of PEO of 4700 g mol^{-1} as steric stabilizer and a ruthenium-based complex as initiator [86, 87] (Scheme 7.40).

The particles generated actually comprise PNB-*g*-PEO and PB-*g*-PEO graft copolymers maintained together through hydrophobic interactions and stabilized by the presence of PEO grafts on their surface. The particle size was found to range from 150 nm to 15 µm depending on the experimental conditions. Such polynorbornene-based particles loaded with active molecules through a pH-sensitive link were shown to release the latter in response to exposure [88]. Cyclooctadiene [89, 90] was also subjected to dispersion ROMP and this study yielded totally different results as compared with the case of NB with respect to both the kinetics of polymerization and the characteristics of the particles formed. In the latter case, part of the PEO chains were trapped in the particle core, reducing the efficiency of its stabilization. The PB particles formed were efficiently hydrogenated using

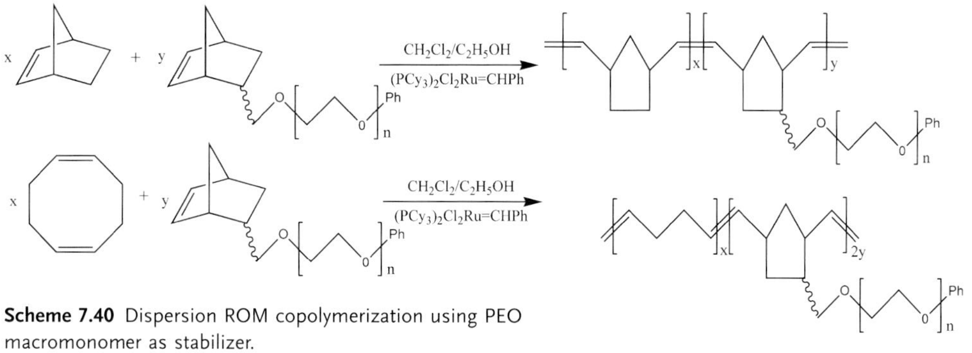

Scheme 7.40 Dispersion ROM copolymerization using PEO macromonomer as stabilizer.

RuCl$_2$(PPh$_3$)$_3$ and fully linear polyethylene-based latexes were obtained with no detrimental effect on particle size and colloidal stability [91].

7.4.3
Suspension ROMP

> *In a typical suspension polymerization process, monomer droplets (50–500 μm) are finely dispersed in an aqueous phase under mechanical stirring. A suspending agent is commonly added to water in order to prevent colloidal degradation. Polymerizations initiated by an oil-soluble initiator occur exclusively within the monomer droplets.*

In 2004, COD was polymerized for the first time by aqueous suspension ROMP by Héroguez and coworkers [92] using RuCl$_2$(PPh$_3$)$_2$(CHPh) complex as initiator. The traditional suspending agent [poly(diallyldimethylammonium chloride)] was ineffective at stabilizing the PB beads formed, which slowly aggregated after discontinuing the stirring. In order to prevent this flocculation, various co-stabilizers based on PEO were tried. To be effective, the co-stabilizer has to be hydrosoluble but also possess a certain hydrophobicity so as not to be completely soluble in the continuous phase and migrate to the particle surface. A stable colloidal suspension of PB particles was obtained in the presence of PB-g-PEO or norbornenyl PS-b-PEO as co-stabilizers: complete consumption of COD was observed in less than 3 h with sizes typically between 20 and 1000 μm. In the same year, Janda's group [93] synthesized cross-linked polynorbornene particles from norbornene, norborn-2-ene-5-methanol and several cross-linkers (Scheme 7.41). Acacia gum was used as suspension agent and first- and second-generation Grubbs catalysts (Scheme 7.42) as initiators. With the first-generation Grubbs catalyst the final particles turned soluble in organic media due to a moderate extent of cross-linking. With the second-generation Grubbs' catalysts a higher extent of cross-linking could be achieved and stable PNB particles isolated with a size of about 50 nm. Through subsequent chemical modification, a

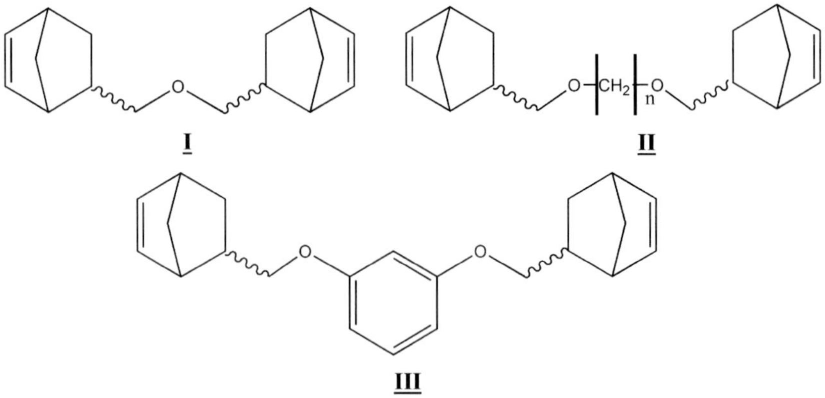

Scheme 7.41 Different cross-linkers used by Janda et al.

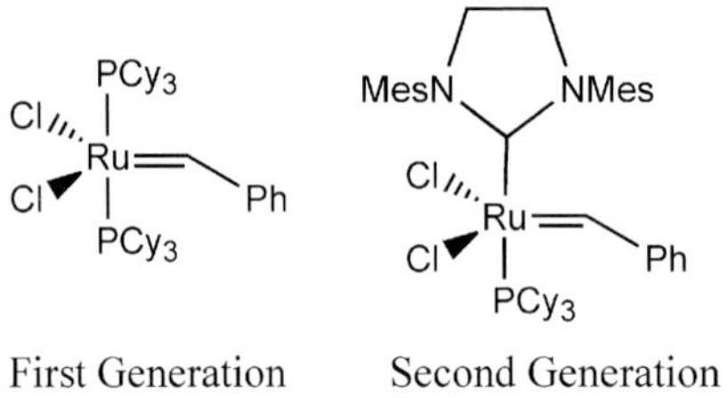

First Generation   Second Generation

Scheme 7.42 Grubbs catalysts.

large variety of functionalized particles could be derivatized for applications in supported chemistry.

7.4.4
Miniemulsion ROMP

In a miniemulsion polymerization process, the monomer is dispersed in a continuous phase – commonly water – through a highly efficient homogenization process such as ultrasonication. A surfactant is dissolved in water to stabilize droplets against coalescence and a lipophilic agent is added to the monomer phase to avoid Ostwald ripening (degradation by molecular diffusion). Polymerization – initiated by either a water-soluble or an oil-soluble compound – occurs exclusively in the monomer droplets. The final polymer particles are thus a perfect copy of the monomer droplets in size and size distribution.

In order to polymerize little-strained cycloolefins such as COD and cyclooctene, Claverie et al. [42] resorted to the miniemulsion process. The method consisted

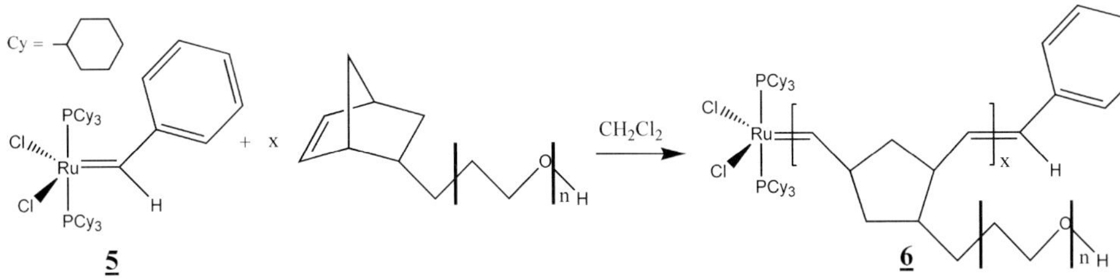

Scheme 7.43 New hydrosoluble PEO-based ruthenium carbenes.

in dispersing the initiator in toluene in the form of miniemulsion droplets. The latter were then contacted with a mixture of monomer and water. Polymerization occurred in the initiator-containing droplets either by diffusion of monomer through the aqueous phase or by collision between the monomer droplets. The size and size distribution of the final particles were different from those of the monomer droplets. However, latexes in the size range 250–650 nm and containing high molar mass polymers were obtained.

In a recent contribution [91, 94], Héroguez and coworkers examined in detail the miniemulsion ROMP of NB in water using both hydrophobic and water-soluble ruthenium-based catalysts. They clearly showed that $RuCl_2(PCy_3)_2(CHPh)$, a well-defined hydrophobic ruthenium carbene, was not appropriate for ROMP of NB in miniemulsion, its use resulting in a large amount of coagulum. In contrast, a more hydrophilic catalyst such as ruthenium trichloride hydrate was found more suitable and stable PNB nanoparticles could be obtained in the size range 200–500 nm depending on the surfactant used (ionic or steric). In addition, the same group demonstrated that the structure of the catalyst influences drastically the rate of polymerization. Based on these observations, hydrosoluble PEO-based ruthenium carbenes (**6**) were employed to trigger the "living" miniemulsion ROMP of norbornene [44] (Scheme 7.43). The precise tuning of the PEO content in the catalyst was essential to obtain stable particles without coagulum.

In a recent contribution to this field, Carrillo and Kane [95] reported the synthesis of nanoparticles of controlled size based on the self-assembly of amphiphilic block copolymers synthesized by the ROMP of a hydrophobic derivative of norbornene with a hydrophilic one. The diameter of the nanoparticles could be varied over a wide range through changes in the composition of the block copolymer. Such particles may serve as novel scaffolds for the multivalent binding of ligands, e.g. sugars, peptides and proteins [96].

7.5
Advanced Materials by ROMP

7.5.1
Liquid Crystalline Polymers

"Living" ROMP was found to be of practical utility in the synthesis of side-chain liquid crystalline polymers (SC-LCPs), which enjoy several applications including high-strength materials (textile fibers) and optical devices. A variety of these SC-LCPs have been synthesized from various cycloolefin monomers carrying various mesogenic groups through different spacers [97, 98].

As well-defined structures with broad functionalities are required in SC-LCPs, most of the ROMPs were initiated with well-defined molybdenum [99–106] or ruthenium [107–109] catalysts. The recent work of the groups of Albouy [110] and Stelzer [98, 111] illustrate the advances made in the domain of SC-LCPS. In particular, Albouy's team described the synthesis of a diblock copolymer A–B consisting of a liquid crystalline block, B, synthesized by ROMP, and an isotropic block, A, obtained by ATRP. The B block was obtained by ROMP of the *exo*-norbornene derivative initiated by a ruthenium initiator; it was terminated by reaction with 4-(2-bromopropionyloxy)-but-2-enyl 2-bromopropionate, which was further used to grow a second polyacrylate block by ATRP of *n*-butyl acrylate (Scheme 7.44). Such a diblock copolymer exhibits microphase separation and a lamellae-type morphology with the SC-LCP block undergoing a nematic–isotropic transition. These hybrid liquid crystalline isotropic materials were studied with a view to developing models for artificial muscles. Depending on the phase of the SC-LCP block – nematic or isotropic – the polymer backbone is elongated along the axis of the nematic director or takes a random conformation. As a re-

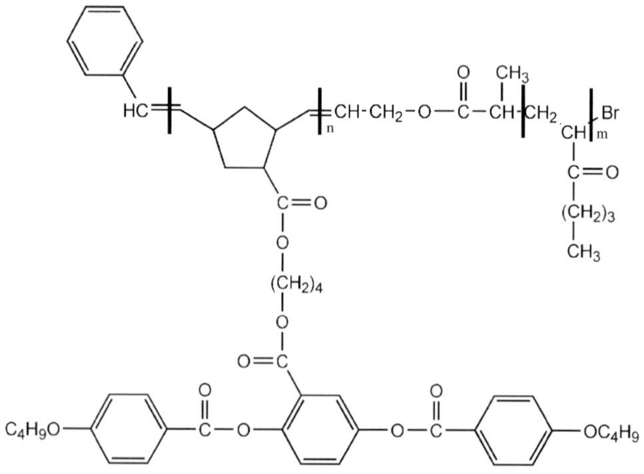

Scheme 7.44 SC-LCP block copolymer obtained by Albouy et al.

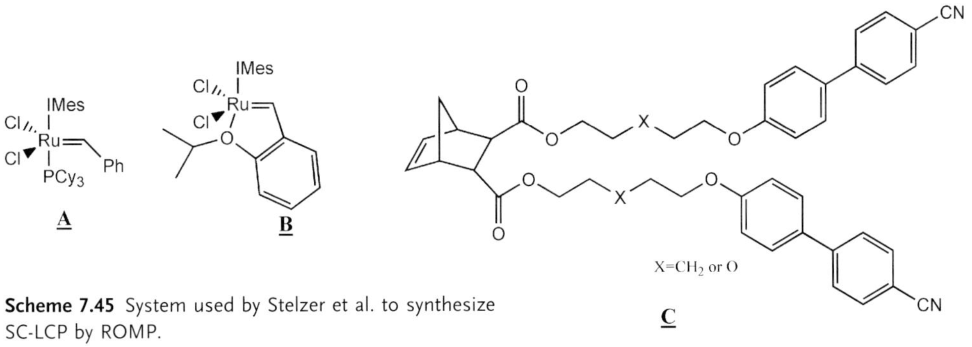

Scheme 7.45 System used by Stelzer et al. to synthesize SC-LCP by ROMP.

sult, reversible contraction and expansion of the elastomer were observed when the LC block underwent the nematic–isotropic phase transition.

The SC-LCP described above contains just one pendant mesogenic group per norbornene unit. Stelzer's group reported the ROMP of norbornene derivatives containing two lateral pendant mesogenic groups, based on {[(4'-cyano-4-biphenyl)yl]oxy}alkyl, per unit (**C**, Scheme 7.45). Despite the incompatibility of the nitrile groups with the ruthenium initiators, it was demonstrated that those containing N-heterocyclic carbene ligands (**A**, **B**) can be used to polymerize cyano-functionalized monomers, but their ROMPs were not "living", affording broadly distributed polymers with polydispersity indexes ranging from 1.6 to 2.4. With the Schrock molybdenum catalyst, the polydispersity decreased to 1.2. However, ruthenium initiators are easier to utilize for a large variety of comonomers.

7.5.2
Conjugated and Electroactive Polymers

Conjugated polymers are characterized by their quasi-infinite π-system that extends over a large number of recurring monomer units. This provides such polymers with directional conductivity. Upon doping, their conductivity can rise by several orders of magnitude and is limited only by the defects in the polymers. When undoped, such polymers behave as semiconductors that can luminesce when exposed to light and more generally function as energy conversion devices. They have therefore found a variety of applications such as light-emitting diodes, transistors, photodetectors and chemical sensing devices with many more exciting possibilities.

The archetype and simplest form of conjugated polymer is polyacetylene. Before Shirakawa showed how to produce thin films of polyacetylene by Ziegler–Natta polymerization, polyacetylene was obtained as an intractable black powder. Through the so-called "Durham route", ROMP offered an alternative elegant method of synthesis of polyacetylene whose chain end structures, molar mass and molar mass distribution could be controlled [112–114]. The "Durham route" actually involved the ROMP of a tricyclic monomer and the formation of a

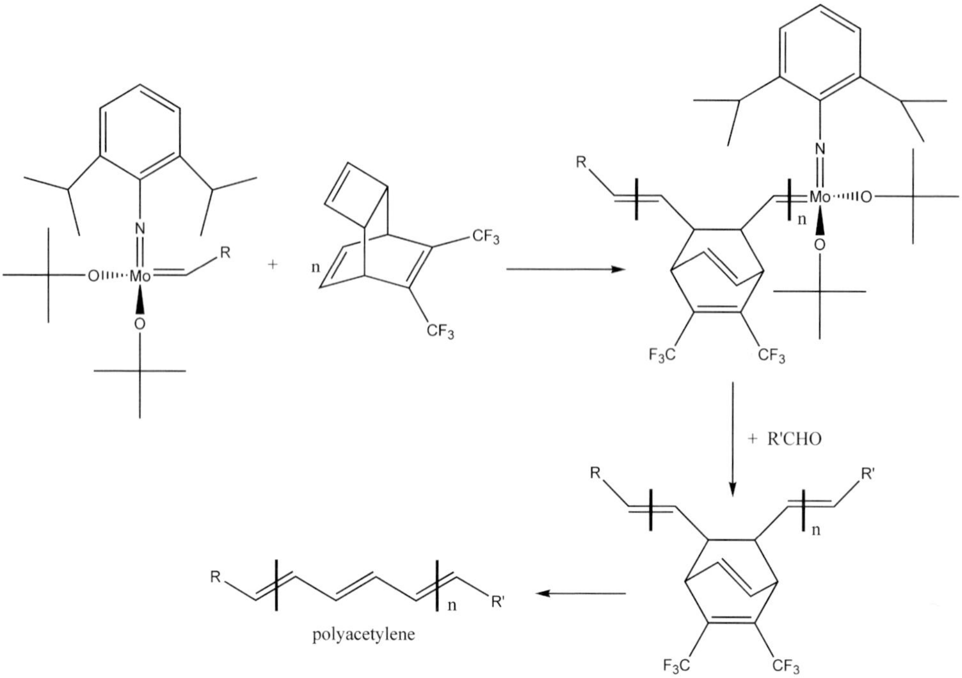

Scheme 7.46 Polyene through the "Durham route".

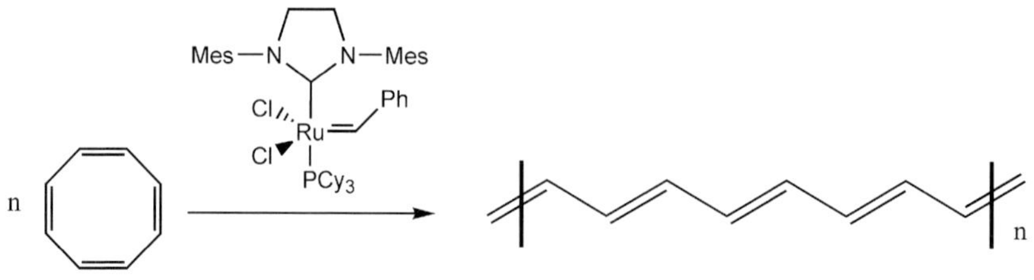

Scheme 7.47 Polyene through the ROMP of COT.

soluble precursor that can subsequently be transformed into polyacetylene (Scheme 7.46).

A second pathway involving the ROMP of 1,3,5,7-cyclooctatetraene (COT) was also proposed as a means to generate polyacetylene. In this case, early transition metal catalysts based on tungsten [115] but also highly active $RuCl_2(CHPh)(PCy_3)(IMesH_2)$ [116, 117] were used to produced films of polyacetylene and soluble telechelic polymers with notable conductivity (Scheme 7.47).

In the domain of electroactive polymers, Schrock's group [118] described the synthesis of polymers and block polymers containing redox-active centers through ROMP of the corresponding monomers. ROMP polymers carrying phenothiazene and ferrocene were synthesized (Scheme 7.48) and the electrochemi-

Scheme 7.48 Phenothiazene and ferrocene ROMP polymers obtained by Schrock et al.

cal independence of their redox centers was demonstrated by solution voltammetry.

7.5.3
Monolithic Supports

Monolithic supports have found many applications in liquid chromatography as a fast and highly efficient separation device or as a support for catalytically active systems [119–121]. They are cross-linked materials with high and defined interconnected porosity in which either interactions (separation) or reactions (catalysis) occur between the solid phase and its surrounding liquid.

Jinner and Buchmeiser [122] developed a ROMP-based methodology that afforded monoliths through the copolymerization of NB with a cross-linker in the presence of porogenic solvents within a column (Scheme 7.49).

Narrowly distributed microglobules were obtained with diameters ranging from 2 to 30 μm, the interglobular void volume and total porosity varying within the range 0–50 and 50–80%, respectively. Because of the "living" character of ROMP reactions, the active ruthenium sites located essentially at the microglobule surface could be used to graft on to the surface of the monolith func-

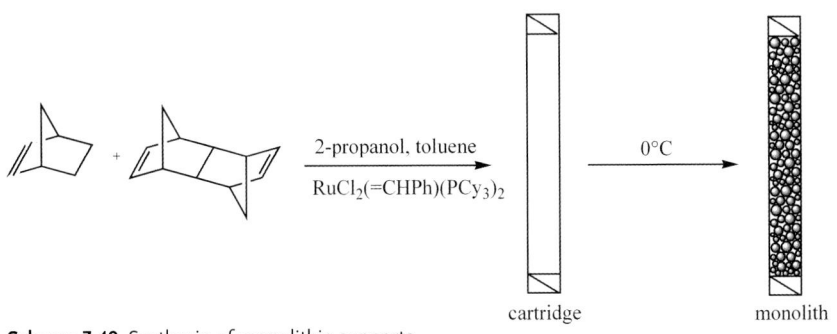

Scheme 7.49 Synthesis of monolithic supports.

tional monomers, which afforded various functionalized monoliths [123, 124]. The utility of these monoliths was demonstrated in different fields. In chromatography they were used to isolate biological compounds such as oligonucleotides, proteins [125] and DNA fragments [126] or polymer chains [127]. These supports also served as carriers of Grubbs-type initiators to catalyze metathesis-based reactions including ring-closing metathesis (RCM), ring-opening cross-metathesis and enyne metathesis [128].

7.5.4
Supported Catalysts

In addition to functionalized monoliths, polymeric beads have also been synthesized [129, 130] and modified using ROMP catalysts and employed as supports [131].

Polymeric multivalent chiral catalysts were synthesized by Brozio et al. [155] by ROMP of oxanorbornene derivatives carrying catalytically active chiral groups (Scheme 7.50). Initiated with $RuCl_2(PCy_3)_2(CHPh)$, ROMPs afforded well-defined polymeric catalysts with polydispersity ranging from 1.1 to 1.4.

Buchmeiser and Wurst [132] supported palladium(II) on a ligand-carrying polymer scaffold obtained by ROM copolymerization of a norbornene derivative with a cross-linker, using a molybdenum catalyst. This supported catalyst was found to be active in the vinylation of aryl iodides and aryl bromides (Heck-type coupling), in the arylation of alkynes and in the amination of aryl bromide (Scheme 7.51).

A new method of "chemical tagging" was reported by Hanson's group [133] in 2002 for homogeneous scavenging. It utilizes 5-norbornene-2-methanol to scavenge a variety of electrophiles that are present in excess. Once tagging is com-

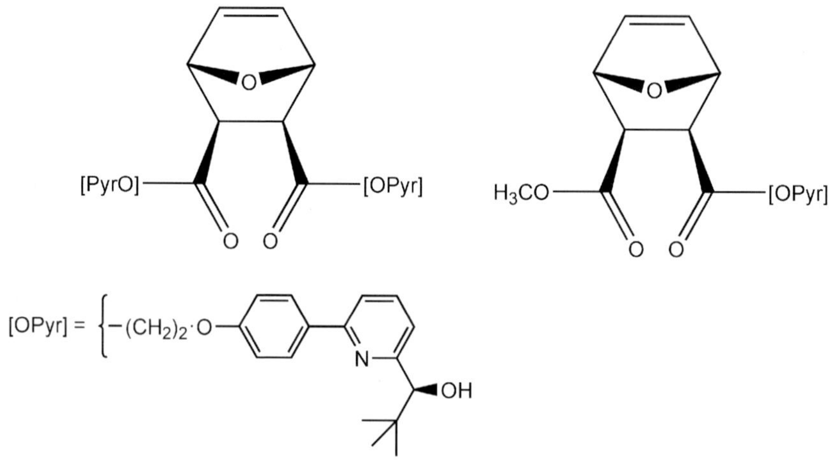

Scheme 7.50 Catalytically active ROMP monomers.

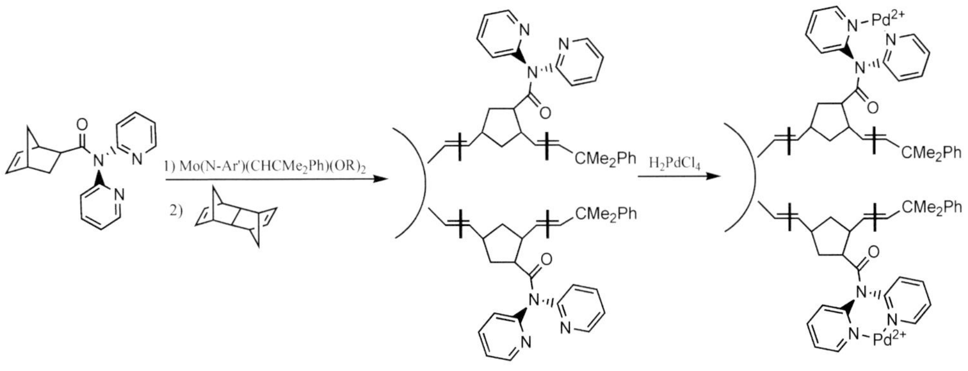

Scheme 7.51 Preparation of palladium-based supported catalyst [Ar′ = 2,6-iPr₂-C₆H₃, R=C(CH₃)(CF₃)₂].

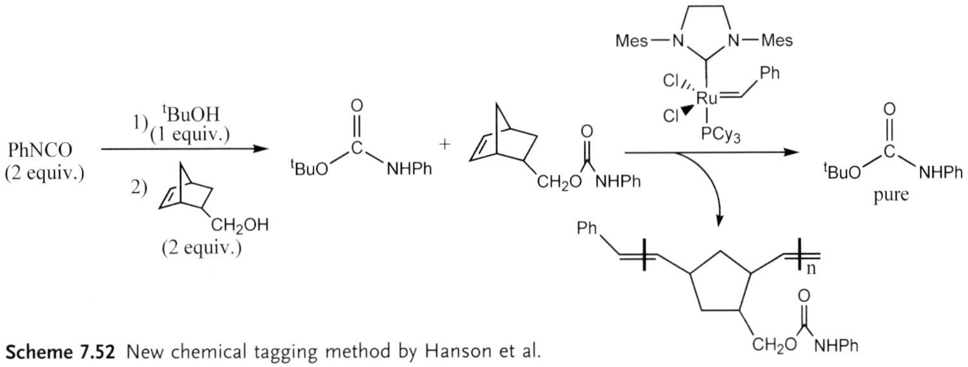

Scheme 7.52 New chemical tagging method by Hanson et al.

plete, the crude reaction mixture is subjected to a rapid and complete ROMP, initiated with a second-generation Grubbs catalyst. This process yields a polymer that precipitates in methanol, affording products in excellent yields and purity (Scheme 7.52).

Polymers with end-functionalized pyridine ligands were synthesized by Nomura et al. [134] to be subsequently used as catalysts for the hydrogenation of PhCH₂CHO in the presence of Ru(acetylacetonato)₃. Such a polymer was prepared by the ROMP of norbornene initiated by Mo(CHCMe₂Ph)(N-2,6-iPr₂C₆H₃)(O-tBu) and deactivated with 4-pyridinecarboxaldehyde (Scheme 7.53). Adding this ω-pyridine polymer increased notably the activity of the hydrogenation of PhCH₂CHO, from 41 mmol of PhCH₂CH₂OH produced/Ru (mmol) without additives to a value of 153. By pouring the reaction mixture into cold methanol, more than 98% of the polymeric ligand was recovered.

An extension of this work described the incorporation of new pyridine moieties for selective hydrogen transfer reduction of ketones [135, 136].

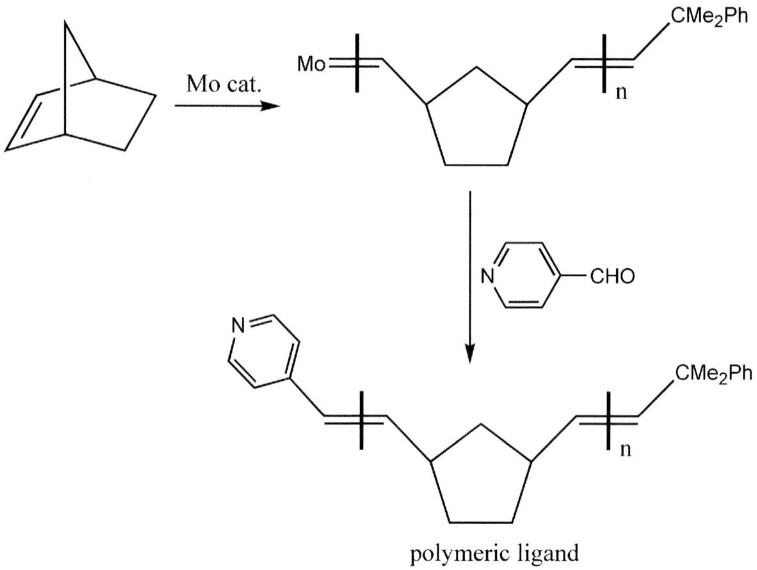

Scheme 7.53 Synthesis of a polymeric ligand for hydrogenation catalysis.

Many other studies have reported the use of polymers bearing ligands for organic synthesis [137–139] and that were prepared by ROMP.

7.5.5
Biological Materials

ROMP has been used in several instances for the preparation of biologically active polymeric materials [22].

Polyvalent carbohydrate-bearing ROMP polymers were prepared by Kiessling and coworkers [140–142]. The 7-oxanorbornene derivative shown in Scheme 7.54 afforded a polymer with protein-binding activity. However, since well-defined ruthenium carbenes – that function even in the presence of polar groups – were discovered only after 1996, ruthenium trichloride was thus employed and ill-defined polymers were produced.

Later publications reported the use of $RuCl_2(CHPh)(PCy_3)_2$, a well-defined Grubbs catalyst, and the formation of polymers with narrow polydispersity. Neoglycopolymers [143, 144] were prepared with high affinity and selective inhibition of P-selectin, a protein that facilitates leukocyte trafficking to the sites of inflammation.

Polyvalent ROMP polymers exhibiting enhanced antibacterial activity against vancomycin-resistant enterococci were also reported by Uemura's group [145] in 1999. They were obtained by ROMP of vancomycin-carrying norbornene ini-

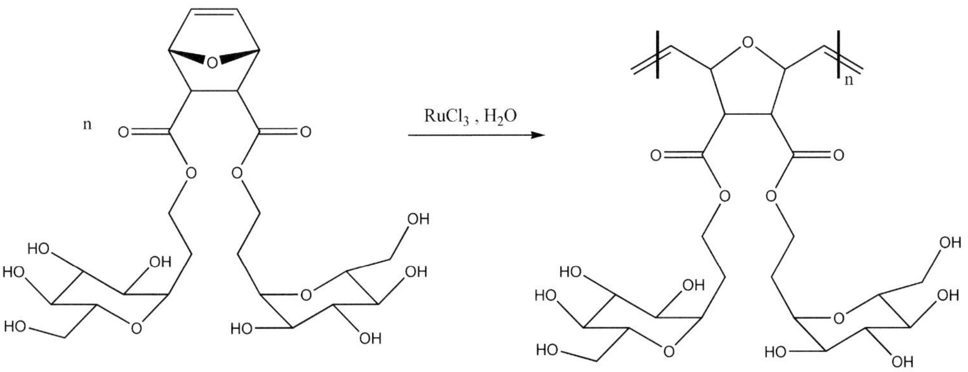

Scheme 7.54 Synthesis of polyvalent ROMP polymer with protein-binding activity.

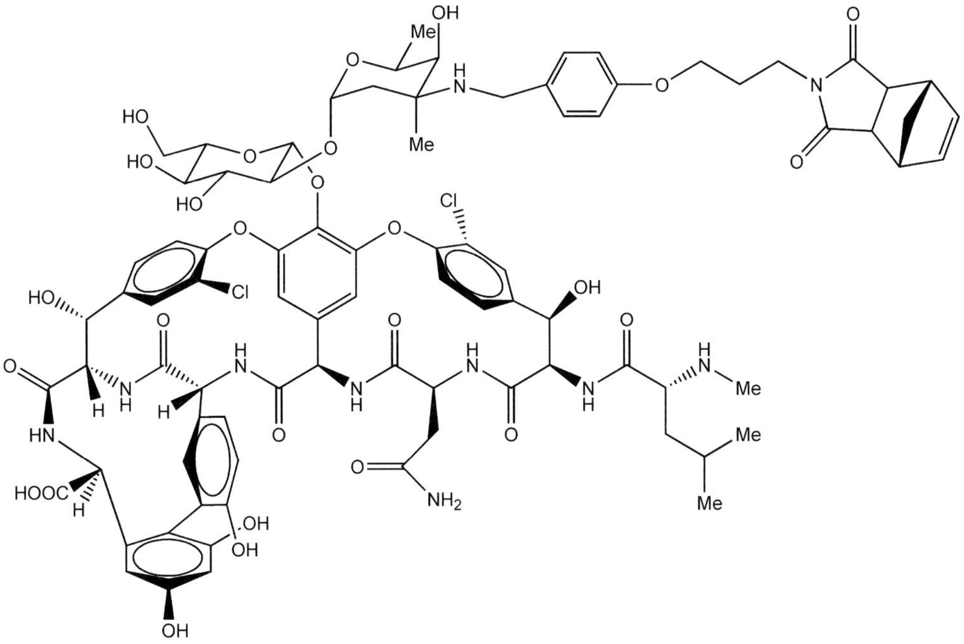

Scheme 7.55 Multivalent ROMP monomer of vancomycin.

tiated with a Grubbs catalyst. Low polymer yields were obtained (4–60%) owing to the low reactivity of the sterically hindered monomer (Scheme 7.55).

Grubbs' group [146] reported in 2000 the synthesis of norbornenyl polymers with pendant glycines, alanines, penta(ethylene oxide)s (EO_5) and cell adhesive sequences glycine–arginine–glycine–aspartic acid (GRGD) and serine–arginine–asparagine (SRN). Copolymer with pendant EO_5 and GRGD groups were

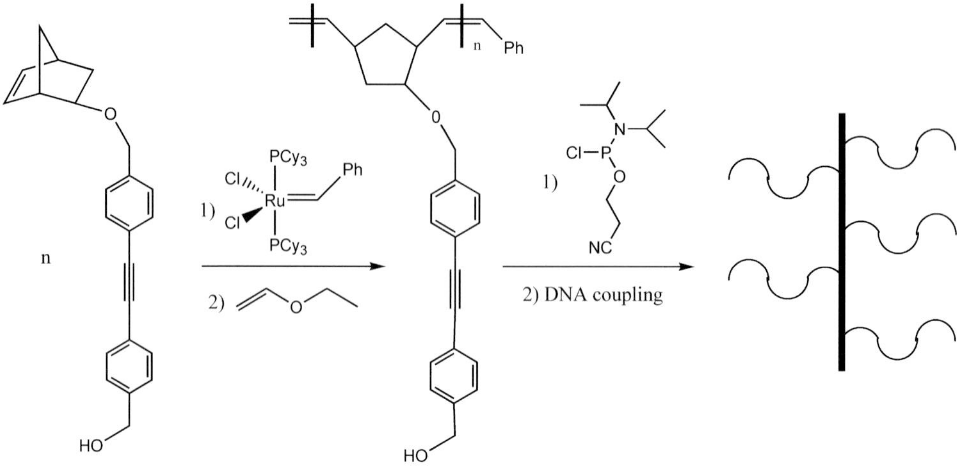

Scheme 7.56 DNA–block copolymer conjugate.

synthesized using RuCl$_2$(CHPh)(PCy$_3$)$_2$ but with extremely low yields and a bimodal molar mass distribution. This problem could be circumvented by using the more active 3,3-dihydroimidazolylidene initiators RuCl$_2$(=CHPh)(PCy$_3$) (DHIMes) and RuCl$_2$(=CH–CH=C(CH$_3$)$_2$(PCp$_3$) (DHIMes). With the latter, a large variety of copolymers were successfully synthesized. Potential applications of these polymers in tumor therapy and in the binding of integrins to proteins were mentioned.

The covalent attachment of DNA to the backbone of a ROMP polymer was reported by Watson et al. in 2001 [147]. The authors aimed at the synthesis of "smart" materials that exhibit the unique property of recognition of DNA borne by a functionalized polymer backbone. The corresponding norbornene derivatives were subjected to ROMP using the Grubbs catalyst RuCl$_2$(CHPh)(PCy$_3$)$_2$ and well-defined polymers were obtained. Subsequent modification of the latter polymers with chlorophosphoramidite followed by coupling with DNA resulted in the desired hybrid product (Scheme 7.56). In an extension of this study, the authors mentioned the synthesis of well-defined electrochemical probes for the detection of DNA [148].

Well-defined amphiphilic ROMP polymers exhibiting lipid membrane disruption activities were designed by Coughlin's group [149] in 2004. Biological applications such as antibacterial activity and gene delivery required the penetration of the cell membranes through favorable interactions followed by aggregation and subsequent disruption. The authors reported the synthesis of a large variety of copolymers through the ROMP of modular monomers using molybdenum- or ruthenium-based initiators (Scheme 7.57). Amphiphilic cationic polymers above certain molar masses were reported to exhibit high membrane disruption activities on lipid vesicles used as rough models of bacterial membranes. An ex-

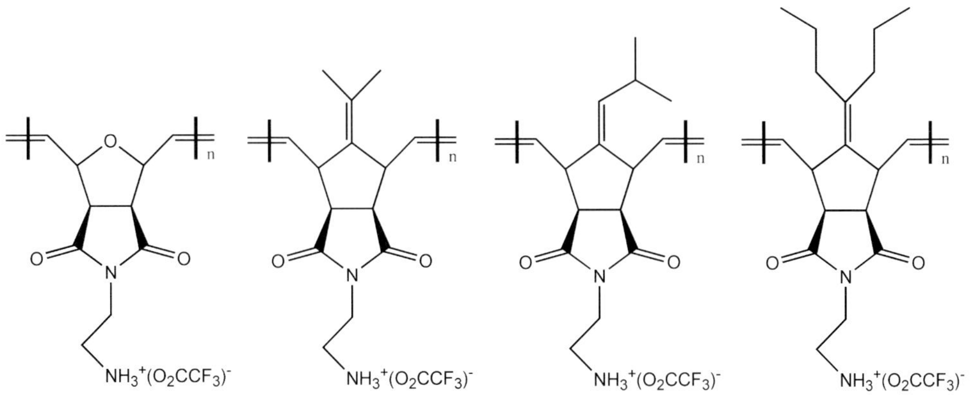

Scheme 7.57 Amphiphilic polynorbornene derivatives.

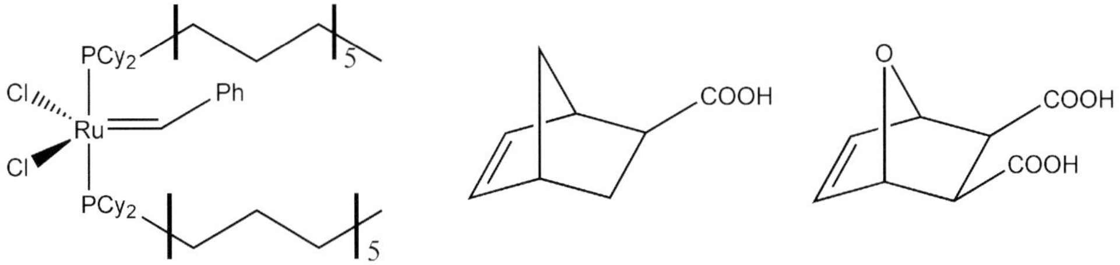

Scheme 7.58 Modified Grubbs catalyst and norbornene derivatives.

tension of this work demonstrated that amphiphilic homopolymers and random copolymers demonstrate both antibacterial and non-hemolytic activities [150].

Sykes' group [151] successfully employed ROMP to synthesize polymeric nodules on the external part of liposomes. The main objective of this work was to mimic the mobility of "living" organisms or objects that move inside cells by polymerizing actin at their surface. The authors investigated the ROMP of norbornene derivatives dissolved in the outer part of liposomes inside which they incorporated – precisely in the bilayer part – a specially designed hydrophobic Grubbs initiator (Scheme 7.58). The growth of polymer nodules from the liposome surfaces occurred, the shapes of the nodules formed being dependent on the monomer hydrophilicity.

Héroguez's group [152] described the synthesis of pH-sensitive biomaterials obtained through the functionalization of polynorbornene nanoparticles by an active molecule (indomethacin). Such particles were obtained by ROMP in dispersion of norbornene with a norbornenyl macromonomer of PEO carrying the drug [153]. Cleavage of the bond linking the active molecule to the particle under low pH induced a controlled release of the bioactive molecule in its native form.

7.5.6
Hybrid Materials/Particles

Using classical "grafting from" and "grafting to" techniques, Buchmeiser [154] modified gold-, silicon- and silica-based surfaces and also other inorganic surfaces by growing/anchoring ROMP-generated polymeric chains.

7.6
Conclusion

In recent years, ROMP has become a mature and valuable tool in macromolecular engineering, standing at the forefront of polymer chemistry. Complex macromolecular architectures, nanostructured particles/devices and well-defined macroscopic materials are accessible by ROMP with the advent of selective catalysts tolerant of polar functions.

References

1 Truett, W. L., Johnson, D. R., Robinson, I. M., Montague, B. A. *J. Am. Chem. Soc.* **1960**, *82*, 2337.
2 Chauvin, Y., Institut Français du Pétrole, Rueil-Malmaison, France.
3 Schrock, R. R., Massachusetts Institute of Technology (MIT), Cambridge, MA, USA.
4 Grubbs, R. H., California Institute of Technology (Caltech), Pasadena, CA, USA.
5 Anderson, A. W., Merckling, N. G. *US Patent 2721189*, **1955**.
6 Eleuterio, H. S. *US Patent 3074918*, **1963**.
7 Peters, E. F., Evering, B. L. Catalysts and their preparation. *US Patent 2963447*, **1960**.
8 Calderon, N., Chen, H. Y., Scott, K. W. *Tetrahedron Lett.* **1967**, *34*, 3327.
9 Calderon, N., Ofstead, E. A., Ward, J. P., Judy, W. A., Scott, K. W. *J. Am. Chem. Soc.* **1968**, *90*, 4133.
10 Calderon, N. *Acc. Chem. Res.* **1972**, *5*, 127.
11 Mol, J. C., Moulijn, J. A., Boelhouwer, C. *J. Chem. Soc., Chem. Commun.* **1968**, 633.
12 Lewandos, G. S., Pettit, R. *J. Am. Chem. Soc.* **1971**, *93*, 7087.
13 Grubbs, R. H., Brunck, T. K. *J. Am. Chem. Soc.* **1972**, *94*, 2538.
14 Fischer, E. O., Maasböl, A. *Angew. Chem. Int. Ed. Engl.* **1964**, *3*, 580.
15 Hérisson, J. L., Chauvin, Y. *Makromol. Chem.* **1971**, *141*, 161.
16 Katz, T. J., McGinnis, J. *J. Am. Chem. Soc.* **1975**, *97*, 1592.
17 Grubbs, R. H., Burk, P. L., Carr, D. D. *J. Am. Chem. Soc.* **1975**, *97*, 3265.
18 Schrock, R. *J. Am. Chem. Soc.* **1974**, *96*, 6796.
19 Schrock, R., Rocklage, S., Wengrovius, J., Rupprecht, G., Fellmann, J. *J. Mol. Catal.* **1980**, *8*, 73.
20 Schrock, R. R. *J. Organomet. Chem.* **1986**, *300*, 249.
21 Nguyen, S. T., Johnson, L. K., Grubbs, R. H. *J. Am. Chem. Soc.* **1992**, *114*, 3974.
22 Grubbs, R. H., *Handbook of Metathesis*. Wiley-VCH, New York, **2003**.
23 Buchmeiser, M. R. *Chem. Rev.* **2000**, *100*, 1565.
24 Ivin, K. J., Mol, J. C., *Olefin Metathesis and Metathesis Polymerization*. Academic Press, London, **1997**.
25 Guillom, L. R., Grubbs, R. H. *Organometallics* **1986**, *5*, 721.
26 Guillom, L. R., Grubbs, R. H. *J. Am. Chem. Soc.* **1986**, *108*, 733.

27 Wallace, K.C., Schrock, R.R. *Macromolecules* **1987**, *20*, 448.
28 Wallace, K.C., Liu, A.H., Dewan, J.C., Schrock, R.R. *J. Am. Chem. Soc.* **1988**, *110*, 4964.
29 Schrock, R., Feldman, J., Cannizzo, L.F., Grubbs, R.H. *Macromolecules* **1987**, *20*, 1169.
30 Wu, Z., Wheeler, D.R., Grubbs, R.H. *J. Am. Chem. Soc.* **1992**, *114*, 146.
31 Schrock, R.R. *Tetrahedron* **1999**, *55*, 8141.
32 Murdzek, J.S., Schrock, R.R. *Macromolecules* **1987**, *20*, 2640.
33 Bazan, G.C., Schrock, R.R., Cho, H.N., Gibson, V.C. *Macromolecules* **1991**, *24*, 4495.
34 Bazan, G.C., Khosravi, E., Schrock, R.R., Feast, W.J., Gibson, V.C., O'Regan, M.B., Thomas, J.K., Davis, W.M. *J. Am. Chem. Soc.* **1990**, *112*, 8378.
35 Bazan, G.C., Oskam, J.H., Cho, H.-N., Park, L.Y., Schrock, R.R. *J. Am. Chem. Soc.* **1991**, *113*, 6899.
36 Schwab, P., Grubbs, R.H., Ziller, J.W. *J. Am. Chem. Soc.* **1996**, *118*, 100.
37 SonBinh, T., Nguyen, T., Grubbs, R. *J. Am. Chem. Soc.* **1993**, *115*, 9858.
38 Lynn, D., Kanaoka, S., Grubbs, R. *J. Am. Chem. Soc.* **1996**, *118*, 784.
39 Wu, Z., Nguyen, S.T., Grubbs, R.H., Ziller, J.W. *J. Am. Chem. Soc.* **1995**, *117*, 5503.
40 Frenzel, U., Weskamp, T., Kohl, F.J., Schattenmann, W.C., Nuyken, O., Hermann, W.A. *J. Organomet. Chem.* **1999**, *586*, 263.
41 Claverie, J.P., Soula, R. *Prog. Polym. Sci.* **2003**, *28*, 619.
42 Claverie, J., Viala, S., Maurel, V., Novat, C. *Macromolecules* **2001**, *34*, 382.
43 Gallivan, J.P., Jordan, J.P., Grubbs, R. *Tetrahedron Lett.* **2005**, *46*, 2577.
44 Quémener, D., Héroguez, V., Gnanou, Y. *J. Polym. Sci. Part A: Polym. Chem.* **2006**, *44*, 2784.
45 Cannizzo, L.F., Grubbs, R.H. *Macromolecules* **1988**, *21*, 1961.
46 Tritto, I., Sacchi, M.C., Grubbs, R.H. *J. Mol. Catal.* **1993**, *82*, 103.
47 Risse, W., Grubbs, R. *Macromolecules* **1989**, *22*, 1558.
48 Risse, W., Grubbs, R. *J. Mol. Catal.* **1991**, *65*, 211.
49 Coca, S., Paik, H.-J., Matyjaszewski, K. *Macromolecules* **1997**, *30*, 6513.
50 Bielawski, C.W., Morita, T., Grubbs, R.H. *Macromolecules* **2000**, *33*, 678.
51 Mahanthappa, M.K., Bates, F.S., Hillmyer, M.A. *Macromolecules* **2005**, *38*, 7890.
52 Demonceau, A., Simal, F., Delfosse, S., Noels, A.F., *Ring Opening Metathesis Polymerization and Related Chemistry*, Vol. 56. Kluwer, Dordrecht, **2002**.
53 Castle, T.C., Hutchings, L.R., Khosravi, E. *Macromolecules* **2004**, *37*, 2035.
54 IUPAC. *IUPAC Compendium of Chemical Terminology.* Blackwell, Oxford, UK **1997**.
55 Nomura, K., Takahashi, S., Imanishi, Y. *Macromolecules* **2001**, *34*, 4712.
56 Feast, W.J., Gibson, V.C., Johnson, A.F., Khosravi, E., Mohsin, M.A. *Polymer* **1994**, *35*, 3542.
57 Feast, W.J., Gibson, V.C., Johnson, A.F., Khosravi, E., Mohsin, M.A. *J. Mol. Catal. A: Chem.* **1997**, *115*, 37.
58 Rizmi, A.C.M., Khosravi, E., Feast, W.J., Mohsin, M.A., Johnson, F. *Polymer* **1998**, *39*, 6605.
59 Héroguez, V., Breunig, S., Gnanou, Y., Fontanille, M. *Macromolecules* **1996**, *29*, 4459.
60 Héroguez, V., Gnanou, Y., Fontanille, M. *Macromol. Rapid Commun.* **1996**, *17*, 137.
61 Héroguez, V., Six, J.-L., Gnanou, Y., Fontanille, M. *Macromol. Chem. Phys.* **1998**, *199*, 1405.
62 Héroguez, V., Amedro, E., Grande, D., Fontanille, M., Gnanou, Y. *Macromolecules* **2000**, *33*, 7241
63 Héroguez, V., Gnanou, Y., Fontanille, M. *Macromolecules* **1997**, *30*, 4791.
64 Desvergne, S., Héroguez, V., Gnanou, Y., Borsali, R. *Macromolecules* **2005**, *38*, 2400.
65 Lesné, T., Héroguez, V., Gnanou, Y., Duplessix, R. *Colloid Polym. Sci.* **2001**, *279*, 190.
66 Allcock, H.R., Denus, C.R.D., Prange, R., Laredo, W.R. *Macromolecules* **2001**, *34*, 2757.
67 Simal, F., Demonceau, A., Noels, A.F. *Tetrahedron Lett.* **1999**, *40*, 5689.

68 Simal, F., Demonceau, A., Noels, A. F. *Angew. Chem. Int. Ed.* **1999**, *38*, 538.
69 Charvet, R., Novak, B. M. *Macromolecules* **2004**, *37*, 8808.
70 Kriegel, R., Rees, W., Weck, M. *Macromolecules* **2004**, *37*, 6644.
71 Mecerreyes, D., Dahan, D., Lecomte, P., Dubois, P., Demonceau, A., Noels, A. F., Jérôme, R. *J. Polym. Sci. Part A: Polym. Chem.* **1999**, *37*, 2447.
72 Rungr, M., Dutta, S., Bowden, N. *Macromolecules* **2006**, *39*, 498.
73 Bazan, G. C., Schrock, R. R. *Macromolecules* **1991**, *24*, 817.
74 Saunders, R. S., Cohen, R. E., Wong, S. J., Schrock, R. R. *Macromolecules* **1992**, *25*, 2055.
75 Dounis, P., Feast, W. J. *Polymer* **1996**, *37*, 2547.
76 Beerens, H., Wang, W., Verdonck, L., Verpoort, F. *J. Mol. Catal. A: Chem.* **2002**, *190*, 1.
77 Méry, D., Astruc, D. *J. Mol. Catal. A: Chem.* **2005**, *277*, 1.
78 Gatard, S., Kahlal, S., Méry, D., Nlate, S., Cloutet, E., Saillard, J.-Y., Astruc, D. *Organometallics* **2004**, *23*, 1313.
79 Murphy, J., Kawasaki, T., Fujiki, M., Nomura, K. *Macromolecules* **2005**, *38*, 1075.
80 Rinehart, R. E., Smith, H. P. *J. Polym. Sci., Polym. Lett.* **1965**, *B3*, 1049.
81 Rinehart, R. *US Patent 3 377 924*, **1968**.
82 Michelotti, F. W., Keaveney, W. P. *J. Polym. Sci.* **1965**, *A3*, 895.
83 Wache, S. *J. Organomet. Chem.* **1995**, *494*, 235.
84 Lynn, D., Mohr, B., Grubbs, R., Henling, L., Day, M. *J. Am. Chem. Soc.* **2000**, *122*, 6601.
85 Lu, S., Quayle, P., Booth, C., Yeattes, H., Padget, J. *Polym. Int.* **1993**, *32*, 1.
86 Héroguez, V., Fontanille, M., Gnanou, Y. *Macromol. Symp.* **2000**, *150*, 269.
87 Chemtob, A., Héroguez, V., Gnanou, Y. *Macromolecules* **2002**, *35*, 9262.
88 Quémener, D., Héroguez, V., Gnanou, Y. *J. Polym. Sci. Part A: Polym. Chem.* **2005**, *43*, 217.
89 Chemtob, A., Héroguez, V., Gnanou, Y. *J. Polym. Sci. Part A: Polym. Chem.* **2004**, *42*, 2705.
90 Chemtob, A., Héroguez, V., Gnanou, Y. *J. Polym. Sci. Part A: Polym. Chem.* **2004**, *42*, 1154.
91 Quémener, D., Chemtob, A., Héroguez, V., Gnanou, Y. *Polymer* **2005**, *46*, 1067–1075.
92 Chemtob, A., Héroguez, V., Gnanou, Y. *Macromolecules* **2004**, *37*, 7619.
93 Lee, B. S., Mahajan, S., Clapham, B., Janda, K. D. *J. Org. Chem.* **2004**, *69*, 3319.
94 Quémener, D., Héroguez, V., Gnanou, Y. *Macromolecules* **2005**, *38*, 7977.
95 Carrillo, A., Kane, R. *J. Polym. Sci. Part A: Polym. Chem.* **2004**, *42*, 3352.
96 Carrillo, A., Yanjarappa, M., Gujraty, K., Kane, R. *J. Polym. Sci. Part A: Polym. Chem.* **2005**, *44*, 928.
97 Pugh, C., Kiste, A. L. *Prog. Polym. Sci.* **1997**, *22*, 601.
98 Trimmel, G., Riegler, S., Fuchs, G., Slugovc, C., Stelzer, F. *Adv. Polym. Sci.* **2005**, *176*, 43.
99 Pugh, C., Dharia, J., Arehart, S. V. *Macromolecules* **1997**, *30*, 4520.
100 Komiya, Z., Schrock, R. R. *Macromolecules* **1993**, *26*, 1387.
101 Pugh, C., Schrock, R. R. *Macromolecules* **1992**, *25*, 6593.
102 Komiya, Z., Pugh, C., Schrock, R. R. *Macromolecules* **1992**, *25*, 6586.
103 Komiya, Z., Pugh, C., Schrock, R. R. *Macromolecules* **1992**, *25*, 3609.
104 Komiya, Z., Schrock, R. R. *Macromolecules* **1993**, *26*, 1393.
105 Arehart, S. V., Pugh, C. *J. Am. Chem. Soc.* **1997**, *119*, 3027.
106 Pugh, C., Bae, J.-Y., Dharia, J., Ge, J. J., Cheng, S. Z. D. *Macromolecules* **1998**, *31*, 5188.
107 Percec, V., Schlueter, D. *Macromolecules* **1997**, *30*, 5783.
108 Maughon, B. R., Weck, M., Mohr, B., Grubbs, R. H. *Macromolecules* **1997**, *30*, 257.
109 Weck, M., Mohr, B., Maughon, B. R., Grubbs, R. H. *Macromolecules* **1997**, *30*, 6430.
110 Li, M.-H., Keller, P., Albouy, P.-A. *Macromolecules* **2003**, *36*, 2284.
111 Demel, S., Riegler, S., Wewerka, K., Schoefberger, W., Slugovc, C., Stelzer, F. *Inorg. Chim. Acta* **2003**, *345*, 363.
112 Dounis, P., Feast, W. J., Widawski, G. *J. Mol. Catal. A: Chem.* **1997**, *115*, 51–60.

113 Edwards, J.H., Feast, W.J. *Polymer* **1980**, *21*, 595.
114 Edwards, J.H., Feast, W.J., Bott, D.C. *Polymer* **1984**, *25*, 395.
115 Klavetter, F.L., Grubbs, R.H. *J. Am. Chem. Soc.* **1988**, *110*, 7807.
116 Scherman, O.A., Rutenberg, I.M., Grubbs, R.H. *J. Am. Chem. Soc.* **2003**, *125*, 8515.
117 Scherman, O.A., Grubbs, R.H. *Synth. Met.* **2001**, *124*, 4314.
118 Albagli, D., Bazan, G., Wrighton, M.S., Schrock, R.R. *J. Am. Chem. Soc.* **1992**, *114*, 4150.
119 Buchmeiser, M.R. *Macromol. Rapid Commun.* **2001**, *22*, 1081.
120 Lubbad, S., Mayr, B., Mayr, M., Buchmeiser, M. *Macromol. Symp.* **2004**, *210*, 1.
121 Buchmeiser, M.R. *J. Mol. Catal. A: Chem.* **2002**, *190*, 145.
122 Sinner, F., Buchmeiser, M.R. *Macromolecules* **2000**, *33*, 5777.
123 Krause, J.O., Lubbad, S.H., Nuyken, O., Buchmeiser, M.R. *Macromol. Rapid Commun.* **2003**, *24*, 875.
124 Lubbad, S., Buchmeiser, M.R. *Macromol. Rapid Commun.* **2003**, *24*, 580.
125 Mayr, B., Tessadri, R., Post, E., Buchmeiser, M.R. *Anal. Chem.* **2001**, *73*, 4071.
126 Lubbad, S., Mayr, B., Christian, G., Huber, M.R.B. *J. Chromatogr. A* **2002**, *959*, 121.
127 Lubbad, S., Buchmeiser, M.R. *Macromol. Rapid Commun.* **2002**, *23*, 617.
128 Mayr, M., Wang, D., Kröll, R., Schuler, N., Prühs, S., Fürstner, A., Buchmeiser, M.R. *Adv. Synth. Catal.* **2005**, *347*, 484.
129 Sinner, F., Buchmeiser, M.R., Tessadri, R., Mupa, M., Wurst, K., Bonn, G. *J. Am. Chem. Soc.* **1998**, *120*, 2790.
130 Buchmeiser, M.R., Atzl, N., Bonn, G. *J. Am. Chem. Soc.* **1997**, *119*, 9166.
131 Buchmeiser, M.R. *Catal. Today* **2005**, 612.
132 Buchmeiser, M.R., Wurst, K. *J. Am. Chem. Soc.* **1999**, *121*, 11101.
133 Moore, J.D., Harned, A.M., Henle, J., Flynn, D.L., Hanson, P.R. *Org. Lett.* **2002**, *4*, 1847–1849.
134 Nomura, K., Ogura, H., Imanishi, Y. *J. Mol. Catal. A: Chem.* **2002**, *185*, 311.
135 Nomura, K., Kuromatsu, Y. *J. Mol. Catal. A: Chem.* in press.
136 Nomura, K., Kuromatsu, Y. *J. Mol. Catal. A: Chem.* **2006**, *245*, 152.
137 Grenz, A., Ceccarelli, S., Bolm, C. *Chem. Commun.* **2001**, 1726.
138 Barrett, A.G.M., Cramp, S.M.J., Hennessy, A., Procopiou, P.A., Roberts, R.S. *Org. Lett.* **2001**, *3*, 271.
139 Bolm, C., Tanyeli, C., Grenz, A., Dinter, C.L. *Adv. Synth. Catal.* **2002**, *344*, 649.
140 Mortell, K.H., Weatherman, R.V., Kiessling, L.L. *J. Am. Chem. Soc.* **1996**, *118*, 2297.
141 Mortell, K.H., Gingras, M., Kiessling, L.L. *J. Am. Chem. Soc.* **1994**, *116*, 12053.
142 Schuster, M.C., Mortell, K.H., Hegeman, A.D., Kiessling, L.L. *J. Mol. Catal. A: Chem.* **1997**, *116*, 209.
143 Gordon, E.J., Sanders, W.J., Kiessling, L.L. *Nature* **1998**, *392*, 30.
144 Manning, D.D., Hu, X., Beck, P., Kiessling, L.L. *J. Am. Chem. Soc.* **1997**, *119*, 3161.
145 Arimoto, H., Nishimura, K., Kinumi, T., Hayakawa, I., Uemura, D. *Chem. Commun.* **1999**, 1361.
146 Maynard, H.D., Okada, S.Y., Grubbs, R.H. *Macromolecules* **2000**, *33*, 6239.
147 Watson, K.J., Park, S.-J., Im, J.-H., Nguyen, S.T., Mirkin, C.A. *J. Am. Chem. Soc.* **2001**, *123*, 5592.
148 Gibbs, J.M., Park, S.-J., Anderson, D.R., Watson, K.J., Mirkin, C.A., Nguyen, S.T. *J. Am. Chem. Soc.* **2005**, *127*, 1170.
149 Ilker, M.F., Schule, H., Coughlin, E.B. *Macromolecules* **2004**, *37*, 694.
150 Ilker, M.F., Nusslein, K., Tew, G.N., Coughlin, E.B. *J. Am. Chem. Soc.* **2004**, *126*, 15870.
151 Jarroux, N., Keller, P., Mingotaud, A.-F., Mingotaud, C., Sykes, C. *J. Am. Chem. Soc.* **2004**, *126*, 15958.
152 Durrieu, M., Héroguez, V., Quémener, D., Baquey, C. *WO2006008386*, **2006**.
153 Héroguez, V., Quémener, D., Durrieu, M. *WO2006008387*, **2006**.
154 Buchmeiser, M.R. *Adv. Polym. Sci.* **2006**, *197*, 137.
155 Bolm, C., Dinter, C., Seger, A., Höcker, H., Brozio, J. *J. Org. Chem.* **1999**, *64*, 5730.

8
Polycondensation

Tsutomu Yokozawa

Polycondensation* is an important method of polymerization that affords polymers containing functional groups and/or aromatic rings in the backbone, which serve as fibers, engineering plastics, electrical materials and so on by virtue of their strong intermolecular forces. Most research had been centered on how to increase the molecular weight of polymers and how to synthesize a variety of condensation polymers. However, polycondensation has developed into a polymerization method that yields not only linear polymers under control of the sequence, molecular weight and molecular weight distribution, but also cyclic and hyperbranched polymers.

8.1
Monomer Reactivity Control (Stoichiometric-imbalanced Polycondensation)

In polycondensation that proceeds in a step-growth polymerization manner, the average degree of polymerization (DP) is expressed as in Eq. (1), where p is the extent of reaction of all of the functional groups of monomers and oligomers in the reaction mixture, according to the basic principle established by Carothers and Flory.

$$DP = \frac{1}{1-p} \qquad (1)$$

However, Eq. (1) is only applicable as long as the exact stoichiometric balance of monomers is maintained. If one of the monomers is used excessively in the

---

* The term *polycondensation* is used in this chapter for polymerization via condensation regardless of whether the polymerization proceeds in a step-growth or a chain-growth polymerization manner according to previous terminology. According to a recent IUPAC recommendation (I. Mita, R. F. T. Stepto, U. W. Suter, *Pure. Appl. Chem.* **1994**, 66, 2483), polymerization via condensation that proceeds in only the step-growth polymerization manner is termed *polycondensation*, but there is no well-known term for polymerization via condensation irrespective of its mechanism.

Macromolecular Engineering. Precise Synthesis, Materials Properties, Applications.
Edited by K. Matyjaszewski, Y. Gnanou, and L. Leibler
Copyright © 2007 WILEY-VCH Verlag GmbH & Co. KGaA, Weinheim
ISBN: 978-3-527-31446-1

polycondensation of AA and BB monomers, Eq. (1) is modified to Eq. (2) in terms of the molar ratio (S, $S>1$) of both of the monomers. Similarly, in polycondensation of an AB monomer, if N_1 mol of a monofunctional compound are added to the polymerization of N_0 mol of the AB monomer, DP is expressed by Eq. (3).

$$DP = \frac{1+S}{1+S-2p} \tag{2}$$

$$DP = \frac{1+\frac{N_1}{N_0}}{(1-p)+\frac{N_1}{N_0}} \tag{3}$$

These equations are based on an assumption that the reactivity of the functional group is not changed during polymerization. If a functional group in the monomer becomes more reactive when the other functional group has reacted, the polycondensation affords a polymer with high molecular weight even when this monomer is used in excess. This is because the AA monomer having the reactivity-enhanced functional groups is liable to react with 2 equiv. of BB to afford BB–AA–BB even though an excess amount of AA is in the reaction mixture. If there is no enhancement of reactivity in the excess monomer AA, both end-groups of the oligomer or polymer would be the AA units to terminate the polymerization when all of BB is consumed. Such stoichiometric-imbalanced polycondensations are desirable, considering that it is sometimes difficult to maintain strict stoichiometric balance between two monomers due to side-reactions and physical loss from the reaction medium by evaporation of the monomers or precipitation of the polymer segments.

8.1.1
Polycondensation of α,α-Dihalogenated Monomers

A well-known example of stoichiometric-imbalanced polycondensation is the polycondensation of dihalomethane and bisphenol derivatives, leading to polyformal [Eq. (4)] [1]. In this polycondensation, a polymer with high molecular weight is obtained even when excess of dihalomethane is used as a solvent. This is attributed to the involvement of an active intermediate formed by dihalomethane and the phenol moiety of the bisphenol, which is more reactive than dihalomethane to react faster with another phenol moiety. Thus, the polymer end-groups are always the phenol moieties, not the halomethyl group even in the presence of excess of dihalomethane, and then polymerization continues without termination with the excess monomer.

8.1 Monomer Reactivity Control (Stoichiometric-imbalanced Polycondensation)

$$\text{CH}_2\text{X}_2 + \text{HO}-\text{C}_6\text{H}_4-\text{C(CH}_3)_2-\text{C}_6\text{H}_4-\text{OH} \xrightarrow{\text{KOH}} [\text{XCH}_2-\text{O}-\text{C}_6\text{H}_4-\text{C(CH}_3)_2-\text{C}_6\text{H}_4-\text{OH}]$$

X = Cl, Br

active intermediate

$$\longrightarrow \ce{+O-C_6H_4-C(CH_3)_2-C_6H_4-OCH_2+}_n \quad (4)$$

A practical problem in this polycondensation is the formation of cyclic oligomers, resulting in a decrease in the average molecular weight of the polymer [2]. This side-reaction is caused by slow dissolution of the potassium salt of bisphenol, which creates high dilution conditions in the reaction mixture. For example, polymerization was carried out at room temperature by addition of the potassium salt of bisphenol A to a solution of dibromomethane in N-methylpyrrolidinone (NMP) over several hours to give polyformal that had a bimodal distribution ($M_n = 46\,900$, $M_w/M_n = 3.1$). In addition, this polyformal contained 23% macrocycles. On the other hand, when bisphenol A and KOH were vigorously stirred in NMP for 2 min, followed by addition of dibromomethane, high molecular weight polymer ($M_n = 250\,100$, $M_w/M_n = 1.3$) was obtained after vigorous stirring for only 5 min. Only miniscule amounts of cyclic oligomers were present.

The production of high molecular weight polyformal by high-intensity mixing can be explained as follows [Eq. (5)]. Since an intermediate bromomethyl ether is much more reactive than dibromomethane, as mentioned above, the intermediate could react with itself to give a cycle under normal stirring conditions where the salt of bisphenol very slowly goes into the solution. Under high-intensity mixing conditions, the concentration of the salt of bisphenol is high so that the intermediate would react preferentially with the salt to give high molecular weight polymer while suppressing the formation of macrocycles.

$$\begin{array}{c}
\text{high conc.} \\
^-\text{O}-\text{Ar}-\text{O}^- \\
^-\text{O}-\text{Ar}-\text{O}\text{mm}\text{OCH}_2-\text{Br} \longrightarrow {}^-\text{O}-\text{Ar}-\text{O}\text{mm}\text{OCH}_2-\text{O}-\text{Ar}-\text{O}^- \\
\text{reactive} \\
\text{low conc.} \qquad\qquad\qquad\qquad\qquad \text{CH}_2\text{Br}_2 \\
^-\text{O}-\text{Ar}-\text{O}^- \qquad\qquad\qquad\qquad\quad {}^-\text{O}-\text{Ar}-\text{O}^- \\
\downarrow \qquad\qquad\qquad\qquad\qquad\qquad \downarrow \\
\text{Ar}-\text{O} \\
\text{O} \quad\quad \text{CH}_2-\text{O} \qquad\qquad\qquad \text{Polyformal with high } M_n
\end{array} \quad (5)$$

The polycondensation of dibromomethane with bisbenzenethiol instead of bisphenol also shows stoichiometric-imbalanced polymerization behavior [Eq. (6)] [3]. When 1.5 equiv. of dibromomethane were reacted with dithiol in NMP at 75 °C for 4 h, polysulfide with the maximum inherent viscosity ($\eta_{\text{inh}} = 0.50$ dL g^{-1}) was obtained. On the basis of the model reaction, a reactive intermediate **1**, in which a bromine in dibromomethane is substituted with the mercapto group of the monomer, is estimated to be 61 times more reactive than dibromomethane.

8 Polycondensation

$$CH_2Br_2 + HS\text{-}Ar\text{-}S\text{-}Ar\text{-}SH \xrightarrow{DBU} [BrCH_2\text{-}S\text{-}Ar\text{-}S\text{-}Ar\text{-}SH]_1$$

$$\longrightarrow \text{-(}CH_2\text{-}S\text{-}Ar\text{-}S\text{-}Ar\text{-}S\text{)}_n \qquad (6)$$

Not only dihalomethane but also 2,2-dichloro-1,3-benzodioxole (**2**) affords a reactive intermediate **3** by the reaction with bisphenol A [Eq. (7)] [4]. Thus, the polycondensation of 1.7 equiv. of **2** with bisphenol A in refluxing dichlorobenzene for 3 h yielded polyorthocarbonate with the highest molecular weight ($M_n = 120\,000$). It should be noted that the theoretical M_n value based on Eq. (2) is only 693 under these stoichiometric-imbalanced conditions. In the model reaction of **2** and phenol at –40 °C, the rate of the second nucleophilic substitution of monochloride formed by the first substitution was 27 times faster than that of the first substitution reaction.

$$\underset{2}{\text{Cl}_2\text{C(OC}_6\text{H}_4\text{O)}} + HO\text{-Ar-}C\text{-Ar-}OH \longrightarrow \underset{3}{[Cl\text{-}C(OC_6H_4O)\text{-}O\text{-Ar-}C\text{-Ar-}OH]_n}$$

$$\longrightarrow \text{-(}O\text{-}C(OC_6H_4O)\text{-}O\text{-Ar-}C\text{-Ar-)}_n \qquad (7)$$

8.1.2
Pd-catalyzed Polycondensation

The above nonstoichiometric polycondensations are derived from the specific reactivity of electrophilic monomers. Catalysts for polycondensation also bring about these polycondensations without strict stoichiometric balance between the monomers. Nomura and coworkers applied palladium-catalyzed allylic substitution (Tsuji–Trost reaction) to the polycondensation between 1,4-diacetoxybut-2-ene (**4**) and diethyl malonate in the presence of a Pd(0) catalyst, resulting in high molecular weight polymer ($M_n = 21\,800$) [Eq. (8)] [5, 6]. Furthermore, the polycondensation by using excess of **4** was also found to afford similar high molecular weight polymers [7]. The observed stoichiometric-imbalanced polymerization behavior is rationalized by involvement of olefin–Pd(0) complex **5**, which leads to cascade bidirectional allylation (Scheme 8.1). Thus, after the first allyla-

$$\underset{4}{AcO\text{-}CH_2\text{-}CH\text{=}CH\text{-}CH_2\text{-}OAc} + \underset{E = CO_2Et}{CH_2E_2} \xrightarrow[\text{base}]{Pd^0} \text{-(}CH(E)_2\text{-}CH_2\text{-}CH\text{=}CH\text{-}CH_2\text{-)}_n \qquad (8)$$

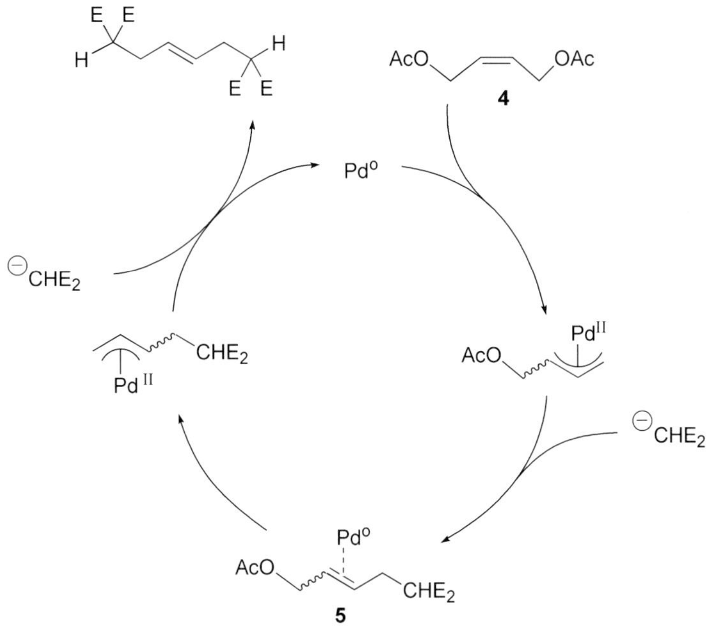

8.1 Monomer Reactivity Control (Stoichiometric-imbalanced Polycondensation)

Scheme 8.1

tion of **4**, **5** selectively forms the allylpalladium(II) complex at the other allylic terminal. Therefore, the polymer end-groups are always the malonic ester moiety even in the presence of excess of **4**, which would not terminate the polycondensation by the attack to both polymer end-groups. This behavior is dependent on the ligand of the Pd(0) catalyst; bis(diphenylphosphino)butane (dppb) is indispensable and use of PPh$_3$ results in the same DP as expected in Carothers' Eq. (2). This dependence of nonstoichiometric polycondensation behavior on a catalyst is interesting in comparison with the polycondensation mentioned in the Section 8.1.1, where that behavior stems only from the structure of the monomers.

The Pd-catalyzed nonstoichiometric polycondensations of other monomers are shown in Table 8.1 [8]. Those electrophilic monomers effect a remote cascade double allylation, where an olefin–Pd(0) complex after the first allylic substitution exchanges the olefin ligand with the remote double bond.

8.1.3
Crystallization Polycondensation

As mentioned at the beginning of Section 8.1, when polycondensation of AB monomer is carried out in the presence of a monofunctional compound, polycondensation is terminated with the monofunctional compound and the DP is estimated by Carothers' Eq. (3). However, Kimura and Kohama reported that

Table 8.1 Stoichiometric-imbalanced polycondensation via the Tsuji-Trost reaction[a)]

Electrophile (X=OCOPh)	Nucleophile (E=CO$_2$Et)	E/N[b)] (DP by Eq. 2)	Polymer (E=CO$_2$Et)	Isolated yield (%)	M_n[c)]	M_w/M_n[c)]	DP[d)]
X⌒⌒=⌒⌒O⌒⌒=⌒⌒X	CH$_2$E$_2$	1.0/1.0 (–)	(structure with EE)	97	16 000	1.53	110
		1.5/1.0 (5)		98	14 000	1.62	99
		1.0/1.5 (5)		–[e)]	800	1.76	6[f)]
	CH$_2$(COMe)$_2$	1.5/1.0 (5)	(structure with MeOC COMe)	82[g)]	13 000	1.39	120
	(indanedione) = CH$_2$E′$_2$	1.5/1.0 (5)	(structure E′E′)	99	14 000	1.52	100
	H–C≡C–CH$_2$ EE (alkyne)	1.5/1.0 (5)	(structure with alkyne EE)	99	16 000	1.47	65
X⌒⌒CH(⌒⌒)X (branched diene)	CH$_2$E$_2$	1.0/1.0 (–)	(copolymer 88/12)	95	32 000	1.57	270
		1.5/1.0 (5)		94	41 000	1.57	340
		2.0/1.0 (3)		96	41 000	1.57	340
		1.0/1.5 (5)		–[e)]	800	1.59	6[f)]

X~~N(Ph)~~X	CH$_2$E$_2$	1.5/1.0 (5)	![structure]	97	13000	1.47	73
X~~(aryl-divinyl)~~X	CH$_2$E$_2$	1.5/1.0 (5)	![structure]	92	10000	1.75	64

a) The polycondensation was catalyzed by 1 mol% of Pd$_2$(dba)$_3$ (5 μmol)–2.0 mol% of 1,4-bis(diphenylphosphino)butane (10 μmol) in the presence of 3.0 mmol of N,O-bis(trimethylsilyl)acetamide in CH$_2$Cl$_2$–DMF (9 : 1, 1.0 mL).
b) The mole ratio of electrophile to nucleophile. The E or N value "1.0" stands for 0.50 mmol; for example, E/N=1.5/1.0, 0.75 mmol of the electrophile and 0.50 mmol of the nucleophile were used.
c) Determined by GPC.
d) Each DP value was calculated using M_n by GPC and the molecular weight of the repeating unit.
e) Not isolated. A crude reaction mixture was analyzed by GPC.
f) The DP value was 5.0±0.5 by ^{1}H NMR.
g) Some amount of an insoluble material was obtained.

the polycondensation of 4-acetoxybenzoic acid (**6**) in the presence of monofunctional compound **7** in liquid paraffin at 320 °C affords poly(4-oxybenzoyl) crystal (whisker) consisting of higher molecular weight polymers compared with the products obtained by the melt polymerization [Eq. (9)] [9]. The *DP* of the aromatic polyester obtained was much higher than that calculated according to Eq. (3). For example, in the polymerization with 30 mol% of **7**, where the calculated *DP* is 3.3, the observed *DP*s were 232 [R in **7**=CH$_3$(CH$_2$)$_9$] and 283 [R in **7**=CH$_3$(CH$_2$)$_{17}$], respectively. This polymerization would proceed through the following mechanisms; oligomers are formed in the solution and when the *DP* of the oligomers exceeds a critical value, they are crystallized. End-free oligomers are preferentially crystallized due to the lower solubility than that of end-capped oligomers. End-capped oligomers are also crystallized and polycondensation proceeds with eliminating the end-capping groups of the oligomers by transesterification just when they are crystallized. Although part of the end-capped oligomers are contained in the crystals, the end-capping groups are excluded by solid-state polycondensation.

$$\text{AcO}-\text{C}_6\text{H}_4-\text{COOH} \; (\mathbf{6}) \; + \; \text{RO}-\text{C}_6\text{H}_4-\text{COOH} \; (\mathbf{7}) \longrightarrow \text{RO}-\text{C}_6\text{H}_4-\text{CO}+\text{O}-\text{C}_6\text{H}_4-\text{CO}+_n\text{OH} \quad (9)$$

Similar polycondensation giving crystals was attained by the reaction of 4-hydroxybenzoic acid (**8**) with 4-ethoxybenzoic acid anhydride (**9**), which immediately affords 4-(4-ethoxybenzoyloxy)benzoic acid (**10**), a monomer like **6**, and monofunctional compound **11** [10]. Polyesters with the *DP* of 38–76 were obtained even if 30–60 mol% of **9** were added to the reaction mixture [Eq. (10)]. This polycondensation is noteworthy not only as stoichiometric-imbalanced polycondensation but also as a valuable method for morphology control of condensation polymers during polymerization (see Section 8.4.3).

$$\text{HO}-\text{C}_6\text{H}_4-\text{COOH} \; (\mathbf{8}) \; + \; \text{EtO}-\text{C}_6\text{H}_4-\text{COOC}-\text{C}_6\text{H}_4-\text{OEt} \; (\mathbf{9}) \longrightarrow$$

$$\text{EtO}-\text{C}_6\text{H}_4-\text{CO}-\text{O}-\text{C}_6\text{H}_4-\text{COOH} \; (\mathbf{10}) \; + \; \text{EtO}-\text{C}_6\text{H}_4-\text{COOH} \; (\mathbf{11}) \longrightarrow \text{EtO}-\text{C}_6\text{H}_4-\text{CO}+\text{O}-\text{C}_6\text{H}_4-\text{CO}+_n\text{OH} \quad (10)$$

8.1.4
Nucleation–Elongation Polycondensation

The closed-system, reversible polymerization of *m*-phenyleneethynylene starter sequences **12** and **13** in solution was driven by the folding energy of the resulting polymers [11, 12]. Zhao and Moore also performed this polymerization under conditions of imbalanced stoichiometry to obtain polymers whose molecular weight was higher than expected from the Flory distribution [13]. For instance, in the reaction of **12** and **13** at a molar ratio of 1:2 in the presence of oxalic acid at room temperature, polymer products were present at equilibrium along with a substantial amount of a monomer but relatively little dimer or trimer. When the polymerization of the corresponding diamine and dialdehyde at the same monomer stoichiometry was carried out in the melt, only low molecular weight oligomers were produced and the degree of polymerization was in good agreement with the prediction from the Flory equation. Because the helical folding of *m*-phenyleneethynylene chains should not take place under these melt conditions, the stoichiometric imbalanced polymerization of **12** and **13** in solution was proposed to proceed in the folding-driven nucleation–elongation mechanism. Thus, an oligomer with a certain length starts to fold in solution and further extending the molecule would result in polymers with increasing folding stability. Excess monomer remains unreacted at equilibrium because it would not take part in producing the folded polymer.

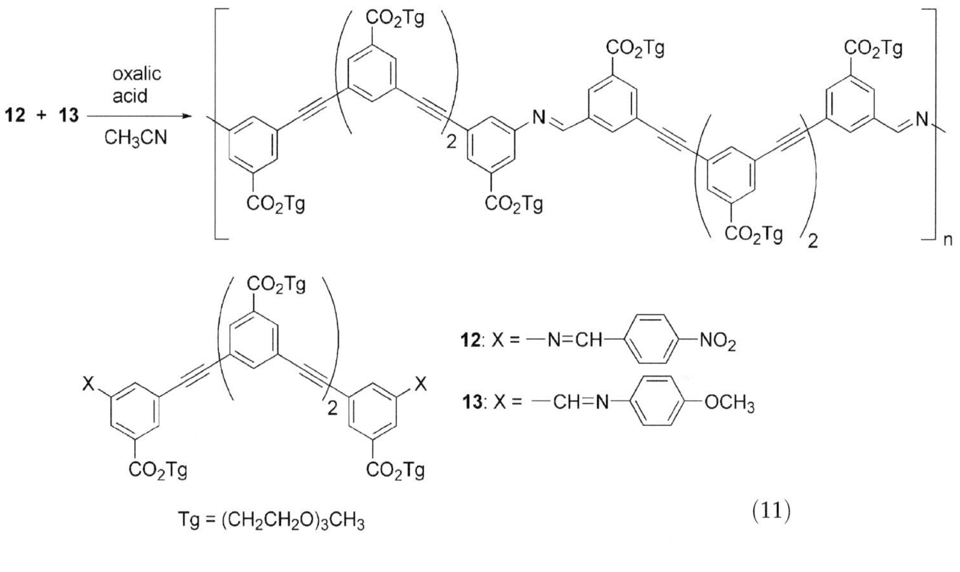

(11)

8.2
Sequence Control

Condensation polymers are often prepared by the reaction between two symmetrical bifunctional monomers (AA + BB). However, polycondensation using unsymmetrical monomers affords polymers having different sequences, which could possess different physical properties according to the different sequences. This means that polymers with different physical properties can be separately synthesized from one unsymmetrical monomer, if the sequence of the monomer units in the polymer is controlled [14].

8.2.1
Sequential Polymers from Symmetrical and Unsymmetrical Monomers

In the polycondensation between symmetrical monomer and unsymmetrical monomers (AA + BB′), there are three probabilities of two adjacent unsymmetrical units in a chain: –BB′–AA–B′B–, –B′B–AA–BB′– and –BB′–AA–BB′– (–B′B–AA–B′B– is indistinguishable). The probability s of a –BB′–AA–BB′– is given by

$$s = [-BB' - AA - BB'-]/([-BB' - AA - B'B-]$$
$$+[-B'B - AA - BB'-] + [-BB' - AA - BB'-])$$

If all units are pointing to the same direction, $s=1$ (head-to-tail) and when the orientation of the units is strictly alternating, $s=0$ (head-to-head or tail-to-tail). If no preference for the different enhancement exists, $s=1/2$ (random chain) [15].

If B is much more reactive than B′ in monomer BB′ and BB′ is slowly added to symmetrical monomer AA, B′B–AA–BB′ is first formed *in situ* and then this intermediate reacts with AA to afford (–AA–B′B–AA–BB′–)$_n$, which is a head-to-head or tail-to-tail polymer ($s=0$). In 1964, Morgan and Kwolek tried to synthesize head-to-head polymers by virtue of preferential acylation to steric hindered diamine [16]. Thus, terephthaloyl chloride or sebacoyl chloride was added dropwise to a solution of *cis*-2,6-dimethylpiperazine in dichloromethane [Eq. (12)]. In this polymerization, the less hindered amine moiety reacted with the acid chloride to form a symmetrical intermediate, which then successively reacted with the acid chloride. In this report, however, there was no characterization of the resulting polymers to support the head-to-head structure. A similar attempt was carried out by using a monomer having two different nucleophiles, amino and hydroxyl groups [Eq. (13)] [17]. The thermal properties of the authentic polymer, which was synthesized by a two-step procedure, and those of the polymers obtained by this method were compared, but no spectral data were shown. There were other preliminary studies, but sequential polymers were not obtained.

(12)

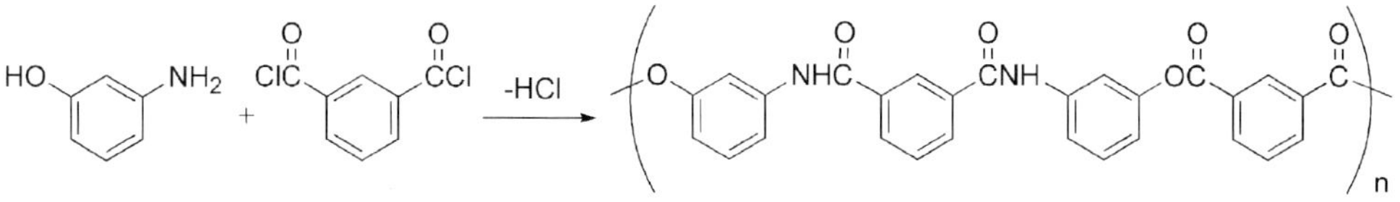

(13)

The first synthesis of a head-to-head polyamide was reported by Pino and co-workers, who used the different aminolysis rates of aliphatic and aromatic 4-nitrophenyl esters [Eq. (14)] [15, 18]. Thus, the polycondensation was carried out at room temperature by slow addition of ethylenediamine to the diester to yield polyamide with $s=0.079$. To improve the regularity, sterically hindered diester was polymerized with ethylenediamine [Eq. (15)]; the value of s was 0 [18].

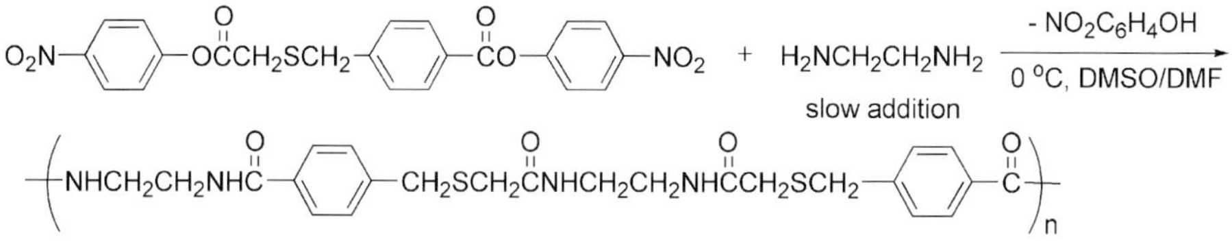

(14)

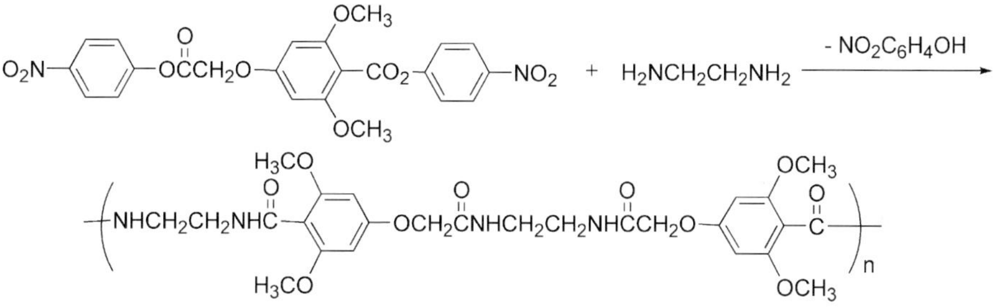

(15)

Ueda et al. synthesized head-to-head polyamide by direct polycondensation using condensing agent **14**, which reacts with the carboxyl group to generate a mixed carboxylic–phosphoric anhydride whose reactivity towards nucleophiles is lower than that of acid chloride, leading to selective amidation. In a similar manner to the above polymerization, a symmetrical monomer, isophthalic acid, was slowly added to a solution of unsymmetrical diamine, **14** and triethylamine to give a polyamide with an inherent viscosity of 0.20 dL g^{-1} [Eq. (16)] [19]. After the synthesis of authentic polyamides with head-to-head, head-to-tail and random orientations and comparison of spectral data between them, the polyamide obtained by direct polycondensation turned out to be the desired head-to-head ($s=0$) sequence. When acid chloride was used for the synthesis of a similar polyamide instead of the direct polycondensation method, the head-to-head regularity was not perfect but $s=0.005$–0.65, depending on the substituents of the diamine and the method of monomer addition [Eq. (17)] [20].

$$(16)$$

$$(17)$$

Since aliphatic and aromatic amines give 10^6-fold different rate constants for aminolysis of an active species, an unsymmetrical diamine, 2-(4-aminophenyl)-ethylamine, was polymerized with isophthalic acid in the presence of 14 and triethylamine, resulting in head-to-head ($s=0$) polyamide with an inherent viscosity of 0.3 dL g^{-1} [Eq. (18)] [21, 22].

$$(18)$$

When head-to-tail polymer is synthesized in polycondensation between AA and BB′ (B is more reactive than B′), the terminal A has to be deactivated when AA–BB′ is formed. If this requirement is satisfied, monomers AA and BB′ are first converted to AA–BB′ and then polymerization of AA–BB′ proceeds to afford head-to-tail polymer (–AA–BB′–)$_n$. However, it is more difficult to synthesize head-to-tail polymers than the head-to-head polymers mentioned above, because many reactions that satisfy the above requirement are not known.

Pino and coworkers tried to synthesize a head-to-tail polyurea by the polycondensation of bis(4-nitrophenyl)carbonate and 2-(4-aminophenyl)ethylamine on the basis of the results that the second substitution of bis(4-nitrophenyl)carbonate with the nucleophile became slower than the first [Eq. (19)] [23]. When the diamine was slowly added to the carbonate, the polyurea with high head-to-tail regularity ($s=0.89$) was obtained. It should be noted that the reverse addition method afforded a polyurea with high head-to-head regularity ($s=0.05$). This observed difference demonstrates that both head-to-tail and head-to-head polymers can be selectively synthesized only by changing the addition method of monomers.

$$\text{(19)}$$

Ueda et al. used a cyclic anhydride instead of a symmetrical dicarboxylic acid for the synthesis of head-to-tail polymer, because it seems to be difficult to carry out the above deactivation system from symmetrical dicarboxylic acids. Thus, succinic anhydride and 2-(4-aminophenyl)ethylamine were mixed at once in the presence of the condensing agent **14** and triethylamine in NMP at room temperature to give the head-to-tail polyamide [Eq. (20)] [24].

$$\text{(20)}$$

The polycondensation of unsymmetrical dihaloaromatic compounds with metal catalysts also has a problem of sequence control, although this polymerization is not classified as the polycondensation of AA and BB′ in this section. For example, the polymerization of 2,5-dibromo-3-alkylthiophene with $Ni(cod)_2$ (cod=1,5-cyclooctadiene) gave a poly(alkylthiophene) with random sequence, whereas the polymerization of the monomer, in which one bromine of the 2,5-dibromo-3-alkylthiophene was exchanged with BrZn [25, 26] or BrMg [27, 28], yielded a head-to-tail poly(3-alkylthiophene) by virtue of a catalytic amount of $Ni(dppe)Cl_2$ [dppe=1,3-bis (diphenylphosphino)ethane] or $Ni(dppp)Cl_2$ [dppp=1,3-bis(diphenylphosphino)propane] [Eq. (21)]. The electrical conductivity of the head-to-tail poly(alkylthiophene) is superior to that of the sequence-random one [29].

$$\text{(21)}$$

8.2.2
Sequential Polymers from Two Unsymmetrical Monomers

It is very difficult to synthesize sequential polymers from two unsymmetrical monomers only by polycondensation (AA' + BB'), because there are four probabilities of diad orientational arrangement: –AA'–BB'–, –AA'–B'B–, –A'A–BB'– and –A'A–B'B–. However, if the two pairs of four A, A', B, B' react selectively, the desired polymer can be synthesized. For example, when A and A' selectively react with B' and B, respectively, sequential polymer $(-AA'-BB'-)_n$ would be obtained. For these distinguishable reactions, condensation and other types of reaction are used. In the polymerization of 4-carboxylphthalic thioanhydride and 4-aminobenzhydrazide in the presence of condensing agent **14** and triethylamine at room temperature, the carboxyl group selectively reacted with the hydrazide group to form an acylhydrazide linkage and then the amino group and the thioanhydride moiety underwent ring-opening addition and condensation to form an imide linkage, resulting in a sequential poly(acylhydrazide–imide) with an inherent viscosity of 0.51 dL g^{-1} [Eq. (22)] [30].

Another example used polyaddition as a reaction distinguishable from polycondensation. Thus, 4-carboxylphenyl acrylate and 4-mercaptoaniline polymerized in the presence of **14** and triethylamine to yield a sequential poly(amide–thioether) with an inherent viscosity of 0.31 dL g^{-1} [Eq. (23)]. In this polymerization, the mercapto group selectively added to the carbon–carbon double bond [31].

$$\text{(22)}$$

$$\text{CH}_2=\overset{\underset{|}{\text{CH}}}{\underset{\text{O}}{\overset{\|}{\text{CO}}}}-\underset{}{\bigcirc}-\overset{\text{O}}{\overset{\|}{\text{COH}}} + \text{H}_2\text{N}-\underset{}{\bigcirc}-\text{SH} \xrightarrow[\text{Et}_3\text{N}]{\mathbf{14}} +\!\!\left(\text{CH}_2\text{CH}_2\overset{\text{O}}{\overset{\|}{\text{CO}}}-\underset{}{\bigcirc}-\overset{\text{O}}{\overset{\|}{\text{CNH}}}-\underset{}{\bigcirc}-\text{S}\right)\!\!_n$$

(23)

8.2.3
Sequential Polymers from Two Symmetrical Monomers and One Unsymmetrical Monomer

Polycondensation of two symmetrical monomers, AA and A′A′ (A is more reactive than A′) and unsymmetrical monomer BB′ (B is more reactive than B′) affords, first, B′B–AA–BB′, which then reacts with A′A′ to yield sequential polymer (–B′B–AA–BB′–A′A′–)$_n$. In this polymer sequence, if B′B–AA and BB′–A′A′ are regarded as monomer units, this polymer can be regarded as a sequential polymer from two unsymmetrical monomers B′B–AA and BB′–A′A′. The direct polycondensation of 2,5-dimethylterephthalic acid, 4,4′-(oxydi-4-phenylene)dibutanoic acid and 4-aminobenzhydrazide was carried out by mixing a pair of dicarboxylic acids and **14** in the presence of triethylamine in NMP–hexamethylphosphoramide (HMPA) at room temperature, followed by addition of 4-aminobenzhydrazide to yield a controlled sequential polymer with an inherent viscosity of 0.40 dL g^{-1} [Eq. (24)]. The ^{13}C NMR spectrum of the polymer obtained was identical with that of the authentic sequential polymer prepared in two steps and completely different from that of the random polymer [32]. A similar sequential polymer was also synthesized by using 2-methoxyisophthalic acid instead of 2,5-dimethylterephthalic acid in the above polycondensation [33].

(24)

One of the dicarboxylic acids can be replaced with an active ester. In the polycondensation of nitroisophthalic acid, bis(2,4,6-trichlorophenyl) isophthalate and 4-aminobenzhydrazide in the presence of **14** and triethylamine, first the benzhydrazide moiety selectively reacted with the carboxyl group activated by **14** and then the aniline moiety and the active ester moiety reacted, yielding a sequence-controlled polyamide with an inherent viscosity of 0.25 dL g^{-1} [Eq. (25)]. A similar polyamide with an inherent viscosity of 0.30 dL g^{-1} was obtained from bis(2,4,6-trichlorophenyl) adipate instead of bis(2,4,6-trichlorophenyl) isophthalate [34].

310 | *8 Polycondensation*

$$\text{(25)}$$

8.2.4
Sequential Polymers from Two Symmetrical Monomers and Two Unsymmetrical Monomers

If four kinds of monomers AA′, A″A″ (reactivity: A > A′ > A″), BB and B′B″ (reactivity: B > B′ > B″) selectively reacted, A′A–BB–AA′ is first formed and then B″B′–A′A–BB–AA′–B′B″ is formed, followed by the reaction with A″A″, resulting in sequential polymer (–B″B′–A′A–BB–AA′–B′B″–A″A″–)$_n$. As described above, this polymer can be regarded as a sequential polymer from three unsymmetrical monomers: B″B′–A′A, BB–AA′ and B′B″–A″A″. The polymerization as shown in Eq. (26) was carried out by mixing **14**, dicarboxylic acid and active ester in NMP at room temperature, followed by addition of two kinds of amines at –15 °C and then stirring at 25–80 °C. The inherent viscosity of the polymer obtained was 0.20 dL g^{-1} and the microstructure was determined by comparing the ^{13}C NMR spectrum of the polymer with that of the authentic sequential polymer [35].

$$\text{(26)}$$

8.3
Molecular Weight and Polydispersity Control

Until fairly recently, controlling molecular weight and polydispersity of a polymer in polycondensation was thought to be essentially impossible, because polycondensation proceeds in a step-growth polymerization manner: the polymeriza-

Artificial polycondensation

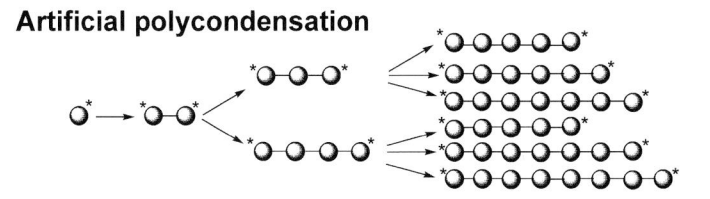

Biological polycondensation

■ : Initiator ○ : Monomer () : Enzyme ○* : Active end

Scheme 8.2

tion is initiated by the reaction of monomers with each other and propagation involves the reaction between both end-groups of all types of oligomers that are formed, in addition to the reaction of the oligomers with the monomer. In contrast, nature synthesizes monodispersed biopolymers such as DNA and polypeptides even by polycondensation. This system involves selective reaction of monomers with the polymer end-group that is activated by an enzyme, that is chain-growth polymerization (Scheme 8.2).

Accordingly, artificial polycondensation is also able to control molecular weight and polydispersity when the polycondensation proceeds in a chain-growth polymerization manner similar to biological polycondensation (the author calls it chain-growth polycondensation). For chain-growth polycondensation, it is necessary that the polymer end-group is always more reactive than the monomer to prevent the monomer from reacting with another monomer, leading to step-growth polymerization. There are three approaches to maintain the reactive polymer end-group during polymerization: transfer of reactive species, derived from the initiator, to the polymer end-group; activation of the polymer end-group by different substituent effects between the monomer and polymer; and transfer of the catalyst to the polymer end-group like biological polycondensation.

8.3.1
Transfer of Reactive Species

Specific elements can transmit a reactive species, stemming from an initiator, with elimination of small molecules. The first category is insertion of the monomer into the terminal M–F bond [Eq. (27)] and the second is transfer of the cationic species [Eq. (28)].

$$\sim\sim\sim M-F + M-X \longrightarrow \sim\sim\sim M\begin{matrix}-F\\ |\\ -M-X\end{matrix} \xrightarrow{-X} \sim\sim\sim M-M-F \qquad (27)$$

$$\sim\sim\sim M^+ + Y-Nu=M-X \longrightarrow \sim\sim\sim M-Nu-\overset{+}{M}-X \xrightarrow{-XY} \sim\sim\sim M-Nu=\overset{+}{M} $$

$$\sim\sim\sim \overset{+}{M}-X + Y-Nu=M-X \longrightarrow \sim\sim\sim M-Nu-\overset{+}{M}-X \xrightarrow{-XY} \sim\sim\sim M=Nu-\overset{+}{M}-X \qquad (28)$$

The polymerization of dimethylsulfoxonium methylide (**15**) initiated by trialkylborane is classified in the first category. Propagation involves insertion of **15** into the terminal C–B bond with elimination of dimethyl sulfoxide (DMSO). The polymerization was carried out in toluene at 70–80 °C, followed by oxidative workup to yield hydroxyl-terminated polymethylene [Eq. (29)]. The M_n values were very close to the calculated values from the feed ratio of **15** to trialkylborane and the M_w/M_n ratio ranged from 1.04 to 1.17. These results are consistent with the character of living polymerization. The insertion mechanism involves initial attack of the ylide on the alkylborane. The borate complex undergoes 1,2-migration of the alkyl group to produce the homologated alkylborane and a molecule of DMSO [Eq. (30)] [36, 37].

$$\underset{\mathbf{15}}{CH_2\overset{+}{S}Me_2} \xrightarrow[70-80\,°C]{Et_3B \quad H_2O_2/NaOH} Et\text{-}(CH_2)_n\text{-}CH_2OH \qquad (29)$$

$$R-CH_2-B\diagup + \underset{\mathbf{15}}{CH_2\overset{+}{S}Me_2} \longrightarrow \begin{matrix}R-CH_2\\ \diagup\\ B\\ Me_2\overset{+}{S}-CH_2\end{matrix} \longrightarrow RCH_2-CH_2-B\diagup + MeSMe \qquad (30)$$

This polyhomologation with alkylboranes is amenable to the synthesis of telechelic polymethylene, because alkylboranes can be prepared by hydroboration of a variety of α-olefins. For example, polymethylene with 4-methoxyphenyl and hydroxy groups at both ends was synthesized from initiator **16** prepared by hydroboration of 4-vinylanisole with $BH_3 \cdot THF$ [Eq. (31)] [36]. Other functional groups including biotin, carbohydrates, primary and secondary amines and dansyl and pyrene fluorescent groups were also introduced as an end-group from the corresponding α-olefins [38].

$$\text{MeO}\text{-}\langle\rangle\text{-}CH=CH_2 \xrightarrow{BH_3\text{-}THF} \left(\text{MeO}\text{-}\langle\rangle\text{-}CH_2CH_2\right)_3\!B$$
$$\mathbf{16}$$

$$\xrightarrow[\text{2) } H_2O_2/NaOH]{\text{1) } n\ \mathbf{15}} \text{MeO}\text{-}\langle\rangle\text{-}CH_2CH_2\text{-}(CH_2)_n\text{OH} \quad (31)$$

When B-thexylboracycloheptane **17**, prepared by the hydroboration of 1,5-heptadiene with thexylborane, was used as an initiator, methylene was inserted into only the C–B bond of boracyclane, not the thexyl–B bond, resulting in ring expansion. The expanded boracycle was treated with sodium cyanide, followed by benzoyl chloride, then peroxide oxidation to yield cyclic ketone [Eq. (32)] [39].

$$\text{(thexyl)BH}_2 + \text{CH}_2=CH(CH_2)_3CH=CH_2 \longrightarrow \mathbf{17} \xrightarrow[\text{2) NaCN}]{\text{1) } n\ \mathbf{15}} \xrightarrow[\text{4) } H_2O_2/NaOH]{\text{3) BzCl}} \text{cyclic ketone}(CH_2)_n \quad (32)$$

Similar polymerization of arsonium ylide **18** initiated by trialkylborane has been reported [40, 41]. However, the reaction of **18** at 0 °C in THF did not lead to polymers substituted on every carbon atom like the polymerization of sulfoxonium ylide **15**, but rather to a polymer, in which the main chain has been elongated by three carbon atoms at a time [Eq. (33)]. Polymers of varying degrees of polymerization were obtained by using varying ratios of ylide **18** to trialkylborane, but the degrees of polymerization were higher than expected from these ratios. The M_w/M_n ratio ranged from 1.21 to 1.58. This suggests that the initiation of the polymerization was not completely efficient, but the propagation was controlled. The polymerization mechanism different from that of the polymerization of ylide **15** involves [1,3]-sigmatropic rearrangement after 1,2-migration of the alkyl group of borate. This is the reason for the elongation of three carbons in this insertion polymerization.

$$R\text{-}B + Ph_3As\text{-}\mathbf{18} \longrightarrow [Ph_3As\cdots B^-R] \xrightarrow{-AsPh_3} \xrightarrow{[1,3]\text{ sigmatropic rearrangement}}$$

$$\xrightarrow{n\ \mathbf{18}} R\text{-}(\ )_n\text{-}B \xrightarrow{H_2O_2/NaOH} R\text{-}(\ )_n\text{-}OH \quad (33)$$

An example of the second category is the polymerization of phosphoranimines **19** initiated with electrophiles, leading to polyphosphazenes. Matyjaszewski and coworkers found that N-silylated phosphoranimine **19a** polymerized with $SbCl_5$

at 100 °C to yield poly[bis(trifluoroethoxy)phosphazene] **20** with M_n of 10 000–50 000 and M_w/M_n of 1.2–2.5 [42]. The molecular weight reached a maximum at partial conversion and leveled off, most likely due to transfer reactions. However, the inverse relationship of molecular weight and SbCl$_5$, the appearance of high molecular weight polymer at partial conversion and the first order in SbCl$_5$ kinetics of the reaction supported that SbCl$_5$ was the true initiator and that polymerization proceeded in a chain-growth manner. The mechanism of polymerization is shown in Eq. (34). The zero order in monomer kinetics of the reaction indicate that a reactive intermediate is formed in fast equilibrium with the monomer, followed by slow unimolecular elimination of trimethylsilyl trifluoroethoxide.

$$\text{Me}_3\text{Si}-\text{N}=\text{P}(\text{OCH}_2\text{CF}_3)_3 \xrightarrow{\text{SbCl}_5} \text{Me}_3\text{Si}-\text{N}=\overset{+}{\text{P}}(\text{OCH}_2\text{CF}_3)_2 \quad \underset{\textbf{19a}}{\rightleftarrows}$$

19a

$$\underset{\substack{|\\ \text{SiMe}_3 \\ \text{reactive intermediate}}}{\text{Me}_3\text{Si}-\text{N}=\text{P}(\text{OCH}_2\text{CF}_3)_2-\overset{+}{\text{N}}-\text{P}(\text{OCH}_2\text{CF}_3)_3} \xrightarrow{\text{slow}} \text{Me}_3\text{Si}-\text{N}=\text{P}(\text{OCH}_2\text{CF}_3)_2-\text{N}=\overset{+}{\text{P}}(\text{OCH}_2\text{CF}_3)_2$$

$$\rightleftarrows \underset{\textbf{20}}{\left[\text{N}=\underset{\overset{|}{\text{OCH}_2\text{CF}_3}}{\overset{\overset{\text{OCH}_2\text{CF}_3}{|}}{\text{P}}}\right]_n} \quad (34)$$

Allcock and coworkers reported that trichloro(trimethylsilyl)phosphoranimine **19b** polymerized with PCl$_5$ at ambient temperature with elimination of trimethylsilyl chloride. The resultant poly(dichlorophosphazene) was treated with an excess of NaOCH$_2$CF$_3$ to give polymer **20** [Eq. (35)]. When the polymerization was carried out in dichloromethane, the molecular weight increased with increasing in the ratio of **19b** to PCl$_5$ (M_w=7000–14 000) and the molecular weight distribution was kept narrow (M_w/M_n=1.04–1.20) [43]. The polymerization in toluene proceeded faster than in dichloromethane to give polymers with higher molecular weights in the region of 10^5 with low polydispersities (<1.3) [44]. Phenyl-substituted monomer **19c**, PhCl$_2$P=NSiMe$_3$, also underwent controlled polymerization until the feed ratio of **19c** to PCl$_5$ was 100. Other related initiators such as SbCl$_5$, TaCl$_5$ and PhPCl$_4$ also appear to initiate the ambient temperature polymerization of **19b** and **c**.

$$\underset{\textbf{19b}}{\text{Me}_3\text{Si}-\text{N}=\text{PCl}_3} \xrightarrow{\text{PCl}_5} \left[\text{N}=\underset{\overset{|}{\text{Cl}}}{\overset{\overset{\text{Cl}}{|}}{\text{P}}}\right]_n \xrightarrow{\text{NaOCH}_2\text{CF}_3} \underset{\textbf{20}}{\left[\text{N}=\underset{\overset{|}{\text{OCH}_2\text{CF}_3}}{\overset{\overset{\text{OCH}_2\text{CF}_3}{|}}{\text{P}}}\right]_n} \quad (35)$$

The polymerization is initiated by the reaction of **19b** with 2 equiv. of PCl$_5$ with elimination of Me$_3$SiCl to generate a salt, with which **19b** successively reacts with elimination of Me$_3$SiCl, resulting in the elongated cation [Eq. (36)].

$$\text{Me}_3\text{Si}-\text{N}=\text{PCl}_3 + 2\,\text{PCl}_5 \longrightarrow [\text{Cl}_3\text{P}{=}\!{=}\text{N}{=}\!{=}\text{PCl}_3]^+\,\text{PCl}_6^-$$
19b

$$\xrightarrow{\textbf{19b}} [\text{Cl}_3\text{P}{=}\!{=}\text{N}{=}\!{=}\text{PCl}_2{=}\!{=}\text{N}{=}\!{=}\text{PCl}_3]^+\,\text{PCl}_6^- \longrightarrow \cdots \longrightarrow {-}\!\!{\left(\text{N}{=}\overset{\text{Cl}}{\underset{\text{Cl}}{\text{P}}}\right)}\!\!{-}_n$$

(36)

Monomer **19b** was synthesized by the reaction of PCl_5 with either $\text{LiN(SiMe}_3)_2$ or $\text{N(SiMe}_3)_3$. These methods give relatively low product yields, because PCl_5 is an initiator for the polymerization of **19b**. To circumvent this concurrent polymerization, a new method for synthesizing **19b** and the subsequent polymerization in one pot has recently been reported [45]. Thus, PCl_3 was reacted with $\text{LiN(SiMe}_3)_2$ to afford $\text{Cl}_2\text{PN(SiMe}_3)_2$, which was oxidized with SO_2Cl_2 to yield **19b**. To the mixture, containing mainly **19b**, Me_3SiCl and LiCl, was added PCl_5 to produce the polyphosphazene. Under the optimized conditions, the derivative polymer **20** showed a relatively narrow molecular weight distribution ($M_w/M_n = 1.24$) even in one-pot reaction from PCl_3 [Eq. (37)].

$$\text{PCl}_3 + \text{LiN(SiMe}_3)_2 \xrightarrow{\text{toluene}} \text{Cl}_2\text{P}-\text{N(SiMe}_3)_2 \xrightarrow{\text{SO}_2\text{Cl}_2} \underset{\textbf{19b}}{\text{Me}_3\text{Si}-\text{N}=\text{PCl}_3}$$

$$\xrightarrow{\text{PCl}_5} {-}\!\!{\left(\text{N}{=}\overset{\text{Cl}}{\underset{\text{Cl}}{\text{P}}}\right)}\!\!{-}_n \xrightarrow{\text{NaOCH}_2\text{CF}_3} \underset{\textbf{20}}{{-}\!\!{\left(\text{N}{=}\overset{\text{OCH}_2\text{CF}_3}{\underset{\text{OCH}_2\text{CF}_3}{\text{P}}}\right)}\!\!{-}_n}$$

(37)

Monomer **19a** also polymerized with an anionic catalyst via a chain-growth process, although this polymerization is not included in the second category of Eq. (28). Montague and Matyjaszewski reported this polymerization before the cationic polymerization mentioned above. The polymerization was carried out with tetrabutylammonium fluoride (TBAF) as an initiator at 95 °C to yield **20** with $M_w/M_n \approx 1.5$ [46]. The proposed polymerization mechanism is as follows. The polymerization is initiated by the abstraction of the silyl group from **19a** with TBAF, followed by the attack of the resultant phosphazene anion on another monomer. Propagation proceeds through the attack of the resultant trifluoroethoxide on the silyl group of the growing polymer chain, producing an anion which can then attack another monomer molecule [Eq. (38)]. Selective desilylation of the polymer end-group can be explained by the possibility that the strength of the N–Si bond on a polymer end-group is weaker than that on a monomer molecule due to long conjugation of the polymer [47].

$$(CF_3CH_2O)_3P=N-SiMe_3 \xrightarrow{TBAF} (CF_3CH_2O)_3P=N^- \xrightarrow{19a} (CF_3CH_2O)_3P=N-\underset{\underset{OCH_2CF_3}{|}}{\overset{\overset{OCH_2CF_3}{|}}{P}}=N-SiMe_3$$
$$\text{19a}$$

$$\xrightarrow{^-OCH_2CF_3} (CF_3CH_2O)_3P=N-\underset{\underset{OCH_2CF_3}{|}}{\overset{\overset{OCH_2CF_3}{|}}{P}}=N \rightleftarrows CF_3CH_2O-\left(\underset{\underset{OCH_2CF_3}{|}}{\overset{\overset{OCH_2CF_3}{|}}{P}}=N\right)_n-SiMe_3$$
$$\text{20} \qquad (38)$$

8.3.2
Different Substituent Effects Between Monomer and Polymer

Even general AB-type monomers, affording polyamide, polyester, polyether and so on, undergo chain-growth polycondensation, if the polymer end-group becomes more reactive than the monomer by virtue of the change in substituent effects between the monomer and polymer (Scheme 8.3). Both the resonance effect and inductive effect of the nucleophilic site on the reactivity of the electrophilic site at the para and meta positions of the monomer are applicable, respectively.

8.3.2.1 Resonance Effect (Polymerization of para-Substituted Monomers)
In the polycondensation of some para-substituted monomers, the polymer end-group was reported to become more reactive than the monomer, but the polycondensation did not precisely control the molecular weight of polymers within narrow molecular weight distributions because of the insolubility of the polymers and the contamination of step-growth polycondensation [48–53]. However, Yokozawa et al. found that the polycondensation of phenyl 4-(octylamino)benzo-

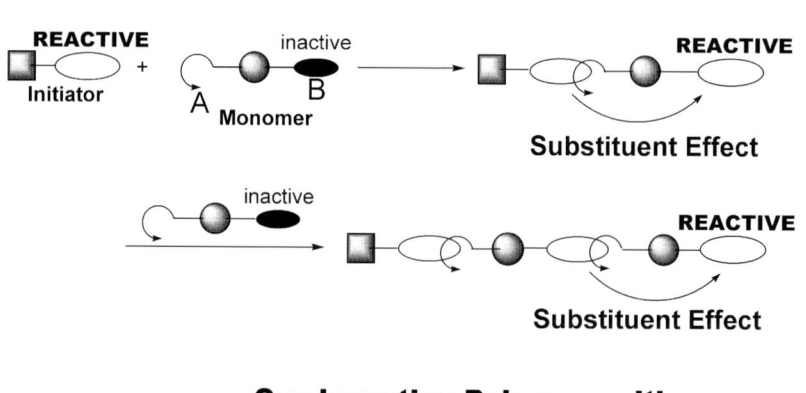

Scheme 8.3

8.3 Molecular Weight and Polydispersity Control

ate (**21a**) proceeded homogeneously in the presence of a base (a combination of N-octyl-N-triethylsilylaniline, CsF and 18-crown-6) and phenyl 4-nitrobenzoate (**22**) as an initiator in THF at ambient temperature to yield well-defined aromatic polyamides with very low polydispersities ($M_w/M_n \leq 1.1$) [Eq. (39)] [54]. The M_n of the polymer was controlled by the feed ratio of monomer **21a** to initiator **22** up to 22000 and the polydispersity was fairly narrow (Fig. 8.1a). Furthermore, the M_n values also increased in proportion to monomer conversion, indicating that this polycondensation proceeded in the manner of chain-growth polymerization (Fig. 8.1b).

$$O_2N-C_6H_4-C(O)-OPh + H-N(C_8H_{17})-C_6H_4-C(O)-OPh \xrightarrow[\text{rt}]{\text{Base}} O_2N-C_6H_4-C(O)-[N(C_8H_{17})-C_6H_4-C(O)]_n-OPh$$

$$\text{Base} = \text{Et}_3\text{Si}-\text{N}(\text{C}_8\text{H}_{17})-\text{Ph} / \text{CsF} / \text{18-crown-6}$$

(39)

This result is explained by the different substituent effects between the monomer and polymer (Scheme 8.4). The base abstracts the proton of the amino group of monomer **21a** to generate the aminyl anion **21a'**. This anion deactivates the phenyl ester moiety of **21a'** by its strong electron-donating ability through the resonance effect, resulting in preventing the monomer from reacting with another one. The anion monomer **21a'** would react with initiator **22**

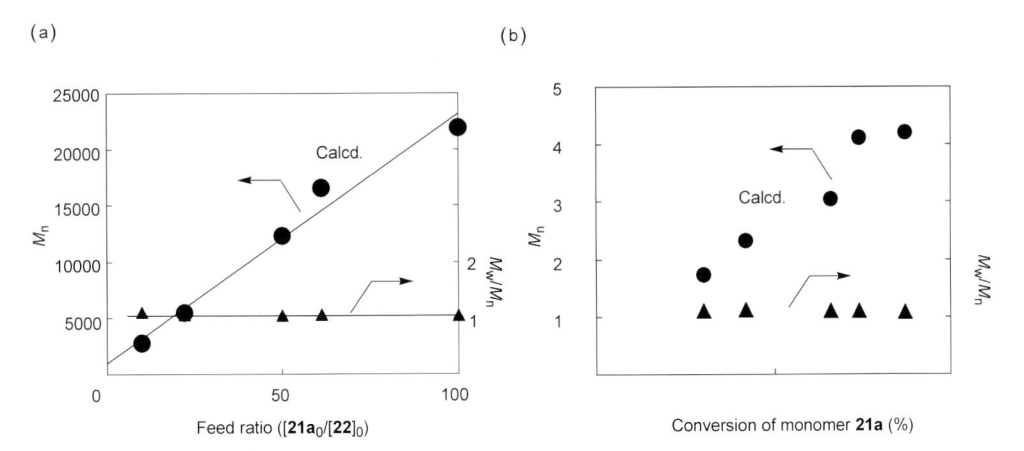

Fig. 8.1 Chain-growth polycondensation of **21a** for aromatic polyamides: (a) M_n and M_w/M_n values of poly-**21a**, obtained with base in the presence of **22** in THF at 25 °C, as a function of the feed ratio of **21a** to **22**. The line indicates the calculated M_n values assuming one polymer chain per unit **22**. (b) M_n and M_w/M_n values of poly-**21a**, obtained with base in the presence of **22** in THF at 25 °C, as a function of monomer conversion. The line indicates the calculated M_n values assuming one polymer chain per unit **22**.

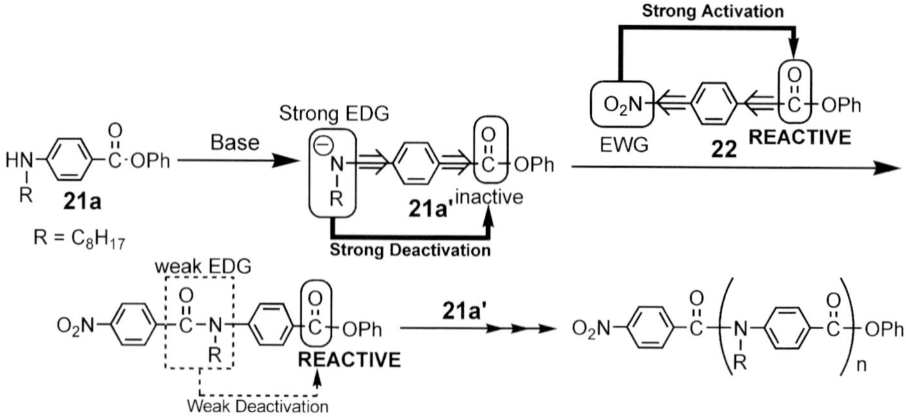

Scheme 8.4 (EDG: Electron-donating group; EWG: Electron-withdrawing group)

having an electron-withdrawing group, because the phenyl ester moiety of **22** is more reactive than that of **21a′**. The amide obtained has a weak electron-donating amide linkage and the phenyl ester moiety of the amide is more reactive than that of **21a′**. Hence the next monomer would selectively react with the phenyl ester moiety of the amide. Growth would continue in a chain polymerization manner by the selective reaction of **21a′** with the polymer terminal phenyl ester moiety.

The above synthetic method for well-defined aromatic polyamides, however, needs a peculiar base, N-octyl-N-triethylsilylaniline, along with CsF and 18-crown-6 and the monomer has a phenyl ester moiety as an electrophilic site, which is not that common compared with a methyl ester or an ethyl ester. Furthermore, it is necessary to separate the obtained polyamide from by-products such as N-octylaniline and phenol by HPLC. For a convenient synthesis, the polycondensation of the corresponding methyl ester monomer **21b** with a commercially available base has been developed [55]. The methyl ester **21b** polymerized with lithium hexamethyldisilazide (LHMDS) in the presence of an initiator in THF at −10 °C [Eq. (40)]. The highly pure polyamide with a defined molecular weight and low polydispersity was obtained after simple treatment of the reaction mixture with aqueous NaOH solution followed by evaporation, because the by-products in this polycondensation after treatment with water are low-boiling methanol, ammonia and hexamethyldisiloxane. The polycondensation of a similar monomer containing a 3-acyl-2-benzothiazolthione as the electrophilic site instead of the methyl ester moiety of **21b** also would make the purification of the polyamide obtained easier; by-products could be washed out with water [56].

$$\text{CH}_3-\text{C}_6\text{H}_4-\overset{\text{O}}{\underset{\|}{\text{C}}}-\text{OPh} + \text{H}-\underset{\underset{\text{C}_8\text{H}_{17}}{|}}{\text{N}}-\text{C}_6\text{H}_4-\overset{\text{O}}{\underset{\|}{\text{C}}}-\text{OMe} \xrightarrow[-10\,°\text{C}]{\text{LHMDS}} \text{CH}_3-\text{C}_6\text{H}_4-\overset{\text{O}}{\underset{\|}{\text{C}}}+\underset{\underset{\text{C}_8\text{H}_{17}}{|}}{\text{N}}-\text{C}_6\text{H}_4-\overset{\text{O}}{\underset{\|}{\text{C}}}+_n \text{OMe}$$

21b

(40)

The synthesis of a well-defined aromatic polyester was more difficult than that of polyamide, because polyester easily undergoes transesterification. The monomer can attack the polymer ester linkage to generate the cleaved chain with the phenoxide moiety and the acyl group at both ends, leading to conventional step-growth polycondensation. Actually, transesterification occurred in the polycondensation of monomer **23**, having an active amide moiety as a good leaving group, even with a weak base such as tertiary amine at room temperature [57]. However, when the polymerization of **23** was carried out at −30 °C with Et$_3$SiH, CsF and 18-crown-6 as a base system, transesterification was almost suppressed and the molecular weight was controlled up to 7300 with low polydispersity ($M_w/M_n \leq 1.3$) [Eq. (41)] [58].

23

R = C$_8$H$_{17}$, (CH$_2$)$_3$OCH$_2$CH$_2$OCH$_3$

(41)

From the perspective of the use of different substituent effects between the monomer and the polymer for chain-growth polycondensation, the synthesis of well-defined polyethers seems to be difficult because the hydroxyl group in a monomer and the ether linkage of a polymer have similar electron-donating ability. However, monomer **24a** bearing a phenoxide moiety underwent chain-growth polycondensation in sulfolane at 150 °C to yield an aromatic polyether with low polydispersity ($M_w/M_n \leq 1.1$). The molecular weight was controlled up to 3500, because the polyether with higher molecular weight than that precipitated during polymerization [59]. Monomer **24b** substituted with an octyl group instead of the propyl group in **24a** also afforded an insoluble polymer in sulfolane when attempts were made to prepare higher molecular weight polymer. In other solvents such as N,N-dimethylimidazolidinone (DMI) and tetraglyme, the polymerization proceeded homogeneously, but both chain-growth and step-growth polymerization took place to give a polyether with broad molecular weight distribution [60] [Eq. (42)].

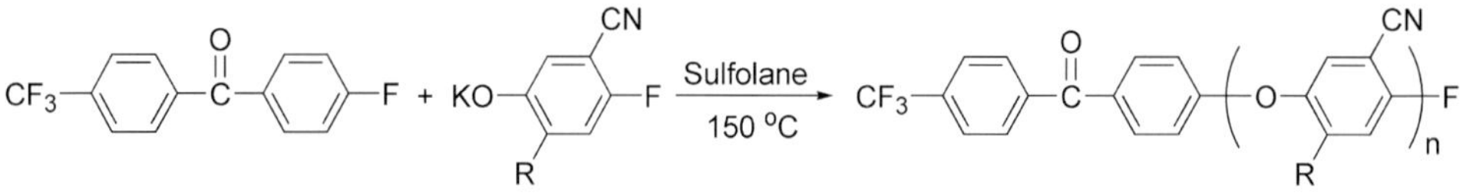

$$\text{CF}_3\text{-C}_6\text{H}_4\text{-CO-C}_6\text{H}_4\text{-F} + \text{KO-C}_6\text{H}_2(\text{CN})(\text{R})\text{-F} \xrightarrow[150\,°\text{C}]{\text{Sulfolane}} \text{CF}_3\text{-C}_6\text{H}_4\text{-CO-C}_6\text{H}_4\text{-[O-C}_6\text{H}_2(\text{CN})(\text{R})]_n\text{-F} \quad (42)$$

24a: R = C₃H₇, **24b**: R = C₈H₁₇

The key to successful chain-growth polycondensation of **24** is the use of phenoxide in the monomer instead of phenol. The phenoxide moiety works as a strong electron-donating group than does the phenol moiety and the ether linkage and the carbon attached to the fluorine in monomer **24** is strongly deactivated to prevent **24** from reacting with itself. Accordingly, **24** reacts selectively with the initiator and the polymer end-group, resulting in chain-growth polycondensation (Scheme 8.5).

An interesting aspect is that the polyether with low polydispersity from chain-growth polycondensation possessed higher crystallinity than that with a broad molecular weight distribution from conventional step-growth polycondensation. The powder X-ray diffraction (XRD) pattern of the former polymer showed a stronger intensity and the differential scanning calorimetry (DSC) profile showed an exothermic peak at 172 °C (cold crystallization) on heating from the glassy state [61]. This implies that the crystallinity of condensation polymers could be controlled by polydispersity.

This polyether synthesis can be applied to the synthesis of a well-defined poly(ether sulfone) by the polycondensation of **25**, which is different from other monomers for chain-growth polycondensation in that the nucleophilic site and electrophilic site are on each benzene ring connected with an electron-withdrawing group, a sulfonyl group [Eq. (43)]. In the polymerization of **25** in the presence of an initiator and 18-crown-6 in sulfolane at 120 °C, the molecular weight was controlled up to 5700 and the molecular weight distribution was <1.5. When the polymerization was carried out at higher feed ratio of the monomer to initiator, both chain-growth and step-growth polycondensation occurred [62]. This undesirable step-growth polycondensation is caused by transetherification of the backbone ether linkage with the monomer and/or fluoride, which is common in the polycondensation for poly(ether sulfone) at high temperature [63, 64].

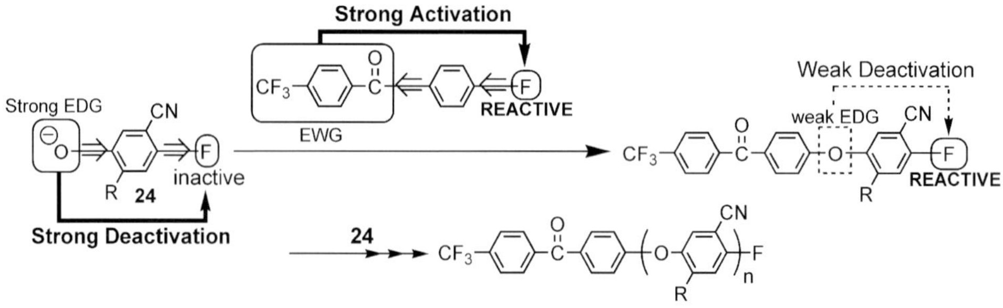

Scheme 8.5

$$CF_3\text{-}C_6H_4\text{-}SO_2\text{-}C_6H_4\text{-}F + KO\text{-}C_6H_4\text{-}SO_2\text{-}C_6H_4\text{-}F \xrightarrow[\text{Sulfolane, 120 °C}]{\text{18-crown-6}}$$

25

$$CF_3\text{-}C_6H_4\text{-}SO_2\text{-}C_6H_4\text{-}(O\text{-}C_6H_4\text{-}SO_2\text{-}C_6H_4)_n\text{-}F \quad (43)$$

8.3.2.2 Inductive Effect (meta-Substituted Monomers)

In the polycondensation of meta-substituted monomers, the inductive effect ($+I$ effect) of the nucleophilic site on the reactivity of the electrophilic site at the meta-position of the monomer is as applicable to chain-growth polycondensation as is the R effect of the para-substituted monomers mentioned above. Thus, ethyl 3-(alkylamino)benzoate (**26**) polymerized in the presence of LHMDS as a base and phenyl 4-methylbenzoate as an initiator in THF at 0 °C to yield N-alkylated poly(m-benzamide)s with well-defined molecular weights and low polydispersities ($M_w/M_n \leq 1.1$) [Eq. (44)]. When the N-alkyl group was octyl, the M_n of the polymer was controlled up to 12 000 by the feed ratio of the monomer to initiator and the polydispersity was kept narrow. In this polymerization, the aminyl anion of deprotonated **26** would deactivate the acyl group at the meta-position through the strong $+I$ effect, resulting in suppression of the self-polycondensation of **26**. The anion of **26** would then selectively react with an initiator and the polymer chain end, the acyl group of which is more reactive than that of the monomer with the aminyl anion and growth would continue in a chain-polymerization manner.

(44)

To support this mechanism, density functional theory (DFT) calculations were performed. The activation energies for the propagation and self-condensation are 21.6 and 27.0 kcal mol^{-1}, respectively. On the basis of the geometries, energies and vibrational frequencies obtained, the theoretical rate constants were then evaluated at 298.15 K and 1 atm. The reaction rate constant (1.1×10^{-3} s^{-1}) for the propagation is 8.6×10^3-fold greater than that for the self-condensation (1.3×10^{-7} s^{-1}) and hence is consistent with the experimental finding that propagation was observed exclusively over self-condensation; that is, chain-growth polycondensation of meta-substituted aminobenzoic ester monomers proceeded [65].

A variety of well-defined poly(m-benzamide)s were synthesized from the corresponding monomers [Eq. (45)]. All these polymers have higher solubility than the para-substituted counterparts [66]. Especially the polyamides having an oligo(ethylene glycol) are soluble in water and show reversible cloud points on heating [67].

$$\text{(structures shown)} \tag{45}$$

R = CH$_3$, C$_2$H$_5$, C$_3$H$_7$, C$_4$H$_9$

OR R = CH$_3$, C$_8$H$_{17}$

CH$_3$ m = 3, 4

8.3.3
Transfer of Catalyst

Polycondensation with a catalyst can involve another mechanism for chain-growth polycondensation, namely the catalyst-transfer mechanism, in which the catalyst activates the polymer end-group, followed by reaction with the monomer and transfer of the catalyst to the elongated polymer end-group, in a similar manner to biological polycondensation. This mechanism has been found in the polycondensation of Grignard thiophene monomer **27a** with an Ni catalyst. The polymerization of **27a** with the Ni catalyst was well known as a regiocontrolled synthetic method for poly(alkylthiophene)s developed by McCullough, but the polymers obtained possessed a broad molecular weight distribution [68]. However, when the polymerization was carried out at room temperature, taking care regarding the exact amount of isopropyl magnesium chloride for generation of monomer **27a** from the corresponding dihalogenated monomer, the M_n values of polymers increased in proportion to monomer conversion while retaining narrow polydispersities and were controlled by the amount of the Ni catalyst; the M_n values were proportional to the [**27a**]$_0$/[Ni catalyst]$_0$ feed ratio [69, 70]. A similar zinc monomer also showed the same polymerization behavior [71]. Furthermore, the M_w/M_n ratios was around 1.1 when the polymerization of **27a** was quenched with hydrochloric acid [72].

After a detailed study of the polymerization of **27a**, the following four important points were clarified: (1) the polymer end-groups are uniform among molecules; one end-group is Br and the other is H (Fig. 8.2); (2) the propagating end-group is a polymer–Ni–Br complex; (3) one Ni molecule forms one polymer chain; and (4) the chain initiator is a dimer of **27 a** formed *in situ*. Based on these results, a catalyst-transfer polycondensation mechanism has been proposed (Scheme 8.6). Thus, Ni(dppp)Cl$_2$ reacts with 2 equiv. of **27 a** and the coupling reaction occurs with concomitant generation of a zero-valent Ni complex. The Ni(0) complex is not stable and it inserts in the intramolecular C–Br bond. Another **27 a** reacts with this Ni, followed by the coupling reaction and transfer of the Ni catalyst to the next C–Br bond. Growth continues in such a way that the Ni catalyst moves to the polymer end-group [73].

In the former chain-growth polycondensation mechanism based on the substituent effect, it would sometimes be difficult to control the molecular weight in the region of very high molecular weights because the reaction between monomers, which leads to self-polycondensation, cannot be completely suppressed in a very low concentration of the propagating group. On the other hand, this catalyst-transfer polycondensation is different in that the monomers essentially do not react with each other but react with only the polymer end-group if the catalyst moves to the polymer end-group. This means that the molecular weight of a polymer can be ideally controlled no matter how high it is.

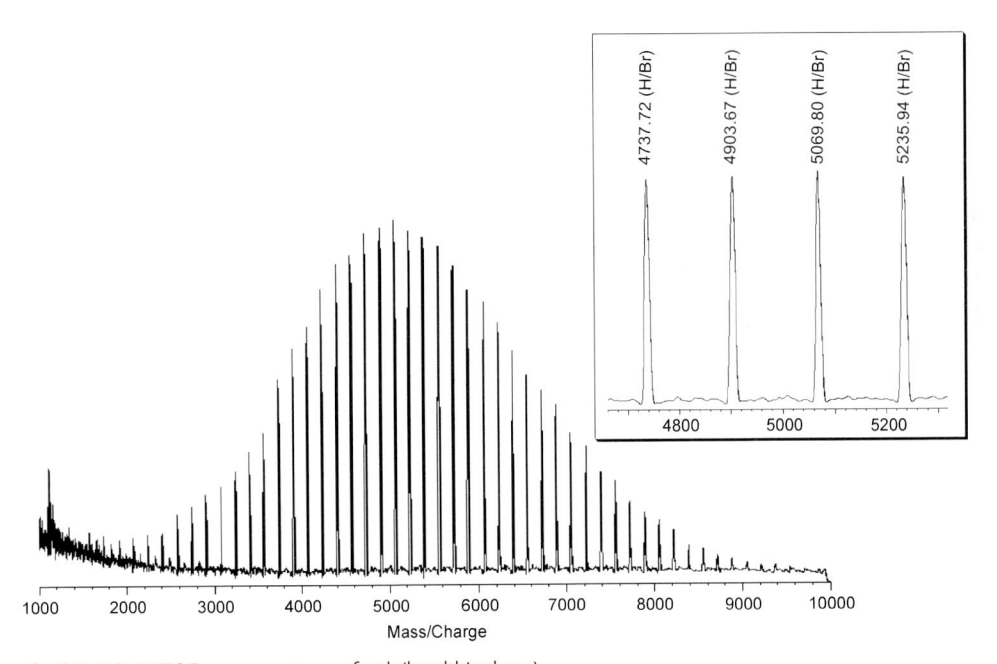

Fig. 8.2 MALDI-TOF mass spectrum of poly(hexylthiophene) obtained by the polymerization of **27 a** with Ni(dppp)Cl$_2$ in THF at ambient temperature. Reproduced with permission from R. Miyakoshi, A. Yokoyama, T. Yokozawa, *J. Am. Chem. Soc.* **2005**, *127*, 17542, Copyright (2005) American Chemical Society.

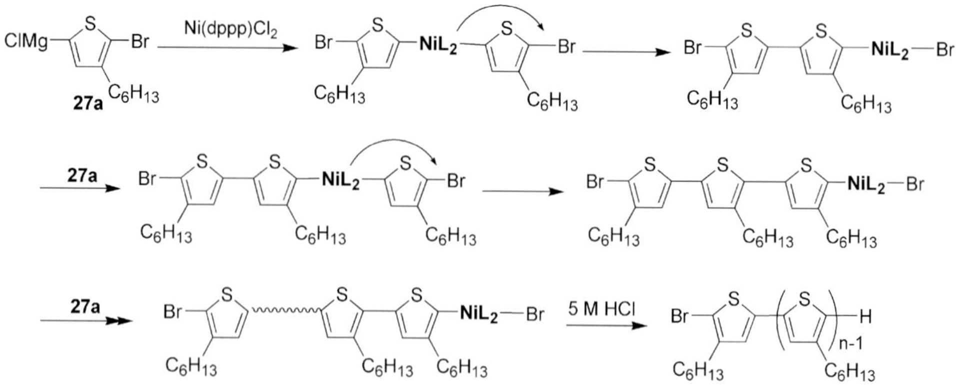

Scheme 8.6

On the basis of the nature of the polythiophene end-group containing the Ni complex, functional groups can be introduced to one or both ends of the polymer by Grignard reagents. Allyl, ethynyl and vinyl Grignard reagents afford monofunctionalized polythiophenes, whereas aryl and alkyl Grignard reagents yield difunctionalized polythiophenes. By utilizing the proper protecting groups, the hydroxyl, formyl and amino groups are also incorporated on to the polymer ends [74].

8.4
Chain Topology and Polymer Morphology Control

8.4.1
Cyclic Polymers

It is thought to be difficult to synthesize cyclic polymers by general polymerization methods such as addition polymerization and ring-opening polymerization. Regarding polycondensation, it also seemed difficult, as indicated by the fact that Carothers and Flory did not take into account cyclization reactions in their theory of step-growth polymerization. However, it was the identification of cyclic polymers that was difficult. The development of matrix-assisted laser desorption/ionization time-of-flight mass spectrometry (MALDI-TOFMS) during the past decade has provided a new, powerful tool for qualitative analyses of cyclic oligomers and polymers. Kricheldorf and Schwarz found in many polycondensations that the percentage and molecular weight of the cyclic polymers increase when the reaction conditions favor high molecular weights. In the absence of side-reactions, all polycondensation products will be cyclic polymers when conversion approaches 100%. Therefore, cyclization limits the degree of polymerization as expressed by Eq. (46), where V_p and V_c are the rates of propagation and cyclization and X is a constant >1.0 [75].

$$DP = \frac{1}{1 - p(1 - \frac{1}{X^\alpha})} \qquad \alpha = V_p/V_c \tag{46}$$

Several examples of producing cyclic polymers are described below. Interfacial hydrolytic polycondensation of bisphenol-A bischloroformate afforded polycarbonate with an M_n of 65 000 under optimized conditions [Eq. (47)]. The MALDI-TOF mass spectrum of this polymer displayed peaks of cyclic oligo- and polycarbonates of up to 18 000 Da and no peaks for linear chains. Furthermore, the optimum mass spectrum of the polymer sample fractionalized by size-exclusion chromatography (SEC) showed the peaks of cyclics of up to 55 000 Da, corresponding to a DP of around 210 [76]. This result indicates that cyclization competes with propagation at any chain length.

$$ \tag{47}$$

Poly(ether sulfone)s are, like polycarbonates, a group of engineering plastics and are synthesized generally by the polycondensation of a bisphenol with 4,4′-dichlorodiphenylsulfone in the presence of K_2CO_3 in DMSO [Eq. (48)]. Commercial poly(ether sulfone)s have M_n in the range 15 000–25 000 and MALDI-TOFMS revealed a moderate content of the cyclic polymers, increasing in parallel with the molecular weights. However, the mass peaks of OH- and Cl-terminated linear chains were detectable, indicating that conversion was far from completion [77]. When more reactive 4,4′-difluorodiphenylsulfone was used instead of the dichloro counterpart and the polymerization conditions were optimized, a polymer with an M_n of 45 000 was obtained. The MALDI-TOF mass spectrum displayed mass peaks of cyclic polymers of up to 13 000 Da and no linear species was detectable [77].

$$ \tag{48}$$

Although rigid-rod poly(p-phenyleneterephthalamide) analogues having alkyl side-chains did not contain cyclic polymers, the polycondensation of silylated m-phenylenediamine and aliphatic dicarboxylic acid chloride afforded predominantly cyclic polyamides [Eq. (49)] [78]. Furthermore, cyclic polymers were also

8 Polycondensation

produced in polycondensations for polyesters, poly(ether ketone)s, polyimides and polyurethanes [75].

$$\text{Me}_3\text{SiHN-C}_6\text{H}_4\text{-NHSiMe}_3 + \text{ClC(O)(CH}_2)_m\text{C(O)Cl} \longrightarrow \text{(-HN-C}_6\text{H}_4\text{-NH-C(O)(CH}_2)_m\text{C(O)-)}_n \tag{49}$$

8.4.2
Hyperbranched Polymers

Hyperbranched polymers have unique properties such as low viscosity, good solubility and multifunctionality. Dendrimers also have similar properties, but hyperbranched polymers are advantageous in that they can be prepared by a one-step polymerization process. From the beginning of research on hyperbranched polymers, polycondensation has been the main synthetic procedure [79–82]. Here AB_2 condensation monomers for hyperbranched polymers are described. For the polymerization of A_2 and B_3, a review is available [83].

8.4.2.1 Polyphenylene

Hyperbranched polyphenylene was prepared by a Suzuki coupling reaction of a monomer shown in Eq. (50). The M_n value determined by GPC was 32 000. It was soluble in organic solvents such as o-dichlorobenzene, tetrachloroethane and THF, although the main chain of the polymer was composed only of rigid aromatic rings [84, 85]. Another hyperbranched polyphenylene was prepared by Diels–Alder reaction of diene, generated by thermolysis of cyclopentadienone and the triple bond [Eq. (50)] [86].

$$\text{(HO)}_2\text{B-C}_6\text{H}_3(\text{X})_2 \qquad \text{(cyclopentadienone derivative with R groups)} \tag{50}$$

X = Br, Cl

R = H, CH$_3$, Ph

8.4.2.2 Polyester

Many kinds of hyperbranched polyesters have been synthesized. Some AB_2 monomers are shown in Eq. (51). Monomer **28** was first used as an AB_2 monomer for the synthesis of aromatic polyester copolymers with an AB monomer by Kricheldorf and Zang [87] and subsequently the successful homopolymerization of **28** was achieved by Fréchet and coworkers [88]. The M_w value of poly-**28**, determined by gel permeation chromatography–multiple angle laser light scattering (GPC–MALLS), was about 80 000 and the degree of branching (*DB*) was calculated to be 0.55–0.60. The hydroxyl-terminated polymers were soluble in common organic solvents and were thermally stable up to 400 °C.

$$\text{(51)}$$

28, **29a**: X = H, **29b**: X = Me$_3$Si, **30**

Monomer **29** gave the same hyperbranched polyester, but a higher reaction temperature was required to obtain a higher molecular weight polymer: the bulk polymerization of **29a** at 250 °C gave a soluble polymer with $M_w > 10^6$, determined by GPC [89]. Monomer **29b** gave a soluble polymer via melt polymerization at 280 °C [90].

Monomer **30** is another type of AB$_2$ monomer for hyperbranched polyester with many terminal carboxyl groups. The product prepared by melt polymerization at 250 °C was insoluble in organic solvents because of intermolecular dehydration between the carboxyl groups. Hydrolysis of the crude product gave a soluble hyperbranched polyester. The M_w value calculated by universal calibration using coupled differential viscometry was 42 900. The *DB* was about 0.5 [91].

Introduction of alkyl chains into the backbone of monomers leads to a decrease in the glass transition temperature of the resulting hyperbranched polymers, which allows melt polycondensation at low temperature. The melt polymerization of **31** at 190 °C in the presence of an organic tin catalyst proceeded efficiently to yield a hyperbranched aromatic-aliphatic polyester [91]. Monomer **32** also gave another hyperbranched aromatic-aliphatic polyester [92].

Hyperbranched polycarbonate was also synthesized by the polymerization of a monomer shown in Eq. (52) in the presence of AgF at 70 °C, followed by hydrolysis of carbonylimidazolide chain ends, leading to the phenol end-groups. The *DB* of polymer was 0.53. This polymer was soluble in common organic solvents and thermally stable up to 350 °C [93].

$$\text{(52)}$$

31, **32**

(52)

8.4.2.3 Polyamide

Kim first reported the synthesis of hyperbranched aromatic polyamides by the polycondensation of **33** and **34** through neutralization of the hydrochloride in the reaction mixture [Eq. (53)] [94]. GPC analysis of the resulting hyperbranched polyamides showed that the molecular weights were in the range 24 000–46 000. Both carboxyl- and amino-terminated polyamides were soluble in amide solvents such as DMF, NMP and DMAc. NMP solution containing more than 40 wt.% of the carboxyl-terminated hyperbranched polyamide exhibited a nematic liquid crystalline phase and did not lose birefringence up to 150 °C [94]. The direct polycondensation of **35–37** with condensation agents, which activate the carboxylic acid group *in situ*, also afforded hyperbranched polyamides [95–98].

Melt polycondensation is also applicable to AB_2 monomers to prepare hyperbranched aromatic polyamides, although it is generally difficult to obtain aromatic linear polyamides with high molecular weight by the molten polycondensation method. The thermal polymerization of **37** at 235 °C proceeded in a stable molten state to give a glassy solid after cooling to room temperature. The M_w value determined by GPC–MALLS was 74 600, which was similar value for the polymer obtained by direct polycondensation [95].

(53)

8.4.2.4 Polyether

Hyperbranched aromatic polyethers have been prepared by nucleophilic aromatic substitution reaction of AB_2 monomers. Miller et al. first reported the synthesis of hyperbranched poly(aryl ether)s by the polycondensation of **38** [Eq. (54)]. The M_w values were in the range 11 300–134 000, and were dependent on the structure of the spacer groups. The polymers were soluble in common organic solvents such as chloroform and THF. Thermogravimetric analysis under nitrogen revealed that the polyethers retained over 95% of their mass up to 500 °C and had high thermal stability [99].

Since certain heterocycles permit nucleophilic aromatic substitution, Hedric and coworkers used quinoxaline monomer **39** as an AB_2 monomer [100]. The polymerization was carried out at 180–220 °C in the presence of potassium carbonate in aprotic solvents. The intrinsic viscosity was about 0.5 dL g^{-1} for both polymers from **39a** and **39b**.

$$(54)$$

Aromatic–aliphatic hyperbranched polyethers were prepared from a variety of monomers [Eq. (55)]. The polymerization of **40** was carried out in the presence of potassium carbonate and 18-crown-6 in acetone to yield a hyperbranched polymer containing C-alkylated methylene units up to 30%, in addition to O-alkylated benzylic methylene units. The M_w value of the polymer determined by GPC– low-angle laser light scattering (LALLS) exceeded 10^5 [101].

The polymerization of **41** was promoted by generation of alkoxide with sodium particles in THF. When the sodium particle was smaller than 1 mm in size, polymer with a low molecular weight was obtained in high yield. In contrast, the use of sodium particles larger than 1 mm led to high molecular weight polymer in low yield [102]. Cyclized components were detected by MALDI-TOFMS [103].

In the polymerization of **42** with sodium hydroxide and phase transfer catalyst in DMSO–water, the methylene group activated by the cyano group served as a difunctional moiety. The M_n determined by ^{1}H NMR spectroscopy was 9100 [104, 105].

$$\tag{55}$$

8.4.2.5 Poly(Ether Ketone) and Poly(Ether Sulfone)

Although the first example of hyperbranched poly(ether ketone)s was the polymer from **38** [Eq. (54)], simple monomers **43** and **44** were also polymerized with potassium carbonate in N-methylpyrrolidone. Soluble hyperbranched poly(ether ketone)s were isolated in good yields and the M_n values were in the range 20 000–55 000 [Eq. (56)]. Interestingly, the DB of poly-**43** was 0.49, whereas that of poly-**44** was only 0.15 [106].

$$\tag{56}$$

Electrophilic aromatic substitution is also useful for the synthesis of hyperbranched poly(ether ketone)s because the aromatic ring connected with the ether linkage is reactive for electrophiles. The polycondensation of **45** or **46** was carried out with phosphorus pentoxide–methanesulfonic acid as a condensing agent to yield soluble hyperbranched polymers. The ammonium salt of the polymer was soluble in water and behaved as a unimolecular micelle [Eq. (57)] [107, 108].

$$\tag{57}$$

Hyperbranched poly(ether sulfone)s were also synthesized by nucleophilic aromatic substitution, similar to the synthesis of poly(ether ketone)s. The polymerization of **47** or **48** was carried out with Cs_2CO_3 and $Mg(OH)_2$ in DMAc at 150–170 °C. MALDI-TOFMS measurements revealed that an intramolecular cyclization competed with propagation, while GPC measurements suggested the formation of hyperbranched poly(ether sulfone) with $M_n > 10^4$ [Eq. (58)] [109–111].

$$X = F, Cl; R = CH_3, Ph \tag{58}$$

8.4.2.6 Poly(Ether Imide)

There are two approaches to poly(ether imide)s: (1) ether bond formation of the monomer containing an imide ring and (2) imide ring formation through amic acid precursors. Moore and coworkers synthesized a hyperbranched aromatic poly(ether imide) with the first approach [112]. Monomer **49** polymerized with CsF at 240 °C to yield a hyperbranched polymer, which was soluble in common organic solvents. The M_n value determined by GPC was 19 200. The DB value determined by ^{1}H NMR spectroscopy was 0.66. The temperature for 10 wt.% loss in air was 530 °C, which represents one of the most thermally stable hyperbranched polymers [Eq. (59)] [112].

Using the second approach, **50** was polymerized in the presence of condensation agents at room temperature, followed by imidization. Chemical imidization in the presence of acetic anhydride and pyridine resulted in soluble polyimide end-capped with an acetamide group. The polymer possessed a DB of 0.48 and an M_w of 188 000, determined by GPC–MALLS. The temperature for 10 wt.% loss in air was 470 °C [113].

$$\tag{59}$$

8.4.2.7 Polyurethane and Polyurea

Generally, urethane and urea linkages are formed by addition of an alcohol or amine to isocyanate. In AB$_2$ monomer, the isocyanate moiety must be protected because it is liable to react with the hydroxyl and amino groups. Since the aryl–aryl carbamate bond is easily cleaved by heating to form an isocyanate moiety, hyperbranched polyurethane was prepared by the polymerization of **51** in the presence of dibutyltin dilaurate as a catalyst in refluxing THF [Eq. (60)] [114]. The end-capped polyurethanes were thermally stable up to 200 °C, whereas the polymers isolated without the end-capping reaction were not stable above 120 °C.

Carbonyl azide is also known as a synthon of isocyanate. The polymerization of monomers **52** and **53** for polyurethane and **54** for polyurea was also investigated [115–117].

(60)

8.4.3
Polymer Morphology Control

Control of polymer morphology including polymer chain alignment is of great importance to exploit the potential expected from the polymer structure. There have not been many reports of morphology control of the polymer during polymerization even in polymerizations other than polycondensation.

As mentioned in Section 8.1.3, wholly aromatic polyesters such as poly(p-oxybenzoyl) (POB) gave whiskers during polymerization [118–125]. The whisker is a microscale single crystal in which the extended polymer chains are aligned along the long axis of the whisker. Kimura's group prepared a POB whisker by the polymerization of p-acetoxybenzoic acid in liquid paraffin with the elimination of acetic acid without stirring [118], and recently prepared a POB nanowhisker, with width and length of 190 nm and 18.6 m, respectively, by changing the temperature during polymerization [126].

A whisker of poly(p-mercaptobenzoyl) (PMB) was also prepared similarly by the polymerization of S-acetyl-4-mercaptobenzoic acid in liquid paraffin at

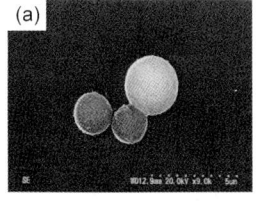

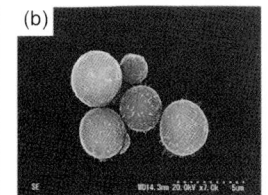

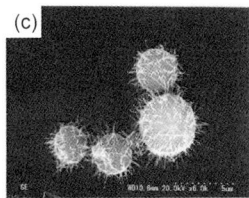

Fig. 8.3 Morphology of the products prepared by the polycondensation of p-acetoxybenzoic acid and 3,5-diacetoxybenzoic acid (10 mol%) in liquid paraffin at 320 °C for (a) 10, (b) 20 and (c) 40 min.

Reproduced with permission from K. Kimura, S. Kohama, S. Kondo, Y. Yamashita, T. Uchida, T. Oohazama, Y. Sakaguchi, *Macromolecules* **2004**, *37*, 1463, Copyright (2004) American Chemical Society.

300 °C [127]. However, the polymerization of the S-propionyl and S-butyryl counterparts yielded PMB microspheres. The microspheres were formed by the liquid–liquid phase separation of oligomers induced by the longer acyl end-groups than the acetyl group of the oligomers due to the reduction in the freezing-point of the oligomers [128].

In the copolymerization of p-acetoxybenzoic acid and 3,5-diacetoxybenzoic acid in liquid paraffin, a microsphere having needle-like crystals on the surface was obtained. This interesting morphology, like pollen and chestnut bur, was formed by the reaction-induced liquid–liquid phase separation and the subsequent crystallization of oligomers during polymerization (Fig. 8.3) [129]. Furthermore, copolymerization at a feed of 10–15 mol% of 3,5-diacetoxybenzoic acid in both monomers yielded a novel network structure comprised of spheres connected by fibrillar crystals. The process of the formation of this morphology is as follows. Co-oligomers were first precipitated at the beginning of the polymerization to form microdroplets. The fibrillar crystals were formed in the coalesced spheres by crystallization of oligomers induced by increase in the molecular weight. The fibrillar crystals connecting the spheres gradually appeared owing to shrinkage of the spheres. The fibrillar crystals grew from the surface of the sphere with crystallization of homo-oligomers of the 4-oxybenzoyl units and finally the network structure was completed (Fig. 8.4) [130].

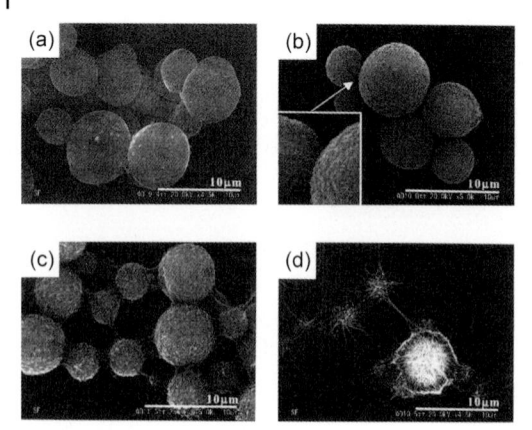

Fig. 8.4 Morphology of the products prepared by the polycondensation of *p*-acetoxybenzoic acid and 3,5-diacetoxybenzoic acid (15 mol%) in Therm S-1000® (mixture of dibenzyltoluene) at 320 °C for (a) 30, (b) 40, (c) 60 and (d) 1200 min. Reproduced with permission from K. Kimura, S. Kohama, S. Kondo, T. Uchida, Y. Yamashita, T. Oohazama, Y. Sakaguchi, *J. Polym. Sci., Part A: Polym. Chem.* **2005**, *43*, 1624, Copyright (2005) Wiley.

8.5
Condensation Polymer Architecture

Development of polycondensation having the nature of living polymerization, as mentioned in Section 8.3, allows condensation polymer architecture to be produced in a similar manner to the architecture that has been attained by living polymerizations of vinyl monomers and cyclic monomers.

8.5.1
Block Copolymers

8.5.1.1 Block Copolymers of Condensation Polymers
By using the polymerization of dimethylsulfoxonium methylide (**15**) initiated by trialkylborane, described in Section 8.3.1, α-hydroxy-ω-(*p*-methoxyphenyl)polymethylene-*b*-polyperdeuteriomethylene was synthesized. Thus, hydroboration of *p*-vinylanisole with $BH_3 \cdot THF$ was first carried out and a solution of **15** was added. Following consumption of ylide, a solution of perdeuterio-**15** was added. After consumption of the second batch of ylide, the tris-polyhomologated alkylborane was oxidized to the terminal alcohol [Eq. (61)] [36]. The polymerization of different arsonium ylides containing **18**, mentioned in Section 8.3.1, initiated by trialkylborane also gave a block copolymer [Eq. (62)] [41].

8.5 Condensation Polymer Architecture

$$\text{MeO-C}_6\text{H}_4\text{-CH=CH}_2 \xrightarrow{\text{BH}_3\text{-THF}} \xrightarrow[\text{15}]{\overset{-}{\text{CH}_2}\overset{+}{\text{SMe}_2}\text{=O}} \xrightarrow{\overset{-}{\text{CD}_2}\overset{+}{\text{S}}(\text{CD}_3)_2\text{=O}} \xrightarrow{\text{H}_2\text{O}_2, \text{NaOH}}$$

MeO–C₆H₄–CH₂CH₂–(CH₂)ₙ–(CD₂)ₘ–OH

(61)

$$\text{Et}_3\text{B} \xrightarrow{\overset{+}{\text{Ph}_3\text{As}}\overset{-}{\diagup}\!\!\overset{R}{\diagdown}} \xrightarrow[\text{18}]{\overset{+}{\text{Ph}_3\text{As}}\overset{-}{\diagup}\!\!\overset{}{\diagdown}} \xrightarrow{\text{H}_2\text{O}_2, \text{NaOH}} \text{Et-[CH=C(R)]}_n\text{-[CH=C(Me)]}_m\text{-OH}$$

(62)

Polyphosphazene block copolymers have also been synthesized. The successive anionic polymerization of N-silylphosphoranimines **19d** and **19a** at 133 °C yielded the block copolymer with an M_w/M_n of 1.4–2.3 [Eq. (63)] [47, 131]. However, due to the presence of two possible leaving groups in **19d**, this approach yielded block copolymers where one of the block segments contained a mixture of side-groups. On the other hand, the cationic polymerization of **19b** with PCl₅ was carried out at ambient temperature, followed by addition of a second phosphoranimine to yield block copolymer with an M_w/M_n of 1.1–1.4, where each block segment had one kind of side-chain [Eq. (64)] [132].

CF₃CH₂O–P(OR')(OR'')=N–SiMe₃ **19d** $\xrightarrow{\text{TBAF}, 133\,°\text{C}}$ CF₃CH₂O–[P(OR')(OR'')=N]ₙ–SiMe₃ $\xrightarrow{(CF_3CH_2O)_3P=N-SiMe_3 \;\; \mathbf{19a}}$

CF₃CH₂O–[P(OR')(OR'')=N]ₙ–[P(OCH₂CF₃)(OCH₂CF₃)=N]ₘ–SiMe₃

R', R'' = CH₃OCH₂CH₂OCH₂CH₂-, CH₃OCH₂CH₂- and/or CF₃CH₂-

(63)

Cl₃P=N–SiMe₃ **19b** $\xrightarrow{\text{PCl}_5}$ [Cl₃P=N–(P(Cl)(Cl)=N)ₙ–PCl₃]⁺ PCl₆⁻ $\xrightarrow[\substack{R=\text{Ph, R'}=\text{Cl}\\R=\text{Me, R'}=\text{Et}\\R=\text{R'}=\text{Me}}]{RR'ClP=N-SiMe_3}$ –(P(Cl)(Cl)=N)ₙ–(P(R)(R')=N)ₘ–

$\xrightarrow{\text{NaOCH}_2\text{CF}_3}$ –(P(OCH₂CF₃)(OCH₂CF₃)=N)ₙ–(P(R)(R')=N)ₘ–

R = Ph, R' = CF₃CH₂O
R = Me, R' = Et
R = R' = Me

(64)

Block copolymers of aromatic polyamides were synthesized by the chain-growth polycondensation of 4-(alkylamino)benzoic acid esters as mentioned in Section 8.3.2.1. An example of a block copolymer of N-alkyl- and N-H-polyamides is shown in Eq. (65) [133]. First, N-octyl monomer **21a** was polymerized and then monomer **21c** having a protecting group on the amino group and a base were added to the reaction mixture. The added **21c** polymerized smoothly from the end of poly-**21a** to yield a block copolymer of poly-**21a** and poly-**21c**. The protecting group was quantitatively removed with trifluoroacetic acid to afford the desired block copolymer of N-alkyl- and N-H-polyamides with narrow polydispersity. The reason why **21c** was used for this block copolymerization was that a monomer having a primary amino group did not polymerize under these polymerization conditions [134]. The block copolymer was arranged in a self-assembly in THF by virtue of intermolecular hydrogen bonding of the N-H-polyamide unit. Scanning electron microscopy (SEM) images showed micrometer-sized bundles and aggregates of flake-like structures.

(65)

Under the conditions for the polycondensation of ethyl 3-(alkylamino)benzoate (**26**) with LHMDS as a base, described in Section 8.3.2.2, a well-defined diblock copolymer of meta- and para-substituted poly(benzamide) was synthesized [Eq. (66)]. Thus, ethyl 3-(octylamino)benzoate (**26a**) was polymerized with 2.2 equiv. of LHMDS at 0 °C to give a prepolymer. A fresh feed of methyl 4-(octylamino)-benzoate (**21b**) was added to the prepolymer in the reaction mixture at the same temperature to yield the block copolymer. It should be noted that excess LHMDS in the polymerization of **26a** as the first step did not react at all with the terminal ester moiety of poly-**26a**, which was able to initiate the polymerization of **21b** as the second step [65].

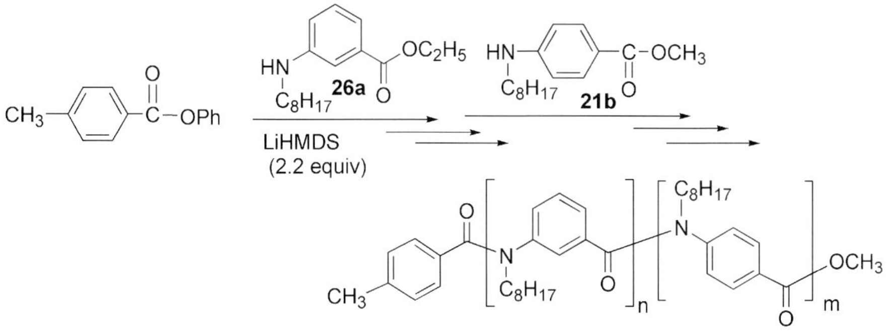

(66)

Chain-growth polycondensation for polyesters also permitted the synthesis of well-defined block polyesters having different side-chains [Eq. (67)] [58].

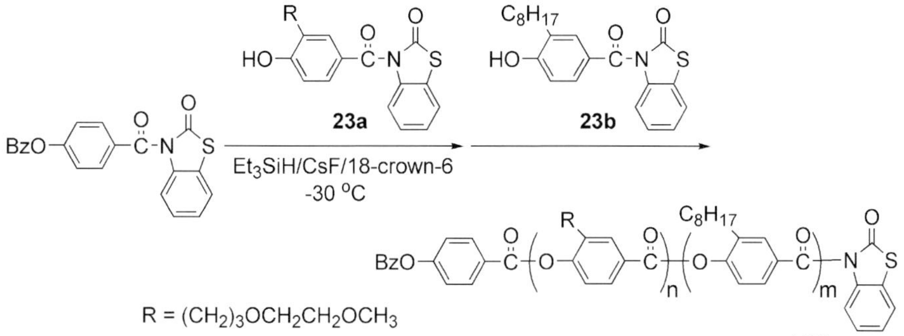

(67)

In the catalyst-transfer polycondensation of Grignard thiophene monomer **27a**, monomer **27b** with a different alkyl side-chain was added to the reaction mixture after consumption of **27a** to yield a block copolymer of polythiophenes [Eq. (68)] [70, 135].

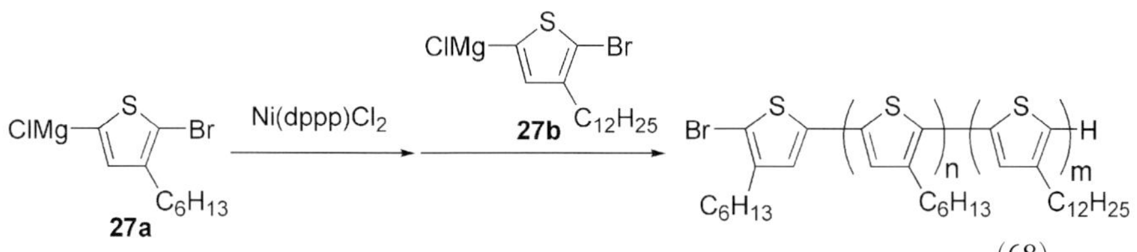

(68)

8.5.1.2 Block Copolymers of Condensation Polymers and Coil Polymers

Some block copolymers of polyphosphazene and coil polymers have been reported. The first example is block copolymers of polyphosphazene and poly(ethylene glycol) (PEG). Amino-terminated PEG was reacted with bromophosphoranimine **55** in the presence of triethylamine to afford PEG–phosphoranimine, which was treated with 2 equiv. of PCl$_5$ at –78 °C, resulting in the formation of macroinitiator **56**. This macroinitiator induced living cationic polymerization of **19b** [see Eq. (35)] to yield the diblock copolymer. Following termination, the chlorine atoms were replaced with trifluoroethoxide groups [Eq. (69)]. When difunctional amino-terminated PEG was used, the triblock copolymer was obtained [136, 137]. The micellar characteristics of this amphiphilic diblock copolymer were studied [138]. Triblock copolymers consisting of PEG and two kinds of polyphosphazenes were also synthesized [139].

$$\text{MeO}\{\text{CH}_2\text{CH}_2\text{O}\}_n\text{CH}_2\text{CH}_2\text{NH}_2 + \text{Br}-\underset{\underset{\text{OCH}_2\text{CF}_3}{|}}{\overset{\overset{\text{OCH}_2\text{CF}_3}{|}}{\text{P}}}=\text{N}-\text{SiMe}_3 \longrightarrow \text{MeO}\{\text{CH}_2\text{CH}_2\text{O}\}_n\text{CH}_2\text{CH}_2\text{NH}-\underset{\underset{\text{OCH}_2\text{CF}_3}{|}}{\overset{\overset{\text{OCH}_2\text{CF}_3}{|}}{\text{P}}}=\text{N}-\text{SiMe}_3$$

55

$$\xrightarrow{\text{PCl}_5} \text{MeO}\{\text{CH}_2\text{CH}_2\text{O}\}_n\text{CH}_2\text{CH}_2\text{NH}-\left[\underset{\underset{\text{OCH}_2\text{CF}_3}{|}}{\overset{\overset{\text{OCH}_2\text{CF}_3}{|}}{\text{P}}}=\text{N}-\text{PCl}_3\right]^+ \text{PCl}_6^- \xrightarrow[\text{19b}]{\text{Cl}_3\text{P}=\text{N}-\text{SiMe}_3} \xrightarrow{\text{NaOCH}_2\text{CF}_3}$$

56

$$\text{MeO}\{\text{CH}_2\text{CH}_2\text{O}\}_n\text{CH}_2\text{CH}_2\text{NH}-\underset{\underset{\text{OCH}_2\text{CF}_3}{|}}{\overset{\overset{\text{OCH}_2\text{CF}_3}{|}}{\text{P}}}-\{\text{N}=\underset{\underset{\text{OCH}_2\text{CF}_3}{|}}{\overset{\overset{\text{OCH}_2\text{CF}_3}{|}}{\text{P}}}\}_m \tag{69}$$

Di- and triblock copolymers of polyphosphazene and polystyrene were prepared by the macroterminator method. Phosphine-terminated polystyrene, prepared by quenching anionic living polystyrene with Ph$_2$PCl, was treated with N$_3$SiMe$_3$ to afford polystyrylphosphoranimine. This species was used as a macromolecular terminator for mono- and difunctional living poly(dichlorophosphazene)s, derived from the cationic polymerization of **19b**, to yield the di- and triblock copolymers, respectively. Following termination, the chlorine atoms were replaced with trifluoroethoxide groups [140] [Eq. (70)]. The monofunctional poly(dichlorophosphazene) was prepared by using nonhalogen phosphoranimines, such as (CF$_3$CH$_2$O)$_3$P=NSiMe$_3$, to generate the initiating species with PCl$_5$ for the living polymerization of **19b** [141]. Diblock copolymers of polystyrene and polyphosphazene with a diethylene glycol monomethyl ether side-chain were also synthesized and the self-association behavior of the block copolymers in aqueous media was investigated [142].

8.5 Condensation Polymer Architecture

$$n\text{-Bu}\!-\!(CH_2CH)_n\!-\!PPh_2 \xrightarrow{N_3SiMe_3} n\text{-Bu}\!-\!(CH_2CH)_n\!-\!PPh_2=NSiMe_3$$
$$\phantom{n\text{-Bu}\!-\!(CH_2CH)_n\!-\!}| |$$
$$\phantom{n\text{-Bu}\!-\!(CH_2CH)_n\!-\!}Ph Ph$$

1) $[Cl_3P=N\text{-}(P=N)_{m-2}\text{-}PCl_3]^+ PCl_6^-$ (with Cl substituents)
2) NaR

1) $[R_3P=N\text{-}(P=N)_{m-1}\text{-}PCl_3]^+ PCl_6^-$ (with Cl substituents)
2) NaR

$R = OCH_2CF_3$

$n\text{-Bu}\!-\!(CH_2CH)_n\!-\!PPh_2=N\!-\!(P=N)_m\!-\!P\!-\!N=PPh_2\!-\!(CHCH_2)_n\!-\!n\text{-Bu}$
(with R substituents, Ph groups)

$R_3P=N\!-\!(P=N)_m\!-\!P\!-\!N=PPh_2\!-\!(CHCH_2)_n\!-\!n\text{-Bu}$
(with R substituents, Ph groups)

(70)

Since the procedure for the formation of a macroinitiator by using **55**, mentioned in the synthesis of block copolymers of polyphosphazene and PEG, allowed the synthesis of monofunctional polyphosphazenes [143], a monoallyl functional polyphosphazene was prepared and underwent hydrosilylation with dihydride-terminated poly(dimethylsiloxane) to produce a polyphosphazene–polysiloxane–polyphosphazene triblock copolymer [Eq. (71)]. When phosphoranimine-terminated polysiloxane was used, polysiloxane–polyphosphazene–polysiloxane triblock copolymer was obtained in a similar manner to that mentioned in the synthesis of block copolymers of polyphosphazene and polystyrene [144, 145].

$$H\!-\!(Si\text{-}O)_m\!-\!Si\!-\!H \; + \; CH_2=CHCH_2NH\!-\!(P=N)_n\!-\!P(OCH_2CF_3)_4 \xrightarrow{\text{Pt catalyst}}$$
(Me, Me substituents on Si; OCH$_2$CF$_3$ substituents on P)

$(CF_3CH_2O)_4P\!-\!(N=P)_n\!-\!NHCH_2CH_2CH_2\!-\!(Si\text{-}O)_m\!-\!Si\!-\!CH_2CH_2CH_2NH\!-\!(P=N)_n\!-\!P(OCH_2CF_3)_4$
(OCH$_2$CF$_3$ substituents on P; Me substituents on Si)

(71)

Block copolymers of aromatic polyamide and conventional coil polymer are prepared by the reaction of the polymer end-group of the polyamide with the living propagating group of the coil polymer. Thus, the phenyl ester moiety of the polyamide reacts with the anionic living end of the coil polymer, whereas the amino group of the polyamide reacts with the cationic living end of the coil polymer (Scheme 8.7).

For example, PEG monomethyl ether was reacted with the polyamide, prepared by the chain-growth polycondensation of **21a**, in the presence of NaH to yield a block copolymer of polyamide and PEG [Eq. (72)]. Excess of PEG was used in this polymer reaction, but unreacted PEG could be washed out with water to isolate the block copolymer [134, 146]. Similar reaction of PEG with a polyamide obtained by the chain-growth polycondensation of **21a** with phenyl terephthalate as a bifunctional initiator gave a PEG–aromatic polyamide–PEG triblock copolymer [147].

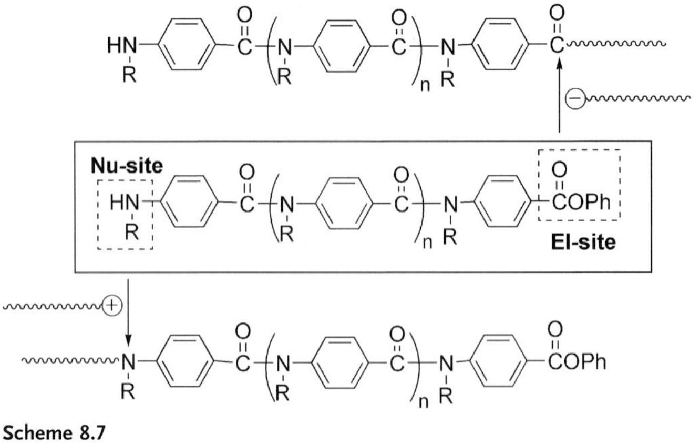

Scheme 8.7

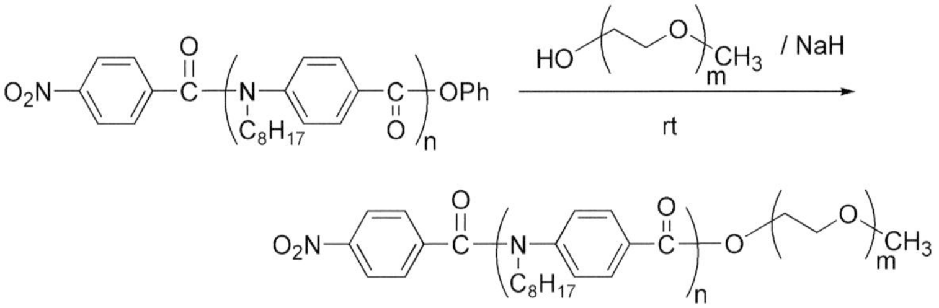

A polyamide having the terminal amino group shown in Scheme 8.7 was prepared by the polymerization of **21a** with an initiator **57** bearing a *tert*-butoxycarbonyl (Boc) group on the amino group, followed by treatment with trifluoroacetic acid to remove the Boc group. The terminal amino group of the polymer reacted with living poly(THF) to yield a block copolymer of polyamide and poly(THF) [Eq. (73)]. When difunctional living poly(THF) initiated by trifluoromethanesulfonic anhydride was reacted with the above polyamide, a polyamide–poly(THF)–polyamide triblock copolymer was produced [148].

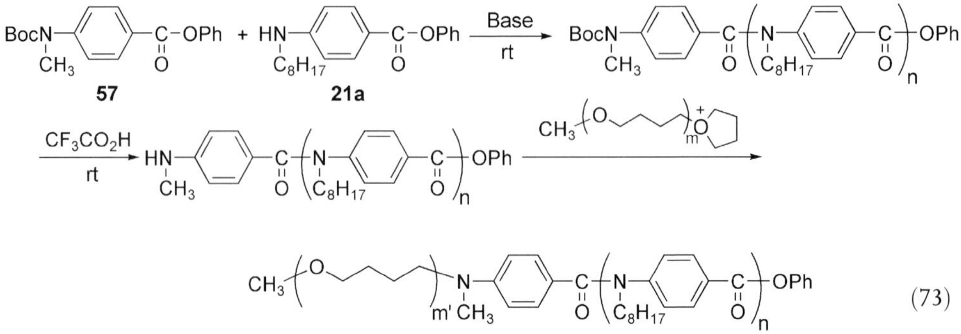

8.5 Condensation Polymer Architecture | 341

Another approach to the block copolymers of aromatic polyamides and coil polymers is the macroinitiator method: the chain-growth polycondensation of **21a** from a macroinitiator derived from coil polymer. A diblock copolymer of polystyrene and a polyamide was synthesized with this approach [Eq. (74)]. First, polystyrene with a terminal carboxyl group was prepared by anionic living polymerization of styrene, followed by quenching with dry-ice, and then the carboxyl group was converted to the phenyl ester by using phenol and a condensation agent. From this terminus, chain-growth polycondensation of **21a** was carried out to yield the desired block copolymer. When low molecular weight macroinitiators were used, block copolymers with low polydispersities were obtained in good yields. When high molecular weight macroinitiators were used, the homopolymer of the polyamide was contaminated. This is probably because the polymer effect of polystyrene decreased the efficiency of initiation from the macroinitiator to induce self-polycondensation of **21a** [149].

(74)

Block copolymers of poly(hexylthiophene) and polystyrene or poly(methyl acrylate) (PMA) were synthesized by the atom transfer radical polymerization (ATRP) of the vinyl monomer from a polythiophene macroinitiator, which was prepared in several steps [150]. After finding catalyst-transfer polycondensation for polythiophene, the block copolymer of polythiophene and PMA was prepared more easily. As mentioned in Section 8.3.3, vinyl-terminated polythiophene was first prepared. The vinyl group was converted to a 2-hydroxyethyl group by hydroboration, followed by esterification with 2-bromopropionyl bromide to give the macroinitiator for ATRP [Eq. (75)] [151]. Allyl-terminated polythiophene, prepared similarly, was used for the synthesis of a block copolymer of poly(hexylthiophene) and polyethylene, in which the ring-opening metathesis polymerization of cyclooctene was carried out in the presence of this polythiophene, followed by hydrogenation [152].

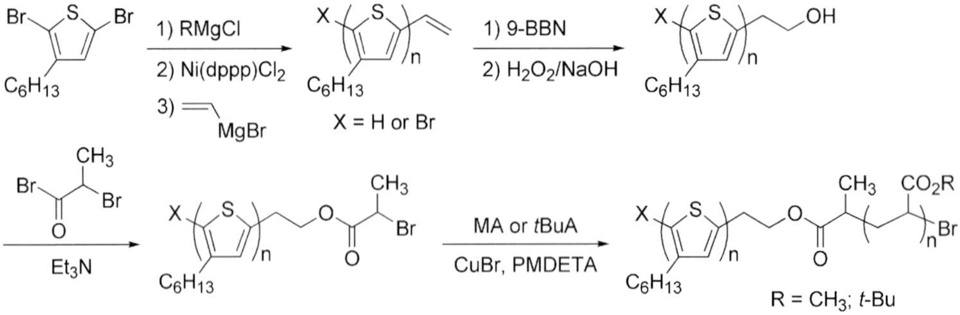

342 | *8 Polycondensation*

(75)

A combination of classical polycondensation and radical living polymerization can also yield block copolymers of condensation polymers and coil polymers, although the condensation polymer unit does not possess a narrow molecular weight distribution. The ABA triblock copolymers of polysulfone (B) with either polystyrene or poly(butyl acrylate) (A) were synthesized by the ATRP of styrene or butyl acrylate from a polysulfone macroinitiator [153, 154] [Eq. (76)]. Telechelic OH polysulfone, prepared by the polycondensation of bisphenol A and 4,4′-difluorosulfone or the dichloro counterpart with bisphenol A in slight excess, was converted to the ATRP macroinitiator by reaction with 2-bromopropionyl bromide or 2-bromoisobutyryl bromide. The M_w/M_n of the macroinitiators was in the range 1.5–1.66. The ATRP of styrene or butyl acrylate from the macroinitiator was carried out at 110 °C in the presence of CuBr and amine ligand in 1,4-dimethoxybenzene or anisole. It is interesting that the M_w/M_ns of the ABA triblock copolymers were narrower than those of the macroinitiators. This was due to the well-defined side blocks prepared by ATRP. The structure and dynamics of the ABA triblock copolymers were also investigated by small-angle X-ray scattering and rheology [154].

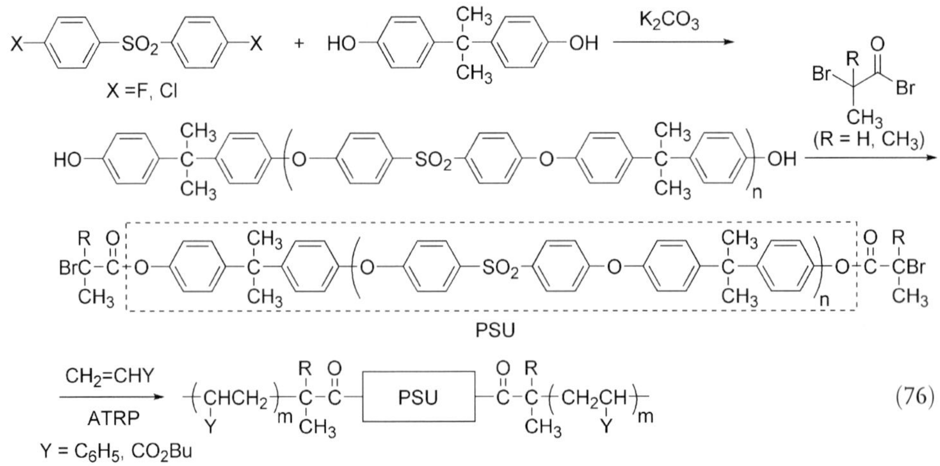

(76)

Telechelic NH$_2$, OH and propargyl polystyrenes were also synthesized by using ATRP, and served as macromonomers for polycondensation and polyaddition [155–158].

8.5.2
Star Polymers

Star polymethylene methanols were synthesized by using the polymerization of dimethylsulfoxonium methylide (**15**) initiated by trialkylborane. For example, polymethylene from **15** was reacted with a,a-dichloromethyl methyl ether, followed by treatment with LiOCEt$_3$ and oxidation with H$_2$O$_2$ and NaOH to give the star polymethylene methanol [Eq. (77)]. The M_n value of the polymer obtained agreed well with the calculated value based on the feed ratio of **15** to triethylborane and the M_w/M_n ratio ranged from 1.02 to 1.13. GPC revealed the absence of linear polymethylene [159]. When 1-boraadamantane·THF (**58**) was used as an initiator for the polymerization of **15**, star polymethylene triol was synthesized (Eq. (78)) [160].

$$BEt_3 \xrightarrow{\underset{15}{CH_2\overset{+}{S}Me_2\ \overset{-}{O}}} \xrightarrow[LiOCEt_3]{CHCl_2OMe} \xrightarrow{H_2O_2,\ NaOH} Et\text{-}(CH_2)_n\text{-}C\text{-}(CH_2)_n\text{-}OH,\ Et\text{-}(CH_2)_n \quad (77)$$

$$\underset{58}{\text{B-THF(adamantane)}} \xrightarrow{\underset{15}{CH_2\overset{+}{S}Me_2\ \overset{-}{O}}} \xrightarrow{H_2O_2/NaOH} HO\text{-}(CH_2)_n\text{-}[adamantane]\text{-}(CH_2)_n\text{-}OH,\ (CH_2)_n\text{-}OH \quad (78)$$

Three-armed star polyphosphazene was also synthesized by the cationic chain-growth polymerization of **19b** from a trifunctional cationic species. This multifunctional initiator was prepared by the reaction of tridentate primary amine with **55**, as mentioned in the synthesis of block copolymers of polyphosphazene and PEG [161].

The chain-growth polycondensation of **21a** for aromatic polyamides was performed from a trifunctional initiator to yield star polyamides with low polydispersities [162]. Polymerization at a higher feed ratio of **21a** to the initiator resulted in not only the three-armed polyamide but also a linear polymer formed by self-polycondensation of **21a**. In the polymerization of **21a** with a monofunctional initiator, no self-polymerization of **21a** takes place as long as the [**21a**]$_0$/[initiator]$_0$ feed ratio is ≤100 [54]. The easy occurrence of the self-polymerization in the polymerization of **21a** with the multifunctional initiator can presumably be ascribed to the low local concentration of the initiator sites in the whole solu-

8.5.3
Graft Polymers

Graft copolymers of polynorbornene, polystyrene or poly(methyl methacrylate) (PMMA) with polyphosphazene have been reported. Monotelechelic polyphosphazene with a norbornene end-group was synthesized through the termination of living poly(dichlorophosphazene) with norbornylphosphoranimine. This material was employed as a macromonomer for the synthesis of a graft copolymer via ring-opening metathesis polymerization of the terminal norbornenyl moiety [Eq. (79)] [163]. With a similar approach, styryl-telechelic polyphosphazene was prepared and underwent radical copolymerization with styrene or MMA to yield a graft copolymer of polystyrene or PMMA with polyphosphazene [Eq. (80)] [140, 164].

(79)

(80)

References

1 For an example: A. S. Hay, F. J. Williams, H. N. Relles, B. M. Boulette, *J. Macromol. Sci. Chem.* **1984**, *A21*, 1065.
2 K. Miyatake, A. R. Hlil, A. S. Hay, *Macromolecules* **2001**, *34*, 4288.
3 H. Iimori, Y. Shibasaki, S. Ando, M. Ueda, *Macromol Symp.* **2003**, *199*, 23.
4 N. Kihara, S. Komatsu, T. Takata, T. Endo, *Macromolecules* **1999**, *32*, 4776.
5 N. Nomura, K. Tsurugi, M. Okada, *J. Am. Chem. Soc.* **1999**, *121*, 7268.
6 N. Nomura, N. Yoshida, K. Tsurugi, K. Aoi, *Macromolecules* **2003**, *36*, 3007.
7 N. Nomura, K. Tsurugi, M. Okada, *Angew. Chem. Int. Ed.* **2001**, *40*, 1932.
8 N. Nomura, K. Tsurugi, T. V. RajanBabu, T. Kondo, *J. Am. Chem. Soc.* **2004**, *126*, 5354.
9 K. Kimura, S. Kohama, Y. Yamashita, *Macromolecules* **2002**, *35*, 7545.
10 K. Kimura, S. Kohama, Y. Yamashita, *Macromolecules* **2003**, *36*, 5045.
11 D. Zhao, J. S. Moore, *J. Am. Chem. Soc.* **2002**, *124*, 9996.
12 D. Zhao, J. S. Moore, *Macromolecules* **2003**, *36*, 2712.
13 D. Zhao, J. S. Moore, *J. Am. Chem. Soc.* **2003**, *125*, 16294.
14 M. Ueda, *Prog. Polym. Sci.* **1999**, *24*, 699.
15 P. Pino, G. P. Lorenzi, U. W. Suter, P. Casartelli, I. Steinman, F. J. Bonner, J. A. Quiroga, *Macromolecules* **1978**, *11*, 624.
16 P. W. Morgan, S. L. Kwolek, *J. Polym. Sci., Part A* **1964**, *2*, 181.
17 J. J. Preston, *J. Polym. Sci., Part A-1* **1970**, *8*, 3135.
18 A. Steinman, U. W. Suter, P. Pino, *Prepr. Int. Symp. Macromol. Chem. (Florence)* **1980**, *2*, 228.
19 M. Ueda, M. Kakuta, T. Morosumi, R. Sato, *Polym. J.* **1991**, *23*, 167.
20 W. R. Meyer, F. T. Gentle, U. W. Suter, *Macromolecules* **1991**, *24*, 642.
21 M. Ueda, T. Morosumi, M. Kakuta, *Polym. J.* **1991**, *23*, 1511.
22 M. Ueda, T. Morosumi, M. Kakuta, R. Sato, *Polym. J.* **1990**, *22*, 733.
23 M. A. Schmucki, P. Pino, U. W. Suter, *Macromolecules* **1985**, *18*, 824.

24 M. Ueda, T. Morosumi, M. Kakuta, J. Sugiyama, *Macromolecules* **1992**, *25*, 6580.
25 T.-A. Chen, R. D. Rieke, *J. Am. Chem. Soc.* **1992**, *114*, 10087.
26 T.-A. Chen, X. Wu, R. D. Rieke, *J. Am. Chem. Soc.* **1995**, *117*, 233.
27 R. D. McCullough, R. D. Lowe, *J. Chem. Soc., Chem. Commun.* **1992**, 70.
28 R. D. McCullough, R. D. Lowe, M. Jayaraman, D. J. Anderson, *J. Org. Chem.* **1993**, *58*, 904.
29 R. Österbacka, C. P. An, X. M. Jiang, Z. V. Vardery, *Science* **2000**, *287*, 839.
30 M. Ueda, H. Sugiyama, *ACS Polym. Prepr.* **1995**, *36*, 717.
31 M. Ueda, T. Okada, *Macromolecules* **1994**, *27*, 3449.
32 M. Ueda, A. Takabayashi, H. Seino, *Macromolecules* **1997**, *30*, 363.
33 O. Haba, H. Seino, K. Aoki, K. Iguchi, M. Ueda, *J. Polym. Sci., Part A: Polym. Chem.* **1998**, *36*, 2309.
34 M. Ueda, H. Sugiyama, *Macromolecules* **1994**, *27*, 240.
35 S. Yu, H. Seino, M. Ueda, *Macromolecules* **1999**, *32*, 1027.
36 K. J. Shea, J. W. Walker, H. Zhu, M. Paz, J. Greaves, *J. Am. Chem. Soc.* **1997**, *119*, 9049.
37 B. B. Busch, M. M. Paz, K. J. Shea, C. L. Staiger, J. M. Stoddard, J. R. Walker, X.-Z. Zhou, H. Zhu, *J. Am. Chem. Soc.* **2002**, *124*, 3636.
38 B. B. Busch, C. L. Staiger, J. M. Stoddard, K. J. Shea, *Macromolecules* **2002**, *35*, 8330.
39 K. J. Shea, S. Y. Lee, B. B. Busch, *J. Org. Chem.* **1998**, *63*, 5746.
40 J.-P.Goddard, P. Lixon, T. L. Gall, C. Mioskowski, *J. Am. Chem. Soc.* **2003**, *125*, 9242.
41 R. Mondière, J.-P. Goddard, G. Carrot, T. L. Gall, C. Mioskowski, *Macromolecules* **2005**, *38*, 663.
42 R. A. Montague, J. B. Green, K. Matyjaszewski, *J. Macromol. Sci. Pure Appl. Chem.* **1995**, *A32*, 1497.
43 C. H. Honeyman, I. Manners, C. T. Morrissey, H. R. Allcock, *J. Am. Chem. Soc.* **1995**, *117*, 7035.
44 H. R. Allcock, S. D. Reeves, C. R. de Denus, C. A. Crane, *Macromolecules* **2001**, *34*, 748.
45 B. Wang, *Macromolecules* **2005**, *38*, 643.
46 R. A. Montague, K. Matyjaszewski, *J. Am. Chem. Soc.* **1990**, *112*, 6721.
47 K. Matyjaszewski, M. K. Moore, M. L. White, *Macromolecules* **1993**, *26*, 6741.
48 R. W. Lenz, C. E. Handlovits, H. A. Smith, *J. Polym. Sci.* **1962**, *58*, 351.
49 A. B. Newton, J. B. Rose, *Polymer* **1972**, *13*, 465.
50 D. B. Hibbert, J. P. B. Sandall, *J. Chem., Soc. Perkin Trans. 2* **1988**, 1739.
51 D. R. Robello, A. Ulman, E. J. Urankar, *Macromolecules* **1993**, *26*, 6718.
52 W. Risse, W. Heitz, *Makromol. Chem.* **1985**, *186*, 1835.
53 V. Percec, J. H. Wang, *J. Polym. Sci., Part A: Polym. Chem.* **1991**, *29*, 63.
54 T. Yokozawa, T. Asai, R. Sugi, S. Ishigooka, S. Hiraoka, *J. Am. Chem. Soc.* **2000**, *122*, 8313.
55 T. Yokozawa, D. Muroya, R. Sugi, A. Yokoyama, *Macromol. Rapid Commun.* **2005**, *26*, 979.
56 Y. Shibasaki, T. Araki, M. Okazaki, M. Ueda, *Polym. J.* **2002**, *34*, 261.
57 A. Yokoyama, K. Iwashita, K. Hirabayashi, A. Aiyama, T. Yokozawa, *Macromolecules* **2003**, *36*, 4328.
58 K. Iwashita, A. Yokoyama, T. Yokozawa, *J. Polym. Sci., Part A: Polym. Chem.* **2005**, *43*, 4109.
59 T. Yokozawa, Y. Suzuki, S. Hiraoka, *J. Am. Chem. Soc.* **2001**, *123*, 9902.
60 Y. Suzuki, S. Hiraoka, A. Yokoyama, T. Yokozawa, *J. Polym. Sci., Part A: Polym. Chem.* **2004**, *42*, 1198.
61 Y. Suzuki, S. Hiraoka, A. Yokoyama, T. Yokozawa, *Macromolecules* **2003**, *36*, 4756.
62 T. Yokozawa, T. Taniguchi, Y. Suzuki, A. Yokoyama, *J. Polym. Sci., Part A: Polym. Chem.* **2002**, *40*, 3460.
63 Y. Wang, K. P. Chan, A. S. Hay, *J. Polym. Sci., Part A: Polym. Chem.* **1996**, *34*, 375.
64 A. Ben-Haida, I. Baxter, H. M. Colquhoun, P. Hodge, F. H. Kohnke, D. J. Williams, *Chem. Commun.* **1997**, 1533.
65 R. Sugi, A. Yokoyama, T. Furuyama, M. Uchiyama, T. Yokozawa, *J. Am. Chem. Soc.* **2005**, *127*, 10172.

66 T. Ohishi, R. Sugi, A. Yokoyama, T. Yokozawa, *J. Polym. Sci., Part A: Polym. Chem.* **2006**, *44*, 4990.
67 R. Sugi, T. Ohishi, A. Yokoyama, T. Yokozawa, *Macromol. Rapid Commun.* **2006**, *27*, 716.
68 R. D. McCullough, *Adv. Mater.* **1998**, *10*, 93.
69 A. Yokoyama, R. Miyakoshi, T. Yokozawa, *Macromolecules* **2004**, *37*, 1169.
70 M. C. Iovu, E. E. Sheina, R. R. Gil, R. D. McCullough, *Macromolecules* **2005**, *38*, 8649.
71 E. E. Sheina, J. Liu, M. C. Iovu, D. W. Laird, R. D. McCullough, *Macromolecules* **2004**, *37*, 3526.
72 R. Miyakoshi, A. Yokoyama, T. Yokozawa, *Macromol. Rapid Commun.* **2004**, *25*, 1663.
73 R. Miyakoshi, A. Yokoyama, T. Yokozawa, *J. Am. Chem. Soc.* **2005**, *127*, 17542.
74 M. Jeffries-El, G. Sauvé, R. D. McCullough, *Macromolecules* **2005**, *38*, 10346.
75 H. R. Kricheldorf, G. Schwarz, *Macromol. Rapid Commun.* **2003**, *24*, 359.
76 H. R. Kricheldorf, S. Böhme, G. Schwarz, C.-L. Schultz, *Macromol. Rapid Commun.* **2002**, *23*, 803.
77 H. R. Kricheldorf, S. Böhme, G. Schwarz, R.-P. Krüger, G. Schultz, *Macromolecules* **2001**, *34*, 8886.
78 H. R. Kricheldorf, S. Böhme, G. Schwarz, *Macromolecules* **2001**, *34*, 8879.
79 Y. H. Kim, *J. Polym. Sci., Part A: Polym. Chem.* **1998**, *36*, 1685.
80 B. Voit, *J. Polym. Sci., Part A: Polym. Chem.* **2000**, *38*, 2505.
81 M. Jikei, M. Kakimoto, *Prog. Polym. Sci.* **2001**, *26*, 1233.
82 M. Jikei, M. Kakimoto, *J. Polym. Sci., Part A: Polym. Chem.* **2004**, *42*, 1293.
83 C. Gao, D. Yan, *Prog. Polym. Sci.* **2004**, *29*, 183.
84 Y. H. Kim, O. W. Webster, *J. Am. Chem. Soc.* **1990**, *112*, 4592.
85 Y. H. Kim, O. W. Webster, *Macromolecules* **1992**, *25*, 5561.
86 F. Morgenroth, K. Müllen, *Tetrahedron* **1997**, *53*, 349.
87 H. R. Kricheldorf, Q.-Z. Zang, G. Schwarz, *Polymer* **1982**, *23*, 1821.
88 C. J. Hawker, R. Lee, J. M. J. Fréchet, *J. Am. Chem. Soc.* **1991**, *113*, 4583.
89 S. R. Turner, B. I. Voit, T. H. Mourey, *Macromolecules* **1993**, *26*, 4617.
90 H. R. Kricheldorf, O. Stöber, D. Lübbers, *Macromolecules* **1995**, *28*, 2118.
91 S. R. Turner, F. Walter, B. I. Voit, T. H. Mourey, *Macromolecules* **1994**, *27*, 1611.
92 P. Kambouris, C. J. Hawker, *J. Chem. Soc., Perkin Trans. 1* **1993**, 2717.
93 D. H. Bolton, K. L. Wooley, *Macromolecules* **1997**, *30*, 1890.
94 Y. H. Kim, *J. Am. Chem. Soc.* **1992**, *114*, 4947.
95 G. Yang, M. Jikei, M. Kakimoto, *Macromolecules* **1998**, *31*, 5964.
96 G. Yang, M. Jikei, M. Kakimoto, *Macromolecules* **1999**, *32*, 2215.
97 S. Russo, A. Boulares, *Macromol. Symp.* **1998**, *128*, 13.
98 O. Haba, H. Tajima, M. Ueda, R. Nagahata, *Chem. Lett.* **1998**, 333.
99 T. M. Miller, T. X. Neenan, E. W. Kwock, S. M. Stein, *J. Am. Chem. Soc.* **1993**, *115*, 356.
100 S. Srinivasan, R. Twieg, J. L. Hedrick, C. W. Hawker, *Macromolecules* **1996**, *29*, 8543.
101 K. E. Uhrich, C. J. Hawker, J. M. J. Fréchet, S. R. Turner, *Macromolecules* **1992**, *25*, 4583.
102 A. Mueller, T. Kowalewski, K. L. Wooley, *Macromolecules* **1998**, *31*, 776.
103 J. K. Gooden, M. L. Gross, A. Mueller, A. D. Stefanescu, K. L. Wooley, *J. Am. Chem. Soc.* **1998**, *120*, 180.
104 R.-H. Jin, Y. Andou, *Macromolecules* **1996**, *29*, 8010.
105 R.-H. Jin, S. Motokucho, Y. Andou, T. Nishikubo, *Macromol. Rapid Commun.* **1998**, *19*, 41.
106 C. J. Hawker, F. Chu, *Macromolecules* **1996**, *29*, 4370.
107 C.-F. Shu, C.-M. Leu, *Macromolecules* **1999**, *32*, 100.
108 C.-F. Shu, C.-M. Leu, F.-Y. Huang, *Polymer* **1999**, *40*, 6591.
109 C. A. Martínez, A. S. Hay, *J. Polym. Sci., Part A: Polym. Chem.* **1997**, *35*, 2015.
110 C. A. Martínez, A. S. Hay, *J. Polym. Sci., Part A: Polym. Chem.* **1997**, *35*, 1781.
111 C. A. Martínez, A. S. Hay, *J. Macromol. Sci., Pure. Appl. Chem.* **1998**, *A35*, 57.

112 D. S. Thompson, L. J. Markoski, J. S. Moore, *Macromolecules* **1999**, *32*, 4764.
113 K. Yamanaka, M. Jikei, M. Kakimoto, *Macromolecules* **2000**, *33*, 1111.
114 R. Spindler, J. M. J. Fréchet, *Macromolecules* **1993**, *26*, 4809.
115 A. Kumar, S. Ramakrishnan, *J. Chem. Soc., Chem. Commun.* **1993**, 1453.
116 A. Kumar, S. Ramakrishnan, *J. Polym. Sci., Part A: Polym. Chem.* **1996**, *34*, 839.
117 A. Kumar, E. W. Meijer, *Chem. Commun.* **1998**, 1629.
118 Y. Yamashita, Y. Kato, S. Endo, K. Kimura, *Macromol. Rapid Commun.* **1988**, *9*, 687.
119 K. Kimura, S. Endo, Y. Kato, Y. Yamashita, *Polymer* **1994**, *35*, 123.
120 K. Kimura, S. Endo, Y. Kato, Y. Yamashita, *Polymer* **1993**, *34*, 1054.
121 K. Kimura, Y. Kato, T. Inaba, Y. Yamashita, *Macromolecules* **1995**, *28*, 255.
122 J. Liu, F. Rybnikar, A. J. East, P. H. Geil, *J. Polym. Sci., Part B: Polym. Phys.* **1993**, *31*, 1923.
123 G. Schwarz, H. R. Kricheldorf, *Macromolecules* **1991**, *24*, 2829.
124 G. Schwarz, H. R. Kricheldorf, *Macromolecules* **1995**, *28*, 3911.
125 J. Liu, F. Rybnikar, P. H. Geil, *J. Polym. Sci., Part B: Polym. Phys.* **1992**, *30*, 1469.
126 K. Kimura, S. Yamazaki, Y. Matsuoka, Y. Yamashita, *Polym. J.* **2005**, *37*, 906.
127 K. Kimura, Y. Yamashita, *Polymer* **1994**, *35*, 3311.
128 K. Kimura, D. Nakajima, K. Kobayashi, S. Kohama, T. Uchida, Y. Yamashita, *Polym. J.* **2005**, *37*, 471.
129 K. Kimura, S. Kohama, S. Kondo, Y. Yamashita, T. Uchida, T. Oohazama, Y. Sakaguchi, *Macromolecules* **2004**, *37*, 1463.
130 K. Kimura, S. Kohama, S. Kondo, T. Uchida, Y. Yamashita, T. Oohazama, Y. Sakaguchi, *J. Polym. Sci., Part A: Polym. Chem.* **2005**, *43*, 1624.
131 M. L. White, K. Matyjaszewki, *Macromol. Chem. Phys.* **1997**, *198*, 665.
132 H. R. Allcock, S. D. Reeves, J. M. Nelson, C. A. Crane, *Macromolecules* **1997**, *30*, 2213.
133 T. Yokozawa, M. Ogawa, A. Sekino, R. Sugi, A. Yokoyama, *J. Am. Chem. Soc.* **2002**, *124*, 15158.
134 T. Yokozawa, M. Ogawa, A. Sekino, R. Sugi, A. Yokoyama, *Macromol. Symp.* **2003**, *199*, 187.
135 R. Miyakoshi, A. Yokoyama, T. Yokozawa, *Polym. Prepr. Jpn.* **2003**, *52*, 1374.
136 J. M. Nelson, A. P. Primrose, T. J. Hartle, H. R. Allcock, *Macromolecules* **1998**, *31*, 947.
137 H. R. Allcock, R. Prange, T. J. Hartle, *Macromolecules* **2001**, *34*, 5470.
138 Y. Chang, R. Prange, H. R. Allcock, S. C. Lee, C. Kim, *Macromolecules* **2002**, *35*, 8556.
139 Y. Chang, E. S. Powell, H. R. Allcock, S. M. Par, C. Kim, *Macromolecules* **2003**, *36*, 2568.
140 R. Prange, S. D. Reeves, H. R. Allcock, *Macromolecules* **2000**, *33*, 5763.
141 H. R. Allcock, E. S. Powell, A. E. Maher, R. L. Prange, C. R. de Denus, *Macromolecules* **2004**, *37*, 3635.
142 H. R. Allcock, E. S. Powell, Y. Chang, C. Kim, *Macromolecules* **2004**, *37*, 7163.
143 H. R. Allcock, J. M. Nelson, R. Prange, C. A. Crane, C. R. de Denus, *Macromolecules* **1999**, *32*, 5736.
144 R. Prange, H. R. Allcock, *Macromolecules* **1999**, *32*, 6390.
145 H. R. Allcock, R. Prange, *Macromolecules* **2001**, *34*, 6858.
146 Y. Izawa, S. Hiraoka, T. Yokozawa, *Polym. Prepr. Jpn.* **2001**, *50*, 215.
147 R. Sugi, Y. Hitaka, A. Sekino, A. Yokoyama, T. Yokozawa, *J. Polym. Sci., Part A: Polym. Chem.* **2003**, *41*, 1341.
148 R. Sugi, A. Yokoyama, T. Yokozawa, *Macromol. Rapid Commun.* **2003**, *24*, 1085.
149 S. Kim, Y. Kakuda, A. Yokoyama, T. Yokozawa, *J. Polym. Sci. Part A: Polym. Chem.*, in press.
150 J. Liu, E. Sheina, T. Kowalewski, R. D. McCullough, *Angew. Chem. Int. Ed.* **2002**, *41*, 329.
151 M. C. Iovu, M. Jeffries-EL, E. E. Sheina, J. R. Cooper, R. D. McCullough, *Polymer* **2005**, *46*, 8582.
152 C. P. Radano, O. A. Scherman, N. Stingelin-Stutzmann, C. Müller,

D. W. Breiby, P. Smith, R. A. J. Janssen, E. W. Meijer, *J. Am. Chem. Soc.* **2005**, *127*, 12502.

153 S. G. Gaynor, K. Matyjaszewski, *Macromolecules* **1997**, *30*, 4241.

154 Y. Zhang, I. S. Chung, J. Huang, K. Matyjaszewski, T. Pakula, *Macromol. Chem. Phys.* **2005**, *206*, 33.

155 K. Matyjaszewski, Y. Nakagawa, S. G. Gaynor, *Macromol. Rapid Commun.* **1997**, *18*, 1057.

156 T. Sarbu, K.-Y. Lin, J. Ell, D. J. Siegwart, J. Spanswick, K. Matyjaszewski, *Macromolecules* **2004**, *37*, 3120.

157 T. Sarbu, K.-Y. Lin, J. Spanswick, R. R. Gil, D. J. Siegwart, K. Matyjaszewski, *Macromolecules* **2004**, *37*, 9694.

158 N. V. Tsarevsky, B. S. Sumerlin, K. Matyjaszewski, *Macromolecules* **2005**, *38*, 3558.

159 K. J. Shea, B. B. Busch, M. M. Paz, *Angew. Chem. Int. Ed.* **1998**, *37*, 1391.

160 C. E. Wagner, K. J. Shea, *Org. Lett.* **2001**, *3*, 3063.

161 J. M. Nelson, H. R. Allcock, *Macromolecules* **1997**, *30*, 1854.

162 R. Sugi, Y. Hitaka, A. Yokoyama, T. Yokozawa, *Macromolecules* **2005**, *38*, 5526.

163 H. R. Allcock, C. R. deDenus, R. Prange, W. R. Laredo *Macromolecules* **2001**, *34*, 2757.

164 H. R. Allcock, E. S. Powell, A. E. Maher, E. B. Berda *Macromolecules* **2004**, *37*, 5824.

9
Supramolecular Polymer Engineering

G. B. W. L. Ligthart, Oren A. Scherman, Rint P. Sijbesma, and E. W. Meijer

9.1
Introduction

The development of polymers in the 20th century has had a dramatic impact on human life. A tremendous accumulation of knowledge in chemistry and physics has spurred the development of innumerous polymeric materials which are currently indispensable. The properties of a polymeric material depend both on the nature of its constituents and on the interactions between them. During the last century, the focus has mainly been on the development of synthetic strategies for polymers consisting of covalently associated monomers. However, with the advent of supramolecular chemistry, the use of non-covalent interactions in polymer chemistry has expanded the range of polymeric materials towards responsive materials. Inspired by the complexity and beauty of the molecular architectures found in nature, supramolecular chemistry uses weak interactions such as hydrogen bonding, hydrophobic and hydrophilic interactions, metal–ligand coordination and ionic interactions to construct complex architectures displaying a certain function. When the covalent bonds that hold together the monomeric repeat units in a macromolecule are replaced by directional non-covalent interactions, supramolecular polymers are obtained (Fig. 9.1). In contrast to covalent polymers, which are irreversibly broken by exerting sufficient force, the formation of supramolecular polymers is reversible and their strength depends on the type of interaction and the chemical environment, such as the solvent or temperature. Through variation of external parameters this reversibility allows for direct control of physical properties. Therefore, supramolecular polymers are expected to reveal new chemical, physical and biomimetic properties. In this chapter, we discuss the scope and limitations of the use of non-covalent interactions in polymeric materials, including a description of the characteristics of the interactions and illustrative examples of their use in polymer chemistry.

Macromolecular Engineering. Precise Synthesis, Materials Properties, Applications.
Edited by K. Matyjaszewski, Y. Gnanou, and L. Leibler
Copyright © 2007 WILEY-VCH Verlag GmbH & Co. KGaA, Weinheim
ISBN: 978-3-527-31446-1

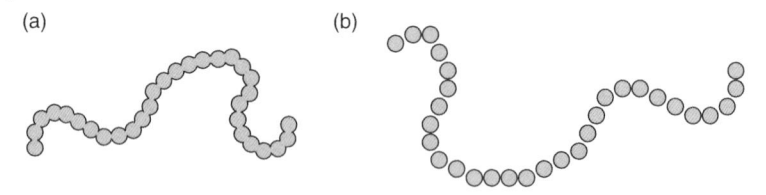

Fig. 9.1 (a) Schematic representation of a covalent polymer; (b) a supramolecular polymer.

9.2
General Aspects of Supramolecular Polymers

Interest in supramolecular polymers has been stimulated to a great extent by the impressive progress made in supramolecular chemistry in general [1] and in the field of synthetic self-assembling molecules in particular [2–5]. The field where supramolecular chemistry and polymer science meet has developed into a vast area of research, ranging from the study of interacting biomacromolecules, such as DNA and proteins, to the self-assembly of large synthetic molecules into well-defined architectures. Examples of the latter include highly organized block-copolymer architectures and self-assembled polymer architectures inspired by the structure of tobacco mosaic virus (TMV) [6]. Non-covalent interactions have also been employed to fold macromolecules (aptly named "foldamers") [7] into well-defined conformations. These examples of using secondary interactions in and between synthetic macromolecules typify the potential of a supramolecular approach to highly organized, functional materials.

In the last two decades, a large number of macromolecular architectures have been disclosed that make use of various non-covalent interactions. Most of the developed structures keep their polymeric properties in solution; it was not until the introduction of multiple hydrogen-bonded supramolecular polymers that systems were obtained that show true polymer materials properties, both in solution and in the solid state [8]. Polymers based on this concept hold promise as a unique class of polymeric materials, because they potentially combine many of the attractive features of conventional polymers with properties that result from the dynamic nature of non-covalent interactions between monomeric units. Structural parameters that determine covalent polymer properties, such as degree of polymerization (DP) and its conformation, are also valid for supramolecular polymers. In addition, dynamic parameters such as lifetime of the chain which are unique to supramolecular polymers are a function of the strength and dissociation rate of the non-covalent interaction, which can be adjusted by external parameters. This dynamic nature results in materials that are able to respond to external stimuli in a way that is not possible for traditional polymers. These aspects of supramolecular polymers have led to a large number of publications bringing together supramolecular chemistry and materials science. It must be stated that some important properties of polymers such as ultimate strength and perfect elasticity require covalent and irreversible bonding

9.2 General Aspects of Supramolecular Polymers

of monomers in a polymer. For applications in which these properties are essential, it is difficult to improve materials properties by introducing supramolecular concepts. However, in many areas these concepts hold enormous potential to improve the properties of polymers in a controlled way.

In order to obtain a fundamental understanding of how to design supramolecular polymers, it is useful to review some of the general aspects. In an earlier review we already proposed a definition for supramolecular polymers [9]. Here, we would like to suggest a slightly modified definition: *Supramolecular polymers are defined as polymeric arrays of monomeric units that are brought together by reversible and highly directional non-covalent interactions, resulting in polymeric properties both in dilute and concentrated solution and in the bulk. The directionality and strength of the supramolecular interaction are important features of systems that can be regarded as polymers and that behave according to well-established theories of polymer physics.* Although the term supramolecular polymers did not include large building blocks bearing recognition motifs and possessing a repetition of chemical fragments in the past, we propose to include these macromolecules in this definition. A major advantage of combining the function of large building blocks with supramolecular interactions can be found in the ease of processing these macromonomers compared with their high molecular weight covalent counterparts.

In general, supramolecular polymers can be classified into two classes: (a) main-chain and (b) side-chain (Fig. 9.2). Main-chain polymers are divided into (i) linear main-chain polymers, (ii) networks and (iii) linear main-chain polymers based on bidirectional units. On the other hand, there are two types of side-chain polymers: (i) polymers bearing binding motifs in the side-chain and (ii) polymers bearing binding motifs in the main-chain.

In a linear main-chain type, supramolecular polymers can either be formed by assembling bi- or multifunctional molecules [(i) and (ii)] or planar structures that have the possibility of assembling on both sides of the plane (iii). In the latter, one structural element is responsible for the formation of the polymer and chain stoppers are difficult to design. Hence the DP is completely governed by the association constant and the concentration. As a result of this structural motif, these supramolecular polymers are rather stiff and resemble rod-like polymers with interesting architectural properties yet exhibit poor mechanical properties in the bulk.

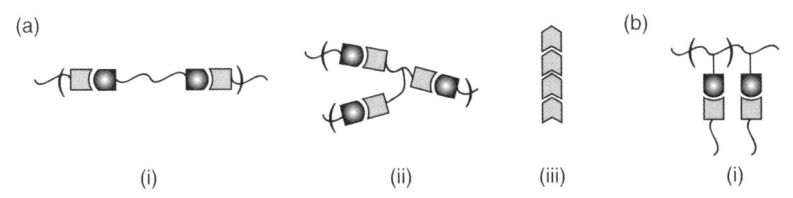

Fig. 9.2 Schematic representation of two classes of supramolecular polymers: (a) main-chain supramolecular polymers and (b) side-chain supramolecular polymers.

Using a directional complementary couple (A–B) or a self-complementary unit (A–A), it is possible to form all known structures of polymers, including linear homo- and copolymers, cross-linked networks and branched structures [10]. In contrast to when self-complementary motifs are applied, the use of complementary motifs allows for the controlled introduction of different structural elements into a supramolecular architecture. Generally, the assembly of bi- or multifunctional monomers can be considered as a step-growth process, with the number-average degree of polymerization (DP) of bifunctional monomers defined by Carothers' equation [11]. If the stepwise polymerization of bifunctional monomers is considered non-cooperative (i.e. equal association constants for each step) and the association constant for the non-covalent interaction is known, the DP can be calculated. The DP is dependent on the concentration of the solution and the association constant and the theoretical relationship is given in Fig. 9.3 [12].

As can be seen from Fig. 9.3, a high association constant between the repeating units is of great importance for obtaining polymers with high molecular weights. In analogy with covalent condensation polymers, the chain length of supramolecular polymers can be tuned by the addition of monofunctional "chain stoppers". This also implies that a high purity of bifunctional monomers is essential, since a small fraction of monofunctional impurity will dramatically decrease the DP. Therefore, as in condensation polymerization, monomer purification is extremely important for obtaining high molecular weight materials.

Supramolecular polymers based on metal coordination bonding can be considered as a separate class since most of the polymers disclosed made use of a strong bonding, in which the reversibility can only be tuned by chemical means. Therefore, these polymers are the closest analogues to conventional macromole-

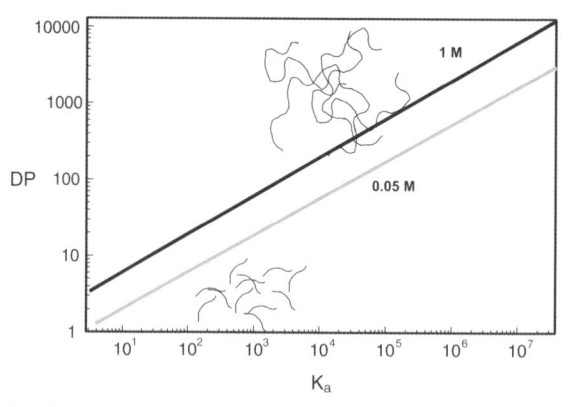

Fig. 9.3 Theoretical relationship between the association constant K_a and DP, using a simple isodesmic association function or "multistage open association" model. Reprinted with permission from Brunsveld, L., Folmer, B. J. B., Meijer, E. W., Sijbesma, R. P. *Chem. Rev.* **2001**, *101*, 4071–4097. Copyright 2001 American Chemical Society.

cules. However, in some cases, an appropriate choice of the metal ion can give rise to reversible bonding. In coordination polymers the DP is similar to that in condensation polymers and achieving exact stoichiometry is also of distinct importance here. Since coordination polymers are not the main focus in this chapter, the reader is referred to Chapter 16 in this volume on inorganic polymers and a number of recent reviews [13–15].

True supramolecular polymers are reversible aggregates that can break and recombine on experimental time-scales. This feature has been investigated in detail by Cates in a physical model, predicting stress relaxation and other viscoelastic properties of entangled "living polymers" as a function of the interaction strength of the monomer end-groups [16–18]. Although this model was developed for worm-like micelles, it was shown by work on ureidopyrimidone-based supramolecular polymers that the model also describes in detail the viscosity behavior of reversible supramolecular polymers [8]. Even though many materials properties of supramolecular polymers are similar to those of traditional polymers, the reversibility can be addressed by external parameters and will lead to unconventional materials properties.

Partly influenced by our own research interests, the focus of this chapter lies in the role of strong and specific non-covalent interactions such as multiple hydrogen bonding and solvophobic interactions in combination with Coulombic interactions to form supramolecular polymers. Special attention is given to bulk properties in these systems. Although crystalline solids in some cases are considered to be supramolecular polymers, we consider these materials and supramolecular polymers to be at opposite ends of the spectrum of molecular materials. In molecular crystals, it is difficult to define a dominant direction of the interactions – crystals are fundamentally three-dimensional – and even when interactions are stronger in one direction than in others, all specific aggregation is lost when these materials are either heated or dissolved. In supramolecular polymers, on the other hand, distinguishable polymeric entities continue to exist in the melt or in (dilute) solution. An interesting intermediate class of materials consists of compounds that form polymers in the liquid crystalline state. In these molecules, cooperativity between a relatively weak non-covalent π-π interaction and excluded volume interactions (which are entropic in nature) can lead to significant degrees of polymerization. However, in the isotropic melt or in solution, most of the polymeric properties are lost. Because supramolecular liquid crystals have played an important role in initiating the field of supramolecular polymers, they will be covered in some detail in this chapter. Polymers that are held together by topological constraints, as in polycatenanes and polyrotaxanes, will not be treated in depth and the reader is referred to other reviews for more detailed information [19, 20]. However, we wish to provide the most illustrative examples of linear interlocked macromolecules. Likewise, supramolecular polymers based on discotic molecules will not be treated in this chapter [21].

9.3
Non-covalent Interactions

The use of each non-covalent interaction has its own set of advantages and limitations and the choice of which interaction to use in designing a supramolecular polymer should be considered carefully. For example, while metal coordination is directional, the strength of the interaction often restricts its dynamic nature and consequently its reversibility. Coulombic interactions between ionic groups, on the other hand, suffer from the fact that these interactions are non-directional in many cases, giving rise to ill-defined aggregation. Solvophobic interactions have the same limitations and are generally weak in solvents besides water. On the other hand, although hydrogen bonding can be strong, binding still remains a challenge in aqueous solutions without the additional aid of hydrophobic interactions.

9.3.1
Hydrogen Bonds

Hydrogen bonds between neutral organic molecules play an important role in determining the three-dimensional structure of chemical and biological systems as a consequence of their specificity and directionality. The most familiar hydrogen bond motifs are the nucleobases found in DNA and RNA. Although hydrogen bonds are not the strongest non-covalent interactions, they hold a prominent place in supramolecular chemistry due to their versatility [22]. Hydrogen bonding arises from a combination of electrostatic, induction, charge-transfer and dispersion forces [23]. Of these effects, electrostatic interactions are generally believed to play the largest role. Hydrogen bonds connect atoms X and Y such as oxygen and nitrogen that have electronegativities larger than that of hydrogen. Generally, the XH group is called the hydrogen bond donor (D) and the Y atom is referred to as the hydrogen bond acceptor. Typically, the energy of a hydrogen bond in the gas phase is 10–80 kJ mol^{-1}. Although this is significantly lower than those of covalent bonds, it is stronger than dipolar or London dispersion force energies (<10 kJ mol^{-1}). The thermodynamic stabilities of hydrogen-bonded complexes in solution are strongly solvent dependent. Stabilities are generally high in apolar solvents without hydrogen bonding properties such as alkanes. In contrast, lower stabilities are displayed in solvents that can act as a hydrogen-bond acceptor or donor themselves such as dimethyl sulfoxide and methanol.

The strong dependence of hydrogen bond strength on electron distribution provides an easy means of influencing the strength. A straightforward way to do so is through electron-donating and -withdrawing substituents. A second approach is the combination of multiple donor and acceptor sites in one molecule. While there are two arrays possible for double hydrogen bond arrays (AA–DD and DA–AD), there are three types of triple hydrogen-bonded dimers and six different quadruple hydrogen-bonded motifs (Fig. 9.4). Although most mo-

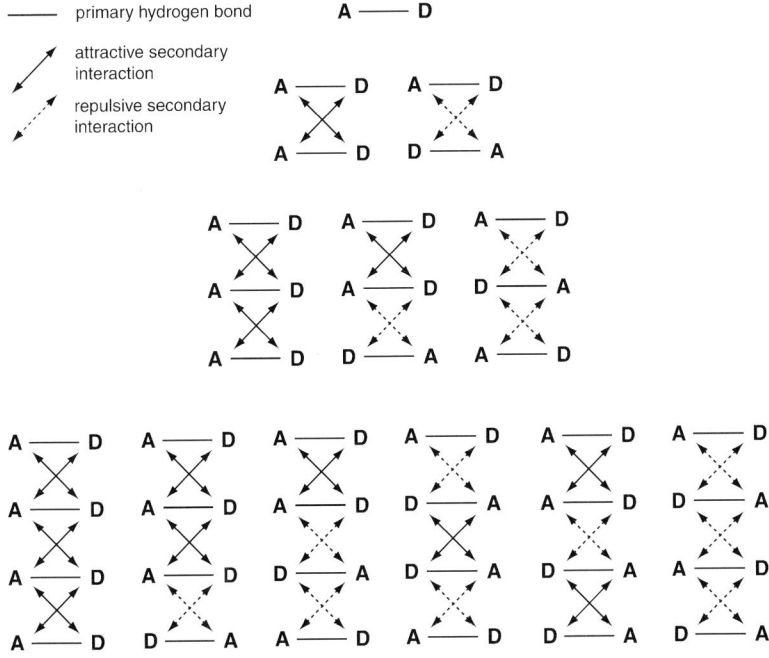

Fig. 9.4 Schematic representation of complexes containing one to four hydrogen bonds together with the secondary interactions hypothesis of Jorgensen and coworkers.

tifs consist of complementary molecules, there are also three self-complementary arrays (DA, AADD and ADAD). The association for these molecules is normally expressed in a dimerization constant (K_{dim}), while binding strength of complementary or heterodimeric complexes is expressed in association constants (K_a). Although the number of primary hydrogen bonds is equal for each double, triple or quadruple hydrogen-bonded unit, these complexes have been found to have very different binding strengths. This observation led Jorgensen and coworkers to propose that secondary electrostatic interactions are important in complex stability [24, 25]. In Fig. 9.4, attractive and repulsive secondary interactions are indicated by solid and dotted double-headed arrows. Early experimental studies on a series of triply hydrogen-bonded complexes by Zimmerman and coworkers provided empirical support of this hypothesis [26, 27]. These effects have been quantified by Sartorius and Schneider for the prediction of multiple hydrogen-bonded complexes in chloroform [28]. Analysis of free energies of complexation for a large number of complexes showed a contribution of ~8 kJ mol^{-1} for each primary interaction, whereas each secondary interaction indicated either an increase or decrease of 2.9 kJ mol^{-1} depending on the interaction being attractive or repulsive. Although good agreement is obtained for small complexes, the validity of this hypothesis has been questioned for complexes with more than three hydrogen bonds in an array [29, 30].

Since the focus of this chapter is on supramolecular polymeric assemblies, for information on well-defined synthetic macromolecular systems the reader is referred to some recent reviews on the use of hydrogen bonds in discrete self-assembly [3, 22, 31–34].

The relationship between the degree of polymerization and the strength of the non-covalent interaction between monomers in a supramolecular polymer (Fig. 9.2) implies that high association constants are required to obtain significant DP. Therefore, either multiple hydrogen bonds must be used or hydrogen bonds should be supported by additional forces, such as excluded volume interactions.

In the last two decades, several reports have described the synthesis of building blocks capable of forming very robust complexes and their association constants have been determined. An illustrative set of complexes is shown in Fig. 9.5. Similarly to Nature, the majority of molecules capable of forming multiple hydrogen bonds, including those in Fig. 9.5, are heterocycles.

In the search for a suitable receptor for barbiturates **1**, Chang and Hamilton developed one of the first strongly associating complexes based on multiple hydrogen bonding [35]. Host **2** shows an inwardly facing cleft capable of making a two triple hydrogen bond arrays with a complementary barbiturate with an association constant of 2.1×10^4 M^{-1} in chloroform. Almost a decade later, in studies aimed at extending hydrogen bonding arrays from three to four linear sites, our group developed the ureido-s-triazine (UTr) dimer **3–3**, which has a dimerization constant (K_{dim}) of 2.0×10^4 M^{-1} in chloroform [36]. As can be seen in the structure, an intramolecular hydrogen bond stabilizes the urea group in the desired conformation. The importance of this interaction becomes clear when the urea is replaced by an amide resulting in a decrease of K_{dim} of nearly two orders of magnitude to 530 M^{-1} in chloroform. Two structurally similar self-complementary molecules **4** and **5** were developed by our group [8, 37, 38] and Zimmerman and coworkers [39, 40], respectively. The ureidopyrimidinone (UPy) **4** could be synthesized very easily in one step from commercially available isocytosines. Crystal structure determination verified the design, as it shows dimers of **4**, which are held together by quadruply hydrogen bonding via a DDAA array [37]. Accurate measurements using a fluorescently tagged UPy derivative determined K_{dim} values of 6×10^7 and 10^8 M^{-1} in chloroform and toluene, respectively [38]. Corbin and Zimmerman had pointed out previously that prototropy (i.e. tautomerism involving proton shifts) can be detrimental to hydrogen-bonded complex formation [39]. Therefore, **5** was designed to display self-complementary quadruple hydrogen bonding arrays irrespective of its tautomeric form. Fluorescence studies analogous to those of Söntjens et al. [38] indicated a lower limit of $K_{dim} > 5 \times 10^8$ M^{-1} in chloroform [40]. Although the previous three dimers form self-complementary arrays, the possibility of complementary multiple hydrogen bond arrays remained largely unexplored until 5 years ago. An early example was presented by Corbin and Zimmerman, who reported the strong selectivity of 2,7-diamido-1,8-naphthyridines **6** for **5** [39]. Using DMSO as a competing cosolvent, they determined a K_a value of 3300 M^{-1} for dimer **5–6** in 5%

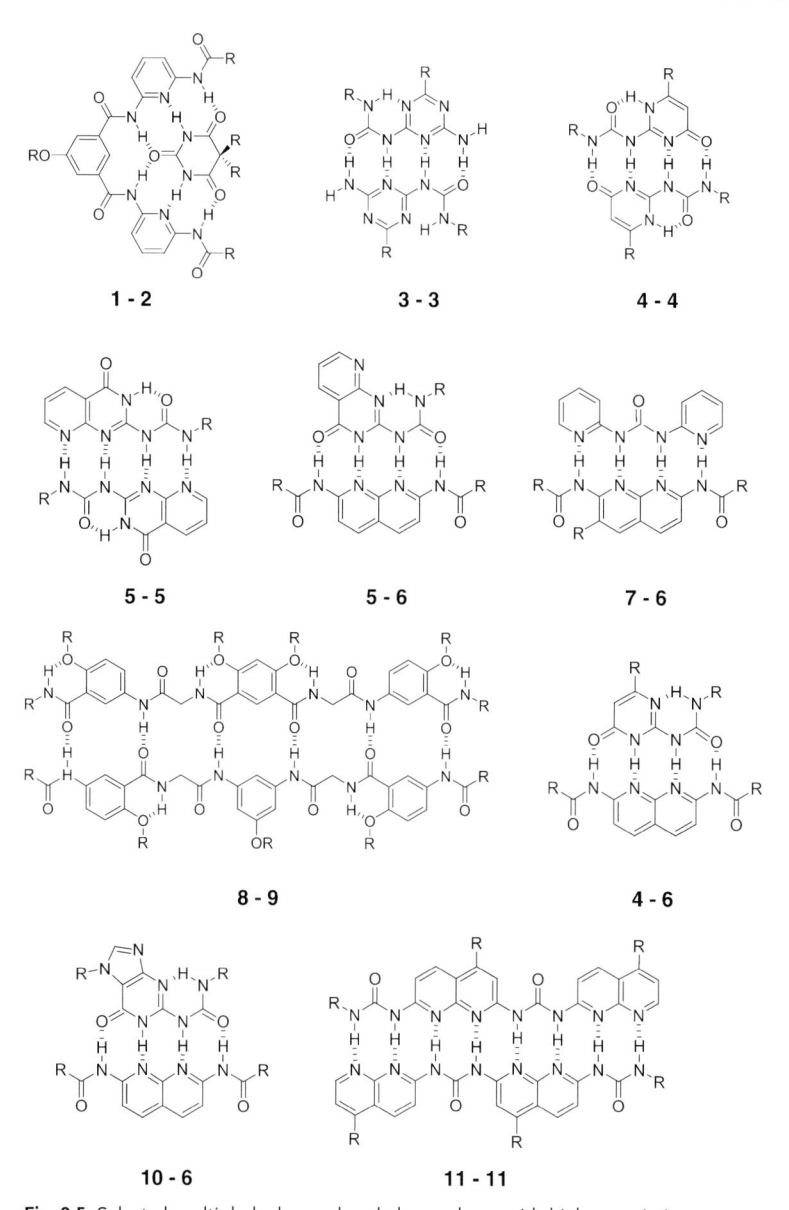

Fig. 9.5 Selected multiple hydrogen-bonded complexes with high association constants.

DMSO-d_6–CDCl$_3$. The importance of the intramolecular hydrogen bond present in **5** is evident when **5** is replaced by *N,N′*-di-2-pyridylurea **7**. These derivatives are known to form inter- and intramolecular hydrogen bonds in solution [41], therefore increasing the energy barrier for formation of heterocomplex **6–7** resulting in K_a values of 2×10^3 M^{-1} in chloroform [42, 43]. Non-heterocyclic compounds **8** and **9** were reported by Gong and coworkers to associate via amide–donor–acceptor interaction reminiscent of those responsible, in part, for the ordered secondary structures of oligopeptides and proteins [44, 45]. Isothermal microcalorimetry was applied to determine an extremely high association constant of 10^9 M^{-1} in chloroform. This strategy of using multiple amide groups not only eliminates repulsive secondary interactions but also its modular approach permits the synthesis of a variety of complementary and self-complementary multiple hydrogen bonding units [46]. More recently, the groups of Li and Zimmerman and ourselves have studied the association behavior of naphthyridine derivatives **6** with complementary tautomeric forms of compounds **4** and **10** [47–50]. Heterodimer **4–6** was shown to exhibit concentration-dependent selectivity with an association constant of 5×10^6 M^{-1} in chloroform [48]. A stronger dimerization of 5×10^7 M^{-1} in chloroform was determined for ureidoguanosine **10** by fluorescence resonance energy transfer experiments [50]. An example of a modular approach in extending linear multiple hydrogen bonding arrays to eight hydrogen bonds in dimer **11–11** was presented recently by Zimmerman and coworkers [51]. NMR dilution experiments in 10% DMSO-d_6–CDCl$_3$ were performed to assess the stability of the dimer indicating a K_a value larger than 5×10^5 M^{-1}.

It is important to note that the hydrogen bonding modules described above are illustrative examples of the types of building blocks that have the potential to be useful in the construction of supramolecular materials. Although there is ample opportunity for developing new multiple hydrogen bonding units, the most important feature for application in supramolecular (co)polymers is the ease of synthesis since molecules **1–11**, with the single exception of **4**, require multi-step syntheses.

9.3.2
Solvophobic and Coulombic Interactions

In addition to directional hydrogen bonding, high association constants are also attainable in the form of inclusion compounds and interlocked structures. Many such compounds typify the term "host–guest" chemistry as the host exhibits an exterior face that interacts with solvent and the interior provides a unique shape and/or environment for the guest molecule. The host molecules are also often referred to as the "wheel" component and the guest molecules that thread them as the "axle" component. This class of complexes is called rotaxanes, a name derived from the Latin *rota* and *axis* meaning wheel and axle, respectively. The non-covalent interactions resulting in these complexes are usually combinations of solvophobic effects and ion–dipole interactions.

Much of the field of supramolecular chemistry finds its origins in the remarkable observations and findings of Pedersen and Cram in their research on crown ethers, spherands, carcerands and, in general, pre-organization. Their studies, and also Lehn's, in the second half of the 20th century allowed for the introduction of many binding motifs for a guest molecule in a specific orienta-

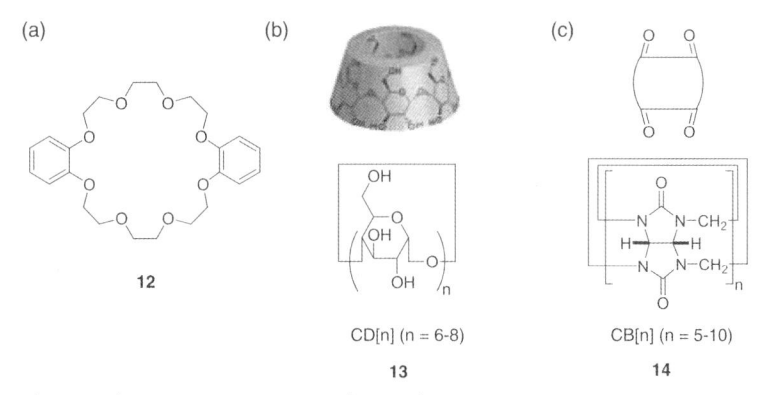

Fig. 9.6 Schematic representation and general structures of wheel components **12–14** used in host–guest chemistry.

tion through synthetic chemistry [1]. In particular, host–guest complexes displaying size and shape selectivity include variations on rotaxanes and catenanes and have led to a large number of papers on the construction of discrete architectures and more recently supramolecular polymers. While linear arrays of hydrogen bonding acceptors and donors allow the tuning of association constants as shown in the previous section, the combination of several attractive interactions in concert has led to remarkably high association constants in rotaxanes in both aqueous and non-polar environments. Typical host compounds include crown ethers such as **12** (K_a values range from 10^2 to 10^4 M^{-1} in polar solvents such as acetonitrile; however, they are strongly dependent on solvent and temperature) [52, 53], the cyclodextrin family **13** (a-, β- and γ-CD; K_a values range up to 10^5 M^{-1} in an aqueous environment) [54, 55] and the cucurbit[n]uril family **14** (CB[5] to CB[10]; K_a values up to 10^{12} M^{-1} in an acidic aqueous environment) [56, 57], as shown in Fig. 9.6. The guest molecules for these hosts range from charged species such as ammonium salts for **12** to a variety of small organics such as adamantyl, ferrocene and *tert*-butyl groups for CD and CB molecules depending on the size of their cavity.

9.4 Supramolecular Polymers

9.4.1 Small Building Blocks

Application of units as associating end-groups in bi- or multifunctional molecules results in the formation of supramolecular polymers with varying DP. The early examples of hydrogen-bonded supramolecular polymers rely on units which associate using single, double or triple hydrogen bonds that all have association constants below 10^3 M^{-1} in chloroform. In isotropic solution, the DP of these polymers is expected to be low. However, in the liquid crystalline state stabilization occurs by excluded volume interactions and the resulting DP is much higher. In the first part of this section, examples of linear supramolecular polymers based on weak hydrogen bonding interactions assisted by liquid crystallinity or phase separation are discussed, followed in the second part by a more thorough discussion of supramolecular polymers based on strong hydrogen bonding and other non-covalent interactions that persist in the isotropic state (melt or solution).

9.4.1.1 Supramolecular Polymers Based on Liquid Crystalline Monomers

The use of directional hydrogen bonds to construct low molecular weight complexes was initially investigated to explain the mesomorphic behavior of carboxylic acids. The first examples of assembly of mesogens based on benzoic acids and pyridines were reported by Kato and Fréchet [58] and Lehn and co-workers [59]. Although the separate components of these assemblies display a narrow liquid crystalline regime or no liquid crystallinity at all, the complexes exhibit thermally stable nematic and smectic (layered) liquid crystalline phases. Shortly thereafter, multisite (≥ 2) mesogens were studied that are considered to be the precursors to main-chain supramolecular polymers [60–62]. The liquid crystalline phase in the supramolecular polymer is stabilized by the increased aspect ratio of aggregates compared with the constituent molecules. There is a strong cooperativity between association and the induction of the liquid crystalline phase, because anisotropy in the liquid crystal strongly enhances the DP of the hydrogen-bonded molecules [63, 64]. Odijk [65, 66], van der Schoot [67] and Ciferri [63] developed a theoretical basis for the relation between chain growth and orientation in supramolecular liquid crystals.

Lehn's group is recognized as the first to develop a supramolecular main-chain polymer based on a ditopic molecule equipped with multiple hydrogen bonding units. These supramolecular polymers are held together by triple hydrogen bonding between diamidopyridines and uracil derivatives (Fig. 9.7a), with an association constant in the order of 10^3 M^{-1} [68, 69]. A 1:1 mixture of the ditopic molecules resulted in aggregates (**15**) that exhibit liquid crystallinity over a broad temperature window, whereas the pure compounds are solids which melt to form isotropic liquids without displaying a liquid crystalline phase. Due to the chirality of the tartaric acid spacer used, the fibers observed by electron microscopy showed biased helicity [70]. The scope of supramolecular polymers was soon expanded by the development of rigid rod polymers (**16**) based on the same triple hydrogen bonding motif [71]. In these polymers, a rigid 9,10-dialkoxyanthracene core connects the hydrogen-bonded groups via a phthalimide group. Due to the increased molecular rigidity, the system is not thermotropic liquid crystalline, but displays a lyotropic liquid crystalline phase that is birefringent. In addition, solutions of these polymers in apolar solvents are highly viscous. By using multifunctional low molecular weight building blocks, St. Pourcain and Griffin were able to obtain materials that exhibit poly-

Fig. 9.7 Liquid crystalline supramolecular polymers: (a) based on triple hydrogen bonding from ditopic chiral monomers (**15**) and from rigid monomers (**16**); (b) based on single hydrogen bonding between multivalent molecules (**17**).

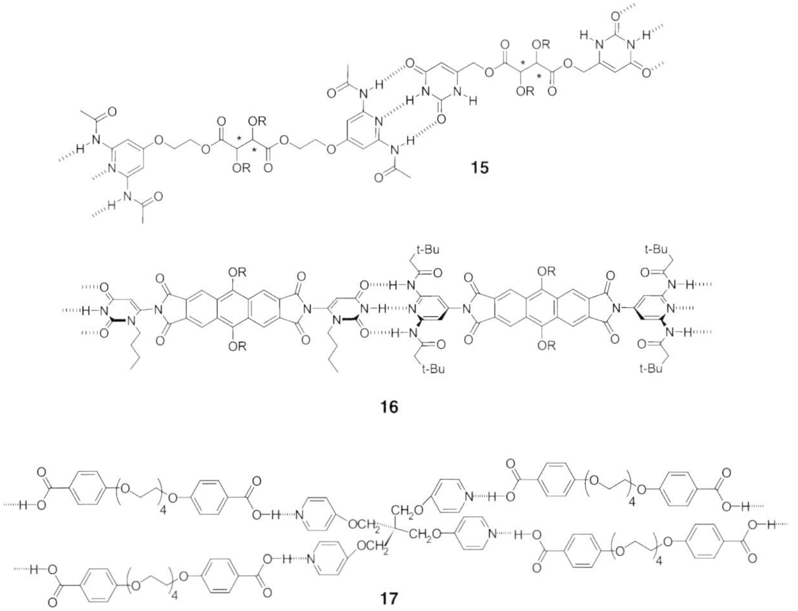

mer-like properties [72]. Hydrogen bonding between pyridine units in a tetrafunctional compound and benzoic acid units in bifunctional compounds (Fig. 9.7b) resulted in the formation of thermoreversible networks (**17**). These materials exhibit properties typical of low molecular weight compounds at high temperature and of polymers at low temperature. Differential scanning calorimetric (DSC) studies on these networks revealed a memory effect, resulting in a consistent decrease in crystallinity as the time during which the material is in the isotropic state increases [73].

9.4.1.2 Supramolecular Polymers in Isotropic Solution

Although chain extension based on single hydrogen bonding can be observed, supramolecular polymers based on non-covalent interactions with $K_a < 10^3 \, \text{M}^{-1}$ are not sufficiently stable to display high DP in solution and in most respects these aggregates resemble small molecules more than polymer chains in solution. Only the triple hydrogen-bonded supramolecular polymers based on bifunctional molecules reported by Lehn and coworkers and multifunctional molecules, such as those reported by St. Pourcain and Griffin [72], displayed some typical polymer properties, such as the capability to be drawn into fibers from the melt.

Up to a decade ago, the number of supramolecular polymers based on very strong non-covalent interactions was relatively small. This was mainly due to the synthetic inaccessibility of most of the monomers. Intriguing architectures such as nanotubes [74] and nanorods [75, 76] are obtained when bifunctional compounds with multiple arrays of hydrogen bonding sites are used. Ghadiri's group has studied nanotubes which self-assemble from bidirectional cyclic peptides (18), composed of an even number of alternating D- and L-amino acids (Fig. 9.8a) [77]. These nanotubes were found to be very robust and turned out to be stable to a wide range of pH and solvents and to physical stress [78]. The cyclic peptides were shown to self-assemble in membranes to form trans-membrane ion channels [79] and the nanotubes have been used in size-selective ion sensing on self-assembled monolayers [80]. Other examples of this subclass of supramolecular main-chain polymers which are based on bidirectional monomers include the polymers based on bisurea motifs developed by Bouteiller and coworkers. These compounds form rigid linear chains in various apolar organic solvents by the cooperative formation of four hydrogen bonds per monomer, as was found by IR spectroscopy, small-angle neutron scattering (SANS), static light scattering and viscometry [81–84]. Figure 9.8b shows a tentative bimolecular wire structure of bisurea molecules which is compatible with the SANS data. More recently, Lehn and coworkers reported the assembly of a bow-shaped molecule into highly ordered sheet-type aggregates via quadruple hydrogen bonds [85]. Although this subclass of supramolecular polymers exhibits many interesting features, they will not be treated in this review.

Independently, the groups of Reinhoudt [75] and Whitesides [76] reported the formation of supramolecular "nanorods" based on the assembly of multiple cyanuric acid–melamine motifs. These hydrogen-bonded polymeric rods are composed of parallel cyanuric acid–melamine rosettes (Fig. 9.8c). The self-assembly of a dimelamine derivative with a dicyanurate derivative results in a distance between the two cyanurate units which is different from the distance between the two melamine units. It was expected that this mismatch would prevent the formation of a closed disk-like assembly and would induce the formation of polymeric entities. Indeed, a 1:1 mixture of dimelamine (19) and dicyanurate (20) resulted in a viscous solution indicating that high molecular weight aggregates were formed. The nature of these and similar aggregates has been investigated by NMR spectroscopy, gel permeation chromatography (GPC) and transmission electron microscopy (TEM). Furthermore, aggregates constructed from compounds similar to 19 and 20 display remarkably high thermodynamic and kinetic stabilities [86, 87].

An important development in the preparation of supramolecular polymers in isotropic solution was the development of the ureidopyrimidinone (UPy) functionality (4), a synthetically accessible quadruple hydrogen bonding unit with a high association constant (6×10^7 M^{-1} in $CDCl_3$) [8, 88]. The UPy unit can be made in a one-step procedure from commercially available isocytosines and isocyanates [37] or amines [89]. Bifunctional monomers (21), possessing two UPy units form very stable and long polymer chains in solution and in the bulk (Fig. 9.9a) [8]. The viscosities of chloroform solutions of 21 are high and exhibit a concentration de-

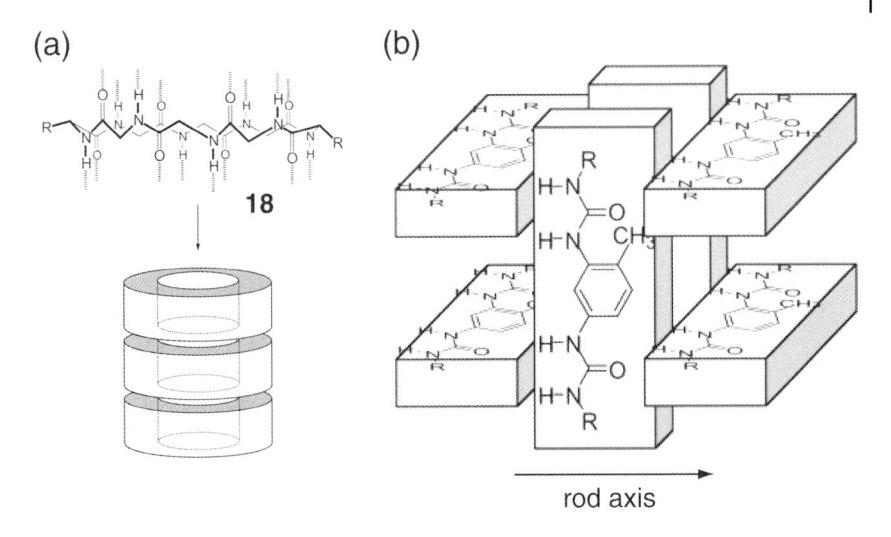

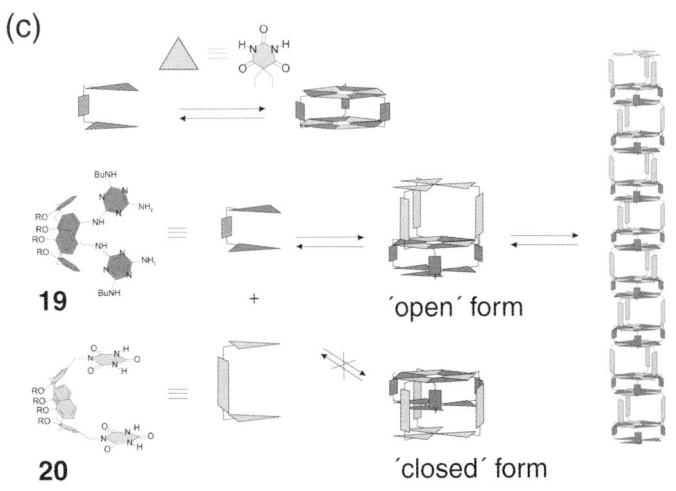

Fig. 9.8 (a) Self-assembled nanotubes based on cyclic peptides (**18**); (b) bimolecular wire structure for bisurea compounds; (c) nano-rods based on the assembly of melamine (**19**) and cyanuric acid (**20**) derivatives. Reprinted with permission from Klok, H. A., Jolliffe, K. A., Schauer, C. L., Prins, L. J., Spatz, J. P., Moller, M., Timmerman, P., Reinhoudt, D. N. *J. Am. Chem. Soc.* **1999**, *121*, 7154–7155. Copyright 1999 American Chemical Society.

pendence that can be attributed to changes in DP. The average DP for highly purified **21** at 40 mM was estimated to be 700, which is consistent with the very high dimerization constant and indicates that no uncontrolled multidirectional gelation occurs. As expected, the presence of monofunctional impurities was shown to lead to a dramatic decrease in DP since they act as chain stoppers and hence limit growth. This was shown by deliberate addition of monofunctional UPy to a solu-

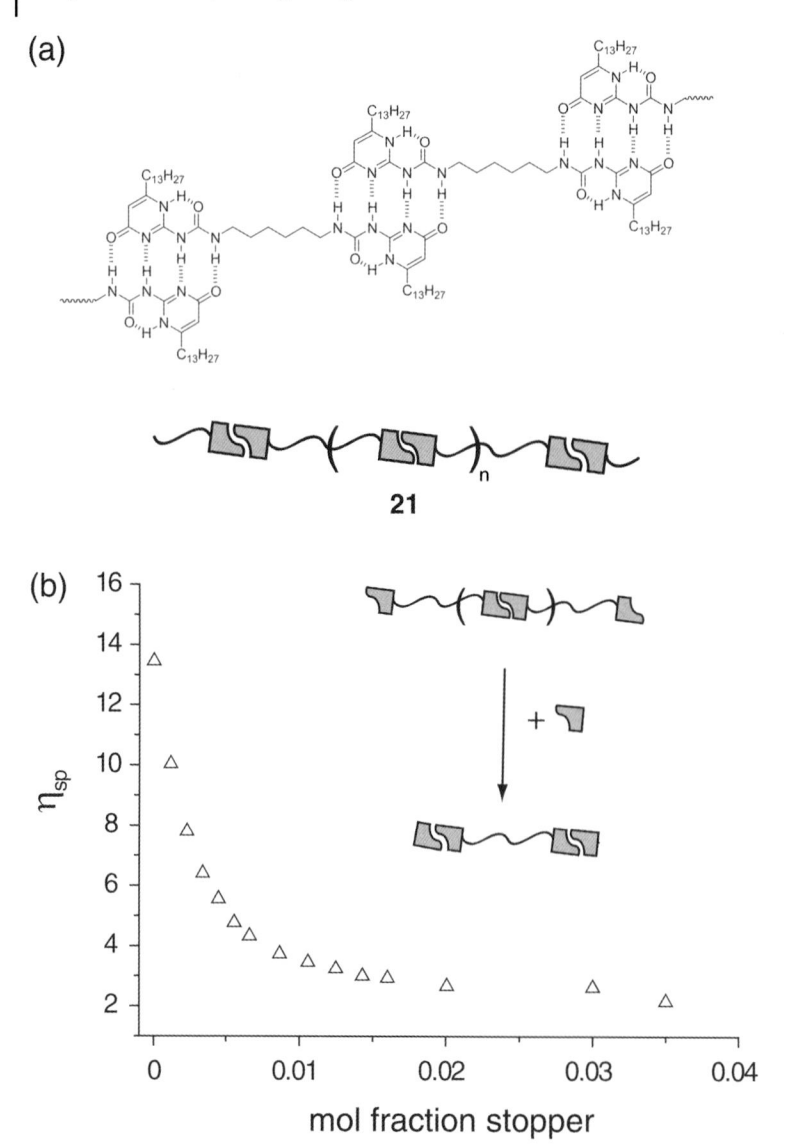

Fig. 9.9 (a) Supramolecular polymer of bifunctional UPy monomer **21**; (b) the effect of a UPy chain stopper on a supramolecular main-chain polymer based on bifunctional UPy monomers.

tion of bifunctional UPy monomers (Fig. 9.9b) and by photogeneration of a monofunctional UPy in a solution of **21** [90].

Although supramolecular polymers based on small bifunctional UPy monomers behave in many ways like conventional polymers, the strong temperature dependence of their mechanical properties really sets them apart from their covalent counterparts. At room temperature, the supramolecular polymers show polymer-like viscoelastic behavior in bulk and solution, whereas at elevated temperatures liquid-like properties are observed. Non-crystallizing UPy derivative **22** was studied using dynamic mechanical analysis (DMTA), rheology and dielectric relaxation spectroscopy in order to obtain information about the bulk viscoelastic properties [91]. One of the salient features of this material, with high relevance for applications, is the extremely high activation energy for viscous flow of 105 kJ mol^{-1} (Fig. 9.10), determined with the Andrade–Eyring equation. This results in a strongly temperature-dependent melt viscosity, which increases processability of these materials at temperatures just above the melting point or T_g. The high activation energy can be attributed to the contribution of three mechanisms to stress relaxation in sheared melts of supramolecular polymers: (i) a mechanism that is shared with covalent polymers, i.e. escape from entanglements by reptation [92]; (ii) supramolecular polymers have enhanced relaxation at higher temperatures because the chains become shorter; and (iii) the supramolecular polymers can lose strain by breaking, followed by recombination of free chain ends without strain [16]. Breaking rates increase with temperature and therefore contribute to the temperature=dependent behavior of supramolecular polymers.

In addition to the work on UPy derivatives, our group also reported studies of bifunctional ureido-s-triazine [36] (UTr) derivatives (**23–26**) [93, 94]. Polymeric aggregates of these derivatives provided with either achiral (**23**) or chiral (**24**) alkyl side-chains are able to self-assemble into columnar structures in dodecane via dimerization and solvophobic stacking of the disk-like dimer as evidenced by SANS [93]. Solutions of **23** in dodecane are highly viscous and show pronounced shear thinning. Furthermore, lateral interactions between the columns cause lyotropic mesophases to form at high concentrations. The combined non-covalent interactions are sufficiently strong to induce formation of columns in

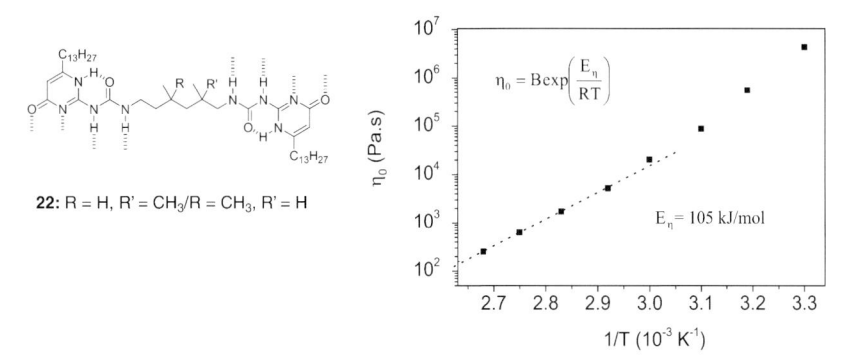

Fig. 9.10 Zero shear viscosity of **22** versus reciprocal temperature; the apparent activation energy is determined using the Andrade–Eyring equation (dotted line).

water by derivatives **25** and **26** equipped with achiral and chiral penta(ethylene oxide) side-chains, respectively [94]. Figure 9.11b shows a schematic representation of the assembly processes. The monomers dimerize into polymeric aggregates in chloroform and are able to organize into columnar disks in dodecane or water. Circular dichroism studies revealed a helical bias in columns formed from monomers **24** and **26**. Moreover, it was shown that introduction of a chiral monofunctional UTr derivative leads to the formation of smaller helixes with biased helicity through cooperative interactions.

Another example which used strong self-complementary binding units was reported by Rebek and coworkers. They described the assembly of ditopic calixarene units functionalized with urea groups on the upper rim (**27**) into polymeric capsules, also termed "polycaps" (Fig. 9.12) [95]. The association between ditopic monomers is based on hydrogen bonding in cooperation with complexation of a solvent molecule or a small guest (**G**). Solutions of these molecules in o-dichlorobenzene show a rheological behavior which is similar to that of polymers, with a strongly concentration-dependent viscosity [96]. The physical integrity of the supramolecular polymers under shear was demonstrated by the observation of strong normal forces in rheometry experiments. Additionally, the "polycaps" can be drawn into fibers with a tensile strength in the order of 10^8 Pa. Networks from tetrafunctional molecules in solution display a stronger elastic component in their rheological behavior and have complicated time-dependent properties, such as shear thickening. As with the linear polycaps, the networks have significant mechanical integrity but the cross-linker imparts a greater elastic component to the viscoelasticity and gels are formed upon mixing tetrafunctional and bifunctional monomers. Furthermore, it was shown that concentrated polycap

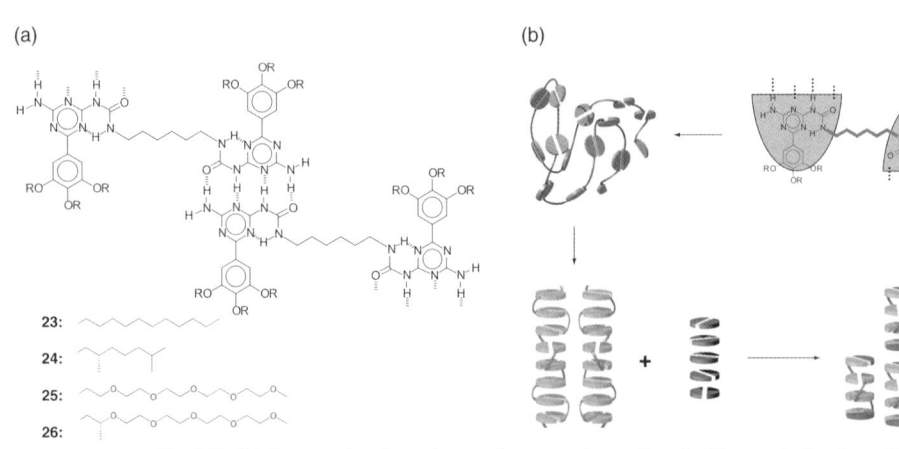

Fig. 9.11 (a) Supramolecular polymer of bifunctional UTr derivatives **23–26**; (b) schematic representation of the formation of columnar aggregates.

Reproduced with permission from Hirschberg, J. H. K. K., Brunsveld, L., Ramzi, A., Vekemans, J. A. J. M., Sijbesma, R. P., Meijer, E. W. *Nature* **2000**, *407*, 167–170.

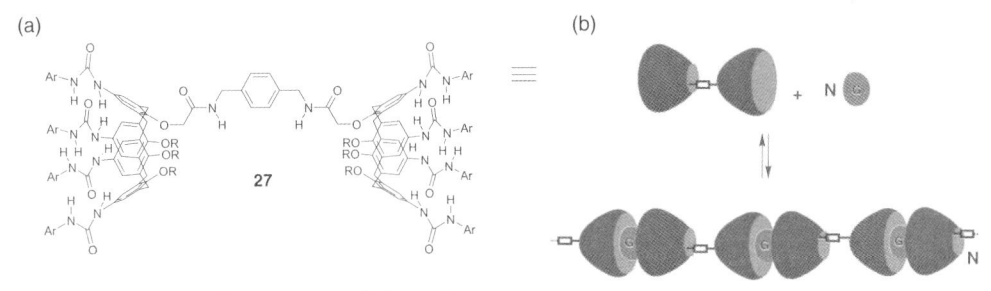

Fig. 9.12 (a) Ditopic calixarene (**27**) and (b) a schematic representation of polycap formation of **27** with guest molecules (**G**). Adapted from Castellano, R. K., Rudkevich, D. M., Rebek, J., Jr. *Proc. Natl. Acad. Sci. USA* **1997**, *94*, 7132–7137. Copyright 1997 National Academy of Sciences, USA.

solutions in chloroform display liquid crystallinity and shear can be used to orient the liquid crystal formulations [97].

Although the aforementioned polymers were prepared by assembly of monomers equipped with self-complementary hydrogen bonding motifs, supramolecular polymers have also been prepared in isotropic solution with complementary motifs A and B. In this way, directionality can be introduced into the supramolecular polymer chain. One of the first examples employing strong complementary hydrogen bonding was reported by Rebek and coworkers, who studied the assembly of bi- and trifunctional polycaps based on the heterodimerization of calixarenes equipped with arylurea and sulfonylurea [98]. The resulting structures were analyzed by ^{1}H NMR and GPC techniques. Inspired by work on multiple hydrogen bond motifs by Hamilton, Lehn's group reported supramolecular polymers based on six-fold hydrogen bonding between a 2,6-bis(diamidopyridine)phenol (**2**) and a double-faced cyanuric acid-type wedge (**28**) in apolar and chlorinated solvents (Fig. 9.13a). ^{1}H NMR spectroscopy, viscometry and electron microscopy (EM) revealed the formation of polymeric aggregates in tetrachloroethane (TCE) and fibers were observed in micrographs of samples of various mixtures. Moreover, a marked influence of stoichiometry and also endcapping and cross-linking upon fiber formation is revealed in solution and by EM. Whereas stoichiometric imbalance and the addition of monofunctional molecules led to a decrease in DP, the introduction of trifunctional molecules led to a cross-linked network (Fig. 9.13b).

The last examples based on the assembly of bifunctional small building blocks furnished with complementary hydrogen bonding motifs are the polymers of Craig and coworkers, which assemble by complementary base pairing between oligonucleotides (Fig. 9.14) [99, 100]. Among several studies of the assembly of various oligonucleotides, polymerization of A–B-type monomer **29** (X=no spacer) was substantiated with UV/visible melting studies. Static and dynamic light scattering (SLS and DLS) and size-exclusion chromatography (SEC) were used to probe the size, structure and dynamics of the system and demon-

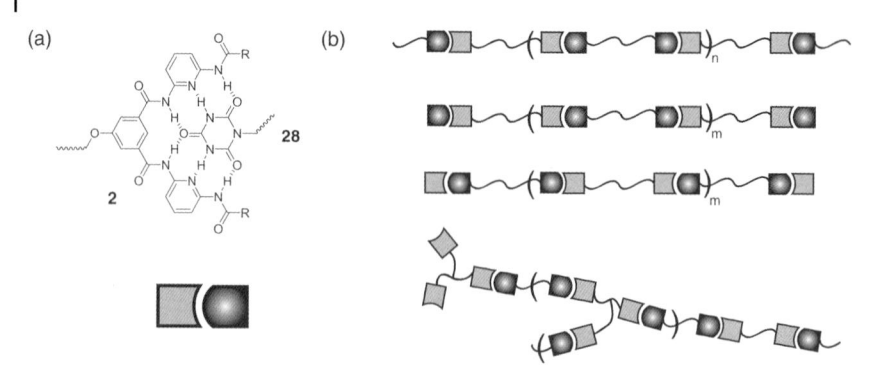

Fig. 9.13 Supramolecular polymerization of hetero-complementary binding units: (a) heterocomplex **2–28**; (b) schematic representation of prepared structures with mono-, bi- and trifunctional building blocks ($m<n$).

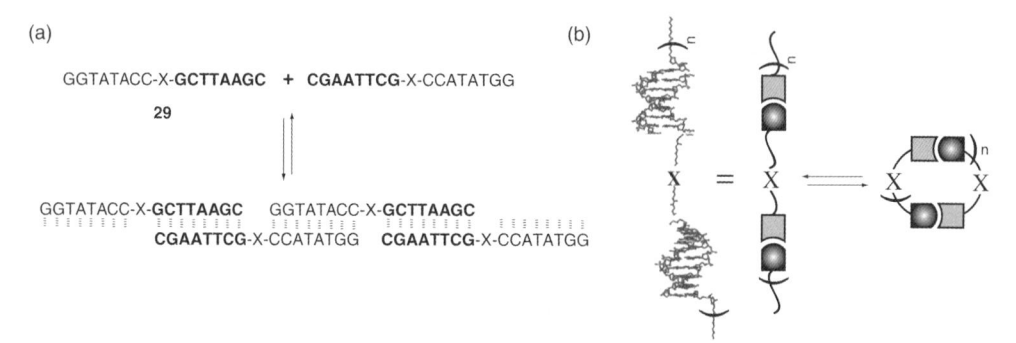

Fig. 9.14 Supramolecular polymerization of complementary oligonucleotides: (a) self-assembly of oligonucleotide **29** (X=no spacer); (b) schematic representation of the dynamic equilibrium in nucleotide based polymers. Reprinted with permission from Xu, J., Fogleman, E.A., Craig, S.L. *Macromolecules* **2004**, *37*, 1863–1870. Copyright 2004 American Chemical Society.

strated that polymers based on **29** resemble rod-like double-stranded DNA. Specifically, viscometry measurements on **29** are in very good agreement with theoretical predictions by Cates since a slope of 3.7 is found for the semidilute regime in a double logarithmic plot of specific viscosity versus concentration. Remarkably, incorporation of short spacers for X within the monomer results in an increased probability of cyclization and increased time-scales of equilibration. This is in contrast to other systems described in Section 9.5 where an increase in spacer length, with X=–$(CH_2)_3$– or –$[(CH_2)_2O]_6$–, results in a decreased tendency to form cyclic aggregates.

Examples of linear supramolecular polymeric materials that utilize host–guest interactions are less abundant. Although several examples of main-chain polyrotaxanes have appeared in the literature, they will not be covered in this chapter, as the polymer chains themselves are all held together by covalent rather than supramolecular interactions. As discussed previously, linear supramolecular polymers can be constructed by combining equivalent amounts of AA and BB complementary units and also by the self-assembly of AB monomers. Examples of AB-type monomers include work by Harada and coworkers on monosubstituted α-cyclodextrins and work on crown ether–ammonium salt oligomers by the groups of Stoddart, Gibson and others. Figure 9.15 illustrates several AB molecules which have appeared in the literature to date.

One of the earlier attempts to make polymeric materials by pseudo-rotaxane inclusion complexes was reported by Cantrill et al. [101]. They reported the synthesis of several dibenzylammonium ion-substituted crown ethers such as **30** and suggested that the interconversion between linear and cyclic daisy-chain arrays might be controlled by factors such as concentration and temperature. However, they concluded that, under a wide range of experimental conditions, most such species existed in the cyclic daisy-chain conformation and fell far short of yielding any linear polymeric assemblies. Among others, Gibson and coworkers also reported the preparation of monosubstituted crown ethers such as **31**. Instead of a dibenzylammonium salt, they employed a methylviologen moiety as the guest [102]. While they suggested that a monocyclic inclusion complex structure was not possible for **31** due to the rigidity of the molecule, only tetrameric species were observed in mass spectrometry experiments. Solution NMR and viscometry experiments, on the other hand, suggested that higher molecular weight species were in fact being formed at high concentrations (>2 M).

An alternative method to linear supramolecular polymers from crown ethers makes use of a two-component system. Here, equimolar amounts of a biscrown ether **33** and a bisdibenzylammonium salt **34** are combined to form an AA–BB linear material (Fig. 9.16). These components self assemble to form a reversible linear supramolecular polymer in solution (acetone–chloroform) with molecular weights up to 20 000 g mol^{-1} (NMR end-group analysis) [103]. Nevertheless, aside from a handful of papers, linear supramolecular polymers from crown ether-containing molecules have not attracted much attention.

In the area of cyclodextrin-based supramolecular polymers, however, Harada and coworkers have reported some progress towards higher molecular weight materials both from AB and AA+BB methods. Recently, they reported the prepa-

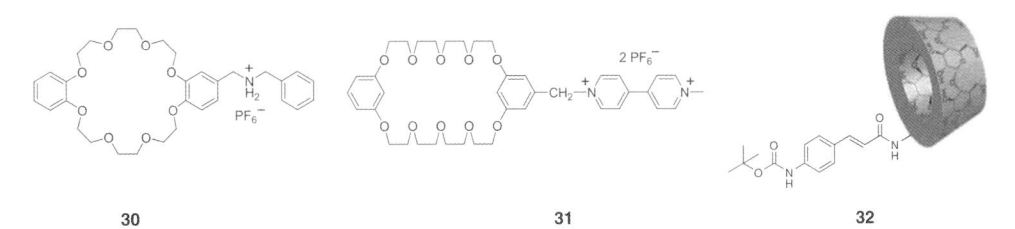

Fig. 9.15 AB-type monomers for supramolecular daisy-chain aggregates.

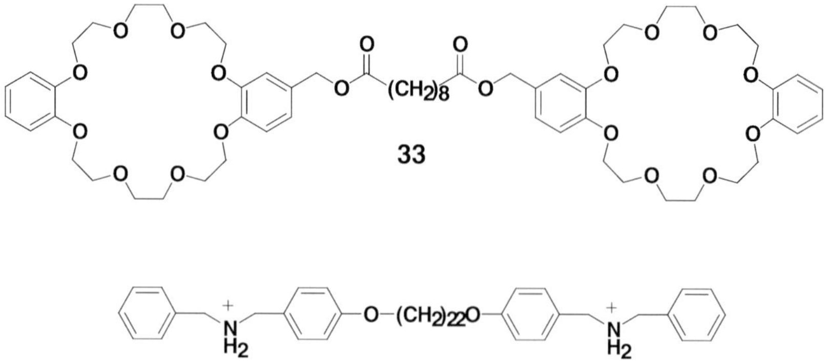

Fig. 9.16 AA ditopic host **33** and BB ditopic guest **34**.

ration of a chiral supramolecular "oligomer" in water through the self-assembly of a monosubstituted α-CD (**32**). A *t*-Boc-protected cinnamoylamino group on the 3-position of an α-CD molecule was used to prepare the AB-type monomer [104]. Both vapor-phase osmometry (VPO) and mass spectrometry experiments suggested a polymeric material with a molecular weight up to 18 000 g mol^{-1}. In addition, circular dichroism experiments showed that the linear supramolecular polymer that formed maintained a specific helicity. The same group also reported the construction of supramolecular polymers by the two-component mixing of bis-CD monomer with a bisadamantyl monomer. The host–guest components were separated by either a rigid alkyl [105] or a flexible oligoethylene glycol linker [106] (Fig. 9.17). Initial results suggest that rigid linkers lead to higher MW linear supramolecular polymers, while the use of a flexible linker increases the amount of cyclic and oligomeric supramolecular structures. In addition, VPO results indicate that molecular weights may be as high as 90–100 kDa. However, the lower K_a values of 10^3–10^4 M^{-1}, the highly dynamic nature of these host–guest interactions and the limitation of these polymerizations to aqueous and polar media may limit the utility of CD inclusion complexes in the construction of supramolecular polymers in the future.

Lastly, the use of cucurbit[*n*]urils (CB[*n*]s) in the construction of linear supramolecular polymers has so far been limited to one paper, by Kim and coworkers [107]. They reported a surface-initiated polymerization stemming from a CB stabilized charge-transfer complex of hydroxynaphthalene with a methylviologen derivative. A CB[8] molecule was employed, capable of sequestering two guest molecules in its cavity forming a charge-transfer complex. A thiolpyridyl derivative was first used to form a self-assembled monolayer on gold with CB[8]. Subsequently, upon introduction of the AB monomer, a directional, linear supramolecular polymer was grown on the surface containing four repeat units on average. Unfortunately, the solubility of CB molecules is currently limited to acidic aqueous environments, which could potentially limit its utility in supramolecular polymeric materials.

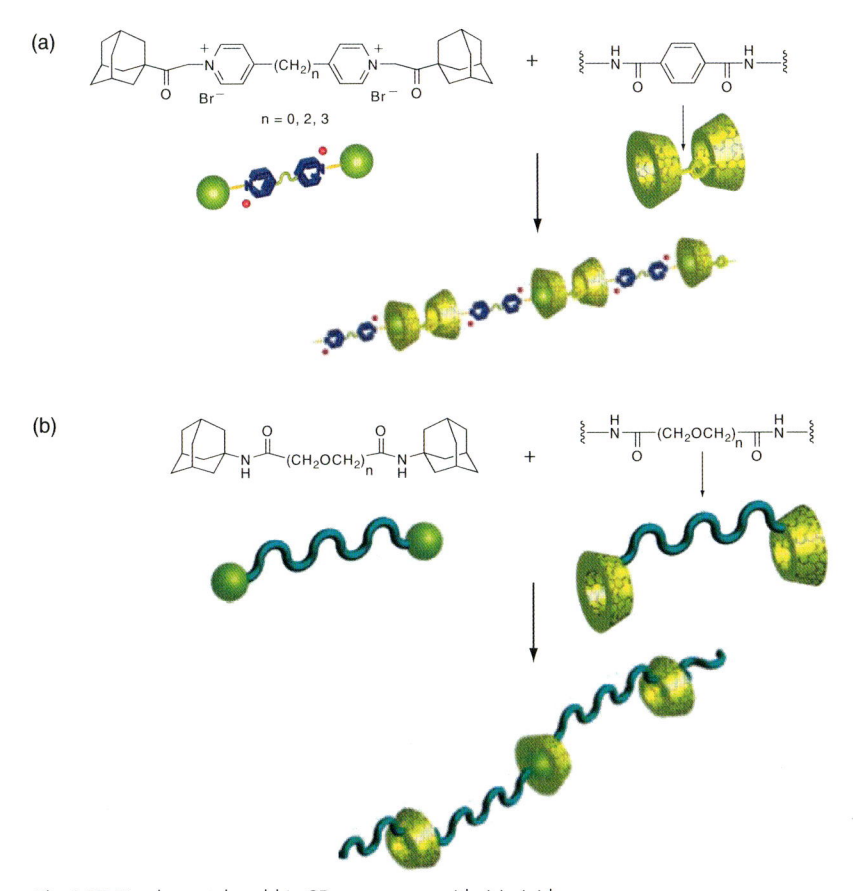

Fig. 9.17 Bisadamantyl and bis-CD monomers with (a) rigid and (b) flexible linkers. Reprinted with permission from Hasegawa, Y., Miyauchi, M., Takashima, Y., Yamaguchi, H., Harada, A. *Macromolecules* **2005**, *38*, 3724–3730. Copyright 2005 American Chemical Society.

9.4.2
Large Building Blocks

Although the supramolecular polymers based on small building blocks possess intriguing new properties, the need for large amounts of material for, e.g., melt-rheological experiments and tensile testing in combination with synthetic barriers do not allow extensive study of the mechanical properties of these materials. Furthermore, small building blocks have an increased tendency to form small discrete assemblies through cyclization (see Section 9.5). Finally, in some cases the recognition unit itself is over several kDa in size and contributes significantly to the observed material properties. Therefore, in order to obtain supramolecular polymers with tunable and macroscopic material properties, the supramolecular functionalities need to be separated by polymeric spacers (or appended to polymer chains). In this way, macromonomers can be elongated or functionalized by reversible interactions. This section will focus on supramolecular polymers based on polymeric building blocks that are equipped with recognition units and their materials properties in solution and in the bulk.

9.4.2.1 Main-chain Supramolecular Polymers

The physical properties of linear polymers and organic molecules are expected to be strongly modified when they are provided with associating end-groups due to the reversible nature of non-covalent interactions. One of the first to acknowledge this were Lenz and coworkers who proposed that the liquid-crystalline behavior and the ability to form elastomeric films of polymeric glycols terminated with diacids could be explained by dimerization of the carboxylic endgroups [108]. Lillya and coworkers have also demonstrated that dimerization of the carboxylic acid moieties in benzoic acid telechelic poly-THF results in a significant improvement in material properties due to the formation of large crystalline domains of the hydrogen-bonded units [109]. A smaller improvement in material properties was observed when polydimethylsiloxanes (PDMS) were provided with benzoic acid groups [110]. Based on detailed FT-IR and NMR spectroscopy and viscometry studies, a quantitative model for the chain length and weight distribution of the functionalized PDMS in solution was described by Bouteiller and coworkers [111–113].

Utilizing a post-polymerization modification route, the groups of Rowan and Long applied nucleobases as recognition motifs to functionalize low MW poly-THF ($M_n = 1.4$ kg mol^{-1}) and polystyrene (PS) ($M_n = 2.6$ kg mol^{-1}) [114–116]. ^{1}H NMR-analysis confirmed the formation of assemblies in a 1:1 mixture of adenine–PS and thymine–PS and thermoreversibility was substantiated in toluene-d_8. In addition, attachment of a nucleobase derivative at the chain ends of bis-amino telechelic poly-THF (Fig. 9.18a) results in a profound change in the properties of the material with an increase of over 100 °C in T_m. While amine-terminated poly-THF is a soft, waxy solid, both adenine- and cytosine-derived telechelic polymers **35** and **36** self-assemble in the bulk to yield materials with sufficient mechanical stability to be melt processed (Fig. 9.18b and c). Both materials show extreme temperature sensitivity, resulting in the formation of very low-viscosity melts. A combination of FT-IR, wide-angle x-ray scattering (WAXS) and rheological experiments demonstrated that the combination of phase segregation between the hard nucleobase components and the soft poly-THF core combined with aromatic amide hydrogen bonding is responsible for the high thermosensitivity of the materials. Interestingly, thymine-functionalized material **37** did not exhibit such properties, which was attributed to its high crystallization temperature.

Although the strength of association between units with such weak non-covalent interactions is low, it was demonstrated that chain extension is a versatile tool to gain a significant improvement in material properties. A modest degree of po-

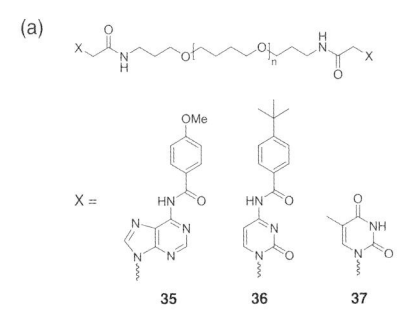

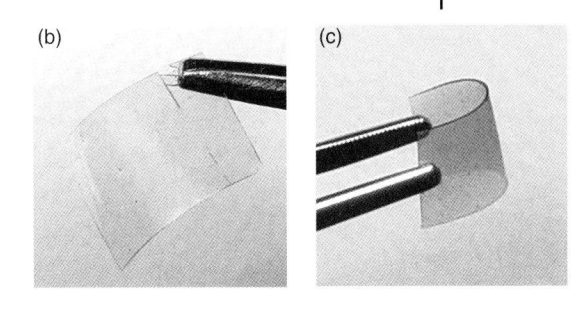

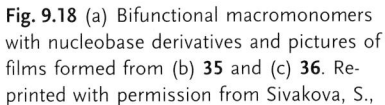

Fig. 9.18 (a) Bifunctional macromonomers with nucleobase derivatives and pictures of films formed from (b) **35** and (c) **36**. Reprinted with permission from Sivakova, S., Bohnsack, D.A., Mackay, M.E., Suwanmala, P., Rowan, S.J. *J. Am. Chem. Soc.* **2005**, *127*, 18202–18211. Copyright 2005 American Chemical Society.

lymerization in combination with physical inter-chain interactions by means of domain formation results in high-MW assemblies with improved material properties. Without domain formation, however, it is generally believed that high association constants are required in order to obtain high-MW polymers.

The quadruple hydrogen bond unit developed by our group has been further employed in the chain extension of telechelic PDMS [117], poly(ethylene–butylenes) (PEB) [118, 119], polyethers [118, 120, 121], polycarbonates [118] and polyesters [118, 122, 123] (Fig. 9.19). Solution viscosity studies and bulk rheological measurements with UPy telechelic oligodimethylsilane **38a** indicated the formation of high-MW PDMS. Likewise, UPy telechelic PDMS **38b** exhibited viscoelastic bulk properties that differed from those of a non-hydrogen bonding PDMS of similar MW. As was demonstrated previously, the purity of the supramolecular materials is extremely important. Therefore, UPy synthon **39** containing a highly reactive isocyanate functionality was designed which is synthetically accessible on a large scale by reaction of commercially available isocytosines in neat diisocyanates. Reaction of this synthon with amino or hydroxy telechelic polymers subsequently allows for an easy work-up procedure to obtain UPy telechelics **40a–e** (Fig. 9.19b).

For example, reaction of hydroxy telechelic PEB ($M_n = 3.5$ kg mol^{-1}, degree of functionalization $F_n = 1.93$) with synthon **39a** on large scale led to supramolecular polymer **40a** with <0.2% residual OH groups [119]. The mechanical properties of the polymer are dramatically different from those of the starting material. Whereas the starting material is a viscous liquid, **40a** is a rubber-like material with a Young's modulus of 5 MPa (Fig. 9.20a and b). Similarly, functionalization of more polar hydroxyl telechelic polymers with synthon **39a** resulted in improved materials properties. Functionalized polyether **40b** displays a rubber plateau in DMTA measurements and a storage modulus of 10 MPa, while the starting material is a viscous liquid. The properties of functionalized polycarbonate **40c** and polyester **40d** and **e** are those of semicrystalline polymers, whereas the starting materials are brittle solids (Fig. 9.20c and d). Especially in-

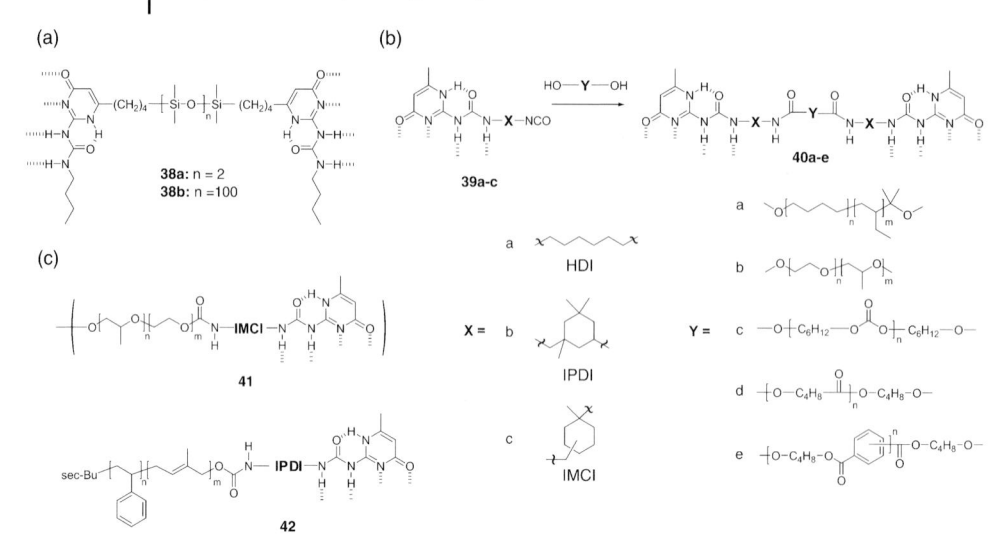

Fig. 9.19 Examples of UPy telechelic polymers: (a) UPy telechelic PDMS **38**; (b) synthesis of various UPy-telechelic supramolecular polymers **40a–e** with UPy-synthons **39a–c** and amine or hydroxy telechelic polymers; (c) trifunctionalized star copolymer of poly(ethylene oxide–propylene oxide) **41** and monofunctional UPy polyisoprene-*block*-polystyrene **42**. HDI, IPDI and IMCI are abbreviations of the commercially available starting diisocyanates.

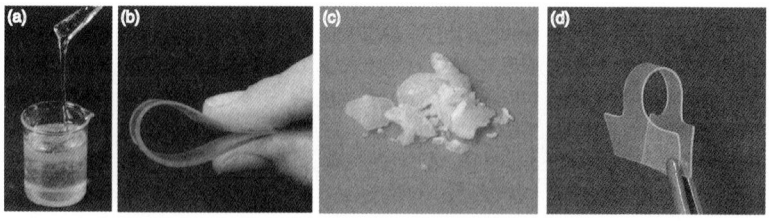

Fig. 9.20 (a) Hydroxy telechelic PEB (M_n = 3.5 kg mol^{-1}); (b) UPy telechelic PEB; (c) hydroxy telechelic PCL (M_n = 2.0 kg mol^{-1}); (d) UPy telechelic PCL. Parts (a) and (b) reproduced with permission from Folmer, B. J. B., Sijbesma, R. P., Versteegen, R. M., van der Rijt, J. A. J., Meijer, E. W. *Adv. Mater.* **2000**, *12*, 874–878. Copyright John Wiley & Sons, Inc.

teresting was polyester material **40d**, since it could easily be processed by different techniques into several scaffolds varying from films and fibers to meshes and grids [122]. An additional notable feature of these polymers is the high activation energy for viscous flow compared with their unfunctionalized counterparts (e.g. 105 kJ mol^{-1} for **40a**). From trifunctional UPy poly(ethylene oxide–propylene oxide) copolymer **41**, supramolecular networks have also been assembled. Solution viscometry measurements, including chain stopper studies, indicated the formation of a reversible network. In addition, due to the formation of reversible cross-links, a higher plateau modulus was observed in dy-

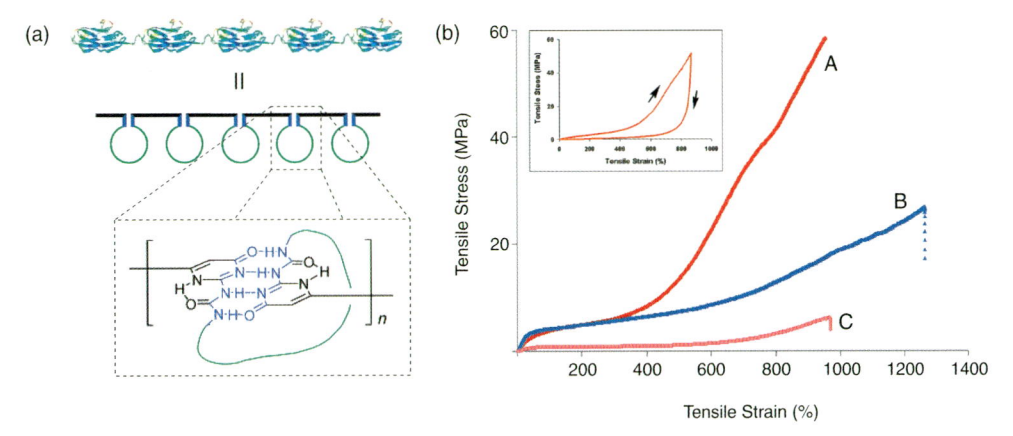

Fig. 9.21 Biomimetic UPy polymer inspired by titin protein: (a) concept of design; (b) stress–strain curves for the UPy-based polymer (A), control polymer (B) and a high-MW polyurethane (C) ($M_n = 10^3$ kg mol^{-1}). A stretching–retraction cycle for A is shown in the inset. Reprinted with permission from Guan, Z. B., Roland, J. T., Bai, J. Z., Ma, S. X., McIntire, T. M., Nguyen, M. *J. Am. Chem. Soc.* **2004**, *126*, 2058–2065. Copyright 2004 American Chemical Society.

namic mechanical analysis. Monofunctional UPy telechelics **42** were synthesized via living anionic polymerization and end-group modification of PS-*block*-PI of variable length with UPy synthon **39 b**.

In summary, the material properties of UPy telechelic polymers were shown to improve dramatically upon functionalization and materials were obtained that combine many of the mechanical properties of conventional macromolecules with the low melt viscosity of organic compounds. An important aspect which sets these materials apart from the previously mentioned supramolecular polymers based on small building blocks is that the viscoelastic properties of UPy telechelic polymers can be attributed to the entanglement of high-MW hydrogen-bonded polymer chains as opposed to physical cross-links by aggregation of the recognition groups into microcrystalline domains.

An elegant supramolecular polymer based on UPy dimerization was reported recently by Guan and coworkers [124]. Inspired by titin, a giant protein of muscle sarcomere that has >100 repeating modules and displays high strength, toughness and elasticity, they synthesized a poly-THF containing UPy functionalities in the main chain (Fig. 9.21a). Analogous to biopolymers, single-molecule nanomechanical properties were studied with atomic force microscopy (AFM) and demonstrated the sequential breaking of UPy dimers. Stress–strain profiles obtained from mechanical analysis of solution-cast films of the UPy-based polymer revealed that the polymer was very elastic, as evidenced by the high strain up to 900% and complete recovery to its original length (Fig. 9.21b). Moreover, the observed hysteresis (inset) demonstrates the great energy dissipation capability of the system, which is an important feature for high toughness.

9.4.2.2 Supramolecular Block Copolymers

Currently, the majority of supramolecular polymers being studied are homopolymers. However, more and more sophisticated applications are being developed that require combinations of properties that cannot be obtained with traditional polymers. As a rule, most polymer blends are immiscible, which results in macrophase separation. The formation of such macrophase morphologies generally results in poor mechanical properties. In block copolymers, however, two or more polymer blocks are linked by means of a strong chemical interaction. The competition between the strong chemical connection and repulsion between different polymer blocks leads to a wealth of interesting microphase structures (typical length scale is about 5–50 nm, depending on the individual length of the blocks). Although non-covalent interactions such as hydrogen bonding play a very important role in the phase segregation in many block copolymer systems such as polyamides, only copolymers based on multiple hydrogen bonding arrays between two end-functionalized blocks will be discussed in this section. The reader is referred to some excellent reviews on the use of multiple weak non-covalent interactions in rod–coil block copolymer systems [125–127].

Recent developments in the field of supramolecular chemistry have shown that both complementary and self-complementary building blocks can lead to large, well-defined structures through self-assembly. Provided that the non-covalent interaction used is strong and the value of χN is sufficiently high, supramolecular copolymers will have the possibility to display microphase separation. Advantages of assembling multi-block copolymers through strong, reversible, non-covalent interactions include a modular approach in synthesis, ease of processing, self-healing properties, facile and selective removal of sacrificial blocks and an extra level of hierarchical control in the self-organization of functional materials [128–131]. It is known that multiple weak hydrogen bonds between two chemically different polymer blocks can result in a homogeneous polymer blend whereas when these interactions are absent the polymers would be immiscible (see examples in the next section on side-chain supramolecular polymers). However, examples using strong non-covalent interactions are scarce. It is generally believed that strong recognition motifs affixed to the polymers will be essential in overcoming the phase separation force and induce the formation of small regular domains. When the non-covalent interaction between two monomers is self-complementary, the block size is not fixed and, depending on the conditions, the structure may vary from a random copolymer through a microstructure to a macrophase separated blend. Using strong, self-complementary ureido-s-triazine units (3) in a supramolecular main-chain rod–coil block copolymer, Hirschberg et al. demonstrated that the presence of kinetically stable microphases in the bulk double in size upon annealing [132]. Therefore, the use of complementary binding motifs is a prerequisite for thermodynamically stable microphase-separated structures. Mixtures of poly(ether ketones) (PEK) and poly(isobutylenes) (PIB) functionalized on the termini with multiple hydrogen bonding arrays were systematically investigated by Binder and coworkers (Fig. 9.22a) [133–135]. They reported that high strength of the hydrogen bond-

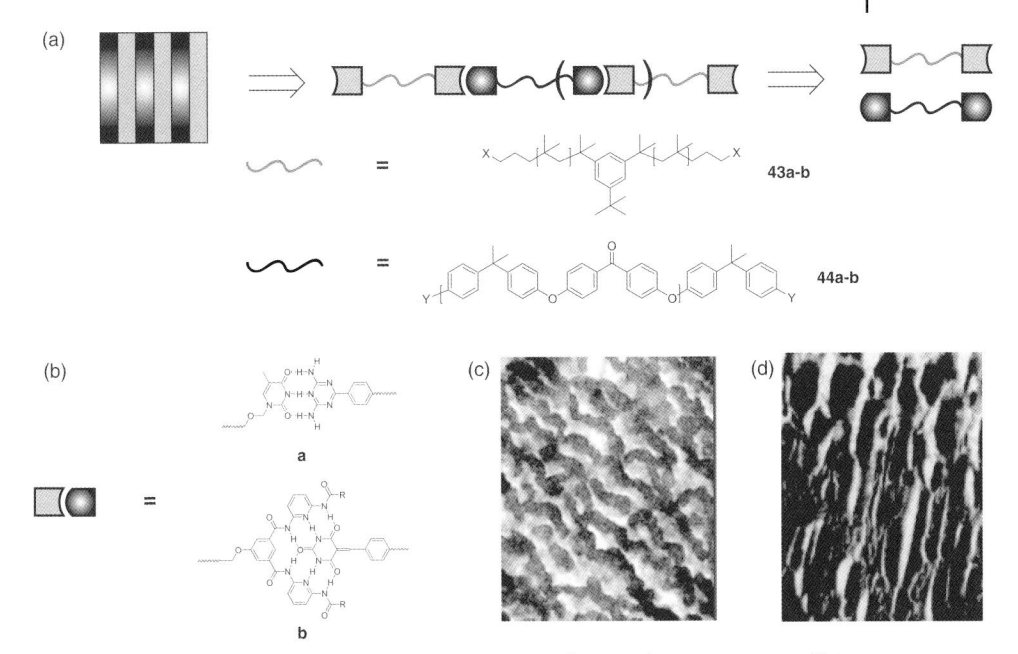

Fig. 9.22 (a) Self-assembly of telechelic polymers [PIB (**43**) and PIK (**44**)] into pseudo-block copolymers; (b) binding motifs **a** and **b**; (c) TEM image of **43a–44a**; (d) TEM image of **43b–44b**. Reproduced with permission from Binder, W. H., Bernstorff, S., Kluger, C., Petraru, L., Kunz, M. J. *Adv. Mater.* **2005**, *17*, 2824–2828. Copyright John Wiley & Sons, Inc.

ing motifs is essential in preventing macrophase separation in the solid state. High association constants ($K_a = 10^3$–10^5 M^{-1}, Fig. 9.22b) lead to the formation of regular arrays of alternating polymers, whereas low K_a values lead to immiscibility of the individual polymers in the bulk [136]. Both SAXS and TEM demonstrated the presence of 14-nm sized PIB and PEK microphases in 1:1 mixtures of **43** and **44**. As expected from the molecular design, sheet-like structures consisting of alternating layers of PEK and PIB were observed (Fig. 9.22c and d). Moreover, clear films of these materials can be obtained, indicating that macrophase separation is prevented. Furthermore, the reversibility of the binding process induced by temperature-mediated breaking of the bonds was revealed by temperature-dependent SAXS measurements. While weak binding motifs such as in **43a–44a** can be broken up to temperatures below the glass transition temperature, the regular structure is retained with strong binding motifs **43b–44b**.

Gong and coworkers reported the microphase separation of supramolecular AB diblock copolymers based on PS and poly(ethylene glycol) (PEG) polymers which were functionalized with a linear complementary six-fold hydrogen bond motif based on oligo-amide complex **8–9** ($K_a > 10^9$ M^{-1}, Fig. 9.23a) [137, 138]. NMR and GPC studies clearly indicated that PS and PEG strands are linked

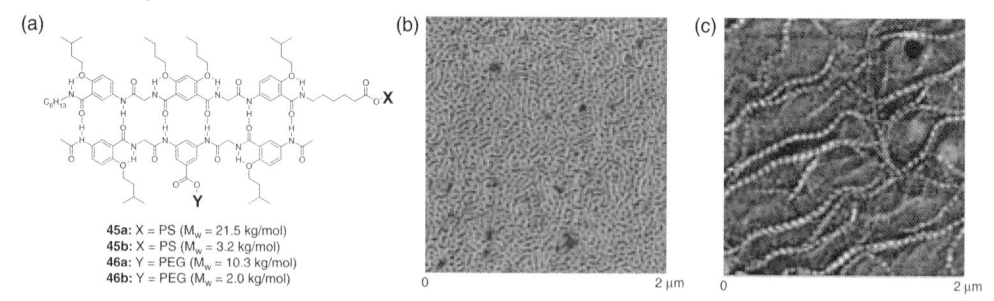

45a: X = PS (M_w = 21.5 kg/mol)
45b: X = PS (M_w = 3.2 kg/mol)
46a: Y = PEG (M_w = 10.3 kg/mol)
46b: Y = PEG (M_w = 2.0 kg/mol)

Fig. 9.23 (a) Supramolecular PS–PEG diblock copolymer functionalized with complementary oligoamide strands and AFM images (height) of (b) **45a–46a** and (c) **45b–46b**. Reproduced with permission of the authors from Gong, B., Yang, X., Kim, W., Ryu, C. Y. *Polym. Prepr.* **2005**, *46*, 1128–1129.

through strong hydrogen bonding. In addition, DSC and AFM measurements suggested the formation of microphase-separated PEG domains when polymers **45a** and **46a** were used (Fig. 9.23b). In contrast, the AFM image of low-MW polymers **45b** and **46b** was completely different (Fig. 9.23c). This was explained by regarding the recognition motif as a third block with comparable MW to the polymers, resulting in a coil–rod–coil triblock type copolymer.

9.4.2.3 Side-chain Supramolecular Polymers

The second generic class of supramolecular polymers is that of side-chain polymers. The general design for this class entails either incorporation of recognition units into a covalent polymer main-chain or as pendant groups on to a covalent polymer. A number of side-chain type polyrotaxanes have been reported recently which exhibit interesting properties in solution. However, the focus in these papers is primarily on changes in solubility upon functionalization with the inclusion complex. Therefore, we intend to cover the use of directional hydrogen bonding arrays in side-chain supramolecular polymers in this section only with a focus on bulk materials properties. Several reviews have appeared recently on this topic [139, 140] so this section is limited to the most illustrative examples of the last two decades. Non-covalent side-chain functionalization strategies were first reported for the synthesis of liquid crystalline (LC) materials (Fig. 9.24a) [141, 142]. Specifically, the benzoic acid–pyridine mesogenic complexes **47** that were first described by Kato and Fréchet are the best studied side-chain supramolecular polymers [143]. These complexes exhibit thermally stable smectic or nematic phases over broad temperature ranges. Additionally, when bifunctional pyridines are used, LC networks **48** are formed [144]. Likewise, mesomorphic molecular assemblies that are based on two components, neither of which is liquid crystalline, can also be obtained through double hydrogen bonding [145, 146]. For example, complex **49** displays a columnar LC phase [146] and supramolecular LC polyamide **50** exhibits a stable and enantio-

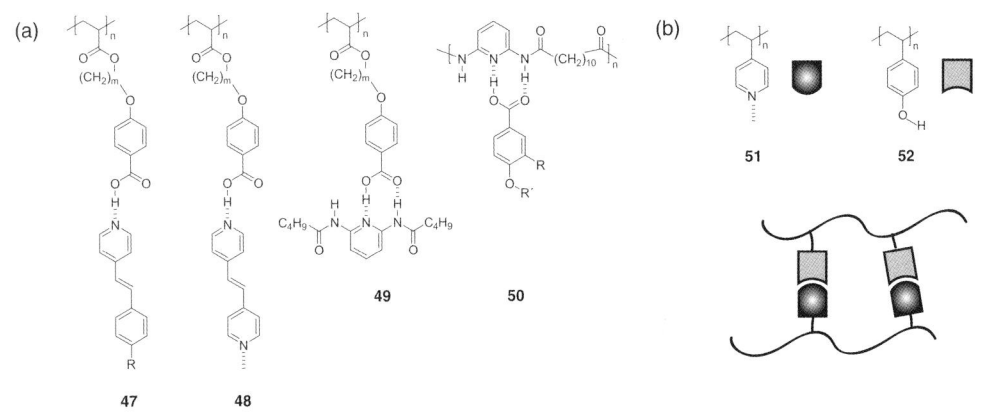

Fig. 9.24 (a) Representative examples of side-chain LC polymers **47–50**; (b) schematic representation of a blend of poly(4-vinylpyridine) (P4VP) **51** and poly(4-hydroxystyrene) **52**.

tropic mesophases [147]. An illustrative example of the effect of multiple weak interactions between two immiscible polymer chains was reported by Meftahi and Fréchet. In these studies, blends were prepared from poly(4-vinylpyridine) (P4VP) **51** and poly(4-hydroxystyrene) **52** (Fig. 9.24b). In contrast to mixtures of **51** and PS, which showed two separate glass-transition temperatures, a single T_g was observed for a 1:1 mixture of **51** and **52** and both polymers were miscible over the whole composition range [148].

Excellent accounts of applying side-chain self-assembly for the design of new polymeric materials have been reported by Ikkala and ten Brinke [127, 131, 149]. An illustrative example is the hierarchical assembly of a comb–coil supramolecular diblock copolymer which is a hydrogen-bonded complex of a PS-*block*-P4VP with a non-mesogenic alkylphenol **53** (Fig. 9.25) [150]. In this manner, a lamellar substructure within a cylindrical mesomorphic architecture is formed (Fig. 9.25b) [151]. In addition, the dimensions and properties of the periods can easily be tailored by modifications of the surfactant. Nanosized objects can be obtained by subsequent selective removal of one of the blocks. A number of materials prepared via this hierarchical self-assembly approach have already been applied for the preparation of photofunctional [152, 153], optical [154], and conductive materials [131, 155, 156].

Stadler and coworkers made an impressive contribution to the field of supramolecular side-chain homopolymers by studying the properties of polybutadienes functionalized with hydrogen-bonded phenylurazole derivatives (**54** and **55**, Fig. 9.26) [157–165]. Due to reversible chain extension and the subsequent formation of small crystalline domains, the functionalized polymers exhibit properties typical of thermoplastic elastomers. At low temperatures, the hydrogen bond interaction contributes to the properties comparable to covalent interactions, whereas at high temperature these interactions disappear and the materials exhibit flow behavior typical of a low molecular weight polymer. The prop-

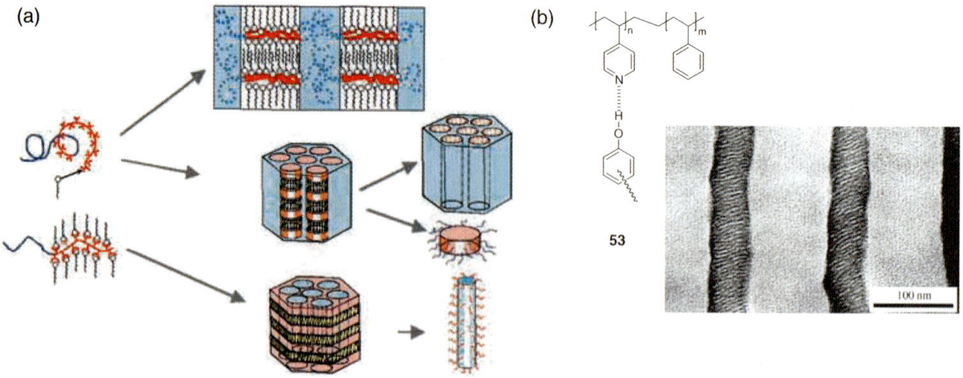

Fig. 9.25 (a) Illustration of hierarchical structure formation by self-assembly of block copolymer-based supramolecular side-chain polymers and subsequent preparation of nanosized architectures by selective removal of a sacrificial block; (b) transmission electron micrograph showing hierarchical self-assembly of PS-*block*-P4VP(PDP)$_{1.0}$ **53** where nominally one pentadecylphenol (PDP) is hydrogen bonded to a pyridine moiety. Reproduced from Ikkala, O., ten Brinke, G. *Chem. Commun.* **2004**, 2131–2137 by permission of The Royal Society of Chemistry.

Fig. 9.26 Network formation by hydrogen bonding of polybutadiene polymers functionalized with phenylurazole derivatives **54** and **55**.

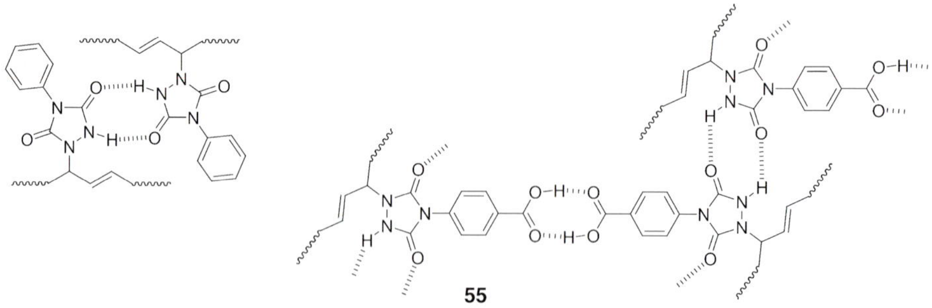

erties of these materials were analyzed by DSC [158], light and x-ray scattering [159], dynamic mechanical analyses [160–162], dielectric spectroscopy [163] and IR spectroscopy [164]. A similar strategy was used by Lange et al., who described the use of triple hydrogen bonding interactions between the alternating copolymer of styrene and maleimide with melamine and 2,4-diamino-6-vinyl-s-triazine [166]. They demonstrated the formation of a polymer network and showed that the presence of the triple hydrogen bonding motif was essential to guarantee complete miscibility of the polymer blend.

Similarly, triple hydrogen bonding moieties have been incorporated into the side-chain of polymers for versatile approaches to polymer functionalization. Especially, Rotello and coworkers have developed "plug and play" polymers [167] based on a post-functionalization strategy of random PS copolymers with either diaminotriazine (**56**), diamidopyridine (**57**) or thymine (**58**) hydrogen bonding groups (Fig. 9.27 a). Initial reports were on the self-assembly of flavin (**59**) on to polymeric scaffolds. Because of an increased intramolecular association of triazines, **57** was more effective in binding **59** [167]. In addition, polymer **56** and thymine-functionalized gold colloids were reported to self-assemble into discrete microspheres that are highly ordered on both the molecular and micrometer scales [168–170]. Other reports describe the formation of recognition-induced polymersomes (RIPs) from **57** and **58** or other bifunctional thymine derivatives (Fig. 9.27 b) [171–175]. These vesicular aggregates were studied with differential interference contrast (DIC), laser confocal scanning microscopy (LCSM) (Fig. 9.27 c), TEM and AFM. Microsphere formation (2–10 μm) was thermally reversible at 50 °C and subsequent reconstruction of the structures was observed on cooling, demonstrating the ability to self-optimize [173, 174]. The strength of this recognition methodology was recently illustrated in an elegant study of the incorporation of thymine-functionalized dendrimer wedges into block copolymer films of **57** analogues. Using dynamic secondary ion mass spectrometry (SIMS), DSC, TEM and x-ray diffraction, the extent of guest-induced phase segregation was analyzed.

Currently, a number of groups are studying the possibility of incorporating multiple recognition groups in order to introduce dynamic multifunctionality into a polymeric material [176, 177]. A system reported by Weck and coworkers

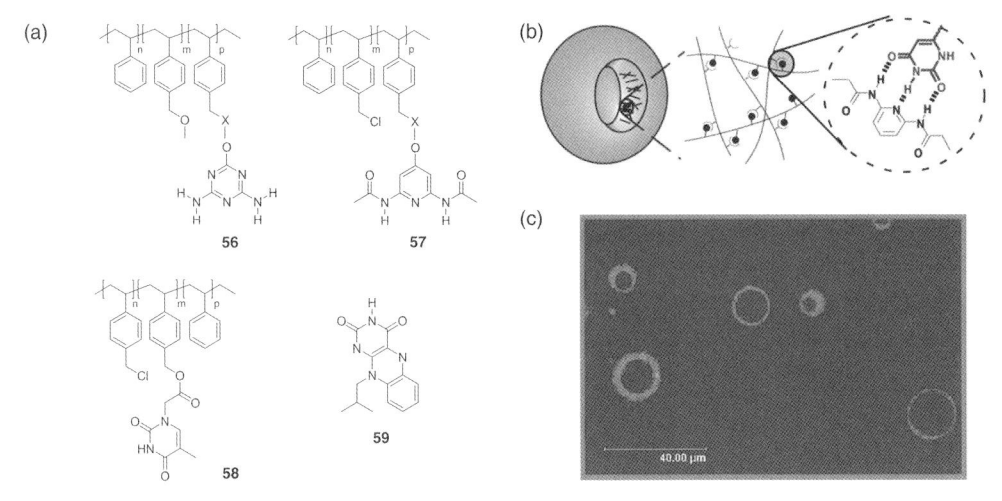

Fig. 9.27 (a) PS copolymers functionalized with triple hydrogen bonding units **56–58** and flavin **59**. (b) Schematic representation of recognition induced polymersomes (RIPs) and the recognition unit. Reproduced with permission from Uzun, O., Xu, H., Jeoung, E., Thibault, R. J., Rotello, V. M. *Chem. Eur. J.* **2005**, *11*, 6916–6920. Copyright John Wiley & Sons, Inc. (c) Laser confocal scanning micrograph of vesicular morphology of RIPs from **57–58** in which **58** was functionalized with Rhodamine B. Reprinted with permission from Uzun, O., Sanyal, A., Nakade, H., Thibault, R. J., Rotello, V. M. *J. Am. Chem. Soc.* **2004**, *126*, 14773–14777. Copyright 2004 American Chemical Society.

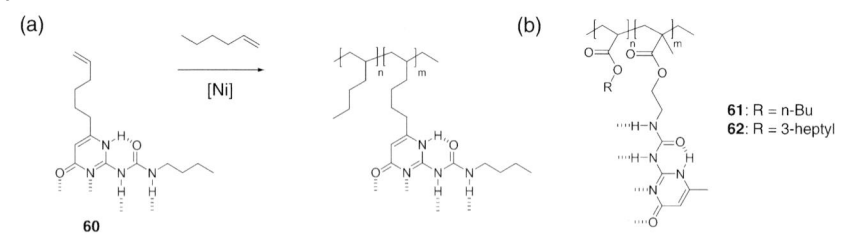

Fig. 9.28 (a) Synthesis of UPy-functionalized poly(1-hexene) copolymer; (b) poly(alkyl methacrylate) copolymer containing pendant UPy units.

is the "universal polymer backbone", which is based on a polymeric scaffold that contains metal–ligand and triple hydrogen bonding complexes [178]. This design was conceptually illustrated by Lehn in 2002 [179].

A different approach to applying self-complementary hydrogen bonding units was taken by the groups of Coates and Long. Coates and coworkers incorporated vinyl-substituted UPy monomer **60** in poly(1-hexene), using a nickel-based polymerization catalyst which is tolerant towards Lewis basic groups (Fig. 9.28a) [180]. With small amounts (<2%) of UPy monomer **60** in the polymer chain, the polyolefins show thermoplastic elastomeric behavior. Long and coworkers described the synthesis and characterization of poly(alkyl methacrylates) containing pendant UPy functionalities (Fig. 9.28b) [181–184]. Interestingly, copolymer **61** exhibited significantly better adhesive properties to glass than a similar unadorned polymer [181]. In addition, these copolymers could easily be spun into fibers (0.1–50 µm) by electrospinning [182]. Tensile experiments on **62** revealed that at low levels of UPy incorporation, copolymers display a high elon-

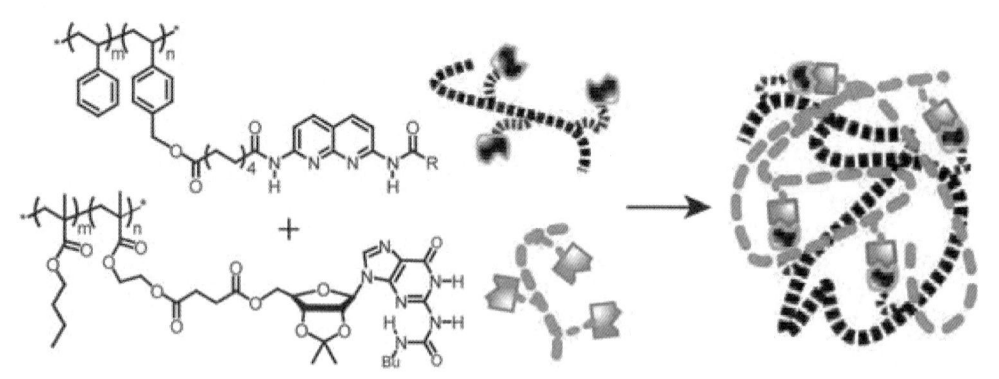

Fig. 9.29 Supramolecular side-chain polymers equipped with complementary quadruple hydrogen bonding motifs (**6** and **10**) and schematic representation of copolymer network formation. Reprinted with permission from Park, T., Zimmerman, S. C., Nakashima, S. *J. Am. Chem. Soc.* **2005**, *127*, 6520–6521. Copyright 2005 American Chemical Society.

gation and low tensile strength [184]. However, with increasing UPy content, the percentage elongation decreased and tensile strength increased. Melt rheological analysis demonstrated that the extent of non-covalent inter-chain interactions largely dictates the polymer physical properties. In contrast to supramolecular main-chain polymers, these materials displayed increased creep resistance, which enhances performance properties.

More recently, Zimmerman and coworkers reported the utility of a complementary quadruple hydrogen bonding complex by the formation of polymer blends of poly(butyl methacrylate) (PBMA) and PS containing a controlled number of units **6** and **10** (Fig. 9.29) [49]. The polymer blends were characterized by DSC, SEC and viscometry which confirmed the formation of reversible networks. Moreover, transparent films could be obtained.

9.4.2.4 Applications Based on Supramolecular UPy Materials

Functional materials possessing good mechanical properties at room temperature which are easy to process are of interest for many applications. In this respect, supramolecular polymers hold much promise, as illustrated by the large number of publications on this topic in the literature over the last decade. More specifically, numerous patent applications involving supramolecular UPy materials [185, 186] have been filed in fields as diverse as coatings [187], adhesives [188], printing [188–193], electronic devices [194] and cosmetics [195–197]. The projected added value from these materials is based on their improved processing in the melt or solution, the ease of synthesis and dynamic nature, which makes the materials responsive to external stimuli. In this section, some illustrative examples of applications based on supramolecular UPy materials are discussed.

The dynamic nature of supramolecular polymers was used by Keizer et al. in polymerization-induced phase separation (PIPS) [121]. In PIPS, a polymer is dissolved in a reactive monomer which is subsequently polymerized (e.g. by UV irradiation), resulting in a certain morphology (Fig. 9.30). This process is currently used to produce high-impact composite materials and avoids the use of solvent and consequently results in the fast and clean production of high-performance materials. The mechanical behavior of materials obtained by fast UV curing of solutions of UPy telechelic polymers in acrylates was comparable to that of high-MW analogues and showed macrophase separation. The enhanced solubility of UPy telechelic polymers makes them suitable candidates for a variety of high-performance materials produced by PIPS. Interestingly, DSM has filed a patent describing the use of supramolecular polymers in coatings for glass-fibers, a process that also requires very short reaction times [187].

A different application in which the differences in phase behavior of supramolecular polymers can be exploited is ink-jet printing. By the ejection of ink droplets through a small orifice, images can be created on a certain substrate. Although the ink needs to display low viscosity before ejection, it needs to be highly viscous when it is printed to the substrate. To this end, the dynamic fea-

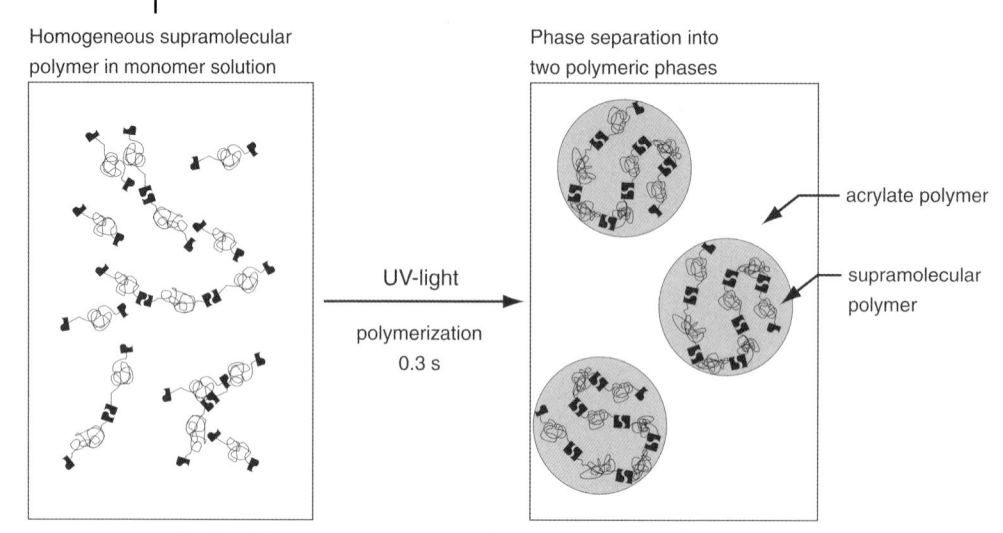

Fig. 9.30 Schematic representation of PIPS using UPy-based supramolecular polymers. Reprinted with permission from Keizer, H. M., Sijbesma, R. P., Jansen, J. F. G. A., Pasternack, G., Meijer, E. W. *Macromolecules* **2003**, *36*, 5602–5606. Copyright 2003 American Chemical Society.

tures of supramolecular polymers seem eminently suitable. Indeed, Xerox and Agfa-Gevaert have filed patents in which supramolecular polymers are used in ink compositions. While Agfa describe UPy-functionalized dyes to improve light fastness [191], Xerox describe the use of multiple UPy units in a low-MW compound as a binder [193]. Mixing of this supramolecular binder with other ingredients at elevated temperatures results in inks that can be used in hot-melt ink printers. In another patent application by Xerox, aqueous-based inks are formulated with supramolecular polymers [192]. In this case, the material properties of the inks are tuned with temperature and the polarity of the solvent. Whereas the ink has a low viscosity before deposition, the viscosity is increased dramatically upon evaporation of the solvent after ink-jetting.

A patent application by Kodak Polychrome Graphics uses the thermosensitivity of supramolecular polymers as a coating to produce a printing plate [190]. The polymers used are prepared by reacting polyfunctional resins with UPy isocyanate **39**. In addition to this polymer, the resulting coating also contains infrared dyes that are able to transform laser light into heat. Because the resins contain several reversible cross-links to reinforce the coating, it is possible to disrupt the non-covalent cross-links by short IR laser illumination and subsequent removal of the exposed areas. This procedure not only ensures a better press-life but also eliminates two processing steps from conventional print plate production.

The dynamic nature of supramolecular polymers also makes them interesting materials for biomedical applications. Dankers et al. demonstrated the use of UPy telechelic oligocaprolactones (**40d**) for a modular approach to bioactive scaf-

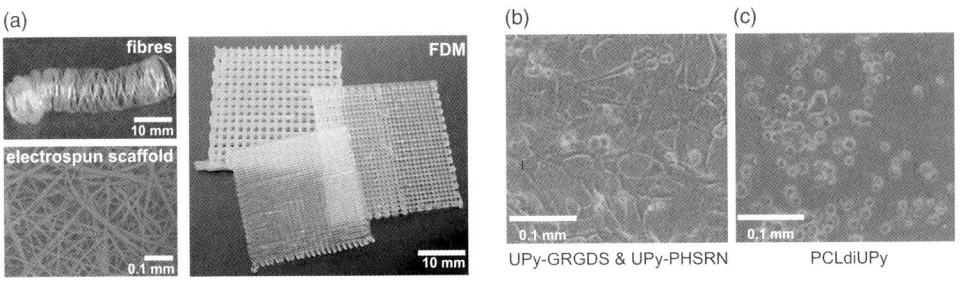

Fig. 9.31 (a) Several scaffolds demonstrating the ease of processing supramolecular UPy materials; (b) cell adhesion and spreading of fibroblast cells on PCldiUPy (**40d**) with UPy–GRGDS and UPy–PHSRN additives; (c) control experiment showing no adhesion to **40d**. Reproduced with permission from Dankers, P.Y.W., Harmsen, M.C., Brouwer, L.A., Van Luyn, M.J.A., Meijer, E.W. *Nat. Mater.* **2005**, *4*, 568–574.

folds for tissue engineering by simply mixing these functional polymers with UPy-modified biomolecules [122]. Various techniques can be used to process these biocompatible and biodegradable supramolecular polymers into a range of different scaffolds (Fig. 9.31a). In addition, *in vitro* experiments indicate strong and specific binding of fibroblasts to the UPy-functionalized bioactive materials containing synergistic UPy peptides [UPy–Gly–Arg–Gly–Asp–Ser (UPy–GRGDS) and UPy–Pro–His–Ser–Arg–Asn (UPy–PHSRN)] (Fig. 9.31b). An even more striking effect is observed *in vivo* where the formation of giant cells at the interface between bioactive material and tissue is triggered.

In summary, over a period of one decade, several applications based on supramolecular polymers with technological relevance have been developed. Due to their dynamic nature, supramolecular polymers are interesting candidates in stimuli-responsive materials and are eminently suitable in a modular approach towards multifunctional materials.

9.5
Ring–Chain Equilibria in Supramolecular Polymers

Cyclization reactions are inevitable side-reactions that occur in any step-growth polymerization. This holds for both covalent and supramolecular polymerizations. The importance of cyclization has been severely under-appreciated and, in the early years of polymer synthesis, the formation of rings was completely ignored. Nevertheless, early work on polycondensation reactions indicated the presence of small rings in polymeric materials. These rings can have a tremendous influence on material properties, since they can act as plasticizers. On the other hand, large rings are interesting because they have properties that are completely different from those of their linear analogues [198]. Considerable effort has been made in the investigation of the cyclization behavior of different condensation polymers by fractionation of equilibrated samples and subsequent analysis by GC or GPC

[199]. The theory of cyclization in polymeric systems has received considerable attention, starting with the influential description of cyclization in condensation polymers by Jacobsen and Stockmayer [200]. The theory, assuming Gaussian chain statistics, states that the probability of ring formation is inversely proportional to $DP^{5/2}$. This implies that larger rings become increasingly less abundant. An important refinement was made by Flory's group, who took directional requirements for cyclization of chain ends into account [201–204]. One of the most important predictions of the theory is the existence of a critical concentration (c^*), below which the equilibrium composition of the system consists entirely of rings. Above this concentration, the concentration of rings stays constant and the remaining material grows as polymer chains. The equilibrium between polymeric and cyclic species also plays an important role in the formation of supramolecular polymers, since it is an inherent phenomenon associated with step-growth polymerization. It is obviously difficult to isolate rings from linear chains in a linear main-chain reversible polymer by fractionation in a way similar to that with their covalent analogues. Recently, a number of groups have studied the ring–chain equilibrium experimentally in supramolecular polymers in more detail. A transition from cyclic to linear species with increasing concentration is expected for supramolecular polymers. Such a transition would be evident from concentration-dependent viscosity measurements. Indeed, an early example by Abed et al. reported the chain to cyclic species transition of benzoic acid-terminated short PDMS monomers upon dilution [111]. Since the average size of cyclic species is only weakly dependent on concentration (c), the specific viscosity (η_{sp}) of such a solution is expected to be roughly proportional to the concentration (i.e. a double logarithmic plot of η_{sp} versus c will have a slope close to 1). At high concentrations, however, viscosity will be dominated by the contribution of polymeric chains, growing in size with increasing concentration. Predictions on worm-like micellar systems by Cates implied that a slope of 3.5–3.7 can be expected for solutions of supramolecular polymers above the overlap concentration [16]. Both predictions were confirmed by studies on bifunctional UPy molecules. Whereas in solutions of bis-UPy **21** in chloroform a concentration-dependent transition [8] could not be observed (a slope of 3.7 was observed, completely in line with Cates' model), the specific viscosity of bis-UPy **63** showed a linear dependence of 1.07 at low concentrations, which increased to 6 for concentrations higher than 0.13 M (Fig. 9.32) [205].

A theoretical treatment of the ring–chain equilibrium in reversible polymers was given by Ercolani et al. [206, 207]. They analyzed the driving force for self-assembly into cyclic oligomers by the introduction of two parameters: (i) K_{inter}, representing an equilibrium constant for the supramolecular interaction which is equal to the K_a and (ii) EM_n, representing the effective molarity of each cyclic n-mer. The latter parameter is defined as the ratio between the intramolecular cyclization constant of a linear n-mer ($K_{intra,n}$) and K_{inter} and represents a cyclization constant corrected for the intrinsic binding strength of the end-groups. According to theory, optimizing the yield of a specific n-mer is equivalent to maximizing EM_n by careful design of the monomer, since this parameter is determined by a combination of steric, enthalpic and template effects. For exam-

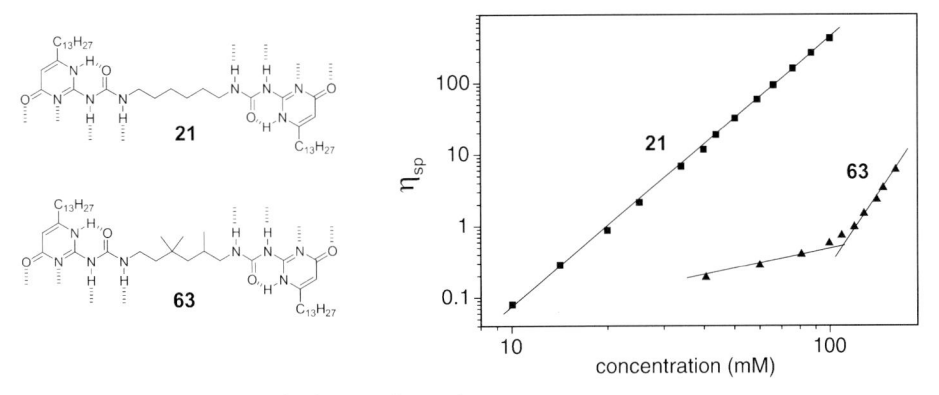

Fig. 9.32 Specific viscosities of solutions of **21** and **63** in chloroform as a function of concentration. Reprinted with permission from Folmer, B.J.B., Sijbesma, R.P., Meijer, E.W. *J. Am. Chem. Soc.* **2001**, *123*, 2093–2094. Copyright 2001 American Chemical Society.

ple, introducing structural bias in a short spacer between two end-groups reduces the number of available conformations and therefore decreases the entropic penalty for cyclization. Molecules **64** and **65** were specifically designed to maximize the number of dimeric cycles [208]. Because mixtures of **64** and **65** give rise to distinct signals of the heterodimer in the ^{1}H NMR spectrum, it was possible to verify the presence of a critical concentration at 7 mM (Fig. 9.33). It was shown that the amount of cyclic dimers increases up to c^* and remains constant followed by incorporation of monomers in oligomeric rings and polymeric species. A systematic study investigating the influence of pre-organization

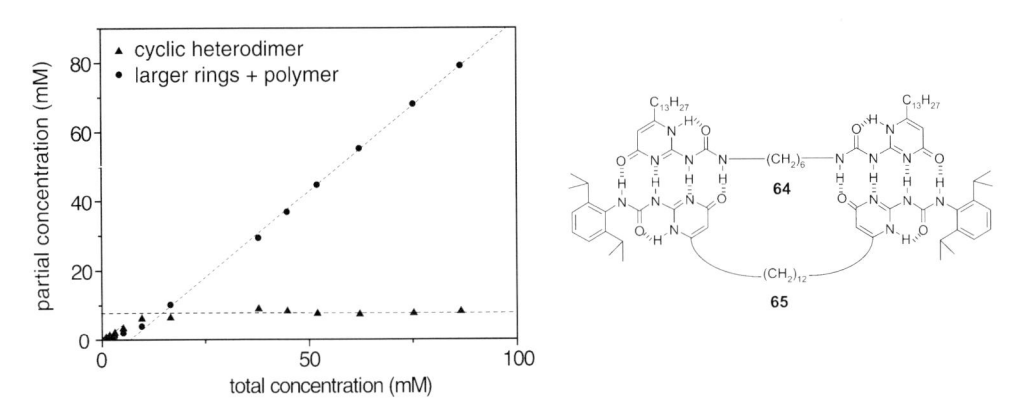

Fig. 9.33 Partial concentrations of cyclic heterodimer and polymer in a 1:1 mixture of **64** and **65** in chloroform. (Reprinted with permission from Söntjens, S.H.M.; Sijbesma, R.P.; van Genderen, M.H.P.; Meijer, E.W. *Macromolecules* **2001**, *34*, 3815–3818. Copyright 2001 American Chemical Society).

between two UPy groups in derivatives **66–69** indicated that critical concentrations can be increased over 0.3 M for **67** when antiparallel association is favored (Table 9.1) [209].

The most frequently used technique to study the size (molecular weight) and size distribution of covalent polymers is SEC. However, when an aggregate is in dynamic equilibrium with its free components, extensive tailing may occur and well-defined peaks can only be obtained when association is extremely high ($K_a \gg 10^8$ M^{-1}) and/or the kinetics are slow [210, 211]. Noteworthy in this case is an interesting study on reversible coordination polymers reported by Paulusse and coworkers, who discussed the equilibrium of cyclic species with high molecular weight palladium coordination polymers by analysis with SEC [212–214].

A more recently developed method to study the size of self-assembled structures is the use of pulse-field-gradient NMR spectroscopy for the determination of relative diffusion constants [215]. Especially when exchange between species is slow on the NMR time-scale, the sizes can be determined separately. The results are generally depicted in a two-dimensional diffusion ordered (DOSY) plot

Table 9.1 Critical concentration of **66–69** in chloroform solution.

Compound	Critical concentration (mM)
66	6
67	>300
68	33
69	95

in which one axis (F1) is the NMR spectrum and the other (F2) represents the information on relative diffusion coefficients. An illustrative example of the presence of cyclic rings and linear chains indicated by DOSY NMR spectroscopy is given in Fig. 9.34a [205]. Depicted is a 0.25 M chloroform solution of bis-UPy **63** in which small cyclic aggregates, polymeric species and the internal reference (β-methylated cyclodextrin) are indicated. From these examples, it is clear that considerable amounts of cyclic species can be formed when small or pre-organized spacers are used between two functional end-groups. Increasing spacer length to polymeric spacers will therefore decrease c^*. However, when the interaction between the end-groups is sufficiently strong, cyclic species are observed up to relatively high concentrations. This was shown to be the case in a system consisting of a bis-UPy derivative and bis(diamidonaphthyridine) derivative with a similar linker length [48]. Even though the bis-UPy derivative has a c^* well below 5 mM, it was shown that only cyclic species were formed by addition of bis(diamidonaphthyridine) up to 40 mM (Fig. 9.34b and c).

Another recent example of a system displaying cyclic species up to a relatively high concentration was reported by our laboratory and consisted of an AB-type molecule with a medium length, flexible spacer of ∼20 atoms [216]. This molecule consisted of one UPy **4** and one diamidonaphthyridine **6** which was capable of forming a monomeric ring. Both solution viscometry and ^{1}H NMR experiments indicated that only rings existed below a c^* of 80 mM with a linear dependence of AB molecule in the specific viscosity versus concentration plot. However, after passing the c^* transition, a viscous chloroform solution was ob-

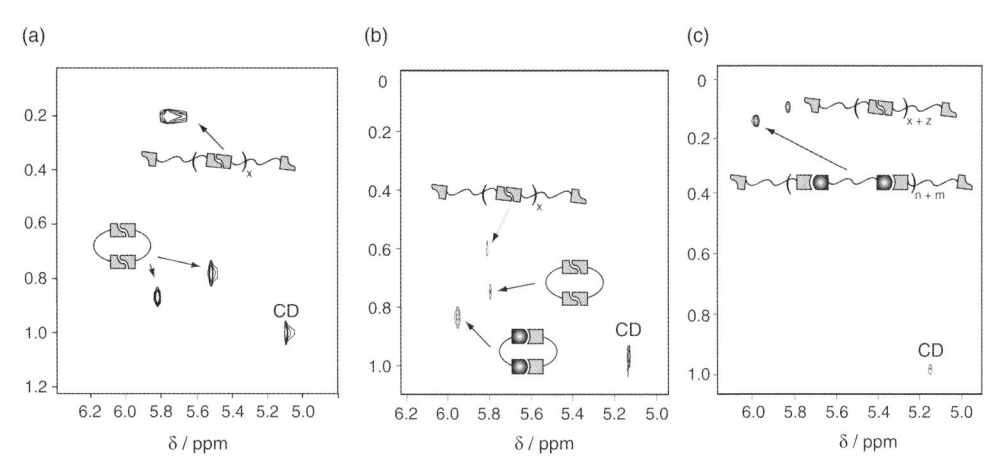

Fig. 9.34 Partial DOSY NMR spectra (CDCl$_3$) of (a) a 250 mM solution of bis-UPy **63**; (b) a solution of a bis-UPy (15 mM) and bis(diamidonaphthyridine) (10 mM); and (c) a solution of a bis-UPy (150 mM) and bis(diamidonaphthyridine) (100 mM). Reprinted with permission from Folmer, B. J. B., Sijbesma, R. P., Meijer, E. W. *J. Am. Chem. Soc.* **2001**, *123*, 2093–2094 and Ligthart, G. B. W. L., Ohkawa, H., Sijbesma, R. P., Meijer, E. W. *J. Am. Chem. Soc.* **2005**, *127*, 810–811. Copyright 2001 and 2005 American Chemical Society.

tained together with a higher slope of 3.3 This served as the first example of a concentration-induced supramolecular ring-opening polymerization of an AB monomer and has potential for applications involving the directional growth of supramolecular polymers from a functional surface.

Many examples are known in the literature in which hydrogen-bonded cyclic assemblies are selectively formed using pre-organized monomers, followed by formation of supramolecular polymers in columnar structures by π-π stacking. For a comprehensive treatment of this topic, the reader is referred to the review literature [3, 22].

9.6
Conclusions and Outlook

This chapter has assembled a number of illustrative examples from the literature on supramolecular polymeric materials, being monomeric units held together by specific and directional non-covalent interactions. The progress in supramolecular chemistry has opened the way to assemble small and large building blocks into high molecular weight assemblies; structures that combine many of the well-known properties of "traditional" macromolecules with the responsive nature of non-covalent interactions. It is certain that this rapidly evolving field has the potential to contribute significantly to materials science. More and more research initiatives are oriented towards supramolecular polymer engineering while suitably strong non-covalent interactions are being developed. Due to the reversibility in the bonding, these supramolecular polymers are under thermodynamic equilibrium and their properties can be adjusted by external stimuli. As demonstrated in this chapter, the foundation for future development to generate tailored polymeric materials has been laid. In our opinion, many challenges remain in fundamental research and in product development of supramolecular polymeric materials. Two fundamental research objectives can be considered most important:

1. The possibility of tuning the properties of a material by simple block copolymer formation arising from the mixing of functional monomers, while maintaining polymeric properties in solution as well as the bulk over a wide range of compositions (Fig. 9.35). This would require molecules which contain both self-complementary and complementary binding motifs.
2. The development of polymeric systems using a number of orthogonal binding arrays that are strong and selective. In contrast to the previous point, the non-covalent interactions employed would not interact with each other. This orthogonal approach would open the way to incorporate multiple functions into supramolecular polymeric systems.

In addition, although a large number of applications are feasible, many engineering opportunities lay ahead leading to, e.g., self-healing materials, hydrogels, sensors, thermoplastic elastomers and other high-performance materials.

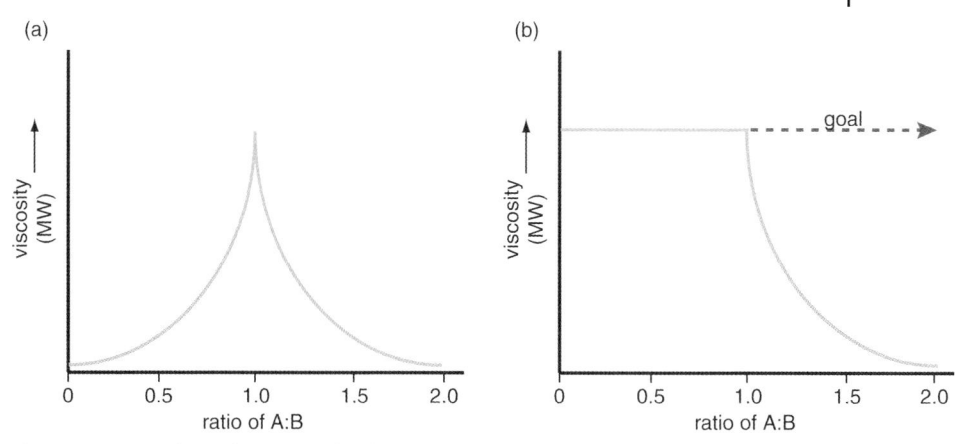

Fig. 9.35 (a) Virtual DP of a supramolecular copolymer versus the ratio of components A and B blocks when A and B are *only* complementary; (b) virtual DP of a supramolecular copolymer versus the ratio of components A and B blocks when A is both self-complementary *and* complementary to B.

In conclusion, with the wealth of materials properties attainable by synthetic and processing techniques firmly established in traditional polymer chemistry together with a fundamental understanding of the principles of supramolecular chemistry, the door is now open for designing new dynamic polymeric materials. We believe that an increasing number of materials based on supramolecular polymers will appear in various products in the coming years. The future, however, will reveal the scope, limitations and applicability of this new class of polymeric materials.

References

1 Lehn, J.-M. *Supramolecular Chemistry.* Wiley-VCH, Weinheim, **1995**.
2 Lindsey, J. S. *New J. Chem.* **1991**, *15*, 153–180.
3 Lawrence, D. S., Jiang, T., Levett, M. *Chem. Rev.* **1995**, *95*, 2229–2260.
4 Hartley, J. H., James, T. D., Ward, C. J. *J. Chem. Soc., Perkin Trans. 1* **2000**, 3155–3184.
5 Elemans, J. A. A. W., Rowan, A. E., Nolte, R. J. M. *J. Mater. Chem.* **2003**, *13*, 2661–2670.
6 Percec, V. *J. Macromol. Sci., Pure Appl. Chem.* **1996**, *A33*, 1479–1496.
7 Hill, D. J., Mio, M. J., Prince, R. B., Hughes, T. S., Moore, J. S. *Chem. Rev.* **2001**, *101*, 3893–4011.
8 Sijbesma, R. P., Beijer, F. H., Brunsveld, L., Folmer, B. J. B., Hirschberg, J. H. K. K., Lange, R. F. M., Lowe, J. K. L., Meijer, E. W. *Science* **1997**, *278*, 1601–1604.
9 Brunsveld, L., Folmer, B. J. B., Meijer, E. W., Sijbesma, R. P. *Chem. Rev.* **2001**, *101*, 4071–4097.
10 Ciferri, A. *Supramolecular Polymers*, 2nd edn. CRC Press, Boca Raton, FL, **2005**.
11 Flory, P. J. *Principles of Polymer Chemistry.* Cornell University Press, Ithaca, NY, **1953**.
12 Ciferri, A. *Macromol. Rapid Commun.* **2002**, *23*, 511–529.

13 Manners, I. *Angew. Chem. Int. Ed. Engl.* **1996**, *35*, 1603–1621.
14 Janiak, C. *J. Chem. Soc., Dalton Trans.* **2003**, 2781–2804.
15 Hofmeier, H., Schubert, U. S. *Chem. Soc. Rev.* **2004**, *33*, 373–399.
16 Cates, M. E. *Macromolecules* **1987**, *20*, 2289–2296.
17 Granek, R., Cates, M. E. *J. Chem. Phys.* **1992**, *96*, 4758–4767.
18 Cates, M. E., Candau, S. J. *J. Phys.: Condens. Matter* **1990**, *2*, 6869–6892.
19 Raymo, F. M., Stoddart, J. F. *Chem. Rev.* **1999**, *99*, 1643–1663.
20 Takata, T., Kihara, N., Furusho, Y. *Adv. Polym. Sci.* **2004**, *171*, 1–75.
21 Demus, D., Goodby, J., Gray, G. W., Spiess, H. W., Vill, V. *Handbook of Liquid Crystals.* Wiley-VCH, Weinheim, **1998**.
22 Prins, L. J., Reinhoudt, D. N., Timmerman, P. *Angew. Chem. Int. Ed.* **2001**, *40*, 2382–2426.
23 Stone, A. J. *The Theory of Intermolecular Forces.* Clarendon Press, Oxford, **1996**.
24 Jorgensen, W. L., Pranata, J. *J. Am. Chem. Soc.* **1990**, *112*, 2008–2010.
25 Pranata, J., Wierschke, S. G., Jorgensen, W. L. *J. Am. Chem. Soc.* **1991**, *113*, 2810–2819.
26 Murray, T. J., Zimmerman, S. C. *J. Am. Chem. Soc.* **1992**, *114*, 4010–4011.
27 Zimmerman, S. C., Murray, T. J. *Philos. Trans. R. Soc. Lond. A* **1993**, *345*, 49–56.
28 Sartorius, J., Schneider, H.-J. *Chem. Eur. J.* **1996**, *2*, 1446–1452.
29 Lukin, O., Leszczynski, J. *J. Phys. Chem. A* **2002**, *106*, 6775–6782.
30 Schneider, H.-J. *J. Phys. Chem. A* **2003**, *107*, 9250–9252.
31 Sherrington, D. C., Taskinen, K. A. *Chem. Soc. Rev.* **2001**, *30*, 83–93.
32 Cooke, G., Rotello, V. M. *Chem. Soc. Rev.* **2002**, *31*, 275–286.
33 Sijbesma, R. P., Meijer, E. W. *Chem. Commun.* **2003**, 5–16.
34 Zimmerman, S. C., Corbin, P. S. *Struct. Bonding (Berlin)* **2000**, *96*, 63–94.
35 Chang, S. K., Hamilton, A. D. *J. Am. Chem. Soc.* **1988**, *110*, 1318–1319.
36 Beijer, F. H., Kooijman, H., Spek, A. L., Sijbesma, R. P., Meijer, E. W. *Angew. Chem. Int. Ed.* **1998**, *37*, 75–78.
37 Beijer, F. H., Sijbesma, R. P., Kooijman, H., Spek, A. L., Meijer, E. W. *J. Am. Chem. Soc.* **1998**, *120*, 6761–6769.
38 Söntjens, S. H. M., Sijbesma, R. P., van Genderen, M. H. P., Meijer, E. W. *J. Am. Chem. Soc.* **2000**, *122*, 7487–7493.
39 Corbin, P. S., Zimmerman, S. C. *J. Am. Chem. Soc.* **1998**, *120*, 9710–9711.
40 Corbin, P. S., Lawless, L. J., Li, Z. T., Ma, Y. G., Witmer, M. J., Zimmerman, S. C. *Proc. Natl. Acad. Sci. USA* **2002**, *99*, 5099–5104.
41 Corbin, P. S., Zimmerman, S. C. *J. Am. Chem. Soc.* **2000**, *122*, 3779–3780.
42 Luning, U., Kuhl, C. *Tetrahedron Lett.* **1998**, *39*, 5735–5738.
43 Corbin, P. S., Zimmerman, S. C., Thiessen, P. A., Hawryluk, N. A., Murray, T. J. *J. Am. Chem. Soc.* **2001**, *123*, 10475–10488.
44 Zeng, H., Miller, R. S., Flowers, R. A., II, Gong, B. *J. Am. Chem. Soc.* **2000**, *122*, 2635–2644.
45 Gong, B. *Synlett* **2001**, 582–589.
46 Gong, B., Yan, Y., Zeng, H., Skrzypczak-Jankunn, E., Kim, Y. W., Zhu, J., Ickes, H. *J. Am. Chem. Soc.* **1999**, *121*, 5607–5608.
47 Wang, X.-Z., Li, X.-Q., Shao, X.-B., Zhao, X., Deng, P., Jiang, X.-K., Li, Z.-T., Chen, Y.-Q. *Chem. Eur. J.* **2003**, *9*, 2904–2913.
48 Ligthart, G. B. W. L., Ohkawa, H., Sijbesma, R. P., Meijer, E. W. *J. Am. Chem. Soc.* **2005**, *127*, 810–811.
49 Park, T., Zimmerman, S. C., Nakashima, S. *J. Am. Chem. Soc.* **2005**, *127*, 6520–6521.
50 Park, T., Todd, E. M., Nakashima, S., Zimmerman, S. C. *J. Am. Chem. Soc.* **2005**, *127*, 18133–18142.
51 Mayer, M. F., Nakashima, S., Zimmerman, S. C. *Org. Lett.* **2005**, *7*, 3005–3008.
52 Loeb, S. J., Wisner, J. A. *Angew. Chem. Int. Ed.* **1998**, *37*, 2838–2840.
53 Loeb, S. J., Tiburcio, J., Vella, S. J. *Org. Lett.* **2005**, *7*, 4923–4926.
54 Connors, K. A. *Chem. Rev.* **1997**, *97*, 1325–1357.
55 Rekharsky, M. V., Inoue, Y. *Chem. Rev.* **1998**, *98*, 1875–1917.
56 Liu, S., Ruspic, C., Mukhopadhyay, P., Chakrabarti, S., Zavalij, P. Y., Isaacs, L. *J. Am. Chem. Soc.* **2005**, *127*, 15959–15967.

57 Lagona, J., Mukhopadhyay, P., Chakrabarti, S., Isaacs, L. *Angew. Chem. Int. Ed.* **2005**, *44*, 4844–4870.
58 Kato, T., Fréchet, J. M. J. *J. Am. Chem. Soc.* **1989**, *111*, 8533–8534.
59 Brienne, M. J., Gabard, J., Lehn, J. M., Stibor, I. *J. Chem. Soc., Chem. Commun.* **1989**, 1868–1870.
60 Kato, T., Wilson, P. G., Fujishima, A., Frèchet, J. M. J. *Chem. Lett.* **1990**, 2003–2006.
61 Alexander, C., Jariwala, C. P., Lee, C. M., Griffin, A. C. *Macromol. Symp.* **1994**, *77*, 283–294.
62 Kihara, H., Kato, T., Uryu, T., Frèchet, J. M. J. *Chem. Mater.* **1996**, *8*, 961–968.
63 Ciferri, A. *Liq. Cryst.* **2004**, *31*, 1487–1493.
64 Kato, T. *Struct. Bonding (Berlin)* **2000**, *96*, 95–146.
65 Odijk, T. *J. Phys. (Paris)* **1987**, *48*, 125–129.
66 Odijk, T. *Curr. Opin. Colloid Interface Sci.* **1996**, *1*, 337–340.
67 van der Schoot, P. *J. Phys. II* **1995**, *5*, 243–248.
68 Fouquey, C., Lehn, J. M., Levelut, A. M. *Adv. Mater.* **1990**, *2*, 254–257.
69 Lehn, J. M. *Macromol. Chem., Macromol. Symp.* **1993**, *69*, 1–17.
70 Gulik-Krzywicki, T., Fouquey, C., Lehn, J. M. *Proc. Natl. Acad. Sci. USA* **1993**, *90*, 163–167.
71 Kotera, M., Lehn, J. M., Vigneron, J. P. *J. Chem. Soc., Chem. Commun.* **1994**, 197–199.
72 St. Pourcain, C. B., Griffin, A. C. *Macromolecules* **1995**, *28*, 4116–4121.
73 Wiegel, K. N., Griffin, A. C., Black, M. S., Schiraldi, D. A. *J. Appl. Polym. Sci.* **2004**, *92*, 3097–3106.
74 Bong, D. T., Clark, T. D., Granja, J. R., Ghadiri, M. R. *Angew. Chem. Int. Ed.* **2001**, *40*, 988–1011.
75 Klok, H. A., Jolliffe, K. A., Schauer, C. L., Prins, L. J., Spatz, J. P., Moller, M., Timmerman, P., Reinhoudt, D. N. *J. Am. Chem. Soc.* **1999**, *121*, 7154–7155.
76 Choi, I. S., Li, X. H., Simanek, E. E., Akaba, R., Whitesides, G. M. *Chem. Mater.* **1999**, *11*, 684–690.
77 Ghadiri, M. R., Granja, J. R., Milligan, R. A., McRee, D. E., Khazanovich, N. *Nature* **1993**, *366*, 324–327.
78 Hartgerink, J. D., Ghadiri, M. R. *New Macromolecular Architecture and Functions*, Springer, Berlin, **1996**.
79 Ghadiri, M. R., Granja, J. R., Buehler, L. K. *Nature* **1994**, *369*, 301–304.
80 Motesharei, K., Ghadiri, M. R. *J. Am. Chem. Soc.* **1997**, *119*, 11306–11312.
81 Boileau, S., Bouteiller, L., Laupretre, F., Lortie, F. *New J. Chem.* **2000**, *24*, 845–848.
82 Lortie, F., Boileau, S., Bouteiller, L., Chassenieux, C., Deme, B., Ducouret, G., Jalabert, M., Laupretre, F., Terech, P. *Langmuir* **2002**, *18*, 7218–7222.
83 Simic, V., Bouteiller, L., Jalabert, M. *J. Am. Chem. Soc.* **2003**, *125*, 13148–13154.
84 van der Gucht, J., Besseling, N. A. M., Knoben, W., Bouteiller, L., Cohen Stuart, M. A. *Phys. Rev. E* **2003**, *67*.
85 Ikeda, M., Nobori, T., Schmutz, M., Lehn, J. M. *Chem. Eur. J.* **2005**, *11*, 662–668.
86 Seto, C. T., Whitesides, G. M. *J. Am. Chem. Soc.* **1993**, *115*, 1330–1340.
87 Mammen, M., Simanek, E. E., Whitesides, G. M. *J. Am. Chem. Soc.* **1996**, *118*, 12614–12623.
88 Söntjens, S. H. M., Sijbesma, R. P., van Genderen, M. H. P., Meijer, E. W. *J. Am. Chem. Soc.* **2000**, *122*, 7487–7493.
89 Keizer, H. M., Sijbesma, R. P., Meijer, E. W. *Eur. J. Org. Chem.* **2004**, 2553–2555.
90 Folmer, B. J. B., Cavini, E., Sijbesma, R. P., Meijer, E. W. *Chem. Commun.* **1998**, 1847–1848.
91 Wubbenhorst, M., van Turnhout, J., Folmer, B. J. B., Sijbesma, R. P., Meijer, E. W. *IEEE Trans. Dielectr. Electr. Insul.* **2001**, *8*, 365–372.
92 Doi, M., Edwards, S. F. *The Theory of Polymer Dynamics*. Clarendon Press, Oxford, **1986**.
93 Hirschberg, J. H. K. K., Brunsveld, L., Ramzi, A., Vekemans, J. A. J. M., Sijbesma, R. P., Meijer, E. W. *Nature* **2000**, *407*, 167–170.
94 Brunsveld, L., Vekemans, J., Hirschberg, J., Sijbesma, R. P., Meijer, E. W. *Proc. Natl. Acad. Sci. USA* **2002**, *99*, 4977–4982.
95 Castellano, R. K., Rudkevich, D. M., Rebek, J., Jr. *Proc. Natl. Acad. Sci. USA* **1997**, *94*, 7132–7137.

96 Castellano, R. K., Clark, R., Craig, S. L., Nuckolls, C., Rebek, J., Jr. *Proc. Natl. Acad. Sci. USA* **2000**, *97*, 12418–12421.
97 Castellano, R. K., Nuckolls, C., Eichhorn, S. H., Wood, M. R., Lovinger, A. J., Rebek, J. *Angew. Chem. Int. Ed.* **1999**, *38*, 2603–2606.
98 Castellano, R. K., Rebek, J., Jr. *J. Am. Chem. Soc.* **1998**, *120*, 3657–3663.
99 Fogleman, E. A., Yount, W. C., Xu, J., Craig, S. L. *Angew. Chem. Int. Ed.* **2002**, *41*, 4026–4028.
100 Xu, J., Fogleman, E. A., Craig, S. L. *Macromolecules* **2004**, *37*, 1863–1870.
101 Cantrill, S. J., Youn, G. J., Stoddart, J. F., Williams, D. J. *J. Org. Chem.* **2001**, *66*, 6857–6872.
102 Yamaguchi, N., Nagvekar, D. S., Gibson, H. W. *Angew. Chem. Int. Ed.* **1998**, *37*, 2361–2364.
103 Gibson, H. W., Yamaguchi, N., Jones, J. W. *J. Am. Chem. Soc.* **2003**, *125*, 3522–3533.
104 Miyauchi, M., Takashima, Y., Yamaguchi, H., Harada, A. *J. Am. Chem. Soc.* **2005**, *127*, 2984–2989.
105 Ohga, K., Takashima, Y., Takahashi, H., Kwaguchi, Y., Yamaguchi, H., Harada, A. *Macromolecules* **2005**, *38*, 5897–5904.
106 Hasegawa, Y., Miyauchi, M., Takashima, Y., Yamaguchi, H., Harada, A. *Macromolecules* **2005**, *38*, 3724–3730.
107 Kim, K., Kim, D., Lee, J. W., Ko, Y. H., Kim, K. *Chem. Commun.* **2004**, 848–849.
108 Hoshino, H., Jin, J. I., Lenz, R. W. *J. Appl. Polym. Sci.* **1984**, *29*, 547–554.
109 Lillya, C. P., Baker, R. J., Hutte, S., Winter, H. H., Lin, Y. G., Shi, J., Dickinson, L. C., Chien, J. C. W. *Macromolecules* **1992**, *25*, 2076–2080.
110 Abed, S., Boileau, S., Bouteiller, L., Lacoudre, N. *Polym. Bull.* **1997**, *39*, 317–324.
111 Abed, S., Boileau, S., Bouteiller, L. *Macromolecules* **2000**, *33*, 8479–8487.
112 Abed, S., Boileau, S., Bouteiller, L. *Polymer* **2001**, *42*, 8613–8619.
113 Duweltz, D., Laupretre, F., Abed, S., Bouteiller, L., Boileau, S. *Polymer* **2003**, *44*, 2295–2302.
114 Yamauchi, K., Lizotte, J. R., Long, T. E. *Macromolecules* **2002**, *35*, 8745–8750.

115 Rowan, S. J., Suwanmala, P., Sivakova, S. *J. Polym. Sci., Part A: Polym. Chem.* **2003**, *41*, 3589–3596.
116 Sivakova, S., Bohnsack, D. A., Mackay, M. E., Suwanmala, P., Rowan, S. J. *J. Am. Chem. Soc.*, 18202–18211.
117 Ky Hirschberg, J. H. K., Beijer, F. H., van Aert, H. A., Magusin, P. C. M. M., Sijbesma, R. P., Meijer, E. W. *Macromolecules* **1999**, *32*, 2696–2705.
118 Folmer, B. J. B., Sijbesma, R. P., Versteegen, R. M., van der Rijt, J. A. J., Meijer, E. W. *Adv. Mater.* **2000**, *12*, 874–878.
119 Keizer, H. M., van Kessel, R., Sijbesma, R. P., Meijer, E. W. *Polymer* **2003**, *44*, 5505–5511.
120 Lange, R. F. M., Van Gurp, M., Meijer, E. W. *J. Polym. Sci., Part A: Polym. Chem.* **1999**, *37*, 3657–3670.
121 Keizer, H. M., Sijbesma, R. P., Jansen, J. F. G. A., Pasternack, G., Meijer, E. W. *Macromolecules* **2003**, *36*, 5602–5606.
122 Dankers, P. Y. W., Harmsen, M. C., Brouwer, L. A., Van Luyn, M. J. A., Meijer, E. W. *Nat. Mater.* **2005**, *4*, 568–574.
123 Yamauchi, K., Kanomata, A., Inoue, T., Long, T. E. *Macromolecules* **2004**, *37*, 3519–3522.
124 Guan, Z. B., Roland, J. T., Bai, J. Z., Ma, S. X., McIntire, T. M., Nguyen, M. *J. Am. Chem. Soc.* **2004**, *126*, 2058–2065.
125 Klok, H.-A., Lecommandoux, S. *Adv. Mater.* **2001**, *13*, 1217–1229.
126 Loos, K., Munoz-Guerra, S. *Supramolecular Polymers*, 2nd edn., Quirk, R. P. (Ed) Taylor & Francis, Boca Raton (FL) **2005**, pp. 393–442.
127 Ten Brinke, G., Ikkala, O. *Chem. Rec.* **2004**, *4*, 219–230.
128 Whitesides, G. M., Grzybowski, B. *Science* **2002**, *295*, 2418–2421.
129 Muthukumar, M., Ober, C. K., Thomas, E. L. *Science* **1997**, *277*, 1225–1232.
130 Stupp, S. I., LeBonheur, V., Walker, K., Li, L. S., Huggins, K. E., Keser, M., Amstutz, A. *Science* **1997**, *276*, 384–389.
131 Ikkala, O., ten Brinke, G. *Chem. Commun.* **2004**, 2131–2137.
132 Hirschberg, J. H. K. K., Ramzi, A., Sijbesma, R. P., Meijer, E. W. *Macromolecules* **2003**, *36*, 1429–1432.

133 Binder, W. H., Kunz, M. J., Ingolic, E. *J. Polym. Sci., Part A: Polym. Chem.* **2003**, *42*, 162–172.
134 Kunz, M. J., Hayn, G., Saf, R., Binder, W. H. *J. Polym. Sci., Part A: Polym. Chem.* **2004**, *42*, 661–674.
135 Binder, W. H., Kunz, M. J., Kluger, C., Hayn, G., Saf, R. *Macromolecules* **2004**, *37*, 1749–1759.
136 Binder, W. H., Bernstorff, S., Kluger, C., Petraru, L., Kunz, M. J. *Adv. Mater.* **2005**, *17*, 2824–2828.
137 Yang, X., Hua, F., Yamato, K., Ruckenstein, E., Gong, B., Kim, W., Ryu, C. Y. *Angew. Chem. Int. Ed.* **2004**, *43*, 6471–6474.
138 Gong, B., Yang, X., Kim, W., Ryu, C. Y. *Polym. Prepr.* **2005**, *46*, 1128–1129.
139 Pollino, J. M., Weck, M. *Chem. Soc. Rev.* **2005**, *34*, 193–207.
140 Kato, T., Mizoshita, N., Kishimoto, K. *Angew. Chem. Int. Ed.* **2006**, *45*, 38–68.
141 Kato, T., Frèchet, J. M. J. *Macromolecules* **1989**, *22*, 3818–3819.
142 Ujiie, S., Iimura, K. *Macromolecules* **1992**, *25*, 3174–3178.
143 Kato, T., Frèchet, J. M. J. *Macromol. Symp.* **1995**, *98*, 311–326.
144 Kato, T., Kihara, H., Kumar, U., Uryu, T., Frèchet, J. M. J. *Angew. Chem.* **1994**, *106*, 1728–1730.
145 Malik, S., Dhal, P. K., Mashelkar, R. A. *Macromolecules* **1995**, *28*, 2159–2164.
146 Kato, T., Nakano, M., Moteki, T., Uryu, T., Ujiie, S. *Macromolecules* **1995**, *28*, 8875–8876.
147 Kato, T., Ihata, O., Ujiie, S., Tokita, M., Watanabe, J. *Macromolecules* **1998**, *31*, 3551–3555.
148 Vivas de Meftahi, M., Frèchet, J. M. J. *Polymer* **1988**, *29*, 477–482.
149 Ikkala, O., ten Brinke, G. *Science* **2002**, *295*, 2407–2409.
150 Ruokolainen, J., ten Brinke, G., Ikkala, O., Torkkeli, M., Serimaa, R. *Macromolecules* **1996**, *29*, 3409–3415.
151 Ruokolainen, J., Saariaho, M., Ikkala, O., ten Brinke, G., Thomas, E. L., Torkkeli, M., Serimaa, R. *Macromolecules* **1999**, *32*, 1152–1158.
152 Osuji, C., Chao, C.-Y., Bita, I., Ober, C. K., Thomas, E. L. *Adv. Funct. Mater.* **2002**, *12*, 753–758.
153 Valkama, S., Kosonen, H., Ruokolainen, J., Haatainen, T., Torkkeli, M., Serimaa, R., ten Brinke, G., Ikkala, O. *Nat. Mater.* **2004**, *3*, 872–876.
154 Chao, C.-Y., Li, X., Ober, C. K., Osuji, C., Thomas, E. L. *Adv. Funct. Mater.* **2004**, *14*, 364–370.
155 Maki-Ontto, R., de Moel, K., Polushkin, E., van Ekenstein, G. A., ten Brinke, G., Ikkala, O. *Adv. Mater.* **2002**, *14*, 357–361.
156 Ruokolainen, J., Makinen, R., Torkkeli, M., Makela, T., Serimaa, R., ten Brinke, G., Ikkala, O. *Science* **1998**, *280*, 557–560.
157 Mueller, M., Dardin, A., Seidel, U., Balsamo, V., Ivan, B., Spiess, H. W., Stadler, R. *Macromolecules* **1996**, *29*, 2577–2583.
158 Hilger, C., Stadler, R., De Lucca Freitas, L. L. *Polymer* **1990**, *31*, 818–823.
159 Bica, C. I. D., Burchard, W., Stadler, R. *Macromol. Chem. Phys.* **1996**, *197*, 3407–3426.
160 De Lucca Freitas, L. L., Stadler, R. *Macromolecules* **1987**, *20*, 2478–2485.
161 De Lucca Freitas, L. L., Stadler, R. *Colloid Polym. Sci.* **1988**, *266*, 1095–1101.
162 Müller, M., Seidel, U., Stadler, R. *Polymer* **1995**, *36*, 3143–3150.
163 Müller, M., Stadler, R., Kremer, F., Williams, G. *Macromolecules* **1995**, *28*, 6942–6949.
164 Seidel, U., Stadler, R., Fuller, G. G. *Macromolecules* **1994**, *27*, 2066–2072.
165 Hilger, C., Stadler, R. *Macromolecules* **1992**, *25*, 6670–6680.
166 Lange, R. F. M., Meijer, E. W. *Macromolecules* **1995**, *28*, 782–783.
167 Ilhan, F., Gray, M., Rotello, V. M. *Macromolecules* **2001**, *34*, 2597–2601.
168 Boal, A. K., Ilhan, F., DeRouchey, J. E., Thurn-Albrecht, T., Russell, T. P., Rotello, V. M. *Nature* **2000**, *404*, 746–748.
169 Frankamp, B. L., Uzun, O., Ilhan, F., Boal, A. K., Rotello, V. M. *J. Am. Chem. Soc.* **2002**, *124*, 892–893.
170 Boal, A. K., Frankamp, B. L., Uzun, O., Tuominen, M. T., Rotello, V. M. *Chem. Mater.* **2004**, *16*, 3252–3256.
171 Ilhan, F., Galow, T. H., Gray, M., Clavier, G., Rotello, V. M. *J. Am. Chem. Soc.* **2000**, *122*, 5895–5896.

172 Thibault, R. J., Jr., Galow, T. H., Turnberg, E. J., Gray, M., Hotchkiss, P. J., Rotello, V. M. *J. Am. Chem. Soc.* **2002**, *124*, 15249–15254.
173 Thibault, R. J., Hotchkiss, P. J., Gray, M., Rotello, V. M. *J. Am. Chem. Soc.* **2003**, *125*, 11249–11252.
174 Uzun, O., Sanyal, A., Nakade, H., Thibault, R. J., Rotello, V. M. *J. Am. Chem. Soc.* **2004**, *126*, 14773–14777.
175 Uzun, O., Xu, H., Jeoung, E., Thibault, R. J., Rotello, V. M. *Chem. Eur. J.* **2005**, *11*, 6916–6920.
176 Gerhardt, W., Crne, M., Weck, M. *Chem. Eur. J.* **2004**, *10*, 6212–6221.
177 Hofmeier, H., Schubert Ulrich, S. *Chem. Commun.* **2005**, 2423–2432.
178 Pollino, J. M., Stubbs, L. P., Weck, M. *J. Am. Chem. Soc.* **2004**, *126*, 563–567.
179 Lehn, J. M. *Polym. Int.* **2002**, *51*, 825–839.
180 Rieth, L. R., Eaton, R. F., Coates, G. W. *Angew. Chem. Int. Ed.* **2001**, *40*, 2153–2156.
181 Yamauchi, K., Lizotte, J. R., Long, T. E. *Macromolecules* **2003**, *36*, 1083–1088.
182 McKee, M. G., Elkins, C. L., Long, T. E. *Polymer* **2004**, *45*, 8705–8715.
183 McKee, M. G., Elkins, C. L., Park, T., Long, T. E. *Macromolecules* **2005**, *38*, 6015–6023.
184 Elkins, C. L., Park, T., McKee, M. G., Long, T. E. *J. Polym. Sci., Part A: Polym. Chem.* **2005**, *43*, 4618–4631.
185 Sijbesma, R. P., Beijer, F. H., Brunsveld, L., Meijer, E. W. (DSM). *US Patent 6320018*, **2001**.
186 Bosman, A. W., Sijbesma, R. P., Meijer, E. W. *Mater. Today* **2004**, *7*, 34–39.
187 Loontjens, J. A., Jansen, J. G. F. A., Plum, B. J. M. (DSM). *EP 1031589*, **2002**.
188 Eling, B. (Huntsman International). *WO 0246260*, **2002**.
189 Ishizuka, Y., Hawakawa, E., Asawa, Y., Pappas, S. P. (Kodak Polychrome Graphis). *WO 02053627*, **2002**.
190 Monk, A., Saraiya, S., Huang, J., Pappas, S. P. (Kodak Polychrome Graphis). *WO 02053626*, **2002**.
191 Locculier, J., Vanmaele, L. L., Meijer, E. W., Fransen, P., Janssen, H. (Agfa Gevaert). *EP 1310533*, **2002**.
192 Luca, D. J., Smith, T. W., McGrane, K. M. (Xerox Corp.). *US Patent Application 2003079644*, **2003**.
193 Goodbrand, H. B. S., Popovic, D., Foucher, D. A., Smith, T. W., McGrane, K. M., (Xerox Corp.). *US Patent Application 20030105185*, **2003**.
194 Hofstraat, J. W., Meijer, E. W., Schenning, A. P. H. J., Hoeben, F. J. M. (Philips Electronics). *WO 111153*, **2004**.
195 Kukulj, D., Goldoni, F. (Unilever). *WO 02092744 A1*, **2002**.
196 Livoreil, A., Mondet, J., Mougin, N. (L'Oreal). *WO 02098377*, **2002**.
197 Eason, M. D., Khoshdel, E., Cooper, J. H., Royles, B. J. L. (Unilever). *WO 03032929*, **2003**.
198 Semlyen, J. A. *Large Ring Molecules*. Wiley, Chichester, **1996**.
199 Semlyen, J. A. *Pure Appl. Chem.* **1981**, *53*, 1797–804.
200 Jacobsen, H., Stockmayer, W. H. *J. Chem. Phys.* **1950**, *18*, 1600–1606.
201 Flory, P. J., Suter, U. W., Mutter, M. *J. Am. Chem. Soc.* **1976**, *98*, 5733–5739.
202 Suter, U. W., Mutter, M., Flory, P. J. *J. Am. Chem. Soc.* **1976**, *98*, 5740–5745.
203 Mutter, M., Suter, U. W., Flory, P. J. *J. Am. Chem. Soc.* **1976**, *98*, 5745–5748.
204 Suter, U. W., Mutter, M. *Makromol. Chem.* **1979**, *180*, 1761–1773.
205 Folmer, B. J. B., Sijbesma, R. P., Meijer, E. W. *J. Am. Chem. Soc.* **2001**, *123*, 2093–2094.
206 Ercolani, G., Mandolini, L., Mencarelli, P., Roelens, S. *J. Am. Chem. Soc.* **1993**, *115*, 3901–3908.
207 Ercolani, G. *J. Phys. Chem. B* **1998**, *102*, 5699–5703.
208 Söntjens, S. H. M., Sijbesma, R. P., van Genderen, M. H. P., Meijer, E. W. *Macromolecules* **2001**, *34*, 3815–3818.
209 ten Cate, A. T., Kooijman, H., Spek, A. L., Sijbesma, R. P., Meijer, E. W. *J. Am. Chem. Soc.* **2004**, *126*, 3801–3808.
210 Seshadri, S., Deming, S. N. *Anal. Chem.* **1984**, *56*, 1567–1572.
211 Zeng, F. W., Zimmerman, S. C., Kolotuchin, S. V., Reichert, D. E. C., Ma, Y. G. *Tetrahedron* **2002**, *58*, 825–843.
212 Paulusse, J. M. J., Sijbesma, R. P. *Chem. Commun.* **2003**, 1494–1495.

213 Paulusse, J. M. J., Sijbesma, R. P. *Angew. Chem. Int. Ed.* **2004**, *43*, 4460–4462.

214 Paulusse, J. M. J., Huijbers, J. P. J., Sijbesma, R. P. *Macromolecules* **2005**, *38*, 6290–6298.

215 Johnson, C. S., Jr. *Prog. Nucl. Magn. Reson. Spectrosc.* **1999**, *34*, 203–256.

216 Scherman, O. A., Ligthart, G. B. W. L., Sijbesma, R. P., Meijer, E. W. *Angew. Chem. Int. Ed.* **2006**, 2072–2076.

10
Polymer Synthesis and Modification by Enzymatic Catalysis

Shiro Kobayashi and Masashi Ohmae

10.1
Introduction

In the field of organic synthesis, enzymes were already in use seven decades ago, as exemplified by lipase catalysis in organic media [1]. This area expanded very actively in the 1980s [2]. The major reason for utilizing enzymatic catalysis is that organic synthesis technology is not adequate to control the stereochemistry and regioselectivity of the products, which are often used further as key intermediates for the total synthesis of complex-structured molecules such as pharmacologically active natural products [3]. For examples, lipase catalysis allowed the synthesis of chiral drugs, liquid crystals, acylated sugar-based surfactants and functional triglycerides. Recent developments in asymmetric synthesis, however, enabled such molecules to be provided in many cases [4].

In the field of polymer synthesis, enzyme-catalyzed polymer synthesis (often referred as *enzymatic polymerization*) was initiated in the late 1980s and has been extensively developed during the last two decades [5–23]. Historically, polymerization catalysts have followed two major directions. The first belongs to the classical catalysts such as acids (Lewis acids, Brønsted acids and various cations), bases (Lewis bases and various anions) and radical-generating species, that have been employed since the 1920s, the early period of polymer science. The second includes transition metal catalysts such as Ziegler–Natta catalysts discovered in the early 1950s and metathesis catalysts recently contributed by Grubbs and Schrock and also rare-earth metal catalysts.

These two major streams of conventional chemical catalysts are still developing towards the goal of precision polymerization and precision polymer synthesis. Enzymatic polymerization can now be regarded as a third stream, although still small compared with the first two [23]. Enzymatic polymerization is defined as "the *in vitro* polymerization of artificial substrate monomers catalyzed by an isolated enzyme via non-biosynthetic (non-metabolic) pathways" [5, 6, 8, 11–14, 16–18, 20, 23]. Therefore, *in vivo* enzymatic reactions via biosynthetic pathways are not included, hence fermentation processes for synthesizing poly(hydroxylalkanoate)s in living cells or *Escherichia coli*-using processes for protein synthesis are not dealt with. Enzymatic catalysis is also employed for polymer modification. Therefore, this chapter overviews polymer synthesis and modification by enzymatic catalysis.

10.2
Characteristics of Enzymatic Catalysis

Enzymatic catalysis *in vivo* is responsible for the production of all the substances necessary for maintaining life and involves several characteristics in general: (i) a high catalytic activity (high turnover number), (ii) reactions under mild conditions with respect to temperature, pressure, solvent, neutral pH, etc., bringing about high energetic efficiency, and (iii) high reaction regulation of regio-, enantio-, chemo-, stereo- and choro-selectivities, giving rise to perfectly structure-controlled products, compared with conventional chemical catalysts [5, 6, 8, 11–14, 16–18, 20, 23]. If these *in vivo* characteristics can be realized for *in vitro* polymer synthesis, we may expect the following outcomes: (i) precise control of polymer structures, (ii) creation of polymers with new structures, (iii) a clean process (without by-products), (iv) a low loading process (energy saving) and (v) product polymers that are biodegradable in many cases. These features are often difficult to achieve by conventional methods, but have been realized in practice in many cases as described below.

Fundamentally important characteristics of *in vivo* enzymatic reactions will be described concerning the following two aspects. First, Fischer in 1894 proposed a "key and lock" theory of the relationship between enzyme and substrate [24]. The theory was based on the findings that via biosynthetic pathways *in vivo* an enzyme induces a reaction of only a specific substrate in a strictly 1:1 fashion like a lock and a key. This phenomenon is nowadays understood in the following way: they recognize each other and form an enzyme–substrate complex as shown in Fig. 10.1. The formation of the complex activates the substrate to lead to a product with perfect structure control. The complex formation is due to supramolecular interactions, mainly of hydrogen bonding.

Second, Pauling pointed out the reason why enzymatic reactions proceed under such mild reaction conditions [25]. An enzyme and a substrate form a complex, which stabilizes a *transition state* and lowers an activation energy substantially compared with a no-enzyme case (Fig. 10.2). The rate enhancement by enzymatic catalysis normally reaches 10^6–10^{12}-fold; however, a recent paper reported even 10^{20}-fold [26]. Figure 10.2 implies that once such a complex is formed, an enzymatic reaction is reversible; a hydrolysis enzyme catalyzing bond cleavage *in vivo* is able to catalyze the bond formation of monomers (formally a reverse reaction of hydrolysis) to produce eventually a polymer by repeating the bond-forming reaction.

Normally, monomers for enzymatic polymerization need to be recognized and activated by the enzyme with formation of an enzyme–monomer complex for

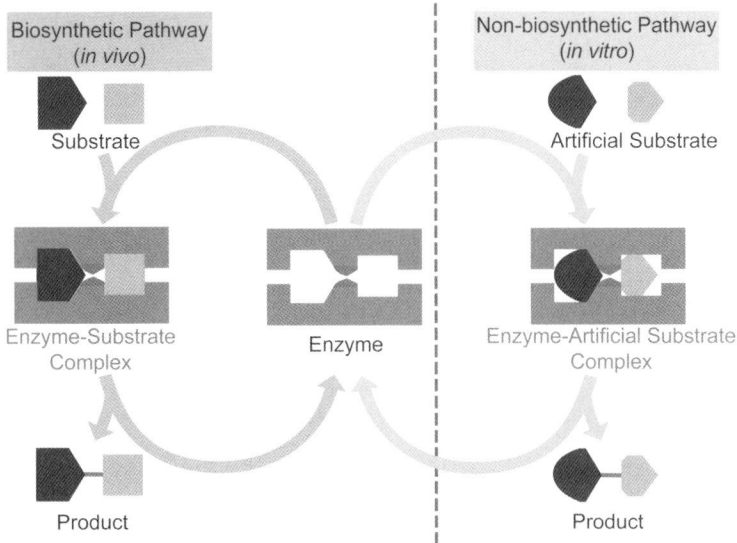

Fig. 10.1 "Key and lock" theory for *in vivo* enzymatic reactions via biosynthetic pathways (left-hand cycle) and for *in vitro* enzymatic reactions via non-biosynthetic pathways (right-hand cycle), showing the importance of the artificial substrate design.

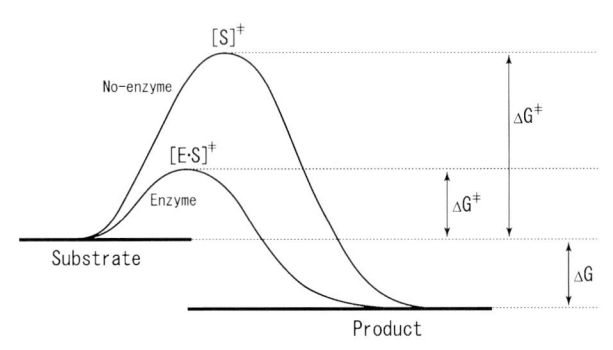

Fig. 10.2 Reaction profile of enzymatic and non-enzymatic reactions, involving the transition state of the enzyme–substrate complex formation for lowering the activation energy.

the polymerization to proceed. Therefore, the key for enzymatic polymerization is the design and synthesis of an artificial monomer structure required for the enzyme to be used, as shown in the right-hand cycle in Fig. 10.1. As a guide to monomer design, a new concept of a *transition-state analogue substrate* (TSAS) monomer was proposed [12–14, 16, 17, 20, 23], which is now widely accepted. This concept of TSAS is also applicable in most cases to enzymatic catalysis of polymer modification.

Table 10.1 Classification of enzymes, examples, typical polymers prepared via *in vitro* enzymatic polymerization (1) and typical polymers used for enzymatic modification (2).

Enzymes	Example enzymes	Typical polymers (1)	Typical polymers (2)
1. Oxidoreductases	Peroxidase Laccase Tyrosinase Glucose oxidase	Polyphenols Polyanilines Vinyl polymers	Polysaccharides Polypeptides (proteins)
2. Transferases	Glycosyltransferase Acyltransferase	Polysaccharides Cyclic oligosaccharides Polyesters	Polysaccharides Polypeptides (proteins)
3. Hydrolases	Glycosidase (cellulase, amylase, chitinase, hyaluronidase) Lipase Peptidase	Polysaccharides Polyesters Polyamides Polycarbonates Poly(amino acid)s	Polysaccharides Polypeptides (proteins)
4. Lyases	Decarboxylase Aldolase Dehydratase		Polysaccharides
5. Isomerases	Racemase Epimerase		
6. Ligases	Acyl CoA synthetase		

All enzymes are classified into six divisions by the Enzyme Commission (EC) (Table 10.1) [8, 11–14, 17, 23]. So far, hydrolase enzymes are most often used for enzymatic polymerization, because they are readily isolable due to their stability and widely availability as commercial products. Polymer synthesis is described below in the order of the enzyme classification.

10.3
Synthesis of Poly(aromatic)s Catalyzed by Oxidoreductases

Representative oxidoreductase enzymes include peroxidases, laccases, bilirubin oxidase and tyrosinase. Peroxidases such as horseradish peroxidase (HRP), soybean peroxidase (SBP), catalase and lignin peroxidase contain an Fe atom in their active site and catalyze an oxidation of a donor to an oxidized donor by the action of hydrogen peroxide, liberating two water molecules. HRP is a single-chain β-type hemoprotein, catalyzing the decomposition of hydrogen peroxide at

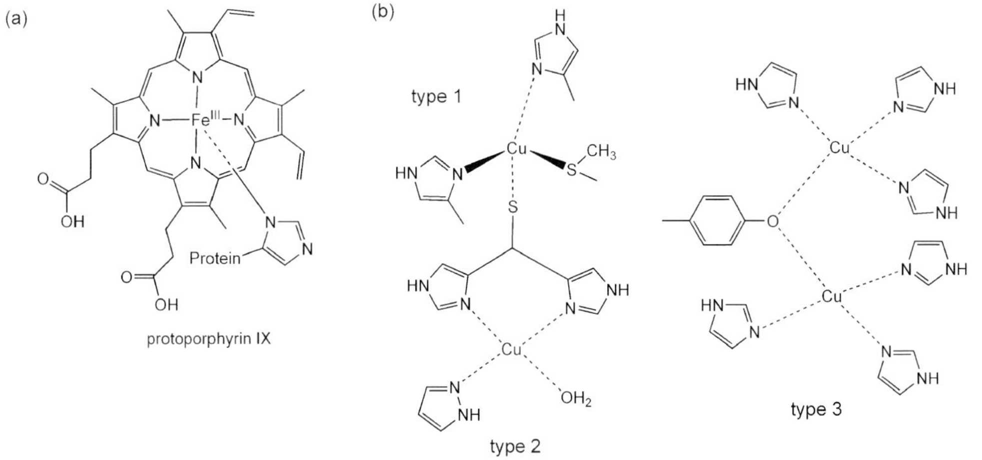

Fig. 10.3 Structures of (a) horseradish peroxidase (HRP) and (b) laccase.

the expense of aromatic proton donors (Fig. 10.3a). Laccases, bilirubin oxidase and tyrosinase, having a Cu active site, catalyze the oxidative coupling of phenols (Fig. 10.3b).

As starting aromatic monomers, phenolic compounds and anilines are typically cited. Here, two categories of phenolic compounds are defined: one is compounds having only one aromatic hydroxyl group and the other is those having more than two hydroxyl groups on the aromatic ring(s) and hence referred as polyphenols. The latter sometimes causes confusion with polymeric (high molecular weight) materials derived from phenolic compounds. In other words, the term "polyphenols" is used for catechol, flavonoid and related compounds and "phenolic polymers" for polymeric materials from phenolic compounds [27].

10.3.1
Synthesis of Polymers from Phenolic Compounds

The following advantages of enzymatic synthesis of phenolic polymers are generally mentioned [28]: (i) the polymerization of phenolic compounds proceeds under mild reaction conditions without using toxic reagents (environmentally benign process); (ii) phenolic monomers having various substituents are polymerized to give a new class of functional poly(aromatic)s; (iii) the structure and solubility of the products can be controlled by changing the reaction conditions; (iv) the procedures for both polymerization and polymer isolation are facile.

Currently, two phenolic polymers are industrially produced, namely, phenol–formaldehyde resin (Bakelite) and poly(2,6-dimethyl-1,4-oxyphenylene). Their high performance properties are well accepted and widely used in practice [29].

10.3.1.1 Polymers from Unsubstituted Phenol

Unsubstituted phenol is the simplest phenol compound and industrially very important. It is desired, however, that the use of formaldehyde is avoided or minimized for environmental reasons. Phenol is a multifunctional monomer for oxidative polymerization, four positions being conceivably reactive, hence conventional methods of using, for example, oxidants of Fe or Cu compounds gave insoluble products with uncontrolled structure. Peroxidase, on the other hand, induced a mild polymerization and produced a polyphenol consisting of phenylene and oxyphenylene units, which shows relatively high thermal stability (Scheme 10.1) [30]. The polymerization of phenol in aqueous buffered alcohol (methanol or ethanol) using HRP or SBP as catalyst at room temperature yielded a powdery phenol polymer which is soluble in N,N-dimethylformamide (DMF) and dimethyl sulfoxide (DMSO). The molecular weight was around 3000–6000 with a distribution of 5.4. This was the first production of a soluble phenol polymer with controlled structure, which is mainly due to the selection of the solvent [30d]. The content of oxyphenylene units increased with increasing amount of methanol in the solvent, varying in the range 32–59%. The solubility increased as the number of oxyphenylene units increased. SBP also catalyzed the polymerization of phenol to produce soluble polyphenol having a controlled structure.

Molecular weight control was more readily achieved by the copolymerization of phenol with 2,4-dimethylphenol, the latter being a monomer with only two reactive positions [31].

The HRP-catalyzed polymerization of phenol was carried out with a dispersion system in 1,4-dioxane–buffer mixtures using poly(ethylene glycol), poly(vinyl alcohol) and poly(methyl vinyl ether) as stabilizers. The dispersion polymerization of phenol leads to the formation of relatively monodisperse polymer particles, typically with a diameter around 250 nm. Substituted phenols also gave such particles. SEM photographs of these particles from phenol and substituted phenols are shown in Fig. 10.4 [32]. Thermal treatment of these particles afforded uniform carbon particles.

By using a template of poly(ethylene glycol) monododecyl ether in water, the HRP-catalyzed polymerization of phenol greatly improved the regioselectivity, producing a phenol polymer with a phenylene unit content close to 90%. The regioselectivity enhancement is speculated to be due to an oriented alignment via hydrogen bonding interaction of phenolic OH groups and ether oxygen atoms of the template. The polymer was obtained in high yields as a precipitate complexed with the template molecules (Fig. 10.5) [33].

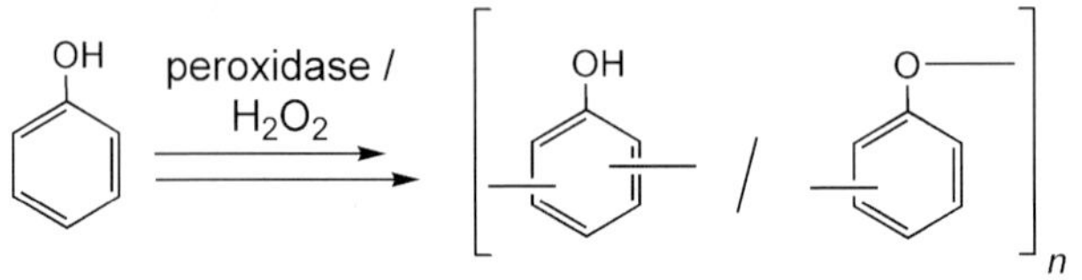

Scheme 10.1

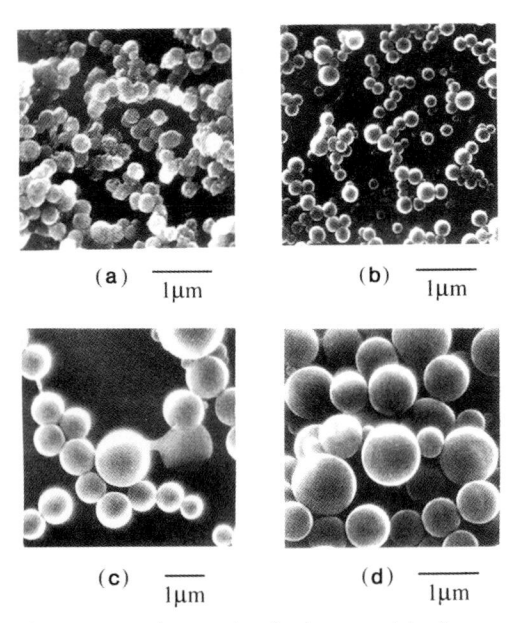

Fig. 10.4 SEM photographs of polymer particles from (a) phenol, (b) *m*-cresol, (c) *p*-cresol and (d) *p*-phenylphenol.

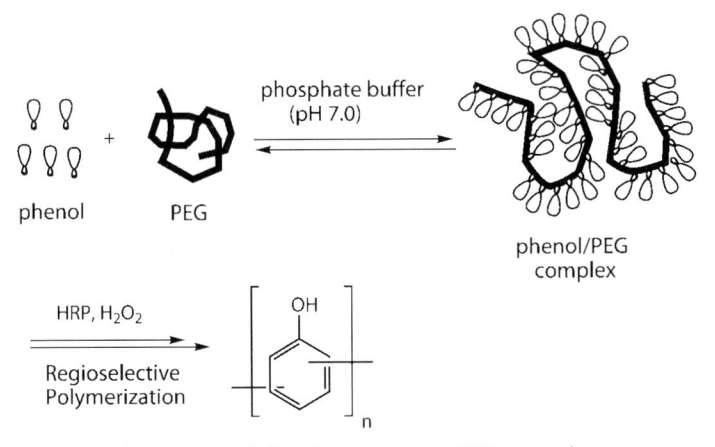

Fig. 10.5 Polymerization of phenol in water using PEG as template.

A novel bienzymatic system (glucose oxidase + HRP) was explored as a catalyst for phenol polymer synthesis (Scheme 10.2) [34]. This system induced the polymerization of phenol in the presence of glucose without the addition of hydrogen peroxide to produce the polymer in a moderate yield. Hydrogen peroxide was formed *in situ* by the oxidation of glucose catalyzed by glucose oxidase, which acted as an oxidizing agent for the polymerization.

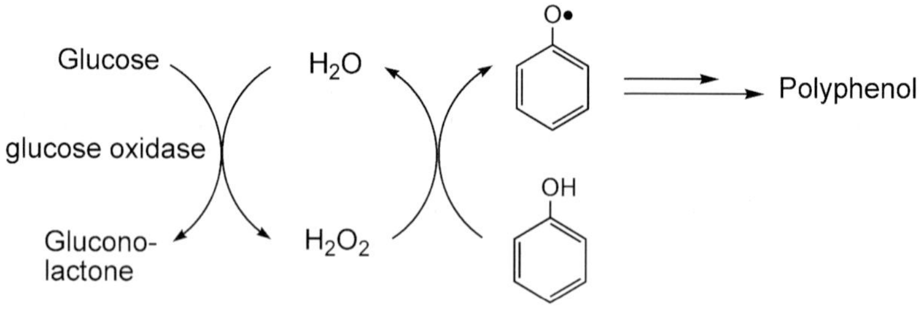

Scheme 10.2

10.3.1.2 Polymers from Substituted Phenols

The HRP-catalyzed polymerization of p-n-alkylphenols in an aqueous 1,4-dioxane solution gave polymers with molecular weights of several thousands, whereby the polymer yield increased as the alkyl chain length increased from 1 to 5 [35]. All cresol isomers were oxidatively polymerized by HRP catalysis, whereas among the o-, m- and p-isopropylphenols, only the p-isomer was polymerized [36]. Poly(p-alkylphenol) products often involve solubility problems; however, a soluble p-ethylphenol with a molecular weight of <1000 was obtained in aqueous DMF [37]. The oxidative polymerization of substituted phenols depends considerably on the structure of the phenols in terms of reaction behavior, product properties, solvent composition and enzyme origin. In the polymerization of para-substituted phenols, the reactive para-position is blocked and hence the product structure becomes simpler. The product structure from p-methylphenol (p-cresol) and p-propylphenol was studied in detail by NMR spectroscopy. The coupling mechanism of p-cresol was discussed with respect to the structure of the dimers produced in the initial stage of polymerization [38].

A natural phenol derivative, glucose–D-hydroquinone (arbutin), was oxidatively polymerized using peroxidase catalyst in a buffer solution, yielding water-soluble polymers with molecular weights ranging 1600 to 3200 (Scheme 10.3) [39]. Acidic deglycosylation of the resulting polymer afforded a soluble poly(hydroquinone), which may have structure with C–C coupling at the ortho-position. Chemoenzymatic synthesis of poly(hydroquinone) from 4-hydroxyphenyl benzoate was reported, the structure being different from that obtained from arbutin [40].

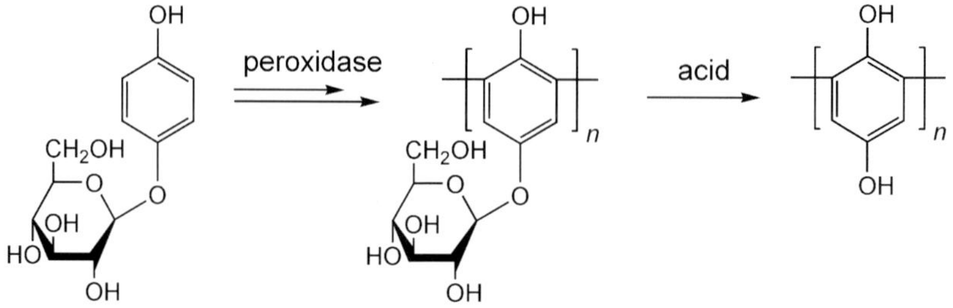

Scheme 10.3

Other examples of natural phenolic monomers are tyrosine ethyl ester and methyl ester, which were polymerized using HRP catalyst in a high concentration of buffer in aqueous solution (Scheme 10.4). The precipitation of tyrosine polymers made from both enantiomeric forms and racemic mixture was observed. The molecular weight was between 1500 and 4000. The polymer structure was estimated to be a mixture of phenylene and oxyphenylene units. The poly(tyrosine ester)s were further converted into a new class of polytyrosine by alkaline hydrolysis of the ester groups. The resulting poly(amino acid) is different from the so-called peptide-type polytyrosine and soluble only in water. Additionally, the enzyme HRP also initiated the oxidative homopolymerization of N-acetyltyrosine and the copolymerization of N-acetyltyrosine and arbutin [41].

Chemoselective reactions are very important for the preparation of functional materials. HRP induced a chemoselective oxidation polymerization of a phenol derivative having a methacryloyl group in aqueous acetone, where only the phenol moiety was involved in the polymerization (Scheme 10.5) [42]. The polymer with $M_n \sim 1700$ was soluble in an organic solvent and cured both thermally and

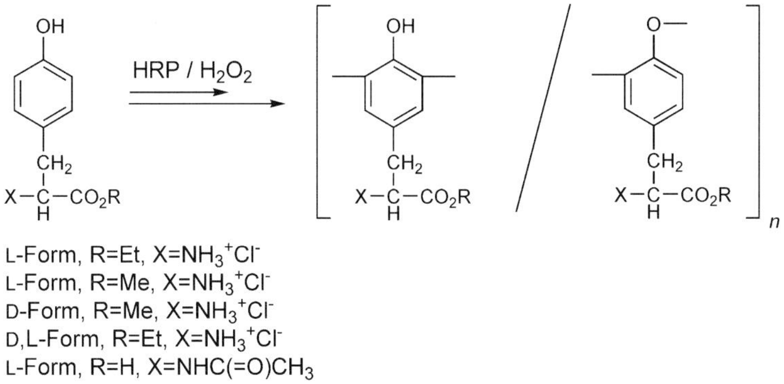

L-Form, R=Et, X=NH$_3^+$Cl$^-$
L-Form, R=Me, X=NH$_3^+$Cl$^-$
D-Form, R=Me, X=NH$_3^+$Cl$^-$
D,L-Form, R=Et, X=NH$_3^+$Cl$^-$
L-Form, R=H, X=NHC(=O)CH$_3$

Scheme 10.4

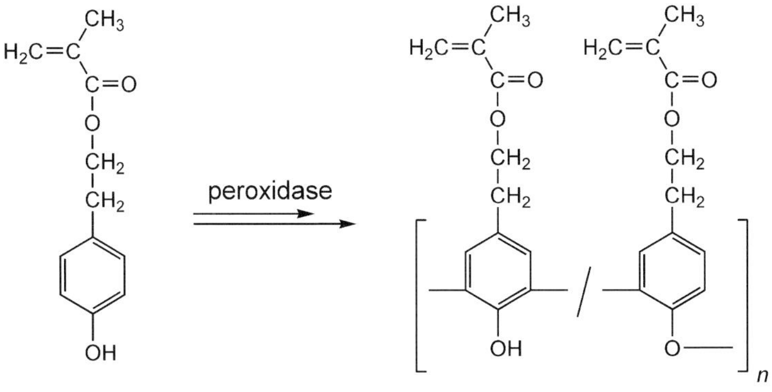

Scheme 10.5

photochemically. The monomer without phenolic OH was polymerized as a vinyl monomer with HRP catalysis.

Other phenols containing vinyl monomers, 4-hydroxy-N-methacryloylanilide, N-methacryloyl-11-aminoundecanoyl-4-hydroxyanilide and 4-hydroxyphenyl-N-maleimide, were chemoselectively polymerized with HRP–H$_2$O$_2$ catalyst. All three monomers were polymerized in the form of their cyclodextrin complexes in an aqueous buffer solution. The resulting phenolic polymers having unreacted double bonds in the side-chain acted as macromonomers, which were subsequently copolymerized with methyl methacrylate or styrene (Scheme 10.6) [43]. The polymers were ultimately cross-linked by radical initiation with AIBN [44].

Polymerization of phenols has been widely studied not only in monophasic solvents but also in interfacial solvents such as micelles, reverse micelles to give p-ethylphenol particles and biphasic and Langmuir trough systems [45]. p-Phenylphenol was polymerized in an aqueous surfactant solution to give a polymer with a relatively narrow molecular weight. The product polymer was not well structurally controlled [46].

HRP and SBP catalyzed the polymerization of bisphenol-A to give a soluble polymer in an aqueous organic solvent [47]. The polymer was composed of a mixture of phenylene and oxyphenylene units and had a molecular weight (M_n) of 1000–4000 ($M_w > 10^4$). It was thermally cured at 150–200 °C to improve the thermal stability. Reaction with an epoxy resin produced an insoluble polymer with high thermal stability, suggesting a possible replacement for the phenol resin without using formaldehyde.

The HRP-catalyzed oxidative polymerization of 4,4′-dihydroxydiphenyl ether in aqueous methanol afforded α,ω-hydroxy-oligo(1,4-oxyphenylene)s in moderate yields [48]. During the reaction, hydroquinone was formed. An unusual mechanism involving the redistribution and/or rearrangement of the quinone–ketal intermediate with liberation of hydroquinone is postulated.

Polymerization of meta-substituted phenols was studied using HRP or SBP catalyst in aqueous methanol, giving rise to soluble polymers with a glass transition temperature (T_g) > 200 °C. HRP effectively catalyzed the polymerization of a phenol monomer having a small substituent, whereas SBP allowed the polymerization of a monomer with a larger substituent in a higher yield, which is

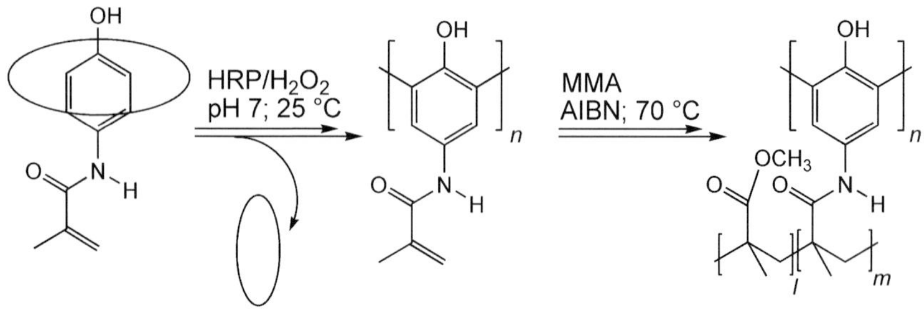

Scheme 10.6

probably due to the larger geometric size of the SBP active site. These enzymatically synthesized polyphenols were applied to positive-type photoresists for printed wire boards because of their high solubility in alkaline solution and high thermal stability [49].

A phenol having an acetylenic group at the meta-position was chemoselectively polymerized by HRP catalyst, giving rise to a phenolic polymer; the acetylenic group did not participate in the polymerization (Scheme 10.7) [50]. The resulting polymer was very reactive and cured at a lower temperature. It was further converted to carbonized polymer in a much higher yield than enzymatically synthesized poly(m-cresol) and is expected to have potential applications as a reactive starting polymer.

Cardanol is the main component derived from cashew nut shell liquid (CNSL) and a phenol derivative having mainly the meta-substituent of a C_{15} unsaturated hydrocarbon chain with one to three double bonds as the major entity. Currently, only a small portion of CNSL obtained in the production of cashew kernel is used in industry, although it has various potential industrial uses such as resins, friction lining materials and surface coatings. Therefore, the development of new applications for cardanol is very attractive for green polymer chemistry [12–14, 17, 23]. A new cross-linkable polymer was synthesized by the SBP- or Fe-salen-catalyzed polymerization of cardanol (Scheme 10.8), where Fe-salen [Fe(II) N,N'-ethylenebis(salicylideneamine)] was regarded as a model complex of peroxidase [51]. The product polymer with molecular weight 2000–4000 was soluble in polar organic solvents and possessed an unsaturated group in the side-chain. Polycardanol was further cross-linked by thermal treatment or by Co naphthenate-catalyzed oxidation, which involves a radical reaction of double bonds. The cross-linked polycardanol showed tough and hard properties as a film with a high gloss surface and is regarded as "artificial urushi" due to a close structural resemblance [52]. In fact, it looks like natural urushi, a traditional Japanese lacquer (see also Section 10.3.2.1). In place of the phenol ring of cardanol, a naphthol ring was introduced; then the cross-linked film became harder [53].

Syringic acid (3,5-dimethoxy-4-hydroxybenzoic acid) derived from plants was polymerized using HRP or SBP catalyst to produce poly(2,6-dimethoxy-1,4-oxyphenylene) having a carboxyl group at the one end and a hydroxy group at the other, with liberation of carbon dioxide and hydrogen during the polymerization. The molecular weight reached 1.5×10^4 with SBP catalyst. The mechanism

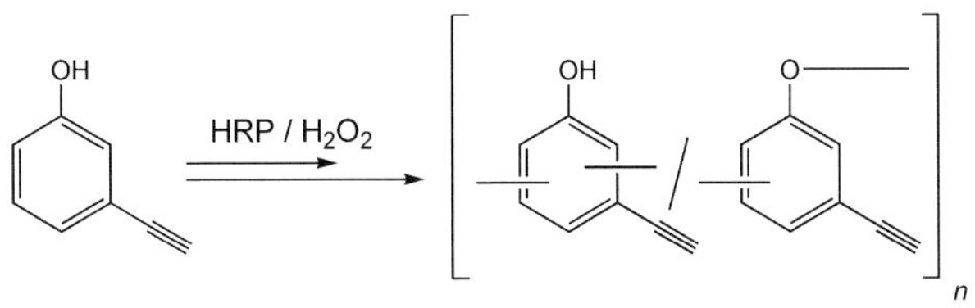

Scheme 10.7

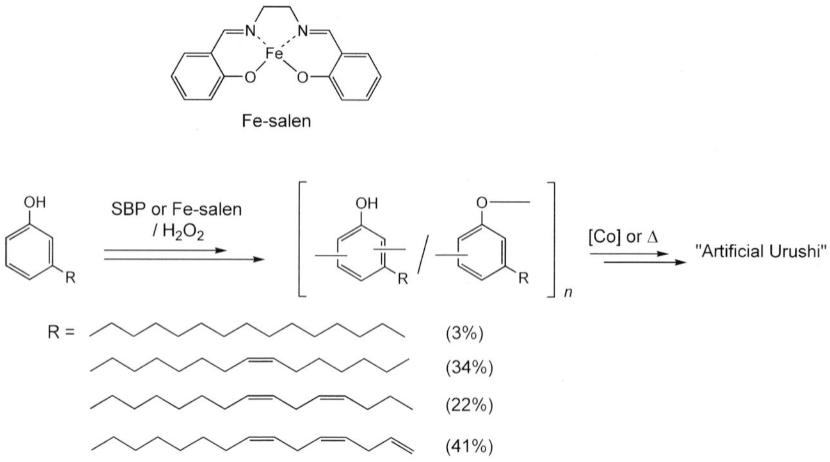

Scheme 10.8

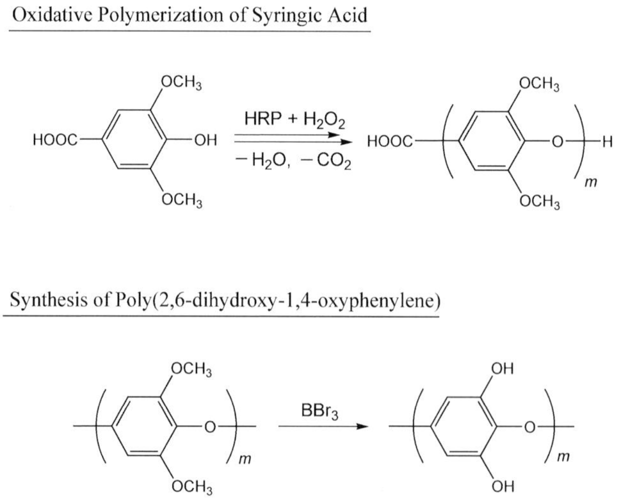

Oxidative Polymerization of Syringic Acid

Synthesis of Poly(2,6-dihydroxy-1,4-oxyphenylene)

Scheme 10.9

of the polymerization was discussed [54]. The demethylation of the product polymer by boron trichloride catalyst gave a new polymer, poly(2,6-dihydroxy-1,4-oxyphenylene) (Scheme 10.9) [55].

Oxidative polymerization of 2-naphthol was carried out in a reverse micellar system to give a polymer in single and interconnected microspheres. The polymer showed the fluorescence characteristics of the 2-naphthol chromophore [56]. The polymerization of 8-hydroxyquinoline-5-sulfonate was studied by *in situ* NMR spectroscopy and the polymerization mechanism was discussed in detail [57].

10.3.1.3 Polymerization of Phenols Catalyzed by Enzyme Model Complexes

Oxidoreductase enzymes contain a catalytic site of a metal moiety. Catalytic functions of the enzymes have been mimicked and new metal complexes created. Such complexes were employed as catalysts for oxidative polymerizations. Regioselective oxidative polymerization of a phenol compound leading to 2,6-unsubstituted poly(1,4-oxyphenylene) (PPO) has not been achieved. This was speculated to be due to the low selectivity in "electrophilic" or "free" phenoxy radical coupling. The regioselective polymerization of 4-phenoxyphenol was achieved for the first time using a tyrosinase model complex as catalyst to give unsubstituted crystalline PPO having melting-points, 171–194 °C (Scheme 10.10) [58].

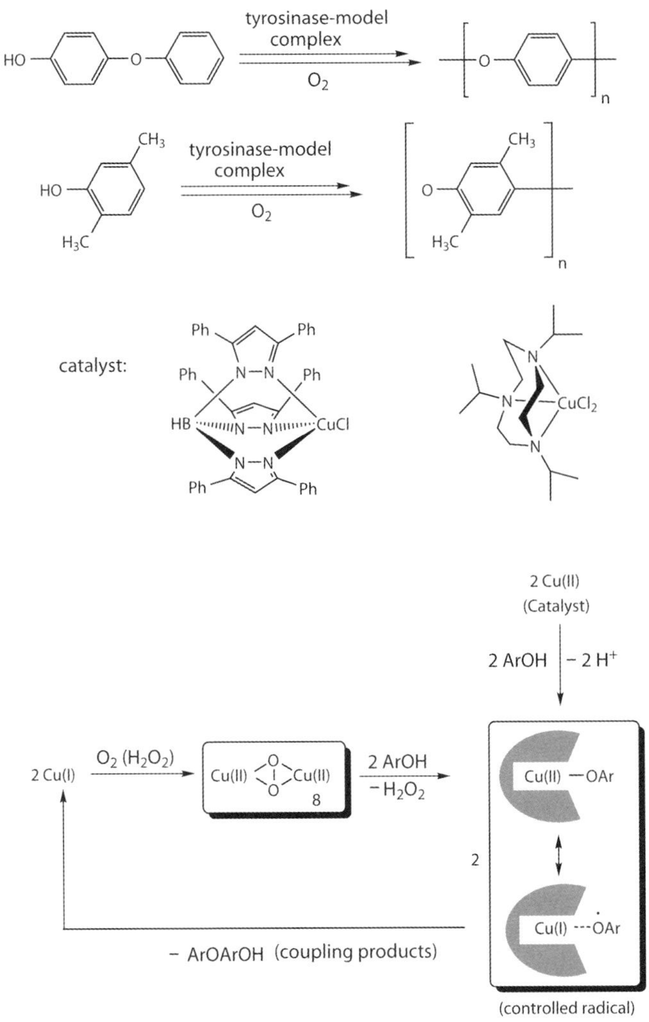

Scheme 10.10

The following may explain the high selectivity. A "nucleophilic" $\mu\text{-}\eta^2:\eta^2$-peroxo dicopper(II) complex is generated as the sole active oxygen complex from the catalyst and abstracts a proton (not a hydrogen atom) from the monomer to give a phenoxo–copper(II) complex, equivalent to a phenoxy radical–copper(I) complex. These intermediates are not "free" radicals, but "controlled" radicals, hence the regioselectivity of the subsequent coupling is regulated. The behaviors in oxidative polymerization of phenol and 4-phenoxyphenol by the tyrosinase model complex catalyst were discussed in detail from a mechanistic viewpoint [59, 60].

The same catalyst induced oxidative polymerization of 2,5-dimethylphenol, giving rise to crystalline poly(2,5-dimethyl-1,4-oxyphenylene) having a melting-point of 275–308 °C [61]. From o- and m-cresols, polymers consisting mainly of 1,4-oxyphenylene units were also formed [62].

Fe-salen catalyzed the oxidative polymerization of p-tert-butylphenol and bisphenol-A, yielding soluble polyphenols [63]. 2,6-Dimethyl- and 2,6-difluorophenols were polymerized using Fe-salen catalyst to give PPO derivatives [64]. The latter showed crystallinity with melting-points higher than 250 °C.

10.3.2
Synthesis of Polymers from Polyphenols

In Section 10.3.1, enzymatic oxidative polymerization of monophenolic compounds was described. This section deals with the enzymatic synthesis of polymers from polyphenols, compounds having more than two hydroxy groups on the aromatic ring(s). In particular, cured phenolic polymers (artificial urushi) and flavonoid polymers are mentioned from the standpoint of the enzymatic synthesis of functional materials [27].

10.3.2.1 Polymers from Catechol Derivatives

Oxidative polymerization of catechol proceeded in the presence of peroxidase catalyst [65], during which an unstable o-quinone intermediate may be formed, following by complicated reaction pathways to give polycatechol. Laccase was also a good catalyst for the oxidative polymerization of catechol [66]. Polymerization using the laccase enzyme conducted batchwise in aqueous acetone gave a polymer with a molecular weight of <1000. Synthesis of polycatechol was demonstrated by multienzymatic processes (Scheme 10.11) [67]. Aromatic compounds were converted to catechol derivatives by the catalytic action of toluene dioxygenase and toluene cis-dihydrodiol dehydrogenase, followed by peroxidase-catalyzed polymerization to give a polymer with a molecular weight of several thousand.

Urushi is a typical Japanese traditional coating material showing excellent toughness and brilliance for a long period. Urushi is a cross-linked coating of "urushiols" obtained from urushi trees, the structure involving a catechol derivative directly linked to unsaturated hydrocarbon chains consisting of a mixture of monoene, dienes and trienes at mainly the 3-position of catechol. Film forma-

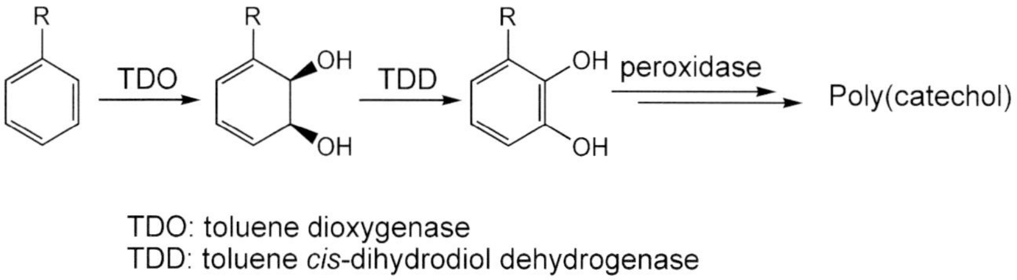

Scheme 10.11

TDO: toluene dioxygenase
TDD: toluene *cis*-dihydrodiol dehydrogenase

tion of urushiols proceeds via laccase catalysis under air at room temperature without involving organic solvents. Urushi is a renewable resource and seems very desirable as a coating material from the environmental standpoint. Urushi is the only example in nature that utilizes enzymatic catalysis for bringing about a practical use.

In vitro enzymatic hardening reaction of catechol derivatives bearing an unsaturated alkenyl group at the 4-position of the catechol ring proceeded using laccase as catalyst to give a cross-linked film showing excellent dynamic viscoelasticity [68]. Fast drying hybrid urushi was developed. Kurome urushi was reacted with silane-coupling agents possessing an amino, epoxy or isocyanate group, resulting in a shorter curing time of urushi [69].

Since natural urushi is very expensive, efforts were aimed at developing a cheaper commodity material. Natural urushi chemistry was mimicked because

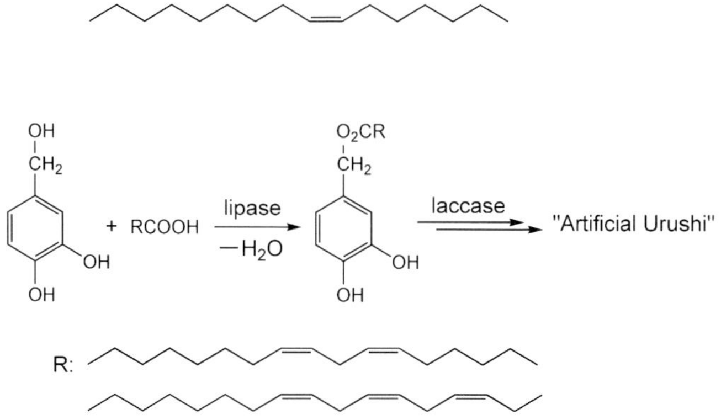

Scheme 10.12

of the difficulty of the chemical synthesis of natural urushiol. New urushiol analogues were developed for the preparation of "artificial urushi" (Scheme 10.12) [70]. The urushiol analogues were cured using laccase catalyst in the presence of acetone powder under mild reaction conditions without using organic solvents, yielding a brilliant film ("artificial urushi") with a high gloss surface (see also Section 10.3.1.2).

Dip pen nanolithography (DPN) is a convenient method to perform surface patterning at the nanometer scale. Nanoscale surface patterning of caffeic acid (3,4-dihydroxycinnamic acid) was achieved on 4-aminothiophenol (p-ATP)-modified gold surfaces by DPN. Then, HRP-catalyzed polymerization was applied on the AFM-patterned caffeic acid features (Scheme 10.13) [71]. In the product polymer, C–O–C coupling, which takes place widely in enzymatic polymerization in solution, was not observed in the surface reaction. The p-ATP monolayer acted as a template for caffeic acid and regioselective polymerization was induced to facilitate exclusive C–C ring coupling due to the participation of the quinone structure from free radical resonance during the reaction.

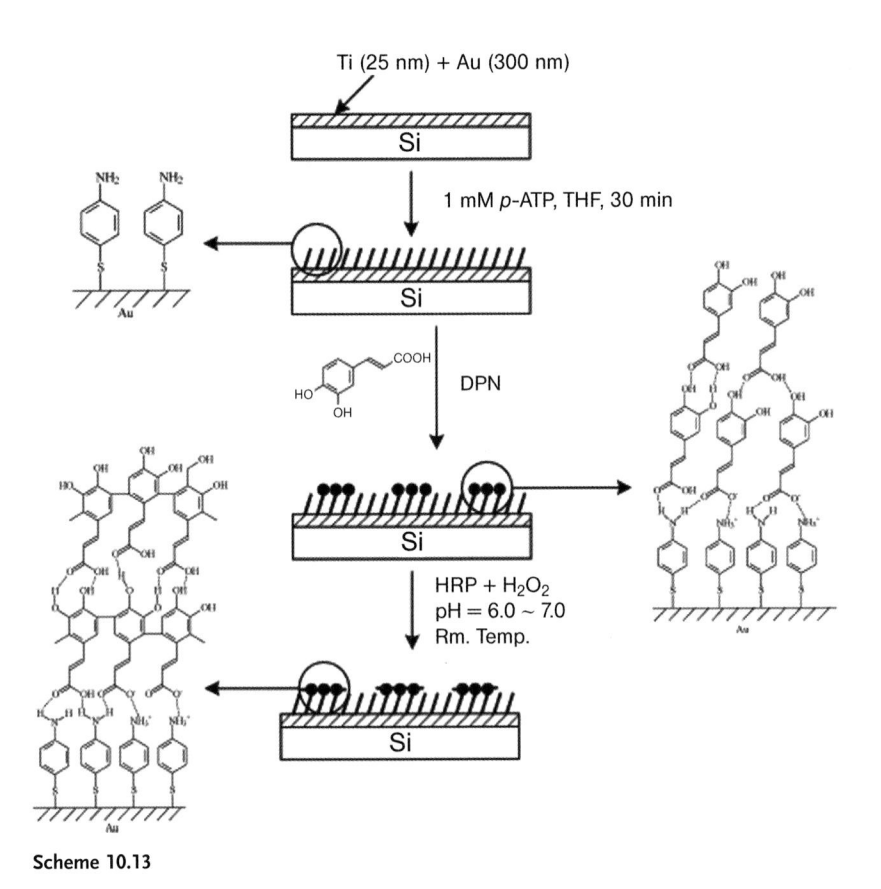

Scheme 10.13

10.3.2.2 Polymers from Flavonoids

Many bioactive polyphenols are found in various plants and used as important materials for human and animal diets. Flavonoids are a broad class of low molecular weight secondary plant polyphenolic compounds, which are benzo-γ-pyrone derivatives with phenolic and pyran rings. Flavonoids are usually subdivided according to their substituents into flavanols, flavones, flavanones, chalcones and anthocyanidines. Recently, there has been increasing interest in flavonoids due to their biological and pharmacological activity induced mainly by antioxidant and anti-carcinogenic properties. In the framework of phenolic polymers, attention is paid to the oxidative polymerization of flavonoids from polymer chemistry.

Green tea is derived from *Camellia sinensis*, an evergreen shrub of the Theaceae family. Most of the polyphenols in green tea are flavanols, commonly known as catechins; the major catechins in green tea are (+)-catechin, (–)-epicatechin, (–)-epigallocatechin, (–)-epicatechin gallate and (–)-epigallocatechin gallate (EGCG) (Fig. 10.6). Numerous biological activities of green tea and red wines have been reported; in particular their anti-cancer property is best known [72].

Polymerization of catechin was carried out aerobically several decades ago by plant polyphenol oxidase to give polymers, whose analytical properties were similar to those obtained by autoxidation with hydrogen peroxide [73] and to tannins extracted from *Acacia catechu*. The main reaction sequence involves an o-quinone intermediate, followed by the coupling of this unstable intermediate. The HRP-catalyzed polymerization of catechin was carried out in an equivolume mixture of 1,4-dioxane and buffer to give a polymer with a molecular weight of 3.0×10^3 in 30% yield [74]. Using methanol as cosolvent improved the polymer yield and molecular weight [75]. In the polymerization of catechin using laccase catalyst, the reaction conditions were examined in detail [76]. A mixture of acetone and acetate buffer (pH 5) was suitable for the efficient synthesis of soluble polycatechin with higher molecular weight. It is interesting that in the ESR spectrum of the enzymatically synthesized polycatechin a singlet peak at $g = 1.982$ was detected, whereas the catechin monomer possessed no peak. Water-soluble oligocatechins were prepared by the HRP-catalyzed polymeriza-

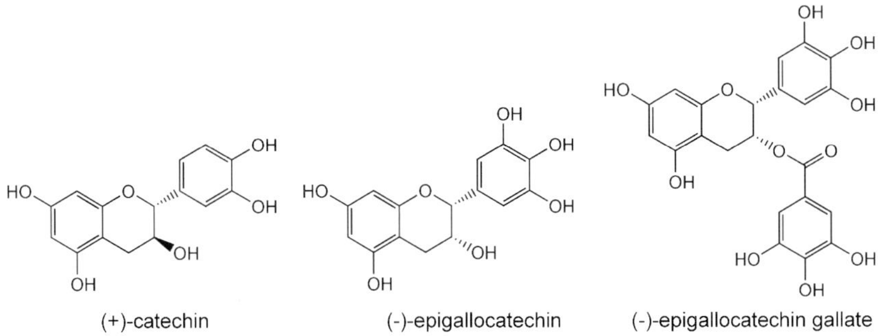

Fig. 10.6 Structures of important components of green tea catechins.

tion of catechin using a polyelectrolyte such as sulfonated polystyrene as template and a surfactant such as sodium dodecylbenzene sulfonate [77].

The polymerization of catechin greatly enhanced the antioxidant property and superoxide anion scavenging activity; the enzymatically obtained polycatechin showed a much higher activity than catechin monomer [75]. Also, poly-EGCG showed excellent inhibition of xanthine oxidase, about 20 times higher than monomeric EGCG, suggesting the possibility of a new pharmacological polymer [78].

10.3.3
Synthesis of Polyaniline and Its Derivatives

Polyaniline and its derivatives belong to one of the most important classes of polymers for use in conducting and optical applications. Well-known methods for the synthesis of polyaniline involve either chemical or electrochemical oxidation of an aniline monomer resulting in polymerization. The reaction conditions are harsh, with extreme pH, high temperature, strong oxidants and highly toxic solvents required. In order to improve the solubility and the processablity of the polymeric products, the synthesized polyaniline is usually treated post-polymerization with fuming sulfuric acid. On the other hand, enzymatic polymerization of aniline and its derivatives provided an alternative method towards the formation of soluble and processable conducting polymers. These reactions are usually carried out at room temperature, in aqueous organic environments at neutral pH. The reaction conditions were greatly improved and the purification process of the final products was simplified compared with the more traditional methods [79].

The conducting property of polyaniline has been extensively studied and is explained from structure analysis as follows. Generally, enzymatically synthesized polyaniline is in the protonated form, which can be converted to the unprotonated base form by treatment with aqueous ammonia solution or other suitable bases. The unprotonated base form of polyaniline consists of reduced base units A and oxidized base units B as repeat units, where the oxidation state of the polymer increases with decreasing values of y ($0 \leq y \leq 1$) (Scheme 10.14). The three extreme possibilities for value of y are 0, 0.5 and 1, corresponding to fully oxidized polyaniline (pernigraniline), half oxidized polyaniline (emeraldine) and fully reduced polyaniline (leucoemeraldine), respectively [80,81].

Peroxidase-catalyzed oxidative polymerization of o-phenylenediamine in a mixture of 1,4-dioxane and phosphate buffer produced a soluble polymer consisting of iminophenylene units (Scheme 10.15) [82]. A new class of polyaromatics was synthesized by peroxidase-catalyzed oxidative copolymerization of phenol derivatives with anilines. In the case of a combination of phenol and o-phenylenediamine, FTIR analysis showed the formation of the corresponding copolymer [83].

In an organic solvent, 1,4-dioxane, HRP catalyzed the enzymatic polymerization of phenylenediamines and aminophenols [84]. The products were soluble in DMF and DMSO. A large molar excess of hydrogen peroxide was necessary to achieve a high polymer yield. Poly(2-aminophenol) and poly(4-aminophenol)

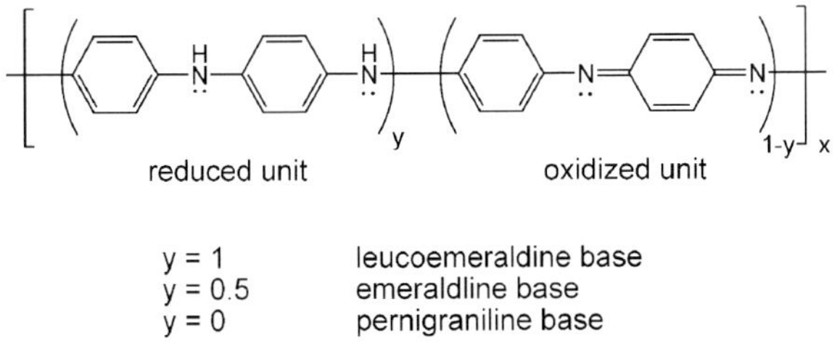

10.3 Synthesis of Poly(aromatic)s Catalyzed by Oxidoreductases

Scheme 10.14

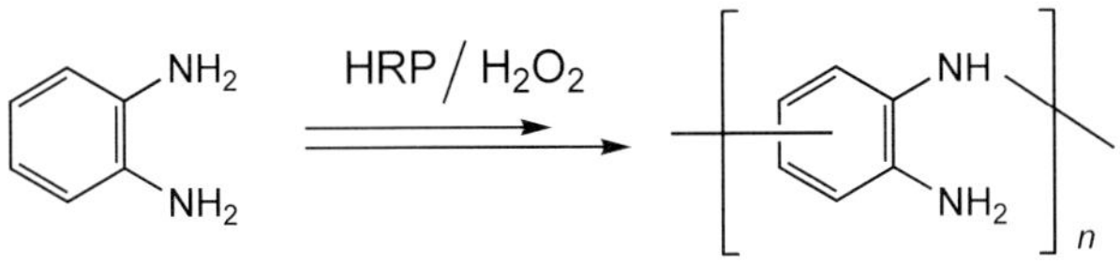

y = 1   leucoemeraldine base
y = 0.5 emeraldline base
y = 0   pernigraniline base

Scheme 10.15

resulting from these reactions showed electroactive properties. Polyazobenzene and its derivatives have applications in optical devices [85]. A novel polyaniline containing azo groups was synthesized by the HRP-catalyzed oxidative coupling of 4,4'-diaminoazobenzene. The polymerization was carried out at pH 6.0 in Tris buffer with 70% yield. The polymer, analyzed by GPC ($M_w = 8.0 \times 10^4$, polydispersity = 4.8), was soluble in DMSO and DMF. Azo groups were detected both in the main-chains and in the side-chains. Photoexcitation studies indicated that cis–trans isomerization of the chromophore may be the result of structural constraints in the polymer [86].

In aqueous solution, water-soluble polyanilines were synthesized by enzymatic templating, chemical or electrochemical methods or copolymerization of aniline with sulfonated aniline [87]. The water-soluble product was electroactive with an average molecular weight of 1.8×10^4. Conductivity was determined to be in the semiconducting region (10^5 S cm^{-1}) at pH 6. The conductivity of poly(2,5-diaminobenzenesulfonate) was about 3–4 orders of magnitude lower because of its branched structure.

Electroactive polyaniline films were synthesized by the catalysis of bilirubin oxidase (a copper-containing oxidoreductase). The polymerization of aniline was carried out on the surface of a solid matrix such as a glass slide, plastic plate or platinum electrode to form homogeneous films [88].

In addition to HRP and bilirubin oxidase, laccases have often been used in the presence of oxygen for the synthesis polyanilines [89]. Also, the hydroxy ferriprotoporphyrin compound hematin was used for the synthesis of various polyanilines [90]. The electrostatic layer-by-layer (ELBL) self-assembly of a polyelectrolyte, poly(dimethyldiallylammonium chloride), and hematin was utilized to construct a nanocomposite film catalyst. The conductive polyaniline was formed not only on the surface of ELBL as a coating, but also in the bulk solution.

The template-assisted synthesis of water-soluble conducting polyaniline was achieved by HRP-catalyzed polymerization in aqueous medium. The aniline complex was formed in the system in the presence of sulfonated polystyrene (SPS), which acted as a polyanionic template. The resulting polymer was complexed to the SPS and exhibited electroactivity. The reduction–oxidation reversibility of the polyaniline–SPS complex was demonstrated [91]. The conductivity of the polyaniline–SPS complex was measured as 0.005 S cm^{-1}. The value increased to 0.15 S cm^{-1} after HCl doping and could be increased further with increase in the aniline to SPS molar ratio [92].

DPN technology was applied to the peroxidase-catalyzed polymerization of 4-aminothiophenol into nanowires on gold surfaces [93]. Reactions were performed in methanol–water (1:1, v/v) with aminothiophenol. The monomer was patterned on to gold by DPN by scratching the surface. The topography of the resulting polymer of aminothiophenol indicates lines 4 mm long with an average width of 210 nm and an average height of 25 nm.

10.4
Synthesis of Vinyl Polymers Catalyzed by Oxidoreductases

In nature there is an only example in biomacromolecular synthesis, to the best of the authors' knowledge, where an enzyme catalyzes C–C bond formation to yield a polymer having a carbon–carbon main-chain structure, that is, natural rubber (NR). The reaction mode of NR synthesis is condensation polymerization, where the rubber transferase enzyme catalyzes the addition of isopentenyl pyrophosphate (monomer) to the propagating rubber molecule end of the pyrophosphate linkage, liberating pyrophosphoric acid in each step to produce NR of *cis*-1,4-polyisoprene *in vivo* [94].

In contrast, *in vitro* polymerization of vinyl-type monomers has been performed by employing mainly oxidoreductase enzymes [95]. The enzymatic polymerization of vinyl monomers was initiated by a reaction system consisting of glucose oxidase, glucose, Fe^{2+} ion and a vinyl monomer in the presence of oxygen, in which glucose oxidase provides hydrogen peroxide, followed by reaction with Fe^{2+} to produce a hydroxyl radical. Then, the radical initiates the polymerization of vinyl monomers [96]. The laccase-catalyzed polymerization of acrylamide in water without any initiator at temperatures ranging from 50 to 80 °C was reported, where the initiation mechanism was not clear. The molecular weight of polyacrylamide was determined as $M_n \approx 10^6$ when carried out for 4 h at 50, 65 and 80 °C, with the highest yield of 70% at 65 °C [97]. Also, the HRP–hydrogen peroxide system induced the polymerization of a methacrylate monomer [42]. Reactions proceeded efficiently at room temperature when using 2,4-pentanedione combined with laccase to produce polyacrylamide in 97% yield ($M_n = 2.3 \times 10^5$) [98]. The β-diketone is considered as a radical generator acting as initiator.

Enzymatic vinyl polymerization using the ternary system HRP–H$_2$O$_2$–β-diketone provided a free radical polymerization system [99]. The effect of eight β-diketones was examined in the polymerization of acrylamide. β-Diketones play an important role in the reactions; on changing the nature of the b-diketone, the molecular weight (5.1×10^3–124×10^3), polydispersity (2.5–4.4) and yield (38–93%) of the product polymers were greatly affected. In the HRP-mediated polymerization of acrylamide in aqueous medium, the effect of surfactant was studied [100]. Chemoselective polymerization of 2-(4-hydroxyphenyl)ethyl methacrylate and polymerization of 2-phenylethyl methacrylate were reported (Scheme 10.5) [42]. Enzymatic polymerization of methyl methacrylate catalyzed by peroxidases including HRP and SBP gave predominantly the formation of syndiotactic polymers [101].

Styrene and its derivatives (4-methylstyrene, 2-vinylnaphthalene and sodium styrenesulfonate) were polymerized with an HRP–H_2O_2–initiator system. On changing the initiator (six types), the molecular weights (2.7×10^4–9.7×10^4) and yields (14–60%) of the polymers were affected [102]. Polymers derived from p-styrenesulfonic acid are used in many applications such as ion-exchange membranes, resins and biomaterials to influence cell adhesion. HRP, hematin and pegylated hematin were used for the polymerization of sodium styrenesulfonate in aqueous systems. Poly(sodium styrenesulfonate) was produced in more than 80% yields. The molecular weight of the polymer increased ($M_n = 6.6 \times 10^4$–16×10^4) with reaction time and also the polydispersity increased from 2.5 to 3.4.

Vitamin C (ascorbic acid)-functionalized polystyrene and poly(methyl methacrylate) were prepared enzymatically for wide applications as antioxidant materials. Two types of starting monomers were synthesized with regioselective lipase catalysis under mild reaction conditions (Fig. 10.7) [103].

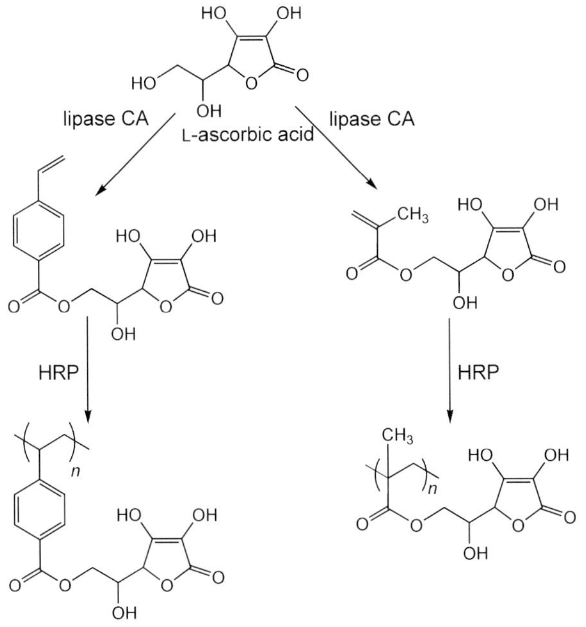

Fig. 10.7 Enzymatic synthesis of vitamin C functionalized polymer.

10.5
Synthesis of Polysaccharides Catalyzed by Hydrolases

Polysaccharides are one of the three major classes of biomacromolecules besides proteins and nucleic acids [5, 8, 12–14, 16–18, 20, 104]. Polysaccharides are produced in major amounts by plants and exist in many cases as glycoconjugates such as glycoproteins and glycolipids in animals. Undoubtedly these biomacromolecules play critical roles in nature and, hence, a number of groups have been attempting to establish facile methods to synthesize such macromolecules. Owing to the development of computer-aided, automated synthesizers [105] and remarkable progresses in genetic engineering [106], the synthesis of proteins and nucleic acids nowadays is relatively easy. In contrast, there have been no universal methods for the production of polysaccharides, owing to the complexity of their chemical structures. Polysaccharides are normally biosynthesized by the catalysis of glycosyltransferases (EC 2.4.x.y), which catalyze exclusively the formation of a glycosidic bond with the corresponding sugar nucleotides as substrates [107]. These enzymes bound to the plasma membrane are unstable and require skill in manipulation; therefore, they are not suitable as catalysts for the *in vitro* production of polysaccharides. This section describes glycoside hydrolases (EC 3.2.1.x) catalyzed polymerizations, which enabled polysaccharides to be produced *in vitro* for the first time. The hydrolases are extracellular enzymes that are more stable and easier to obtain than glycosyltransferases. It should be noted that appropriate design of the substrate monomers and control of polymerization conditions are essential for the synthesis of polysaccharides by hydrolases, which catalyze *in vivo* glycosidic bond cleavage of polysaccharide chains. Here, two families of monomers, sugar fluorides and sugars having a 4,5-dihydrooxazole (oxazoline) ring, are used as activated substrate monomers for polycondensation and ring-opening polyaddition, respectively [8, 12–14, 16, 17, 20, 104].

10.5.1
Synthesis of Polysaccharides via Polycondensation

10.5.1.1 Cellulose and Its Derivatives
Cellulose, a linear polymer of D-glucose (Glc) linking through a (1 → 4)-β-glycoside, is the most abundant organic material on Earth; hundreds of billions of tons of cellulose are photosynthesized annually with carbon dioxide fixation. Cellulose is an important renewable resource utilized for housing, clothing and also industrial products such as papers and fibers. In the history of polymer science, such an important substance as cellulose has been a target molecule for many prominent workers with respect to its structure, molecular weight and derivatization reactions since the 1920s; cellulose has a very symbolic existence for polymer scientists [108]. However, no successful synthesis had been reported since the first attempt at chemical synthesis in 1941 [109].

In 1991, the first *in vitro* synthesis of cellulose was reported [16, 20, 110]. A chemically prepared β-cellobiosyl fluoride (β-CF) was successfully polymerized as an activated sugar monomer by the catalysis of cellulase from *Trichoderma viride* (CelTV), providing synthetic cellulose as white crystals in good yields (Scheme 10.16). The monomer having a fluorine atom with β-orientation was designed on the basis of the mechanism of catalysis of CelTV belonging to the glycoside hydrolase (GH) family 5 [111]. CelTV is a retaining enzyme involving a double-displacement mechanism [112]. Furthermore, addition of an organic cosolvent such as acetonitrile (CH_3CN:buffer = 5:1) greatly suppressed the hydrolysis of the monomer and also the product by the enzyme, leading preferentially to polymerization.

Figure 10.8 illustrates the postulated reaction mechanisms of CelTV catalysis in the hydrolysis of cellulose (a) and in the polymerization of monomer (b) [20, 104].

It is generally accepted that GH5 cellulases have two carboxylic acid residues involved in the catalysis. In the hydrolysis of cellulose (Fig. 10.8a), one residue pulls (protonation) the glycosidic oxygen and the other pushes the C-1 carbon in the general acid–base mode to assist the glycosidic bond cleavage (stage a). Then, a highly reactive intermediate (or transition state) of a glycosyl–carboxylate is formed (stage b). This intermediate may have another possible structure of an oxocarbenium carboxylate. A water molecule attacks the C-1 of the intermediate from the β-side to complete the hydrolysis with formation of the hydrolyzate having a β-configuration (from stage b to c). In the polymerization (Fig. 10.8b), the monomer β-CF is readily recognized and activated at the donor site via a general acid–base mode to cleave the C–F bond to form a highly reactive glycosyl–carboxylate (from stage a to b). The 4-hydroxy group of another monomer molecule or the growing chain end attacks the C-1 of the intermediate from the β-side to form a β-glycoside (from stages b to c). The product moves to the right for the next monomer entering the donor site. The repetition of stage a–c is a polycondensation with liberation of an HF molecule, leading to the production of synthetic cellulose. β-CF is considered as a transition-state analogue substrate (TSAS) monomer, because of the structural similarity involved from stage a to b in both reactions [20]. From polymer chemistry, this is an "activated monomer mechanism", in contrast to an "active chain-end mechanism" as normally observed in many vinyl polymerizations.

The importance of the fluorine atom in the monomer has mainly two reasons: the size of fluorine is close to that of oxygen and the fluoride anion is an excellent leaving group. This finding initiated an enzymatic approach to the synthesis of polysaccharides via polycondensation of sugar fluoride monomers. In

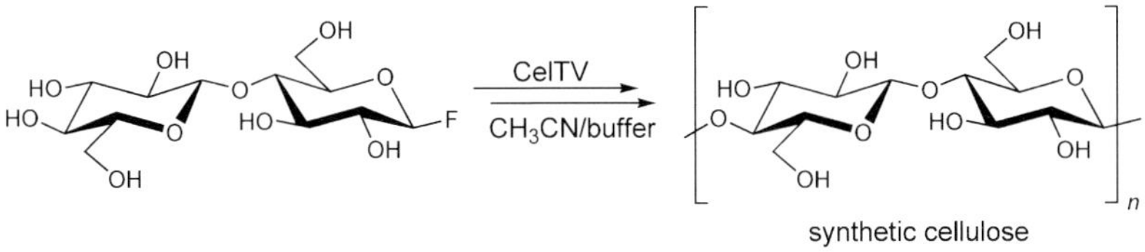

Scheme 10.16

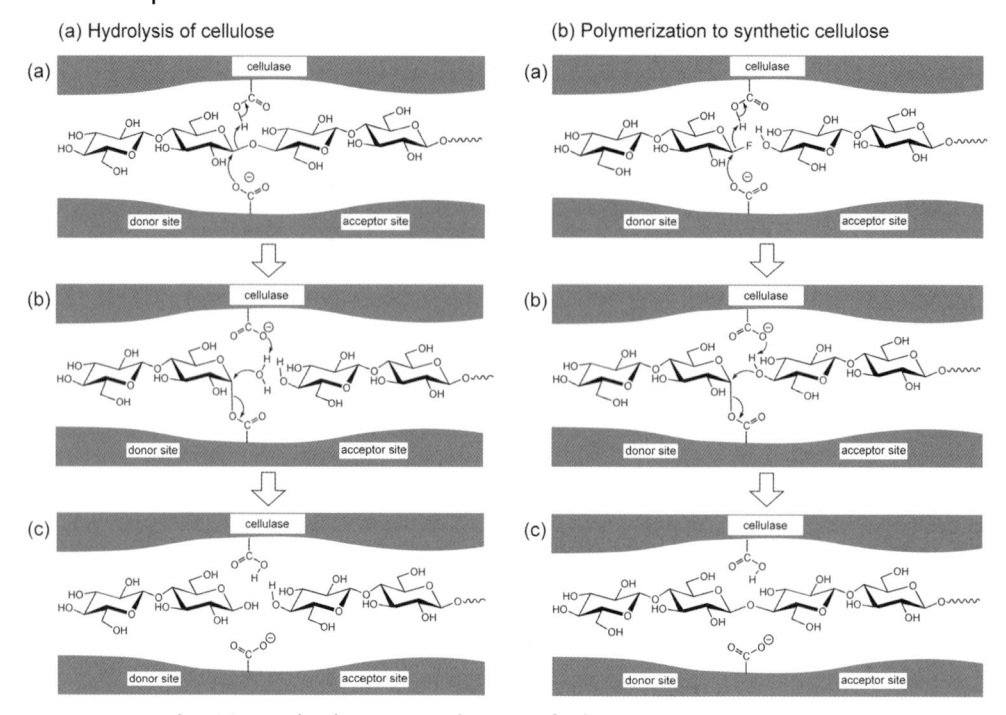

Fig. 10.8 Postulated reaction mechanisms of CelTV:
(a) hydrolysis of cellulose and (b) polycondensation of β-CF monomer to synthetic cellulose.

fact, sugar fluoride monomers have been used extensively for oligo- and polysaccharide synthesis [113].

Cellulose shows typically two allomorphs of high-ordered structure: cellulose I consists of a parallel molecular chain packing, which it was believed only Nature produces and is thermodynamically metastable, and cellulose II allomorph is of anti-parallel crystal form and thermodynamically stable. Once cellulose I has been converted to cellulose II, it never returns to cellulose I. The *in vitro* formation of synthetic cellulose I allomorph, therefore, has been a challenging problem. The challenge was taken up for the first time by the purification of enzyme; polycondensation of the monomer using a partially purified CelTV provided a microfibril of synthetic cellulose I [114]. On the other hand, polycondensation using a crude cellulase mixture led to the formation of the synthetic cellulose II spherulites (Fig. 10.9) [115a,b].

This situation was explained by micelle formation for a macromolecular chain-aligned structure leading to cellulose I (Fig. 10.10) and a novel concept, "choroselective polymerization" (meaning space-selective polymerization), was proposed (Fig. 10.9) [116].

A recent study revealed a mechanism for the formation of self-assembled cellulose II allomorph during enzymatic polymerization by means of *in situ* and

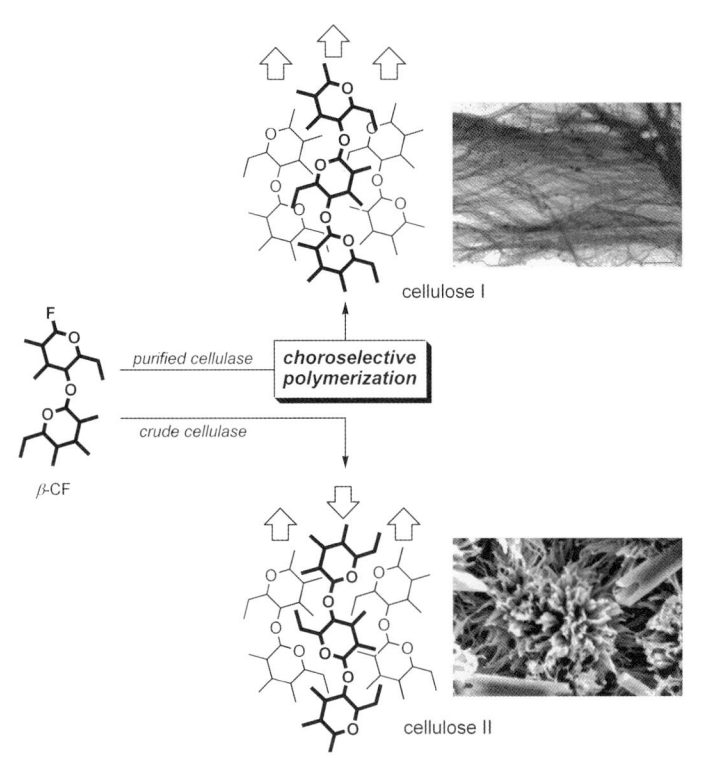

Fig. 10.9 Two allomorphs of cellulose I (parallel) via choroselective polymerization and cellulose II (antiparallel).

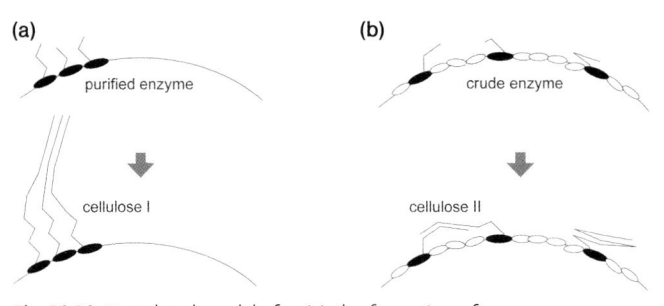

Fig. 10.10 Postulated models for (a) the formation of synthetic cellulose I by partially purified CelTV and (b) synthetic cellulose II by a crude enzyme mixture.

time-resolved small-angle neutron scattering (SANS) [115c]. The SANS profiles exhibited changes in the scattered intensity expressed by a characteristic power law behavior of q^{-a}, which was estimated as q^{-4} and $q^{-3.7}$ before and after polymerization, respectively (Fig. 10.11a). The results strongly suggest the following: β-CF is recognized and activated at the active site of the enzyme located on the

(a)

[Graph showing SANS profiles: $\frac{d\Sigma}{d\Omega}(q)$ (cm^{-1}) vs q (nm^{-1}), with legend: Synthetic cellulose (After polymerization), Enzyme solution (Before polymerization), Solvent. Slopes labeled $q^{-3.7}$ and q^{-4}, with equation $\frac{d\Sigma}{d\Omega}(q) = Aq^{-(6-D_s)}$]

(b)

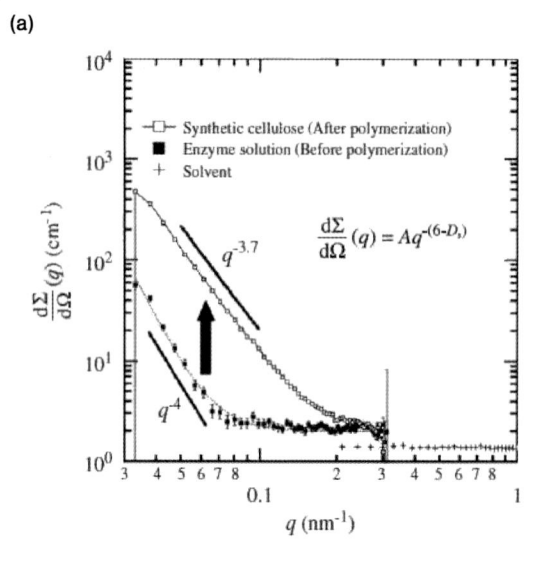

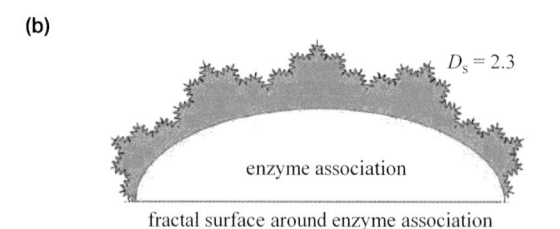

[Schematic: dome shape labeled "enzyme association" with fractal surface, $D_s = 2.3$; caption "fractal surface around enzyme association"]

Fig. 10.11 (a) SANS profiles before and after polymerization of β-CF catalyzed by CelTV. (b) Schematic illustration of the self-assembly of synthetic cellulose during polymerization.

micelle with a smooth interface. Then, a large number of cellulose chains are produced from each active site by the enzyme-catalyzed polycondensation, leading to the formation of an aggregated structure with a rough surface (the fractal dimension $D_s = 2.3$; Fig. 10.11b). Thus, single-step polymerization to synthetic cellulose catalyzed by CelTV allowed direct observations for the detailed investigation of the assembling mechanism.

Cellooligosaccharides were prepared using β-lactosyl fluoride via a step-by-step chain elongation by the combined use of CelTV and β-galactosidase [117], which catalyzes glycosidation of the fluoride to the non-reducing 4-hydroxy group of the acceptor and cutting-off a β-galactoside from a newly added chain end (Scheme 10.17).

Furthermore, a chemically modified monomer having a 6-O-methyl group was successfully polymerized by CelTV, giving rise to an alternating 6-O-methyl-

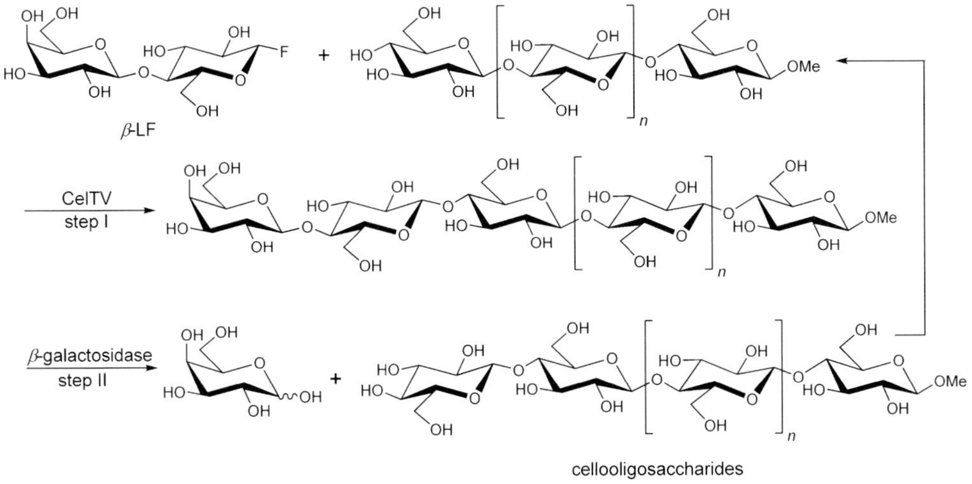

Scheme 10.17

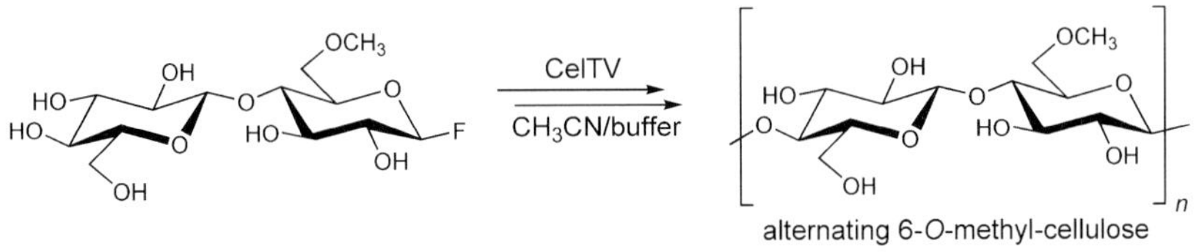

Scheme 10.18

cellulose due to the catalytic tolerance of the enzyme in substrate recognition at the donor and acceptor sites (Scheme 10.18) [118].

10.5.1.2 Xylan

Naturally occurring xylan existing mostly in the hemicellulose form is an important biomacromolecule composed of a $(1 \rightarrow 4)$-β-linked D-xylose (Xyl) mainchain, which bears side-chains such as arabinose and 4-O-methylglucuronic acid (Fig. 10.12a). As in the case of cellulose synthesis, β-xylobiosyl fluoride monomer was designed for the catalysis of CelTV containing xylanase. The enzyme catalyzed the polycondensation of monomer, providing synthetic xylan in good yields having an average DP of 23 without side-chains with perfect control of regioselectivity and stereochemistry (Fig. 10.12b) [119].

10.5.1.3 Amylose Oligomers

Amylose is a water-soluble $(1 \rightarrow 4)$-α-Glc polymer produced *in vivo* as a substance for energy storage. Maltose, a disaccharide component of amylose, was

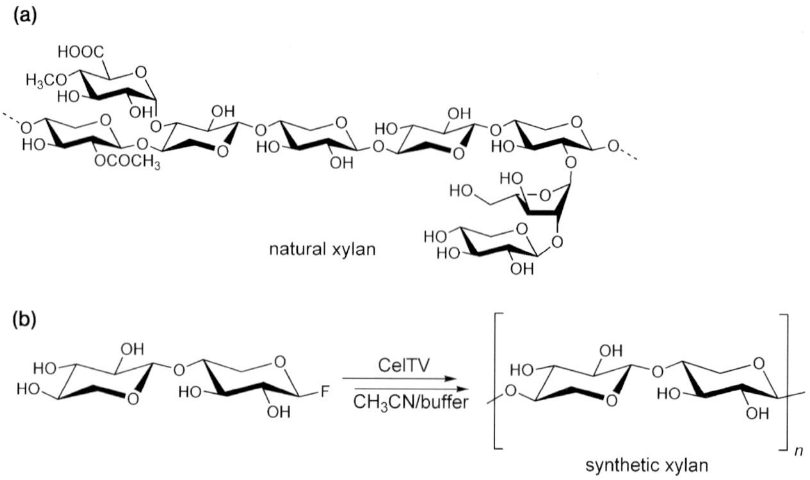

Fig. 10.12 Structures of (a) natural xylan and (b) synthetic xylan.

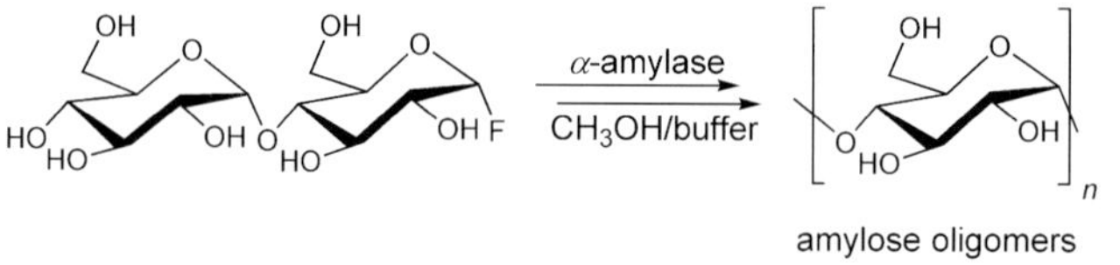

Scheme 10.19

converted to α-maltosyl fluoride as monomer for the catalysis of α-amylase (EC 3.2.1.1), which belongs to a retaining enzyme producing maltooligosaccharides with α-anomer [111]. The monomer readily disappeared on addition of the enzyme (within 1 h) to give a mixture of maltooligosaccharides (amylose oligomers) (Scheme 10.19) [120]. The product contained oligomers consisting not only of even- but also odd-numbered Glc units, such as maltotriose, maltopentaose and maltoheptaose. These results indicate that hydrolysis and/or rearrangement of the product amylose occurs by the catalysis of α-amylase.

It is reported that potato phosphorylase (EC 2.4.1.1) belonging to the glycosyltransferase family 35 catalyzes the polymerization of α-glucosyl phosphate to produce amylose polymer in the presence of maltooligomers as essential glycosyl primers [121]. This reaction, however, belongs to a metabolic pathway.

10.5.1.4 Hybrid Polysaccharides

Transferases strictly discriminate their favorable substrates, whereas hydrolases sometimes show tolerance in their substrate recognition. Taking advantage of this characteristic of hydrolases, molecularly hybrid-type polysaccharides were

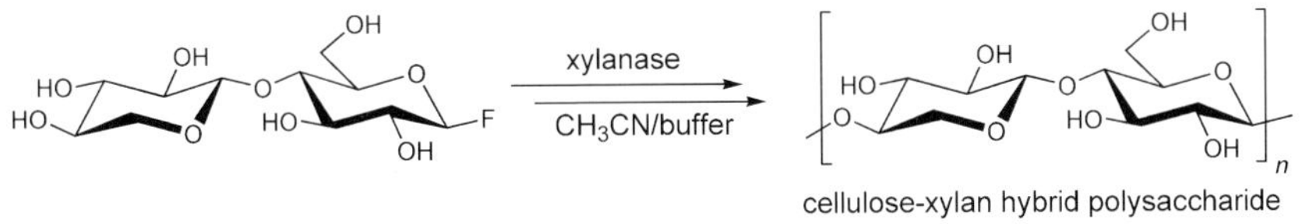

10.5 Synthesis of Polysaccharides Catalyzed by Hydrolases

Scheme 10.20

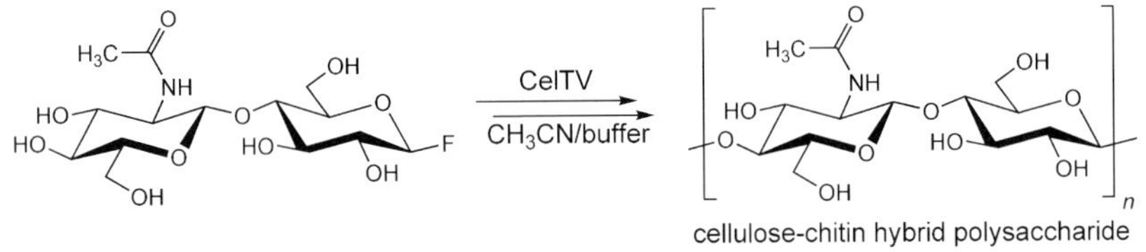

Scheme 10.21

prepared via enzymatic polymerization. A novel β-fluoride monomer of Xylβ(1 → 4)Glc was recognized and catalyzed by xylanase from *Trichoderma viride*, providing a cellulose–xylan hybrid polysaccharide in good yields [122]. (Scheme 10.20).

Recently, an important hybrid polysaccharide having an alternating structure of (1 → 4)-β-linked N-acetyl-D-glucosamine (GlcNAc) and Glc, that is, a cellulose-chitin hybrid polysaccharide, was synthesized via two modes of hydrolase-catalyzed polymerizations [123]. A newly designed β-fluoride monomer of GlcNAcβ(1 → 4)Glc was polymerized to a cellulose–chitin hybrid polysaccharide in good yields via CelTV-catalyzed polycondensation with perfect structure control (Scheme 10.21).

10.5.1.5 Oligo- and Polysaccharide Synthesis by Mutated Enzymes

During the past decade, the number of reports that deal with synthesis of oligo- and polysaccharides by utilizing genetically engineered glycoside hydrolases as catalysts has increased [124]. These mutants keep a folded protein structure and commonly lack the catalytic nucleophile, that is, the amino acid residue (Asp or Glu) is replaced with another residue unable to induce a nucleophilic attack (e.g. Ala or Gly), causing almost suppressed hydrolysis activity of the enzyme. Interestingly, the glycosynthases act as inverting enzymes requiring activated sugar substrates with an opposite anomer configuration to that in the product formed, although their original function is retained. For example, the Glu358Ala mutant of the *Agrobacterium* sp. β-glucosidase/galactosidase (AbgGlu358Ala) catalyzed transglycosylation of α-galactosyl fluoride to form a β-galactoside (Fig. 10.13) [113], whereas the wild-type enzyme employs a β-galactoside donor for the catalysis to form another β-galactoside [125]. The glycosynthase recognizes α-galactosyl fluoride, an activated substrate mimicking the glycosyl–enzyme intermediate in the hydrolysis,

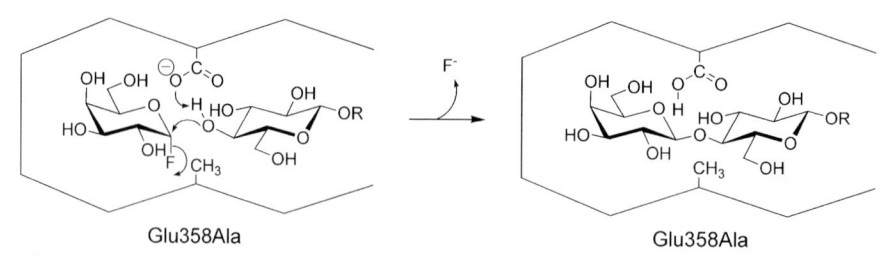

Fig. 10.13 Postulated reaction mechanisms of the mutant glycosynthase AbgGlu358Ala.

Fig. 10.14 Schematic representation of EGII expressed by *Saccaromyces cerevisiae* and the mutant EGII$_{core}$.

where the fluoride is directly attacked by a hydroxy group of a sugar chain from the β-side, resulting in the formation of a β-glycosidic linkage. This important finding led to the successful synthesis of (1 → 4)-β- and (1 → 3)-β-glucans [126].

Recently, another approach to the synthesis of cellulose was advanced utilizing the EGII$_{core}$ mutant, which is the endoglucanase II lacking the cellulose-binding domain (CBD) (Fig. 10.14) [127]. The CBD is responsible for loosening the molecular packing of the cellulose chain for easy access of the catalytic domain to the insoluble substrate [128]. The EGII$_{core}$ catalyzed the polycondensation of β-CF to give synthetic cellulose, whereas it suppressed the hydrolysis of the product cellulose.

10.5.2
Synthesis of Polysaccharides via Ring-opening Polyaddition

Naturally occurring polysaccharides often contain 2-acetamido-2-deoxy sugars such as GlcNAc and *N*-acetyl-D-galactosamine (GalNAc), which in many case exhibit important biological activities [129]. Chitin is the most abundant polysaccharide consisting of (1 → 4)-β-linked GlcNAc in the animal world, which is distributed widely in the skeletons of invertebrates. Hyaluronan (hyaluronic acid; HA), chondroitin (Ch) and chondroitin sulfate (ChS) are the major class of glycosaminoglycans existing in extracellular matrices, which include GlcNAc or

GalNAc in their repeating disaccharides. This section deals with the synthesis of these polysaccharides including 2-acetamido-2-deoxy sugars via hydrolase-catalyzed ring-opening polyaddition of sugar dihydrooxazole (oxazoline) monomers.

10.5.2.1 Chitin and its Derivatives

Chitin and its derivatives are frequently utilized for biocompatible, biodegradable and bioactive materials such as immuno-adjuvant substances [130], inhibitors of metastases of tumor cells [131], wound-healing materials [132], additives for cosmetics [133] and drug carriers [134]. The synthesis of such an important biopolymer of chitin was first reported in 1996, utilizing chitinase from *Bacillus* sp. (ChiBs) as catalyst [135]. ChiBs belonging to GH18 involves a "substrate-assisted mechanism" during the catalysis [111, 136]. An N,N-diacetylchitobiose [GlcNAcβ(1 → 4)GlcNAc] oxazoline derivative (ChiNAc-oxa) was recognized and catalyzed by ChiBs as a TSAS monomer, giving rise to synthetic chitin in quantitative yields via ring-opening polyaddition with perfect control of regioselectivity and stereochemistry (Scheme 10.22). The polymerization proceeded under weak alkaline conditions such as at pH 10.6. Normally, ChiBs has a pH optimum around at 7–8 for hydrolysis of chitin; therefore, the hydrolysis activity was suppressed at pH 10.6, resulting in polymerization exclusively. In addition, 6-deoxyfluorinated chitin derivatives have been successfully synthesized via ChiBs-catalyzed polymerization of 6-deoxyfluorinated oxazoline monomers [137].

Notably, synthetic chitin was obtained as white crystals of α-chitin allomorph, which shows antiparallel molecular chain packing normally found in naturally occurring chitin. Furthermore, the formation of chitin spherulite was observed during the polymerization [138]. Figure 10.15 illustrates the postulated reaction mechanisms of ChiBs in (a) the hydrolysis of chitin and (b) polymerization to synthetic chitin.

In the hydrolysis, the glycosidic oxygen is protonated by the carboxylic acid immediately after recognition in stage 1. Then, the acetamido oxygen of the saccharide unit at the donor site attacks the neighboring C-1 carbon to form the corresponding oxazolinium ion stabilized by another carboxylate, leading to scission of the glycosidic linkage (stage 2). Nucleophilic attack by a water molecule from the β-side opens the oxazolinium ring to accomplish the hydrolysis reac-

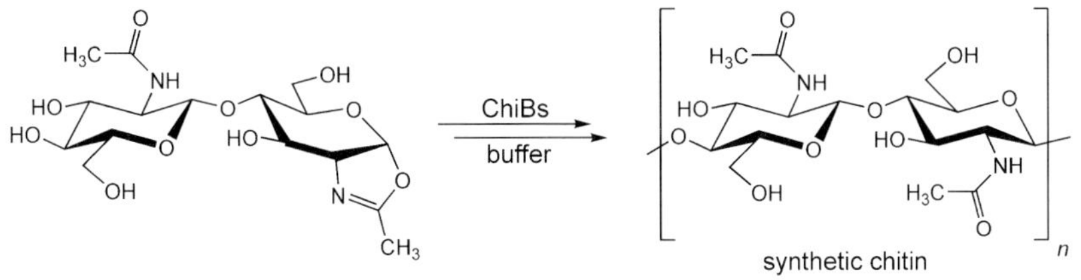

Scheme 10.22

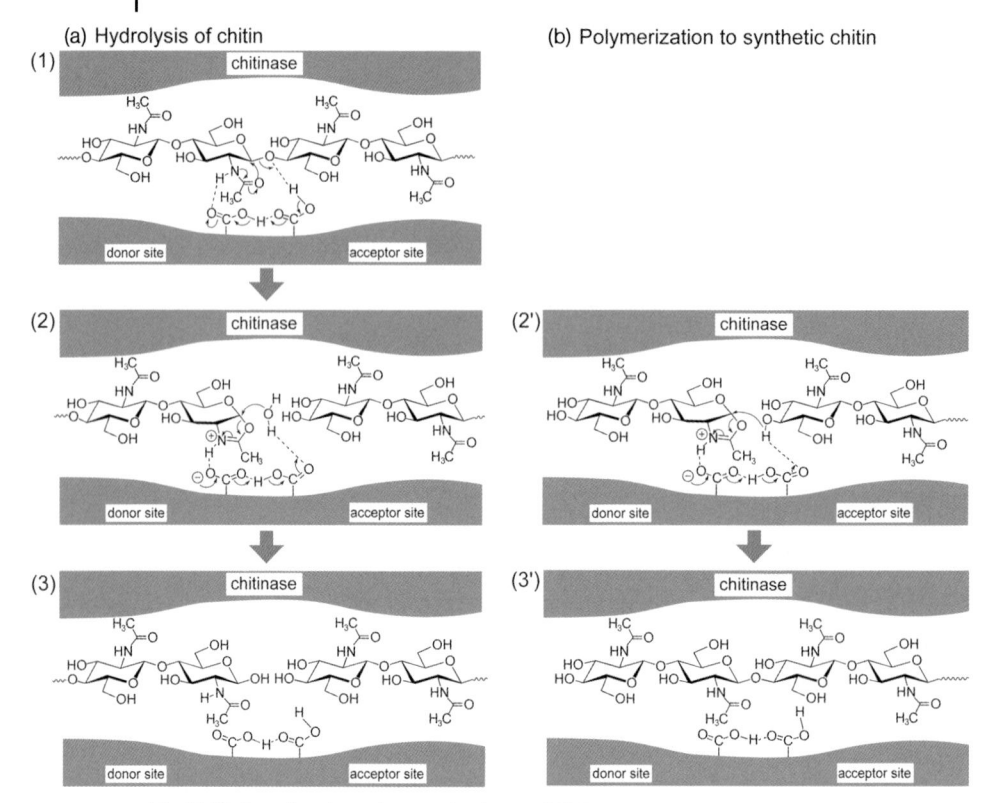

Fig. 10.15 Postulated reaction mechanisms of ChiBs: (a) hydrolysis of chitin and (b) ring-opening polyaddition of oxazoline monomer to synthetic chitin.

tion, giving rise to the hydrolyzate having a β-configuration (stage 3). In the polymerization, the oxazoline monomer is readily recognized at the donor site as a TSAS and is immediately protonated by the carboxylic acid to form the corresponding oxazolinium ion. The protonation step on to nitrogen must have a small barrier. The hydroxy group at the C-4 of another monomer or the growing chain end attacks the C-1 of the oxazolinium from the β-side (stage 2′), resulting in the formation of a (1 → 4)-β-glycosidic linkage (stage 3′). Repetition of these reactions is a ring-opening polyaddition, leading to the formation of synthetic chitin with perfect control of regioselectivity and stereochemistry. The key point is the structural similarity of the transition state (or the intermediate), which is commonly involved in both stages 2 and 2′. ChiNAc-oxa monomer is very close to this moiety, showing the importance of the concept of TSAS monomer [104, 135–137].

On the basis of the polymerization mechanism, the mutant enzyme derived from *Bacillus circulans* WL-12 chitinase A1 [139] was employed for chitin synthesis. The amino acid residue E204 in the enzyme is assumed to be the catalytic

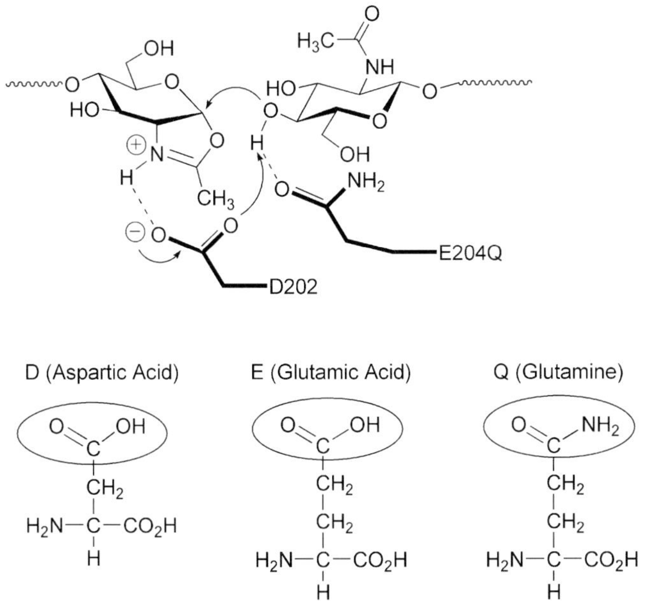

Fig. 10.16 Polymerization of oxazoline monomer by mutant chitinase.

acid/base and D202 serves as a stabilizer for the oxazolinium ion during the hydrolysis. The mutant E204Q, where the glutamic acid at the 204th amino acid residue was replaced by glutamine, exclusively exhibited catalytic activity for the formation of a (1 → 4)-β-glycosidic linkage, suppressing the hydrolysis of the product [140], although the total catalytic activity was much reduced (Fig. 10.16). This result shows the mechanistic difference between hydrolysis and polymerization; the latter inherently does not need the second carboxyl group.

Chitin oligomers were prepared through a step-by-step chain elongation catalyzed by ChiBs using an *N*-acetyllactosamine (LacNAc) oxazoline derivative as glycosyl donor (Scheme 10.23) [141], as observed above for cellooligomers. A LacNAc oxazoline derivative was effectively recognized and catalyzed by ChiBs, providing chitin oligomers bearing a (1 → 4)-β-galactoside at the non-reducing end. The terminal Gal residue is regarded as a protecting group against chain elongation by the enzyme; therefore, once it is removed by β-galactosidase, an unprotected GlcNAc residue is formed at the non-reducing end, which can serve as a glycosyl acceptor for the next glycosylation.

Further, ChiBs catalyzed the oligomerization of ChiNAc-oxa monomers with a 3-*O*-methyl group, affording a variety of 3-*O*-methylchitooligomers [142]. A ChiNAc-oxa monomer with a 6'-*O*-carboxymethyl group was subjected to the chitinase-catalyzed reaction; chitinase from *Streptomyces griseus*, one of the different kinds of GH18 chitinases, efficiently induced the production of an alternating 6-*O*-carboxymethylchitotetraose (Scheme 10.24) [143].

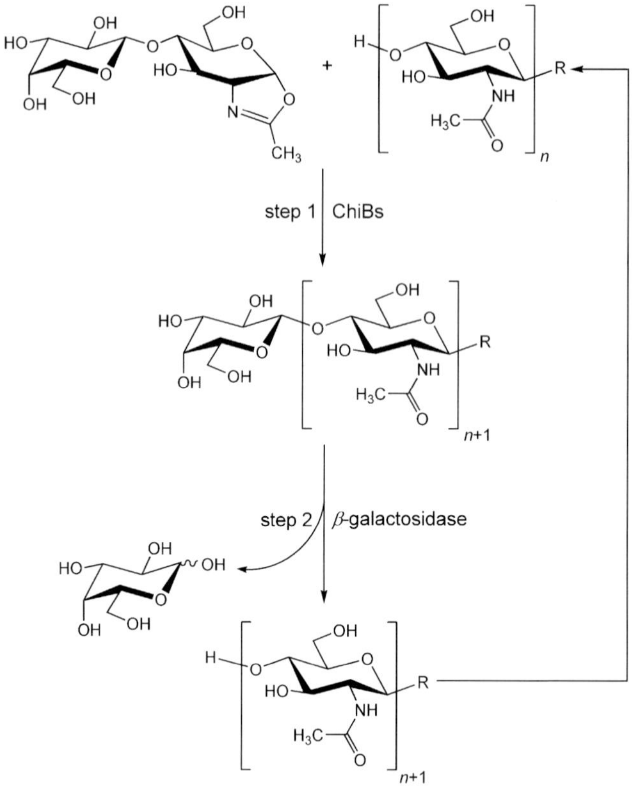

434 | *10 Polymer Synthesis and Modification by Enzymatic Catalysis*

Scheme 10.23

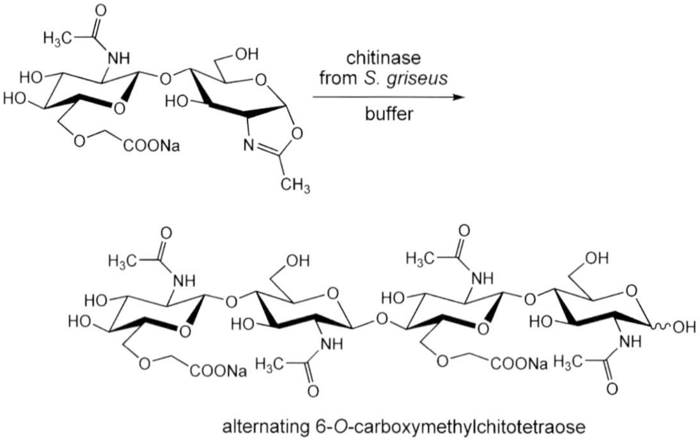

Scheme 10.24

alternating 6-O-carboxymethylchitotetraose

10.5.2.2 Glycosaminoglycans

Glycosaminoglycans (GAGs) are one of the classes of naturally occurring bioactive polysaccharides consisting of alternating connection of a hexosamine (D-glucosamine or D-galactosamine) and a uronic acid (D-glucuronic acid or L-iduronic acid) [129b]. Figure 10.17 illustrates the structures of six polysaccharides organizing GAG families. In the extracellular matrices (ECMs) and on the cell surface, GAGs construct ECMs through association with collagens, fibronectins and so forth, providing scaffolds for cellular proliferation and differentiation in morphogenesis and regeneration of tissues [144]. Thus, GAGs play critical roles in living organisms and are frequently used as therapeutic materials and as food supplements [145]. However, investigation of GAG functions at a molecular level is currently challenging.

Hyaluronidase (HAase; EC 3.2.1.35) is one of the *endo-β-N*-acetylglucosaminidases responsible for metabolism of HA, Ch and ChS [146], which belongs to GH56 [111]. The enzyme catalyzes the hydrolysis of such GAGs via a substrate-assisted mechanism similarly to GH18 chitinase [147]. On the basis of the TSAS concept, molecular design of a monomer structure bearing an oxazoline moiety allowed the production of HA, Ch and ChS via ring-opening polyaddition by HAase catalysis.

Hyaluronan and its derivatives

As illustrated in Fig. 10.17, HA is composed of a heterodisaccharide repeat, that is, $(1 \rightarrow 4)$-β-linked *N*-acetylhyalobiuronate [GlcAβ$(1 \rightarrow 3)$GlcNAc; MeHAbu] [148]. Therefore, a MeHAbu oxazoline derivative was designed as a TSAS mono-

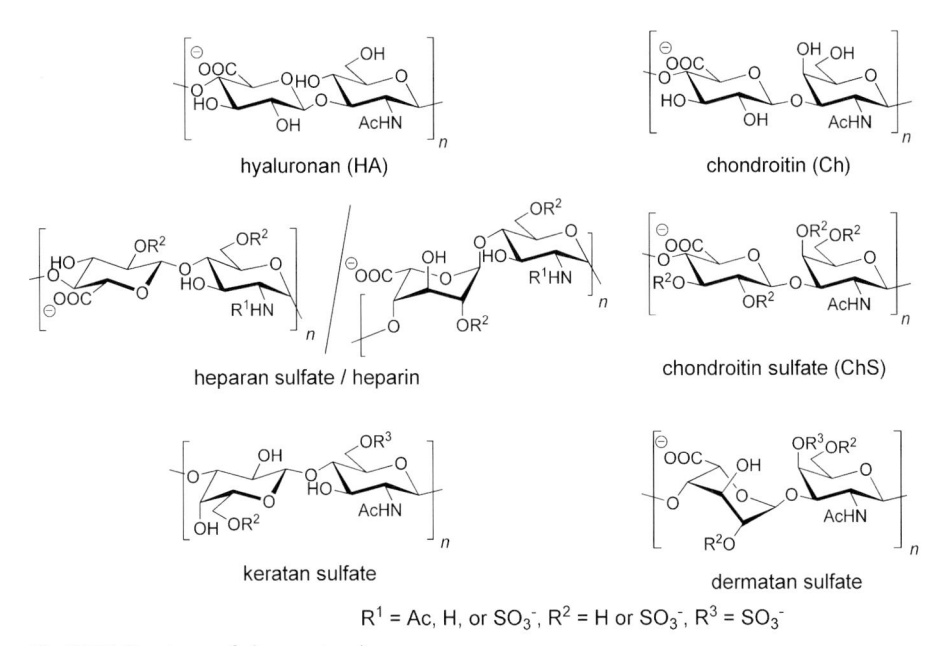

Fig. 10.17 Structures of glycosaminoglycans.

R^1 = Ac, H, or SO_3^-, R^2 = H or SO_3^-, R^3 = SO_3^-

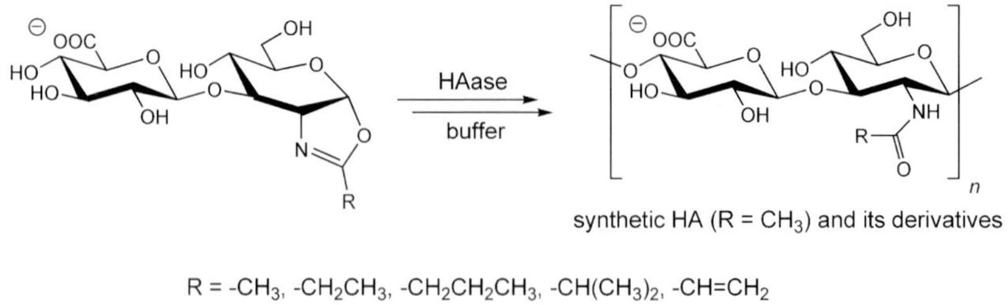

Scheme 10.25

R = -CH$_3$, -CH$_2$CH$_3$, -CH$_2$CH$_2$CH$_3$, -CH(CH$_3$)$_2$, -CH=CH$_2$

synthetic HA (R = CH$_3$) and its derivatives

mer for HA synthesis. The monomer was efficiently recognized and catalyzed by HAase with perfect structure control, providing for the first time *in vitro* synthetic HA with a relatively high molecular weight ($M_n \approx 20\,000$) in good yields (Scheme 10.25) [149]. Furthermore, 2-substituted oxazolines of GlcAβ(1 → 3)GlcNAc such as 2-ethyl, 2-*n*-propyl and 2-vinyl derivatives were also polymerized effectively by HAase catalysis, giving rise to unnatural HA derivatives with the corresponding *N*-acyl group. In particular, an *N*-vinyl (acrylamido) HA derivative has potential for utilization as a macromonomer, telechelic and so forth, leading to application in medical materials.

These results motivated us to carry out HAase-catalyzed copolymerization between various 2-substituted HAbu oxazoline monomers, giving rise to some unnatural HA derivatives [150].

Chondroitin and its derivatives

Ch exists predominantly as a carbohydrate part of proteoglycans in *Caenorhabditis elegans* [151] or in higher organisms as a precursor of ChS, mainly found in cartilage, cornea and brain matrices [152]. A number of reports have described the biological functions of Ch and ChS, for example, maintaining cartilage elasticity [153], regulation of tissue morphogenesis and promotion of neurite outgrowth [154] and neuronal migration [155], which are frequently associated with the sulfation patterns [156]. Such important biomacromolecules of Ch and ChS have been synthesized for the first time via ring-opening polyaddition catalyzed by HAase [157, 158].

Ch is composed of a heterodisaccharide repeating unit, GlcAβ(1 → 3)GalNAc (*N*-acetylchondrosine; MeChd), which is connected through a (1 → 4)-β-glycosidic linkage (Fig. 10.17). Therefore, a MeChd oxazoline derivative was designed for the synthesis of Ch via HAase-catalyzed ring-opening polyaddition (Scheme 10.26) [157]. In addition to the MeChd monomer, polymerization of various oxazoline monomers with 2-ethyl, 2-*n*-propyl, 2-isopropyl, 2-vinyl and 2-phenyl substituents was investigated. Monomers with 2-methyl, 2-ethyl and 2-vinyl substituents were efficiently catalyzed by HAase at pH 7.5, providing synthetic Ch and its derivatives bearing the corresponding *N*-acyl groups with perfect structure control. The molecular weights (M_n) of these products reached several thousand, which correspond to that of naturally occurring Ch. Monomers with

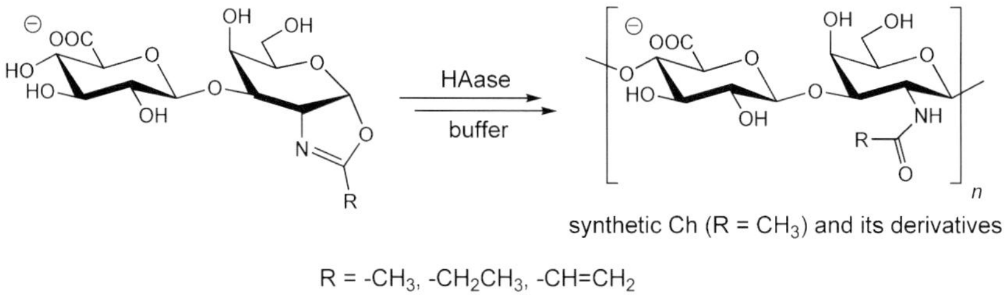

Scheme 10.26

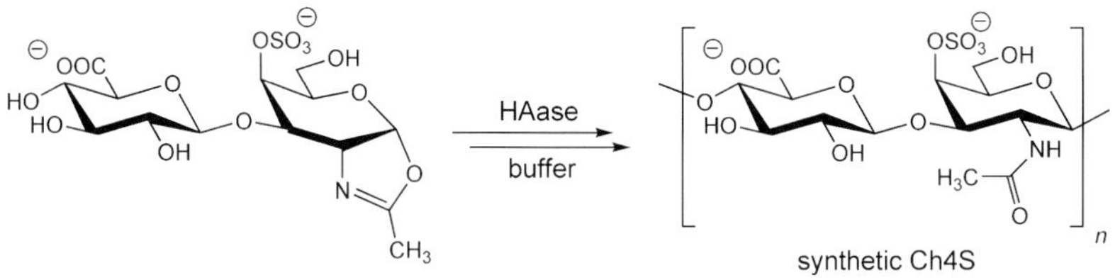

R = -CH₃, -CH₂CH₃, -CH=CH₂

Scheme 10.27

2-isopropyl and 2-phenyl substituents were not catalyzed by the enzyme, probably due to the steric bulkiness of these substituents.

HAase-catalyzed polymerization of a MeChd oxazoline with a sulfate group at the C-4 of GalNAc successfully progressed at pH 7.5 and 0–40 °C, giving rise to synthetic Ch4S having sulfate groups exclusively at the C-4 of the GalNAc unit (Scheme 10.27) [158]. However, the MeChd monomers with a sulfate group at the C-6 and both the C-4 and C-6 positions were not polymerized at all. These results suggest that HAase distinguishes sulfate groups at different positions on the GAG chains. Synthetic Ch4S is expected to have potentials for pharmaceutical and medical applications due to its defined new structure.

10.5.2.3 Unnatural Hybrid Polysaccharides

As observed in Section 10.5.1.4, ChiBs similarly exhibits substrate recognition, resulting in the formation of hybrid polysaccharides. A cellulose–chitin hybrid polysaccharide can also be synthesized via ChiBs-catalyzed ring-opening polyaddition of a Glcβ(1→4)GlcNAc oxazoline monomer, which has an oppositely aligned sugar structure to that of a β-fluoride monomer of GlcNAcβ(1→4)Glc employed in cellulase-catalyzed polycondensation [123]. ChiBs recognized the oxazoline monomer and catalyzed the ring-opening polyaddition at pH 9.0 and 30 °C, providing a cellulose–chitin hybrid polysaccharide having a reversed unit structure compared with that produced by CelTV-catalyzed polycondensation (Scheme 10.28).

ChiBs induced the production of a chitin–chitosan hybrid polysaccharide via ring-opening polyaddition. A newly designed oxazoline monomer having

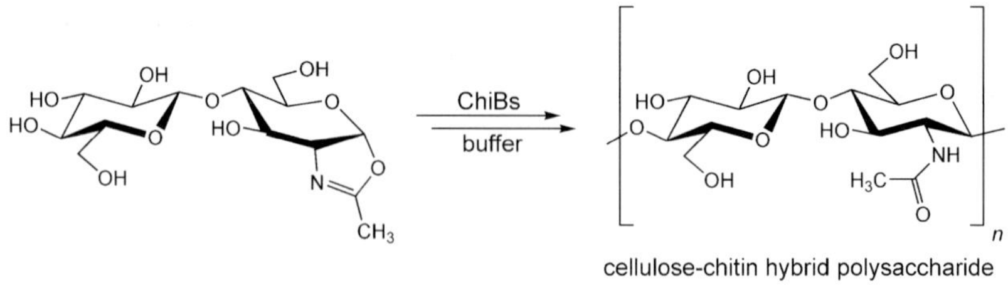

Scheme 10.28

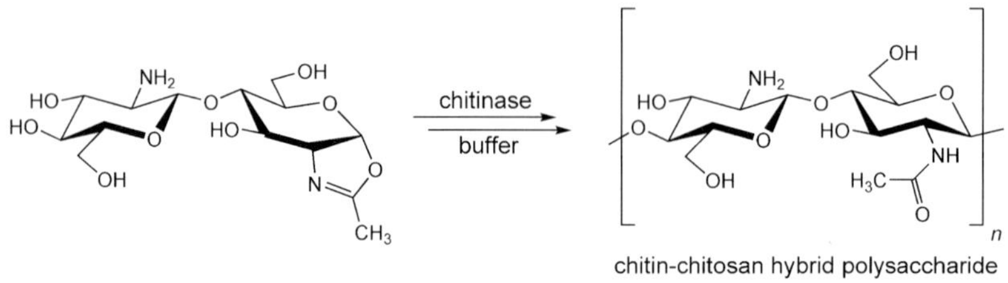

Scheme 10.29

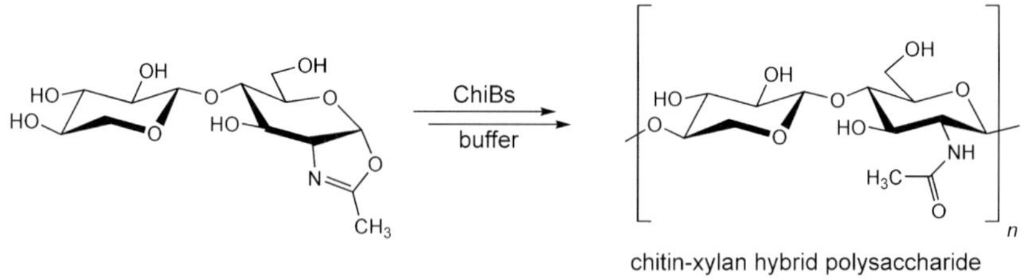

Scheme 10.30

(1 → 4)-β-D-glucosaminide at the non-reducing end [GlcNβ(1 → 4)GlcNAc-oxa] was successfully recognized and catalyzed by ChiBs, giving rise to a hybrid polysaccharide in good yields (Scheme 10.29) [159]. Chitinase from *Serratia marcescens* also catalyzed the polymerization with higher efficiency, providing a hybrid polysaccharide with higher molecular weight in higher yields. The product hybrid showed excellent solubility in water; the reaction proceeded homogenously.

Another water-soluble hybrid polysaccharide was produced by the catalysis of ChiBs. The enzyme efficiently catalyzed ring-opening polyaddition of a novel oxazoline monomer bearing (1 → 4)-β-D-xyloside at the non-reducing end [Xylβ(1→4)GlcNAc-oxa] in a regio- and stereoselective manner, affording a chitin–xylan hybrid polysaccharide in good yields (Scheme 10.30) [160]. The molecular weight reached several thousand as determined by MALDI-TOF mass spectrometry.

10.6
Synthesis of Polyesters Catalyzed by Hydrolases, Mainly by Lipases

Among the enzymatic syntheses of polymers, lipase-catalyzed polymerization is one of the most actively studied fields [5, 6, 8–15, 17–23]. Lipase is an enzyme that catalyzes the hydrolysis of fatty acid esters *in vivo* with bond cleavage, yet lipase catalyzes polymerization reactions *in vitro* with the bond formation, when the lipase catalyst and substrate monomer are appropriately combined in the reaction. In fact, polyester synthesis has been found to proceed via various polymerization modes. Further, lipases are sometimes stable in organic solvents and can be used widely as catalysts for esterifications and transesterifications in organic syntheses [161, 162]. Three reaction modes of lipase-catalyzed polymerization are typically shown in Scheme 10.31.

10.6.1
Polyesters via Ring-opening Polymerization

10.6.1.1 Ring-opening Polymerization of Lactones

Discovery of lipase catalyst
In 1993, lipase-catalyzed ring-opening polymerization was reported for the first time on ε-caprolactone (ε-CL; seven-membered) and δ-valerolactone (δ-VL; six-membered) by two independent groups [163, 164]. Since then, various lactones of different ring sizes and of other cyclic esters have been extensively examined

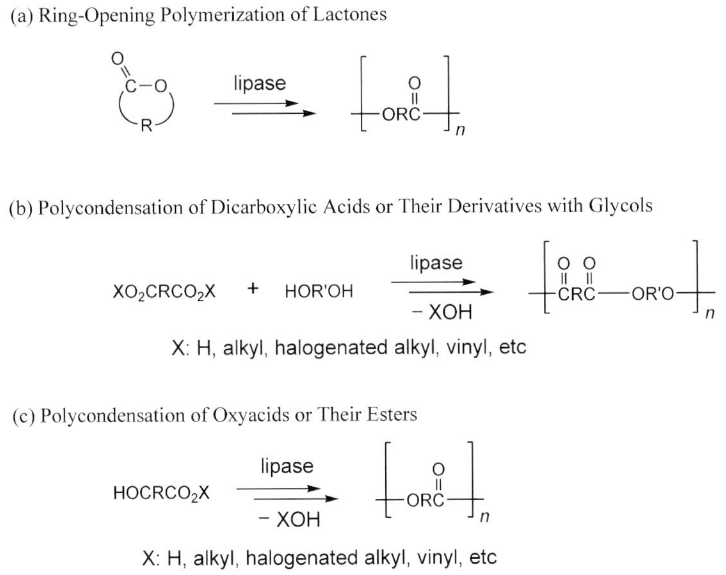

(a) Ring-Opening Polymerization of Lactones

$$\text{lactone} \xrightarrow{\text{lipase}} [-ORC(O)-]_n$$

(b) Polycondensation of Dicarboxylic Acids or Their Derivatives with Glycols

$$XO_2CRCO_2X + HOR'OH \xrightarrow[-XOH]{\text{lipase}} [-CRC(O)-OR'O-]_n$$

X: H, alkyl, halogenated alkyl, vinyl, etc

(c) Polycondensation of Oxyacids or Their Esters

$$HOCRCO_2X \xrightarrow[-XOH]{\text{lipase}} [-ORC(O)-]_n$$

X: H, alkyl, halogenated alkyl, vinyl, etc

Scheme 10.31

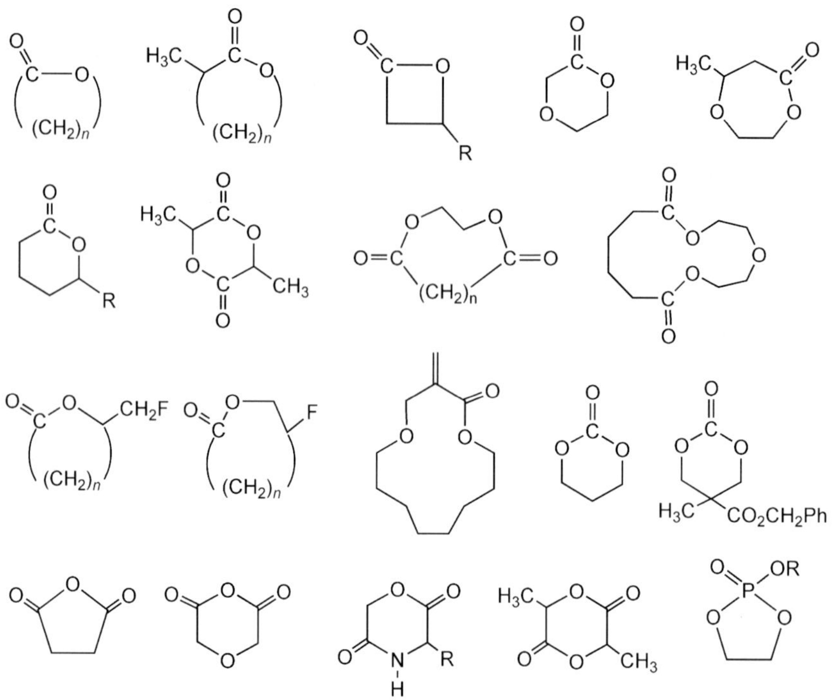

Fig. 10.18 Cyclic monomers for lipase-catalyzed polyester synthesis.

(Fig. 10.18) [11–15, 17, 21, 23, 164, 165]. Lipase acted as an efficient catalyst for the ring-opening polymerization of 4–17-membered non-substituted lactones (Scheme 10.32) [166, 167]. The polymerization can be performed in bulk, in an organic solvent or in other solvents. The polymer obtained in bulk has the terminal structure of an alcohol at one end and a carboxylic acid at the other and hence it belongs to telechelics.

Among lactone monomers, ε-CL has been most extensively studied [163–179]. ε-CL was polymerized rapidly by various lipases of different origin. Among them, *Candida antarctica* (lipase CA, the immobilized form often as Novozyme 435) was the most effective for the polymerization of ε-CL [180, 181]. The catalyst amount could be reduced to less than 1 wt.%, compared with the previously

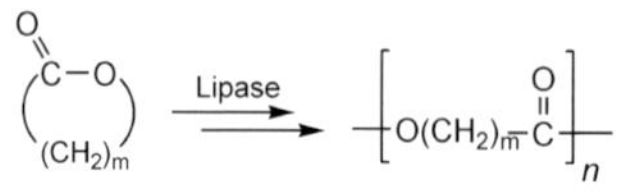

m=2 (4-membered) : β-PL   m=10 (12-membered) : UDL
m=4 (6-membered) : δ-VL   m=11 (13-membered) : DDL
m=5 (7-membered) : ε-CL   m=14 (16-membered) : PDL
m=7 (9-membered) : 8-OL   m=15 (17-membered) : HDL

Scheme 10.32

reported systems of 10–50% of lipase from other origins. The molecular weight (M_n) of poly(ε-CL) reached $>8.9\times10^4$ [171, 182]. During the polymerization of ε-CL, extensive degradation and polymerization occurred simultaneously [181].

Reaction mechanism and ring-opening polymerizability
Lipase-catalyzed polymerization of lactones is explained by considering the following reactions as the principal reaction course involving an acyl-enzyme intermediate (Scheme 10.33) [12–15, 17, 183]. Lipase-catalyzed hydrolysis of an ester is generally accepted to proceed via a similar intermediate [184]. The catalytic site of lipase is known to be a serine residue. The key step is the reaction of lactone with lipase involving the ring opening of lactone to give an acyl-enzyme intermediate (enzyme-activated monomer, EM). The initiation is a nucleophilic attack of water, which is probably contained in the enzyme, on the acyl carbon of the intermediate to produce ω-hydroxycarboxylic acid ($n=1$), the shortest propagating species. In the propagation stage, EM is nucleophilically attacked by the terminal hydroxyl group of a propagating polymer to produce a one-monomer-unit elongated polymer chain. Thus, the polymerization proceeds via an "activated monomer mechanism".

β-Propiolactone (β-PL; four-membered) and substituted β-PL were polymerized by lipase catalysis in bulk, yielding a mixture of linear and cyclic oligomers to polymers with molecular weights up to 2×10^4 [185]. α-Methyl-β-PL gave a polymer with a similar structure to poly(lactic acid). Nine-membered lactone (8-octanolide, 8-OL) was also polymerized by lipase catalyst, producing a polymer with a molecular weight of 1.6×10^4 at 75 °C after 10 days [186]. Concerning macrolides, 11-undecanolide (UDL; 12-membered), 12-dodecanolide (DDL; 13-membered), 15-pentadecanolide (PDL; 16-membered) and 16-hexadecanolide (HDL; 17-membered) were enzymatically polymerized [166, 183, 187]. Polyesters with molecular weights $>2\times10^4$ were readily obtained from these macrolides under appropriate reaction conditions. The lipase-catalyzed polymerization of UDL, DDL and PDL proceeded even in an aqueous medium [188].

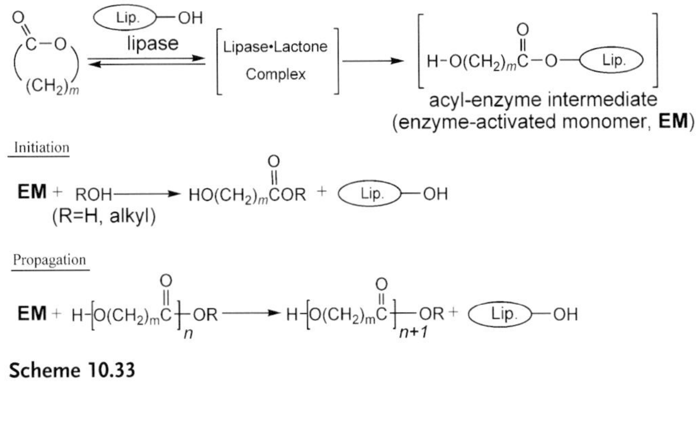

Scheme 10.33

The reactivity of cyclic compounds generally depends on the ring size; small- and medium-sized compounds show higher reactivity towards ring-opening polymerization owing to the higher strain in their rings in comparison with large-sized macrolides. Table 10.2 summarizes the dipole moments and reactivities of unsubstituted lactones with different ring sizes. The dipole moments of the monomers are taken as an indication of their ring strain. The values for the macrolides are lower than that of ε-CL and close to that of an acyclic fatty acid ester (butyl caproate). The rate constants of the macrolides in alkaline hydrolysis and anionic polymerization are much smaller than those of ε-CL. These data imply that the macrolides have much lower ring strain and, hence, show less anionic reactivity and polymerizability than ε-CL.

All the lipase-catalyzed ring-opening polymerizations obeyed Michaelis–Menten kinetics, where the formation of an acyl-enzyme intermediate was supposed to be the rate-determining step of the overall reaction, because the reactivity of the intermediate must be high and the following steps are very rapid [166, 167]. Kinetic analysis was performed based on the Michaelis–Menten equation for lipase PF (*Pseudomonas fluorescens* lipase)-catalyzed polymerizations. The relative rate of the enzymatic polymerization was thus derived from semi-quantitative rate data (Table 10.2). With lipase PF, the larger the ring size, the higher is the polymerization rate. (When lipase of a different origin is used, this reactivity may turn out to be different.) On the other hand, the Zn-catalyzed anionic polymerizability of the lactones showed the reverse direction with a large difference;

Table 10.2 Comparison of dipole moments and reactivities of various unsubstituted lactones.

Lactone (ring-size)	Dipole moment (C m)	Rate constant		Relative rate of polymerization	
		Alkaline hydrolysis [a] (L mol^{-1} s^{-1}×10^4)	Propagation [b] (s^{-1}×10^3)	Enzymatic polymerization [c]	Anionic polymerization [d]
δ-VL (6)	4.22	55 000	–	–	2500
ε-CL (7)	4.45	2 550	120	0.10	330
8-OL (9)	–	–	–	–	21
UDL (12)	1.86	3.3	2.2	0.13	0.9
DDL (13)	1.86	6.0	15	0.19	1.0
PDL (16)	1.86	6.5	–	0.74	0.9
HDL (17)	–	–	–	1.00	1.0
Butyl caproate	1.75	8.4	–	–	–

a) Alkaline: NaOH. Measured in 1,4-dioxane–water at 0 °C.
b) Measured using NaOMe initiator in THF at 0 °C.
c) Kinetics of the polymerization were studied using lipase PF as catalyst in the presence of 1-octanol in diisopropyl ether (10 mL) at 60 °C.
d) [Zn(Oct)$_2$]$_0$ = [BuOH]$_0$ = 0.28 mol L^{-1} with bulk polymerization carried out at 100 °C [189].

10.6 Synthesis of Polyesters Catalyzed by Hydrolases, Mainly by Lipases

the polymerizability is governed by the ring strain [189]. The higher reactivity of the macrolides with a lipase catalyst was explained by the higher rate of the formation of the acyl-enzyme intermediate.

Lipase CA-catalyzed copolymerization of a racemic lactone (methyl substituted) with an achiral lactone (unsubstituted) was investigated (Scheme 10.34) [190a]. The reaction induced enantioselective copolymerization of the racemic monomer. From the viewpoint of propagation, there are four different diads for a binary copolymer. For example, the copolymerization of β-butyrolactone (β-BL,

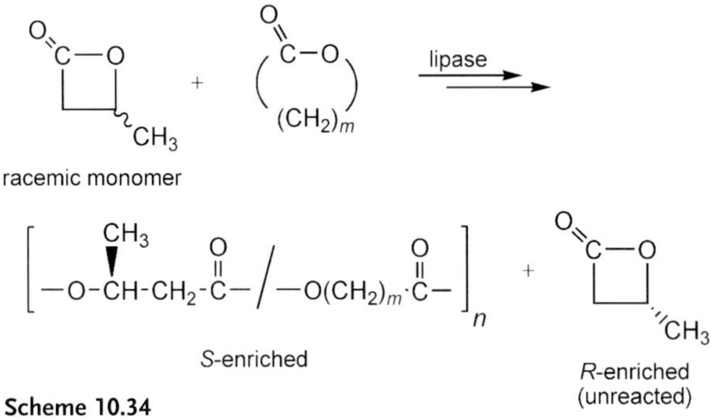

Scheme 10.34

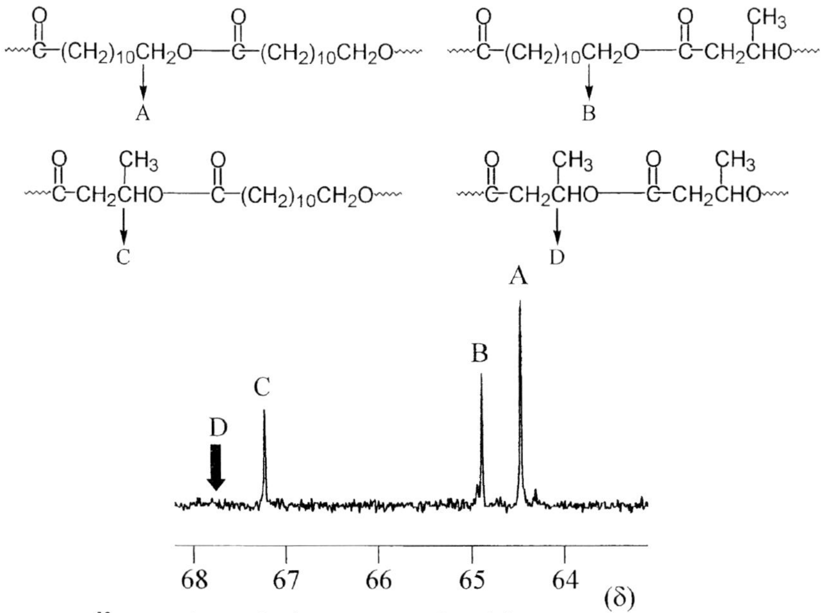

Fig. 10.19 ^{13}C NMR: four peaks due to C–O signals and their diad sequences of the copolymer from β-butyrolactone and 12-dodecanolide.

four-membered) with 12-dodecanolide involves four modes, which was proven by the ^{13}C NMR spectrum (Fig. 10.19). Four peaks (A, B, C and D) were assigned to the respective diad and their intensity ratio was approximately 10:6:4:0. The corresponding four elementary propagations are given in Fig. 10.19, showing the steps where an acyl-enzyme intermediate (or transition-state) is attacked by a nucleophile of the propagating end of the hydroxyl group. If the rate-determining step of the lipase-catalyzed ring-opening polymerization is the formation of an acyl-enzyme intermediate because the subsequent reactions are very fast due to the high reactivity of the intermediate, the formation of the diad A and C and of the diad B and D should be the same in amount; however, the results are different; they are 10:4 and 6:0, respectively. This suggests that the reaction steps of Fig. 10.20 involve a propagating end of sterically two different nucleophile structures (primary and secondary alcohols) and, hence, the structure of the nucleophile greatly affects the overall polymerization.

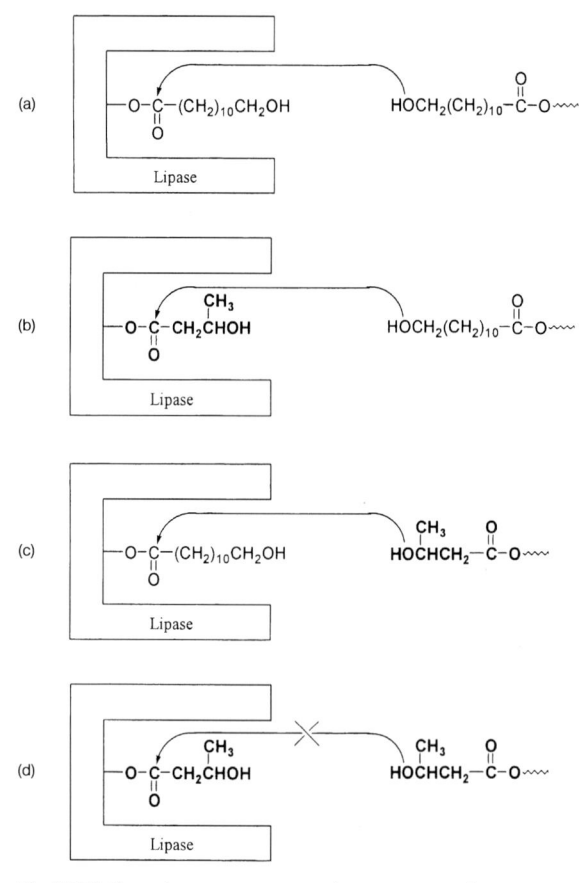

Fig. 10.20 Four elementary propagations corresponding to reactions giving A, B, C and D diads.

Similar results were obtained for the copolymerization of ε-CL (six-membered) with 12-dodecanolide and also other binary copolymerizations [190a].

The above observations led to a mechanism of the lipase-catalyzed ring-opening polymerization in which the formation of the acyl-enzyme intermediate (acylation step) and/or the subsequent reaction of the intermediate (deacylation step) are operative depending on the nature of the monomer. In particular, the deacylation step becomes more strongly operative when the propagating alcohol end is a sterically bulky nucleophile, the step of which is rate determining. Enantioselection is therefore induced possibly at both the acylation and deacylation steps (Fig. 10.21) [190b].

A mechanism of the important deacylating step was described previously, from the lipase CA-catalyzed ring-opening polymerization of 4-alkyl-substituted ε-CL where methyl, ethyl and propyl groups were employed as an alkyl group [191]. The following were observed: (1) the polymerization rate was greatly affected by the substituent size with relative values of none/methyl/ethyl/propyl = 140/70/14/1 and (2) the enantioselectivity changed from S for methyl and ethyl to R for propyl, and decreased greatly with increasing substituent size. On the other hand, in the hydrolysis of these monomers by the same enzyme: (1) the relative rate was none/methyl/ethyl/propyl = 2.5/5.8/5.2/1 and (2) the enantioselectivity was S in all cases. From these observations, it was suggested that the deacylation step must be rate determining for the polymerization of these monomers (analogously shown in Fig. 10.21). However, in the hydrolysis, the rate-determining step is most likely the formation of the acyl-enzyme intermediate, because water is a very small nucleophile whereas a propagating chain alcohol end is not [191].

In the lipase PF-catalyzed copolymerizations between unsubstituted lactones, typically the copolymerization between PDL and ε-CL produced a copolymer

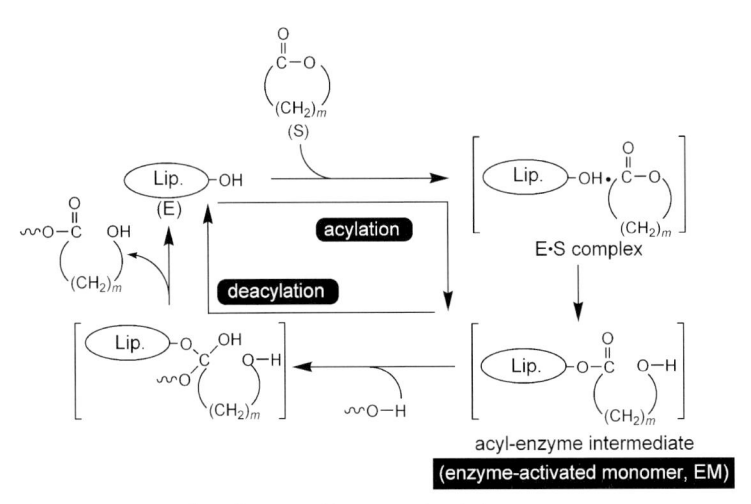

Fig. 10.21 Proposed mechanism of lipase-catalyzed ring-opening polymerization of lactones involving the acylation step and/or the deacylation step as the rate-determining step.

having the diad sequence distribution: PDL–*PDL*=0.81, PDL–*ε-CL*=0.19, ε-CL–*PDL*=0.78, ε-CL–*ε-CL*=0.22, where the italics denote the monomer unit of propagating end [187b]. The result shows that both ε-CL and PDL monomers have comparable polymerizability. Also, with the same catalyst the copolymerization of ε-CL with δ-VL gave a similar result [163b]. These data imply that in both copolymerizations the formation of an acyl-enzyme intermediate mainly governs the overall reaction. This is probably because the structure of propagating ends is very similar in four elementary propagations; all of the propagating ends are primary alcohols and the deacylation step is comparable in rate for the copolymerization.

It should be noted that a transesterification by lipase catalyst often takes place under severe reaction conditions. For example, when the ring-opening polymerization of a macrolide (DDL or PDL) was carried out in the presence of an aliphatic polyester [poly(ε-CL) or poly(1,4-butylene adipate)] at 75 °C, a polyester copolymer from the cyclic monomer and the polyester was produced [192]. Also, lipase CA-catalyzed intermolecular transesterification between two different polyesters took place to give a polyester copolymer composed of the two polyesters repeating units [173, 193]. Copolymerization of ε-CL with D,L-lactide was carried out with lipase CA catalyst and the product copolyester was structurally well characterized [194].

Enantioselective and chemoselective polymerizations

Lipase PC catalyzed the enantioselective polymerization of 3-methyl-4-oxa-6-hexanolide (MOHEL) (Scheme 10.35) [166]. The initial reaction rate of the S-isomer was seven times larger than that of the R-isomer, indicating that the enantioselective polymerization of MOHEL was induced effectively. (S)-MOHEL was also polymerized by lipase PF, whereas no polymerization of the R-isomer took place.

α-Methyl-β-propiolactone was enantioselectively polymerized by lipase PC to produce an optically active S-enriched polymer with enantiomeric excess (*ee*) up to 50% [185c]. Lipase CA-catalyzed enantioselective polymerization of fluorinated lactones with a ring size from 10 to 14 was reported. Interestingly, the corresponding oxyacid gave an optically inactive polyester [195]. A highly S-enriched substituted poly(ε-CL) was prepared using seven-membered substituted ε-CLs. Racemic 4-methylcaprolactone [(R,S)-4-MeCL] and 4-ethylcaprolactone [(R,S)-β-4-EtCL] were enantioselectively polymerized by lipase to produce the corresponding polyesters showing *ee* higher than 95% with molecular weight about

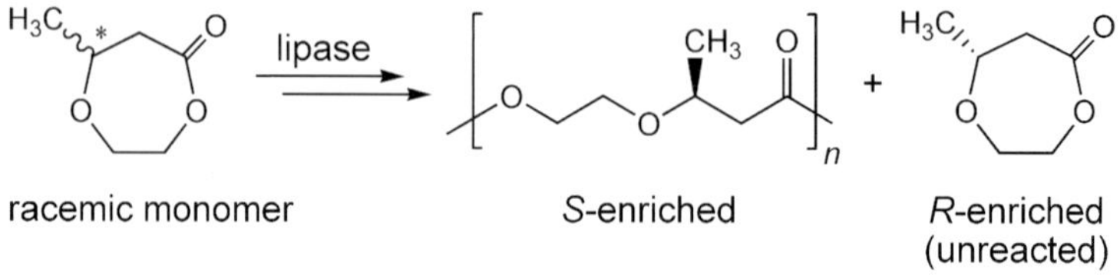

Scheme 10.35

5000 [196]. The enantioselective polymerization of (*R*,*S*)-β-BL was observed by using the natural poly(hydroxybutyrate) depolymerase and compared with that of the lipase catalysis, which resulted in a lower selectivity [197].

Ring-opening polymerization of methyl-substituted ε-CLs was examined with lipase CA catalyst. 3-Methyl- and 4-methyl-ε-BLs were polymerized moderately fast but 6-methyl-ε-BL did not polymerize. The importance of the stereochemistry of enantiomeric secondary alcohols involved in the propagating end is suggested from the polymerization results for 6-methyl-ε-BL [198].

In the lipase CA-catalyzed copolymerization of β-butyrolactone (β-BL) with DDL, (*S*)-β-BL was preferentially reacted to give the *S*-enriched optically active copolymer with *ee* of the β-BL unit = 69% (Scheme 10.34) [190]. δ-Caprolactone (six-membered) was also enantioselectively copolymerized with achiral lactones by the lipase catalyst to give the *R*-enriched optically active polyesters reaching to 76% *ee*.

Chiral oligoesters were produced by a new method of ring-opening polymerization of 6-methyl-ε-BL with iterative tandem catalysis [199]. Iterative tandem catalysis means a polymerization in which the chain growth is effected by a combination of two different catalytic processes that are both compatible and complementary. With the combination of lipase CA catalysis for ring-opening polymerization of racemic 6-methyl-ε-BL and Ru-catalyzed racemization of propagating secondary alcohol in one pot, optically pure oligoesters were obtained; the principle is shown in Scheme 10.36. First, lipase CA catalyzes the ring-opening of the *S*-monomer enantioselectively to give the benzyl alcohol adduct, but the *S*-alcohol is less favored to react with the monomer and hence Ru-catalyzed racemization takes place to give the racemic alcohol. Then, the favored *R*-alcohol is selectively consumed by reaction with the monomer. This situation repeats and ends up with the production of *R*-oligoesters from the racemic monomer.

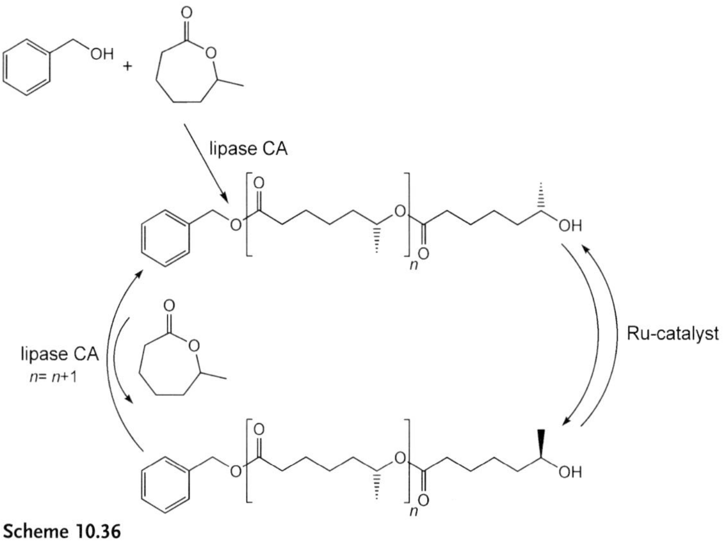

Scheme 10.36

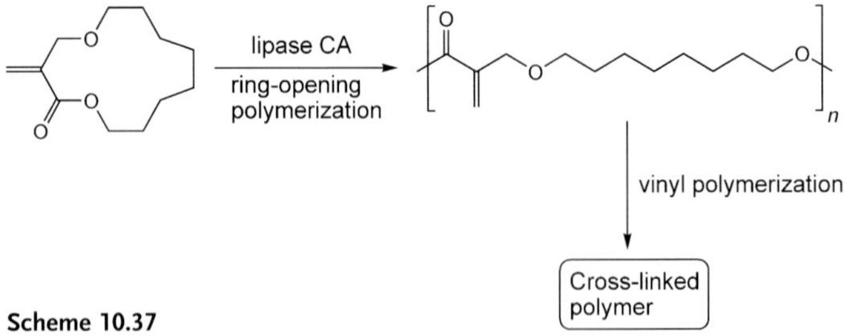

Scheme 10.37

The lipase catalyst chemoselectively induced the ring-opening polymerization of 2-methylene-4-oxa-12-dodecanolide, a cyclic derivative of methyl methacrylate, yielding a polyester having a reactive *exo*-methylene group in the main chain. This type of polymer cannot be obtained using a conventional chemical initiator. The chemoselective polymerization of α-methylenemacrolides having various groups in the ring, e.g. aromatic, ether and amine groups, was carried out enzymatically, anionically and radically. The lipase-catalyzed polymerization selectively afforded polyesters through the ring-opening process, whereas anionic and radical initiators induced vinyl polymerization. The polyester was further cross-linked via reaction through the vinylene group to produce a polymer gel (Scheme 10.37) [200, 201].

Synthesis of functionalized polyesters

The control of the polymer terminal structure is very important for the synthesis of end-functionalized polymers, typically macromonomers and telechelics. Various methodologies for synthesis of these polymers have been developed; however, they often required elaborate and time-consuming procedures. Lipase catalysis, on the other hand, provided a novel methodology for a single-step synthesis of end-functionalized polyesters by facile procedures.

As shown in Scheme 10.38, a nucleophile such as water and an alcohol can act as initiating species in the ring-opening polymerization of lactones. The lipase CA-catalyzed polymerization of lactones in the presence of functional alco-

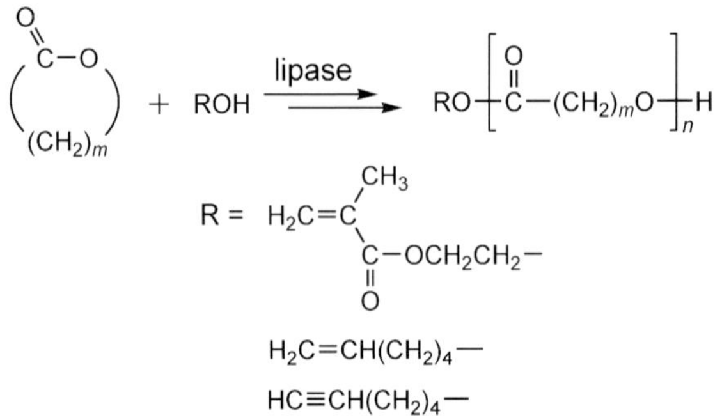

Scheme 10.38

hols produced end-functionalized polyesters ("initiator method"). In the polymerization of DDL employing 2-hydroxyethyl methacrylate as initiator, the methacryloyl group was quantitatively introduced at the polymer terminus, yielding a methacryl-type polyester macromonomer [202]. A similar methodology was employed to synthesize ω-alkenyl- and alkynyl-type macromonomers by using 5-hexen-1-ol and 5-hexyn-1-ol as initiators. Alkyl glucopyranosides [203, 204] and a first-generation dendrimer having six primary alcohols [205] initiated the polymerization of ε-CL in the presence of lipase CA, where the regioselective initiation (monoacylation of the initiator) took place. Thus, the primary hydroxy group was regioselectively acylated to give a polymer bearing the sugar moiety at the polymer terminus. A recent paper reported the lipase-catalyzed ring-opening polymerization of a seven-membered lactone to give end-functionalized and triblock polyesters [206].

A single-step, convenient production of end-functionalized polyesters was developed by lipase-catalyzed polymerization of DDL in the presence of vinyl esters (see also Section 10.6.2.1. for vinyl esters) [207]. The vinyl ester acted as terminator during the polymerization ("terminator method"). In using vinyl methacrylate as terminator, the methacryl group was quantitatively introduced at the polymer terminus to give a methacryl-type polyester macromonomer (Scheme 10.39). Polymerization in the presence of vinyl 10-undecanoate produced the ω-alkenyl-type macromonomer. This system can be applied to the synthesis of telechelics having a carboxylic acid group at both ends by the addition of divinyl

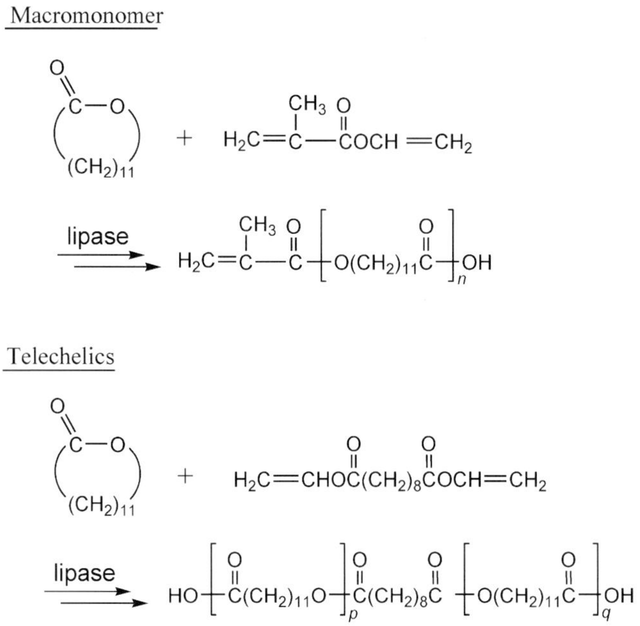

Macromonomer

Telechelics

Scheme 10.39

sebacate to the reaction mixture. The vinyl ester method was applied for the synthesis of an amphiphilic macromonomer of methacryl and sugar group-containing polycaprolactones [208]. The methacryl-type polyester macromonomer was in fact radically polymerized to afford a polymer of brush structure [209].

A new approach to a biodegradable polyester system was made by the lipase-catalyzed ring-opening polymerization of ε-CL and p-dioxanone monomers initiated from an alcohol attached on a gold surface [210]. It is considered that these polyesters can be used as biocompatible/biodegradable polymers for coating materials in biomedical areas such as passive or active coatings of stents. This method would be beneficial in applications where the minimization of harmful species is critical. A similar surface-initiated polymerization was reported for the *in situ* solid-phase synthesis of biocompatible poly(3-hydroxybutyrate) [211].

Enzymatic ring-opening polymerization was applied to synthesize hyperbranched aliphatic copolyesters by copolymerization of ε-CL with 2,2-bis(hydroxymethyl)butyric acid, catalyzed by lipase CA. In preparing the AB$_2$ polyesters, the degree of branching and the density of functional end-groups were controlled [212].

A novel chemoenzymatic method was developed by the combination of enzymatic ring-opening polymerization (eROP) of lactones and atom transfer radical polymerization (ATRP) [213]. The method allows a versatile, one-pot synthesis of block copolymers consisting from a polyester chain and a vinyl polymer chain by using a bifunctional initiator, whose reaction routes are shown in Scheme 10.40 [214]. For example, an initiator having both a hydroxy group for eROP and a bromide group for ATRP can be used. Lipase CA (Novozyme 435)-catalyzed ROP of ε-CL (reaction: 5 g of ε-CL with 0.5 g of the initiator and 0.5 g of the lipase, 60 °C in toluene for 3 h) to give 3.1 g of poly(ε-CL). Then poly(ε-CL) having a bromide group was isolated and used for the Cu-catalyzed radical polymerization of styrene at 85 °C in 1,4-dioxane to give poly(ε-CL-*b*-St) in high

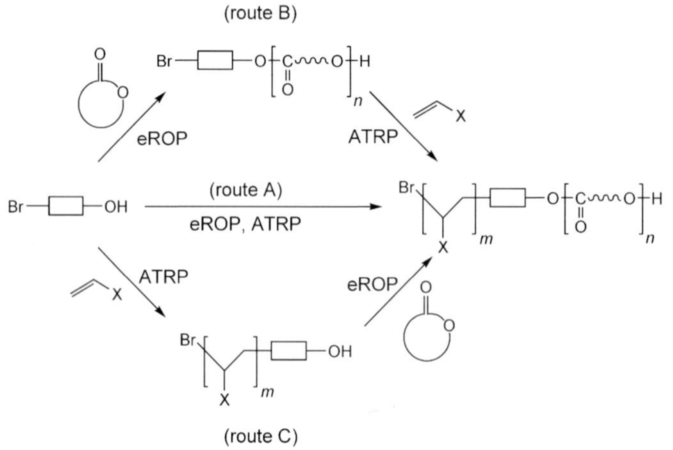

Scheme 10.40

yield, in which the ε-CL chain exhibited $M_n = 5.8 \times 10^3$ and the St chain 15×10^3, via route B [214a]. A similar principle was applied for the synthesis of such block copolymers by using a nitroxide-mediated radical process instead of a metal-catalyzed radical process via route B [214b]. Branched polymers with or without a poly-MMA chain from poly(ε-CL) macromonomer were produced by using the chemoenzymatic technique [214c].

Approach to environmental problems: chemical recycling and reaction solvents
It is demanded worldwide that recycling of polymeric materials is applied as one of the possible ways to combat to environmental problems (energy, natural resources, carbon dioxide emission and so forth) to maintain a green sustainable society. Polymer scientists are encouraged to contribute to overcoming these problems by pursuing "green polymer chemistry" [12, 13, 17, 20]. Among recycling methods, chemical recycling is the most desirable because starting materials can be reproduced. However, industrial examples of chemical recycling are limited, for example, alcoholysis of poly(ethylene terephthalate) and poly(butylene telephthalate) and their processes normally consume much energy. Biodegradable polymers are expected to provide an alternative to traditional non-biodegradable polymers. These polymers are subjected to degradation by living organisms, whereas the degradation products are not directly converted to the original polymers. Thus, a new concept of chemical recycling of polymers using a lipase catalyst was proposed [215, 216]. The principle is simply stated: lipase catalysis of lactone polymerization gives cyclic oligomers including monomer in a dilute solution and gives higher molecular weight polymers in a concentrated solution, which can be degraded to cyclic oligomers again in a dilute solution by the same catalyst; this cycle can be performed repeatedly. The amount of water in the system must always be controlled.

The lipase CA-catalyzed degradation of poly(ε-CL) with molecular weight 4×10^4 was readily induced in toluene at 60 °C to give oligomers with molecular weight <500. After the removal of the solvent from the reaction mixture, the residual oligomer was polymerized using the same catalyst. This approach provides a basic concept that the degradation–polymerization repetition could be controlled by presence or absence of the solvent, suggesting a new methodology of recycling plastics (Scheme 10.41) [216].

The same concept was applied to poly(lactic acid) (PLA), which is one of the most promising green plastics [217]. PLA could be repeatedly chemically recycled with lipase catalysis through polymerizable cyclic oligomers having a molecular weight of a few hundred. Poly(L-LA) having an M_w of 12×10^4 was transformed into cyclic oligomers by lipase CA at 100 °C. Similar results were obtained by the lipase-catalyzed degradation of poly(D,L-LA).

To approach environmental problems, reaction solvents are seriously considered. An organic solvent is often used; in addition, however, water, supercritical carbon dioxide (scCO$_2$) and an ionic liquid are typically regarded as green solvents. Macrolides, such as UDL and DDL, were polymerized by lipase even in water to produce the corresponding polyesters [188].

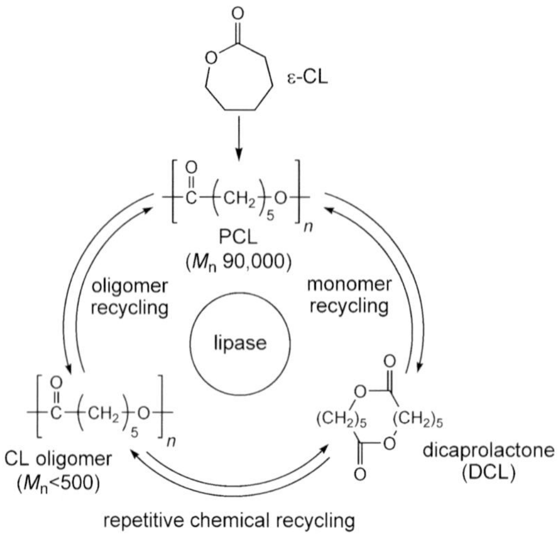

Scheme 10.41

scCO$_2$ is inexpensive, inert, non-toxic and non-flammable and possesses potential for polymer synthesis and recycling. It was used for the first time for the lipase CA-catalyzed polymerization of ε-CL to produce poly(ε-CL) with molecular weight $M_w \approx 1.1 \times 10^4$ in high yields [218] and later the highest molecular weight (M_w) reached 7.4×10^4 [219]. The enzyme and scCO$_2$ were repeatedly used for the polymerization. The lipase-catalyzed degradation of poly(ε-CL) in the presence of acetone produced oligomers with molecular weights of <500 in scCO$_2$ [220, 221]. The ε-CL oligomer produced was again polymerized with lipase CA to yield poly(ε-CL) having $M_n > 8.0 \times 10^4$ [221]. Chemoenzymatic synthesis combining eROP and ATRP was performed in a simultaneous one-pot system using ε-CL and MMA in scCO$_2$ via route A in Scheme 10.40 and the processes of routes B and C also worked well in producing poly(ε-CL-b-MMA) [222]. Semifluorinated block copolymers containing poly(ε-CL) and poly(fluorooctyl methacrylate) were synthesized in scCO$_2$. The first eROP step from a hydroxyalkyl bromide-type bifunctional initiator afforded a poly(ε-CL) macroinitiator and then polymerization of the methacrylate monomer proceeded via ATRP from the macroinitiator, giving rise to a block copolymer with M_n in the range $(13-30) \times 10^3$.

There are few reports of enzymatic ring-opening polymerization in ionic liquids. The lipase-catalyzed polymerization of ε-CL was carried out in ionic solvents, such as 1-butyl-3-methylimidazolium salts [223].

10.6.1.2 Ring-opening Polymerization of Other Cyclic Monomers

A cyclic acid anhydride was subjected to lipase-catalyzed reaction with glycols to produce a polyester, liberating a water molecule, which is a new type of "ring-opening poly(addition–condensation)" (Scheme 10.42) [224]. Enzymatic synthe-

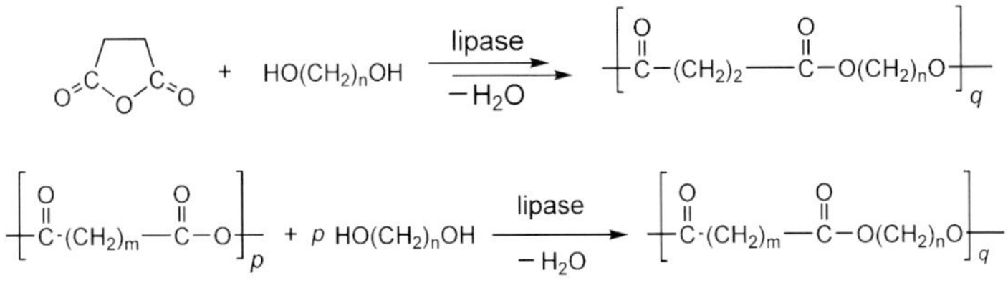

Scheme 10.42

sis of polyesters was also achieved by the reaction of polyanhydrides and glycols. The reaction of poly(azelaic anhydride) and 1,8-octanediol took place with lipase CA catalysis, giving rise to the corresponding polyester with a molecular weight of several thousand [225].

A poly(ester–ether) was synthesized by the ring-opening polymerization of 1,5-dioxepan-2-one, an ε-CL derivative, with lipase CA catalyst, at 60 °C for 4 h with 10 wt.% of the catalyst with respect to the monomer, which gave a polymer of high molecular weight up to $M_n = 5.6 \times 10^4$ and $M_w = 11.2 \times 10^4$ [226]. The polymerization behaved like a living system and the monomer was consumed much faster than ε-CL.

A polycarbonate was prepared via the lipase-catalyzed ring-opening polymerization of a six-membered cyclic carbonate, 1,3-dioxan-2-one (trimethylene carbonate, TMC), reported first in 1997 (Scheme 10.43) [227]. The reactions were clean and no elimination of carbon dioxide by the enzymatic polymerization of TMC was detected, in contrast to the other processes. TMCs with or without a methyl substituent were polymerized in the presence of lipase between 60 and 100 °C to yield the corresponding polycarbonates [228]. Polycarbonates having a pendant carboxyl group or hydroxy group were obtained from homopolymerizations or copolymerizations of substituted TMC monomers [229]. Lipase CA also catalyzed the ring-opening polymerization of macrocyclic dicarbonates to yield the corresponding polycarbonates. The dicarbonates copolymerized with lactones, affording carbonate/ester copolymers [230].

Lipase-catalyzed ring-opening polymerization of morpholine-2,5-dione derivatives (cyclic depsipeptides) was extensively studied to produce poly(ester–amide)s [231]. Four kinds of lipase were used as catalysts for the reaction at 100–130 °C for 72 h. The polymer was obtained with 20–80% conversions, with molecular weights ranging from 3500 to 30 000.

Interestingly, a cyclic phosphate (ethylene isopropyl phosphate) was polymerized via the enzyme-catalyzed ring-opening method (Scheme 10.44) [232]. The

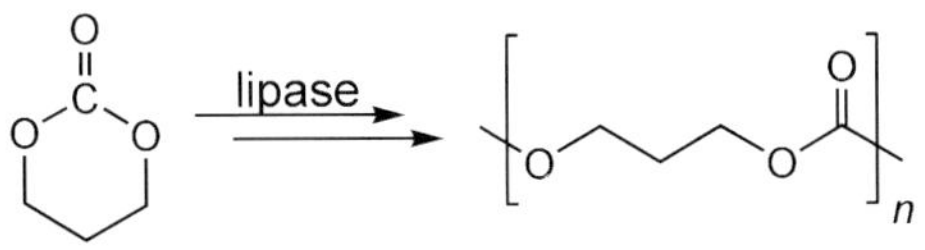

Scheme 10.43

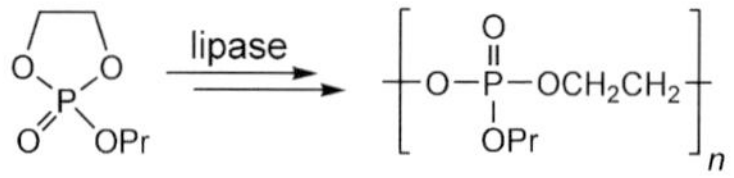

Scheme 10.44

ring-opening polymerization of ethylene isobutyl phosphate was also performed using an immobilized lipase at 70 °C to yield the corresponding polyphosphate with M_n values ranging from 1.6×10^3 to 5.8×10^3. A new enzymatic ring-opening copolymerization of ethyl ethylene phosphate and TMC was performed in bulk at 100 °C using lipase catalyst to yield random copolymers having molecular weights from 3.2×10^3 to 10×10^3.

10.6.2
Polyesters via Polycondensation

Polyester synthesis by lipase catalysis via polycondensation (condensation polymerization) arises because some of the isolated lipases can act as catalysts for the reverse reactions, esterifications and transesterifications (Scheme 10.45) [3, 233]. By utilizing lipase catalysis, functional aliphatic polyesters have been synthesized via mainly two polymerization modes (Scheme 10.31 b and c).

10.6.2.1 Polycondensation of Dicarboxylic Acids and Their Derivatives with Glycols

Lipase catalyzes the dehydration polycondensation using free carboxylic acids, for example, the polycondensation of adipic acid and 1,4-butanediol in diisopropyl ether gave a polyester with a degree of polymerization (DP) of 20 [234]. Higher molecular weight polyesters were obtained enzymatically by the polymerization of sebacic acid and 1,4-butanediol under vacuum [235]. In lipase-catalyzed polymerization in hydrophobic solvents of high boiling-points such as diphenyl ether and veratrole, the molecular weight of polyesters obtained from various combinations of diacids and glycols were $>4 \times 10^4$. An increase in the molecular weight of aromatic polyesters was also observed by polymerization under vacuum [236].

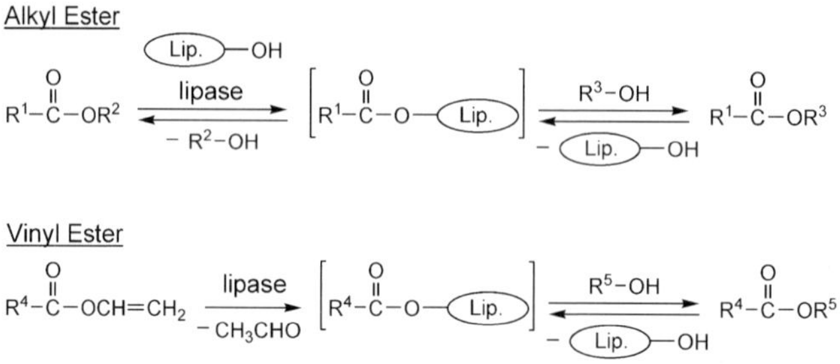

Scheme 10.45

In a solvent-free system reported in 1998, lipase CA efficiently catalyzed the polycondensation of dicarboxylic acids and glycols under mild reaction conditions, despite the heterogeneous mixture of the monomers and catalyst [237]. The methylene chain length of the monomers greatly affected the polymer yield and molecular weight. A polymer with molecular weight $>1\times10^4$ was obtained by reaction under reduced pressure. In the polymerization of adipic acid and 1,6-hexanediol, the loss of the enzymatic activity was small during the polymerization, whereas less than half of the activity remained when using glycols with a methylene chain length <4 [238]. A scale-up experiment produced a polyester from the same combination in >200 kg yield. This solvent-free system claimed high potential as an environmentally friendly synthetic process for polymeric materials owing to the mild reaction conditions without using organic solvents and toxic catalysts.

A dehydration reaction is generally performed in non-aqueous media. Since the product water of the dehydration is in equilibrium with the starting materials, the solvent water disfavors the dehydration to proceed in an aqueous medium due to the law of mass action. Nevertheless, it was found that lipase catalysis provided dehydration polymerization of a dicarboxylic acid and glycol in water [166, 239]. The view of dehydration in an aqueous medium is a new aspect in organic chemistry and the importance of water solvent was discussed above in Section 10.6.1.1 under "Approach to environmental problems: chemical recycling and reaction solvents". Lipase CA and other lipases were active for the polymerization of sebacic acid and 1,8-octanediol. In the polymerization of an a,ω-dicarboxylic acid and glycol, the polymerization behavior depended greatly on the methylene chain length of the monomers. The polymer was obtained in good yields from 1,10-decanediol, whereas no polymer formation was observed from 1,6-hexanediol, suggesting that a combination of monomers with appropriate hydrophobicity is needed for polymer production.

Normally, transesterification of alkyl esters is slow due to the reversible nature of the reaction, as shown in Scheme 10.45. Therefore, it is often difficult to obtain high molecular weight polyesters by lipase-catalyzed polycondensation of dialkyl esters with glycols. Lipase CA-catalyzed polycondensation of dimethyl succinate and 1,6-hexanediol reached equilibrium between the starting materials and polymer. Adsorption of methanol by molecular sieves or elimination of methanol by nitrogen bubbling shifted the thermodynamic equilibrium. Polyesters with molecular weights of several thousand were prepared from a,ω-alkylene dicarboxylic acid dialkyl esters and, whatever the monomer structure, cyclic oligomers were formed [240]. In the polymerization of dimethyl terephthalate and diethylene glycol catalyzed by lipase CA in toluene, unique macrocyclic oligomers were formed [241]. The molecular weight was greatly enhanced under vacuum to remove the alcohols formed, leading the equilibrium towards the product polymer [242]; a polyester with molecular weight of 2×10^4 was obtained by the lipase-catalyzed polymerization of sebacic acid and 1,4-butanediol in diphenyl ether or veratrole under reduced pressure.

Protease was also effective as a catalyst for aromatic polyester synthesis, catalyzing the oligomerization of esters of terephthalic acid with 1,4-butanediol

[243]. Recently, room-temperature ionic liquids have attracted much attention as green designer solvents. It was first found that, interestingly, ionic liquids acted as a good medium for the lipase-catalyzed production of polyesters [244]. The polycondensation of diethyl adipate and 1,4-butanediol using lipase CA as catalyst proceeded efficiently in 1-butyl-3-methylimidazolinium tetrafluoroborate or hexafluorophosphate under reduced pressure. The polymerization of diethyl sebacate and 1,4-butanediol in 1-butyl-3-methylimidazolinium hexafluorophosphate took place even at room temperature in the presence of lipase PC [245].

Transesterifications using lipase catalysts are often very slow owing to the reversible nature of the reactions, as discussed above. To shift the equilibrium to the product polymer, activated esters were used. Examples of activated esters of alcohol components are 2-chloroethanol, 2,2,2-trifluoroethanol and 2,2,2-trichloroethanol, which increase the electrophilicity (reactivity) of the acyl carbonyl and avoid significant alcoholysis of the products by decreasing the nucleophilicity of the leaving alcohols.

Polycondensation of bis(2,2,2-trichloroethyl) glutarate and 1,4-butanediol proceeded with porcine pancreas lipase (PPL) catalyst at room temperature in diethyl ether to produce the polyesters with a molecular weight of 8.2×10^3 [246]. In the reaction of bis(2-chloroethyl) succinate and 1,4-butanediol catalyzed by lipase PF, only oligomeric products were formed [247]. This may be due to the low enzymatic reactivity of the succinate substrate. The lipase-catalyzed synthesis of polyesters was achieved in a supercritical fluid. The polymerization of bis(2,2,2-trichloroethyl) adipate and 1,4-butanediol using PPL catalyst proceeded in a supercritical fluoroform solvent to give the polymer with a molecular weight of several thousand [248]. In the PPL-catalyzed polymerization of bis(2,2,2-trifluoroethyl) glutarate with 1,4-butanediol in 1,2-dimethoxybenzene, the periodic vacuum method increased the molecular weight to nearly 4×10^4 [249].

A perfect irreversible procedure for the lipase-catalyzed acylation using vinyl esters was developed, where a leaving group of vinyl alcohol tautomerizes to acetaldehyde. In this case, the reaction with the vinyl ester proceeds much faster to produce the desired compound in higher yields in comparison with the alkyl esters. The synthesis of polyesters using divinyl esters was reported for the first time in 1994 (see also Section 10.6.1.1 under "Synthesis of functionalized polyesters" for vinyl esters) [250]. In the lipase PF-catalyzed polymerization of divinyl adipate and 1,4-butanediol in diisopropyl ether at 45 °C, a polyester with a molecular weight of 6.7×10^3 was formed, whereas adipic acid and diethyl adipate did not afford polymeric materials under the similar reaction conditions (Scheme 10.45). The reaction mechanism is considered to be similar to that of the ring-opening polymerization of lactones to involve an acyl-enzyme intermediate (enzyme-activated monomer, EM), liberating acetaldehyde. The reaction of EM with the glycol produces a 1:1 adduct of both monomers. The propagation stage starts from the nucleophilic attack of the hydroxy group on the acyl-enzyme intermediate formed from the vinyl ester group of the monomer and the 1:1 adduct and then the propagation steps follow this reaction cycle repeatedly.

Lipase catalysts showed high catalytic activity towards the polymerization of divinyl adipate and divinyl sebacate with α,ω-glycols with different chain lengths [251]. A combination of divinyl adipate and 1,4-butanediol afforded a polymer with a molecular weight of 2.1×10^4. During the lipase-catalyzed polymerization of divinyl esters and glycols, there was competition between the enzymatic transesterification and hydrolysis of the vinyl end-group, resulting in limitation of the polymer growth [252]. A batch-stirred reactor was developed to minimize temperature and mass-transfer effects [253]. Using this reactor, poly(1,4-butylene adipate) with a molecular weight of 2.3×10^4 was obtained within 1 h at 60 °C.

A combinatorial approach was applied for the biocatalytic production of polyesters [254]. A library of polyesters was synthesized in 96 deep-well plates from a combination of divinyl esters and glycols with lipases of different origin. In this screening, lipase CA was the most active biocatalyst for polyester production. As acyl acceptor, 2,2,2-trifluoroethyl esters and vinyl esters were examined and the former produced the polymer of highest molecular weight.

Supercritical carbon dioxide ($scCO_2$) was employed as solvent for the polycondensation of divinyl adipate and 1,4-butanediol [218]. Quantitative consumption of both monomers was achieved to give a polyester with a molecular weight of 3.9×10^3, indicating that $scCO_2$ was a good medium for enzymatic polycondensation.

Aromatic polyesters were efficiently synthesized from aromatic diacid divinyl esters. Lipase CA induced the polymerization of divinyl esters of isophthalic acid, terephthalic acid and p-phenylenediacetic acid with glycols to give polyesters containing an aromatic moiety in the main-chain with the highest molecular weight (7.2×10^3) [255]. Enzymatic polymerization of divinyl esters and aromatic diols also afforded aromatic polyesters [256].

Lipase-catalyzed copolymerization of divinyl esters, glycols and lactones produced ester copolymers with molecular weight $>1\times10^4$ (Scheme 10.46) [257]. Lipases showed high catalytic activity for this copolymerization, involving both polycondensation and ring-opening polymerization simultaneously in a one-pot system to produce ester copolymers. The results also suggest the frequent occurrence of transesterification between the resulting polyesters during the polymerization [258].

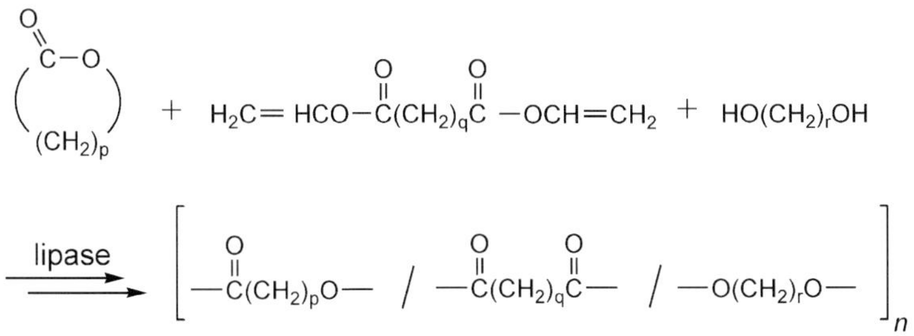

Scheme 10.46

10.6.2.2 Polycondensation of Oxyacid Derivatives

An early paper in 1985 reported a lipase-catalyzed polymerization of an oxyacid monomer, 10-hydroxydecanoic acid, in benzene using poly(ethylene glycol) (PEG)-modified lipase soluble in the medium [259]. The DP value of the product was >5. PEG-modified esterase induced the oligomerization of glycolic acid, the shortest oxyacid. The lipase-catalyzed polymerization of lactic acid gave only a low molecular weight polymer even under a variety of the reaction conditions. The addition of a small amount of dicarboxylic acid or cyclic anhydrides enhanced the molecular weight (Scheme 10.47) [260].

The polymerization of ricinoleic acid proceeded using lipases as catalyst at 35 °C in water, hydrocarbons or benzene to give a polymer with a molecular weight around 1×10^3 [261]. These lipases also induced the polymerization of 12-hydroxyoctadecanoic acid, 16-hydroxyhexadecanoic acid and 12-hydroxydodecanoic acid. In the lipase CA-catalyzed oligomerization of cholic acid, a hydroxy group at the 3-position was regioselectively acylated to give the oligoester with a molecular weight of 920 [262].

Polyesters of relatively high molecular weight were produced enzymatically from 10-hydroxydecanoic acid and 11-hydroxyundecanoic acid using a large amount of lipase CR catalyst (10 weight-fold with respect to the monomer). In the case of 11-hydroxyundecanoic acid, the corresponding polymer with a molecular weight of 2.2×10^4 was obtained in the presence of activated molecular sieves [263]. Lipase CA also polymerized hydrophobic oxyacids efficiently [264]. The DP value was >100 in the polymerization of l6-hydroxyhexadecanoic acid, 12-hydroxydodecanoic acid or 10-hydroxydecanoic acid under vacuum at a high temperature (90 °C) for 24 h, whereas a polyester with a lower molecular weight was formed from 6-hydroxyhexanoic acid under similar reaction conditions. This difference is probably due to the different strengths of the lipase–substrate interaction as shown in the lipase-catalyzed ring-opening polymerization of lactones of different ring size [167, 265].

Lipase-catalyzed polymerization of methyl 6-hydroxyhexanoate, an oxyacid ester, was reported [164]. A polymer with DP up to 100 was synthesized by reaction in hexane at 70 °C for more than 50 days. The PPL-catalyzed polymerization of methyl 5-hydroxypentanoate for 60 days produced a polymer with a DP of 29. Various hydroxy esters, ethyl esters of 3- and 4-hydroxybutyric acids, 5- and 6-hydroxyhexanoic acids, 5-hydroxidodecanoic acid and 15-hydroxypentadecanoic acid, were polymerized by *Pseudomonas* sp. lipase at 45 °C to give the corresponding polyesters with molecular weights of several thousand [266].

An ester–thioester copolymer was synthesized enzymatically [267]. The lipase CA-catalyzed copolymerization of ε-CL with 11-mercaptoundecanoic acid or 3-

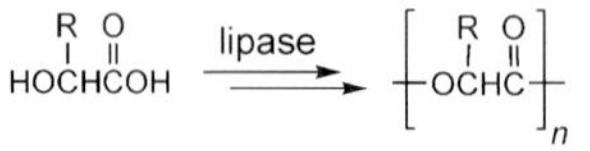

$$\text{HOCHCOH} \xrightarrow{\text{lipase}} \text{[OCHC]}_n$$
(with R and O substituents)

Scheme 10.47

mercaptopropionic acid under reduced pressure produced a polymer with a molecular weight $>2\times10^4$. Transesterification between poly(ε-CL) and 11-mercaptoundecanoic acid or 3-mercaptopropionic acid also took place.

10.6.2.3 Synthesis of Functional Polyesters

Lipase-catalyzed enantioselective polymerization was first reported in 1989; the reaction of bis(2,2,2-trichloroethyl) *trans*-3,4-epoxyadipate with 1,4-butanediol in diethyl ether gave a highly optically active polyester [268]. The molar ratio of the diester to the diol was adjusted to 2:1 to produce the (–)-polymer with enantiomeric purity >96%. An optically active oligoester was prepared enantioselectively from racemic 10-hydroxyundecanoic acid using lipase CR catalyst. The resulting oligomer was enriched in the *S*-enantiomer to a level of 60% *ee* and the residual monomer was recovered with 33% *ee* favoring the antipode [269]. Optically active oligomers (DP<6) were also synthesized from racemic ε-substituted ε-hydroxy esters using PPL catalyst [270]. The enantioselectivity increased as a function of bulkiness of the monomer substituent.

Polyesters having a sugar moiety in the main-chain were synthesized via protease catalysis. In the polycondensation of sucrose with bis(2,2,2-trifluoroethyl) adipate catalyzed by an alkaline protease showing esterase activity, the regioselective acylation of sucrose at the 6- and 1'-positions was claimed to yield a sucrose-containing polyester [271]. The reaction proceeded slowly; the molecular weight reached $>1\times10^4$ after 7 days. A two-step synthesis of sugar-containing polyesters with lipase CA catalyst was reported [272]. Lipase CA catalyzed the condensation of sucrose with an excess of divinyl adipate to produce sucrose 6,6'-*O*-divinyl adipate, which was reacted with a,ω-alkylene diols with the same catalyst, yielding polyesters containing a sucrose unit in the main-chain. This method conveniently afforded sugar-containing polyesters with relatively high molecular weight. Similarly, a trehalose-containing polyester was obtained from trehalose 6,6'-*O*-divinyl adipate with lipase CA catalysis.

Lipase CA catalysis produced reduced sugar-containing polyesters regioselectively from divinyl sebacate and sorbitol, in which sorbitol was exclusively acylated at the primary alcohol at the 1- and 6-positions (Scheme 10.48) [273]. Mannitol and *meso*-erythritol were also regioselectively polymerized with divinyl sebacate. The enzymatic formation of a high molecular weight sorbitol-containing polyester was confirmed by the combinatorial approach [272].

The lipase CA-catalyzed polycondensation of adipic acid and sorbitol also took place in bulk [274]. In polymerization at 90 °C, the molecular weight reached 1×10^4; however, the regioselectivity decreased (85%), probably due to the high temperature and/or the bulk conditions. These data suggest that the divinyl ester is a suitable monomer for regioselective synthesis. The copolymerization of adipic acid, sorbitol and 1,8-octanediol enhanced the molecular weight of sugar-containing polyesters. The melting and crystallization temperatures and the melting enthalpy decreased with increasing sorbitol content, due to the polyol units along the polyester chain that disrupt crystallinity [275].

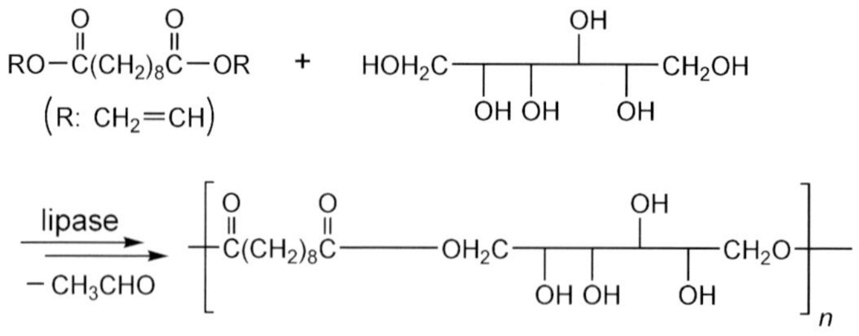

460 10 Polymer Synthesis and Modification by Enzymatic Catalysis

Scheme 10.48

In the following the synthesis of some examples of reactive polyesters is described. Polyols such as glycerol were regioselectively polymerized at a primary hydroxy group with divinyl esters using a lipase catalyst to produce polyesters having a secondary hydroxy group in the main-chain [276]. NMR analysis of the polymer obtained from divinyl sebacate and glycerol using lipase CA catalyst at 60 °C in bulk showed that 1,3-diglyceride is the main unit and a small amount of a branching unit (triglyceride) is present. The regioselectivity of the acylation between primary and secondary hydroxy groups was 74:26. In polymerization at 45 °C, the regioselectivity was perfectly controlled to give a linear polymer consisting exclusively of 1,3-glyceride units [277].

A new cross-linkable polyester was obtained by the lipase-catalyzed polymerization of divinyl sebacate and glycerol in the presence of unsaturated higher fatty acids derived from renewable plant oils (Scheme 10.49) [278]. Polymerization under reduced pressure improved the polymer yield and molecular weight. The curing of the polymer obtained using linoleic or linolenic acid proceeded with a cobalt naphthenate catalyst or thermal treatment to give a cross-linked transparent film. Biodegradability of the film obtained was verified by measurement of the biochemical oxygen demand (BOD).

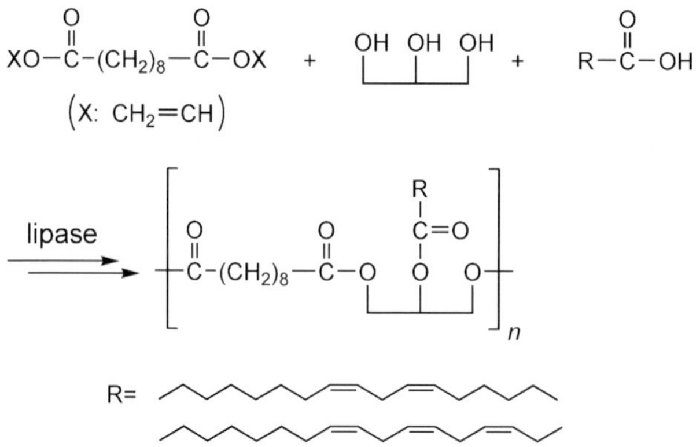

Scheme 10.49

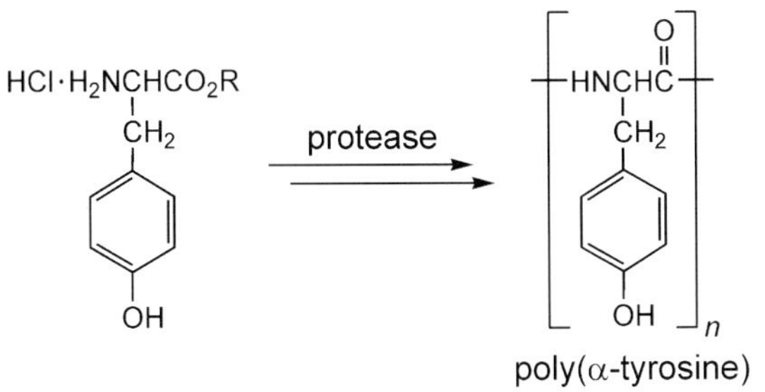

Scheme 10.50

Long-chain unsaturated α,ω-dicarboxylic acid methyl esters and their epoxidized derivatives were polymerized with 1,3-propanediol or 1,4-butanediol in the presence of lipase CA to produce reactive polyesters [279]. All the resulting polymers possessed melting-points in the range 23–75 °C. Epoxide-containing polyesters were synthesized enzymatically via two routes using unsaturated fatty acids as the starting substrate [280]. Curing of the resulting polymers proceeded thermally, yielding transparent polymeric films with a high gloss surface. Pencil scratch hardness of the film was improved compared with that of the cured film obtained from the polyester having an unsaturated fatty acid in the side-chain. The film obtained showed good biodegradability in an activated sludge.

Chemoenzymatic synthesis of alkyds (oil-based polyester resins) was demonstrated. PPL-catalyzed transesterification of triglycerides with an excess of 1,4-cyclohexanedimethanol mainly produced 2-monoglycerides, followed by thermal polymerization with phthalic anhydride to give the alkyd resins with molecular weights of several thousands [281].

Structural control of the polymer terminus has been extensively studied since terminal-functionalized polymers, typically macromonomers and telechelics, are often used as prepolymers for the synthesis of functional polymers. The enzymatic polymerization of 12-hydroxydodecanoic acid in the presence of 11-methacryloylaminoundecanoic acid conveniently produced the methacrylamide-type polyester macromonomer [282].

As reference, peptide synthesis is cited; L-tyrosine ester and L-glutamic acid ester were used for protease-catalyzed polymerization and copolymerization to give corresponding poly(amino acid)s with an α-peptide linkage having molecular weights of several thousands (Scheme 10.50) [283].

10.7
Modification of Polymers by Enzymatic Catalysis

Isolated enzymes show effective catalytic activity for many reactions of small molecular substrates and can often be used also for those of polymer substrates. These reactions permit the modification of polymers in various ways. The main reason for this is to improve the polymer properties in order to add value for specific applications. Many examples of modifications have involved water-soluble polymers, such as polysaccharides and proteins. Modification reactions include esterification, oxidation, reduction, acylation, amidation, glycosylation and polymerization. Polymer properties such as solubility, (bio)compatibility, gel (formation) properties, surface properties, barrier properties, thermal stability, interfacial properties, cross-linking ability and other chemical and physical properties can be modified. The modified polymers can be used for various value-added purposes such as gel materials and drug delivery systems. Furthermore, enzymatic modifications often have advantages over chemical processes by being environmentally friendly, particularly non-toxic, can take place under mild reaction conditions and have the potential for targeting reactive groups of molecules that react chemo-, regio- or stereoselectively.

10.7.1
Modification of Polysaccharides

Cellulose is the most abundant renewable resource and is inexpensive. Cellulose acetate is the most important organic acid ester of cellulose and is produced commercially by a heterogeneous solution process, in which solid cellulose is reacted with acetic anhydride with sulfuric acid catalysis in acetic acid solution. Early enzymatic modification studies on polysaccharides are represented by the lipase-catalyzed acylation of cellulose acetate with various fatty acids in an organic media [284]. Immobilized lipase CA catalyzed the acylation of cellulose acetate [degree of substitution (DS) = 2.4] with lauric and oleic acids in acetonitrile. The final conversion of both fatty acids was about 35% after 96 h of incubation at 50 °C, the amounts of cellulose acetate, fatty acid and enzyme all being 150 mg in 5 mL of acetonitrile. The total ester bonds were increased by 2% after the esterification, suggesting that the acylation took place either in the free hydroxy groups of the cellulose acetate by direct esterification or in the acetylated hydroxy groups by transesterification (Fig. 10.22).

Acylation of cellulose was carried out in homogeneous solution processes, for example, in DMSO–paraformaldehyde or in LiCl–dimethylacetamide, or under heterogeneous conditions [285]. Carboxymethylcellulose was acetylated using a lipase catalyst with vinyl acetate in an aqueous homogeneous solution. Lipase A12 was found to be the most active among 12 lipases tested; the acetyl group content in the product was 1.48 wt% after 24 h at 40 °C [286]. A similar reaction was conducted on a solid cellulose suspension; however, the acetyl content was very low, ~0.16 wt% after 48 h. Heterogeneous enzymatic esterification of solid cellulose generally gave poor results [287]. Also, lipase-catalyzed modification of hydroxyethylcellulose (HEC) in film or powder form afforded HEC-g-poly(ε-CL) with a low DS [288].

Starch (amylose) is biocompatible, biodegradable and non-toxic, hence it is often used for biomedical purposes including drug carriers. Depending on the mode of usage, regioselective modification is required. Starch is soluble in water

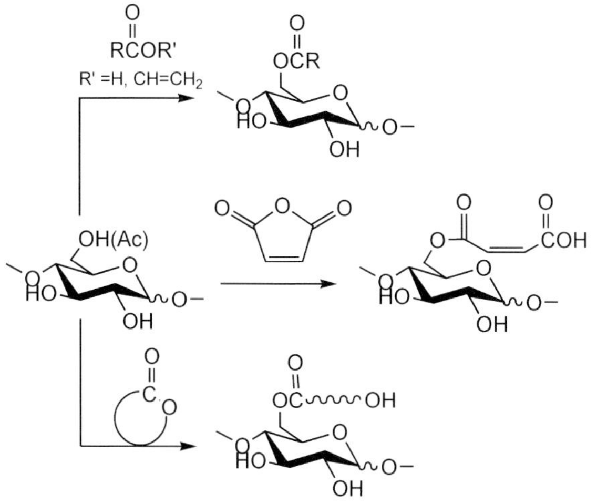

Fig. 10.22 Lipase-catalyzed acylations of cellulose, starch and their derivatives.

and in polar aprotic solvents, where regioselective acylation is very difficult owing to problems such as the low catalytic activity of lipase. A new system was developed for selective acylation [289]. Starch nanoparticles were incorporated into reverse micelles stabilized by AOT [sodium bis(2-ethylhexyl)sulfosuccinate]. AOT-coated starch particles and their solubility in non-polar solvents such as isooctane allowed their diffusion to sites within the pores of the physically immobilized lipase CA catalyst at which esterification of starch took place. Acylating agents were vinyl esters of fatty acids differing the chain length, maleic anhydride and ε-CL. These acylations occurred regioselectively at the 6-OH group; motivations for these modifications were (i) increasing starch hydrophobicity, (ii) introducing both a carboxylate side-chain and sites for radical cross-linking and (iii) initiating polyester graft polymerization (Fig. 10.22). In the first case, up to 30% of vinyl stearate was involved in the acylation of starch, giving a DS of 0.8 (40 °C for 48 h). Infrared microspectroscopic analysis showed that AOT-coated starch nanoparticles can diffuse into the outer 50 μm shell of catalyst beads, suggesting that catalyst lipase, starch and the acylating agent are all in close proximity on the bead surface, promoting esterification.

Chitin and chitosan are polysaccharides from animals. They are often used for biomedical applications due to their good biocompatibility, bioabsorbability and non-toxic nature. Chitosan was cross-linked with tyrosinase catalysis. Tyrosinases are ubiquitous oxidative enzymes capable of converting low molecular weight phenols or accessible tyrosyl residues of proteins such as gelatin into quinones. These quinones are reactive and can undergo non-enzymatic reactions with a variety of nucleophiles. Tyrosinases are responsible for reactions leading to melanin production and the enzymatic browning of foods. The primary amino groups of chitosan react with the quinone to give a gelatin-grafted polymer or cross-linked gels under mild reaction conditions (Scheme 10.51)

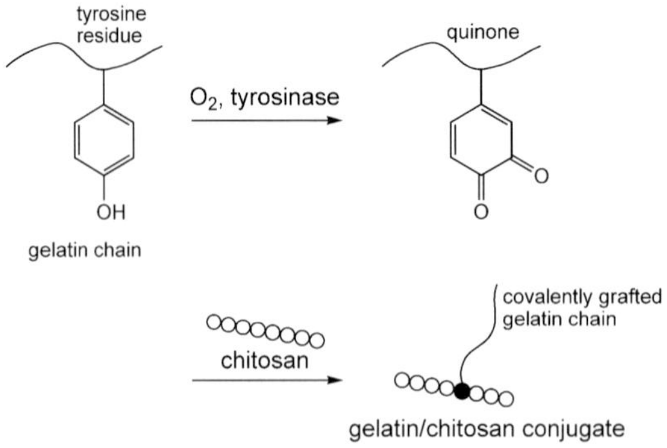

Scheme 10.51

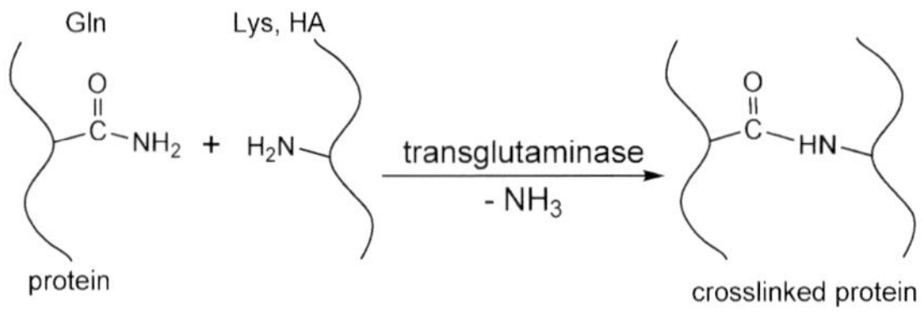

Scheme 10.52

[290]. The chemistry and physics of tyrosinase-catalyzed chitosan–gelatin gels are still uncertain.

An enzymatic modification of chitosan was applied for the synthesis of a chitosan–catechin conjugate. The formation of a Michael-type adduct and/or Schiff base was proposed during the polyphenol oxidase-catalyzed conjugation of catechin with chitosan [291]. Rheological measurements demonstrated that the resulting conjugate behaves as an associative thickener.

Transglutaminase (TGase, amine γ-glutaminyl transferase) is known to catalyze transamidation reactions between glutamine residues and lysine residues that lead to the formation of an isopeptide side-chain bridge (Scheme 10.52) [290–293]. TGase was employed as the catalyst of gel formation from gelatin, from a gelatin–hyaluronan (HA) mixture and from an HA derivative bearing lysinyl substituents [292]. The presence of HA in a gelatin gel makes the network more hydrophilic and bioactive due to the ability of HA to promote cell regeneration. Such mixed gels are expected to provide good materials to replace fleshy tissues.

10.7.2
Modification of Other Polymers

A protein, typically gelatin, has been widely employed as a starting material for various gels or conjugates [290–294]. The gel materials from TGase catalyzed-cross-linking of gelatin were characterized in terms of rheologically estimated times, equilibrium swelling in water and in phosphate buffer and rigidity modulus (Scheme 10.53) [292]. The rate of TGase-catalyzed cross-linking reactions was examined for several oligopeptide substrates including PEG chain-containing substrates that serve for tissue engineering and surgical adhesive applications [293]. The specificities of lysine peptides varied by several orders of magnitude, whereas those of glutamine peptides varied only modestly. By selecting the substrate sequence structure, 3,4-dihydroxyphenylalanine (DOPA), which confers adhesive strength between the protein and the surface, was incorporated into synthetic polymer gels for medical applications. Besides lysine residue as acyl acceptors, primary alkylamines can also be used as substrates, which allows the selective alkylation of proteins via their accessible glutamine residues [295].

A gelatin–catechin conjugate was prepared by the laccase-catalyzed oxidation of catechin in the presence of gelatin [294]. The conjugate showed good scavenging activity against superoxide anion radicals and an amplified inhibition effect on human low-density lipoprotein (LDL) oxidation. Poly(ε-lysine) is a biopolymer produced from culture filtrates of *Streptomyces albulus* that shows good antimicrobial activity against Gram-positive and -negative bacteria and therefore is widely used as an additive in the food industry. A new inhibitor against disease-related enzymes, collagenase, hyaluronidase and xanthine oxidase, was developed by the conjugation of catechin on poly(ε-lysine) by using laccase as catalyst [296].

The conjugation of catechin was conducted on a synthetic polymer, polyallylamine, using laccase as catalyst under air [297]. During the conjugation, the reaction mixture turned brown and a new peak at 430 nm was observed in the UV–visible spectrum. The reaction rate was maximum at pH 7. Conjugation hardly occurred in the absence of laccase, indicating that the reaction proceeded via enzyme catalysis. The resulting polyallylamine–catechin conjugate showed more lasting antioxidant activity against LDL peroxidation induced by 2,2′-azobis(2-amidinopropane) dihydrochloride (AAPH), compared with unconjugated catechin.

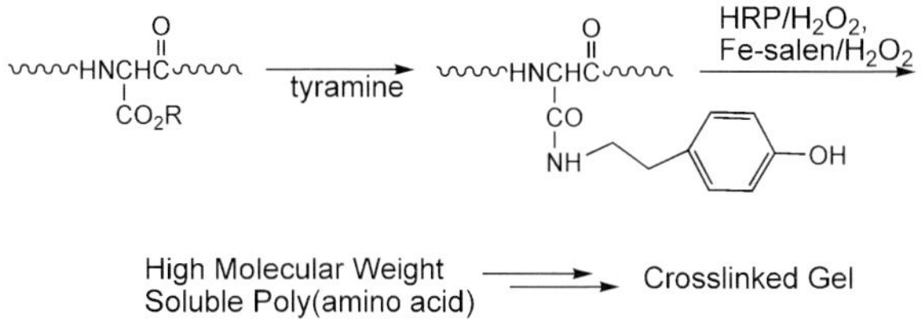

Scheme 10.53

Catechin-immobilizing polymer particles were prepared by laccase-catalyzed oxidation of catechin in the presence of amine-containing porous polymer particles [298]. The resulting particles showed good scavenging activity towards stable free 1,1-diphenyl-2-picrylhydrazyl radical and 2,2′-azinobis(3-ethylbenzothiazoline-6-sulfonate) radical cation. These particles may be applied in packed column systems to remove radical species such as reactive oxygen closely related to various diseases.

An intermolecular oxidative coupling of a phenol moiety of a poly(amino acid) was examined [299]. Fe-salen-catalyzed oxidative coupling of poly(a-tyrosine) did not give a higher molecular weight coupling product, probably due to steric factors of the phenol moiety. Tyramine was introduced to polyglutamine and polyasparagine as a phenol moiety having a spacer (Scheme 10.53). The tyramine-containing poly(amino acid)s underwent HRP- or Fe-salen-catalyzed oxidative coupling to produce a high molecular weight poly(amino acid); for example, polyglutamine having a tyramine moiety (50 unit % with respect to glutamine units) showed a weight-average molecular weight of 2.7×10^6 and was soluble in DMF. In the Mark–Houwink–Sakurada equation ($[\eta] = KM^a$), the a value of the polymer was 0.5, which is lower than that of a linear polymer (0.6–0.8), suggesting a branched structure of the poly(amino acid) with an ultra-high molecular weight. Further prolonged reactions resulted in gel formation. Also, an ultra-high molecular weight poly(m-cresol) was obtained by the Fe-salen-catalyzed oxidative coupling of poly(m-cresol) starting polymer. The molecular weight of the product polymer reached 6×10^6 without formation of insoluble gel [300].

10.8
Conclusion

At the beginning of the 1990s, research on the enzyme-catalyzed synthesis of polymers actually took off and has been developed extensively since then. The main reasons for this increasing research activity may stem from two aspects: (i) catalysis specificities: the enzymatic reactions take place under mild conditions and are excellent in controlling regio-, chemo-, stereo- and choroselectivities; and (ii) environmental problems: a trend to shift from petroleum-based resources to bio-based renewable resources is much in demand because of the increase in atmospheric carbon dioxide causing global warming and of the rising price of crude petroleum. Further, a variety of enzymes have recently become more readily available at a lower cost than before and enzymes are non-toxic. Over the years, enzymatic reactions have attracted the attention from many scientists in the fields of biochemistry, organic chemistry, medicinal chemistry, polymer chemistry, material sciences and, of course, enzymology. Their reaction mechanisms are still one of central important problems in these fields.

Here, the enzymatic approach and conventional chemical approach are compared in terms of advantages and disadvantages as an aid in designing production processes in the future.

- *Advantages of enzymatic method:*
  - Perfect control of regioselectivity, chemoselectivity and stereochemistry.
  - No need for protection and deprotection.
  - Many reactions: clean (small amounts of by-products), mild conditions (energy-saving, environmentally desirable).
  - Catalyst enzymes: non-toxic.
- *Disadvantages of enzymatic method:*
  - Limitation of substrates.
  - Availability and cost of enzymes.
  - Reactions: often slow despite high turnover.
- *Advantages of chemical method:*
  - Variety of substrates.
  - Wide selection of reactions and catalysts.
  - Reactions: normally fast.
- *Disadvantages of chemical method:*
  - Difficulty sometimes in control of regioselectivity, chemoselectivity and stereochemistry.
  - Tedious procedures for protection and deprotection depending on the reaction.
  - Reactions: often harsh conditions; high temperature or pressure (environmentally undesirable).
  - Catalysts: often using toxic metals or compounds, too strong acids or bases (environmentally undesirable).

In considering the above properties of enzymes as catalysts, some polymer production processes are expected to be commercialized in the future with enzymatic catalysis alone and/or chemo-enzymatic processes ("green polymer chemistry") to contribute to the chemical industry for a sustainable society.

References

1 (a) E. A. Sym, *Biochem. Z.* **1933**, *258*, 304–324; (b) E. A. Sym, *Enzymologia* **1936**, *1*, 156–160.
2 (a) A. Zaks, A. M. Klibanov, *Science*, **1984**, *224*, 1249–1251; (b) A. Zaks, A. M. Klibanov, *Proc. Natl. Acad. Sci. USA* **1985**, *82*, 3192–3196.
3 (a) G. M. Whitesides, C.-H. Wong, *Angew. Chem. Int. Ed. Engl.* **1985**, *24*, 617–638; (b) J. B. Jones, *Tetrahedron* **1986**, *42*, 3351–3403; (c) C.-H. Wong, G. M. Whitesides, *Enzymes in Synthetic Organic Chemistry*. Pergamon Press, Oxford, **1994**.
4 R. Noyori, *Asymmetric Catalysis in Organic Synthesis*. Wiley, New York, **1994**.
5 S. Kobayashi, S. Shoda, H. Uyama, *Adv. Polym. Sci.* **1995**, *121*, 1–30.
6 S. Kobayashi, S. Shoda, H. Uyama, Enzymatic polymerization, in *The Polymeric Materials Encyclopedia*, J. C. Salamone, ed. CRC Press, Boca Raton, FL, **1996**, 2102–2107.
7 M. Yalpani, ed., *Biomedical Functions and Biotechnology of Natural and Artificial Polymers*. ATL Press, Mount Prospect, IL, **1996**.
8 S. Kobayashi, S. Shoda, H. Uyama, Enzymatic catalysis, in *Catalysis in*

Precision Polymerization, S. Kobayashi, ed. Wiley, Chichester, **1997**, 417–441.
9 H. Ritter, in *Desk Reference of Functional Polymers, Synthesis and Applications*, R. Arshady, ed. American Chemical Society, Washington, DC, **1997**, 103–113.
10 R. A. Gross, D. L. Kaplan, G. Swift, eds., *Enzymes in Polymer Synthesis*. ACS Symposium Series, Vol. 684. American Chemical Society, Washington, DC, **1998**.
11 S. Kobayashi, H. Uyama, Biocatalytical routes to polymers, in *Material Science and Technology Series, Synthesis of Polymers*, A. D. Schlueter, ed. Vol. 54. Wiley-VCH, Weinheim, **1999**, 549–569.
12 S. Kobayashi, *J. Polym. Sci. Part A: Polym. Chem.* **1999**, *37*, 3041–3056.
13 S. Kobayashi, H. Uyama, M. Ohmae, *Bull. Chem. Soc. Jpn.* **2001**, *74*, 613–635.
14 S. Kobayashi, H. Uyama, S. Kimura, *Chem. Rev.* **2001**, *101*, 3793–3818.
15 R. A. Gross, A. Kumar, B. Karla, *Chem. Rev.* **2001**, *101*, 2097–2124.
16 S. Kobayashi, J. Sakamoto, S. Kimura, *Prog. Polym. Sci.* **2001**, *26*, 1525–1560.
17 S. Kobayashi, H. Uyama, Enzymatic polymerization, in *Encyclopedia of Polymer Science and Technology*, 3rd edn., J. I. Kroschwitz, ed. Wiley, New York, **2002**, on the website.
18 S. Kobayashi, H. Uyama, *Macromol. Chem. Phys.* **2003**, *204*, 235–256.
19 R. A. Gross, H. N. Cheng, eds. *Biocatalysis in Polymer Science*. ASC Symposium Series, Vol. 840. American Chemical Society, Washington, DC, **2003**.
20 S. Kobayashi, *J. Polym. Sci. Part A: Polym. Chem.* **2005**, *43*, 693–710.
21 I. K. Varma, A.-C. Albertsson, R. Rajkhova, R. K. Srivastava, *Prog. Polym. Sci.* **2005**, *30*, 949–981.
22 H. N. Cheng, R. A. Gross, *Polymer Biocatalysis and Biomaterials*, ASC Symposium Series, Vol. 900. American Chemical Society, Washington, DC, **2005**.
23 S. Kobayashi, H. Ritter, D. Kaplan, eds. Enzyme-catalyzed Synthesis of Polymers. Special Issue. *Adv. Polym. Sci.* **2006**, *194*.
24 E. Fischer, *Chem. Ber.* **1894**, *27*, 2985–2993.
25 (a) L. Pauling, *Chem. Eng. News* **1946**, *24*, 1375–1377; (b) P. A. Kollman, B. Kuhn, O. Donini, M. Perakyla, R. Santon, D. Bakowies, *Acc. Chem. Res.* **2001**, *34*, 72–79.
26 S. Borman, *Chem. Eng. News* **2004**, February 23, 35.
27 (a) M. Reihmann, H. Ritter, *Adv. Polym. Sci.* **2006**, *194*, 1–49; (b) H. Uyama, S. Kobayashi, *Adv. Polym. Sci.* **2006**, *194*, 51–67.
28 H. Uyama, S. Kobayashi, *CHEMTECH* **1999**, *29*, 22–28.
29 A. S. Hay, *J. Polym. Sci. Part A: Polym. Chem.* **1998**, *36*, 505–517.
30 (a) J. S. Dordick, M. A. Marletta, A. M. Klibanov, *Biotechnol. Bioeng.* **1987**, *30*, 31–36; (b) H. Uyama, H. Kurioka, I. Kaneko, S. Kobayashi, *Chem. Lett.* **1994**, 423–426; (c) H. Uyama, K. Kurioka, J. Sugihara, S. Kobayashi, *Bull. Chem. Soc. Jpn.* **1996**, *69*, 189–193; (d) T. Oguchi, S. Tawaki, H. Uyama, S. Kobayashi, *Macromol. Rapid Commun.* **1999**, *20*, 401–403; (e) T. Oguchi, S. Tawaki, H. Uyama, S. Kobayashi, *Bull. Chem. Soc. Jpn.* **2000**, *73*, 1389–1396.
31 N. Mita, S. Tawaki, H. Uyama, S. Kobayashi, *Polym. J.* **2001**, *33*, 374–376.
32 (a) H. Uyama, H. Kurioka, S. Kobayashi, *Chem. Lett.* **1995**, 795–796; (b) H. Kurioka, H. Uyama, S. Kobayashi, *Polym. J.* **1998**, *30*, 526–529.
33 (a) Y. J. Kim, H. Uyama, S. Kobayashi, *Macromolecules* **2003**, *36*, 5058–5060; (b) Y. J. Kim, H. Uyama, S. Kobayashi, *Macromol. Biosci.* **2004**, *4*, 497–502.
34 H. Uyama, H. Kurioka, S. Kobayashi, *Polym. J.* **1997**, *29*, 190–192.
35 (a) H. Kurioka, I. Komatsu, H. Uyama, S. Kobayashi, *Macromol. Rapid Commun.* **1994**, *15*, 507–510; (b) H. Uyama, H. Kurioka, J. Sugihara, I. Komatsu, S. Kobayashi, *J. Polym. Sci. Polym. Chem. Ed.* **1997**, *35*, 1453–1459.
36 H. Uyama, H. Kurioka, J. Sugihara, I. Komatsu, S. Kobayashi, *Bull. Chem. Soc. Jpn.* **1995**, *68*, 3209–3214.
37 M. S. Ayyagari, K. A. Marx, S. K. Tripathy, J. A. Akkara, D. L. Kaplan, *Macromolecules* **1996**, *28*, 5192–5197.
38 (a) S. K. Sahoo, W. Liu, A. L. Samuelson, L. A. Kumar, A. L. Cholli, *Macromolecules* **2002**, *35*, 9990–9998; (b) X. Wu, W. Liu,

R. Nagarayan, J. Kumar, L. A. Samuelson, A. L. Cholli, *Macromolecules* **2004**, *37*, 2322–2324.

39 P. Wang, D. Martin, S. Parida, D. G. Rethwisch, J. S. Dordick, *J. Am. Chem. Soc.* **1995**, *117*, 12885–12886.

40 H. Tonami, H. Uyama, S. Kobayashi, K. Rettig, H. Ritter, *Macromol. Chem. Phys.* **1999**, *200*, 1998–2002

41 T. Fukuoka, Y. Tachibana, H. Tonami, H. Uyama, S. Kobayashi, *Biomacromolecules* **2002**, *3*, 768–774.

42 H. Uyama, C. Lohavisavapanich, R. Ikeda, S. Kobayashi, *Macromolecules* **1998**, *31*, 554–556.

43 (a) M. H. Reihmann, H. Ritter, *Macromol. Chem. Phys.* **2000**, *201*, 798–804; (b) M. H. Reihmann, H. Ritter, *Macromol. Chem. Phys.* **2000**, *201*, 1593–1597.

44 Y. Pang, H. Ritter, M. Tabatabai, *Macromolecules* **2003**, *36*, 7090–7093.

45 N. S. Kommareddi, M. Tata, C. Karayigitoglu, V. T. John, G. L. McPherson, M. F. Herman, C. J. O'Connor, Y. S. Lee, J. A. Akkara, D. L. Kaplan, *Biochem. Biotechnol.* **1995**, *51/52*, 241–252.

46 P. Xu, J. Kumar, L. Samuelson, A. L. Cholli, *Biomacromolecules* **2002**, *3*, 889–893.

47 S. Kobayashi, H. Uyama, T. Ushiwata, T. Uchiyama, J. Sugihara, H. Kurioka, *Macromol. Chem. Phys.* **1998**, *199*, 777–782.

48 T. Fukuoka, H. Tonami, N. Maruichi, H. Uyama, S. Kobayashi, H. Higashimura, *Macromolecules* **2000**, *33*, 9152–9155.

49 (a) H. Tonami, H. Uyama, S. Kobayashi, M. Kubota, *Macromol. Chem. Phys.* **1999**, *200*, 2365–2371; (b) J. Kadota, T. Fukuoka, H. Uyama, K. Hasegawa, S. Kobayashi, *Macromol. Rapid Commun.* **2004**, *25*, 441–444.

50 H. Tonami, H. Uyama, S. Kobayashi, T. Fujita, Y. Taguchi, K. Osada, *Biomacromolecules* **2000**, *1*, 149–151.

51 (a) R. Ikeda, H. Tanaka, H. Uyama, S. Kobayashi, *Macromol. Rapid Commun.* **2000**, *21*, 496–499; (b) R. Ikeda, H. Tanaka, H. Uyama, S. Kobayashi, *Polym. J.* **2000**, *32*, 589–593; (c) R. Ikeda, T. Tsujimoto, H. Tanaka, H. Oyabu, H. Uyama, S. Kobayashi, *Proc. Jpn. Acad. Ser. B*, **2000**, *76*, 155–160; (d) R. Ikeda, H. Tanaka, H. Uyama, S. Kobayashi, *Polymer* **2002**, *43*, 3475–3481.

52 S. Kobayashi, H. Uyama, R. Ikeda, *Chem. Eur. J.* **2001**, *7*, 4755–4760.

53 (a) T. Tsujimoto, R. Ikeda, H. Uyama, S. Kobayashi, *Macromol. Chem. Phys.* **2001**, *202*, 3420–3425; (b)T. Tsujimoto, H. Uyama, S. Kobayashi, *Macromolecules* **2004**, *37*, 1777–1782.

54 R. Ikeda, J. Sugihara, H. Uyama, S. Kobayashi, *Polym. Int.* **1998**, *47*, 295–301.

55 R. Ikeda, H. Uyama, S. Kobayashi, *Polym. Bull.* **1997**, *38*, 273–277.

56 R. S. Premachandran, S. Banerjee, X.-K Wu, V. T. John, G. L. McPherson, J. Akkara, M. Ayyagari, D. Kaplan, *Macromolecules*, **1996**, *29*, 6452–6460.

57 K. S. Alva, L. Samuelson, J. Kumar, S. Tripathy, A. L. Cholli, *J. Appl. Polym. Sci.* **1998**, *70*, 1257–1264.

58 (a) H. Higashimura, K. Fujisawa, Y. Moro-oka, M. Kubota, A. Shiga, A. Terahara, H. Uyama, S. Kobayashi, *J. Am. Chem. Soc.* **1998**, *120*, 8529–8530; (b) H. Higashimura, M. Kubota, A. Shiga, K. Fujisawa, Y. Moro-oka, H. Uyama, S. Kobayashi, *Macromolecules* **2000**, *33*, 1986–1995; (c) H. Higashimura, K. Fujisawa, S. Namekawa, M. Kubota, A. Shiga, Y. Moro-oka, H. Uyama, S. Kobayashi, *J. Polym. Sci., Polym. Chem. Ed.* **2000**, *38*, 4792–4804; (d) H. Higashimura, M. Kubota, A. Shiga, M. Kodera, H. Uyama, S. Kobayashi, *J. Mol. Catal. A: Chem.* **2000**, *161*, 801–806.

59 S. Kobayashi, H. Higashimura, *Prog. Polym. Sci.* **2003**, *28*, 1015–1048.

60 H. Higashimura, K. Fujisawa, M. Kubota, S. Kobayashi, *J. Polym. Sci. Part A: Polym. Chem.* **2005**, *43*, 1955–1962.

61 (a) H. Higashimura, K. Fujisawa, Y. Moro-oka, S. Namekawa, M. Kubota, A. Shiga, H. Uyama, S. Kobayashi, *Macromol. Rapid Commun.* **2000**, *21*, 1121–1124; (b) S. Kobayashi, H. Uyama, H. Tonami, T. Oguchi, H. Higashimura, R. Ikeda, M. Kubota, *Macromol. Symp.* **2001**, *175*, 1–10.

62 H. Higashimura, K. Fujisawa, Y. Moro-oka, M. Kubota, A. Shiga, H. Uyama, S. Kobayashi, *Appl. Catal. A*,

2000, *194/195*, 427–433;
(b) H. Higashimura, K. Fujisawa, Y. Moro-oka, M. Kubota, A. Shiga, H. Uyama, S. Kobayashi, *J. Mol. Catal. A*. 2000, *155*, 201–207.

63 H. Tonami, H. Uyama, H. Higashimura, T. Oguchi, S. Kobayashi, *Polym. Bull.* 1999, *42*, 125–129.

64 (a) H. Tonami, H. Uyama, S. Kobayashi, H. Higashimura, T. Oguchi, *J. Macromol. Sci. Pure Appl. Chem.* 1999, *A36*, 719–730; (b) R. Ikeda, H. Tanaka, H. Uyama, S. Kobayashi, *Macromolecules* 2000, *33*, 6648–6652.

65 S. Dubey, D. Singh, R. A. Misra, *Enzyme Microb. Technol.* 1998, *23*, 432–437.

66 N. Aktas, N. Sahiner, Ö. Kantolu, B. Salih, A. Tanyolac, *J. Polym. Environ.* 2003, *11*, 123–128.

67 G. Ward, R. E. Parales, C. G. Dosoretz, *Environ. Sci. Technol.* 2004, *38*, 4753–4757.

68 M. Terada, H. Oyabu, Y. Aso, *J. Jpn. Soc. Colour. Mater.* 1994, *66*, 681.

69 K. Nagase, R. Lu, T. Miyakoshi, *Chem. Lett.* 2004, *33*, 90–91.

70 (a) S. Kobayashi, R. Ikeda, H. Oyabu, H. Tanaka, H. Uyama, *Chem. Lett.* 2000, 1214–1215; (b) R. Ikeda, T. Tsujimoto, H. Tanaka, H. Oyabu, H. Uyama, S. Kobayashi, *Proc. Acad. Jpn.* 2000, *76B*, 155–160; (c) R. Ikeda, H. Tanaka, H. Oyabu, H. Uyama, S. Kobayashi, *Bull. Chem. Soc. Jpn.* 2001, *74*, 1067–1073.

71 P. Xu, H. Uyama, J. E. Whitten, S. Kobayashi, D. L. Kaplan, *J. Am. Chem. Soc.* 2005, *127*, 11745–11753.

72 J. Jankun, S. H. Selman, R. Swiercz, E. Skrzypczak-Jankun, *Nature* 1997, *387*, 561–561.

73 (a) D. E. Hathway, J. W. T. Seakins, *Biochemistry* 1957, *67*, 239–245; (b) D. E. Hathway, J. W. T. Seakins, *Nature* 1955, *176*, 218.

74 L. Mejias, M. H. Reihmann, S. Sepulveda-Boza, H. Ritter, *Macromol. Biosci.* 2002, *2*, 24–32.

75 M. Kurisawa, J. E. Chung, Y. J. Kim, H. Uyama, S. Kobayashi, *Biomacromolecules* 2003, *4*, 469–471.

76 M. Kurisawa, J. E. Chung, H. Uyama, S. Kobayashi, *Macromol. Biosci.* 2003, *3*, 758–764.

77 F. F. Bruno, S. Nagarajan, R. Nagarajan, J. Kumar, L. A. Samuelson, *J. Macromol. Sci. Part A: Pure Appl. Chem.* 2005, *42*, 1547–1554.

78 M. Kurisawa, J. E. Chung, H. Uyama, S. Kobayashi, *Chem. Commun.* 2004, 294–295.

79 P. Xu, A. Singh, D. L. Kaplan, *Adv. Polym. Sci.* 2006, *194*, 69–94.

80 S. K. Sahoo, R. Nagarajan, L. A. Samuelson, J. Kumar, A. L. Cholli, S. K. Tripathy, *J. Macromol. Sci. Pure Appl. Chem.* 2001, *A38*, 1315–1328.

81 A. G. MacDiarmid, J. C. Chiang, A. F. Richter, A. J. Epstein, *Synth. Met..* 1987, *18*, 285–290.

82 S. Kobayashi, I. Kaneko, H. Uyama, *Chem. Lett.* 1992, 393–394.

83 H. Uyama, H. Kurioka, I. Kaneko, S. Kobayashi, *Macromol. Rep.* 1994, *A31*, 421–427.

84 (a) J. Shan, S. Cao, *Polym. Adv. Technol.* 2000, *11*, 288–293; (b) J. Shan, L. Han, F. Bai, S. Cao, *Polym. Adv. Technol.* 2003, *14*, 330–336.

85 S. K. Tripathy, D. Y. Kim, L. Li, N. K. Viswanathan, S. Balasubramanian, W. Liu, P. Wu, S. Bian, L. Samuelson, J. Kumar, *Synth. Met.* 1999, *102*, 893–896.

86 K. S. Alva, T. S. Lee, J. Kumar, S. K. Tripathy, *Chem. Mater.* 1998, *10*, 1270–1275.

87 K. Shridhara, J. Kumar, K. A. Marx, S. K. Tripathy, *Macromolecules* 1997, *30*, 4024–4029.

88 M. Aizawa, L. Wang, H. Shinohara, Y. Ikariyama, *J. Biotechnol.* 1990, *14*, 301–309.

89 E. I. Solomon, U. M. Sundaram, T. E. Machonkin, *Chem. Rev.* 1996, *96*, 2563–2605.

90 (a) B. C. Ku, S. Y. Lee, W. Liu, J. A. He, J. Kumar, F. F. Bruno, L. A. Samuelson, *J. Macromol. Sci. Pure. Appl. Chem.* 2003, *A40*, 1335–1346; (b) S. Roy, J. M. Fortier, R. Nagarajan, S. Tripathy, J. Kumar, S. L. A. amuelson, F. F. Bruno, *Biomacromolecules* 2002, *3*, 937–941.

91 L. A. Samuelson, A. Anagnostopoulos, K. S. Alva, J. Kumar, S. K. Tripathy, *Macromolecules*, 1998, *31*, 4376–4378.

92 W. Liu, J. Kumar, S. Tripathy, K. J. Senecal, L. Samuelson, *J. Am. Chem. Soc.* **1999**, *121*, 71–78.

93 P. Xu, D. L. Kaplan, *Adv. Mater.* **2004**, *16*, 628–628.

94 Y. Tanaka, *Prog. Polym. Sci.* **1989**, *14*, 339–371.

95 A. Singh, D. L. Kaplan, *Adv. Polym. Sci.* **2006**, *194*, 211–224.

96 H. Iwata, Y. Hata, T. Matsuda, Y. Ikada, *J. Polym. Sci. Part A: Polym. Chem.* **1991**, *29*, 1217–1218.

97 R. Ikeda, H. Tanaka, H. Uyama, S. Kobayashi, *Macromol. Rapid. Commun.* **1998**, *19*, 423–425.

98 O. Emery, T. Lalot, M. Brigodiot, E. Marechal, *J. Polym. Sci. Part A: Polym. Chem.* **1997**, *35*, 3331–3333.

99 D. Teixeira, T. Lalot, M. Brigodiot, E. Marechal, *Macromolecules* **1999**, *32*, 70–72.

100 Kalra, R. A. Gross, *Green Chem.* **2002**, *4*, 174–178.

101 B. Kalra, R. Gross, *Biomacromolecules*, **2000**, *1*, 501–505.

102 (a) A. Singh, D. Ma, D. L. Kaplan, *Biomacromolecules* **2000**, *1*, 592–596; (b) A. Singh, S. Roy, L. Samuelson, F. Bruno, R. Nagarajan, J. Kumar, V. John, D. L. Kaplan, *J. Macromol. Sci. Pure Appl. Chem.* **2001**, *A38*, 1219–1230.

103 (a) A. Singh, D. L. Kaplan, *Adv. Mater.* **2003**, *15*, 1291–1291; (b) A. Singh, D. L. Kaplan, *J. Macromol. Sci. Pure Appl. Chem.* **2004**, *41*, 1377–1386.

104 S. Kobayashi, M. Ohmae, *Adv. Polym. Sci.* **2006**, *194*, 159–210.

105 E. Atherton, R. C. Sheppard, eds., *Solid-phase Peptide Synthesis: a Practical Approach*. Oxford University Press, Oxford, **1989**.

106 B. Alberts, A. Johnson, J. Lewis, M. Raff, K. Roberts, P. Walter, eds., Manipulating proteins, DNA and RNA, in *Molecular Biology of the Cell*, 4th edn. Garland Science, New York, **2002**, *8*, 469–546.

107 (a) R. Perrin, C. Wilkerson, K. Keegstra, *Plant Mol. Biol.* **2001**, *47*, 115–130; (b) H. Merzendorfer, *J. Comp. Physiol. B* **2006**, *176*, 1–15; (c) N. Taniguchi, K. Honke, M. Fukuda, eds., *Handbook of Glycosyltransferase and Related Genes*. Springer, Tokyo, **2002**.

108 H. Mark, *Cellul. Chem. Technol.* **1980**, *14*, 569–581.

109 H. H. Schlubach, L. Luehrs, *Liebigs Ann. Chem.* **1941**, *547*, 73–85.

110 S. Kobayashi, K. Kashiwa, T. Kawasaki, S. Shoda, *J. Am. Chem. Soc.* **1991**, *113*, 3079–3084.

111 http://www.cazy.org/CAZY/index.html. and references cited therein.

112 H.-L. Lai, L. G. Butler, B. Axelrod, *Biochem. Biophys. Res. Commun.* **1974**, *60*, 635–640.

113 For examples: (a) S. Fort, V. Boyer, L. Greffe, G. J. Davies, O. Moraz, L. Christiansen, M. Schuelein, S. Cottaz, H. Driguez, *J. Am. Chem. Soc.* **2000**, *122*, 5429–5437; (b) L. F. Mackenzie, Q. Wang, R. A. J. Warren, S. G. Withers, *J. Am. Chem. Soc.* **1998**, *120*, 5583–5584.

114 J. H. Lee, R. M. Brown, Jr., S. Kuga, S. Shoda, S. Kobayashi, *Proc. Natl. Acad. Sci. USA* **1994**, *91*, 7425–7429.

115 (a) S. Kobayashi, S. Shoda, J. H. Lee, K. Okuda, R. M. Brown, Jr., S. Kuga, *Macromol. Chem. Phys.* **1994**, *195*, 1319–1326; (b) S. Kobayashi, L. J. Hobson, J. Sakamoto, S. Kimura, J. Sugiyama, T. Imai, T. Itoh, *Biomacromolecules* **2000**, *1*, 168–173; (c) T. Hashimoto, H. Tanaka, S. Koizumi, K. Kurosaki, M. Ohmae, S. Kobayashi, *Biomacromolecules* **2006**, *7*, 2479–2482.

116 S. Kobayashi, E. Okamoto, X. Wen, S. Shoda, *J. Macromol. Sci. Pure Appl. Chem.* **1996**, *A33*, 1375–1384.

117 (a) S. Kobayashi, T. Kawasaki, K. Obata, S. Shoda, *Chem. Lett.* **1993**, 685–686; (b) S. Shoda, T. Kawasaki, K. Obata, S. Kobayashi, *Carbohydr. Res.* **1993**, *249*, 127–137; (c) O. Karthaus, S. Shoda, H. Takano, K. Obata, S. Kobayashi, *J. Chem. Soc., Perkin Trans. 1* **1994**, 1851–1857.

118 E. Okamoto, T. Kiyosada, S. Shoda, S. Kobayashi, *Cellulose* **1997**, *4*, 161–172.

119 S. Kobayashi, X. Wen, S. Shoda, *Macromolecules* **1996**, *29*, 2698–2700.

120 S. Kobayashi, J. Shimada, K. Kashiwa, S. Shoda, *Macromolecules* **1992**, *25*, 3237–3241.

121 G. Ziegast, B. Pfannemüller, *Carbohydr. Res.* **1987**, *160*, 185–204.

122 M. Fujita, S. Shoda, S. Kobayashi, *J. Am. Chem. Soc.* **1998**, *120*, 6411–6412.

123 S. Kobayashi, A. Makino, H. Matsumoto, S. Kunii, M. Ohmae, T. Kiyosada, K. Makiguchi, A. Matsumoto, M. Horie, S. Shoda, *Biomacromolecules* **2006**, *7*, 1644–1656.

124 G. Perugino, B. Cobucci-Ponzano, M. Rossi, M. Moracci, *Adv. Synth. Catal.* **2005**, *347*, 941–950.

125 S. G. Withers, R. A. J. Warren, I. P. Street, K. Rupitz, J. B. Kempton, R. Aebersold, *J. Am. Chem. Soc.* **1990**, *112*, 5887–5889.

126 M. Hrmova, T. Imai, S. J. Rutten, J. K. Fairweather, L. Pelosi, V. Bulone, H. Driguez, G. B. Fincher, *J. Biol. Chem.* **2002**, *277*, 30102–30111.

127 I. Nakamura, H. Yoneda, T. Maeda, A. Makino, M. Ohmae, J. Sugiyama, M. Ueda, S. Kobayashi, S. Kimura, *Macromol. Biosci.* **2005**, *5*, 623–628.

128 B. Henrissat, *Cellulose* **1994**, *1*, 169–196.

129 (a) R. A. A. Muzzarelli, C. Muzzarelli, *Adv. Polym. Sci.* **2005**, *186*, 151–209; (b) R. Raman, V. Sasisekharan, R. Sasisekharan, *Chem. Biol.* **2005**, *12*, 267–277; (c) K. R. Taylor, R. L. Gallo, *FASEB J.* **2006**, *20*, 9–22.

130 H. Hasegawa, T. Ichinohe, P. Strong, I. Watanabe, S. Ito, S.-I. Tamura, H. Takahashi, H. Sawa, J. Chiba, T. Kurata, T. Sata, *J. Med. Virol.* **2005**, *75*, 130–136.

131 I. Saiki, *Jpn. J. Pharmacol.* **1997**, *75*, 215–242.

132 S. B. Lee, Y. H. Kim, M. S. Chong, Y. M. Lee, *Biomaterials* **2004**, *25*, 2309–2317.

133 L. Chen, Y. Du, H. Wu, L. Xiao, *J. Appl. Polym. Chem.* **2002**, *83*, 1233–1241.

134 F. L. Mi, S. S. Shyu, Y. M. Lin, Y. B. Wu, C. K. Peng, Y. H. Tsai, *Biomaterials* **2003**, *24*, 5023–5036.

135 S. Kobayashi, T. Kiyosada, S. Shoda, *J. Am. Chem. Soc.* **1996**, *118*, 13113–13114.

136 I. Tews, A. C. T. van Scheltinga, A. Perrakis, K. S. Wilson, B. W. Dijkstra, *J. Am. Chem. Soc.* **1997**, *119*, 7954–7959.

137 (a) A. Makino, J. Sakamoto, M. Ohmae, S. Kobayashi, *Chem. Lett.* **2006**, *35*, 160–161; (b) A. Makino, M. Ohmae, S. Kobayashi, *Macromol. Biosci.* **2006**, *6*, 862–872.

138 J. Sakamoto, J. Sugiyama, S. Kimura, T. Imai, T. Itoh, T. Watanabe, S. Kobayashi, *Macromolecules* **2000**, *33*, 4155–4160, 4982.

139 (a) T. Watanabe, K. Suzuki, W. Oyanagi, K. Ohnishi, H. Tanaka, *J. Biol. Chem.* **1990**, *265*, 15659–15665; (b) T. Watanabe, K. Kobori, K. Miyashita, T. Fujii, H. Sakai, M. Uchida, H. Tanaka, *J. Biol. Chem.* **1993**, *268*, 18567–18572.

140 J. Sakamoto, T. Watanabe, Y. Ariga, S. Kobayashi, *Chem. Lett.* **2001**, *30*, 1180–1181.

141 (a) S. Shoda, M. Fujita, C. Lohavisavapanichi, Y. Misawa, K. Ushizaki, Y. Tawata, M. Kuriyama, M. Kohri, H. Kuwata, T. Watanabe, *Helv. Chim. Acta* **2002**, *85*, 3919–3936; (b) S. Shoda, T. Kiyosada, H. Mori, S. Kobayashi, *Heterocycles* **2000**, *52*, 599–602.

142 J. Sakamoto, S. Kobayashi, *Chem. Lett.* **2004**, *33*, 698–699.

143 H. Ochiai, M. Ohmae, S. Kobayashi, *Chem. Lett.* **2004**, *33*, 694–695.

144 M. Bernfield, M. Götte, P. W. Park, O. Reizes, M. L. Fitzgerald, J. Lincecum, M. Zako, *Annu. Rev. Biochem.* **1999**, *68*, 729–777.

145 (a) K. A. Johnson, D. A. Hulse, R. C. Hart, D. Kochevar, Q. Chu, *Osteoarthritis Cart.* **2001**, *9*, 14–21; (b) R. Duncan, *Nat. Rev. Drug Discov.* **2003**, *2*, 347–360; (c) T. E. McAlindon, M. P. LaValley, J. P. Gulin, D. T. Felson, *JAMA* **2000**, *283*, 1469–1475.

146 R. Stern, M. J. Jedrzejas, *Chem. Rev.* **2006**, *106*, 818–839.

147 Z. Marković-Housley, G. Miglierini, L. Soldatova, P. J. Rizkallah, U. Müller, T. Schirmer, *Structure* **2000**, *8*, 1025–1035.

148 L. Lapčík, Jr., L. Lapčík, S. De Smedt, J. Demeester, *Chem. Rev.* **1998**, *98*, 2663–2684.

149 (a) S. Kobayashi, H. Morii, R. Itoh, S. Kimura, M. Ohmae, *J. Am. Chem. Soc.* **2001**, *123*, 11825–11826; (b) H. Ochiai, T. Mori, M. Ohmae, S. Kobayashi, *Biomacromolecules* **2005**, *6*, 1068–1084.

150 H. Ochiai, T. Mori, M. Ohmae, S. Kobayashi, *Polym. Prepr. Jpn.* **2004**, *53*, 1882.

151 S. Mizuguchi, T. Uyama, H. Kitagawa, K. H. Nomura, K. Dejima, K. Gengyo-Ando, S. Mitani, K. Sugahara, K. Nomura, *Nature* **2003**, *423*, 443–448.

152 (a) T. Laabs, D. Carulli, H. M. Geller, J. W. Fawcett, *Curr. Opin. Neurobiol.* **2005**, *15*, 116–120.

153 H. Watanabe, K. Kimata, S. Line, D. Strong, L.-y. Gao, C. A. Kozak, Y. Yamada, *Nat. Genet.* **1994**, *7*, 154–157.

154 A. M. Clement, S. Nadanaka, K. Masayama, C. Mandl, K. Sugahara, A. Faissner, *J. Biol. Chem.* **1998**, *273*, 28444–28453.

155 N. Maeda, M. Noda, *J. Cell. Biol.* **1998**, *142*, 203–216.

156 (a) H. Habuchi, O. Habuchi, K. Kimata, *Glycoconj. J.* **2004**, *21*, 47–52; (b) M. Kusche-Gullberg, L. Kjellén, *Curr. Opin. Struct. Biol.* **2003**, *13*, 605–611.

157 S. Kobayashi, S. Fujikawa, M. Ohmae, *J. Am. Chem. Soc.* **2003**, *125*, 14357–14369.

158 S. Fujikawa, M. Ohmae, S. Kobayashi, *Biomacromolecules* **2005**, *6*, 2935–2942.

159 A. Makino, K. Kurosaki, M. Ohmae, S. Kobayashi, *Biomacromolecules* **2006**, *7*, 950–957.

160 S. Kobayashi, A. Makino, N. Tachibana, M. Ohmae, *Macromol. Rapid Commun.* **2006**, *27*, 781–786.

161 A. M. Klibanov, *Acc. Chem. Res.*, **1990**, *23*, 114–120.

162 E. Santaniello, P. Ferraboschi, P. Grisenti, A. Manzocchi, *Chem. Rev.* **1992**, *92*, 1071–1140.

163 (a) H. Uyama, S. Kobayashi, *Chem. Lett.* **1993**, 1149–1152; (b) H. Uyama, K. Takeya, S. Kobayashi, *Proc. Jpn. Acad. Ser. B* **1993**, *69*, 203–207.

164 D. Knani, A. L. Gutman, D. H. Kohn, *J. Polym. Sci. Polym. Chem. Ed.* **1993**, *31*, 1221–1232.

165 (a) S. Kobayashi, H. Uyama, in *Advances in Biochemical Engineering/Biotechnology*, Vol. 71, Biopolyesters, W. Babel, A. Steinbüchel, eds. Springer, Heidelberg, **2001**, 241–262; (b) S. Kobayashi, H. Uyama, in *Handbook of Biopolymers*, Vol. 3a, Polyesters I, Y. Doi, A. Steinbuechel, eds. Wiley-VCH, Weinheim, **2002**, 374–400; (c) S. Kobayashi, H. Uyama, *Curr. Org. Chem.* **2002**, *6*, 209–222; (d) A.-C. Albertsson, I. K. Varma, *Biomacromolecule* **2003**, *4*, 1466–1486.

166 S. Kobayashi, H. Uyama, S. Namekawa, *Polym. Degrad. Stabil.* **1998**, *59*, 195–201.

167 S. Kobayashi, H. Uyama, *Macromol. Symp.* **1999**, *144*, 237–246.

168 (a) S. Matsumura, *Macromol. Biosci.* **2002**, *2*, 105–126; (b) S. Matsumura, *Adv. Polym. Sci.* **2006**, *194*, 95–132.

169 R. T. MacDonald, P. S. Kulapura, Y. Y. Svirkin, R. A. Gross, D. L. Kaplan, J. Akkara, G. Swift, S. Wolk, *Macromolecules* **1995**, *28*, 73–78.

170 L. A. Henderson, Y. Y. Svirkin, R. A. Gross, D. L. Kaplan, G. Swift, *Macromolecules* **1996**, *29*, 7759–7766.

171 S. Matsumura, H. Ebata, K. Toshima, *Macromol. Rapid. Commun.* **2000**, *21*, 860–863.

172 H. Ebata, K. Toshima, S. Matsumura, *Chem. Lett.* **2001**, 798–799.

173 A. Kumar, B. Kalra, A. Dekhterman, R. A. Gross, *Macromolecules* **2000**, *33*, 6303–6309.

174 Y. Mei, A. Kumar, R. Gross, *Macromolelcules* **2002**, *35*, 5444–5448.

175 A. Cordova, T. Iversen, K. Hult, *Polymer*, **1999**, *40*, 6709–6721.

176 H. Dong, S. Cao, Z. Li, S. Han, D. You, J. Shen, *J. Polym. Sci. Part A: Polym. Chem.* **1999**, *37*, 1265–1275.

177 S. Divakar, *J. Macromol. Sci. Pure Appl. Chem.* **2004**, *A41*, 537–546.

178 A. Kumar, R. A. Gross, *Biomacromolecules* **2000**, *1*, 133–138.

179 S. Kobayashi, K. Takeya, S. Suda, H. Uyama, *Macromol. Chem. Phys.* **1998**, *199*, 1729–1736.

180 H. Uyama, S. Suda, H. Kikuchi, S. Kobayashi, *Chem. Lett.* **1997**, 1109–1110.

181 G. Sivalingam, G. Madras, *Biomacromolecules* **2004**, *5*, 603–609.

182 H. Ebata, K. Toshima, S. Matsumura, *Biomacromolecules* **2000**, *1*, 511–514.

183 H. Uyama, K. Takeya, S. Kobayashi, *Bull. Chem. Soc. Jpn.* **1995**, *68*, 56–61.

184 E. M. Anderson, *Biocatal. Biotransform.* **1998**, *16*, 181–204.
185 (a) S. Matsumura, H. Beppu, K. Nakamura, S. Osanai, K. Toshima, *Chem. Lett.* **1996**, 795–796; (b) S. Namekawa, H. Uyama, S. Kobayashi, *Polym. J.* **1996**, *28*, 730–731; (c) Y. Y. Svirkin, J. Xu, R. A. Gross, D. L. Kaplan, G. Swift, *Macromolecules* **1996**, *29*, 4591–4597; (d) G. A. R. Nobes., R. J. Kazlauskas, R. H. Marchessault, *Macromolecules* **1996**, *29*, 4829–4833.
186 S. Kobayashi, H. Uyama, S. Namekawa, H. Hayakawa, *Macromolecules* **1998**, *31*, 5655–5659.
187 (a) H. Uyama, K. Takeya, N. Hoshi, S. Kobayashi, *Macromolecules* **1995**, *28*, 7046–7050; (b) H. Uyama, H. Kikuchi, K. Takeya, S. Kobayashi, *Acta Polym.* **1996**, *47*, 357–360; (c) K. S. Bisht, L. A. Henderson, R. A. Gross, D. L. Kaplan, G. Swift, *Macromolecules* **1997**, *30*, 2705–2711; (d) S. Namekawa, H. Uyama, S. Kobayashi, *Proc. Jpn. Acad.* **1998**, *74B*, 65–68; (e) A. Kumar, B. Kalra, A. Dekhterman, R. A. Gross, *Macromolecules* **2000**, *33*, 6303–6309.
188 S. Namekawa, H. Uyama, S. Kobayashi, *Polym. J.* **1998**, *30*, 269–271.
189 A. Duda, A. Kowalski, S. Penczek, H. Uyama, S. Kobayashi, *Macromolecules* **2002**, *35*, 4266–4270.
190 (a) H. Kikuchi, H. Uyama, S. Kobayashi, *Macromolecules* **2000**, *33*, 8971–8975; (b) S. Kobayashi, *Macromol. Symp.* **2006**, *240*, 178–185.
191 J. W. Peeters, O. van Leeuwen, A. R. A. Palmans, E. Meijer, *Macromolecules* **2005**, *38*, 5587–5592.
192 S. Namekawa, H. Uyama, S. Kobayashi, *Macromol. Chem. Phys.* **2001**, *202*, 801–806.
193 T. Takamoto, P. Kerep, H. Uyama, S. Kobayashi, *Macromol. Biosci.* **2001**, *1*, 223–227.
194 J. Wahlberg, P. V. Persson, T. Olsson, E. Hedenstrom, T. Iversen, *Biomacromolecules* **2003**, *4*, 1068–1071.
195 M. Runge, D. O'Hagan, G. Haufe, *J. Polym. Sci. Polym. Chem. Ed.* **2000**, *38*, 2004–2012.
196 T. F. Al-Azemi, L. Kondaveti, K. S. Bisht, *Macromolecules*, **2002**, *35*, 3380–3386.
197 Y. Suzuki, S. Taguchi, T. Hisano, K. Toshima, S. Matsumura, Y. Doi, *Biomacromolecules* **2003**, *4*, 537–543.
198 J. Peeters, M. Veld, F. Scheijen, A. Heise, A. R. A. Palmans, E. W. Meijer, *Biomacromolecules* **2004**, *5*, 1862–1868.
199 B. A. C. van As, J. van Buijtenen, A. Heise, Q. B. Broxterman, G. K. M.Verzijl, A. R. A. Palmans, E. W. Meijer, *J. Am. Chem. Soc.* **2005**, *127*, 9964–9965.
200 H. Uyama, S. Kobayashi, M. Morita, S. Habaue, Y. Okamoto, *Macromolecules* **2001**, *34*, 6554–6556.
201 S. Habaue, M. Asai, M. Morita, Y. Okammoto, H. Uyama, S. Kobayashi, *Polymer* **2003**, *44*, 5195–5200.
202 H. Uyama, S. Suda, S. Kobayashi, *Acta Polym.* **1998**, *49*, 700–703.
203 K. S. Bisht, F. Deng, R. A. Gross, D. L. Kaplan, G. Swift, *J. Am. Chem. Soc.* **1998**, *120*, 1363–1367.
204 A. Córdova, T. Iversen, K. Hult, *Macromolecules* **1998**, *31*, 1040–1045.
205 A. Córdova, A. Hult, K. Hult, H. Ihre, T. Iversen, E. Malmström, *J. Am. Chem. Soc.* **1998**, *120*, 13521–13522.
206 R. K. Srivastava, A.-C. Albertsson, *Macromolecules* **2006**, *39*, 46–54.
207 (a) H. Uyama, H. Kikuchi, S. Kobayashi, *Chem. Lett.* **1995**, 1047–1048; (b) H. Uyama, H. Kikuchi, S. Kobayashi, *Bull. Chem. Soc. Jpn.* **1997**, *70*, 1691–1695.
208 (a) A. Gordova, *Biomacromolecules* **2001**, *2*, 1347–1351; (b) J. Li, W. Xie, H. N. Cheng, R. G. Nickol, P. G. Wang, *Macromolecules* **1999**, *32*, 2789–2792; (c) R. Kumar, R. A. Gross, *J. Am. Chem. Soc.* **2002**, *124*, 1850–1851.
209 B. Kalra, A. Kumar, R. A. Gross, M. Baiardo, M. Scandola, *Macromolecules* **2004**, *37*, 1243–1250.
210 K. R. Yoon, K.-B. Lee, Y. S. Chi, W. S. Yun, S.-W. Joo, I. S. Choi, *Adv. Mater.* **2003**, *15*, 2063–2066.
211 Y.-R. Kim, H.-J. Paik, C. K. Ober, G. W. Coates, C. A. Batt, *Biomacromolecules* **2004**, *5*, 889–894.
212 S. Skaria, M. Smet, H. Frey, *Macromol. Rapid Commun.* **2002**, *23*, 292–296.

213 For example: K. Matyjaszewski, J. Xia, *Chem. Rev.* **2001**, *101*, 2921–2990.
214 (a) U. Meyer, A. R. A. Palmans, T. Loontjens, A. Heise, *Macromolecules* **2002**, *35*, 2873–2875; (b) B. A. C. van As, P. Thomassen, B. Kalra, R. A. Gross, E. W. Meijer, A. R. A. Palmans, A. Heise, *Macromolecules* **2004**, *37*, 8973–8977; (c) J. W. Peeters, A. R. A. Palmans, E. W. Meijer, C. E. Koning, A. Heise, *Macromol. Rapid Commun.* **2005**, *26*, 684–689; (d) M. de Geus, J. W. Peeters, M. Wolffs, T. Hermans, A. R. A. Palmans, C. E. Koning, A. Heise, *Macromolecules* **2005**, *38*, 4220–4225; (e) K. Sha, D. Li, S. Wang, L. Qin, J. Wang, *Polym. Bull.* **2005**, *55*, 349–355; (f) M. Geus, L. Schormans, A. R. A. Palmans, C. E. Koning, A. Heise, *J. Polym. Sci. Polym. Chem.* **2006**, *44*, 4290–4297.
215 S. Kobayashi, H. Uyama, T. Takamoto, *Biomacromolecules* **2000**, *1*, 3–5.
216 (a) H. Ebata, K. Toshima, S. Matsumura, *Biomacromolecules* **2000**, *1*, 511–514; (b) S. Sugihara, K. Toshima, S. Matsumura, *Macromol. Rapid Commun.* **2006**, *27*, 203–297.
217 (a) Y. Takahashi, K. Okajima, K. Toshima, S. Matsumura, *Macromol. Biosci.* **2004**, *4*, 346–353; (b) Y. Osanai, K. Toshima, S. Matsumura, *Macromol. Biosci.* **2004**, *4*, 936–942.
218 T. Takamoto, H. Uyama, S. Kobayashi, *e-Polymers* **2001**, *4*, 1–6.
219 F. C. Loeker, C. J. Duxbury, R. Kumar, W. Gao, R. A. Gross, S. M. Howdle, *Macromolecules* **2004**, *37*, 2450–2453.
220 T. Takamoto, H. Uyama, S. Kobayashi, *Macromol. Biosci.* **2001**, *1*, 215–218.
221 S. Matsumura, H. Ebata, R. Kondo, K. Toshima, *Macromol. Rapid Commun.* **2001**, *22*, 1325–1329.
222 (a) C. J. Duxbury, W. X. Wang, M. de Geus, A. Heise, S. M. Howdle, *J. Am. Chem. Soc.* **2005**, *127*, 2384–2385; (b) S. Villarroya, J. Zhou, C. J. Duxbury, A. Heise, S. M. Howdle, *Macromolecules* **2006**, *39*, 633–640.
223 H. Uyama, T. Takamoto, S. Kobayashi, *Polym. J.* **2002**, *34*, 94–96.
224 S. Kobayashi, H. Uyama, *Makromol. Chem. Rapid Commun.* **1993**, *14*, 841–844.
225 H. Uyama, S. Wada, S. Kobayashi, *Chem. Lett.* **1999**, 893–894.
226 R. K. Srivastava, A.-C. Albertsson, *J. Polym. Sci. Part A: Polym. Chem.* **2005**, *43*, 4206–4216.
227 (a) S. Kobayashi, H. Kikuchi, H. Uyama, *Macromol. Rapid Commun.* **1997**, *18*, 575–579; (b) S. Matsumura, K. Tsukada, K. Toshima, *Macromolecules* **1997**, *30*, 3122–3124; (c) K. S. Bisht, Y. Y. Svirkin, L. A. Henderson, R. A. Gross, D. L. Kaplan, G. Swift, *Macromolecules* **1997**, *30*, 7735–7742.
228 F. Deng, R. A. Gross, *Int. J. Biol. Macromol.* **1999**, *25*, 153–159.
229 (a) S. Matsumura, K. Tsukada, K. Toshima, *Int. J. Biol. Macromol.* **1999**, *25*, 161–167; (b) T. F. Al-Azemi, K. S. Bisht, *Macromolecules*, **1999**, *32*, 6536–6540; (c) T. F. Al-Azemi, J. P. Harmon, K. S. Bisht, S. Kirpal, *Biomacromolecule* **2000**, *1*, 493–500; (d) T. F. Al-Azemi, K. S. Bisht, *J. Polym. Sci. Part A: Polym. Chem.* **2002**, *40*, 1267–1274.
230 S. Namekawa, H. Uyama, S. Kobayashi, H. R. Kricheldorf, *Macromol. Chem. Phys.* **2000**, *201*, 261–264.
231 (a) Y. Feng, J. Knufermann, D. Klee, H. Hoecker, *Macromol. Rapid Comunn.***1999**, *20*, 88–90; (b) Y. Feng, D. Klee, H. Keul, H. Hoecker, *Macromol. Chem. Phys.* **2000**, *201*, 2670–2675; (c) Y. Feng, D. Klee, H. Hoecker, *Macromol. Biosci.* **2004**, *4*, 587–590.
232 J. Wen, R. X. Zhuo, *Macromol. Rapid. Commun.* **1998**, *19*. 641–642.
233 H. Uyama, S. Kobayashi, *Adv. Polym. Sci.* **2006**, *194*, 133–158.
234 F. Binns, S. M. Roberts, A. Taylor, C. F. Williams, *J. Chem. Soc., Perkin. Trans. 1* **1993**, 899–904.
235 Z. L. Wang, K. Hiltunen, P. Orava, J. Seppälä, Y. Y. Linko, *J. Macromol. Sci. Pure. Appl. Chem.* **1996**, *A33*, 599–612.
236 (a) X. Wu, Y. Y. Linko, J. Seppälä, *Ann. N. Y. Acad. Sci.* **1998**, *864*, 399–404; (b) A. Mahapatro, B. Kalra, A. Kumar,

R. A. Gross, *Biomacromolecules* **2003**, *4*, 544–551.

237 (a) H. Uyama, K. Inada, S. Kobayashi, *Chem. Lett.* **1998**, 1285–1286; (b) F. Binns, P. Harffey, S. M. Roberts, A. Taylor, *J. Polym. Sci. Part A: Polym. Chem.* **1998**, *36*, 2069–2079.

238 (a) F. Binns, P. Harffey, S. M. Roberts, A. Taylor, *J. Chem. Soc., Perkin. Trans. 1* **1999**, 2671–2676; (b) H. Uyama, K. Inada, S. Kobayashi, *Polym. J.* **2000**, *32*, 440–443.

239 (a) S. Kobayashi, H. Uyama, S. Suda, S. Namekawa, *Chem. Lett.* **1997**, 105; (b) S. Suda, H. Uyama, S. Kobayashi, *Proc. Jpn. Acad.* **1999**, *75B*, 201–206.

240 C. Berkane, G. Mezoul T. Lalot, M. Brigodiot, E. Maréchal, *Macromolecules* **1997**, *30*, 7729–7734.

241 A. Lavalette, T. Lalot, M. Brigodiot, E. Maréchal, *Biomacromolecules* **2002**, *3*, 225–228.

242 Y.-Y. Linko, J. Seppälä, *CHEMTECH* **1996**, *26*, 25–31.

243 H. G. Park, H. N. Chang, J. S. Dordick, *Biocatalysis* **1994**, *11*, 263–271.

244 H. Uyama, T. Takamoto, S. Kobayashi, *Polym. J.* **2002**, *34*, 94–96.

245 S. J. Nara, J. R. Harjani, M. M. Salunkhe, A. T. Mane, P. P. Wadgaonkar, *Tetrahedron Lett.* **2003**, *44*, 1371–1373.

246 J. S. Wallace, C. J. Morrow, *J. Polym. Sci. Part A: Polym. Chem.* **1989**, *27*, 3271–3284.

247 Y.-Y. Linko, Z.-L Wang, J. Seppälä, *Biocatalysis* **1994**, *8*, 269–282.

248 A. K. Chaudhary, E. J. Beckman, A. J. Russell, *J. Am. Chem. Soc.* **1995**, *117*, 3728–3733.

249 E. M. Brazwell, D. Y. Filos, C. J. Morrow, *J. Polym. Sci. Part A: Polym. Chem.* **1995**, *33*, 89–95.

250 H. Uyama, S. Kobayashi, *Chem. Lett.* **1994**, 1687–1690.

251 H. Uyama, S. Yaguchi, S. Kobayashi, *J. Polym. Sci. Part A: Polym. Chem.* **1999**, *37*, 2737–2745.

252 A. K. Chaundhary, E. J. Beckman, A. J. Russell, *Biotechnol. Bioeng.* **1997**, *55*, 227–239.

253 B. J. Kline, S. S. Lele, P. J. Lenart, E. J. Beckman, A. J. Russell, *Biotechnol. Bioeng.* **2000**, *67*, 424–434.

254 D. Y. Kim, J. S. Dordick, *Biotechnol. Bioeng.* **2001**, *76*, 200–206.

255 H. Uyama, S. Yaguchi, S. Kobayashi, *Polym. J.* **1999**, *31*, 380–383.

256 R. L. Rodney, B. T. Allinson, E. J. Beckman, A. J. Russell, *Biotechnol. Bioeng.* **1999**, *65*, 485–489.

257 S. Namekawa, H. Uyama, S. Kobayashi, *Biomacromolecules* **2000**, *1*, 335–338.

258 T. Takamoto, P. Kerep, H. Uyama, S. Kobayashi, *Macromol. Biosci.* **2001**, *1*, 223–227.

259 A. Ajima, T. Yoshimoto, K. Takahashi, Y. Tamaura, Y. Saito, Y. Inada, *Biotechnol. Lett.* **1985**, *7*, 303–306.

260 (a) Y. Ohya, T. Sugitou, T. Ouchi, *J. Macromol. Sci. Pure. Appl. Chem.* **1995**, *A32*, 179–190; (b) K. R. Kiran, S. Divakar, J. World, *Microbiol. Biotechnol.* **2003**, *19*, 859–865.

261 S. Matsumura, J. Takahashi, *Makromol. Chem. Rapid. Commun.* **1986**, *7*, 369–373.

262 O. Noll, H. Ritter, *Macromol. Rapid. Commun.* **1996**, *17*, 553–557.

263 (a) D. O'Hagan, N. A. Zaidi, *J. Chem. Soc., Perkin. Trans. 1* **1993**, 2389–2390; (b) D. O'Hagan, N. A. Zaidi, *Polymer*, **1994**, *35*, 3576–3578.

264 A. Mahapatro, A. Kumar, R. A. Gross. *Biomacromolecules* **2004**, *5*, 62–68.

265 S. Namekawa, H. Uyama, S. Kobayashi, *Int. J. Biol. Macromol.* **1999**, *25*, 145–151.

266 H. Dong, H. D. Wang, S. G. Cao, J. C. Shen, *Biotechnol. Lett.* **1998**, *20*, 905–908.

267 S. Iwata, K. Toshima, S. Matsumura, *Macromol. Rapid. Commun.* **2003**, *24*, 467–471.

268 J. S. Wallace, C. J. Morrow, *J. Polym. Sci. Part A: Polym. Chem.* **1989**, *27*, 2553–2567.

269 D. O'Hagan, A. H. Parker, *Polym. Bull.* **1998**, *41*, 519–524.

270 D. Knani, D. H. Kohn, *J. Polym. Sci. Part A: Polym. Chem.* **1993**, *31*, 2887–2897.

271 D. R. Patil, D. G. Rethwisch, J. S. Dordick, *Biotechnol. Bioeng.* **1991**, *37*, 639–646.

272 O. J. Park, D. Y. Kim, J. S. Dordick, *J. Polym. Chem. Part A: Polym. Chem. Ed.* **2000**, *70*, 208–216.

273 H. Uyama, E. Klegraf, S. Wada, S. Kobayashi, *Chem. Lett.* **2000**, 800–801.

274 A. Kumar, A. S. Kulshrestha, W. Gao, R. A. Gross, *Macromolecules* **2003**, *36*, 8219–8221.

275 H. Fu, A. S. Kulshrestha, W. Gao, R. A. Gross, M. Baiardo, M. Scandola, *Macromolecules* **2003**, *36*, 9804–9808.

276 B. J. Kline, E. J. Beckman, A. J. Russell, *J. Am. Chem. Soc.* **1998**, *120*, 9475–9480.

277 (a) H. Uyama, K. Inada, S. Kobayashi, *Macromol. Rapid Commun.* **1999**, *20*, 171–174; (b) H. Uyama, K. Inada, S. Kobayashi, *Macromol. Biosci.* **2001**, *1*, 40–44.

278 (a) T. Tsujimoto, H. Uyama, S. Kobayashi, *Biomacromolecules* **2001**, *2*, 29–31; (b) T. Tsujimoto, H. Uyama, S. Kobayashi, *Macromol. Biosci.* **2002**, *2*, 329–335.

279 S. Warwel, C. Demes, G. Steinke, *J. Polym. Sci. Part A: Polym. Chem.* **2001**, *39*, 1601–1609.

280 H. Uyama, M. Kuwabara, T. Tsujimoto, S. Kobayashi, *Biomacromolecules* **2003**, *4*, 211–215.

281 G. S. Kumar, A. Ghogare, D. Mukesh, *J. Appl. Polym. Sci.* **1997**, *63*, 35–45.

282 (a) K. Pavel, H. Ritter, *Makromol. Chem.* **1991**, *192*, 1941–949; (b) O. Noll, H. Ritter, *Macromol. Rapid. Commun.* **1997**, *18*, 53–58.

283 H. Uyama, T. Fukuoka, I. Komatsu, T. Watanabe, S. Kobayashi, *Biomacromolecules* **2002**, *3*, 318–323.

284 V. Sereti, H. Stamatis, E. Koukios, F. N. Kolisis, *J. Biotechnol.* **1998**, *66*, 219–223.

285 (a) B. Tosh, C. N. Saikia, *Trends Carbohydr. Chem.* **1999**, *4*, 55–67; (b) O. A. El Seoud, G. A. Marson, G. T. Ciacco, E. Frollini, *Macromol. Chem. Phys.* **2000**, *201*, 882–889.

286 K. Yang, Y.-J. Wang, *Biotechnol. Prog.* **2003**, *19*, 1664–1671.

287 J. Xie, Y. L. Hsieh, *J. Polym. Sci. Part A: Polym. Chem.* **2001**, *39*, 1931–1939.

288 J. Li, W. Xie, H. N. Cheng, R. G. Nickol, P. G. Wang, *Macromolecules* **1999**, *32*, 2789–2792.

289 S. Chakraborty, B. Sahoo, I. Teraoka, L. M. Miller, R. A. Gross, *Macromolecules* **2005**, *38*, 61–68.

290 (a) T. Chen, H. D. Embree, L. Q. Wu, G. F. Payne, *Biopolymer* **2002**, *64*, 292–302; (b) T. Chen, H. D. Embree, E. M. Brown, M. M. Taylor, G. F. Payne, *Biomaterials* **2003**, *24*, 2831–2841.

291 L. Q. Wu, H. D. Embree, B. M. Balgley, P. J. Smith, G. F. Payne, *Environ. Sci. Technol.* **2002**, *36*, 3446–3454.

292 V. Crescenzi, A. Francescangeli, A. Taglienti, *Biomacromolecules* **2002**, *3*, 1384–1391.

293 B-H. Hu, P. B. Messersmith, *J. Am. Chem. Soc.* **2003**, *125*, 14298–14299.

294 J. E. Chung, M. Kurisawa, H. Uyama, S. Kobayashi, *Biotechnol. Lett.* **2003**, *25*, 1993–1997.

295 T. Otsuka, A. Sawa, R. Kawabata, N. Nio, M. Motoki, *J. Agric. Food Chem.* **2000**, *48*, 6230–6233.

296 N. Ihara, S. Schmitz, M. Kurisawa, J. E. Chung, H. Uyama, S. Kobayashi, *Biomacromolecules* **2004**, *5*, 1633–1636.

297 J. E. Chung, M. Kurisawa, Y. Tachibana, H. Uyama, S. Kobayashi, *Chem. Lett.* **2003**, *32*, 620–621.

298 N. Ihara, Y. Tachibana, J. E. Chung, M. Kurisawa, H. Uyama, S. Kobayashi, *Chem. Lett.* **2003**, *32*, 816–817.

299 (a) T. Fukuoka, H. Uyama, T. Kakuchi, S. Kobayashi, *Macromol. Rapid Commun.* **2002**, *23*, 698–702; (b) T. Fukuoka, H. Uyama, S. Kobayashi, *Biomacromolecules* **2004**, *5*, 977–983; (c) T. Fukuoka, H. Uyama, S. Kobayashi, *Macromolecules* **2004**, *37*, 8481–8484.

300 T. Fukuoka, H. Uyama, S. Kobayashi, *Macromolecules* **2003**, *36*, 8213–8215.

11
Biosynthesis of Protein-based Polymeric Materials

Robin S. Farmer, Manoj B. Charati, and Kristi L. Kiick

Advances in polymer synthetic methods, including anionic, ring-opening, coordination, enzymatic catalysis and other polymerization methods highlighted in these volumes, have been crucial in the development of new polymeric materials with controlled tacticity, block and graft structure, topology and chemical reactivity. The development of these methods has been crucial for the expansion of polymeric materials in applications such as drug delivery, tissue engineering, soft lithography and biomineralization. Over the last 15 years, biosynthetic methods for the production of protein-based polymers have continued to be prominent, as the genetically directed nature of the biosynthetic method allows for absolute control over polymer length and amino acid sequence and, consequently, also over the folded structure of the protein polymer. These synthetic strategies have therefore afforded new materials with well-controlled properties and a variety of macromolecules have been produced; some mimic natural molecules such as silk, elastin and collagen and others are based on artificial amino acid sequences designed to exhibit specific architectural features.

All of the information for directing the biosynthesis of proteins and polypeptides is encoded in the genetic material of the chromosomes, i.e. double-stranded DNA. DNA sequences are converted to their corresponding protein sequences at the ribosome via two general steps of high fidelity (error rates less than one in 10^4) mediated by Watson–Crick base pairing [1–3]. First, a specified nucleic acid sequence is copied into a corresponding mRNA sequence (*transcription*). Second, these mRNA fragments are used as templates to direct the synthesis of proteins (*translation*). The translation of the mRNA template into a protein proceeds with the involvement of a second class of RNA molecules – transfer RNA (tRNA) – that acts as an adaptor between the mRNA and the amino acids of the protein (Fig. 11.1); each tRNA molecule carries a specific amino acid to the ribosome, thereby ensuring the fidelity of the synthesis of proteins with precisely defined structures and functions. It is the control of sequence that is paramount, as it is the amino acid sequence of a protein that contains the information necessary to direct the proper folding of a protein into the specific secondary (e.g. α-helix, β-sheet) and higher order structures that control its

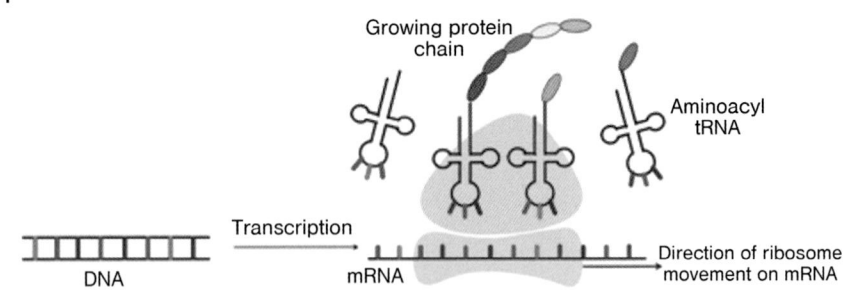

Fig. 11.1 Schematic of protein biosynthesis.

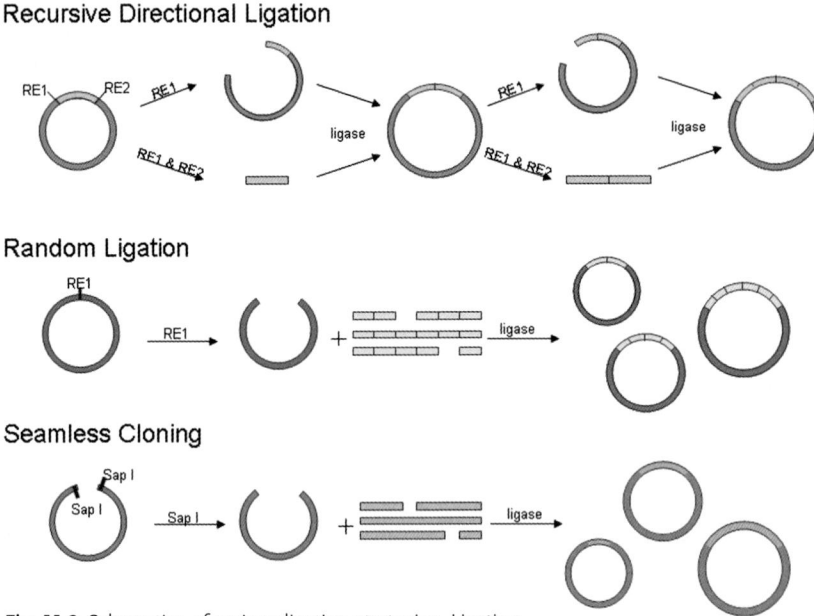

Recursive Directional Ligation

Random Ligation

Seamless Cloning

Fig. 11.2 Schematics of various ligation strategies. Ligation results in a collection of expression plasmids with varying gene lengths. Only a single expression plasmid is transformed into an *E. coli* cell, ensuring that a single expression host encodes a protein polymer of a distinct and precise length.

function. This sequence control has a unique and powerful advantage in polymer synthetic approaches and has been exploited for the precise specification of macromolecular sequence, structure and properties, at the angstrom length scale, in advanced protein-based polymeric materials.

The DNA fragments that encode a desired amino acid sequence are obtained via one of two strategies: either extraction from an organism that produces the sequence or solid-phase techniques. In both cases, the desired DNA fragment is then incorporated (ligated) into a circular plasmid DNA and the expression host, most commonly *Escherichia coli*, is transformed with that plasmid. The target protein polymer can then be produced by cultures of the expression host. The direct cloning of cDNA fragments from organisms has been used widely for the production of natural proteins. The use of solid-phase techniques, however, permits the production of oligonucleotides that encode polypeptides that either mimic natural proteins or are novel sequences not found in nature. Because many natural structural proteins are composed of repetitive amino acid sequences, the production of protein-based polymers via genetic methods has been focused primarily on methods in which multiple copies of a DNA fragment are ligated to form a repetitive gene. A variety of ligation methods, including random ligation, recursive directional ligation [4] and seamless cloning [5], have been employed to incorporate oligonucleotide sequences into expression plasmids for the production of protein-based polymers with repetitive amino acid sequences and of multiblock protein-based copolymers (Fig. 11.2). Expression plasmids containing the multimeric DNA are transformed into the desired expression host, most commonly *E. coli*. During protein expression, host cells are grown to a desired density, at which point protein expression is thermally or chemically induced. Protein-based polymers of many different compositions have been expressed at high levels, accumulating intracellularly, and can be isolated from cellular proteins via cell lysis, precipitation and chromatographic techniques.

11.1
Protein Polymers That Mimic Natural Proteins

11.1.1
Silk

Naturally occurring silks from silkworms and spiders have mechanical properties that are superior to those of most synthetic fibers and, because of their high toughness and elastic moduli, natural silk proteins are desirable materials for lightweight and high-performance fiber applications. Although silk has attractive materials properties, it is not readily made in bulk quantities as production is dependent on the insects or spiders. This has motivated its production via other expression hosts to expand the possible applications for silk-like materials; silk's repetitive amino acid sequences have facilitated the production of silk-mimetic artificial proteins via genetic methods. Attempts to design polymers with mechanical properties similar to those of naturally occurring silks have been focused primarily on silk protein sequences from the *Nephila clavipes* spider, the *Bombyx mori* silkworm and the *Samia cynthia ricini* silkworm.

The *N. clavipes* dragline silk has two consensus sequences, [AGQGGYGGLG-SQGAGRGGLGGQGAGA$_7$GG] and [(GPGGYGPGQQ)$_3$GPSGPGSA$_{10}$], spidroin I and spidroin II, respectively. *B. mori* silk has a glycine- and alanine-rich consensus sequence, GAGAGS, while the *S. c. ricini* silkworm has a conserved

a) Silk-like Polypeptides

b) Elastin-like Polypeptides

c) Collagen-like Polypeptides

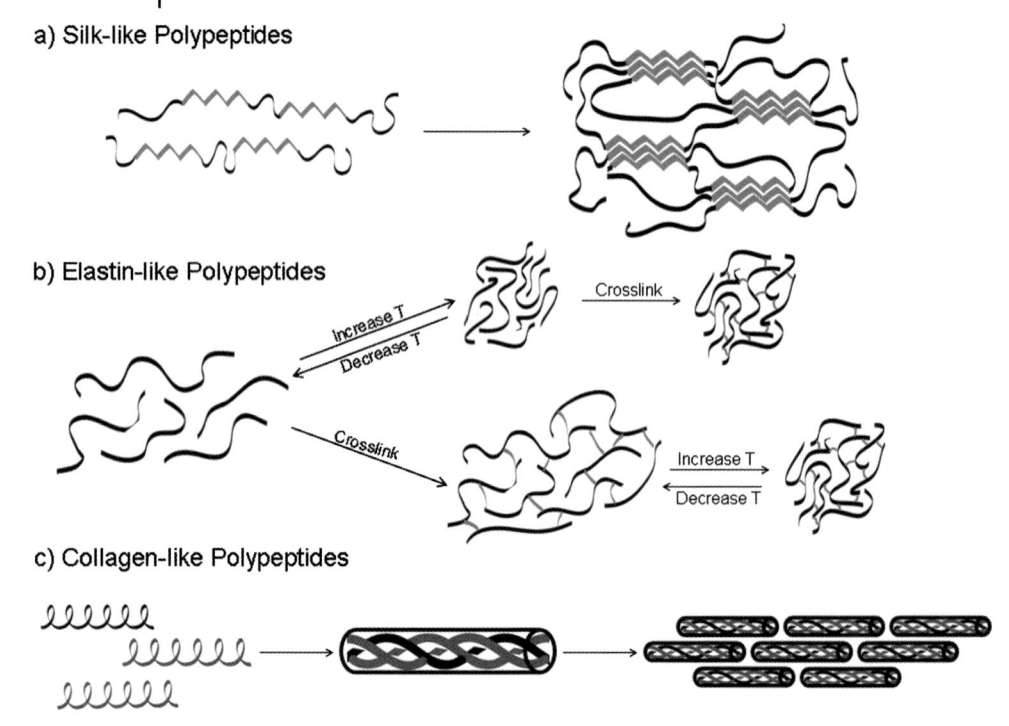

Fig. 11.3 Schematics of (a) silk-like polypeptides, (b) elastin-like polypeptides and (c) collagen-like polypeptides.

stretch of polyalanine. The polyalanine- and alanine-rich segments in the proteins from both spiders and silkworms have been indicated to play an important role in the superior mechanical properties of the silks, as they form β-sheet, "cross-linked" regions interconnected by a disordered structure (Fig. 11.3a) [6]. There are various types of spider silk, including dragline silk, which is used to support the spider, and capture silk, which is used in the web spiral. Many protein polymers with sequences based on the dragline silk from *N. clavipes* have been produced, although there have also been limited studies on flagelliform silk found in capture silk [7] and silk from other spiders [8].

The expression of silk proteins, whether natural or mimetic, has proven to be a difficult task. Low yields and protein truncation have complicated the production of recombinant silks because of poor codon matching of the gene with the bacterial expression host and depletion of the aminoacyl-tRNA pool of certain amino acids due to the highly repetitive sequence [9]. The yields of various silk analogues range from 2 to 15 mg of protein polymer per liter of culture, which is too low to be useful in many materials applications. The yields, however, have been increased via improved purification procedures [10] and there are limited reports of high expression yields under specific conditions employing *Pichia pastoris* or high cell density production from *E. coli* [11, 12].

As mentioned above, most spider silk analogue proteins have sequences based on the spidroin I and spidroin II consensus sequences [13, 14]. The dragline silk from other spiders has also been investigated, albeit to a lesser extent than the dragline silk of *N. clavipes*. The *Araneus diadematus* spider, for example, has two major dragline silk proteins, ADF-3 and ADF-4. These two proteins have similar sequences – GPGGX repetitive regions followed by a polyalanine region – but the two proteins exhibit different solubilities and behave differently during assembly. Because of the higher solubility of the analogues of *A. diadematus* dragline silks, they show higher expression levels than analogues of the *N. clavipes* dragline silk [8], which may offer improved opportunities for the application of silk-like protein polymers in multiple applications.

Although *E. coli* has been the prominent expression host for spider silk analogues and recombinant spider silk, a variety of other hosts have been used to express these proteins. The use of expression hosts such as yeast has increased expression yields and decreased the truncation of the protein polymer sequences [15]. Transgenic plants, such as tobacco and potato plants, are also viable options as expression hosts for the silk-like proteins [16, 17]. For example, the DP1 silk analogue based on *N. clavipes* dragline silk has been produced in the model plant *Arabidopsis thaliana*. Artificial genes encoding DP1 with molecular weights of 64 and 127 kDa were cloned into the plant genome and the resultant proteins accumulated in both the seeds and the leaves, with higher levels of accumulation observed in the seeds. Fusion of a signal peptide and retention peptide to the N- and C-termini, respectively, of the DP1 sequence has been shown to guide the silk protein polymer to specific sites in the cells [18], thereby increasing the yields of purified protein up to 10 times greater than those reported for transgenic plants without protein targeting. These results are extremely promising for the industrial application of spider silk analogues, as these high yields make *Arabidopsis* plants, and the seeds in particular, a feasible option as an expression host for silk protein polymers.

Recombinant spider silk proteins have also been expressed in mammalian cells and insect cells in an attempt to overcome the low yields and truncation seen in *E. coli* expression. Partial cDNA clones of dragline silk from the *N. clavipes* and the *A. diadematus* were transfected into bovine mammary epithelial alveolar cells and baby hamster kidney cells. The silk proteins were expressed and secreted in the media with yields of approximately 25–50 mg L^{-1} of culture. In addition, *Araneus diadematus* dragline silks, ADF-3 and ADF-4, have also been expressed from an insect cell line derived from the fall armyworm (*Spodoptera frugiperda*) [19].

Although a great deal of research has been conducted on spider silk, silk from the cocoons of silkworms has also been of great interest. Silk from two types of silkworms, *B. mori*, which contains (GAGAGS)$_n$ and (GAGAGVGY)$_n$ consensus sequences, and *Samia cynthia ricini*, which consists of a polyalanine consensus sequence, have been the primary focus of research on silkworm silk. Asakura et al. have combined the polyalanine sequence from *S. c. ricini* with the (GAGAGVGY)$_n$ sequence from *B. mori* to form a chimeric silkworm silk that is soluble in 8M urea and adopts a helical structure, in contrast to the β-sheet forming *S. c. ricini* silk [20]. Although most research has centered on the fibroin protein of silkworm silk, sericin, a second protein found in *B. mori* silk, binds two fibroin fibers to form the silk fiber. The consensus sequence of the sericin protein of *B. mori* silk – (SSTGSSSNTDSNSNSVGSSTSGGSSTYGYSSNSRD-GSV)$_n$ – has been expressed from an *E. coli* expression host with yields ranging from 15 to 36 mg per liter of culture [21]. The protein polymer displayed a mixture of random coil and β-sheet conformations after purification, but with dialysis against deionized water, the protein polymer adopts a β-sheet conformation.

Given that recombinant silk and silk analogues from either spiders or silkworms can usually only be solubilized in harsh solvents such as formic acid and HFIP, modifications to the consensus sequences have been employed as a method to improve the processability of the engineered protein polymers. Bulky side-groups, such as those in methionine residues, have been incorporated into silk sequences to control the formation of the β-sheet structure through a triggering process. When the residue is in its reduced form, the protein can form a β-sheet structure, and when the residue is oxidized, there is a decrease in the β-sheet content [22–24]. Similar triggering behavior was achieved via incorporation of a serine residue that can be phosphorylated [25]. In addition, consensus sequences of spider silk and silkworm silk have been encoded in various combinations with each other or with artificial amino acid sequences in an attempt to increase the solubility of the protein polymers [20, 26, 27].

The biological function of silk analogues has also been modified, via inclusion of cell-binding domains in the amino acid sequence; the resulting protein polymers are targeted as a mechanically strong scaffold for tissue engineering applications. Sequences from the silkworm silks of *B. mori* and *S.c. ricini* have been employed in these studies and the amino acid sequence RGD from fibronectin has been included to promote cell attachment to the resulting materials [28, 29]. In particular, a hybrid silk protein combining the polyalanine region of *S.c. ricini* with the cell adhesive site RGD was expressed in *E. coli*. The resultant protein polymer showed higher levels of cell adhesion and cell growth than collagen and *S.c. ricini* fibroin, indicating the potential of this new hybrid protein as a tissue engineering scaffold.

11.1.2
Elastin

Elastin is a structural protein that is found in connective tissues, the lungs and blood vessels. The protein gives tissues extensibility and recoil, which allows them to return to their original state after expansion. Elastin is expressed as tropoelastin, which, after hydroxylation and cross-linking, assembles into an insoluble mature elastin network. The main component of elastin is a hydrophobic domain which contains valine and glycine in abundance, in addition to proline, alanine, leucine and isoleucine [30]. In mammals, the consensus repeat sequence VPGVG mediates the mechanical properties of the protein and most elastin-like protein polymers contain variations of this sequence. Interest in recapitulating the highly elastic behavior of the natural elastins, coupled with the ease of expression of the elastin-like protein polymers, has driven the enormous expansion of research on these polypeptides over the last several years. Most ELPs have been expressed in high yields from *E. coli* and the development of other expression systems has therefore not been as widely pursued as in the silk-like polypeptides.

The useful thermally responsive behavior of ELPs has also motivated their extensive study and application. Polypeptides of this sequence exhibit an inverse temperature transition, in which the polymer phase separates from solution with an increase in temperature, upon its collapse into a β-spiral structure and concomitant liberation of water associated with the polypeptide chain (Fig. 11.3 b). This thermally tiggered behavior is conserved for all protein polymers of the general sequence [VPGXG]$_n$, where X can be any amino acid except proline, as determined by Urry and coworkers [31–34]. The temperature at which the protein polymer undergoes the phase transition, T_t, can be controlled via changes in the identity of X, which offers many promising opportunities for various applications including drug delivery, surface modification, nanoparticle formation, actuation and purification.

The thermal properties of the general sequence [VPGXG]$_n$ can also be tuned via incorporation of varying ratios of amino acids in position X. Chilkoti and coworkers have demonstrated the utility of modifying ratios of valine, glycine or alanine for the production of ELPs with physiologically relevant T_ts for polymer-mediated targeting of drugs to hyperthermic areas in the body such as tumors [4, 35–37]. The transition behavior of the ELPs can also be controlled via the use of copolymers containing blocks of (VPGVG)- and the (VPGXG)-based sequences. Based on these studies, a predictive model to determine the T_t for a solution of an ELP with a given amino acid composition (V:G:A in the X position), molecular weight and concentration has been developed, allowing the design of specific ELPs for specific applications [37]. Tirrell and coworkers have produced ELPs with lower T_ts for applications as vascular graft replacement materials. ELPs of the sequence MG[LDCS5(GVPGI)$_x$]$_y$LE, where $x:y$ is 40:3 or 20:5, were expressed in *E. coli* [38] with yields of 40–60 mg L^{-1}. The lower transition temperatures (22–28 °C) of these VPGIG-based ELPs facilitate the use of these polypeptides under physiological conditions [39] and the inclusion of the cell-binding domain of CS5 has been useful for promoting endothelial cell attachment to these materials.

Because the phase separation of ELPs at elevated temperatures is completely reversible upon cooling, methods to cross-link the phase-separated coacervate have been developed (Fig. 11.3 b). One approach for creating insoluble scaffolds for tissue engineering applications employs the cross-linking of the protein polymers through γ-irradiation or chemical reaction to form a hydrogel irreversibly. Controlled extents of cross-linking are possible for the ELPs via these strategies. For example, (IPGVG)$_{260}$ and (VPGVG)$_{251}$ have been cross-linked via γ-irradiation and (IPGVG)$_{260}$ has been cross-linked via radical reactions initiated by a dicumyl peroxide initiator. As expected, the moduli of the hydrogels increases with increase in cross-linking density [40, 41].

Functional amino acid residues have also been incorporated into the (VPGXG) sequences to introduce sites for chemical cross-linking. The two most predominantly used residues have been lysine and glutamic acid and there have been numerous sequences designed in which multiple repeats of (VPGVG) and [VPG(E/K)G] have been interdispersed. For example, Urry and coworkers have expressed the ELP sequence [(GVGVP)$_2$(GXGVP)(GVGVP)$_3$]$_x$ in *E. coli*, where X is either lysine or glutamic acid [41, 42]. The incorporation of a combination of lysine and valine residues in specific positions of the general (VPGXG)$_x$ sequence has also been widely employed and provides protein polymers in which cross-linking density is controlled via control of the distance between the lysine positions [43–48]. For example, McMillan et al. reported the use of a seamless cloning strategy to synthesize the sequence [(VPGVG)$_4$(VPGKG)]$_x$, which can be cross-linked via reaction of 85% of the lysine residues with bis(sulfosuccinimidyl) suberate to form a hydrogel [43]. The density of lysine residues also affects the hydrogel formation outside of considerations of cross-linking density. Specifically, sequences with higher lysine density form hydrogels over a wider range of protein polymer concentrations than sequences with lower lysine density [47].

Elastin-like polypeptides containing both lysine and glutamic acid residues have also been cross-linked to form hydrogels. The protein polymer sequence (VPGVG VPGKG VGPVG VPGVG VPGEG VPGIG)$_x$ has been expressed in *E. coli* [49] and the uncross-linked protein polymer exhibits expected inverse temperature transition behavior with transition temperatures decreasing from 63 to 35 °C with increasing polymer concentration from 1 to 50 mg mL^{-1}. Polymers that contain both lysine and glutamic acid residues may also offer advantages in the design of elastin-like polypeptide-based hydrogels, as cross-linking can be mediated through the nucleophilic reaction of the amine-containing side-chain, and the glutamic acid side-chain can be modified with ligands or other moieties that yield inverse temperature transition behavior that is responsive to chemical, rather than just thermal, stimuli. For example, Glu-containing ELPs have been modified not only with flavin adenine dinucleotide (FAD) [42], to control thermally responsive behavior via biologically relevant oxidation/reduction reactions, but also with metal-binding ligands [50], which can be employed for protein purification.

11.1.2.1 Elastin-like Polypeptide Copolymers

Other approaches for manipulating hydrogel formation or mechanical properties of ELP materials focus on the use of copolymers of (VPGVG)$_x$ motifs. For example, BAB triblock copolymers have been produced by Conticello and coworkers, in which the B block, [(IPAVG)$_4$(VPAVG)]$_n$, is hydrophobic and plastic and the A block, [(VPGVG)$_4$(VPGEG)]$_m$, is hydrophilic and elastic [51, 52]. Both blocks exhibit inverse transition temperatures, but block B has a lower T_t (near room temperature), with its mechanical behavior more plastic under physiological conditions. These protein polymers can be cast into films and, depending on the solution from which the film is cast, the films exhibit different mechanical properties based on solvent-induced differences in the microstructure of the copolymer. Young's moduli ranging from 0.03 to 35 MPa and elongation to break values of 250–1300% can be obtained via choice of processing conditions [51, 52]. In addition, nanoparticles can be formed from triblock copolymers with similarly designed blocks, and also by diblock ELPs with blocks of differing hydrophobicity [53–55]. The assembly process can be manipulated via sequence variations that allow triggered assembly (into nanoparticles or networks) on the basis of changes in pH, temperature and ionic strength.

ELP sequences that combine exon sequences from human elastin in an arrangement different to natural elastin have also been reported by Woodhouse and coworkers [56]. Sequences combining (PGVGVA)$_x$ and (VPGVG)$_x$ with elastin-based lysine-rich cross-linking domains have been utilized to understand the conversion of tropoelastin into insoluble β-fibril networks. These polypeptides form coacervates above the transition temperature and further incubation at temperatures greater than T_t results in a fibrillar structure, which displays mechanical behavior commensurate with natural elastin [57–59]. These polypeptides can be used not only as model systems to understand the formation of fibrils in elastin, but also as well-defined self-assembling biomaterials for applications in tissue engineering.

11.1.2.2 Silk–Elastin-like Polypeptides

Silk analogues and elastin mimetic polypeptides have been combined on the genetic level to form silk–elastin-like polypeptides (SELPs) that exhibit the mechanical properties of silk and the inverse temperature transitions of elastin. This class of protein polymers has been expressed from *E. coli* and has great potential for use in biomedical applications. The SELPs contain both crystalline and flexible domains and spontaneously form hydrogels via changes in temperature or solvent; the lack of a need for chemical or irradiation cross-linking affords this group of protein polymers a processing advantage over the gels produced from strictly elastin-based macromolecules. Cappello et al. have reported various SELP sequences which contain varying numbers of repeats of the *B. mori* silk sequence, (GAGAGS), in combination with (VPGVG) and (VPGKG) elastin-like sequences [60–62]. Association of the silk blocks in these protein polymers is indicated to be the driving force for observed gel formation, as an increase in the number of silk blocks in a repeat unit leads to gelation. The swelling ratio for specific hydrogels has been shown to be independent of temperature, pH and ionic strength, but is dependent on the protein polymer concentration, as the concentration controls the number of silk domains available to form physical cross-links. In particular, the polymer ProLastin 47K with the sequence [(GVGVP)$_4$GKGVP(GVGVP)$_3$(GAGAGS)$_4$]$_{12}$, has been shown to have useful gelation and swelling properties for drug delivery applications [61, 63, 64].

Because these protein polymers form hydrogels under physiologically relevant conditions, they are prime candidates for tissue engineering and drug delivery applications. Megeed et al. reported that the resorption and biocompatibility of the SELP hydrogels are controlled by the number of silk-like repeats in the SELP [65]. Upon implantation, the hydrogels of lower silk content caused a mild immune response before being resorbed, whereas the hydrogels of high silk content triggered the formation of healing tissue around the implant and little or no resorption. SELP hydrogels have also been investigated as drug delivery vehicles for therapeutic molecules, including theophylline, vitamin B_{12} and cytochrome c, and plasmid DNA [64–67]. The release of these molecules from ProLastin hydrogels depended on the cross-linking density and the size of the solute molecule; a more porous network permits smaller molecules to diffuse faster from the matrix as predicted [64].

11.1.2.3 Applications of ELPs

ELPs are candidates for use in a wide variety of applications because of their thermally responsive behavior and assembly. Many applications for ELPs involve the attachment of another molecule to the protein polymer for targeting or binding; for example, therapeutic molecules have been attached to ELP sequences to target cancers or tumors. ELPs that exhibit transition temperatures near 40 °C have been produced and have been shown to accumulate preferentially in heated tumors [68–71]. Elastin-like polypeptides of the general sequence [VPGXG]$_x$ (where X = V, A or G in ratios of 5:3:2 or 1:8:9) exhibit transition temperatures of 40 and 42 °C, respectively and have been used for transporting the anticancer drug doxorubicin to tumor sites [69, 70]. The cytotoxicity of the drug in association with ELP was found to be similar to that of the free drug, although via a distinct mechanism. These two protein polymer sequences have more recently been combined with peptide sequences, including a c-Myc inhibitory peptide sequence that can stop the production of c-Myc, an oncogene which is overexpressed in tumors and known to play a role in tumor proliferation [71]. The protein polymers, with a T_t near the hyperthermic temperature of the tumor, show high levels of uptake and the presence of the protein polymer causes a redistribution of the c-Myc into the cytoplasm, inhibiting cell transcription activity and therefore cell proliferation. The delivery of drugs to specific sites through the use of ELP sequences shows significant promise for tumor treatment methods that reduce accumulation of drugs in non-malignant areas.

Peptide-based cell-binding sequences can also be included in the ELP sequence on the genetic level. The addition of the CS5 cell-binding domain of fibronectin (GEEIQIGHIPREDVDYHYP) to the ELP sequence yields a protein polymer that is a candidate for vascular tissue engineering applications. The CS5 cell binding sequence has been included in the ELP sequence (VPGIG)$_n$ to produce ELPs capable of binding endothelial cells [38, 72]. The protein polymer can be easily expressed from *E. coli* and cell adhesion studies, performed on polypeptide-coated glass slides, showed HUVEC (human umbilical vein endothelial cell) attachment only on the ELP–CS5 films and not on ELP controls. Another ELP–CS5 protein polymer, containing the elastin-based amino acid sequence [(VPGIG)$_2$(VPGKG)(VPGIG)$_2$]$_x$, was designed for the production of protein polymer films through chemical cross-linking of the lysine residues with bis(sulfosuccinimidyl) suberate [73]. Enhanced mechanical properties were observed for films with higher cross-link densities and the moduli of the films ranged from 0.7 to 0.97 MPa, commensurate with the moduli of native elastin, 0.6 MPa [74]. In cell attachment studies, HUVECs spread on the ELP-based surfaces exhibit similar behavior to those spread on fibronectin and approximately 60% of the cells attached to the protein polymer film can withstand physiologically relevant shear stresses [75]. In addition, Girotti et al. have incorporated a target sequence for elastase – VGVAPG – into a similar ELP–CS5 protein polymer, permitting the enzymatic degradation of the protein polymer when employed under conditions relevant for tissue engineering applications [76].

ELPs are also being used in applications involving purification of a variety of molecules and macromolecules. The isolation of small molecules such as heavy metals [77, 78] and organophosphorus compounds [79] from solution, and the isolation of macromolecular species such as antibodies [80, 81], DNA [82] and proteins [83, 84] have all been simplified and enhanced through the use of ELP sequences. Specifically, the attachment of binding motifs to ELPs subsequently allows isolation of target molecules from solution upon heating of the solution above the transition temperature of the ELP–target molecule complex. The ELPs can be cycled through the purification process multiple times with little to no decrease in the efficiency of target compound removal. Similarly, proteins with ELP fusion tags can be purified via association with free ELPs at elevated temperatures, with protein isolation yields that are comparable to those in other purification processes [84].

11.1.3
Collagen

Collagen is a family of fibrous structural proteins found in the extracellular matrix of connective tissues. Although there are more than 20 varieties of collagen, all share the general amino acid sequence Gly–X–Y, where X is generally proline and Y is hydroxyproline. Collagen is expressed as a propeptide, an individual polypeptide chain which adopts a polyproline-like helical structure (Fig. 11.3c). Tropocollagen fibers are formed by the assembly of three propeptide chains into a triple helix [85] and further hierarchical assembly produces high-strength collagen fibers found in connective tissues such as tendons, ligaments and cartilage. Biologically, collagen fibers aid in cell attachment, development and proliferation, and also wound healing and tissue remodeling. The structural, mechanical and biological properties of collagen make it an attractive material for use in biomedical applications, foods, cosmetics and therapeutics, and the production of recombinant collagen and collagen-like polypeptides via protein engineering strategies offers the option of tailoring the primary sequence of collagen for these applications.

Most research involving the expression of collagen proteins has focused on the production of recombinant collagen, in which the gene encoding the collagen sequence is transferred from the target mammal to an expression host, avoiding the need for isolation of collagen from animal sources. The expression of various mammalian collagens in yeast expression hosts has been widely investigated. The expression of propeptides, the precursors to native collagen, provides yields of 3 to 15 g L^{-1} of the propeptide from culture [86–93]. Insects and mammalian cells have also been used as expression hosts for both propeptides and collagen [94–101]. When the protein polymer is coexpressed with prolyl 4-hydroxylase or the culture is supplemented with ascorbate, the melting temperatures and secondary structure of the protein polymer are commensurate with those of the natural protein [96, 97, 99].

The interest in producing repetitive collagen-like polypeptides has increased significantly since the initial attempts in 1989 [102]. Most work has focused on the use of yeast as the expression host for the collagen analogues [90, 93, 103]. Kajino et al. investigated the modification, on the DNA level, of two sequences found in human collagen to provide matching of the codons in the gene encoding the target protein with the preferred codon usage of the yeast expression host [103]. Tandem repeats of the sequences (GESGREGAPGAAEGSPGRDGSP-GAKGDRGET) and (GPAGPPGAPGAPGAPGPVGPAGKSGDRGET) were expressed in *Bacillus brevis* and after 6 days of expression, 0.5 g L^{-1} of protein was secreted from the cells and collected from the supernate via ammonium sulfate precipitation. Solutions of the artificial collagen protein showed thermoreversible viscosity changes similar to native collagen. In addition, a fully synthetic gelatin-based protein polymer with the general sequence (GXY)$_n$ has been expressed in the yeast expression host *P. pastoris* [90]. In efforts to improve expression from *E. coli*, a synthetic gelatin-like polypeptide was designed with the hydrophilic glutamine included in the Y position. The sequence (GAPGAPGSQ-GAPGLQ)$_x$ can be expressed from *E. coli* with yields of 100–200 mg per liter of culture [104]. The high expression yields of collagen or gelatin analogues that can be achieved using either yeast or *E. coli* expression hosts are promising for future synthesis and application of collagen-like polypeptides.

Amino acid sequences of collagen found in the extracellular matrix contain the tripeptide sequence GER, which is important for cell binding and angiogenesis [105]. A collagen-like polypeptide containing both the cell adhesion sequences and the cross-linking sequences found in natural collagen has been designed to mimic the collagen found in the extracellular matrix [106]. The amino acid sequence [GERGDLGPQGIAGQRGVV(GER)$_3$GAS]$_8$ has been expressed from *E. coli* and the purified protein polymers display circular dichroism (CD) spectra consistent with those expected from the collagen triple helix. Cell adhesion studies show that the attachment of mouse fibroblast cells to the collagen-like polypeptide exceeds that observed for cells on native collagen surfaces.

11.1.4
Other Naturally Occurring Proteins

11.1.4.1 Resilin

Resilin is an elastomeric protein mainly found in the specialized regions of the cuticle in most insects and was first discovered four decades ago by Weis-Fogh during a study of flight systems in locusts [107]. Resilin is a highly elastomeric protein and is present in most, if not all, insects, where it has been adapted for flight [107, 108], jumping [109, 110] and sensory mechanisms. It confers long-range elasticity to the cuticle and probably functions as both an energy store [111, 112] and a damper of vibrations in insect flight. Resilin has also been found in sound-producing organs of some insects such as cicadas [113] and moths [114], where it withstands frequencies of vibration up to 4 kHz. Like elastin, resilin has a high content of glycine (35–40%) and proline (7–10%) [115], but in contrast to elastin, resilin is cross-linked in its soluble domain via the formation of dityrosine and trityrosine cross-links [111].

Andersen and coworkers identified a gene from *Drosophila melanogaster*, CG15290, likely to be a precursor for *Drosophila* resilin [116]. In a recent report, Elvin and coworkers cloned the N-terminal domain of the CG15290 gene, dominated by 17 putative repeat motifs of GGRPSDSYGAPGGGN, and expressed this gene in *E. coli* as a soluble protein [117], with a yield of 15 mg L^{-1}. An Ru(II)-mediated photochemical method was used to cross-link the resilin to form dityrosine linkages. This synthetic resilin could be stretched to three times its original length and was found to have a resilience of 97%, superior to that of elastin and other synthetic rubbers. It is likely that this elasticity arises from entropic restoring forces, although two different mechanisms of chain deformation and entropic recovery have been proposed. One mechanism attributes the elasticity to the reduction in conformational entropy of the highly randomly coiled, thermally agitated chains upon strain [118, 119] and the other suggests that the presence of glycine and proline residues introduces type II β-turns, with the repetition of this conformational unit giving rise to an easily deformable β-spiral conformation [116, 120]. The expression of recombinant forms of resilin will therefore be certain to result not only in new elastomeric materials but also in a more detailed understanding of elastic behavior in rubber-like polypeptides and proteins.

11.1.4.2 Mussel Adhesive Plaque Protein and Glutenin

The production of polypeptide-based mimics of other natural proteins, such as the mussel adhesive plaque protein (MAP), has also been investigated. To date, recombinant versions of MAP have been expressed in *E. coli*, but yields have been low due to poor codon matching and depletion of the tRNA pools of certain amino acids during expression. Expression remains of continued interest, however, as the production of polypeptides that can mimic the adhesive and mechanical properties of this protein may prove beneficial in a variety of applications [121, 122]. Wheat glutenin has also been studied, with specific goals of probing the viscoelastic properties of the protein [123]. Glutenin is comprised of numerous subunits, which, in addition to their importance to the viscoelastic behavior of the protein, have interesting structural characteristics. The identification and production of additional repetitive proteins with useful structural, mechanical, thermal and/or biological properties promise new generations of protein-based materials for specialized applications in both biomedical and materials arenas.

11.2
Protein Polymers of *De Novo* Design

11.2.1
β-Sheet-forming Protein Polymers

In addition to studies intended to produce repetitive polypeptides that mimic naturally occurring proteins, the utility of genetically directed polymer synthesis has also been extended to the production of completely artificial proteins. The use of genetically directed methods to produce artificial amino acid sequences of *de novo* design offers important advantages for the systematic study of structure–function relationships and the design of materials with novel properties for specific applications in nanotechnology and biotechnology.

In original investigations of this kind by Tirrell and coworkers, a protein polymer inspired by the sequence and structure of *B. mori* silk, $[(AG)_xEG]_y$, was expressed from *E. coli* and was shown to form well-defined lamellar crystals with β-sheet structure [124, 125]. Variations in x resulted in alterations in the length of the β-sheet strands between the glutamic acid residues in order to alter crystal thickness. Other variations, in which the glutamic acid residue was replaced by other natural amino acids, yielded protein polymers that form β-sheet structures with controlled lamellar spacing that was directly related to the size of the amino acid encoded in place of E [126, 127]. In total, these investigations demonstrated the utility of the $[(AG)_xXG]_y$ sequences in the formation of lamellar β-sheet crystals of specified architecture and surface functionality. Topilina et al. have recently reported the incorporation of tyrosine, glutamic acid, histidine and lysine residues at the X position of the polypeptide sequence. This polypeptide, $[(GA)_3GY(GA)_3GE(GA)_3GH(GA)_3GK]_y$, displays the X residues at the turn position of the β-sheet, with hydrophobic groups and charged residues displayed on opposite sides [128]. A bilayer β-sheet, fibril self-assembles through hydrophobic interactions and these fibrils were characterized via atomic force microscopy (AFM) and transmission electron microscopy (TEM). Recently, $[(AG)_3EG]_{10/20}$ sequences have also been used as a middle block in a triblock polymer with poly(ethylene glycol) (PEG) end-blocks [129]. The polypeptide domain of the triblock copolymer mediates the formation of β-sheet structures, while the PEG end-blocks control the higher order assembly of the macromole-

a) PEG–[(AG)₃EG]ₓ–PEG

b) Coiled-Coil Hydrogel

c) Artificial Helical Protein Polymers

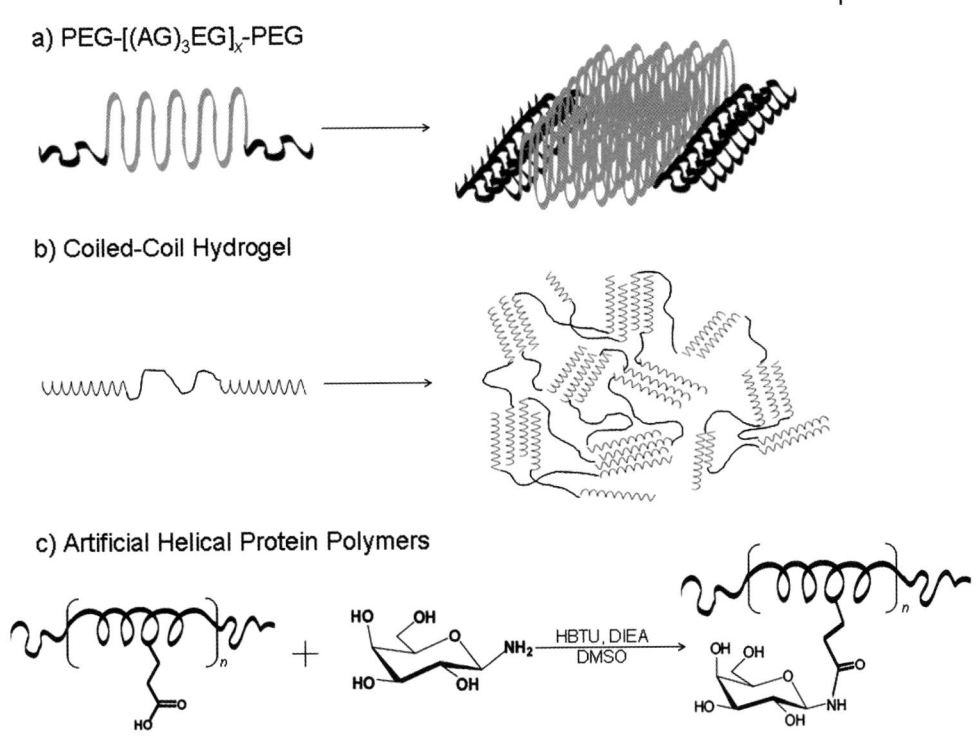

Fig. 11.4 Schematics of (a) PEG–[(AG)₃EG]ₓ–PEG protein polymers, (b) coiled coil triblock copolymer network and (c) alanine-rich α-helical protein polymers.

cules (Fig. 11.4a). The attachment of the PEG chains, via reaction between terminal cysteine residues on the protein polymer and the maleimide function on the PEG molecules, allows the polypeptide domain to form β-sheet fibrils, but inhibits the further aggregation of the fibrils into lamellar crystals. These fibrils, formed through the assembly of antiparallel β-sheet structures, are formed on the micron length scale as observed via TEM.

11.2.2
Liquid Crystals

While the sequence control of genetically directed polymer synthetic methods permits manipulation of folded structures as shown above, the molecular weight control offers unique opportunities for controlled assembly. The utility of such precision in controlling assembly has been demonstrated via the production of monodisperse benzyl ester derivatives of poly(L-glutamic acid) (PLGA) via a combination of biosynthetic and chemical methods [130, 131]. PLGAs of varying molecular weights have been expressed in *E. coli* and the purified protein polymers are chemically modified to form the poly(γ-benzyl-α-L-glutamate) (PBLG) derivatives. Films of the protein polymer prepared from trifluoroacetic acid exhibit smectic-like ordering, with X-ray diffraction reflections corresponding directly to the lengths of the monodisperse molecules. Such smectic-like ordering is not possible in chemically produced PBLGs owing to their polydispersity.

11.2.3
Coiled Coils

Coiled coil motifs are found in transcription factors and related helical motifs can be found in structural proteins including keratin. These motifs consist of a heptad repeat sequence, *abcdefg*, in which hydrophobic residues occupy positions *a* and *d*, whereas charged residues tend to occupy positions *e* and *g*. Under specific solution conditions, these sequences will form amphipathic helices that assemble via hydrophobic interactions. By controlling the sequence and length of the protein polymer, the assembly and stability of these molecules can be manipulated.

In original investigations employing coiled coils in protein polymer assembly, BAB triblock copolymers were produced, with the coiled coils as the B blocks and a random coil protein polymer as the middle block (Fig. 11.4b). These triblock copolymers are driven to form hydrogels through the association of the coiled coil sequences [132], which is a result of four chain bundles forming cross-links, as suggested via small-angle X-ray scattering and analytical ultracentrifugation [133]. The formation of hydrogels is pH dependent; at low pH values the association of the coiled coil and the protonation of the charged midblock cause the precipitation of the protein polymer. The hydrogel can be disrupted by either increasing the pH, which also induces swelling of the charged midblock, or increasing the temperature, but the dissociation of the molecules is fully reversible and the gel re-forms at lower pH and temperature. The transient nature of the cross-links, coupled with their small aggregation number, however, causes these gels to erode readily. Shen et al. have incorporated cysteine residues into the helical domains of the triblocks, both internally and terminally, in order to enhance the stability of the hydrogels through the formation of disulfide bonds [134]. The cysteine residues do not adversely affect the helical structure of the coiled coil sequences and the disulfide bonds stabilize the hydrogel, as monitored via tracer bead release studies. Hydrogels formed via the assembly of triblock copolypeptides in which the end-blocks are dissimilar coiled coil domains that do not associate with each other, have also been reported by Shen et al. [135]. This change in network topology, which minimizes the formation of loops that erode from the network, causes marked improvements in the moduli and erosion profiles of these hydrogel materials, further indicating the potential of genetically directed strategies for the explicit design of hydrogels with applications in biology and medicine.

11.2.4
Helical Protein Polymers

Protein engineering techniques have recently been applied to the production of polyelectrolytes with charged functional groups regularly displayed along an alanine-rich α-helical backbone; these macromolecules can direct the electrostatic assembly of nanostructured materials [136]. When the sequence $(A_3E)_n$, expressed from an *E. coli* host, is mixed with cationic lipids, a polyelectrolyte–lipid complex forms, which consists of peptides adsorbed on the membrane surface. In other studies by Kiick and coworkers, a series of alanine-rich protein polymers, of the general sequence $[(AAAQ)_y(AAAE)(AAAQ)_y]_x$, have been designed to permit variation in the number and spacing of glutamic acid residues via changes in the x and y values [137, 138]. These protein polymers can be produced in *E. coli* with yields ranging from 10 to 40 mg L^{-1} of culture depending on the molecular weight of a given protein polymer. The protein polymers are highly helical under ambient conditions, as monitored via CD spectroscopy and via modification of the glutamic acid residues with saccharides, the protein polymers can be employed in systematic studies of protein–saccharide binding events (Fig. 11.4c). The attachment of saccharides does not affect the conformational behavior of the protein polymers and the resulting glycopolymers are promising for use as toxin inhibitors [139].

11.2.5
Non-structured Protein Polymers

Although the protein polymers discussed up to this point have been designed specifically to adopt β-sheet or α-helical structures, there have also been studies aimed at the synthesis of protein polymers that have no regular secondary structure. For example, the sequence $(GKGSAQA)_3$, which, by design, has an unordered secondary structure, has been produced in *E. coli* [140]. Two sequences, $(GAGQGEA)_{n=18-72}$ and $(GAGQGSA)_{n=12-48}$, have been used as drag-tags for DNA oligonucleotides during capillary electrophoresis experiments [141]. The conjugation of the polypeptide sequence increases the elution time of the DNA oligonucleotides and this elution time can be tuned by varying the molecular weight of the drag-tag. Non-structured block copolypeptides, with an anchor domain comprising poly(glutamic acid) and a tail domain comprising polyproline, have also been used for specific attachment of the anchor domain to particles with extension of the tail domain used to control particle–particle interactions [142].

The production of protein polymers via genetically directed strategies has demonstrated its unique promise for the synthesis of polymeric materials with precisely specified and controlled physicochemical, conformational, mechanical, assembly and biological properties. In addition, the derivatization of biosynthetically produced protein polymers with chemically synthesized polymers has also proven beneficial in controlling assembly and may find use in nanotechnology and device applications. As mentioned previously, van Hest and coworkers have integrated the polypeptide sequence $[(AG)_3EG]_n$ with chemically synthesized PEG to form PEG–polypeptide–PEG triblock copolymers in which the β-sheet assembly can be modulated [129, 143]. These studies suggest important opportunities for the production of advanced macromolecular systems via the integration of biological and chemical strategies.

11.3
Proteins Containing Non-natural Amino Acids

Ingenious modification of protein polymers to increase their biological and chemical versatility has been a long sought-after goal, as the chemical versatility offered by conventional *de novo* protein engineering techniques that employ the repertoire of the 20 natural amino acids is limited to the functionality of amines, carboxylic acids and thiols. Incorporation of non-natural amino acids into protein- and polypeptide-based materials produced *in vivo* is a rapidly expanding approach that has broadened the repertoire of amino acids considerably and has added a new dimension to the synthesis of protein-based polymeric materials, primarily as a result of its ability to permit chemical modification at multiple sites in a protein and to effect significant changes in polypeptide physical properties. Several strategies have evolved for the incorporation of a variety of functional groups [144–146], including alkenes, alkynes, halides, ketones and azides, and these novel amino acids can then be selectively modified via desired chemical approaches (Table 11.1). Selected useful methods that have been recently developed are described below.

11.3.1
Synthetic Methodologies

11.3.1.1 Chemical Synthesis
Chemical synthetic approaches, such as solid-phase peptide synthesis (SPPS) [147, 148] and solution-based polymerizations of NCA monomers [149], have been widely employed to synthesize polypeptidic molecules functionalized with non-natural amino acids. Such techniques are straightforward and the number of non-natural amino acids that can be incorporated in this fashion is essentially infinite. However, the efficiency of the coupling steps in SPPS limits the length of the peptide to approximately 50 amino acids and the NCA-based polymerizations offer limited sequence and molecular weight control. To overcome the size limitations of SPPS, efficient strategies have been developed to ligate synthetic peptides together to make larger polypeptides. Native chemical ligation allows the direct coupling of peptide fragments to form a native peptide linkage [150–152]. In this method, a peptide with a thioester at its C-terminus is ligated to a second peptide with an N-terminal cysteine residue through a transthioesterification reaction to form a native peptide bond at the ligation site. It has also

Table 11.1 Incorporation strategies and applications for non-natural amino acids.

Amino acid analogue	aaRS and expression machinery	Application
Homopropargylglycine	WT MetRS	Cu(I)-catalyzed cycloaddition, Sonagashira coupling
Azidohomoalanine	WT MetRS	Staudinger ligation, Cu(I)-catalyzed cycloaddition
Homoallylglycine	WT MetRS	Heck reaction
Photo-methionine	WT MetRS	Study protein–protein interactions, photocrosslinking reactions
Trifluoroleucine	WT LeuRS	Altering surface properties, enhanced thermal stability and hydrophobicity
Trifluoroisoleucine	WT IsoRS	Altering surface properties, enhanced thermal stability and hydrophobicity
Hexafluoroleucine	Overexpression of WT LeuRS	Altering surface properties, ^{19}F NMR
Photo-leucine	WT LeuRS	Study protein–protein interactions, photocrosslinking reactions
p-Fluorophenylalanine	Overexpression of WT PheRS	Altering surface properties, ^{19}F NMR
p-Bromophenylalanine	A294G PheRS[a]	Heck reaction, Sonagashira coupling, structure determination via XRD
p-Iodophenylalanine	A294G PheRS[a]	Heck reaction
p-Azidophenylalanine	A294G PheRS[a]	Photocrosslinking reactions
p-Ethynylphenylalanine	A294G PheRS[a]	Cu(I)-catalyzed cycloaddition
p-Acetylphenylalanine	T251G/A294G PheRS[a]	Side-chain can be modified with hydrazine and hydroxylamine reagents
Benzofuranylalanine	A294G PheRS[a]	Photocrosslinking reactions
4-Hydroxyproline	Expression of ProRS under hyperosmotic concentrations of NaCl	Impart thermal stability and promote proper protein folding
Azetidine-2-carboxylic acid	WT ProRS	Study polymer folding
Dehydroproline	WT ProRS	Can react with H_2O_2 and Br_2
4-Fluoroproline	WT ProRS	Studying thermodynamics of protein folding

a) All the mutant aaRS are overexpressed.

been shown that an auxiliary sulfhydryl group, which can be removed after ligation, allows ligation at a non-cysteine residue [153]. However, the method is impractical for the synthesis of peptides or proteins of poor solubility or high molecular weight.

11.3.1.2 In Vitro Suppression Strategies

A second general approach for non-natural amino acid incorporation is the use of suppression-based methods. In these methods, a suppressor tRNA is chemically modified with a non-cognate amino acid and, when added to a suitably programmed ribosome-mediated, *in vitro* protein synthetic system can insert different non-natural amino acids in response to either a *nonsense codon* (one of the three stop codons encoded in mRNA), a *missense codon* (alterations of one sense codon to another so that different amino acids are determined) or a *frameshift* mutation. The groups of Schultz [154], Sisido [155] and Chamberlin [156] have adopted suppression-based methodology to incorporate many non-natural amino acids in proteins using amber and/or frameshift suppressor tRNAs.

A wide variety of non-natural amino acids can be incorporated into polypeptides via suppression-based methods because the discriminatory control of aaRS is evaded. However, yields of proteins synthesized via *in vitro* translational protocols are often very low [157] due to the difficulty of obtaining sufficient amounts of misacylated tRNAs. Thus, purified cell-free translation systems have been developed; these systems permit the synthesis of peptides, containing non-natural amino acids, in response to several neighboring and randomly chosen sense codons [158, 159]. A purified translation system could in theory allow all 64 codons to be recognized by synthetic aminoacyl-tRNAs because there are no contaminating aminoacyl-tRNA synthetases that would otherwise proofread and recharge the tRNAs with the natural amino acids. Although the method is currently limited to the synthesis of short peptides in small yields (30–50% compared with peptides with natural amino acids), this method may some day make it possible to use different synthetic aminoacyl-tRNAs as substrates for all 64 codons in the genetic code, allowing the synthesis of polypeptide molecules with versatile side-chain and backbone chemistries.

11.3.1.3 In Vivo Suppression Strategies

The use in materials studies of the *in vitro* translation and purified translation systems described above has been curbed by very low yields. Therefore, *in vivo* strategies for site-specific incorporation of non-natural amino acids, which employ heterologous expression systems (an expression host in which both the suppressor tRNA and the desired aaRS are imported from a foreign organism), have been developed. The essential requirement for the success of this method is orthogonality; the introduced aaRS must recognize *only* the non-natural amino acid and catalyze attachment of this amino acid *exclusively* to the introduced suppressor tRNA. In addition, the wild-type aaRS of the expression host must not charge the non-natural amino acid or the introduced suppressor tRNA. These methods have been used for site-specific incorporation of a variety of different non-natural amino acids with azide, acetylene, electrophilic and ketone-containing side-chains. In a recent report [160], Schultz and coworkers produced a heterologous expression host also capable of producing the non-natural amino acid targeted for incorporation. The *E. coli* expression host was equipped with an aaRS and suppressor tRNA pair, from *Streptomyces venezuele*, that is capable of selectively charging *p*-aminophenylalanine. In addition to carrying the heterologous aaRS–tRNA pair, the host was also modified to carry the genes responsible for the production of *p*-aminophenylalanine in *S. venezuele*. The modified *E. coli* host was capable of producing *p*-aminophenylalanine from a basic carbon source and incorporating it into proteins with fidelity and efficiency similar to those for natural amino acids. This approach highlights the versatility of biosynthetic pathways, but like other suppression methods, is currently limited to the single-site incorporation of non-natural amino acids that can be made via identified metabolic pathways.

11.3.2
Multisite Incorporation of Non-natural Amino Acids into Protein Polymers *In Vivo*

Although applications for site-specific incorporation of a single non-natural amino acid are numerous (biosynthetic probes for protein folding studies, enzyme activity assays, etc.), multisite incorporation of non-natural amino acids in the synthesis of protein polymer materials is uniquely suited for permitting chemical modification at multiple sites and for effecting significant changes in the physical properties of the protein polymers. In multisite incorporation methods, multiple amino acid residues of one kind are replaced by a non-natural amino acid during protein biosynthesis. Experimentally, the host cells are starved for a specific natural amino acid and the synthesis of a specific protein is induced, with concomitant addition of the non-natural amino acid to the culture medium. These strategies may also provide long-term opportunities for the incorporation of multiple types of non-natural amino acids at specified multiple positions in a protein, which would further expand the versatility of the approach.

The incorporation of a desired non-natural amino acid into a target protein polymer without harmful effects on the expression host requires that the non-natural analogue must meet numerous criteria. The basic requirements include that the analogue is easily transported across the cell membrane into the amino acid pool of the expression host and does not disrupt metabolic processes once inside the cell. The non-natural analogue must be a good substrate for elongation factor (EF-Tu) for efficient delivery of the analogue-charged tRNA to the ribosomal A site [161]. Most importantly, the non-natural analogue must be recognized by the aaRS and form a stable aminoacyl-tRNA. As mentioned in the Introduction, aaRSs are pivotal participants in translation, catalyzing accurate biosynthesis of aminoacyl-tRNAs (the key molecules of translation) and thereby safeguarding the fidelity of protein biosynthesis.

Typically, the aaRS enzymatically link amino acids to the 3′-end of the cognate tRNA in a two-step process:

$$aaRS + aa + ATP \rightleftharpoons [aaRS:aa \sim AMP] + PP_i$$

$$[aaRS:aa \sim APMP] + tRNA^{aa} \rightleftharpoons aaRS + AMP + aa \sim tRNA^{aa}$$

where ATP is adenosine triphosphate, AMP is adenosine monophosphate, aaRS: aa~AMP is the aminoacyladenylate complex with the enzyme and aa~tRNAaa is the aminoacyl-tRNA. An amino acid is first activated via the linking of its carboxyl group directly to an AMP moiety to form a highly reactive adenylated amino acid. In the second reaction, the AMP-linked carboxyl group is transferred to the hydroxyl group on the sugar at the 3′-end of the tRNA to form the aminoacyltRNA, which then carries the amino acid to the appropriate site in the ribosome. Via the use of ATP–PP$_i$ assays, Tirrell and coworkers [162–164] observed a good correlation between the rate of activation of a non-natural amino acid by an aaRS and incorporation of the analogue during protein biosynthesis, indicating that the initial recognition of an analogue by the aaRS is the key step in non-natural amino acid incorporation into proteins *in vivo*. Therefore, strategies that manipulate the activities of an aaRS offer significant promise for the engineering of protein polymers via incorporation of non-natural amino acids.

A key experimental strategy for incorporation of non-natural amino acids into protein polymers *in vivo* is the minimization of the concentration of the natural amino acid in the culture medium during protein biosynthesis. Such minimization is required owing to the high specificity and fidelity exhibited by the aaRS towards natural amino acids and is achieved via the use of a metabolically engineered class of bacterial expression hosts that cannot produce a given natural amino acid (*auxotrophs*). The experimental procedure for incorporation of non-natural analogues using an auxotrophic strain involves the growth of the host strain in a medium supplemented with the natural amino acid until a sufficient number of cells are grown. These cells now act as a "factory" for protein polymer synthesis and are centrifuged, washed and resuspended in a minimal media containing the non-natural amino acid of interest. Protein expression is then induced and, after isolation of the target protein, the extent of analogue incorporation can be determined via amino acid analysis, tryptic digest mass spectrometry, Edman degradation analysis and NMR spectroscopy. The expression system can be further optimized via controlling the activity of the translation components such as the aaRS.

The incorporation of non-natural amino acids into proteins of an auxotrophic expression host equipped with the wild-type (unmodified) biosynthetic apparatus has been known for decades [165–167]. Such incorporation requires that a wild-type aaRS is capable of charging an appropriate tRNA with the non-natural amino acid of interest. Since the earliest studies, several additional fluorinated [167–169], unsaturated [170], electroactive [171] and other [166, 172] non-natural amino acids have been incorporated into protein polymers via the use of the wild-type biosynthetic apparatus in an auxotrophic expression host. For example, in the early 1990s, Tirrell and coworkers reported the incorporation of selenomethionine in place of methionine into a β-sheet-forming polypeptide [(GA)$_3$GM]$_9$ [172]. In addition to selenomethionine, telluromethionine, norleucine, trifluoromethionine and ethionine have all been shown to have translational activity in the wild-type biosynthetic apparatus [173, 174]. Methionine analogues 1–4 (Fig. 11.5) have been shown to replace methionine *in vivo* in these conventional bacterial expression hosts [175–177]. In more recent reports, Tirrell and coworkers have also reported replacement of 92% of the leucine sites with trifluoroleucine (**15**) in coiled-coil polypetides [178] and 85% substitution of isoleucine by trifluoroleucine in murine interleukin-2 [162].

Additional strategies involving the engineering of the activity of aaRS have been employed to permit incorporation of a larger set of analogues. In the simplest of these strategies, overexpression of a wild-type aaRS in an expression host has been shown to permit protein biosynthesis with non-natural amino acid that are poor substrates for a given aaRS [179–182]. For example, Kiick et al. demonstrated that the overexpression of wild-type methionyl tRNA synthetase (MetRS) permits the replacement of 92% of the methionine residues in mDHFR by *trans*-crotylglycine (Tcg, **10**) with protein yields of approximately 12 mg L^{-1} [181]. The analogues **5–9**, which are activated up to 340 000 times more slowly than methionine, can also be incorporated into protein polymers via similar strategies [179], although the engineered expression host must be supplemented with large quantities of the analogue (500 mg L^{-1}). Similarly, in studies by Tang and Tirrell, hexafluoroleucine (**17**) (Hfl) was incorporated in place of leucine in coiled-coil polypeptides with levels of incorporation of 74%, via the use of a leucine auxotroph equipped with extra copies of the leucyl tRNA synthetase (LeuRS) [182].

Further mutation of the aaRS activity has also been exploited as an additional approach to expand the chemical diversity of the side-chains that can be incorporated into protein polymers. Specifically, mutations at the active site of the aaRS can be performed in order to relax the substrate specificity and thereby permit aminoacylation of tRNAs with a broader range of non-natural amino acids. Expression of such mutant aaRS in an auxotrophic strain has been shown to permit incorporation of a wide range of non-natural amino acids with varied chemical functionality [183–185]. For example, an engineered mutant form of phenylalanyl tRNA synthetase (PheRS), which has an Ala294Gly mutation at the binding pocket (A294GPheRS) [186], was shown to have relaxed substrate specificity towards *p*-chlorophenylalanine (**19**) [187]. Overexpression of this mutant PheRS in *E. coli* has permitted the multisite incorporation of *p*-bromo- (**22**) [183], *p*-iodo- (**23**), *p*-azido- (**24**), *p*-ethynyl- (**25**) and *p*-cyanophenylalanine (**26**) [184]. A double mutant of PheRS having both Ala294Gly and Thr251 mutations consequently has a larger PheRS binding pocket and further relaxed substrate specificity. An *E. coli* expression host equipped with this double mutant PheRS was shown to incorporate *p*-acetylphenylalanine (**27**) during biosynthesis of the target protein mDHFR [185]. Given that several aaRS also exhibit editing func-

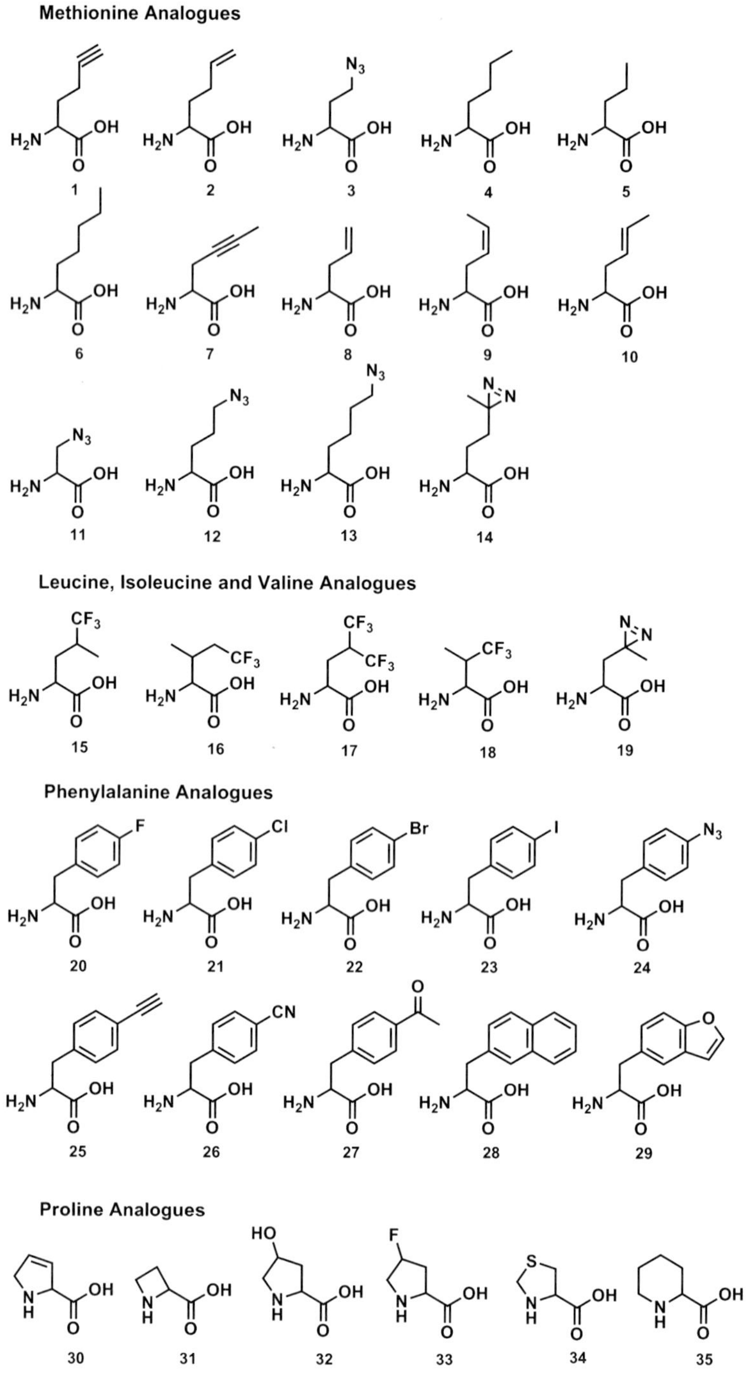
Methionine Analogues
1 2 3 4 5
6 7 8 9 10
11 12 13 14
Leucine, Isoleucine and Valine Analogues
15 16 17 18 19
Phenylalanine Analogues
20 21 22 23 24
25 26 27 28 29
Proline Analogues
30 31 32 33 34 35

Fig. 11.5 Non-natural amino acids.

tions that hydrolyze incorrectly aminoacylated tRNAs, mutation of the editing activity of a given aaRS has also been useful/required for the incorporation of an expanded set of non-natural amino acids [164, 188–190].

All of the above strategies provide efficient incorporation of non-natural amino acids at multiple sites in a protein and proteins that contain both the natural amino acid and the non-natural analogue can be synthesized via control of the ratio of natural to non-natural amino acid in the culture medium. This approach, however, does not exert precise control over the placement of the natural and the non-natural amino acids in the protein polymer chain, as the non-natural analogue must share codons with the natural amino acid. Given that there are 20 amino acids encoded by 61 codons, all amino acids (with the exception of tryptophan and methionine) have more than one codon. Reassignment of the degenerate codons to a non-natural analogue therefore offers the potential for expanding the useful combinations of natural and non-natural building blocks for polypeptide synthesis *in vivo*. For example, Kwon et al. [191] employed a yeast tRNA^{Phe} and a mutant yeast PheRS (T251G) to reassign UUU, one of the two degenerate phenylalanine codons, to the non-natural amino acid 2-naphthylalanine (**28**). *E. coli* expression hosts equipped with this yeast PheRS/tRNA^{Phe} pair were used to produce mDHFR from cultures supplemented with 2-naphthylalanine. Analysis of tryptic digest fragments of mDHFR via matrix-assisted laser desorption/ionization mass spectrometry (MALDI-MS) shows that 2-naphthylalanine was incorporated at positions encoded by UUU and phenylalanine was inserted in positions encoded by UUC. This method offers enormous potential in protein polymer engineering, which will be realized as additional heterologous aaRS–tRNA pairs, which charge specific tRNAs with desired non-natural amino acids, are developed.

11.3.3
Types of Chemically Novel Amino Acids Incorporated into Protein Polymers

In this section, different amino acid analogues that have been incorporated into protein polymers and their impact are discussed; a partial list of these analogues and their applications are given in Table 11.1. The non-natural analogues have been classified according to the chemical functionality of the side-chain.

11.3.3.1 Halide-functionalized Side-chains
Polymers containing aryl bromide or aryl iodide functional groups can be chemically modified via useful transformations such as Heck reactions and Sonagashira coupling. Additionally, fluorinated amino acids are used to probe changes in local environments using ^{19}F NMR spectroscopy [192, 193] and present potential tools for *in vivo* ^{19}F magnetic resonance imaging. Bromination can be used for structure determination using X-ray diffraction (XRD) [194, 195]. Protein polymers that contain amino acid analogues with halide functionality are therefore attractive candidates for generating novel biomaterials.

Earliest investigations of the incorporation of fluorinated analogues demonstrated that they could reduce the surface energy of crystals of the protein polymer [169]. More recently, fluorinated amino acids have been used to increase the hydrophobicity, and hence the stability, of hydrophobically stabilized polypeptide structures. As mentioned above, trifluoroleucine (Tfl, **15**) has been incorporated into coiled coil proteins via the use of a leucine auxotroph equipped with the wild-type biosynthetic apparatus [178]. About 92% substitution of leucine by Tfl was observed and the incorporation of the fluorinated amino acid increased the melting temperature of the coiled coil from 54 to 67 °C without significantly perturbing the coiled coil structure. Similarly, trifluoroisoleucine (Tfi, **16**) was shown to support the synthesis of murine interleukin-2 (mIL-2) in wild-type expression hosts, with levels of replacement of 93% [162]. The Tfi-containing mIL-2 was shown to have an equivalent maximum proliferative response when compared with wild-type mIL-2.

Additional fluorinated amino acids have been incorporated into proteins via the overexpression of wild-type and mutant aaRS. For example, hexafluoroleucine (Hfl, **17**) was incorporated in coiled coil proteins via overexpression of wild-type LeuRS [182]. Coiled coil proteins synthesized from such expression hosts exhibited levels of Hfl incorporation of 74% as determined via MALDI-MS and amino acid analysis. The coiled coil proteins containing Hfl maintain their coiled coil conformation and exhibited a melting transition of 76 °C, which is 22 °C higher compared with leucine-containing controls. In a recent example that broadly highlights the versatility of the biosynthetic apparatus, the 2S,3R-stereoisomer of 4,4,4-trifluorovaline (**18**) was incorporated in place of either valine or isoleucine, under conditions in which the expression host was equipped with extra copies of ValRS or IleRS, respectively, and grown on cultures depleted of Val or Ile [196]. These results clearly indicate the versatility of the incorporation of fluorinated amino acids and that such incorporation can improve the thermal stability of protein–protein interfaces and structures without compromising activity. The methods may therefore be generally useful in the engineering of stable, protein-based polymeric structures.

Mutant aaRS have also been used to permit the incorporation of other halogenated amino acids into proteins *in vivo*. As mentioned above, the mutant PheRS (A294G) has been employed for incorporation of *p*-bromophenylalanine (**22**) into mDHFR produced in a phenylalanine auxotrophic expression host [183]. The *p*-bromophenylalanine was incorporated in mDHFR at levels of replacement of 88% and the yields obtained were approximately 70% of the yield of mDHFR obtained from cells grown with phenylalanine ($\sim$20–25 mg L^{-1}). In addition, *p*-iodophenylalanine was incorporated in mDHFR using a similar strategy and was shown to replace 45% of the phenylalanine residues [184]. The UV spectrum of mDHFR containing *p*-iodophenylalanine (**23**) is similar in peak intensity and position with the free *p*-iodophenylalanine, indicating that the side-chain is not modified by the bacterial host during protein biosynthesis. As mentioned earlier, aromatic halides can undergo metal-catalyzed reactions, cyanation and amidation. Performing such chemistries on proteins containing these amino acids would allow new strategies for side-chain modifications, immobilization of proteins on surfaces and synthesis of cross-linked polymers.

11.3.3.2 Azide-functionalized Side-chains

Azides have been shown to be viable chemical reporters for labeling biomolecules in essentially any biological locale and are useful in multiple chemical transformations and in photocross-linking events [197–199]. This versatile functional group is abiotic in animals, absent from nearly all naturally occurring species, bioorthogonal and can survive the cellular metabolism. The alkylazides are particularly useful in aqueous compatible chemistries such as Staudinger ligation and click chemistry transformations (Fig. 11.6), and have also been employed in the chemical synthesis of functionalized polymers [199–203]. In first reports of azide incorporation into proteins, Kiick et al. demonstrated the use of azidohomoalanine (Aha, **3**) as a methionine surrogate and its selective modification via Staudinger ligation [175]. About 95% of the methionine residues were replaced with Aha when mDHFR was synthesized by expression hosts equipped with the wild-type biosynthetic apparatus in media supplemented with Aha. Selective tagging of Aha was performed in the presence of other cellular proteins via Staudinger ligation with a triarylphosphine–FLAG conjugate, demonstrating the selectivity of incorporation and modification. Link and Tirrell reported incorporation of Aha, via a similar strategy to that mentioned above, into a mutated outer membrane protein C (OmpC). The azide groups in this case were selectively biotinylated via click chemistry protocols that employ copper-mediated [3+2] azide–alkyne cycloaddition [204]. Subsequent incubation with a streptavidin-functionalized fluorescent dye allowed the differentiation, via flow cytometry, of cells decorated with Aha from cells containing methionine. With the aid of this technique,

Link et al. were also able to detect the incorporation into OmpC of methionine analogues such as azidoalanine (**11**), azidonorvaline (**12**) and azidonorleucine (**13**) [205], which were previously shown to be inefficient substrates [175]. Thus, the incorporation of azido-functionalized amino acids into proteins represents a promising route for developing novel bioorthogonal chemical reporters [204, 205] and for synthesizing chemically diverse protein-based polymeric materials.

Arylazides are also interesting targets as they allow intramolecular photoactivated cross-linking upon exposure to light and are among the most commonly used photocross-linking agents (Fig. 11.6). Recent investigations have demonstrated that *p*-azidophenylalanine (**24**) can replace 63% of the phenylalanine residues in mDHFR when mDHFR is produced in an expression host equipped with the Ala294Gly PheRS mutant [184]. Artificial protein polymers containing *p*-azidophenylalanine have been utilized as scaffolds to immobilize recombinant proteins on surfaces for protein microarray production [206]. Specifically, *p*-azidophenylalanine was incorporated into a protein polymer scaffold that contains two separate domains: a protein capture domain, ZE, which mediates capture of

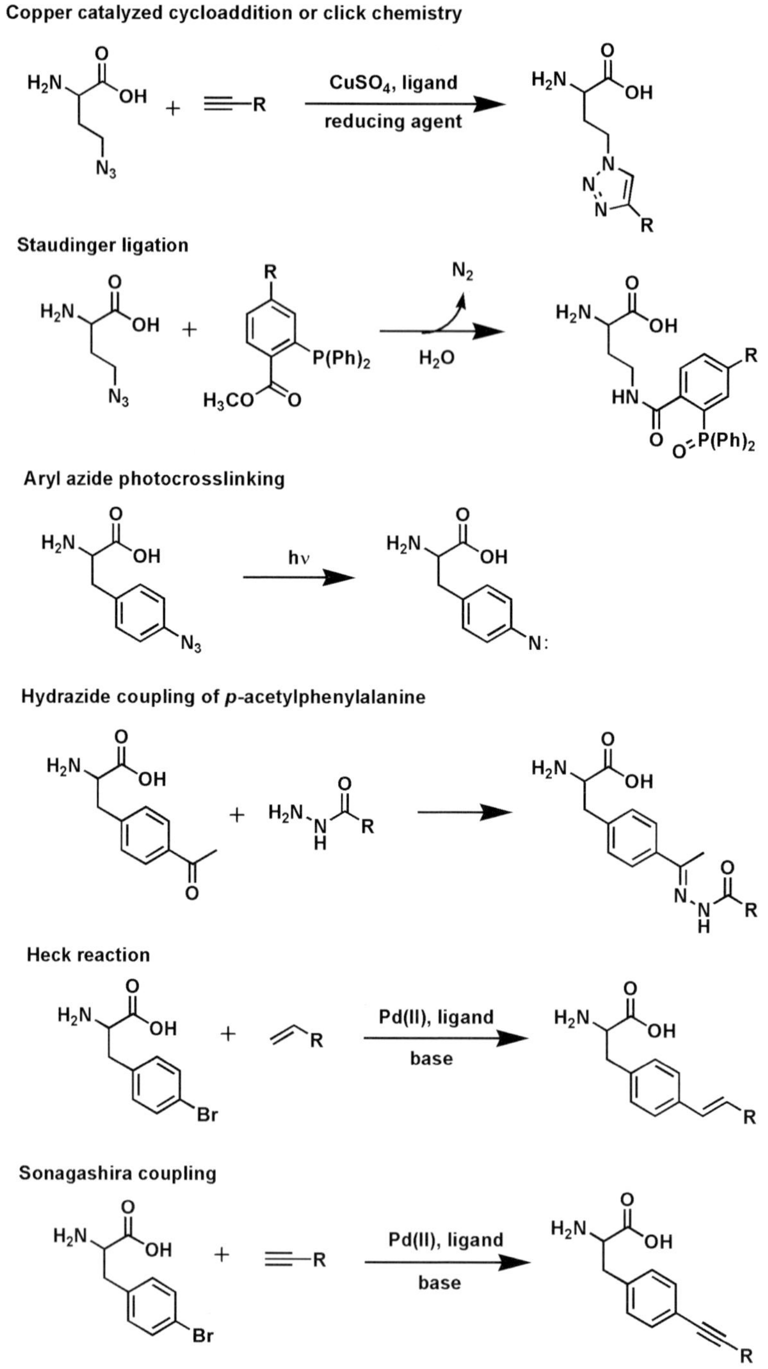

Fig. 11.6 Useful chemistries available via the incorporation of non-natural amino acids.

a protein target (ZR) via heterodimeric coiled coil formation, and an elastin-mimetic surface anchor that contains sites for *p*-azidophenylalanine incorporation. The surface anchor was covalently attached to a glass substrate through photodecomposition of *p*-azidophenylalanine to form uniform protein films. Target proteins that contain a ZR fusion tag can be immobilized on these films from a crude cell lysate and can be detected via the use of protein microarrays.

11.3.3.3 Ketone-functionalized Side-chains

Ketones can also be used as bioorthogonal chemical reporters and these mild electrophiles are smart choices for modifying biosynthetically derived protein polymers, as they are essentially inert to the reactive moieties normally found in proteins, lipids and other biomolecules. In addition, the ketone side-chain can be selectively modified with hydrazide (Fig. 11.6), hydroxylamino and thiosemicarbazide reagents under physiological conditions [207–210]. In 2002, Datta et al. reported successful experiments for the incorporation of *p*-acetylphenylalanine (**27**) into mDHFR *in vivo* [185]. In these studies, a computational approach was used to develop a PheRS mutant with two active site mutations. Overexpression of this PheRS double mutant (Ala294Gly and Thr251Gly) allowed the incorporation of *p*-acetylphenylalanine in mDHFR, at levels of replacement of 80%. The modified mDHFR has been shown to be active towards hydrazine reagents without chain degradation.

11.3.3.4 Alkyne- and Alkene-functionalized Side-chains

Amino acids containing alkene and alkyne side-chains have long been attractive choices for incorporation into protein polymers owing to the powerful and versatile chemistries they offer. For example, homoallylglycine (Hag, **2**) has been used to effect structural self-assembly and self-organization of hydrogen-bonded peptide dimers via ruthenium-catalyzed metathesis reactions [211] and homopropargylglycine (Hpg, **1**) has been used for fluorescence labeling of protein polymers via Cu(I)-catalyzed cycloaddition [212]. Protein polymers containing terminal alkene and alkyne side-chains can also be derivatized using Heck coupling or Sonagashira reaction (Fig. 11.6) with aryl halides respectively.

Hag and Hpg have been incorporated, as methionine surrogates, in mDHFR via the use of the wild-type biosynthetic apparatus. Hag replaces methionine at a level of approximately 92% [176], whereas Hpg was shown to replace methionine in mDHFR at a level of 88% as determined via amino acid analysis and Edman degradation, with no reduction in protein yields [177]. The incorporation of ethynylphenylalanine (Eth, **25**) into mDHFR has also been demonstrated, with measured levels of incorporation of approximately 55% [184]. In a recent report, Beatty et al. expressed the barstar protein in *E. coli* cultures supplemented with Hpg or Eth [212]. The incorporation of either of these analogues permitted the fluorescence labeling, via the Cu(I)-catalyzed cycloaddition between alkynes and azides, of the newly synthesized barstar protein with fluorescent coumarins containing a terminal azide. Similar labeling procedures can be employed for identification of enzyme inhibitors and labeling the surface of *E. coli* [205]. The copper-mediated reaction has also been used to tag azides installed within virus particles [213], nucleic acids [214] and proteins from complex tissue lysates [215] with virtually no background labeling. It should be noted that the same reaction can be conducted with the alkyne as the chemical reporter.

11.3.3.5 Photoreactive Side-chains

Photocross-linking agents offer advantages over chemical cross-linking agents, since they generate highly reactive intermediates for very efficient, although nonspecific, cross-linking [216], in reactions that are simple to monitor and that produce no byproducts. Non-natural amino acids containing photoreactive groups have therefore been employed to study protein–protein interactions and protein–nucleic acid interactions and to attach protein polymers covalently to glass substrates for uniform film production. In recent studies, Bentin et al. used a mutant PheRS (Ala294Gly) to incorporate benzofuranylalanine (**29**) into mDHFR [217]. The level of incorporation was found to be approximately 90% via mass spectrometric analysis. Benzofuranylalanine undergoes [2+2] photocycloaddition with a variety of alkenes and therefore may be very useful for modification and cross-linking of protein polymers.

Thiele and coworkers [218] have demonstrated the incorporation of two new photoreactive amino acids, photo-methionine (**14**) and photo-leucine (**19**), into mammalian proteins. Their similarity to the natural amino acids methionine and leucine allows their efficient incorporation into proteins *in vivo* by employing unmodified mammalian translation machinery. Activation of the analogue by ultraviolet light induces covalent cross-linking of interacting proteins, which can be detected via Western blot analysis. These methods have proven advantageous for the investigation of large protein complexes that regulate cholesterol biosynthesis in the endoplasmic reticulum and may be similarly effective in protein polymer synthesis and assembly approaches.

11.3.3.6 Unsaturated and Structural Amino Acid Analogues

A variety of proline analogues have been incorporated into proteins for chemical modification purposes and also for modification and investigation of protein folding behavior. Dehydroproline (Dhp, **30**) and azetidine-2-carboxylic acid (Aza, **31**) were among the earliest proline analogues to be systematically studied and were incorporated into the protein polymer [(AG)$_x$PEG]$_y$ in cultures of an *E. coli* proline auxotroph equipped with the wild-type prolyl tRNA synthetase. The Dhp was easily incorporated at levels of 100%, whereas the Aze was incorporated at levels of 40% as determined via amino acid analysis and ^{1}H NMR spectroscopy. Dhp was shown to react readily with H_2O_2 and Br_2 to produce hydroxylated and brominated forms of Dhp [170] while the smaller side-chain Aze increased chain flexibility as suggested by the presence of β-sheet structure in the Aze-containing [(AG)$_x$PEG]$_y$ protein polymers [219].

4-Hydroxyproline (Hyp, **32**) residues have an essential role in providing the collagen triple helices with thermal stability. The denaturation temperature of a non-hydroxylated type I collagen is only 24 °C, whereas a triple helix consisting of fully hydroxylated collagen polypeptide chains is stable up to 39 °C [220, 221]. In mammalian systems, the hydroxylation of proline is a post-translational modification that permits folding and improved stability of collagen triple helix. In prokaryotic systems, however, these enzymes are not present and therefore the modification of proline has been accomplished via the introduction of the gene encoding the enzyme prolyl 4-hydroxylase in the expression host. The direct incorporation of Hyp in place of proline would obviate the need for the introduction of such additional enzymatic pathways. Buechter et al. have expressed the a_1 fragment of human type I collagen in *E. coli* proline auxotroph, under conditions in which the auxotroph is grown in proline-depleted cultures that contain Hyp and hyperosmotic concentrations of sodium chloride (500 mM) [222]. CD spectroscopic studies on the Hyp-containing collagen indicate that the multisite substitution of Hyp in place of proline does not destabilize the collagen triple helix. Developing the capability to incorporate Hyp selectively in the Y-positions of the collagen GXY repeat should offer additional strategies for producing biomimetic collagen-like polypeptides without the need for co-expression of hydroxylating enzymes.

The versatility of proline analogue incorporation and its utility in unraveling protein folding questions have been demonstrated in recent studies by Conticello and coworkers [223]. In these studies, 4-fluoroproline (**33**) analogues were incorporated into elastin-like polypeptides (ELPs) in order to study the effect of amino acid substitution on the protein polymer structure [223]. In particular, (2S,4S)- and (2S,4R)-4-fluoroproline were incorporated as proline surrogates, via the use of the unmodified translational machinery, to probe stereoelectronic effects on the thermodynamics of protein polymer conformational transitions. The incorporation of (2S,4R)-4-fluoroproline in the polypeptide MGH$_{10}$S$_2$GHID$_4$KHM [(VPGVG)$_4$VPGIG]$_{16}$V was found to reduce the temperature at which the transition to the type II β-turn structure occurs, suggesting, consistent with computational modeling, that the (2S,4R)-4-fluoroproline stabilizes this type of turn structure [223]. In contrast, incorporation of (2S,4S)-4-fluoroproline destabilizes the type II -turn structure and therefore increases T_t. These elegant studies offer insights into the mechanisms of elastic behavior in important protein-based materials and can therefore offer additional criteria for the design of novel elastomeric materials with desired properties for particular applications. Proline analogues such as 1,3-thiazolidine-4-carboxylic acid (**34**) and piperidine-2-carboxylic acid (**35**) were also found to be translationally active under modified expression conditions [224]. 1,3-Thiazolidine-4-carboxylic acid (**34**) was found to be susceptible to intracellular oxidative degradation in *E. coli*; therefore an *E. coli* proline auxotroph, containing mutations in the genes of the enzymes responsible for the degradation (*putA* and *proC*), was employed for the synthesis of elastin mimetic polypeptides containing **34**. The modified proline auxotroph was grown in proline-depleted cultures containing **34** and hyperosmotic concentrations of sucrose (800 mM) and was also engineered for overexpression of the wild-type prolyl-tRNA synthetase (ProRS). In these experiments, **34** was shown to replace proline in elastin mimetic polymers at a level of 20%, with polypeptide yields of 27 mg L^{-1}. The proline analogue piperidine-2-carboxylic acid (**35**) was reported to be an ineffective substrate for ProRS, although a ProRS mutant enzyme (Cys443Gly) was found, via ATP–PP$_i$ exchange assays, to activate proline analogues such as **35**. Overexpression of this mutant ProRS in a proline auxotroph and hyperosmotic concentrations of sodium chloride (600 mM) allowed the incorporation of **35** in elastin mimetic polymers. About 13% of proline residues were replaced by **35** and the yield of the protein polymer was reported to be 16 mg L^{-1}. These investigations further highlight the versatility of bacterial expression hosts for the production of chemically and structurally novel polypeptides.

11.4
Prospects for Protein-based Polymers

Control over monomer arrangement at the molecular level, coupled with the diversity of amino acid sequences that provide useful materials properties, has created enormous interest in the genetic engineering of biomaterials, with exciting academic and commercial applications. Protein polymers with designs based on natural fibrous proteins have been engineered to optimize biological and mechanical properties. Modulation of the properties of these polymers can be achieved through precise changes in sequence and length to produce protein polymers with controlled chain folding architectures, liquid crystalline phases and self-assembled hydrogels. In addition, the design of artificial protein polymers can offer opportunities for the production of novel materials that display functional groups in a manner that controls materials assembly, mediates interactions with biological targets and yields polymers with interesting properties.

Protein-based polymers are environmentally clean over their entire life cycle from production to disposal; they can be produced from renewable resources and they are biodegradable. Commercial viability, however, requires a cost of production that would begin to rival that of petroleum-based polymers. The cost associated with the protein polymer synthesis depends on the cost of the materials used in the synthesis and the scale of the synthesis. Many of these protein polymers can be produced using technologically advanced cell biosystems, including microbial and mammalian systems, as well as transgenic plants (tobacco and potato) and animals (goat). For example, the advent of genetic engineering methods provides the means to produce recombinant proteins in bacteria and yeast, making them productive bioreactors. However, very long and repetitive proteins are difficult to produce in bacterial systems because the repetitive nature of the genes often leads to genetic instability due to recombination events. The alternative approach is to produce recombinant proteins in higher eukaryotic organisms such as plants and animals, which generally possess the capability to make long repetitive proteins.

A variety of companies have implemented the commercial production of protein-based polymers and numerous patents have been granted, with topics focused on expression of artificial proteins, specific application needs addressed by the unique properties of these protein polymers and the incorporation of non-natural amino acids. Since 1998, Protein Polymer Technologies, Inc. (PPTI) has been a pioneer in protein polymer design and synthesis and has extensive patent literature in the area of silk-based and elastin-based polymeric materials. The high molecular weight, genetically engineered biomaterials are processed into products with properties tailored to specific needs; over 50 protein polymer sequences have been designed by PPTI for commercial application and some can be produced in kilogram quantities. Targeted products include urethral bulking agents for the treatment of stress urinary incontinence, dermal augmentation products for cosmetic and reconstructive surgery, tissue adhesives and sealants, and scaffolds for wound healing and tissue engineering. In addition to medical applications, industrial applications are also of interest for protein polymers and Genencor International, Inc. obtained a worldwide exclusive license to develop PPTI's protein polymers in both industrial and personal care applications. Nexia Biotechnologies, Inc. manufactures complex recombinant proteins with industrial and medical applications; its lead product, BioSteel, is based on recombinant spider-silk proteins and production of these proteins from transgenic goats is under continued development. Targeted applications for these materials include medical sutures, surgical meshes and artificial ligaments, and also materials applications such as technical sporting gear (e.g. biodegradable fishing lines), soft body armor and composites. In addition, products based on collagen are also of increasing interest, as the use of recombinant human collagen in biomaterials applications would eliminate the potential for immune reactions or the possibility of a transfer of pathogens from animal material. Meristem Therapeutics has been developing technology using transgenic tobacco plants to produce human recombinant collagen, in commercial quantities, for use as biological glue.

Despite the challenges for the commercial implementation of protein polymers, their enormous potential for research and technological advancement has not been quenched. In addition to the protein polymers mentioned above, progress in protein design will permit structure-based protein engineering to control finely the physical, chemical and biological properties of new protein polymers. Technological growth in expression systems such as mammalian cells offers several advantages, including proper protein folding, assembly and post-translational modification and will expand the architectures and functionality of protein-based polymers. Hence the biosynthesis of protein polymers remains a powerful technique for the creation of the next generation of functional materials.

References

1 J. Parker, *Microbiol. Rev.* **1989**, *53*, 273.
2 R. Sankaranarayanan, D. Moras, *Acta Biochim. Pol.* **2001**, *48*, 323.
3 S. Cusack, *Nat. Struct. Mol. Biol.* **1995**, *2*, 824.
4 D. E. Meyer, A. Chilkoti, *Biomacromolecules* **2002**, *3*, 357.
5 N. L. Goeden-Wood, V. P. Conticello, S. J. Muller, J. D. Keasling, *Biomacromolecules* **2002**, *3*, 874.
6 J. D. van Beek, S. Hess, F. Vollrath, B. H. Meier, *Proc. Natl. Acad. Sci. USA* **2002**, *99*, 10266.
7 Y. T. Zhou, S. X. Wu, V. P. Conticello, *Biomacromolecules* **2001**, *2*, 111.
8 D. Huemmerich, C. W. Helsen, S. Quedzuweit, J. Oschmann, R. Rudolph, T. Scheibel, *Biochemistry* **2004**, *43*, 13604.
9 S. Arcidiacono, C. Mello, D. Kaplan, S. Cheley, H. Bayley, *Appl. Microbiol. Biotechnol.* **1998**, *49*, 31.
10 C. M. Mello, J. W. Soares, S. Arcidiacono, M. M. Butlers, *Biomacromolecules* **2004**, *5*, 1849.
11 J. P. O'Brien, S. R. Fahnestock, Y. Termonia, K. C. H. Gardner, *Adv. Mater.* **1998**, *10*, 1185.
12 A. Panitch, K. Matsuki, E. J. Cantor, S. J. Cooper, E. D. T. Atkins, M. J. Fournier, T. L. Mason, D. A. Tirrell, *Macromolecules* **1997**, *30*, 42.
13 J. T. Prince, K. P. McGrath, C. M. Digirolamo, D. L. Kaplan, *Biochemistry* **1995**, *34*, 10879.
14 S. R. Fahnestock, S. L. Irwin, *Appl. Microbiol. Biotechnol.* **1997**, *47*, 23.
15 S. R. Fahnestock, L. A. Bedzyk, *Appl. Microbiol. Biotechnol.* **1997**, *47*, 33.
16 J. Scheller, K. H. Guhrs, F. Grosse, U. Conrad, *Nat. Biotechnol.* **2001**, *19*, 573.
17 E. S. Piruzian, V. G. Bogush, K. V. Sidoruk, I. V. Goldenkova, K. A. Mysiychuk, V. G. Debabov, *Mol. Biol.* **2003**, *37*, 554.
18 J. J. Yang, L. A. Barr, S. R. Fahnestock, Z. B. Liu, *Transgenic Res.* **2005**, *14*, 313.
19 D. Huemmerich, T. Scheibel, F. Vollrath, S. Cohen, U. Gat, S. Ittah, *Curr. Biol.* **2004**, *14*, 2070.
20 T. Asakura, K. Nitta, M. Y. Yang, J. M. Yao, Y. Nakazawa, D. L. Kaplan, *Biomacromolecules* **2003**, *4*, 815.

21 J. Huang, R. Valluzzi, E. Bini, B. Vernaglia, D. L. Kaplan, *J. Biol. Chem.* **2003**, *278*, 46117.
22 R. Valluzzi, S. Szela, P. Avtges, D. Kirschner, D. Kaplan, *J. Phys. Chem. B* **1999**, *103*, 11382.
23 S. Szela, P. Avtges, R. Valluzzi, S. Winkler, D. Wilson, D. Kirschner, D. L. Kaplan, *Biomacromolecules* **2000**, *1*, 534.
24 S. Winkler, S. Szela, P. Avtges, R. Valluzzi, D. A. Kirschner, D. Kaplan, *Int. J. Biol. Macromol.* **1999**, *24*, 265.
25 S. Winkler, D. Wilson, D. L. Kaplan, *Biochemistry* **2000**, *39*, 12739.
26 M. Y. Yang, T. Asakura, *J. Biochem.* **2005**, *137*, 721.
27 Y. Qu, S. C. Payne, R. P. Apkarian, V. P. Conticello, *J. Am. Chem. Soc.* **2000**, *122*, 5014.
28 J. P. Anderson, J. Cappello, D. C. Martin, *Biopolymers* **1994**, *34*, 1049.
29 T. Asakura, C. Tanaka, M. Y. Yang, J. M. Yao, M. Kurokawa, *Biomaterials* **2004**, *25*, 617.
30 Z. Indik, H. Yeh, N. Ornstein-Goldstein, P. Sheppard, N. Anderson, J. C. Rosenbloom, L. Peltonen, J. Rosenbloom, *Proc. Natl. Acad. Sci. USA* **1987**, *84*, 5680.
31 D. T. McPherson, C. Morrow, D. S. Minehan, J. G. Wu, E. Hunter, D. W. Urry, *Biotechnol. Prog.* **1992**, *8*, 347.
32 D. W. Urry, D. C. Gowda, T. M. Parker, C. H. Luan, M. C. Reid, C. M. Harris, A. Pattanaik, R. D. Harris, *Biopolymers* **1992**, *32*, 1243.
33 C. H. Luan, D. W. Urry, *J. Phys. Chem. B* **1991**, *95*, 7896.
34 D. W. Urry, C. H. Luan, T. M. Parker, D. C. Gowda, K. U. Prasad, M. C. Reid, A. Safavy, *J. Am. Chem. Soc.* **1991**, *113*, 4346.
35 A. Chilkoti, M. R. Dreher, D. E. Meyer, *Adv. Drug Deliv. Rev.* **2002**, *54*, 1093.
36 H. Betre, L. A. Setton, D. E. Meyer, A. Chilkoti, *Biomacromolecules* **2002**, *3*, 910.
37 D. E. Meyer, A. Chilkoti, *Biomacromolecules* **2004**, *5*, 846.
38 A. Panitch, T. Yamaoka, M. J. Fournier, T. L. Mason, D. A. Tirrell, *Macromolecules* **1999**, *32*, 1701.

39 T. Yamaoka, T. Tamura, Y. Seto, T. Tada, S. Kunugi, D. A. Tirrell, *Biomacromolecules* **2003**, *4*, 1680.
40 J. Lee, C. W. Macosko, D. W. Urry, *Macromolecules* **2001**, *34*, 5968.
41 J. Lee, C. W. Macosko, D. W. Urry, *Biomacromolecules* **2001**, *2*, 170.
42 D. W. Urry, S. Q. Peng, L. C. Hayes, D. McPherson, J. Xu, T. C. Woods, D. C. Gowda, A. Pattanaik, *Biotechnol. Bioeng.* **1998**, *58*, 175.
43 R. A. McMillan, T. A. T. Lee, V. P. Conticello, *Macromolecules* **1999**, *32*, 3643.
44 R. A. McMillan, V. P. Conticello, *Macromolecules* **2000**, *33*, 4809.
45 L. Huang, R. A. McMillan, R. P. Apkarian, B. Pourdeyhimi, V. P. Conticello, E. L. Chaikof, *Macromolecules* **2000**, *33*, 2989.
46 M. Hong, D. Isailovic, R. A. McMillan, V. P. Conticello, *Biopolymers* **2003**, *70*, 158.
47 K. Trabbic-Carlson, L. A. Setton, A. Chilkoti, *Biomacromolecules* **2003**, *4*, 572.
48 X. L. Yao, V. P. Conticello, M. Hong, *Magn. Reson. Chem.* **2004**, *42*, 267.
49 A. Junger, D. Kaufmann, T. Scheibel, R. Weberskirch, *Macromol. Biosci.* **2005**, *5*, 494.
50 H. Stiborova, J. Kostal, A. Mulchandani, W. Chen, *Biotechnol. Bioeng.* **2003**, *82*, 605.
51 K. Nagapudi, W. T. Brinkman, J. Leisen, B. S. Thomas, E. R. Wright, C. Haller, X. Y. Wu, R. P. Apkarian, V. P. Conticello, E. L. Chaikof, *Macromolecules* **2005**, *38*, 345.
52 K. Nagapudi, W. T. Brinkman, B. S. Thomas, J. O. Park, M. Srinivasarao, E. Wright, V. P. Conticello, E. L. Chaikof, *Biomaterials* **2005**, *26*, 4695.
53 E. R. Wright, V. P. Conticello, *Adv. Drug Deliv. Rev.* **2002**, *54*, 1057.
54 E. R. Wright, R. A. McMillan, A. Cooper, R. P. Apkarian, V. P. Conticello, *Adv. Funct. Mater.* **2002**, *12*, 149.
55 T. A. T. Lee, A. Cooper, R. P. Apkarian, V. P. Conticello, *Adv. Mater.* **2000**, *12*, 1105.
56 C. M. Bellingham, K. A. Woodhouse, P. Robson, S. J. Rothstein, F. W. Keeley, *Biochim. Biophys. Acta* **2001**, *1550*, 6.
57 C. M. Bellingham, M. A. Lillie, J. M. Gosline, G. M. Wright, B. C. Starcher, A. J. Bailey, K. A. Woodhouse, F. W. Keeley, *Biomaterials* **2003**, *70*, 445.
58 G. C. Yang, K. A. Woodhouse, C. M. Yip, *J. Am. Chem. Soc.* **2002**, *124*, 10648.
59 F. W. Keeley, C. M. Bellingham, K. A. Woodhouse, *Philos. Trans. R. Soc. London, Ser. B* **2002**, *357*, 185.
60 J. Cappello, J. W. Crissman, M. Crissman, F. A. Ferrari, G. Textor, O. Wallis, J. R. Whitledge, X. Zhou, D. Burman, L. Aukerman, E. R. Stedronsky, *J. Control. Release* **1998**, *53*, 105.
61 J. Cappello, E. R. Stedronsky (to Protein Polymer Technologies, Inc.), *US Patent 6 380 154*, **2002**.
62 E. R. Stedronsky, J. Cappello (to Protein Polymer Technologies, Inc.), *US Patent 6 423 333*, **2003**.
63 A. A. Dinerman, J. Cappello, H. Ghandehari, S. W. Hoag, *Biomaterials* **2002**, *23*, 4203.
64 A. A. Dinerman, J. Cappello, H. Ghandehari, S. W. Hoag, *J. Control. Release* **2002**, *82*, 277.
65 Z. Megeed, J. Cappello, H. Ghandehari, *Adv. Drug Deliv. Rev.* **2002**, *54*, 1075.
66 Z. Megeed, M. Haider, D. Q. Li, B. W. O'Malley, J. Cappello, H. Ghandehari, *J. Control. Release* **2004**, *94*, 433.
67 Z. Megeed, J. Cappello, H. Ghandehari, *Pharm. Res.* **2002**, *19*, 954.
68 D. E. Meyer, B. C. Shin, G. A. Kong, M. W. Dewhirst, A. Chilkoti, *J. Control. Release* **2001**, *74*, 213.
69 M. R. Dreher, D. Raucher, N. Balu, O. M. Colvin, S. M. Ludeman, A. Chilkoti, *J. Control. Release* **2003**, *91*, 31.
70 D. Y. Furgeson, M. R. Dreher, A. Chilkoti, *J. Control. Release* **2006**, *110*, 362.
71 G. L. Bidwell, D. Raucher, *Mol. Cancer Ther.* **2005**, *4*, 1076.
72 E. R. Welsh, D. A. Tirrell, *Biomacromolecules* **2000**, *1*, 23.
73 K. Di Zio, D. A. Tirrell, *Macromolecules* **2003**, *36*, 1553.
74 Y. C. Fung, *Biomechanics: Mechanical Properties of Living Tissues*, 2nd edn, Springer, New York, **1993**.
75 S. C. Heilshorn, K. A. DiZio, E. R. Welsh, D. A. Tirrell, *Biomaterials* **2003**, *24*, 4245.

76 A. Girotti, J. Reguera, J. C. Rodriguez-Cabello, F. J. Arias, M. Alonso, A. M. Testera, *J. Mater. Sci. Moder. Medicine* **2004**, *15*, 479.

77 J. Kostal, A. Mulchandani, W. Chen, *Macromolecules* **2001**, *34*, 2257.

78 J. Kostal, A. Mulchandani, K. E. Gropp, W. Chen, *Environ. Sci. Technol.* **2003**, *37*, 4457.

79 M. Shimazu, A. Mulchandani, W. Chen, *Biotechnol. Bioeng.* **2003**, *81*, 74.

80 J. Y. Kim, A. Mulchandani, W. Chen, *Biotechnol. Bioeng.* **2005**, *90*, 373.

81 J. Y. Kim, S. O'Malley, A. Mulchandani, W. Chen, *Anal. Chem.* **2005**, *77*, 2318.

82 J. Kostal, A. Mulchandani, W. Chen, *Biotechnol. Bioeng.* **2004**, *85*, 293.

83 N. Nath, A. Chilkoti, *Anal. Chem.* **2003**, *75*, 709.

84 K. Trabbic-Carlson, L. Liu, B. Kim, A. Chilkoti, *Protein Sci.* **2004**, *13*, 3274.

85 D. J. Prockop, K. I. Kivirikko, *Annu. Rev. Biochem.* **1995**, *64*, 403.

86 M. W. T. Werten, T. J. van den Bosch, R. D. Wind, H. Mooibroek, F. A. de Wolf, *Yeast* **1999**, *15*, 1087.

87 P. D. Toman, G. Chisholm, H. McMullin, L. M. Gieren, D. R. Olsen, R. J. Kovach, S. D. Leigh, B. E. Fong, R. Chang, G. A. Daniels, R. A. Berg, R. A. Hitzeman, *J. Biol. Chem.* **2000**, *275*, 23303.

88 M. Nokelainen, H. M. Tu, A. Vuorela, H. Notbohm, K. I. Kivirikko, J. Myllyharju, *Yeast* **2001**, *18*, 797.

89 D. R. Olsen, S. D. Leigh, R. Chang, H. McMullin, W. Ong, E. Tai, G. Chisholm, D. E. Birk, R. A. Berg, R. A. Hitzeman, P. D. Toman, *J. Biol. Chem.* **2001**, *276*, 24038.

90 M. W. T. Werten, W. H. Wisselink, T. J. J. van den Bosch, E. C. de Bruin, F. A. de Wolf, *Protein Eng.* **2001**, *14*, 447.

91 O. Pakkanen, E. R. Hamalainen, K. I. Kivirikko, J. Myllyharju, *J. Biol. Chem.* **2003**, *278*, 32478.

92 C. Yang, P. Hillas, J. Tang, J. Balan, H. Notbohm, J. Polarek, *J. Biomed. Mater. Res.* **2004**, *69B*, 18.

93 D. Olsen, J. Jiang, R. Chang, R. Duffy, M. Sakaguchi, S. Leigh, R. Lundgard, J. Ju, F. Buschman, V. Truong-Le, B. Pham, J. W. Polarek, *Protein Expr. Purif.* **2005**, *40*, 346.

94 A. Fertala, W. B. Han, F. K. Ko, *J. Biomed. Mater. Res.* **2001**, *57*, 48.

95 M. Nokelainen, A. Lamberg, T. Helaakoski, J. Myllyharju, T. Pihlajaniemi, K. I. Kivirikko, *Matrix Biol.* **1996**, *15*, 194.

96 A. Lamberg, T. Helaakoski, J. Myllyharju, S. Peltonen, H. Notbohm, T. Pihlajaniemi, K. I. Kivirikko, *J. Biol. Chem.* **1996**, *271*, 11988.

97 J. Myllyharju, A. Lamberg, H. Notbohm, P. P. Fietzek, T. Pihlajaniemi, K. I. Kivirikko, *J. Biol. Chem.* **1997**, *272*, 21824.

98 W. V. Arnold, A. L. Sieron, A. Fertala, H. P. Bachinger, D. Mechling, D. J. Prockop, *Matrix Biol.* **1997**, *16*, 105.

99 A. Fichard, E. Tillet, F. Delacoux, F. Ruggiero, *J. Biol. Chem.* **1997**, *272*, 30083.

100 D. C. A. John, R. Watson, A. J. Kind, A. R. Scott, K. E. Kadler, N. J. Bulleid, *Nat. Biotechnol.* **1999**, *17*, 385.

101 P. D. Toman, F. Pieper, N. Sakai, C. Karatzas, E. Platenburg, I. de Wit, C. Samuel, A. Dekker, G. A. Daniels, R. A. Berg, G. J. Platenburg, *Transgenic Res.* **1999**, *8*, 415.

102 I. Goldberg, A. J. Salerno, T. Patterson, J. I. Williams, *Gene* **1989**, *80*, 305.

103 T. Kajino, H. Takahashi, M. Hirai, Y. Yamada, *Appl. Environ. Microbiol.* **2000**, *66*, 304.

104 J. Yin, J. H. Lin, T. Y. Li, D. I. C. Wang, *J. Biotechnol.* **2003**, *100*, 181.

105 C. G. Knight, L. F. Morton, D. J. Onley, A. R. Peachey, A. J. Messent, P. A. Smethurst, D. S. Tuckwell, R. W. Farndale, M. J. Barnes, *J. Biol. Chem.* **1998**, *273*, 33287.

106 J. M. Yao, S. Yanagisawa, T. Asakura, *J. Biochem.* **2004**, *136*, 643.

107 T. Weis-Fogh, *J. Exp. Biol.* **1960**, *37*, 889.

108 S. N. Gorb, *Naturwissenschaften* **1999**, *86*, 552.

109 M. Rothschild, J. Schlein, *Philos. Trans. R. Soc. London, Ser. B* **1975**, *271*, 457.

110 M. Burrows, *Nature* **2003**, *424*, 509.

111 S. O. Andersen, *Biochim. Biophys. Acta* **1964**, *93*, 213.

112 J. Gosline, M. Lillie, E. Carrington, P. Guerette, C. Ortlepp, K. Savage, *Philos. Trans. R. Soc. London, Ser. B* **2002**, *357*, 121.

113 D. Young, H. C. Bennetclark, *J. Exp. Biol.* **1995**, *198*, 1001.
114 N. Skals, A. Surlykke, *J. Exp. Biol.* **1999**, *202*, 2937.
115 K. Bailey, T. Weis-Fogh, *Biochim. Biophys. Acta* **1961**, *48*, 452.
116 D. H. Ardell, S. O. Andersen, *Insect Biochem. Mol. Biol.* **2001**, *31*, 965.
117 C. M. Elvin, A. G. Carr, M. G. Huson, J. M. Maxwell, R. D. Pearson, T. Vuocolo, N. E. Liyou, D. C. C. Wong, D. J. Merritt, N. E. Dixon, *Nature* **2005**, *437*, 999.
118 T. Weis-Fogh, *J. Mol. Biol.* **1961**, *3*, 520.
119 T. Weis-Fogh, *J. Mol. Biol.* **1961**, *3*, 648.
120 D. W. Urry, C. H. Lvan, S. Q. Peng in *Molecular Biology and Pathology of Elastic Tissues*, CIBA Foundation Symp., **1995**, *192*, 4.
121 D. S. Hwang, H. J. Yoo, J. H. Jun, W. K. Moon, H. J. Cha, *Appl. Environ. Microbiol.* **2004**, *70*, 3352.
122 Y. J. Wang, X. Zheng, L. H. Zhang, Y. Ohta, *Process Biochem.* **2004**, *39*, 659.
123 K. A. Feeney, N. Wellner, S. M. Gilbert, N. G. Halford, A. S. Tatham, P. R. Shewry, P. S. Belton, *Biopolymers* **2003**, *72*, 123.
124 Y. Deguchi, M. J. Fournier, T. L. Mason, D. A. Tirrell, *J. Macromol. Sci.* **1994**, *A31*, 1691.
125 M. T. Krejchi, E. D. T. Atkins, A. J. Waddon, M. J. Fournier, T. L. Mason, D. A. Tirrell, *Science* **1994**, *265*, 1427.
126 J. X. Wang, A. D. Parkhe, D. A. Tirrell, L. K. Thompson, *Macromolecules* **1996**, *29*, 1548.
127 E. J. Cantor, E. D. T. Atkins, S. J. Cooper, M. J. Fournier, T. L. Mason, D. A. Tirrell, *J. Biochem.* **1997**, *122*, 217.
128 N. I. Topilina, S. Higashiya, N. Rana, V. V. Ermolenkov, C. Kossow, A. Carlsen, S. C. Ngo, C. C. Wells, E. T. Eisenbraun, K. A. Dunn, I. K. Lednev, R. E. Geer, A. E. Koaloyeros, J. T. Welsh, *Biomacromolecules* **2006**, *7*, 1104.
129 J. M. Smeenk, M. B. J. Otten, J. Thies, D. A. Tirrell, H. G. Stunnenberg, J. C. M. van Hest, *Angew. Chem. Int. Ed.* **2005**, *44*, 1968.
130 G. Zhang, M. J. Fournier, T. L. Mason, D. A. Tirrell, *Macromolecules* **1992**, *25*, 3601.
131 S. M. Yu, V. Conticello, G. Zhang, C. Kayser, M. J. Fournier, T. L. Mason, D. A. Tirrell, *Nature* **1997**, *389*, 187.
132 W. A. Petka, J. L. Hardin, K. P. McGrath, D. Wirtz, D. A. Tirrell, *Science* **1998**, *281*, 389.
133 S. B. Kennedy, K. Littrell, P. Thiyagarajan, D. A. Tirrell, T. P. Russell, *Macromolecules* **2005**, *38*, 7470.
134 W. Shen, R. G. H. Lammertink, J. K. Sakata, J. A. Kornfield, D. A. Tirrell, *Macromolecules* **2005**, *38*, 3909.
135 W. Shen, K. Zhang, J. A. Kornfield, D. A. Tirrell, *Nat. Mater.* **2006**, *5*, 153.
136 I. Koltover, S. Sahu, N. Davis, *Angew. Chem. Int. Ed.* **2004**, *43*, 4034.
137 R. S. Farmer, K. L. Kiick, *Biomacromolecules* **2005**, *6*, 1531.
138 R. S. Farmer, L. M. Argust, J. D. Sharp, K. L. Kiick, *Macromolecules* **2005**, *39*, 162.
139 Y. Wang, K. L. Kiick, *J. Am. Chem. Soc.* **2005**, *127*, 16392.
140 J. I. Won, A. E. Barron, *Macromolecules* **2002**, *35*, 8281.
141 J. I. Won, R. J. Meagher, A. E. Barron, *Electrophoresis* **2005**, *26*, 2138.
142 A. Tulpar, D. B. Henderson, M. Mao, B. Caba, R. M. Davis, K. E. van Cott, W. A. Ducker, *Langmuir* **2005**, *21*, 1497.
143 J. M. Smeenk, L. Ayres, H. G. Stunnenberg, J. C. M. van Hest, *Macromol. Symp.* **2005**, *225*, 1.
144 N. Budisa, *Angew. Chem. Int. Ed.* **2004**, *43*, 6426.
145 N. Budisa, *Engineering the Genetic Code: Expanding the Amino Acid Repertoire for the Design of Novel Proteins*, Wiley-VCH, Weinheim, **2005**.
146 L. Wang, P. G. Schultz, *Angew. Chem. Int. Ed.* **2005**, *44*, 34.
147 R. B. Merrifield, *Pure Appl. Chem.* **1978**, *50*, 643.
148 W. C. Chan, P. D. White (Eds.), *Fmoc Solid Phase Peptide Synthesis: a Practical Approach*, **2000**. Oxford University Press, Oxford.
149 T. J. Deming, *Nature* **1997**, *390*, 386.
150 T. W. Muir, *Structure* **1995**, *3*, 649.
151 P. E. Dawson, T. W. Muir, I. Clark-Lewis, S. B. H. Kent, *Science* **1994**, *266*, 776.
152 P. E. Dawson, S. B. H. Kent, *Annu. Rev. Biochem.* **2000**, *69*, 923.

153 L. E. Canne, S. J. Bark, S. B. H. Kent, *J. Am. Chem. Soc.* **1996**, *118*, 5891.
154 C. J. Noren, S. J. Anthonycahill, M. C. Griffith, P. G. Schultz, *Science* **1989**, *244*, 182.
155 T. Hohsaka, Y. Ashizuka, H. Sasaki, H. Murakami, M. Sisido, *J. Am. Chem. Soc.* **1999**, *121*, 12194.
156 J. D. Bain, E. S. Diala, C. G. Glabe, T. A. Dix, A. R. Chamberlin, *J. Am. Chem. Soc.* **1989**, *111*, 8013.
157 D. Mendel, V. W. Cornish, P. G. Schultz, *Annu. Rev. Biophys. Biomol.* **1995**, *24*, 435.
158 Y. Shimizu, A. Inoue, Y. Tomari, T. Suzuki, T. Yokogawa, K. Nishikawa, T. Ueda, *Nat. Biotechnol.* **2001**, *19*, 751.
159 A. C. Forster, Z. Tan, M. N. L. Nalam, H. Lin, H. Qu, V. W. Cornish, S. C. Blacklow, *Proc. Natl. Acad. Sci. USA* **2003**, *100*, 6353.
160 R. A. Mehl, J. C. Anderson, S. W. Santoro, L. Wang, A. B. Martin, D. S. King, D. M. Horn, P. G. Schultz, *J. Am. Chem. Soc.* **2003**, *125*, 935.
161 H. Asahara, O. C. Uhlenbeck, *Proc. Natl. Acad. Sci. USA* **2002**, *99*, 3499.
162 P. Wang, Y. Tang, D. A. Tirrell, *J. Am. Chem. Soc.* **2003**, *125*, 6900.
163 P. Wang, N. Vaidehi, D. A. Tirrell, W. A. Goddard, *J. Am. Chem. Soc.* **2002**, *124*, 14442.
164 Y. Tang, D. A. Tirrell, *Biochemistry* **2002**, *41*, 10635.
165 D. B. Cowie, G. N. Cohen, *Biochim. Biophys. Acta* **1957**, *26*, 252.
166 T. W. Tuve, H. H. Williams, *J. Am. Chem. Soc.* **1957**, *79*, 5830.
167 M. H. Richmond, *J. Mol. Biol.* **1963**, *6*, 284.
168 D. B. Cowie, G. N. Cohen, E. T. Bolton, H. Derobichonszulmajster, *Biochim. Biophys. Acta* **1959**, *34*, 39.
169 E. Yoshikawa, M. J. Fournier, T. L. Mason, D. A. Tirrell, *Macromolecules* **1994**, *27*, 5471.
170 T. J. Deming, M. J. Fournier, T. L. Mason, D. A. Tirrell, *J. Macromol. Sci.* **1997**, *A34*, 2143.
171 S. Kothakota, T. L. Mason, D. A. Tirrell, M. J. Fournier, *J. Am. Chem. Soc.* **1995**, *117*, 536.
172 M. J. Dougherty, S. Kothakota, T. L. Mason, D. A. Tirrell, M. J. Fournier, *Macromolecules* **1993**, *26*, 1779.
173 N. Budisa, B. Steipe, P. Demange, C. Eckerskorn, J. Kellermann, H. Huber, *Eur. J. Biochem.* **1995**, *230*, 788.
174 H. Duewel, E. Daub, V. Robinson, J. F. Honek, *Biochemistry* **1997**, *36*, 3404.
175 K. L. Kiick, E. Saxon, D. A. Tirrell, C. R. Bertozzi, *Proc. Natl. Acad. Sci. USA* **2002**, *99*, 19.
176 J. C. M. van Hest, D. A. Tirrell, *FEBS Lett.* **1998**, *428*, 68.
177 J. C. van Hest, K. L. Kiick, D. A. Tirrell, *J. Am. Chem. Soc.* **2000**, *122*, 1282.
178 Y. Tang, G. Ghirlanda, W. A. Petka, T. Nakajima, W. F. DeGrado, D. A. Tirrell, *Angew. Chem. Int. Ed.* **2001**, *40*, 1494.
179 K. L. Kiick, R. Weberskirch, D. A. Tirrell, *FEBS Lett.* **2001**, *502*, 25.
180 K. L. Kiick, D. A. Tirrell, *Tetrahedron* **2000**, *56*, 9487.
181 K. L. Kiick, J. C. M. van Hest, D. A. Tirrell, *Angew. Chem. Int. Ed.* **2000**, *39*, 2148.
182 Y. Tang, D. A. Tirrell, *J. Am. Chem. Soc.* **2001**, *123*, 11089.
183 N. Sharma, R. Furter, P. Kast, D. A. Tirrell, *FEBS Lett.* **2000**, *467*, 37.
184 K. Kirshenbaum, I. S. Carrico, D. A. Tirrell, *ChemBioChem* **2002**, *3*, 235.
185 D. Datta, P. Wang, I. S. Carrico, S. L. Mayo, D. A. Tirrell, *J. Am. Chem. Soc.* **2002**, *124*, 5652.
186 P. Kast, H. Hennecke, *J. Mol. Biol.* **1991**, *222*, 99.
187 M. Ibba, P. Kast, H. Hennecke, *Biochemistry* **1994**, *33*, 7107.
188 V. Doring, H. D. Mootz, L. A. Nangle, T. L. Hendrickson, V. de Crecy-Lagard, P. Schimmel, P. Marliere, *Science* **2001**, *292*, 501.
189 R. S. Mursinna, T. L. Lincecum, S. A. Martinis, *Biochemistry* **2001**, *40*, 5376.
190 R. S. Mursinna, S. A. Martinis, *J. Am. Chem. Soc.* **2002**, *124*, 7286.
191 I. Kwon, K. Kirshenbaum, D. A. Tirrell, *J. Am. Chem. Soc.* **2003**, *125*, 7512.
192 M. H. J. Seifert, D. Ksiazek, M. K. Azim, P. Smialowski, N. Budisa, T. A. Holak, *J. Am. Chem. Soc.* **2002**, *124*, 7932.

193 J. G. Bann, J. Pinkner, S. J. Hultgren, C. Frieden, *Proc. Natl. Acad. Sci. USA* **2002**, *99*, 709.

194 S. A. Shah, A. T. Brunger, *J. Mol. Biol.* **1999**, *285*, 1577.

195 C. C. Correll, B. Freeborn, P. B. Moore, T. A. Steitz, *Cell* **1997**, *91*, 705.

196 P. Wang, A. Fichera, K. Kumar, D. A. Tirrell, *Angew. Chem. Int. Ed.* **2004**, *43*, 3664.

197 R. J. Griffin, *Prog. Med. Chem.* **1994**, *31*, 121.

198 T. Saegusa, Y. Ito, T. Shimizu, *J. Org. Chem.* **1970**, *35*, 2979.

199 H. C. Hang, C. R. Bertozzi, *Acc. Chem. Res.* **2001**, *34*, 727.

200 W. G. Lewis, L. G. Green, F. Grynszpan, Z. Radic, P. R. Carlier, P. Taylor, M. G. Finn, K. B. Sharpless, *Angew. Chem. Int. Ed.* **2002**, *41*, 1053.

201 B. Parrish, R. B. Breitenkamp, T. Emrick, *J. Am. Chem. Soc.* **2005**, *127*, 7404.

202 P. Wu, M. Malkoch, J. N. Hunt, R. Vestberg, E. Kaltgrad, M. G. Finn, V. V. Fokin, K. B. Sharpless, C. J. Hawker, *Chem. Commun.* **2005**, 5775.

203 M. Malkoch, R. J. Thibault, E. Drockenmuller, M. Messerschmidt, B. Voit, T. P. Russell, C. J. Hawker, *J. Am. Chem. Soc.* **2005**, *127*, 14942.

204 A. J. Link, D. A. Tirrell, *J. Am. Chem. Soc.* **2003**, *125*, 11164.

205 A. J. Link, M. K. S. Vink, D. A. Tirrell, *J. Am. Chem. Soc.* **2004**, *126*, 10598.

206 K. C. Zhang, M. R. Diehl, D. A. Tirrell, *J. Am. Chem. Soc.* **2005**, *127*, 10136.

207 J. Shao, J. P. Tam, *J. Am. Chem. Soc.* **1995**, *117*, 3893.

208 L. K. Mahal, K. J. Yarema, C. R. Bertozzi, *Science* **1997**, *276*, 1125.

209 K. J. Yarema, L. K. Mahal, R. E. Bruehl, E. C. Rodriguez, C. R. Bertozzi, *J. Biol. Chem.* **1998**, *273*, 31168.

210 R. Sadamoto, K. Niikura, T. Ueda, K. Monde, N. Fukuhara, S. I. Nishimura, *J. Am. Chem. Soc.* **2004**, *126*, 3755.

211 T. D. Clark, M. R. Ghadiri, *J. Am. Chem. Soc.* **1995**, *117*, 12364.

212 K. E. Beatty, F. Xie, Q. Wang, D. A. Tirrell, *J. Am. Chem. Soc.* **2005**, *127*, 14150.

213 Q. Wang, T. R. Chan, R. Hilgraf, V. V. Fokin, K. B. Sharpless, M. G. Finn, *J. Am. Chem. Soc.* **2003**, *125*, 3192.

214 T. S. Seo, X. Bai, H. Ruparel, Z. Li, N. J. Turro, J. Ju, *Proc. Natl. Acad. Sci. USA* **2004**, *101*, 5488.

215 A. E. Speers, B. F. Cravatt, *ChemBioChem* **2004**, *5*, 41.

216 V. Chowdhry, F. H. Westheimer, *Annu. Rev. Biochem.* **1979**, *48*, 293.

217 T. Bentin, R. Hamzavi, J. Salomonsson, H. Roy, M. Ibba, P. E. Nielsen, *J. Biol. Chem.* **2004**, *279*, 19839.

218 M. Suchanek, A. Radzikowska, C. Thiele, *Nat. Methods* **2005**, *2*, 261.

219 T. J. Deming, M. J. Fournier, T. L. Mason, D. A. Tirrell, *Macromolecules* **1996**, *29*, 1442.

220 R. A. Berg, D. J. Prockop, *Biochem. Biophys. Res. Commun.* **1973**, *52*, 115.

221 K. Mizuno, T. Hayashi, D. H. Peyton, H. P. Bachinger, *J. Biol. Chem.* **2004**, *279*, 38072.

222 D. D. Buechter, D. N. Paolella, B. S. Leslie, M. S. Brown, K. A. Mehos, E. A. Gruskin, *J. Biol. Chem.* **2003**, *278*, 645.

223 W. Kim, R. A. McMillan, J. P. Snyder, V. P. Conticello, *J. Am. Chem. Soc.* **2005**, *127*, 18121.

224 W. Kim, A. George, M. Evans, V. P. Conticello, *Chem. Bio. Chem.* **2004**, *5*, 928.

12
Macromolecular Engineering of Polypeptides Using the Ring-opening Polymerization of α-Amino Acid *N*-Carboxyanhydrides

Harm-Anton Klok and Timothy J. Deming

12.1
Introduction

Proteins can be regarded as the ultimate polymers. Proteins are linear polypeptides that have high molecular weights, yet possess uniform chain lengths and precisely defined monomer sequences (primary structure) [1]. Nature masters this synthetic challenge using a unique pathway that involves polymerization of activated monomers (aminoacyl-tRNA) on a messenger RNA template. Chain length uniformity and defined monomer sequences are two important characteristics that allow folding and organization of linear polypeptide polymers into hierarchically ordered tertiary and quaternary protein structures. Many of the exquisite properties of proteins are related to their ability to form hierarchically organized assemblies.

There is increasing interest in both natural and *de novo* designed polypeptides for a variety of applications in biomedicine, biomaterials and diagnostics [2–4]. The development of DNA recombinant technologies allows the biological synthesis of high molecular weight, perfectly monodisperse and sequence-specific peptide polymers [5]. However, the yields of protein that can be isolated using this technique can vary greatly depending on the primary structure of the protein of interest and optimization of process conditions to improve expression yields and isolation/purification procedures can be cumbersome. Furthermore, non-natural amino acids can only be incorporated to a limited extent. Chemical synthesis represents an attractive alternative. Two main strategies can be distinguished. The first is solid-phase peptide synthesis (SPPS) [6, 7]. Although SPPS allows accurate control over monomer sequence and affords perfectly monodisperse peptides, this technique is only practical for the synthesis of relatively short peptides containing up to 30–40 amino acids and also cannot be infinitely scaled up without great expense. The most facile chemical route for the preparation of polypeptides is the polymerization of α-amino acid *N*-carboxyanhydrides (NCAs) (Scheme 12.1) [8]. Polymerization of NCAs yields neither perfectly monodisperse nor sequence-specific polypeptides. The method, however, uses

Macromolecular Engineering. Precise Synthesis, Materials Properties, Applications.
Edited by K. Matyjaszewski, Y. Gnanou, and L. Leibler
Copyright © 2007 WILEY-VCH Verlag GmbH & Co. KGaA, Weinheim
ISBN: 978-3-527-31446-1

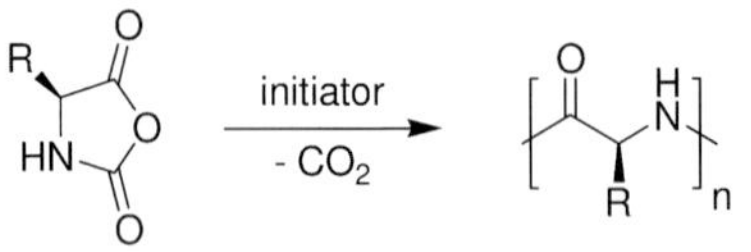

Scheme 12.1 Synthesis of polypeptides via the ring-opening polymerization of α-amino acid N-carboxyanhydrides.

simple starting materials and allows the preparation of high molecular weight polypeptides in good yields. Another attractive feature of NCA polymerization is that it does not affect the stereochemistry at the chiral center of the amino acid building blocks.

The history of NCA polymerization dates back to the early 20th century. NCAs were first described in 1906 by Hermann Leuchs, who discovered that heating N-ethoxycarbonyl and N-methoxycarbonyl α-amino acid chlorides under reduced pressure resulted in the formation of the corresponding NCAs [9–11]. Upon exposure to water, Leuchs observed that the NCAs reacted and were transformed into an insoluble mass. The concept of polymers not being established at that time, Leuchs described this product as cyclic oligopeptides. It was not until the 1920s, when the existence of macromolecules became widely accepted, that systematic studies on the synthesis and polymerization of NCAs were started [12–17]. Since then, NCA polymerization has matured and has become the most commonly applied technique for the large-scale (i.e. multi-gram) synthesis of high molecular weight polypeptides and polypeptide-based block copolymers. During the last 5–10 years, significant advances have been made in the polymerization of NCAs, which has led to enhanced control over composition, functionality and topology of the resulting polypeptides. This chapter will first give an overview of the different approaches, conventional and more recent, that are available for the polymerization of NCAs. Subsequently, the feasibility of using these different NCA polymerization methods to synthesize polypeptides and peptide hybrid polymers with increasingly complex architectures will be discussed.

12.2
Polymerization of α-Amino Acid N-Carboxyanhydrides

12.2.1
Conventional Methods

The most important initiators for NCA ring-opening polymerization are nucleophiles and bases, such as primary amines, tertiary amines and alkoxide and hydroxide ions [8, 18, 19]. Other initiating systems, such as temperature, metal salts and organometallic compounds, have also been used, but these have not found widespread application [8]. Polymerization of NCAs using nucleophilic or

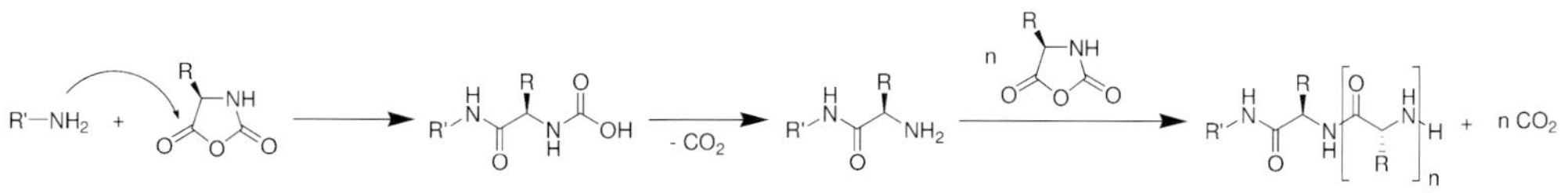

Scheme 12.2 Polymerization of NCAs via the amine mechanism.

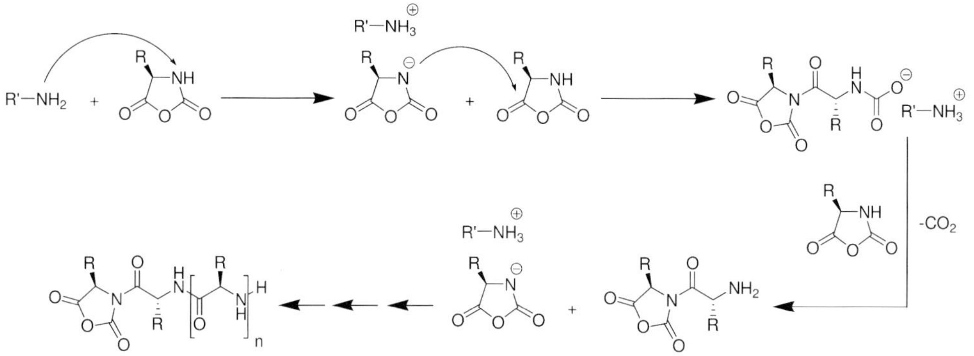

Scheme 12.3 Polymerization of NCAs via the activated monomer mechanism.

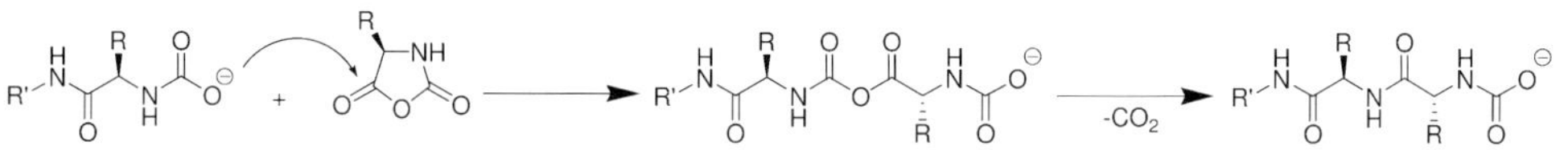

Scheme 12.4 Chain growth during the NCA polymerization via the carbamate mechanism.

basic initiators is thought to proceed along two main pathways, the amine mechanism (Scheme 12.2) and the activated-monomer mechanism (Scheme 12.3) [8, 18, 19]. The amine mechanism is a nucleophilic ring-opening chain-growth process. The activated-monomer mechanism, on the other hand, starts with deprotonation of an NCA, which subsequently acts as the nucleophile that initiates chain growth. If decarboxylation is slow, chain growth may proceed along a third pathway, the carbamate mechanism (Scheme 12.4). The preferred pathway of polymerization for a given initiator is determined by its nucleophilicity/basicity ratio. Generally, NCA polymerizations initiated by primary amines, which have a high nucleophilicity/basicity ratio, predominantly follow the primary amine mechanism. NCA polymerizations initiated by tertiary amines or alkoxide/hydroxide ions, which are more basic than nucleophilic, mainly proceed along the activated-monomer mechanism. This classification, however, is by no means unambiguous and it is important to note that a given polymerization can switch back and forth many times between the different mechanisms during the course of the reaction. In other words, a propagation step for one mechanism is a side-reaction for the other and vice versa.

The existence of multiple, competitive mechanisms, which may act simultaneously during the course of a polymerization reaction, restricts control over the molecular weight and the molecular weight distribution and also hampers the formation of well-defined block copolymers. Using very nucleophilic or basic initiators, NCA polymerization can be biased towards the amine and activated-monomer mechanism, respectively. These initiators allow a certain level of control over the polymerization and can produce polypeptides with molecular weights which roughly correlate with the monomer to initiator ratio. In addition to the existence of multiple competing propagating pathways, conventional NCA polymerizations are also susceptible to termination due to reaction of the amine end with an ester side-chain, attack of DMF by the amine end or chain transfer to monomer. Another aspect which may complicate the polymerization of certain NCAs is the build-up of secondary structure during chain growth, which may lead to precipitation during polymerization.

12.2.2
Transition Metal-mediated NCA Polymerization

It was mentioned before that the inherent problem of using any of the traditional initiators for the NCA ring-opening polymerization is that the amino group at the growing peptide's N-terminus can act both as a nucleophile and as a base. As a result, chain growth may proceed both via the amine and the activated-monomer mechanism. One approach to prevent chain growth via multiple, competing pathways is the use of transition metal complexes to control the reactivity of the growing polymer chain end. Transition metal complexes have been explored as initiators for the NCA ring-opening polymerization since the 1960s [20–23]. Unfortunately, most of these complexes were characterized by relatively low initiator efficiencies and did not permit the synthesis of polypeptides with precise control of molecular weight and narrow molecular weight distributions with high yields. This situation changed dramatically, however, with the discovery of zerovalent nickel and cobalt complexes, which were found to be able to promote very efficient "living" polymerization of NCAs [24, 25]. The putative mechanism for the polymerization of NCAs using these initiators, bipy-Ni(COD) and (PMe$_3$)$_4$Co, is outlined in Scheme 12.5 [19]. These transition metal initiators were found to offer unprecedented control over molecular weight and allowed the high-yield synthesis of polypeptides with molecular weights between 500 and 500 000 Da and polydispersities < 1.20.

12.2.3
Other NCA Polymerization Methods

The use of transition metal complexes to control the reactivity of the growing polymer chain end is described above. An interesting alternative approach, which allows chain growth to take place preferentially via the amine mechanism and suppresses the activated-monomer pathway, is to use primary amine hydro-

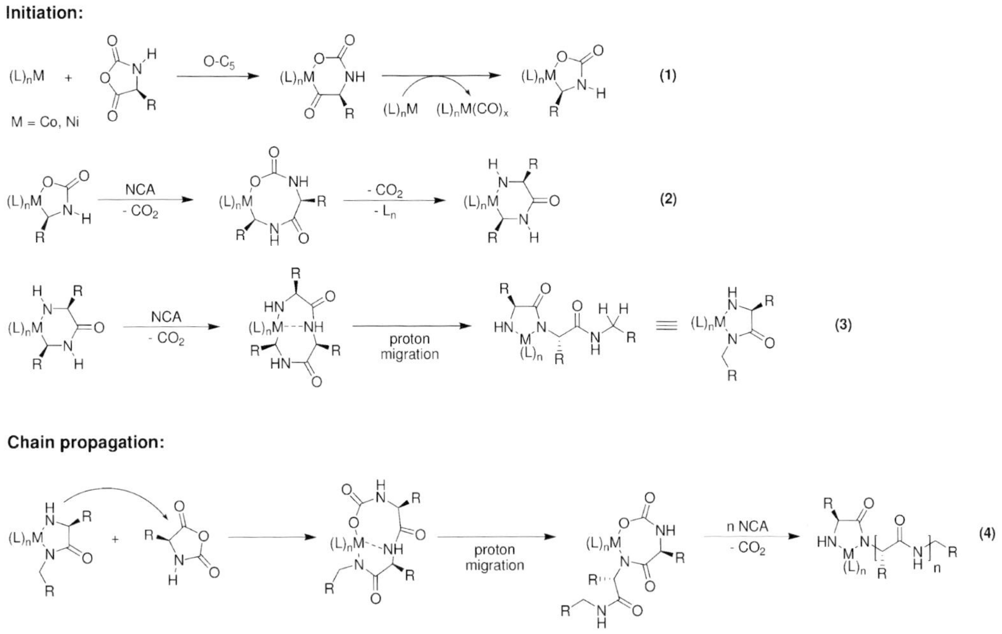

Scheme 12.5 Transition metal-mediated NCA polymerization [19].

chlorides as initiators for the NCA ring-opening polymerization [26]. The use of primary amine hydrochlorides as initiators for the NCA ring-opening polymerization is based on early work by Knobler and coworkers [27, 28], who investigated stoichiometric reactions between primary amine hydrochlorides and NCAs. These investigations revealed that this reaction proceeds smoothly and without the formation of polypeptides. When anionically prepared polystyrene macroinitiators with a primary amine hydrochloride end-group were used for the ring-opening polymerization of ε-benzyloxycarbonyl-L-lysine NCA, hybrid block copolymers with very narrow polydispersities (<1.03) were obtained [26]. These narrow polydispersities, which are close to a Poisson distribution, suggest that primary amine hydrochlorides promote polymerization of NCAs in a well-controlled fashion. The mechanism proposed for the polymerization of NCAs in the presence of primary amine hydrochlorides is outlined in Scheme 12.6. The primary amine hydrochloride is regarded as a dormant species, which is in equilibrium with the corresponding free amine and $H^+(Cl^-)$. While the primary amine end-group is responsible for chain growth via the amine mechanism, the liberated $H^+(Cl^-)$ serves to protonate any NCA anion that may be present, thereby preventing polymerization via the activated-monomer mechanism. The mechanism outlined in Scheme 12.6 is reminiscent of the persistent radical effect, which forms the basis of all controlled radical polymerization techniques and which is also based on a fast and reversible deactivation of a reactive species [29, 30].

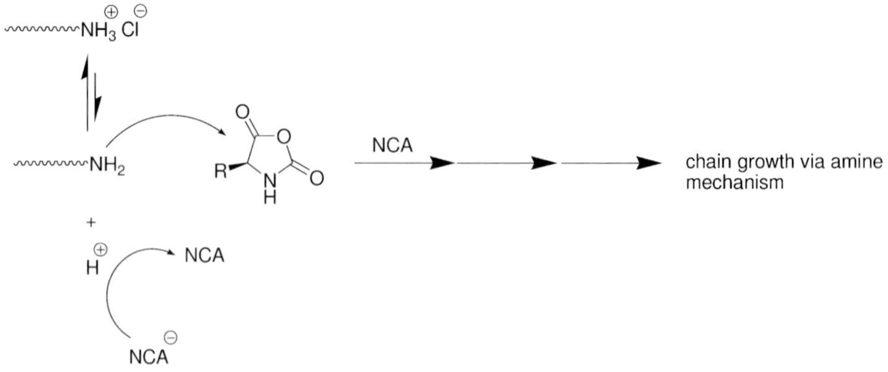

Scheme 12.6 Polymerization of NCAs using primary amine hydrochloride initiators [26].

In 2004, the group of Hadjichristidis demonstrated that polypeptide homopolymers and block copolypeptides with molecular weights corresponding to the monomer to initiator ratio and with narrow molecular weight distributions could also be prepared using high-vacuum techniques, which have traditionally been used for the anionic polymerization of vinyl monomers [31]. High-vacuum techniques allow careful purification of all reagents and can provide a reaction environment that is virtually free of any impurities that can interfere with polymerization [32]. Under these conditions, using monomers purified via high-vacuum techniques and thoroughly dried solvents, primary amine-initiated NCA polymerizations can produce polypeptide homo- and block copolymers with precise control of molecular weight and composition and narrow polydispersities [31]. These primary amine-initiated NCA ring-opening polymerizations fulfill all the requirements to be considered truly "living" polymerizations. For example, poly(benzyl-L-glutamate) homopolymers with a number-average degree of polymerization of ~ 440 ($M_n \approx 100\,000$ Da) could be produced with a polydispersity of 1.18. These results suggest that the side-reactions that are normally observed in amine-initiated NCA polymerizations are a consequence of impurities. However, the main side-reactions in amine-initiated NCA polymerizations (termination by reaction of the amine end with an ester side-chain, attack of DMF by the amine end or chain transfer to monomer) are not due to adventitious impurities such as water, but rather involve reactions with monomer, solvent or polymer. As a result, it seems very unlikely to the authors that the lack of control in amine-initiated NCA polymerizations is due to the presence of (traces of) impurities. Instead, the observed polymerization control under high-vacuum conditions may be due to a catalytic effect of impurities on side-reactions with monomer, solvent or polymer. An interesting hypothesis, which would require further studies, would be that polar species such as water can bind to monomer or the propagating chain end and thus affect their reactivity.

12.3
Block Copolymers

12.3.1
Block Copolypeptides

12.3.1.1 Conventional NCA Polymerization

For the examination of model protein–protein interactions and the assembly of novel three-dimensional structures, block copolypeptides are required which have structural domains (i.e. amino acid sequences) whose size and composition can be precisely adjusted. Such materials have proven elusive using conventional techniques. Strong base-initiated NCA polymerizations are very fast. These polymerizations are poorly understood and block copolymers cannot be prepared. Primary amine-initiated NCA polymerizations are also not free of side-reactions. Even after fractionation of the crude preparations, the resulting polypeptides are relatively ill-defined, which may complicate unequivocal evaluation of their properties and potential applications. Nevertheless, there are many reports on the preparation of block copolypeptides using conventional primary amine initiators [33]. Examples include many hydrophilic-hydrophobic and hydrophilic-hydrophobic-hydrophilic di- and triblock copolypeptides (where hydrophilic residues were glutamate and lysine and hydrophobic residues were leucine [34, 35], valine [36], isoleucine [37], phenylalanine [38] and alanine [39]) prepared to study conformations of the hydrophobic domain in aqueous solution. These conformational preferences of different amino acid residues were used as the basis for early models for predicting protein conformations from sequence. Consequently, these copolypeptides were studied under conditions favoring isolated single chains (i.e. high dilution) and self-assembly of the polymers was not investigated. These copolymers were often subjected to only limited characterization (e.g. amino acid compositional analysis) and, as such, their structures and the presence of homopolymer contaminants were not conclusively determined. Some copolymers, which had been subjected to chromatography, showed polymodal molecular weight distributions containing substantial high and low molecular weight fractions [38]. The compositions of these copolymers were found to be very different from the initial monomer feed compositions and varied widely for different molecular weight fractions. It appears that most, if not all, block copolypeptides prepared using amine initiators have structures different than predicted by monomer feed compositions and likely have considerable homopolymer contamination due to the side-reactions described above.

12.3.1.2 Controlled NCA Polymerization

Polypeptide block copolymers prepared via transition metal-mediated NCA polymerization are well defined, with the sequence and composition of block segments controlled by order and quantity of monomer, respectively, added to initiating species. These block copolypeptides can be prepared with the same level of control as found in anionic and controlled radical polymerizations of vinyl monomers, which greatly extends the potential of polypeptide materials. The unique chemistry of these initiators and NCA monomers also allows NCA monomers to be polymerized in any order, which is a challenge in most vinyl copolymerizations and the robust chain-ends allow the preparation of copolypeptides with many block domains (e.g. >4). The self-assembly of these block copolypeptides has also been investigated, e.g. to direct the biomimetic synthesis of ordered silica structures [40], to form polymeric vesicular membranes [41, 42] or to prepare self-assembled polypeptide hydrogels [43]. Furthermore, poly(L-lysine)-b-poly(L-cysteine) block copolypeptides have been used to generate hollow, organic–inorganic hybrid microspheres composed of a thin inner layer of gold nanoparticles surrounded by a thick layer of silica nanoparticles [44]. Using the same procedure, hollow spheres could also be prepared, which consisted of a thick inner layer of core-shell CdSe/CdS nanoparticles and a thinner silica nanoparticle outer layer [45]. The latter spheres are of interest, since they allowed for microcavity lasing without the use of additional mirrors, substrate spheres or gratings.

12.3.2
Hybrid Block Copolymers

12.3.2.1 Conventional NCA Polymerization

The first peptide–synthetic hybrid block copolymers were reported by the groups of Gallot [46] and Yamashita [47] in the 1970s. Since then, these initial studies have been followed by numerous other publications on the preparation of AB- and ABA-type peptide–synthetic hybrid block copolymers (A=polypeptide; B=synthetic polymer) [48–50]. Most of these block copolymers have been prepared in a two-step process and use a synthetic polymer with either one or two primary amine end-groups as a macroinitiator for the NCA ring-opening polymerization. The main advantage of this approach is that, by using any of the available controlled/living polymerization techniques, most synthetic polymers can be prepared with controlled chain lengths, low polydispersities and a high degree of amine end functionalization. However, as expected for a conventional primary amine-initiated NCA polymerization, this strategy results in block copolymers with a relatively high polydispersity in the peptide segment and is accompanied by side-reactions, which can lead to the formation of, e.g., homopolypeptide impurities. In spite of these limitations, the ring-opening polymerization of NCAs using primary amine end-functionalized synthetic polymers provides reasonably well-defined samples when carefully executed and has been successfully used to prepare a broad range of peptide–synthetic hybrid block copolymers. Table 12.1 gives an overview of the different block copolymers that have been prepared using this methodology [48–50].

The synthetic polymer blocks are typically prepared by addition polymerization of conventional vinyl monomers such as styrene or butadiene, and also by ring-opening polymerization in case of, e.g., ethylene oxide or ε-caprolactone.

12.3 Block Copolymers

Table 12.1 Examples of polypeptide hybrid block copolymers prepared from macromolecular amine initiators [48–50]: block architectures are primarily AB diblocks or ABA triblocks, where the polypeptide segment is the A domain and the macroinitiator is the B domain.

Amine macroinitiator	Polypeptide segments (*architecture*) [a]
Polystyrene	PBLG (*AB*), PZLL (*AB*), PMDG (*ABA*)
Polybutadiene	PZLL(*AB,ABA*), PBLG (*AB,ABA*), PBL/DG (*ABA*), PML/DG (*ABA*)
Polyisoprene	PBLG (*ABA*)
Polydimethylsiloxane	PBLG (*AB, ABA*), poly(L/D-Phe) (*AB*)
Poly(ethylene glycol)	PZLL (*AB,ABA*), PBLG (*AB,ABA*), poly(L-Pro) (*ABA*), PBLA (*AB,ABA*)
Poly(propylene oxide)	PBLG (*ABA*)
Poly(2-methyloxazoline)	PBLG (*AB*), poly(L-Phe) (*AB*) [51]
Poly(2-phenyloxazoline)	PBLG (*AB*), poly(L-Phe) (*AB*) [51]
Poly(methyl methacrylate)	PZLL (*AB*), PBLG (*AB*), PMLG (*AB*)
Poly(methyl acrylate)	PBLG (*AB*)
Polyoctenamer	PBLG (*ABA*)
Polyethylene	PBLG (*ABA*)
Polyferrocenylsilane	PBLG (*AB*) [52]
Poly(9,9-dihexylfluorene)	PBLG (*ABA*) [53]
Poly(ε-caprolactone)	PBLG (*ABA*), poly(L-Phe)(*ABA*), poly(Gly)(*ABA*), poly(L-Ala)(*ABA*) [54]

a) Abbreviations: PBLA, poly(β-benzyl-L-aspartate);
PBLG, poly(γ-benzyl-L-glutamate);
PBL/DG, poly(γ-benzyl-DL-glutamate);
PMDG, poly(γ-methyl-D-glutamate);
PMLG, poly(γ-methyl-L-glutamate);
PML/DG, poly(γ-methyl-DL-glutamate);
PZLL, poly(ε-benzyloxycarbonyl-L-lysine).

The polypeptide segments are mainly based on ε-benzyloxycarbonyl-L-lysine and γ-benzyl-L-glutamate, since these can form α-helical polypeptides with good solubility characteristics. ε-Benzyloxycarbonyl-L-lysine and γ-benzyl-L-glutamate are also easily deprotected to afford polypeptide segments that are both water soluble and conformationally responsive to stimuli such as pH or temperature [55]. The self-assembly properties of peptide–synthetic hybrid block copolymers, both in the solid state and in solution, have been extensively discussed in several recent reviews [50, 56].

Uchida et al. used a slightly different approach for the preparation of peptide–synthetic hybrid block copolymers, which starts with the synthesis of an isocyanate end-capped polyurethane prepolymer [57]. The polyurethane was mixed with γ-methyl-L-glutamate N-carboxyanhydride and hydrazine initiator in order to yield a polypeptide *in situ* that is capped with primary amine groups at both chain ends. Reaction of the difunctional polypeptide with the isocyanate-functionalized polyurethane afforded (AB)$_n$-type peptide–synthetic hybrid multiblock copolymers.

12.3.2.2 Controlled NCA Polymerizations

The feasibility of primary amine hydrochlorides to act as initiators for the controlled polymerization of NCAs was demonstrated using polystyrene-based macroinitiators. Dimitrov and Schlaad showed that in this way well-defined polystyrene-*b*-poly(ε-benzyloxycarbonyl-L-lysine) (PS–PZLL) diblock copolymers with narrow polydispersities could be prepared in good yields [26]. More recently, this technique was used by Lutz and coworkers to prepare poly(ethylene glycol)-*b*-poly(γ-benzyl-L-glutamate) (PEG–PBLG) and poly(ethylene glycol)-*b*-poly(β-benzyl-L-aspartate) (PEG–PBLA) diblock copolymers [58]. Although the chain length distributions of the resulting diblock copolymers were very narrow ($M_w/M_n < 1.05$), the NCA polymerization was rather slow and the experimentally determined peptide block lengths were significantly smaller than those expected based on the NCA to macroinitiator ratio. The authors also noted that in some cases the final polymers contained small quantities of low molecular weight PBLG or PBLA homopolypeptides, which, however, could be removed by selective precipitation in cold DMF.

As mentioned earlier, zerovalent nickel and cobalt complexes are very efficient initiators that can promote controlled NCA polymerization. A drawback of this method, however, is that the active propagating species are generated *in situ*, where the C-terminal end of the polypeptide is derived from the first NCA monomer. As a consequence, this method does not allow the attachment of a synthetic polymer to the carboxyl chain end and is not suitable for the synthesis of peptide–synthetic hybrid block copolymers. This limitation could be overcome, however, by the direct synthesis of the amido amidate metallacycle propagating species [59]. Scheme 12.7 shows different examples of initiators that were synthesized from N-alloc-α-amino acids, bidentate ligands and Ni(COD)$_2$. These complexes allowed controlled polymerization of NCAs, which also provided additional support for their validity as polymerization intermediates. More importantly, the initiating ligand was quantitatively incorporated as a C-terminal group, which allowed the introduction a broad variety of functional end-groups. Interestingly, the direct synthesis of functional amido amidate metallacycles is not restricted to low molecular weight substituents, but can also be extended to polymeric macroinitiators [60]. This is outlined in Scheme 12.8, which shows the transformation of α,ω-amino polyoctenamers obtained via ADMET into metallacycle substituted macroinitiators. These macroinitiators were used to polymerize γ-benzyl-L-glutamate N-carboxyanhydride. Selective hydrogenation of the polyoctenamer double bonds subsequently afforded poly(γ-benzyl-L-glutamate)-*b*-polyethylene-*b*-poly(γ-benzyl-L-glutamate) triblock copolymers. The strategy illustrated in Scheme 12.8 was also successfully used for the synthesis of poly(γ-benzyl-L-glutamate)-*b*-poly(ethylene glycol)-*b*-poly(γ-benzyl-L-glutamate) (PBLG–PEG–PBLG) and poly(γ-benzyl-L-glutamate)-*b*-polydimethylsiloxane-*b*-poly(γ-ben-

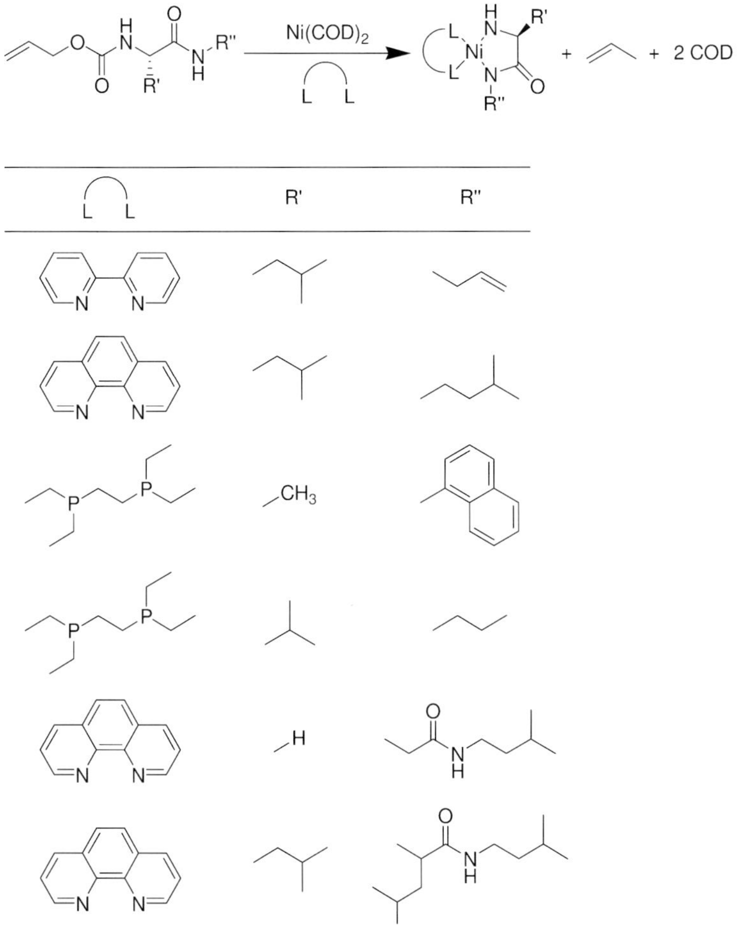

Scheme 12.7 Synthesis of amido amidate metallacycles from
N-alloc-α-amino acids [59].

zyl-L-glutamate) (PBLG–PDMS–PBLG) triblock copolymers using appropriate, commercially available α,ω-bisamino-functionalized PEG and PDMS precursors [61]. In a similar fashion, poly(methyl acrylate)-b-poly(γ-benzyl-L-glutamate) (PMA–PBLG) diblock copolymers were also synthesized using an amine-terminated PMA macroinitiator, which was obtained via atom transfer radical polymerization (ATRP) [62].

Electrophiles such as isocyanates can act as chain-terminating agents in NCA polymerizations since they can react with the propagating amine chain end. Following this rationale, Deming and coworkers discovered that reactive nickelacycle peptide chain ends could be quantitatively capped by reaction with excess isocyanate, isothiocyanate or acid chloride [61]. This method was subsequently used to prepare PEG-capped CABAC-type hybrid pentablock copolymers, where

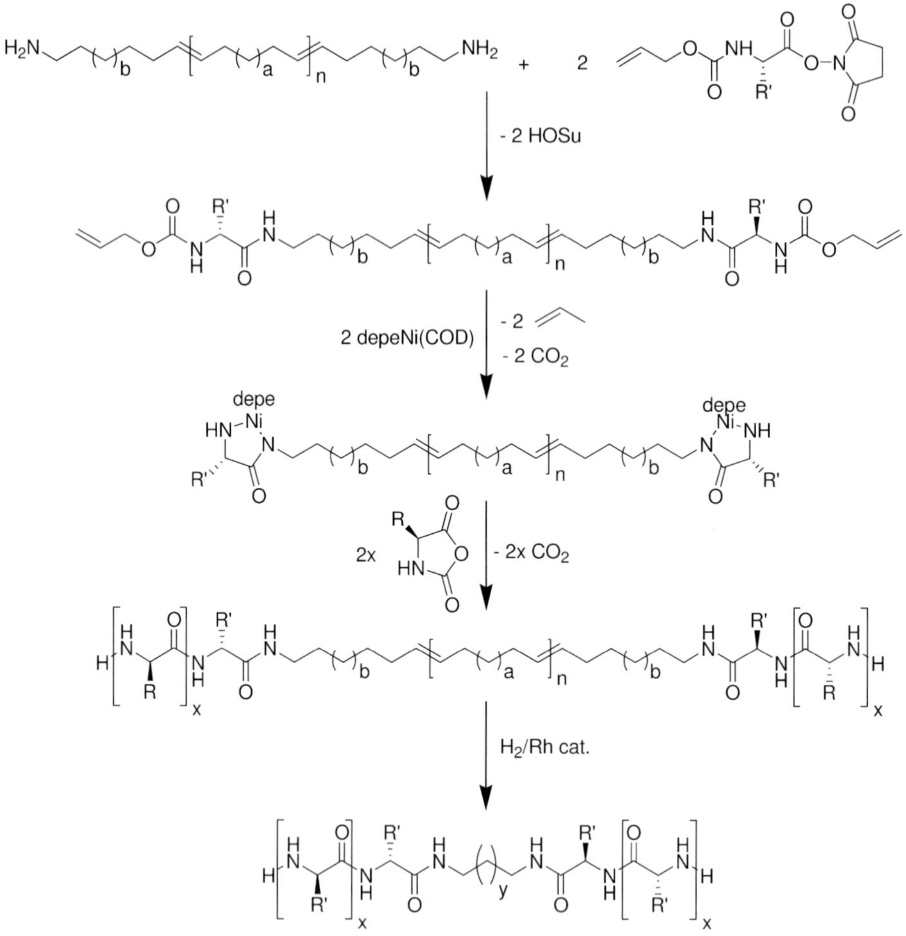

Scheme 12.8 Synthesis of polypeptide-b-polyethylene-b-polypeptide triblock copolymers [60].

A = PBLG, B = polyoctenamer, PDMS or PEG and C = PEG. Since excess PEG was used to end-cap the living polypeptide chains, the pentablock copolymers required purification, which was achieved by repeated precipitation from THF into methanol.

Recently, Kros et al. reported a one-pot, tandem approach for the preparation of poly(γ-benzyl-L-glutamate)-b-polyisocyanide rod-rod type peptide–synthetic hybrid diblock copolymers (Scheme 12.9) [63]. The remarkable feature of this synthesis is that it is a one-pot process in which a nickel active species is successively involved in the polymerization of a NCA and an isocyanide monomer. Upon addition of (S)-(−)-α-methylbenzyl isocyanide (MBI) or L-isocyanoalanyl-L-alanine methyl ester (L,L-IAA) to the poly(γ-benzyl-L-glutamate) with the active nickel chain end, the authors observed a color change indicating an alteration

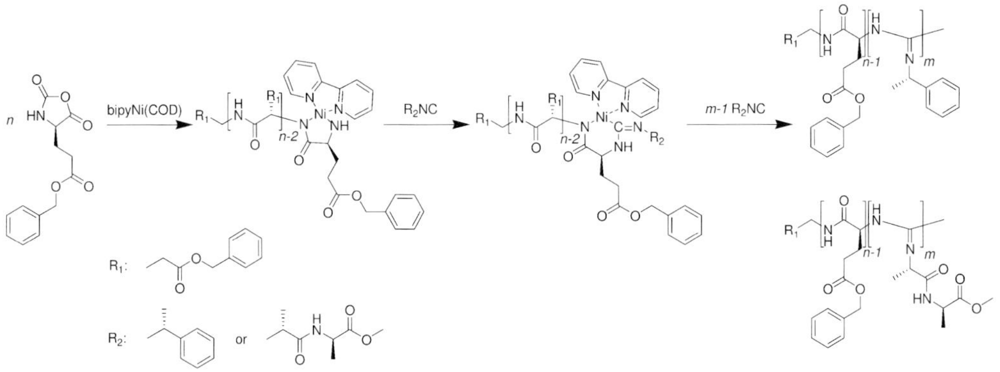

Scheme 12.9 Synthesis of poly(γ-benzyl-L-glutamate)-b-poly-isocyanide peptide–synthetic hybrid diblock copolymers [63].

in the coordination sphere of the metal from the metallacycle state that mediates the NCA polymerization to the iminoacyl species that catalyzes the isocyanide polymerization. Using this strategy, a series of diblock copolymers containing 122–225 γ-benzyl-L-glutamate and 130–200 isocyanide repeat units was prepared.

12.4
Star Polypeptides

12.4.1
Conventional NCA Polymerization

Principally, star polymers can be prepared via two main synthetic routes, the so-called arm-first and core-first methodologies. In the first case, "living" or end-reactive polymer chains are coupled to a multifunctional core. In the second case, a multifunctional core molecule is used to initiate the polymerization of the arms. Most of the star polypeptides reported so far have been obtained following the core-first strategy and were prepared using conventional primary amine-initiated NCA polymerization. In this way, Daly and coworkers successfully prepared a series of three-, four-, six- and nine-arm poly(γ-stearyl-L-glutamate) star polypeptides [64, 65]. Inoue and coworkers used hexakis(4-aminophenoxy)cyclo-triphosphazene as initiator for the synthesis of six-arm poly(β-benzyl-L-aspartate) and poly(β-benzyl-L-glutamate) stars [66, 67]. They used the ability of 5-[4-(di-methylamino)phenyl]-2,4-pentadienal (DMAPP) to react selectively with aromatic primary amines with the formation of the corresponding Schiff bases to quantify the efficiency of the initiation step. For the poly(β-benzyl-L-aspartate) stars, complete consumption of all aromatic primary amine initiating groups, i.e. the formation of well-defined six-arm star polymers, was observed at sufficiently high monomer to initiator ratios (≥100) and high conversions [66]. In a subsequent paper, Inoue et al. reported the synthesis of oligo(ethylene glycol)-

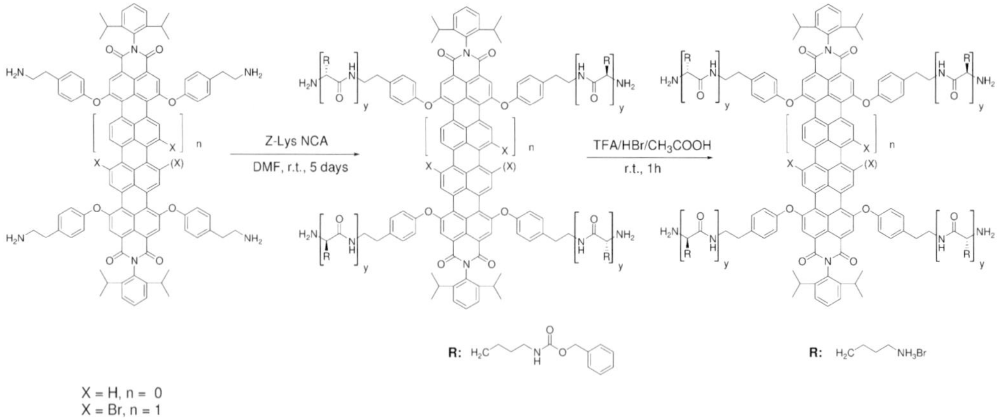

Scheme 12.10 Synthesis of four-arm star polypeptides from rylene-based multifunctional primary amine initiators [69, 70].

modified six-arm poly(L-glutamate) star polymers. These polymers were prepared via transesterification of the benzyl ester groups of six-arm star poly(γ-benzyl-L-glutamate) stars with di- or triethylene glycol monomethyl ether in the presence of *p*-toluenesulfonic acid [68]. Depending on the reaction time and temperature, degrees of substitution of 52–63% could be achieved. Teflon films of these ethylene glycol-substituted six-arm polypeptide star polymers were found to be excellent enantioselective membranes for amino acids.

Klok and coworkers have prepared several water-soluble, fluorescent, and near-infrared-absorbing four-arm star polypeptides based on L-lysine and L-glutamic acid (Scheme 12.10) [69, 70]. The synthesis of these star polymers started with the ring-opening polymerization of γ-benzyl-L-glutamate N-carboxyanhydride or ε-benzyloxycarbonyl-L-lysine N-carboxyanhydride using perylene-, terrylene- or quaterrylene-based multifunctional primary amine initiators. Subsequent removal of the side-chain protective groups afforded the corresponding water-soluble star polypeptides. The experimentally determined peptide arm lengths were in good agreement with those expected based on the monomer to initiator ratio. Furthermore, for the lower molecular weight star polymers, ^{1}H NMR spectra indicated complete consumption of all initiator groups, which provided additional evidence for the structural integrity of the star polypeptides.

So far, only star polypeptides containing a relatively small number of peptide arms have been discussed. Star polypeptides with a much larger number of arms can be obtained when, e.g., primary amine-functionalized poly(amidoamine) (PAMAM) or poly(propylenimine) dendrimers are used as macroinitiators for the NCA ring-opening polymerization. Higashi and coworkers, for example, used a third-generation amine-terminated PAMAM dendrimer to prepare a 32-arm poly(γ-benzyl-L-glutamate) star polymer [71, 72]. ^{1}H and ^{13}C NMR spectroscopy were used to validate the structure of the star polymer. From the ^{13}C NMR spectra, it was concluded that the primary amine initiator groups at the periphery of the PAMAM dendrimer were quantitatively consumed during the NCA polymerization. Okada and coworkers followed an identical strategy to prepare sugar-substituted many-arm star polypeptides [73]. In this case, generation 3, 4 and 5 amine-terminated PAMAM dendrimers were used as initiators for the ring-opening polymerization of O-(tetra-O-acetyl-β-D-glucopyranosyl)-L-serine N-carboxyanhydride and O-(2-acetamido–3,4,6-tri-O-acetyl-2-deoxy-β-D-glucopyranosyl)-L-serine N-carboxyanhydride. Since the polymerizations were carried out using relatively small monomer to initiator ratios, star polypeptides were obtained that contained only up to five amino acid residues per arm [73]. The same authors also reported the synthesis of 64-arm polysarcosine star polymers via ring-opening polymerization of sarcosine N-carboxyanhydride using a poly(propylenimine) dendrimer as macroinitiator [74].

12.4.2
Controlled NCA Polymerizations

A first example of the use of a controlled NCA polymerization technique for the preparation of star polypeptides was reported by Hadjichristidis and coworkers [75]. Using high-vacuum techniques, they first prepared several linear poly-(γ-benzyl-L-glutamate), poly(ε-benzyloxycarbonyl-L-lysine), poly(γ-benzyl-L-glutamate)-b-poly(ε-benzyloxycarbonyl-L-lysine) and poly(ε-benzyloxycarbonyl-L-lysine)-b-poly(γ-benzyl-L-glutamate) precursors, which were subsequently coupled to a trifunctional core, triphenylmethane 4,4′,4″-triisocyanate. An excess of the linear precursor and reaction times of up to 4 weeks were needed to drive the coupling reaction to completion. Although the excess of the linear precursor could be removed by an additional precipitation step and well-defined star polypeptide could be obtained, this illustrates the main drawbacks of the arm-first method as compared with the core-first method that has been most widely used for the synthesis of star polypeptides.

12.5
Graft and Hyperbranched Polypeptides

All graft and hyperbranched polypeptides that have been reported so far have been prepared using conventional NCA polymerization techniques. The first graft polypeptides were published in 1956 by Sela et al. [76]. Using poly(L-lysine) or poly(DL-ornithine) as multifunctional initiator for the ring-opening polymerization of a number of different NCAs, they prepared a variety of graft polypeptides, which they termed multichain polyamino acids. Since the NCA polymerizations were carried out in aqueous dioxane, the synthesis of the graft polypeptides was accompanied by the formation of short linear polypeptides. These by-products, however, could be removed by dialysis. For the graft polymerization, poly(L-lysine) or poly(DL-ornithine) initiators with degrees of polymerization between 20 and 200 were used. The number of amino acid residues per graft was determined via chromatographic analysis of the hydrolyzed product or via end-group titrations. Depending on the length of the poly(L-lysine) or poly(DL-ornithine) initiator, the number of amino acids per graft varied from three to 25. In a subsequent paper, Sela et al. used this procedure to prepare a family of multi-chain copolymers with grafts containing L-tyrosine, L-glutamic acid and L-alanine residues [77]. The interest in these polymers was due to their potential use as synthetic polypeptide antigens. In this case, in addition to poly(L-lysine) homopolymer, also copolymers of DL-alanine and L-lysine were used as multifunctional initiators for the NCA ring-opening polymerization. Furthermore, the grafts of the multichain copolymers were not only simple homopolypeptides but, in most cases, block-type sequences. Multichain polymers with block copolypeptide grafts were prepared by utilizing the appropriate multichain polyamino acid precursors as the initiator for a subsequent NCA ring-opening polymerization step.

Since, in the examples discussed above, the polymerizations were carried out in aqueous media, the formation of linear polypeptide byproducts is inevitable. To overcome this problem, attempts have been made to synthesize multichain polypeptides in anhydrous, polar aprotic solvents such as DMF and DMSO. Sakamoto and coworkers, for example, have prepared different multichain polypeptides using random copolymers of L-lysine and γ-methyl-L-glutamate as initiator for the ring-opening polymerization of ε-benzyloxycarbonyl-L-lysine N-carboxyanhydride, γ-benzyl-L-glutamate N-carboxyanhydride and β-benzyl-L-aspartate N-carboxyanhydride [78–80]. These random copolymers typically had degrees of polymerizations ranging from 82 to 118 and contained 12–36 lysine residues. The NCA polymerizations were carried out in DMF containing 3 or 9% (v/v) DMSO. However, also under these conditions, the NCA polymerization generated linear homopolypeptide byproducts, which had to be removed by reprecipitation in diethyl ether [poly(ε-benzyloxycarbonyl-L-lysine)], methanol [poly(γ-benzyl-L-glutamate)] or acetone [poly(β-benzyl-L-aspartate)]. The number-average degree of polymerization of the polypeptide grafts was estimated by osmometric molecular weight determination and amino acid analysis and varied from 20 to 60 amino acids, depending on the initiator and the relative amounts of monomer and initiator that were used. The findings by Sakamoto and co-workers are in agreement with earlier observations by Yaron and Berger, who also identified linear homopolypeptide byproducts when the NCA polymerization was carried out in dry dioxane or DMF [81]. Tewksbury and Stahmann, in contrast, reported that the synthesis of multichain poly(amino acids) from poly(L-lysine) and DL-phenylalanine N-carboxyanhydride, L-leucine N-carboxyanhydride or benzyl-L-glutamate N-carboxyanhydride in anhydrous DMSO was not accompanied by the formation of linear byproducts [82].

These contradictory observations are characteristic for primary amine-initiated NCA polymerizations, where the success of the polymerization, i.e. the homogeneity of the reaction product, can vary significantly depending on the type of monomer and the precise polymerization conditions that are used.

Hudecz has used modified versions of the multichain polypeptides pioneered by Sela and coworkers for the preparation of antitumor drug conjugates, boron conjugates for boron neutron capture therapy and synthetic antigens [83]. The graft polypeptides studied by Hudecz are based on a poly(L-lysine) backbone that contains short side-chains composed of about three DL-alanine residues and one additional amino acid (X), which is placed either at the end of the grafts or at the position next to the poly(L-lysine) backbone. The degree of polymerization of the poly(L-lysine) backbone of these constructs was either 80–120 or 400–500. While the short oligo(DL-alanine) branches were introduced via ring-opening polymerization of the corresponding NCA, the single residue X was attached using standard peptide coupling chemistry. These short peptide grafts were attached using the procedure developed by Sela et al. [76]. Using carbodiimide-mediated coupling chemistry, the amino groups of the N-terminal residues of the peptide grafts were modified with methotrexate or a daunomycin derivative to yield antitumor drug conjugates. Boron-modified conjugates that are of potential interest for boron neutron capture therapy were prepared by coupling the polyhedral borane $Cs_2B_{12}H_{11}SH$ to the N-termini of the peptide grafts. The attachment of the borane to the polypeptide was achieved using heterobifunctional coupling agents such as N-succinimidyl-3-(2-pyridyldithio)propionate (SPDP) or m-maleimidobenzoyl-N-hydroxysuccinimide ester (MBS).

More complex, higher branched polypeptide architectures can be obtained when functional groups in the side-chains of the peptide grafts are used to initiate a subsequent NCA ring-opening polymerization step. Repetition of this graft-on-graft strategy leads to highly branched, so-called dendritic-graft, polypeptides [84]. This is illustrated in Scheme 12.11, which shows the synthesis of dendritic-graft poly-

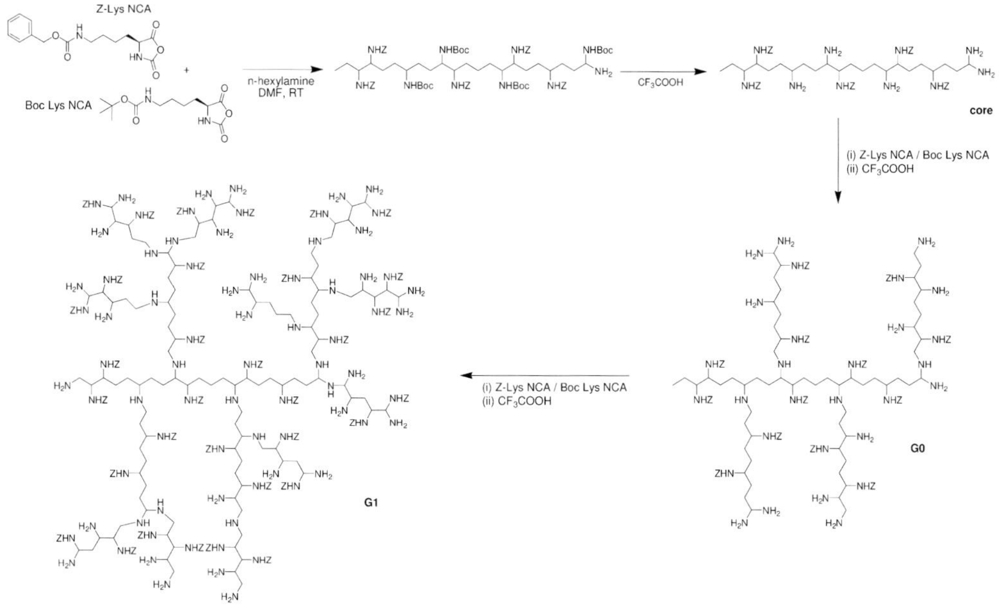

Scheme 12.11 Synthesis of dendritic-graft polypeptides [84].

lysine via a repetitive sequence of ring-opening copolymerization and deprotection steps using two orthogonally protected L-lysine NCA derivatives. Following this strategy, dendritic-graft polylysines containing up to ~160 amino acids, corresponding to a number-average molecular weight of ~40 kDa, could be prepared in just four ring-opening copolymerization–deprotection cycles [84].

In addition to the graft-on-graft strategy discussed above, highly branched polypeptides can also be prepared via an iterative sequence of NCA ring-opening polymerization and end-functionalization reactions as illustrated in Scheme 12.12 [85]. Each NCA ring-opening polymerization step is followed by an end-functionalization reaction with an appropriate $N^{\alpha},N^{\varepsilon}$-diprotected lysine derivative. Deprotection of the lysine amine groups doubles the number of end-groups that can be used to initiate a subsequent NCA ring-opening polymerization step and leads to branching of the polymer architecture. The strategy outlined in Scheme 12.12 has been used to prepare highly branched polylysines with number-average molecular weights of up to 33 kDa in only a small number of reaction steps. Birchall and North used a related approach to prepare highly branched block copolypeptides [86]. In this case, however, relatively hydrophobic water-insoluble amino acids such as alanine, leucine and phenylalanine were used, which made it necessary to keep the polymer chains relatively short (~5–10 amino acid repeat units) in order to avoid solubility problems.

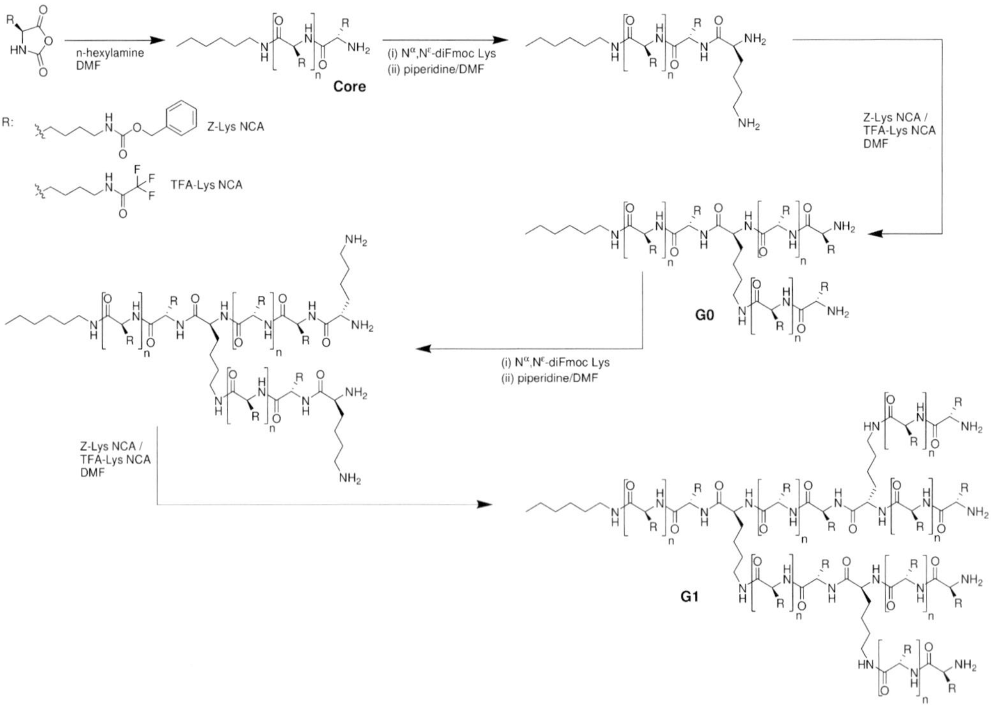

Scheme 12.12 Synthesis of highly branched polypeptides [85].

In addition to the pure polypeptide graft and hyperbranched polymers discussed above, hybrid systems have also been reported. Maeda and Inoue prepared a series of four different poly(γ-benzyl-L-glutamate)-based hybrid graft copolymers by copolymerization of styrene end-functionalized PBLG macromonomers with styrene or methyl methacrylate [87]. The PBLG macromonomers had a number-average degree of polymerization of ∼10 and were obtained via ring-opening polymerization of γ-benzyl-L-glutamate N-carboxyanhydride using N-methyl-N-(4-vinylphenethyl)ethylenediamine as the initiator. Graft copolymerization was carried out in benzene with AIBN and afforded copolymers with GPC number-average molecular weights between 15 000 and 40 000 g mol^{-1}. The number of PBLG grafts in the copolymers was relatively small and varied between 0.5 and 2.5 mol%. Schmidt and coworkers, in contrast, attempted to prepare hybrid graft copolymers in which each backbone repeat unit is functionalized with a polypeptide graft [88]. Such densely grafted copolymers are also referred to as bottlebrushes. Two strategies were explored for the synthesis of polypeptide bottlebrushes. The first approach, the so-called "grafting-through" approach, involved the homopolymerization of N-terminal methacrylate-functionalized PZLL or PBLG macromonomers. The macromonomers were synthesized via n-hexylamine-initiated polymerization of the corresponding NCAs following by modification of the N-terminal amine group with methacryloyl chloride. The number-average degree of polymerization of the macromonomers varied between ∼10 and ∼40. From the four macromonomers that were prepared, however, only one could be successfully polymerized to afford a relatively low molecular weight graft copolymer ($M_n \approx 10 000$ g mol^{-1}). A more successful approach to preparing synthetic polymers densely grafted with polypeptide arms is the so-called "grafting-from" strategy, which uses a synthetic polymer that is substituted with a primary amine group on each repeat unit as a macroinitiator for the NCA ring-opening polymerization. This strategy is illustrated in Scheme 12.13. Schmidt et al. used a poly[N-(3-aminopropyl)methacrylamide] macroinitiator with a weight-average molecular weight according to light scattering of 240 000 g mol^{-1} and a GPC polydispersity of $M_w/M_n = 2.5$. Attempts to characterize the resulting PBLG and PZLL graft polymers and also the corresponding side-chain deprotected L-glutamic acid and L-lysine analogues with GPC and light-scattering were not successful. Individual graft copolymer molecules, however, could be visualized with atomic force microscopy (AFM). From the AFM experiments, the length of the polypeptide grafts was estimated as about five repeat units.

12.6
Summary and Conclusions

Since its initial discovery by Leuchs about 100 years ago, NCA ring-opening polymerization has grown to become the most important method for the large-scale synthesis of polypeptides and peptide–synthetic hybrid polymers. Traditional primary amine- or strong base-initiated NCA polymerizations may pro-

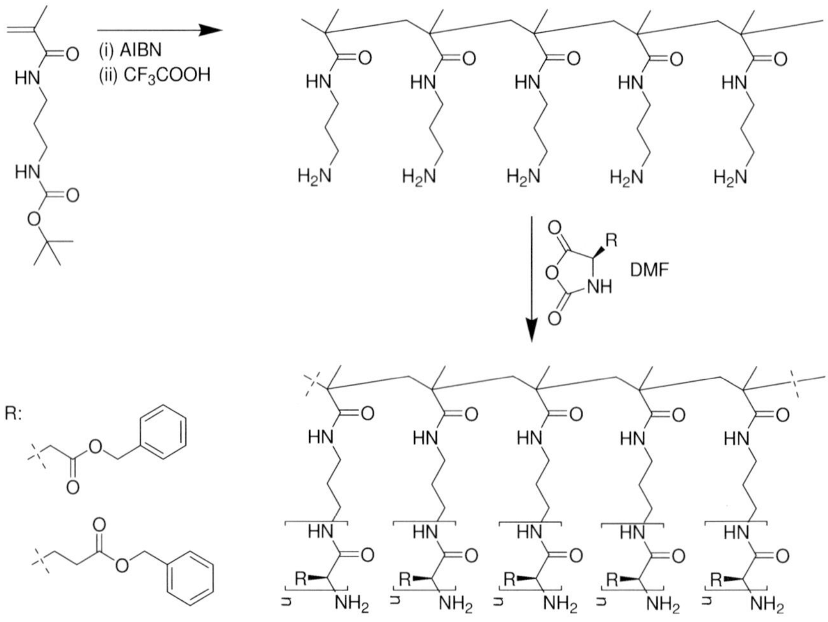

Scheme 12.13 Synthesis of polypeptide bottlebrushes [88].

ceed along multiple coexisting reaction pathways, which limits control over polymer molecular weight, architecture and end-group functionality. Very often, block copolymers or more complex polypeptide architectures prepared via these traditional methods contain homopolypeptide contaminations, which need to be removed in additional purification steps. The recent development of a number of techniques that allow controlled NCA polymerization and can produce narrowly distributed polypeptide polymers with predictable molecular weights represents a major step forward. In particular, the use of transition metal complexes has been proven to be a very versatile strategy to control the reactivity of the growing polymer chain ends and has been successfully used to prepare a broad range of well-defined polypeptide homo- and block copolymers. Block copolypeptides prepared using transition metal-based initiators have been used, for example, to prepare hydrogels and to direct the biomimetic synthesis of ordered silica structures. There is, however, also a need for further improvement. In particular, synthetic methods that facilitate the preparation of peptide–synthetic hybrid polymers or which allow easier access to more complex polypeptide architectures could pave the way for new structures, functions and possible applications. Currently, the preparation of peptide–synthetic polymer hybrid polymers and the synthesis of more complex branched polypeptide architectures is based on multistep protocols and mostly uses conventional amine- or base-initiated NCA polymerizations. The recent discovery that isocyanides and NCAs can be sequentially polymerized to form hybrid block copolymers in a one-pot, transition metal-mediated tandem process represents a significant improvement that overcomes many of the drawbacks of the more traditional approaches. In view of the current vibrancy in the field, there seems little doubt that other exciting discoveries will follow.

References

1. C. Branden, J. Tooze, *Introduction to Protein Structure*, 2nd edn, Garland Publishing, New York and London, **1999**.
2. R. Fairman, K. S. Akerfeldt, *Curr. Opin. Struct. Biol.* **2005**, *15*, 453.
3. M. P. Lutolf, J. A. Hubbell, *Nat. Biotechnol.* **2005**, *23*, 47.
4. X. J. Zhao, S. G. Zhang, *Trends Biotechnol.* **2004**, *22*, 470.
5. J. C. M. van Hest, D. A. Tirrell, *Chem. Commun.* **2001**, 1987.
6. G. B. Fields, R. L. Noble, *Int. J. Pept. Protein Res.* **1990**, *35*, 161.
7. W. C. Chan, P. D. White (Eds.), *Fmoc Solid Phase Peptide Synthesis*, Oxford University Press, Oxford, **2000**.
8. H. R. Kricheldorf, *α-Aminoacid-N-carboxyanhydrides and Related Heterocycles – Synthesis, Properties, Peptide Synthesis, Polymerization*, Springer, Berlin, **1987**.
9. H. Leuchs, *Ber. Dtsch. Chem. Ges.* **1906**, *39*, 857.
10. H. Leuchs, W. Manasse, *Ber. Dtsch. Chem. Ges.* **1907**, *40*, 3235.
11. H. Leuchs, W. Geiger, *Ber. Dtsch. Chem. Ges.* **1908**, *41*, 1721.
12. F. Wessely, *Hoppe-Seylers Z. Physiol. Chem.* **1925**, *146*, 72.
13. F. Sigmund, F. Wessely, *Hoppe-Seylers Z. Physiol. Chem.* **1926**, *157*, 91.
14. F. Wessely, F. Sigmund, *Hoppe-Seylers Z. Physiol. Chem.* **1926**, *159*, 102.
15. F. Wessely, M. John, *Hoppe-Seylers Z. Physiol. Chem.* **1927**, *170*, 38.
16. F. Wessely, J. Mayer, *Monatsh. Chem.* **1928**, *50*, 439.
17. For an early review, see E. Katchalski, *Adv. Protein Chem.* **1951**, *6*, 123.
18. T. J. Deming, *Adv. Mater.* **1997**, *9*, 299.
19. T. J. Deming, *J. Polym. Sci., Part A: Polym. Chem.* **2000**, *38*, 3011.
20. S. Freireich, D. Gertner, A. Zilkha, *Eur. Polym. J.* **1974**, *10*, 439; T. Makino, S. Inoue, T. Tsuruta, *Makromol. Chem.* **1971**, *150*, 137.
21. K. Matsuura, S. Inoue, T. Tsuruta, *Makromol. Chem.* **1967**, *103*, 140.
22. S. Yamashita, H. Tani, *Macromolecules* **1974**, *7*, 406.
23. S. Yamashita, K. Waki, N. Yamawaki, H. Tani, *Macromolecules* **1974**, *7*, 410.
24. T. J. Deming, *Nature* **1997**, *390*, 386.
25. T. J. Deming, *Macromolecules* **1999**, *32*, 4500.
26. I. Dimitrov, H. Schlaad, *Chem. Commun.* **2003**, 2944.
27. Y. Knobler, S. Bittner, M. Frankel, *J. Chem. Soc.* **1964**, 3941.
28. Y. Knobler, S. Bittner, D. Virov, M. Frankel, *J. Chem. Soc.* **1969**, 1821.
29. H. Fischer, *Chem. Rev.* **2001**, *101*, 3581.
30. K. Matyjaszewski, J. Xia, *Chem. Rev.* **2001**, *101*, 2921.
31. T. Aliferis, H. Iatrou, N. Hadjichristidis, *Biomacromolecules* **2004**, *5*, 1653.
32. N. Hadjichristidis, H. Iatrou, S. Pispas, M. Pitsikalis, *J. Polym. Sci., Part A: Polym. Chem.* **2000**, *38*, 3111.
33. F. Uralil, T. Hayashi, J. M. Anderson, A. Hiltner, *Polym. Eng. Sci.* **1977**, *17*, 515.
34. H. E. Auer, P. Doty, *Biochemistry* **1966**, *5*, 1708.
35. S. E. Ostroy, N. Lotan, R. T. Ingwall, H. A. Scheraga, *Biopolymers* **1970**, *9*, 749.
36. R. E. Epand, H. A. Scheraga, *Biopolymers* **1968**, *6*, 1551.
37. S. Kubota, G. D. Fasman, *Biopolymers* **1975**, *14*, 605.
38. F. Cardinaux, J. C. Howard, G. T. Taylor, H. A. Scheraga, *Biopolymers* **1977**, *16*, 2005.
39. R. T. Ingwall, H. A. Scheraga, N. Lotan, A. Berger, E. Katchalski, *Biopolymers* **1968**, *6*, 331.
40. J. N. Cha, G. D. Stucky, D. E. Morse, T. J. Deming, *Nature* **2000**, *403*, 289.
41. E. G. Bellomo, M. D. Wyrsta, L. Pakstis, D. J. Pochan, T. J. Deming, *Nat. Mater.* **2004**, *3*, 244.
42. E. P. Holowka, D. J. Pochan, T. J. Deming, *J. Am. Chem. Soc.* **2005**, *127*, 12423.

43 A. P. Novak, V. Breedveld, L. Pakstis, B. Ozbas, D. J. Pine, D. Pochan, T. J. Deming, *Nature* **2002**, *417*, 424.
44 M. S. Wong, J. N. Cha, K.-S. Choi, T. J. Deming, G. D. Stucky, *Nano Lett.* **2002**, *2*, 583.
45 J. N. Cha, M. H. Bartl, M. S. Wong, A. Popitsch, T. J. Deming, G. D. Stucky, *Nano Lett.* **2003**, *3*, 907.
46 B. Perly, A. Douy, B. Gallot, *C. R. Acad. Sci., Ser. C* **1974**, *279*, 1109.
47 Y. Yamashita, Y. Iwaya, K. Ito, *Makromol. Chem.* **1975**, *176*, 1207.
48 B. Gallot, *Prog. Polym. Sci.* **1996**, *21*, 1035.
49 T. J. Deming, *Adv. Polym. Sci.* **2006**, *202*, 1.
50 H.-A. Klok, S. Lecommandoux, *Adv. Polym. Sci.* **2006**, *202*, 75.
51 K. Tsutsumiuchi, K. Aoi, M. Okada, *Macromolecules* **1997**, *30*, 4013.
52 K. T. Kim, G. W. M. Vandermeulen, M. A. Winnik, I. Manners, *Macromolecules* **2005**, *38*, 4958.
53 X. Kong, S. A. Jehneke, *Macromolecules* **2004**, *37*, 8180.
54 H. R. Kricheldorf, K. Hauser, *Biomacromolecules* **2001**, *2*, 1110.
55 G. D. Fasman, *Poly α-Amino Acids*, Marcel Dekker, New York, **1967**.
56 H. Schlaad, *Adv. Polym. Sci.* **2006**, 202.
57 S. Uchida, T. Oohori, M. Suzuki, H. Shirai, *J. Polym. Sci., Part A: Polym. Chem.* **1999**, *37*, 383
58 J.-F. Lutz, D. Schütt, S. Kubowicz, *Macromol. Rapid Commun.* **2005**, *26*, 23
59 S. A. Curtin, T. J. Deming, *J. Am. Chem. Soc.* **1999**, *121*, 7427.
60 K. R. Brzezinska, T. J. Deming, *Macromolecules* **2001**, *34*, 4348.
61 K. R. Brzezinska, S. A. Curtin, T. J. Deming, *Macromolecules* **2002**, *35*, 2970.
62 K. R. Brzezinska, T. J. Deming, *Macromol. Biosci.* **2004**, *4*, 566.
63 A. Kros, W. Jesse, G. A. Metselaar, J. J. L. M. Cornelissen, *Angew. Chem. Int. Ed.* **2005**, *44*, 4349.
64 W. H. Daly, D. Poche, P. S. Russo, I. Negulescu, *Polym. Prepr.* **1992**, *33*, 188.
65 W. H. Daly, D. S. Poche, P. S. Russo, I. I. Negulescu, *J. Macromol. Sci., Pure Appl. Chem.* **1994**, *A31*, 795.

66 K. Inoue, H. Sakai, S. Ochi, T. Itaya, T. Tanigaki, *J. Am. Chem. Soc.* **1994**, *116*, 10783.
67 K. Inoue, S. Horibe, M. Fukae, T. Muraki, E. Ihara, H. Kayama, *Macromol Biosci.* **2003**, *3*, 26.
68 K. Inoue, A. Miyahara, T. Itaya, *J. Am. Chem. Soc.* **1997**, *119*, 6191.
69 H.-A. Klok, J. Rodriguez-Hernandez, S. Becker, K. Müllen, *J. Polym. Sci., Part A: Polym. Chem.* **2001**, *39*, 1572.
70 J. Rodriguez-Hernandez, J. Qu, E. Reuther, H.-A. Klok, K. Müllen, *Polym. Bull.* **2004**, *52*, 57.
71 N. Higashi, T. Koga, N. Niwa, M. Niwa, *Chem. Commun.* **2000**, 361.
72 N. Higashi, T. Koga, M. Niwa, *Adv. Mater.* **2000**, *12*, 1373.
73 K. Aoi, K. Tsutsumiuchi, A. Yamamoto, M. Okada, *Tetrahedron* **1997**, *53*, 15415.
74 K. Aoi, T. Hatanaka, K. Tsutsumiuchi, M. Okada, T. Imae, *Macromol. Rapid Commun.* **1999**, *20*, 378.
75 T. Aliferis, H. Iatrou, N. Hadjichristidis, *J. Polym. Sci., Part A: Polym. Chem.* **2005**, *43*, 4670.
76 M. Sela, E. Katchalski, M. Gehatia, *J. Am. Chem. Soc.* **1956**, *78*, 747.
77 M. Sela, S. Fuchs, R. Arnon, *Biochem. J.* **1962**, *85*, 223.
78 M. Sakamoto, Y. Kuroyanagi, *J. Polym. Sci., Polym Chem. Ed.* **1978**, *16*, 1107.
79 M. Sakamoto, Y. Kuroyanagi, R. Sakamoto, *J. Polym. Sci., Polym. Chem. Ed.* **1978**, *16*, 2001.
80 M. Sakamoto, Y. Kuroyanagi, *J. Polym. Sci., Polym. Chem. Ed.* **1979**, *17*, 2577.
81 A. Yaron, A. Berger, *Biochim. Biophys. Acta* **1965**, *107*, 307.
82 D. A. Tewksbury, M. A. Stahmann, *Arch. Biochem. Biophys.* **1964**, *105*, 527.
83 F. Hudecz, *Anti-Cancer Drugs* **1995**, *6*, 171.
84 H.-A. Klok, J. Rodriguez-Hernandez, *Macromolecules* **2002**, *35*, 8718.
85 J. Rodriguez-Hernandez, M. Gatti, H.-A. Klok, *Biomacromolecules* **2003**, *4*, 249.
86 A. C. Birchall, M. North, *Chem. Commun.* **1998**, 1335.
87 M. Maeda, S. Inoue, *Makromol. Chem., Rapid Commun.* **1981**, *2*, 537.
88 B. Zheng, K. Fischer, M. Schmidt, *Macromol. Chem. Phys.* **2005**, *206*, 157.

13
Segmented Copolymers by Mechanistic Transformations

M. Atilla Tasdelen and Yusuf Yagci

13.1
Introduction

Segmented copolymers are macromolecules composed of two or more long, contiguous, chemically dissimilar sequences of monomer repeat units. Segmented copolymers have attracted considerable interest in their synthesis, properties and solid-state morphologies because of their versatile properties. In most cases, the corresponding homopolymers do not form a homogeneous phase. However, a linear arrangement of blocks in segmented copolymers by chemical bonds results in the realization of a stable structure with two phases separated. Each segment exerts its character or function on the bulk of the copolymers. For example, multiblock thermoplastic elastomers (TPE) are segmented copolymers with alternately soft and hard segments in their structure. One of the phases, the hard one, is rigid and can participate in intermolecular interactions, whereas the soft part provides flexibility and usually has a glass-transition temperature below room temperature. Among the different segmented copolymers, polycondensates, namely the polyurethanes [1–3] and poly(ether ester)s [4–7], are the most widely studied copolymers. Polycondensate-type segmented copolymers are usually synthesized by combining blocks or segments of two dissimilar homopolymers along the chain backbone. Generally, the synthesis is achieved in two ways: (1) transesterification of homopolymers [8] (Scheme 13.1; for convenience, only one ester group is represented) or (2) coupling reactions of antagonist groups present in the prepolymer chain ends (Scheme 13.2: (a) [AB] and (b) [ABA] block copolymers and (c) [AB]$_n$ multiblock copolymers) [9, 10].

It should be pointed out that the final morphologies observed in these materials are complex as the sequences of the resulting copolymers are irregular.

Although a significant proportion of commercial polymers is prepared by polycondensation techniques, even today the problems associated with control over the structure and molar mass in these processes have not been fully solved. However, in chain growth processes there are several methods allowing

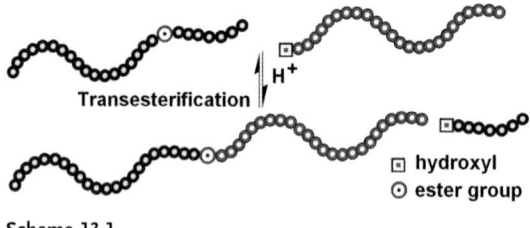

Scheme 13.1

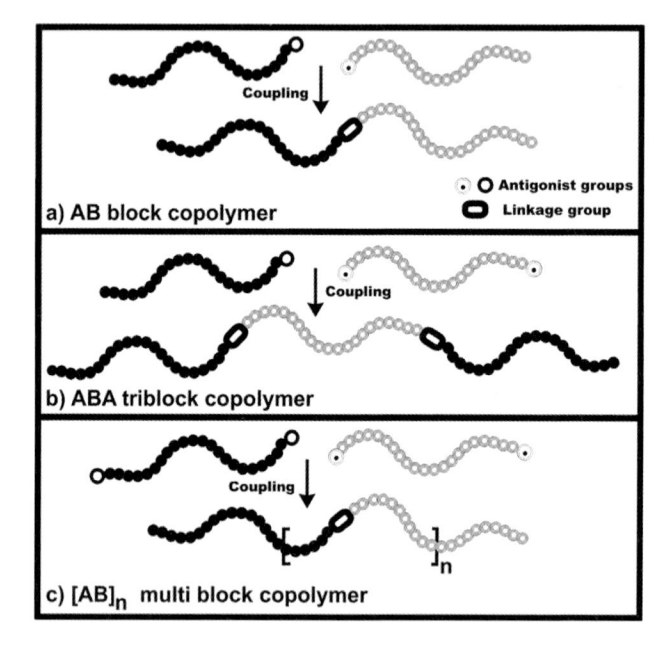

- ☐ hydroxyl
- ⊙ ester group

Transesterification | H⁺

Scheme 13.2

a) AB block copolymer

- ⊙ ○ Antigonist groups
- ▭ Linkage group

b) ABA triblock copolymer

c) [AB]$_n$ multi block copolymer

the formation of polymers with controlled structure and polydispersity. For example, living ionic polymerization is an elegant method for the controlled synthesis of segmented copolymers [11, 12]. Numerous studies have investigated the characteristics of micro-phase ordered AB diblock, ABA triblock and (AB)$_n$ multiblock copolymers prepared by living sequential anionic polymerization and fundamental relationships between the molecular architecture and block mass in these systems have been established [13]. However, in addition to high purity requirements, this technique is limited to ionically polymerizable monomers and excludes monomers that polymerize by other mechanisms. In fact, there are some limitations even for the ionically polymerizable monomers. Whether the block copolymerization of two ionically polymerizable monomers can or cannot be carried out is critically dependent on the structure and relative reactivity of the ionic species and the monomers. For example, only a few monomers are suitable for the preparation of block copolymers by anionic polymerization. A similar situation is valid for cationic polymerization. Furthermore, the synthesis of block copolymers between structurally different polymers, i.e. condensation and vinyl polymers, by a single polymerization method is difficult due to the nature the respective polymerization mechanisms.

In recent years, a high level of control over molar mass, polydispersity, end-groups and architecture has become a high priority for polymerization methods. The rapid development of metallocene polymerization of olefins and controlled radical polymerization strongly reflects this trend. In order to extend the range of polymers for the synthesis of segmented copolymers, a mechanistic transformation approach was postulated by which the polymerization mechanism could be changed from one to another which is suitable for the respective monomers. The pioneering work on the mechanistic transformation dates back about three decades ago to the studies of Richards and coworkers [14–16], who established the versatility of the method.

Although some review articles related to mechanistic transformation approach have already been published [17–20], there is growing interest in the area. Many new studies regarding the synthesis of segmented copolymers using particularly newly developed polymerization techniques in combination with the already existing polymerization modes have been reported recently. Currently three approaches generally appear to be used and the major processes will be discussed in this chapter:

1. *Direct transformation reactions:* The transformation of a polymerization mechanism is carried out at the end of the first block segment in the polymerization mixture, which means that the species which initiated the polymerization mechanism of the first monomer by one mechanism was transformed to another mechanism by a redox process without termination and isolation (Scheme 13.3).
2. *Indirect transformation:* This type of transformation usually requires multi-step reactions (Scheme 13.4). The stable but potentially reactive functional group for the second polymerization mode is introduced to the chain ends either in the initiation or the termination steps of the polymerization of the first mono-

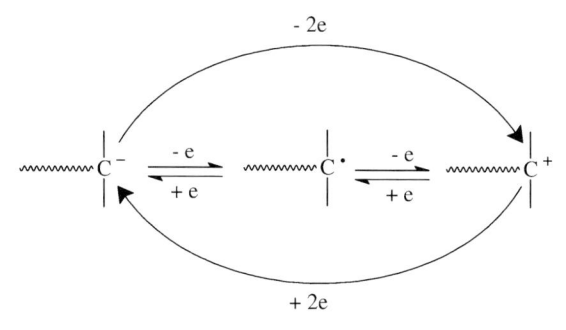

Scheme 13.3

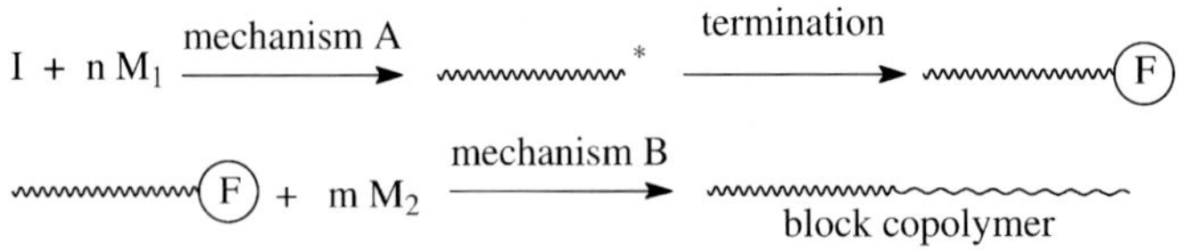

13 Segmented Copolymers by Mechanistic Transformations

$$I + n\,M_1 \xrightarrow{\text{mechanism A}} \sim\sim\sim^{*} \xrightarrow{\text{termination}} \sim\sim\sim\,F$$

$$\sim\sim\sim\,F + m\,M_2 \xrightarrow{\text{mechanism B}} \sim\sim\sim\sim\sim \text{ block copolymer}$$

Scheme 13.4

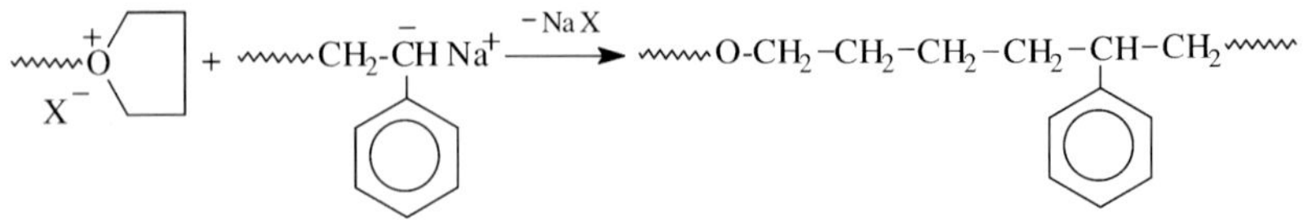

$$\sim\sim\sim\overset{+}{O}\underset{X^-}{\diagup} + \sim\sim\sim CH_2\text{-}\overset{-}{CH}\,Na^+ \xrightarrow{-\,NaX} \sim\sim\sim O\text{-}CH_2\text{-}CH_2\text{-}CH_2\text{-}CH_2\text{-}CH\text{-}CH_2\sim\sim\sim$$

Scheme 13.5

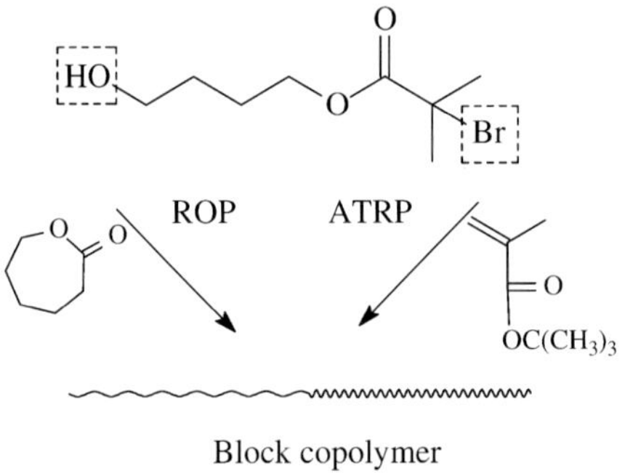

Scheme 13.6

mer. The polymer is isolated and purified and finally the functional groups are converted to another kind of species by external stimulation such as photoirradiation, heating or chemical reaction.

3. *Coupling reactions and concurrent polymerizations:* The coupling reactions in which the mutually reactive polymerizing ends are terminated with each other, i.e. cationic and anionic polymerization [20] (Scheme 13.5), can also yield block copolymers of monomers polymerizable with different modes [20, 21].

In the case of concurrent polymerizations, the polymerizations of two monomers, each polymerizing by different chemistries, for example ring-opening polymerization (ROP) and atom transfer radical polymerization (ATRP) [22], are initiated simultaneously by a molecule bearing initiating centers for the respective polymerization (Scheme 13.6). However, in both cases the polarity of the propagating sites remains the same for the individual steps and mechanistic transformation is not involved.

13.2
Direct Transformations

In this case, propagating active centers are transformed directly to another active center with different polarity. This transfer occurs through an electron transfer as shown in Scheme 13.7 for the transformation involving anionic and cationic systems.

There has been a lack of interest in direct transformations, mainly because of the short lifetime of propagating sites, particularly radicals. The active center must have a lifetime sufficient to permit transformation. Furthermore, a thermodynamic limitation for a successful redox process may result from unsuitable redox potentials of the propagating species and oxidant and reductant. The first reported development for the direct transformation was by Endo and coworkers [23, 24]. This process involves the reduction of the cationic propagating end of polytetrahydrofuran (PTHF) to the anionic one by samarium iodide-hexamethylphosphoramide (SmI_2-HMPA). The two electron reductions of propagating oxonium ion proceeded quantitatively to give PTHF with terminal organosamarium moieties. The transformed anionic species reacted with *tert*-butyl methacrylate (tBMA), ε-caprolactone (CL) [25] and δ-valerolactone (VL) [26] to yield block copolymers of THF with the respective monomers, as shown in Scheme 13.8. This synthetic procedure offers novel block copolymers of cationically and anionically polymerizable monomers.

In another approach, the direct transformation of a propagating radical into a propagating cation took place. The transformation of the radicals to corresponding cations was achieved in the presence of electron transfer agents such as onium salts [27]. This transformation reaction was confirmed by an ESR study of model radicals and by the preparation of block copolymers of *p*-methoxysty-

Scheme 13.7

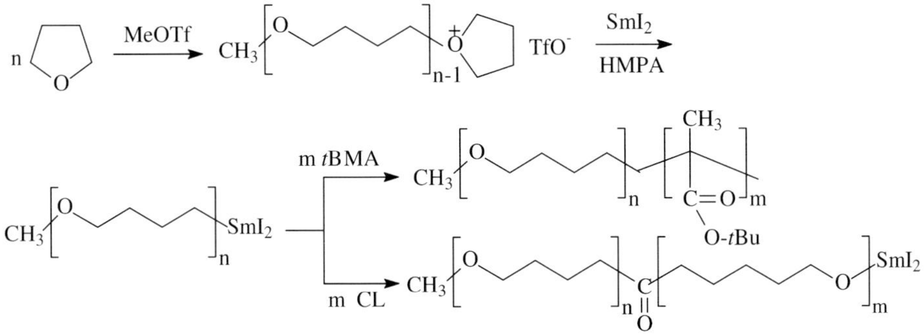

Scheme 13.8

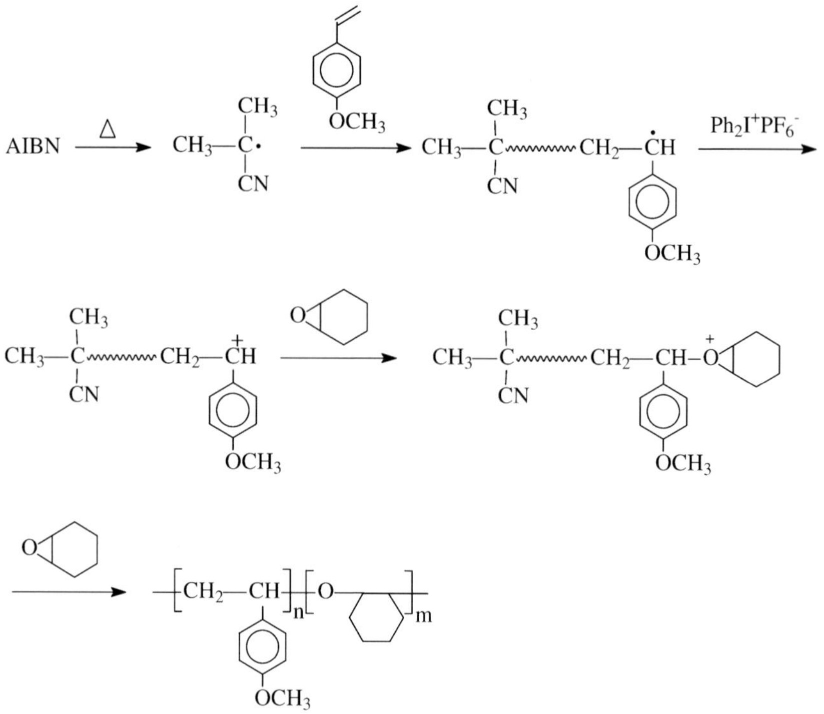

Scheme 13.9

rene (MeSt) and cyclohexene oxide (CHO) according to the reactions shown in Scheme 13.9.

13.3
Indirect Transformation

The most popular and best documented method is indirect transformation, which uses various polymerization modes. Although indirect transformation involves several multi-step paths leading to the transformation of active centers, it is much more convenient to achieve than direct transformation. In the following section, the methods will be classified according to the nature of the propagating centers involved in the transformation polymerization.

13.3.1
Transformations Involving Condensation Polymerization

In contrast to block copolymers synthesized by addition polymerization, which have mainly AB- or ABA-type di- and triblock structures, block copolymers synthesized by step polymerization are $(AB)_n$ multiblock copolymers [28]. The syntheses of AB or ABA block copolymers by condensation or coupling cannot be considered as true polymerization reactions. The synthesis of block copolymers between step-growth and addition polymers is even more difficult due to the respective polymerization mechanisms. While condensation polymers are formed by multi-step coupling of two species, chain growth in addition polymerization occurs by rapid addition of monomer to a small number of active centers. Many methods have been developed to bridge these differences. These methods involve either the chain-growth or the step-growth polymer being prepared first. The synthesis of the addition polymer first requires that the polymer possesses one or two functional groups at either of the chain ends. Using monofunctional chain polymers as macromonomers in a step-growth polymerization results in ABA block copolymers with A=chain and B=step-growth polymer; difunctional polymers result in $(AB)_n$ multiblock copolymers. The preparation of telechelic polymers [29], which can be considered as the elementary units of the block copolymers, can be accomplished by using living polymerizations or through functional transfer agents. However, as usual in step polymerization, the major drawbacks arise from (i) the control of stoichiometry and therefore the control of functionality and purity of starting reactive oligomers and (ii) the control of side-reactions, which lead to undesired structures and a decrease in molecular weight, especially at high temperatures. These drawbacks essentially result in incomplete block copolymer formation.

13.3.1.1 Condensation Polymerization to Conventional Radical Polymerization

Transformations through azo initiators
The structurally suitable azo initiators can be classified as transformation agents since they are able to combine step-growth and radical polymerizations.

Bifunctional azo initiators with groups enabling them to participate in condensation or addition reactions can be classified as condensation-radical and addition-radical transfer agents, respectively (Scheme 13.10).

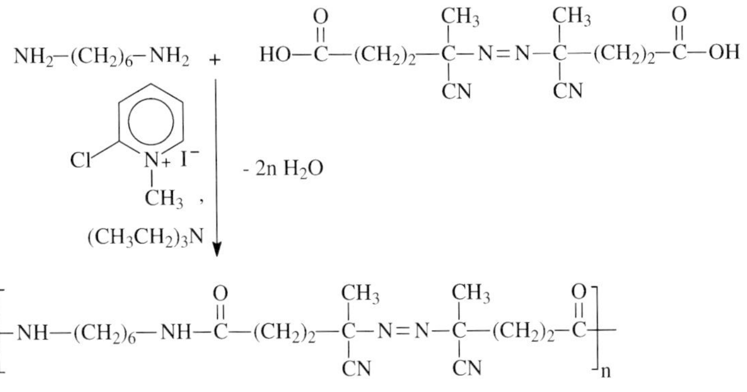

Scheme 13.10

Table 13.1 Bifunctional azo initiators used as transformation agents.

Azo functionality	Polycondensate segment	Radical segment[a)]	Ref.
[HO–...–NC–N=]$_2$	Polyester Polyamide Polyurethane	PSt, PB, PMMA, PSt and PIP PEO PSt PBA, PEA, PMMA	30–34 35, 36 37
[HO–...–NC–N=]$_2$	Polyurethane Polyester Polyamide	PMMA, PSt PMMA, PSt PMMA and PEO, PPO	38, 39 40 41
[Cl–...–NC–N=]$_2$	Polyester Polyacrylate Polyamide Polyether	PEO, PSt PMMA, PDMS, PAa, PCL, C, Cac, PVP PSt, PMMA PSt, PMMA, PBMA, PHEMA, PVAc PSt	42–51 52–54 55 56
[Cl$_3$C–...–NC–N=]$_2$	Polyamide	PSt, PMMA	57
[N–...–O–...–NC–N=]$_2$	Polyester	PSt, PMMA	58
[dibenzazepine–...–NC–N=]$_2$	Poly(dibenzazepine)	PSt, PMMA, PMA	59
[Ph–...–Ph–O–...–NC–N=]$_2$ (RO)	Polyester	PSt, PMMA	60–63
[Ph–...–N–O–...–NC–N=]$_2$	Polyamide	PSt	64

Table 13.1 (continued)

Azo functionality	Polycondensate segment	Radical segment[a]	Ref.
[structure: –NH–C(=O)–C(CH₃)₂–N=N–C(CH₃)₂–]ₙ	Polyamide	PSt, PMMA	65–68

a) PSt, polystyrene; PB, polybutadiene; PMMA, poly(methyl methacrylate); PEA, poly(ethyl acrylate); PBA, poly(butyl acrylate); PEO, poly(ethylene oxide); PDMS, polydimethylsiloxane; PAa, polyacrylamide; PCL, poly(ε-caprolactone); C, cellulose; Cac, cellulose acetate; PVP, poly(vinylpyrrolidone); PBMA, poly(butyl methacrylate); PHEMA, poly(hydroxyethyl methacrylate); PVAc, poly(vinyl acetate); PMA, poly(methyl acrylate); PIP, polyisoprene; PPO, poly(propylene oxide).

They are able to combine radically polymerizable monomers with low molecular weight compounds reacting in polycondensation or addition reactions, thus generating block copolymers. In a large number of publications, the acid or acid chloride of the well-known radical initiator azobisisobutyronitrile (AIBN) was used to condense monomer or preformed polymer with alcohol, amine, acid chloride or acid end-groups. Regarding polyaddition, mostly dialcohols containing one central azo group were reacted with diisocyanates. In Table 13.1, block copolymers prepared by the mechanistic transformation from condensation polymerization to conventional polymerization using azo compounds are listed.

Transformations through photoinitiators
Like azo compounds, certain photoinitiators with suitable functional groups may be used as transformation agents in segmented block copolymer synthesis [69]. In this case, photolabile groups are incorporated in the polycondensate chain. Irradiation of these polymers at appropriate wavelengths in the presence of vinyl monomers gives block copolymers. Benzoin derivatives, acyloxime esters, N-nitroso compounds and alkyl halides in conjunction with manganese decacarbonyl [Mn(CO)$_{10}$] have been successfully used as photoinitiating systems (Table 13.2).

13.3.1.2 Condensation Polymerization to Controlled Radical Polymerization

Block copolymers resulting from the above-described conventional radical polymerization have an overall structure that depends on the mode of the termination by either disproportionation and/or combination of the vinyl monomer. The polymers of structure AB, ABA or (AB)$_n$ can be formed according to the distribution of functional groups and termination of the polymerization of the particular monomer involved. Also, by the nature of a free radical polymeriza-

Table 13.2 Bifunctional photoinitiators used as transformation agents.

Photoactive group	Polycondensate segment	Radical segment[a]	Ref.
Ph−C(=O)−C(H)(OCH₃)−Ph	Polycarbonate	PMMA, PEMA, PAN	70, 71
Ph−C(=O)−C(=N−OH)(CH₃)	Polyester	PSt	72
−N(H)−R−C(=O)−O−N=N−CH₂−R′−C(=O)−	Polyamide	Polyolefin	73, 74
$Mn_2(CO)_{10}$	Polyamide	PSt, PMMA	75

[a] PMMA, poly(methyl methacrylate); PEMA, poly(ethyl methacrylate); PAN, polyacrylonitrile; PSt, polystyrene.

tion, the vinyl blocks are ill defined and homopolymer formation arising from the transfer reactions cannot be avoided. The development of controlled radical polymerizations has promised to overcome these problems [76–78]. Among them, ATRP has been used [79] in combination with step growth polymerization to prepare well-defined block copolymers of polysulfone with vinyl monomers. In the transformation approach, polysulfone with activated halogen groups was used as a macroinitiator for the ATRP. Polysulfone with phenolic end-groups was first prepared and treated with 2-bromopropionyl bromide in the presence of pyridine to yield polymers with halogen groups. The ABA block copolymers of polysulfone (B) with either styrene (St) or butyl acrylate were prepared by applying ATRP [79, 80]. Interestingly, the molecular weight distributions of the two copolymers were narrower than the initial polysulfone macroinitiator due to the well-defined side blocks prepared by ATRP. The initiating 2-bromopropionyloxy end-groups can be incorporated in other polycondensates. Shen et al. [81] prepared polycarbonates possessing end-chain and side-chain 2-bromopropionyloxy moieties by condensation polymerization of bisphenol A and triphosgene in the presence of functional terminator and comonomer, respectively. The subsequent ATRP of St yielded ABA type block copolymers and graft copolymers (Scheme 13.11). Other examples of materials prepared from the condensation polymerization to controlled radical polymerization are given in Table 13.3.

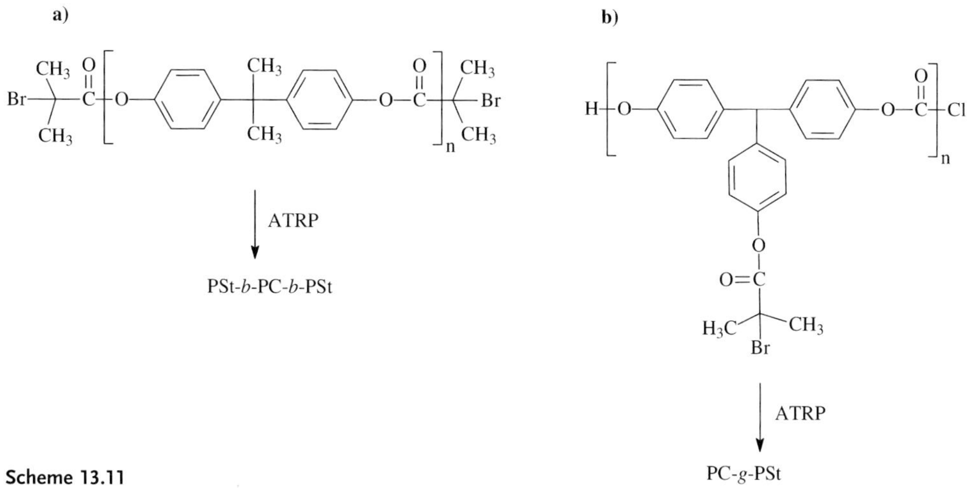

Scheme 13.11

Table 13.3 Examples of the transformation of condensation polymerization into controlled radical polymerization.

Polymer	Controlled radical polymerization	Ref.
poly(St-b-sulfone-b-St), poly(BA-b-sulfone-b-BA)	ATRP	79, 80
poly(St-b-dimethyladipate-b-St)	ATRP	82
poly(St-b-bisphenol A carbonate-b-St) or poly(MMA-b-bisphenol A carbonate-b-MMA)	ATRP	83
poly(10-hydroxydecanoic acid-b-styrene)	ATRP	84
poly(oxadiazole-b-ester)	ATRP	85
poly(St-b-fluoroaromatic ethers)	ATRP	86
poly(p-phenylenevinylene-b-St)	NMP	87
poly(St-b-ethylene adipate-b-St)	NMP	88

13.3.1.3 Macrocyclic Polymerization to Condensation Polymerization

It has been shown by Kricheldorf and coworkers [89–93] that cyclic tin compounds allow a ring-opening polymerization of lactones via an insertion mechanism. The so-called macrocyclic polymerization proceeds according to a living pattern and the polymers are obtained with almost quantitative conversion (Scheme 13.12).

It was also demonstrated by these authors that the macrocyclic polymers can be used as difunctional monomers in step-growth polymerization using diacid chlorides [94, 95] (Scheme 13.13).

Various polycondensations of poly(ε-caprolactone) (PCL), poly(δ-valerolactone) (PVL) and poly(β-D,L-butyrolactone) with terephthaloyl chloride, sebacoyl chloride, isophthaloyl chloride, 4,4'-biphenyldicarbonyl chloride and 4,4'-phenylenebisacryloyl chloride were conducted to yield multiblock segmented copolymers.

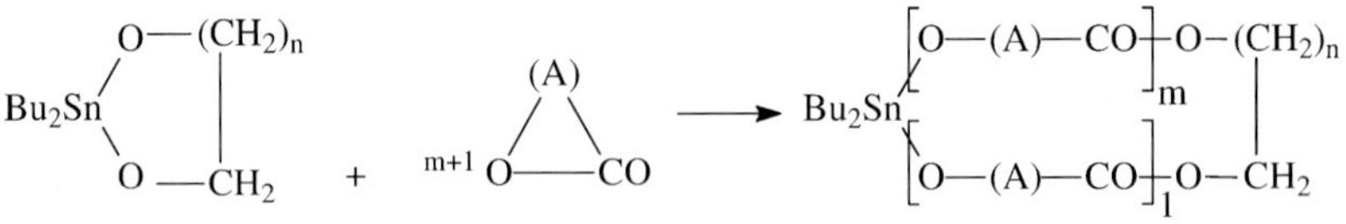

Scheme 13.12

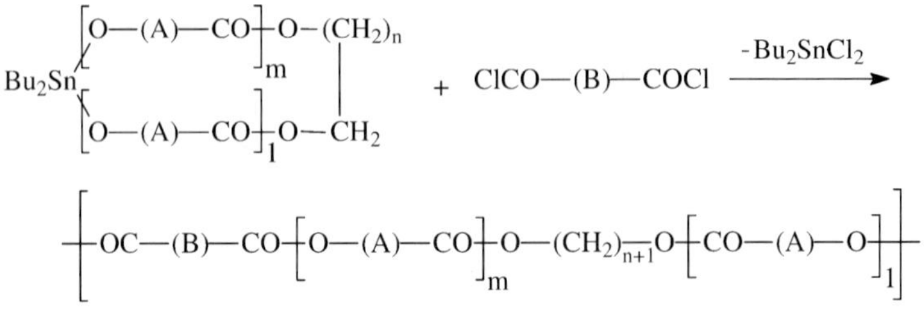

Scheme 13.13

More recently, a similar approach was applied to silicon-mediated polycondensation [96]. In this case, tin-containing macrocyclic polymers of CL were reacted *in situ* with an excess of isophthaloyl chloride and subsequently was polycondensed with silylated bisphenol A.

Thermotropic multiblock copolyesters were also prepared from silylated methylhydroquinone and 1,10-bis(4-chloroformyl)phenoxydecane. The isolated block copolymers showed a two-phase melt with an isotropic phase containing the polylactone blocks and a nematic phase consisting of aromatic blocks. The combination of macrocyclic polymerization with ring-opening polycondensation is a versatile method to prepare multiblock copolymers of polyesters in a manner such that the chemical structure and the lengths of block can be varied independently.

13.3.1.4 Condensation Polymerization to Anionic Coordination-Insertion Polymerization

Lee et al. [97] recently reported synthesis biodegradable poly(*trans*-4-hydroxy-*N*-benzyloxycarbonyl-L-proline)-*β*-poly(*ε*-caprolactone) copolymers by transforming condensation polymerization into anionic coordination-insertion polymerization. First, hydroxyl end functional poly(*trans*-4-hydroxy-*N*-benzyloxycarbonyl-L-proline) was prepared with 1-hexanol initiating the melt bulk polycondensation of *trans*-4-hydroxy-*N*-benzyloxycarbonyl-L-proline. The hydroxyl groups were then used as the initiation sites with organocatalyst D,L-lactic acid for the ring-opening polymerization of CL to produce a block copolymer which is capable of forming micelles in the aqueous phase.

13.3.1.5 Transformations Involving Suzuki and Yamamoto Polycondensations

Poly(p-phenylene)s (PPP) are typical conjugated, electroluminescent polymers for light-emitting devices in combination with excellent mechanical properties and thermal and thermo-oxidative stability. Current methodologies for the direct synthesis of derivatized PPP are primarily based on nickel- and palladium-mediated cross-coupling reactions due largely to their preservation of regio-chemistry and nearly quantitative yields. The key structural factor in describing the supramolecular ordering of PPP is their anisotropic shape, which follows from a rod-like architecture that differentiates them from flexible polymers [98]. Unfortunately, PPPs are insoluble in many organic solvents, which limits their processability. Therefore, attachment of conformationally mobile alkyl side-chains to the backbone is important because it allows the controlled synthesis of soluble and processable PPPs with high molecular weight. On combining a stiff, insoluble, rod-like polymer such as PPP with a soft coil, e.g. PSt, it is possible to form a new polymer with novel and interesting properties [99]. We have recently focused on the synthesis of PPP-type graft copolymers [100, 101], that can present nanostructures between a conductive and an insulating polymer, by using the transformation approach in which controlled polymerizations (ATRP [102–106], anionic [105–108] and cationic ring-opening polymerization [103]) are combined with metal-catalyzed Suzuki or Yamamoto polycondensation, specific to obtaining soluble, high molecular weight, conjugated polymers. In this approach, dual functional initiators having both bromine benzyl groups useful for ATRP of St or and bromine atoms or boronic acid functionalities directly

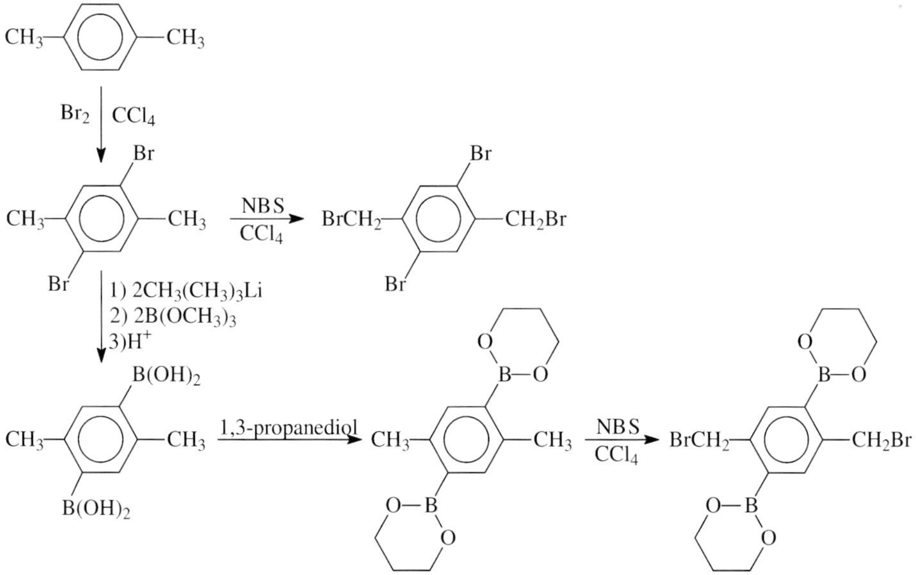

Scheme 13.14

Scheme 13.15

linked to a benzene ring, useful for Suzuki- or Yamamoto-type reactions, were designed and synthesized (Scheme 13.14).

Benzyl bromides are efficient initiators for both ATRP (Scheme 13.15). On the other hand, aryl halides do not initiate these types of reactions.

Using them in Suzuki polycondensation or Yamamoto coupling, new segmented graft copolymers of PPPs with PSt side-chains were obtained. The overall synthetic strategy as outlined in Fig. 13.1 was successfully applied to prepare a wide range of comb-like polymers possessing side-chains of PCL, PTHF, polyoxazoline or their block copolymers.

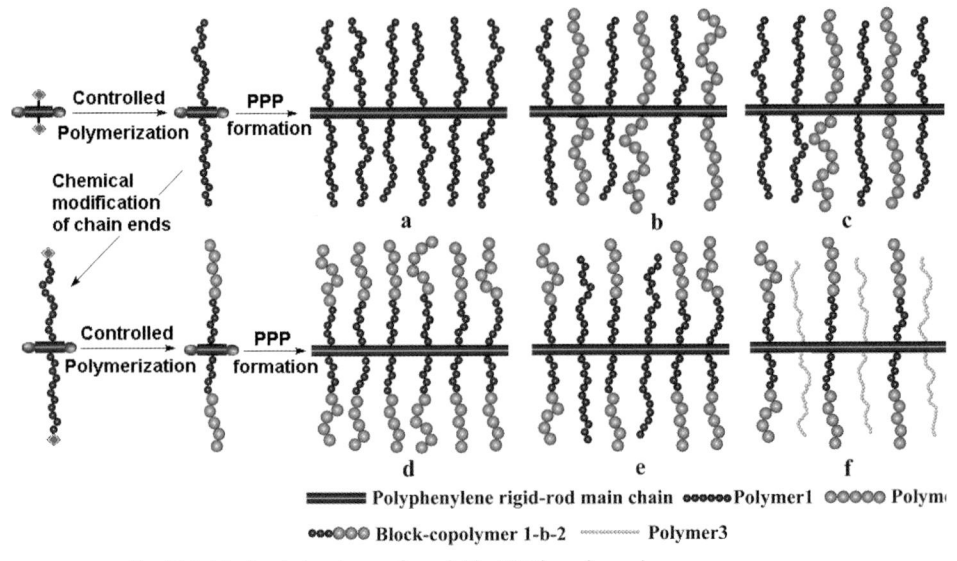

Fig. 13.1 Idealized structures of comb-like PPP-based copolymers.

13.3.2
Transformation of Anionic Polymerization to Radical Transformation

13.3.2.1 Anionic Polymerization to Conventional Radical Polymerization

This approach is one of the most widely used processes in transformation technology. In principle, a wide range of segmented block copolymers of anionically polymerizable monomers and free radically polymerizable monomers are accessible with this route. Various examples of this transformation have been surveyed in previous reviews [109, 110] and are summarized in Table 13.4.

The most recent example of this transformation concerns the preparation of a block copolymer composed of crystalline polyethylene oxide (PEO) and non-crystalline PSt by combination of anionic polymerization and photoinduced charge-transfer polymerization [132]. In this process, the amino group of *p*-aminophenol was protected and then reacted with metallic potassium. The phenoxy anion thus formed initiated the polymerization of ethylene oxide.

The PEO prepolymer with Schiff base end-group was deprotected by acidolysis with acetic acid (Scheme 13.16). Then the recovered amino group in conjunction with benzophenone formed a charge-transfer complex under UV irradiation to initiate free radical polymerization of St (Scheme 13.17).

Pure block copolymers of ethylene oxide (EO) and St were thus formed since only macroradicals with aniline end-groups are capable of initiating the poly-

Table 13.4 Block copolymers prepared by anion to conventional radical polymerization transformation.

Block copolymer [a]	Ref.
PSt-*b*-PMMA, PSt-*b*-PVC, PSt-Vinyl	111–122
PIP-*b*-PVC	123
PB-*b*-PMMA, PB-*b*-PMMA-*b*-PIB	124
PAN-*b*-Vinyl	125–127
PSt-*b*-PTHF-*b*-PMMA	128
PPp-*b*-PMMA	129
PCL-*b*-PHPMAm star block copolymer	130
PEO-*b*-PAN	131

[a] PSt, polystyrene; PB, polybutadiene; PMMA, poly(methyl methacrylate); PIP, polyisoprene; PVC, poly(vinyl chloride); PIB, polyisobutene; PAN, polyacrylonitrile; PTHF, polytetrahydrofuran; PEO, poly(ethylene oxide); PPp, polypeptide; PCL, poly(ε-caprolactone); PHPMAm, poly[N-(2-hydroxypropyl)methacrylamide].

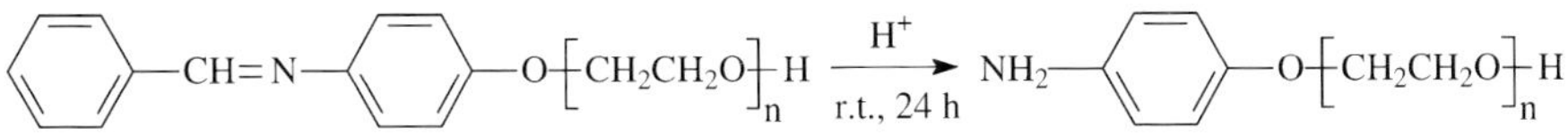

Scheme 13.16

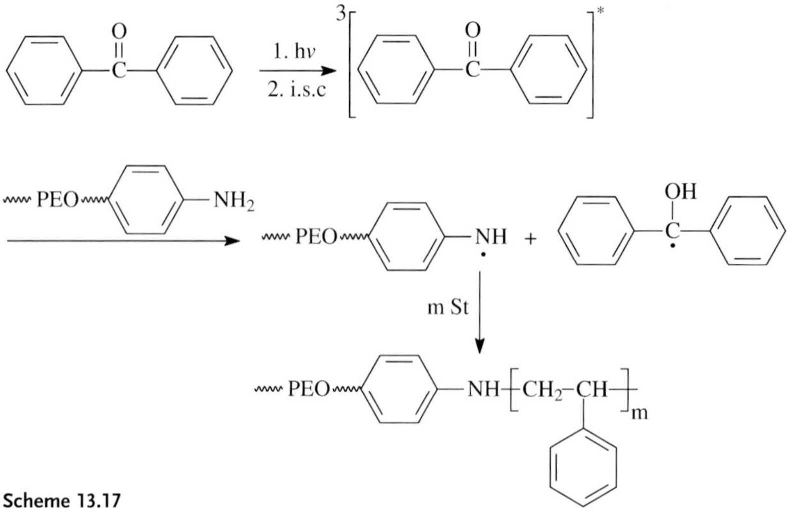

Scheme 13.17

merization. The same group also reported the preparation of PMMA-*b*-PSt-*b*-PMMA and PEO-*b*-PMMA-*b*-PSt triblock segmented copolymers by using a combination of anionic and photoinduced charge-transfer radical polymerization [133, 134].

13.3.2.2 Anionic Polymerization to Controlled Radical Polymerization

Following the pioneering work of Rizzardo and coworkers, special attention has recently been focused on the use of stable nitroxyl radicals such as 2,2,6,6-tetramethylpiperidine-1-oxyl (TEMPO) in order to achieve living conditions in conventional radical polymerization [135–139]. In principle, these nitroxide-mediated polymerizations (NMP) involve reversible termination of the polymer radical with TEMPO and chain growth during the lifetime of the polymeric radical as shown in Scheme 13.18.

Besides organo tin compounds described in Section 13.3.1.3, several other new initiators can be used to synthesize designed polymers based on PCL [140–145]. Among them, metallic alkoxides are particularly useful for introducing functional groups selectively at one chain end of PCL [146–148]. Yoshida and

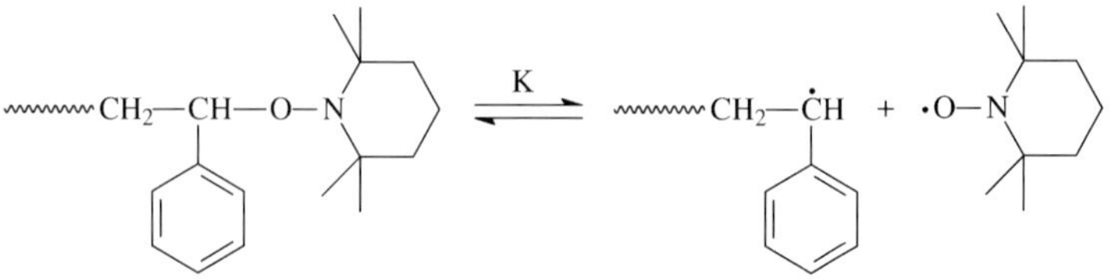

Scheme 13.18

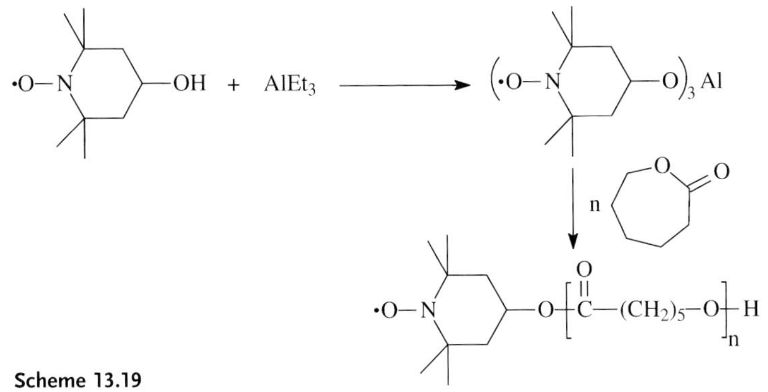

Scheme 13.19

Osagawa [149] reported the stable radical functionalization of PCL by using a specially designed aluminum alkoxide initiator. For this purpose, aluminum tri(4-oxy-TEMPO), which was prepared by the reaction of triethylaluminum with three equimolar amounts of 4-hydroxy-TEMPO, was used as an initiator for the anionic polymerization of CL (Scheme 13.19).

PCL with the TEMPO moiety behaved as a polymeric counter radical for the polymerization of St, resulting in the quantitative formation of PCL-*b*-PSt. The radical polymerization was found to proceed in accordance with a living mechanism without undesirable side-reactions (Scheme 13.20).

The thermal analysis of the block copolymer indicated that the components of PCL and PSt were completely immiscible and microphase separated. The incorporation of TEMPO moiety into PEO chain ends in the radical form was also achieved [150]. In this case, TEMPO-Na was used as an initiator in the living anionic polymerization of EO (Scheme 13.21) under conditions such that the stable nitroxyl radical at the end of the PEO chain could not be destroyed. Again, the resulting PEO with a TEMPO moiety acted as a macromolecular radical trapper in NMP of St to give PEO-*b*-PSt with narrow polydispersity. It was found that PEO with high molecular weight is less efficient at trapping chain ends and so can enhance the polymerization rate.

Polybutadiene-*b*-polystyrene (PB-*b*-PSt) [151–153], polydimethylsiloxane-*b*-polystyrene (PDMS-*b*-PSt) [154], poly(ethylene oxide)-*b*-polystyrene (PEO-*b*-PSt) [155] and poly(ethylene oxide)-*b*-poly(4-vinylpyridine) (PEO-*b*-PVP) [156] copolymers were synthesized by terminating the corresponding living anionic polymerization with a suitable TEMPO derivative and subsequent NMP.

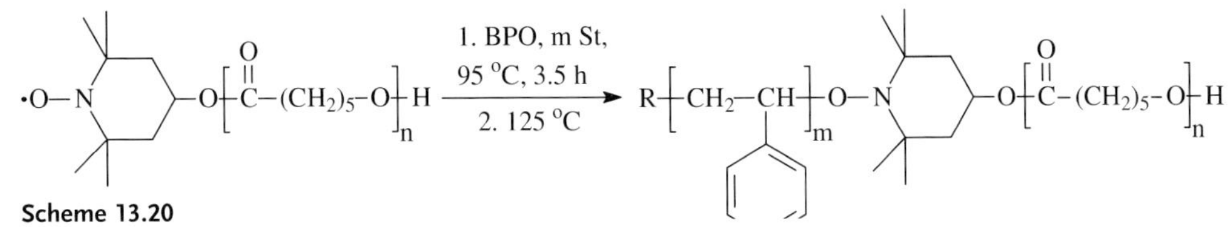

Scheme 13.20

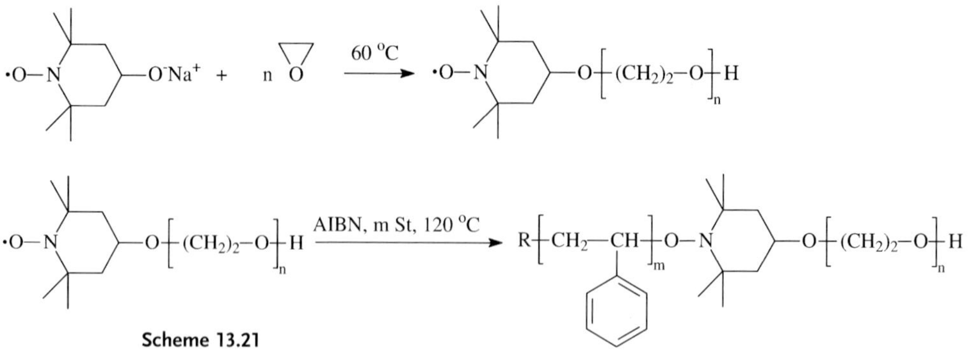

Scheme 13.21

Stable nitroxyl radicals can also be incorporated into polymers as side groups. Endo et al. [157] copolymerized nitroxyl radical containing epoxide with glycidyl phenyl ether anionically using potassium *tert*-butoxide as initiator (Scheme 13.22). The ratio of the nitroxyl radical moiety in the resulting copolymer can be controlled by the feed ratio. Subsequent polymerization of St in the presence of this polymer is expected to yield graft copolymers.

An interesting variation of this approach was reported by Yagci and coworkers [158], who demonstrated that a stable TEMPO radical can undergo a one-electron redox reaction with potassium naphthalene. Although the TEMPO alcoholate thus formed does not initiate the polymerization of St, the polymerization of ethylene oxide was readily accomplished. PEO obtained in this way possessed TEMPO terminal units and was subsequently used as an initiator for NMP of St to give block copolymers (Scheme 13.23).

ATRP is the most widely used controlled radical polymerization in anion to radical transformation methodology. The first example was reported by Acar and Matyjaszewski [159] and utilized for the preparation of AB- and ABA-type block copolymers. The macroinitiators, PSt and polystyrene-*b*-polyisoprene (PSt-*b*-PIP) containing 2-bromoisobutyryl end-groups, were prepared by living anionic polymerization and a suitable termination agent. These polymers were then used as macroinitiators for ATRP to prepare block copolymers with methyl acrylate (PSt-*b*-PMA), butyl acrylate (PSt-*b*-PBA), methyl methacrylate (PSt-*b*-

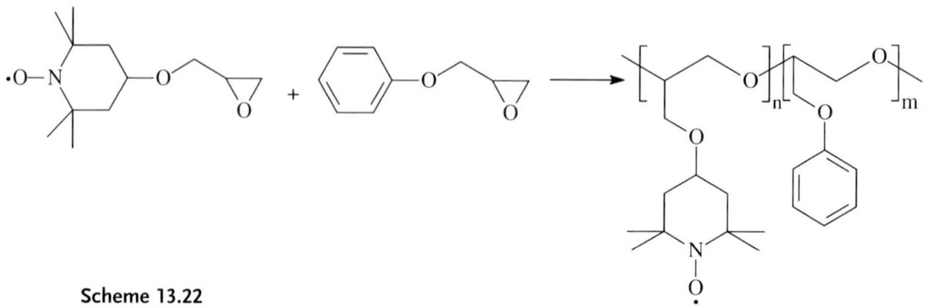

Scheme 13.22

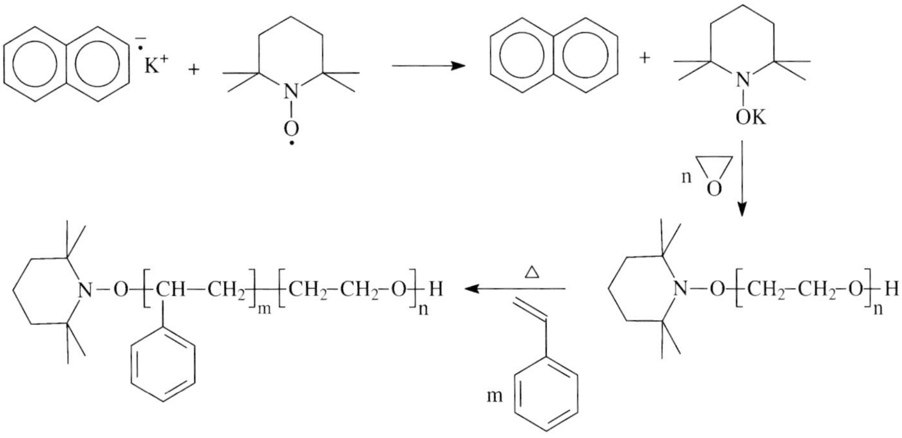

Scheme 13.23

PMMA), a mixture of styrene and acrylonitrile [PSt-*b*-P(St-*r*-AN)] and also chain extension with St (PSt-*b*-PSt and PSt-*b*-PIP-*b*-PSt) (Scheme 13.24).

Other examples of materials prepared from the anionic polymerization to ATRP are given in Table 13.5. It can be seen that the transformation approach involving the combination of living anionic polymerization and ATRP allowed the preparation of segmented copolymers with an exciting range of structural variety. In this way, multiblock copolymers possessing soft segments and glassy segments, graft terpolymers, comb-like block copolymers, star and dendrimer-like architectures and polymer-ceramic hybrid materials were successfully prepared. A very interesting application concerns the incorporation of a fluorescent dye at the junction point of PMMA-*b*-PBA copolymer. The overall process is depicted in Scheme 13.25 [161].

Recently, reversible addition-fragmentation transfer polymerization (RAFT), which was proven to be another controlled radical polymerization method, has also been used in this transformation. The mechanism involves the chain trans-

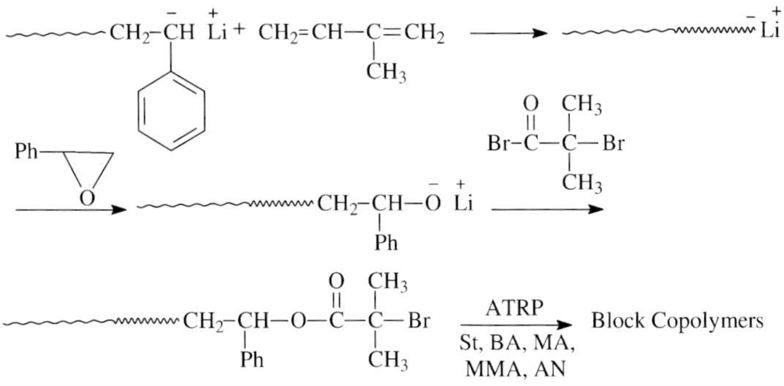

Scheme 13.24

Table 13.5 Block copolymers prepared by anionic to ATRP transformation.

Anionic segment (A)[a]	ATRP segment (B)[a]	Ref.
PSt	PVP	160
PIP-b-PSt	PSt	161
PMA	PMMA	161
PIP	PSt	162
PSt-PB, PSt	PMA, PMMA	163
PFS	PMMA	164
PEO	PSt, PHMA, (AB$_2$, AB$_3$, A$_2$B, A$_2$B$_2$) PSt and (three-arm) PSt-b-PtBMA star, block copolymers	165–168
PMMA	PSt-b-PtBA, PSt-b-PMMA, PBA-b-PSt	169, 170
PE-co-PBu	PSt, PAcSt	171
PE-co-PP-b-PEO	PHMA	172
PCL	PSt, PDMAEMA, PODMA block; PHEMA, PEGMA, PMMA, PtBA, PMAA star; PMMA, PHEMA star, block copolymers	173–183
PDMS	PSt, POEGMA	184, 185
PLL	PSt, PMMA, PtBA dendrimer based star, PMMA, PtBA, PBzA (ABA) triblock, copolymer	186–188

[a] PSt, polystyrene; PVP, poly(vinylpyrrolidone); PIP, polyisoprene; PMA, poly(methyl acrylate); PMMA, poly(methyl methacrylate); PB, polybutadiene; PFS, poly(ferrocenyldimethylsilanes); PEO, poly(ethylene oxide); PtBA, poly(tert-butyl acrylate); PCL, poly(ε-caprolactone); PE, polyethylene; PBu, polybutene, PAcSt, poly(4-acetoxystyrene); PP, polypropylene; PHMA, poly(hexyl methacrylate); PtBMA, poly(tert-butyl methacrylate); PDMAEMA, poly[(dimethylamino)ethyl methacrylate]; PMAA, poly(methacrylic acid); PODMA, poly(n-octadecyl methacrylate); PHEMA, poly(2-hydroxyethyl methacrylate); PEGMA, poly[(ethylene glycol)methacrylate]; POEGMA, poly[oligo(ethylene glycol) methyl ether methacrylate]; PDMS, poly(dimethylsiloxane); PLL, polylactide; PBzA, poly(benzyl acrylate).

fer of active species such as the radicals stemming from the decomposition of the initiator and propagating radicals to the chain-transfer agents (RAFT agents), forming an unreactive adduct radical, followed by fast fragmentation to a polymeric RAFT agent and a new radical. The radical initiates the polymerization. The equilibrium is established by subsequent chain trans-ferfragmentation steps. It was shown that PEO containing a xanthate end-group can be used as a macro RAFT agent in the polymerization of N-vinylformamide (NVF) to yield polyethylene-b-poly(N-vinylformamide) (PEO-b-PNVF) (Scheme 13.26) [189].

In another case, hydroxy functionalities of PEOs were converted to dithiobenzoyl groups and used as macro RAFT agents in the RAFT polymerization of N-isopropylacrylamide (NIPAa) (Scheme 13.27). Depending on the functionality of the initial polymers, AB- and ABA-type block copolymers with well-defined structures were prepared [190, 191].

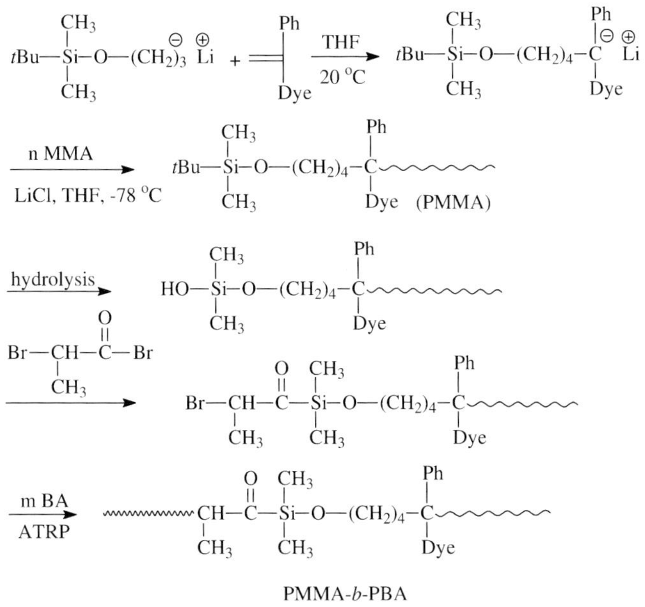

Scheme 13.25

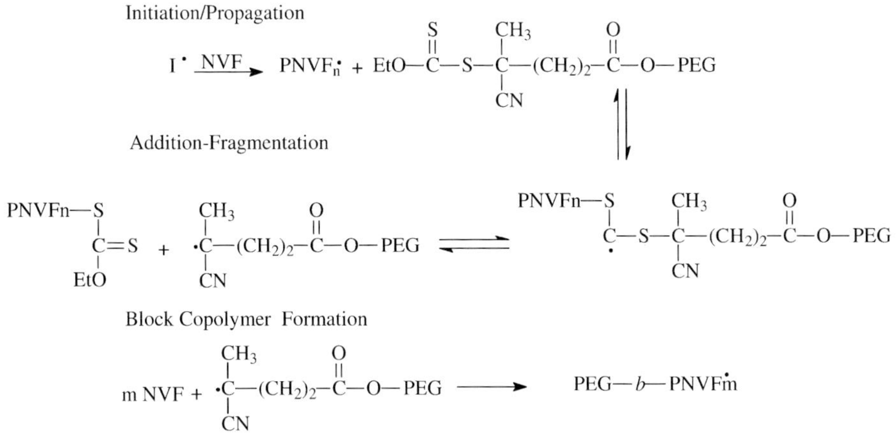

Scheme 13.26

Obviously, the most important step of these transformations is the modification of the chain end into a good leaving group. In order to obtain quantitatively functionalized macro RAFT agents or ATRP initiators, modifications of living anionic PB with diphenylethylene, St and haloalkanes have recently been investigated [192, 193]. Matyjaszewski and coworkers described the versatility of combining living anionic polymerization with RAFT to prepare segmented graft ter-

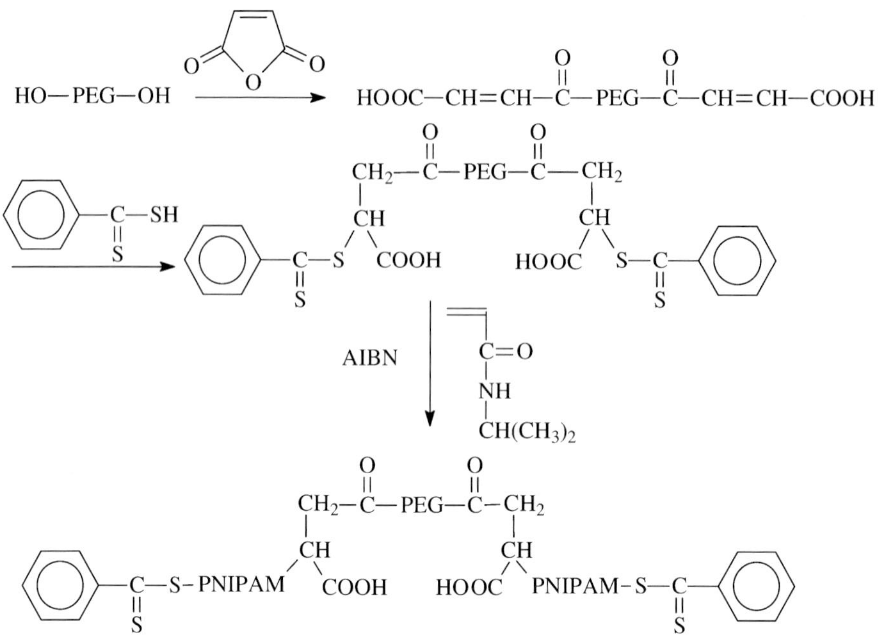

Scheme 13.27

polymers with a controlled molecular structure. Anionically prepared polylactide (PLL) and poly(dimethylsiloxane) (PDMS) macromonomers were used in the RAFT polymerization of alkyl methacrylates [194].

A conceptually different transformation reaction was applied for the preparation of PCL-b-(PMMA-co-PSt)-b-PCL by using the iniferter technique in the controlled radical polymerization step. Substituted tetraphenylethanes represent a class of thermal iniferters applicable to the radical polymerization of many

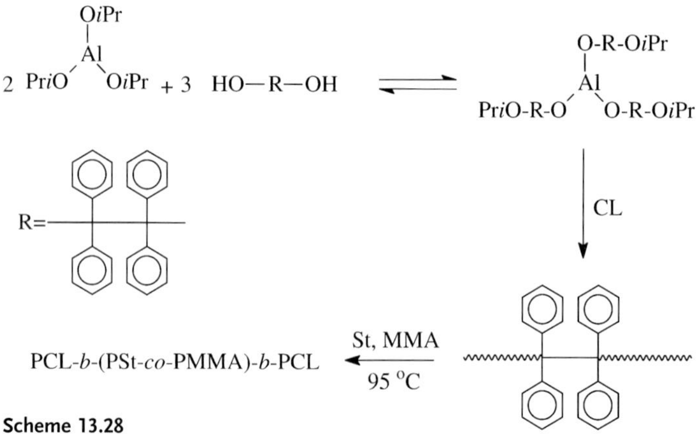

Scheme 13.28

monomers in a controlled manner. The initiation of anionic coordination polymerization of CL by aluminum triisopropoxide in the presence of benzpinacol leads to the formation of polymers possessing an iniferter structure in the middle of the chain [195].

The benzpinacolate groups incorporated into the polymer chain then initiate the polymerization of St and MMA via a controlled radical mechanism at 95 °C to yield the desired block copolymers (Scheme 13.28).

13.3.3
Transformation of Cationic Polymerization to Radical Polymerization

13.3.3.1 Cationic Polymerization to Conventional Radical Transformation

Low molecular weight azo compounds have frequently been used in cationic polymerizations to produce azo-containing polymers. Thus, combination of ionically and radically polymerizable monomers into block copolymers has been achieved. Azo compounds were used in all steps of cationic polymerization: to initiate the polymerization without any loss of azo function, they were applied as initiator, monomers and terminating agents.

For initiating cationic chain polymerizations, the reaction of chlorine-terminated azo compounds with various silver salts has been thoroughly investigated. 4,4′-Azobis(4-cyanopentanoyl chloride) (ACPC), a compound frequently used in condensation-type reactions discussed previously, was reacted with Ag^+X^-, X^- being BF_4^- or SbF_6^- [196–198]. This reaction results in two oxycarbenium cations, being very suitable initiating sites for the cationic polymerization of tetrahydrofuran (THF) (Scheme 13.29). Thus, PTHF with M_n between 3×10^3 and 4×10^3 containing exactly one central azo link per molecule was synthesized. Furthermore, N-vinylcarbazole (NVC), n-butyl vinyl ether (NBVE) and CHO have been polymerized following this procedure, although CHO yielded low monomer conversions [198].

Azo-containing PTHF have been used for preparing novel liquid crystalline block copolymers [199–202]. The liquid crystalline blocks were obtained by polymerizing different acrylates containing substituted biphenyl mesogenes by

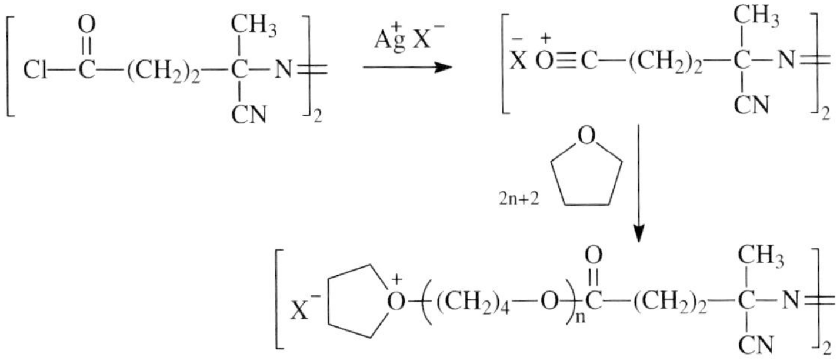

Scheme 13.29

means of azo functionalized prepolymer. Moreover, a PTHF-based azo initiator was employed for the copolymerization of St and divinylbenzene, giving macroporous beads with good swelling properties [203].

Azo-containing poly(cyclohexene oxide) (PCHO) may also be synthesized by a promoted cationic polymerization [204]. Subsequent transformation to radical polymerization was used for the preparation of new classes of liquid crystalline (LC) block copolymers comprising a semicrystalline block, PCHO and LC block of different structures (Scheme 13.30). The block copolymers obtained are essentially microphase separated systems and form smectic mesophases, analogous to the corresponding LC homopolymers.

Living ends of cationic polymerization of THF initiated by triflic anhydride or azo-oxocarbenium salt may be terminated with pyridinium N-oxide derivatives as shown in Scheme 13.31 for the azo initiator [197].

PTHF with terminal N-alkoxy pyridinium ions obtained from the triflic initiator can participate in photoinduced electron transfer to yield photosensitizer radical cation and macroradical. The latter readily initiates polymerization of MMA resulting in the formation of MMA and THF block copolymers with quantitative yields [204] (Scheme 13.32).

Possessing both thermally labile azo sites and photolytically decomposable pyridinium ions, the resulting polymer from the azo oxocarbenium salt is a difunctional initiator being a suitable precursor for ABC triblock copolymers [197].

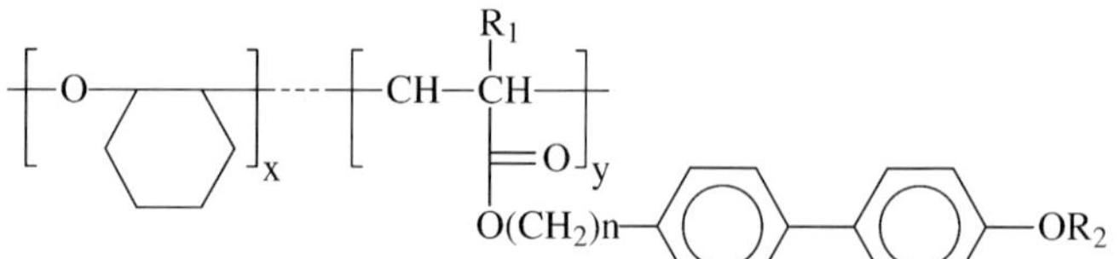

Scheme 13.30

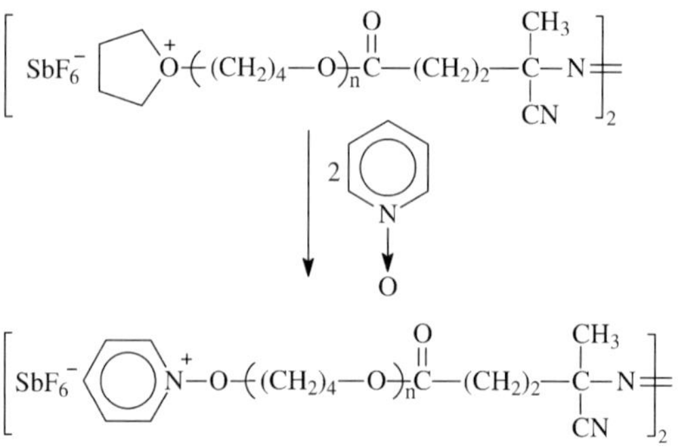

Scheme 13.31

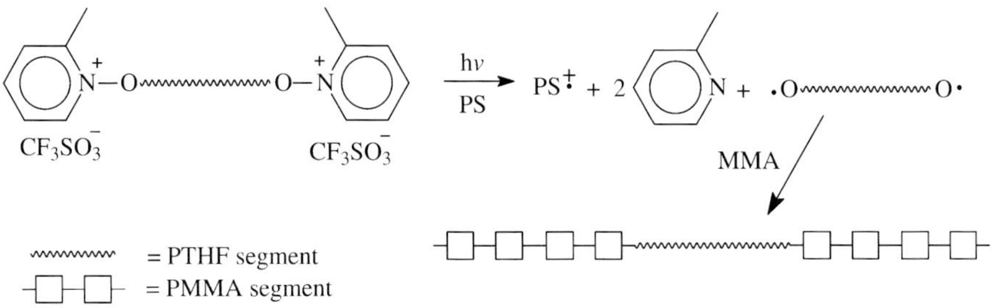

Scheme 13.32

$$\text{PTHF-N}^+\text{(pyridinium)-O}\cdots\text{O-N}^+\text{(pyridinium)} \xrightarrow{h\nu, \text{PS}} \text{PS}^{\ddagger} + 2\,\text{N(pyridinium)} + {}^\bullet\text{O}\cdots\text{O}^\bullet$$

CF$_3$SO$_3^-$ CF$_3$SO$_3^-$

↓ MMA

PMMA—PTHF—PMMA block copolymer

∿∿∿ = PTHF segment
▫▫▫ = PMMA segment

An interesting pathway in synthesizing polymeric azo initiators is the living polymerization of an azo-containing monomer. Nuyken et al. [205] investigated a system consisting of monomer, HI and a coinitiator [tetrabutylammonium perchlorate (TBAP)]. Since an α,ω-divinyl ether containing one central azo function was used for reaction with the initiator in the very first polymerization step, polymeric initiators (M_n between 1.3×10^3 and 1.6×10^4) with one thermolabile function per repeating unit were obtained. If the azo-containing divinyl ether were used not only in the initiation step but also in the subsequent polymerization, insoluble cross-linked polymer would be formed due to the monomer's bifunctionality [206]. The macro azo initiator prepared was used for thermally polymerizing MMA. It was found that the polymerization rates of the macro azo initiator are identical with those of AIBN.

Azo initiator-involved transformation provides an excellent methodology for the synthesis of multicomponent copolymers containing a polyisobutene (PIB) segment [207]. First isobutene was polymerized cationically using a difunctional organic tertiary ether azo initiator and a Lewis acid initiator combination to produce azo group-containing PIB. In the second step, the functional PIB was used as a macroinitiator for the radical polymerization of vinyl monomers such as St and MMA to yield multicomponent block copolymers (Scheme 13.34).

In addition to being used as initiators and monomers, azo compounds may also be employed to terminate cationic polymerization. Thus, the living cationic

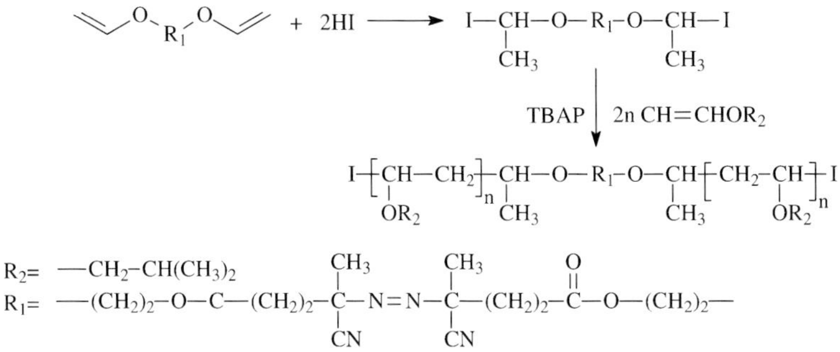

Scheme 13.33

$$\text{CH}_2=\text{CH–O–R}_1\text{–O–CH=CH}_2 + 2\text{HI} \longrightarrow \text{I–CH(CH}_3\text{)–O–R}_1\text{–O–CH(CH}_3\text{)–I}$$

↓ TBAP, $2n$ CH=CHOR$_2$

$$\text{I–[CH(OR}_2\text{)–CH}_2\text{]}_n\text{–CH(CH}_3\text{)–O–R}_1\text{–O–CH(CH}_3\text{)–[CH}_2\text{–CH(OR}_2\text{)]}_n\text{–I}$$

R$_2$ = —CH$_2$–CH(CH$_3$)$_2$
R$_1$ = —(CH$_2$)$_2$–O–C(=O)–(CH$_2$)$_2$–C(CH$_3$)(CN)–N=N–C(CH$_3$)(CN)–(CH$_2$)$_2$–C(=O)–O–(CH$_2$)$_2$—

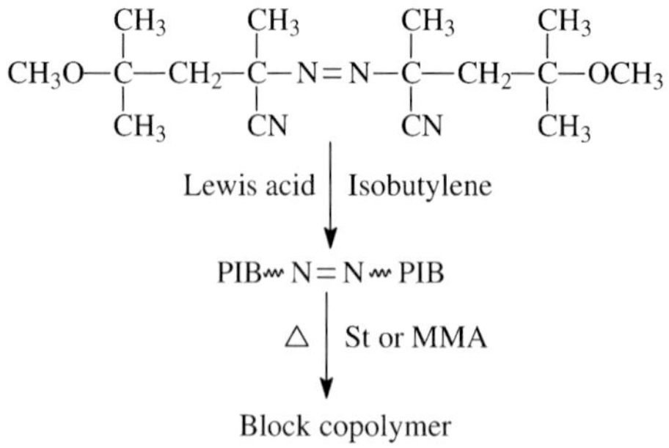

Scheme 13.34

$$CH_3O-\underset{\underset{CH_3}{|}}{\overset{\overset{CH_3}{|}}{C}}-CH_2-\underset{\underset{CN}{|}}{\overset{\overset{CH_3}{|}}{C}}-N=N-\underset{\underset{CN}{|}}{\overset{\overset{CH_3}{|}}{C}}-CH_2-\underset{\underset{CH_3}{|}}{\overset{\overset{CH_3}{|}}{C}}-OCH_3$$

Lewis acid | Isobutylene

PIB~N=N~PIB

Δ | St or MMA

Block copolymer

Scheme 13.34

polymerization of isobutyl vinyl ether (IBVE) initiated by the above-mentioned system HI-coinitiator was terminated by an azo-containing alcohol in the presence of ammonia [208] (Scheme 13.35).

Another method to terminate cationic polymerization was described by D'Haese et al. [209]. They reacted a living PTHF chain with either azetidinium or thiolanium, thus stabilizing the cationic sites. In a second reaction step, the PTHF was treated with the sodium salt of the azo initiator disodium 4,4′-azobis(4-cyanopentanoate), yielding PTHF-based polymeric initiators with either one azo group in the middle of the backbone or several azo groups at regular distances in the chain (Scheme 13.36). Both initiator types were successfully used for the thermal synthesis of PSt and PMMA blocks.

Polymers containing labile peroxide groups can initiate vinyl polymerization to obtain block copolymers. In this way, linear and star block copolymers consisting of PTHF as cationic segment and PSt and PMMA as radical segments were prepared [210, 211].

Cation to radical transformation was also carried out by preparing PTHF via cationic polymerization followed by functionalization (with a terminal bromine atom [212, 213] or photoactive group [214]) and used for the radical polymerization of MMA and/or St. To incorporate a terminal bromine functionality into the cationically prepared living PTHF chain, the polymerization was terminated by addition of lithium bromoacetate (Scheme 13.37). Bromomethyl groups acti-

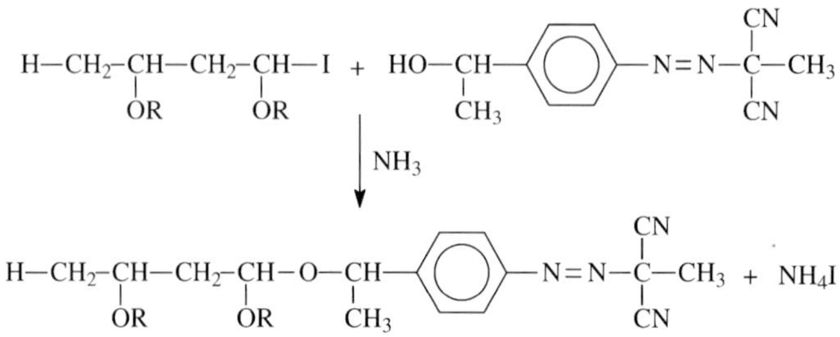

Scheme 13.35

H—CH$_2$—CH—CH$_2$—CH—I + HO—CH—⟨⟩—N=N—C(CN)(CN)—CH$_3$
    | | |
    OR   OR   CH$_3$

↓ NH$_3$

H—CH$_2$—CH—CH$_2$—CH—O—CH—⟨⟩—N=N—C(CN)(CN)—CH$_3$ + NH$_4$I
    | | |
    OR   OR   CH$_3$

Scheme 13.35

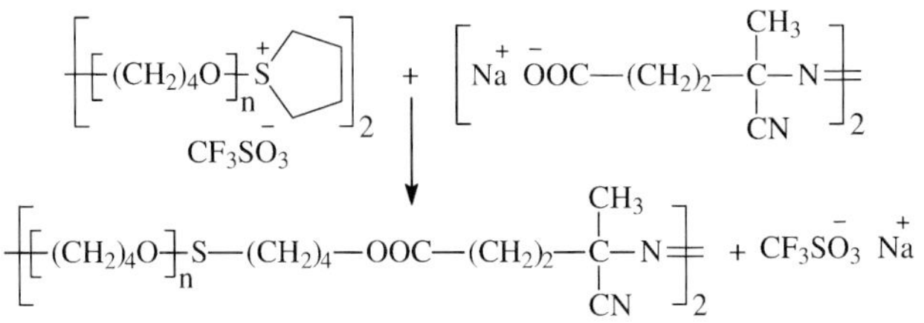

Scheme 13.36

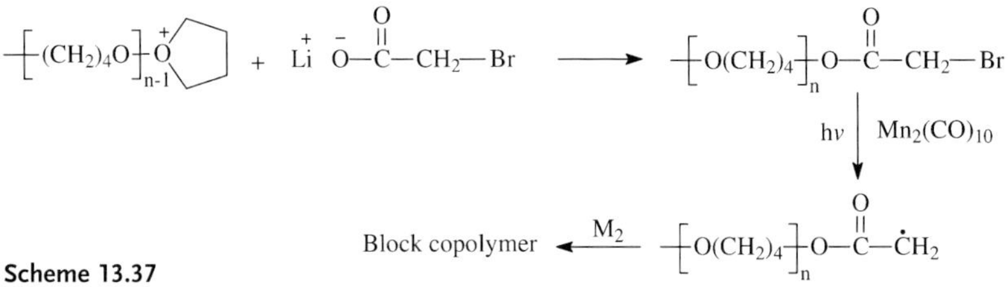

Scheme 13.37

vated by proximity to a carbonyl group are reactive in free radical generation using a metal carbonyl-halide system and were used to effect a transformation from cationic to free radical propagation.

It was shown that terminally brominated PTHF with a molecular weight of 20 000 can act as an effective macroinitiator in conjunction with the manganese compound (photolyzed at 436 nm) and that the kinetics of polymerization for St and MMA were closely similar to those for initiation from low molecular weight initiators [213].

Another elegant transformation between cationic and initer or iniferter radical polymerization has also been demonstrated [128, 215, 216]. This report concerns

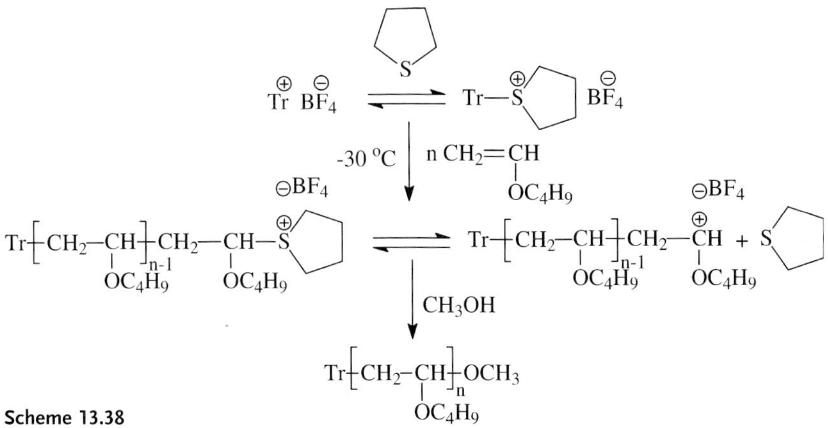

Scheme 13.38

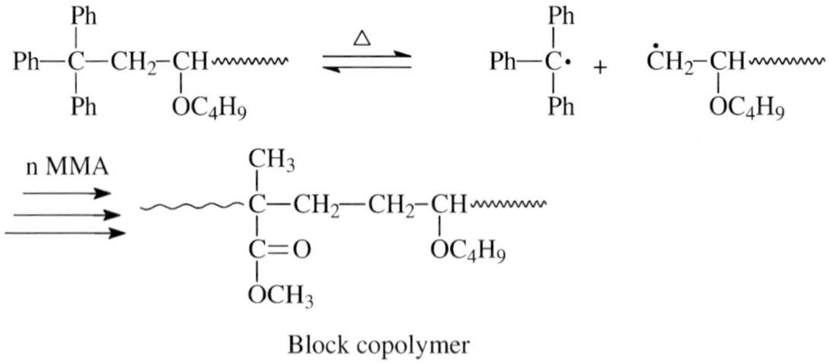

Scheme 13.39

the synthesis of polymer with a terminal trityl group or dithiocarbamate, using the corresponding cationic initiator and its further use in conventional radical polymerization for the preparation of block copolymers. For this purpose, polymerization of BVE was carried out with a combined triphenylcarboniumtetrafluoroborate-thiolane initiator system. Thiolane was used to prevent both hydride abstraction of trityl cation and chain transfer reactions dominating vinyl ether polymerization [217].

Trityl-terminated PBVE (Tr-PBVE) was subsequently used to initiate free radical polymerization of MMA via the initer mechanism [218, 219] (Scheme 13.39).

13.3.3.2 Cationic Polymerization to Controlled Radical Transformation

Yoshida and Sugita [220, 221] described the synthesis of PTHF possessing a nitroxy radical, by terminating the polymerization of living PTHF with sodium 4-oxy-TEMPO. The polymer obtained in this way acted as a counter radical in the polymerization of St in the presence of a free radical initiator to yield PSt-b-PTHF (Scheme 13.40).

NMP was also extended to azo-containing polymeric initiators obtained by cationic polymerization [222]. In this case, ω-alkoxyamine PTHF was obtained and upon heating at 125 °C, polymeric and stable nitroxyl radicals were formed. In

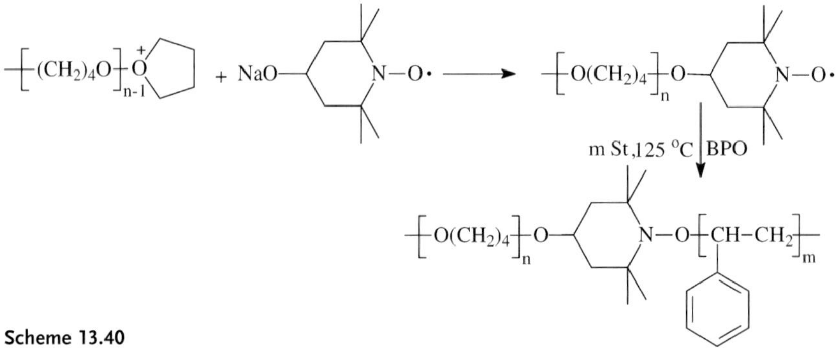

Scheme 13.40

the presence of St, the block copolymers produced had a controlled molecular weight, since termination reactions were minimized and the equilibrium between dormant and active species allowed controlled growth (Scheme 13.41).

An alternative approach to this type of transformation was also reported [223]. The living propagating chain end was quenched with previously prepared sodium 2,2,6,6-tetramethylpiperidin-1-oxylate according to the reactions shown in Scheme 13.42. In the subsequent step, radical polymerization of St was carried out with alkoxyamine-terminated PTHF. Although the increase in conversion with polymerization time was observed and block copolymers with polydispersities close to those of the prepolymers were readily formed, the initiation efficiency of ω-alkoxyamine PTHF was rather poor. This was attributed to the relatively slow decomposition and initiation of alkoxyamine attached to unsubstituted methylene groups. It has been reported that alkoxyamines containing an unsubstituted carbon atom are very slow to decompose and an α-methyl group

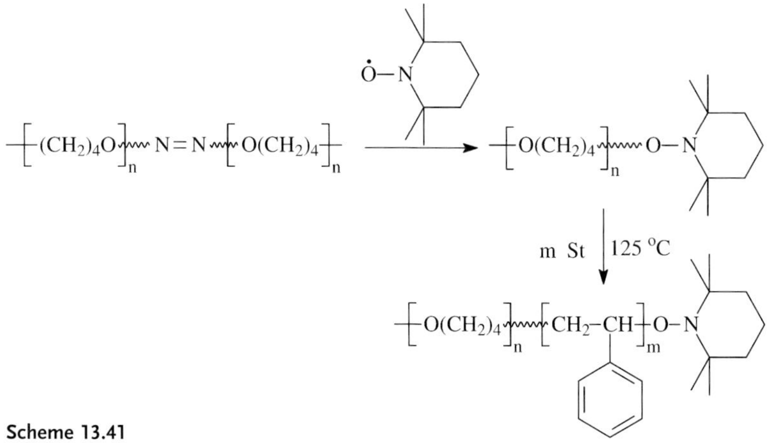

Scheme 13.41

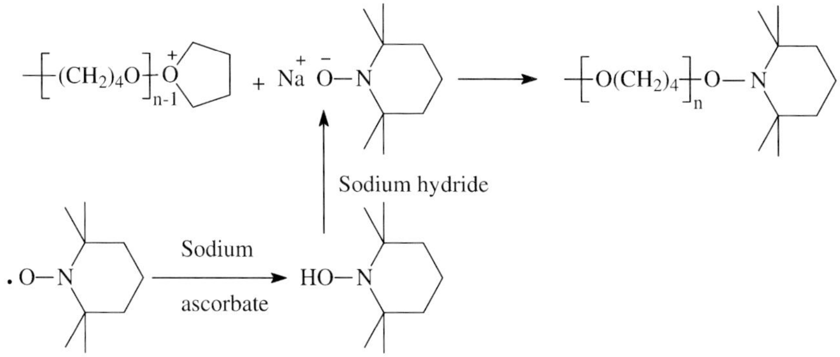

Scheme 13.42

is essential for the conventional radical polymerization to proceed with a truly living character [224].

Cation to ATRP to form AB- and ABA-type block copolymers has also been performed [225]. One or two bromopropionyl end-groups were introduced to PTHF by using a functional initiator and termination approaches in the ring-opening polymerization of tetrahydrofuran, respectively (Scheme 13.43).

Bromo-functionalized PTHFs obtained in this way were used as initiators in ATRP of St, MMA and methyl acrylate (MA) to yield AB- and ABA-type block copolymers. Notably, in the case of St and MA, the formation of triblock copolymers was significantly slower.

It was also reported [226] that PSt with chlorine termini, synthesized by living cationic polymerization, without any additional reaction, was an efficient macroinitiator for living ATRP of St, MMA and MA (Scheme 13.44).

With some variations in the initiator design, more complex structures such as block, graft and miktoarm star-block copolymers having PTHF [227–232] chains as the cationic segment were synthesized.

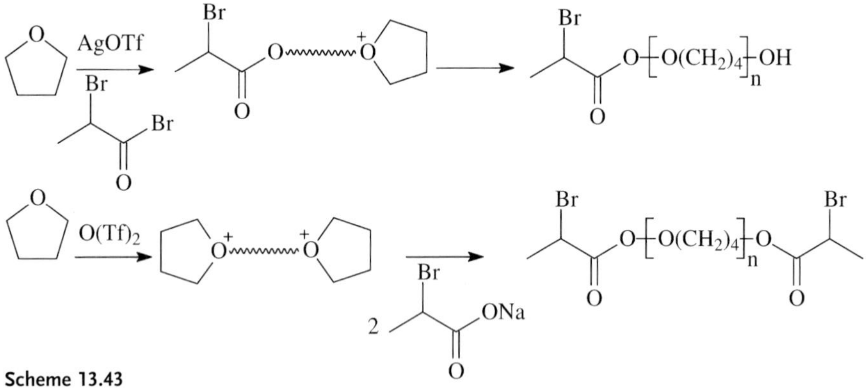

Scheme 13.43

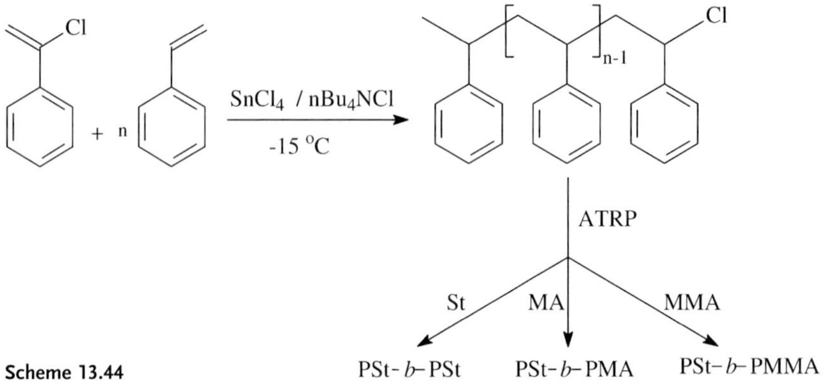

Scheme 13.44

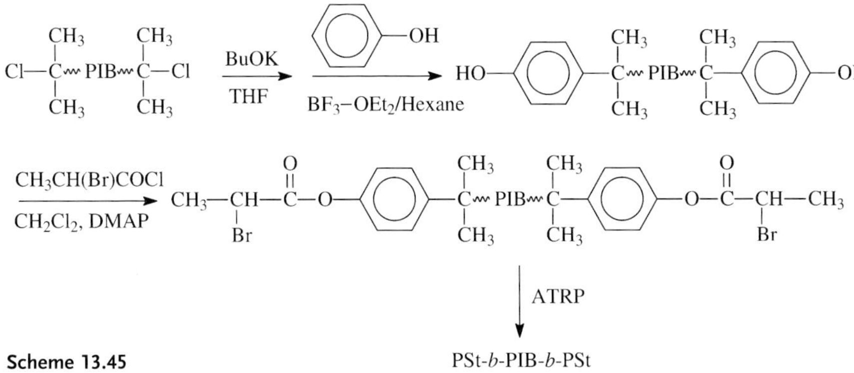

Scheme 13.45

Cationic to ATRP transformation was also used in the synthesis of triblock copolymers consisting of PIB as the middle sequence. These materials are particularly useful as thermoplastic elastomers. In this case, a few units of St were added to living difunctional PIB after the isobutene had reacted. The isolated PIBs could act as bifunctional macroinitiators for ATRP [233]. A similar strategy was used by Batsberg et al. for the synthesis of block copolymers of isobutene with p-acetoxystyrene or styrene (PIB-b-PAcSt) or (PIB-b-PSt) [234].

In a further study [235], chlorine end-groups of PIB were quantitatively converted to bromo ester groups to facilitate ATRP from end-positioned activated ester groups (Scheme 13.45). In this way, the capping with short blocks of PSt observed in the earlier method could be avoided. A similar strategy gave rise to polyisobutene-b-poly(methyl methacrylate) (PIB-b-PMMA) [236].

The concept was further extended to the preparation of PtBA-b-PIB-b-PSt (ABC) triblock [237] and amphiphilic pentablock copolymers based on PIB [238]. Using the same combination, segmented graft copolymers can also be obtained. For example, Hong et al. [239] prepared PIB-g-PMMA and PIB-g-PSt by using partially brominated PIB-co-Pp-MeSt as a macroinitiator in ATRP of the respective monomers (Scheme 13.46).

The phase and dynamic mechanical behaviors were strongly affected by the composition and/or the side-chain architectures. Moreover, the properties of the segmented graft copolymers were controlled in a wide range, leading to toughened glassy polymers or elastomers.

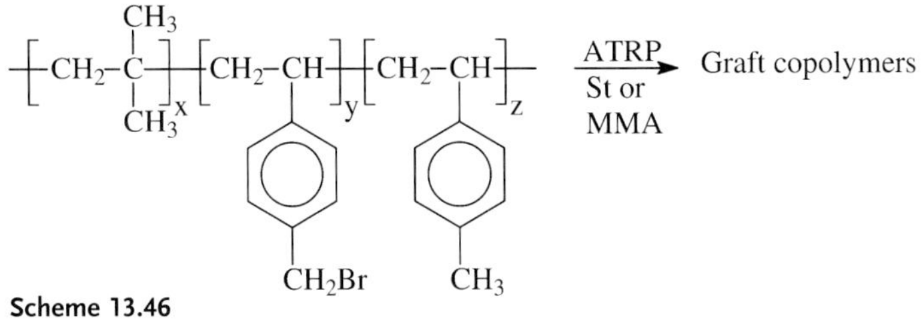

Scheme 13.46

13.3.4
Transformation of Radical Polymerization to Anionic Polymerization

13.3.4.1 Conventional Radical Polymerization to Anionic Polymerization

The basic strategy for this type of transformation is the use of amino or hydroxy telechelics as macroinitiators for the anionic polymerization of the amino acid N-carboxy anhydrides (AA-NCAs), D,L-lactide (LL) and CL [240–242]. Thus any radical method yielding polymers with terminal amino group(s) is expected to be a convenient pathway to this particular transformation. The overall strategy is outlined in Scheme 13.47.

Several examples of this type of transformation yielding polypeptide blocks and grafts have been reported (Table 13.6).

Radical to anionic transformation was also carried out by the radical copolymerization of AcSt and subsequently the anionic polymerization of MMA by treating the copolymer with BuLi. The conversion of the acetate group to hydroxyl group in the copolymer was accomplished in 90% yield. The grafting efficiency of the reaction was not stated [245].

Another example [246], the synthesis of an St graft propylene oxide (PO) copolymer, is presented in Scheme 13.48.

The synthesis of PSt-g-PEO has been demonstrated by Rempp et al. [247] and Ito et al. [248] using macromonomer techniques. However, Jannasch and Wesslen [249] have also demonstrated the synthesis of graft copolymers containing PEO side-chains on a PSt backbone using transformation concept. The St copolymers containing between 5 and 15 mol% acrylamide (Aa) and methacrylamide

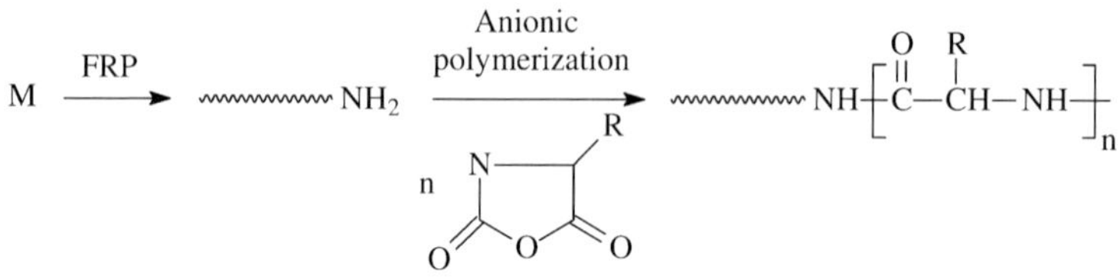

Scheme 13.47

Table 13.6 Synthesis of polypeptide block and graft copolymers by radical to anionic transformation.

Amine functionalization	Amine group	Copolymer [a]	Ref.
Bifunctional radical initiator	Terminal	PPp-b-PMMA-b-PPp	240
Chain transfer agent	Terminal	PSt-b-PPp and PMMA-b-PPp	243
Copolymerization with amine monomer	Side-chain	PBMA-g-PPp	244

a) PPp, polypeptide; PSt, polystyrene; PMMA, poly(methyl methacrylate); PBMA, poly(butyl methacrylate).

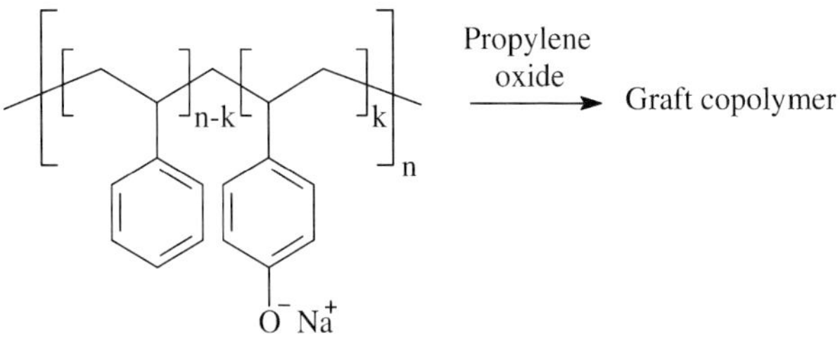

Scheme 13.48

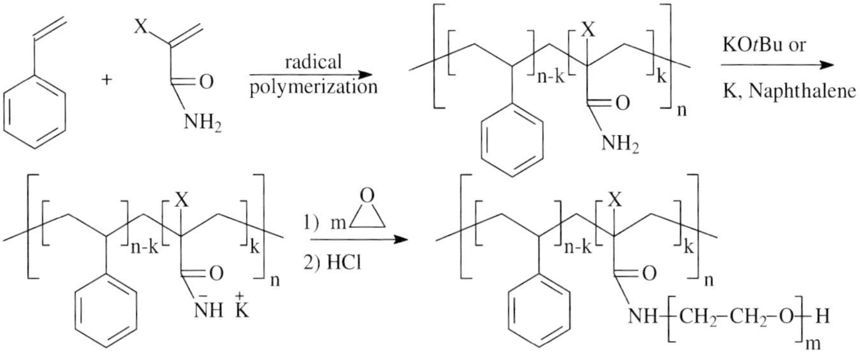

Scheme 13.49

(MAa) were synthesized by a free radical mechanism using AIBN initiator. The amide groups in the copolymers were ionized by using potassium *tert*-butoxide or potassium naphthalene and grafting was achieved by utilizing the amine anion as initiating sites for the polymerization of EO in 2-ethoxyethyl ether at 65 °C. The grafting efficiency was very high. The general route for the preparation of the graft copolymers is shown in Scheme 13.49.

13.3.4.2 Controlled Radical Polymerization to Anionic Polymerization

The most widely applied controlled radical polymerization method for this particular transformation is ATRP. This is mainly because hydroxyl and amino groups, potential initiating sites for the ring-opening anionic polymerization of certain monomers, are compatible with the ATRP of vinyl monomers. Examples of such transformations are given in Table 13.7 and the general concept is illustrated in Scheme 13.50 with the example of the combination ATRP of vinyl monomers with the ring-opening polymerization of lactides [250].

Table 13.7 Block copolymers prepared by ATRP to anion transformation.

ATRP segment	Anionic segment [a]	Ref.
Poly(acrylate ethyl lactose octaacetate)	Poly(benzyl-L-glutamate)	251
Poly(acrylate ethyl lactose)	Poly(l-alanine N-carboxyanhydride)	252
PMMA	PLL, PCL	253, 254
Polyacrylates	PCL	255
PSt	PEO, PLL, PPp, γ-benzyl-L-glutamate N-carboxyanhydride or PtBA	256–259

a) PMMA, poly(methyl methacrylate); PCL, poly(ε-caprolactone); PLL, polylactide; PPp, polypeptide; PSt, polystyrene; PEO, poly(ethylene oxide).

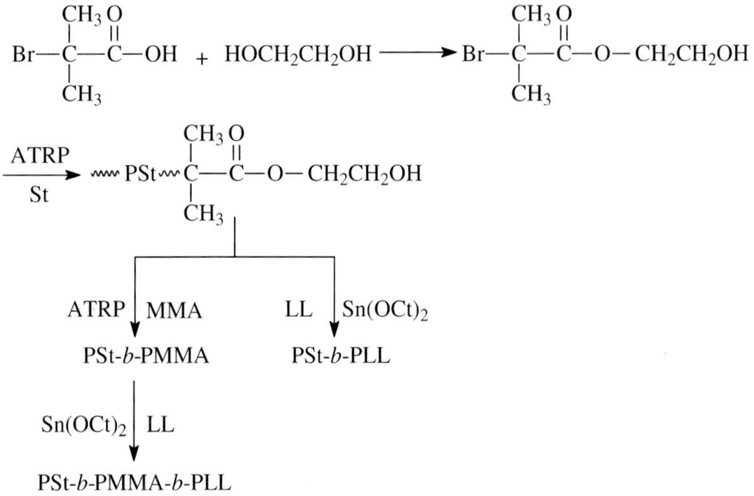

$$Br-\underset{\underset{CH_3}{|}}{\overset{\overset{CH_3}{|}}{C}}-\overset{\overset{O}{\|}}{C}-OH + HOCH_2CH_2OH \longrightarrow Br-\underset{\underset{CH_3}{|}}{\overset{\overset{CH_3}{|}}{C}}-\overset{\overset{O}{\|}}{C}-O-CH_2CH_2OH$$

$$\xrightarrow[St]{ATRP} \sim\sim PSt\sim\sim \underset{\underset{CH_3}{|}}{\overset{\overset{CH_3}{|}}{C}}-\overset{\overset{O}{\|}}{C}-O-CH_2CH_2OH$$

ATRP | MMA   LL | Sn(OCt)$_2$

PSt-b-PMMA   PSt-b-PLL

Sn(OCt)$_2$ | LL

PSt-b-PMMA-b-PLL
Scheme 13.50

13.3.5
Transformation of Radical Polymerization to Cationic Polymerization

So far, most of the applied radical to cationic transformation polymerizations are based on mainly using redox or transfer agent to functionalize the polymers in the course of the radical process. The functionalized polymers, usually alkyl halides or acyl halides, were employed to generate a carbocation which is capable of initiating copolymerization by reacting with silver salts such as AgClO$_4$, AgPF$_6$ or AgSbF$_6$. Table 13.8 furnishes some examples pertaining to such combinations.

An interesting variation of this theme is accomplished with dual initiation controlled by a different polymerization mechanism, i.e. radical and cationic. Yagci et al. [271] reported the preparation of a block copolymer of St and CHO

Table 13.8 Examples of the transformation of conventional radical polymerization into cationic polymerization.

Block copolymer [a]	Ref.
Polyvinyl-b-PTHF	260
PSt-b-PTHF, PSt-b-PNVC, PSt-b-PNVP	117
PTHF-b-PSt-b-PTHF	261
PSt-g-PTHF	262, 263
PMMA-b-PBVE	264
PMeSt-b-PCHO	265–267
Pt-BuSiSt-b-PCHO	268
PNVC-b-PCHO	269
PMeSt-b-PCHO, PMeSt-b-PECO	270

a) PTHF, polytetrahydrofuran; PSt, polystyrene; PNVC, poly(N-vinylcarbazole); PNVP, poly(N-vinylpyrrolidone); PBVE, poly(butyl vinyl ether); PMeSt, poly(methylstyrene); PCHO, poly(cyclohexene oxide); Pt-BuSiSt, poly[4-(tert-butyldimethylsiloxy)styrene]; PECO, poly(1,2,5,6-diepoxycyclooctane).

by a combination of radical and cationic polymerization. PSt bearing photolabile benzoin groups was prepared by the thermal decomposition of an azobenzoin initiator in the presence of St monomer [60, 61, 272] (Scheme 13.51).

The polymeric alkoxy radicals formed upon irradiation of this polymer were oxidized, by means of pyridinium salts, to the corresponding carbocation capable of initiating the cationic polymerization. Thus block copolymers of ABA type were formed, A and B being PCHO and PSt, respectively (Scheme 13.52).

Kobayashi et al. [273] polymerized 2-methyl-2-oxazoline (MeOZO) or 2-ethyl-2-oxazoline (EtOZO) by using tosylated or bromoacylated copolymer of ethylene and vinyl acetate. The acetate moieties or bromoacylate copolymer were converted to hydroxyl groups by saponification, which were subsequently tosylated or bromoacetylated (Scheme 13.53).

The random copolymer of St and p-AcSt was prepared by radical copolymerization and its acetoxy groups were partially converted to sodium phenoxide groups by alkaline hydrolysis. The sodium phenoxide groups of the copolymer initiated the polymerization of propylene oxide to generate a hydroxyl end-group

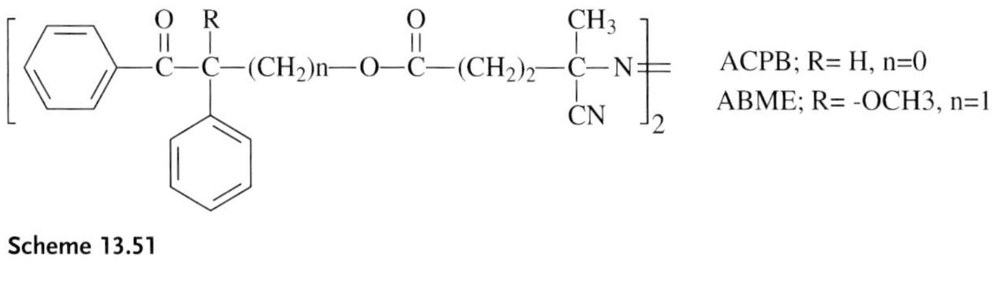

ACPB; R= H, n=0
ABME; R= -OCH3, n=1

Scheme 13.51

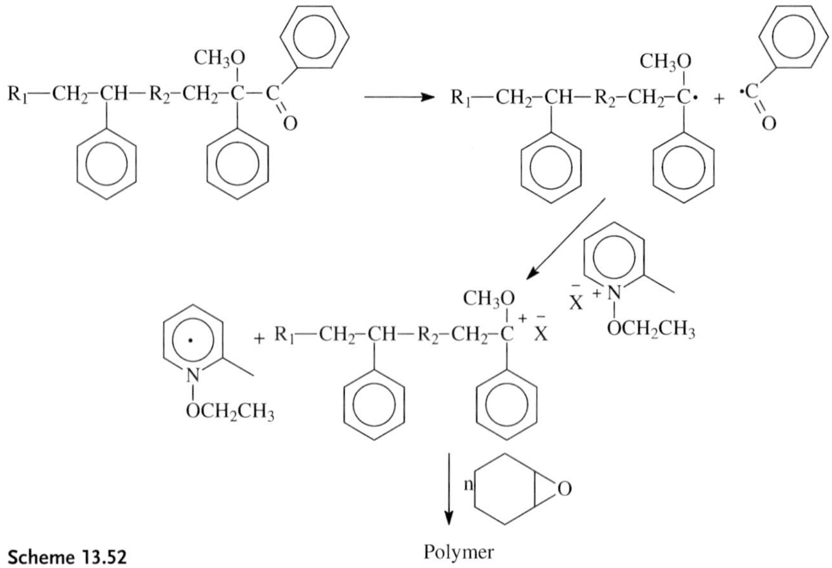

13 Segmented Copolymers by Mechanistic Transformations

Scheme 13.52

at each propylene oxide branch. The hydroxyl groups were converted to tosylate groups from which cationic polymerization of MeOZO was initiated [246] to yield poly(St-g-[PO-b-MeOZO]) (Scheme 13.54).

Polymer-bound sulfonium salts have been used for the preparation of graft copolymers [274, 275] (Scheme 13.55). In the first step, a sulfonium containing St-based monomer was prepared. This monomer was polymerized or copolymerized with St in the presence of AIBN at 60°C, thus yielding sulfonium-modified PSt with variations in the concentration of sulfonium containing units. Heating the macroinitiator in the presence of cationically polymerizable monomer makes the latter grafted on to PSt backbones. This scheme was followed using a bicyclo orthoester as cationically polymerizable monomer. The

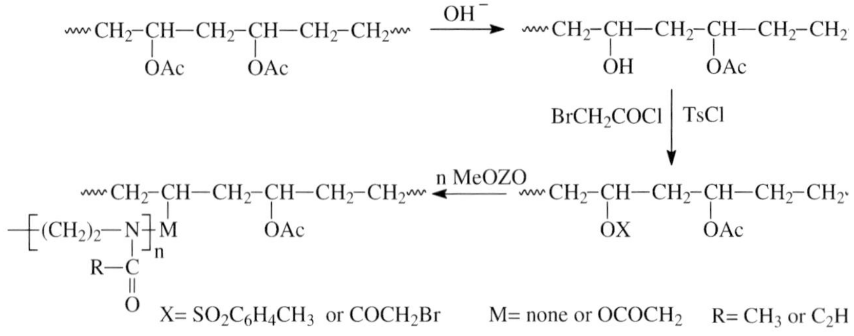

Scheme 13.53

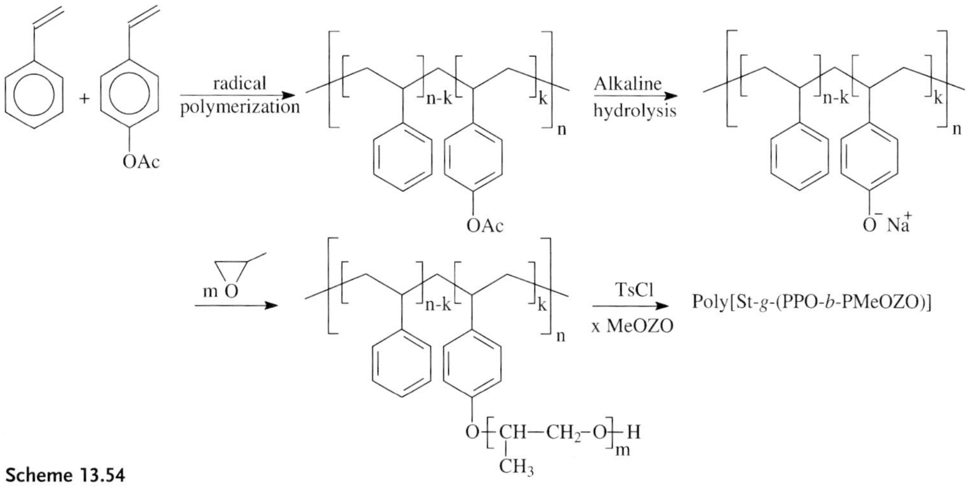

13.3 Indirect Transformation

Scheme 13.54

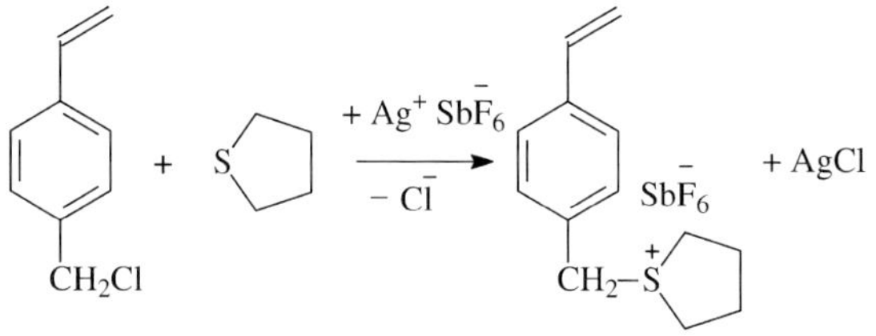

Scheme 13.55

Table 13.9 Transformations of controlled radical polymerization to cationic polymerization.

Controlled radical polymerization	Cationic polymerization	Type of segmented copolymer[a]	Ref.
ATRP	Carbocationic	PIB-b-PSt-b-PMMA-PSt-b-PIB	276, 277
ATRP	Ring opening	S-(PSt)$_2$-(PDOP)$_2$ miktoarm star	278
ATRP	Vinyl	PMVE-b-PtBu, PMVE-b-PAA	279
ATRP	Ring opening	PSt-b-PTHF, PTHF-b-PSt-b-PTHF	280
ATRP	Ring opening	PTMO-b-PSt and PTMO-b-PSt-b-PMMA	281
ATRP	Promoted cationic	PSt-b-PCHO	282
DPE	Promoted cationic	PSt-b-PCHO and PMMA-b-PCHO	283
DPE	Carbocationic	PMMA-b-PSt-g-PIB	284, 285
NMP	Ring opening	PSt-g-PEI comb polymer	286

a) PIB, polyisobutene; PMMA, poly(methyl methacrylate); PSt, polystyrene; PDOP, polydioxapane, PMVE, poly(methyl vinyl ether); PtBu, poly(tert-butyl acrylate); PTHF, polytetrahydrofuran; PTMO, poly(trimethylene oxide); PCHO, poly(cyclohexene oxide); PEI, poly(ethylenimine).

onium salt-carrying monomer has also been copolymerized radically with an unsaturated spiro orthoester and a third vinyl monomer (MMA or AN).

Parallel to the recent advances in controlled radical polymerizations, many transformations of ATRP, NMP and RAFT to cationic polymerization have been reported (Table 13.9).

13.3.6
Transformations Involving Anionic and Cationic Polymerizations

Anion to cation or reverse transformation reactions were successfully employed to prepare block copolymers. The particular advantage of these reformations is that both anionic and cationic blocks can be prepared under living polymerization conditions. In this connection, prominence must be given to pioneering work of Richards and coworkers [14–16], demonstrating the great versatility of the transformation reactions. They prepared bromine-terminated PSt by direct reaction of excess bromine or xylene dibromide with living PSt (Scheme 13.56).

It was found that competing Wurtz coupling reactions (Scheme 13.57) may be prevented [287] using a Grignard intermediate. The desired bromo-functionalized PSt was obtained at up to 95% efficiency.

Bromo-functionalized polymer was employed to prepare block copolymer upon generating carbocations by reacting suitable silver salts (Scheme 13.58).

Although block copolymers with narrow polydispersity were obtained, quantitative transformation efficiency was not achieved even at low temperatures due to the β-proton elimination reactions (Scheme 13.59). Termination by β-proton

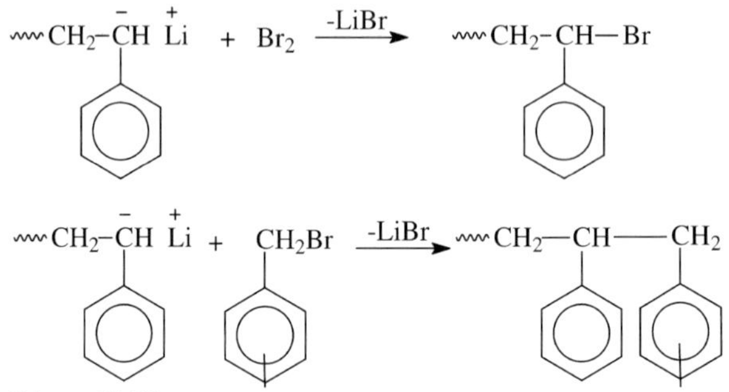

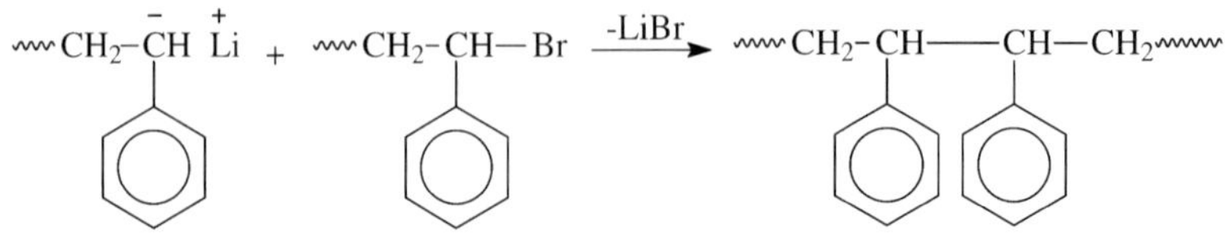

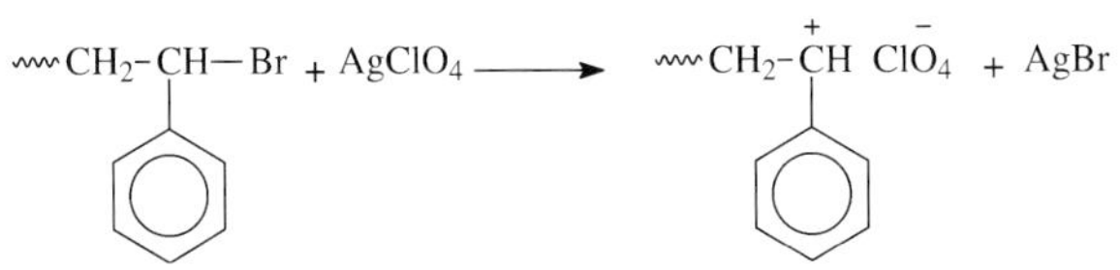

Scheme 13.58

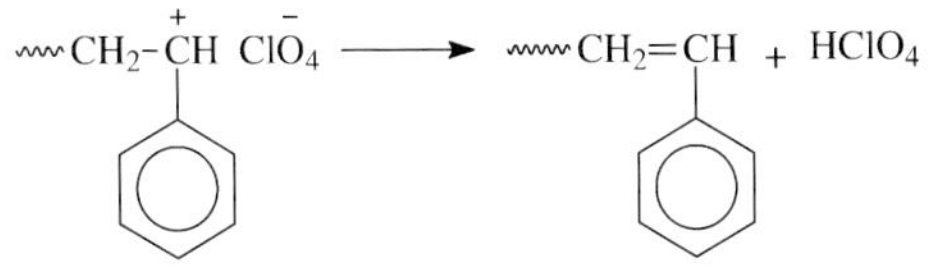

$$\sim\sim CH_2-CH(Ph)-Br + AgClO_4 \longrightarrow \sim\sim CH_2-\overset{+}{C}H(Ph)\ \overset{-}{ClO_4} + AgBr$$

Scheme 13.59

$$\sim\sim CH_2-\overset{+}{C}H(Ph)\ \overset{-}{ClO_4} \longrightarrow \sim\sim CH_2=CH(Ph) + HClO_4$$

elimination may be avoided by using xylene dibromide in the halogenation process. However, it did not improve the overall transformation efficiency.

A conceptually similar approach was used by Muhlbach and Schulz [288] to prepare styrene and 1-azabicyclo[4.2.0]octane block copolymer. By end-capping living PSt with ethylene oxide and subsequent reaction with bromoacetyl bromide, a polymer with bromoacetyl groups was obtained. This polymer together with AgClO$_4$ acted as a macroinitiator for living polymerization of the cyclic monomer (Scheme 13.60). Very little homopolymer was formed. Two distinct glass transitions at 10 and 94 °C were observed with the block copolymer corresponding to poly(1-azabicyclo[4.2.0]octane) and PSt sequences, respectively, which indicates that both blocks are incompatible and phase separated.

Cationic polymerization of cyclic amines is well known [289–291]. Low molecular weight initiators such as ethyl tosylate induce polymerization of cyclic amines such as 1-*tert*-butylaziridine. The idea of using a macroinitiator having a tosylate end-group to polymerize cyclic amines prompted Kazama et al. [292] to polymerize 1-*tert*-butylaziridine using PDMS having a terminal tosylate group. However, no polymerization occurred when macroinitiator was used. Neverthe-

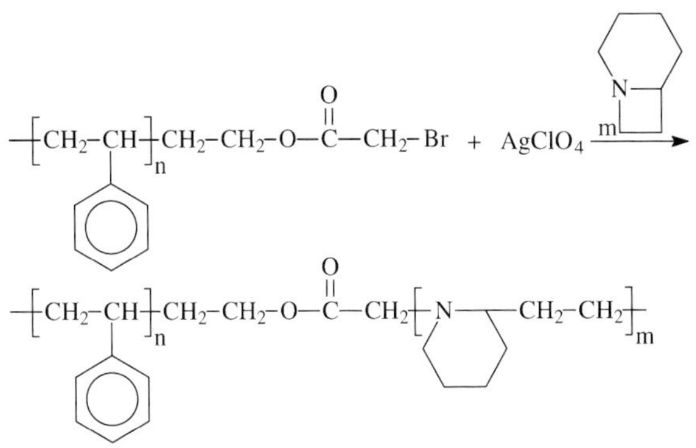

Scheme 13.60

less, this clearly indicates the initiative behind the transformation reaction between anionic and cationic polymerizations.

Anionic polymerization is one of the best methods to prepare end-functionalized polymers. The haloalkyl groups at one end of the vinyl polymers were prepared by anionic polymerization followed by termination of the living anion with an excess of 1,2-dichloroethane or 1,4-dibromobutane [293]. These polymers served as macroinitiators for the polymerization of MeOZO and aromatic vinyl monomers (Scheme 13.61).

Block copolymers consisting of poly(N-acylethylenimine) and poly(ethylene oxide) chains were prepared by initiating the polymerization of 2-methyl-2-oxazoline and 2-ethyl-2-oxazoline with α,ω-ditosylated or mesylated PEO [294] (Scheme 13.62). The blocking efficiency was close to 100%.

Simionescu et al. [295] used poly(ethylene oxide adipate) having tosylate groups at both ends as macroinitiators for the cationic polymerization of MeOZO to produce ABA-type block copolymers. The concept was further explored [296] and block copolymers consisting of PPO and PMeOZO were prepared by using poly(propylene oxide)-p-nitrobenzene sulfonate as a macroinitiator for the cationic polymerization of MeOZO (Scheme 13.63). As the conversion to the sulfonate functionality was quantitative, the polymerization of MeOZO by the macroinitiator produced a mixture of AB- and ABA-type block copolymer.

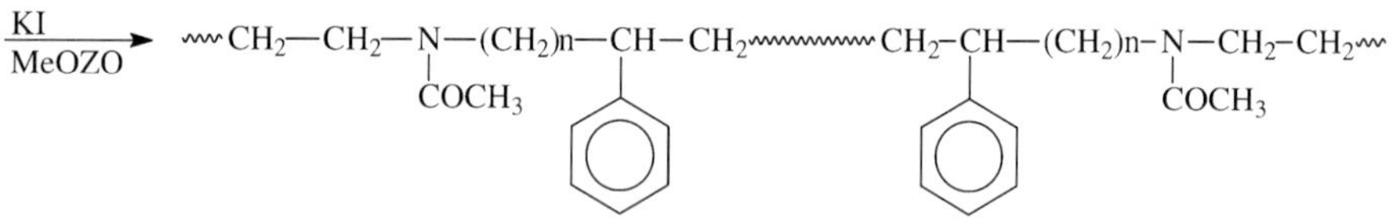

Scheme 13.61

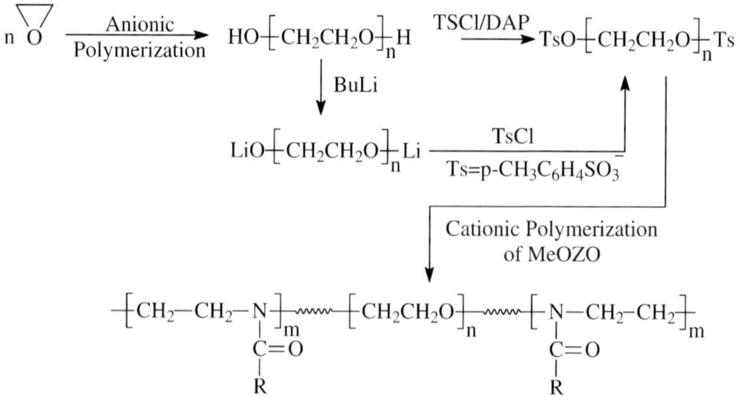

Scheme 13.62

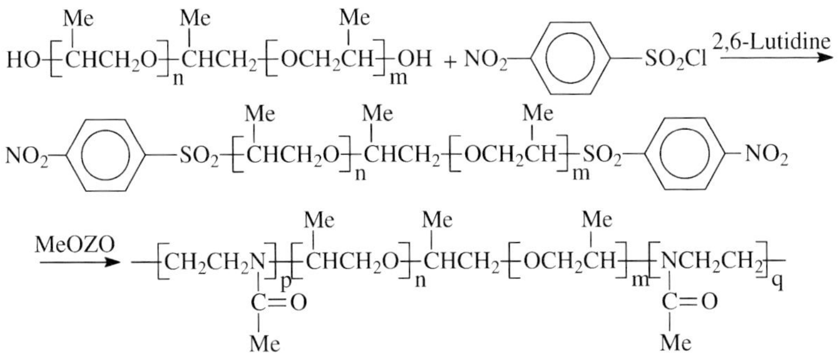

Scheme 13.63

An interesting application of new polymer architectures by combination of anionic and cationic polymerization was described by Deffieux and Schappacher [297, 298]. In this study, quantitative formation of comb-like polymers was obtained by grafting polystyryllithium on to poly(chloroethyl vinyl ether) (PCEVE) (Scheme 13.64).

The advantage of the system arises from the fact that since the PCEVE backbone and PSt grafts can be prepared by both living cationic and anionic polymerization, it is possible to synthesize graft copolymers possessing both a backbone with controlled dimensions and adjustable number of branches having precise length and narrow molecular weight distribution. Star and hyperbranched polymers were also prepared by following the same strategy.

Many examples of active site transformations from cation to anion to prepare block and graft copolymers have been reported (Table 13.10).

A report [311] on this type of transformation illustrates the versatility of the method to prepare new and unique polymer architectures. Living PIB chains were quantitatively captured with 1,1-diphenylethylene, leading to diphenylmethoxy and diphenylvinyl end-groups (Scheme 13.65).

The stable macroanions obtained by the subsequent metalation of the endgroups, according to the reactions shown in Scheme 13.66, were used to initiate living anionic polymerization *tert*-butyl methacrylate (*t*BMA) yielding with PIB-*b*-P*t*BMA block copolymers with almost quantitative efficiency [312].

The hydrolysis of the ester groups of the acrylate segment further allowed the preparation of amphiphilic polyisobutene-*b*-poly(methacrylic acid) (PIB-*b*-

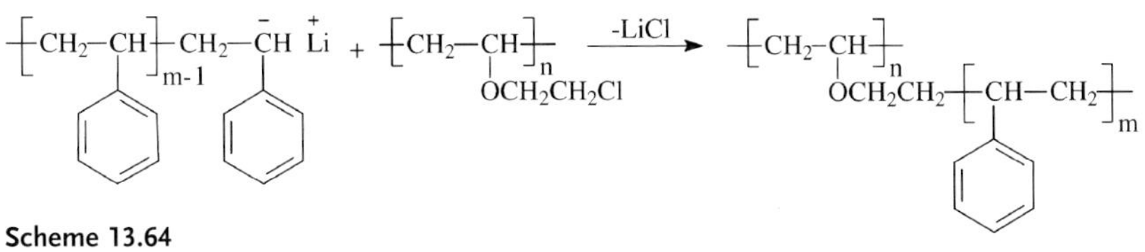

Scheme 13.64

Table 13.10 Block copolymers via cation to anion transformation reactions.

Block copolymer[a]	Ref.
PIB-b-PB	299
PIB-b-PMMA	300
PIB-b-PS	301
PIB-b-PMMA	302
PIB-b-PCL	303
PIB-b-MeSt-g-PPV	304
PIBVE-b-PCL	305
PIB-b-PtBMA, PIB-b-PMAA	306
PEtOZO-b-PLL, PEtOZO-b-PCL	307
PTHF-b-PMMA	308, 309
PTHF-b-PSt, PTHF-b-PIP	310

a) PIB, polyisobutene; PB, polybutadiene; PMMA, poly(methyl methacrylate); PS, polysiloxane; PCL, poly(ε-caprolactone); PMeSt, poly(methylstyrene); PPV, polypivalactone; PIBVE, poly(isobutyl vinyl ether); PtBMA, poly(tert-butyl methacrylate); PMAA, poly(methacrylic acid); PEtOZO, poly(2-ethyl-2-oxazoline); PLL, polylactide; PTHF, polytetrahydrofuran; PIP, polyisoprene.

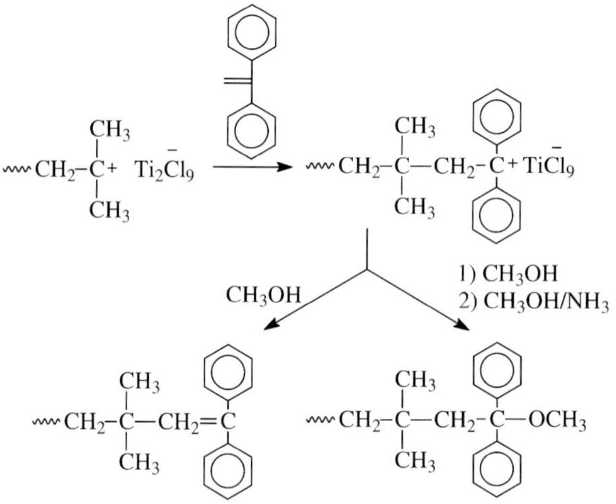

Scheme 13.65

PMAAc). A series of linear and star copolymers consisting of PIB and PMMA were also prepared [312].

The process was further improved by replacing diphenylethylene with thiophene in the end-capping process [313]. The advantage of this modification was related to the quantitative functionalization of living PIB with thiophene and the possibility of metalation of the thiophene end-groups with n-BuLi. This is

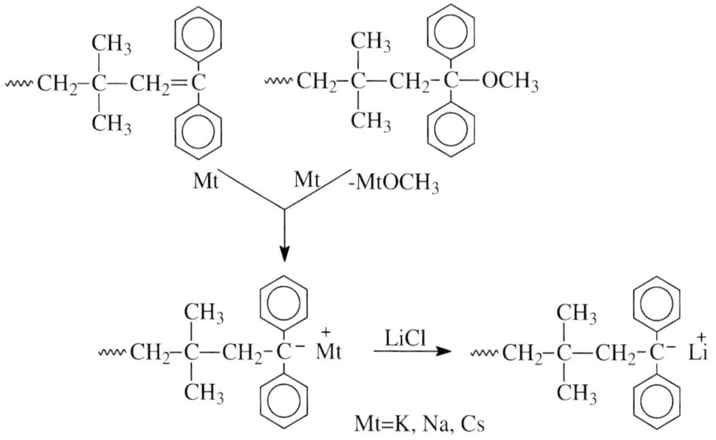

Scheme 13.66

an important improvement for industrial processes since lithiation by *n*-BuLi is much more convenient than metalation with Na/K alloy.

Graft copolymers consisting of a PMMA backbone and poly(isobutyl vinyl ether) (PIVBE) side-chains were also prepared by combined anionic and cationic polymerizations based on the homopolymerization of a bifunctional monomer [314].

13.3.7
Transformations Involving Activated Monomer Polymerization

Polymerization by the activated monomer mechanism opened up promising pathways for transformation reactions. Kubisa and coworkers [315–317] reported that cationic polymerization of oxiranes may proceed in the presence of hydroxyl-containing compounds by an activated monomer (AM) mechanism according to the mechanism shown in Scheme 13.67.

Thus propagation involves the reaction of a protonated (activated) monomer molecule with a nucleophilic site in the neutral growing macromolecule. Yagci et al. [318] adapted this procedure to transformation reactions. 4,4′-Azobis(4-cyanopentanol) was used in AM polymerization of epichlorohydrin (ECH) to produce polymers with an azo linkage in the main chain. Polymerization was conducted under typical conditions, i.e. by slow addition of ECH to the solution of initiator containing catalyst. Reaction was considerably slower than in the presence of simple diols (e.g. ethylene oxide) and only 28% conversion was obtained

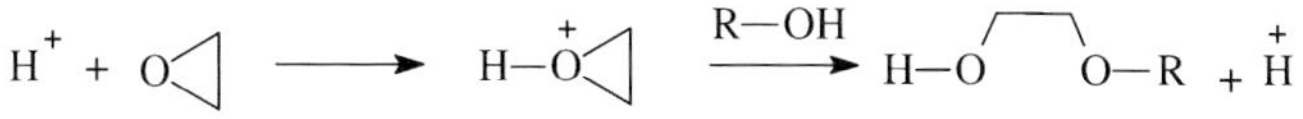

Scheme 13.67

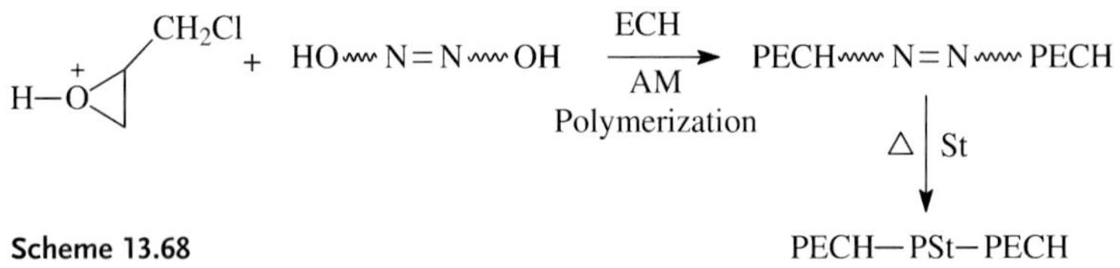

Scheme 13.68

under conditions sufficient to reach complete conversion in the polymerization initiated by ethylene oxide. Poly(epichlorohydrin) (PECH) prepared in this way was used in the polymerization of St to produce block polymer (Scheme 13.68).

The reverse mode of this transformation is also accessible. For instance, Steward [319] has shown that by using a free radically prepared hydroxy-terminated polybutadiene as a macro-catalyst in AM polymerization of ECH, block copolymers are produced in quantitative yield.

Polymers with a variety of photochromophoric groups have been extensively used as precursors for block and graft copolymers [320]. More recently, benzoin derivatives containing hydroxy groups were used as initiators of AM polymerization of ECH. The resulting polymers contain photoactive benzoin terminal groups [321].

As shown for radical to cationic transformation (see above), the polymerization of acrylates and cyclic ethers can be photochemically initiated by a benzoin moiety incorporated in polymers (Scheme 13.69). In the latter case, organic oxidants such as onium salts are essential for converting originally formed radicals to cations to afford cationic polymerization of CHO [322].

It is clear that various transformation reactions can be realized by using suitable selected hydroxy compounds in AM polymerization.

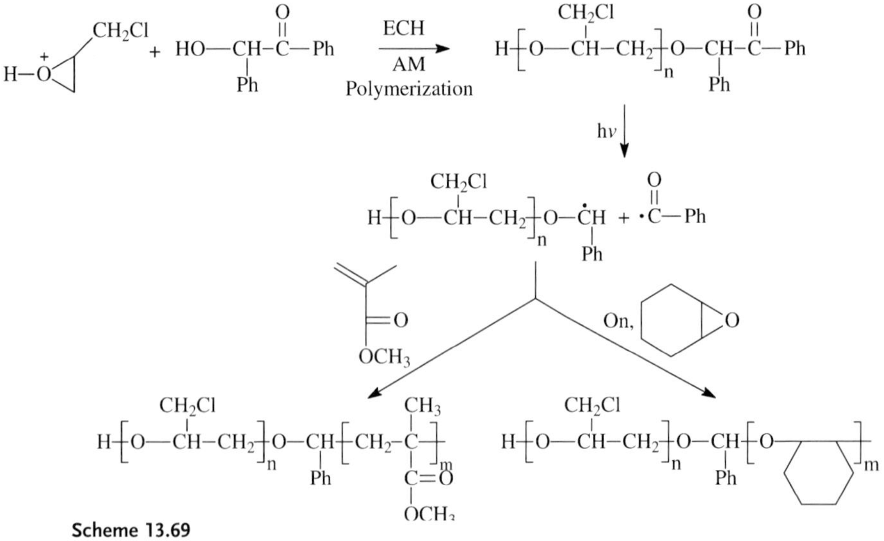

Scheme 13.69

13.3.8
Transformations Involving Metathesis Polymerization

Ring-opening metathesis polymerization is convenient route to prepare well-defined polymers [323–325] and initiated by Ti, Mo, W, Ta, Re and Ru complexes [326–331]. However, this type of polymerization is limited to highly strained cycloolefins such as norbornene, norbornadiene, dicyclopentadiene and other strained polycyclics for which ring opening is essentially irreversible [332]. It therefore seems that the transformation reactions involving metathesis polymerization allow extension of the range of attainable block copolymers. The first reported example [333, 334] involves block copolymerization of St and cyclopentene. When polystyryl anion was used in conjunction with tungsten hexachloride, the propagating anion was transformed to covalent species and propagating centers for the polymerization of the second monomer might have a bridged structure as shown in Scheme 13.70.

The same catalyst system was re-examined, but instead of a copolymer, only dimers and oligomers of the polystyryl co-catalyst were isolated [335]. Using different catalysts, namely ruthenium and molybdenum initiators, various block and graft copolymers were also prepared via anionic to ring-opening metathesis polymerization [336–338].

Risse et al. [339] and Tritto et al. [340] reported two independent transformation reactions for block copolymer synthesis. The first report [339, 341] involves changing the mechanism from living metathesis polymerization of cycloalkene to group transfer polymerization of silyl vinyl ether. Second, they prepared block copolymers of norbornene and ethylene by transforming metathesis polymerization to Ziegler–Natta polymerization [342]. The reverse transformation is also possible [343].

More recently, Matyjaszewski and coworkers [344] and others [345, 346] reported general methods of transformation of living metathesis polymerization into ATRP and anionic polymerization for the preparation of block copolymers. In this approach, polynorbornene and poly(dicyclopentadiene) with terminal bromide were synthesized by end-capping of the corresponding living chain ends with benzyl bromide (Scheme 13.71).

These polymers served as efficient macroinitiators for the homogeneous ATRP of St and MA according to the mechanism described earlier (see above).

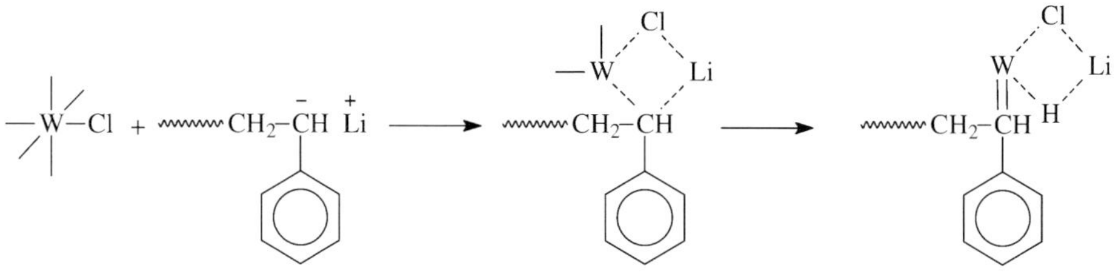

Scheme 13.70

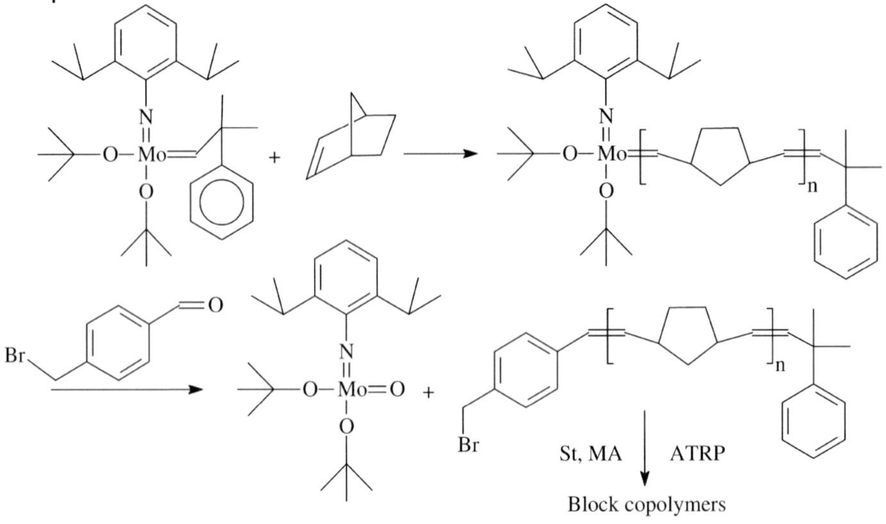

Scheme 13.71

Several other examples [347–349] of this type of transformation, including those used for the preparation of liquid crystalline block [350] and graft copolymers [351], have been reported.

13.3.9
Transformations Involving Ziegler–Natta Polymerization

Ziegler–Natta polymerization is well known and involves a two-stage process. In the first stage, an aluminum alkyl such as trialkylaluminum is reacted with $TiCl_4$ in order to give the active β-$TiCl_3$. Alkyl radicals, also produced in this reaction, were terminated by coupling and give inert products. Subsequent alkylation of β-$TiCl_3$ then occurs to generate the titanium species capable of initiating polymerization of olefins such as ethylene (Scheme 13.72).

As the polymerization results in the incorporation of alkyl ligand in the final product, polymeric aluminum compounds may conveniently be employed instead of small molecule analogues to affect an anion to Ziegler–Natta transformation process yielding novel block copolymers [352–355]. Following this strat-

$TiCl_4 + R_3Al \longrightarrow RTiCl_3 + R_2AlCl \longrightarrow \beta\text{—}TiCl_3 + R\cdot$

$\beta\text{-}TiCl_3 + R_3Al \longrightarrow RTiCl_2 + R_2AlCl$

$RTiCl_3 + CH_2{=}CH_2 \begin{array}{c} \xrightarrow{\text{Coordination}} RTiCl_2C_2H_4 \\ \xrightarrow{\text{Insertion}} RCH_2CH_2TiCl_2 \end{array}$

Scheme 13.72

$$3 \;\text{\textasciitilde\textasciitilde\textasciitilde}^{\ominus} \text{Li}^{\oplus} + \text{AlCl}_3 \longrightarrow (\text{\textasciitilde\textasciitilde\textasciitilde})_3 \text{Al} + 3\;\text{LiCl}$$

Scheme 13.73

egy, aluminum compounds have been successfully synthesized by successive alkylation of aluminum halide with living anionic polymers as shown in Scheme 13.73. The method has been used to prepare block copolymers of St with ethylene and acetylene. Notably, extremely low transformation efficiencies were obtained. However, purified block copolymer of the latter has potential use as an electroactive polymer [356].

Another useful active site transformation from Ziegler–Natta to radical has been achieved by Agnuri et al. [357]. Alkenic-vinylic-type block copolymers containing crystalline and amorphous sequences were prepared by using this transformation. Ziegler–Natta polymerization was induced with diethylzinc as the transition complex and peroxy groups were incorporated into crystalline polymer through oxidation of the carbon-zinc bond. Thermolysis of the macroinitiator formed in this way resulted in the generation of a pair of radicals. Both radicals are capable of initiating the polymerization of vinyl monomers such as MMA. Therefore, block copolymer formation was accompanied by the formation of homopolymer arising from the ethoxy radicals (Scheme 13.74).

Doi and coworkers [358, 359] have demonstrated the living coordination polymerization of propylene (P) by a soluble Ziegler–Natta catalyst composed of vanadium acetylacetonate and diethylaluminum chloride. The living polypropylene chain end can be transformed to iodide by treatment with a solution of iodine in toluene. In combination with $AgClO_4$, the polymer containing iodide generates a carbocation which initiates cationic polymerization of THF at $0\,°C$ as shown in Scheme 13.75. However, the blocking efficiency was negligible at $20\,°C$.

Mulhaupt and coworkers have reported several studies related to the preparation of block copolymers from thiol, maleic acid and hydroxy functional polypropylene prepared using a metallocene catalyst [360, 361].

A facile and inexpensive reaction process for the preparation of polypropylene (PP)-based graft copolymers containing an isotactic PP main chain and several

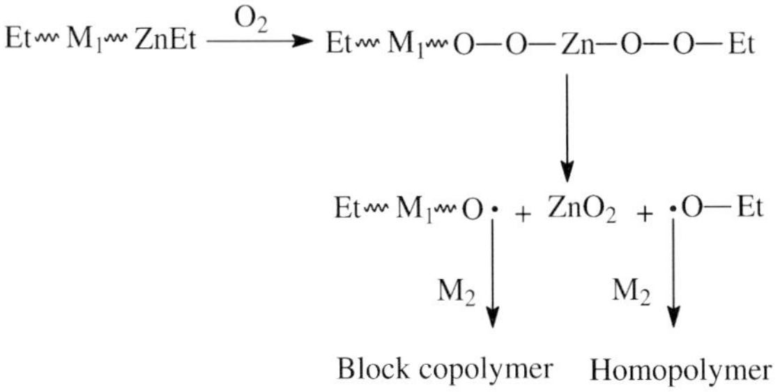

$$\text{Et\textasciitilde\textasciitilde} M_1 \text{\textasciitilde\textasciitilde} \text{ZnEt} \xrightarrow{O_2} \text{Et\textasciitilde\textasciitilde} M_1 \text{\textasciitilde\textasciitilde} O-O-Zn-O-O-Et$$

$$\downarrow$$

$$\text{Et\textasciitilde\textasciitilde} M_1 \text{\textasciitilde\textasciitilde} O\cdot + ZnO_2 + \cdot O-Et$$

$$\quad\quad M_2 \downarrow \quad\quad\quad M_2 \downarrow$$

$$\quad\text{Block copolymer}\quad\text{Homopolymer}$$

Scheme 13.74

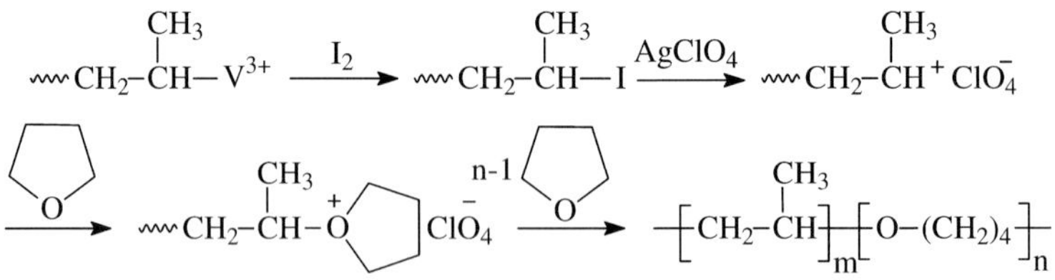

Scheme 13.75

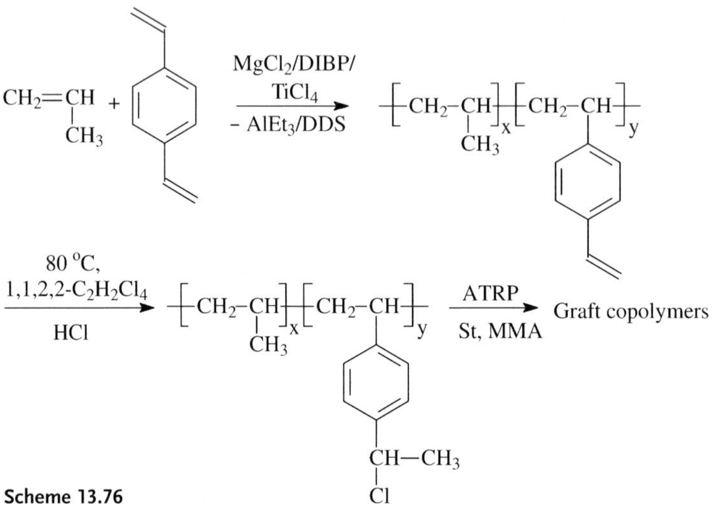

Scheme 13.76

functional polymer side-chains was recently described by Chung and coworkers [362, 363] (Scheme 13.76). In this case, Ziegler–Natta to ATRP transformation was applied.

The same group also reported the transformation of metallocene-mediated olefin polymerization to anionic polymerization by a novel consecutive chain-transfer reaction for the preparation of polypropylene-based block copolymers [364].

The metallocene–ATRP route was successfully followed by Matsugi et al. to produce polyethylene-b-poly(methyl methacrylate) (PE-b-PMMA). The block copolymers obtained exhibited unique morphological features that depended on the content of the PMMA segment. Moreover, the block copolymers effectively compatibilized the corresponding homopolymer blend at the nanometer level [365].

A relatively new coordination olefin polymerization method, degenerative transfer coordination polymerization was recently combined with ATRP to prepare block and graft copolymers with linear PE segments [366–369].

13.3.10
Transformations Involving Group Transfer Polymerization

Group transfer polymerization is another [370, 371] polymerization for producing well-defined acrylic polymers by silyl keten acetal activation using nucleophilic or electrophilic catalysts. This type of polymerization involves the generation of keten acetal at the chain end for each monomer unit addition (Scheme 13.77).

In contrast to anionic living polymerization, group transfer polymerization can be performed at room temperature or above. The preparation of block copolymers by utilizing only the group transfer polymerization mechanism is limited to MA-type polymers. Both batch process and sequential monomer addition techniques were used. Block and graft copolymers of methacrylates with nonacrylic monomers using group transfer polymerization in conjunction with other polymerization routes can be realized via the transformation approach. In principle, any polymer independent of polymerization mode, possessing terminal or side-chain silyl keten acetal groups, would act as macroinitiators for group transfer polymerization. Ester groups were readily converted to silyl keten acetal macroinitiators by sequential treatment with lithium diisopropylamide (LDA) and chlorotrimethylsilane [372, 373] (Scheme 13.78).

Typical examples of this procedure were reported by Ruth et al. [373] and Verma et al. [21]. They prepared PIB or poly(alkyl vinyl ether) oligomer by living cationic polymerization. The hydroxyl end-groups were first converted to ester groups and subsequently the above-described chemistry was applied to afford the silyl keten acetal macroinitiator. Reaction of the polymeric silyl keten acetal with MMA in the presence of tetrabutylammonium dibenzoate (TBADB) produced a block copolymer (Scheme 13.79).

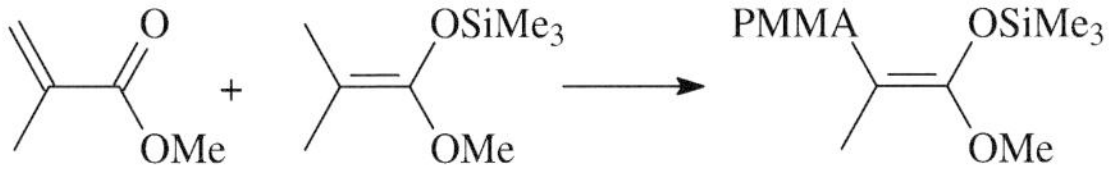

Scheme 13.77

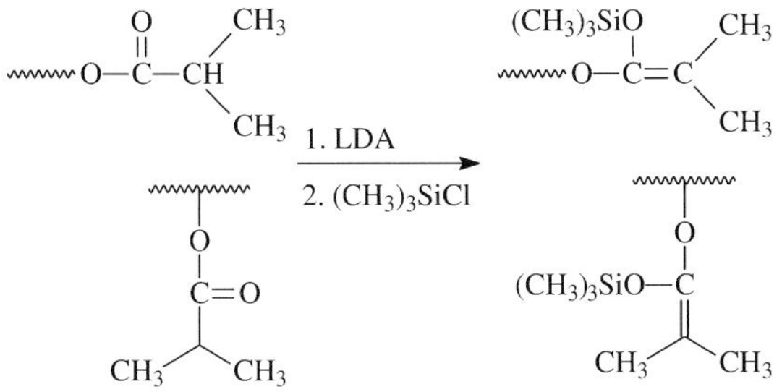

Scheme 13.78

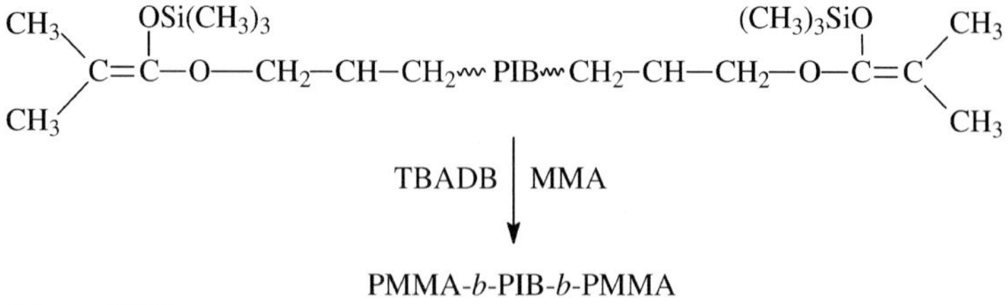

$$\underset{CH_3}{\overset{CH_3}{>}}C=\underset{}{\overset{OSi(CH_3)_3}{C}}-O-CH_2-CH-CH_2\text{\textasciitilde\textasciitilde}PIB\text{\textasciitilde\textasciitilde}CH_2-CH-CH_2-O-\underset{}{\overset{(CH_3)_3SiO}{C}}=C\underset{CH_3}{\overset{CH_3}{<}}$$

$$\xrightarrow{\text{TBADB} \mid \text{MMA}}$$

PMMA-*b*-PIB-*b*-PMMA

Scheme 13.79

Although MMA conversion was quantitative, a significant amount of unreacted PIB was reported by the authors, so contamination problems exist. The coupling reactions of the two corresponding living homopolymers also yielded block copolymers with limited success [374]. Coupling efficiencies up to 80% could be obtained. However, both studies confirmed the potential use of transformation reactions for the preparation of thermoplastic elastomers.

Jenkins et al. [372] applied this approach to prepare St, MMA graft copolymers. In this case, PSt, containing a minor portion of residues with functional groups, served as macroinitiator for the subsequent group transfer polymerization.

PMMA–Br samples prepared by group transfer polymerization and subsequent bromination were used to initiate sequential conventional free radical [375] and ATRP [169] to yield triblock copolymers.

13.4
Coupling Reactions and Concurrent Polymerizations

As stated in the Introduction, the direct coupling of preformed living blocks (usually cation and anion or group transfer) also permits the formation of block copolymers (Table 13.11). Typical example of such a coupling process between oppositely charged macroions is presented in Scheme 13.80 for the preparation of polystyrene-*b*-poly(ethyl vinyl ether) (PSt-*b*-PEVE).

Table 13.11 Summary of coupling reactions.

Type of coupling	Block copolymer [a]	Ref.
Anionic-cationic	PSt-*b*-PTHF	20
Anionic-cationic	PSt-*b*-PEVE	376
Anionic-cationic	PMMA-*b*-PTHF	377
Anionic-cationic	Poly(glycopeptide)-*b*-PMeOZO	378
Group transfer-cationic	PMMA-*b*-PBVE	21

a) PSt, polystyrene; PTHF, polytetrahydrofuran; PEVE, poly(ethyl vinyl ether); PB, polybutadiene; PMMA, poly(methyl methacrylate); PEO, poly(ethylene oxide); PCL, poly(ε-caprolactone); PVP, poly(vinylpyrrolidone); PBA, poly(butyl acrylate); PIP, polyisoprene; PMA, poly(methyl acrylate); PFS, poly(ferrocenyldimethylsilanes).

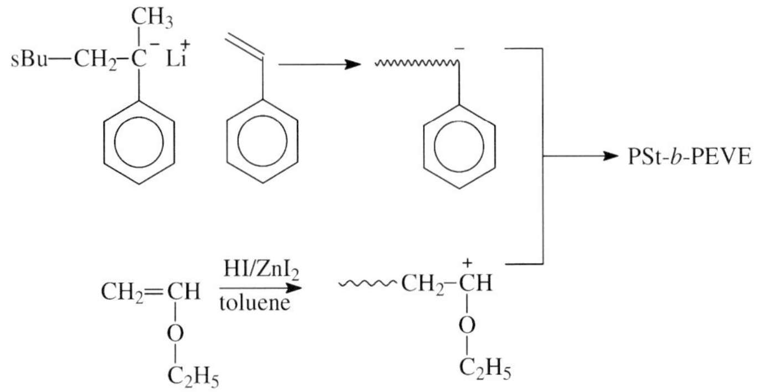

Scheme 13.80

Another synthetic scheme to produce block copolymers comprised of monomers that are polymerizable by different polymerization mechanisms is concurrent polymerizations in one step. In this concept, a single initiator (also called as bifunctional, dual or double-head initiator) is used to perform two mechanistically distinct polymerizations, without the need for intermediate transformation or activation steps. Sogah and coworkers first reported the synthesis of multifunctional initiators possessing initiating sites for different types of polymerization and their use in the synthesis of block and graft copolymers [286, 379].

This concept was further developed by Hawker et al. [380], who performed dual living polymerizations from a single initiating molecule without a requirement for additional reaction. The compatibility of either NMP or ATRP with living ring-opening polymerization of ε-caprolactone was demonstrated by the synthesis of a variety of well-defined block copolymers. The basic strategy followed for the dual polymerization is demonstrated in Scheme 13.81.

It is interesting that block copolymers with low molecular weight distributions were prepared with either sequences, i.e. living radical polymerization or living ring-opening polymerization first.

Similarly, hydroxy-functionalized ATRP initiators can be used as bifunctional initiators for the polymerization of both CL and a variety of vinyl monomers (Scheme 13.82). The novel block copolymers obtained possessed low polydispersities and controllable molecular weights for both of the blocks.

Lim et al. used a palladium complex for the cationic polymerization of THF and ring-opening metathesis polymerization of norbornene [381]. It was also demonstrated that even condensation and chain polymerizations can be performed simultaneously in one step. This was achieved by the use of unimolecular compounds which can simultaneously act both as an initiator for chain polymerization and as an end-capper for condensation polymerization. The method introduced provides a simple way to combine ring-opening polymerization, NMP or ATRP with a condensation polymerization to yield interesting and useful block copolymers [382].

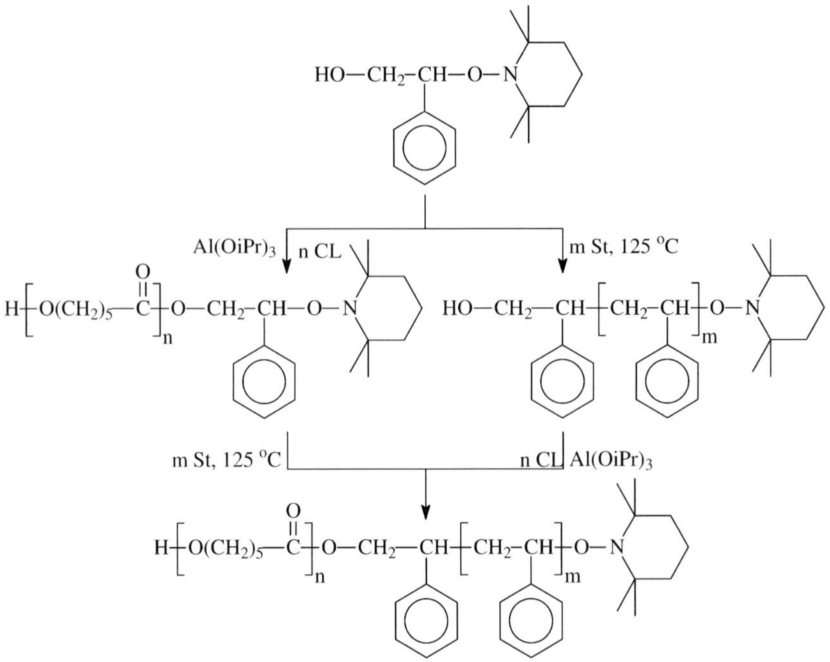

Scheme 13.81

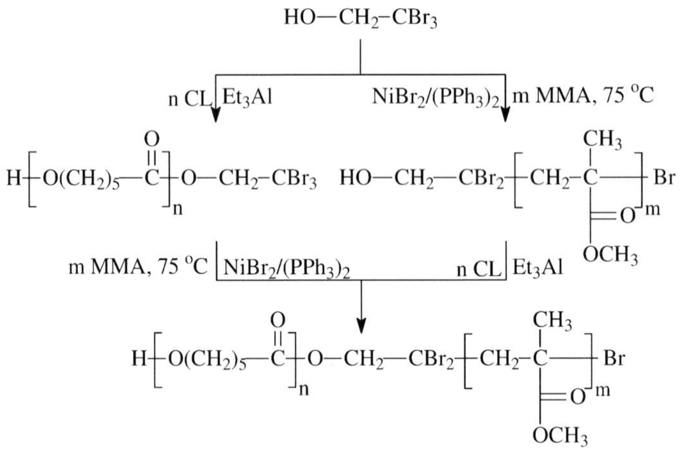

Scheme 13.82

13.5
Conclusions

Using transformation reactions, i.e. combining different polymerization mechanisms, novel polymeric materials may be synthesized from new and existing monomers. As shown in this chapter, a full range of possible block and graft copolymers built from monomers with different chemical structures are accessible through transformation reactions. Transformations can be achieved not only between different polymerization methods, but also within the same mechanism using different initiating systems. For example, ATRP is combined with conventional, photoinitiated and NMP, radical polymerization methods. It should be pointed out that the transformation within the same polymerization process is not limited only to free radical polymerization. For example, transformations involving vinyl and ring-opening anionic or cationic polymerizations are also possible (Table 13.12).

As stated previously and with some representative examples, the transformation approach can be applied to the preparation of segmented graft copolymer via the macromonomer technique. In this case, macromonomers were prepared with one mode of polymerization and (co)polymerized with another mechanism. Recent examples of such transformations have been reported for various polymerization mechanisms including ATRP, FRP, AROP and RAFT [415–423].

It is also possible to combine several polymerization methods to form more complex structures. ABC-type miktoarm star polymers can be prepared with a core-out method via a combination of ROP, NMP and ATRP [424]. It is clear that the transformation reactions will continue to attract interest in the near future because of the possibility of the various newly developed "living"/controlled polymerization mechanisms. It should be possible to design and synthesize materials having precise structures with desired properties by combination of such mechanisms.

Table 13.12 Some examples of polymers obtained by transformation involving the same polymerization mechanisms.

Transformation [a]	Ref.
FRP-FRP	383–391
FRP-ATRP	392–397
ATRP-FRP	398–400
NMP-ATRP	401–403
Cobalt mediated ATRP	404
DT-ATRP	366
Anionic vinyl-AROP	405–411
Cationic vinyl-CROP	412–414

a) FRP, conventional free radical polymerization; DT, degenerative transfer polymerization; AROP, anionic ring-opening polymerization; CROP, cationic ring-opening polymerization.

References

1 Clough, S. B., Schneider, N. S., King, A. O. *J. Macromol. Sci. Rev.* **1968**, *2*, 641.
2 Seymour, R. W., Estes, G. M., Cooper, S. L. *Macromolecules* **1970**, *3*, 579.
3 Xu, M., MacKnight, W. J., Chen, C. H. Y., Thomas, E. L. *Polymer* **1983**, *24*, 1329.
4 Hoeschele, G. K., Witisepe, W. K. *Angew. Makromol. Chem.* **1973**, *29/30*, 267.
5 Lilaonitkul, A., West, J. C., Cooper, S. L. *J. Macromol. Sci. Phys.* **1976**, *12*, 563.
6 Ma, D., Zhang, G., Huang, Z., Luo, X. *J. Polym. Sci. Part A: Polym. Chem.* **1998**, *36*, 2961.
7 Saiani, A., Rochas, C., Eeckhaut, G., Daunch, W. A., Leenslag, J.-W., Higgins, J. S. *Macromolecules* **2004**, *37*, 1411.
8 Lenz, R. W., Go, S. *J. Polym. Sci. Polym. Chem. Ed.* **1974**, *12*, 1.
9 Iwakura, Y., Taneda, Y., Uchida, S. *J. Appl. Polym. Sci.* **1961**, *5*, 108.
10 Garcia-Gaitan, B., Perez-Gonzalez, M. del P., Martinez-Richa, A., Luna-Barcenas, G., Nuno-Donlucas, S. M. *J. Polym. Sci. Part A: Polym. Chem.* **2004**, *42*, 4448.
11 Spontak, R. J., Smith, S. D., Satkowski, M. M., Ashraf, A., Zielinski, J. M. in *Polymer Solutions Blends and Interfaces*, Noda, I., Rubingh, D. N. (Eds). Elsevier, Amsterdam, **1992**, pp. 65–88.
12 Hadjichristidis, N., Pispas, S. S., Floudas, G. *Block Copolymers: Synthetic Strategies Physical Properties and Applications* Chapter 6. Wiley, Hoboken, NJ, **2003**, p. 91.
13 Spontak, R. J., Smith, S. D. *J. Polym. Sci. Part B: Polym. Phys.* **2001**, *39*, 947–955.
14 Burgess, F. J., Cunliffe, A. V., MacCallum, J. R., Richards, D. H. *Polymer* **1977**, *18*, 719–725.
15 Burgess, F. J., Cunliffe, A. V., MacCallum, J. R., Richards, D. H. *Polymer* **1977**, *18*, 726–732.
16 Burgess, F. J., Cunliffe, A. V., Dawkins, J. V., Richards, D. H. *Polymer* **1977**, *18*, 733–740.
17 Hadjichristidis, N., Pitsikalis, M., Iatrou, H. *Adv. Polym. Sci.* **2005**, *189*, 1.
18 Yagci, Y. in *Advanced Functional Molecules and Polymers*, Nalwa, H. S. (Ed.). Gordon & Breach, New York, **2001**, Vol. 1, p. 233.
19 Hadjichristidis, N., Pispas, S., Floudas, G. *Block Copolymers: Synthetic Strategies Physical Properties and Applications.* Wiley, Hoboken, NJ, **2003**.
20 Richards, D. H., Kingston, S. B., Souel, T. *Polymer* **1978**, *19*, 68–72.
21 Verma, A., Nielsen, A., McGrath, J. E., Riffle, J. S. *Polym. Bull.* **1990**, *23*, 563–570.
22 Nasser-Eddine, M., Delaite, C., Hurtrez, G., Dumas, P. *Eur. Polym. J.* **2005**, *41*, 313–318.
23 Nomura, R., Endo, T. *Macromolecules* **1994**, *27*, 5523–5526.
24 Nomura, R., Narita, M., Endo, T. *Macromolecules* **1994**, *27*, 7011–7014.
25 Nomura, R., Endo, T. *Macromolecules* **1995**, *28*, 1754–1757.
26 Nomura, R., Shibasaki, Y., Endo, T. *Polym. Bull.* **1996**, *37*, 597–601.
27 Guo, H. Q., Kajiwara, A., Morishima, Y., Kamachi, M. *Macromolecules* **1996**, *29*, 2354–2358.
28 Fradet, A. in *The Polymeric Materials Encyclopedia*, Salamone, J. C. (Ed.). CRC Press, Boca Raton, FL, **1996**, p. 797.
29 Yagci, Y., Nuyken, O., Graubner, V. in *Encyclopedia of Polymer Science and Technology*, 3rd edn, Kroschwitz, J. I. (Ed.). Wiley, New York, **2005**, Vol. 12, pp. 57–130.
30 Matsukawa, K., Ueda, A., Inoue, H. *J. Polym. Sci. Part A: Polym. Chem.* **1990**, *28*, 2107–2114.
31 Ueda, A., Nagai, S. *J. Polym. Sci. Part A: Polym. Chem.* **1986**, *24*, 405–418.
32 Simon, J., Bajpai, A. *J. Appl. Polym. Sci.* **2001**, *82*, 2922–2933.
33 Terada, H., Haneda, Y., Ueda, A., Nagai, S. *J. Macromol. Sci. Pure Appl. Chem.* **1994**, *A31*, 173–178.
34 Ueda, A., Nagai, S. *Kobunshi Ronbunshu* **1986**, *43*, 97–103.
35 Nuyken, O., Dauth, J., Pekruhn, W. *Angew. Makromol. Chem.* **1991**, *190*, 81–98.
36 Ueda, A., Nagai, S. *J. Polym. Sci. Part A: Polym. Chem.* **1984**, *22*, 1611–1621.
37 Kinoshita, H., Ooka, M., Tanaka, N., Araki, T. *Kobunshi Ronbunshu* **1993**, *50*, 147–157.
38 Cheikhalard, T., Massardier, V., Tighzert, L., Pascault, J. P. *J. Appl. Polym. Sci.* **1998**, *70*, 613–627.

39 Kinoshita, H., Tanaka, N., Araki, T., Syoji, A., Ooka, M. *Makromol. Chem. Macromol. Chem. Phys.* **1993**, *194*, 829–839.
40 Takahashi, H., Ueda, A., Nagai, S. *J. Polym. Sci. Part A: Polym. Chem.* **1997**, *35*, 69–76.
41 Yuruk, H., Ulupinar, S. *Angew. Makromol. Chem.* **1993**, *213*, 197–206.
42 Hazer, B. *Makromol. Chem. Macromol. Chem. Phys.* **1992**, *193*, 1081–1086.
43 Haneda, Y., Terada, H., Yoshida, M., Ueda, A., Nagai, S. *J. Polym. Sci. Part A: Polym. Chem.* **1994**, *32*, 2641–2652.
44 Ueda, A., Shimada, M., Agari, Y., Nagai, S. *Kobunshi Ronbunshu* **1994**, *51*, 453–458.
45 Ikeda, I., Hirose, T., Suzuki, K. *Sen-I Gakkaishi* **1997**, *53*, 111–114.
46 Chang, T.C., Chen, H.B., Chen, Y.C., Ho, S.Y. *J. Polym. Sci. Part A: Polym. Chem.* **1996**, *34*, 2613–2620.
47 Hazer, B., Erdem, B., Lenz, R.W. *J. Polym. Sci. Part A: Polym. Chem.* **1994**, *32*, 1739–1746.
48 Hepuzer, Y., Guvener, U., Yagci, Y. *Polym. Bull.* **2004**, *52*, 17–23.
49 Nagamune, T., Ueda, A., Nagai, S. *J. Appl. Polym. Sci.* **1996**, *62*, 359–365.
50 Cakmak, I., Hazer, B., Yagci, Y. *Eur. Polym. J.* **1991**, *27*, 101–103.
51 Eroglu, M.S., Hazer, B., Baysal, B.M. *J. Appl. Polym. Sci.* **1998**, *68*, 1149–1157.
52 Ahn, T.O., Kim, J.H., Lee, J.C., Jeong, H.M., Park, J.Y. *J. Polym. Sci. Part A: Polym. Chem.* **1993**, *31*, 435–441.
53 Ahn, T.O., Kim, J.H., Jeong, H.M., Lee, S.W., Park, L.S. *J. Polym. Sci. Part B: Polym. Phys.* **1994**, *32*, 21–28.
54 Ohishi, H., Ohwaki, T., Nish, T. *J. Polym. Sci. Part A: Polym. Chem.* **1998**, *36*, 2839–2847.
55 Hirano, T., Amano, R., Fujii, T., Onimura, K., Tsutsumi, H., Oishi, T. *Polym. J.* **1999**, *31*, 864–871.
56 Alli, A., Hazer, B., Menceloglu, Y., Suzer, S. *Eur. Polym. J.* **2006**, *42*, 740–750.
57 Yagci, Y., Muller, M., Schnabel, W. *J. Macromol. Sci. Chem.* **1991**, *A28*, 37–46.
58 Yagci, Y., Hizal, G., Tunca, U. *Polym. Commun.* **1990**, *31*, 7–10.
59 Bamford, C.H., Ledwith, A., Yagci, Y. *Polymer* **1978**, *19*, 354–356.
60 Onen, A., Yagci, Y. *J. Macromol. Sci. Chem.* **1990**, *A27*, 743–753.
61 Onen, A., Yagci, Y. *Angew. Makromol. Chem.* **1990**, *181*, 191–197.
62 Yagci, Y., Onen, A. *J. Macromol. Sci. Chem.* **1991**, *A28*, 129–141.
63 Hepuzer, Y., Bektas, M., Denizligil, S., Onen, A., Yagci, Y. *J. Macromol. Sci. Pure Appl. Chem.* **1993**, *A30*, 111–115.
64 Onen, A., Denizligil, S., Yagci, Y. *Macromolecules* **1995**, *28*, 5375–5377.
65 Denizligil, S., Yagci, Y. *Polym. Bull.* **1989**, *22*, 547–551.
66 Yagci, Y., Denizligil, S. *Eur. Polym. J.* **1991**, *27*, 1401–1404.
67 Yagci, Y., Denizligil, S., Bicak, N., Atay, T. *Angew. Makromol. Chem.* **1992**, *195*, 89–95.
68 Denizligil, S., Bicak, N., Yagci, Y. *J. Macromol. Sci. Pure Appl. Chem.* **1992**, *29*, 293–301.
69 Yagci, Y., Mishra, M.K. in *Macromolecular Design: Concept and Practise*, Chapter 7. Mishra, M.K. (Ed.). Polymer Frontiers International, New York, **1994**.
70 Smets, G. *Polym. J.* **1985**, 17, 153–165.
71 Doi, T., Smets, G. *Macromolecules* **1989**, *22*, 25–29.
72 Lanza, E., Berghman, H., Smets, G. *J. Polym. Sci. Part B: Polym. Phys.* **1973**, *11*, 95–108.
73 Craubner, H. *J. Polym. Sci. Part A: Polym. Chem.* **1982**, *20*, 1935–1939.
74 Craubner, H. *J. Polym. Sci. Part A: Polym. Chem.* **1980**, *18*, 2011–2020.
75 Bamford, C.H., Middleton, I.P., Allamee, K.G., Paprotny, J. *Br. Polym. J.* **1987**, *19*, 269–274.
76 Matyjaszewski, K., Gaynor, S.G., Greszta, D., Mardare, D., Shigemoto, T., Wang, J.S. *Macromol. Symp.* **1995**, *95*, 217–231.
77 Hawker, C.J. *Trends Polym. Sci.* **1996**, *4*, 183–188.
78 Sawamoto, M., Kamigaito, M. *Trends Polym. Sci.* **1996**, *4*, 371–377.
79 Gaynor, S.G., Matyjaszewski, K. *Macromolecules* **1997**, *30*, 4241–4243.
80 Zhang, Y., Chung, I.S., Huang, J.Y., Matyjaszewski, K., Pakula, T. *Macromol. Chem. Phys.* **2005**, *206*, 33–42.
81 Shen, D.W., Shi, Y., Fu, Z.F., Yang, W.T., Cai, X.P., Lin, R.X., Zhang, D.S. *Polym. Bull.* **2003**, *49*, 321–328.

82 Gaynor, S.G., Edelman, S.Z., Matyjaszewski, K. *Polym. Prepr.* **1997**, *38*, 703–704.

83 Mennicken, M., Nagelsdiek, R., Keul, H., Höcker, H. *Macromol. Chem. Phys.* **2004**, *205*, 143–153.

84 Li, D., Sha, K., Li, Y., Liu, X., Wang, W., Wang, S., Xu, Y., Ai, P., Wu, M., Wang, J. *Polym. Bull.* **2006**, *56*, 111–117.

85 Tzanetos, N.P., Kallitsis, J.K. *J. Polym. Sci. Part A: Polym. Chem.* **2005**, *43*, 1049–1061.

86 Huang, X.Y., Lu, G.L., Peng, D., Zhang, S., Qing, F.L. *Macromolecules* **2005**, *38*, 7299–7305.

87 Stalmach, U., de Boer, B., Videlot, C., van Hutten, P.F., Hadziioannou, G. *J. Am. Chem. Soc.* **2000**, *122*, 5464–5472.

88 Yoshida, E., Nakamura, M. *Polym. J.* **1998**, *30*, 915–920.

89 Kricheldorf, H.R., Lee, S.R. *Macromolecules* **1995**, *28*, 6718–6725.

90 Kricheldorf, H.R., Lee, S.R., Bush, S. *Macromolecules* **1996**, *29*, 1375–1381.

91 Kricheldorf, H.R., Lee, S.R. *Macromolecules* **1996**, *29*, 8689–8695.

92 Kricheldorf, H.R., Eggerstedt, S. *Macromol. Chem. Phys.* **1998**, *199*, 283–290.

93 Kricheldorf, H.R., Lee, S.R., Schittenhelm, N. *Macromol. Chem. Phys.* **1998**, *199*, 273–282.

94 Kricheldorf, H.R., Eggerstedt, S. *J. Polym. Sci. Part A: Polym. Chem.* **1998**, *36*, 1373–1378.

95 Kricheldorf, H.R., Hauser, K. *Macromolecules* **1998**, *31*, 614–620.

96 Kricheldorf, H.R., Eggerstedt, S. *Macromolecules* **1998**, *31*, 6403–6408.

97 Lee, R.S., Li, H.R., Yang, J.M., Tsai, F.Y. *Polymer* **2005**, *46*, 10718–10726.

98 Berresheim, A.J., Muller, M., Mullen, K. *Chem. Rev.* **1999**, *99*, 1747–1785.

99 Francois, B., Widawski, G., Rawiso, M., Cesar, B. *Synth. Met.* **1995**, *69*, 463–466.

100 Cianga, I., Hepuzer, Y., Yagci, Y. *Macromol. Symp.* **2002**, *183*, 145–157.

101 Cianga, I., Yagci, Y. *Prog. Polym. Sci.* **2004**, *29*, 387–399.

102 Cianga, I., Yagci, Y. *Eur. Polym. J.* **2002**, *38*, 695–703.

103 Cianga, I., Hepuzer, Y., Yagci, Y. *Polymer* **2002**, *43*, 2141–2149.

104 Cianga, I., Yagci, Y. *Polym. Bull.* **2001**, *47*, 17–24.

105 Yurteri, S., Cianga, A., Demirel, A.L., Yagci, Y. *J. Polym. Sci. Part A: Polym. Chem.* **2005**, *43*, 879–896.

106 Yurteri, S., Cianga, I., Yagci, Y. *Design. Monom. Polym.* **2005**, *8*, 61–74.

107 Yurteri, S., Cianga, I., Degirmenci, M., Yagci, Y. *Polym. Int.* **2004**, *53*, 1219–1225.

108 Demirel, A.L., Yurteri, S., Cianga, I., Yagci, Y. *Macromolecules* **2005**, *38*, 6402–6410.

109 Schue, F. in *Comprehensive Polymer Science*, 2nd edn, Allen, G., Bevington, J.C. (Eds). Pergamon Press, Oxford, **1989**, Vol. 6, p. 10.

110 Yagci, Y., Mishra, M.K. in *The Encyclopedia of Polymeric Materials*, Salamone, J.C. (Ed.). CRC Press, Boca Raton, **1996**, Vol. 1, p. 789.

111 Bamford, C.H., Eastmond, G.C., Woo, J., Richards, D.H. *Polymer* **1982**, *23*, 643–645.

112 Nicolova-Anankova, Z., Palacin, F., Raviola, F., Riess, G. *Eur. Polym. J.* **1975**, *11*, 301–304.

113 Abadie, M.J.M., Schue, F., Souel, T. *Polymer* **1981**, *22*, 1076–1080.

114 Hazer, B., Cakmak, I., Kucukyavuz, S., Nugay, T. *Eur. Polym. J.* **1992**, *28*, 1295–1297.

115 Vinchon, Y., Reeb, R., Riess, G. *Eur. Polym. J.* **1976**, *12*, 317–321.

116 Riess, G., Reeb, R. *Polym. Prepr.* **1980**, *179*, 112.

117 Abadie, M.J.M., Ourahmoune, D., Mendjel, H. *Eur. Polym. J.* **1990**, *26*, 515–520.

118 Cunliffe, A.V., Hayes, G.F., Richards, D.H. *J. Polym. Sci. Part C: Polym. Lett.* **1976**, *14*, 483–488.

119 Lindsell, W.E., Service, D.M., Soutar, I., Richards, D.H. *Br. Polym. J.* **1987**, *19*, 255–262.

120 Eastmond, G.C., Parr, K.J., Woo, J. *Polymer* **1988**, *29*, 950–957.

121 Reeb, R., Vinchon, Y., Riess, G., Catala, J.M., Brossas, J. *Bull. Soc. Chim. Fr.* **1975**, 2717–2721.

122 Catala, J.M., Riess, G., Brossas, J. *Makromol. Chem.* **1977**, *178*, 1249.

123 Souel, T., Schue, F., Abadie, M., Richards, D. H. *Polymer* **1977**, *18*, 1292–1294.
124 Ren, Q., Zhang, H. J., Zhang, X. K., Huang, B. T. *J. Polym. Sci. Part A: Polym. Chem.* **1993**, *31*, 847–851.
125 Yagci, Y., Menceloglu, Y. Z., Baysal, B. M., Gungor, A. *Polym. Bull.* **1989**, *21*, 259–263.
126 Jenkins, A. D., Lappert, M. F., Srivasta, R. C. *Eur. Polym. J.* **1971**, *7*, 289.
127 Billingh, N., Boxall, L. M., Jenkins, A. D. *Eur. Polym. J.* **1972**, *8*, 1045.
128 Eastmond, G. C., Woo, J. *Polymer* **1990**, *31*, 358–361.
129 Vlasov, G. P., Rudkovskaya, G. D., Ovsyannikova, L. A. *Makromol. Chem.* **1982**, *183*, 2635–2644.
130 Lele, B. S., Leroux, J. C. *Polymer* **2002**, *43*, 5595–5605.
131 Sui, K. Y., Gu, L. X. *J. Appl. Polym. Sci.*, **2003**, *89*, 1753–1759.
132 Huang, J. L., Huang, X. Y., Zhang, S. *Macromolecules* **1995**, *28*, 4421–4425.
133 Lu, Z. J., Wan, D. C., Huang, J. L. *J. Appl. Polym. Sci.* **1999**, *74*, 2072–2076.
134 Huang, X. Y., Chen, S., Huang, J. L. *J. Polym. Sci. Part A: Polym. Chem.* **1999**, *37*, 825–833.
135 Solomon, D. H., Rizzardo, E., Cacioli, P. *US Patent 4 581 429*, **1986**.
136 Rizzardo, E. *Chem. Aust.* **1987**, *54*, 32.
137 Georges, M. K., Veregin, R. P., Kazmaier, P. M., Hamer, G. K. *Polym. Mater. Sci. Eng.* **1993**, *6*, 68.
138 Georges, M. K., Veregin, R. P. N., Kazmaier, P. M., Hamer, G. K. *Macromolecules* **1993**, *26*, 2987–2988.
139 Veregin, R. P. N., Georges, M. K., Kazmaier, P. M., Hamer, G. K. *Macromolecules* **1993**, *26*, 5316–5320.
140 Ouhadi, T., Hamitou, A., Jerome, R., Teyssie, P. *Macromolecules* **1976**, *9*, 927–931.
141 Kricheldorf, H. R., Mang, T., Jonte, J. M. *Macromolecules* **1984**, *17*, 2173–2181.
142 Endo, M., Aida, T., Inoue, S. *Macromolecules* **1987**, *20*, 2982–2988.
143 Shen, Y. Q., Shen, Z. Q., Shen, J. L., Zhang, Y. F., Yao, K. M. *Macromolecules* **1996**, *29*, 3441–3446.
144 Stevels, W. M., Ankone, M. J. K., Dijkstra, P. J., Feijen, J. *Macromolecules* **1996**, *29*, 8296–8303.
145 Hamitou, A., Jerome, R., Hubert, A. J., Teyssie, P. *Macromolecules* **1973**, *6*, 651–652.
146 Ouhadi, T., Stevens, C., Teyssie, P. *J. Appl. Polym. Sci.* **1976**, *20*, 2963–2970.
147 Vion, J. M., Jerome, R., Teyssie, P., Aubin, M., Prudhomme, R. E. *Macromolecules* **1986**, *19*, 1828–1838.
148 Ropson, N., Dubois, P., Jerome, R., Teyssie, P. *J. Polym. Sci. Part A: Polym. Chem.* **1997**, *35*, 183–192.
149 Yoshida, E., Osagawa, Y. *Macromolecules* **1998**, *31*, 1446–1453.
150 Hua, F. J., Yang, Y. L. *Polymer* **2001**, *42*, 1361–1368.
151 Yoshida, E., Ishizone, T., Hirao, A., Nakahama, S., Takata, T., Endo, T. *Macromolecules* **1994**, *27*, 3119–3124.
152 Kobatake, S., Harwood, H. J., Quirk, R. P., Priddy, D. B. *Macromolecules* **1998**, *31*, 3735–3739.
153 Miura, Y., Hirota, K., Moto, H., Yamada, B. *Macromolecules*, **1999**, *32*, 8356–8362.
154 Morgan, A. M., Pollack, S. K., Beshah, K. *Macromolecules* **2002**, *35*, 4238–4246.
155 Wang, Y. B., Chen, S., Huang, J. L. *Macromolecules* **1999**, *32*, 2480–2483.
156 Lu, G. Q., Jia, Z. F., Yi, W., Huang, J. L. *J. Polym. Sci. Part A: Polym. Chem.* **2002**, *40*, 4404–4409.
157 Endo, T., Takuma, K., Takata, T., Hirose, C. *Macromolecules* **1993**, *26*, 3227–3229.
158 Cianga, I., Senyo, T., Ito, K., Yagci, Y. *Macromol. Rapid Commun.* **2004**, *25*, 1697–1702.
159 Acar, M. H., Matyjaszewski, K. *Macromol. Chem. Phys.* **1999**, *200*, 1094–1100.
160 Ramakrishnan, A., Dhamodharan, R. *J. Macromol. Sci. Pure Appl. Chem.* **2000**, *37*, 621–631.
161 Tong, J. D., Ni, S. R., Winnik, M. A. *Macromolecules* **2000**, *33*, 1482–1486.
162 Tong, J. D., Zhou, C. L., Ni, S. R., Winnik, M. A. *Macromolecules* **2001**, *34*, 696–705.
163 Liu, B., Liu, F., Luo, N., Ying, S. K., Liu, Q. *Chin. J. Polym. Sci.* **2000**, *18*, 39–43.
164 Korczagin, I., Hempenius, M. A., Vancso, G. J. *Macromolecules* **2004**, *37*, 1686–1690.

165 Angot, B., Taton, D., Gnanou, Y. *Macromolecules* **2000**, *33*, 5418–5426.
166 Mahajan, S., Renker, S., Simon, P. F. W., Gutmann, J. S., Jain, A., Gruner, S. M., Fetters, L. J., Coates, G. W., Wiesner, U. *Macromol. Chem. Phys.* **2003**, *204*, 1047–1055.
167 Peleshanko, S., Jeong, J., Shevchenko, V. V., Genson, K. L., Pikus, Y., Ornatska, M., Petrash, S., Tsukruk, V. V. *Macromolecules* **2004**, *37*, 7497–7506.
168 Nasser-Eddine, M., Reutenauer, S., Delaite, C., Hurtrez, G., Dumas, P. *J. Polym. Sci. Part A: Polym. Chem.* **2004**, *42*, 1745–1751.
169 Masar, B., Vlcek, P., Kriz, J. *J. Appl. Polym. Sci.* **2001**, *81*, 3514–3522.
170 Masar, B., Janata, M., Vlcek, P., Policka, P., Toman, L. *Macromol. Symp.* **2002**, *183*, 139–144.
171 Jankova, K., Kops, J., Chen, X. Y., Batsberg, W. *Macromol. Rapid Commun.* **1999**, *20*, 219–223.
172 Mahajan, S., Cho, B. K., Allgaier, A., Fetters, L. J., Coates, G. W., Wiesner, U. *Macromol. Rapid Commun.* **2004**, *25*, 1889–1894.
173 Chen, J., Zhang, H. L., Chen, J. F., Wang, X. Z., Wang, X. Y. *J. Macromol. Sci. Pure Appl. Chem.* **2005**, *A42*, 1247–1257.
174 Sha, K., Qin, L., Li, D. S., Liu, X. T., Wang, J. Y. *Polym. Bull.* **2005**, *54*, 19.
175 Jakubowski, W., Lutz, J.-F., Slomkowski, S., Matyjaszewski, K. *J. Polym. Sci. Part A: Polym. Chem.* **2005**, *43*, 1498–1510.
176 Slomkowski, S., Gadzinowski, M., Sosnowski, S., De Vita, C., Pucci, A., Ciardelli, F., Jakubowski, W., Matyjaszewski, K. *Macromol. Symp.* **2005**, *226*, 239–252.
177 Zheng, G., Stöver, H. D. H. *Macromolecules* **2003**, *36*, 7439–7445.
178 Chen, Y. M., Wulff, G. *Macromol. Rapid Commun.* **2002**, *23*, 59–63.
179 Magbitang, T., Lee, V. Y., Connor, E. F., Sundberg, L. K., Kim, H. C., Volksen, W., Hawker, C. J., Miller, R. D., Hedrick, J. L. *Macromol. Symp.* **2004**, *215*, 295–305.
180 Yuan, W. Z., Huang, X. B., Tang, X. Z. *Polym. Bull.* **2005**, *55*, 225–233.
181 Johnson, R. M., Fraser, C. L. *Macromolecules* **2004**, *37*, 2718–2727.
182 Jia, Z. F., Zhou, Y. F., Yan, D. Y. *J. Polym. Sci. Part A: Polym. Chem.* **2005**, *43*, 6534–6544.
183 Hedrick, J. L., Trollsås, M., Hawker, C. J., Atthoff, B., Claesson, H., Heise, A. *Macromolecules* **1998**, *31*, 8691–8705.
184 Miller, P. J., Matyjaszewski, K. *Macromolecules* **1999**, *32*, 8760–8767.
185 Kurjata, J., Chojnowski, J., Yeoh, C.-T., Rossi, N. A. A., Holder, S. J. *Polymer* **2004**, *45*, 6111–6121.
186 Zhao, Y., Shuai, X., Chen, C., Xi, F. *Chem. Commun.* **2004**, 1608–1609.
187 Zhao, Y. L., Shuai, X. T., Chen, C. F., Xi, F. *Macromolecules* **2004**, *37*, 8854–8862.
188 Messman, J. M., Scheuer, A. D., Storey, R. F., *Polymer* **2005**, *46*, 3628–3638.
189 Shi, L. J., Chapman, T. M., Beckman, E. J. *Macromolecules* **2003**, *36*, 2563–2567.
190 Hong, C.-Y., You, Y-Z., Pan, C.-Y. *J. Polym. Sci. Part A: Polym. Chem.* **2004**, *42*, 4873–4881.
191 Rzayev, J., Hillmyer, M. A. *Macromolecules* **2005**, *38*, 3–5.
192 Donkers, E. H. D., Willemse, R. X. E., Klumperman, B. *J. Polym. Sci. Part A: Polym. Chem.* **2005**, *43*, 2536–2545.
193 Jagur-Grodzinski, J. *J. Polym. Sci. Part A: Polym. Chem.* **2002**, *40*, 2116–2133.
194 Lutz, J. F., Jahed, N., Matyjaszewski, K. *J. Polym. Sci. Part A: Polym. Chem.* **2004**, *42*, 1939–1952.
195 Guo, Z. R., Wan, D. C., Huang, J. L., *Macromol. Rapid Commun.* **2001**, *22*, 367–371.
196 Yagci, Y., *Polym. Commun.* **1986**, *27*, 21–22.
197 Denizligil, S., Baskan, A., Yagci, Y. *Macromol. Rapid Commun.* **1995**, *16*, 387–391.
198 Yagci, Y. *Polym. Commun.* **1985**, *26*, 7–8.
199 Chiellini, E., Galli, G., Serhatli, E. I., Yagci, Y., Kaus, M., Angeloni, A. S. *Ferroelectrics* **1993**, *148*, 311.
200 Chiellini, E., Galli, G., Angeloni, A. S., Laus, M., Bignozzi, M. C., Yagci, Y., Serhatli, E. I. *Macromol. Symp.* **1994**, *77*, 349–358.
201 Serhatli, I. E., Galli, G., Yagci, Y., Chiellini, E. *Polym. Bull.* **1995**, *34*, 539–546.

202 Hepuzer, Y., Serhatli, I. E., Yagci, Y., Galli, G., Chiellini, E. *Macromol. Chem. Phys.* **2001**, *202*, 2247–2252.
203 Akar, A., Aydogan, A. C., Talinli, N., Yagci, Y. *Polym. Bull.* **1986**, *15*, 293–296.
204 Hizal, G., Yagci, Y., Schnabel, W. *Polymer* **1994**, *35*, 2428–2431.
205 Nuyken, O., Kroner, H., Aechtner, S. *Makromol. Chem. Rapid Commun.* **1988**, *9*, 671–679.
206 Vollmert, B., Bolte, H. *Makromol. Chem.* **1960**, *36*, 17.
207 Mishra, M. K. in *Recent Advances in Macromolecular Engineering*, Mishra, M. K., Nuyken, O., Kobayashi, S., Yagci, Y., Sar, B. (Eds). Plenum Press, New York, **1995**, p. 143.
208 Nuyken, O., Kroner, H., Aechtner, S. *Makromol. Chem. Macromol. Symp.* **1990**, *32*, 181–197.
209 D'Haese, F., Goethals, E. J., Tezuka, Y., Imai, K. *Makromol. Chem. Rapid Commun.* **1986**, *7*, 165–170.
210 Savaskan, S., Hazer, B. *Angew. Makromol. Chem.* **1996**, *239*, 13–26.
211 Arslan, H., Hazer, B. *Eur. Polym. J.* **1999**, *35*, 1451–1455.
212 Macit, H., Hazer, B. *J. Appl. Polym. Sci.* **2004**, *93*, 219–226.
213 Niwa, M., Higashi, N. *Macromolecules* **1989**, *22*, 1000.
214 Yagci, Y., Ledwith, A. *J. Polym. Sci. Part A: Polym. Chem.*, **1988**, *26*, 1911–1918.
215 Acar, M. H., Gulkanat, A., Seyren, S., Hizal, G. *Polymer* **2000**, *41*, 6709–6713.
216 Acar, M. H., Kucukoner, M. *Polymer* **1997**, *38*, 2829–2833.
217 Goethals, E. J., Haucourt, N. H., Verheyen, A. M., Habimana, J. *Makromol. Chem. Rapid Commun.* **1990**, *11*, 623–627.
218 Demircioglu, P., Acar, M. H., Yagci, Y. *J. Appl. Polym. Sci.* **1992**, *46*, 1639–1643.
219 Lin, C. H., Matyjaszewski, K. *Polym. Prepr.* **1993**, *206*, 56.
220 Yoshida, E., Sugita, A. *Macromolecules* **1996**, *29*, 6422–6426.
221 Yoshida, E., Sugita, A. *J. Polym. Sci. Polym. Chem.* **1998**, *36*, 2059–2068.
222 Yagci, Y., Duz, A. B., Onen, A. *Polymer* **1997**, *38*, 2861–2863.
223 Denizligil, S., Yagci, Y. *Design. Monom. Polym.* **1998**, *1*, 121.
224 Hawker, C. J., Barclay, G. G., Orellana, A., Dao, J., Devonport, W. *Macromolecules* **1996**, *29*, 5245–5254.
225 Kajiwara, A., Matyjaszewski, K. *Macromolecules* **1998**, *31*, 3489–3493.
226 Coca, S., Matyjaszewski, K. *Macromolecules* **1997**, *30*, 2808–2810.
227 Xu, Y. J., Pan, C. Y. *Macromolecules* **2000**, *33*, 4750–4756.
228 Feng, X. S., Pan, C. Y. *Macromolecules* **2002**, *35*, 2084–2089.
229 Guo, Y. M., Pan, C. Y. *Polymer* **2001**, *42*, 2863–2869.
230 Bernaerts, K. V., Schacht, E. H., Goethals, E. J., Du Prez, F. E. *J. Polym. Sci. Part A: Polym. Chem.* **2003**, *41*, 3206–3217.
231 Lu, J., Liang, H., Zhang, W., Cheng, Q. *J. Polym. Sci. Part A: Polym. Chem.* **2003**, *41*, 1237–1242.
232 Lu, J., Liang, H., Li, A. L., Cheng, Q. *Eur. Polym. J.* **2004**, *40*, 397–402.
233 Coca, S., Matyjaszewski, K. *J. Polym. Sci. Part A: Polym. Chem.* **1997**, *35*, 3595–3601.
234 Chen, X. Y., Ivan, B., Kops, J., Batsberg, W. *Macromol. Rapid Commun.* **1998**, *19*, 585–589.
235 Jankova, K., Kops, J., Chen, X., Gao, B., Batsberg, W. *Polym. Bull.* **1998**, *41*, 639.
236 Keszler, B., Fenyvesi, G., Kennedy, J. P. *J. Polym. Sci. Part A: Polym. Chem.* **2000**, *38*, 706–714.
237 Breland, L. K., Murphy, J. C., Storey, R. F. *Polymer* **2006**, *47*, 1852–1860.
238 Storey, R. F., Scheuer, A. D., Achord, B. C. *Polymer* **2005**, *46*, 2141–2152.
239 Hong, S. C., Pakula, T., Matyjaszewski, K. *Macromol. Chem. Phys.* **2001**, *202*, 3392–3402.
240 Vlasov, G. P., Rudkovskaya, G. D., Ovsyannikova, L. A. *Makromol. Chem.-Macromol. Chem. Phys.* **1982**, *183*, 2635–2644.
241 Chung, T. W., Cho, K. Y., Lee, H. C., Nah, J. W., Yeo, J. H., Akaike, T., Cho, C. S. *Polymer* **2004**, *45*, 1591–1597.
242 Luo, L. B., Ranger, M., Lessard, D. G., Le Garrec, D., Gori, S., Leroux, J. C., Rimmer, S., Smith, D. *Macromolecules* **2004**, *37*, 4008–4013.

243 Tanaka, M., Mori, A., Imanishi, Y., Bamford, C. H. *Int. J. Biol. Macromol.* **1985**, *7*, 173–181.
244 Higuchi, S., Mozawa, T., Maeda, M., Inoue, S. *Macromolecules* **1986**, *19*, 2263–2267.
245 Mulvaney, J. E., Ottaviani, R. O. *J. Polym. Sci. Part A: Polym. Chem.* **1982**, *20*, 1941–1942.
246 Ishikawa, S., Ishizu, K., Fukutomi, T. *Polym. Bull.* **1986**, *16*, 223–228.
247 Rempp, P., Lutz, P., Masson, P., Chaumont, P., Franta, E. *Makromol. Chem. Macromol. Chem. Phys.* **1985**, 47–66.
248 Ito, K., Hashimura, K., Itsuno, S., Yamada, E. *Macromolecules* **1991**, *24*, 3977–3981.
249 Jannasch, P., Wesslen, B. *J. Polym. Sci. Part A: Polym. Chem.* **1993**, *31*, 1519–1529.
250 Tao, L., Luan, B., Pan, C. Y. *Polymer* **2003**, *44*, 1013–1020.
251 Dong, C. M., Faucher, K. M., Chaikof, E. L. *J. Polym. Sci. Part A: Polym. Chem.* **2004**, *42*, 5754–5765.
252 Dong, C. M., Sun, X. L., Faucher, K. M., Apkarian, R. P., Chaikof, E. L. *BioMacromolecules* **2004**, *5*, 224–231.
253 Ydens, I., Degee, P., Dubois, P., Libiszowski, J., Duda, A., Penczek, S. *Macromol. Chem. Phys.* **2003**, *204*, 171–179.
254 Shinoda, H., Matyjaszewski, K. *Macromolecules* **2001**, *34*, 6243–6248.
255 Mecerreyes, D., Atthoff, B., Boduch, K. A., Trollsas, M., Hedrick, J. L. *Macromolecules* **1999**, *32*, 5175–5182.
256 Francis, R., Taton, D., Logan, J. L., Masse, P., Gnanou, Y., Duran, R. S. *Macromolecules* **2003**, *36*, 8253–8259.
257 Han, D.-H., Pan, C. Y. *J. Polym. Sci. Part A: Polym. Chem.* **2006**, *44*, 2794–2801.
258 Abraham, S., Ha, C.-S., Kim, I. *J. Polym. Sci. Part A: Polym. Chem.* **2006**, *44*, 2774–2783.
259 Babin, J., Leroy, C., Lecommandoux, S., Borsali, R., Gnanou, Y., Taton, D. *Chem. Commun.* **2005**, 1993–1995.
260 Nicora, C., Borsini, G., Ratti, L. *J. Polym. Sci. Part C Polym. Lett.* **1966**, *4*, 151.
261 Hizal, G., Tasdemir, H., Yagci, Y. *Polymer* **1990**, *31*, 1803–1806.
262 Cai, G. F., Yan, D. Y. *Makromol. Chem. Macromol. Chem. Phys.* **1987**, *188*, 1005–1015.
263 Okada, M., Sumitomo, H., Kakezawa, T. *Makromol. Chem.* **1972**, *162*, 285.
264 Braun, H., Yagci Y., Nuyken, O. *Eur. Polym. J.* **2002**, *38*, 151–156.
265 Guo, H. Q., Kajiwara A., Morishima, Y., Kamachi, M. *Macromolecules* **1996**, *29*, 2354–2358.
266 Kamachi, M., Guo, H. Q., Kajiwara. A. *Macromol. Symp.* **1997**, *118*, 149–161.
267 Kamachi, M., Guo, H. Q., Kajiwara. A. *Macromol. Chem. Phys.* **2002**, *203*, 991–997.
268 Hashidzume, A., Kurokawa, M., Morishima, Y. *Macromolecules* **2002**, *35*, 5326–5330.
269 Guo, H. Q., Kajiwara, A., Morishima, Y., Kamachi, M. *Polym. Adv. Techn.* **1997**, *8*, 196–202.
270 Guo, H. Q., Fang, Z., Kajiwara, A., Kamachi, M. *Acta Polym. Sci.* **2002**, *5*, 555–559.
271 Yagci, Y., Onen, A., Schnabel, W. *Macromolecules* **1991**, *24*, 4620–4623.
272 Yagci, Y., Onen, A. *J. Macromol. Sci. Chem.* **1991**, *A28*, 25–29.
273 Kobayashi, S., Shimano, Y., Saegusa, T. *Polym. J.* **1991**, *23*, 1307–1315.
274 Uno, H., Endo, T. *Chem. Lett.* **1986**, 1869–1870.
275 Uno, H., Takata, T., Endo, T. *J. Polym. Sci. Part A: Polym. Chem.* **1989**, *27*, 1675–1685.
276 Toman, L., Janata, M., Spevacek, J., Vlcek, P., Latalova, P., Masar, B., Sikora, A. *J. Polym. Sci. Part A: Polym. Chem.* **2004**, *42*, 6098–6108.
277 Toman, L., Janata, M., Spevacek, J., Vlcek, P., Latalova, P., Sikora, A., Masar, B. *J. Polym. Sci. Part A: Polym. Chem.* **2005**, *43*, 3823–3830.
278 Guo, Y. M., Xu, J., Pan, C. Y. *J. Polym. Sci. Part A: Polym. Chem.* **2001**, *39*, 437–445.
279 Bernaerts, K. V., Du Prez, F. E. *Polymer* **2005**, *46*, 8469–8482.
280 Xu, Y. J., Pan, C. Y. *J. Polym. Sci. Part A: Polym. Chem.* **2000**, *38*, 337–344.
281 Luan, B., Yuan, Q., Pan, C. Y., *Macromol. Chem. Phys.* **2004**, *205*, 2097–2104.

282 Duz, A. B., Yagci, Y. *Polym. J.* **1999**, *35*, 2031–2038.
283 Tasdelen, M. A., Degirmenci, M., Yagci, Y., Nuyken, O. *Polym. Bull.* **2003**, *50*, 131–138.
284 Wieland, P. C., Schafer, M., Nuyken, O. *Macromol. Rapid Commun.* **2002**, *23*, 809–813.
285 Schafer, M., Wieland, P. C., Nuyken, O. *J. Polym. Sci. Part A: Polym. Chem.* **2002**, *40*, 3725–3733.
286 Puts, R. D., Sogah, D. Y. *Macromolecules* **1997**, *30*, 7050–7055.
287 Burgess, F. J., Richards, D. H. *Polymer* **1976**, *17*, 1020–1023.
288 Muhlbach, K., Schulz, R. C. *Makromol. Chem. Macromol. Chem. Phys.* **1989**, *190*, 2551–2562.
289 Munir, A., Goethals, E. J. *Makromol. Chem. Rapid Commun.* **1981**, *2*, 693–697.
290 Schacht, E. H., Goethals, E. J. *Makromol. Chem. Macromol. Chem. Phys.* **1973**, *167*, 155–169.
291 Muhlbach, K., Schulz, R. C. *Makromol. Chem. Macromol. Chem. Phys.* **1988**, *189*, 1267–1277.
292 Kazama, H., Tezuka, Y., Imai, K. *Makromol. Chem. Macromol. Chem. Phys.* **1988**, *189*, 985–992.
293 Morishima, Y., Tanaka, T., Nozakura, S. I. *Polym. Bull.* **1981**, *5*, 19–24.
294 Miyamoto, M., Sano, Y., Saegusa, T., Kobayashi, S. *Eur. Polym. J.* **1983**, *19*, 955–961.
295 Simionescu, C. I., Rabia, I., Crisan, Z. *Polym. Bull.* **1982**, *7*, 217–222.
296 Miyamoto, M., Aoi, K., Yamanaka, H., Saegusa, T. *Polym. J.* **1992**, *24*, 405–409.
297 Schappacher, M., Deffieux, A. *Macromol. Chem. Phys.* **1997**, *198*, 3953–3961.
298 Deffieux, A., Schappacher, M. *Macromol. Symp.* **1998**, *132*, 45–55.
299 Nemes, S., Kennedy, J. P. *J. Macromol. Sci. Chem.* **1991**, *A28*, 311–328.
300 Kennedy, J. P., Price, J. L. *Polym. Prepr.* **1991**, *201*, 23-PMSE.
301 Wilczek, L., Mishra, M. K., Kennedy, J. P. *J. Macromol. Sci. Chem.* **1987**, *A24*, 1033–1049.
302 Kitayama, T., Nishiura, T., Hatada, K. *Polym. Bull.* **1991**, *26*, 513–520.
303 Wondraczek, R. H., Kennedy, J. P. *J. Polym. Sci. Part A: Polym. Chem.* **1982**, *20*, 173–190.
304 Harris, J. F., Sharkey, W. H. *Macromolecules* **1986**, *19*, 2903–2908.
305 Verma, A., Glagola, M., Prasad, A., Marand, H., Riffle, J. S. *Makromol. Chem. Makromol. Symp.* **1992**, *54/55*, 95–106.
306 Schuch, H., Klingler, J., Rossmanith, P., Frechen, T., Gerst M., Feldthusen, J., Muller, A. H. E. *Macromolecules* **2000**, *33*, 1734–1740.
307 Lee, S. C., Chang, Y. K., Yoon, J. S., Kim, C., Kwon, I. C., Kim, Y. H. *Macromolecules* **1999**, *32*, 1847–1852.
308 Liu, L. H., Jiang, B. Z., Zhou, E. L. *Polymer* **1996**, *37*, 3937–3943.
309 Tseng, S. S., Zhang, H. Z., Feng, X. D. *Polym. Bull.* **1982**, *8*, 219–224.
310 Abadie, M. J. M., Schué, F., Souel, T., Hartley, D. B., Richards, D. H. *Polymer* **1982**, *23*, 445–451.
311 Feldthusen, J., Ivan, B., Muller, A. H. E. *Macromolecules* **1997**, *30*, 6989–6993.
312 Feldthusen, J., Ivan, B., Muller, A. H. E. *Macromolecules* **1998**, *31*, 578–585.
313 Martinez-Castro, N., Lanzendorfer, M. G., Muller, A. H. E., Cho, J. C., Acar, M. H., Faust, R. *Macromolecules* **2003**, *36*, 6985–6994.
314 Zhang, H. M., Ruckenstein, E. *Macromolecules* **1998**, *31*, 746–752.
315 Brzezinska, K., Szymanski, R., Kubisa, P., Penczek, S. *Makromol. Chem. Rapid Commun.* **1986**, *7*, 1–4.
316 Penczek, S., Kubisa, P., Szymanski, R. *Makromol. Chem. Macromol. Symp.* **1986**, *3*, 203–220.
317 Kubisa, P. *Makromol. Chem. Macromol. Symp.* **1988**, *13*, 203–210.
318 Yagci, Y., Serhatli, I. E., Kubisa, P., Biedron, T. *Macromolecules* **1993**, *26*, 2397–2399.
319 Steward, M. J. in *New Methods of Polymer Synthesis*, Ebdon, J. R. (Ed.). Blackie, New York, **1991**, p. 107.
320 Yagci, Y., Schnabel, W. *Prog. Polym. Sci.* **1990**, *15*, 551–601.
321 Yagci, Y., Hepuzer, Y., Onen, A., Serhatli, I. E., Kubisa, P., Biedron, T. *Polym. Bull.* **1994**, *33*, 411–416.

322 Hepuzer, Y., Yagci, Y., Biedron, T., Kubisa, P. *Angew. Makromol. Chem.* **1996**, *237*, 163–171.
323 Grubbs, R. H., Tumas, W. *Science* **1989**, *243*, 907–915.
324 Ivin, K. J., Mol, J. C. *Olefin Metathesis and Metathesis Polymerization*. Academic Press, San Diego, **1997**.
325 Breslow, D. S. *Prog. Polym. Sci.* **1993**, *18*, 1141–1195.
326 Gilliom, L. R., Grubbs, R. H., *J. Am. Chem. Soc.* **1986**, *108*, 733–742.
327 Bazan, G. C., Schrock, R. R., Cho, H. N., Gibson, V. C. *Macromolecules* **1991**, *24*, 4495–4502.
328 Schrock, R. R., Feldman, J., Cannizzo, L. F., Grubbs, R. H. *Macromolecules* **1987**, *20*, 1169–1172.
329 Wallace, K. C., Liu, A. H., Dewan, J. C., Schrock, R. R. *J. Am. Chem. Soc.* **1988**, *110*, 4964–4977.
330 Toreki, R., Schrock, R. R. *J. Am. Chem. Soc.* **1990**, *112*, 2448–2449.
331 Nguyen, S. T., Johnson, L. K., Grubbs, R. H., Ziller, J. W. *J. Am. Chem. Soc.* **1992**, *114*, 3974–3975.
332 Novak, B. M., Risse, W., Grubbs, R. H. *Adv. Polym. Sci.* **1992**, *102*, 47–72.
333 Amass, A. J., Gregory, D. *Br. Polym. J.* **1987**, *19*, 263–268.
334 Amass, A. J., Bas, S., Gregory, D., Mathew, M. C. *Makromol. Chem. Macromol. Chem. Phys.* **1985**, *186*, 325–330.
335 Thorncsanyi, E., Kessler, M., Schwartau, M., Semm, U. *J. Mol. Catal.* **1988**, *46*, 385–394.
336 Feast, W. J., Gibson, V. C., Johnson, A. F., Khosravi, E., Mohsin, M. A. *J. Mol. Catal. A Chem.* **1997**, *115*, 37–42.
337 Castle, T. C., Hutchings, L. R., Khosravi, E. *Macromolecules* **2004**, *37*, 2035–2040.
338 Khosravi, E., Hutchings, L. R., Kujawa-Welten, M. *Design. Monom. Polym.* **2004**, *7*, 619–632.
339 Risse, W., Grubbs, R. H. *J. Mol. Catal.* **1991**, *65*, 211–217.
340 Tritto, I., Sacchi, M. C., Grubbs, R. H. *Polym. Prepr.* **1994**, *207*, 348.
341 Risse, W., Grubbs, R. H. *Macromolecules* **1989**, *22*, 1558–1562.
342 Tritto, I., Sacchi, M. C., Grubbs, R. H. *J. Mol. Catal.* **1993**, *82*, 103–111.
343 Manivannan, R., Sundarajan, G., Kaminsky, W. *Macromol. Rapid Commun.* **2000**, *21*, 968–972.
344 Coca, S., Paik, H. J., Matyjaszewski, K. *Macromolecules* **1997**, *30*, 6513–6516.
345 Katayama, H., Fukuse, Y., Nobuto, Y., Akamatsu, K., Ozawa, F., *Macromolecules* **2003**, *36*, 7020–7026.
346 Liaw, D. J., Huang, C. C., Ju, J. Y. Presented at the IUPAC World Polymer Congress MACRO 2004, Paris, **2004**.
347 Bielawski, C. W., Morita, T., Grubbs, R. H. *Macromolecules* **2000**, *33*, 678–680.
348 Katayama, H., Yonezawa, F., Nagao, M., Ozawa, F. *Macromolecules* **2002**, *35*, 1133–1136.
349 Kriegel, R. M., Rees, W. S., Weck, M. *Macromolecules* **2004**, *37*, 6644–6649.
350 Li, M. H., Keller, P., Albouy, P. A. *Macromolecules* **2003**, *36*, 2284–2292.
351 Charvet, R., Novak, B. M. *Macromolecules* **2004**, *37*, 8808–8811.
352 Aldissi, M. *J. Chem. Soc. Chem. Commun.* **1984**, 1347–1348.
353 Aldissi, M., Bishop, A. R. *Polymer* **1985**, *26*, 622–624.
354 Aldissi, M. *Synth. Met.* **1986**, *13*, 87–100.
355 Stowell, J. A., Amass, A. J., Beevers, M. S., Farren, T. R. *Polymer* **1989**, *30*, 195–201.
356 Aldissi, M., Hou, M., Farrell, J. *Synth. Met.* **1987**, *17*, 229–234.
357 Agnuri, E., Favier, G., Laputte, R., Philardeau, T., Rideau, J. in *Symposium on Block and Graft Copolymerization: Preprints*. GFP, Mulhouse, **1972**, p. 55.
358 Doi, Y., Watanabe, Y., Ueki, S., Soga, K. *Makromol. Chem. Rapid Commun.* **1983**, *4*, 533–537.
359 Doi, Y., Keii, T. *Adv. Polym. Sci.* **1986**, *73*, 201–248.
360 Mulhaupt, R., Duschek, T., Rieger, B. *Makromol. Chem. Macromol. Symp.* **1991**, *48*, 317–332.
361 Mulhaupt, R., Duschek, T., Rosch, J. *Polym. Adv. Tech.* **1993**, *4*, 465.
362 Cao, C. G., Zou, J. F., Dong, J. Y., Hu, Y. L., Chung, T. C. *J. Polym. Sci. Part A: Polym. Chem.* **2005**, *43*, 429–437.
363 Zou, J. F., Cao, C. G., Dong, J. Y., Hu, Y. L., Chung, T. C. *Macromol. Rapid Commun.* **2004**, *25*, 1797–1804.

364 Chung, T. C., Dong, J. Y, *J. Am. Chem. Soc.* **2001**, *123*, 4871–4876.
365 Matsugi, T., Kojoh, S., Kawahara, N., Matsuo, S., Kaneko, H., Kashiwa, N., *J. Polym. Sci. Part A: Polym. Chem.* **2003**, *41*, 3965–3973.
366 Kaneyoshi, H., Inoue, Y., Matyjaszewski, K. *Macromolecules* **2005**, *38*, 5425–5435.
367 Inoue, Y., Matyjaszewski, K. *J. Polym. Sci. Part A: Polym. Chem.* **2004**, *42*, 496–504.
368 Inoue, Y., Matsugi, T., Kashiwa, N., Matyjaszewski, K. *Macromolecules* **2004**, *37*, 3651–3658.
369 Hong, S. C., Jia, S., Teodorescu, M., Kowalewski, T., Matyjaszewski, K., Gottfried, A. C., Brookhart, M. *J. Polym. Sci. Part A: Polym. Chem.* **2002**, *40*, 2736–2749.
370 Webster, O. W., Hertler, W. R., Sogah, D. Y., Farnham, W. B., Rajanbabu, T. V. *J. Am. Chem. Soc.* **1983**, *105*, 5706–5708.
371 Eastmond G. C., Webster, O. W. In *New Methods of Polymer Synthesis*, Ebdon, J. R. (Ed.). Blackie, New York, **1991**, p. 22.
372 Jenkins, A. D., Tsartolia, E., Walton, D. R. M., Horskajenkins, J., Kratochvil, P., Stejskal, J. *Makromol. Chem. Macromol. Chem. Phys.* **1990**, *191*, 2511–2520.
373 Ruth, W. G., Moore, C. G., Brittain, W. J., Si, J. S., Kennedy, J. P. *Polym. Prepr.* **1993**, *205*, 61.
374 Takacs, A., Faust, R. *Macromolecules* **1995**, *28*, 7266–7270.
375 Eastmond, G. C., Grigor J. *Makromol. Chem. Rapid* **1986**, *7*, 375–379.
376 Creutz, S., Vandooren, C., Jerome, R., Teyssie, P. *Polym. Bull.* **1994**, *33*, 21–28.
377 Wang, G. J., Yang, D. Y., Muller, A. H. E. *Chem. J. Chin. Univ.* **2001**, *22*, 157–159.
378 Tsutsumiuchi, K., Aoi, K., Okada, M. *Macromol. Rapid Commun.* **1995**, *16*, 749–755.
379 Weimer, M. W., Scherman, O. A., Sogah, D. Y. *Macromolecules* **1998**, *31*, 8425–8428.
380 Hawker, C. J., Hedrick, J. L., Malmstrom, E. E., Trollsas, M., Mecerreyes, D., Moineau, G., Dubois, P., Jerome, R. *Macromolecules* **1998**, *31*, 213–219.
381 Lim, N. K., Arndtsen, B. A. *Macromolecules* **2000**, *33*, 2305–2307.
382 Klaerner, G., Trollsas, M., Heise, A., Husemann, M., Atthoff, B., Hawker, C. J., Hedrick, J. L., Miller, R. D. *Macromolecules* **1999**, *32*, 8227–8229.
383 Simionescu, C., Sik, K. G., Comanita, E., Dumitriu, S. *Eur. Polym. J.* **1984**, *20*, 467–470.
384 Shaikh, A. S., Comanita, E., Sumitriu, B., Simionescu, C. *Angew. Makromol. Chem.* **1981**, *100*, 147–158.
385 Dumitriu, S., Shaikh, A. S., Comanita, E., Simionescu, C. I. *Eur. Polym. J.* **1983**, *19*, 263–266.
386 Piirma, I., Chou, L. H. *J. Appl. Polym. Sci.* **1979**, *24*, 2051–2070.
387 Sik, G., Dumitriu, S., Comanita, E., Simionescu, C. *Polym. Bull.* **1984**, *12* 419–425.
388 Onen, A., Denizligil, S., Yagci, Y. *Angew. Makromol. Chem.* **1994**, *217*, 79–89.
389 Erim, M., Erciyes, A. T., Serhatli, E. I., Yagci, Y. *Polym. Bull.* **1992**, *27*, 361–366.
390 Erciyes, A. T., Erim, M., Hazer, B., Yagci, Y. *Angew. Makromol. Chem.* **1992**, *200*, 163–171.
391 Semsarzadeh, M. A., Mirzaei, A., Vasheghani-Farahani, E., Haghighi, M. N. *Eur. Polym. J.* **2003**, *39*, 2193–2201.
392 Destarac, M., Pees, B., Boutevin, B. *Macromol. Chem. Phys.* **2000**, *201*, 1189–1199.
393 Destarac, M., Matyjaszewski, K., Silverman, E. *Macromolecules* **2000**, *33*, 4613–4615.
394 Paik, H. J., Teodorescu, M., Xia, J. H., Matyjaszewski, K. *Macromolecules* **1999**, *32*, 7023–7031.
395 Erel, I., Cianga, I., Serhatli, E., Yagci, Y. *Eur. Polym. J.* **2002**, *38*, 1409–1415.
396 Matyjaszewski, K., Beers, K. L., Kern, A., Gaynor, S. G. *J. Polym. Sci. Part A: Polym. Chem.* **1998**, *36*, 823–830.
397 Roos, S. G., Mueller, A. H. E., Matyjaszewski, K. *Macromolecules* **1999**, *32*, 8331–8335.
398 Degirmenci, M., Cianga, I., Yagci, Y. *Macromol. Chem. Phys.* **2002**, *203*, 1279–1284.
399 Yagci, Y., Degirmenci, M. *ACS Symp. Ser.* **2003**, *854*, 383–393.
400 Li, H., Zhang, Y. M., Xue, M. Z., Liu, Y. G. *Polym. J.* **2005**, *37*, 841–846.

401 Miura, Y., Narumi, A., Matsuya, S., Satoh, T., Duan, Q., Kaga, H., Kakuchi, T. *J. Polym. Sci. Part A: Polym. Chem.* **2005**, *43*, 4271–4279.

402 Durmaz, H., Aras, S., Hizal, G., Tunca, U. *Design. Monom. Polym.* **2005**, *8*, 203–210.

403 Celik, C., Hizal, G., Tunca, U. *J. Polym. Sci. Part A: Polym. Chem.* **2003**, *41*, 2542–2548.

404 Debuigne, A., Caille, J.-R., Willet, N., Jerome, R. *Macromolecules* **2005**, *38*, 9488–9496.

405 Heuschen, J., Jerome, R., Teyssie, P. *Macromolecules* **1981**, *14*, 242–246.

406 Wang, Y. B., Hillmyer, M. A. *Macromolecules* **2000**, *33*, 7395–7403.

407 Wang Y. B., Hillmyer, M. A. *J. Polym. Sci. Part A: Polym. Chem.* **2001**, *39*, 2755–2766.

408 Schmidt, S. C., Hillmyer, M. A. *Macromolecules* **1999**, *32*, 4794–4801.

409 Coulembier, O., Degee, P., Cammas-Marion, S., Guerin, P., Dubois, P. *Macromolecules* **2002**, *35*, 9896–9903.

410 Vangeyte, P., Jerome, R. *J. Polym. Sci. Part A: Polym. Chem.* **2004**, *42*, 1132–1142.

411 Mecerreyes, D., Dubois, P., Jerome, R., Hedrick, J. L. *Macromol. Chem. Phys.* **1999**, *200*, 156–165.

412 Liu, Q., Konas, M., Davis, R. M., Riffle, J. S. *J. Polym. Sci. Part A: Polym. Chem.* **1993**, *31*, 1709–1717.

413 Kennedy, J. P., Kurian, J. *Polym. Bull.* **1990**, *23*, 259–263.

414 Gadkari, J. P., Kennedy, J. P. *J. Appl. Polym. Sci. Appl. Polym. Symp.* **1989**, *44*, 19–34.

415 Matyjaszewski, K., Beers, K. L., Kern, A., Gaynor, S. G. *J. Polym. Sci. Part A: Polym. Chem.* **1998**, *36*, 823–830.

416 Roos, S. G., Mueller, A. H. E., Matyjaszewski, K. *Macromolecules* **1999**, *32*, 8331–8335.

417 Shinoda, H., Matyjaszewski, K. *Macromolecules* **2001**, *34*, 6243–6248.

418 Shinoda, H., Matyjaszewski, K. *Macromol. Rapid Commun.* **2001**, *22*, 1176–1181.

419 Shinoda, H., Miller, P. J., Matyjaszewski, K. *Macromolecules* **2001**, *34*, 3186–3194.

420 Borner, H. G., Matyjaszewski, K. *Macromol. Symp.* **2002**, *177*, 1–15.

421 Neugebauer, D., Zhang, Y., Pakula, T., Matyjaszewski, K. *Polymer* **2003**, *44*, 6863–6871.

422 Neugebauer, D., Theis, M., Pakula, T., Wegner, G., Matyjaszewski, K. *Macromolecules* **2006**, *39*, 584–593.

423 Shinoda, H., Matyjaszewski, K., Okrasa, L., Mierzwa, M., Pakula, T. *Macromolecules* **2003**, *36*, 4772–4778.

424 Tunca, U., Ozyurek, Z., Erdogan, T., Hizal, G. *J. Polym. Sci. Part A: Polym. Chem.* **2004**, *42*, 4228–4236.

14
Polymerizations in Aqueous Dispersed Media

Bernadette Charleux and François Ganachaud

14.1
Introduction

Heterogeneous polymerization methods offer many invaluable practical advantages and are, for instance, the most important industrial processes for the production of synthetic polymers via free radical polymerization. A variety of systems can be used, which differ from each other in the initial state of the polymerization mixture, in the mechanism of particle formation, in the size of the final polymer particles and in the kinetics of polymerization. The details of these differences have been well reviewed in the literature [1, 2]. Polymerization in dispersed media is a very broad field that cannot be entirely covered in the context of this chapter. We shall consider here only systems where polymerization takes place mainly in stabilized particles, with the purpose of forming well-defined polymer dispersions. It will concern suspension, miniemulsion and emulsion polymerizations in water as a dispersing agent and also surfactant templated polymerizations (in microemulsions and vesicles). In the specific case of dispersion polymerization, organic media and carbon dioxide could also be considered, but these systems will not be presented here. All processes have been essentially developed for free radical polymerization; nevertheless, other polymerization techniques are now applicable in such aqueous dispersed media and we shall focus on the most recent results, paying special attention to the polymerization methods that afford precise control over the macromolecular structure. These methods are primarily controlled/living free radical polymerizations (CRP) [3–6] and to a lesser extent anionic and cationic polymerization [7], ring-opening metathesis polymerization [8, 9], catalytic polymerization of ethylene and dienes and polycondensation/polyaddition reactions [10]. The characteristics of all of them are fully described in other chapters of this book. For the sake of simplicity, only open literature is cited here, although patents were sometimes the first to describe innovative processes, such as the ionic polymerization or the polycondensation of silicon-based monomers in aqueous dispersions.

14.2
Aqueous Suspension Polymerization

14.2.1
Conventional Free Radical Polymerization

In suspension polymerization [2, 11], a water-insoluble monomer is dispersed by vigorous stirring in the continuous aqueous phase as micrometer- to millimeter-range liquid droplets. An oil-soluble radical initiator, mainly a peroxide, is employed to initiate polymerization inside the monomer droplets. A stabilizer or a mixture of stabilizers or dispersants, which may be water-soluble polymers, such as poly(N-vinylpyrrolidone) or poly(vinyl alcohol-co-vinyl acetate) or insoluble inorganic salts such as carbonates, silicates or phosphates are added to prevent coalescence of the monomer droplets and sticking of the partially polymerized particles during the course of polymerization. The stabilizer can improve the dispersion stability and maintain the particle size by increasing the viscosity of the aqueous phase, although most often it acts by forming a film at the droplet–particle surface, thus preventing coalescence by steric repulsion. The diameters of the particles obtained from suspension polymerization are usually in the range 20–2000 µm, depending on the stirring rate, volume ratio of the monomer to water, concentration of the stabilizer, viscosity of both phases and the design of the reaction vessel.

From the kinetic viewpoint, each particle behaves as an isolated microreactor and the polymerization rate follows the same principles as those of bulk polymerization. Steady-state conditions apply and the local radical concentration is therefore the same as in a homogeneous system. Further, the molar masses and molar mass distributions are similar to those obtained in conventional free radical polymerization. Consequently, the droplet size and the amount of stabilizer do not affect the polymerization rate. The continuous aqueous phase serves only to decrease the viscosity and to dissipate heat generated by the polymerization.

An important property, directly related to the target application, is the morphology of the individual particles. Particles with the polymer soluble (or swellable) in its own monomer have a smooth surface and a relatively homogeneous texture, such as polystyrene and poly(methyl methacrylate). When the polymer is not soluble in its own monomer, such as poly(vinyl chloride) and polyacrylonitrile, the particles have a rough surface and a porous morphology. Cross-linked polystyrene beads with controlled degree of porosity and controlled pore structure are obtained by suspension polymerization using a diluent that properly dissolves the styrene and divinylbenzene comonomers but that is a poor swelling agent for the copolymer. These beads are of particular interest in the production of ion-exchange resins and polymer supports.

14.2.2
Controlled/Living Free Radical Polymerization

Suspension polymerization is certainly the easiest heterogeneous process to apply to controlled/living free radical polymerization as it exhibits the same fundamental kinetic principles as a homogeneous system. However, the number of examples to be found in the open literature is small. For industrial applications of suspension polymerization, one might guess that the control over molar mass and (co)polymer architecture is not of major interest. Nevertheless, in the case of cross-linked particles, for instance, CRP would provide a much better design of the polymer network (homogeneous distribution of the cross-links) with a smaller amount of cross-linker, but, to the best of our knowledge, this feature has never been investigated.

14.2.2.1 Nitroxide-mediated Controlled Free Radical Polymerization (NMP)

The first publication [12] describing NMP in suspension concerned the copolymerization of styrene and butadiene. The reaction, initiated by benzoyl peroxide in the presence of 2,2,6,6-tetramethylpiperidine-l-oxyl (TEMPO), yielded a copolymer with a much narrower molar mass distribution than that prepared in the absence of nitroxide. This example was given in order to demonstrate that controlled free radical polymerization could be performed in suspension besides solution or bulk processes. After this preliminary work, more examples became available, mainly devoted to the homopolymerization and copolymerization of styrene in the presence of TEMPO as a control agent [13, 14]. With this nitroxide, the polymerization temperature is necessarily higher than 100 °C, which is above the boiling-point of water and also the glass transition temperature (T_g) of the (co)polymers. This means that the polymerization had to be carried out under a slight pressure, which is actually not unusual in industrial systems. Stability of the soft particles during synthesis was, however, a challenge. The polymerizations were carried out using sodium dodecyl sulfate, poly(vinyl alcohol) and stearic acid as a stabilizing system. Polymer beads were obtained with diameters in the range 1–3 mm. The tendency towards agglomeration was overcome by using a high stirring speed. As expected, the rate of polymerization and the control over molar mass and molar mass distribution were not found to be different from those in bulk and followed the same rules.

14.2.2.2 Atom-transfer Radical Polymerization (ATRP)

Very few articles have been published since the first reports showing that ATRP could be conducted in a living fashion under the conditions of suspension polymerization [3, 15]. ATRP has, for instance, been applied to the production of poly(methyl methacrylate) spherical beads with an oil-soluble copper-based catalyst and poly(N-vinylpyrrolidone) or poly(vinyl alcohol) as a stabilizer [16, 17]. In some cases, NaCl was added to the aqueous phase to decrease the diffusion of the catalyst towards water. The control over molar mass was good although the polydispersity index was larger than expected. Treatment of the polymer beads with an acidic aqueous solution allowed elimination of most of the copper catalyst, which might be an advantage of the process [16]. The method was also successfully used for the production of polymer microcapsules with a cross-linked amphiphilic copolymer wall and an organic liquid core, which were not achievable via conventional free radical suspension polymerization [18, 19].

14.2.2.3 Other Controlled Free Radical Polymerization Methods

In addition to NMP and ATRP, the reversible addition–fragmentation chain transfer (RAFT) methodology was used for the controlled free radical suspension polymerization of methyl methacrylate [20]. Poly(vinyl acetate) was also prepared in a controlled manner using cobalt-mediated polymerization [21]. In both cases the suspension polymerization led to well-defined homopolymers, as millimeter size beads.

14.2.3
Ring-opening Metathesis Polymerization (ROMP)

The ring-opening metathesis polymerization of olefins in aqueous dispersed media has been described in well-documented, recent review articles [8] and elsewhere in this book. The advent of ROMP in aqueous dispersed systems has been made possible with the development of a new class of carbene initiators tolerant to water, based on a group VIII metal such as ruthenium. An example of ROMP of 1,5-cyclooctadiene using $(Cy_3P)_2Cl_2Ru=CHPh$ as a catalyst in aqueous suspension was described by Chemtob et al. [22]: the polymerization leads to beads of 1,4-polybutadiene with diameter in the region of 20 µm, stabilized by an amphiphilic norbornenyl-ended polystyrene-b-poly(ethylene oxide) macromonomer or a poly(butadiene-g-ethylene oxide) graft copolymer.

14.2.4
Ionic Polymerizations

In this chapter devoted to ionic polymerization methods, we shall consider equally all processes analogous to an aqueous suspension polymerization, since it is mainly the interface which controls the reactions rather than the size or hydrophobicity of the particles.

14.2.4.1 Oil-in-Water Processes

The cationic or anionic polymerization of heterocycles or vinyl monomers in direct suspension according to the interfacial polymerization process where fast transfer and/or termination reactions to water are observed has hardly been described in the literature (a noticeable exception concerns alkyl cyanoacrylates, for which anionic polymerization in suspension is living even in acidic media, to give particles of typically a few micrometers stabilized by cyclodextrin or Pluronic surfactants [23, 24]). Some tricks were then used to facilitate the polymerization inside the large monomer droplets or to reduce the extent of the transfer/termination reactions to water with respect to propagation.

One option is to choose a well-defined organic initiator/activating catalyst system, rather than to implement initiation via proton or hydroxide. For instance, in the cationic, surfactant-free, suspension polymerization of styrene with $B(C_6F_5)_3$ as a water-tolerant Lewis acid [25], the remarkably high chain-end functionality of the chains formed was due to an exclusive initiation by the 4-methoxy-α-methylbenzyl alcohol–Lewis acid couple on the one hand and a very fast termination reaction to water on the other, which prevents any other side-reaction from occurring. Unfortunately, all efforts to reactivate the thus-generated carbinol chain end by a strong Lewis acid have failed so far.

A second approach consists in polymerizing monomers under the activated monomer mechanism [26] to promote the polymerization inside the monomer droplets. For instance, the cationic polymerization of multifunctional epoxide monomers proceeds in the presence of various superacids, either introduced in water [27] or produced by photoactivation [28, 29]. The numerous alcohol functions generated by a side-reaction, namely hydrolysis of the oxirane ring, initiate polymerization competitively with water molecules entering the droplets/particles. Falk and Crivello [28, 29] made use of organic porogens and alcohol group post-derivatization to prepare highly porous and hyper-functionalized polymer beads devoted to heterogeneous catalysis. In anionic polymerization processes, only N-carboxyanhydrides (NCAs) were successfully polymerized in aqueous suspension through the activated monomer mechanism, as reported in two patents and one paper [30]. Triethylamine was used as a water-soluble base, while toluene or dichloromethane ensured a sufficient growth of the chains by avoiding early precipitation of the oligopeptides exhibiting a β-sheet conformation (a major drawback in "conventional" aqueous-based polymerization of NCA [7]); after evaporation of the organic solvent, the submicrometer particles formed (in the presence of nonionic surfactants) or the larger beads (without surfactant) were logically found to be porous.

A third way consists in using dispersions of a metallic catalyst. For instance, the coordinated anionic polymerization of episulfides was performed in toluene–water dispersions [31, 32]. Zinc or cadmium catalysts were thoroughly hydrolyzed to give dihydroxy aggregates in the presence of monomer, which coordinated through the sulfur atom to add to the polymer chains; the molar masses were found to be much higher than those obtained with emulsion processes (see the section 14.4.5). The so-called "pearl polymerization" process [32], i.e. the generation of a polymer shell around the finely grounded catalyst colloids (diameter below 1 µm), generated shelf-stable suspensions.

14.2.4.2 Water-in-Oil Processes

Although the inverse suspension polymerization systems did not lead to well-designed aqueous polymer dispersions, numerous studies allowed some understanding of the process of ionic polymerization in aqueous media. For the record, the conjunction of a proton donor, such as mineral acids or sulfonic acids and ytterbium (or another water-tolerant metal such as zinc or copper) triflate in very large quantities promoted the polymerization of alkoxystyrene in suspension, with or without solvents and surfactants [33–37]. Cauvin et al. [38] showed recently that the system was an inverse suspension, in which the water droplets were stabilized against Ostwald ripening by the large load of salt generated from the Lewis acid dissolution in water. Further, $HYb(OTf)_4$, a superacid, was undoubtedly the true initiator of the interfacial polymerization reaction (a rapid description of this chemistry is given in the section 14.3.5.1). Recently, isobutene was dispersed in water-based mixtures of eutectic salts and polymerized using a diborane Lewis acid at $-80\,°C$ to generate, in less than 10 s, high molar mass polymers [39].

14.2.5
Polycondensation/Polyaddition

A decade ago, Arshady reviewed the three different techniques, i.e. suspension, dispersion/precipitation and interfacial polycondensation processes, that one may use to prepare polymer beads directly in aqueous dispersion [10]. He also described the various polymers generated in these studies, e.g. formaldehyde-based resins, polyurethanes, polyamides and polyesters. Since 1993, numerous studies have been published mainly on interfacial polycondensation processes, with a view to generating polymers, by means of phase-transfer catalysis, with or without organic solvent, or microcapsules, through a reaction of an organic monomer with a water-soluble one.

14.2.5.1 Phase-transfer Catalysis

There have been many reviews on the use of interfacial suspension polymerization to produce various industrially important aliphatic and aromatic polycondensates, including polyamides, polyesters and polycarbonates (e.g. [40–42]). The principle is to react an organic (possibly water-sensitive) monomer, dissolved in toluene or dichloromethane, with a water-soluble monomer at the solvent/water interface. A phase-transfer catalysis (PTC) agent, such as a tetrabutylammonium hydroxide or a proton acceptor, such as triethylenamine (when the reaction releases protons) is generally added. The advantage of working in the presence of water is twofold: (i) side-reactions that commonly arise in melt processes, such as transesterification, are avoided and (ii) linear chains rather than macrocycles are formed, the latter being obtained quantitatively in solution processes.

The PTC agent was purposely chosen to be symmetrical so as to avoid curvature of the interface and discrepancies in terms of solubility of the two monomers right at the interface. Casassa et al., however, showed in seminal studies [43, 44] that adding a surfactant, if its nature and concentration were correctly chosen, could help to increase the average molar mass of the polycondensate, whatever the chemistry involved. Like PTC agents, cationic surfactants can bring hydroxide ions close to the interface to trap the unwanted acid released in some processes, without any need for external additives. In addition, the increase in interfacial area and the possibility of better swelling of the interface with monomer are advantageous in terms of enhanced kinetics, for instance when using water-sensitive monomers, but also on increasing the polymer chain length. A limitation on such an improvement occurs with too high surfactant concentrations, which would generate direct and inverse micelles or too small droplets, capable of capturing too much water at their inner interface and thus of preventing efficient polycondensation. Since then, different groups have made use of surfactants to prepare, among others, polysulfonates [45–48] and poly(thio)-esters [49, 50]. The most revealing example, in our opinion, comes from the work of Saam and coworkers on polycondensation reactions between hydroxyl-based monomers and carboxylic acid- or aldehyde-based monomers in aqueous suspension, in the presence of sulfonate surfactants or strong acids [51] (such work was recently applied by Kobayashi's group for simple esterification reactions [52] and reproduced in part by another Japanese team [53]). The main requirements of this process are (i) total water insolubility of the two monomers, (ii) surface activity of the catalyst and possibly of the diol and (iii) the necessity to work at low temperature, in order to avoid penetration of water into the droplets. Polymers with molar masses of typically 2000 g mol^{-1} were generated in these processes, which could be slightly increased by off-stoichiometric formulations (to counterbalance the difference in water solubility of the monomers) or increasing the surfactant/catalyst content [dodecylbenzenesulfonic acid (DBSA)]; surprisingly, low polydispersities (generally around 1.5) were also found [53].

Another trend forced by environmental issues consists in preparing polycondensates in water via a "green" chemistry process, i.e. using neither solvents nor surfactants (precipitation process). In the case of aromatic polymer synthesis, catalysis is performed by heating the monomer–water mixture, possibly in the presence of an organic base, such as an aromatic amine. The authors showed that activated aromatic monomers involved in the polycondensation reactions, such as halides [54], bisoxazolines [55] or sulfides [56], are little prone to hydrolysis. Even though the conversions were hardly quantitative, the molar masses thus achieved were comparable to those obtained in N-methylpyrrolidone or dimethyl sulfoxide (DMSO) solution processes. In the same vein, very interesting and, it is hoped, valuable work on the synthesis of polyimides is currently under way by an Australian team [57, 58]. Various polyimides were prepared from a dianhydride (previously hydrolyzed at ambient pressure and reflux of water) and diamine, under pressure and at less than 200 °C. Such mild conditions may render the water-based process competitive with solution or melt processes, currently used in industry. Examples of aliphatic polyamides prepared directly in water also exist, starting from the carboxylate salts: lithium bromoundecanoate monomer formed polycondensate of up to 23 monomer units with better thermal resistance than polymers prepared in polar solvents, which present defects due to side-reactions [59]. Amino acid derivatives activated by a carbodiimide react with ethylenediamine to produce amorphous and biodegradable polypeptides in reasonable yields with molar masses as large (up to 3×10^5 g mol^{-1}) as those obtained by NCA solution polymerization under strictly anhydrous conditions [60]. Another recent paper described the preparation of poly(hydroxyurethane) polymers starting from water-insoluble cyclic carbonates and hexamethylenediamines [61].

14.2.5.2 Preparation of Microcapsules

Microcapsules have long been studied for their remarkable ability to protect or slowly release active ingredients from a tailored formulation (for complete information on microcapsule synthesis and use, see the series edited by Arshady [62]). Among various techniques of encapsulation, polycondensation/polyaddition processes were revealed to be facile routes to the generation of capsules with a large size range (from a few μm to mm scale), thin membranes (typically a few percent of the capsule diameter) and high content of encapsulated material (typically above 75%). The most striking application of these is self-healing materials [63], i.e. an epoxy matrix loaded with a metathesis catalyst and encapsulated monomer, which under mechanical stress is released and polymerizes through ring-opening metathesis polymerization to fill the crack.

Microcapsules are prepared mostly by starting from the oil to be encapsulated, a water-insoluble monomer and possibly a water-soluble one. The chemistry of these systems is usually very complex. For instance, in the synthesis of polyurethanes, the isocyanate group may be hydrolyzed to a reactive amine [64, 65] or react with the alcohol groups of the poly(vinyl alcohol) dispersant [66] or of the nonionic surfactant [64]. In polyamide chemistry, numerous reactions, including transamidification reactions, probably occur [67]. Also, microcapsules are subjected to irreversible coagulation through inter-wall cross-linking, a feature that cannot be considered in conventional free radical polymerization where the chains are intrinsically dead. There is no need to differentiate the processes used, since (direct or inverse) suspension and precipitation processes are all controlled by the chemistry taking place at the oil/water interface, not by the nature of each phase. On the other hand, the wall formation of the final capsules is closely related to the kinetics [68] and mechanisms of polycondensation, the cross-linking reactions and/or the degree of crystallinity of the resulting polymer.

The shell formation may proceed according to three different processes, as reported in a variety of literature articles: (i) if both monomers are soluble in the organic phase, then the capsule wall grows slowly by precipitation of the polymer; this process occurs for most polyurea systems, where the bare isocyanate reacts with an amine (possibly formed by hydrolysis of another isocyanate group); (ii) if monomers are soluble in the aqueous phase and in the organic phase, either one or both monomers have to diffuse through the polymer packed at the interface to ensure efficient wall formation with time; in this instance, the organic phase must swell the membrane sufficiently, but still favors precipitation of the polymers – this is typically the case for polyamide or polyurethane chemistries; (iii) the third process entails the generation of a prepolymer, which then further reacts at the interface; this is the case, for instance, for surfactant-like molecules, such as poly(ethylene glycol) macromonomers and urea–formaldehyde prepolymers [69].

The nature of the monomers used, whether they bear aromatic groups or long alkyl chains [70], affects the crystalline content of the materials through abundant hydrogen bonding interactions whenever urea, urethane or amide groups are involved. Similarly, the use of trifunctional monomers modifies the membrane brittleness [71]. Both factors have a deep impact on the roughness and porosity of the membrane capsule, which is of utmost importance in slow-release applications. Shrinking of the polymer through crystallization or intense cross-linking opens larger pores than a smooth auto-repairing surface [72]. It seems that these factors may be adapted only through a specific trial-and-error approach.

Other examples include the generation of polyepoxide microcapsules in an inverse suspension system, i.e. entailing an aqueous core, in which magnetite and vitamin B_6 were encapsulated [73]. The originality of this work lies in the 10-fold swelling ratio of the membrane simply by decreasing the pH, thus protonating the secondary amino groups. The kinetics of wall formation through precipitation/aggregation of small polymer colloids were followed kinetically while preparing poly(urea–formaldehyde) capsules, thus accounting for the very high rugosity observed by scanning electron microscopy [66].

14.3
Aqueous Miniemulsion Polymerization

14.3.1
Conventional Free Radical Polymerization

Historically, miniemulsion polymerization [74] is younger than emulsion polymerization. It was first proposed by Ugelstadt et al. [75] and studied in detail later by the groups of El-Aasser [76–79], Schork [80], Landfester [81, 82] and Asua [83].

Although the final latex does not differ from that formed in an emulsion polymerization (same particle size, same solids content, same surfactant, …), the key principle of miniemulsion polymerization is that the monomer droplets are the major loci of nucleation and subsequent propagation. This, however, does not make miniemulsion polymerization really similar to suspension polymerization. There are actually many differences: first, the initial monomer-in-water emulsion is specially prepared so as to exhibit submicrometer-size droplets (50–500 nm), second, a conventional surfactant is used as a stabilizer, and third, a water-soluble initiator can be employed rather than exclusively an oil-soluble initiator. To ensure droplet nucleation, in contrast to emulsion polymerization the surfactant should not form micelles [concentration below the critical micelle concentration (CMC) in the presence of the organic phase]; it should reside at the monomer/water interface where it stabilizes the droplets against coalescence. Typical surfactants bear an anionic head such as the very common sodium dodecyl sulfate, but cationic surfactants such as cetyltrimethylammonium chloride or bromide or nonionic species based on poly(ethylene oxide) have sometimes been employed. Moreover, the emulsion degradation via molecular diffusion of monomer from the small droplets toward the largest ones (so-called Ostwald ripening) is prevented by the addition of an (ultra)hydrophobe to the organic phase. Typical hydrophobes are hexadecanol and hexadecane, the roles of which differ slightly. The function of the long-chain alcohol can be twofold: limiting the coalescence by forming a barrier at the surface of the droplets by combination with the surfactant and/or preventing the Ostwald ripening by building up an osmotic pressure within the monomer droplets, which counterbalances the Laplace pressure. The ultrahydrophobe behaves exclusively according to the second mechanism. Other hydrophobes, such as silanes, perfluorinated alkanes, polymers and even very hydrophobic comonomers, initiators and chain transfer agents have also been used. In particular, the presence of a few percent of added polymer has been shown to enhance droplet nucleation. The choice of a suitable hydrophobe is therefore a key issue for successful miniemulsion polymerization. The so-formed unstable emulsion is then subjected to a very strong shear force, which can be ultrasonication at the laboratory scale. For large volumes, however, mechanical homogenizers are preferred.

Once formed, the metastable miniemulsion is polymerized. With an oil-soluble initiator, all droplets immediately behave as independent nano-reactors; they become the locus of all reactions normally taking place in a free radical polymerization process. With a water-soluble initiator, initiation and the first propagation events take place in the water phase in a very similar way as for emulsion polymerization. Simply because the monomer droplets are of very small size and exhibit a large surface area and because no micelles are present, they are the single objects for oligoradical capture. Radical entry is therefore the initial polymerization step inside the droplets. Ideally, whatever the type of initiator, the concept of miniemulsion polymerization aims at producing a latex which is a 1:1 copy of the original droplets, thereby achieving direct control over the number of particles, avoiding the complex nucleation step that exists in conventional emulsion polymerization. With a water-soluble radical initiator, homogeneous nucleation (see Section 14.4.1) is still possible depending on the monomer, but can be limited by adjusting the experimental parameters.

An ideal miniemulsion polymerization with a water-soluble radical initiator displays different kinetics from an emulsion polymerization, as studied for styrene by Bechthold and Landfester [84]. A particle formation (droplet nucleation) stage is initially observed accompanied by an increasing polymerization rate and an increasing average number of radicals per particle, $\tilde{n}$ (interval I). After the polymerization rate has reached a maximum and $\bar{n}$ has reached a value of 0.5, this period is immediately followed by *interval III* (same definition as in emulsion polymerization) with a steadily decreasing rate and a constant value of $\bar{n}=0.5$. During this stage, the polymerization follows first-order kinetics with respect to monomer as in bulk, solution or suspension polymerizations. *Interval II* in emulsion polymerization, characterized by a constant polymerization rate, is missing because the transport of monomer is negligible: there is no period of constant monomer concentration in the polymerization locus but an exponential decay. A last stage with increasing rate and $\bar{n}$ can sometimes be observed, typical of the gel effect. The polymerization rate is greatly affected by the droplet size and hence by the surfactant concentration.

Technically, miniemulsion polymerization offers some unique advantages over emulsion polymerization. This process is, for instance, especially useful for polymerizations involving a very water-insoluble ingredient, which would transport slowly through the aqueous phase in an emulsion polymerization and would therefore not be incorporated into the latex particles. It is also a very powerful process for encapsulating various materials (such as dye, pigment, inorganic nanoparticles or preformed polymer such as polyester) in the final particles or for conducting copolymerizations of monomers with very different water solubilities.

Although miniemulsion polymerization is a well-described process, following the aforementioned experimental rules, here we shall also consider microsuspension polymerization as an analogous process; in the latter system, submicrometer droplets are likely prepared by ultrasonication, except that no ultrahydrophobe is added to the monomer phase.

14.3.2
Controlled/Living Free Radical Polymerization

Although much less often used for industrial applications than the traditional emulsion polymerization, it can nevertheless be stated that miniemulsion polymerization was the first successful approach to achieve CRP in an aqueous dispersed system, hence affording "living" latexes with good control over both the molecular characteristics of the (co)polymers and the colloidal properties of the particles [3–6].

14.3.2.1 Nitroxide-mediated Controlled Free Radical Polymerization

As in bulk NMP, the initiating system is either two-component (conventional radical initiator and free nitroxide) or single-component (preformed molecule with an alkoxyamine group). In the latter case, the initiator can also be a macroalkoxyamine, i.e. a polymer with an alkoxyamine chain end, usually prepared by NMP of the corresponding monomer in a preliminary step. An additional complexity in miniemulsion polymerization comes from the possibility for the initiator to be either oil or water soluble. In all cases, however, the nitroxide has to

Fig. 14.1 Structure of the nitroxide SG1.

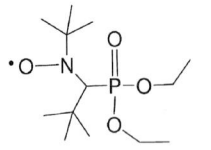

be sufficiently compatible with the organic phase to ensure a living polymerization process in the monomer droplets/polymer particles. Two main nitroxides have been used in miniemulsion polymerization systems: TEMPO (or TEMPO derivatives) and N-tert-butyl-N-(1-diethylphosphono-2,2-dimethylpropyl)-N-oxyl (SG1) (Fig. 14.1). With TEMPO, most studies were devoted to styrene [5], whereas with SG1 both styrene and n-butyl acrylate have been polymerized and copolymerized [85–87]. Because of the kinetic features of NMP and particularly the polymerization temperature, which is usually above 100 °C, aqueous miniemulsion polymerizations were carried out under pressure. The surfactant has to be stable enough at high temperature and, for this reason, molecules with a sulfonate hydrophilic group instead of a sulfate have been preferred. The hydrophobe was usually hexadecane and, most of the time, ultrasonication was applied to create the submicrometer monomer droplets.

The use of oil-soluble radical initiators (benzoyl peroxide, azobisisobutyronitrile, thermal autopolymerization in the case of styrene) in conjunction with TEMPO or SG1 or oil-soluble preformed alkoxyamines (molecular or macromolecular, based on either TEMPO or SG1) is certainly the most straightforward method to apply NMP in miniemulsion polymerization processes. None of the results observed were actually fundamentally different from those found in bulk polymerization. The main reasons are the choice of similar reagents together with the possibility for all of the polymerization events to take place in the same locus, i.e. the monomer droplet/polymer particle behaving as an individual nano-reactor. As in homogeneous systems, the polymerization kinetics are governed by the activation–deactivation equilibrium and by the persistent radical effect and are hence independent of the particle size and surfactant concentration [88]. This characteristic is consequently a fundamental difference between conventional radical polymerization in miniemulsion polymerization and NMP. Moreover, control over molar mass according to the monomer/initiator molar ratio and narrow molar mass distribution are not different features from those also commonly observed in bulk. The exit of the primary radicals from the monomer droplet/polymer particle is not a major event because the droplet size is large enough that initiation is usually favored over exit and potential termination reactions in the aqueous phase [89]. The partition coefficient of the nitroxide between the oil phase and the aqueous phase should remain in favor of the oil phase, otherwise fast polymerization together with large polydispersity indexes can be observed. For high precision in macromolecular engineering, preformed alkoxyamines are preferred over two-component initiators, because the latter suffer from low initiator efficiency and imprecise control over the concentration of free nitroxide and hence over the polymerization kinetics.

With water-soluble radical initiators (two-component systems with potassium persulfate, potassium persulfate–sodium metabisulfite in conjunction with TEMPO or SG1; monocomponent initiators with a water-soluble hydroxycarbonylprop-2-yl sodium salt initiating radical attached to the SG1 [90]), the quality of control depends on the events taking place in the aqueous phase together with the extent of mass transfer from the aqueous phase to the oil phase. When degradation of the nitroxide takes place in the aqueous phase (owing, for instance, to a strong pH decrease with persulfate) or when preferential irreversible termination reactions occur before efficient entry of the oligoradicals in the monomer droplets, the quality of molar mass control can be dramatically affected [3, 6].

When all parameters have been well adjusted to achieve good control over molar mass and molar mass distribution along with high chain-end functionality, it is very easy to add a new monomer after consumption of the first one and then design block copolymers [3, 6]. It should be kept in mind, however, that, in NMP, 100% conversions are never reached due to the persistent radical effect [91], which induces a continuous release of free nitroxide throughout the polymerization, concomitantly with the production of dead chains; this kinetic feature leads to a continuous reduction in the polymerization rate. Consequently, a compromise between high conversion and high chain livingness has to be found. Since in a latex it is sometimes difficult to eliminate the residual monomer, a random copolymerization takes place in the second step. Again, the best conditions have to be found for acceptable purity of the second block in relation to the target properties of the diblock copolymer. In parallel with block copolymers, other types of architectures have been developed using NMP, such as random/gradient copolymers [86] and gels, for which the polymerization in miniemulsion was shown to affect the cross-link density with respect to a bulk process [92].

Finally, in most of the examples from the literature dealing with NMP, the miniemulsion process was able to afford stable latexes with a fair amount of surfactant. In none of these studies, however, did the latex exhibit a very narrow particle size distribution. Degradation of the colloidal properties was observed during the polymerization, most probably because of the high temperature and the long polymerization times (significantly longer than in conventional free radical polymerizations) [3–6].

14.3.2.2 Atom-transfer Radical Polymerization

ATRP conducted in water-borne systems and particularly in a miniemulsion polymerization process has been reviewed recently [3, 6, 93]. The polymerization follows the same kinetic principles as NMP, but with a difference of importance for miniemulsion polymerization: the requirement for a metallic complex activator to produce radicals, instead of the thermal dissociation of the dormant chain end. Consequently, the solubilities of both the activator and the deactivator have to be taken into account. Most, if not all, of the studies in miniemulsion polymerization have been performed with a copper–complex catalyst. The important criterion to fulfill should then be high solubility in the monomer phase of both the Cu(I) and Cu(II) complexes, which emphasizes the crucial role of the ligand. Indeed, only ligands with a sufficiently high hydrophobicity are efficient. Because all organic species are initially located in the monomer droplets and no transport is required, their water solubility is not an issue. Additionally, selection of a suitable surfactant is a nontrivial task; indeed, only nonionic and cationic surfactants that do not interact with the catalyst led to living polymerizations, but only with the former were stable latexes achieved. With respect to NMP, a wider variety of monomers have been successfully polymerized via ATRP in miniemulsion, namely styrene, acrylates and methacrylates, and the polymerization temperature was much lower, typically below 100 °C. Ultrasonication has always been the typical homogenization method, together with the addition of hexadecane to prevent Ostwald ripening.

Miniemulsion polymerization was applied to direct and reverse ATRP processes, with sometimes the need for a high surfactant concentration to maintain the colloidal stability of the latexes, a feature that depends, however, on the catalyst activity: a high activity allows the temperature to be decreased, which is in favor of better stability. Technically, reverse ATRP was preferred owing to the poor stability in air of the Cu(I) activator needed in direct ATRP, but the method invariably suffered from low initiator efficiency. Moreover, reverse ATRP requires high deactivator concentrations and does not allow the design of complex architectures. The most remarkable results in miniemulsion polymerization were obtained when novel initiation methods were applied, namely simultaneous reverse and normal initiation (SR&NI) [94] and activator generated by electron transfer (AGET) [95]. Both methods entail the invaluable advantage of using an alkyl halide initiator as in direct ATRP together with Cu(II) deactivator as in reverse ATRP. Such components are then perfectly suited for the miniemulsion polymerization technique. With a proper choice of the ligand, the methods allow the metal concentration to be drastically reduced. Moreover, AGET can also be carried out in the presence of air. For the polymerization to start, Cu(II) has to be converted into Cu(I), either via dissociation of a conventional oil-soluble or water-soluble radical initiator followed by deactivation of the primary radicals or oligoradicals by Cu(II) (SR&NI) or via *in situ* reduction of Cu(II) into Cu(I) by the water-soluble ascorbic acid. With SR&NI [96, 97], employing very hydrophobic catalysts in low concentration (0.2 equivalent based on the alkyl halide initiator), good control over the polymer growth was achieved with molar masses being mainly controlled by the initial concentration of the alkyl halide initiator.

Direct ATRP in miniemulsion has been used to generate diblock copolymer particles in a two-step method [98]. An additional advantage of ATRP with SR&NI [99] and AGET [100] concerns the possibility of synthesizing complex (co)polymer architectures, under more convenient and more efficient experimental conditions. Complex architectures such as triblock copolymers or stars result from the possible use of an alkyl halide initiator or a macroinitiator with a functionality larger than 1. This is illustrated by the synthesis of three-arm star and star-block copolymers. Moreover, AGET has the unique advantage of avoiding the presence of an additional conventional initiator, hence leading to well-defined block copolymers without residual homopolymer.

14.3.2.3 Control via Reversible Chain Transfer

Because the method is based on a chain transfer reaction, with the simple addition of a chain transfer agent to an otherwise identical polymerization recipe, virtually any kind of process commonly applied to free radical polymerization can also be applied to CRP via reversible chain transfer reaction. In general, the chain transfer agents are oil-soluble molecules. It is important to remember, however, that the method requires the use of an initiator, able to produce radicals throughout the polymerization reaction. Consequently, in a multiphase system, careful attention should be paid to the loci of radical generation and growth and to their continuous ability to react with the chain transfer agents, in order to achieve good quality of control over molar mass and molar mass distribution. In particular, a secondary nucleation has to be prevented, as it might lead to the formation of particles free of any chain transfer agent, where chains would grow in a non-controlled manner. In miniemulsion polymerizations, two main methods have actually been employed [3, 6, 101]: iodine-transfer polymerization (ITP) and RAFT (generic acronym employed here for all methods based on a reversible addition–fragmentation chain transfer reaction, for which the chain transfer agents are unsaturated polymethacrylates or most commonly thiocarbonylthio compounds). All of them operate at low temperature, i.e. below 100 °C.

The miniemulsion polymerization process is particularly convenient when the added chain transfer agent is too hydrophobic to diffuse through the aqueous phase. Consequently, thanks to its initial location within the monomer droplets, the respective concentrations of monomer and control agent are the same in all the polymerization loci. This was well illustrated in ITP when the very hydrophobic $C_6F_{13}I$ was used as a chain transfer agent [102, 103]: the molar mass and molar mass distribution were identical with those observed in bulk. In addition, the colloidal properties of the latexes were not altered with respect to those achieved in a conventional free radical polymerization; this was explained by the low chain transfer constant of $C_6F_{13}I$ along with its insolubility in the water phase; as a consequence, whatever the type of initiator (oil-soluble or water-soluble), the first chain transfer events take place in the monomer droplets and lead to sufficiently long oligomers to be trapped inside.

When the chain transfer agent is much more reactive (a dithio ester, for instance), two types of difficulty have been encountered, however: rate retardation and colloidal instability. The rate retardation is a fairly common characteristic of the RAFT method in bulk or solution, when very effective RAFT agents such as dithiobenzoates are used, but it was somehow enhanced in dispersed systems, owing to the ability of the primary radicals created upon transfer to exit the particles and potentially terminate in the aqueous phase [104]. This situation was, of course, dependent on the hydrophilicity of the leaving radical of the RAFT agent. Rate retardation was less pronounced when a macromolecular chain transfer agent was used [105] (oligomer or polymer synthesized in a preliminary bulk polymerization in the presence of a RAFT agent). The colloidal instability resulted in the formation of a colored organic top layer during the polymerization, which contained part of the monomer and living oligomers. The existence of a high concentration of very short oligomers was suspected to modify completely the thermodynamic properties of the dispersed phase and to result in a superswelling state [106], favoring the formation of very large droplets and leading to destabilization of the system. To prevent this effect, nonionic were preferred to ionic surfactants and the concentration of the hydrophobe (hexadecane or Kraton) was increased [107].

In conditions where the stability issues are overcome, it is possible to synthesize well-defined homopolymers and block copolymers from a variety of monomers, with good prediction over molar mass and narrow molar mass distribution, using miniemulsion polymerization via RAFT [3, 6, 80, 101]. Moreover, particles with target morphologies such as capsules with a liquid core and a well-defined polystyrene shell have been prepared using this method [108].

The field of controlled free radical miniemulsion polymerization via RAFT is very active and the number of publications shows constant growth. The most recent developments concern, for instance, mechanistic understanding [109–111] and process implementation [112, 113].

14.3.2.4 Other Controlled Free Radical Polymerization Methods

Cobalt-mediated miniemulsion polymerization allowed well-defined poly(vinyl acetate) to be prepared at low temperature (0–30 °C) with a fairly fast rate, yielding latexes with small particles (diameters ~ 100 nm) and good stability [114]. It is one of the best methods (besides RAFT using a xanthate as a chain transfer agent) to produce well-defined poly(vinyl acetate) and its successful implementation in an aqueous dispersed system is an important step.

14.3.3
Ring-opening Metathesis Polymerization

As stated for aqueous suspension polymerization, the development of ROMP catalysts tolerant to water allowed the application of the polymerization of norbornene [115, 116], cyclooctene [116] and cyclooctadiene [116] in a miniemulsion process, to yield stable latexes with submicrometer particles. In the earlier paper, however, although the term miniemulsion was employed [116], the process was not exactly of the miniemulsion type: first, droplets of the hydrophobic $(Cy_3P)_2Cl_2Ru=CHPh$ initiator in an organic solvent were formed employing the techniques of emulsification and stabilization usually applied in miniemulsion polymerization and only then was the monomer added. Consequently, monomer transport was necessary to ensure polymerization in the catalyst droplets. The results were, however, not completely satisfactory from both the colloidal and kinetic viewpoints. More recent results of ring-opening metathesis aqueous miniemulsion polymerization of norbornene have been published [115]. With the oil-soluble initiator $(Cy_3P)_2Cl_2Ru=CHPh$, using two different procedures of introduction (adding the initiator solution to a miniemulsion of norbornene or adding norbornene to an aqueous miniemulsion of the initiator in toluene solution), the miniemulsion polymerization did not result in the formation of a stable latex. In contrast, with a water-soluble initiator based on $RuCl_3$ and an alcohol, stable particles with diameters in the range 200–500 nm were achieved in the presence of sodium dodecyl sulfate or poly(styrene)-b-poly(ethylene oxide) stabilizer and hexadecane. Conversion of norbornene ranged from 10 to 90% after 15 h of reaction, depending on the experimental conditions.

For more detailed information, the reader may refer to review articles [8, 9] and to other chapters of this book.

14.3.4
Catalytic Polymerization of Ethylene and Butadiene

The polymerization of olefins in aqueous dispersed systems is a new and challenging domain [8, 117, 118]. Late transition metal complexes are less water-sensitive than the early transition metal Ziegler or metallocene catalysts and therefore allow the polymerization of ethylene to be conducted in the presence of water [8, 119]. With a lipophilic catalyst based on nickel(II), the miniemulsion technique was applied to disperse the catalyst solution in the aqueous phase, in the presence of a surfactant and a hydrophobe. Then, ethylene was continuously fed into the reactor under pressure. As described earlier for ROMP, this method differs substantially from the traditional miniemulsion polymerization, in which a monomer-in-water miniemulsion is initially formed. The method nevertheless led to polyethylene latexes, but those were unstable in the long term. This drawback was explained by the high crystallinity of the polymer, responsible for the generation of a much larger number of particles than the initial droplets owing to the expulsion of the crystallites from the organic phase, leading to the formation of irregular, non-spherical, submicrometer particles [120]. When ethylene was copolymerized with a long-chain α-olefin, crystallinity decreased and a stable, 30% solids latex could be recovered, with spherical particles exhibiting a smooth surface [121]. It was moreover found that the process allowed a higher incorporation of the comonomer than in the case of organic phase copolymerization.

In contrast to ethylene, the aqueous phase catalytic polymerization of butadiene is much older [8], but did not always lead to stable latexes. The recent synthesis of 1,2-polybutadiene stable latexes with a controlled microstructure was reported by Monteil et al. [122], using a miniemulsion of a cobalt(I)–allyl complex catalyst organic solution in water, with sodium dodecyl sulfate as a surfactant and hexadecane as a hydrophobe. Stable latexes with particle diameters of $\sim$200 nm were obtained.

14.3.5
Ionic Polymerization

The advent of miniemulsion systems allowed significant progress to be made in the ionic polymerization of cyclosiloxanes and organic monomers in aqueous media. Indeed, in most cationic and some anionic catalyzed processes, polymerization occurs compulsorily at the droplet interface, which is favored by the large specific surface area offered by a miniemulsion system. On the other hand, monomers, for which the polymer chain end is reactivable, can be equally polymerized in miniemulsion (still giving low molar mass polymers, see below) or (preferentially) in other processes, including emulsion polymerization and polymerization in spontaneously formed monomer-in-water emulsions (for a recent review, see [123]).

14.3.5.1 Irreversible Deactivation of the Chain Ends by Water

Indifferently, heterocycles, such as phenyl glycidyl ether (PGE) [124] and vinyl monomers, for instance *p*-methoxystyrene (pMOS) [125, 126], were polymerized by ionic polymerization in miniemulsion, using anionic and cationic processes, respectively. The main criteria for the choice of the monomers are threefold: (i) hydrophobicity, to generate a stable enough microsuspension or miniemulsion without the need for a hydrophobe; (ii) stability towards hydrolysis at "extreme" pH; and (iii) a fairly fast polymerization rate in bulk systems, with conventional acid or base initiators.

The choice of the surfactant is crucial in such a polymerization process, for which all steps (initiation, propagation, termination) are located at the interface (Fig. 14.2). The surfactant both draws the catalyst (H^+ or OH^-) towards the in-

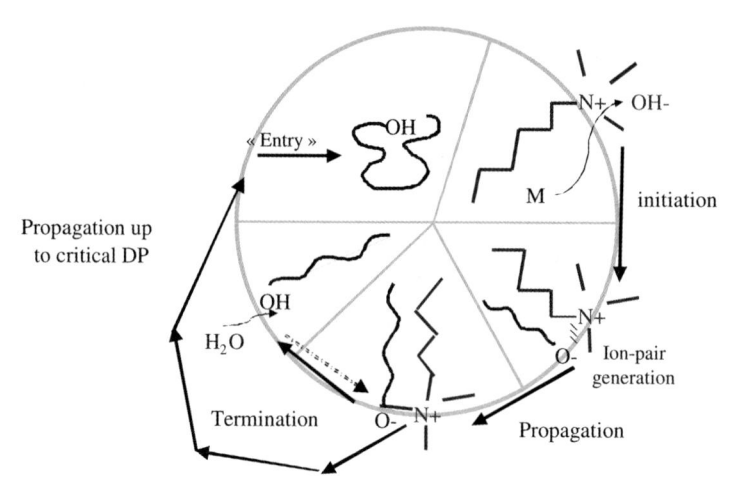

Fig. 14.2 Mechanism of interfacial (an)ionic polymerization in miniemulsion.

terface and, after initiation, pairs with the propagating chain ends to generate a bulky, and thus highly reactive, ion pair. Once chains have been generated, they grow at a fast rate until the oxyanion or carbocation undergoes chain transfer to water, which kills the chains irreversibly. Note that here the final molar mass is only a function of the ratio of propagation to termination rate constants and can sometimes reach several tens of thousands of g mol^{-1} (see, for instance, the cationic polymerization of tetramethylcyclotetrasiloxane or D_4^H [137]). Surfactants bearing ammonium hydroxide and phenylsulfonic acid heads are preferred for carrying out anionic and cationic polymerizations, respectively. The nature of the alkyl side-chains is also important, since it should ensure good stabilization of the monomer droplets after sonication and limit the content of interfacial water. A superacid–nonionic surfactant couple was also shown to work in the case of pMOS [126].

For irreversibly deactivated chains prepared from PGE [124] and pMOS [125], the average degree of polymerization (*DP*) of the oligomers is directly related to the "polarity" of the interface, which changes slightly with conversion. Indeed, oligo(PGE) and oligo(pMOS) bear two and one OH end-groups, respectively, thus acting as cosurfactants in the system and evacuating, to some extent, water from the interface. Typical molar masses ranged from 500 [124] to 1000 g mol^{-1} [125] in these systems. The colloidal state of the miniemulsion may be changed, however; for instance, by adding hexadecane or by opting for a dialkylated surfactant [126], smaller droplet sizes could be achieved, hence increasing the monomer/droplet interfacial area and leading to faster kinetics. In contrast to microsuspension systems, the hydrophobe added in the miniemulsion process also permits the dispersions to be better stabilized once the monomer has been totally consumed (side-reactions such as chain scissions occur in the case of pMOS and possible diffusion of the low molar mass species thus created might affect the particle size distribution and latex stability [125]). Two recent studies made use of ytterbium Lewis acid complexes to try to control the polymerization [127, 128]; however, both failed to produce a living polymerization system as the ytterbium salt catalyst instantaneously separated into inactive ions; interfacial polymerization is again only induced by some protonic acids added to the system.

14.3.5.2 Reversible Deactivation of the Chain Ends by Water

The fact that monomers exhibiting living polymerization may stop propagating at relatively low *DP* has been understood recently, mainly from studies on cyclosiloxanes. The so-called "critical *DP* limit" (Fig. 14.2) corresponds to the point where deactivated chains are no longer surface-active and prefer the core of the particle rather than the interface. These chains do not easily come back to the particle surface, where they might be activated again, and propagation simply ceases. Such a limit *a priori* does not apply to systems where termination reactions are very slow compared with propagation, for instance alkyl cyanoacrylates polymerized in a spontaneous emulsification system gives chains of $\geq 10^4$ g mol^{-1} [123].

Anionic and cationic homopolymerizations of cyclosiloxanes in miniemulsion have been extensively studied, first by researchers in industry [129, 130] and more recently by academic groups [131–139], because of their clear industrial interest, inherent reactivity and facile preparation of monomer-in-water miniemulsions. Working in aqueous media has been motivated by the peculiar kinetics (compared with bulk polymerization) of the numerous reactions involved in those systems, namely initiation, propagation and opposite back-biting reactions, termination and opposite chain-end reactivation, redistribution, polycondensation and opposite hydrolysis reaction (both last reactions treated in a following section 14.3.6) [131]. Similar catalyst/surfactants to those reported before were used, principally fatty sulfonic acid and quaternary ammonium [132]. In the specific model studies of anionic polymerizations of octamethylcyclotetrasiloxane or D_4 [134], the basic features of ionic emulsion polymerization were respected, e.g. slow initiation, fast propagation and back-biting and reversible termination, all of them taking place at the interface. Chains were limited to molar masses of about 2500 g mol^{-1}, the critical DP limit for chains propagating on both ends, before increasing after 70% conversion, thanks to polycondensation reactions. The overall ratio of rings to linear chains was found to be slightly smaller than that found in the bulk polymerization thermodynamic equilibrium.

Functional polycyclosiloxanes were also prepared via anionic polymerization of rings with three $SiO(CH_3)R$ units (with $R=CH_2CH_2CF_3$ [135] or phenyl [133]) and four $SiO(CH_3)R$ units (with R=vinyl [136] or phenyl [133]). Similar kinetic features were found, except for a faster polymerization for the highly strained monomers (sometimes to a point where conversion goes up to 100% before decreasing to the equilibrium state) and a larger final ring content (however, remaining below that observed in bulk). More recently, D_4^H was also polymerized using a cationic process and different surfactant mixtures [137–139]. Good control of the polymerization was achieved using DBSA and a nonionic surfactant, although the extensive redistribution of D_4^H into larger rings was not clearly explained. The best system in terms of rate of conversion and control over the linear chain molecular characteristics was nevertheless achieved by employing fatty phosphonic acids. One should finally note reports on copolymerization studies with a view to preparing functionalized random [140–143] and/or block copolymers [136].

Apart from cyclosiloxanes, most of the conventional monomers that can be polymerized anionically and for which chain ends are reactivable have hardly been tested in miniemulsion polymerization, since pre-emulsification is not a criterion for polymerization to proceed. One notable exception concerns the polymerization of n-butyl cyanoacrylate (BCA) in miniemulsion [144], where a sulfonic acid surfactant was used both as a stabilizer and to bring protons to the monomer/water interface, which prevents too fast polymerization.

14.3.6
Polycondensation/Polyaddition Reactions

As for ionic polymerization systems, polycondensation and polyaddition reactions in miniemulsion have attracted interest recently, since they allowed nano- (in place of micro-) spheres or capsules to be prepared. Miniemulsion polymerization can also be considered here as a model of a dispersed medium for studying the kinetics and mechanisms of complex chemistry processes.

The polycondensation of bis-silanol polydimethylsiloxane (PDMS) using acid [130, 145] and base [146] catalysts was very different in terms of ring formation and final molar masses. Anionic catalysis favors intramolecular back-biting, so that rings quickly build in the particles. In contrast, cationic catalysis favors chain-end reaction, including end-biting, and therefore rings are very few, even after long polycondensation times. The difference between the two processes lies in the content and nature of rings present in the oil droplets, which sets the content of intraparticular water and thus the equilibrium reaction between polycondensation and hydrolysis reactions. Other key factors are the monomer/water ratio and the temperature, which again affect the water content inside the droplets [145]. A recent article [147] proposed to cross-link bis-silanol polydimethylsiloxane by tetraethoxysilane in microsuspension particles, using a sodium dodecyl sulfate–HCl catalytic complex.

Hydrolysis–condensations of alkoxysilanes (including trifunctional monomers) have hardly been reported in the open literature, since seminal work done by Dow Corning [129]. Dimethyldiethoxysilane (DMDES) [148] was added drop-by-drop to a water–ammonia mixture, the latter being the catalyst of the reaction. The species formed were mainly D_4 and small oligomers (particularly when adding methanol), i.e. polymerization was not favored. The stability of these emulsions, where no surfactant was added in the recipe, were recently explained by the high content of hydroxyl groups (arising from hydrolysis reactions) at the droplet/water interface [149]. The control of the particle size was, in a second approach, exerted by introducing surfactants in the recipe, among which silicone-based surfactants led to particles of about 25 nm in diameter [150]. However, only a mixture of di- and trifunctional alkoxysilanes was able to produce resins of various shapes and cross-linking densities [151]. Further studies [152, 153] took advantage of interfacial catalysis by an ammonium hydroxide surfactant to enhance the kinetics by decreasing the droplet size, in addition to preparing high molar mass silicones. Cross-linked nanocapsules were also recently achieved by first preparing polydimethylsiloxane dispersions by a two-step process of hydrolysis–condensation of either DMDES alone or with the trifunctional monomer, then removing linear PDMS chains by solvent extraction [154]. Further, the instantaneous hydrolysis of dichlorodimethylsilane, in the presence of an electrosteric surfactant (to resist the high load of ions produced), first generates cyclosiloxanes that later polymerize according to a conventional cationic polymerization process [155].

Independent of silicone chemistry, three other monomer systems have been applied for polycondensation-polyaddition reactions in miniemulsion, namely epoxide–amine [156], isocyanate–alcohol [157–160] and an esterification system [161] similar to that reported by Saam et al. [51]. In all cases, strictly water-insoluble components were chosen to ensure efficient reaction. No clear differences could be found in terms of the influence of particle size on polymerization rate and conversion, e.g. compared with the suspension polymerization processes. Various studies on polyurethane chemistry showed that an aliphatic diisocyanate, typically isophorone diisocyanate (IPDI), hydrolyzes relatively slowly under these conditions, despite the large water content in the small droplets. Furthermore, polyurethanes with molar masses of up to 7×10^4 g mol^{-1} were obtained by adding excess IPDI over diol (twofold) and incorporating an organic tin catalyst in the recipe [158].

14.4
Aqueous Emulsion Polymerization

14.4.1
Conventional Free Radical Polymerization

Emulsion polymerization is by far the most studied heterogeneous polymerization process and has been the subject of a multitude of books [1, 74, 162–167] and articles since 1947 [168–170]. This comes mainly from its huge industrial interest and also its intrinsic complexity. Like miniemulsion polymerization described earlier, emulsion polymerization leads to particles of submicrometer size that remain suspended in the aqueous medium, thus forming a latex. The particles are stabilized against flocculation and coalescence by a surfactant, which is adsorbed on their surface and provides stabilization by either an electrostatic effect (anionic and cationic surfactants), steric effect (nonionic surfactants) or electrosteric effect (amphiphilic polyelectrolytes). Emulsion polymerization differs, however, from miniemulsion in the quality of the initial monomer-in-water emulsion. In *ab initio* emulsion polymerization, an unstable emulsion of the monomer in water is formed by stirring the mixture in the presence of a surfactant, at a concentration usually larger than the CMC. The monomer is partitioned between different phases: the large monomer droplets (diameter >1 μm) formed by stirring and stabilized by the surfactant adsorbed at the interface, the continuous water phase (saturation concentration), which also dissolves the initiator, and free surfactant molecules and the micelles (when [surfactant]>CMC).

The polymerization is started in the aqueous phase by the introduction of a water-soluble radical initiator (generally a negatively charged compound such as persulfate), which forms oligoradicals upon initiation and subsequent polymerization with the dissolved monomer molecules. When the oligoradicals become surface-active, they are captured by the monomer-swollen micelles of surfactant and generate particles by the so-called micellar nucleation mechanism. Only part of the micelles are actually nucleated, the others serving as surfactant reservoirs to stabilize the interfaces created by nucleation and particle growth. Nucleation ceases when all micelles have been consumed. Depending, however, on the surfactant concentration and the water solubility of the monomer, formation of the particles might also follow a so-called homogeneous nucleation mechanism, for instance in the case of methyl methacrylate and vinyl acetate. In that situation, the oligoradicals grow in the water-phase by monomer addition until they become insoluble and precipitate; colloidal stability of such primary particles is ensured by the charged fragment of the initiator at the chain end and might be enhanced by the limited coagulation of several nuclei in order to increase the charge surface density (homogeneous–coagulative nucleation). Nucleation ceases when capture of the oligoradicals by the already existing particles prevails. In all cases, droplet nucleation is and should be negligible. The monomer is actually not polymerized in the droplets but in the particles only, which swell with monomer as soon as they are formed. Consequently, the droplets act only as a monomer reservoir and provide monomer(s) to the polymerization loci via molecular diffusion through the water phase. An accurate understanding of the particle nucleation mechanisms is not only interesting for fundamental knowledge, but also is particularly useful for predicting the final number of particles, N_p, to quantify the effect of the various parameters such as the type of surfactant and its concentration, the type of initiator and its concentration, the nature of the monomer(s) and the temperature and to predict the occurrence of continuous or secondary nucleation. The nucleation step (or particle formation stage) is usually referred to as *interval I* of an *ab initio* emulsion polymerization. During *interval I*, both the number of particles and the polymerization rate increase. The duration of *interval I* varies within the range 2–10% conversion and its end corresponds to the stabilization of the value of N_p. Polymerizations starting with *interval I* are called "*ab initio* emulsion polymerizations". When the emulsion polymerization is started with preformed polymer particles, it is called "*seeded* emulsion polymerization" and begins directly with *interval II*. Detailed descriptions of the nucleation step can be found in several well-documented books (e.g. [1, 162]).

Interval II corresponds to the main step of particle growth: propagation is permitted by the capture of oligoradicals generated in the aqueous phase and polymerized monomer is continuously replaced by new monomer molecules supplied from the droplets via diffusion through the water phase. During *interval II*, both the number of particles and the monomer concentration inside the particles remain constant (the latter corresponds to the maximum concentration of monomer in a polymer particle, as a result of a thermodynamic equilibrium described by Morton's equation [171]). Accordingly, the polymerization rate is relatively constant, as it is proportional to those values and also to the average number of radicals per particle. Propagation thus obeys a zero internal order with respect to monomer concentration during *interval II* and not a first order as observed in homogeneous polymerizations and suspension polymerization. *Interval II* may extend from 5–10% to 30–70% conversion, ending with the disappearance of the monomer droplets. A complete description of the kinetics of particle growth can be found in Gilbert's book [1] with the presentation of all models that have been developed. The unique feature of kinetics in emulsion polymerization results from the compartmentalization of the propagating radicals within separate particles. The polymerization kinetics are not simply depicted by initiation, propagation and termination as in a homogeneous radical polymerization. The situation is complicated by the existence of at least two phases for these events to take place. Continuous entry of oligoradicals from the water phase into the particles, exit of propagating radicals from the particles via transfer reactions to small molecules, transportation of monomer, etc., all have to be considered. A direct consequence of the compartmentalization of the propagating radicals is a much faster polymerization rate than in a homogeneous system, owing to a much higher overall concentration of propagating radicals.

Interval III is characterized by a reduction in the polymerization rate, resulting from the decreasing monomer concentration inside the polymer particles. Hence first-order kinetics are expected in this final stage until a gel effect takes over and the polymerization rate can start to increase again.

The molar masses of the polymers obtained from emulsion polymerization (usually in the order of 10^6 g mol^{-1}) are significantly larger than those obtained from bulk polymerization, due to a longer lifetime of the propagating radicals resulting from compartmentalization. This is basically due to the fact that it is not the rate of chain termination that determines the lifetime of each growing chain, but the rate of entry of new radicals into the particles. It is therefore possible in an emulsion polymerization to increase simultaneously the molar mass and the polymerization rate by increasing N_p, whereas in bulk or suspension polymerization the molar mass decreases when the rate is increased (via an increase in the initiator concentration). The molar mass distribution is predicted to be broader in general compared with homogeneous systems.

Various processes other than a simple batch system can be easily used and offer advantages for the control of the polymerization (kinetics, number of particles, heat control, latex stability, copolymer composition, etc.). For instance, a semi-batch process with starved monomer feed allows control of the propagation rate and of copolymer composition, avoiding the composition drift observed in batch systems. A similar process is also used to create particles with unique morphologies, the core–shell structure being the most common.

Numerous applications exist for polymers prepared in emulsion systems (use of the latex itself after formulation or of the polymer after coagulation or film formation) and their number will probably increase in the future owing to environmental regulations aimed at limiting the use of volatile organic compounds in industry. For instance, emulsion polymerization has been the dominant process used for the synthesis of poly(vinyl acetate), polychloroprene, poly(acrylic ester) copolymers and styrene–butadiene-based synthetic rubbers.

14.4.2
Controlled/Living Free Radical Polymerization

For industrial applications, emulsion polymerization is considered to be the preferential process to produce latexes, as it is much simpler and more convenient for large scales than miniemulsion polymerization. Applying CRP in such a process and particularly in *ab initio* emulsion polymerization has therefore been a target of major importance. However, unexpected difficulties were encountered and delayed the success of the method. For all three techniques, NMP, ATRP and RAFT, the main reason was related to the complexity of the nucleation step. The most recent and best achievements actually overcame those difficulties by essentially varying both the chemical nature of the control agent and the process employed [3, 6].

14.4.2.1 Nitroxide-mediated Controlled Free Radical Polymerization

Very few articles have described the use of TEMPO- and SG1-mediated CRP, entirely in *ab initio* emulsion polymerization. The first ones were related to the use of a two-component initiating system with a water-soluble radical initiator in a batch process, but the results were poor regarding the latex stability, the colloidal characteristics of the particles and the quality of control over molar mass and molar mass distribution [3, 6]. Similar results and in particular poor stability were observed with water-soluble alkoxyamines when they were employed as initiators in a batch emulsion polymerization process. The explanation invoked was the prevalence of nucleation of the large droplets, owing to the partitioning of the long-lived oligoalkoxyamines between water and monomer. In a better approach, a seed latex of preformed TEMPO-based polystyrene alkoxyamine was prepared via a microprecipitation process [172]: an acetone solution of the oligoalkoxyamine was slowly added to a water phase containing poly(vinyl alcohol) as a stabilizer according to the solvent shifting spontaneous emulsification technique [123]; after elimination of acetone, the seed particles obtained were swollen with styrene and the polymerization was performed by heating the suspension to the desired temperature. The results were fairly satisfactory from the macromolecular and colloidal viewpoints, but the need for an organic cosolvent cannot be considered fully acceptable for industrial production. The application of water-soluble SG1-based mono- and difunctional alkoxyamines in a multistep process actually gave the best results so far [173–175]. To favor micellar nucleation instead of monomer droplet nucleation, the first step of the process was the preparation of a very low solids content seed latex, with a low monomer concentration with respect to that of the surfactant. In the second step, a "one-shot" or continuous monomer addition (*n*-butyl acrylate and/or styrene) was performed, to reach solids contents up to ~ 30 wt%. Well-defined homopolymers and also di- and triblock copolymers, depending on the initiator, were obtained. With the monofunctional alkoxyamine initiator, the latexes were stable but exhibited rather large diameters and broad particle size distributions.

The unique structure of the dialkoxyamine initiator and especially its two negative charges have permitted the average diameters to be dramatically decreased and narrow particle size distributions to be obtained. With a living SG1-capped poly(sodium acrylate) macroinitiator, surfactant-free emulsion polymerizations of styrene and n-butyl acrylate were successfully achieved in *ab initio* batch conditions, but with a low initiator efficiency owing to termination reactions in the aqueous phase. The method led to amphiphilic diblock copolymer nanoparticles, analogous to the so-called crew-cut micelles, with a narrow particle size distribution [176, 177].

14.4.2.2 Atom-transfer Radical Polymerization

Direct ATRP is still difficult to achieve under true emulsion polymerization conditions, owing to the requirements for a water-soluble initiator and mass transfer of living chains from the aqueous phase to the oil phase, while maintaining good quality of control. With an oil-soluble initiator, the system was often referred to as a microsuspension polymerization and the mechanistic aspects of nucleation still have to be elucidated [3, 6, 178]. At present, only the reverse ATRP method affords suitable conditions for emulsion polymerization to be carried out. Most studies were focused on the polymerization of n-butyl methacrylate, using Cu(II) in conjunction with dialkylbipyridine ligand as the deactivator and a nonionic surfactant. The water-soluble azo initiators led to better control than persulfate and the most effective temperature was 70–90 °C. Well-controlled polymerizations were usually achieved, exhibiting a linear increase in molar mass with monomer conversion and low polydispersity index, but the initiator efficiency invariably remained low, as a consequence of extensive termination reactions in the water phase. From the colloidal viewpoint, the latexes were very stable with particle diameters ranging from 150 to 300 nm, depending on the experimental conditions.

In addition to the production of latex particles containing well-defined (co)-polymer chains, ATRP has also been used to initiate polymer chains at the surface of latex particles. The latter were produced in conventional free radical emulsion polymerization and in a final stage an organic comonomer containing a halide group was introduced to create a functional shell with attached ATRP initiator. Water-soluble chains were grafted from the particles by surface-initiated aqueous ATRP and allowed hairy particles to be formed [179–181].

14.4.2.3 Control via Reversible Addition–Fragmentation Chain Transfer

In the early days of the reversible addition–fragmentation chain transfer method, unsaturated polymethacrylates were generated via a catalytic chain transfer reaction in free radical emulsion polymerization. In a second step, a methacrylic monomer was added to the latex under starve–feed conditions and the polymerization proceeded with reversible addition–fragmentation chain transfer, leading to chain extension and hence to the formation of a diblock copolymer [3, 182].

This was one of the first examples of the generation of block copolymers via emulsion polymerization.

In *ab initio* batch emulsion polymerization, fast diffusion of the oil-soluble chain transfer agent from the monomer droplets to the polymer particles is a key parameter, in order to allow molar masses to be well controlled [3, 6, 101, 183]. Most of the xanthates studied so far exhibited sufficient water solubility and diffusion was not the rate-determining step of their consumption. Because of a low chain transfer constant, semi-continuous addition in starved monomer conditions were sometimes applied in order to favor the transfer reaction over propagation [184]. With a fluorinated xanthate agent of high chain transfer constant, *ab initio* emulsion polymerization could nevertheless be carried out with good control over molar mass and a relatively narrow molar mass distribution, in well-defined latex particles [185]. Although this was not a major effect in bulk polymerization, in emulsion polymerization the rate was found to be retarded when the amount of xanthate was increased. This result was assigned to an enhanced exit of the primary radicals formed upon transfer in the particles, along with an unexpectedly slow rate of entry of the oligoradicals from the aqueous phase [186]. With the RAFT agents exhibiting a low chain transfer constant, the colloidal properties of the latexes are generally not an issue. In a seeded emulsion polymerization process, which avoids the difficulties with the nucleation step, well-defined latexes of block copolymers were synthesized [187]. With the more reactive dithio ester RAFT agents, in contrast, severe rate retardation and stability problems were encountered with phase separation leading to a loss of part of the control agent and hence to poor control of the polymer characteristics [3, 6]. Again, droplet nucleation had to be avoided. Most of the difficulties encountered were overcome by utilizing an amphipathic asymmetric trithiocarbonate RAFT agent in semi-continuous monomer addition. The polymerization was first started with acrylic acid in alkaline conditions to form *in situ* a water-soluble oligomeric RAFT agent. A hydrophobic acrylate monomer was then added at a slow feeding rate to form amphiphilic diblock copolymers able to self-assemble into micelles where polymerization further took place by chain extension. All the characteristics of a controlled polymerization were observed together with good colloidal properties of the surfactant-free latexes formed [188].

In a different way, polyelectrolytes with a dithio ester chain-end functionality can be employed as reactive hydrophilic species in surfactant-free batch emulsion polymerization in order to impart electrosteric stabilization to the latex [189]. Such species, which are synthesized in a single polymerization step, via a RAFT method, are a new alternative to the former water-soluble macromonomers, with the advantages of being much easier to prepare and of being able to react in the early stage of the emulsion polymerization reaction. In such a situation, control over all the characteristics of the polymer formed is not a major target, but the method offers a new tool towards surfactant-free emulsion polymerization, which is of high industrial relevance.

14.4.3
Ring-opening Metathesis Polymerization

An example of ROMP of norbornene in emulsion was given by Claverie et al. [116], using ruthenium-based water-soluble initiators with tris(3-sulfonatophenyl)phosphine ligands. The polymerization yielded high molar mass polynorbornene in high solids content latexes with submicrometer particles and a relatively narrow particle size distribution. A homogeneous–coagulative nucleation mechanism was suspected, even in the presence of micelles of surfactant.

14.4.4
Catalytic Polymerization of Ethylene

In the various articles by Mecking and coworkers devoted to the aqueous polymerization of ethylene, the nickel(II)-based catalyst was hydrophilic, which allowed an emulsion polymerization process to take place in the presence of a surfactant [8, 117, 118, 190]. The nucleation mechanism was supposed to be similar to that usually observed in free radical emulsion polymerization, with capture of the insoluble growing chains by the micelles. Stable polyethylene latexes exhibiting submicrometer particles were formed by this process. As in miniemulsion polymerization, the semicrystalline structure of polyethylene induces the formation of non-spherical particles, some of them consisting of a single lamella [191].

14.4.5
Ionic Polymerization

Cyanoacrylic esters [192–197], methylidene malonates [198–200] and episulfides [201, 202] can all be reacted via an anionic polymerization method in an aqueous emulsion process, since the polymer chain ends are reactivable in these conditions. Polymerization proceeds without the help of an ionic surfactant, conventionally acting as a provider of catalyst and a bulky counterion of the active center; also, the critical DP is rapidly reached through propagation–reversible termination steps.

Cyanoacrylic esters were added drop-by-drop to an aqueous phase loaded with nonionic surfactants, whether molecular (Tween, Triton) or macromolecular (Pluronics, Dextran). The pH was set at low values (typically 2) to avoid any aggregation due to too fast polymerization. The size of the particles was mainly due to the nature of the surfactant, nonionic surfactants producing particles diameters as small as 30 nm. Other parameters included the monomer concentration and the pH, without any change of the molar masses, typically set around 2500 g mol^{-1} [203]. Methylidene malonates are less reactive than cyanoacrylic esters, thus allowing their polymerization at reasonable pH (around 6). For this particular monomer, it was shown that the growing oligomers were able to form a cyclic trimer, the favored partitioning of which in water degraded the colloidal stability of nanospheres. This drawback was then overcome by implementing a spontaneous emulsification process [123]. Note that 1,1-disubstituted electron-deficient olefins [204], such as vinylidene cyanide, could likely be polymerized in emulsion [205].

The emulsion polymerization of propylene sulfide was initiated by using an initiator (an alkylated dithiol)–organic base {e.g. 1,8-diazabicyclo[5.4.0]undec-7-ene (DBU)} pair [201]. Polymerization was living, as demonstrated by the quantitative end-functionalization of the propagating sulfide anion by iodoacetamide, throughout the course of polymerization. Nevertheless, the final conversions were around 60% and the chain lengths were limited to 60 monomer units. These two facts are the signature of the critical DP effect. To overcome the problem of instability due to coalescence [poly(propylene sulfide) is an oil at room temperature], the authors then made use of a tetrafunctional initiator to obtain cross-linked polymers; the particles obtained were to be tested in biological applications [202]. Another recent example of ring-opening anionic emulsion polymerization concerns the polymerization of NCA in water–DMSO mixture, initiated simultaneously by a conventional butylamine and an amino-terminated poly(ethylene glycol) [206]. Self-assembly of the block copolymers prepared from the latter initiator generated submicrometer particles in which conventional initiation by the butylamine could occur.

Finally, the only example of cationic emulsion polymerization was recently published [207] using pMOS as the monomer, an ytterbium Lewis acid surfactant as the catalyst and suitable initiators in place of the previously chosen protonic acids [127, 128]. Since polymerization proceeded inside the droplets, the distribution of which was fairly broad at the beginning of the process but narrowed with conversion, the molar masses were much larger than those usually obtained by interfacial polymerization; such an effect was ascribed to both the low content of water inside the particles and the elimination of the critical DP limit.

14.5
Polymerization in Surfactant Templates

14.5.1
Microemulsion Polymerization

14.5.1.1 Conventional Free Radical Polymerization

The formation of microemulsions, i.e. droplets exhibiting diameters as small as 5–10 nm, requires a surfactant–cosurfactant pair, generally in a ratio close to unity [208]. These are thermodynamically stable, since their overall surface tension is almost nil. Globular microemulsions have been found suitable templates to polymerize monomers with a view to generating nanolatexes, which exhibit outstanding specific surface areas; either hydrophobic monomers, e.g. styrene or methyl methacrylate, or hydrophilic monomers, e.g. acrylamide, were polymerized in direct (oil-in-water) and inverse (water-in-oil) microemulsion systems, respectively (polymerization in bicontinuous systems is also a simple way of preparing ultraporous polymer films or bulk materials, of use in membrane applications). The promising advent of generating particles with diameters between typically 5 and 50 nm have inspired many workers, whose studies have been reviewed in several seminal papers (e.g. [74, 209–211]). Here, only the main features of the direct microemulsion polymerization process are summarized.

The translation of droplets into particles by polymerization generally does not occur according to a 1:1 copy, as ideally observed in miniemulsion processes. Indeed, the conversion of monomer into polymer moves, in the phase diagram, the ideal dispersion formulation towards thermodynamically unstable domains. The size of the final particles is generally larger than expected, since apart from droplet nucleation, micellar and possibly homogeneous nucleation renders the physico-chemistry and the polymerization chemistry complex. Antonietti et al. have shown that there was a direct linear relationship between the ratio of oil and surfactants and the size of the droplets after polymerization over a wide surfactant content range [212]. In a series of three papers [213–215] making use of the most advanced techniques to characterize the polymer nanolatexes, such as neutron scattering, Kaler and coworkers proposed a theoretical description and mathematical translation of the microemulsion polymerization process of different industrial monomers, including acrylates and styrene. The curve representing the rate of polymerization, and also the final large molar mass distribution, can both be accurately fitted according to a theory entailing a competition between micellar and homogeneous nucleations, also taking into account the rate of monomer transfer between the different loci (water/micelle/droplet/particle).

In recent advances it was proposed to starve–feed the monomer into a pre-existing microemulsion to obtain small nanolatexes (particle size ranging between 20 and 80 nm) with both lower surfactant content (<10 wt%) and much higher final solids content (up to 40 wt%) [216]. Such results arose from a preliminary study in which the authors tried to polymerize a Winsor I system, i.e. composed of a monomer phase on top of a microemulsion. Other studies reported the generation of large solids content nanolatexes by feeding the monomer through a hollow-fiber apparatus filled with a microemulsion [217].

14.5.1.2 Controlled Free Radical Polymerization

There are almost no examples of CRP conducted in microemulsion systems *per se*. Because the method offers the same advantages as miniemulsion, but without the requirement for high shearing, it was applied as a first step to create a low solids content latex, further used as a seed for chain extension and particle growth. Consequently, the technique was essentially regarded as a seeded emulsion polymerization. It was applied to nitroxide-mediated CRP [173–175] and to ATRP with the AGET technique [218, 219] and was an important achievement in the domain of latex formation via CRP.

14.5.1.3 Ionic Polymerization

Alkyl cyanoacrylates were polymerized by inverse (micro)emulsion processes to generate water-filled nanocapsules typically 80–250 nm in size and with membrane thicknesses of <20 nm [220–222]. These colloids, meant to encapsulate aqueous drugs, were found particularly difficult to redisperse in water; several centrifugation–redispersion steps with an aqueous phase containing mixtures of surfactants were needed. Cyclosiloxanes such as D_4 have been polymerized by starve–feed microemulsion systems [223], a study that was inspired by numerous patents on the subject. Both cationic and anionic polymerization processes operated, using fairly low contents of an ionic surfactant or a mixture of an ionic with a nonionic surfactant. Particles with diameters as small as 25 nm and a narrow particle size distribution were produced, with an oil content as high as 40 wt%. Other papers reported the anionic polymerization of D_4 in microemulsion, using nonionic surfactants and cosurfactants such as ethylene glycol or aminoethanol [224, 225]. Thanks to the large specific surface area of these 30-nm sized particles, initiators such as KOH or potassium silanolate could be used with no detrimental decrease in the polymerization rate, the kinetics of which followed the Morgen–Kaler theory.

14.5.1.4 Polycondensation

To our knowledge, polycondensation reactions in microemulsions were mainly conducted in inverse systems. Jong and Saam [226] showed that polyesterification catalyzed in a closed vessel by sulfonic acid surfactants generated thermodynamically stable water-in-oil microemulsions, with a rather poor result in term of molar mass enhancement, compared with direct suspension systems. Bischoff and Sigwalt [227] studied the condensation of low molar mass bis-silanol polydimethylsiloxanes catalyzed by triflic acid to produce micrometer-sized water droplets, stabilized by the initial surface active silicone oligomers, until their mass grew to the point where they no longer prevented coalescence. IPDI–propylene glycol prepolymers were synthesized in a bicontinuous inverse microemulsion stabilized by AOT [228].

14.5.2 Polymerization in Vesicles

Capsules containing an embedded water core are among the most difficult to prepare and the approach consisting of using vesicles as templates seems the most promising (for reviews, see [229, 230]). Polymerizations carried out *inside* the vesicle bilayer were mainly of radical type. Although some authors claimed the synthesis of the expected capsules, most often a "parachute-like" structure, attributed to polymer chains demixing as a bead trapped in the bilayer, was observed [231]. Predictable nanocapsule synthesis therefore required complex chemistry to be carried out in order to prepare tailored surfactants bearing either a polymerizable group or an alkyl/fluorinated tail.

Catanionic vesicles were also used as templates for the cross-linking reaction of tetramethylcyclotetrasiloxane *inside* their bilayers, leading to non-porous, impermeable, highly cross-linked water-filled hollow spheres of diameter about 100 nm [232].

14.6
Conclusions

Almost all polymerization techniques have been used in water dispersion processes in recent years. This has led to rejuvenation of the field, which was almost entirely devoted to conventional free radical polymerization. The application of living polymerization methods, in particular controlled/living free radical polymerization, opens the door to new systems. Diblock, triblock and star-block copolymers have been achieved in aqueous dispersed systems, with similar characteristics to those synthesized via homogeneous polymerization methods. From the point of view of the macromolecular architectures, CRP in aqueous dispersed systems presents the advantage of the process, which is solvent-free, environmentally friendly and hence might be attractive for industrial production. In ionic polymerizations or polycondensation reactions, aqueous dispersion processes favor the generation of linear chains over (macro)cyclic species; these are functional oligomers or telomers, depending on the initiator species, with an end-functionality very close to 1 or 2, respectively. The morphology of the nano- or micro-colloids is definitely a major feature of aqueous dispersed processes. The development of well-defined capsules, of core–shell particles made of diblock copolymers, the ability to fine-tune the location of the cross-links in the particle volume and to design particles with different types of polymers originating from different polymerization chemistries, and so on, has only been possible thanks to the conjunction of the physico-chemistry of emulsions and the chemistry of polymerizations.

References

1 R.G. Gilbert, *Emulsion Polymerization: a Mechanistic Approach*, Academic Press, London, **1995**.
2 R. Arshady, *Colloid Polym. Sci.* **1992**, *270*, 717–732.
3 J. Qiu, B. Charleux, K. Matyjaszewski, *Prog. Polym. Sci.* **2001**, *26*, 2083–2134.
4 M.F. Cunningham, *Prog. Polym. Sci.* **2002**, *27*, 1039–1067.
5 M.F. Cunningham, *C.R. Chim.* **2003**, *6*, 1351–1374.
6 M. Monteiro, B. Charleux, in *Chemistry and Technology of Emulsion Polymerization*, A. van Herk (Ed.), Blackwell, Oxford, **2005**, pp. 111–139.
7 F. Ganachaud, in *Les Latex Synthétiques: Élaboration, Propriétés, Applications*, J.C. Daniel, C. Pichot (Eds.), Lavoisier, Paris, **2006**, pp. 949–980.
8 J.P. Claverie, R. Soula, *Prog. Polym. Sci.* **2003**, *28*, 619–662.
9 D. Quémener, A. Chemtob, V. Héroguez, Y. Gnanou, *Polymer* **2005**, *46*, 1067–1075.
10 R. Arshady, M.H. George, *Polym. Eng. Sci.* **1993**, *33*, 865–876.

11 J.V. Dawkins, in *Comprehensive Polymer Science. The Synthesis, Characterization Reactions and Applications of Polymers*, Vol. 4, G. Allen (Ed.), Pergamon Press, New York, **1989**, pp. 231–241.
12 M.K. Georges, R.P.N. Veregin, P.M. Kazmaier, G.K. Hamer, *Macromolecules* **1993**, *26*, 2987–2988.
13 G. Schmidt-Naake, M. Drache, C. Taube, *Angew. Makromol. Chem.* **1999**, *265*, 62–68.
14 C. Taube, G. Schmidt-Naake, *Macromol. Mater. Eng.* **2000**, *279*, 26–33.
15 Y. Fuji, T. Ando, M. Kamigaito, M. Sawamoto, *Macromolecules* **2002**, *35*, 2949–2954.
16 N. Bicak, M. Gazi, U. Tunca, I. Kucukkaya, *J. Polym. Sci., Part A: Polym. Chem.* **2004**, *42*, 1362–1366.
17 C. Zhu, F. Sun, M. Zhang, J. Jin, *Polymer* **2004**, *45*, 1141–1146.
18 M.M. Ali, H.D.H. Stöver, *Macromolecules* **2003**, *36*, 1793–1801.
19 M.M. Ali, H.D.H. Stöver, *J. Polym. Sci., Part A: Polym. Chem.* **2006**, *44*, 156–171.
20 J.D. Biasutti, T.P. Davis, F.P. Lucien, J.P.A. Heuts, *J. Polym. Sci., Part A: Polym. Chem.* **2005**, *43*, 2001–2012.
21 A. Debuigne, J.R. Caille, C. Detrembleur, R. Jerome, *Angew. Chem. Int. Ed.* **2005**, *44*, 3439–3442.
22 A. Chemtob, V. Héroguez, Y. Gnanou, *Macromolecules* **2004**, *37*, 7619–7627.
23 S.J. Douglas, S.S. Davis, S.R. Holding, *Br. Polym. J.* **1985**, *17*, 339–342.
24 A. Tuncel, H. Çiçek, M. Hayran, E. Pişkin, *J. Biomed. Mater. Res.* **1995**, *29*, 721–728.
25 S. Kostjuk, F. Ganachaud, *Macromolecules* **2006**, *39*, 3110–3113.
26 P. Kubisa, S. Penczek, *Prog. Polym. Sci.* **1999**, *24*, 1409–1437.
27 F.H. Walker, J.B. Dickenson, C.R. Hegedus, F.R. Pepe, *Prog. Org. Coat.* **2002**, *45*, 291–303.
28 B. Falk, J.V. Crivello, *Chem. Mater.* **2004**, *16*, 5033–5041.
29 B. Falk, J.V. Crivello, *J. Appl. Polym. Sci.* **2005**, *97*, 1574–1585.
30 K. Goto, Y. Yamakawa, Y. Yoshida, T. Hayashi, *Seita Zairyo* **1998**, *16*, 145–151 (in Japanese).
31 G.P. Belonovskaja, Y.P. Kuztenov, B.A. Dolgoplosk, V.E. Eskin, *Eur. Polym. J.* **1979**, *15*, 187–195.
32 Y.B. Amerik, L.A. Shirokova, I.M. Toltchinsky, B.A. Krentsel, *Makromol. Chem.* **1984**, *185*, 899–904.
33 K. Satoh, M. Kamigaito, M. Sawamoto, *Macromolecules* **1999**, *32*, 3827–3832.
34 K. Satoh, M. Kamigaito, M. Sawamoto, *Macromolecules* **2000**, *33*, 4660–4666.
35 K. Satoh, M. Kamigaito, M. Sawamoto, *Macromolecules* **2000**, *33*, 5836–5840.
36 K. Satoh, M. Kamigaito, M. Sawamoto, *J. Polym. Sci., Part A: Polym. Chem.* **2000**, *38*, 2728–2733.
37 R.F. Storey, A.D. Scheuer, *J. Macromol. Sci. Pure Appl. Chem.* **2004**, *A41*, 257–266.
38 S. Cauvin, F. Ganachaud, V. Touchard, P. Hémery, F. Leising, *Macromolecules* **2004**, *37*, 3214–3221.
39 S.P. Lewis, L.D. Henderson, B.D. Chandler, M. Parvez, W.E. Piers, S. Collins, *J. Am. Chem. Soc.* **2005**, *127*, 46–47.
40 Y. Imai, *J. Macromol. Sci. Chem.* **1981**, *A15*, 833–852.
41 S. Boileau, in *New Methods for Polymer Synthesis*, W.J. Mijs (Ed.), Plenum Press, New York, **1992**, pp. 179–210.
42 L.H. Tagle, in *Handbook of Phase Transfer Catalysis*, Y. Sasson, R. Neumann (Eds.), Chapman and Hall, London, **1997**, pp. 200–243.
43 E.Z. Casassa, *J. Macromol. Sci. Chem.* **1981**, *A15*, 787–797.
44 E.Z. Casassa, D.-Y. Chao, M. Henson, *J. Macromol. Sci. Chem.* **1981**, *A15*, 799–813.
45 F.D. Karia, P.H. Parsania, *Eur. Polym. J.* **1998**, *35*, 121–125.
46 K.M. Rajkotia, M.M. Kamani, P.H. Parsania, *Polymer* **1997**, *38*, 715–719.
47 B.G. Manwar, S.H. Kavthia, P.H. Parsania, *Eur. Polym. J.* **2003**, *40*, 315–321.
48 Y.V. Patel, P.H. Parsania, *J. Macromol. Sci. Pure Appl. Chem.* **2002**, *A39*, 145–154.
49 W. Podkoscielny, D. Wdowicka, *J. Appl. Polym. Sci.* **1991**, *43*, 2213–2217.
50 W. Podkoscielny, S. Szubinska, *J. Appl. Polym. Sci.* **1986**, *32*, 3277–3297.
51 M. Baile, Y.J. Chou, J.C. Saam, *Polym. Bull.* **1990**, *23*, 251–257.

52 K. Manabe, S. Iimura, X. M. Sun, S. Kobayashi, *J. Am. Chem. Soc.* **2002**, *124*, 11971–11978.
53 H. Tanaka, T. Kurihashi, *Polym. J.* **2003**, *35*, 359–363.
54 Y. Yamada, A. Kameyama, T. Nishikubo, *Polym. J.* **1997**, *29*, 899–903.
55 T. Nishikubo, A. Kameyama, A. Kaneko, Y. Yamada, *J. Polym. Sci., Part A: Polym. Chem.* **1997**, *35*, 2711–2717.
56 K. Kimura, H. Sato, A. Kameyama, T. Nishikubo, *J. Polym. Sci., Part A: Polym. Chem.* **2000**, *38*, 3399–3404.
57 J. Chiefari, B. Dao, A. M. Groth, J. H. Hodgkin, *High Perform. Polym.* **2003**, *15*, 269–279.
58 J. Chiefari, B. Dao, A. M. Groth, J. H. Hodgkin, *High Perform. Polym.* **2006**, *18*, 31–44.
59 Y. Shigetomi, T. Kojima, *J. Appl. Polym. Sci.* **2003**, *89*, 130–134.
60 A. Okamura, T. Hirai, M. Tanihara, T. Yamaoka, *Polymer* **2002**, *43*, 3549–3554.
61 B. Ochiai, Y. Satoh, T. Endo, *Green Chem.* **2005**, *7*, 765–767.
62 R. Arshady (Ed.), *Microspheres, Microcapsules and Liposomes*, Vols. 1 and 2, Citus, London, **1999**.
63 S. R. White, N. R. Sottos, P. H. Geubelle, J. S. Moore, M. R. Kessler, S. R. Sriram, E. N. Brown, S. Viswanathan, *Nature* **2001**, *409*, 794–797.
64 J. S. Cho, A. Kwon, C. G. Cho, *Colloid Polym. Sci.* **2002**, *280*, 260–266.
65 L. Zhuo, C. Shuilin, Z. Shizhou, *Int. J. Polym. Mater.* **2004**, *53*, 21–31.
66 K. Mizuno, Y. Taguchi, M. Tanaka, *J. Chem. Eng. Jpn.* **2005**, *38*, 45–48.
67 N. Zydowicz, P. Chaumont, M. L. Soto-Portas, *J. Membr. Sci.* **2001**, *189*, 41–58.
68 M. Kubo, Y. Harada, T. Kawakatsu, T. Yonemoto, *J. Chem. Eng. Jpn.* **2001**, *34*, 12, 1506–1515.
69 E. N. Brown, M. R. Kessler, N. R. Sottos, S. R. White, *J. Microencapsulation* **2003**, *20*, 719–730.
70 K. Hong, S. Park, *J. Appl. Polym. Sci.* **2000**, *78*, 894–898.
71 Y. Frère, L. Danicher, P. Gramain, *Eur. Polym. J.* **1998**, *34*, 193–199.
72 L. Danicher, P. Gramain, Y. Frère, A. Le Calvé, *React. Funct. Polym.* **1999**, *42*, 111–125.
73 N. Yamazaki, Y. Z. Du, M. Nagai, S. Omi, *Colloids Surf. Biointerfaces* **2003**, *29*, 159–169.
74 P. A. Lovell, M. S. El-Aasser, *Emulsion Polymerization and Emulsion Polymers*, Wiley, Chichester, **1997**.
75 J. Ugelstadt, M. S. El-Aasser, J. W. Vanderhoff, *J. Polym. Sci. Polym. Lett. Ed.* **1973**, *11*, 503–513.
76 Y. T. Choi, M. S. El-Aasser, E. D. Sudol, J. W. Vanderhoff, *J. Polym. Sci., Part A: Polym. Chem.* **1985**, *23*, 2973–2987.
77 C. M. Miller, E. D. Sudol, C. A. Silebi, M. S. El-Aasser, *Macromolecules* **1995**, *28*, 2754–2764.
78 C. M. Miller, E. D. Sudol, C. A. Silebi, M. S. El-Aasser, *Macromolecules* **1995**, *28*, 2765–2771.
79 C. M. Miller, E. D. Sudol, C. A. Silebi, M. S. El-Aasser, *Macromolecules* **1995**, *28*, 2772–2780.
80 F. J. Schork, Y. Luo, W. Smulders, J. P. Russum, A. Butté, K. Fontenot, *Adv. Polym. Sci.* **2005**, *175*, 129–255.
81 K. Landfester, *Macromol. Rapid Commun.* **2001**, *22*, 896–936.
82 M. Antonietti, K. Landfester, *Prog. Polym. Sci.* **2002**, *27*, 689–757.
83 J. M. Asua, *Prog. Polym. Sci.* **2002**, *27*, 1283–1346.
84 N. Bechthold, K. Landfester, *Macromolecules* **2000**, *33*, 4682–4689.
85 C. Farcet, M. Lansalot, B. Charleux, R. Pirri, J. P. Vairon, *Macromolecules* **2000**, *33*, 8559–8570.
86 C. Farcet, B. Charleux, R. Pirri, *Macromol. Symp.* **2002**, *182*, 249–260.
87 C. Farcet, J. Nicolas, B. Charleux, *J. Polym. Sci., Part A: Polym. Chem.* **2002**, *40*, 4410–4420.
88 B. Charleux, *Macromolecules* **2000**, *33*, 5358–5365.
89 B. Charleux, *ACS Symp. Ser.* **2003**, *854*, 438–451.
90 J. Nicolas, B. Charleux, O. Guerret, S. Magnet, *Macromolecules* **2004**, *37*, 4453–4463.
91 H. Fischer, *Chem. Rev.* **2001**, *101*, 3581–3610.
92 P. B. Zetterlund, M. N. Alam, H. Minami, M. Okubo, *Macromol. Rapid Commun.* **2005**, *26*, 955–960.

93 M. Li, K. Matyjaszewski, *J. Polym. Sci., Part A: Polym. Chem.* **2003**, *41*, 3606–3614.
94 J. Gromada, K. Matyjaszewski, *Macromolecules* **2001**, *34*, 7664–7671.
95 W. Jakubowski, K. Matyjaszewski, *Macromolecules* **2005**, *38*, 4139–4146.
96 M. Li, K. Matyjaszewski, *J. Polym. Sci., Part A: Polym. Chem.* **2003**, *41*, 3606–3614.
97 M. Li, K. Min, K. Matyjaszewski, *Macromolecules* **2004**, *37*, 2106–2112.
98 Y. Kagawa, H. Minami, M. Okubo, J. Zhou, *Polymer* **2005**, *46*, 1045–1049.
99 M. Li, N. M. Jahed, K. Min, K. Matyjaszewski, *Macromolecules* **2004**, *37*, 2434–2441.
100 K. Min, H. Gao, K. Matyjaszewski, *J. Am. Chem. Soc.* **2005**, *127*, 3825–3830.
101 J. B. McLeary, B. Klumperman, *Soft Matter* **2006**, *2*, 45–53.
102 A. Butté, G. Storti, M. Morbidelli, *Macromolecules* **2000**, *33*, 3485–3487.
103 M. Lansalot, C. Farcet, B. Charleux, J.-P. Vairon, R. Pirri, *Macromolecules* **1999**, *32*, 7354–7360.
104 M. J. Monteiro, M. Hodgson, H. de Brouwer, *J. Polym. Sci., Part A: Polym. Chem.* **2000**, *38*, 3864–3874.
105 M. Lansalot, T. P. Davis, J. P. A. Heuts, *Macromolecules* **2002**, *35*, 7582–7591.
106 Y. Luo, J. G. Tsavalas, F. J. Schork, *Macromolecules* **2001**, *34*, 5501–5507.
107 H. de Brouwer, M. J. Monteiro, J. G. Tsavalas, F. J. Schork, *Macromolecules* **2000**, *33*, 9239–9246.
108 A. J. P. van Zyl, R. F. P. Boscha, J. B. McLeary, R. D. Sanderson, B. Klumperman, *Polymer* **2005**, *46*, 3607–3615.
109 Y. Luo, R. Wang, L. Yang, B. Yu, B. Li, S. Zhu, *Macromolecules* **2006**, *39*, 1328–1337.
110 L. Yang, Y. Luo, B. Li, *Polymer* **2006**, *47*, 751–762.
111 H. Matahwa, J. B. McLeary, R. D. Sanderson, *J. Polym. Sci., Part A: Polym. Chem.* **2006**, *44*, 427–442.
112 W. Smulders, C. W. Jones, F. J. Schork, *Macromolecules* **2004**, *37*, 9345–9354.
113 J. P. Russum, C. W. Jones, F. J. Schork, *Macromol. Rapid Commun.* **2004**, *25*, 1064–1068.
114 C. Detrembleur, A. Debuigne, R. Bryaskova, B. Charleux, R. Jerome, *Macromol. Rapid Commun.* **2006**, *27*, 37–41.
115 D. Quémener, V. Héroguez, Y. Gnanou, *Macromolecules* **2005**, *38*, 7977–7982.
116 J. P. Claverie, S. Viala, V. Maurel, C. Novat, *Macromolecules* **2001**, *34*, 382–388.
117 S. Mecking, A. Held, F. M. Bauers, *Angew. Chem. Int. Ed.* **2002**, *41*, 544–561.
118 S. Mecking, J. P. Claverie, In *Late Transition Metal Polymerization Catalysis*; B. Rieger, L. S. Baugh, S. Kacker, S. Striegler (Eds.), Wiley-VCH, Weinheim, **2003**, p. 231.
119 A. Bastero, S. Mecking, *Macromolecules* **2005**, *38*, 220–222.
120 R. Soula, C. Novat, A. Tomov, R. Spitz, J. Claverie, X. Drujon, J. Malinge, T. Saudemont, *Macromolecules* **2001**, *34*, 2022–2026.
121 R. Soula, B. Saillard, R. Spitz, J. Claverie, M. F. Llaurro, C. Monnet, *Macromolecules* **2002**, *35*, 1513–1523.
122 V. Monteil, A. Bastero, S. Mecking, *Macromolecules* **2005**, *38*, 5393–5399.
123 F. Ganachaud, J. L. Katz, *ChemPhysChem* **2005**, *6*, 209–216.
124 C. Maitre, F. Ganachaud, O. Ferreira, J. F. Lutz, Y. Paintoux, P. Hémery, *Macromolecules* **2000**, *33*, 7730–7736.
125 S. Cauvin, A. Sadoun, R. Dos Santos, J. Belleney, F. Ganachaud, P. Hémery, *Macromolecules* **2002**, *35*, 7919–7927.
126 S. Cauvin, R. Dos Santos, F. Ganachaud, *e-polymers* **2003**, No. 50.
127 V. Touchard, C. Graillat, C. Boisson, F. D'Agosto, R. Spitz, *Macromolecules* **2004**, *37*, 3136–3142.
128 S. Cauvin, F. Ganachaud, *Macromol. Symp.* **2004**, *215*, 179–189.
129 D. R. Weyenberg, D. E. Findlay, J. Cekada, A. E. Bey, *J. Polym. Sci. Part C* **1969**, *27*, 27.
130 D. Graiver, D. J. Huebner, J. C. Saam, *Rubber Chem. Technol.* **1983**, *56*, 918.
131 A. De Gunzbourg, J.-C. Favier, P. Hémery, *Polym. Int.* **1994**, *35*, 179–188.
132 A. De Gunzbourg, S. Maisonnier, J.-C. Favier, C. Maitre, M. Masure, P. Hémery, *Macromol. Symp.* **1998**, *132*, 359.
133 J.-R. Caille, D. Teyssié, L. Bouteiller, R. Bischoff, S. Boileau, *Macromol. Symp.* **2000**, *153*, 161–166.

134 M. Barrère, F. Ganachaud, D. Bendejacq, M.-A. Dourges, C. Maitre, P. Hémery, *Polymer* **2001**, *42*, 7239–7246.
135 M. Barrère, C. Maitre, M.-A. Dourges, P. Hémery, *Macromolecules* **2001**, *34*, 7276–7280.
136 C. Ivanenko, C. Maitre, F. Ganachaud, P. Hémery, *e-polymers* **2003**, No. 10.
137 S. Maisonnier, J.-C. Favier, M. Masure, P. Hémery, *Polym. Int.* **1999**, *48*, 159–164.
138 G. Palaprat, F. Ganachaud, M. Mauzac, P. Hémery, *Polymer* **2005**, *46*, 11213–11218.
139 B. Yactine, F. Ganachaud, O. Senhaji, B. Boutevin, *Macromolecules* **2005**, *38*, 2230–2236.
140 X. Z. Kong, E. Ruckenstein, *J. Appl. Polym. Sci.* **1999**, *73*, 2235–2245.
141 W.-D. He, C.-T. Cao, C.-Y. Pan, *J. Appl. Polym. Sci.* **1996**, *61*, 383–388.
142 M. Okaniwa, *Polymer* **2000**, *41*, 453–460.
143 M. Lin, F. Chu, E. Bourgeat-Lami, A. Guyot, *J. Dispersion Sci. Technol.* **2004**, *25*, 827–835.
144 C. Limouzin, A. Cavvigia, F. Ganachaud, P. Hémery, *Macromolecules* **2003**, *36*, 667–674.
145 J. C. Saam, D. J. Huebner, *J. Polym. Sci., Part A: Polym. Chem.* **1982**, *20*, 3351.
146 M. Barrère, C. Maitre, F. Ganachaud, P. Hémery, *Macromol. Symp.* **2000**, *151*, 359–364.
147 Z. Gao, J. Schulze Nahrup, J. E. Mark, A. Saki, *J. Appl. Polym. Sci.* **2003**, *90*, 658–666.
148 T. M. Obey, B. Vincent, *J. Colloid Interface Sci.* **1994**, *163*, 454.
149 B. Neumann, B. Vincent, R. Krustev, H.-J. Müller, *Langmuir* **2004**, *20*, 4336–4344.
150 K. R. Anderson, T. M. Obey, B. Vincent, *Langmuir* **1994**, *10*, 2493–2494.
151 M. I. Goller, T. M. Obey, D. O. H. Teare, B. Vincent, M. R. Wegener, *Colloids Surf. A* **1997**, *183*, 123–124.
152 F. Baumann, B. Deubzer, M. Geck, J. Dauth, M. Schmidt, *Macromolecules* **1997**, *30*, 7568–7573.
153 N. Jungmann, M. Schmidt, M. Maskos, J. Weis, J. Ebenhoch, *Macromolecules* **2002**, *35*, 6851–6857.
154 H. Wang, P. Chen, X. Zheng, *J. Mater. Chem.* **2004**, *14*, 1648–1651.
155 G. Palaprat, F. Ganachaud, *C. R. Chim.* **2003**, *6*, 1385–1392.
156 K. Landfester, F. Tiarks, H. P. Hentze, M. Antonietti, *Macromol. Chem. Phys.* **2000**, *201*, 1–5.
157 F. Tiarks, K. Landfester, M. Antonietti, *J. Polym. Sci., Part A: Polym. Chem.* **2001**, *39*, 2520–2524.
158 M. Barrère, K. Landfester, *Macromolecules* **2003**, *36*, 5119–5125.
159 L. Torini, J. F. Argilier, N. Zydowicz, *Macromolecules* **2005**, *38*, 3225–3236.
160 K. Landfester, U. Pawelzik, M. Antonietti, *Polymer* **2005**, *46*, 9892–9898.
161 M. Barrère, K. Landfester, *Polymer* **2003**, *44*, 2833–2841.
162 E. S. Daniels, E. D. Sudol, M. S. El-Aasser (Eds.), *Polymer Latexes. Preparation, Characterization and Applications.* ACS Symposium Series, Vol. 492. American Chemical Society, Washington, DC, **1992**.
163 J. M. Asua (Ed.), *Polymeric Dispersions: Principles and Applications.* NATO ASI Series. Kluwer, Dordrecht, **1996**.
164 R. M. Fitch, *Polymer Colloids: a Comprehensive Introduction.* Academic Press, London, **1997**.
165 A. M. Van Herk, *Chemistry and Technology of Emulsion Polymerization*, Blackwell, Oxford, **2005**.
166 M. Nomura, H. Tobita, K. Suzuki, *Adv. Polym. Sci.* **2005**, *175*, 1–128.
167 A. M. Van Herk, M. Monteiro, in *Handbook of Radical Polymerization*, K. Matyjaszewski, T. P. Davis (Eds.), Wiley, New York, **2002**, pp. 301–331.
168 W. D. Harkins, *J. Am. Chem. Soc.* **1947**, *69*, 1428–1444.
169 W. V. Smith, R. H. Ewart, *J. Chem. Phys.* **1948**, *16*, 592–599.
170 J. L. Gardon, *J. Polym. Sci., Part A-1* **1968**, *6*, 623–641.
171 M. Morton, S. Kaizerman, M. W. Altier, *J. Colloid Sci.* **1954**, *9*, 300–312.
172 A. R. Szkurhan, M. K. Georges, *Macromolecules* **2004**, *37*, 4776–4782.
173 J. Nicolas, B. Charleux, O. Guerret, S. Magnet, *Angew. Chem. Int. Ed.* **2004**, *43*, 6186–6189.

174 J. Nicolas, B. Charleux, O. Guerret, S. Magnet, *Macromolecules* **2005**, *38*, 9963–9973.
175 J. Nicolas, B. Charleux, S. Magnet, *J. Polym. Sci., Part A: Polym. Chem.* **2006**, *44*, 4142–4153.
176 G. Delaittre, J. Nicolas, C. Lefay, M. Save, B. Charleux, *Chem. Commun.* **2005**, 614–616.
177 G. Delaittre, J. Nicolas, C. Lefay, M. Save, B. Charleux, *Soft Matter* **2006**, *2*, 223–231.
178 H. Eslami, S. Zhu, *Polymer* **2005**, *46*, 5484–5493.
179 M. Manuszak-Guerrini, B. Charleux, J.-P. Vairon, *Macromol. Rapid Commun.* **2000**, *21*, 669–674.
180 J. N. Kizhakkedathu, D. Goodman, D. E. Brooks, *ACS Symp. Ser.* **2003**, *854*, 316–330.
181 J. N. Kizhakkedathu, D. Goodman, D. E. Brooks, *Macromolecules* **2003**, *36*, 591–598.
182 J. Krstina, G. Moad, E. Rizzardo, C. L. Winzor, *Macromolecules* **1995**, *28*, 5381–5385.
183 S. Nozari, K. Tauer, A. M. I. Ali, *Macromolecules* **2005**, *38*, 10449–10454.
184 D. Charmot, P. Corpart, H. Adam, S. Z. Zard, T. Biadatti, G. Bouhadir, *Macromol. Symp.* **2000**, *150*, 23–32.
185 M. J. Monteiro, M. M. Adamy, B. J. Leeuwen, A. M. van Herk, M. Destarac, *Macromolecules* **2005**, *38*, 1538–1541.
186 W. Smulders, R. G. Gilbert, M. J. Monteiro, *Macromolecules* **2003**, *36*, 4309–4318.
187 W. Smulders, M. J. Monteiro, *Macromolecules* **2004**, *37*, 4474–4483.
188 C. J. Ferguson, R. J. Hughes, D. Nguyen, B. T. T. Pham, R. G. Gilbert, A. K. Serelis, C. H. Such, B. S. Hawkett, *Macromolecules* **2005**, *38*, 2191–2204.
189 M. Manguian, M. Save, B. Charleux, *Macromol. Rapid Commun.* **2006**, *27*, 399–404.
190 F. M. Bauers, S. Mecking, *Macromolecules* **2001**, *34*, 1165–1171.
191 F. M. Bauers, R. Thomann, S. Mecking, *J. Am. Chem. Soc.* **2003**, *125*, 8838–8840.
192 P. Couvreur, B. Kante, M. Roland, P. Guiot, P. Bauduin, P. Speiser, *J. Pharm. Pharmacol.* **1979**, *31*, 331–332.
193 N. Behan, C. Birkinshaw, *Macromol. Rapid Commun.* **2000**, *21*, 884–886.
194 N. Behan, C. Birkinshaw, N. Clarke, *Biomaterials* **2001**, *22*, 1335–1344.
195 S. J. Douglas, L. Illum, S. S. Davis, J. Kreuter, *J. Colloid Interface Sci.* **1984**, *101*, 149–158.
196 S. J. Douglas, L. Illum, S. S. Davis, J. Kreuter, *J. Colloid Interface Sci.* **1984**, *103*, 154–163.
197 S. J. Douglas, S. S. Davis, S. R. Holding, *Br. Polym. J.* **1985**, *17*, 339–342.
198 J. L. De Keyser, J. H. Poupaert, P. Dumont, *J. Pharm. Sci.* **1991**, *80*, 67–70.
199 F. Lescure, C. Seguin, P. Breton, P. Bourrinet, D. Roy, P. Couvreur, *Pharm. Res.* **1994**, *11*, 1270–1277.
200 P. Breton, X. Guillon, D. Roy, G. Riess, N. Bru, C. Roques-Carmes, *Biomaterials* **1998**, *19*, 271–281.
201 A. Rehor, N. Tirelli, J. A. Hubbell, *Macromolecules* **2002**, *35*, 8688–8693.
202 A. Rehor, J. A. Hubbell, N. Tirelli, *Langmuir* **2005**, *21*, 411–417.
203 A. Bootz, T. Russ, F. Gores, M. Karas, J. Kreuter, *Eur. J. Pharm. Biopharm.* **2005**, *60*, 391–399.
204 P. Klemarczyk, *Polymer* **1998**, *39*, 173–181.
205 H. Gilbert, F. F. Miller, S. J. Averill, R. F. Smidt, F. D. Stewart, H. L. Trumbull, *J. Am. Chem. Soc.* **1954**, *76*, 1074–1076.
206 M. Matsusaki, T. Waku, T. Kaneko, T. Kida, M. Akashi, *Langmuir* **2006**, *22*, 1396–1399.
207 S. Cauvin, F. Ganachaud, M. Moreau, P. Hémery, *Chem. Commun.* **2005**, 2713–2715.
208 J. Klier, *Adv. Mater.* **2000**, *12*, 1751–1757.
209 P. Y. Chow, L. M. Gan, *Adv. Polym. Sci.* **2005**, *175*, 257–298.
210 F. M. Pavel, *J. Dispers. Sci. Technol.* **2004**, *25*, 1–16.
211 C. Larpent, *Surf. Sci.* **2003**, *115*, 145–187.
212 M. Antonietti, R. Basten, S. Lohmann, *Macromol. Chem. Phys.* **1995**, *196*, 441–466.

213 C. C. Co, R. de Vries, E. W. Kaler, *Macromolecules* **2001**, *34*, 3224–3232.
214 R. de Vries, C. C. Co, E. W. Kaler, *Macromolecules* **2001**, *34*, 3233–3244.
215 C. C. Co, P. Cotts, S. Burauer, R. de Vries, E. W. Kaler, *Macromolecules* **2001**, *34*, 3245–3254.
216 X.-J. Xu, L. M. Gan, *Curr. Opin. Colloid Interface Sci.* **2005**, *10*, 239–244.
217 X. J. Xu, K. S. Siow, M. K. Wong, L. M. Gan, *Langmuir* **2001**, *17*, 4519–4524.
218 K. Min, K. Matyjaszewski, *Macromolecules* **2005**, *38*, 8131–8134.
219 K. Min, H. Gao, K. Matyjaszewski, *J. Am. Chem. Soc.* **2006**, *128*, 10521–10526.
220 S. Watnasirichaikul, N. M. Davies, T. Rades, I. G. Tucker, *Pharm. Res.* **2000**, *17*, 684–689.
221 S. Watnasirichaikul, I. G. Tucker, T. Rades, N. M. Davies, *Int. J. Pharm.* **2002**, *235*, 237–246.
222 K. Krauel, N. M. Davies, S. Hooka, T. Rades, *J. Control. Release* **2005**, *106*, 76–87.
223 M. Barrère, S. Capitao da Silva, F. Ganachaud, *Langmuir* **2002**, *18*, 941.
224 D. Zhang, X. Jiang, C. Yang, *J. Appl. Polym. Sci.* **2003**, *89*, 3587–3593.
225 D. Zhang, X. Jiang, C. Yang, *J. Appl. Polym. Sci.* **2005**, *98*, 347–352.
226 L. Jong, J. C. Saam, *ACS Symp. Ser.* **1996**, *624*, 332–349.
227 R. Bischoff, P. Sigwalt, *Polym. Int.* **1995**, *36*, 57–71.
228 J. Texter, P. Ziemer, *Macromolecules* **2004**, *37*, 5841–5843.
229 D. H. W. Hubert, M. Jung, A. L. German, *Adv. Mater.* **2000**, *12*, 1291–1294.
230 V. T. John, B. Simmons, G. L. McPherson, A. Bose, *Curr. Opinion Colloid Interface Sci.* **2002**, *7*, 288.
231 M. Jung, D. H. W. Hubert, A. M. van Herk, A. L. German, *Macromol. Symp.* **2000**, *151*, 393–398.
232 M. Kepczynski, F. Ganachaud, P. Hémery, *Adv. Mater.* **2004**, *16*, 1861–1863.

15
Polymerization Under Light and Other External Stimuli

Jean Pierre Fouassier, Xavier Allonas, and Jacques Lalevée

15.1
Introduction

The initiation step of a polymerization reaction is usually a thermal process. Other stimuli, however, can be used, such as light, electron beam, X-rays, γ-rays, plasma, microwaves and pressure. Among them, UV, visible and near-infrared radiation represent a very elegant and powerful excitation approach to start a polymerization reaction. In fact, this basic process has induced a lot of fruitful developments in industrial areas referred to as "UV curing" (or "radiation curing", which also includes electron beam curing) and laser imaging where traditional applications sit alongside newly emerging high-tech sectors combining photochemistry, organic and polymer chemistry, physics, optics, electronics, telecommunications, medicine, etc. During the last two decades, several books have appeared in this field of growing interest, discussing the mechanistic aspects of the reactions encountered and detailing the applications [1–26]. The present chapter is intended to give an overview focused on the basic principles, the photochemical reactions involved, the properties of the cured polymer films, the reactivity–efficiency relationships, the design of many versatile, fascinating and original properties of the polymerized materials and the existing applications in everyday life.

15.2
Background

15.2.1
Photopolymerization Reactions

Photopolymerization (see, e.g., [1]) is a general word to represent polymerization which starts under light exposure. A photopolymer refers to a polymer sensitive to light and generally devoted to the imaging science and electronic areas.

Macromolecular Engineering. Precise Synthesis, Materials Properties, Applications.
Edited by K. Matyjaszewski, Y. Gnanou, and L. Leibler
Copyright © 2007 WILEY-VCH Verlag GmbH & Co. KGaA, Weinheim
ISBN: 978-3-527-31446-1

Moreover, this word is often used to represent a polymer requiring light to be formed. A photoinduced polymerization reaction is considered as a chain reaction similar to that encountered in thermal polymerization reactions: the only and striking difference lies in the fact that the initiation step is governed by a photochemical event. In a photocross-linking reaction, the formation of a cross-link between two macromolecular chains requires the absorption of a photon. Photocuring is a practical word and refers to the use of light to induce this rapid conversion of formulated reactive monomeric liquids to solids. When using multifunctional monomers or oligomers, the photoinduced polymerization reaction does not proceed to form a linear polymer because it also leads to a cross-linking reaction, thereby creating a polymer network. In both reactions (photoinduced polymerization and photocross-linking), the liquid or viscous tacky matrix is converted into a tack-free polymer film.

15.2.2
Light Sources

Typical sources of light are typically [27] conventional light sources such as mercury or xenon vapor lamps, laser beams and, recently introduced, light-emitted diodes (LEDs) [28]. Excimer lamps can find applications [29]. Sunlight is also a very convenient and inexpensive source of light but the applications are rather restricted in the radiation curing field.

A light source is characterized by an emission spectrum which can appear continuous in wavelength λ (for the Sun), composed of a set of particular wavelengths (for an Hg lamp) or consisting of only one wavelength (for a laser). The intensity of the source is defined by the number of photons delivered at a given wavelength per unit surface area and time unit (I_0). A photon corresponds to the energy quantum $h\nu$, where h is Planck's constant and ν the frequency of the radiation. The quantity $I_0 h\nu$ represents the amount of energy in J cm^{-2} s^{-1}, i.e. the power density (W cm^{-2}) emitted by the light source. With conventional sources, the light is delivered continuously as a function of time, whereas lasers can also emit the light as a pulse (in that case, very high power densities can be attained). The reciprocity law states that the phenomenon produced upon light absorption should be dependent only on the amount of energy absorbed, whatever I_0 and the time distribution are; if it is not the case, there is a reciprocity failure [30].

Laser beams present specific characteristics: monochromatic light, high energy concentration on to a small surface, high spatial resolution, spectral selectivity, narrow bandwidth of the emission, very short exposure times, allowing scanning of the film surface by the laser spot, and easy focusing. Laser-induced polymerization reactions are mostly encountered in laser imaging technology where the excitation light has to scan a surface or needs specific properties (for example, in holographic recording).

15.2.3
Absorption of Light

A molecule exhibits an absorption spectrum in the UV, visible or very near-infrared wavelength range as a function of its chemical structure [31]. This absorption is due to electronic transitions between a ground state (singlet state S_0) and an excited state (singlet state S_1, S_2, ...). The S_1 state converts more or less efficiently into an usually reactive T_1 triplet state. Deactivation of the excited states occurs through well-known processes [31].

The absorption spectrum of a molecule reflects the probability of absorbing photons at given wavelengths through the molar extinction coefficient ε. The amount of light absorbed I_{abs} is written as

$$I_{abs} = I_0[1 - \exp(-\varepsilon l c)]$$

where l is the optical path length and c the molar concentration. As a consequence, a high absorption is achieved by using molecules having high ε values at the emission wavelengths of the lamp.

The rate of absorption, defined as

$$-d[S_0]/dt = d[S_1]/dt = I_{abs}$$

governs the concentration of the S_1 excited state and thereby the concentration of the reactive species. As most of the photoinitiators work from their lowest triplet state, the intersystem crossing process must be fast.

15.2.4
Initiation Step

Monomers, oligomers and prepolymers usually absorb the photons delivered by the lamps in the UV range [32]. This absorption, however, does not allow one to produce reactive species with a high yield. In the polymerizable matrix, it is therefore necessary to add a photoinitiator PI or a photosensitizer PS. A photoinitiator is employed either alone or with a co-initiator, absorbs the light and directly leads to reactive species. A photosensitizer absorbs the light and transfers the energy to the photoinitiator. The reactive species can correspond to radicals, cations, anions, radical ions, etc.

15.3
Photoinitiators and Photosensitizers

This point is largely developed in [1, 21] and recent reviews have covered several aspects (e.g. [33, 34]).

15.3.1
Photoinitiators

15.3.1.1 Direct Production of Reactive Species

The reactive species are dependent on the starting molecule and, accordingly, the usual type of polymerization reactions (radical, cationic, anionic) can be encountered. The efficiency of their production will obviously lead to an efficient initiation step (as evaluated from the number of starting chains per photon absorbed).

15.3.1.2 Radical Photoinitiators

Most efficient photoinitiators developed so far work via a homolytic bond cleavage in the triplet state (type I photoinitiator) and lead to the generation of two radicals. In most cases, the cleavage takes place at a C–C or a C–P bond at the α-position of a carbonyl group (in benzoin ethers, hydroxyalkylacetophenones, phosphine oxides and amino ketones) [21, 33, 35, 36] or at a C–S bond at the β-position of a carbonyl group (sulfonyl ketones) [37].

Another approach consists in using a two-component system based on a photoinitiator and a co-initiator (type II photoinitiator). The co-initiator is a hydrogen donor (alcohol, amine, thiol, etc.) and the process is either a direct hydrogen transfer or an electron transfer followed by a proton transfer; in both cases, two radicals are formed: one on the hydrogen/electron donor (hydroxyisopropyl, aminoalkyl or thiyl radical, etc.) and the other on the photoinitiator (ketyl type radical, etc.) [1, 21, 38, 39].

The hydrogen donor can also be a polymer chain possessing labile hydrogens: the hydrogen abstraction process creates a macromolecular radical which will induce a cross-linking reaction or, in an appropriate system, a graft reaction [40].

Generally, all of these compounds absorb in the UV or the near-UV range. Many attempts have been made to shift the absorption towards the visible region of the spectrum through the introduction of well-selected substituents. However, photoinitiators sensitive to visible light are usually based on dyes or metal complexes. Titanocene derivatives exhibit an excellent absorption around 500 nm and cleave very efficiently into two radicals [21]. Various available structures (see, e.g., [34, 41, 42]) allow the whole of the visible region to be covered.

15.3.1.3 Cationic Photoinitiators

Useful cationic photoinitiators belong to three main classes:
1. *Diazonium salts.* These generate a Lewis acid which is able to start an epoxy ring-opening polymerization reaction [43].
2. *Onium salts.* These represent the main class and consist of a cationic moiety [where a positively charged central atom such as iodine (iodonium salts) or sulfur (sulfonium salts) is linked to aromatic groups] and a counteranion.

The introduction of long-wavelength absorbing chromophores allows the absorption to be shifted to the visible region. The decomposition process [1, 21, 44] involves a heterolytic or homolytic cleavage of a C–I or C–S bond followed by an in-cage or, in the presence of a hydrogen donor, an out-of-cage process, yielding a strong acid and a radical. A lot of studies have focused on the search for (i) new onium salts and (ii) new anions. The progress of the cationic polymerization reaction is obviously strongly affected by the nucleophilicity of this anion. A new anion (pentafluorophenylborate) has recently been proposed to overcome this problem [45]: high rates of polymerization are attained both in the usual epoxide cationic organic matrices and in epoxysilicones.

3. *Metallocene derivatives.* One example is concerned with ferrocenium salt derivatives, which absorb in the visible range and yield a Lewis acid [21].

15.3.1.4 Anionic Photoinitiators

This kind of photoinitiation process has seldom been described. Anionic photoinitiators [46] for the polymerization of cyanoacrylates are based on:

1. inorganic complexes able to release negatively charged nucleophiles. Such compounds include pentacarbonyl complexes, Schiff bases, ferrocenes and phosphazene bases;
2. ferrocene, leading to a ground-state complex with the monomer followed by an electron transfer upon excitation.

15.3.1.5 Photoacid and Photobase Generators

Such nonionic compounds are sometimes designed as photolatent compounds [21, 47]. They are used in photocross-linking reactions working through a proton- or a base-catalyzed process for the design of chemically amplified photoresists. They involve, for example, iminosulfonates, which form a sulfonic acid [48], or O-acyloximes, which generate an amine [49]. The acid or base is photochemically produced in a first step. In a second step, a thermal process favors the cross-linking reactions of suitable monomers such as epoxides (acid or base) or isocyanates (base).

15.3.1.6 Absorption Spectra

Most photoinitiators absorb in the UV region. The absorption spectra of specific photoinitiators spread into visible wavelengths. It is obvious that a good matching of these absorption spectra with the emission spectrum of the light source is necessary when

1. the idea is to recover the maximum number of photons emitted in the visible region;
2. a well-defined and monochromatic excitation is imposed (when using lasers);
3. a pigmented medium has to be polymerized. The high light absorption by the pigment has a detrimental effect on the light absorption by the photoini-

tiator. As a consequence, the design of a photosensitizer that absorbs in the spectral window offered by the pigment is necessary [1, 21]; this often requires the design of complex photoinitiating systems [34];
4. the photoinitiator presents an absorption only at short wavelengths (e.g. iodonium salts).

15.3.1.7 Excited State Reactivity
In addition to the role of the usual photophysical processes in the excited states, other deactivation pathways can seriously affect the photoinitiation step. Among these, the quenching by
- oxygen in radical photopolymerization
- water in cationic photopolymerization
- monomer (e.g. the problem of ketone–double bond interaction)
- other additives such as UV absorbers, light stabilizers, the environment (e.g. the presence of phenolic compounds in wood coating photopolymerization reactions [50])

might play a significant role. The production of the reactive species will therefore be in competition with all these processes [1].

15.3.2
Photosensitizers

15.3.2.1 Processes
The photosensitized decomposition of a photoinitiator PI in the presence of a photosensitizer PS can be described as resulting from two main processes largely encountered in photochemistry [31]: energy transfer or electron transfer. In the first process, the energy transfer occurs from the PS excited state to the PI excited state (of the same spin multiplicity): the same reactive species that would result from direct excitation of PI are generated. The energetic requirement is that the PS excited state energy must be higher than that of PI. In an electron transfer process, a radical ion pair is formed which undergoes further reactions leading to reactive species. The driving force of this process is the free enthalpy change of the reaction, which must obviously be negative.

15.3.2.2 Examples

Radical photoinitiator/photosensitizer systems
Such an example is provided in a well-known system where an energy transfer occurs between a thioxanthone derivative and a morpholino ketone derivative [1, 21]. More complex systems can be designed: a very general picture has been established many times in three-component combinations ([33] and references cited therein). In that case, the system is primarily working through an electron

transfer between a ketone and an amine [51]. The role of the third additive (such as an iodonium salt) is to scavenge the ketyl-type radical which is known as a terminating agent of the growing macromolecular chains.

Cationic photoinitiator/photosensitizer systems

Various photosensitizers can help to photolyze cationic photoinitiators [1, 21]: ketones, hydrocarbons, phenothiazines, dyes, radicals, dye/onium salt intra-ion pairs, etc. Another approach proposed recently [52] suggests using an addition/fragmentation reaction involving a radical source and a cationic salt (these two species can also be linked in a single molecule): careful selection of the radical source allows one to tune the absorption to higher wavelengths.

Photoacid and photobase generator/photosensitizer systems

An example of a photosensitized production of a base is shown in [53]: decomposition of an O-acyloxime derivative occurs through energy transfer when benzophenone is used and electron transfer in the presence of naphthoquinone.

15.3.3
Properties of Photoinitiators and Photosensitizers

In the radiation curing area, photoinitiators and photosensitizers must exhibit specific properties:
- final properties of the polymerized material: no formation (or as low as possible) of yellowing photolysis products, etc.;
- fast polymerization (well-adapted spectral range absorption, low sensitivity to oxygen, high molar extinction coefficients, excellent photochemical reactivity, high chemical reactivity of the reactive species, etc.);
- handling (solubility in the formulation, compatibility, etc.);
- marketing and safety (low price, good shelf stability, absence of odor, no toxicity, low content of extractable compounds, etc.).

These requirements have spurred numerous studies aimed at the development of photoinitiator or photosensitizer systems having improved properties (1, 21, 54–56), such as polymeric, copolymerizable, hydrophilic or water-soluble photoinitiators, red-shifted absorbing photoinitiators, surface-active photoinitiators, liquid photoinitiators and chemically bound photoinitiators/photosensitizers.

15.4
Monomers and Oligomers

A photoinitiated polymerization reaction must transform a liquid medium into a solid polymer upon exposure to light with a high rate and high final percentage conversion. In industrial applications (for coatings and varnishes), the formulation is based on

1. an unsaturated multifunctional oligomer (referred to as a resin) for film formation. The backbone of the polymer network formed after irradiation will be strongly dependent on the nature and structure of this oligomer;
2. a monomer that behaves as a diluent to facilitate the formulation handling. It readily participates in the polymerization process;
3. a photoinitiating system;
4. additives (inhibitors, light stabilizers, fillers, pigments, formulation agents, etc.).

15.4.1
Various Systems

Light-curable multifunctional monomer and oligomer formulations for radical and cationic polymerization reactions are increasingly being developed since the final properties of the cured coating are governed by the chemistry of the formulation. The development of systems for anionic polymerization is more difficult, as is finding suitable matrixes for photobase-catalyzed cross-linking reactions.

15.4.1.1 Radical Monomers and Oligomers
The most widely used systems on an industrial scale are based on radical monomers and oligomers such as the early developed unsaturated polyester/styrene resins and more recently mono- and multifunctional acrylates. The oligomer backbone (epoxy, polyurethane, polyester, etc.) allows the design of tailor-made properties.

15.4.1.2 Cationic Monomers and Oligomers
Typical cationic monomers and oligomers for industrial applications in UV curing are usually based on epoxides and vinyl ethers. Oxetanes are alternatives to epoxides and exhibit the same performance capabilities with higher reactivity.

15.4.2
Current Developments

New developments include, for example [1, 6, 10, 21–26]
1. the synthesis of novel or modified known systems;
2. the proposal of particular promising systems such as water-borne UV-curable formulations or powder coatings;
3. the search for new, high-performance or high-tech properties;
4. the use of specific possibilities such as dual cure and hybrid cure.

15.4.3
General Properties

Depending on the applications of the polymerized formulations for coatings in UV technologies, numerous specific and very different properties can be searched for, for example:

1. physical and mechanical properties: hardness, flexibility, abrasion, scratch resistance, adhesion, heat resistance, tensile strength, flexural strength, modulus, elongation at break, impact resistance, glass transition temperature, surface properties (wetting, hydrophilicity/hydrophobicity), thermal conductivity, specific heat, thermal diffusivity, low shrinkage, stiffness, barrier properties, moisture permeability, etc.;
2. chemical properties: resistance to solvents, water, acids and bases, etc.;
3. optical properties: matting, gloss, transparency, color change, yellowing, blistering, etc.;
4. usage properties: low odor, low skin irritation, low volatility, weatherability, wear properties, anti-soiling properties, antibacterial properties, superabsorbancy, gas barrier properties, etc.

Extensive research work (see also in [23–26]) has sought to meet these requirements through the continuous search for new systems, for example:

- reactive monomers avoiding acrylates and leading to less odor and low irritation (vinyl ethers which can undergo radical-initiated copolymerization with acrylate functions, etc.) [57];
- new monomers and oligomers providing new properties, e.g.:
  - hyperbranched acrylates (decrease in viscosity) [58];
  - calixarene derivatives containing acrylate groups (high transition temperature, excellent thermal stability) [59];
  - fluoroalkyl propenyl ethers [60] (high thermal stability, excellent chemical resistance, low dielectric constant, refractive index);
  - silsesquioxanes having oxetanyl groups (high modulus at high temperature and fairly high thermal stability [61]);
  - hybrid acrylate–mineral particle nanocomposites (heat resistance, strength, stiffness, barrier properties, mechanical properties) [62];
  - liquid polybutadiene prepolymers (resistance to water, acids and bases, low moisture permeability, etc.), epoxyacrylates (better flexibility), polyester oligomers (abrasion resistance), acrylated polyesters (superior wear properties and enhanced chemical and moisture resistance), liquid crystalline monomers and oligomers [1];
  - low-viscosity systems for spray applications, low skin-irritating monomers, low-volatility organic raw materials, etc.

15.5
Brief Overview of Applications in UV Curing

UV curing is growing rapidly because of the wide range of applications in various industrial sectors, such as [1, 6, 10, 21, 23–26]:
1. coatings, varnishes and paints for many applications on a large variety of substrates (optical fibers, wood, plastics, metal, papers, etc.) in very different sectors [flooring, packaging, release coating, powder coating on medium-density fiber (MDF) wood panels, automotive, pipe linings, etc.];
2. adhesives (laminating, pressure sensitive, hot melt, etc.);
3. graphic arts (drying of inks, ink jets, manufacture of printing plates, etc.);
4. 3D machining (or 3D photopolymerization or stereolithography giving the possibility of making objects for prototyping applications;
5. microelectronics (photoresists for printed circuits, integrated circuits, very large and ultra-large-scale integrated circuits, soldering resists, mask repairs, encapsulants, conductive screen ink, metal conductor layers, etc.);
6. laser imaging (laser direct imaging for writing complex relief patterns for the manufacture of microcircuits or patterning selective areas in microelectronic packaging; computer-to-plate technology, a similar process used in graphic arts to reproduce a document directly (stored in the computer) on a printing plate by using a laser beam controlled by the computer;
7. dentistry and medicine (restorative and preventive denture relining, wound dressing, ophthalmic lenses, glasses, artificial eyelenses, drug microencapsulation, etc.);
8. optics (holographic recording and information storage, computer-generated holograms, embossed holograms, manufacture of optical elements such as diffraction gratings, mirrors, lenses, waveguides, array illuminators, display applications, etc.).

Growing applications are expected in thick pigmented coatings, multi-layer coatings, antistatic coatings, good weatherability or scratch-resistant coatings, coatings with good adhesion on metal, spray coatings, coil coatings, silicone coatings, water-reducible coatings, coatings for exterior applications, optical fiber coatings, UV-cured powders, pressure-sensitive adhesives (PSA), solvent-free adhesives, hot melt adhesives, laminating adhesives, optically clear adhesives, lenticular films, flexographic inks, UV inks, overprint varnishes for food packaging, UV inkjet inks, metal decorating, wood protection products, automotive uses and refinishing (plastic parts, metal, etc.), stereolithography, 3D microfabrication with two photon-photosensitive systems, nanocomposites, photolatent base technology, composites, polymeric microcapsules, biomaterials (microchips, etc.). Major efforts have been made in the development of curing equipment (excimer lamps, spot lamps, LEDs, radiofrequency excited lamps, tunable UV lamps, small portable lamps, higher intensity UV light sources, visible light sources, laser diodes, etc.). Curing under an inert gas such as carbon dioxide has recently attracted strong interest. The finding of efficient photosensitive systems under visible light can allow one to explore the possibility of sun curing.

Many innovations are currently making headlines. Driving forces include performance (high surface quality for coatings, application versatility, enhanced product durability, etc.), economy (energy saving, low-temperature operation, small space requirements, fast cure speed) and ecology (almost no volatile organic compounds, etc.). UV curing represents a green technology for the 21st century.

15.6
Photochemical/Chemical Reactivity and Final Properties

In the usual industrial applications in the radiation curing area, the targets of the research are the investigation of how to:
1. increase further the level of performance (cure speed, high percentage conversion, etc.);
2. design improved physical, chemical, optical properties for classical applications such as in coating manufacture;
3. limit the oxygen inhibition in radical processes;
4. achieve better photostability of the cured coating, etc.

In other high-tech sectors, the search for specific new properties and new functionalities (from a judicious choice of the systems and the experimental conditions) well adapted to potential applications is the driving factor of the research work. According to these objectives, the photopolymerization reaction can present many different aspects which will be briefly mentioned below.

15.6.1
Different Aspects of Photopolymerization Reactions

15.6.1.1 Examples
Photopolymerization reactions can be encountered in many experimental conditions and/or in many different chemical monomers, oligomers or polymer media, allowing extensive possibilities of (potential) applications inside (see above) and outside the UV curing area (see below). The following list is not exhaustive:
- usual film radical or cationic photopolymerization under conventional light sources;
- photopolymerization of various particular systems under lasers [63] (optical resins, as self-written waveguides, optical recording media, surface relief gratings, microlens arrays; anisotropic materials as photoalignment layers, gradient structure materials, computer-to-plate systems, photoresists in microelectronics, etc.);
- photopolymerization of interpenetrated networks (light has the unique advantage of triggering the reaction so that the two networks can be formed successively through concurrent cationic/radical photopolymerization) [64];

- surface photopolymerization of adsorbed or oriented molecules (photopolymerization of styrene from self-assembled monolayers on gold form films as thin as 7–190 nm [65]);
- gas-phase photopolymerization (the photopolymerization of acrolein on to a metallic substrate takes place at the surface absorbed layer [66]);
- photopolymerization in aerosols (for forming microcapsules [67]);
- *in situ* photopolymerization in microfluidic devices (production of polymer microstructures such as fibers, tubes, spheres for biomedical applications [68]);
- photo-oriented polymerization (self-focusing laser excitation controls the polymerization in a hybrid SiO_2 glass waveguide by a high degree of spatial localization, thereby leading to optical fibers on chips [69]);
- alternating copolymerization (using the charge-transfer absorption properties of substituted fluoroethylene and vinyl ethers under light [70]);
- photopolymerization in ionic liquids (in the case of MMA, this leads to a large increase in the propagation rate constant, attributed to the polarity increase, and a decrease in the termination rate constant, ascribed to the increased viscosity [71]);
- photopolymerization of composites [72] (glass fibers, composites, prepreg, carbon-reinforced composites, glass-fiber mats, silica-filled composites, etc.);
- photopolymerization of nanocomposites (incorporation of zeolites or clay derivatives in a photopolymerizable formulation allows better properties to be obtained for the final material [73]);
- surface graft photopolymerization: this allows one to modify the surface of a polymer material [74]. It has been applied to the wettability and adhesivity of low-density polyethylene, which are considerably improved by grafting hydrophilic monomers;
- dendritic photopolymerization (the dynamic viscosity of dendritic acrylates is lower than that of the corresponding linear oligomers, which facilitates the formulation handling; the cure speed is faster and the tensile strengths increase [75, 76]);
- gradient photopolymerization (this can be done by the creation of a double bond conversion gradient using the attenuation of the excitation light through the film thickness; the polymerized material exhibits a difference in structures on the top and bottom of a 5-mm thick sample [77]);
- controlled polymerization (a photoiniferter behaves as a photoinitiator, a transfer agent and a chain terminator; well-designed iniferters would give polymers bearing controlled end-groups; in the presence of N,N-alkyl dithiocarbamate, a living radical polymerization can be obtained provided that a photochemical method is used to allow reversibility of the termination reaction, which cannot be possible through a thermal process) [78, 79];
- template photopolymerization (used for the development of nano-structured organic–inorganic composite materials that can be reliably and controllably synthesized [73]; the use of photopolymerizable lyotropic liquid crystals allows one to template specific structures on to the organic and inorganic phases);

- sol–gel photopolymerization of organic–inorganic hybrid resins for the production of hybrid sol–gel (HSG) materials (synthesis of cross-linked glass-like matrices bearing Si–O–Si linkages that involves acid- or base-catalyzed hydrolysis of polyalkoxysilanes and where the presence of an acrylate monomer allows photoinduced densification of the gel, thereby giving unique electrical, optoelectronic and mechanical properties) [80, 81];
- photopolymerization of hydrogels (these systems, polymerized using photoinitiator-free donor–acceptor monomer couples, are of interest in controlled-release applications [82]);
- topochemical photopolymerization (this occurs within the solid state of organic crystals which are made up by packing molecules into a three-dimensional regular lattice; stereospecifically regulated polymers are formed [83]);
- photopolymerization in multilayers (this can allow the preparation of extended and highly organized ultrathin layers of polymers by the solid-state photopolymerization of multilayers built from monomolecular films of monomers using the Langmuir–Blodgett technique [84]);
- photopolymerization in (micro)heterogeneous media [85–87] (microemulsion, inverse emulsion, emulsion, dispersion, etc.);
- photochemical block copolymerization [88];
- two-photon photopolymerization [this reaction does not use the sequential absorption of two photons where the absorption is governed by the same one-photon selection rules; a two-photon absorption process relates to the simultaneous absorption of two photons via a virtual state. The great advantage of such a two-photon photopolymerization (of acrylates, epoxides, thiol–enes, etc.) is to generate polymeric microstructures because of the three-dimensional spatial resolution inherent to the process] [89–91];
- pulsed laser-induced polymerization (see below);
- photopolymerization with evanescent waves (this behaves as a stereolithographic method and allows one to produce nano-objects – 1 μm in-plane resolution and 30 nm out-of-plane resolution – usable in imaging, holographic recording or pixel-by-pixel addressing set-ups [92]);
- spatially controlled photopolymerization (usable for the manufacture of optical elements such as microlenses [93], holographic recording, etc.);
- *in vivo* light-induced polymerization [94];
- photopolymerization under a magnetic field (which shows a faster rate of radical monomer conversion that is connected with the role of the radical pair formed after excitation of the photoinitiator) [95].

15.6.1.2 Functional Properties

Through the above-mentioned reactions, many very interesting and varied properties can be attained for the polymers obtained or new functionalities for the polymer materials:

- formation of anisotropic polymers (formed from monomers containing mesogenic groups having a polymerizable group and a specific group leading to hydrogen bonding; the anisotropic network shows selective reflection [96]);
- large refractive index change allowing photo-optical control [obtained by photodimerization of poly(vinyl cinnamate)] [97];
- photoalignment of side-chain by using liquid crystalline polymers (thin films of photocross-linkable side-chain liquid crystalline copolymers irradiated with linearly polarized light that induces alignment of both mesogenic groups with the electric field vector [98]);
- optically induced birefringence in photoresponsive polymers (in a photopolymer containing both mesogenic groups and photoisomerizable/photodimerizable groups, reorientation effects induce an optical anisotropy under linearly polarized light whereas the mesogenic moieties and the dimerization process have an effect on the birefringence [99]);
- chiro-optical behavior (when irradiated with circularly polarized light, an achiral liquid crystalline azo polymer exhibits photoinduced chirality [100]);
- optical properties of photopolymers [101] (in the manufacture of various optical elements: volume phase gratings by photopatterning of hybrid sol–gel glasses, quadratic optical properties of doped photopolymers, diffraction gratings, etc.);
- electro-optical properties due to the uniform orientations obtained in a polymerization-induced phase separation process of monomer molecules in liquid crystal droplets (dispersed in a thermoplastic matrix) under a magnetic field [102];
- high-T_g polymers by exploiting the heterogeneity of cross-linked photopolymers (heterogeneous networks with broad distribution times allow a T_g to be reached – depending on the curing temperature – higher than that obtained in a more homogeneous network [103]);
- core hair-type microgels with various hair lengths (their viscosities and dispersibilities are particularly suitable for applications in water-developable screen printing plates [104]);
- nanogels with specific properties (the photocross-linking of well-adapted graft terpolymers can give a nanogel possessing a temperature-responsive core and pH-responsive arms [105]);
- metallo-polymers [through photopolymerization of metal-containing (liquid crystalline, etc.) monomers as a form of metal carboxylates bearing acrylate functions; 5% of zinc can be incorporated] [106];
- polymeric nanoparticles (by photopolymerization in a water-in-oil microemulsion [107]);
- monolithic polymers with controlled porous properties (monoliths containing hydrophilic, hydrophobic or ionizable functionalities are prepared by free radical polymerization, in spatially defined positions using a photolithographic set-up, on microfluidic ships [108]);
- expanding monomers (these UV-polymerizable systems are excellent candidates for the design of low-shrinkage materials [109]).

15.6.1.3 Some Typical Reactions of Industrial Interest in the UV Curing Area

Film photopolymerization of acrylates

The radical polymerization of multifunctional monomers in film is very fast (in the seconds time range under mercury lamps in industrial conditions), sensitive to the presence of oxygen and presents a low dark effect. Under continuous laser sources, the exposure time can drop to 1 ms or less.

Oxygen inhibition was found to affect significantly the top surface layer (on about 1 µm) in terms of physical properties and network structure, as confirmed by a recent investigation using advanced surface characterization techniques [110]. In films up to 100 µm, oxygen leads to a decrease in the reaction rate and spatial anisotropy of the polymer yield [111]. Many efforts have been made to overcome the O_2 inhibition by using various strategies [1]: irradiation under high-intensity lamps and high cure dosages, incorporation of oxygen barriers, use of sandwiching materials, design of oxygen-insensitive formulations, benefit of the oxygen scavenging effect of amines [112] or phosphorus compounds [113], inerting conditions leading to oxygen-free systems (under nitrogen; more recently, curing under a CO_2 atmosphere [114] should allow a promising future). The introduction of less sensitive monomers such as allyl-functionalized derivatives has been made.

The addition of UV absorbers (UVAs) and sterically hindered amine light stabilizers (HALSs) reduces the curing speed to a small extent [1]. These compounds are very useful, however, to photostabilize the coating and to avoid the rapid photodegradation of the polymer material (which is known to occur because of the generation of peroxyl radicals, peroxides and hydroperoxides upon further exposure of the cured coating to light). During the coating life, the UVAs absorb the incident light (and thereby avoid the polymer matrix from absorbing) and relax after excitation by non-radiative processes. HALSs scavenge oxygen through well-established processes, react with radicals and behave as a chain-breaking agent by scavenging the peroxyl radicals [1].

Outstanding properties can be achieved if the matrix is properly selected [1]: high T_g and low shrinkage for isobornyl acrylates, high flexibility of an ethoxylated trimethylolpropane triacrylate, high tensile strengths of polyether acrylates, hardness and flexibility of carbamate acrylate monomers, etc. Adhesion is improved by surface treatment with chemicals or plasmas, the use of adhesion promoters, chemical modification of monomers and oligomers, adequate formulation and selection of existing compounds, etc. [1].

Photopolymerization of thick, clear coatings (cm range) can be achieved if a suitable photoinitiating system is used: the penetration of the light into the depth of the film is favored if good bleaching occurs. The photopolymerization of thick pigmented coatings appears to be a real challenge because of the strong absorption of the pigment. Suitable high-performance photoinitiating systems have been proposed and allow the curing of paints usable on industrial lines [34].

The substrate can have a specific influence on the cure speed; for example, phenolic compounds sharply decrease the photopolymerization efficiency in wood coatings [50, 115].

Photopolymerization of epoxides and vinyl ethers

Cationic polymerization in films is fast, insensitive to oxygen, sensitive to moisture and water and presents an important dark post effect. The range of available monomers has been greatly expanded. Epoxides and vinyl ethers are widely used and give coatings with high thermal capability, excellent adhesion and good chemical resistance. Cyclic ethers such as oxetanes [116] can be an alternative to epoxides for achieving faster curing speeds in industrial lines.

The reaction rate can be increased through the use of additional initiation processes originating from the aryl free radical formed in the photodecomposition of iodonium salts [117]. This radical enhances the reaction efficiency through the generation of an additional initiation step: proper selection of cationic monomers possessing readily abstractable hydrogen atoms (alkyl vinyl ethers, epoxides bearing acetal groups, etc.) should help in the design of high-performance systems.

Epoxy-containing silicone monomers can be polymerized using cationic photoinitiators. High cure speeds are attained as in acrylate-modified silicones. The final properties correspond to those of a silicone polymer (excellent temperature, chemical and environmental resistance and good electrical characteristics).

Organic–inorganic hybrid resins based on a siloxane structure bearing pendent epoxy groups display high reactivity in photoinitiated cationic ring-opening polymerization. The mechanical properties are related to the length and type of the spacer group between the epoxy group and the siloxane chain [118].

Thiol–ene photopolymerization

Thiol–ene polymerization of suitable systems in films is insensitive to oxygen and the coating formed exhibits good adhesion to the substrate [1]. The reaction concerns the addition of a thiol to a double bond (vinyl, allyl, acrylate, methacrylate, etc.) and in recent years has seen a revival of interest [119].

As known, thiol–vinyl ether or –allyl ether polymerization [120] shows some interesting features: a very fast process, low or even no oxygen inhibition effect and formation of highly cross-linked networks with good adhesion, physical and mechanical properties.

The polymerization reaction in photoinitiator-free conditions was recently revisited; this reaction could be helpful if a non-UV-absorbing coating must be obtained because in that case no photoinitiator decomposition products are present in the cured coating. The addition of a UV photoinitiator allows the cure speed to be increased. Several papers have recently been devoted to the study of the propagation mechanism of the polymerization reaction and the formation of the polymer network [121]. The photoinduced polymerization of thiol–acrylate systems has also attracted particular attention: the high reactivity and the fact that under UV light the presence of a photoinitiator is not necessary make these systems very attractive. The polymer network structure is still under study and the role of the mechanism of the polymerization propagation is emphasized. All these systems are of interest in the fields of coatings, adhesives, laser imaging, optics (the thiol moieties allowing a high refractive index of the cured system), etc. The problem of the odor of thiols is nowadays solved by using new high molecular weight compounds and is not important in the case of a grafted system. Liquid crystalline thiol–enes [122] and the photopolymerization of pigmented thiol–enes [123] have also been studied.

Photopolymerization of water-borne light-curable systems
The basic idea is to develop solventless systems to conform to new environmental regulations [1, 5, 6, 124]. Water-based formulations are less viscous than the conventional acrylate mixtures. When used on porous substrates such as wood, the formulation will swell the wood fibers and increase the wood–resin interface, resulting in better adhesion. These systems are available as water-soluble media (urethane acrylates with low molecular weights, water-soluble polyether acrylates, etc.), emulsions (ethylene–vinyl acetate–methacrylate, methacrylate–butyl acrylate–glycidyl methacrylate, etc.), dispersions (urethane acrylates, epoxy acrylates or polyester acrylates stabilized with an added surfactant or functionalized with a surfactant moiety, etc.). Reactive monomer/oligomer cross-linkers can be added. Tack-free clear coatings can be cured using the photopolymerization of an aqueous polyurethane dispersion in the presence of a radical photoinitiator; however, obtaining good surface properties (hardness, chemical resistance) requires drying the coating before light exposure. Clear water-borne latexes for outdoor applications cross-link under sunlight via an oxidative mechanism. Sunlight curing of pigmented water-borne paints under air has been successfully achieved using a latex and a cross-linker [125].

Photopolymerization of powder formulations
Powder coatings are very attractive compared with liquid finishes since they are VOC free and easier to transport and to handle [5, 6]. Conventional systems are thermally cross-linked at high temperature (200 °C). The interest in using UV powder coatings is derived from attributes such as easy handling, no solvent, small-sized equipment, fast drying and a two-step process. The powder is applied on the substrate and exposed to an IR source to achieve coalescence of the solid particles. Then, the cross-linking reaction is carried out under UV (or visible) light in a very short time and at a lower temperature (100–120 °C), which allows these coatings to be used on heat-sensitive substrates such as wood (for medium-density fiberboard applications). The nature of the resin governs the final properties of the cured coating: alkyds (adhesion, brightness, flexibility), epoxides (mechanical properties, resistance to solvents and corrosion, insulating properties), polyurethanes (light fastness), unsaturated polyesters (hardness), etc. Commonly encountered systems are based on polyesters (with maleate or fumarate unsaturations) and polyurethane–vinyl ethers which copolymerize according to a radical process. A combination of amorphous polyesters and functionalized polyesters (with allyl ethers) has recently been proposed: allyl ethers are cheaper and available with a wide range of functionalities and structures, thereby allowing tuning of the final properties [126].

Charge-transfer photopolymerization

The polymerization reactions of maleimide–vinyl ether show some interesting features [1, 5, 6, 127–130]: very high cure speed (similar to that obtained with acrylates) and no or low sensitivity to oxygen. Using vinyl ethers appears to be an environment friendly alternative to acrylate monomers, which can smell and cause skin and eye irritation.

These systems can be used in photoinitiator-free formulations: in that case, a charge-transfer polymerization complex is formed between an acceptor molecule (maleimide) and a donor molecule (vinyl ether). After excitation, the generated intermediates from the complex might be a biradical, a zwitterion or an ion pair. A primary hydrogen abstraction process has been demonstrated to take place in the initiation step: it generates free radicals which can react with the monomer or participate in the propagation step of the polymerization through a chain-transfer reaction. The polymerization efficiency increases sharply in the presence of a UV or visible radical photoinitiator. Several papers have also recently been devoted to the study of the propagation of the polymerization reaction and the formation of the polymer network, the kinetics of the reaction and the possible applications.

In maleate–vinyl ethers, an alternate copolymer is formed. This is explained on the basis of a homopolymerization through a charge-transfer complex rather than a cross propagation.

One drawback of such polymerizable systems is the high cost and the availability of polyfunctional maleimides. Photoinitiator-free and UV/visible photoinitiator-containing maleimides and allyl ethers are less expensive, less irritating than maleimide–vinyl ethers and also exhibit fairly good efficiency. The fast curing of a maleimide–allyl or –vinyl ether film formulation can be achieved within 1 s in the presence of a four-component photoinitiating system under low light intensity [131].

Dual cure polymerization

The dual cure technique is a two-step process [1, 5, 6, 132], usually consisting in the combination of a UV irradiation and a thermal drying process; for example, this can be helpful to cure shadow areas. A typical dual cure system involves a two-pack material, typically based on a polyisocyanate and an acrylate. The acrylate is first cured upon light exposure and the subsequent dark reaction of the isocyanates with the hydroxyl groups of the acrylates allows the coating to be fully cured, thereby producing a tack-free film.

Dual cure can also refer to the use of two consecutive UV exposures (pregelification low intensity curing/high intensity curing or short-wavelength curing/conventional lamp curing) or a UV curing process followed by air drying [1].

Hybrid cure polymerization

The hybrid cure technique [1] corresponds to the photopolymerization of a blend of different monomers/oligomers which react through a different mechanism [133].

For example, a blend of acrylates and vinyl monomers in the presence of radical and cationic photoinitiators will lead to a different network compared with those obtained in separate radical or/and cationic formulations. Cross-linked copolymers combining the properties of the two homopolymer networks and interpenetrating networks exhibiting new properties can be formed. The starting point of the reaction can be driven by a selective and separate excitation of the two photoinitiators depending on the wavelengths used: the possibility of this two-step procedure (thanks to the photochemical initiation process) allows the two networks to be generated separately (and upon request).

Other examples [1] include the use of a light exposure and a dark process (thermal decomposition of peroxides, acid curing, "remote curing process", etc.).

15.6.2
Kinetics and Efficiency of the Photopolymerization Reaction

15.6.2.1 Overall Processes

A photoinduced polymerization reaction consists of a photochemical part and a chemical part. The initiation quantum yield ϕ_i, which represents the number of starting chains per photon absorbed, governs the initiation rate R_i, the polymerization rate R_p and the polymerization quantum yield ϕ_m, defined as the number of monomer units polymerized per photon absorbed:

$$R_i = \phi_i I_{abs}$$

$$R_p = K(R_i)^{0.5}$$

$$\phi_m = R_p / I_{abs}$$

This picture, which is well established in solution polymerization, holds true in the fast bulk polymerization of multifunctional monomers at low conversion and has been widely used in the literature.

15.6.2.2 Monitoring of the Photopolymerization Reaction

The monitoring of the film polymerization reaction can be done using different techniques, e.g. real-time IR spectroscopy, IR attenuated total reflection or near-IR reflection spectroscopy, differential scanning calorimetry, *in situ* microwave dielectric analysis, fluorescence emission of a probe sensitive to the viscosity, optical pyrometry, dilatometry and shrinkage measurement, gel point determination, Raman spectroscopy and photoacoustic calorimetry. The most widely used method is real-time FTIR spectroscopy [134], which allows the evolution of the monomer conversion to be recorded as a function of time. The corresponding curves for a radical polymerization exhibit a characteristic S shape because of the oxygen inhibition at very short exposure times or/and autoacceleration effects. The slope of the curve yields the rate of polymerization R_p at each conversion time. The slope at the origin (or the apparent linear part) is usually referred to as R_p in most publications where the idea is to evaluate the relative efficiency of various polymerizable formulations.

15.6.2.3 Kinetics of Photopolymerization

The investigation of the fast photopolymerization kinetics of multifunctional monomers is a rather hard and complex task which has not yet led to a clear scheme [135]. It is admitted that the usual expression for R_p for a chain-free radical in a quasi-stationary regime can be successfully applied only in the early stages of such a polymerization. A network is rapidly formed, even at low conversion, and the viscosity increases; as a consequence, the rate constants of propagation k_p and termination k_t will change as a function of the conversion σ in addition to the initial quantum yield, which is dependent on the cage escape reactions of the radicals. The following relationship (based on the reaction diffusion concept [136]) has been verified recently [135] in the photopolymerization of difunctional/monofunctional acrylate coatings having T_g of –5 and +63 °C:

$$k_t/[k_p(1-\sigma)] = \text{constant}$$

in the range $\sigma = 0.2-1$, which means that, for $\sigma > 0.2$, k_p and k_t are not constant. For monofunctional monomers, these rate constants can be considered to be almost unaffected until $\sigma < 0.8$.

Various aspects of the polymerization kinetics have been studied: kinetics and structural evolution in cross-linking polymerizations [137], sequential photoinduced living graft polymerizations [138], effect of chain length-dependent termination [139], modeling thermal and optical effects [140], autoacceleration effects [141], kinetic model for radical trapping [142], determination of copolymerization ratios [143], heat and mass transfer effects in thick films [144], bulk photopolymerization at a high degree of conversion [145], living photopolymerization [130], etc. Other polymerization kinetics have been explored for thiol–enes and thiol–acrylates [146, 147], cationic monomers [44], acrylamide in inverse emulsion [85], etc.

There is interest in the polymerization of new monomers able to lead to faster cure speeds, such as novel methacrylic monomers [147] and highly reactive fast monomers [148]. For example, a direct correlation between the hydrogen bonding/dipole moment and the rate of polymerization has been proposed [149]; the hydrogen bonding affects the tacticity of the polymer formed; the propagation and termination steps are influenced by the hydrogen bonding and the dipole moment, respectively.

Frontal photopolymerization [150] and cure depth in photopolymerization [151] are of prime importance for the polymerization of thick coatings.

Pulsed laser polymerization is particularly useful for evaluating the rate constants of propagation and termination [152, 153].

15.6.2.4 Photochemical and Chemical Reactivity

Photoinitiation quantum yield
The yield of the different primary reactions that occurs before the initiating step is governed by the photophysical and photochemical processes involved in the excited states, for example in a radical polymerization: cleavage reactions, electron transfer reactions, hydrogen abstraction reactions, quenching of excited states by monomer, oxygen, light stabilizers or other additives, energy transfer reactions, secondary reactions, etc. As a consequence, the initiation quantum yield ϕ_i appears strongly dependent on the kinetics of the excited state processes [1].

Investigation of the excited state processes
The investigation of the mechanisms involved in photoinitiating systems can be carried out using time-resolved laser spectroscopy [1] based on CIDNP-NMR or CIDEP-ESR techniques in real time [154], time-resolved laser absorption spectroscopy on both nanosecond [1, 37, 39, 155] and picosecond time-scales [35, 156, 157], laser-induced step scan FT-IR [158], real-time photoconductivity [159], laser-induced photothermal techniques [53, 160], etc. All these methods provide the usual information on the rate constants, the quantum yields of cleavage or electron transfer, the side-reactions of the radicals formed, the quenching of the radicals by the monomer, the interaction of O_2 with the excited species, etc. [1, 21, 33]. Other generally unknown (or rather scarce) data such as intersystem crossing quantum yields, triplet state energy levels, quantum yields of dissociation of cleavable photoinitiators, bond dissociation energies (BDE) of amines, formation enthalpies of the initiating radicals and growing polymer radicals, interaction rate constants between initiating radicals and monomer double bonds, addition reaction enthalpy and propagation and termination rate constants of the polymerization reaction also become accessible [161–165]. Full diagrams of the excited state processes are or will be known. Structure–excited state reactivity–practical efficiency relationships can be discussed.

The introduction of molecular modeling calculations to study the excited state processes from a theoretical point of view might be a way to design new, powerful photoinitiator, photosensitizer or complex photoinitiating systems. Recent examples include the cleavage process of photoinitiators [161], the decomposition of photobases [49], the bond dissociation of amines [162] and the energy transfer process [163], which is important when discussing the ability of various donor–acceptor couples to initiate a photocross-linking or a photopolymerization reaction.

Access to the rate constants of the chemical reactions
The addition reaction of an initiating radical to a radical monomer unit can be studied with time-resolved laser spectroscopy, pulsed radiolysis or laser-induced photocalorimetry techniques. The theoretical reactivity of selected carbon-centered radicals (such as aminoalkyl, methyl and benzoyl) towards double bonds also representative of widespread monomers (vinyl ether, vinyl acetate, acrylonitrile, methyl acrylate, etc.) was recently examined in detail by using molecular

orbital calculations. The observed reactivity in different radical–double bond systems is strongly influenced by the reaction exothermicity, demonstrating that the energy barrier is governed in a large part by the enthalpy term; the polar effect computed from the transition state structures can drastically enhance the reactivity [164, 166].

The investigation of the reactivity of acrylate radicals produced after the initiation reaction is particularly important. However, these radicals are difficult to observe because of their absorption in the UV region. A recent paper [167] shows that it is now possible to follow the initiating radical R and the first monomer radical RM due to the addition of R to the monomer double bond M. The interaction of RM with the inhibitors of polymerization (oxygen, stabilizers, spin traps) can be investigated in detail. In addition, to the recombination of RM and its addition to a new monomer unit can also be determined; the measured rate constants might be representative of the reactivity of the polymerization propagating radicals at low conversion, that is, the rate constants of propagation k_p and termination k_t of the radical polymerization reaction become accessible, much more easily than by using conventional pulsed laser polymerization (PLP) methods [167].

Solution and bulk reactivity

The reactivity in bulk compared with that in model conditions in solution deserves to be deeply investigated. Differences will likely originate from changes in the rate constants of the processes involved so that the balance between the competitive reactions is affected. Although these effects must be carefully taken into account, an interesting lesson of a recent study [168] is the evidence of an acceptable correlation of the relative polymerization efficiency in viscous media [epoxyacrylate–hexane diol diacrylate (HDDA) or HDDA-poly(methyl methacrylate) (PMMA)] with the dissociation quantum yield of selected photoinitiators. This correlation can remain valid in less viscous media (in BA–PMMA) but does not work well in fluid (in butyl acrylate, methyl acrylate or methyl methacrylate solutions). A broad knowledge of the photophysics and photochemistry of the photoinitiators in solution can thus serve as a good basis for the prediction of the polymerization efficiency in film experiments.

The investigation of excited state processes in real time in photopolymer layers is also promising [41, 169]. The same holds true for the study of the phenomena that occur under high-intensity laser light excitation or two-photon excitation.

15.7
Electron Beam, Microwave, Gamma Rays, Plasma and Pressure Stimuli Compared with Temperature and Light

High-energy radiation – X-rays (0.01–0.5 nm), γ-rays (^{60}Co, 10 pm, photon energy around 1 MeV) – is unselectively absorbed by molecules. The interaction with the electrons of the substrate (organic molecules, macromolecules) induces an ionization process and the excess energy is converted into kinetic energy for the ejected electrons. Inelastic collisions between these electrons and the substrate cause an excitation or an ionization process so that excited species and cation radicals are produced. Fragmentation reactions can then occur and the radicals generated are able to initiate a radical polymerization or lead to chemical changes (chain scission and/or photocross-linking) according to the nature of the material. Applications of X-rays can be found in the microlithography sector for very large-scale integration chip production [170] or in solid-state polymerization [171]. The use of γ-rays in the polymerization area is still encountered in biomaterial applications and in various other areas (core–shell latex [172], polymerization in ionic liquids [173], etc.).

In electron beam (EB) technology (typically energy range 50–500 keV), electrons having a high kinetic energy are directly produced. EB curing is well known in the industrial applications of radiation curing [23–26], where good radiation penetration is necessary or an absence of any other organic molecules (such as the photolysis products of the photoinitiator in UV curing) is required. The main advantages are a higher energy efficiency and a higher curing speed, an excellent depth of cure, an ability to cure a coating sandwiched between several layers containing material, a decreased number of toxic by-products, improved coating adhesion and physical properties (e.g. thermal, chemical, barrier, impact), etc. This leads to materials with improved dimensional stability, reduced stress cracking, reduced solvent and water permeability and improved thermomechanical properties. Examples of applications are encountered in coatings, inks, adhesives, plastics, fiber optics, paper, fabrics, food packaging and microlithography (integrated circuits). Others include graft, surface, multilayer, solid-state and emulsion polymerization applications [174].

Plasma polymerization is an interesting tool [175, 176] for modifying the surface properties of a material; it can either polymerize a suitable organic vapor or create a polymer film that cannot be obtained under the usual polymerization conditions. A discharge applied to an organic gas or a vapor in well-defined conditions produces a low-pressure and low-temperature plasma: ionization occurs and high-energy electrons are produced. In the case of a monomer, the electrons will collide with hydrocarbon molecules to produce positive ions (which are accelerated to the cathode and produce secondary electrons), excited molecular or atomic fragments and radicals. Excited atoms emit photons and create a glow. The plasma, which is electrically neutral, contains electrons, ions and radicals. The polymer deposition occurs on to the substrate in contact with the glow. The chemical reactions encountered are complex but, generally, the polymerization proceeds according to a free radical mechanism. The plasma polymer chains are highly branched and randomly terminated with a high degree of cross-linking. They lead to a cross-linking gradient over the polymer layer thickness and the extent of cross-linking can be adjusted by the plasma energy. All these result in interesting properties (excellent coating adhesion on almost any substrates; chemical, mechanical and thermal stability; high barrier effect, etc.) and applications (such as scratch-resistant coatings, corrosion-protective coatings, anti-bonding and anti-soiling coatings, barrier layers and pretreatment of metal surfaces).

Under microwave exposure [177], compared with conventional heating, a large number of reactions exhibit a significant increase in speed (at least because a high temperature is obtained in the microwave oven) and in yield (the excitation can circumvent the formation of by-products or the exhaustion of catalysts; the shortness of the heating periods keeps the undesired reactions to a minimum). The reduction of the side-reactions improves the purity and properties of the polymer. The selective excitation of ions or zerovalent metals has pronounced effects when using metal catalysis or ionic species. The kinds of polymerization reactions observed are (i) a step-growth polymerization (the most widely investigated) in polyamides, polyethers, polyesters, systems involving C–C coupling reactions and phase-transfer catalyzed systems and (ii) ring-opening polymerizations, free radical and controlled radical polymerizations and emulsion polymerizations. Applications can be found in the preparation of dental materials, the modification of polymers and curing processes.

An external pressure applied to a solid medium is able to finely tune the distances and the orientations of the molecules. In unsaturated bond-containing systems, chemical reactions can occur. The physical properties of a solvent and the reaction rates and also the thermodynamic and kinetic parameters are modified by the pressure, which is therefore a very interesting activation mode of reactions where volume changes are concerned. New and specific properties of the polymer formed are noted, for example: high-pressure polymerization of acetylene [178] above 3 GPa in the orthorhombic crystal phase governs the final conformation of the polymer; polyphenylacetylene prepared at 12 GPa is formed by conjugated segments having both cis and trans orientations; and pressure-induced polymerization of single-wall carbon nanotubes leads to a super-hard material with a very high bulk modulus.

15.8
Conclusion

This chapter has aimed to show the current status of light-induced polymerization reactions. The traditional UV curing area and the emerging new possibilities represent a field of growing importance in an industrial context. On the other hand, the development of potential high-tech applications of photopolymerizable materials, mainly based on the clever design of tailor-made monomers, oligomers and photosensitive systems, will continue to open up fascinating new fields of research.

References

1 J.P. Fouassier, *Photoinitiation, Photopolymerization, Photocuring*, Hanser, Munich, **1995**.

2 S.P. Pappas, *UV-Curing: Science and Technology*, Technology Marketing Corp., Stamford CT, **1986**; Plenum Press, New York, **1992**.

3 J.P. Fouassier, J.F. Rabek (Eds.), *Radiation Curing in Polymer Science and Technology*, Chapman and Hall, London, **1993**.

4 C.G. Roffey, *Photopolymerization of Surface Coatings*, Wiley, New York, **1982**.

5 R. Holman (Ed.), *UV and EB Chemistry*, Sita Technology, London, **1999**.

6 S. Davidson, *Exploring the Science, Technology and Application of UV and EB Curing*, Sita Technology, London, **1999**.

7 A.B. Scranton, A. Bowman, R.W. Peiffer (Eds.), *Photopolymerization: Fundamentals and Applications*, ACS Symposium Series 673, American Chemical Society, Washington, DC, **1997**.

8 J.P. Fouassier, J.F. Rabek (Eds.), *Lasers in Polymer Science and Technology: Applications*, CRC Press, Boca Raton, FL, **1990**.

9 V. Krongauz, A. Trifunac (Eds.), *Photoresponsive Polymers*, Chapman and Hall, New York, **1994**.

10 D.C. Neckers, *UV and EB at the Millennium*, Sita Technology, London, **1999**.

11 K.D. Belfied, J.V. Crivello (Eds.), *Photoinitiated Polymerization*, ACS Symposium Series 847, American Chemical Society, Washington, DC, **2003**.

12 J.P. Fouassier (Ed.), *Photosensitive Systems for Photopolymerization Reactions*, Trends in Photochemistry and Photobiology, Vol. 5, Research Trends, Trivandrum, **1999**.

13 J.P. Fouassier, *Light Induced Polymerization Reactions*, Trends in Photochemistry and Photobiology, Vol. 7, Research Trends, Trivandrum, **2001**.

14 J.G. Drobny, *Radiation Technology for Polymers*, CRC Press LLC, Boca Raton, FL, **2003**.

15 N.S. Allen, M. Edge, I.R. Bellobono, E. Selli (Eds.), *Current Trends in Polymer Photochemistry*, Ellis Horwood, New York, **1995**.

16 R. Holman, P. Olding (Eds.) *Chemistry and Technology of UV and EB Formulation for Coatings, Inks and Paints*, Vols. I–VIII, Wiley, Chichester and Sita Technology, London, **1997**.

17 C.E. Hoyle, J.F. Kinstle (Eds.), *Radiation Curing of Polymeric Materials*, ACS Symposium Series 417, American Chemical Society, Washington, DC, **1990**.

18 H. Bottcher, *Technical Applications of Photochemistry*, Deutscher Verlag für Grundstoffindustrie, Leipzig, **1991**.

19 J.F. Rabek, *Mechanisms of Photophysical and Photochemical Reactions in Polymers: Theory and Practical Applications*, Wiley, New York, **1987**.

20 J.P. Fouassier (Ed.), *Photochemistry and UV Curing: new trends*, Research Signpost, Trivandrum, **2006**.

21 K. Dietliker, *A Compilation of Photoinitiators Commercially Available for UV Today*, Sita Technology, London, **2002**.

22 J.P. Fouassier, *Photoinitiated Polymerization: Theory and Applications*, Rapra Review Reports, vol. 9, issue 4, Rapra Technology, Shawbury, **1998**.

23 *Technical Conference Proceedings*, RadTech International North America, Chevy Chase, **1998**, **2000**, **2002**, **2004**; http://www.radtech.org/.

24 *Conference Proceedings*, RadTech Europe Association, Hannover, **1999**, **2001**, **2003**, **2005**; http://www.radtech-europe.com/.

25 *Technical Conference Proceedings*, RadTech Asia, Tokyo, **2003**; http://www.radtechjapan.org/.

26 *Technical Conference Proceedings*, RadTech China, Beijing, **2005**; http://www.radtechchina.com/.

27 A. Braun, M.T. Maurette, E. Oliveros, *Technologie Photochimique*, Presses Polytechniques Romandes, Lausanne, **1986**.

28 M.R. Zonca, Jr., B. Falk, J.V. Crivello, *J. Macromol. Sci., Pure Appl. Chem.* **2004**, *41*, 741–756.

29 T. Oppenlaender, *Trends Chem. Eng.* **2003**, *8*, 123–126.

30 J.W. Martin, J.W. Chin, T. Nguyen, *Prog. Org. Coat.* **2003**, *47*, 292–311.

31 N.J. Turro, *Modern Molecular Photochemistry*, Benjamin, New York, **1978**, 1990.

32 T. Scherzer, *J. Polym. Sci., Part A: Polym. Chem.* **2004**, *42*, 4, 894–901.
33 J. P. Fouassier, *Recent Res. Dev. Photochem. Photobiol.* **2000**, *4*, 51–74; *Curr. Trends Polym. Sci.* **1999**, *4*, 163–183; *Recent Res. Dev. Polym. Sci.* **2000**, *4*, 131–145.
34 J. P. Fouassier, X. Allonas, D. Burget, *Prog. Org. Coat.* **2003**, *47*, 16–36.
35 S. Jockusch, I. V. Koptyug, P. F. McGarry, G. W. Sluggett, N. J. Turro, D. M. Watkins, *J. Am. Chem. Soc.* **1997**, *119*, 11495–11499.
36 M. Spichty, N. J. Turro, G. Rist, J. L. Birbaum, K. Dietliker, J. P. Wolf, G. Gescheidt, *J. Photochem. Photobiol. A: Chem.* **2001**, *142*, 209–213.
37 X. Allonas, C. Grotzinger, J. Lalevée, J. P. Fouassier, M. Visconti, *Eur. Polym. J.* **2001**, *37*, 897–906.
38 J. Paczkowski, D. C. Neckers, *Electron Transfer Chem.* **2001**, *5*, 516–585.
39 X. Allonas, J. P. Fouassier, L. Angiolini, D. Caretti, *Helv. Chim. Acta* **2001**, *84*, 2577–2588.
40 Q. Wu, B. Qu, *Polym. Eng. Sci.* **2001**, *41*, 1220–1226.
41 T. Uurano, H. Ito, T. Yamaoka, *Polym. Adv. Technol.* **1999**, *10*, 321–328.
42 J. Yang, D. C. Neckers, *J. Polym. Sci., Part A: Polym. Chem.* **2004**, *42*, 3836–3841.
43 U. Muller, A. Utterodt, W. Morke, B. Deubzer, C. Herzig, in *Photoinitiated Polymerization* (K. D. Belfield, J. Crivello, Eds.), ACS Symposium Series 847, American Chemical Society, Washington, DC, **2003**, pp. 202–212.
44 J. V. Crivello, J. Ma, F. Jiang, H. Hua, J. Ahn, R. A. Ortiz, *Macromol. Symp.* **2004**, 165–177.
45 F. Castellanos, J. P. Fouassier, C. Priou, J. Cavezzan, *J. Appl. Polym. Sci.* **1996**, *60*, 705–713.
46 C. Kutal, Y. Yamaguchi, W. Dei, C. Sanderson, in *Photoinitiated Polymerization* (K. D. Belfield, J. Crivello, Eds.), ACS Symposium Series 847, American Chemical Society, Washington, DC, **2003**, Ch. 29, p. 332–350.
47 M. Shirai, K. Suyama, M. Tsunooka, *Trends Photochem. Photobiol.* **1999**, *5*, 169–180.
48 J. Lalevée, X. Allonas, J. P. Fouassier, M. Shirai, M. Tsunooka, *Chem. Lett.* **2003**, *32*, 178–179.
49 J. Lalevée, X. Allonas, J. P. Fouassier, H. Tachi, A. Izumitani, M. Shirai, M. Tsunooka, *J. Photochem. Photobiol., A: Chem.* **2002**, *151*, 27–37.
50 M. Dossot, H. Obeid, X. Allonas, P. Jacques, J. P. Fouassier, A. Merlin, *J. Appl. Polym. Sci.* **2004**, *92*, 1154–1164.
51 X. Allonas, J. P. Fouassier, M. Kaji, Y. Murakami, *Photochem. Photobiol. Sci.* **2003**, *2*, 224–229.
52 Y. Yagci, A. Onen, in *Photoinitiated Polymerization* (K. D. Belfield, J. Crivello, Eds.), ACS Symposium Series 847, American Chemical Society, Washington, DC, **2003**, Ch. 16, p. 187–201.
53 X. Allonas, J. Lalevée, J. P. Fouassier, H. Tachi, M. Shirai, M. Tsunooka, *Chem. Lett.* **2000**, 1090–1091.
54 L. Angiolini, D. Caretti, C. Carlini, E. Corelli, E. Salatelli, *Polymer* **1999**, *40*, 7197–7207.
55 M. Visconti, M. Cattaneo, *Prog. Org. Coat.* **2000**, *40*, 243–251.
56 R. Liska, *J. Polym. Sci., Part A: Polym. Chem.* **2002**, *40*, 1504–1518.
57 C. Decker, D. Decker, *J. Macromol. Sci. Pure Appl. Chem.* **1997**, *A34*, 605–625.
58 C. Devadoss, P. Bharathi, J. S. Moore, *Macromolecules* **1998**, *31*, 8091–8099.
59 T. Nishikubo, A. Kameyana, H. Kudo, in *Photoinitiated Polymerization* (K. D. Belfield, J. Crivello, Eds.), ACS Symposium Series 847, American Chemical Society, Washington, DC, **2003**, Ch. 31, p. 363–377.
60 S. Y. Shim, D. H. Suh, in *Photoinitiated Polymerization* (K. D. Belfield, J. Crivello, Eds.), ACS Symposium Series 847, American Chemical Society, Washington, DC, **2003**, Ch. 24, p. 277–284.
61 H. Suzuki, S. Tajima, H. Sasaki, in *Photoinitiated Polymerization* (K. D. Belfield, J. Crivello, Eds.), ACS Symposium Series 847, American Chemical Society, Washington, DC, **2003**, Ch. 27, p. 306–316.
62 L. Keller, C. Decker, K. Zahouily, S. Benfahri, J. M. Le Meins, J. Miehe-Brendle, *Polymer* **2004**, *45*, 7437–7447.

63 C. Crouxte-Barghorn, S. Calixto, D. J. Lougnot, *Proc. SPIE* **1997**, *2998*, 222–231.

64 C. Decker, in *Photoinitiated Polymerization* (K. D. Belfield, J. Crivello, Eds.), ACS Symposium Series 847, American Chemical Society, Washington, DC, **2003**, *8*, 92–104.

65 R. Paul, R. Schmidt, J. Feng, D. J. Dyer, *J. Polym. Sci., Part A: Polym. Chem.* **2002**, *40*, 3284–3291.

66 F. C. Chen, H. S. Lackritz, *Macromolecules* **1997**, *30*, 5986–5996.

67 C. Esen, T. Kaiser, M. A. Borchers, G. Schweiger, *Colloid Polym. Sci.* **1997**, *275*, 131–137.

68 S. Xu, Z. Nie, M. Seo, P. Lewis, E. Kumacheva, H. A. Stone, P. Garstecki, D. B. Weibel, I. Gitlin, G. M. Whitesides, *Angew. Chem. Int. Ed.* **2004**, *43*, 2–6.

69 K. Saravanamuttu, M. P. Andrews, *Polym. Mater. Sci. Eng.* **1999**, *81*, 477–478.

70 M. Gaboyard, Y. Hervaud, B. Boutevin, *Polym. Int.* **2002**, *51*, 577–584.

71 S. Harrisson, S. R. Mackenzie, D. M. Haddleton, *Macromolecules* **2003**, *36*, 14, 5072–5075.

72 V. Sipani, L. S. Coons, B. Rangarajan, A. B. Scranton, *RadTech Rep.* **2003**, *17* (3), 22–26.

73 C. A. Guymon, J. H. Norton, in *Technical Conference Proceedings, RadTech 2002, The Premier UV/EB Conference and Exhibition*, Indianapolis, IN, **2002**, April 28–May 1, pp. 338–348.

74 N. Luo, J. B. Hutchison, K. S. Anseth, C. N. Bowman, *J. Polym. Sci., Part A: Polym. Chem.* **2002**, *40*, 1885–1891.

75 A. Momotake, T. Arai, *Polymer* **2004**, *45*, 5369–5390.

76 A. Asif, W. F. Shi, *Polym. Adv. Technol.* **2004**, *15*, 669–675.

77 N. Desilles, L. Lecamp, P. Lebaudy, C. Bunel, *Polymer* **2003**, *44*, 6159–6165.

78 T. Otsu, *J. Polym. Sci., Polym. Chem.* **2000**, *38*, 2121–2127.

79 J. Kwak, P. Lacroix-Desmazes, J. J. Robin, B. Boutevin, N. Torres, *Polymer* **2003**, *44*, 5119–5125.

80 O. Soppera, P. J. Moreira, P. V. Marques, A. P. Leite, *Proc. SPIE* **2004**, *5451*, 173–181.

81 W.-S. Kim, R. Houbertz, T.-H. Lee, B.-S. Bae, *J. Polym. Sci., Polym. Chem.* **2004**, *42*, 10, 1979–1986.

82 S. Jonsson, V. Kalyanaraman, K. Lindgren, S. Swami, L.-T. Ng, *Polym. Prepr.* **2003**, *44*, 7–8.

83 M. Hasegawa, C.-M. Chung, K. Kinbara, *J. Photosci.* **1997**, *4*, 147–159.

84 T. Seki, K. Tanaka, K. Ichimura, *Polym. J., Tokyo* **1998**, *30*, 646–652.

85 L. Liu, W. Yang, *J. Polym. Sci., Polym. Phys.* **2004**, *42*, 846–852.

86 J. K. Oh, J. Wu, M. A. Minnik, G. P. Craun, J. Rademacher, R. Farwaha, *J. Polym. Sci., Polym. Chem.* **2002**, *40*, 1594–1607.

87 S. E. Shim, H. Jung, H. Lee, J. Biswas, S. Choe, *Polymer* **2003**, *44*, 5563–5572.

88 M. Degirmenci, G. Hizal, Y. Yagci, *Macromolecules* **2002**, *35*, 8265–8270.

89 K. D. Belfield, X. Ren, E. W. van Stryland, D. J. Hagan, V. Dubikovsky, E. J. Miesak, *J. Am. Chem. Soc.* **2000**, *122*, 1217–1218.

90 H. B. Sun, S. Kawata, *Adv. Polym. Sci.* **2004**, *170*, 169–273.

91 Y.-X. Yan, X.-T. Tao, Y.-H. Sun, C.-K. Wang, G.-B. Xu, W.-T. Yu, H.-P. Zhao, J.-X. Yang, X.-Q. Yu, Y.-Z. Wu, X. Zhao, M.-H. Jiang, *New J. Chem.* **2005**, *29*, 479–484.

92 C. Ecoffet, A. Espanet, D. J. Lougnot, *Adv. Mater.* **1998**, *10*, 411–414.

93 C. Croutxe-Barghorn, O. Soppera, D. J. Lougnot, *Appl. Surf. Sci.* **2000**, *168*, 89–91.

94 J. H. de Groot, K. Dillingham, H. Deuring, H. J. Haitjema, F. J. van Beijma, K. Hodd, S. Norrby, *Biomacromolecules* **2001**, *2*, 1271–1278.

95 I. V. Khudyakov, N. Arsu, S. Jockusch, N. J. Turro, *Designed Monom. Polym.* **2003**, *6*, 91–95.

96 S. Greenfield, M. J. Goulding, M. Verrall, O. L. Parri, G. W. Gray, D. Coates, D. Sherrington, *Br. Pat. Appl. 2305925*, **1997**.

97 S. Murase, K. Kinoshita, K. Horie, S. Morino, *Macromolecules* **1997**, *30*, 8088–8090.

98 N. Kawatsuki, C. Suehiro, T. Yamamoto, *Macromolecules* **1998**, *31*, 5984–5990.

99 D. H. Choi, H. T. Hong, K. J. Cho, J. H. Kim, *Korea Polym. J.* **1999**, *7*, 189–195.
100 D. K. Hore, A. L. Natansohn, P. L. Rochon, *J. Phys. Chem. B* **2003**, *107*, 11, 2506–2518.
101 M. Feuillade, C. Croutxe-Barghorn, L. Mager, C. Carré, A. Fort, *Chem. Phys. Lett.* **2004**, *398*, 151–156.
102 F. P. Nicoletta, G. De Filpo, D. Cupelli, M. Macchione, G. Chidichimo, *Appl. Phys. Lett.* **2001**, *79*, 4325–4329.
103 H. Lu, L. G. Lovell, C. N. Bowman, *Macromolecules* **2001**, *34*, 8021–8025.
104 T. Takahashi, H. Watanabe, N. Miyagawa, S. Takahara, T. Yamaoka, *Polym. Adv. Technol.* **2002**, *13*, 33–39.
105 D. Kuckling, C. D. Vo, S. E. Wohlrab, *Langmuir* **2002**, *18*, 4263–4269.
106 L. Marcot, P. Maldivi, J.-C. Marchon, D. Guillon, M. Ibn-Elhaj, D. J. Broer, T. Mol, *Chem. Mater.* **1997**, *9*, 2051–2058.
107 E. F. Craparo, G. Cavallaro, M. L. Bondi, G. Giammona, *Macromol. Chem. Phys.* **2004**, *205*, 1955–1964.
108 C. Yu, M. Xu, F. Svec, J. M. J. Frechet, *J. Polym. Sci., Polym. Chem.* **2002**, *40*, 755–769.
109 C. C. Chappelow, C. S. Pinzino, L. Jeang, C. D. Harris, A. J. Holder, J. D. Eick, *J. Appl. Polym. Sci.* **2000**, *76*, 1715–1724.
110 H. Cao, E. Currie, M. Tilley, Y. C. Jean, in *Photoinitiated Polymerization* (K. D. Belfield, J. Crivello, Eds.), ACS Symposium Series 847, American Chemical Society, Washington, DC, **2003**, Ch. 13, pp.152–164.
111 V. V. Krongauz, C. P. Chawla, J. Dupré, *Photoinitiated Polymerization* (K. D. Belfield, J. Crivello, Eds.), ACS Symposium Series 847, American Chemical Society, Washington, DC, **2003**, Ch. 14, pp. 165–177.
112 R. Sato, T. Kurihara, M. Takeishi, *Polym. J.* **1998**, *30*, 158–160.
113 W. Kuang, C. E. Hoyle, K. Viswanathan, S. Jonsson, in *Technical Conference Proceedings, RadTech* **2002**, pp. 292–299.
114 K. Studer, C. Decker, E. Beck, R. Schwalm, *Prog. Org. Coat.* **2003**, *48*, 92–100.
115 S. Yin, A. Merlin, A. Pizzi, X. Deglise, B. George, M. Sylla, *J. Appl. Polym. Sci.* **2004**, *92*, 3499–3507.
116 H. Kato, H. Sasaki, in *Photoinitiated Polymerization* (K. D. Belfield, J. Crivello, Eds.), ACS Symposium Series 847, American Chemical Society, Washington, DC, **2003**, Ch. 25, pp. 285–295.
117 J. V. Crivello, in *Photoinitiated Polymerization* (K. D. Belfield, J. Crivello, Eds.), ACS Symposium Series 847, American Chemical Society, Washington, DC, **2003**, Ch. 15, pp. 178–186.
118 K. Y. Song, R. Ghoshal, J. V. Crivello, in *Photoinitiated Polymerization* (K. D. Belfield, J. Crivello, Eds.), ACS Symposium Series 847, American Chemical Society, Washington, DC, **2003**, Ch. 22, pp. 253–265.
119 C. E. Hoyle, T. Y. Lee, T. Roper, *J. Polym. Sci., Polym. Chem.* **2004**, *42*, 5301–5338.
120 Carlsson, A. Harden, S. Lundmark, A. Manea, N. Rehnberg, L. Svensson, in *Photoinitiated Polymerization* (K. D. Belfield, J. Crivello, Eds.), ACS Symposium Series 847, American Chemical Society, Washington, DC, **2003**, Ch. 6, pp. 65–75.
121 S. K. Reddy, N. B. Cramer, A. K. O'Brien, T. Cross, R. Raj, C. N. Bowman, *Macromol. Symp.* **2004**, *206*, 361–374.
122 H. T. A. Wilderbeek, M. G. M. van der Meer, C. W. M. Bastiaansen, D. J. Broer, *J. Phys. Chem. B* **2002**, *106*, 12874–12883.
123 T. M. Roper, T. Kwee, T. Y. Lee, C. A. Guymon, C. E. Hoyle, *Polymer* **2004**, *45*, 2921–2929.
124 C. Decker, F. Masson, R. Schwalm, *J. Coat. Technol.* **2004**, *1*, 127—130.
125 C. Bibaut-Renaud, D. Burget, J. P. Fouassier, C. G. Varelas, J. Thomatos, G. Tsagaropoulos, L. O. Ryrfors, O. J. Karlsonn, *J. Polym. Sci., Polym. Chem.* **2002**, *40*, 3171–3181.
126 D. Burget, C. Mallein, F. Mauguière-Guyonnet, J. P. Fouassier, C. G. Varelas, S. Apostolatos, G. Charalampoulou, M. Manea, S. Lundmark, in *Proceedings of RadTech Europe 2003 Conference*, Berlin, **2003**, 56–62.
127 C. E. Hoyle, S. C. Clark, K. Viswanathan, S. Jonsson, *Photochem. Photobiol. Sci.* **2003**, *2*, 1074–1079.
128 C. K. Nguyen, T. B. Cavitt, C. E. Hoyle, V. Kalyanaraman, S. Jonsson, in *Photo-*

initiated Polymerization (K.D. Belfield, J. Crivello, Eds.), ACS Symposium Series 847, American Chemical Society, Washington, DC, **2003**, Ch. 3, pp. 27–40.

129 C. Decker, C. Bianchi, S. Jönsson, *Polymer* **2004**, *45*, 5803–5811.

130 N.B. Cramer, S.K. Reddy, M. Cole, C. Hoyle, C.N. Bowman, *J. Polym. Sci., Polym. Chem.* **2004**, *22*, 5817–5826.

131 D. Burget, C. Mallein, J.P. Fouassier, *Polymer* **2003**, *44*, 7671–7678.

132 C. Decker, F. Masson, R. Schwalm, *Macromol. Mater. Eng.* **2003**, *288*, 17–28.

133 J.-D. Choe, J.W. Hong, *J. Appl. Polym. Sci.* **2004**, *93*, 1473–1483.

134 T. Scherzer, U. Decker, *Nucl. Instrum. Methods Phys. Res., Sect. B* **1999**, *151*, 306–312.

135 I.V. Khudyakov, M.B. Purvis, N.J. Turro, in *Photoinitiated Polymerization* (K.D. Belfield, J. Crivello, Eds.), ACS Symposium Series 847, American Chemical Society, Washington, DC, **2003**, Ch. 10, pp. 113–127.

136 G.V. Schultz, *Z. Phys. Chem.* **1956**, *8*, 290–293.

137 C.N. Bowman, K.S. Anseth, *Polym. Mater. Sci. Eng.* **1997**, *77*, 375–376.

138 H. Ma, R.H. Davis, C.N. Bowman, *Macromolecules* **2000**, *33*, 331–335.

139 K.A. Berchtold, B. Hacioglu, L. Lovell, J. Nie, C.N. Bowman, *Macromolecules* **2001**, *34*, 5103–5111.

140 A.K. O'Brien, C.N. Bowman, *Macromolecules* **2003**, *36*, 7777–7782.

141 M.D. Goodner, C.N. Bowman, *Macromolecules* **1999**, *32*, 6552–6559.

142 M. Wen, A.V. McCormick, *Macromolecules* **2000**, *33*, 9247–9254.

143 J.F.G.A. Jansen, E.E.J.E. Houben, P.H.G. Tummers, D. Wienke, J. Hoffman, *Macromolecules* **2004**, *37*, 2275–2286.

144 M.D. Goodner, C.N. Bowman, *Chem. Eng. Sci.* **2002**, *57*, 887–900.

145 U. Schmelmer, A. Paul, A. Kueller, R. Jordan, A. Goelzhaeuser, M. Grunze, A. Ulman, *Macromol. Symp.* **2004**, *217*, 223–230.

146 N.B. Cramer, C.N. Bowman, *J. Polym. Sci., Polym. Chem.* **2001**, *39*, 3311–3319.

147 L.-T. Ng, S. Swami, S. Jonsson, *Radiat. Phys. Chem.* **2004**, *69*, 321–328.

148 J.F.G.A. Jansen, A.A. Dias, M. Dorschu, B. Coussens, *Macromolecules* **2003**, *36*, 3861–3873.

149 J.F.G.A. Jansen, A.A. Dias, M. Dorschu, B. Coussens, *Macromolecules* **2002**, *35*, 7529–7531.

150 M. Belk, K.G. Kostarev, V. Volpert, T.M. Yudina, *J. Phys. Chem.* **2003**, *107*, 10292–10299.

151 J.H. Lee, R.K. Prud'Homme, I.A. Aksay, *J. Mater. Res.* **2001**, *16*, 3536–3544.

152 M. Buback, M. Egorov, A. Feldermann, *Macromolecules* **2004**, *37*, 1768–1776.

153 J.M. Asua, S. Bauermann, M. Buback, P. Castignolles, B. Charleux, R.G. Gilbert, R.A. Hutchinson, J.R. Leiza, A.N. Nikitin, J.P. Vairon, A.M. van Herk, *Macromol. Chem. Phys.* **2004**, *205*, 2151–2156.

154 I. Gatlik, P. Rzadek, G. Gescheidt, G. Rist, B. Hellrung, J. Wirz, K. Dietliker, G. Hug, M. Kunz, J.-P. Wolf, *J. Am. Chem. Soc.* **1999**, *121*, 8332–8336.

155 W. Schnabel, in *Lasers in Polymer Science and Technology*, Vol. II (J.P. Fouassier, J.F. Rabek, Eds.), CRC Press, Boca Raton, FL, **2000**, pp. 95–144.

156 F. Morlet-Savary, C. Ley, P. Jacques, J.P. Fouassier, *J. Phys. Chem.* **2001**, *105*, 11026–11033.

157 U. Kolczak, G. Rist, K. Dietliker, J. Wirz, *J. Am. Chem. Soc.* **1996**, *118*, 6477–6482.

158 C.S. Colley, D.C. Grills, N.A. Besley, S. Jockusch, P. Matousek, A.W. Parker, M. Towrie, N.J. Turro, P.M.W. Gill, M.W. George, *J. Am. Chem. Soc.* **2002**, *124*, 14952–14959.

159 X. Allonas, C. Ley, R. Gaume, P. Jacques, J.P. Fouassier, *Trends Photochem. Photobiol.* **1999**, *5*, 93–102.

160 X. Allonas, C. Ley, C. Bibaut, P. Jacques, J.P. Fouassier, *Chem. Phys. Lett.* **2000**, *322*, 483–490.

161 X. Allonas, J. Lalevée, J.P. Fouassier, *J. Photochem. Photobiol., A: Chem.* **2003**, *159*, 127–133.

162 J. Lalevée, X. Allonas, J.P. Fouassier, *J. Am. Chem. Soc.* **2002**, *124*, 9613–9621.

163 J. Lalevée, X. Allonas, F. Louerat, J. P. Fouassier, *J. Phys. Chem.* **2002**, *106*, 6702–6709.
164 H. Fischer, L. Radom, *Angew. Chem.* **2001**, *40*, 1340–1348.
165 X. Allonas, J. Lalevée, J. P. Fouassier, *Photoinitiated Polymerization* (K. D. Belfield, J. Crivello, Eds.), ACS Symposium Series 847, American Chemical Society, Washington, DC, **2003**, Ch. 12, pp. 140–151.
166 J. Lalevée, S. Genêt, X. Allonas, J. P. Fouassier, *J. Am. Chem. Soc.* **2003**, *125*, 9377–9380; J. Lalevée, X. Allonas, J. P. Fouassier, *J. Org. Chem.* **2005**, *70*, 814–819.
167 J. Lalevée, X. Allonas, J. P. Fouassier, *Chem. Phys. Lett.* **2005**, *401*, 483–486.
168 J. Lalevée, X. Allonas, S. Safi, J. P. Fouassier, *Macromolecules* **2006**, *39*, 1872–1879.
169 T. Urano, H. Ito, K. Takahama, T. Yamaoka, *Polym. Adv. Technol.* **1999**, *10*, 201–205.
170 K. Emoto, *Eur. Pat. Appl. 1398648*, **2004**.
171 J. E. Hernandez, J. E. Whitten, *Polym. Mater. Sci. Eng.* **2000**, *31*, 238–245.
172 M. Zou, Z. Zhang, X. Shen, J. Nie, X. Ge, *Radiat. Phys. Chem.* **2005**, *74*, 323–330.
173 W. Liu, G. Wu, D. Long, G. Zhang, *Polymer* **2005**, *46*, 8403–8409.
174 J. Pusch, V. Herk, M. Alex, *Macromolecules* **2005**, *38*, 8694–8700.
175 A. J. Beck, J. D. Whittle, N. A. Bullett, P. Eves, S. MacNeil, S. A. McArthur, G. Alexander, *Plasma Process. Polym.* **2005**, *2*, 641–649.
176 L. Francesch, E. Garreta, M. Balcells, E. R. Edelman, S. Borros, *Plasma Process. Polym.* **2005**, *2*, 605–611.
177 F. Wiesbrock, R. Hoogenboom, U. Schubert, *Macromol. Rapid Commun.* **2004**, *25*, 1739–1764.
178 M. Santoro, L. Ciabini, R. Bini, V. Schettino, *J. Raman Spectrosc.* **2003**, *34*, 557–566.

16
Inorganic Polymers with Precise Structures

David A. Rider and Ian Manners

16.1
Metal-containing Polymers

16.1.1
Introduction

Many of the valuable physical and chemical properties and functions of solid-state and biological materials can be attributed to the presence of metallic elements. Examples include magnetic materials used in data storage, superconductors, electrochromic materials and catalysts, including metalloenzymes. It has long been recognized that the incorporation of metal atoms into synthetic polymer chains may also lead to desirable properties and thereby generate a new and versatile class of functional materials capable of enhanced processability. However, until recently, synthetic difficulties associated with the creation of macromolecular chains possessing metal atoms as a key structural component have held back progress in the field.

Over the past decade or so, these synthetic obstacles have been, in part, overcome through many creative procedures to prepare new materials [1–9]. The new approaches that are now available have led to macromolecular structures in which metals are not only incorporated via the use of traditional covalent bonds but also by potentially reversible coordination interactions. Of comparable importance to these synthetic developments, these new materials have shown that they can possess a diverse range of interesting and useful properties and potential applications that complement those of organic macromolecules [1–9].

The additional challenge of preparing precise metallopolymer structures has become a key goal since the early 1990s. Metallopolymers with precisely controlled chain length or with block, star or dendritic architectures are attracting rapidly expanding attention as a result of their properties and potential applications.

Metal-containing block copolymers, in particular, represent an area of rapidly growing interest as a result of their self-assembly into phase-separated nanodo-

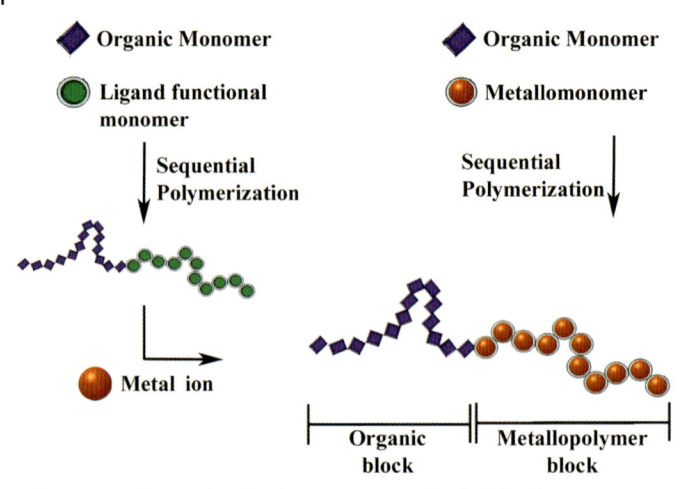

Scheme 16.1 Approaches for the preparation of well-defined metal-containing block copolymers.

main structures in thin films and micelles in block-selective solvents [1, 8]. In principle, two different synthetic approaches can be envisaged for the preparation of these materials (Scheme 16.1). In the following sections, the use of these two approaches is reviewed in terms of the mechanisms involved.

16.1.2
Chain Growth Polymerizations

Chain growth polymerizations represent the most efficient method for preparing polymers with controlled architectures. These processes are well illustrated by addition polymerizations, which typically involve unsaturated molecules as monomers where the thermodynamic driving force for polymerization involves the formation of new σ-bonds from π-bonds. These polymerizations (shown in Scheme 16.2) are initiated by a reactive intermediate (e.g. a free radical, cation or anion) that adds to a monomer to which its reactivity is transferred. Ideally, this process repeats with further monomer additions until all the monomers are consumed. The polymer chain ceases to grow from chain termination or chain transfer steps where the reactive center at the chain end is deactivated or transferred to another molecule or site. By virtue of the highly reactive chain ends, which readily and efficiently react with monomers, high molecular weight macromolecules are formed at low monomer conversion compared with step growth polymerizations.

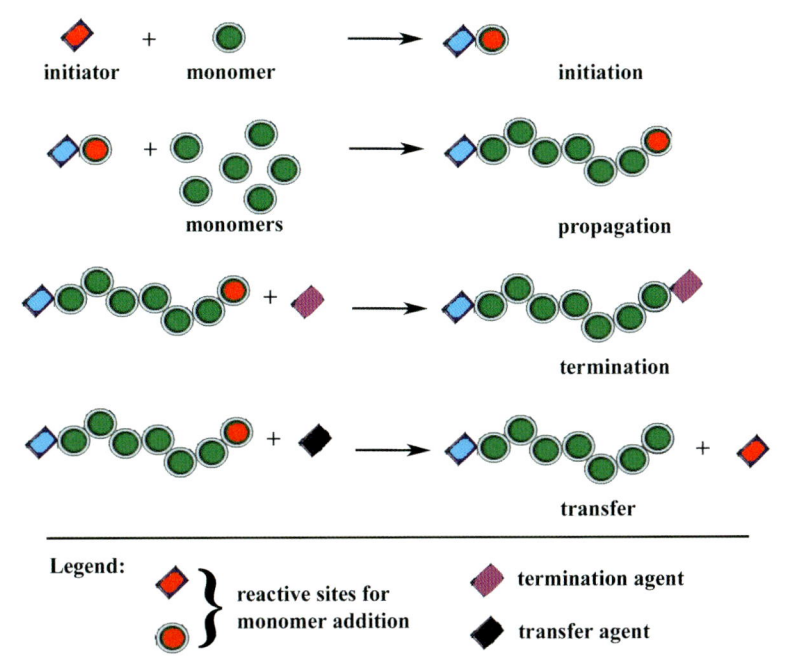

Scheme 16.2 Chain growth polymerization. Initiation, propagation, chain termination and chain transfer shown from top to bottom.

16.1.3
Living Polymerizations and Controlled Polymerizations

The term "living" has been assigned to polymerizations that lack chain transfer and chain termination steps [10]. In living polymerizations, when the rate of initiation is fast compared with the rate of chain propagation, the resulting polymer has a narrow molecular weight distribution [the polydispersity index (PDI) is close to 1.00]. A "controlled" polymerization, however, has a slightly more vague definition [10, 11]. In this case, a certain structural aspect of the polymer chain produced and/or a kinetic feature of the polymerization is subject to control. With this in mind, we provide an overview of living and controlled routes to metallopolymers within the following survey. Specifically, we cover (i) nitroxide-mediated radical polymerization, (ii) atom-transfer radical polymerization, (iii) reversible addition fragmentation termination polymerization, (iv) cationic polymerization, (v) anionic polymerization and (vi) metathesis polymerization. In addition, the use of these techniques to prepare metal-containing block copolymers, metallo-linked block copolymers and star-homo and star-block copolymers is outlined. We then present an overview of some exciting applications of these unique materials. Lastly, we outline polymers based on main group elements.

16.2
Use of Nitroxide-mediated Radical Polymerization

16.2.1
Introduction

Nitroxide (NO)-mediated radical polymerization is a controlled polymerization process where bimolecular termination is suppressed. This is achieved by the addition of persistent nitroxide radicals ($X^\bullet$) which reversibly couple with propagating chains ($\sim P_n^\bullet$) to form a dormant species ($\sim P_n$–X). Shown in Scheme 16.3b are some typical nitroxides used for this process, namely 1-(1-phenylethoxy)-2,2,6,6-tetramethylpiperidine (PhEt-TEMPO) and 2,2,5-trimethyl-3-(1-phenylethoxy)-4-phenyl-3-azahexane (PhEt-TIPNO). For these initiators and their respective dormant polymer analogues, homolytic cleavage of the C–O bond results in a nitroxide radical ($X^\bullet$) with stability due to steric protection from the substituents on the nitrogen atom. The carbon-based radical counterpart is much more reactive, however, and readily adds to the monomer [12]. Scheme 16.3a shows the nitroxide-mediated polymerization mechanism where k_a and k_{da} are rate constants for the radical chain-end activation and deactivation reactions, respectively. NO-mediated polymerizations are sometimes slow (1–3 days) due to small $k_a:k_{da}$ ratios and occur at elevated temperatures (125–140 °C).

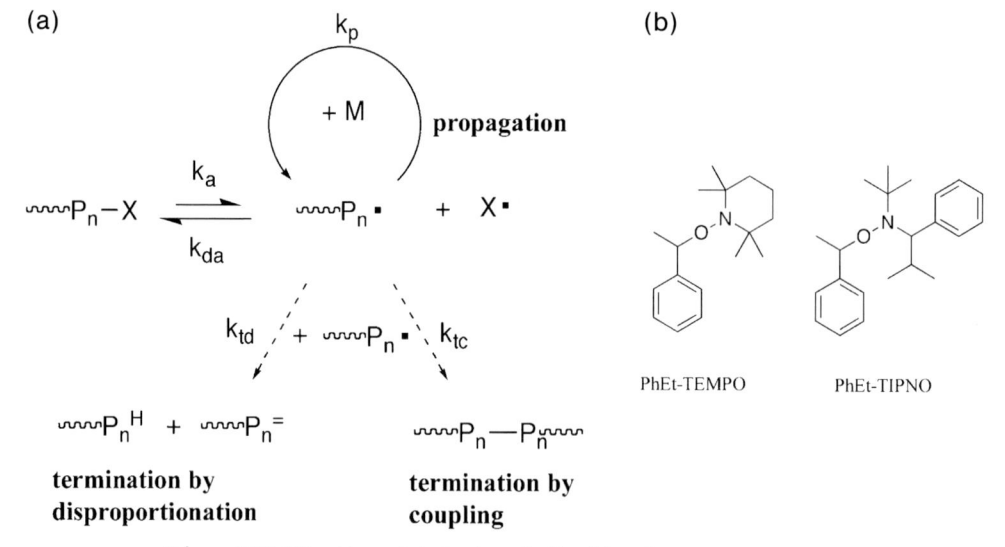

Scheme 16.3 Nitroxide-mediated polymerization (a) and some nitroxide-mediated polymerization initiators (b).

16.2.2
Nitroxide-mediated Radical Polymerization Routes to Ligand Functional Homopolymers

Nitroxide-mediated radical polymerization is a tremendous accomplishment and is also experimentally straightforward. Mainly styrenyl- and acrylate-type monomers undergo reproducible, controlled polymerizations with predictable molecular weights and low polydispersity indices [13]. The following sections highlight the controlled polymerization routes of some styrenyl monomers as representative examples.

Tew et al. have developed the NO-mediated random copolymerization of styrene and a terpyridine-containing styrenyl monomer, which is available from condensation of 5-aminopentyl 4'-(2,2',6',2''-terpyridinyl) ether with 4-vinylbenzoic acid [14]. As shown in Scheme 16.4, this styrene-based terpyridine monomer can be copolymerized with styrene via thermal initiation with PhEt-TIPNO to afford a random copolymer containing 10 mol% of terpyridine substituents ($M_n = 32\,100$, PDI = 1.33). Although no metallization has been reported with this specific material, an analogous route to poly(methyl methacrylate) (PMMA)-based block-random copolymers with pendant terpyridine units was subsequently investigated and these results will be discussed in Section 16.3.3.

Recently, Grubbs and coworkers have developed NO-mediated polymerization routes to alkyne-functional homopolymers [15]. Shown in Scheme 16.5 is the polymerization of 4-(phenylethynyl)styrene (PES) initiated and mediated by PhEt-TIPNO to afford poly[4-(phenylethynyl)styrene] (PPES, $M_n = 5400–10\,800$, PDI = 1.15–1.21). After approximately 48 h, conversions greater than 83% are observed with near-linear plots for M_n versus conversion and for $\ln(1 - \text{conver-}$

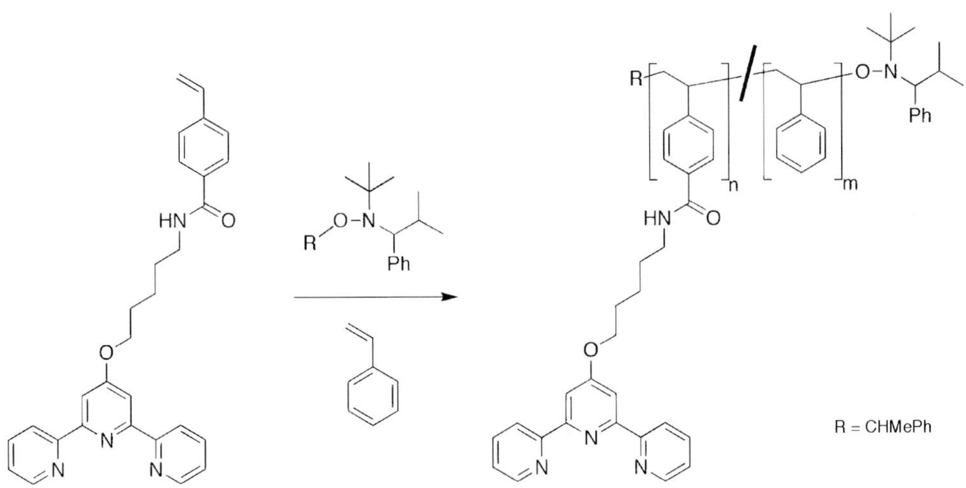

Scheme 16.4 NO-mediated random copolymerization of a styrenylterpyridine monomer.

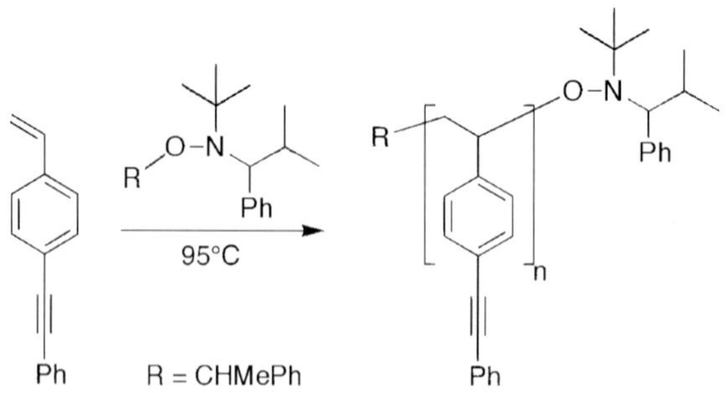

Scheme 16.5 NO-mediated synthesis of alkyne-functional polystyrene.

sion) versus time. Slight deviations from well-controlled polymerizations occur at higher conversion, which are thought to arise from side-reactions of the nitroxide with the monomer alkyne groups. Again, limited information regarding the homopolymer metallization is available but subsequent application of these techniques to metallized PPES block copolymers will be described in Section 16.2.3.

16.2.3
Nitroxide-mediated Radical Polymerization Routes to Ligand Functional Block Copolymers

Poly(vinylpyridine) (PVP) block copolymers are readily synthesized via NO-mediated polymerization but are usually made by anionic polymerization [16, 17]. Their subsequent complexation with transition metals is therefore described in the anionic polymerization section (see Sections 16.6.2 and 16.6.3).

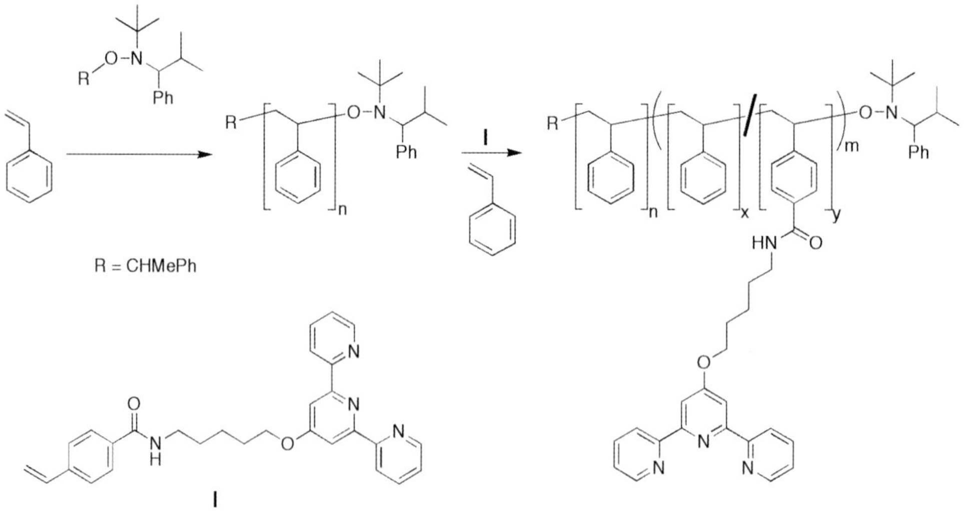

Scheme 16.6 NO-mediated polymerization of styrenylterpyridine to block-random copolymers.

Tew has recently extended his group's work to the synthesis of styrene-based block copolymers with a terpyridine-containing segment (Scheme 16.6) [18]. Using a polystyrene (PS)-based TIPNO macroinitiator ($M_n = 44\,400$, PDI = 1.13) for the NO-mediated random copolymerization of styrene and styrenylterpyridine (Scheme 16.6, **I**), monomers at 125 °C afforded the block random material in moderate yields ($M_n = 67\,200$, PDI = 1.40). Broader molecular weight distributions were reported to arise from the monomer reactivity differences.

More recently, Grubbs and coworkers have used NO-mediated polymerization routes to prepare alkyne-functional block copolymers that could be subsequently complexed with organometallic moieties [15, 18, 19]. Shown in Scheme 16.7 is the polymerization of 4-(phenylethynyl)styrene (PES) initiated and mediated by a PS–TEMPO macroinitiator (e.g. $M_n = 19\,200$, PDI = 1.08), which produced PS-b-poly[4-(phenylethynyl)styrene]s ($M_n = 22\,700$–$32\,700$, PDI = 1.09–1.19). After approximately 17 h of polymerization, conversions greater than 43% were observed with near-linear plots for M_n versus conversion and for $\ln(1 - \text{conversion})$ versus time. A slight deviation from well-controlled polymerization above this conversion is thought to arise from side-reactions of the nitroxide with the monomer alkynyl substituents. To introduce a transition metal on to the alkynyl polymer side-chains, the poly[4-(phenylethynyl)styrene] block copolymers were reacted with $Co_2(CO)_8$ under reflux in toluene to induce cluster formation. The resulting metalloblock copolymers are film-forming materials which are easily

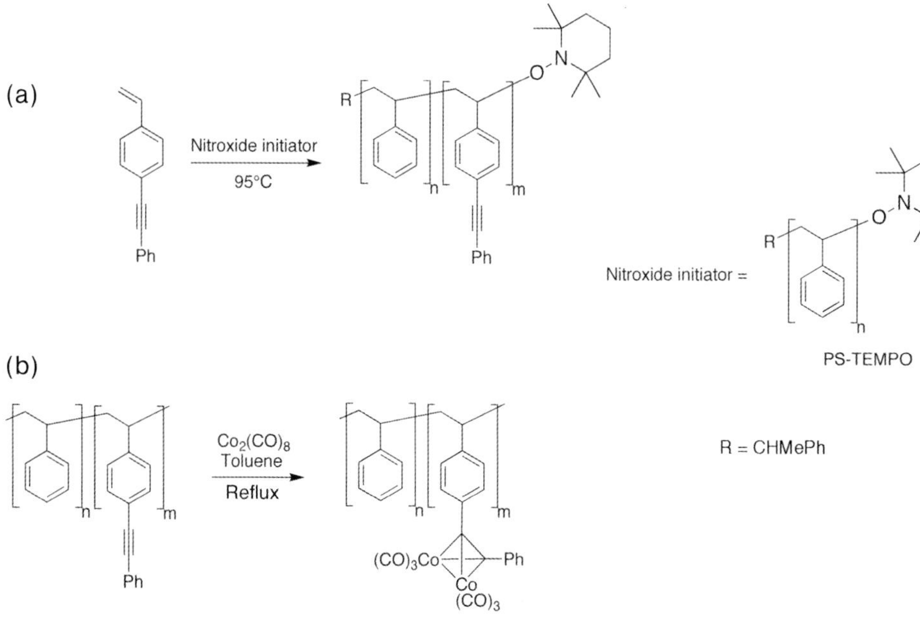

Scheme 16.7 NO-mediated synthesis of alkyne-functional polystyrene block copolymers (a) and subsequent clusterization with $Co_2(CO)_8$ (b).

handled in solution. The degree of alkyne clusterization was estimated to be ca. 91% using elemental analysis techniques. These organic–organocobalt block copolymers were of sufficient immiscibility to permit phase separation, leading to arrays of nanometer-sized metal-rich domains in bulk films.

16.2.4
Nitroxide-mediated Radical Polymerization of Metallomonomers

Frey and coworkers investigated the NO-mediated homopolymerization of vinylferrocene and its random copolymerization with styrene using the radical initiator AIBN at 120 °C and 2,2,6,6-tetramethyl-1-piperidinyl-1-oxy (TEMPO) as the mediating agent [20]. After 24–72 h, the molecular weights of isolated products at approximately 40% conversion were in the range 1300–3600 for poly(vinylferrocene) (PVFc, PDI = 1.24–1.84). For random copolymers (Scheme 16.8), it was found that for an increasing target fraction of vinylferrocene (i) the macromolecular weights obtained were increasingly lower than theoretical estimates, (ii) the maximum conversion decreased and (iii) the polydispersity indices increased. Moreover, the polydispersity indices were also reported to increase with near-complete conversion. The lack of a well-controlled polymerization for vinylferrocene was attributed its high constant for chain transfer, where electron transfer from ferrocenyl groups to the radical chain end occurs.

16.2.5
Nitroxide-mediated Radical Polymerization of Metallomonomers to Prepare Block Copolymers

The nitroxide-mediated radical synthesis of metal-containing block copolymers has also mainly focused on vinylferrocene as monomer. Unfortunately, the NO-mediated synthesis of poly(vinylferrocene) block copolymers is hindered by the good chain transfer agent characteristics of the metallomonomer and the intramolecular electron transfer from the organometallic chain to the chain end which terminates chain growth. Nevertheless, Frey and coworkers have investigated the limits to PVFc-based block copolymers. By using TEMPO-terminated polystyrene (M_n = 3900, PDI = 1.21) as a macromolecular initiator at 120 °C, chain growth of vinylferrocene was conducted [20]. It was found that only very short PVFc segments (M_n = 500–700) could be grown from the PS chain with a maxi-

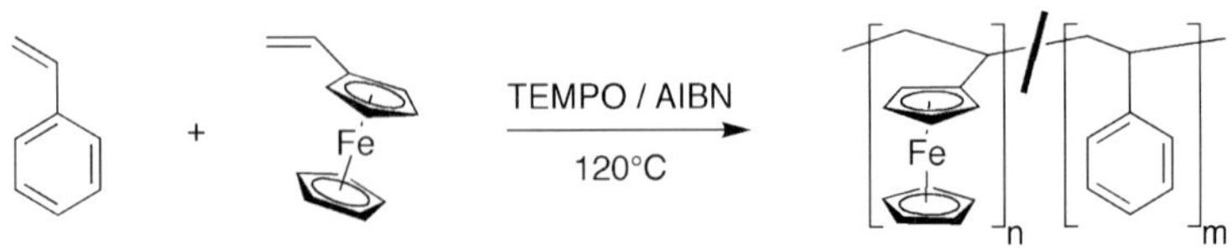

Scheme 16.8 NO-mediated radical polymerization of styrene and vinylferrocene.

mum conversion of 32%. The polydispersity index of the isolated block copolymers was 1.11 and the vinylferrocene content for the entire block copolymers was 3 mol%. This limited chain extension of PVFc was attributed to the slow dissociation of TEMPO from the PVFc chain end thus significantly lowering the concentration of active chain extension sites. Alternatively, using the same type of macroinitiator ($M_n=8900$, PDI=1.17) and extending a random copolymer segment comprised of styrene and vinylferrocene, the conversion and vinylferrocene content were shown to increase to 48% and to 17 mol%, respectively. The molecular weights for the random copolymer segments were tunable in the range $M_n=2000–8700$, where the block copolymer polydispersity indices were typically 1.17–1.29.

16.3
Use of Atom-transfer Radical Polymerization (ATRP)

16.3.1
Introduction

Another and perhaps even more powerful method by which the controlled radical polymerization of unsaturated monomers can be accomplished is atom-transfer radical polymerization (ATRP) [21]. Many of the concepts for mediation of radical concentration are analogous to those for NO-mediated polymerization, but in this case a transition metal complex is used as the agent for equilibrating dormant and active radical chain ends. Shown in Scheme 16.9 is a general schematic for ATRP. Usually, fast initiation of a coinitiator (typically an organic halide, R–X, X=Br) with a copper(I) halide (Y=L_xCuBr, L_x=ligand set) results, where halogen transfer from R–Br to the metal center occurs to give a free radi-

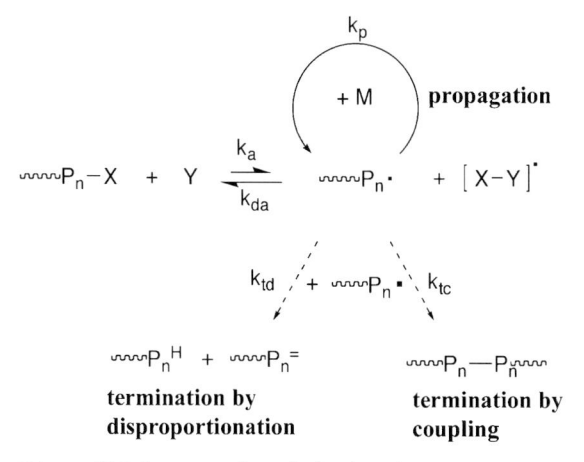

Scheme 16.9 Atom-transfer radical polymerization.

cal (R•) and a persistent copper(II) metalloradical (L_xCuBr_2) [12]. As in NO-mediated polymerizations, the free-radical site rapidly and irreversibly adds monomer for chain growth, while the metalloradical rapidly and reversibly deactivates radical chain ends ($\sim P_n^\bullet$) through halogen transfer producing dormant polymer chains ($\sim P_n$–X). Through this mechanism, termination reactions are minimized by a dramatically reduced radical concentration. Linear plots of molecular weight versus conversion and $\ln(1 - \text{conversion})$ versus time and low polydispersity indices have been reported, confirming controlled polymerization. In many cases, however, broadened molecular weight distributions can arise due to uncontrolled chain termination from (i) the slow production of the persistent metalloradical at the low conversion stage and (ii) a reduced polymerization rate at the high conversion stage.

In the following sections we outline some recent examples of the use of ATRP to afford ligand functional or transition metal-containing homopolymers and block copolymers. The following cited examples do not represent an exhaustive summary of all such systems as successful routes to poly(4-vinylpyridine) [22], poly(dimethylaminoethyl methacrylate) [23], poly(acrylic acid) and others have also been demonstrated using ATRP [24, 25].

16.3.2
Atom-transfer Radical Polymerization to Ligand Functional Homopolymers

Side-chain terpyridine polymers prepared using ATRP have been investigated by several groups as the terpyridine ligand is known to chelate to many transition metals. Kallitsis and coworkers investigated the ATRP of a terpyridine-containing monomer (Scheme 16.10, complex II) [26]. Homopolymers of II were produced with a bipyridine–Cu^{I}Br system at 110 °C where the monomer:bipyridine:Cu^{I}Br mole ratio was 10:2:2. Molecular weights were 1800–3000 with polydispersity indices of 1.14–1.38. Molecular weight versus conversion and $\ln([M]_0/[M])$ versus time were shown to be linear up to 94% conversion, suggesting well-controlled radical chain growth. Subsequent ligation of the polymeric terpyridine side-chains to an Ru(II) center (Scheme 16.10, complex III) was conducted under reductive conditions (THF–EtOH with N-ethylmorpholine). Gel permeation chromatography using a UV-visible detector with a UV-visible monitor set at 496 nm showed a multimodal molecular weight distribution for the metallopolymers. This was proposed to arise from polymer–polymer cross-linking via Ru(II) centers.

16.3.3
Atom-transfer Radical Polymerization Routes to Ligand Functional Block Copolymers

Tew et al. employed an indirect route to afford terpyridine-bearing polymethacrylamides (PMAM$_{\text{terpy}}$) block copolymers [14]. Scheme 16.11 shows the synthesis of poly(methyl methacrylate) (PMMA) block copolymers from a PMMA

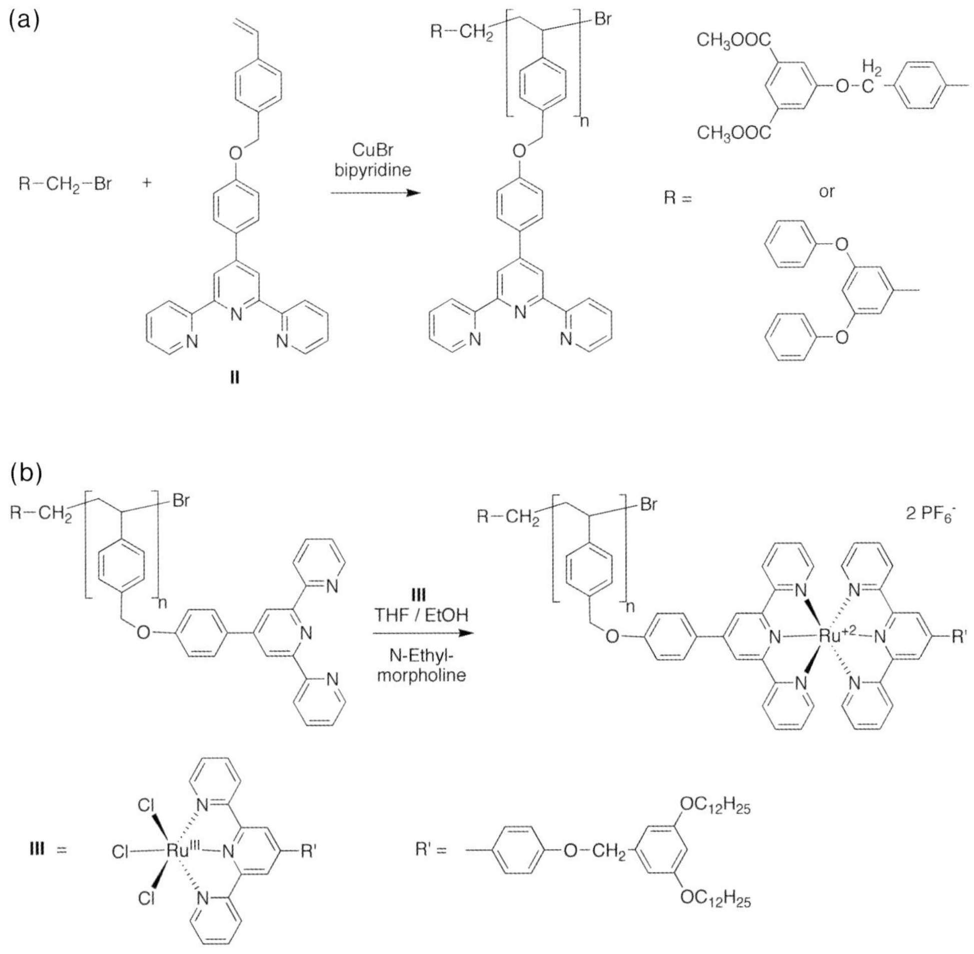

Scheme 16.10 Atom-transfer radical polymerization of **II**, a styrenylterpyridine monomer (a) and subsequent metallization with **III** (b).

macroinitiator followed by block selective macromolecular modification to afford the desired PMMA-b-PMAM$_{terpy}$. The terpyridine derivatization was found to proceed quantitatively as determined by ^{1}H-NMR spectroscopy. An analogous procedure was also successful for the preparation of PS-b-PMAM$_{terpy}$, poly(butyl methacrylate)-b-PMAM$_{terpy}$ and poly(ethylene glycol methacrylate)-b-PMAM$_{terpy}$ with their corresponding macroinitiators. In all cases the precursor block copolymers were reported to be available with controlled molecular weights and relatively narrow molecular weight distributions. On introduction of Cu(II) nitrate to solutions of the PMMA-b-PMAM$_{terpy}$, the viscosity was observed to increase rapidly, which was thought to be due to intermolecular cross-linking via the

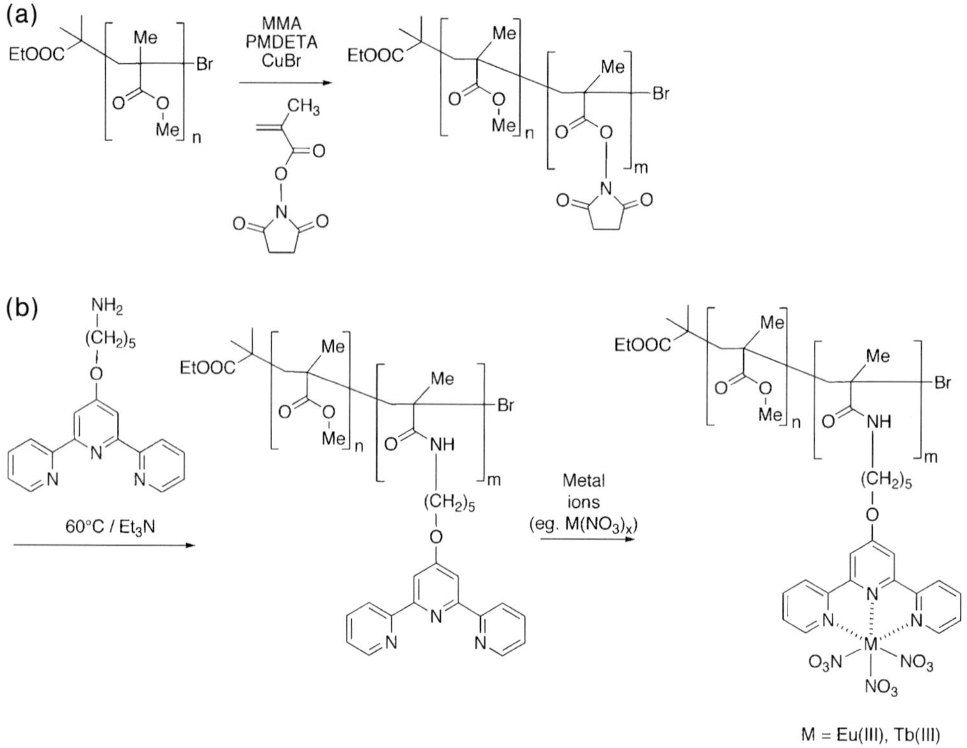

Scheme 16.11 Atom-transfer radical polymerization of MMA and N-methacryloxysuccinimide (a) followed by subsequent terpyridine derivatization and metallization (b).

Cu(II) ions. For high terpyridine content copolymers, precipitation was observed, affording highly insoluble materials, probably due to a high degree of cross-linking. Additionally, lanthanide-based polymers could be synthesized by the same method whereby the same terpyridine block copolymers were reacted with europium nitrate, terbium nitrate or an equimolar mixture of these lanthanoidal nitrates. The resulting mixed lanthanide-containing block copolymers exhibited interesting luminescent properties [27].

16.3.4
Atom-transfer Radical Polymerization of Metallomonomers

Due to the excellent tolerance of ATRP to different functional groups, including polar and protic monomers, well-controlled polymerization of metallomonomers has been achieved. Chan and coworkers successfully employed ATRP to polymerize rhenium-containing monomers (Scheme 16.12, **IV** and **V**) that were readily available from esterification of methacryloyl chloride and diiminetricarbonylrhenium(I) chloride complexes with pendant alcohol groups [28]. These

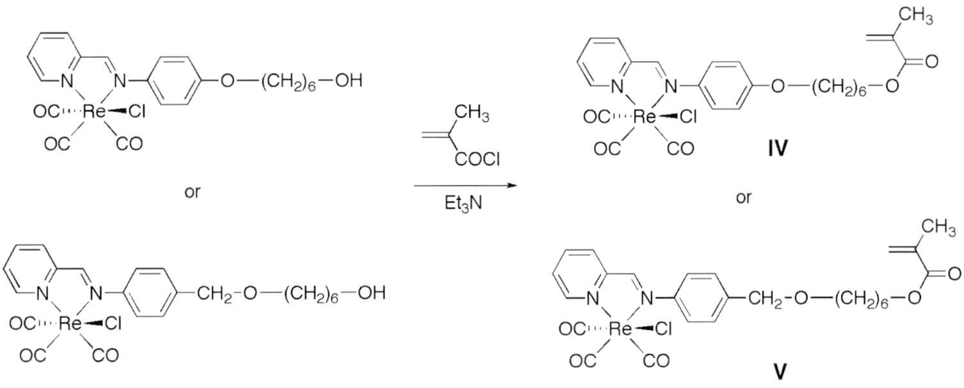

Scheme 16.12 Synthesis of Re-containing methacrylate monomers.

monomers were polymerized at 100 °C with copper(I) bromide, 1,1,4,7,7-pentamethyldiethylenetriamine (PMDETA) (a ligand for copper) and methyl 2-bromopropionate (M:R–X:CuBr:L_x mole ratio = 30:1:2:2) in xylene or dioxane for 24–48 h. Metallo-homopolymers of **IV** and **V** with molecular weights of 16 300 (PDI = 1.18) and 15 800 (PDI = 1.23) were isolated, respectively. As expected for ATRP, linear plots of molecular weight versus conversion (up to 70%) and of ln([M_0]/[M]) versus time were observed for **IV**. For the polymerization of **V**, however, precipitation of the growing homopolymer was also reported. Random copolymerization with methyl methacrylate helped circumvent this problem. Unfortunately, however, this appeared to compromise the "livingness" of the polymerization, which was thought to result from differing reactivity of the monomers with the growing radical chain ends.

16.3.5
Use of Atom-transfer Radical Polymerization of Metallomonomers to Prepare Block Copolymers

Using the techniques explored above, Chan and coworkers have extended their studies to the ATRP of the rhenium-containing monomer **V** using an ATRP-generated macroinitiator [28]. A difunctional initiator (Scheme 16.13, isolated from reaction of 1,4-butanediol with 2-bromoisobutyryl bromide) was used with copper(I) bromide, 1,1,4,7,7-pentamethyldiethylenetriamine (PMDETA), methyl methacrylate (MMA) and xylene for 45 min at 75 °C to produce the PMMA-based macroinitiator **VI** (M_n = 8600, PDI = 1.15). Once isolated, macroinitiator **VI** was used to polymerize the rhenium diimine methacrylate monomer **V** using similar conditions to those used for homopolymerizations (see Section 16.3.4). The block copolymerization proceeded slowly, as seen for the homopolymer, and eventually precipitation was observed after prolonged reaction times. Nevertheless, the isolated material had a molecular weight and a polydispersity index

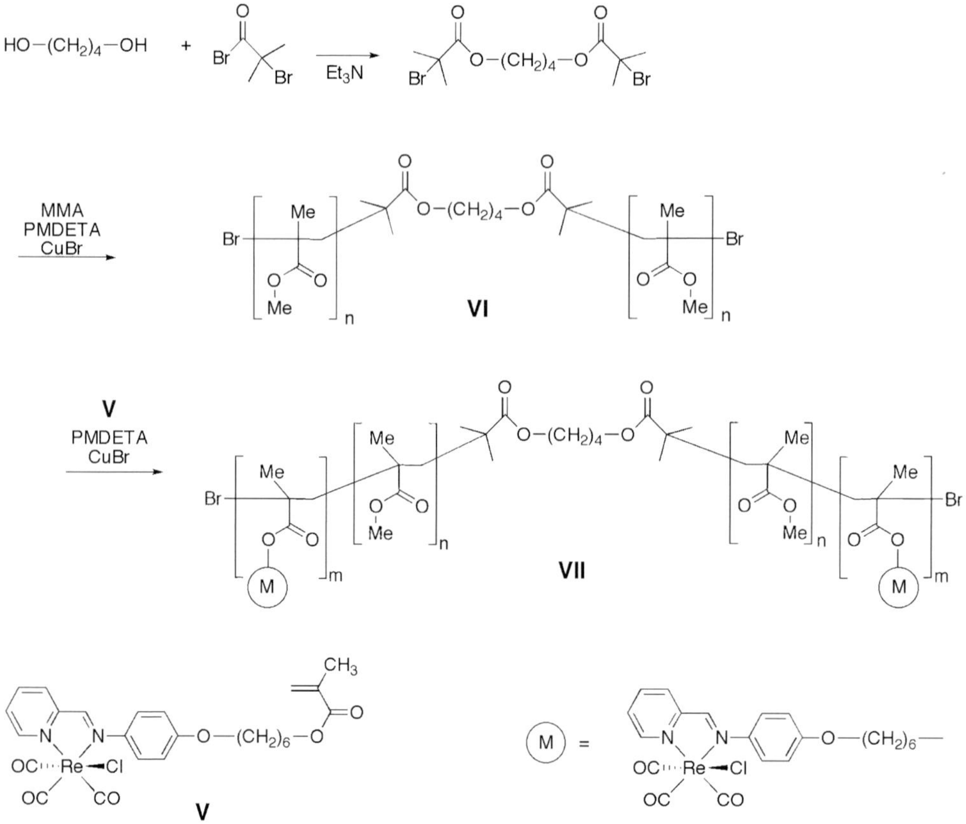

Scheme 16.13 Synthesis of Re-containing ABA triblock copolymer.

of 18 300 and 1.3, respectively. Deviation from well-controlled conditions was thought to be a result of inefficient initiation by the macroinitiator.

16.4
Use of Reversible Addition Fragmentation Termination (RAFT) Polymerization

16.4.1
Introduction

To mediate an active chain end of a free-radical polymerization, a chain transfer agent such as a dithioester (e.g. cumyl dithiobenzoate) can also be used (Scheme 16.14 a, R=Me$_2$CPh R′=Ph). A controlled polymerization results since the transferred end-group in the polymeric dithioester is as labile as in the dithioester initiator species [29]. The resulting equilibrium between dormant

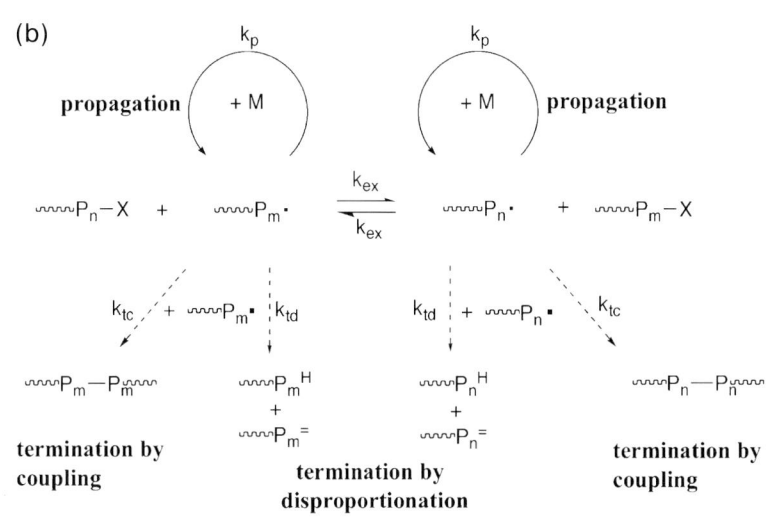

Scheme 16.14 Initiation (a), propagation and termination (b) of RAFT polymerization.

and propagating chains allows for well-controlled radical polymerization (Scheme 16.14b) [12]. Narrow molecular weight distributed polymers are afforded when irreversible bimolecular termination reactions are very slow and initiation is rapid.

16.4.2
Reversible Addition Fragmentation Termination Polymerization to Prepare Ligand Functional Block Copolymers

In addition to the ATRP polymerization of terpyridine-containing methacrylates, recent advances in the RAFT polymerization of these monomers has been reported [18]. Firstly, by using a RAFT chain transfer agent, MMA and AIBN, a PMMA-based macroinitiator was synthesized and isolated ($M_n = 19\,000$, PDI = 1.13). This macro RAFT chain transfer agent was then used to polymerize both MMA and styrenylterpyridine (**VIII**) monomers (Scheme 16.15) in a random fashion. The resulting block random copolymer was isolated in a high yield ($M_n = 37\,300$, PDI = 1.22). Elemental analysis and ^{1}H-NMR spectroscopy showed a 3 mol% terpyridine content at 95% conversion of second block monomers, suggesting a slower tendency for incorporation of terpyridinyl monomers into the growing polymer chain end. Details regarding the subsequent use of

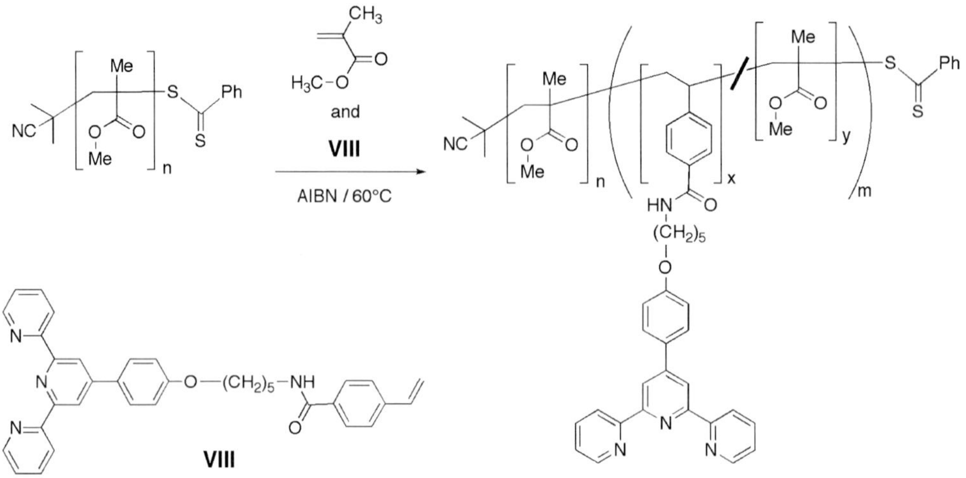

Scheme 16.15 Use of reversible addition fragmentation termination polymerization to form block random copolymers with pendant terpyridine moieties.

these ligand functional block random copolymers for metallization were not available but results similar to those found for analogous ATRP-derived terpyridine-containing block copolymers (see Section 16.3.3) can be envisioned.

16.5
Use of Living Cationic Polymerization

Living cationic polymerizations are more difficult to accomplish than their anionic polymerization counterparts (see Section 16.6) and have rarely been used to prepare well-defined metal-containing polymers. For unsaturated monomers, initiation usually involves Brønsted or Lewis acids, the latter of which often require protogens (e.g. water, hydrogen halide, carboxylic acid) or a cationogen (e.g. *tert*-butyl chloride, Ph_3CCl). Ideally, initiation generates a monomer-based cationic species with a non-nucleophilic anion. Once initiated, the growing chain ends are cationic by virtue of the repeated insertion reactions of monomer. Usually, these reactive cationic centers require stabilization (e.g. with electron-donating groups on the monomer) to limit chain transfer and chain termination reactions whilst maintaining sufficient reactivity for initiation and propagation. For this reason, limited examples of well-controlled polymerizations are known and typically these involve strict reaction conditions. Moreover, high molecular weight polymers are not often produced, as chain transfer to monomer is usually significant. Proton transfer from the growing polymer chain to monomer is a well-known chain transfer pathway, for example, whereas chain termination can occur via combination with the counterion. Some typical mono-

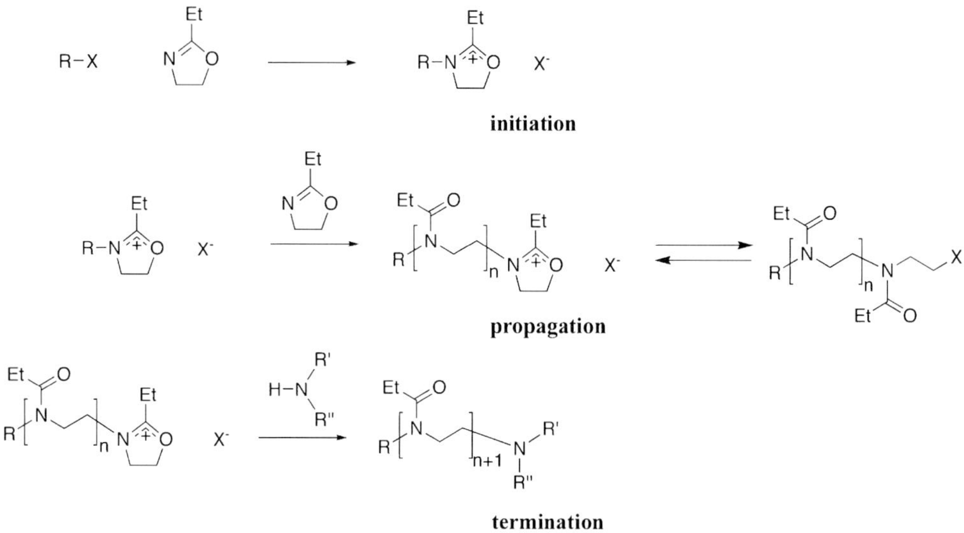

Scheme 16.16 Mechanism for living ring-opening cationic polymerization of 2-ethyl-2-oxazoline.

mers for cationic polymerization are isobutene, vinyl ethers and styrenyl monomers. Shown in Scheme 16.16 is the ring-opening polymerization (ROP) of 2-ethyl-2-oxazoline as an example of a cationic polymerization mechanism. Initiation involves reaction of the generic halide (R–X) with 2-ethyl-2-oxazoline to generate the oxazolinium ion and its halide counterion (X$^-$). Propagation proceeds with nucleophilic attack of the monomer on the oxazolinium species. The growing chain end is also known to interconvert between this oxazolinium ion and the product of its anion–cation recombination where only the former can further propagate. Deliberate termination is accomplished by addition of a secondary amine (HNR′R″).

Fraser's group has investigated the cationic polymerization of 2-ethyl-2-oxazoline (EOX) using di-, tetra- and hexafunctional ruthenium tris(bipyridine) metalloinitiators (Scheme 16.17, $x=2, 1, 0$, respectively) [30]. The resulting star-like 6-star-poly(2-ethyl-2-oxazoline) [star-Ru(PEOX)$_{6-2x}$] is a rare example of well-controlled cationic polymerization to afford a transition metal-containing polymer system. Depicted in Scheme 16.17 is the procedure used for generating the star-Ru(PEOX)$_{6-2x}$. It was found that an Ru center was preferable for these metalloinitiators as the resulting star polymers were more resilient against arm dismemberment [e.g. loss of bipy(PEOX)$_2$] during molecular weight determination by GPC. For polymerization, the Ru metalloinitiator and EOX were dissolved in acetonitrile and then heated to 80 °C for 1.5 days. For deliberate termination, a small amount of dipropylamine was added after cooling to room temperature. The star-Ru(PEOX)$_{6-2x}$ products were isolated in moderate yields as red–orange glassy solids (for $x=0$, $M_n=6300$–22 700, PDI$=1.09$–1.24). Near-linear plots of

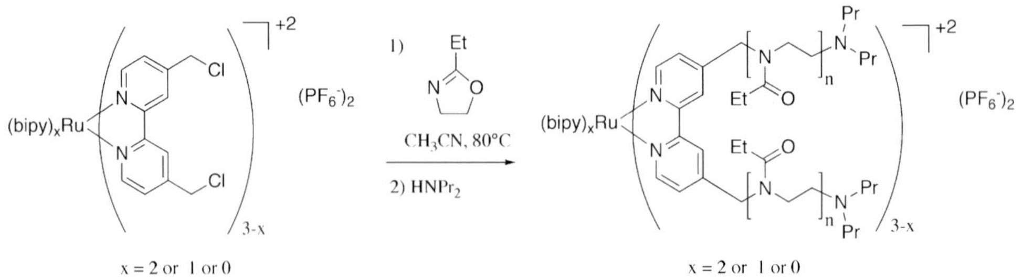

Scheme 16.17 Cationic polymerization of 2-ethyl-2-oxazoline from metalloinitiator.

molecular weight versus conversion and $\ln([M]_0/[M])$ versus time were found for conversions up to 60%. Also reported was the tendency of the Fe(II)-centered metalloinitiator to afford polymeric systems of higher molecular weights (up to 120 000).

16.6
Use of Living Anionic Polymerization

16.6.1
Introduction

In anionic polymerization, chain propagation involves the reaction of monomers with anionic chain ends. Typically, ionic metal alkyls or amides, alkoxides, hydroxides, cyanides, phosphines, amines or Grignard reagents are used as initiators. The range of solvents is also restricted to aliphatic and aromatic hydrocarbons and ethers where halogenated and ester/ketone-containing solvents readily react with the anionic chain end [12]. Monomers for anionic polymerization typically possess vinylic groups or are strained rings with electrophilic centers. The thermodynamic driving force for anionic polymerization of these species is the conversion of π-bonds to σ-bonds and the release of monomeric ring strain, respectively. Additional tuning of monomer reactivity is possible through stabilization of anionic centers by introducing electron-withdrawing substituents near the carbanion site. For this reason, very strong nucleophiles such as alkyllithium species are required for the initiation of 1,3-butadiene or styrene, which have relatively weak electron-withdrawing groups. Dianionic chain propagation is also known to occur with certain initiators such as with sodium or potassium naphthalide where two reactive chain ends per polymer are created. The choice of the monomer polymerization sequence for block copolymers is often important to ensure the effective initiation of the following blocks. Deliberate addition of terminating agents (alcohol or water) quenches the living anionic chain ends and allows for the isolation of the polymers.

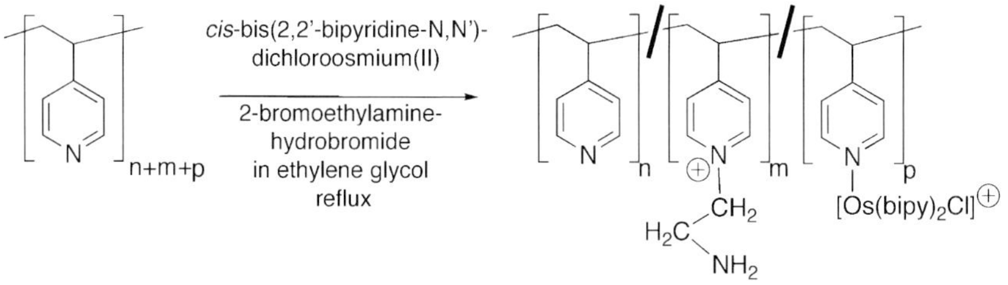

Scheme 16.18 Synthesis of the polycationic osmium-containing poly(4-vinylpyridine).

16.6.2
Living Anionic Polymerization Routes to Ligand Functional Homopolymers

Side-chain transition metal-containing homopolymers of poly(4-vinylpyridine) (P4VP) have been of interest in fields such as electron mediators for the detection of enzymatic redox processes. For metallization [31, 32], the pyridinyl substituents can be partially complexed with cis-bis(2,2′-bipyridine)-N,N′-dichloroosmium whereas remaining substituents can be further quaternized with 2-bromoethylamine hydrobromide. The resulting polycationic osmium-containing poly(4-vinylpyridine) is shown in Scheme 16.18.

16.6.3
Living Anionic Polymerization Routes to Ligand Functional Block Copolymers

The anionic polymerization of vinylpyridine has permitted the synthesis of numerous block copolymers with simple or elaborate architectures. Many poly(vinylpyridine) (PVP) diblock copolymers are now commercially available [33]. Presented here are a few examples of PVP metallization. Two samples of polystyrene-b-poly(4-vinylpyridine) (PS-b-P4VP, $M_n = 40\,900$, PDI = 1.13 and $M_n = 26\,000$, PDI = 1.16) prepared from anionic polymerization routes have been used to pre-

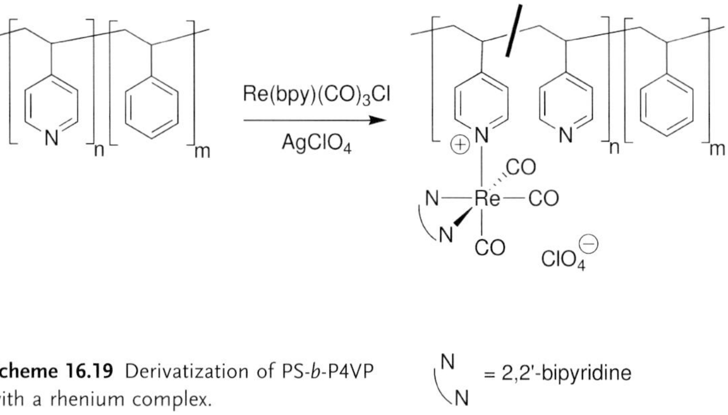

Scheme 16.19 Derivatization of PS-b-P4VP with a rhenium complex.

pare rhenium-substituted block copolymers ($M_n = 53\,900$ and $30\,800$) where the P4VP segment was 13 and 21 mol% metallized, respectively (Scheme 16.19) [34]. Although no molecular weight distribution data were reported, the formation of spherical and cylindrical core–shell structures in block-selective solvents was described. Similarly, analogous treatments of PS-b-P4VP with palladium acetate were found to attach Pd complexes to the P4VP blocks [35]. Comparable approaches have also afforded electrostatically bound $\overline{AuCl_4}$ ions when pyridine groups of PS-b-P2VP are protonated by treatment with $HAuCl_4$ [36, 37].

Poly(butadiene) (PB) block copolymers have also been used as polymers with coordination sites for transition metal complexes [38]. PS-b-PB-b-PS was reacted with various metal complexes such as $Fe_3(CO)_{12}$, $[Rh(\mu\text{-}Cl)(CO)_2]_2$ and $PdCl_2(NCMe)_2$ to produce Fe-, Rh- and Pd-containing block copolymers. Complications due to intermolecular cross-linking and insolubility on drying were reported with some of these systems.

Polystyrene-b-poly(vinyltriphenylphosphine) has also been studied with respect to potential coordination to metal complexes [39]. Upon introduction of Pd salts for polymer metal loading via the pendant phosphine substituents, complex aggregation resulted. This was likely due to intermolecular cross-linking as discussed above.

16.6.4
Living Anionic Polymerization of Metallomonomers

The living anionic polymerization of vinylferrocene has been studied by Nuyken and coworkers [40–42]. The use of alkyllithium initiators in THF at low temperatures (–70 to –30 °C) overcame the challenging initiation of vinylferrocene as a result of lithium ion complexation with the solvent. A decreased polymerization rate was found with polymerization temperatures below –45 °C, resulting in low conversions of prolonged polymerizations due to the increasing influence of moisture impurities. For n-butyllithium-initiated polymerization in THF at –45 °C, vinylferrocene polymerized with (i) very fast initiation, (ii) predictable molecular weights (dependent on [M]:[I] ratio), (iii) linear molecular weight dependence with conversion, (iv) low PDI and (v) further chain extension with sequential vinylferrocene addition. Polyvinylferrocene was produced with molecular weights in the range 1900–10 000 with polydispersity indices of 1.08–1.27 (Scheme 16.20).

The anionic ROP of ring-strained metallocenophanes was first reported in 1994 [43, 44]. The silicon-bridged [1]ferrocenophanes were shown to undergo polymerization in the presence of an anionic initiator such as n-butyllithium. When appropriately pure monomer and solvents were used, a living anionic ROP was achieved. The polymerization involves the formation of an anionically propagated polymer, the living end of which can subsequently be terminated with various capping agents to afford the polyferrocenylsilanes (Scheme 16.21). Since the initiation process is rapid and no significant chain transfer or uncontrolled termination occurs, polyferrocenylsilanes with very low polydispersity in-

Scheme 16.20 Living anionic polymerization of vinylferrocene.

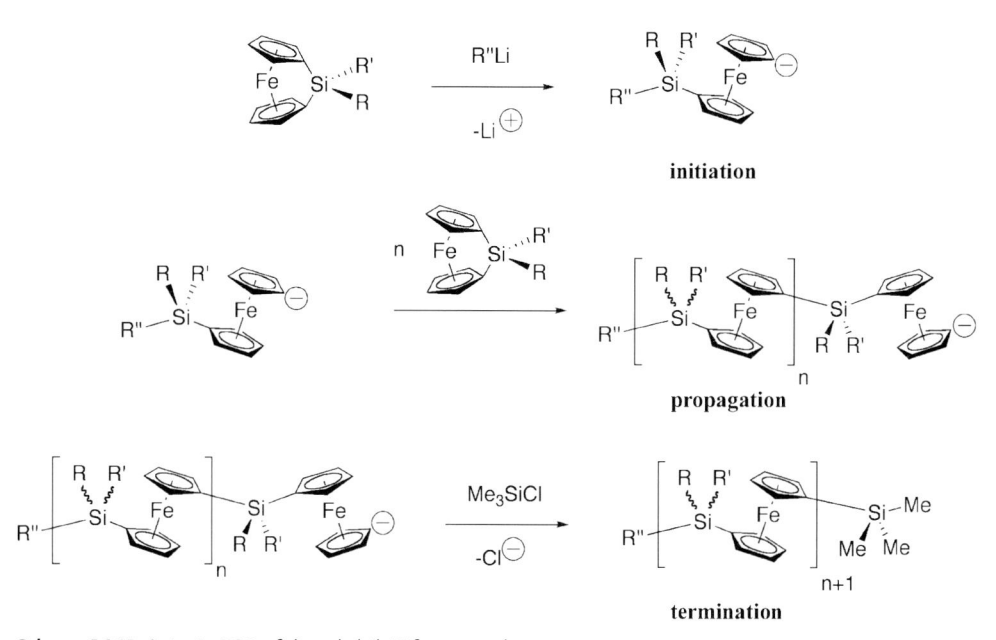

Scheme 16.21 Anionic ROP of dimethylsila[1]ferrocenophane (R = R' = Me).

dices are accessible (PDI values are typically <1.10). As expected for a living polymerization, the molecular weights of the resulting polymers (up to 120 000) are dependent on the monomer-to-initiator ratio. Moreover, this living anionic ROP appears to be very general for silicon-bridged [1] ferrocenophanes and therefore has been extended to a range of different ring-strained monomers. The living anionic ROP of phosphorus-bridged [1]ferrocenophanes has also been achieved with controlled molecular weights (up to 36 000) and narrow molecular weight distributions (PDI = 1.08–1.25) [45, 46].

The living *photolytic* anionic ROP of silicon-bridged [1]ferrocenophanes was also successfully accomplished with the anionic η^5-initiator Na[C_5H_5] [47]. Variation of the ratio of monomer to Na[C_5H_5] from 25:1 to 200:1 under UV–visible irradiation at 5 °C, followed by chain termination with water, afforded samples of polyferrocenylsilane with controlled molecular weights and narrow polydispersities (PDI = 1.04–1.21). Shown in Scheme 16.22 are the synthetic procedures

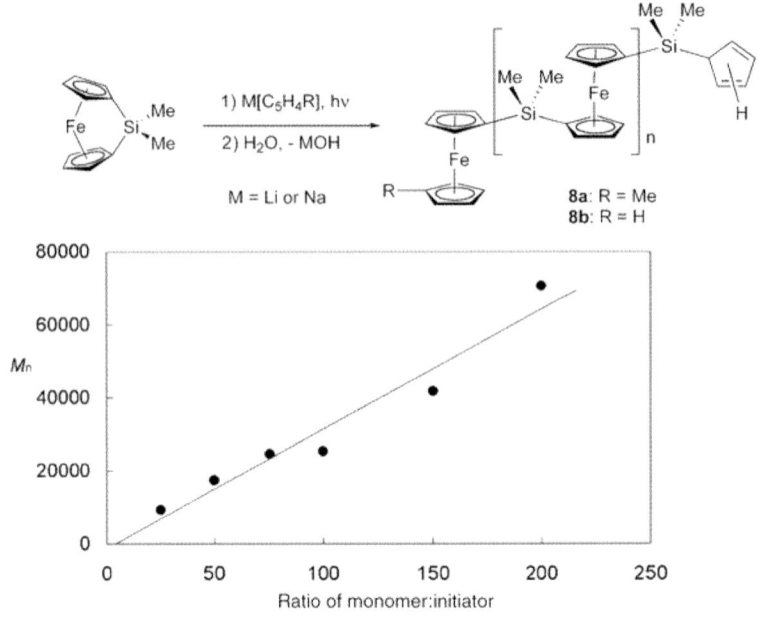

Scheme 16.22 Living photolytic ROP of sila[1]ferrocenophane. Reproduced with permission from [47].

and the plot for molecular weights ranging from $M_n = 9300$ to $70\,000$ versus monomer:initiator.

16.6.5
Living Anionic Polymerization of Metallomonomers to Prepare Block Copolymers

Poly(vinylferrocene) (PVFc) block copolymers have also been prepared from the anionic polymerization techniques outlined in Section 16.6.4. Living anionic poly(vinylferrocene) was used to initiate the anionic polymerization of styrene or methyl methacrylate at $-70\,°C$ [40–42]. The lower temperature polymerization of the second block was required in order to minimize undesirable side-reactions. Alternatively, sequential polymerization of propylene sulfide at $0\,°C$ was also possible with initiation from living PVFc. In all cases no contamination of poly(vinylferrocene) homopolymer was reported, further confirming the living anionic polymerization of vinylferrocene. The isolated diblock copolymers PVFc-b-PS, PVFc-b-PMMA and PVFc-b-poly(propylene sulfide) had molecular weights of $30\,000$ (PDI = 1.13), $25\,500$ (PDI = 1.17) and $11\,900$ (PDI = 1.11), respectively. Additionally, an A–B–A block copolymer was synthesized via the initiation and polymerization of styrene using the dianionic initiator potassium naphthalide in THF at $-70\,°C$. Once the chain growth of dianionic living polystyrene was complete, the temperature was increased to -30 to $0\,°C$, at which point the vinylfer-

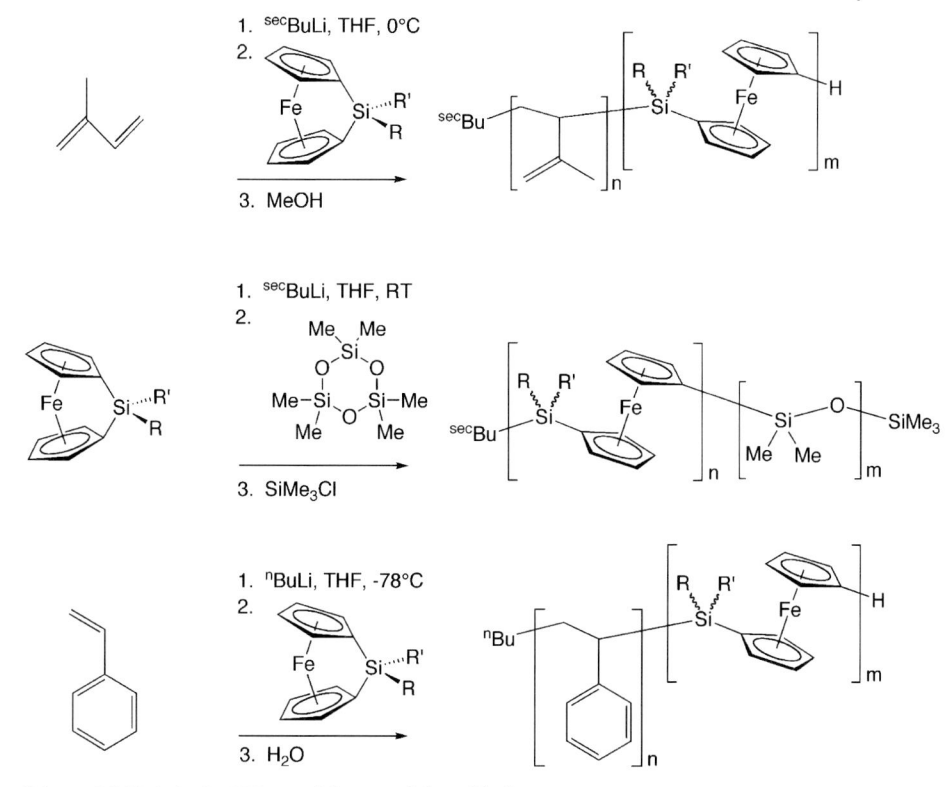

Scheme 16.23 Anionic ROP to polyferrocenylsilane block copolymers.

rocene was introduced. Molecular weights of 7500–25 000 (PDI = 1.19–1.20) were observed for the PVFc-b-PS-b-PVFc.

Polyferrocenylsilane (PFS) block copolymers are readily available via ROP of strained silicon-bridged [1]ferrocenophane monomers [48]. The prototypical PFS block copolymers contained inorganic [polydimethylsiloxane (PDMS)] and organic (polystyrene and polyisoprene) coblocks, respectively (Scheme 16.23). The living anionic ROP of phosphorus-bridged [1]ferrocenophanes for block copolymer synthesis is also possible. A range of di-, tri- and even pentablock materials with crystalline or amorphous polyferrocenes are now available where either inorganic or organic block are easily incorporated [49–51].

16.7
Use of Metathesis Polymerization

16.7.1
Introduction

The ring-opening metathesis polymerization (ROMP) of cycloalkenes requires coordination initiators in catalytic amounts to afford the corresponding polyalkenes [52]. The thermodynamic driving force for ROP is the release of monomeric ring strain, thus limiting ROMP to unsaturated ring-strained species. The reaction mechanism for polymerization is analogous to the transalkylidenation of two alkenes brought about by metal–alkylidene complexes of W, Mo, Rh or Ru. Well-controlled polymerizations result from using stable, isolable metal–carbene initiators such as those developed extensively by Schrock and Grubbs. Shown in Scheme 16.24 is the mechanism for the ROMP of norbornene [12]. Initiation involves complexation of the cycloalkene to the metal–alkylidene followed by formation of a metallocyclobutane by sequential π- and σ-bond cleavage in the ring-strained monomer. The new regenerated metal–alkylidene can extend the chain via repeated metathesis of the cycloalkene monomer. Deliberate termination is usually executed by addition of an aldehyde or vinylic ether for Schrock's or Grubbs' catalysts, respectively. High molecular weight polymers from numerous cycloalkene monomers (only oligomers produced from strainless cyclohexene) have been reported owing to the living nature of the metal–alkylidene chain end. Additionally, ROMP of cycloalkyne monomers is known to proceed by alkyne metathesis, where metal–alkylidyne catalysts are used instead.

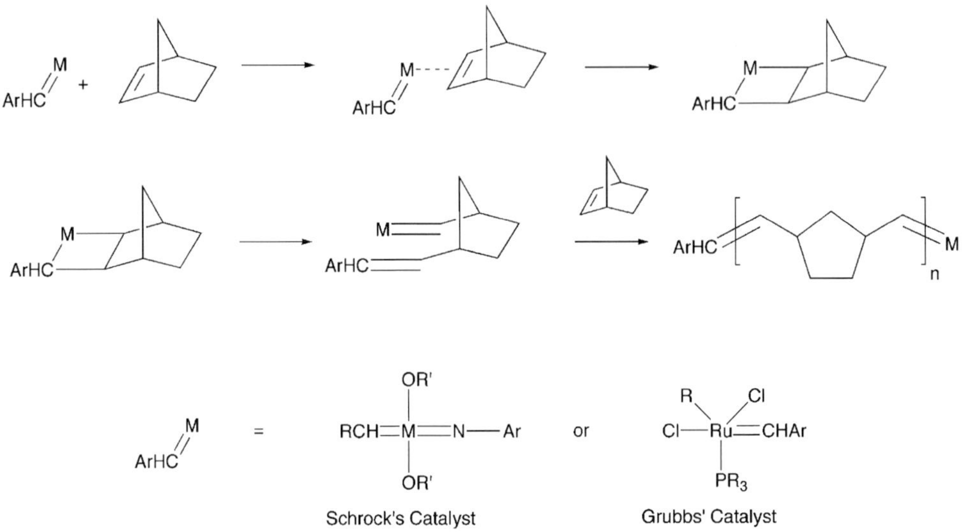

Scheme 16.24 ROMP mechanism for poly(norbornene).

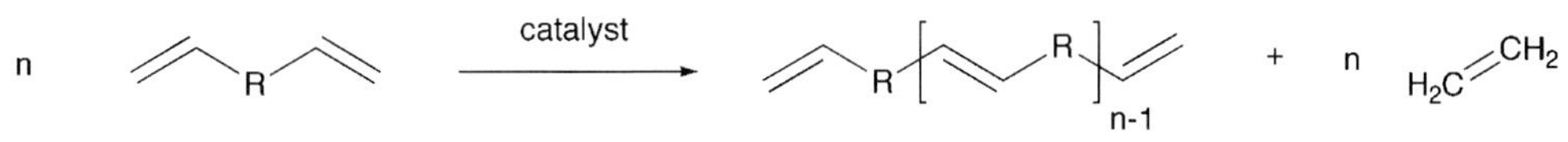

Scheme 16.25 Acyclic metathesis polymerization of a dienyl monomer.

The acyclic metathesis polymerization (ADMET) of unsaturated alkenyl and alkynyl monomers is also feasible with metal–alkylidene and metal–alkylidyne complexes, respectively [12]. Shown in Scheme 16.25 is the ADMET polymerization of a monomer where similar transalkylidenation of two alkenes affords a polymer. In this case the thermodynamic driving force for polymerization involves the elimination of a volatile small molecule (ethylene) and formation internal rather than terminal C=C bonds.

16.7.2
Metathesis Polymerization Routes to Ligand Functional Block Copolymers

In the 1990s, Schrock's group reported on the sequential ROMP of methyltetracyclododecane (MTD) and phosphine-functionalized norbornenes to afford the phosphino block copolymers P(MTD)-b-P(NORPHOS) and P(MTD)-b-P(NBE-R) [53, 54]. The phosphine-containing monomers {NORPHOS = 2,3-bis(diphenylphosphino)bicyclo[2.2.1]hept-5-ene and NBE-R = dioctylphosphine derivatives of 5-norbornene-2-methanol} are shown in Scheme 16.26. By using a molybdenum-based Schrock-type catalyst, the resulting polyene block copolymer, P(MTD)$_{300}$-b-P(NORPHOS)$_{20}$ was isolated (M_n = 81 000, PDI = 1.08). Silver and gold complexes were used for metallization as shown in Scheme 16.26. Based on the rapid and quantitative substitution of Ag-bound COD ligands with PPh$_3$ on Ag model complexes, it was assumed that similar substitution was permitted with the block copolymer. In the case of Au, the reaction required to be driven to completion by drying of the sample and thus the evolution of displaced PMe$_3$ ligands. The metallized phosphine-containing block copolymers were not further characterized in terms of molecular weight but the Au- and Ag-containing P(MTD)$_{300}$-b-P(NORPHOS)$_{20}$ were found to undergo solid-state self-assembly when allowed to dry over several days in the absence of light. In the case of the analogous P(MTD)$_x$-b-P(NBE-R)$_y$ ligand functional diblock copolymers [Scheme 16.26, R = P(O)(oct)$_2$, O(CH$_2$)$_5$P(oct)$_2$, O(CH$_2$)$_5$P(O)(oct)$_2$], bimodal molecular weight distributions were observed for these materials.

16.7.3
Metathesis Polymerization of Metallomonomers

A successful ROMP of a ferrocenylnorbornene has been reported [55]. Using a molybdenum alkylidene as a ROMP initiator, norbornyl monomers were polymerized and, following cleavage of the metal catalyst from the chain end with

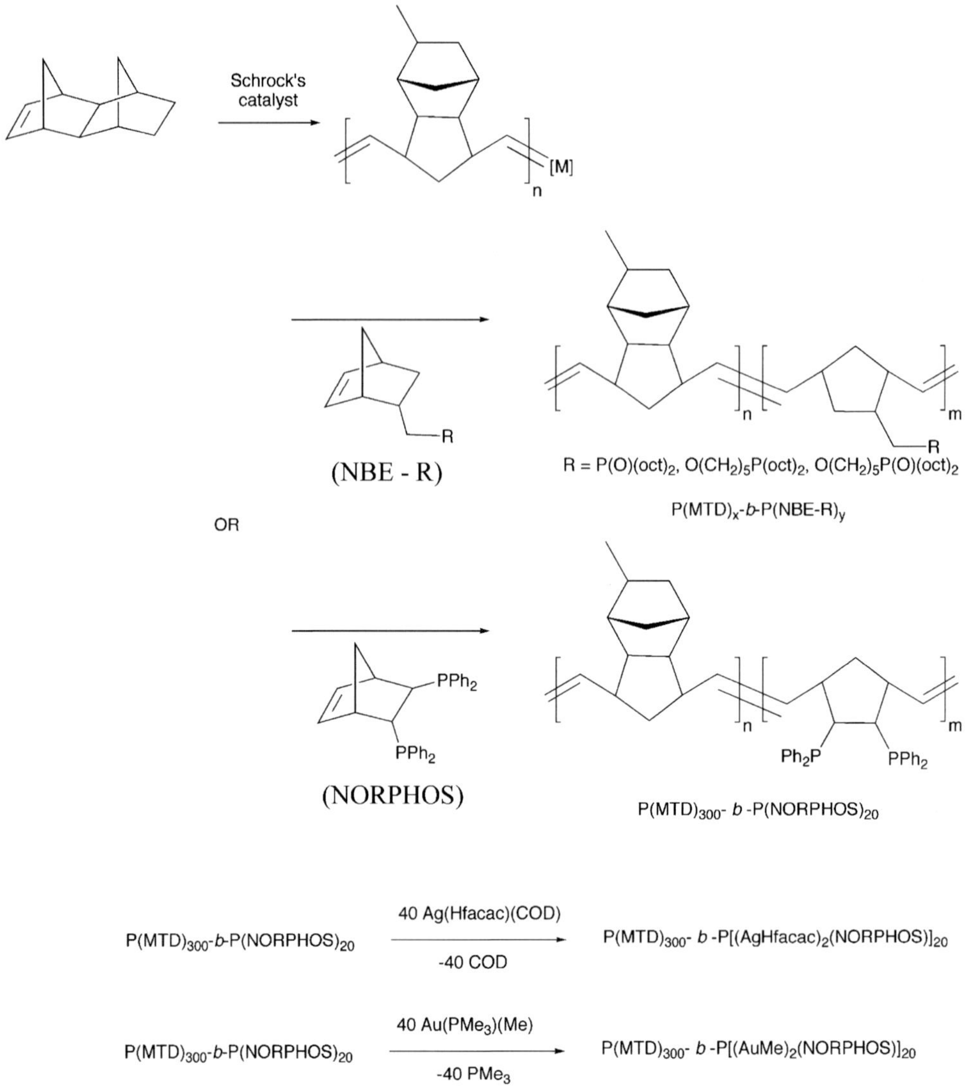

Scheme 16.26 ROMP of methyltetracyclododecane and phosphine-functionalized norbornenes.

trimethylsilylbenzaldehyde, a polynorbornene with pendant ferrocenyl sidechains was isolated (Scheme 16.27). The molecular weights were controlled between 5100 and 9000 with a polydispersity of 1.13.

Abd-El-Aziz et al. have investigated the ROMP of norbornenes bearing arenyliron moieties [56, 57]. It was found that an iron(III)-bearing norbornenyl monomer could be polymerized using a Grubbs-type initiator. After polymerization for 2 h, ethyl vinyl ether was introduced to cleave the catalyst from the polymer,

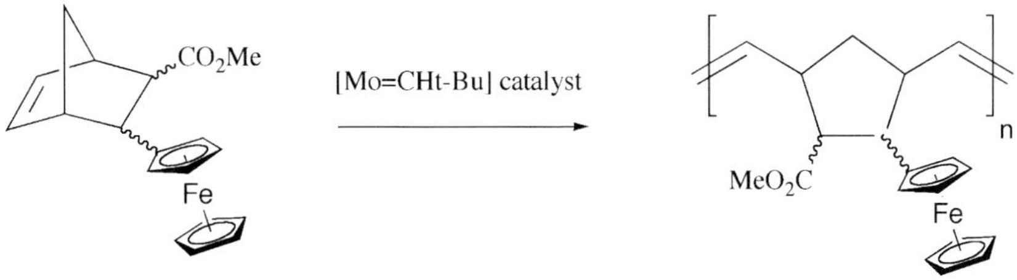

Scheme 16.27 ROMP route to polynorbornene with pendant ferrocenyl groups.

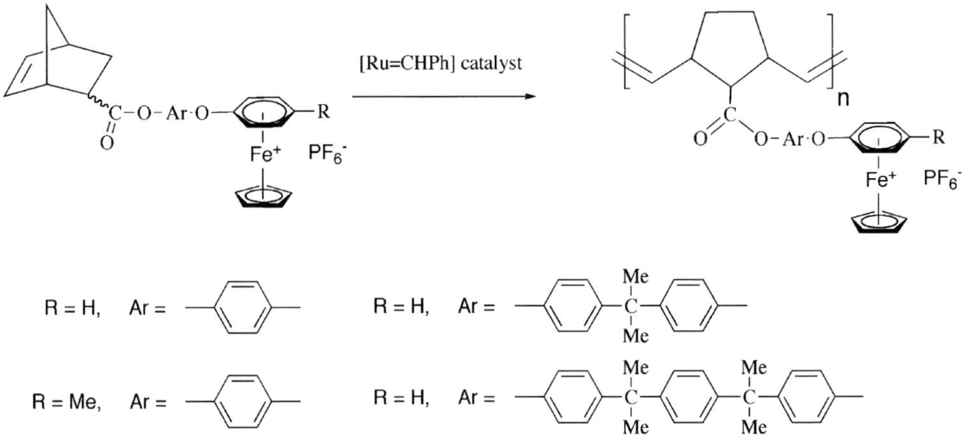

Scheme 16.28 ROMP route to polynorbornene with pendant cationic organoiron groups.

which was then deliberately precipitated from solution for purification. The resulting side-chain organoiron(III) poly(norbornene) material was thus isolated (Scheme 16.28). Attempted direct molecular weight estimation by GPC with CH_2Cl_2 as eluent failed and therefore indirect GPC estimates were made based on the photolyzed polymer, where cleavage of the iron(III) from the arene (O–C_6H_4–R) was induced. The molecular weights were 11500–18200 (PDI=1.4–2.7). This, in combination with unreported thorough investigations into molecular weight control based on monomer:initiator ratios, suggests that only a moderately controlled polymerization is possible.

In 1992, Schrock and coworkers investigated the ROMP of Pt- and Pd-containing norbornene derivatives [58]. By treating a solution of the Pt- or Pd-containing norbornyl monomers with catalytic amounts of a Mo catalyst, polynorbornenes with pendant Pt and Pd complexes (P[NBE-Pd] and P[NBE-Pt], respectively) were afforded in 15 min (Scheme 16.29). At this point, ^{1}H-NMR spectra showed that the conversion was nearly quantitative as no signals attributable to monomeric Pd or Pt norbornenes were present. GPC analysis of P[NBE-Pd] and

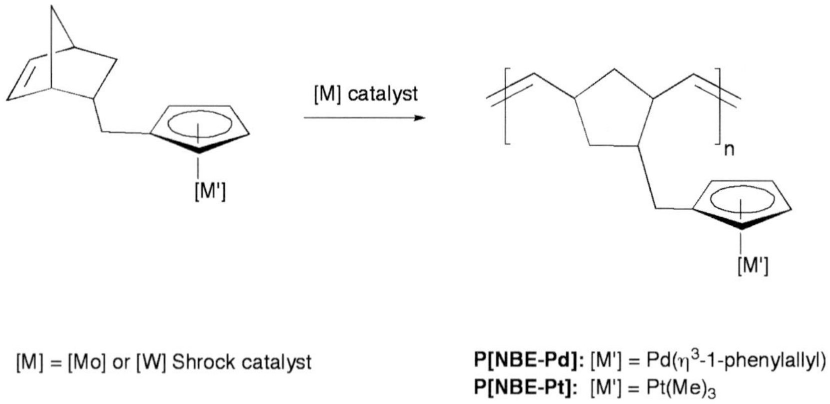
[M] catalyst
[M']
n
[M']
[M] = [Mo] or [W] Shrock catalyst
P[NBE-Pd]: [M'] = Pd(η3-1-phenylallyl)
P[NBE-Pt]: [M'] = Pt(Me)3

Scheme 16.29 ROMP of Pt- and Pd-containing norbornene derivatives.

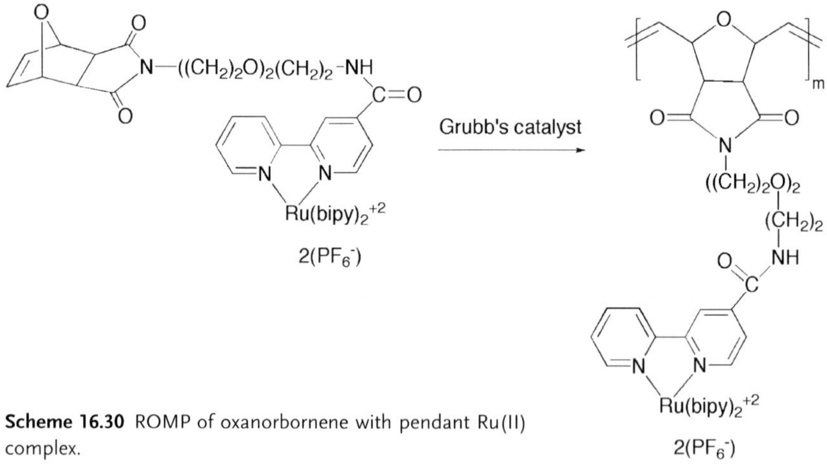
N—((CH2)2O)2(CH2)2-NH
C=O
N
N
Ru(bipy)2+2
2(PF6-)
Grubb's catalyst
m
N
((CH2)2O)2
(CH2)2
NH
Ru(bipy)2+2
2(PF6-)

Scheme 16.30 ROMP of oxanorbornene with pendant Ru(II) complex.

P[NBE-Pt] found molecular weights of 22 500 (PDI = 1.16) and 17 400 (PDI = 1.13), values that were close to the molecular weights expected based on the initial monomer: initiator ratios. In addition to these diagnostics, the ability to synthesize diblock copolymers with these monomers (see Section 16.7.4) was further evidence for well-controlled polymerization conditions.

Chen and Sleiman have recently shown that the ROMP of oxanorbornene with pendant Ru(II) complexes proceeds under controlled polymerization conditions (Scheme 16.30) [59]. The corresponding Ru-containing polyoxanorbornenes were isolated in high yields with conversions up to 90%. ^{1}H-NMR monitoring confirmed slow termination reactions as metal alkylidene signals were shown to persist during the complete 20 min of polymerization and up to an additional 12 h. By lowering the polymerization temperature to 0 °C, conversion could also be

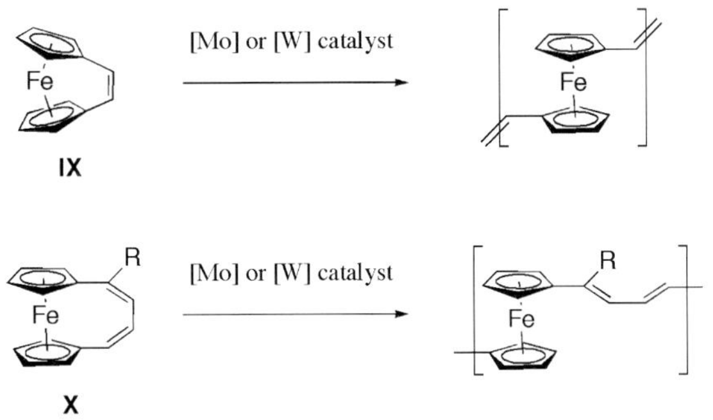

Scheme 16.31 ROMP of [2]- and [4]ferrocenophanes.

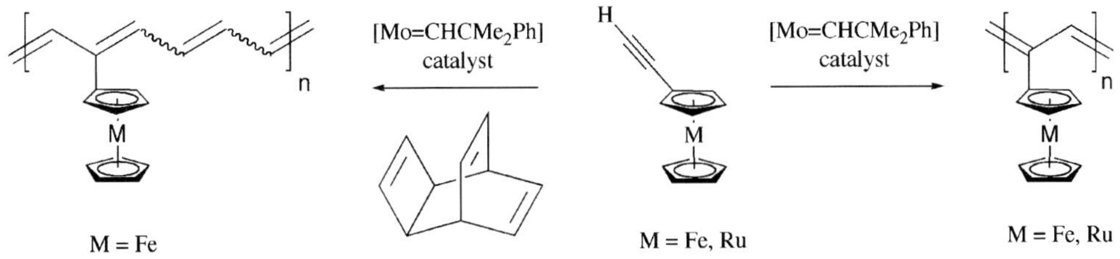

Scheme 16.32 Alkyne metathesis polymerization to polyacetylenes with metallocene side-chains.

monitored. It was found that monomer conversion versus time plots were linear up to 60% conversion, permitting the synthesis of block copolymers (see Section 16.7.4). These results indicate that the ROMP of the oxanorbornene with pendant Ru(II) complexes proceeds as a living process. Unfortunately, molecular weight estimates using GPC were not available as the polymer was thought to interact strongly with the size-exclusion media in the GPC system.

The ROMP of [2]- and [4]ferrocenophanes with unsaturated –C=C– and –C=C–C=C– bridges, respectively, has been accomplished to obtain π-conjugated polyferrocenylenedivinylenes (Scheme 16.31, **IX** and **X**, R=H) [60, 61]. Unfortunately, the resulting polymers were highly insoluble unless copolymerized with norbornene or *sec*-butylcyclooctatetraene, respectively. However, Lee and coworkers were able to circumvent this solubility issue by the introduction of an alkyl substituent in the monomeric [4]ferrocenophane bridge (Scheme 16.31, **X**, R=*tert*-butyl) [62]. With varied monomer to initiator ratios, soluble polyferrocenylenedivinylenes were obtained with varied molecular weights ($M_n \approx 25\,000$–$300\,000$, PDI=1.6–2.3). Studies on the kinetics of the ROMP of [4]ferrocenophanes have not yet been reported, so it is difficult to assess the degree of control of this polymerization.

Buchmeiser and coworkers have reported the polymerization of ethynylmetallocenes using metathesis polymerization techniques [63–65]. Employing a Schrock-type initiator for the acyclic alkyne metathesis polymerization of ethynylferrocene and ethynylruthenocene, polyacetylenes with side-chain ferrocenyl or ruthenocenyl groups were obtained (Scheme 16.32). E conformations for backbone double bonds were confirmed to result from head-to-tail monomer addition. Soluble ferrocenyl-containing polyacetylenes with conjugation up to 50 double bonds was also observed. The authors reported a living polymerization for this system.

16.7.4
Use of Metathesis Polymerization of Metallomonomers to Prepare Block Copolymers

The success achieved with homopolymerization of metallomonomers using ROMP has provided well-controlled routes to metal-containing block copolymers. The versatility of this approach is demonstrated by the variety of metal atoms, namely Sn, Pb, Zn, Pd and Pt, that can be incorporated in the sidegroups of polynorbornenes. Shown in Scheme 16.33 is the representative example of the sequential ROMP of norbornene and an Sn-containing derivative of norbornene [66]. The ROMP catalyst chosen was a tungsten alkylidene, with which norbornene was first polymerized followed by the norbornyltin monomer. Cleavage of the catalyst with 1,3-pentadiene was used to afford the organotin block copolymers.

In 1992, Schrock's group reported the ROMP of norbornenes with pendant platinum and palladium complexes [58]. In the same report, the block copolymerization with methyltetracyclododecane (MTD) was discussed. It was found that the Pd- and Pt-containing block copolymers P[MTD]-b-P[NBE-Pd] and P[MTD]-b-P[NBE-Pt] could be isolated with quantitative conversion of the metallomonomers (Scheme 16.34). Tuned block ratios were easily achieved by using varied MTD:metallomonomer ratios. No GPC molecular weight estimates were

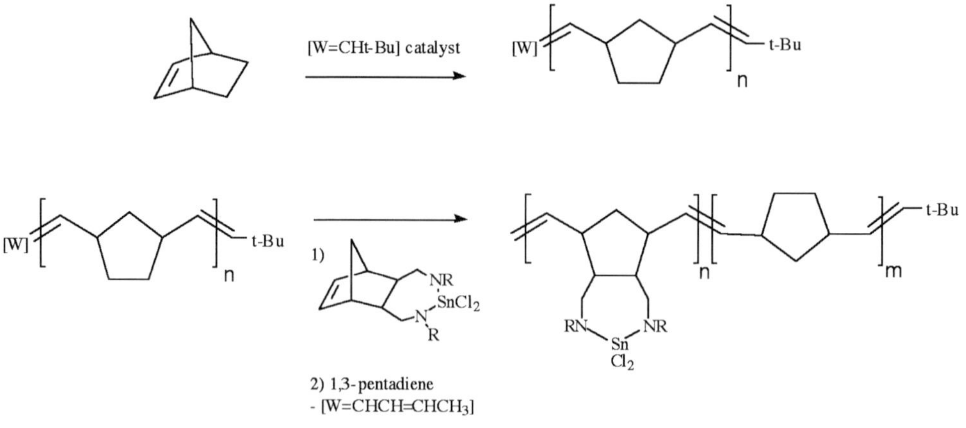

Scheme 16.33 ROMP route to tin-containing polynorbornene block copolymer.

reported for the diblock copolymers but GPC analysis of aliquots taken prior to delivery of the metallomonomer confirmed narrowly distributed P[MTD] (M_n = 23 200–41 800, PDI = 1.04–1.07). These estimates, in addition to those reported for the P[NBE-Pd] and P[NBE-Pt] homopolymers, suggest that well-defined molecular weights were feasible. These Pd- and Pt-containing block copolymers were also observed to undergo solid-state phase separation when dried from solution.

In 2004, Sleiman and coworkers reported the polymerization of oxanorbornenes with pendant Ru(II)bipy$_2$ units. The study was extended to diblock copolymers with oxabicyclo[2.2.1]hept-5-ene-2,3-dicarboximide (N-butyloxanorbornene, Scheme 16.35) [59]. By first polymerizing the N-butyloxanorbornene with a Grubbs-type catalyst (monomer:initiator = 15:1) followed by delivery of the Ru-containing oxanorbornenes, an Ru-containing diblock copolymer was isolated in high yields. Conversion of the metallomonomer was shown to be nearly 98% after approximately 2 h and no homopolymer of the first block was observed. ^{1}H-NMR spectroscopy was used to estimate the molecular weight (M_n = 28 000) as accurate GPC molecular weight estimates were hindered by undesired polymer interactions with the size-exclusion media in the GPC system.

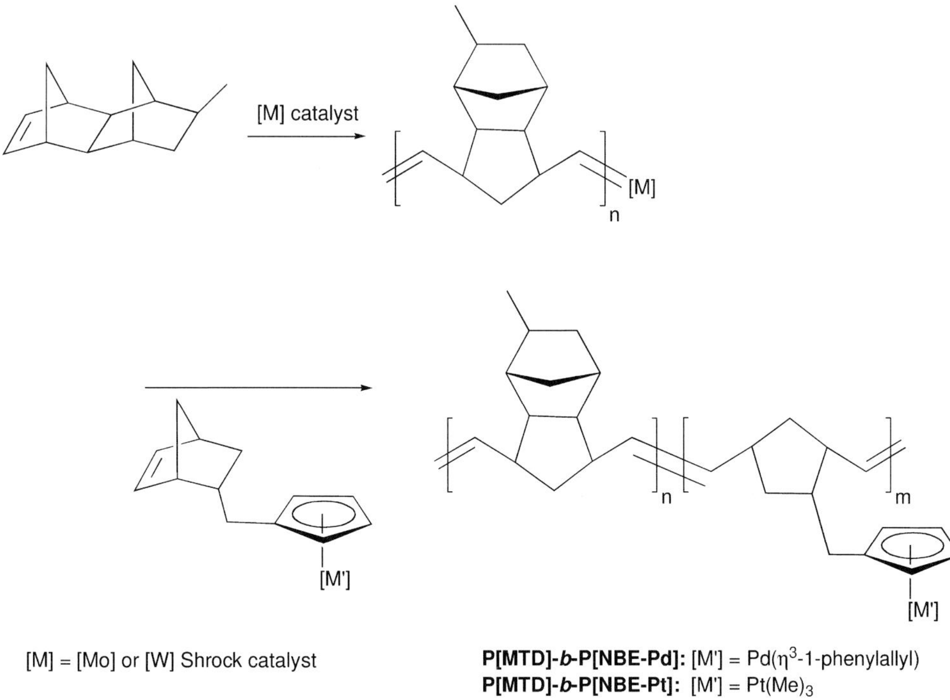

[M] = [Mo] or [W] Shrock catalyst

P[MTD]-*b*-P[NBE-Pd]: [M'] = Pd(η^3-1-phenylallyl)
P[MTD]-*b*-P[NBE-Pt]: [M'] = Pt(Me)$_3$

Scheme 16.34 ROMP of methyltetracyclododecane and norbornenes with pendant platinum and palladium complexes.

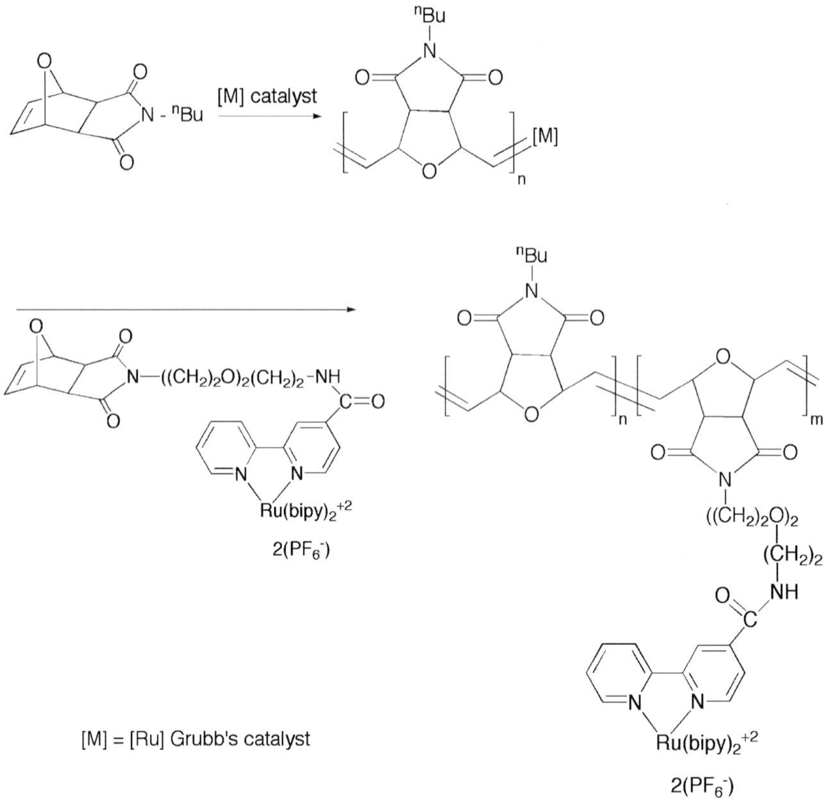

Scheme 16.35 Sequential ROMP of oxanorbornene and oxanorbornenes with pendant Ru(II)bipy₂ groups.

16.8
Indirect Sequential Polymerization Routes to Metal-containing Block Copolymers

Many significant synthetic challenges exist for the synthesis and efficient metallation of well-defined ligand-functionalized polymers and for the controlled polymerization of metallomonomers (Scheme 16.1). Therefore, novel synthetic strategies are being developed in order to circumvent the side-reactions that often prevent the preparation of desired materials using these methods. Often the use of a combination of potentially controlled polymerization procedures is more successful in the preparation of transition metal-containing block copolymers.

Diblock copolymers of poly(dimethylaminoethyl methacrylate) (PDMAEMA) and poly(methyl methacrylate) (PMMA) with polymetallocenes have represented a synthetic target. However, the presence of carbonyl groups in the monomer and polymer leads to side-reactions when living anionic polymerization strategies are used. Success with a non-sequential polymerization approach has been demonstrated by homopolymer-based macroinitiators for the synthesis of PFS-*b*-

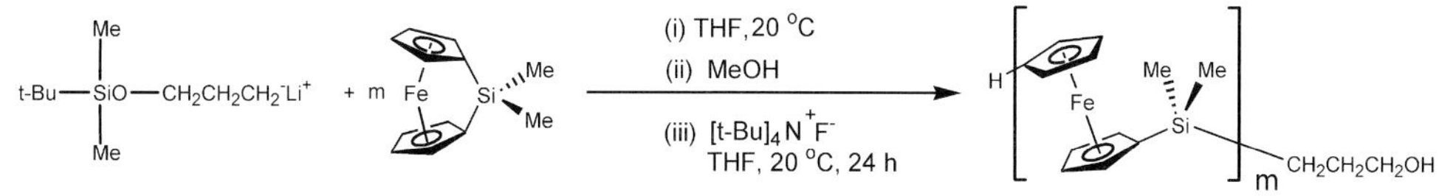

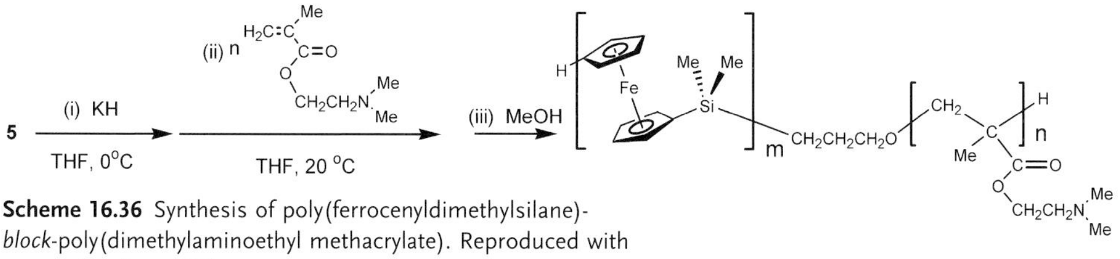

Scheme 16.36 Synthesis of poly(ferrocenyldimethylsilane)-block-poly(dimethylaminoethyl methacrylate). Reproduced with permission from [67].

PDMAEMA and PFS-*b*-PMMA. Additionally, an approach where the anionic polymer chain end is mediated by a carbanion pump strategy allows the successful synthesis of PFS-*b*-PMMA without having to isolate homopolymer-based macroinitiators.

The first PFS-*b*-polymethacrylate copolymer was reported in 2002 [67]. The procedure involved a two-step anionic polymerization (Scheme 16.36). Hydroxy-terminated polyferrocenylsilane (PFS-OH) was first synthesized using *tert*-butyldimethylsilyloxy-1-propyllithium as an initiator with protected alcohol functionality. Once isolated, PFS-OH was deprotonated using potassium hydride to afford the alkoxy chain end that initiates dimethylaminoethyl methacrylate (DMAEMA) for anionic polymerization. The PFS-*b*-PDMAEMA block copolymer was obtained in high yield ($M_n = 11\,000$, PDI = 1.3) and with a high polyferrocenylsilane content (PFS:PDMAEMA = 1:5).

Vancso and coworkers recently investigated the anionic ROP of dimethylsila-[1]ferrocenophane followed by end-group transformations for the isolation of a polyferrocenylsilane-based ATRP macroinitiator for MMA polymerization [68]. By quenching the living anionic polyferrocenylsilane chain end with siloxypropylchlorosilane, a homopolymer bearing a protected alcohol chain end was generated (Scheme 16.37). Subsequent chain end hydrolysis and derivatization with 2-bromoisobutyric anhydride afforded a PFS macroinitiator for the ATRP of MMA. The macroinitiators were isolated as narrow distributed molecular weight homopolymers ($M_n = 8000$–$12\,000$, PDI = 1.02–1.07). Following introduction of MMA monomer in the presence of an ATRP Ru-based catalyst [*p*-cymeneruthenium(II)chloride–tricyclohexylphosphine], PFS-*b*-PMMA diblock copolymers were produced. The metal-containing diblock copolymers were available with tunable polyferrocenylsilane content (between 7 and 26 wt%) via variations in

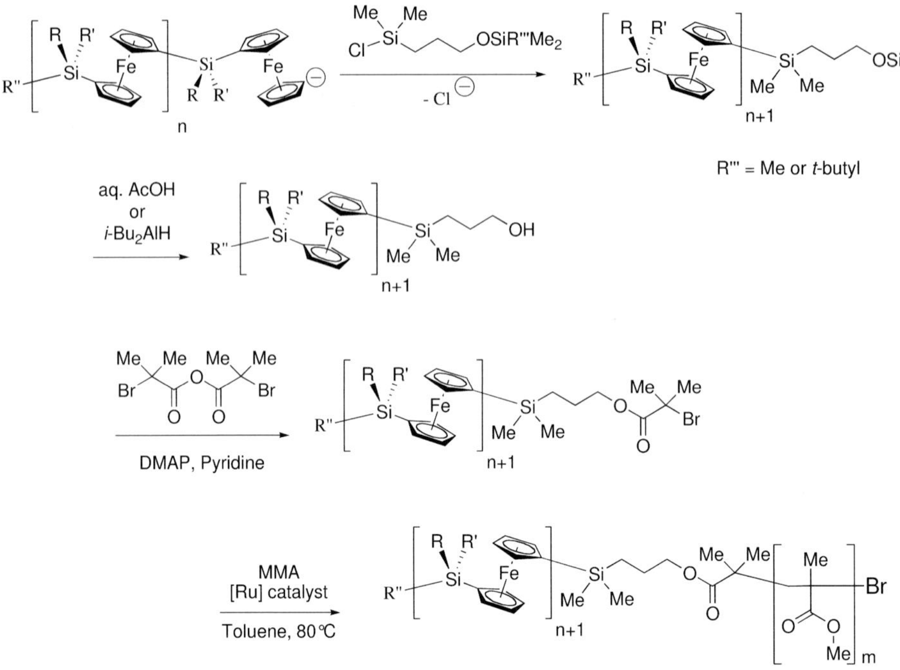

Scheme 16.37 Synthesis of poly(ferrocenyldimethylsilane)-block-poly(methyl methacrylate) via a chain end switch from anionic polymerization to ATRP.

the macroinitiator concentration with respect to monomer concentration. Molecular weights were also varied from 40 500 to 101 000 (PDI = 1.06–1.18) by maintaining constant monomer to macroinitiator ratios. The MMA conversion was 45–60% in all cases.

Recently, Kloninger and Rehahn successfully synthesized PFS-b-PMMA without the need for homopolymer isolation and various chain-end transformations [69]. To circumvent the undesired reaction of anionic PFS chain ends with the carbonyl functionality of MMA, the mediation of the reactivity of the living anionic PFS chain end was investigated. A living PFS chain end mediated by the addition of non-propagating diphenylethylene (DPE) was thus targeted. First, to ensure optimal coupling of DPE, living PFS chain ends were used for the ring-opening reaction of dimethylsilacyclobutane (Scheme 16.38). From matrix-assisted laser desorption/ionization time-of-flight mass spectrometric (MALDI-TOF-MS) analysis of oligomeric model homopolymers for PFS, it was found that 1–3 units of the silacyclobutane are incorporated in the efficient chain end functionalization process using DPE (>95% by ^{1}H-NMR). Having sufficiently reduced the reactivity of the chain via incorporation of DPE, subsequent anionic polymerization of MMA was completed. Unfortunately, approximately 5% of the chain ends were reported to die due to undesired spontaneous chain termina-

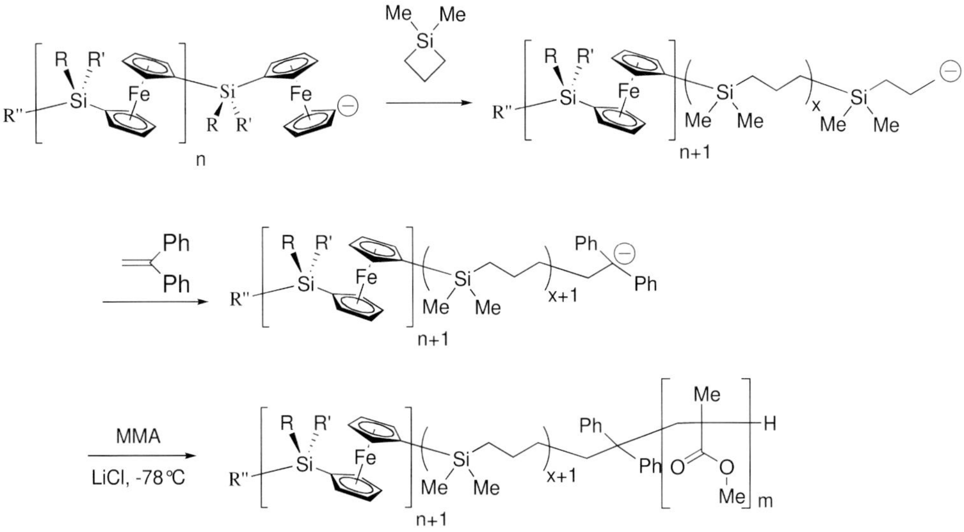

Scheme 16.38 Synthesis of poly(ferrocenyldimethylsilane)-block-poly(methyl methacrylate) via chain end mediation using DPE.

tion or homocoupling. Nevertheless, following selective precipitation PFS-b-PMMA diblock copolymers were isolated with tunable molecular weights ($M_n = 27\,100$–$84\,500$, PDI = 1.03–1.08) and poly(ferrocenylsilane) content (27–47 wt%).

16.9
Routes to Metallo-linked Block Copolymers

Homopolymer–homopolymer coupling strategies are well known and an interesting field has emerged when transition metal atoms are used as linking agents for the generation of block copolymers. Often the resulting materials have many of the characteristics of all-organic block copolymers, yet they incorporate a polymer backbone metal center at the junction point, which introduces unique properties and opportunities.

Lohmeijer and Schubert have reported the preparation of terpyridine (terpy) end-group functionalized polystyrene (PS) and poly(ethylene oxide) (PEO) from a nitroxide-mediated terpy functionalized initiator and OH-terminated PEO transformation, respectively (Scheme 16.39a and b) [70]. These polymer segments act as the building blocks for the coupling reaction between terpy moieties and a transition metal, namely ruthenium(II). The AB block copolymerization process involves two straightforward steps (Scheme 16.39c). First, the terpy end-group-functionalized PS or PEO can be complexed with $RuCl_3$ to afford selectively a monocomplex (PS–$RuCl_3$ or PEO–$RuCl_3$). Second, the monocomplex

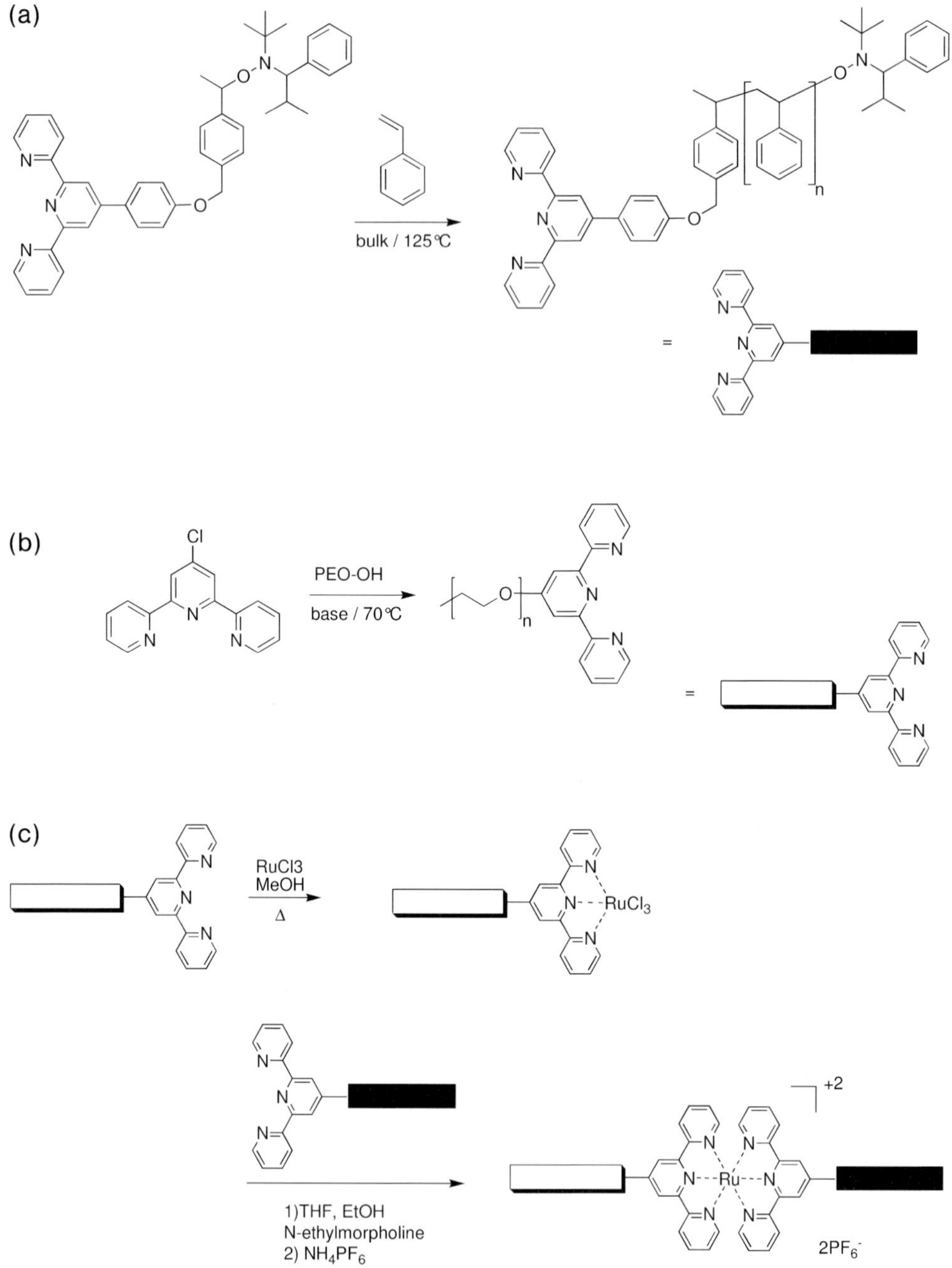

Scheme 16.39 Synthesis of terpyridine-functionalized polystyrene (a), terpyridine-functionalized poly(ethylene oxide) (b) and ruthenium (II)-connected PEO-b-PS diblock copolymer (c).

is reacted with an additional terpy-containing polymer segment under reducing conditions, forming the diblock copolymer linked by the transition metal element. Molecular weight estimates for these AB metal-linked block copolymers from GPC were not possible as the ruthenium center was reported to interact unfavorably with the size-exclusion media in the GPC system. Although the metal content in these materials is relatively low, important features are introduced with the redox-active link. First, the link permits copolymer characterization and manipulation. Second, the linkage is easily cleaved, permitting the formation of nanoporous thin films (see Section 16.11). Recently, Zhou and Harruna have extended this methodology to include terpy-containing RAFT initiators for the controlled polymerization of styrene (M_n=4500–13900, PDI=1.04–1.18) and N-isopropylacrylamide (M_n=1900–2900, PDI=1.06–1.11) [71].

16.10
Routes to Metal-centered Star- and Star-block Copolymers

An interesting divergent polymer chain growth was demonstrated by Haddleton and coworkers [72] in which a transition metal-centered six-arm macroinitiator for ATRP was synthesized. This was used to create a six-arm PMMA star polymer using Cu-catalyzed ATRP and a divergent approach. A series of different transition metal macroinitiators [Fe(II), Ru(II) and Zn(II) based] were compared with respect to MMA polymerization (Scheme 16.40). The resulting six-arm transition metal-centered 6-*star*-PMMAs were isolated in moderate yields with molecular weights in the ranges M_n=4900–32400 (PDI=1.24–1.67), M_n=1000–5500 (PDI=1.24–1.36) and M_n=16900–173900 (PDI=1.01–1.20) for Fe-, Ru- and Zn-based metalloinitiators, respectively. For Zn, Fe and Ru metal centers, linear plots of conversion versus time were reported but only in the case of the Zn metalloinitiator could a linear plot of molecular weight versus conversion be constructed.

As an elegant example of the use of a convergent approach to a metallo-star polymer, we discuss work done by Harruna's group. In 2004, it was reported that a controlled route to tris(2,2′-bipyridine)ruthenium(II)-centered polystyrenes was achieved [73]. For this purpose, a 4,4′-bis(dithioester)bipyridine was synthesized and used to initiate living radical polymerization of styrene. Linear plots for ln([M_0]/[M]) versus time and molecular weight versus conversion were reported with conversions in excess of 60%. The bipyridine-centered polystyrene was isolated in moderate yields and narrow molecular weight distributions (M_n=8500–21500, PDI=1.01–1.27). These homopolymers could subsequently be functionalized with Ru(II)(bipy)$_2$Cl$_2$ in the presence of Ag[PF$_6$] to afford the Ru(II)-centered 6-*star*-polystyrene. Ru(II)-containing diblock copolymers were also targeted [74]. In this case, however, Fraser's group synthesized 4,4′-polycaprolactone-bipyridine (PCL-bipy; M_n=11700–32000, PDI=1.08) and 4,4′-polystyrene-bipyridine (PS-bipy; M_n=8400–25400, PDI=1.09–1.14) with terminal ATRP groups from anionic ROP and ATRP, respectively. Subsequent chain

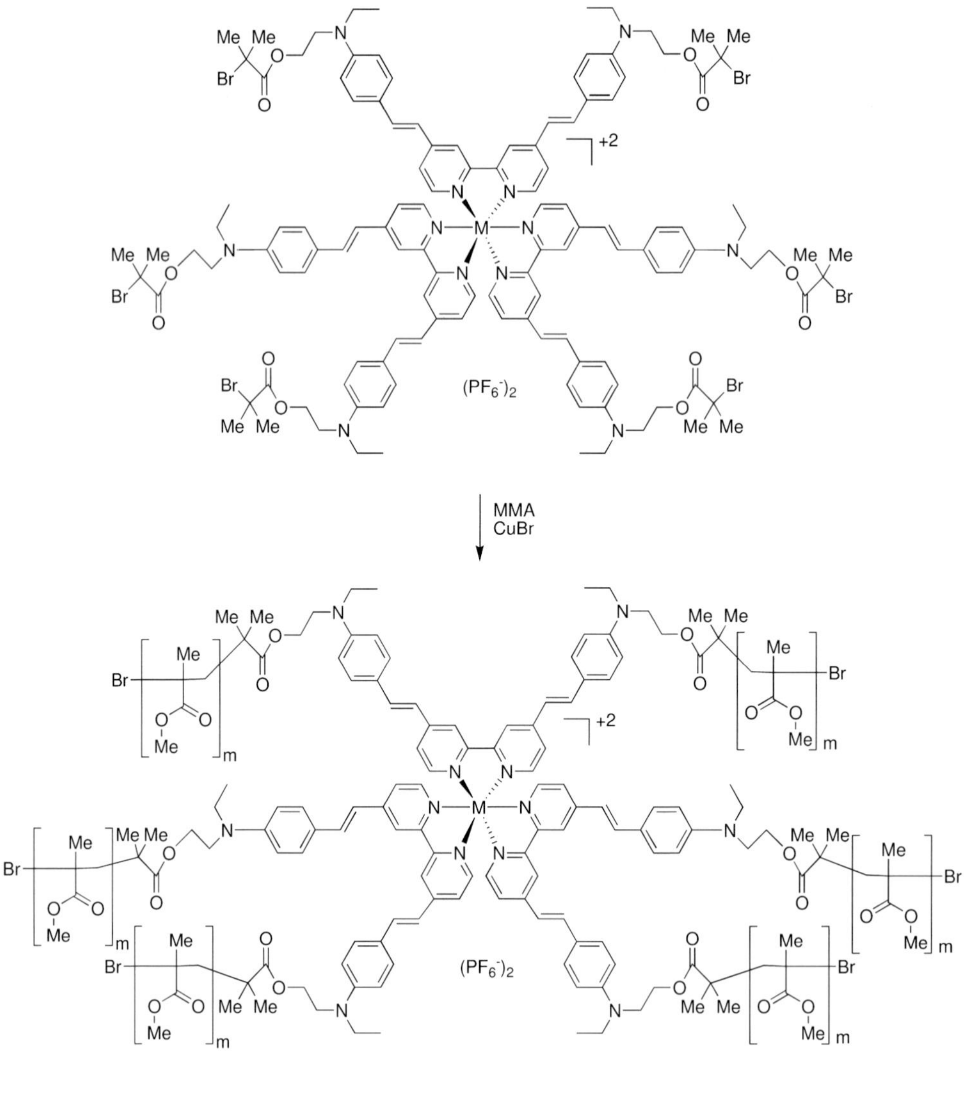

Scheme 16.40 ATRP from a metalloinitiator with divergent chain growth.

growth of PMMA afforded the bipyridine-centered PMMA-b-PCL-b-PMMA ($M_n = 33\,800$–$174\,400$, PDI = 1.08–1.37) and PMMA-b-PS-b-PMMA ($M_n = 55\,500$–$127\,200$, PDI = 1.34) triblock macro bipyridinyl ligands. Linear plots of ln(conversion) versus time and molecular weight versus conversion were reported with conversions in excess of 85%. The bipyridinyl ligands in the triblock material

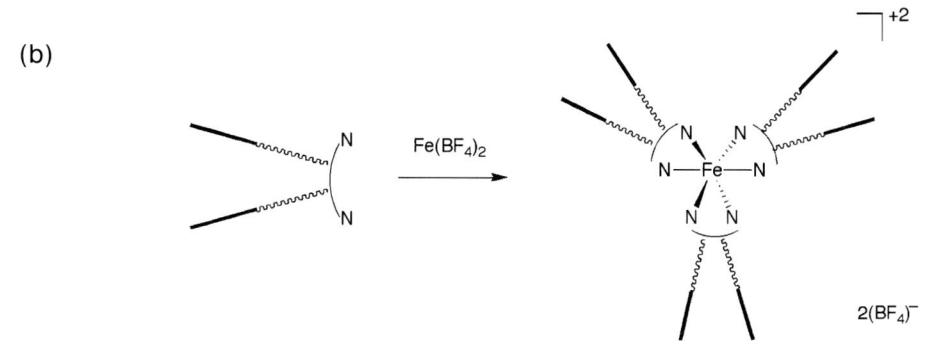

Scheme 16.41 Anionic ROP and ATRP to multiblock macroligands, convergent growth.

could then be further reacted with Fe[BF$_4$]$_2$ to afford the Fe-centered 6-*star*-[polystyrene-*block*-poly(methyl methacrylate)] copolymers (Scheme 16.41 b).

16.11
Applications of Metal-containing Polymers with Precise Structures

The successful integration of transition metals into polymeric frameworks has permitted a range of novel applications. In this section, we provide a brief overview of some recent and exciting applications of metal-containing block copolymers. Each example is intended to illustrate the use of some of the unique properties that result from metallopolymers with well-defined structures such as redox-induced lability, catalytic properties and plasma etch resistance.

In 2005, Schubert and coworkers used supramolecular diblock copolymers for the generation of a nanoporous thin film [75]. The diblock copolymer studied was a PS$_{375}$–[Ru]-PEO$_{225}$ with a Ru(II) junction point {where [Ru] denotes bis-(terpyridine)Ru complex; see Section 16.9.1}. It was found that by casting a thin film, ca. 74 nm thick, spontaneous vertical orientation of cylindrical microdomains of PEO occurred within a PS matrix (AFM phase image shown in Scheme 16.42 c). The center-to-center distance was 63 nm whereas the cylindrical diameter was 33 nm. The film was then exposed to deep UV radiation so as to stabilize the PS matrix via cross-linking. Next, a simple 1-h immersion in an acidic (pH = 1) aqueous solution of Ce(SO$_4$)$_2$ followed by a water rinse and N$_2$ drying afforded a nanoporous PS film. The mechanism for removal of the PEO cylindrical domains involves oxidation of the Ru(II) junction by the dissolved Ce(SO$_4$)$_2$. This involved (i) swelling of the PEO phase so as to permit oxidation of the Ru(II) by the PEO-infiltrated Ce(IV) ions, (ii) formation of cleaved copolymer chains made up of PS and PEO segments and (iii) selective dissolution of the PEO and Ru(III) species into the aqueous phase. The resulting film (AFM phase image shown in Scheme 16.42 d) exhibits a phase contrast that is opposite to that of the precursor film, evidencing its nanoporosity. The generation of a porous PS thin film was further confirmed by X-ray photoelectron spectroscopy and X-ray reflectivity experiments on both PS$_{375}$–[Ru]-PEO$_{225}$ and nanoporous PS films.

To illustrate another application, we discuss the Lewis base-catalyzed Michael addition studied by Balsara and coworkers [76]. The representative reaction studied was the Michael addition of ethyl 2-oxycyclopentanecarboxylate and methyl vinyl ketone. It is known that Lewis acids such as iron(III) and scandium(III) ions catalyze this reaction under mild reaction conditions. First, poly(vinylferrocene)-*block*-polyisoprene, synthesized by anionic polymerization techniques, was oxidized using silver triflate (AgOTf) to give the poly(vinyl ferroceniumtriflate)-*block*-polyisoprene. Second, insoluble cross-linked disks were generated by press molding the polymer and using an electron beam-induced polyisoprene cross-linking reaction. The resulting material with a phase-separated lamellar morphology was characterized by TEM (Scheme 16.43). Third, by swelling the

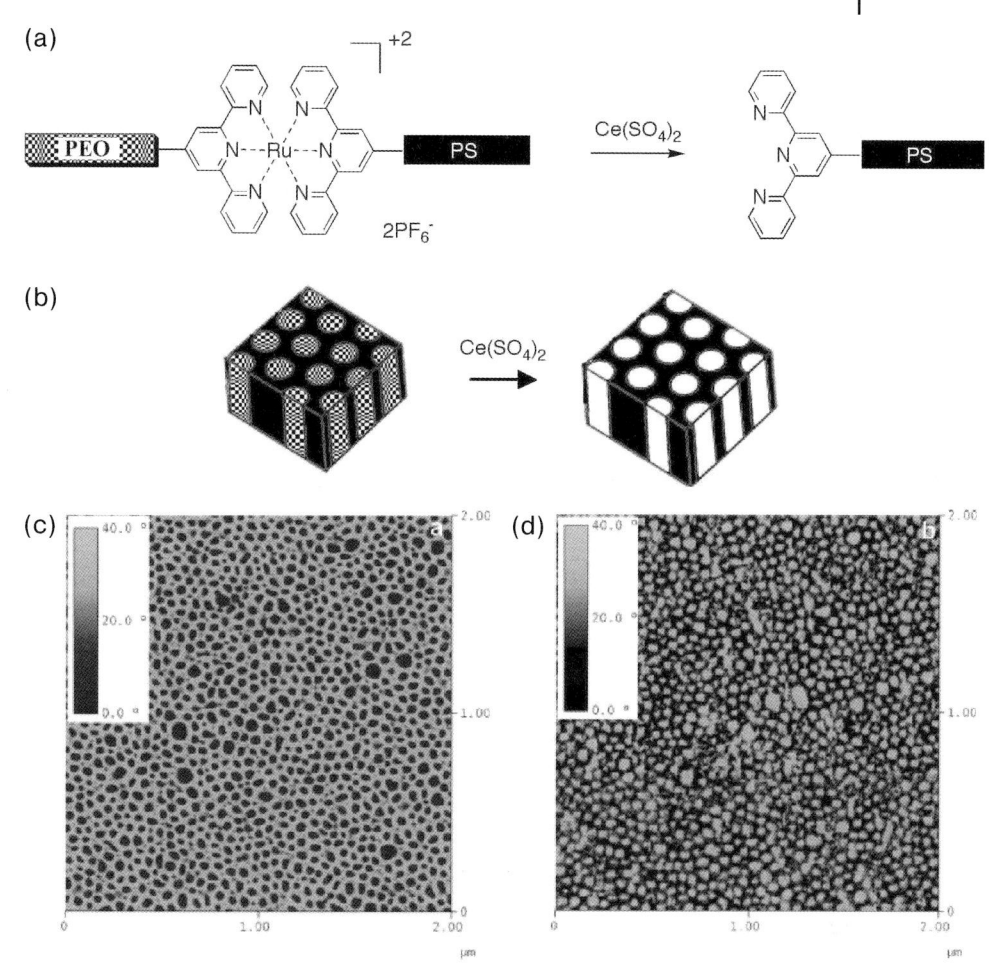

Scheme 16.42 (a, b) General schematic for the generation of nanoporous PS film. Atomic force microscopy (AFM) phase images of (c) PS$_{375}$–[Ru]–PEO$_{225}$ thin film and (d) nanoporous PS film from oxidation-induced cleavage of the [Ru] junction. Reproduced with permission from [75].

cross-linked diblock copolymer film in dichloromethane–methanol (6:1), washing to remove all free polymer chains was conducted. Lastly, by introducing the reactants for the Lewis acid-catalyzed Michael addition into the vessel with the solvent-swollen diblock copolymer gel, the catalytic activity could be monitored. It was found that the polymer gel network with supported ferrocenium catalysts were comparable in efficiency to dissolved ferrocenium and poly(vinylferrocenium) catalysts. Perhaps most importantly, the use of a disk-shaped polymeric gel catalyst avoided tedious separation of the product from catalyst material.

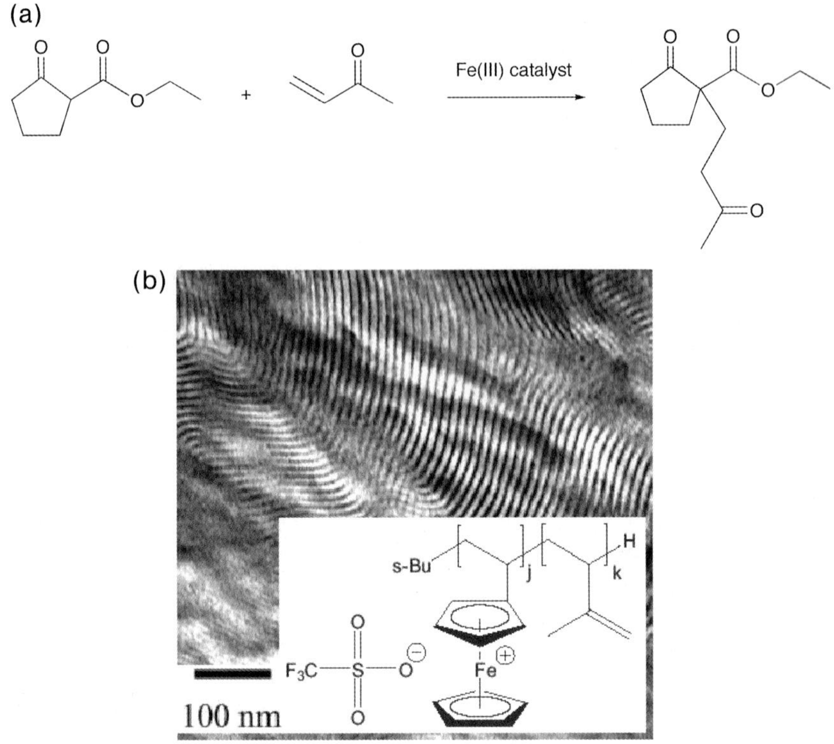

Scheme 16.43 (a) Fe(III)-catalyzed Michael addition of ethyl 2-oxycyclopentanecarboxylate and methyl vinyl ketone. (b) TEM image of bulk section of cross-linked poly(vinyl ferroceniumtriflate)-*block*-polyisoprene. Reproduced with permission from [76].

As the interest in nanolithography grows, efficient routes to nanotemplates are required. An emerging method involves films of phase-separated diblock copolymers [77]. As an exciting application of transition metal-containing block copolymers, we discuss the lithographic template produced using PS-*b*-PFS. Ross and coworkers demonstrated that by spin coating a film of a PS-*b*-PFS, comprising plasma etch-resistant PFS spheres in a PS matrix, direct transfer of the pattern to underlying layers was possible [78]. Using oxygen-reactive ion etching (RIE), much of the PS matrix is eliminated so as to permit the CF_3H etch of an underlying silica layer (Scheme 16.44c). Further CF_4 and O_2 RIE followed by ashing permits the etching of tungsten and isolation of an array of tungsten posts. Lastly, an ion beam etch was used to transfer the pattern into the cobalt layer. The end result is an array of tungsten-capped cobalt dots (Scheme 16.44f) that have dimensions on the nanometer scale whose pattern has been inherited from the PS-*b*-PFS film.

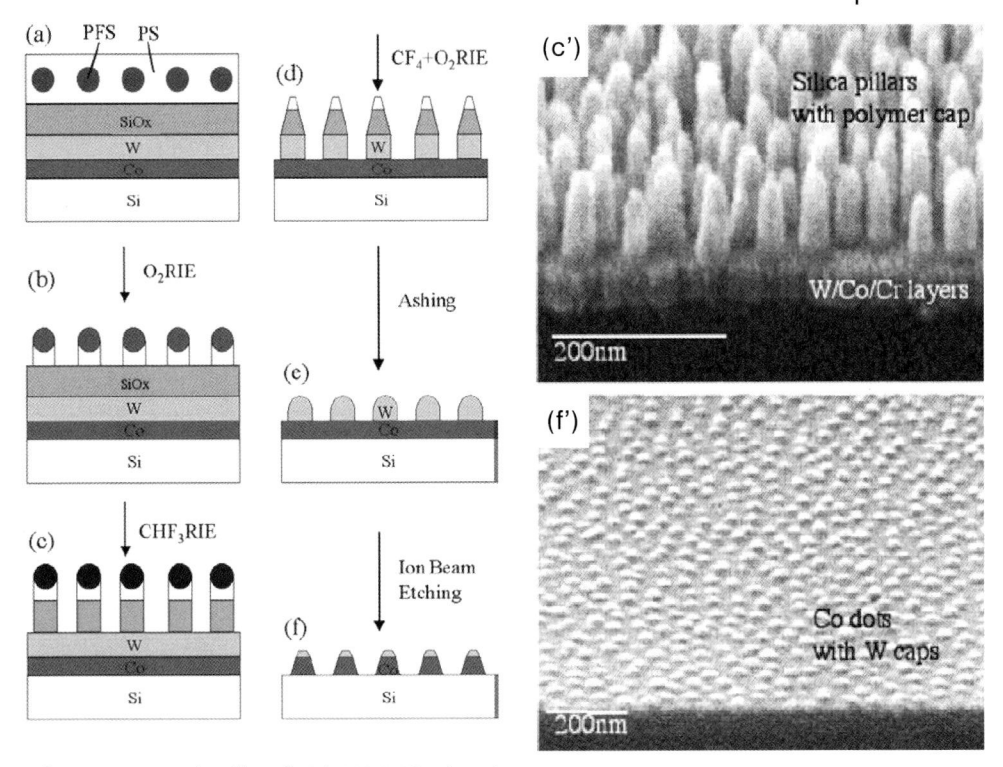

16.11 Applications of Metal-containing Polymers with Precise Structures

Scheme 16.44 (a) Thin film of PS-b-PFS (PFS spheres) on a silicon substrate with Co, W and SiO$_x$ layers. (b) Oxygen-reactive ion etch of PS-b-PFS. (c) CHF$_3$-reactive ion etch of SiO$_x$ layer [SEM of resulting film in (c')]. (d) CF$_4$+O$_2$ reactive ion etch of PS-b-PFS-derived lithographic mask and W layer. (e) Ashing to remove all material above W. (f) Ion beam etching to produce cobalt dots with W caps [SEM of resulting film in (f')] (reproduced with permission from [78]).

An additional application utilizing the catalytic properties of transition metal particles afforded from polyferrocenylsilane has recently been demonstrated [79, 80]. By combining bottom-up and top-down approaches, the groups of Manners and Winnik in collaboration with Agilent Technologies were able to pattern high-quality, small-diameter single-walled carbon nanotubes (SWCNTs) using PS-b-PFEMS (Scheme 16.45, PFEMS = poly(ferrocenylethylmethylsilane)) [81]. The approach first involved the generation of low-depth wells in a photoresist medium using conventional UV photolithography. On this substrate was deposited phase-separating PS-b-PFEMS, which has been shown to afford vertically orienting cylindrical PFEMS domains in a PS matrix. By simply removing the underlying photoresist material using a PS-b-PFEMS non-solvent, an array of catalytically active islands of phase-separated material was generated. Subse-

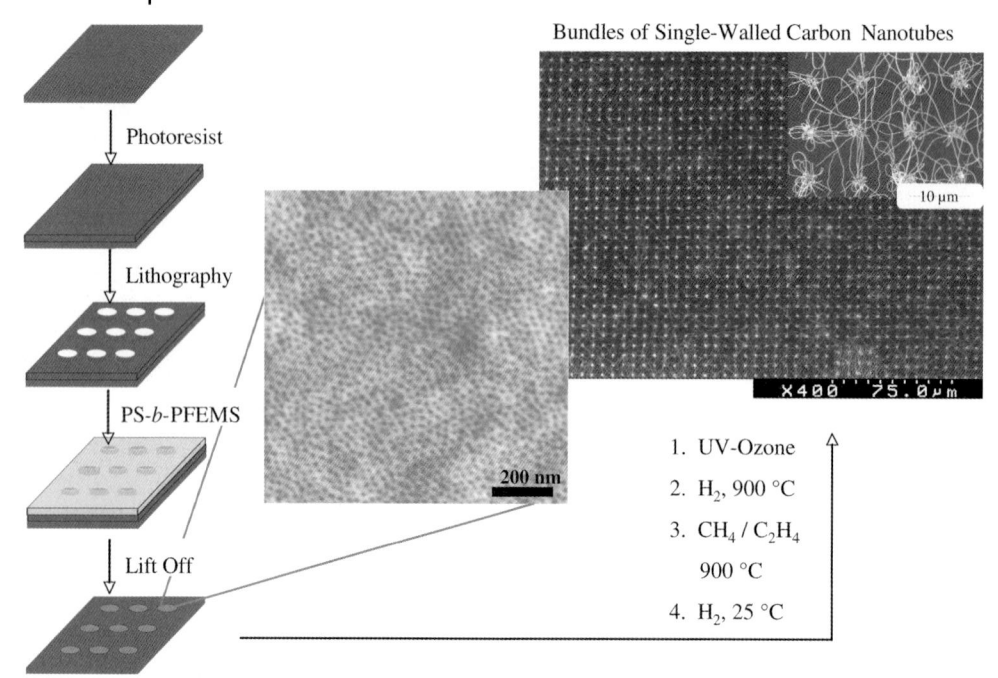

Scheme 16.45 Carbon nanotube growth from patterned PS-b-PFEMS film (reproduced with permission from [81]).

quent UV ozonation and chemical vapor deposition of SWCNTs from methane–ethylene at 900 °C afforded a patterned array of small-diameter SWCNTs.

A new type of nanotextured Ag surface was generated by Lu et al. from a metal-containing diblock copolymer and was found to exhibit high and uniform surface-enhanced Raman spectroscopy (SERS) activity [82]. The metal-containing diblock copolymer, PS-b-PFEMS, was first used for thin-film self-assembly into hexagonally close-packed standing cylindrical PFEMS-based nanostructures surrounded by a PS matrix. UV ozonation was used to selectively remove the PS matrix and convert PFEMS into iron-containing silicon oxide cylinders. A silver-nanotextured surface was then generated by sputtering a thin layer of silver over these inorganic cylinders (Scheme 16.46). Uniformly enhanced Raman signals on the silver-nanotextured surfaces have been observed for absorbed benzenethiol, with enhancement factors up to 10^6 (Scheme 16.46c). This metal-containing block copolymer template approach provides a simple and straightforward method to fabricate SERS-active substrates with great manufacturing potential. Furthermore, since the size and spacing of the cylinders can be adjusted by tailoring the polymer chain lengths, the electromagnetic field can potentially be tuned to achieve even higher SERS activity. More importantly, this block copolymer approach has established the potential for reproducible fabrication of uniformly enhancing SERS-active substrates.

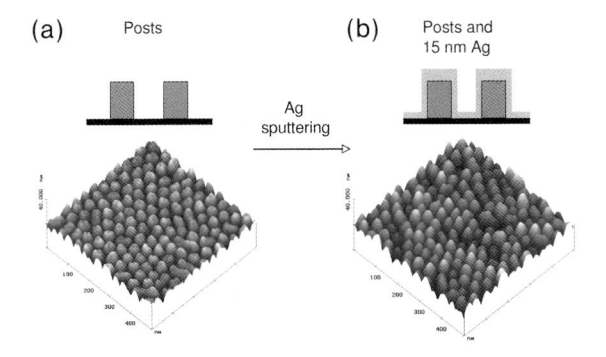

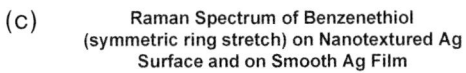

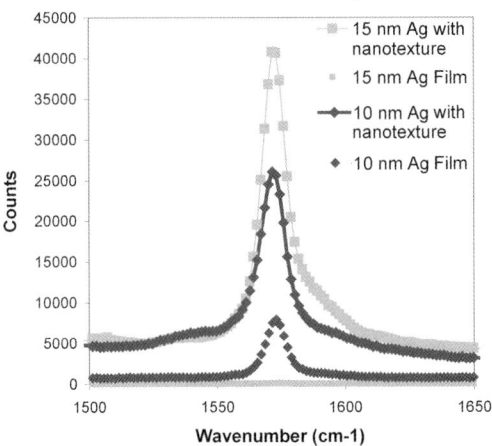

Scheme 16.46 Generation of nanotextured Ag film [AFM height image in (a)] from sputtering on UV ozone-etched PS-b-PFEMS film [AFM height image in (b)]. Surface-enhanced Raman spectra (c) of benzenethiol on nanotextured Ag films [(◆) 10 and (■) 15 nm] and control experiments with smooth Ag films [(◆) 10 and (■) 15 nm] (reproduced with permission from [82]).

16.12
Polymers Based on Main Group Elements

16.12.1
Introduction

Polymers containing main group elements have also attracted considerable attention as they offer opportunities to access new properties and functions [83, 84]. For example, main group elements can introduce low-temperature flexibility, flame retardancy, high-temperature and high-oxidative stability and unusual electronic properties which arise from phenomena such as σ-delocalization. The ability to prepare precise structures has been realized relatively recently. In this section, we survey the developments which have permitted chain length and architectural control.

16.12.2
Polysiloxanes

Polysiloxanes, first developed in the 1930s and 1940s, represent a billion dollar global industry [85, 86]. Indeed, these are the only class of inorganic polymers which can be considered as commodity materials. The exceptional properties of polysiloxanes are a direct result of their inorganic backbone of silicon and oxygen atoms and have resulted in their widespread use as high-performance elastomers and fluids, surface modifiers, adhesives, biomedical materials and materials for soft contact lenses and artificial skin [87–91]. The polysiloxane backbone contains long Si–O bonds (1.64 Å compared with 1.54 Å for a C–C bond) and lacks substituents on every other skeletal atom (oxygen). The wide bond angle at oxygen (Si–O–Si 143° compared with C–C–C 109°) imparts unique dynamic flexibility and therefore these materials retain elasticity at very low temperatures (T_g=–123 °C for polydimethylsiloxane and –137 °C for polymethylhydrosiloxane). The strong Si–O bonds (bond energies: Si–O ca. 450 and C–C ca. 348 kJ mol^{-1}) bring unique oxidative, thermooxidative and UV radiation stability to polysiloxanes.

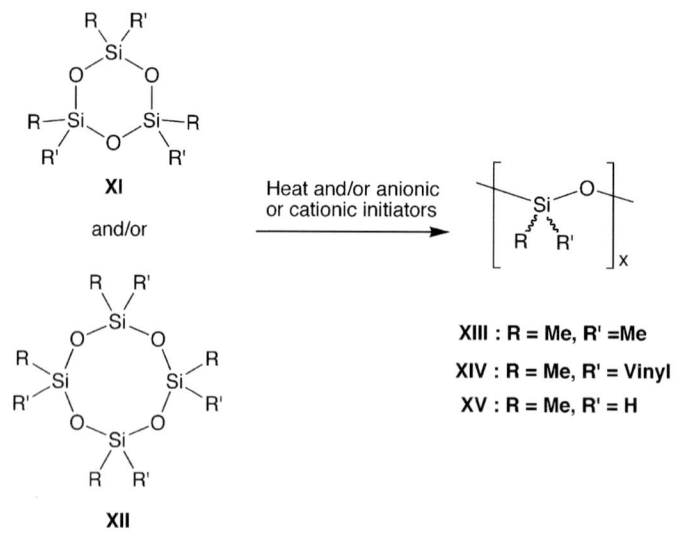

XIII : R = Me, R' = Me
XIV : R = Me, R' = Vinyl
XV : R = Me, R' = H

Scheme 16.47 ROP of hexamethylcyclotrisiloxane (**XI**) and octamethylcyclotetrasiloxane (**XII**).

The main methods of synthesis involve polycondensation and anionic or cationic ring-opening routes [see Scheme 16.47; **XIII** = polydimethylsiloxane (PDMS)]. Cross-linking for elastomer applications can be achieved by a variety of techniques including heating with peroxides or transition metal-catalyzed hydrosilylations for Si–vinyl (Scheme 16.47; **XIV**) and Si–H (Scheme 16.47; **XV**) functionalized polymers.

The living anionic ROP of strained cyclic siloxanes such as **XI** = $(Me_2SiO)_3$ can be achieved using organolithium (e.g. BuLi) or siloxanolate (e.g. $[Me_3SiO]^-$) initiators is very important as this method permits molecular weight control and access to block copolymers with organic monomers [92]. A wide range of multi-block copolymers of PDMS exist and we illustrate these with two diblock copolymers prepared via sequential anionic polymerization and one example of a diblock copolymer synthesized via a non-sequential procedure (Scheme 16.48).

Polystyrene-*block*-polydimethylsiloxane (PS-*b*-PDMS), an organic–inorganic block copolymer, was first prepared in the early 1970s by researchers at Dow Corning via anionic polymerization [93]. Living polystyrenyllithium was prepared and used to initiate the ROP of hexamethylcyclotrisiloxane (Scheme 16.48a). Propagation of the polysiloxane segment required addition of promoters such as tetrahydrofuran or diglyme. Deliberate termination using acetic acid or chlorodimethylvinylsilane was conducted. Polydispersity indices approaching unity and predictable molecular weights were confirmed by GPC analysis; moreover, diblock copolymers with minor homopolymer impurities (<6 wt%) were found to have tunable inorganic segment contents in the range of 30–60 wt%. Selected materials from this series were also found to undergo solid-state self-assembly in thin films.

The use of the above procedures for the generation of organometallic-*block*-inorganic diblock copolymers was investigated by Manners and coworkers in the mid-1990s [48, 94]. By initiating the anionic ROP of dimethylsila[1]ferrocenophane with alkyllithium species, living PFS segments were produced (Scheme 16.48b). Addition of hexamethylcyclotrisiloxane yielded a PDMS block. Deliberate termination with $ClSiMe_3$ was conducted and PFS-*b*-PDMS diblock copolymers were isolated. Unimodal and low polydispersity index GPC traces confirmed the diblock copolymer structure. Since this early report, PFS-*b*-PDMS materials have been further confirmed to have predictable molecular weights (M_n up to 10^5, PDI < 1.1) and tunable organometallic or inorganic content [95–97].

The utility of the well-controlled polymerization of hexamethylcyclotrisiloxane has been demonstrated by the non-sequential synthesis of polydimethylsiloxane-*block*-poly(2-ethyl-2-oxazoline) (PDMS-*b*-PEOX) [98]. A series of monofunctional benzyl chloride-terminated PDMS homopolymers (M_n = 500–20 000; low PDI) were prepared following the procedure outlined in Scheme 16.48c. Once isolated, these materials were activated for the cationic initiation of 2-ethyl-2-oxazoline using NaI. Following cationic ROP of 2-ethyl-2-oxazoline monomers at 110 °C, deliberate termination was executed with KOH–MeOH. GPC traces for homopolymers and diblock copolymers were shown for purified PDMS-*b*-PEOX samples. Narrow molecular weight distribution curves were also observed and

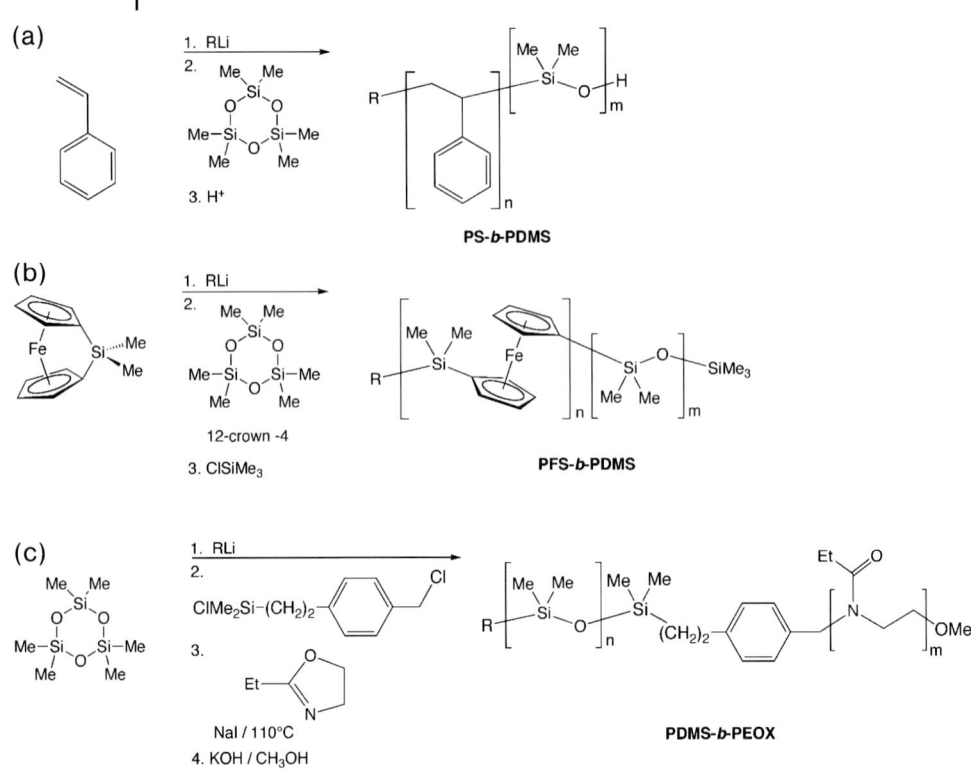

Scheme 16.48 PDMS block copolymers from (i) sequential anionic polymerization of styrene and dimethylsila[1]ferrocenophane with hexamethylcyclotrisiloxane [(a) and (b), respectively] and (ii) combination of anionic ROP of hexamethylcyclotrisiloxane and cationic polymerization of 2-ethyl-2-oxazoline (c).

theoretical compositions closely matched those of isolated materials (tunable from 10 to 66 wt% PDMS). ^{1}H-NMR spectra obtained in block-selective solvents also suggested that these materials self-assemble into micellar structures.

16.12.3
Polysilanes

Soluble polysilanes, first reported in the late 1970s, have attracted intense interest from both fundamental and applied perspectives [87, 99–101]. Unique electronic and optical properties exist for polysilanes as a result of the delocalization of σ-electrons in the backbone silicon atoms. As a consequence of this delocalization, the σ–σ* transition occurs at 300–400 nm in high polymers and therefore appreciable electrical conductivity results following doping. In addition, many of the polymers are thermochromic as the conformations adopted by the

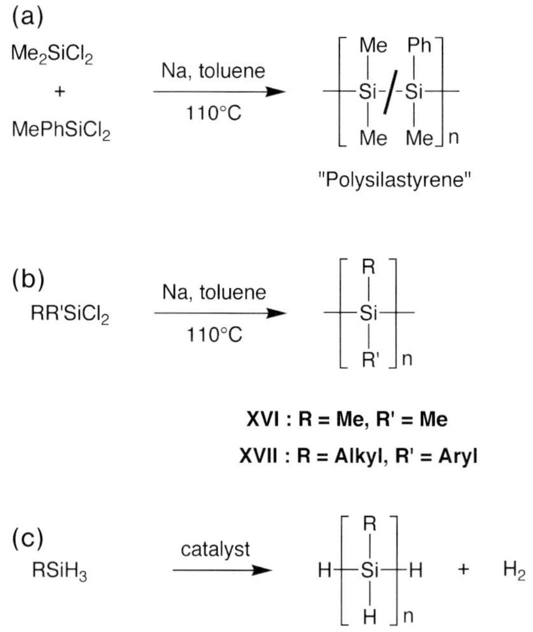

Scheme 16.49 Wurtz coupling polymerization of organodichlorosilanes (a, b) and dehydrocoupling polymerization of organotrihydrosilanes (c).

(a)
Me_2SiCl_2 + $MePhSiCl_2$ $\xrightarrow{\text{Na, toluene}, 110°C}$ $\left[\begin{array}{c} Me \\ | \\ -Si- \\ | \\ Me \end{array} \middle/ \begin{array}{c} Ph \\ | \\ -Si- \\ | \\ Me \end{array} \right]_n$

"Polysilastyrene"

(b)
$RR'SiCl_2$ $\xrightarrow{\text{Na, toluene}, 110°C}$ $\left[\begin{array}{c} R \\ | \\ -Si- \\ | \\ R' \end{array} \right]_n$

XVI : R = Me, R' = Me
XVII : R = Alkyl, R' = Aryl

(c)
$RSiH_3$ $\xrightarrow{\text{catalyst}}$ $H\left[\begin{array}{c} R \\ | \\ -Si- \\ | \\ H \end{array} \right]_n H$ + H_2

polymer change with temperature, which alters the degree of σ delocalization along the main chain. Due to their low-energy σ–σ* transitions, polysilanes are photosensitive and have attracted considerable attention as photoresist materials in microlithography [87, 99, 100]. These polymers also function as thermal precursors to silicon carbide fibers, as hole transport layers in electroluminescent devices and as photoconductors when doped with C_{60} [102].

The first soluble polysilane, then termed "polysilastyrene" (Scheme 16.49a), was reported by West and coworkers in 1978. The material was prepared by the treatment of a mixture of organodichlorosilanes with sodium metal. Polydimethylsilane (Scheme 16.49b, **XVI**) had been previously prepared as a highly crystalline insoluble material [87, 99, 100]. The introduction of phenyl groups in the random copolymer reduces the crystallinity and allows the material to be soluble and easily processed. The use of this Wurtz coupling route has allowed the preparation of a range of high molecular weight polymers ($M_n > 10^5$) with alkyl or aryl substituents at silicon (Scheme 16.49b, **XVII**). Although improvements in this process have been reported, the harsh conditions for this reaction tend to limit the side-groups to nonfunctionalized alkyl and aryl units. The dehydrogenative coupling polymerization process (Scheme 16.49c) catalyzed by early transition metals (usually titanocene or zirconocene derivatives) is potentially very attractive; however, the molecular weights of the polysilanes formed to date are generally fairly low ($M_n < 8000$) [103, 104]. In addition, Hsiao and Waymouth have shown that a variety of new side-groups can be introduced by using a side-group derivatization approach [105].

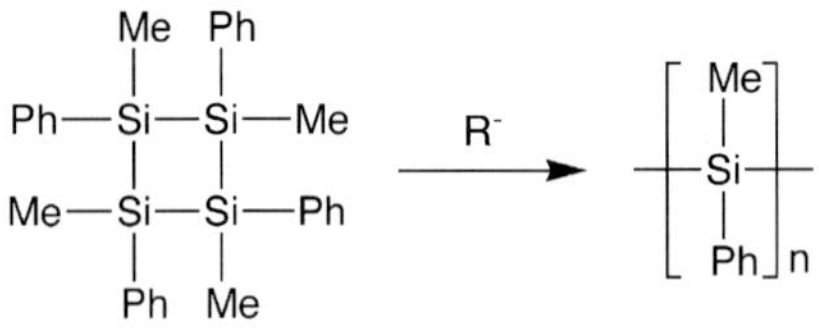

Scheme 16.50 Anionic ROP of 1,2,3,4-tetramethyl-1,2,3,4-tetraphenylcyclotetrasilane.

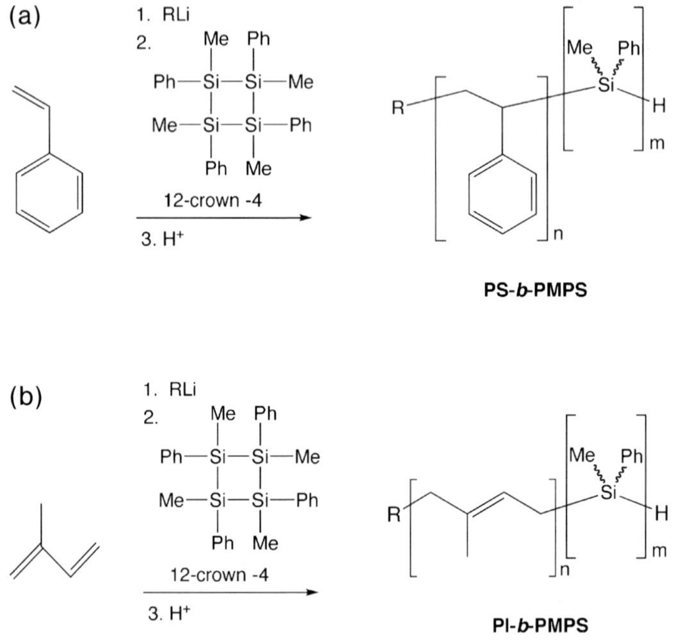

Scheme 16.51 Sequential anionic polymerization of (a) styrene and (b) isoprene with 1,2,3,4-tetramethyl-1,2,3,4-tetraphenylcyclotetrasilane.

In 1991, Matyjaszewski and coworkers reported the anionic ROP route to polysilanes (Scheme 16.50) [106, 107]. The key to this approach was to replace phenyl groups from octaphenylcyclotetrasilane by smaller methyl substituents (by a two-step process) as the former is too sterically crowded to undergo ROP.

The anionic ROP of 1,2,3,4-tetramethyl-1,2,3,4-tetraphenylcyclotetrasilane [106] remains the best route for well-controlled polymerization to polysilanes. The methodology listed above has been extended to the synthesis of organic-*block*-inorganic block copolymers for isoprene and styrene monomers (Scheme 16.51) [108]. For polystyrene-*block*-polymethylphenylsilane (PS-*b*-PMPS) and polyisoprene-*block*-polymethylphenylsilane (PI-*b*-PMPS), the sequential polymerization of styrene and isoprene blocks was performed first followed by the poly-

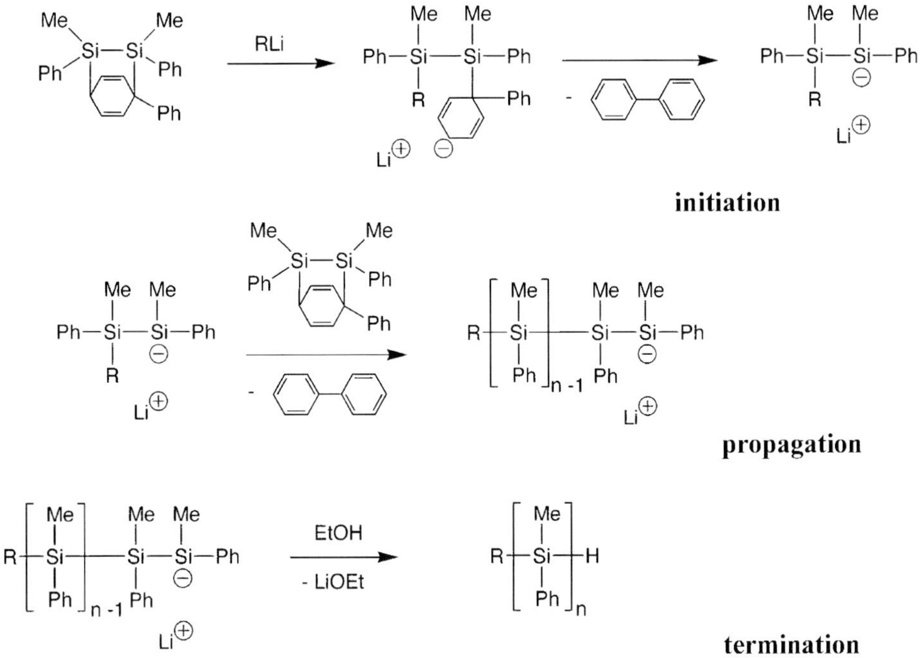

Scheme 16.52 Living anionic ROP of a disilabicyclooctadiene.

merization of the cyclotetrasilane monomer. The living chain ends of PS and PI were of appropriate nucleophilicity for the ring opening of the cyclotetrasilane monomer although 12-crown-4 is required for enhancing the reactivity of the silyl anions for propagation. GPC of the isolated diblock copolymers showed efficient shifting of PS or PI homopolymer traces to higher molecular weights upon addition of the cyclotetrasilane monomer and 12-crown-4. Typical isolated (PS-b-PMPS) and (PI-b-PMPS) were found to have tunable molecular weights with low polydispersity indices (e.g. M_n for PS-b-PMPS = 9700–26 400, PDI = 1.2–1.3; M_n for PS = 2400–5400 and M_n for PI-b-PMPS = 13 400, PDI = 1.3; M_n for PI = 3700). Additionally, it was found that PS-b-PMPS could self-assemble in the solid state into nanoscopic PMPS-based cylinders in a PS matrix or in the solution state into micelles using the PS-selective 1,4-dioxane as a solvent [109].

Sakurai and coworkers described another living route to polysilanes through anionic polymerization of disilabicyclooctadienes, which function as masked disilenes (Scheme 16.52) [110, 111]. Block copolymers available by this method have also been prepared and the formation of micelles with photodegradable polysilane cores have been prepared and studied [112]. Dendritic polysilanes with extensively σ-delocalized structures have also recently been prepared [113].

16.12.4
Polyphosphazenes

Polyphosphazenes have a phosphorus–nitrogen backbone. The key developments concerning this polymer system took place in the mid-1960s, when it was shown that if the ROP of pure [NPCl$_2$]$_3$ is carried out carefully, uncross-linked polydichlorophosphazene [NPCl$_2$]$_n$ is formed, which is soluble in organic solvents [114]. This permitted replacement of the halogen substituents by reaction with nucleophiles to yield hydrolytically stable polyorganophosphazenes (Scheme 16.53). This approach permits the immense structural diversity of polyorganophosphazenes and allows the properties to be tuned and specific properties to be introduced. The phosphazene backbone possesses a unique range of unusual properties. For example, it is extremely flexible, thermally and oxidatively stable, optically transparent from 220 nm to the near-IR region and it imparts flame-retardant properties. Some of the most important polymers are fluoroalkoxy derivatives and amorphous copolymers are very useful as flame-retardant and hydrocarbon solvent- and oil-resistant elastomers and have found aerospace and automotive applications. Polymers such as the comb polymer poly[bis(methoxyethoxyethoxy)-phosphazene] (PMEEP) are of considerable interest as components of polymeric electrolytes in battery technology. Polyphosphazenes are also of interest as biomedical materials and bioinert, bioactive, membrane-forming and bioerodable materials and hydrogels have been prepared [114].

The reaction of polydichlorophosphazene with organometallic reagents such as Grignard or organolithium reagents generally leads to chain cleavage and substitution and does not provide a satisfactory route to polymers with only alkyl and aryl side-groups bound by direct P–C bonds. In the early 1980s, a solution to this problem was provided by the discovery of a thermal condensation route to poly(alkyl/arylphosphazene)s from phosphoranimines [115]. The polymerization is in fact a chain-growth reaction and allows access to high molecular weight polyphosphazenes such as polydimethylphosphazene and polymethylphenylphosphazene (Scheme 16.54). However, detailed understanding of the mechanism of this interesting polymerization is not yet available.

Although several polyphosphazenes have been commercialized, recent work has focused on the development of cheaper and more convenient methods for making these materials. In addition, the thermal ROP and condensation routes (Schemes

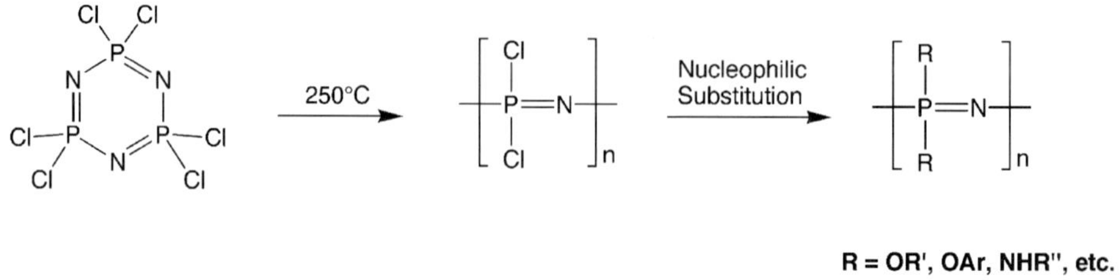

R = OR', OAr, NHR'', etc.

Scheme 16.53 Thermal ROP route to polyphosphazenes.

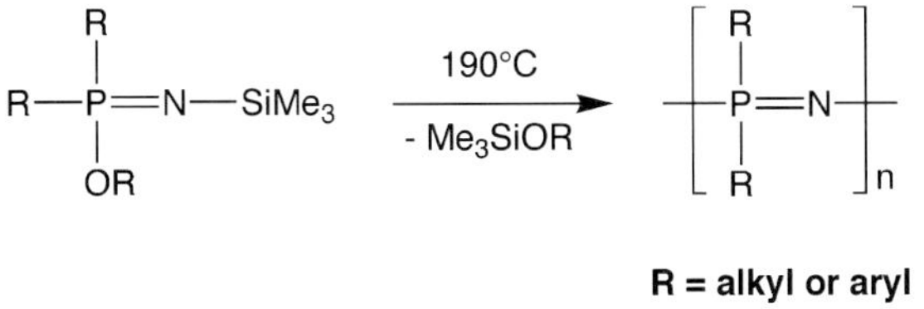

Scheme 16.54 Thermal condensation route to polyphosphazenes.

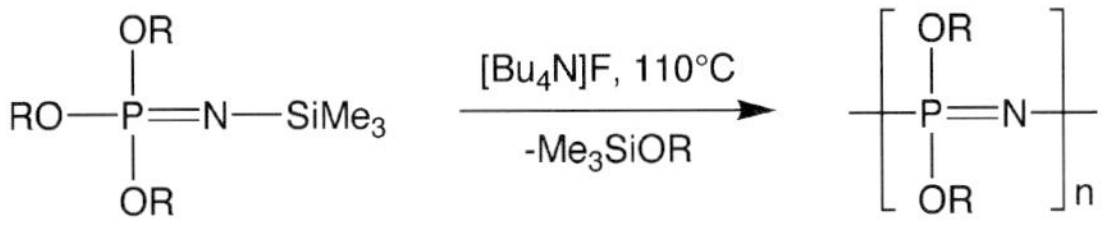

Scheme 16.55 Synthesis of polyphosphazenes via the anionic polymerization of phosphoranimines.

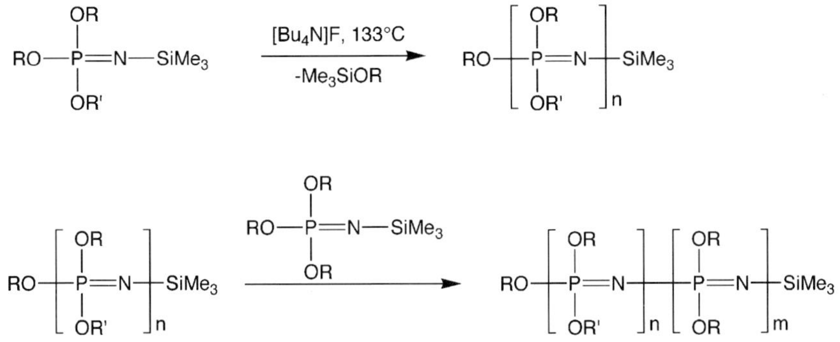

Scheme 16.56 Synthesis of polyphosphazene block co-polymers via the sequential living anionic polymerization of phosphoranimines.

R = -CH$_2$CF$_3$
R' = -CH$_2$CH$_2$OCH$_3$ or
-CH$_2$CH$_2$OCH$_2$CH$_2$OCH$_3$

16.53 and 16.54) do not permit convenient molecular weight control and broad molecular weight distributions result. Attention has therefore also been focused on the development of new routes which allow control of chain length and architecture.

Another condensation approach provides an alternative, direct route to fluoroalkoxyphosphazene polymers and related derivatives. This process shows characteristics of a living polymerization and allows control of molecular weight, narrow molecular weight distributions and access to block copolymers (Schemes 16.55 and 16.56) [116–120]. The mechanism is believed to involve attack of a fluoride anion at the silicon center to create a living phosphazene anion, which subsequently attacks the silicon center of other monomers in the chain propagation step.

The development of thermal condensation routes to polydichlorophosphazene have also been reported. For example a promising thermal condensation route from monomeric Cl$_3$P=N–P(O)Cl$_2$ at 200 °C yields polymers with broad molecular weight distributions and involves the elimination of POCl$_3$ [121].

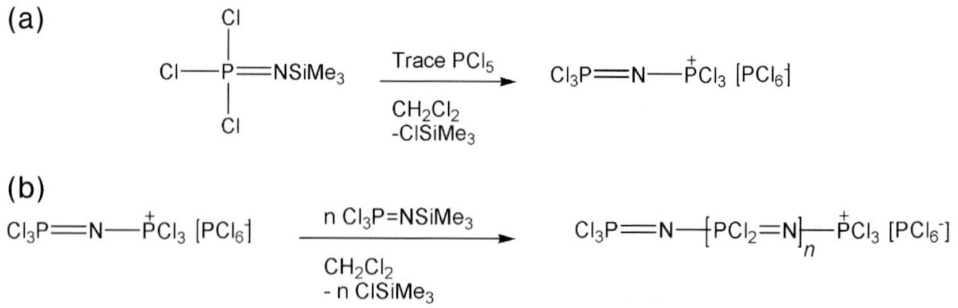

Scheme 16.57 Proposed mechanism for the living cationic chain polymerization of N-silyltrichlorophosphoranimine. (a) Initiation and (b) propagation.

(a) $Cl_2(Cl)P=NSiMe_3 \xrightarrow[CH_2Cl_2]{\text{Trace } PCl_5, -ClSiMe_3} Cl_3P=N-\overset{+}{P}Cl_3 \ [PCl_6^-]$

(b) $Cl_3P=N-\overset{+}{P}Cl_3 \ [PCl_6^-] \xrightarrow[CH_2Cl_2]{n \ Cl_3P=NSiMe_3, -n \ ClSiMe_3} Cl_3P=N-[PCl_2=N]_n-\overset{+}{P}Cl_3 \ [PCl_6^-]$

In 1995, a synthesis of polydichlorophosphazene that operates at room temperature and allows molecular weight control and narrow molecular weight distributions (M_n values of up to ca. 50 000) was described [122]. The monomer $Cl_3P=NSiMe_3$ is also now conveniently available from PCl_3 in high yield in a one-pot process [123]. The mechanism involves cationic chain growth (Scheme 16.57) and has been demonstrated for model oligomer compounds [124, 125]. Block copolymers with organic or inorganic blocks are available by using end-functionalized polymers as initiators or by sequential addition of monomers [126, 127].

The series of developments that have emerged since 1990 offer the prospect of improved routes to phosphazene polymers that may well facilitate more rapid and extensive commercialization in addition to allowing access to new polymer architectures with new applications.

16.13
Conclusions and Outlook

Metal-containing polymers have long represented an area of considerable but untapped potential. The structural diversity of the materials now available is impressive, as is the range of function. An inexhaustive list of uses includes applications as catalysts, electrode mediators, sensors and stimuli responsive gels; as photonic, conductive, photoconductive and luminescent materials; as precursors to magnetic ceramics and nanopatterned surfaces; and as bioactive materials and metalloenzyme models [1–9]. In particular, despite the synthetic challenges, increasing focus is being applied to the area of metallomacromolecules with precise structures, as illustrated by the area of metal-containing block copolymers. Similar considerations exist for the area of polymers based on main group elements where useful thermophysical properties and unusual electronic characteristics can be introduced. Many future advances can be expected in the near future in the blossoming field of inorganic polymers with precise structures and controlled architectures.

References

1 Review: R. P. Kingsborough, T. M. Swager, *Prog. Inorg. Chem.* **1999**, *48*, 123.
2 Review: I. Manners, *Science* **2001**, *294*, 1664.
3 Review: G. R. Newkome, E. F. He, C. N. Moorefield, *Chem. Rev.* **1999**, *99*, 1689.
4 Review: P. Nguyen, P. Gómez-Elipe, I. Manners, *Chem. Rev.* **1999**, *99*, 1515.
5 Review: U. S. Schubert, C. Eschbaumer, *Angew. Chem. Int. Ed.* **2002**, *41*, 2893.
6 Review: B. J. Holliday, T. M. Swager, *Chem. Commun.* **2005**, 23.
7 R. D. Archer, *Inorganic and Organometallic Polymers*, Wiley-VCH, Weinheim, **2001**.
8 I. Manners, *Synthetic Metal-containing Polymers*, Wiley-VCH, Weinheim, **2004**.
9 V. Chandrasekhar, *Inorganic and Organometallic Polymers*, Springer, New York, **2005**.
10 S. Penczek, *J. Polym. Sci., Part A: Polym. Chem.* **2002**, *40*, 1665.
11 K. Matyjaszewski, *Macromolecules* **1993**, *26*, 1787.
12 G. Odian, *Principles of Polymerization*, Wiley-VCH, New York, **2005**.
13 C. J. Hawker, A. W. Bosman, E. Harth, *Chem. Rev.* **2001**, *101*, 3661.
14 G. N. Tew, K. A. Aamer, R. Shunmugam, *Polymer* **2005**, *46*, 8440.
15 L. B. Sessions, L. A. Mîinea, K. D. Ericson, D. S. Glueck, R. B. Grubbs, *Macromolecules* **2005**, *38*, 2116.
16 J. Bohrisch, U. Wendler, W. Jaeger, *Macromol. Rapid Commun.* **1997**, *18*, 975.
17 M. Baumann, G. Schmidt-Naake, *Macromol. Chem. Phys.* **2000**, *201*, 2751.
18 K. A. Aamer, G. N. Tew, *Macromolecules* **2004**, *37*, 1990.
19 R. B. Grubbs, *J. Polym. Sci., Part A: Polym. Chem.* **2005**, *43*, 4323.
20 M. Baumert, J. Fröhlich, M. Stieger, H. Frey, R. Mülhaupt, H. Plenio, *Macromol. Rapid Commun.* **1999**, *20*, 203.
21 K. Matyjaszewski, J. Xia, *Chem. Rev.* **2001**, *101*, 2921.
22 J. Xia, X. Zhang, K. Matyjaszewski, *Macromolecules* **1999**, *32*, 3531.
23 X. Zhang, J. Xia, K. Matyjaszewski, *Macromolecules* **1998**, *31*, 5167.
24 E. J. Ashford, V. Naldi, R. O'Dell, N. C. Billingham, S. P. Armes, *Chem. Commun.* **1999**, 1285.
25 K. A. Davis, B. Charleux, K. Matyjaszewski, *J. Polym. Sci., Part A: Polym. Chem.* **2000**, *38*, 2274.
26 N. P. Tzanetos, A. K. Andreopoulou, J. K. Kallitsis, *J. Polym. Sci., Part A: Polym. Chem.* **2005**, *43*, 4838.
27 R. Shunmugam, G. N. Tew, *J. Am. Chem. Soc.* **2005**, *127*, 13567.
28 C. W. Tse, L. S. M. Lam, K. Y. K. Man, W. T. Wong, W. K. Chan, *J. Polym. Sci., Part A: Polym. Chem.* **2005**, *43*, 1292.
29 C. Barner-Kowollik, T. P. Davis, J. P. A. Heuts, M. H. Stenzel, P. Vana, M. Whittaker, *J. Polym. Sci., Part A: Polym. Chem.* **2003**, *41*, 365.
30 J. E. McAlvin, C. L. Fraser, *Macromolecules* **1999**, *32*, 6925.
31 P. A. Lay, A. M. Sargeson, H. Taube, *Inorg. Synth.* **1986**, *24*, 291.
32 B. A. Gregg, A. Heller, *J. Phys. Chem.* **1991**, *95*, 5970.
33 Polymer Source Inc., www.polymersource.com.
34 S. Hou, W. K. Chan, *Macromol. Rapid Commun.* **1999**, *20*, 440.
35 S. Klingelhöfer, W. Heitz, A. Greiner, S. Oestreich, S. Förster, M. Antonietti, *J. Am. Chem. Soc.* **1997**, *119*, 10116.
36 J. P. Spatz, T. Herzog, S. Mössmer, P. Ziemann, M. Möller, *Adv. Mater.* **1999**, *11*, 149.
37 J. P. Spatz, S. Mössmer, C. Hartmann, M. Möller, T. Herzog, M. Krieger, H.-G. Boyen, P. Ziemann, *Langmuir* **2000**, *16*, 407.
38 L. M. Bronstein, M. V. Seregina, O. A. Platonova, Y. A. Kabachii, D. M. Chernyshov, M. G. Ezernitskaya, L. V. Dubrovina, T. P. Bragina, P. M. Valetsky, *Macromol. Chem. Phys.* **1998**, *199*, 1357.
39 D. M. Chernyshov, L. M. Bronstein, H. Börner, B. Berton, M. Antonietti, *Chem. Mater.* **2000**, *12*, 114.
40 O. Nuyken, V. Burkhardt, C. Hübsch, *Macromol. Chem. Phys.* **1997**, *198*, 3353.
41 O. Nuyken, V. Burkhardt, T. Pöhlmann, M. Herberhold, *Makromol. Chem. Macromol. Symp.* **1991**, *44*, 195.

42 O. Nuyken, V. Burkhardt, T. Pöhlmann, M. Herberhold, F. J. Litterst, C. Hübsch, *Macromolecular Systems: Microscopic Interactions and Macroscopic Properties*, Wiley-VCH, Weinheim, **2000**, pp. 305–324.

43 R. Rulkens, A. J. Lough, I. Manners, *J. Am. Chem. Soc.* **1994**, *116*, 797.

44 R. Rulkens, Y. Ni, I. Manners, *J. Am. Chem. Soc.* **1994**, *116*, 12121.

45 C. H. Honeyman, T. J. Peckham, J. A. Massey, I. Manners, *J. Chem. Soc., Chem. Commun.* **1996**, 2589.

46 T. J. Peckham, J. A. Massey, C. H. Honeyman, I. Manners, *Macromolecules* **1999**, *32*, 2830.

47 M. Tanabe, I. Manners, *J. Am. Chem. Soc.* **2004**, *126*, 11434.

48 Y. Ni, R. Rulkens, I. Manners, *J. Am. Chem. Soc.* **1996**, *118*, 4102.

49 K. Kulbaba, I. Manners, *Macromol. Rapid Commun.* **2001**, *22*, 711.

50 X. S. Wang, M. A. Winnik, I. Manners, *Macromolecules* **2002**, *35*, 9146.

51 D. A. Rider, K. A. Cavicchi, K. N. Power-Billard, T. P. Russell, I. Manners, *Macromolecules* **2005**, *38*, 6931.

52 M. E. Piotti, *Curr. Opin. Solid State Mater. Sci.* **1999**, *4*, 539.

53 Y. N. C. Chan, R. R. Schrock, R. E. Cohen, *Chem. Mater.* **1992**, *4*, 24.

54 D. E. Fogg, L. H. Radzilowski, R. Blanski, R. R. Schrock, E. L. Thomas, *Macromolecules* **1997**, *30*, 417.

55 D. Albagli, G. Bazan, M. S. Wrighton, R. R. Schrock, *J. Am. Chem. Soc.* **1992**, *114*, 4150.

56 A. S. Abd-El-Aziz, L. J. May, J. A. Hurd, R. M. Okasha, *J. Polym. Sci., Part A: Polym. Chem.* **2001**, *39*, 2716.

57 A. S. Abd-El-Aziz, *Macromol. Rapid Commun.* **2002**, *23*, 995.

58 Y. N. C. Chan, G. S. W. Craig, R. R. Schrock, R. E. Cohen, *Chem. Mater.* **1992**, *4*, 885.

59 B. Chen, H. F. Sleiman, *Macromolecules* **2004**, *37*, 5866.

60 M. A. Buretea, T. D. Tilley, *Organometallics* **1997**, *16*, 1507.

61 C. E. Stanton, T. R. Lee, R. H. Grubbs, N. S. Lewis, J. K. Pudelski, M. R. Callstrom, M. S. Erickson, M. L. McLaughlin, *Macromolecules* **1995**, *28*, 8713.

62 R. W. Heo, F. B. Somoza, T. R. Lee, *J. Am. Chem. Soc.* **1998**, *120*, 1621.

63 M. Buchmeiser, R. R. Schrock, *Macromolecules* **1995**, *28*, 6642.

64 M. R. Buchmeiser, *Macromolecules* **1997**, *30*, 2274.

65 M. R. Buchmeiser, N. Schuler, G. Kaltenhauser, K.-H. Ongania, I. Lagoja, K. Wurst, H. Schottenberger, *Macromolecules* **1998**, *31*, 3175.

66 C. C. Cummins, M. D. Beachy, R. R. Schrock, M. G. Vale, V. Sankaran, R. E. Cohen, *Chem. Mater.* **1991**, *3*, 1153.

67 X. S. Wang, M. A. Winnik, I. Manners, *Macromol. Rapid Commun.* **2002**, *23*, 210.

68 I. Korczagin, M. A. Hempenius, G. J. Vancso, *Macromolecules* **2004**, *37*, 1686.

69 C. Kloninger, M. Rehahn, *Macromolecules* **2004**, *37*, 1720.

70 B. G. G. Lohmeijer, U. S. Schubert, *J. Polym. Sci., Part A: Polym. Chem.* **2004**, *42*, 4016.

71 G. Zhou, I. I. Harruna, *Macromolecules* **2005**, *38*, 4114.

72 L. Viau, M. Even, O. Maury, D. M. Haddleton, H. Le Bozec, *Compt. Rend. Chim.* **2005**, *8*, 1298.

73 G. Zhou, I. I. Harruna, *Macromolecules* **2004**, *37*, 7132.

74 R. M. Johnson, C. L. Fraser, *Macromolecules* **2004**, *37*, 2718.

75 C. A. Fustin, B. G. G. Lohmeijer, A.-S. Duwez, A. M. Jonas, U. S. Schubert, J.-F. Gohy, *Adv. Mater.* **2005**, *17*, 1162.

76 D. A. Durkee, H. B. Eitouni, E. D. Gomez, M. W. Ellsworth, A. T. Bell, N. P. Balsara, *Adv. Mater.* **2005**, *17*, 2003.

77 C. J. Hawker, T. P. Russell, *MRS Bull.* **2005**, *30*, 952.

78 J. Y. Cheng, C. A. Ross, V. Z.-H. Chan, E. L. Thomas, R. G. H. Lammertink, G. J. Vancso, *Adv. Mater.* **2001**, *13*, 1174.

79 S. Lastella, Y. J. Jung, H. Yang, R. Vajtai, P. M. Ajayan, C. Y. Ryu, D. A. Rider, I. Manners, *J. Mater. Chem.* **2004**, *14*, 1791.

80 C. Hinderling, Y. Keles, T. Stöckli, H. F. Knapp, T. De Los Arcos, P. Oelhafen, I. Korczagin, M. A. Hempenius, G. J. Vancso, R. Pugin, H. Heinzelmann, *Adv. Mater.* **2004**, *16*, 876.

81 J. Q. Lu, T. E. Kopley, N. Moll, D. Roitman, D. Chamberlin, Q. Fu, J. Liu, T. P.

Russell, D. A. Rider, M. A. Winnik, I. Manners, *Chem. Mater.* **2005**, *17*, 2227.

82 J. Q. Lu, D. Chamberlin, D. A. Rider, I. Manners, T. P. Russell, *Nanotechnology* **2006**, *17*, 5792.

83 I. Manners, *Angew. Chem., Int. Ed. Engl.* **1996**, *35*, 1602.

84 J. E. Mark, H. R. Allcock, J. West, *Inorganic Polymers*, Oxford University Press, Oxford, **2005**.

85 J. A. Semlyen, S. J. Clarson (Eds.), *Siloxane Polymers*, Prentice Hall, Englewood Cliffs, NJ, **1991**.

86 E. G. Rochow, *Silicon and Silicones*, Springer, Heidelberg, **1987**.

87 J. E. Mark, H. R. Allcock, J. West, *Inorganic Polymers*, Prentice Hall, Englewood Cliffs, NJ, **1992**.

88 B. Arkles, *CHEMTECH* **1983**, *13*, 542.

89 *Chem. Eng. News* 11 December **1995**, 10.

90 P. MacDonald, N. Plavac, W. Peters, S. Lugowski, D. Smith, *Anal. Chem.* **1995**, *67*, 3799.

91 D. M. Gott, J. J. B. Tinkler, *Silicone Implants and Connective Tissue Disease: Evaluation of Evidence for an Association between the Implantation of Silicones and Connective Tissue Disease*, Medical Devices Agency, London, **1994**.

92 J. Chojnowski, *J. Inorg. Organomet. Polym.* **1991**, *1*, 299.

93 J. C. Saam, D. J. Gordon, S. Lindsey, *Macromolecules* **1970**, *3*, 1.

94 R. Rulkens, Y. Ni, I. Manners, *J. Am. Chem. Soc.* **1994**, *116*, 12121.

95 R. Resendes, J. A. Massey, H. Dorn, K. N. Power, M. A. Winnik, I. Manners, *Angew. Chem. Int. Ed.* **1999**, *38*, 2570.

96 R. Resendes, J. A. Massey, K. Temple, L. Cao, K. N. Power-Billard, M. A. Winnik, I. Manners, *Chem. Eur. J.* **2001**, *7*, 2414.

97 X. S. Wang, M. A. Winnik, I. Manners, *Macromolecules* **2002**, *35*, 9146.

98 Q. Liu, G. R. Wilson, R. M. Davis, J. S. Riffle, *Polymer* **1993**, *34*, 3030.

99 R. West, *J. Organomet. Chem.* **1986**, *300*, 327.

100 R. D. Miller, J. Michl, *Chem. Rev.* **1989**, *89*, 1359.

101 K. Matyjaszewski, *J. Inorg. Organomet. Polym.* **1991**, *1*, 463.

102 Y. Wang, R. West, C. H. Yuan, *J. Am. Chem. Soc.* **1993**, *115*, 3844.

103 C. T. Aitken, J. F. Harrod, E. Samuel, *J. Organomet. Chem.* **1985**, *279*, C11.

104 C. T. Aitken, J. F. Harrod, E. Samuel, *J. Am. Chem. Soc.* **1986**, *108*, 4059.

105 Y.-L. Hsiao, R. M. Waymouth, *J. Am. Chem. Soc.* **1994**, *116*, 9779.

106 M. Cypryk, Y. Gupta, K. Matyjaszewski, *J. Am. Chem. Soc.* **1991**, *113*, 1046.

107 E. Fossum, K. Matyjaszewski, *Macromolecules* **1995**, *28*, 1618.

108 K. Matyjaszewski, P. J. Miller, E. Fossum, Y. Nakagawa, *Appl. Organomet. Chem.* **1998**, *12*, 667.

109 E. Fossum, K. Matyjaszewski, S. S. Sheiko, M. Möller, *Macromolecules* **1997**, *30*, 1765.

110 K. Sakamoto, K. Obata, H. Hirata, M. Nakajima, H. Sakurai, *J. Am. Chem. Soc.* **1989**, *111*, 7641.

111 K. Sakamoto, M. Yoshida, H. Sakurai, *Polymer* **1994**, *35*, 4990.

112 T. Sanji, Y. Nakatsuka, S. Ohnishi, H. Sakurai, *Macromolecules* **2000**, *33*, 8524.

113 J. B. Lambert, J. L. Pflug, C. L. Stern, *Angew. Chem. Int. Ed. Engl.* **1995**, *34*, 98.

114 H. R. Allcock, *Chemistry and Applications of Polyphosphazenes*, Wiley, Hoboken, NJ, **2002**.

115 R. H. Neilson, P. Wisian-Neilson, *Chem. Rev.* **1988**, *88*, 541.

116 R. A. Montague, K. Matyjaszewski, *J. Am. Chem. Soc.* **1990**, *112*, 6721.

117 K. Matyjaszewski, *J. Inorg. Organomet. Polym.* **1992**, *2*, 5.

118 K. Matyjaszewski, R. Montague, J. Dauth, O. Nuyken, *J. Polym. Sci., Part A: Polym. Chem.* **1992**, *30*, 813.

119 K. Matyjaszewski, M. K. Moore, M. L. White, *Macromolecules* **1993**, *26*, 6741.

120 K. Matyjaszewski, U. Franz, R. A. Montague, M. L. White, *Polymer* **1994**, *35*, 5005.

121 G. D'Halluin, R. De Jaeger, J. P. Chambrette, P. Potin, *Macromolecules* **1992**, *25*, 1254.

122 C. H. Honeyman, I. Manners, C. T. Morrissey, H. R. Allcock, *J. Am. Chem. Soc.* **1995**, *117*, 7035.

123 B. Wang, E. Rivard, I. Manners, *Inorg. Chem.* **2002**, *41*, 1690.

124 H. R. Allcock, C. A. Crane, C. T. Morrissey, J. M. Nelson, S. D. Reeves, C. H.

Honeyman, I. Manners, *Macromolecules* **1996**, *29*, 7740.
125 E. Rivard, A. J. Lough, I. Manners, *Inorg. Chem.* **2004**, *43*, 2765.
126 H. R. Allcock, S. D. Reeves, J. T. Nelson, I. Manners, *Macromolecules* **2000**, *33*, 3999.
127 H. R. Allcock, S. D. Reeves, J. M. Nelson, C. A. Crane, I. Manners, *Macromolecules* **1997**, *30*, 2213.